The South American Camelids

An Expanded and Corrected Edition

Monograph 64

Cotsen Institute of Archaeology Monographs

CONTRIBUTIONS IN FIELD RESEARCH AND CURRENT ISSUES IN ARCHAEOLOGICAL METHOD AND THEORY

To Ana

THE
SOUTH AMERICAN CAMELIDS

BY
DUCCIO BONAVIA
Translated by Javier Flores Espinoza

COTSEN INSTITUTE OF ARCHAEOLOGY
UNIVERSITY OF CALIFORNIA, LOS ANGELES
2008

THE COTSEN INSTITUTE OF ARCHAEOLOGY at UCLA is a research unit at the University of California, Los Angeles that promotes the comprehensive and interdisciplinary study of the human past. Established in 1973, the Cotsen Institute is a unique resource that provides an opportunity for faculty, staff, graduate students, research associates, volunteers and the general public to gather together in their explorations of ancient human societies.

Former President and CEO of Neutrogena Corporation Lloyd E. Cotsen has been associated with UCLA for more than 30 years as a volunteer and donor and maintains a special interest in archaeology. Lloyd E. Cotsen has been an advisor and supporter of the Institute since 1980. In 1999, The UCLA Institute of Archaeology changed its name to the Cotsen Institute of Archaeology at UCLA to honor the longtime support of Lloyd E. Cotsen.

Cotsen Institute Publications specializes in producing high-quality data monographs in several different series, including Monumenta Archaeologica, Monographs, and Perspectives in California Archaeology, as well as innovative ideas in the Cotsen Advanced Seminar Series and the Ideas, Debates and Perspectives Series. Through the generosity of Lloyd E. Cotsen, our publications are subsidized, producing superb volumes at an affordable price.

This book is set in Janson Text
Edited and produced by Leyba Associates, Santa Fe, New Mexico
Cover design by William Morosi
Index by Robert and Cynthia Swanson

Library of Congress Cataloging-in-Publication Data

Bonavia, Duccio, 1935–
  [Camélidos sudamericanos. English]
  The South American camelids / by Duccio Bonavia.
      p. cm. — (Monograph ; 64)
  Includes bibliographical references and index.
  ISBN 978-1-931745-40-6 (pbk. : alk. paper) —
  ISBN 978-1-931745-41-3 (cloth : alk. paper)
  1. *Lama* (Genus)—South America. I. Title.
  QL737.U5413B65 2008
  636.2'96098—dc22
                        200802665

Contents

LIST OF ILLUSTRATIONS

LIST OF TABLES

Acknowledgments from the First Spanish Edition

Duccio Bonavia

The writing of any book requires the help of many individuals and institutions, and acknowledging their help is a matter of conscience. Although this may appear to be an exaggeration, this is indeed the case because one always worries about not having thanked everyone adequately and about having inadvertently overlooked someone. On the other hand, listing everyone is almost impossible. This book has not overcome these difficulties, and I hereby apologize to anyone whom I may have inadvertently left out.

This study would not have been possible without the crucial involvement of the Institut Français d'Études Andines in Lima. I began collecting the data when the Institute gave me the opportunity to be Associate Research Investigator for one year (1990). My deepest thanks goes to the Institute for their help in publishing the first edition of this book (1996). I also extend my thanks to Christian de Muizon, Director of the Institute while the manuscript was being prepared, and I thank Georges Pratlong, who took over this position as the book went to press.

I extend my gratitude to all the staff of the Institut Français d'Études Andines, who enthusiastically cooperated in preparing and publishing this study. I am particularly grateful to Zaida Lanning de Sánchez for her invaluable help in the search for data in the Institute's library and in solving various bibliographical problems.

The Universidad Peruana Cayetano Heredia gave me a sabbatical year in 1994, during which I devoted all of my attention to the preparation of this book; the Universidad later helped with its publication. Raúl Ishiyama C., then Head of the Biology Department, provided all the help needed. Jasmín Zelaya Barrientos of the University library helped me solve bibliographical quandaries.

My thanks also go to Conservation International, whose Carlos F. Ponce helped with the edition.

There is another kind of support which is just as crucial and difficult to acknowledge. I refer to the moral support, the motivation, and the comfort that I received from my family and friends. This is priceless. In this regard I must single out the enduring presence of my wife, the exemplary support of my children and son-in-law, and my mother's tenderness. And I cannot forget the loyal and genuine friendship of Antonio Muñoz Nájar.

While preparing the manuscript I had a very serious coronary illness, and this manuscript would never have made it to publication had it not been for the intervention of a group of physicians. Here I would like to mention Augusto Yi Chu, Miguel Sánchez-Palacios Paiva, Waldo Fernández, Jaime Ulloa Schiantarelli, Régulo Agusti Campos, Raúl Ames Enríquez, and all of the staff in the Servicios de Cardiología y Cirugía Cardiovascular of the Guillermo Almenara Public Hospital in Lima.

In writing this book I had to cover topics that fall within other specialties, and I was fully aware of the risks this involves. I was fortunate in having the help of good friends who made the task easier. David S. Webb kindly read the chapters on taxonomy and phylogeny, as well as that on paleontology, and made invaluable suggestions. For the later section I had the help of Christian de Muizon, while for the physiological aspects I was assisted by Carlos Monge C. and Ramiro Castro de la Mata.

I am most thankful to Jorge Flores Ochoa and Sabine Dedenbach-Salazar Sáenz. Their studies were an inspiration in more than one way. Their research in this area is of major importance, and in some cases all I did was expand their database to strengthen what was not only their position but also mine. We have covered several topics from the same perspective, and thus I have made a real effort to be precise in my citations. I apologize for any instance where I have not clearly specified my source of information.

Many colleagues made suggestions on various points throughout the years that I spent developing this work. I would like to single out Ramiro Castro de la Mata, Franklin Pease G. Y., Peter Dressendörfer, and Jean Guffroy. Others who kindly provided me with unpublished data from their personal archives include Karen Olsen Bruhns, Cristóbal Campana, Claude Chauchat, Lorenzo Huertas, José A. Ochatoma Paravicino, Elizabeth J. Reitz, María Rostworowski de Diez Canseco, John H. Rowe, Víctor F. Vásquez Sánchez, and Jorge Zevallos Quiñones. My thanks to them all.

Research in Peru is a difficult endeavor due to the lack of good libraries. Two individuals who opened their libraries to me were Ramiro Castro de la Mata and Franklin Pease G. Y. In addition, many colleagues sent me copies of papers and books, and this help was invaluable. I must mention Karen Olsen Bruhns, Gonzalo Castro, Hilda Codina, Peter Dressendörfer, Thomas Fisher, Jorge Flores Ochoa, César Gálvez Mora, Jean Guffroy, Roswith Hartmann, Olaf Holm, Patricia Majluf, Ramiro Matos Mendieta, George R. Miller, Gordon McEwan, Laura Laurencich Minelli, Omar Ortiz Troncoso, Elizabeth J. Reitz, Anne Rowe, Helaine Silverman, Rubén Stehberg,

Douglas H. Ubelaker, Víctor F. Vásquez Sánchez, D. David Webb, and Elizabeth S. Wing. A special acknowledgment to the Corporación Nacional Forestal de Chile, and to Mario Alvarado Eva, who kindly provided me the data I needed on the guanaco in Chile. And Bruna Bonavia Fisher carried out a difficult bibliographical search in Canada which was essential for the completion of this book.

Hermilio Rosas Lanoire gave me all the help possible to analyze and document the ceramic collections of the Museo Nacional de Antropología y Arqueología of Lima during his tenure as Director. I later had the same cooperation from Fernando Rosas, who also held the position of Director in what is now the Museo Nacional de Antropología, Arqueología e Historia de Lima.

Some colleagues kindly provided photographs or helped me get them, including Cristóbal Campana, Jorge Chávez Salas, Daniel Goodwin, Danièle Lavallée, Craig Morris, Sergio Purin, Michael Snarskis, Hernán Torres, and Wilder Trejo. I also had the help of several institutions which provided photographs from their archives, including the American Museum of Natural History of New York, the Rijksmuseum voor Volkenkunde of Leyden, and the Facultad de Ciencias Forestales of the Universidad Nacional Agraria-La Molina in Lima, Peru.

Several institutions kindly gave permission to reproduce some illustrations from their publications. I must mention the Smithsonian Institution (Washington), The Carnegie Museum of Natural History (Pittsburgh), the Society for American Archaeology (Washington), and the Pontificia Universidad Católica del Perú.

I cannot end these acknowledgements without naming Leopoldo Chiappo and Luis León Herrera, who assisted me more than once in solving doubts of a linguistic nature.

ACKNOWLEDGMENTS FOR THE ENGLISH EDITION

DUCCIO BONAVIA

I decided to prepare a second edition of this book once the Spanish edition was sold out, in response to the insistent requests of my colleagues. However, given the problems such an endeavor would have in Peru, I decided to try to publish it abroad in English. In general a scientific text becomes outdated quickly, and my book is no exception. I have therefore updated it, adding the most recent information I was able to get. I am fully aware that, because of the time that elapses between the moment the manuscript is handed to the editor and its final publication, a book of this nature may yet fall again out of date. To this we must add the time it takes not just to translate it, but also to secure its financing. Even so, I believe that no significant changes have arisen in the topics discussed. That said, this updating also allowed me to correct the inadvertent errors that always find their way into studies of this kind.

The data required for such a study as this are often unpublished, and one must resort to the help of colleagues. In this regard I thank Lidio Valdez Cárdenas, Juan Castañeda, María Benavides, Eduardo Frank, Michael Hink, Maurizio Tosi, and Martina Gerken.

It is difficult to obtain statistical data on the number of camelids present throughout the world. It would have been impossible to present the figures that appear in the book without the help of the following institutions and individuals. I must thank Dar Wassink from the Alpaca Registry Inc. in the United States, the Canadian Llama and Alpaca Registry, Jorge Herrera Hidalgo, and Rony Garibay Suárez of the Consejo Nacional de Camélidos Sudamericanos of Peru,

Martine Lavelatte from the French Ministère de l'Agriculture et de la Pèche, Chris Tuckwel from Australia, Ángel María Martínez Churiaque in Spain, Jean-Jacques Lauvergne in France, Paul Rose in England, Carlo Renieri from the Università degli Studi di Camerino in Italy, John Eliasov and Boris Krasnov de Ramon from the Science Center of Israel, and Alexander Grobman, who put me in touch with them. Finally, Elmo León Canales and Javier Delgado provided me with statistics on camelids in Europe.

Aurelio Bonavia, Lawrence Kaplan, Joyce Marcus, and Kent Flannery helped me obtain some publications which were unavailable in Peru. Zaida Lanning de Sánchez, from the Institut Français d'Études Andins, and Jazmín Zelaya Barrientos, from the Universidad Peruana Cayetano Heredia, helped me search for the bibliographical data in the libraries of their institutions.

A special acknowledgment is due to Ramiro Castro de la Mata, who was my advisor in managing the information on camelid physiology, and let me use his library, thus greatly aiding my study. I would also like to thank Jean Noel Martínez, who provided me with paleontological data.

I cannot forget Jorge Castillo Yui, who helped me print out the many drafts of the translation as these were corrected.

Bruna Bonavia-Fisher, Thomas Fisher, Aurelio Bonavia, Tom Dillehay, Lawrence Kaplan, Elizabeth Reitz, Laura Johnson-Kelly, John Topic, Frank Salomon, and Carmen Arellano revised and suggested translations for technical terms. Hernán Torres helped me find the funds for the translation.

It is not easy getting the funds to translate a book of this kind, as few institutions really understand the goals that are being pursued. My gratitude to the two institutions that supported me will therefore be insufficient: World Wildlife Fund Inc., and the US Fish and Wildlife Service. In the latter case I note the great help given by Frank F. Rivera-Milan in solving the many problems that arise when one works at long distances.

Translations are always difficult. The Italian aphorism "traduttore traditore," which holds a great truth, always hangs over the head of the translator like the sword of Damocles. However, in this case the saying proved false. Javier Flores not only prepared a faithful translation, correctly interpreting what I wanted to say, but also made several suggestions to improve this book. Furthermore, he patiently accepted many changes and additions which were due to my obsessive desire for perfection. No thanks can suffice in this regard.

I left my thanks to Gonzalo Castro and Ramiro Matos Mendieta to the end in order to give them special meaning. To a great extent they actually are responsible for the completion of this book. Both are old friends. Gonzalo first became acquainted with archaeological sites with me when he was a child and then as my student in the university, and we have always retained a warm friendship despite the distance. Ramiro in turn was my fellow student in the university and is now a distinguished colleague. Both have been sympathetic to my concerns ,and both realized what this work has meant to my professional life. They not only helped me from the very beginning, but also encouraged me through the hard times. It was they who throughout all these years sought and made the essential contacts with institutions and individuals needed to complete this task. If the truth be told, this edition would have been impossible without them. Words fail me in expressing my thanks.

FOREWORD

ALBERTO REX GONZÁLEZ

The study of South American camelids, with its multiple aspects and problems, is of interest to archaeologists, ethnographers, ethnozoologists, social anthropologists, ethnologists, ethnohistorians, and veterinarians alike. This animal family essentially formed the basis of the subsistence economy for both hunter-gatherers in early hunting societies and the more complex agricultural societies that followed. Thus, scholars from different disciplines who study pre-Columbian cultures must have abundant, detailed knowledge of the different camelid species. In one way or another, they will be faced with the problem of identifying camelids through their remains, as well as with understanding their behavior in relationship to the humans who domesticated and used them in the past, and still do so profitably throughout the subcontinent.

It was for these reasons that the Spanish edition of this book was welcomed by researchers and scholars. It received favorable reviews, and I expect that the English-language edition will greatly expand the book's readership. As happens with other essential elements associated with pottery-making agriculturalists on the American continent, such as the great variety of cultigens that such societies manipulate, analytical treatises on animals that provide a clear, comprehensive, and easily accessible overview of the animals are rare. Nor are there published studies that can be used to avoid having to dedicate long hours to acquiring the zoological basics.

A personal experience may prove instructive in this regard. One of my first archaeological field experiences was the excavation of Intihuasi Cave in the central Argentinian sierra (González 1963). This cave had been occupied since 6000 BC by groups of hunter-gatherers whose major food was guanaco meat. The stratigraphy of the cave showed that three different cultural traditions were superimposed. For millennia, the humans in the cave had relied almost exclusively on consuming guanaco meat, and the bone remains of these animals were scattered in the archaeological deposits. Although their identification was made by a distinguished paleontologist (R. Pascual), I personally tried to study the descriptive information on the dominant species, which proved a laborious task, particularly because there was no easily accessible monograph on these mammals. My own experience, therefore, has convinced me of the importance of publishing this book by Duccio Bonavia, which its author has modestly subtitled "An Introduction to Their Study." The South American Camelids actually is a bona fide general, almost exhaustive treatise on the many perspectives and diverse aspects of this field that goes far beyond the articles and studies published previously.

Duccio Bonavia is more than qualified to prepare a study of this scope. A graduate of the Universidad Nacional Mayor de San Marcos in Lima, Peru, he was a professor at the Universidad Peruana Cayetano Heredia in Lima. After graduating he worked with the renowned French prehistorian François Bordes. Upon returning to Peru, he dedicated himself to carrying out archaeological studies in Peru, where he distinguished himself with his multifaceted vision, which led him to inquire about the various aspects of the archaeological record, from lithic technology to prehispanic art, from paleobotany to agriculture, from paleogeography to ecology, medicine, and paleoeschatology, and finally, with this book,to paleozoology.

In the ethnobotanical field, during his student years Bonavia collaborated with the renowned Professor Paul C. Mangelsdorf, a maize specialist, from whom he received invaluable instruction. Bonavia's interest in this area led to his spending a season at the Botanical Museum of Harvard University. His studies and discoveries have left their mark on our knowledge of the major cultigens of South America, including the varieties of maize known today in Peru..

The seminal studies published by Bonavia either as sole author or as co-author on maize races in preceramic times are landmarks in the study of prehispanic cultigens. I thus believe that it was not by chance that Bonavia later studied camelids, for these two subjects complement one another and are essential for a clear picture of the development of the subsistence economy of the native peoples of South America.

Moreover, Bonavia's study is not just a compilation and synthesis of others' work. It goes far beyond this and tries to clear up the misunderstandings and mistakes that have sometimes been made by researchers, such as the widespread belief that the domestication of the llama brought about the end of its ability to produce milk. This mistake is also found in the Spanish Royal Academy's important *Diccionario de la Lengua Española*, the ultimate reference for the Spanish language, which until the twentieth edition credited the Andean llama with producing milk for human consumption.

Beyond these minor corrections, Bonavia analyzed fundamental problems and resolved issues that have been misinterpreted by most scholars. An example of such misinterpretations is assuming that the llama was an archetypical high-altitude animal, which, if true, would have helped early Andean cultures adapt to the high-altitude environments in the cordillera. In this regard, one chapter in this book specifically scrutinizes the physiology of these animals in conjunction with studies on human adaptation to high altitude. Studies of these animals' phys-

iology have often been carried out in the same laboratories and directed by the same researchers who studied human physiology. These studies, which preceded Bonavia's work, support his hypothesis, because the physiology of these animals has shown that these camelids are not specialized for high altitudes. Bonavia tracked down paleontological, archaeological, and historical evidence, and his results show that this conclusion is true. In fact, every line of evidence seems to indicate that the fossil remains of these animals were dispersed in the South American lowlands, and that it was the first hunters-gatherers to reach the Andean area who pushed the wild species into the high-altitude areas. But then, throughout the prehispanic era, camelids lived in all the environmental zones of the Andes, from sea level to the highest altitudes. It was the Europeans who finally marginalized them in these areas when they introduced Old World animals, and it was then that the myth that camelids were high-altitude animals was born—a myth with longlasting consequences.

In each chapter on the specialized topics (taxonomy, physiology), Bonavia sought the collaboration of renowned specialists, which speaks to his commitment to obtain the most accurate information on the complex problems he addresses in his study. This is clearly evident in the historical data in Chapter 1, which deals with the taxonomy and phylogeny of camelids. A great number of camelid fossils dating to the Pleistocene and Holocene have been found in Argentina. The fact that this country has a long tradition of paleontological studies (by F. Ameghino, R. Pascual, D. López Aranguren, F. Tonni, C. Rusconi, and L. Kraglievich, among others) meant that the studies sometimes were complicated, with several confusing synonyms extant in the literature on species, subspecies, and genera. Examining these contributions thus was arduous and discouraging. At the time, the specialists themselves had doubts over these classifications, and even more over the phylogeny of the main species. During my first trip to the United States I consulted the eminent paleontologist G. Simpson because I wanted to clear up the problems that existed in classifying and distinguishing the different camelid species when we found just a few bone remains in our sites. Simpson explained that such confusion was widespread and almost unavoidable.

Today, the conclusions are far more precise, owing to new studies on phylogeny and comparative anatomy, but the classification still has not been cleared up definitively. However, Bonavia's contribution in this regard is substantial, for it is the first time that the synonyms for genera and species created by paleontologists and that have confused nonspecialists have been assembled in a single work.

Another essential topic addressed in this book is the knowledge base concerning camelid reproduction for the management of llama herds in the Andes or for guanaco hunting strategies in Patagonia and the Andean valleys.

Also of great interest is the chapter on diseases common to the camelids, such as mange, which not only affects the reproductive capacity of herds but also spread to humans, causing an unbearable itch. Scratching the itch produces scabies that are easily ulcerated and become infected, with unpredictable results, particularly in children. This can become an endemic problem, as it has in the Chaco region. Addressing this disease from a historical perspective, Bonavia has shown that the original camelid populations suffered from a different type of mange than that introduced by the Europeans when they imported Old World animals, and that the mange introduced with European livestock was one of the causes of the great morbidity these animals sustained in Colonial times.

The chapter on camelid species includes figures and statistics that show the same care exhibited throughout the book and will be of practical use to those in charge of protecting these animals, usually anthropologists, as in the case of the province of Catamarca in Argentina. In the region of Laguna Blanca in this same province, the vicuñas are almost extinct, owing to overhunting for their wool, which is used by local blanket weavers and has high economic value. The provincial authorities passed special laws and established an organization that was placed under the direction of an anthropologist, and in just a few years the "great basin of Laguna Blanca" saw the recovery of large herds of vicuñas, like those I saw half a century ago. Although the guanaco

abounds in Patagonia, in some zones it is almost extinct as a result of the killing of *chulengos*, animals whose hides were used to prepare excellent blankets.

The historical bent of archaeologists suggests they will also enjoy reading the history of the biological species detailed in this book. The American and Asian camelids, for example, had common ancestors that populated the North American plains, and then spread to Asia and South America. The early history of these groups underscores Bonavia's call for comparative studies of camels, llamas, and alpacas and should be borne in mind by specialists.

Another important chapter discusses the possible centers of domestication of these animals. After reviewing many sources, Bonavia believes that one of the major centers must have been the puna and intermediate zones, and adds that these centers are not necessarily restricted to a single zone; indeed, further research should be directed toward the intermediate zones.

The domestication of camelids, a process that varies according to the wild species, cultural frontiers, and the ecosystem in which the process of domestication took place, is important not just for its significance to the development of the autochthonous cultures of South America but also for our understanding of the general process of animal domestication as a major evolutionary achievement to be addressed in any theory of culture. In the specific case of camelids, as Bonavia shows, there are interesting comparisons to be made with the same phenomenon as it occurred with other animals in other parts of the globe.

Bonavia also compiled the historical information available on camelid distribution at the time of the Spanish conquest, a problem of renewed interest in light of potential corroboration from recent archaeological finds. The problem still stands, however, for some regions. Bonavia examines in detail the recollections and descriptions of Ulrich Schmidel, a German soldier-chronicler who recounts the discovery of the Río de la Plata and Paraguay and who took part in the first settlement of Buenos Aires. Schmidel's narrative records the presence of llamas in the zone and includes graphic examples of this camelid, and he swears he used one of them to "ride" for some distance.

However, excavation of the urban center and possible sites of the earliest settlement of Buenos Aires was begun in the last decade, and Dr. Daniel Schavelzon, who directed these excavations, reports he has thus far not found a single camelid bone, domesticated or wild, even though several sources mention the presence of the guanaco throughout most of the province of Buenos Aires. This discrepancy means we must keep this problem in mind.

An interesting chapter analyzes the camelid remains found in Peruvian archaeological sites. These animals evidently were present throughout Peru from preceramic times to the Inca Empire. However, the data are so scattered that they gave rise to a completely distorted view of reality, and it is only now that we can analyze and understand the exact role of these animals. Bonavia also shows—and this is most important—how little this field has been researched. The Peruvian example can be generalized to neighboring countries.

Bonavia warns other scholars about the significant problems that arise in trying to identify the different camelid species solely from bones, which are often recovered at archaeological sites in a fragmentary state. This analysis is not limited to Peru alone but applies to other geographic areas as well, such as the Araucanian area of Chile, where there is a specific type of llama that was very important for the Mapuche economy and whose presence in this culture raises some very difficult problems, among them accurately establishing the moment when the displacement of the ancestors of the Chilihueque to the regions of its Mapuche habitat took place. This problem is intimately connected to the cultural development of a large part of northern Chile and northwestern Argentina, a problem that concerns anthropologists working in both countries.

Listing all of the points covered by Bonavia would take too long; suffice it to say, the book is most comprehensive. All that remains to be noted is the excellent and copious bibliography, which includes more than eleven hundred entries transcribed with a probity and accuracy easily detectable in the numerous references in the text that correct mistakes in the citation of dates or in literal transcriptions made by some scholars. It should also be pointed out that Bonavia is the only one who has taken the trouble of including a bibliography of references mentioned by the authors cited, which is a notable aid for scholars interested in this subject. To this we should add the numerous tables that classify numerically the analytical topics in the text, and the photographs and figures of live camelids, as well as archaeological reproductions. Indices of names and places complete and facilitate the use of this invaluable volume.

Bonavia not only synthesizes and summarizes the studies of scholars who come to the same conclusions, he also includes the work of those whose research has led them to different views. An example is some paleontologists' belief that wild representatives of today's domesticated species, such as the llama and the alpaca, were already present in the Pleistocene. If supported, this hypothesis would substantially change the phylogenetic deductions pertaining to the origins of various species. This is a problem of importance to paleontologists, but also one that archaeologists cannot ignore.

In sum, owing to the invaluable analytical data amassed in one volume and the high scientific standard of the presentation, this book is a landmark in the study of camelids and a must in the personal library of all students of New World cultures, past and present.

Note to the Reader

To avoid misunderstandings, I would like to draw the reader's attention to some details in this book.

First, it will be noticed that imported animals are often labeled "European." I am well aware that this adjective is not correct, because most of the animals brought to America were not native to Europe but were introduced into Europe much earlier. I use the term for practical reasons and to avoid long disquisitions.

I often had to use ancient measures for weights and distances, which were then converted to our decimal system to facilitate comparisons and to simplify reading. To make the conversions I used the handbook by Llerena Landa (1957). This is indicated in some cases but not in others, to avoid having a mass of citations

Finally, the reader should be aware that the bibliography has two sections, which is unusual. I have personally verified, wherever possible, the citations that did not specify the original source, to save other scholars from having to do so themselves. It so happens that sometimes one has to resort to the original sources to check and expand the data. To save the reader from this annoyance, I introduced the second section in the bibliography. The first section (15.1), entitled "References Cited by the Author," includes all the titles I read and cited. This section tries to be as exhaustive as was possible, and I assume full responsibility for any mistakes or omissions. The second section (15.2), entitled "References Used by the Authors Cited," includes all the bibliographical items that each author cited; these are items I did not believe were essential reading, or items I was unable to get. The bibliographical data were taken directly from the bibliographies of the authors who used them and have not been modified, except in

some cases, as indicated. It will be seen that even the journal name abbreviations, which I preferred not to use in my own bibliography, have been retained. I have tried to include all references cited by the authors I mentioned. Should one be missing, it is an inadvertent omission.

In the case of authors with two or more studies published in one year, only one of which was used while the remaining one was cited by another author mentioned in the text, I added letters from the alphabet to the year of publication. Therefore, if in the first section there appears the letter b and not a, this means that the latter will be found in the second section, and vice versa. An example is Hesse, whose 1982a study appears in the first bibliography section, while the 1982b study is included in the second section.

In the second section the reader will find a few authors not cited in the text. These are studies that I believe may be useful, most of which are included in Universidad Nacional Mayor de San Marcos et al. (1985) and in Muñoz de Linares and García Sánchez (1986).

I duly noted in the text whenever an author used the same source as I did but in a different edition, with the date of the one I used indicated.

I believe that if both sections of the bibliography, plus some studies that have recently appeared and that I was unable to get, as well as some bibliographies cited here (see section 15.2), are all combined, we have an initial database for an even larger general study of camelids that remains to be done. This will naturally entail a most detailed cross-checking of the data, because not all authors were careful in the use of information. It should be recalled that for the historical subject, there is an important bibliography in Dedenbach Salazar (1990:248–320).

Llamas in Churcampa (Department of Huancavelica, province of Tayacaja, district of Churcampa, 3113 masl). *Photo by Duccio Bonavia.*

INTRODUCTION

[T]he herders of the *ayllus* of Paratía continue to tend their alpacas, weaving their cloth and traveling over the routes that their fathers and their fathers' fathers have followed for generations, driving trains of llamas loaded with textiles and animal products. But for how long? We do not know. This question could only be answered by the Spirit of the Windy Cliffs, *Apu Wayra Qaqa*, and the Spirit of Philinko, *Apu Philinko*.

(Flores Ochoa 1979b:119, 1967:102)

Despite the significant role that South American camelids have had, and still have, in Andean societies, and despite the fact that their role was already understood at the time of the Spanish conquest, this is still a subject with far too many inaccurate assumptions and stereotyped judgments, and in which there still are great gaps in our knowledge. For instance, it is amazing that so serious and well written a book as *Las Armas de la Conquista*, by Alberto Mario Salas (1950), mentions horses and completely ignores the llama as a vehicle of transport for the Inca armies, and even for the Spanish forces. Or that the twentieth edition of the Spanish Royal Academy's *Diccionario de la Lengua Española* still claims, under the heading "llama," that "its milk is used" (1984: 848). Fortunately, this has been corrected in subsequent editions (i.e., starting with the twenty-first, 1992:vol. II, 1279). Some specialists are aware of this (e.g., Gade 1977:113; Franklin 1982: 458), but awareness is still not at the level required.

Besides, it is interesting to note that a serious and difficult problem is posed in this field, precisely the same one Andean archaeology has to face: that is, heuristics. It will be seen throughout this book that the sources have not always been used as they should be and that citations are often not the most complete. There can be no doubt, however, that, as Franklin (1982:458) noted, "[o]ver the past two decades [Franklin wrote in 1982] there has been a serious attempt to systematically investigate and discover the true facts."

Nor should we go to the other extreme and start with the assumption that because of the "tremendous amount of post-Hispanic disturbance of both herders and camelid environmental zones," some aspects will never be elucidated, and therefore claim, as Kent (1987:172) did, that "[d]ocuments are of little help because the Spaniards, who had never seen the New World camelids before, gave them names such as '*ovejas indias*' or '*carneros de la tierra*', indiscriminately assigning either of these names to whatever camelid happened to be around." This certainly is an overstatement. There is no question that there are difficulties, and these difficulties are great, but the information that is still there to be found is considerable and involves many disciplines. And as we shall see, the writings of the Spaniards who came to the New World hold considerable invaluable data that let us expand the scope of our knowledge of camelids enormously. The point is that these materials have not been properly examined.

There is much to be done in the archaeological field. There are scattered data that I have tried to assemble, but except in the case of a few isolated efforts that are always limited in time and in geographic space, these are incidental and are lost in the broader context. It must also be pointed out that the study of animal remains in Peruvian archaeology is of relatively recent interest, so that a great deal of the data on this subject lies forgotten among the refuse of archaeological excavations. The isolated efforts I just mentioned were those of Horkheimer, who was interested in the subject, but he focused on Mochica llamas, and ultimately raised many questions but did not provide any answers (see Horkheimer 1958). My own interest dates back to the same time and had its origin in the many trips I made with Horkheimer along the north coast in the 1950s, and particularly in the observations made in the long explorations of looted cemeteries in the Peruvian coast and its abandoned and destroyed ruins, undertaken with Ernesto Tabío in the 1950s and 1960s. There we saw many camelid remains, but it is only now that I am able to crystallize my thoughts in this regard, although in a fragmentary fashion. Finally, the interest of the Shimadas must be noted: since the late 1970s they have been analyzing materials that bear witness to the presence of camelids on the north coast of Peru. They emphasize the lack of archaeological data regarding this topic and the potential the different sources have to expand the subject (see Izumi Shimada 1982:161).

It was only in the 1970s that a serious and systematic study was began that sought to classify and analyze the faunal remains obtained in archaeological studies, mainly by Elizabeth Wing and her students, and to a lesser extent by Jane Wheeler. These studies will be covered in depth. However, it must be pointed out that these materials had been obtained in projects that had other goals, and in which the camelid issue was secondary. As far as I know, no specific archaeological project has ever been organized to study these animals. Nevertheless, in Appendices A and B (especially prepared for this new edition), we have short reports on the fauna MacNeish excavated in a series of caves in the Ayacucho Basin.

Terminology is one of the aspects that reflects the ignorance there is on this issue. The term camelid is actually almost unknown in common Peruvian parlance. The term auchenid is always used for the llama and its congeners, even though, as has been pointed out more than once, this term

is incorrect and must not be used, as will be seen below in the discussion of taxonomy. As it happens, the term Auchenia was used in the scientific vocabulary for more than a century before it was declared taxonomically unsound (Franklin 1982:464). Chances are that no matter what scientists do to eliminate this word, they will "not manage to halt the popular usage of a word with which llamas, alpacas, huanacos and vicuñas have been called in Peru for over a century and a half" (A:1973).

The serious issue here is that this mistake is still made in scientific circles. Thus, Armando Cardozo (1954), one of the most cited authors in the literature on camelids—unfairly in my opinion, as shall be seen throughout this book—in one of his publications uses the word "Auquénidos" in its title, and although it is explained in the text that this term "has become vulgarized" (Cardozo 1954:17), at no point is it pointed out that it must not be used. And the same term is still used in universities, and often appears in publications.

Aside from the scientific terminology, there are also problems and confusion surrounding the common names (relating to use) of these animals. This is a most complex subject in which the magical world of the Andean Indians intervenes, and where the meanings vary according to whether the animals are domestic or wild ones, or according to sex, the quality and color of the fiber, or several other points. This is a topic to which attention must be drawn, but one beyond the scope of this book. Interested readers can find ample data in a study by Flores Ochoa (1978; see also Dedenbach Salazar Sáenz 1993).

In this context of inaccurate assumptions, perhaps the one most emphasized and deeply rooted is that Andean camelids are "high-altitude animals"—therefore denying the possibility that they were able to live on the coast and in environments other than high-altitude zones.

This is a mistake that dates to Colonial times and has been systematically repeated, without any attempt to ascertain to what extent it is true. In the sixteenth century Father Acosta (1954:137) wrote: "All this livestock likes a cold weather and therefore it lives in the sierra and dies on the plains [i.e., the coast] with the heat." And in the seventeenth century Pedro de León Portocarrero (see Lohmann 1967), better known as the Anonymous Portuguese (1958:80), also said that "the rams of Peru, and these and the vicuñas are not raised on the plains [i.e., the coast])." Of course, there also are authors who in addition to making a mistake introduce fantasies, such as Neveu-Lemaire and Grandidier (1911:40), who wrote that the llama "[w]as created for the upper regions of the Andes, where the air is fresh and pure; it is heavily loaded and sent to the virgin forests, where a humid heat holds sway, or instead to the hot sands of the coast, where they find no more than a strange food, [and] millions of their kind die of exhaustion." This assertion aside—which clearly is not just an exaggeration but also an example of what can be said if reality is not known—claims of this kind have been repeated to the present day. Thus, when mentioning the llama, Mason (1962:44, 50, 58, 1964:30, 39, 47) systematically repeats in his famous handbook that "it is a native of the high lands," while Lumbreras (1993:291; claims that "[t]he four species live perfectly well in the upper parts of the Andean territory." Interestingly enough, even individuals well acquainted with the subject make mistakes of this kind. When mentioning the camelid family, Wheeler (1985a: 28) says that it groups the "animals par excellence of the Andean puna."

Other authors believe that the "adaptation" of camelids consists in being able to live in a non-high-altitude environment. A good example is given by Lumbreras (1974:102), who clearly says that "although . . . [it] is a highland animal, it appears able to adapt to coastal conditions."

The issue of camelids' potential to live in coastal zones or outside a high-altitude habitat has divided those who study the subject. It is worth citing both those who reject this possibility and those who accept and defend it to see what arguments were used, and to understand why I decided to find out more about this subject.

In the first half of the nineteenth century, Johan Jacob von Tschudi was one of the scholars who most concerned himself with this topic, and he is certainly one of its pioneers; his contribution has been taken to be "the first scientific study of the llama" (Custred 1969:121).[1] His work is of

major significance even though it has some mistakes. His ideas are worth knowing because they have been considered *verba magistri*, as we shall see.

Tschudi wrote:

These animals have never been acclimatized in the Pacific littoral, where they came and went as beasts of burden. The reports in the oldest chroniclers must be taken with much caution. It is true that llamas were also raised in the warm valleys west of the coastal cordillera, but it was exclusively in the highest and coldest parts, called "*cabeceras*," where the climate is the appropriate one and food easy [to get]. At present they have almost vanished from these locales, and the fact that their remains have been found in the tombs and ruins of Ancón, located on the coast close to Lima, is in no way proof that they ever lived permanently on the coast.

The Indians who came carrying the corpses of their relatives to bury them at Ancón brought these llamas from high in the sierra, and [these] were either buried alive together with the corpses, or dead and dismembered to be placed beside them as provisions of food. (Tschudi 1885:94, 1891:96–97, 1918:206, 1969:124–125)

Interestingly enough, Troll (1958:28), one of the most serious geographers and someone who was acquainted with the Andean area like few others, took the words of Tschudi literally, for in one of his classic studies he said the following: "According to the opinion of J. J. von Tschudi, [which was] based on a critical study of historical sources and to whose accurate knowledge of the country we undoubtedly owe the most valuable data on the natural and cultural history of the llama (57), these too never acclimatized on the coastal Peruvian deserts, but went there every now and then as beasts of burden" (n. 57 in Troll gives the studies by Tschudi in 1844–1846, 1885, 1891). We thus find that although it is true that Tschudi was a most serious scholar and that he certainly had read the main historical sources, on this point he did not base his claims on solid arguments; however, these were taken axiomatically by Troll.

There is no question that Gilmore (1950) is the author of an important study of the South American fauna one of the sources most used by archaeologists. He also does not believe that camelids may have lived on the coast, and states that "[t]he Coastal herds [in the text it is not clear whether he means the llama, the guanaco, or both] in Inca times probably never bred well or had no pastures and were maintained from Highland stock; at least they were not integrated well into the Coastal cultures, because they disappeared soon after the fall of the Inca Empire" (Gilmore 1950:433). He adds that "the coastal Peruvian aborigines are said to have had high mountain breeding grounds for their llamas; from this it can be inferred that llamas did not breed well on the Peruvian coasts, but this itself may have been a secondary result caused by the absence of a forage crop and pasture for any but the work llamas (castrated?)" (Gilmore 1950:432). We will, however, see that Gilmore accepts the presence of herds of llamas on the coast. But it is interesting that it never occurred to him that the causes of camelids' vanishing on the coast with the arrival of the Europeans were not specifically ecological ones. And it is likewise interesting that he, like other authors we meet later on, considers that that possible lack of food for these animals was a serious problem, whereas I do not consider this a real problem.

This position has, in one way or another, been held by most of the authors who, without denying the presence of llamas on the coast, believe they were not raised there but were instead brought from the highlands. So it is with Bushnell (1963:29) and Menzel (1977:42). Flannery et al. (1989:115–116) summarize this situation. They said: "Significantly, when llama trains reached sea level they often found the *yunga*, or coastal natives, as eager for the llamas themselves as for the products they bore. . . . Coastal people bought llamas for food or for sacrifice, the result being that many coastal archaeological sites have camelid bones in their refuse middens, even when no camelids were raised locally." They then add:

Llamas are found at high altitudes today because that is where they interfere the least with agriculture and where Spanish culture has made the fewest inroads. They did very well at lower elevations in prehispanic times. There are even records of herds kept by highland immigrants at coastal oases (Diez [1567–1568] 1964), but as Murra (1965:188) points out, ". . . special arrangements would have to be made to feed such coastal animals." Guanacos and llamas do well over the whole area between the puna and the *chaupi yunga* (coastal piedmont), but the hot coastal desert itself is outside their normal range.

Not all authors agreed with these claims, and some claimed that camelids were raised on the coast.

Horkheimer, as was already mentioned, was one of the advocates of this position. At first he was not so emphatic (1958:26–27), and believed that the disappearance of llamas on the coast may have been due to the "great plague (mange) of 1544–45," which was mentioned by Garcilaso. His major argument was the evidence noted by Tello (1942:607) of a considerable number of llama bones in cemeteries close to the lomas. Horkheimer thought that this was "proof that in prehispanic Peru the llama was kept at lower[-altitude] levels than today, at least temporarily" (Horkheimer1958:27). However, in a later study he was far more emphatic, for he wrote: "That animal may have lived on the coast itself where its bones have been found, sometimes in great number. The old argument that the auchenids do not procreate on the plains is null, for [some] beautiful llamas and vicuñas were recently born in Puruchuco, close to Lima, whose parents had lived before on the coastal site for several months" (Horkheimer 1961:31–32).

Another author who held this position was Lanning (1967a:17). He believed that "[t]hough it is often affirmed that these animals [he specifically referred to llamas] bred well only in the highlands, there is no good evidence to this effect. Small herds are successfully bred on the coast today. Certainly the large numbers of their bones found in coastal midden sites, and the extensive use of their wool on the coast even at periods when there is no evidence of trade with the highlands, imply that considerable herds were raised on the coast in ancient times."

For María Rostworowski (1981:61), "[t]he presence of llamas on the coast is proven," because she considers that all of the ethnohistorical data she has found are conclusive proof. In a later study, Rostworowski (1988a) dedicates almost an entire chapter to the presence of camelids on the coast, repeating her 1981 data but adding some more recent information for preceramic times.

The Shimadas have become interested in this subject but only on the north coast, where they have tried to collect documentation, which I will use later. At first their position was not too clear. In 1981 they claimed, discussing all the available data collected on llamas, that these data were "not enough proof for their being raised and herded on the coast, nor for the presence of a different coastal species" (Shimada and Shimada 1981:63). However, the following year Izumi Shimada (1982:161) published an English version of the same text, but with different phraseology. It reads: "there has been no serious attempt to document possible herding and breeding on the coast," which is altogether different. Either this is a contradiction or there is a problem in the translation, because the Spanish version came out before the English one and one can assume it is the original one. I believe that it is a mistake in the translation, because Izumi Shimada (1982:163) himself says of the llama that "their herding and breeding was possible on the coast and most likely practiced." This is consistent with a later study (Shimada and Shimada 1985) entirely dedicated to the subject that makes significant contributions, but even so the discussion is most superficial, more research is needed, and not all of the physiological data used are the most appropriate data. However, it must be pointed out that the Shimadas distinguish between llama and alpaca; whereas the llama may have "adapted" to the coast, they do not believe the same may have been true of the alpaca. Thus, in a reply to a comment made by some colleagues (see Lange Topic et al. 1987), the Shimadas claim they only suggested that the alpacas were tolerant when they were on the north coast. "Whe [sic] made no claim that alpacas were adapted to and thrived in the coastal environment, a claim that would carry strong evolutionary implications" (Shimada and Shimada 1987:837).

However, Lange Topic et al. (1987:834) consider that in general terms, the evidence presented by Shimada and Shimada (1987) is conclusive, and that they have "demonstrate[d] that the llama can be considered a coastal staple, to be included in the roster of items necessary for self-sufficiency, probably at the community level and perhaps at the household level. Additional studies are needed to uncover details of llama herding, distribution, ownership, and use. Alpacas, though, must be considered exotic to the coast, and their wool falls into the category of prestige good, procured for the benefit of the elite segments of society who oversaw its use and redistribution."

Sumar apparently also accepts this position. In the introduction to a bibliography of camelids (see Universidad Nacional Mayor de San Marcos et al. 1985) that leaves much to be desired because of its incompleteness and many mistakes, Sumar says, "The current distribution of the llama and the alpaca is circumscribed to ecological levels above 3000 masl, and this has given rise to the mistaken belief that these animals may only thrive and reproduce at high altitudes. However, the archaeozoology and ethnohistory of pre-Columbian America have shown that camelids were intensively raised and used on the coast of Peru and other regions of South America" (Sumar 1985:12). This is no more than an assertion, because Sumar does not support his position. In fact, in an article full of mistakes, Sumar (1992:84) subsequently repeated exactly the same words, merely adding "at sea level" after South America, and gives as his only source Shimada, with no date or other bibliographical evidence.

Flores Ochoa seems to have changed his mind on this topic over time. When referring to the wealth of livestock, he noted that "it was not exclusive to the altiplano, for there is evidence of great herds in other zones, and even special herds on the coast" (1977b:23). However, when discussing the work of Gilmore (1950), Rostworowski (1977), and Murra (1978), and what they claim about the presence of llamas on the coast, Flores Ochoa later said, "Perhaps it was temporary to collect guano, to carry goods for exchange, to be sacrificed at the site for ceremonial reasons, or to provide fresh meat" (Flores Ochoa 1982:63).

It should be noted that some scholars claim that the presence of camelids on the coast was essentially due to transhumance. For instance, bearing in mind that the dry season at high altitudes coincides with the wet season in the lomas, Tello (1942:19) deemed that the resources on the coastal lomas were emergency sites for highland herders. Thus, he stated that "[w]hen [the number of] the auchenid livestock was greater, as happened in antiquity, this migration must have likewise been greater, and that would explain the abundance of llama skeletons found in the cemeteries close to the lomas." Murra (1975:118–119) took a similar stance, adding some ethnohistorical data to the archaeological information. Cardich (1976:34), however, is skeptical regarding transhumance, but his arguments are not too convincing.

From this brief summary of different authors' positions regarding the possibility that there may have been breeding and raising of camelids on the Pacific coast, it is evident that little has actually been contributed. In general, these positions are based in most cases on few data, but no effort was made to search for new evidence. In truth, it must be pointed out that the Shimadas are the only ones who have broached the subject with a critical eye and have tried to indicate the problems, both existing and pending. First of all, they noted that although there apparently is a repetitive or redundant characterization of the coastal valleys, it is valid only at a most general level (Izumi Shimada 1982:183). This is something we should remember after we review the existing evidence. Second, some points must be elucidated in order to estimate the carrying capacity llamas had, essentially as a means of transport, on the coast. For example, the nutritional value of coastal forage must be determined. Our knowledge of coastal herds, their breeding and management, is still scant. Finally, we need to analyze the changes that took place throughout time in the coastal environment as a result of natural and human factors, such as coastal uplift, activities related to dunes, deforestation to make charcoal, and the burning of fields, all of which may have influenced these animals (Shimada and Shimada 1985:20). These are important suggestions that may be answered in these pages, but clearly they are just some of the many questions that remain and that the above-mentioned authors have not fully considered. Among these questions, the European presence in America since the sixteenth century is of major importance. I must note that continental uplift is the only point on which I disagree with Shimada and Shimada (1985). I discussed this before (see Bonavia 1991:37–49) and agree with Sébrier that it is difficult to make an interpretation in tectonic terms as regards the coast, for which reason he proposes an interpretation based on isostasy. However, the essential point for the subject that interests us is the climate—in other words, whether climatic changes have occurred that have affected the survivability of camelids. I do not see how the continental uplift posited by Shimada and Shimada (1985) could have had a role in it. In any case, all the evidence thus far available tends to show that since the end of the Pleistocene, the Peruvian coast, with the exception of the far north, has been subject to the same conditions of aridity as it is at present, with the sole exception of the vegetated river valleys, where an increase in river water took place during the last deglaciation and led to a rise in groundwater levels, which led to an increase in vegetation on the margins of the valleys (see Bonavia 1991). But in any case this would be an argument for the presence of camelids on the coast.

At the beginning I mentioned my own interest in this subject. It is therefore worth summarizing my point of view and the contribution I made up to the moment I began the research for this book.

In the 1970s, while finishing my study of the preceramic period at Huarmey, I reviewed the literature on the subject and published a critical review of the data on camelids of that time period (see Bonavia 1982a:392–394). From these data, from my ample experience exploring archaeological sites, and from the long discussions I had at the time with José Whittembury and Carlos Monge C., I became acquainted with the physiology of camelids; I likewise became convinced that there was an inaccurate conception of their distribution, particularly on the coast, and that these in no way were high-altitude animals. I then posited as a working hypothesis that these animals were perhaps originally distributed throughout all of the Andes, and that their ascent to high altitudes was not a natural phenomenon, that is, that they were instead forced to do so when the first hunters began to decimate them. I then wrote, "Their flight to high-altitude [areas] would not be . . . anything but a defensive mechanism" (Bonavia 1982a:394–395). These ideas were repeated in a book I completed in the 1990s (Bonavia 1991:73–74, 115–116). I was aware, however, that this would remain a working hypothesis until the existing information, both in the archaeological and the historical database, was collected and essentially checked. Here I present the data I managed to collect, and at the end I evaluate them to see what conclusions may be drawn.

I do not intend to say that the idea that forms the basis of this study is original. Others were also aware of it but for some reason did not publish their data. In this regard I had a constant source of inspiration in the work of Jorge Flores Ochoa, and I fully agree that to establish the changes in the current limits of camelid distribution with respect to the ancient limits, one must "look for information that can let us determine with some accuracy the distribution alpacas and llamas had at the time that the Spanish invasion took place; then show what the process of depopulation was like in the native pasture zones due to the pressure sheep and cattle raisers exerted, and still exert; [and] how [these] became new elements in the biotic community to compete with the native [ones] for favorable pastures and environments" (Flores Ochoa 1982:65).

So, although my original intention was to cast some light on the changing ecology of these animals and to try to explain it, I realized that in my search for data, additional materials had been gathered on related issues I had not thought of discussing. I felt I had to provide this information to my colleagues, so that it may be used by those who want to continue investigating the life and customs of camelids and their relationship with humans. The subject is so vast and complex that it cannot be studied by a single individual, and one who moreover has dedicated his life to archaeology. A multidisciplinary study dedicated exclusively to the study of camelids sensu latu must one day be organized. I will be greatly satisfied and will feel my labors well compensated if the ideas here outlined and the data collected help begin a study like the one I envision.

NOTE

1 The first edition dates to 1885. Some slight modifications were introduced to the same text on the llama in a later study published in 1891. In 1918 Urteaga and Romero first published the Spanish translation of Germán Torres Calderón, which is good, and which will be used here. However, in 1969 another translation was published that is awful, full of mistakes and omissions, and pretends to be the account "is now published for the first time in Spanish" (Custred 1969:121), ignoring the abovementioned 1918 edition by Urteaga and Romero. Since this last Spanish edition is the most accessible one and may lead to error, every time Tschudi is mentioned I will refer to all editions so that the reader may compare them.

CHAPTER 1

THE TAXONOMY AND PHYLOGENY OF THE SOUTH AMERICAN CAMELIDS

with Christian de Muizon[*]

We do not intend to dwell on taxonomic details, but it is worth summarizing some general data because these are not only useful, especially for archaeologists, but also not easy to obtain, and they clear up some of the mistakes that appear in the literature.

We were unable to get data on the very first studies of South American camelids. In Colonial times, the Crown sent the naturalist Hernández to study the natural resources of New Castille, and Hernández was probably the first scientist to describe the vicuña (Ridout 1942:401). For López Aranguren (1930a:16), however, it was Lund (1837–1843) who was the first to discuss camelids. Lund based his statements on remains found in Brazil.

Taxonomic nomenclature has changed through time, and, as we shall see, there are still some disagreements. In 1758 Linnaeus classified the llama (*Camelus glama*) and the alpaca (*Camelus pacos*), placing them in the same genus with Old World camels. The guanaco and the vicuña, the two wild New World species, were classified, respectively, as *Camelus guanicoe* by Müller in 1776 and *Camelus vicugna* by Molina in 1782. In 1775 Frisch suggested that all four South American camelids should be placed in a separate genus, *Lama*. However, the International Commission on Zoological Nomenclature met in 1953 and a year later issued Statement 258, which declared

the work of Frisch (1775, *Das Natur-System der vierfüssigen Thiere*) taxonomically invalid "because the author did not apply the principles of binomial nomenclature" (Hemming and Noakes 1958b:[First installment, Titles 1–58] Title No. 8, 2).[1] The Commission met again in 1955 and in 1956 issued Statement 39, which accepts the genus *Lama* that Cuvier (1800–1805) used in Volume 1 of his five-volume *Leçons d'Anatomie Comparé: Recueillies et publiées sous ses yeux par C. Dumeril* (Hemming and Noakes 1958a:[direction 32, 4] Title No. 18, 5).

In 1804 Tiedemann proposed the term *Lacma*, but this should be considered the oldest unused synonym; it has since been discarded and is completely unknown in the literature. In 1811 Illiger used the term *Auchenia*, which could not be used as it had already been defined by *Lama*; moreover, it was a homonym of *Auquenia*, which Thumberg had created in 1789 for a beetle genus. This was made clear in 1827, when Lesson published "the first taxonomically correct separate genus designation for the New World Camelidae, *Lama*" (Novoa and Wheeler 1984:116). However, as Franklin (1982:464) points out, this term "was used for more than a century until [it was] declared a taxonomically invalid name." Furthermore, the term "*auquénido*" is still common in South America as a synonym of camelid, particularly in Peru, and is even used in the scientific literature, although it should be discarded. In 1842 Lesson was the first to identify the vicuña as a separate genus, but this was not accepted until Miller restated it in 1924 with osteological data.

In 1940 Cabrera and Yepes established two genera, *Lama*, which included the llama, guanaco, and alpaca, and *Vicugna* which includes the vicuña. In 1952–1953 Herre placed them all in the genus *Lama*, a position adopted by most European scholars, while most North and South American scholars continued to use the two genera. Novoa and Wheeler (1984) have upheld the one-genus position, as does Kent (1987). Even so, Novoa and Wheeler (1984:116) admit that "[s]ufficient systematic research is not yet available, however, to determine if the correct classification of the llama, alpaca, guanaco and vicuña is at the species or subspecies level." Despite this, Wheeler pointed out in a 1991 review of the classification and nomenclature of camelids that Miller's 1924 study was incomplete, as he had not studied llama and alpaca incisors, which was why he left these species in the genus *Lama* together with the guanaco, "implying by default that both animals are descendants of the guanaco." Wheeler points out that alpaca incisors are similar to those of the vicuña, whereas llama incisors are similar to those of the guanaco. She quotes her own studies (Wheeler 1982[b], 1984a, 1984b[c], 1986 [this reference is wrong and should be 1985]) (Wheeler 1991:14). It was for this reason that Wheeler changed her position vis-à-vis her 1984 study (Novoa and Wheeler 1984) and stated that "the taxonomic classification of the four South American camelids could be as follows: the genus *Lama* (Cuvier 1800), with the wild guanaco *L. guanicoe* (Müller 1776), the domestic llama *L. glama* (Linnaeus 1758), the genus *Vicugna* (Molina) (Miller 1924) with the wild vicuña *V. vicugna* (Molina 1782), and the domestic alpaca *V. pacos* (Linnaeus 1758). The validity of this classification requires additional morphological and biochemical studies, but the osteoarchaeological remains from sites occupied six thousand years ago, when the process of domestication began, prove that the alpaca is a domesticated vicuña and the llama a domesticated guanaco (Wheeler 1984a and b[c], 1986 [a mistake, it should be 1985]; Wing 1977[a], 1986)" (Wheeler 1991:14–15). Molecular analy-

ses based on the mtDNA sequence (Stanley et al. 1994) later made Wheeler change her position once more. In 1995 she accepted two genera—*Lama*, with three species (*glama*, *pacos*, and *guanicoe*), and *Vicugna*, with one species (*vicugna*) (Wheeler 1995).

Franklin (1982:464) believes the subdivision into two genera posited by Cabrera and Yepes in 1940 (1960:73) should still be maintained. In this regard, Koford (1957:154–155) wrote,

> The phylogenetic classification of the lamoids seems to be indeterminate, at least on the basis of structural criteria. Of the group, the llama and guanaco are the most similar, and the vicuña, although resembling the alpaca in some characteristics, is the most distinct. It has been variously classified as specifically, subgenerically, or generically separate from the other lamoids. The characteristic, unique among living artiodactyls, that best supports generic separation of the vicuña is its peculiar lower incisors. In *Vicugna* these are very long with parallel sides, enamel on but one face, and open root. (Miller 1924:2).

Koford then noted that another characteristic distinguishing the vicuña from other lamoids is some kind of long-haired bib hanging from its chest.

The four forms were also described by Edgardo Pires-Ferreira (1979 and 1981–1982) as *Lama glama* subspecies and variants of *L. g. glama*, incorrectly using the domestic form according to Novoa and Wheeler (1984). This study is lacking in scientific seriousness (Hernando de Macedo, pers. commun., 23 September 1991) and should not be considered further.

The South American wild camelid subspecies and the variants of the domestic species have been described. Four guanaco subspecies are known: *Lama guanicoe cacsilensis* (Lomberg 1913) in southern Peru, Bolivia, and northeastern Chile, which, as we shall see, has been questioned by several scholars; *Lama guanicoe huanacus* (Molina 1782), on the western slopes of the Chilean Andes; *Lama guanicoe voglii* (Krumbiegel 1944), in the drylands and plains of Argentina and perhaps Uruguay, north of 32° S latitude; and *Lama guanicoe guanicoe* (Müller 1776), south of 35° S latitude (or 38° according to Torres 1922a), in Patagonia and Tierra del Fuego. The subspecies that occupies the continent farther north, in northern Peru (8° S latitude), is unidentified. According to Wheeler (1991:17), today only *L. g. guanicoe* and *L. g. cacsilensis* are accepted, but she points out that "there must be other subspecies or geographical races still unknown to science."

There also are two vicuña subspecies that have the same color, according to most authors: *Vicugna vicugna mensalis* (Thomas 1917), found in southern Peru, western Bolivia, and northern Chile, and the Argentinian vicuña, *Vicugna vicugna vicugna* (Molina 1782), found on the eastern Andean slopes south of 18° S latitude. According to Torres (1992a:32), this subspecies is larger and lighter in color than the other subspecies.

Two phenotypic varieties of the alpaca have also been distinguished in terms of their wool and body size. *Huacaya* has short wavy wool hairs, and *suri* has long wavy hairs. Two phenotypic varieties of the llama are essentially distinguishable, the *chaku* (also *ch'aku*) or woolly llamas, and the *ccara* (or *q'ara*) llamas, which have short hair. Even so, Wheeler (1991:28) indicates that "it is possible that others still remain unknown." For the data presented here, the reader should turn to the studies of Thomas (1891), Krumbiegel (1944), Gilmore (1950), Franklin (1975; 1982:464–465), Novoa and Wheeler (1984:116), Torres (1992a: 32–33), and Wheeler (1991). There is a folk taxonomy that mainly concerns the crossing of different camelids. Interested readers should see the study by Flores Ochoa (1977b).

One of the reasons for this confusion is a fact that has been asserted frequently in the scientific literature—the possibility that crosses among these four camelids took place. It has been said that these do not occur, or that their offspring are sterile when they do try to breed. This is not so. There is much evidence that these crossings take place between pure and hybrid kin. Besides, all four New World camelid species have the same 2n = 74 karyotype (see Novoa and Wheeler 1984: 116, who relied on the studies made by Gray 1972 [sic]; Fernández Baca 1971; Hsu and Benirschke 1967, 1974; see also Wheeler [1991:33]. We should bear in mind that the date for Gray is given as 1972 in Novoa

and Wheeler [1984], whereas Wheeler [1991] gives 1954, which is the correct date.) According to Sumar (1988:23),[2] the only exception is the cross "between two wild species (vicuña × guanaco)," which apparently has not been reported in the literature (to support his statement, Sumar cites León, but without giving a date [this must be his dissertation; see León 1932b], Fernández-Baca 1971, and MacDonagh 1940). Wheeler (1991:36) considers that most of these crossings were carried out by man and do not occur naturally. Kent (1987: 172) discusses this point (see below).

It is now clear that the solution to many of these problems will come with genetic studies. However, DNA and molecular data are wholly unknown for the Camelidae family (Larramendy et al. 1984:95). According to Larramendy et al. (1984: 93), the data available in the literature indicate that the chromosomal makeup of the Old World camelids is similar to that of the New World ones (n = 74). Even so, these scholars indicate that all of these studies (they refer to those of Taylor et al. 1968) were carried out using conventional cytogenetic techniques to determine chromosome morphology and number. Today these techniques are not considered adequate to establish the degree of chromosome homology between species.

> The remarkable phenotypic divergence, the acquisition of different adaptive characteristics, and the distribution pattern exhibited by the members of the Camelidae family have led to a systematic subdivision of the taxon into six species (Pieters 1954; Walker 1964). So far, the data for the biological characterization of each species are scant and there are no comparative interspecies studies, nor any phenotype-genotype correlation studies that would provide a basis for this hierarchical distribution. (Larramendy et al. 1984:93)

These scholars have established the chromosomal makeup of the llama and the guanaco and measured the degree of chromosomal affinity between both. According to them, "a modal number of 2n = 74 chromosomes was established for the llama and the guanaco, with 74 bibranchial elements for the females and one single telocentric element for males. . . . These results indicate that at the level of conventional cytogenetic techniques, the llama and guanaco share a karyotype of similar characteristics. The individualization of different pairs of homologous chromosomes and the preparation of karyotypes with G banding was attained through an analysis of 15 stained metaphases per species. A comparison of the various karyotypes allowed the establishment of a G-banding pattern and the ideograph for each species.... This analysis showed that both patterns coincide, thus establishing a 100% chromosomal homology between the llama and the guanaco" (Larramendy et al. 1984:94–95). These same researchers noted that the DNA of llamas and that of vicuñas exhibit similar behavior and have similar makeup values (1984.:96). In conclusion, they wrote,

> A remarkable morphometric similarity between llama and guanaco chromosomes was established through the use of cytogenetic techniques. The karyotypic comparison, aided by G banding, likewise revealed that the similarity between both species reaches to the level of chromosome segments. The karyotypic uniformity observed therefore indicates that the phenotypic differentiation reached by these species was not accompanied by rough chromosome rearrangements. If any chromosomal differences are present, these would lie below the resolution level of the techniques currently used.

> A variable number of DNA fractions with a high content of G-C has been detected in several species of artiodactyls, which form bands of satellite DNA when subjected to ultracentrifugation in cesium chloride gradients (Curtain et al. 1973; Kurent et al. 1973). The remarkable species-specificity of these DNAs has led to their being widely used as molecular markers in phylogenetic studies. In camelids it has yet to be established whether the thermostable fractions described correspond to satellite DNA, and whether they are universally present throughout the group or are an element of differentiation between some of its members. (Larramendy et al. 1984:96)

According to Sumar (1988:23–24), other researchers have studied this same problem, and he in fact cites Bunch and Foote (there are mistakes and gaps in the bibliography, just as in most of Sumar's studies; in this case, no date is given), who, according to Sumar, have found a similarity between llama and Bactrian camel karyotypes; the band patterns of individual chromosomes are apparently similar. This seems to agree with the results obtained by Larramendy et al. (1984:96). Sumar goes further and "wonders about the possibility of a cross between Old and New World camels." I am not a specialist, but, in view of the study by Larramendy et al., it seems that the problem is far more complex than it might appear at first sight, particularly if the conclusions of the aforementioned study are read carefully. However, newspaper articles (Anonymous 1998a, 1998b, 1998c) report the successful crossing of a camel and a llama. The details of how pregnancy was achieved are unknown. All that is noted is that "two years of research were needed to achieve conception through artificial insemination" (Anonymous 1998a:A1). This new experiment was carried out in the Dubai Camelid Reproduction Centre, United Arab Emirates, under the direction of Julian Skidmore. The newly born "looks more like his father the camel, has small ears and a large tail just like camels, but with two-toed feet like the llamas" (Anonymous 1998a:A1). From the available data, it is difficult to speculate on the practical consequences of this experiment.

Going back once more to the problem of crosses between the various camelid forms, and given the doubts and mistakes that have arisen, I believe it is worthwhile mentioning actual evidence. For example, crossing the alpaca and the vicuña (paco-vicuña) has been known "indisputably since 1840, when it was carried out by Juan Pablo Cabrera, a priest in Carabaya (Puno), who performed the operation with industrial goals in mind" (León 1939:102). In 1864, when Raimondi was studying the province of Carabaya, he noted that "[f]rom Crucero (the provincial capital) I set out for the town of Macusani and passed through the hamlet of Acoyani, where I was able to see a small number of graceful paco-vicuñas, the remains of the herd grown thanks to the care of the priest Cabrera, who seeking to combine in one individual the fine wool of the vicuña and the abundant wool of the alpaca, crossed these two animals and obtained the hybrids called paco-vicuñas, thus achieving his objective to a great extent" (Raimondi 1874:177). Hoffstetter (in a personal communication to Larry Marshall et al. 1984:21) also verified that *Lama* and *Vicugna* are interfertile. Kent (1987: 172) is also emphatic in this regard: "All of the camelids interbreed occasionally, and most crosses occur without any human intervention. Most of the hybrids are fertile (Fernández-Baca 1971, 1978)." Interestingly, this researcher believes that "[t]his success may be relatively recent, and the result of deliberate and intense human selection during the early twentieth century (see Kent 1982[a]:23–29)" (Kent 1987:172). Flores Ochoa (1982:66) also reports these crosses and points out that "[a]t present they can take place . . . between the domestic species. The most common crosses are the so-called *paqo-vicuña*, an offspring of the alpaca and the vicuña; the *wari*, a cross of llama and alpaca with two variants called *paqowari* and *llamawari*, according to whether the phenotypic or external characteristics of the alpaca or llama prevail. It is possible that llamas and alpacas breed with the guanaco in natural conditions" (see also Wheeler 1995:289–290).

I believe that an effort to summarize camelid fossil remains, with their synonyms, will be of interest to nonspecialists, even though the terminological complexities often confuse those who are not familiar with the subject. Cardozo (1975a) wrote a summary, but it was disorganized, had mistakes, and had not been proofread, which is why I decided not to use it and do not recommend it to others. Instead, I rely mainly on Webb (1974) and Hoffstetter (1952).

The animals I am discussing belong to the Order Artiodactyl, Suborder Tylopoda, Family Camelidae, Subfamily Camelinae, Tribe Lamini (Camelini, the other tribe, encompasses the Old World camels).

Lama Cuvier (1800)

This is a synonym of *Auchenia* Illiger (1811); nec Thunberg (1789), *Mesolama* Ameghino (1884), *Stilauchenia* Ameghino (1889), and *Neoauchenia* Ameghino (1891).

Both Webb (1974:173) and Hoffstetter (1952: 314) accept three species: *L. glama* Linné [Linnaeus] 1758, llama, found in the Pleistocene of Argentina and Bolivia and as a domesticate in the Holocene of Argentina, Chile, Bolivia, Peru, and Ecuador. *L. guanicoe* Müller 1776, guanaco or huanaco, found in the Pleistocene of Argentina and, in the Holocene, in the south and west of South America as far as Peru. *L. pacos* Linné [Linnaeus] 1758, alpaca, found in the Pleistocene of Argentina and as a domestic animal in the Holocene in Bolivia and Peru.

According to Cabrera (1931), both researchers give two extinct species: *L. owenii* and *L. angustimaxila*.

Lama owenii
(H. Gervais & Ameghino 1880)

Found in Late Pleistocene deposits in Argentina, it has as synonyms *Palaeolama owenii* H. Gervais & Ameghino 1880, *Auchenia weddelli* P. Gervais 1855, in Ameghino (1889, in part), *Palaeolama leptognatha* Ameghino 1889, *Stilauchenia owenii* H. Gervais & Ameghino 1880, in Ameghino 1889, *Lama gigantea* López Aranguren 1930(b).

Lama angustimaxila
(Ameghino 1884)

Found from the Early Pleistocene to the Sub-Recent of Argentina and Bolivia, it has three synonyms: *Mesolama angustimaxila* Ameghino (1884), *Palaeolama weddelli* P. Gervais (1855) in Boule & Thévenin (1920, in part), and *Lama angustimaxila* (Ameghino 1884), in López Aranguren (1930[b]).

Vicugna Gray (1872)

There is only one species, *V. vicugna*, found in sites ranging from the Middle Pleistocene to the Sub-Recent of Argentina and Bolivia. Four synonyms are accepted for it: *Auchenia* (H. Gervais & Ameghino 1880, in part), *Palaeolama* (Ameghino 1889, in part), *Hemiauchenia* (Ameghino 1891, in part), and *Lama* (López Aranguren 1930[b], in part).

Eulamaops Ameghino (1889)

Known only from the Late Argentinian Pleistocene, it has only one species, *E. parallelus*. *Eulamaops* has three synonyms: *Auchenia parallela*

Ameghino 1884, *Eulamaops parallelus* Ameghino 1889, and *Palaeolama brevirostris* Rusconi 1930(b). However, Hoffstetter (1952:315) has some doubts about the latter synonym.

Palaeolama
P. Gervais (1867)

This genus is found in lands that range from the Middle to the Late Pleistocene of Bolivia, Ecuador, and Peru and north to Florida. In this case Webb and Hoffstetter disagree over the synonyms. They agree only about *Protauchenia* Branco 1883 (Webb 1974:175; Hoffstetter 1952:316). Webb (1974:175) in turn gives the following as synonyms: *Tanupolama* (Simpson 1928), *?Tanupolama* (Simpson 1932), *Palaeolama* (Hoffstetter 1952), *Tanupolama* (Bader 1957), and *Palaeolama* (*Astyolama*) (Churcher 1965). Meanwhile, Hoffstetter (1952:316) gives as synonyms *Auchenia* (Lund 1843; P. Gervais 1855; Liais 1872; Winge 1906), *Lama* (López Aranguren 1930[b], Cabrera 1931), *Palaeolama* (P. Gervais 1867), *Palaeolama* (H. Gervais & Ameghino 1889; Ameghino 1889; Boule & Thevenin 1920; Rusconi 1930[b], 1931), and *Hemiauchenia* (H. Gervais & Ameghino 1880).

Hoffstetter also indicates that the only form of *Palaeolama* from the pampas has also received the following names: *Palaeolama weddelli* P. Gervais (1867), *Palaeolama weddelli* (P. Gervais 1855) in part in H. Gervais & Ameghino (1880), *Palaeolama major* H. Gervais & Ameghino (1880), *Hemiauchenia paradoxa* H. Gervais & Ameghino (1880), *Palaeolama leptognatha* Ameghino (1889), *Hemiauchenia leptognatha* Ameghino (1889), *Hemiauchenia paradoxa* H. Gervais & Ameghino (1880) in Boule & Thévenin (1920), *Palaeolama weddelli* (P. Gervais 1855) in López Aranguren (1930[a]), *Hemiauchenia major* (Liais 1872) in López Aranguren (1930[a]), *Lama major* (H. Gervais & Ameghino 1880) in López Aranguren (1930[b]), *Hemiauchenia paradoxa* and the subspecies *elongata* Rusconi (1930[b]), *Palaeolama weddelli* (P. Gervais 1855) in part in Rusconi (1930[b], 1931), and *Palaeolama weddelli* (P. Gervais 1855) in part in Cabrera (1931, 1935). Hoffstetter concludes (1952:317) that the valid name for the Argentinian species is *Palaeolama paradoxa* (H. Gervais & Ameghino 1880).

There are remarkable disagreements over the *Palaeolama* species. Cabrera (1931:114) accepted only one species. Hoffstetter (1952: 317–320) for his part accepted six (and a seventh with reservations), while Webb (1974:175–196) only considers three.

The only species accepted by Cabrera (1931: 114) is *P. weddelli* (P. Gervais), while the species approved by Hoffstetter (1952:317–320) are *P. paradoxa* (H. Gervais & Ameghino 1880), *P. weddelli* (P. Gervais 1855), *P. major* (E. Liais 1872), *P. reissi* (Branco 1883), *P. crassa*, and *P. aequatorialis*. Hoffstetter (1952:320) has not passed judgment on *P. brevirostris* Rusconi (1930[b]).

Like Hoffstetter, Webb (1974:175–176) accepts *P. weddelli* and *P. aequatorialis*, but adds *P. mirifica* (Simpson 1929), which is not mentioned by Hoffstetter. However, Webb disagrees with Hoffstetter's remaining species. Thus, the *P. paradoxa* accepted by Hoffstetter (1952) is for Webb (1974: 198) a synonym of *Hemiauchenia paradoxa*, *P. major* is a synonym of *Hemiauchenia major* (Webb 1974: 199), *P. reissi* is a synonym of *P. weddelli* (Webb 1974:176), and *P. crassa* is likewise a synonym of *P. weddelli* (Webb 1974:177). He does not mention *P. brevirostris*, which Hoffstetter did. (To prevent confusion I must point out that Webb gives the date for Cabrera as 1932 [1974:248, repeatedly quoted throughout the text], whereas it is 1931.)

Hemiauchenia
H. Gervais & Ameghino (1880)

This genus is found from the Middle to the Late Pleistocene in North America, and from the Uquian to the Lujanian (i.e., throughout the Late Pleistocene) in southeastern South America.

A great number of synonyms are acknowledged for this genus: *Auchenia* (Lund 1842, in part), *Auchenia* (P. Gervais 1869, in part), *Auchenia* (Liais 1872, in part), *Auchenia* (Cope 1878, in part), *Palaeolama* (H. Gervais & Ameghino 1880, in part), *Hemiauchenia* (H. Gervais & Ameghino 1880), *Holomeniscus* (Cope 1884, in part), *Holomeniscus* (Cope 1893, in part), *Camelops* (Wortman 1898, in part), *Camelus* (Wortman 1898, in part), *Procamelus* (Hay 1921, in part), *Lama* (Meriam & Stock 1925, in part), *Tanupolama* (Stock 1928), *Hemiauchenia* (López Aranguren 1930[a]), *Palaeolama* (López Aranguren

1930[a], in part), *Palaeolama* (Cabrera 1931 and 1935, in part), *Lama* (Kraglievich 1946, in part), *Palaeolama* (Hoffstetter 1952, in part), *Palaeolama* (*Palaeolama*) (Churcher 1965, in part).

Six species are acknowledged for *Hemiauchenia*, each with a great many synonyms. They are listed below.

Hemiauchenia paradoxa H. Gervais & Ameghino (1880). It is known from the Uquian or Chapadmalalan (Early Pleistocene) through Lujanian (Late Pleistocene) in Argentina. It is a synonym of *Palaeolama weddelli* (P. Gervais 1855) in P. Gervais (1867, in part), *Hemiauchenia paradoxa* H. Gervais & Ameghino (1880), *Palaeolama weddelli* (H. Gervais & Ameghino 1880), *Palaeolama major* H. Gervais & Ameghino (1880), nec *Auchenia major* Liais (1872), *Palaeolama leptognatha* Ameghino (1889), *Hemiauchenia paradoxa elongata* Rusconi (1930[b]), *Palaeolama weddelli* (P. Gervais 1855) in Rusconi (1930[b]), *Palaeolama weddelli* (P. Gervais 1855) in López Aranguren (1930[a]), *Hemiauchenia major* (H. Gervais & Ameghino 1880) in López Aranguren (1930[a]), *Palaeolama weddelli parodii* Rusconi (1936), *Lama* sp. Kraglievich (1946), and *Palaeolama paradoxa* (H. Gervais & Ameghino 1880) in Hoffstetter (1952).

Hemiauchenia major (Liais 1872). It comes from Late Pleistocene deposits in the caves of Lagoa Santa, Minas Gerais, in Brazil. Its synonyms are *Auchenia minor* Lund (1843, *nomen nudum*), *Auchenia major* Liais (1872), *Auchenia major* (Lund [sic]) in Winge (1906), *Hemiauchenia major* (Liais 1872), in López Aranguren (1930[a]), *Palaeolama weddelli* Cabrera (1931, in part) and *Palaeolama* major (Liais-Winge 1906) in Hoffstetter (1952).

Hemiauchenia vera (Matthew). This is essentially a species that belongs to the Hemphillian deposits of the United States. It has five synonyms: *Pliauchenia humphreysiana* Cope in Wortman (1898), *Pliauchenia vera* (Matthew and Osborn 1909), *Tanupolama vera* (Matthew) in J. T. Gregory (1939), *Tanupolama vera* (Matthew) in Hibbard (1963), and *Tanupolama vera* (Matthew) in Webb (1965).

Hemiauchenia blancoensis (Meade 1945). This genus also comes from the North American Blancan deposits. It has only three synonyms: *Tanupolama* cf. *T. longurio* (Hay 1921) in Gazin (1942),

Tanupolama blancoensis Meade (1945), and *Tanupolama blancoensis* Meade (1945) in Hibbard and Riggs (1949).

Hemiauchenia seymourensis Hibbard & Dalquest (1962) is a North American genus like the previous one from the late Kansas strata. A single synonym is ascribed to it: *Tanupolama seymourensis* Hibbard & Dalquest (1962).

Hemiauchenia macrocephala (Cope 1893). It is also found only in North America and belongs to the Middle Pleistocene. A great number of synonyms are attributed to it: *Holomeniscus macrocephalus* Cope (1893), *Camelus americanus* Wortman (1898), *Camelops vitakerianus* (Cope 1878) in Wortman (1898 in part), *Lama stevensi* Merriam & Stock (1925), *Tanupolama stevensi* (Merriam & Stock 1928) in Stock (1928), *Proauchenia americana* (Wortman 1898) in Frick (1929), *Lama* (?) *hollomani* Hay & Cook (1930), *Tanupolama hollomani* (Hay & Cook 1930) in Meade (1953), *Tanupolama macrocephala* (Cope 1893) in Hibbard and Dalquest (1962), and *Tanupolama stevensi* (Merriam & Stock 1925) in W. E. Miller (1968).

To sum up: I find that the arrangement of the Lamini tribe can be formally summarized as follows: *Pliauchenia* Cope (1875) in the Early and Middle North American Pliocene; *Hemiauchenia* H. Gervais & Ameghino (1880) (= *Tanupolama* Stock 1928), from the Middle Pliocene to the Late Pleistocene of North America, and in South America from the Early to the Late South American Pleistocene; *Palaeolama* P. Gervais (1869) (= *Protoauchenia* Branco 1883; *Palaeolama* [*Astylolama*] Churcher 1965), from the Middle to Late Pleistocene of southern North America and northern South America; *Eulamaops* Ameghino (1889), in the South American Late Pleistocene; *Vicugna* Gray (1872) from the Middle to the Recent Pleistocene of South America, and finally *Lama* Cuvier (1800), which is found in the Early to Recent Pleistocene in South America (see Webb 1974:211).

Owing to disagreements among taxonomists and paleontologists over camelid systematics, one of the most controversial points is the origin of the modern forms of these animals. Several scholars have summarized this problem. Some have focused on discussing the domestic forms (llama and alpaca), while others have considered the problem in toto, that is, including wild and domestic forms. In the first group we have Wheeler (1984a:405; 1984b; 1985b:78) and Franklin (1982:464), in the second group Kent (1987:171–172; 1988a:26–28). I believe the latter is the clearest and best-documented summary.

As Kent (1987:171) has pointed out, a result of this confusion is the multiplication of possible phylogenies, which can be arranged into four major groups (see Figure 1.1):

1. A wild guanaco is the ancestor of the modern guanaco, the modern llama, and the modern alpaca. Scientists who accept this position are Thomas (1891), Cabrera (1922), Cook (1925), Mann (1930), Strooks (1937), Gilmore (1950), Herre (1952, 1961, 1968), Fallet (1961), Zeuner (1963), Jungius (1971), Herre and Röhrs (1977), and Otte and Venero (1979). It should be noted that Hemmer (1975, 1976) agrees as far as the llama is concerned but disagrees on the alpaca (see below, third point). It is worth pointing out that Flannery et al. (1989:89) report that Wing (1977a) accepts that "the llama's wild ancestor was the guanaco." However, Wing's position is not as conclusive as has been said, for she wrote, "The guanaco is wild, and *may be* close to the wild ancestor of both domestic forms" (i.e., the llama and the alpaca; Wing 1977a:847; emphasis added).

2. The wild guanaco is the ancestor of the modern forms of the guanaco and llama, while the wild vicuña is the ancestor of the present-day vicuña and alpaca. This position is defended by Burmeister (1879), Ameghino (1889), Antonius (1922), Latcham (1922), Krumbiegel (1952), Steinbacher (1953), Capurro et al. (1960), Wheeler (1985b), Flannery et al. (1989), and Bustinza (1970a). Otte and Venero (1979) in turn categorically deny that the vicuña was the ancestor of the alpaca.

3. A wild guanaco is the ancestor of the modern guanaco and llama forms, while a cross of llama and vicuña would have produced the modern alpaca. This is the position held by Hemmer (1975).

4. The present-day llama descends from a wild, extinct llama type, while present-day alpacas descend from a wild, extinct alpaca type. In

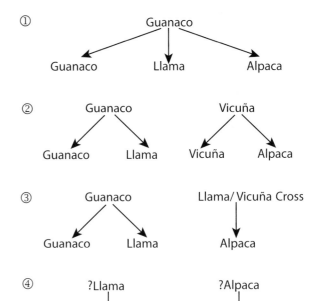

FIGURE 1.1. Hypotheses explaining the phylogeny of domesticated South American camelids.

other words, the ancestors of the llama and the alpaca were wild forms of the same respective type whose fossil remains have been found in sites dating to the Middle and Recent Pleistocene (see below). This was proposed by López Aranguren (1930b). However, she accepts that the fossil remains of *Lama pacos* are few, and considers instead that the *Lama gracilis* species "could well have been the ancestor of *pacos*, for I have not found fossil remains of the alpaca, save for a subfossil which Ameghino classified as *A. lujanensis*" [López Aranguren 1930b:116; see also 118–119]), Cabrera (1931) and Cabrera and Yépez (1940). However, Tonni and Laza (1976) have rejected this possibility, and so has Wheeler (1995:282).

Wheeler (1993:17–19, and 1995:281–282, Fig. 2) prepared an overview presenting the proposed camelid phylogenies. Although she does expand the database used by Kent (1987:171), the essential statements have not changed, and the four major hypotheses remain, as summarized in Figure 1.1.

Stanley et al. (1994) discuss the origins of the llama and the alpaca on the basis of a molecular evolution analysis supported by the study of mitochondrial DNA. They conclude that hybridization cannot have taken place in the wild species, whereas it did take place in the domestic ones. The authors suggest that this hybridization happened after the Spanish conquest, and that this leads to a misinterpretation of the data obtained in living animals. In this regard, they wrote:

> [W]e can conclude . . . that hybridization among the domestic camelids has occurred at some time in the past. Therefore, although we can eliminate the hypothesis that no hybridization has occurred, three interpretations could be made from the data for the alpaca. It is possible that the alpaca did indeed originate as a domesticated form of the vicuña but that hybridization has occurred since, resulting in the mix of genotypes observed. Alternatively, the alpaca may have originated as a llama-vicuña cross. In this case, either genotype could originally have been present in the alpaca, depending upon the female in the cross. A male llama crossed with a female vicuña may be more likely (based on size alone), but either of these crosses may have occurred, together with backcrossing to llama (or guanaco) in the F2 or subsequent generations. Finally, we cannot eliminate the possibility that the alpaca is a second domesticated form of guanaco, although incisor morphology would not support this conclusion (Wheeler 1991). Data from the mitochondrial genome alone are therefore insufficient to allow an unambiguous determination of the relationship of the domestic animals to their wild ancestors. (Stanley et al. 1994:4)

We are still far from a solution to this problem, and this confirms to a great extent (points 1, 2, and 3) the position of Kent (1987:171–172) presented above.

A later study by Kessler et al. (1996) restates the results from Stanley et al. (1994). Kessler et al. (1996:269, 271–272) wrote: "The results support the hypothesis that llamas were originally derived from the guanaco since the animals studied share the same cytoplasm. In the alpaca population, on the other hand, since mitochondrial DNA is maternally inherited, this would not exclude crossbreeding events between all four groups in the DNA studied." They conclude, "Our results show that the mtDNA's [sic] found in guanacos is identical to llama mtDNA. Some mtDNA haplotypes observed in the alpaca were also detected in the vicuña. In addition the alpaca contains mtDNA haplotypes identical to guanaco mtDNA. These data are in agreement with results presented by Stanley et al., 1994, which imply that alpacas are of mixed maternal origin."

It is worth pointing out that Tschudi (1846) posited that the four camelids represent independent species, a position later accepted by Pocock (1923). Although I have followed Kent (1987) in this synthesis, I have also augmented the database.

From what has been presented, and still following Kent (1987:172), I conclude that present-day domestic llamas descend either from the guanaco or from a llama-like form that lived in the wild in the Pleistocene. In the case of the alpacas, these descend either from the guanaco; or from a similar wild Pleistocene form; or from a llama, which in turn descends from a guanaco, so that the alpaca is the outcome of a cross between a vicuña and a llama; or the alpaca descends from the vicuña.[3]

NOTES

*   Christian de Muizon assisted the author with the taxonomic nomenclature.

1   Novoa and Wheeler (1984:116) give 1950 as the year when the International Commission on Zoological Nomenclature made its decision. This is a mistake, for Wheeler (1991:14) herself later gives the correct date. There is also another slip that must be cleared up. Wheeler (1991:14) states that the work done by Frisch was rejected "because the principles of binomial nomenclature were lacking," whereas Statement 258 reads *ad litteram* "because the author *did not apply* the principles" (emphasis added). I also believe that publication of the decisions of the International Commission on Zoological Nomenclature should be credited to Hemming and Noakes, and not just to the former, as Wheeler did (1991:14, 44).

2   The article by Sumar should be used with due caution because his citations are incomplete. Besides, with regard to the subject under consideration, he clearly

copied almost to the letter Novoa and Wheeler (1984), who figure in his bibliography but are not mentioned in this case. I cite Sumar's words because the references he mentions are hard to find.

3 Kessler et al. (1996:269) quite rightly note that "[t]he evolution of domesticated South American Camelids, Llama, and Alpaca, is still a matter of controversy." New data from Kadwell et al. (2001) suggest that the alpaca is descended from the vicuña and should be reclassified as *Vicugna pacos*.

CHAPTER 2

NOTES ON THE BIOLOGY OF THE
THE SOUTH AMERICAN CAMELIDS

with the collaboration of Carlos Monge C.*

2.1 INTRODUCTION

This chapter does not attempt to provide a comprehensive review of all the biological characteristics of camelids. Instead, we mention only those aspects we believe are relevant to the discussions that occur throughout the book and of interest to researchers who study these animals mainly from a cultural perspective.

As is well known, camelids live in Central Asia, in western South America, and in parts of Africa, from the Mediterranean to Senegal to the Atlantic and Indian Ocean (*Enciclopedia de los Animales* 1970:289). As we saw in Chapter 1, they are divided into two tribes, the Lamini and the Camelini. Both tribes have the essential ruminating processes, but several characteristics distinguish them from the Suborder Pecora (ruminants), including their stomach morphology, the absence of horns or antlers, the presence of true canines separated from the premolars by a diastema (or gap), the anatomy of the hind legs, which allows camelids to rest on their belly with knees bent and paws pushed backward, and the presence of callous pads ending in claws instead of hooves (Wheeler 1991:12). The following discussion considers only South American camelids.

The wild and domestic camelids are clearly the most important and largest native mammal herbivores of South America. As Franklin (1982:457)

notes, their ecological domain, their uniqueness, and their major contribution to mankind in the past and the present are unmatched.

Although the South American mammalian fauna is extremely rich in rodents and bats, big herbivores are missing. For this reason, South America is one of the best examples of an ecologically "unsaturated" fauna: it has few large mammals (Keast 1972; Simpson 1962). There are only about 19 species of wild native ungulates (artiodactyls and perissodactyls: three tapirs, three peccaries, two camelids, and 11 deer), a very small number in comparison with the 93 species in Africa (five horses, two rhinoceroses, three suids, two hippos, one tragulid, two giraffes, and 78 bovids). This low number is surprising, because South America is about 60% the size of Africa, and about 20%–25% of the continent offers habitat preferred by the great herbivores, that is, woodlands, grasslands, steppes, or scrub. Less than 3% of South American mammalian species are terrestrial herbivores of medium to large size.

Looking at a map, one would expect to find great concentrations of herbivores in the pasturelands of the pampa, the *páramo*, or the puna, or in the steppe/scrubland of Patagonia. However, only four wild ungulates (two cervids and two camelids) occur in all the land comprised by the altiplano, the pampas, and Patagonia. Camelids are the most important of these ungulates because of their distribution, number, size, and cultural and commercial value.

Camelids have long been a dry-environment species. Both the wild and domestic South American camelids are the dominant large herbivores on the high plains of the Andes and the dry montane slopes, as well as in the dry lowlands of the steppes and plains.

It is worth bearing in mind that pastoral habitats like the savannas and the deserts probably are more widely distributed in South America than in any other vegetation area (Goodland 1966). Almost all of Argentina and Uruguay, at least half of Bolivia, Chile, and Venezuela, about a fifth of Brazil and Colombia, all of the coast, the western highlands, and the puna of Peru, the highlands of Ecuador, and a large part of the Guyanas are desert, tundra, meadow, pampa, prairie, scrub steppe, or savanna. Hershkovitz (1972) included most of these lands in what he called the Patagonian Subregion, which comprises the *páramo* and the southern parts of Ecuador, the desert coast and the puna regions of Peru, the alpine highlands of Bolivia, all of the Chilean alpine highlands, the pastoral plains of Uruguay, all of the Argentinian plains except the northeastern subtropical forest, and the pastoral areas of the Argentinian, Bolivian, and Paraguayan Chaco.

The mammalian fauna of the Patagonian Subregion is comparatively poor and mostly insular. The characteristic mammals are relics of marsupials, edentates, and caviomorph and cricetine rodents. The latter two groups constitute 52–76 genera.

Throughout the Tertiary, the American continent had a rich and diversified fauna of endemic ungulates and great nonungulate herbivores. In the Early Pleistocene there was a fauna with big herbivores that, according to Keast (1972), was as rich in its major categories as that of present-day Africa. Keast believes that the pampas have changed since the Pleistocene (or Pliocene) in such a way as to favor the invasion and dominance of caviomorph and cricetine rodents rather than ungulates. The camelids are the only native ungulates throughout most of the area, and they represent the specialized pastoral forms of this Patagonian Subregion (Hershkovitz 1972; see Franklin 1982:457, 471).

2.2 HABITAT

In the Central Andes, camelids have long been associated with the puna. These are high-altitude grasslands, between 3900 and 4600 meters above sea level (masl). Whole series of plant formations develop there (see Tosi 1960; ONERN 1976). Known as classic puna, these grasslands essentially lie across the highlands that extend from central Peru to western Bolivia, northwestern Argentina, and northern Chile. The puna is a unique biological zone because of its high altitude, its proximity to the equator, and the fact that it extends from the tree line to the snowline. Its topography is quite varied and includes great plains, wide open valleys, flattened cordilleras, rolling hills, rocky bluffs, and jagged peaks.

The puna is a cold, dry, and often windy environment. Precipitation occurs mostly in summer, and at night the temperature almost always falls below freezing. There are two seasons on this alpine prairie where no trees grow: a temperate season with rainfall in December–April, and a cold, dry season from May to November. From the data obtained at Pampa Galeras (see Franklin 1978, 1983), we know that the typical mean annual temperature ranges between 5.6°C and 5.5°C, with an annual rainfall ranging between 400 and 700 mm. The humidity appears as rain and hail, and only occasionally as snow. The cloud cover from late spring to the summer storms has a profound effect on the minimum night temperatures, which in turn affect the length of the growing season and, finally, the scant production of plants in the puna (Franklin 1978, 1983). There are several comprehensive studies on plants, communities, and vegetation types (see Franklin 1978, 1983; Tosi 1960; ONERN 1976; Tovar 1973).

The dry and windy Andean slopes of Patagonia are also an important habitat for the South American camelids. Patagonia is a high plateau with bush and prairies approximately south of 45° S latitude. Its flat and undulating plains rise among the hills of the desert and the eastern slopes of the Andes. (All of these data are from Franklin 1982:457, 471, 473.)

The association of the camelids with the puna has long been noted. In the eighteenth century, Molina (1782:Bk. IV, 311), when seeking to explain the differences between Andean camelids and Old World camels, pointed out that one of the most important differences was "[they] can live in the cordilleras among the snow," and naively attributed this to the abundant fat Andean camelids have between the skin and the flesh and to a "prodigious" amount of blood in the veins, which would give them the necessary warmth to survive.

However, no study has been done to determine whether this habitat was always an "ideal" place for these animals, to the point where Troll's statement (1958:29) that "in a biological sense, llamas and alpacas are members of the puna biotype" has turned into a consensus gentium (e.g., Murra 1975:118) and the puna has been defined as "the optimum habitat of the domestic camelids" (Browman 1974a:194). The fact that these animals now live at high altitudes does not mean this was always the case or that the high-altitude zones were always their "optimal habitat." Some physiologists have discussed this matter, but anthropologists and archaeologists seem to be unaware of their data. This situation has given rise to a distorted view of camelids, whose adaptation to high altitude has not been understood. As Flores Ochoa (1975a:7) correctly noted, "[t]he Andean camelids are magnificently adapted to high altitudes through anatomic and physiological characteristics . . . that enable them to withstand the cold of the puna, make good use of the tough grass in this area, and turn its high cellulose content into high-quality fibers and meat. Birth, for instance, takes place in the rainy season, when the best conditions for the survival of the offspring are present owing to the abundance of pastures that allow the mothers to produce milk." However, this genotypic adaptation does not prevent these animals from leaving this habitat. Such an erroneous idea, which has been accepted almost dogmatically, has ensured that the evidence from areas where by definition "high-altitude" camelids could not live (e.g., the coast) was not considered. A typical example of this interpretation appears in Cardich (1976:34, repeated verbatim in 1987b:21–22), who claims that

> [a]ccording to important contemporary studies, we can note that the Andean

camelids do not usually come down to lower zones, not just because their favorite pastures disappear, but also because they have a physiological specialization for life at high altitudes, especially the alpaca (*Lama paco* [sic]), the vicuña (*Vicugna vicugna)*, the llama (*Lama glama*) and, probably to a lesser degree, the guanaco (*Lama guanicoe*). Thus, for example, even when the llama has lived at sea level for several generations, it still exhibits the characteristics of "adaptation to hypoxia (high altitudes): for example, a high concentration of hemoglobin in the red blood cells, a longer lasting survival of erythrocytes, and a high hemoglobin affinity for oxygen" (Kreuzer 1966, quoted in Jensen 1974:16 [It should be noted that the original study was done by Hall (1937) and was taken up by Kreuzer (1966)]). This characteristic is validated by the fact that the coastal climate is harmful for the llama (Maccagno 1932:43, in Jensen 1974:17), so we can reject the massive seasonal displacement of camelid herds to coastal regions. We can also add another minor inconvenience, that a relatively extensive semidesert area, poor in pastures, lies between the lomas and the puna.

These generalizations are made from second-hand data, without any real understanding of the phenomenon.

It is worth noting here a fact that reinforces my own argument. A volume in the *Handbook of South American Indians*, one of the most widely read handbooks in the Andean archaeological literature, was published in the 1950s, yet an important claim made there has gone unnoticed. In this volume, Gilmore (1950:432) said of the llama that it "does not seem to be a specialized high-altitude breed." After analyzing several aspects of this animal, among them its special physiology, Gilmore noted that one factor that ensured the continuing importance of the llama was "its tolerance to altitudes" (1950:436, 438). I have the impression that Gilmore realized the problem but did not expand on this point. And even though Wheeler (1977) does not give a proper explanation of this phenomenon, she does use the correct terms and thus shows she is aware of it. Thus, when discussing the highlands in the Department of Huánuco above 4000 masl, she defines them as "the primary habitat of the Camelidae," but then indicates that "hemoglobin studies (Aste Salazar 1964) may also provide evidence concerning the location of the Camelidae *prior to their migration into the high punas* around 10,000–8000 BC" (Wheeler 1977:3, 15; emphasis added). The reference to Aste is not relevant because he studied sheep and camelids and made no reference to the adaptation problems of the camelids, because his assumption was that camelids were high-altitude animals.

I believe it was Carlos Monge C. who correctly stated the problem. Monge wrote:

> The international biological, physiological and medical literature mentions some Andean animals as of high-altitude. Thanks to the studies done in mammals (humans included), birds and batrachia, our group . . . considers that this denomination must be revised within the context of an altitudinal gradient. For example, physiologists believe that the South American camelids, Andean birds and others animals are high-altitude animals, whereas *they actually occupy the Andean gradient from sea level to altitudes of 5000 m or more. . . .* A high affinity of hemoglobin for oxygen and lack of an erythemic response to environmental hypoxia are taken as an indication of genetic adaptations to high altitude. These characteristics are absent both in the European animals introduced into the American continent during the Spanish conquest as well as in native animals whose ancestral natural habitat was not a high-altitude environment. A remarkable exception occurs in the Asian and African camels, which despite their ancestral habitat at sea level exhibit as high a hemoglobin affinity as their relatives, the South American camelids. We intend to approach this problem with two hypotheses: the first assumes that those animals that have a high-altitude genetic mark acquired it through an evolu-

tionary period in high-altitude Andean zones. The second hypothesis is related to the concept of pre-adaptation and has to do with a mutational change that precedes the invasion of an ecological niche and facilitates it. . . . The cases of camels and South American camelids favor the concept of pre-adaptation. (1989:8; emphasis added)

Monge then concludes that "[w]e can consider . . . camelids [as a case of] pre-adaptation or, alternatively, as an instance of animals that, having acquired the genetic mark of high affinity at high altitudes, then migrated to sea level, where the acquired characteristics do not disturb the normal physiological and reproductive development of the species, even though the selective pressure (hypoxia) does not exist." And in the 1990s a significant truth was stated: "the commonly held belief that they cannot survive at low elevation is a heritage of the Spanish conquest" (Wheeler et al. 1992:470). I return to this discussion after presenting and analyzing the evidence collected.

2.3 MORPHOLOGY AND PHYSIOLOGY

The relevant morphological and physiological aspects of the Andean camelids are worth reviewing briefly in relation to the habitat these species live in.

From a morphological point of view, the llama has a high cardiac weight with respect to its body weight, similar to other mammals, while the vicuña has a considerably higher weight (Jürgens et al. 1988; Jürgens 1989). The capillary density in the heart and the soleus muscle is greater, thus facilitating oxygen diffusion to the tissues. It should be noted that vicuñas exhibit these characteristics to a greater degree than other camelids (Jürgens et al. 1988). Studies of the trunk of the pulmonary artery, and the aorta, including their histology, extensibility, and collagen and elastin content, do not show much difference between llamas, dogs, ovines, cattle, and pigs living at altitudes of 4330 masl (Heath et al. 1968). However, in a subsequent study conducted on llamas living at 4720 masl, the same team did not find the right cardiac hypertrophy or hypertrophy of the muscular layer of the pulmonary trunk artery, which are considered functional advantages acquired through evolution (Heath et al. 1974).

On the other hand, the width of the arteries of the pulmonary muscles and the weight of the two ventricles in three high-altitude llamas and one alpaca did not differ from those of a llama and two guanacos raised at sea level. This can be explained by the modest hypertensive pulmonary response of these animals when living at high altitudes as compared with members of the same species born and raised at sea level. These properties were attributed to adaptive evolution (Harris et al. 1982). It was found that the small pulmonary arteries of the llama have thinner walls than those of other domestic animals acclimatized to the Peruvian Andes (Heath et al. 1969). From a functional point of view, it was found that the llama responds to altitude with a moderate pulmonary hypertension (Banchero et al. 1971a).

The histological study conducted on high-altitude human beings, cattle, guinea pigs, and rabbits showed that the carotid bodies were enlarged. The llama did not show this enlargement (Heath et al. 1974).

On the other hand, study of the alveolar dimensions of the llama's lungs did not show any difference from those of other mammals of a similar size. Nor was it possible to show a change in this morphological parameter in other mammals raised at high altitudes (Tenney and Remmers 1966). For other morphological studies of the cardiovascular system, see Heath et al. 1976 and Williams et al. 1981.

It is worth emphasizing that South American camelids have some blood properties in common with Asian and African camels. These properties include very small, elliptical red blood cells with a high hemoglobin concentration and a hematocrit (percentage of red blood cells in the blood) that is the lowest among mammals. The low hematocrit gives rise to a fluid, low-viscosity blood (Whittembury et al. 1968). Unlike in sea-level mammals, the percentage of red corpuscles in South American camelids does not increase under conditions of a lower concentration of atmospheric oxygen, as happens at high altitudes (Hall et al. 1936; Reynafarge et al. 1968; Miller and Banchero 1971; Banchero et al. 1971b). Stud-

ies of red blood cell formation in llamas, alpacas, and vicuñas have shown that these animals have a greater output and destruction of red blood cells than animals native to high altitudes, but that the total globular mass is less. No secretion of erythropoietin (the substance that regulates the formation of red corpuscles) was found in normal conditions at 4200 masl, but it did appear after profuse bleeding in one alpaca (Reynafarge 1966; Reynafarge et al. 1968). Increased secretion of erythropoietin was also verified in these animals after acute exposure to high altitudes (Scaro and Aggio 1966). In addition, their hemoglobin has a high affinity for oxygen (there are studies on the special properties of the red corpuscles of South American camelids; see Jain and Keeton 1974 and Smith et al. 1979). The fragility of the osmotic dilution is much increased, a property that is similar in camels (Gurmendi 1966). It is worth recalling that Viault (1890) first described the high number of red corpuscles of the llama.

A study of blood types in the guanaco, llama, and their hybrids led to the conclusion that these camelids exhibit a polymorphism in their red cells and that natural isoantibodies are absent. In blood tests these can be typed with rabbit antisera, as in cattle, or by using lectins and electrophoresis in starch gels. Although these studies can be used to classify these species, Miller et al. (1985) concluded that more tests were needed to establish their significance for the taxonomic classification of New World camelids. A later study by Penedo et al. (1988) of genetic variation in llamas and alpacas suggested that these animals were subspecies rather than separate species.

A property of camelid hemoglobin that has been intensively studied is its high affinity for oxygen, a property that is shared by other families with a genotypic adaptation to high altitudes, such as Andean birds and batrachia (Winslow and Monge 1987; Hall 1936, 1937; Sillau et al. 1976; Banchero et al. 1971b; Banchero 1973; Bartels et al. 1963; Bauer et al. 1980; Chiodi 1962, 1971; Jürgens et al. 1988; Braunitzer 1979, 1980; Braunitzer et al. 1977a, 1977b, 1980; Meschia et al. 1960; Van Nice et al. 1980; Petschow et al. 1977; Monge M. and Monge C. 1968; Villavicencio et al. 1970). 2,3-DPG is an allosteric regulator of hemoglobin affinity that is absent in ovines and bovines but present in llamas and alpacas (Reynafarge and Rosenmann 1971).

Based on studies of the molecular structure of hemoglobin, Poyart et al. (1992) suggested that the ancestor of the four existing camelid species had a high-affinity hemoglobin. Based on the amino acid sequence in the α and β chains of the hemoglobin, Poyart et al. proposed a family tree in which a common ancestor gave rise to the dromedary camel, on the one hand, and to an ancestor of the South American camelids on the other. The latter ancestor would have given rise to the vicuña, the llama, the alpaca, and the guanaco. Although this proposed relationship does not clarify the tree derived from paleontological data, we find this general family tree interesting.

Unlike humans native to high altitudes, who lose their ability to increase their lung capacity significantly when subjected to acute hypoxia, high-altitude llamas do retain this response (Brooks and Tenney 1968). Llama tissue has a very high capacity for oxygen extraction in comparison with high-altitude humans (Banchero et al. 1971a). This capacity has also been confirmed in the fetuses of llamas (Benavides et al. 1989) and alpacas (Sillau et al. 1976). The distance that oxygen must diffuse between capillaries in the llama's placenta is the shortest in all domestic ungulates studied (Stevens et al. 1980). However, although the llama's placenta has six chorionic layers, the estimated difference in oxygen throughout the placenta is small, indicating low resistance to the diffusion of oxygen (Meschia et al. 1960).

A study of llamas undergoing a graduated exercise showed that they require high circulatory flows in conditions of intense exercise because they have a low oxygen capacity. These flows carry substrates at a speed compatible with the required high oxygen consumption (Hochachka et al. 1987). On the other hand, biochemical studies done on alpacas show that concentrations of lactic dehydrogenase, glucose-6-phosphate dehydrogenase, and myoglobin are high in animals living at high altitudes and fall when the animals move down to sea level (Reynafarge 1971). The study of the chemical structure of alpaca DNA showed no difference between the genes that code for the same molecules in other vertebrates (Melgar et al. 1971). A study of llamas' renal function (Becker et al. 1955) is worth reiterating here, because its results do not agree with what would be expected in an animal with a low hematocrit.

The marked physiological differences between the South American camelids and the mammals introduced in South America during the Spanish conquest allow a study of genotype and phenotype adaptations (Monge and Whittembury 1976). It should be noted that from the point of view of man's capacity to live at high altitudes, humans do not present a genotypic adaptation because their physiological capacity to adjust to high altitudes is reversed when they move down to sea level, and it is not transmitted to their offspring.

I shall now analyze the differences between the physiological adaptations of South American camelids and high-altitude natives, emphasizing the fact that camelids do not display the symptoms of nonadaptation to high altitudes indicative of humans' limited capacity to live at great altitudes (Winslow and Monge 1987).

Unlike in native high-altitude humans, in South American camelids the walls of the small pulmonary arteries are small and lacking in muscle formation. They also respond to high-altitude hypoxia with a moderate pulmonary hypertension. Camelids do not show an increase in the size of the right heart. Their blood has a low viscosity, with levels of hemoglobin close to those in man, but they do not respond with polycythemia at high altitudes, and their hemoglobin has a high affinity for oxygen. Camelids do not show an increase in the size of the carotid body and do not develop a low respiratory rate in response to acute hypoxia. Their tissues have more capillaries per surface area, and there is a short diffusion distance for oxygen in the placenta capillaries. Camelids respond to exercise with a remarkable increase in circulatory flow and with a high capacity to extract oxygen from the blood flowing in the tissues.

For scientists studying adaptive physiology (e.g., Monge and León Velarde 1991), the described properties seem to give South American camelids a great capacity to live at high altitudes. High-altitude natives have a limited capacity for life at those altitudes compared with the domestic animals introduced into South America some 500 years ago. The study of South American camelids thus becomes an important tool for the interpretation of life in high mountains, including aspects of livestock health and yields that indirectly affect the well-being of Andean populations. For more information see Bonavia 1995, and Bonavia and Monge 1996 and 1997.

Jensen (1974), who extensively reviewed the literature on high-altitude adaptation, indiscriminately adopted unacceptable positions regarding camelids, claiming "it is possible that the respiratory adaptations (including chemical alterations in the blood) to the hypoxic effects of the blood tend to restrict the species to high altitudes; in other words, is incomplete adaptation of the respiratory system the best way to adapt for species that move between different altitudes?" (Jensen 1974:17–18). He based these claims on data taken from the literature which indicated that llamas had trouble surviving at sea level, and that alpacas fared even worse. At the same time, he accepted that at high altitudes, llamas would sustain genetic modifications to their respiratory system as a result of hypoxic effects. By combining these interpretations, Jensen came up with an evolutionary explanation for the incorrect claim that llamas thrive only at high altitudes, and suggested that those animals that occupy a range of Andean environments might do so through their incomplete respiratory adaptation. The latter point is mere fantasy. Following this line of reasoning, he discussed the guanaco as an animal with no specialized respiratory apparatus for high altitudes and which instead can live perfectly well at sea level. It is worth noting that no specific studies of the respiratory capacity of the guanaco existed, so these claims were mere speculation. The guanaco does not differ substantially from other camelids with respect to blood characteristics.

The reader should be warned of the risk entailed in reading any review of the physiological literature on South American camelids prepared by those without any physiological background whatsoever; much of it is speculative and based on data that include serious conceptual and factual errors.

The fact that camels living at sea level have similar blood characteristics to South American camelids is of great significance from the perspective of evolutionary physiology. Some researchers consider that certain selected mutations that arise in a given habitat can be favorable to survival in another habitat, thus allowing migration to another habitat without additional natural selection. This capacity is (improperly) called pre-adaptation (Monge C. 1989).[1] Pre-adaptation perhaps took place in ancestral camels, whose natural capacities enabled them to occupy desert habitats in Asia and Africa and high-altitude habitats in South America. In the specific case of high-oxygen-affinity hemoglobin, we believe that Asian and African camels might have acquired this characteristic independently after migrating from America, while passing through the high mountains of Asia. This hypothesis is based on the observation that this property can be acquired in a brief evolutionary period, as described by León Velarde et al. (1991).

It is also worth noting that llamas, alpacas, and guanacos do not alter their body temperature in the face of severe dehydration, as dromedaries do. In this regard, the poikilothermal property acquired by dromedaries in their adaptation to an intense desert life in North Africa is not shared by South American camelids (Bligh et al. 1975: 707). It is not known whether the Asian camel is also poikilothermal.

Eduardo N. Frank (in letters dated 19 November 1999 and 29 February 2000) writes that Eva Güttler's dissertation (1986), written on Aguada Blanca in the province of Catamarca, Argentina, shows that the rectal body temperature measured in 34 llamas early in the morning and early in the afternoon ranged on average between 35.9°C and 39.8°C, while the environmental temperature was between 3.5°C and 43°C. Eduardo N. Frank and Carlos M. Nuevo Freire obtained similar data but in semidesert grassland conditions. However, this variation places the llamas studied within the definition of homeothermy given by Bligh and Johnson (1973; see Bligh et al. 1975:705–706). The same thing happens with the vicuña, according to Bligh et al. (1975:707).

2.4 REPRODUCTION

Long pregnancies, low fertility, and a single conceptus at a time undoubtedly are some of the factors that have strongly influenced the cultural aspects of camelids.

Male and female camelids reach puberty at 12 months of age, but the reproductive age is usually delayed until two years in females and three in males. These animals tend toward seasonal reproduction, but when males are kept separate from females, the latter are sexually active all year long (Fernández Baca 1971). Pregnancies in llamas take 348 days, according to Novoa and Wheeler (1984: 117), 352 according to Sumar (1975:300), and 348–368 according to Wheeler (1991:31). In alpacas, pregnancies last 342 days, according to Novoa and Wheeler (1984:117), 343 according to Sumar (1975:300), and 342–345 days according to Wheeler (1991:31). For the guanaco and vicuña, pregnancy lasts about 11 months, according to Novoa and Wheeler (1984), although Wheeler gives a period of 345–360 days for the guanaco (Wheeler 1991:20, but she attributes this to Franklin [1982], who gives different figures: "13 to 15 months" [Franklin 1982:482]), and between 330 and 350 days for the vicuña (Wheeler 1991:25), which accords well with the data in Franklin (1992:477), who gives 11 months (± 1–2 weeks).

It is known that all domestic animals have low fertility on the puna, and the same is true for camelids. The veterinarians from the Instituto de Investigaciones Tropicales y de Altura (IVITA) consider a fertility rate of 50% typical for llamas in the puna. This means that on average, each fertile female will be able to bear only one offspring per year throughout all of her fertile lifetime. This is not just a problem in females, as male fertility is also low even in llamas born and raised on the puna.

Flannery et al. (1989), who studied the Ayacucho herders in the Yanahuaccra-Toqtoqasa region, confirm that every year, each female llama has about one out of two chances of giving birth, and that a female cannot expect to have a birth in two successive years. Since the females become juveniles at about three years of age (although this differs from what Fernández Baca [1971] notes, as was already indicated), even a high-reproductive female cannot have more than four to six offspring over the course of her lifetime. These herders say that a llama can live for about 20 years, though most do not make it. Besides, their usefulness begins to decline when they are about 15, once their teeth are worn to the point where they cannot graze effectively and the animal loses

weight until it loses its strength. According to Flannery et al., the Ayacucho herders carefully control their animals so as to detect this critical moment, and the llamas that reach it are eaten before they begin to lose weight. In females, these are animals that no longer can nurse their offspring, and in the case of males, these are animals no longer able to carry loads. Almost all herders agree they would rather eat their old llamas than sell them, because they have a strong feeling for their herds. Some herders pointed out that bad luck can strike the other animals should one be sold to a stranger (Flannery et al. 1989:99).

It is assumed that low fertility is due to poor pastures, problems with the climate and altitude, and an increase in the sublethal genetic load resulting from selective breeding for recessive traits. It seems that the right uterine horn is usually atrophied in camelids, thus reducing the possibility of gamete fertilization (Maccagno 1932: 24; Browman 1974a:191). Even more interesting, Fernández Baca (1971:19) found some hints of differential fertility at different altitudes.

The low fertility problem was correctly noted by Flannery et al. (1989:214), who state that in past times, guanacos used to migrate extensively in search of pastures, whereas today, llamas are penned in *kanchas*, where their communal dung piles become fertilizer for potato fields, instead of marking the limits of their land. Today it is the herders who have to migrate to find fresh pastures. Herders establish several huts (*estancias*) in the upper and lower *sallqa* (i.e., on the puna) and move their herds on a seasonal basis between them. The animals must be kept far away from the *chakras* during the agricultural season, at higher altitudes. This is achieved by taking the llamas to higher altitudes for longer periods of time than their ancestors, the guanacos, would have gone. This high-altitude move lowered fertility, something that characterizes all domestic mammals taken to the *sallqa*.

I believe that the human intervention mentioned by Flannery et al. (1989), which brought about the low fertility of camelids, is not the only one that has taken place. Another, earlier one took place long before the domestication of these animals, although in different circumstances. This point is discussed later in the chapter.

Births are seasonal when the animals live in herds, between December and March, so the offspring are born in the rainy season, when there are better pastures. However, births can take place throughout the year, and births occur between sunrise and early afternoon, an adaptation to the environment that increases the survival rate of the newly born, as this avoids giving birth at night, when the temperatures are very low and frosts occur (Novoa and Wheeler 1984:117).

However, it is worth noting that there is a high fetal mortality within the first 30 days (Fernández Baca 1971) that, according to Novoa and Wheeler (1984:117), can reach an average of 40%. According to Wheeler (1985b:71), the death rates at birth are relatively low among wild camelids. Raedecke (1979:199) notes a mortality of 4.40% among males and 4.39% for females among newborn guanacos in Tierra del Fuego. Wheeler (1985b:71) says that Franklin (1978:42) reported a mortality of 10%–30% in the first four months among vicuñas in the Pampa Galeras Reserve. Among guanacos, death is usually due to hunger, while most of the vicuña deaths are due to predators. There appears to be no support for the claim that disease or abortion in the last months of pregnancy contributes to morbidity in either species. Mortality is quite marked among young llamas and alpacas, contrary to what occurs among wild camelids. Fernández Baca (1971:29) indicates a mean annual mortality of 50% during the first 40 days of life. Enterotoxemia is the major cause of death (see Wheeler 1985b:71–72; 1984c:79). On the other hand, it is known that mortality among the alpaca herds at the Estación de Camélidos de La Raya (4100 masl) is 30% in the first year of life, even though the animals are kept under "ideal conditions" (Miller 1981:21).

2.5 Nutrition and Other Characteristics

From a cultural perspective, nutrition is a most important aspect and one I am particularly interested in. The early evolution of these camelids is related to the divergence between creodonts and condylarths. According to Cardozo (1975a:63), the volume of some compartments of the stomach increased to slow the rate of digestion. This is appropriate for animals that consume rough fodder, which must be processed to yield its nutrients. In ruminants, the rate of digestion is determined by the folds of the rumen. This effect was achieved in camelids with the development of glandular pouches. According to Cardozo (1975a: 63), Langer (1974:306) assumes three evolutionary models in artiodactyls, but in all of them the camelids appear independently, though closer to the ruminants. In fact, the South American camelids have essentially the same digestive process as the ruminants, although there are some differences with the advanced ruminants (Novoa and Wheeler 1984:116–117). At the same time there are some important differences, as San Martín et al. (1989:98, 108) noted, that influence selectivity. For instance, according to work by Vallenas (1965), the morphology of the stomach is different in advanced ruminants and camelids. The latter consume less fodder (Reiner et al. 1987), have slower passage of food, and have a longer retention time for stomach contents than ovines do (San Martín 1987). These characteristics allow more complete digestion of dry fodder (Van Soest 1982).

The major anatomical difference between camelids' stomach and the stomachs of true ruminants is motility. San Martín and Bryant (1987:12; 1988:84–85) explain that the stomach of South American camelids has more continuous activity than is observed in advanced ruminants. Engelhardt and Rubsamen (1979) indicate that the anatomy of the camelid digestive system is very different from that of advanced ruminants, so any analogy between the two species would be difficult to prove. In fact, "[i]t has been suggested (Vallenas and Stevens 1971a, 1971b) that the greater digestive efficiency of South American camelids over other ruminants may occur because of their more continuously active forestomach and more frequent cycles of rumination, which, combined with glandular pouch secretions, may provide for more adequate and efficient mixing, maceration, and perhaps even absorption" (Franklin 1982:482).

This greater retention time permits the digestion of food with a high lignocellulose (woody cell wall) content (like that found in the high-altitude native pastures; San Martín and Bryant 1988:92). We believe it is worth emphasizing this point and noting the time camelids take to digest their food because it has not been considered by archaeologists, although it has an important role in the interpretation of camelid excreta. San Martín and Bryant (1988:15) point out that studies comparing South American camelids and advanced ruminants show that camelids retain food in the digestive tract for a longer time. Whereas an ovine retains food for 40.9–43.2 hours, the alpaca does so for 50.3 hours and the llama for 62.3 hours (San Martín and Bryant used the studies by Flórez 1973 and San Martín 1987). This shows, as San Martín and Bryant (1987:25) note, that South American camelids are more efficient than ovines at digesting medium- and low-quality foods.

An important element in the digestive system of camelids is a specialized dentition, which, as we shall see later, is an important archaeological factor for the identification of remains from these animals. Some even believe the dentition can be used to distinguish wild from domestic animals. The teeth are characterized by continual growth (but according to Wing [1973], this happens only in the vicuña). Besides, the harelip, so typical of these animals, allows them to eat the tough grass of the puna with ease (see Novoa and Wheeler 1984:117; Flores Ochoa 1975b:300). (Interested readers can find a complete description of the camelids' digestive system and its structure and function in San Martín and Bryant 1987.)

The truth, as Sumar (1988:28) correctly notes, is that our knowledge of the digestive physiology of camelids is still deficient, notwithstanding all the work done to date. But, as San Martín and Bryant (1987:56–57) insist, there is no doubt that South American camelids are well adapted to areas where the amount of fodder is limited and the nutrients are highly diluted, hard-to-digest carbohydrates. These are precisely the characteristics of the present vegetation in the altiplano, where long droughts occur throughout the year (there usually are four dry months) and cycles with dry years are frequent. San Martín and Bryant conclude that "[u]nder these circumstances, South American camelids are the most suitable species to use the scant and fibrous vegetation present in the altiplano region due to their selective characteristics, reduced consumption,

the longer time the rate of digestion takes in their digestive tract, as well as being physically adapted to survive in high-altitude zones (Reynafarge 1975 [it should be Reynafarge et al.])."

There is another problem intimately connected with eating and that, from a cultural perspective, is worth elucidating: the capacity of camelids to make long journeys with scant food and water. Much has been written about this, but not always on a sound basis. For example, Murra (1978:88) says: "In the chroniclers we do not find details on the possibility of going without food and water for long, much like camels, as has been noted by contemporary observers." As we shall see, there are references in some chronicles (e.g., Zárate 1968:176) and travelers' accounts. Tschudi (1885:107; 1891: 107; 1918:227; 1969:137), for example, wrote: "The llamas do not graze once the sun sets, and they can even go without food for two or three days. The fact that they only look for their food by day is no doubt what determined the need to make them travel no more than short distances of about 20 km." Much has clearly been exaggerated in this regard, but there is some truth to it. It is a fact that lamoids and cameloids exhibit a specialized adaptation to an arid or semiarid environment, but only the latter can withstand long periods without water (see Novoa and Wheeler 1984:117). Gilmore (1950:430) explains that the mechanism for storing water in true camels apparently lies in the metabolism of the hump, the subcutaneous fat, and the corporal carbohydrates, and there might be a parallelism in the lamoids, even without the hump. The use of water in the organism probably takes place during the long, forced marches the llama can make without water, or is known to have done. From a physiological perspective, camels and lamoids are probably well adapted to thrive in environments with scant vegetation and to derive many carbohydrates from cellulose through an enzymatic or bacterial action in the rumen of their complex stomach. Carbohydrates produce more water than grease in the transformation of metabolic energy. Gilmore believes this advantage is only relative, because the intake of a strictly hydrocarbonated diet would yield about 15% more water than a diet of pure fat. Since an herbivore eats grasses with a given amount of lipids and cannot live on a fatty diet, the major factor in the diet, as far as the intake of water is concerned, actually depends more on the dryness of the pastures than on its chemical composition.

Raedecke (1976:93) extensively studied the guanaco and wrote as follows in this regard:

> Many myths about the frugal consumption of water and the storage capacity of the camelid family are still believed. It is said that camels store water in their humps. The camel is actually an efficient water user, but even so it has to drink it every given number of days. Many people in Tierra del Fuego still believe that the guanaco comes to the sea to drink salty water, but this was not observed in the three years of this study. A small population of guanacos lives in a great salty plain but this is a matter of survival, not a search for salty water as the area is inaccessible and there are no sheep. The great amount of water the guanaco needs is probably supplied by the vegetation it consumes, and this because the area is often wet and rainy, and there is much water available in the pastures.

Another most important aspect of the biology of these animals, and one that has most moved conservationists to defend this family, is the very special anatomical characteristic of their feet, which exhibit great advantages over those of the animals introduced after the Spanish invasion. The ability of the llama to walk along the Inca roads without causing any damage is proverbial (see Manrique 1985; Usselmann 1987). This evidently is a great advantage for these animals, because, as Sumar (1992:93) explains, their small third phalanx does not have the hoof typical of other ruminants and has nails covering the dorsal surface. Their pedal characteristics give camelids a broad support surface that does not harm the road, as happens with other hoofed animals. For this reason, according to Sumar (1992:93), the llama and the alpaca are commonly said to behave like "ladies with the soil." This characteristic was also used to classify the camelids in the Suborder Tylopoda, distinguishing them from other hoofed ruminants, the Pecora.

For Koford (1957:162), the lamoids are probably more efficient than Old World camels in walking and running on slopes and over rocks.

A most important aspect of the camelids as far as human-animal interaction is concerned, and one I believe was crucial during the process of domestication, is the sense of territoriality these animals have, although some species exhibit it more than others. This is indicated by delimiting the territory of the group through the accumulation of excrement at certain points, defecating and urinating on common piles. This is a most particular characteristic of the behavior of South American camelids. It caught the attention of Darwin (1921:239; 1969:172), who wrote as follows: "The guanacos have one singular habit, which is to me quite inexplicable; namely, that on successive days they drop their dung in the same defined heap. I saw one of these heaps which was eight feet in diameter, and was composed of a large quantity." Raedecke (1976:112) observed that several of these common piles of dung were used year after year and were very big, almost 3 m in diameter and 20–30 cm thick. According to Raedecke, Franklin (1973; this is actually a mimeographed "progress report") established the dry weight of the piles of vicuña excrement, which was 45 kg. Koford (1957: 169–170) noticed that alpacas and llamas used the same piles as the vicuñas. A typical big mound is 30 cm thick in its center and has a diameter of 4.50 m. These piles of dung are a most important resource for the economy of the pampas Indians, who use them as fuel to cook.

2.6 Diseases: Sarna

Disease is another subject that must be discussed, because some diseases were of great significance to human history, particularly from the moment domestication took place. Sarna is particularly important; it was mentioned extensively by the Spanish chroniclers and had major consequences for the economy of the Andean people.

Flannery et al. (1989:102–103, 113–114) provide a good review of the problem. They show that camelids actually have two diseases that bear the same name: *qarachi* in Quechua and *sarna* in

Spanish (mange in English). Both are caused by mites 200–450 microns in size.

It is suspected that *Sarcoptes scabiei*, the mite that causes the itch, evolved with mankind for several thousand years and was then transferred to domestic animals. Today there are different varieties for man, dogs, sheep, pigs, cattle, goats, Old World camels, South American camelids, rabbits, and horses. These varieties can infest mammals that are not their specific host without causing a permanent infestation (Baker and Wharton 1952:363).

The *Sarcoptes* female can penetrate the epidermis in less than three minutes and continue digging at an average rate of 2–3 mm per day until it has made a burrow up to 3 cm long in the horny outer layer of the skin (Belding 1942:765). Here she lays her eggs, which become adults in 8–15 days and begin digging their own burrow. The initial infestation does not itch, but after a month or so the host becomes sensitive, the itch becomes unbearable, and the hair begins to fall off. The density of llama herds is the significant factor in spreading this infection. This is what happens with sheep in a corral or soldiers in a crowded barracks. This condition was quite common in the history of the Old World and was cyclical. Since this disease is associated with humans, is very old, and has a Quechua name, Flannery et al. (1989:102–103, 113–114) suspect that it must have arrived in the Andes with its first inhabitants, who then passed it on to the llama. One specific variety found on sheep, *S. scabiei* var. *ovis*, was probably introduced by Europeans. Although these scholars do not mention it, the variety that attacks camelids must be *S. scabiei* var. *auchenia*, as noted by Cunazza (1976a:152–153), based on work by Guerrero (1971).

According to Flannery et al. (1989), psoroptic mange is present in Peru today, most likely introduced by the Spaniards. Sheep scabies (*Psoroptes equi* var. *ovis*, as noted by Cunazza [1976a:161], who in turn cites Plaza 1973) has a historically clear relation with Old World sheep and is much less likely to have arrived in the Andes with the first humans than scabies. For this reason, Flannery et al. (1989) suggest that the first outbreak of *Psoroptes* in llamas and alpacas happened in 1544–1545 as a result of the introduc-

tion of European sheep into the Andes. They also posit that the loss of llamas in prehispanic times was perhaps considerably lower than in the early Colonial period, and perhaps even lower than it is now. In this regard, the data in Cunazza (1976:162) are interesting, for he showed, based on the studies by Dennler de la Tour (1954), that guanaco sarna is not transmitted to sheep.

Flannery et al. (1989:113–114) conclude that the antiquity of the *Sarcoptes* mite, its long association with humans, and the presence of a Quechua word for mange suggest that it existed before the conquest. On the other hand, it seems that *Psoroptes* was not present prior to the introduction of sheep by Europeans. Flannery et al. recall the famous passage in Garcilaso (discussed below) and note that the description of the "sheep scabies" or psoroptic mange he gives is almost perfect. They also note that we can assume it was an introduced mite against which the camelids had not developed resistance, to judge from its virulence and the time it occurred. Flannery et al. end their discussion by noting that although *Sarcoptes* might have been a problem in ancient times, it is unlikely that prehispanic herds ever had to face an epidemic like that caused by *Psoroptes* in 1544–1545.

Often forgotten is the fact that humans are susceptible to infestation by nonhuman varieties of *S. scabiei* that normally affect other species, including camelids (see Faust et al. 1979:615–616).

2.7 SELECTION AND MANAGEMENT OF HERDS

These biological aspects (and others that I probably have not mentioned) cannot be ignored if one seeks to understand the significance of camelids within the context of Andean culture. Herre (1969:118) insisted on this point, especially on man's effect on these animals. The llama was essentially raised as a beast of burden, while the alpaca was raised instead for its wool. Here we have a good example of the close interaction between biology and culture, and of the need for specialists from both fields to collaborate in all discussions. It is worth bearing in mind that according to the data Sumar (1988:24) collected, there is an area located at a considerable altitude that extends over four South American countries

(Peru, Bolivia, northwestern Argentina, and northern Chile) and comprises more than 25 million hectares with 350,000 families who depend solely on camelid herding for their living. If some species, such as the llama and the alpaca, have survived to this day, it is only because they are crucial for the Andean Indians.

For Sumar (1988:24), "[b]reeding and herd management practiced on the high Andean grassland are determined by a mosaic of traditional and Hispanic techniques, which to many outsiders seem backward or irrational. The systems of livestock production may seem inefficient and unproductive, and it is hard to explain why the local herders resist technological improvements. However, the herders of the Peruvian highlands are responsive to programs of change, if these are properly conceived and directed."

2.8 A BRIEF DESCRIPTION OF THE SOUTH AMERICAN CAMELIDS

In this section I briefly describe the four South American camelids, noting their major characteristics. However, it should be kept in mind that these four lamoids resemble each other in structure but differ in size, range, pelage, temperament, and usefulness to humans (Koford 1957: 154). I shall begin with the guanaco and the vicuña, two allopatric[2] species and the large dominant herbivores of the South American undisturbed environments. They are the only wild ungulates living in the continent's deserts, in the high plateau scrublands, and in the grasslands (Franklin 1983:573).

2.8.1 The Guanaco (*Lama guanicoe*)

Of the living species, the guanaco is for Cardozo (1975a:76) the one that usually has a smaller endostyle[3] than the other *Lama* species. This could mean that it is the oldest species in its genus (see Figure 2.1).

When discussing the taxonomy of the Camelidae, I noted the position of the guanaco, so there is no need to repeat it. Instead, I will discuss its subspecies, which were mentioned in Chapter 1. Four subspecies have been described: *Lama guanicoe guanicoe*, found in Patagonia and Tierra del Fuego, south of 35° S latitude; *L. g. huanacus*,

found in Chile; *L. g. cacsilensis*, found in the puna of southern Peru and Bolivia; and *L. g. voglii*, found in Argentina north of 32° S latitude. The southern subspecies is considered to be the biggest South American ungulate but it is always smaller than the domestic llama, even though *cacsilensis* falls within the size range of the next smallest species, the alpaca (Novoa and Wheeler 1984: 121; Franklin 1982).

Wheeler (1995:276) noted the possibility that there is an undescribed guanaco species on the Peruvian coast, for which she claims there are few data available. She attributes this idea to Ponce del Prado and Otte (1984); however, that study makes no comment on this point. I do not know the source of Wheeler's information. As far as I have been able to determine, *cacsilensis* alone of the subspecies listed here has problems and has been questioned.

Lama guanicoe cacsilensis, described by Lönnberg in 1913, presumably lived between 4500 and 4600 masl (close to Caxile in the Ñuñoa area, Department of Puno). It should be borne in mind that all other subspecies live at altitudes that extend from sea level to 4260 masl (Wheeler 1984c:78; 1985b:69). According to Wheeler

FIGURE 2.1. Guanaco (*Lama guanicoe*). *Hernán Torres, courtesy of the author.*

(1984a:401; 1984c:78; 1985b:69) (who received a personal communication from Carl Edelstam), this subspecies was apparently described from just one specimen, but has never been verified. Wheeler's position is ambiguous, as we shall see. This datum is confirmed by León (1939:100), citing Lönnberg in a footnote, who says, "with a cranium and a skin found by an expedition from the Stockholm Museum, but he did not see the live animal, nor has anyone ever found another one since then."

Miller and Gill (1990:58–59) explain that the measurements of the guanaco were taken from Tierra del Fuego animals, where the species seems to have reached a great size—larger, in fact, than the northern animals. What made Lönnberg create *L. g. cacsilensis* was the discovery of an exceptionally small guanaco cranium in the area of Ñuñoa, with some characteristics similar to the vicuña and which, according to Lönnberg, was osteometrically different from its bigger cousins in Patagonia. However, neither the existence nor the exact size of the *cacsilensis* subspecies nor its geographic distribution is secure. Its taxonomic validity was questioned by Osgood (1916), Allen (1942), and more recently by Franklin (1982: 464).

Miller and Gill (1990:59) are emphatic:

> The exact size of the putative sub-species is uncertain because Lönnberg published only cranial measurements of the type specimen and because of the scarcity of living Peruvian guanacos as well as of their skeletons in museum collections. The one Peruvian skeleton that we have been able to examine, from the Museum of Paleontology at the University of California, Berkeley, is smaller than all of the Tierra del Fuego specimens. . . . Of perhaps most [sic] importance is the fact that Lönnberg himself believed the *cacsilensis* subspecies to . . . constitute a local race confined to a certain district only and . . . that Peru must be inhabited by a large Guanaco as well. . . (Lönnberg 1913:8).

According to Miller and Gill, Lönnberg used as comparative material a male guanaco, studied by Tschudi in 1840, whose measurements are bigger than those of the average guanaco from Patagonia, to judge from the data presented by Franklin (1982:465). However, they believe that a passage from Cieza de León (1973:244 [1984:294]) on the guanaco that says these were very big can be added to Lönnberg's argument. For Miller and Gill (1990:59), this means that the guanaco Cieza was familiar with was at least as big as the llama.

From the data I have assembled, I can say that Cabrera accepts the *cacsilensis* subspecies (this was taken from Grimwood 1969:71; however, Grimwood does not give the date of Cabrera's study, which is 1961), and he even determined its northern limits in the mountainous area of southern Peru, specifically in Cuzco and Ica. Wheeler's stance on this issue has changed over time. She had already said once of the *cacsilensis* subspecies that it "has not been verified" (1984a:401; 1984c:78; 1985b:69), but then noted that "[t]he Peruvian guanaco *L. g. cacsilensis* is poorly known because nowadays this animal is extremely rare in Peru . . . and has almost completely vanished from the Junín region" (1985a:28–29). In her 1991 article, Wheeler accepted the existence of this breed with no reservations (Wheeler 1991:17). Curiously enough, she still accepted it later (e.g., Wheeler 1993:19), but noted that "[t]his animal . . . is . . . virtually unknown to science" (Wheeler 1995:277). However, Pascual and Odreman (1973:34–35) report that Augusto Cardich told them that "this species lived there until just a few years ago," a point that should be verified with more evidence.

Of all the South American camelids, the guanaco is the one that occupies the most extensive area and most varied habitats. Its habitat requirements are far more flexible than those of the vicuña. Thus, we find it on the western slopes and spurs of the Andes in northern Peru, in parts of Bolivia and central Chile, in the coastal lomas, on the dry slopes of the southern Andes, and throughout Patagonia and Tierra del Fuego, including the island of Navarino—in other words, from approximately 8° S latitude to 53° S latitude. The guanacos were far more abundant on the cold scrub prairies of Patagonia. The prairie extends from sea level to 4250 masl, and the guanaco occupies the deserts, semideserts, scrub, grassland zones, savannas, bushy zones, high-altitude pam-

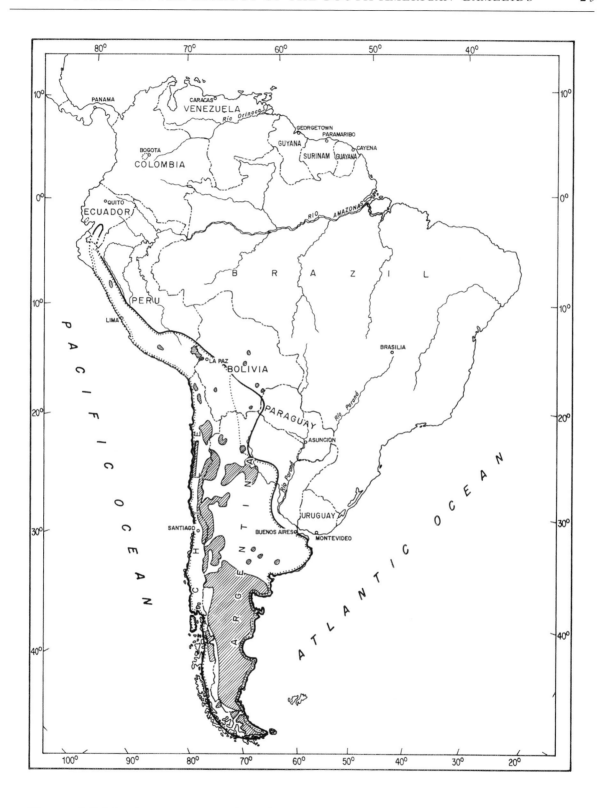

FIGURE 2.2. Map showing current and past distribution of the guanaco. *Prepared by the author and drawn by Osvaldo Saldaña.*

    ....   Approximate distribution in pre-Columbian times.

    ⋯   Approximate distribution during the conquest and Colonial times.

    ▨   Present distribution.

pas, plateaus, foothills, and mountains (Figure 2.2). Interestingly, these animals have adapted very well, even in Europe, and reproduce without many problems if well tended (*Enciclopedia de los Animales* 1970:302).

The guanaco avoids steep slopes with cliffs and rocks. Despite its preference for dry and open habitats, its adaptability allows it to live even in the wet and rainy coastal woodlands of Tierra del Fuego. The guanaco thus lives in extreme habitats: in the Peruvian-Atacameño desert, one of the driest in the world, and in the wet Fuego archipelago, which has high humidity all year long. Morrison (1966) has described bare skin patches on the sides of the guanaco that act as "thermal windows" to disperse excess body heat in warm zones (the data were taken from Franklin 1975[:191]; 1982[:474, 482]; and 1983[:605], who in turn relied on the studies by Prichard 1902; Dennler de la Tour 1954; Matthews 1971; Miller et al. 1973; Franklin 1975, 1982; Raedecke 1979). According to Wheeler (1985a:29), there are also important data in Neveu-Lemaire and Grandidier (1911:45–49). Gilmore (1950:450), Novoa and Wheeler (1984:121), and Torres (1992a:33) can also be used.

The wide distribution of the guanaco and its dominant role in the most varied kinds of dry environments are the result of its flexible behavior and social organization. For these reasons, the guanaco has a wider distribution than the vicuña.

It is clear that the physiology of the guanaco is insufficiently known, yet it holds the secret to its adaptability to such varied environments. Its water- and energy-efficient metabolism appears to be one of the major adaptations that enable it to live successfully in such a great variety of dry environments, with the exception of the pastures in the puna of the central Peruvian altiplano (Franklin 1982:481; 1983:605, 620). The guanaco is both a grazing and a browsing animal, and we know from the studies Raedecke (1979) made that it can adapt to different diets, for example, when it competes with sheep. Raedecke (1980) believes the guanaco was able to maintain a generalized diet without the competitive pressure of other herbivores because it was the only ungulate living on the prairies of Patagonia since the late Pleistocene, and of course before cattle and sheep were introduced into this environment.

Although the reason why the guanaco occupies such extremely arid environments has not been studied, it is surely related to its ability to live for long periods without drinking water, especially when fodder is moist and provides enough water. The guanaco needs to drink water periodically. These animals have been seen drinking briny or salty water, including seawater in puddles left behind by the receding tides (Franklin 1982:482 [in turn based on Vaughan 1978]; 1983:605 [based on Musters 1871 and a personal communication from Payne]). It was for these reasons that Raedecke (1976:24–25) classified the guanaco as a typical animal of the steppe that can be found in the transitional zone of the pampa.

The studies by Raedecke (1976:109) show the guanaco is a diurnal animal, active only in the daytime and sleeping at night. It is also highly territorial (Franklin 1982:482). Another important characteristic of these animals is that their "gallop . . . is faster than that of horses, especially if they are climbing hills" (Arbocco Arce 1974:11). They are also skilled swimmers (Arbocco Arce 1974:10; Gilmore 1950:450), as Darwin (1921:238; 1969:172) had already noted: "The guanacos readily take to the water: several times at Port Valdes they were seen swimming from island to island."

Sumar (1988:26) says that "the guanaco exists only in the wild state." This is not entirely true. Cooper (1946a:143) found data showing that the ancient Tehuelche tamed young guanacos and used them as decoys to hunt their kin (Cooper's database comes from Pigafetta 1906, 1:52; Mori [1535] 1889:320). Pigafetta (1927:59) entered the following into his diary while in the Port of San Julián on 19 May 1520: "They brought four animals of those I have mentioned [guanacos], tied with some kind of halter, but they were small and of the kind used to trap the big ones, for which they tie the small ones to a shrub; the big ones come to play with them and the men, hidden in the thicket, kill them with arrows." Another important testimony is by Molina (1782:Bk. IV, 320), who not only claimed that the guanaco was easily tamed but also claimed to have seen it personally. Darwin (1921:238 [1969:171]) made a similar statement: "These animals are very easily domesticated, and I have seen some thus kept in northern Patagonia near a house, though not under any restraint." Franklin (1981:67) likewise confirms that Patagonians used to have the guanaco as a domestic animal. Gilmore (1950:450) believes that the taming of the guanaco could hold the key to the process of domestication, because the wild specimens might be the ancestral stock of the domestic llama.

Another source reports that when young, the guanaco behaves well and is trusting and affectionate, following its owner like a dog and letting humans get close to it because it is as tame as a lamb. However, with age the guanaco becomes ever more unfriendly and rebellious, finally losing all of its good disposition toward humans (*Enciclopedia de los Animales* 1970:302). However, this does not seem to be entirely true, to judge from the experience of the Franklin family (Franklin 1981), who domesticated two guanacos without having any major problem. Cardozo (1975a:107) also says that "There have been and are attempts at taming and domestication, but their numerical impact is insignificant. It [the guanaco] is therefore still considered a non-domestic species."

Guanacos can live in sedentary and migratory groups (see San Martín and Bryant 1987:5; Raedecke 1979; Franklin 1982; 1983:605; Novoa and Wheeler 1984:121). Flannery et al. (1989:92) note that one of the keys to understanding the wide geographic dispersal of the guanaco is its flexible social organization. Responding to local environmental conditions, the guanaco can be sedentary when resources are plentiful and migratory when conditions worsen. However, Franklin (1983:620–621) emphasized two factors that favored the migratory habits of the guanaco. The first is the winter snow, and therefore the almost complete lack of dry fodder at this time of the year, and the second is the possibility of finding alternative grazing sites in other places. Franklin (1983:612) also pointed out that in wintertime there are migratory groups of guanacos with 176 individuals, and he believes that in ancient times there must have been migratory herds with thousands of animals. (I recommend the book by Flannery et al. [1989] to readers interested in the social organization of these animals.)

It is usually stated that the guanaco has the same body size as the llama (San Martín and Bryant 1987:5), but the situation is far more complex. Although not confirmed, Franklin (1982:464) suggests that from north to south, the size of the guanaco's body increases. What remains constant in all subspecies is the reddish brown color. Raedecke (1976:104) is far more definite in this regard, because he states that the guanacos in Magallanes are far bigger than the northern South American populations. This seems to be supported by Molina (1782:Bk. IV, 318), who claimed to have seen guanacos the size "of a good horse." Torres (1992a:33) believes the size of these animals ranges between 1.2 and 1.75 m long, including the head and the body, while its height ranges between 0.90 m and 1.0 m. They weigh between 48 and 140 kg. However, these data differ from those given by Wheeler (1991:17–18), who cites numerous sources. For the *L. g. guanicoe* of Tierra del Fuego and Patagonia, Wheeler gives a range of 1.10–1.15 m and 1.20 m for the height at shoulder in adult animals. The length from the tip of the nose to the base of the tail[4] ranges from 1.67 m to 1.85 m, 1.91 m, and 2.10 m. In the case of *L. g. cacsilensis* (from Calipuy), the length ranges between 0.90 m and 1.10 m. The average weight of adult forms of *L. g. guanicoe* is 120–130 kg, and 96 kg for *L. g. cacsilensis* (Wheeler's data come from Cabrera and Yepes 1960, Franklin 1982, Raedecke 1979, Mac Donagh 1949, Dennler de la Tour 1954a [1954 in my bibliography], Miller et al. 1973, and Kostritsky and Vilches 1974). The measurements taken by Franklin (1981:63) in the Torres del Paine National Park, Chile were 1.52 m high and a weight of 113.4 kg.

Guanaco females are sexually active and able to reproduce when they are 14 months old (Raedecke 1976:111), but "[t]here is a long-standing discussion about the rate of reproduction in the literature. . . . Walker et al. (1964) claim that the guanaco reproduces every other year and has one offspring; Housse (1930) claims that it can have up to three a year; Cabrera and Yepes (1960), England et al. (1969), Cardozo (1954), and others agree on one offspring a year. This discrepancy shows the general lack of reliable data on the reproductive biology of the species" (Raedecke 1976:125). "Several authors claim that pregnancies in guanacos take 10 to 11 months (Walker et al. 1964; Schmidt 1973; Cardozo 1954; Cabrera and Yepes 1940; Strass 1916 [he does not appear in the bibliography]). This also seems to be true among the guanacos of Tierra del Fuego, despite the scant data in this regard collected during this study" (Raedecke 1976:129). However, Raedecke (1976:130) himself notes that after studying reports of births in 22 zoological gardens, Schmidt (1973) discovered that guanaco births are extremely limited.

Flannery et al. (1989:94) were also interested in this subject. They believe that guanacos of both sexes are capable of reproducing in their second year, but most cannot. The females in a wild state usually do not have enough food, and males less than five years old do not have access to the females. Besides, according to Raedecke's studies (1979), there is a small decline in the fertility of the older guanacos once the younger ones begin to mate, at least throughout the 12th year. As a result, the female does not give birth before its third year and usually does not give birth every year, but can remain fertile until a relatively old age. Flannery et al. (1989:94) demonstrated this among the Ayacucho llamas they studied, but I wonder whether it holds true for the guanaco.

It seems that the food used by the guanaco changes from year to year and from one season to another. The availability of certain kinds of plants changes with the season, the climate, the year, location, landscape conditions, and several other factors, which obviously affect the eating habits of the animal. It seems that in some areas, such as Magallanes, food is the major limiting factor in determining the number of guanacos (Raedecke 1976:82). However, Raedecke (1976) admits that there are no data on their feeding habits.

San Martín and Bryant (1987:40–41) recently wrote on the subject, primarily using research done by Ortega (1985). For them, the guanaco shows a strong preference for the Poaceae. Scrub plants were eaten only when these were covered by snow. San Martín and Bryant also cite the study done by Bahamonte et al. (1986) on the feces deposited all year long. The results show that scrub plants are the major component of the diet (30%), followed by herbaceous plants (15%), the Poaceae (15%), and plants similar to the latter (6%). These results show the extensive use the guanaco makes of trees and scrub plants in wintertime, a good adaptive strategy for areas where snow covers the vegetation. However, we should bear in mind that

the study was made with guanacos in the province of Neuquén, Argentina.

I have already discussed the distribution of the guanaco in general terms and the various habitats it occupies (see above). Tschudi (1885:93–94; 1891:96; 1918:205; 1969:124) also addressed this topic, noting that "The most dispersed South American auchenid is the guanaco, which extends from Central Peru almost to Tierra del Fuego." At present, most scholars believe that the area occupied by the guanaco extends from 8° S latitude in Peru, in the Department of La Libertad, across the coastal and highland Peruvian mountains to western Bolivia, northern and central Chile, and then up to the eastern Andes as far as the Argentinian pampas. From here they extend northward to the sierra of Curámadal and La Ventana, in the province of Buenos Aires, and in ancient times perhaps to the Paraguay River, and to the south along the Andean cordillera up to the Navarino Isles in Tierra del Fuego (see Gilmore 1950:124; Novoa and Wheeler 1984:121 and Fig. 14.3).

However, when we study the distribution of the guanaco at the time of European arrival, we see that Cieza de León noted its presence in the Ecuadorian sierra, as Dedenbach Salazar (1990: 96) correctly noted.

I believe that Grimwood (1969:71–72) has made the best study and the most complete monitoring of these animals in Peru. However, we must not forget that his study dates from 1969. We know the guanaco is found on the coast in northern Peru and extends up to 4000 masl. The northern limit seems to lie around 8° S latitude. Cieza confirms the presence of these animals in the Huamachuco zone, as discussed in Chapter 6 (Cieza de León 1984:Pt. I, Chap. LXXXI, 36). However, it is worth recalling that according to Cieza, there were also guanacos in the heights of Huancabamba, in the Department of Piura (Cieza de León 1984:Pt. I, Chap. LVIII, 185). The guanaco is in danger of extinction in several areas, and there are no reliable figures on their real numbers. There once was a considerable number of these camelids in several areas on the southern coast. The guanaco was mostly found all year long on the upper part of the slopes of the Andean spurs, going down to the lomas in winter. The only evidence Grimwood found of survivors in this region were small groups at 3000 masl south of Matucana (12°00′ S latitude); more than 3000 masl inland in Palpa (14°20′ S latitude), where three groups were detected; at 3500 masl north of the Pampa Galeras Reserve (14°30′ S latitude); and finally at 3700 masl close to Parinacochas (15°20′ S latitude). In each case the groups had fewer than a dozen individuals.

Guanacos are no longer seen in the lomas of Lachay (11°20′ S latitude) or in the Ancón area (12°00′ S latitude), where they could be found until 1950. And what was once an important population in the lower part of the Lurín River (12°15′ S latitude) "has almost surely disappeared," according to Grimwood. In fact, from data I have gathered, this population no longer exists. On the other hand, it is possible that some individuals occasionally wander down to the lomas of Atiquipa and Taymara (15°50′ S latitude), because there seems to be a small population inland, and groups of 10–12 individuals are still regularly seen in the lomas of Morosoma (18°00′ S latitude). This was confirmed by Valdez (1996:100), who reports that in 1987, the people of Atiquipa and the Acarí Valley (in the province of Caravelí, Department of Arequipa) saw herds of guanacos on the lomas.

In the Calipuy National Reserve, in the district and province of Santiago de Chuco (8°30′ S latitude) in the Department of La Libertad (between 1000 and 4000 masl), a small population still survives because it was protected. In 1965 it was estimated that there were 1,000 animals. That year an epidemic broke out, and 400–500 are believed to have survived. For Franklin (1975:195, 201), Grimwood's estimate was still valid in 1975, and this was said to be the biggest and northernmost population left in Peru. However, in 1992 Hoces (1992:Table 10, 54; see my Table 12.2) claimed that there were still 1,000 guanacos in this reserve. I find this figure striking, as it agrees with the one given by Grimwood (1968, 1969) for 1965. Is the figure for 1992 correct? It seems that it is not, because 538 animals were counted in 1996 (according to data from the Consejo Nacional de Camélidos Sudamericanos, Ministerio de Agricultura). In 1992 a total of 134 animals were reported in the Department of Ayacucho within the area of the Pampa Galeras National

Reserve (Table 12.2), yet the 1996 census said there were 1,167 (according to data from the Consejo Nacional de Camélidos Sudamericanos, Ministerio de Agricultura).

There are few data for other areas. Some specimens have been reported locally at an altitude of 3500–4500 masl on both sides of the Pachachaca Valley in the Department of Apurímac, and also more to the east, close to Pachacona, Huachircas, and Antabamba. The presence of guanacos was also reported in the Chivay region in the Department of Arequipa. However, the presence of 148 animals was reported in 1992 in the Salinas y Aguada Blanca National Reserve, which is in the Departments of Arequipa and Moquegua (Table 12.2), and 1,203 were reported in the 1996 census (according to data from the Consejo Nacional de Camélidos Sudamericanos, Ministerio de Agricultura). Grimwood was unable to obtain information on their presence in the Departments of Puno, Cuzco, Junín, Huánuco, and Ancash. There apparently are no data for these departments save for Puno, which had 30 animals in 1992 (Table 12.2) and 71 in 1996 (according to data from the Consejo Nacional de Camélidos Sudamericanos, Ministerio de Agricultura). However, according to the 1996 census, there were 211 guanacos in the Department of Huancavelica, 516 in Ica, and 95 in Tacna (according to data obtained from the Consejo Nacional de Camélidos Sudamericanos of Peru's Ministerio de Agricultura).

Although it extends over a larger area than the vicuña, the guanaco is hard to track down and is believed to be heading toward extinction. It seems unlikely, Grimwood (1969) noted, that the population ever reached 5,000 individuals. Hoces (1992:53, Table 10, 54) later reported 1,347 in 1992 (see Table 2.1), which evidently entails dramatic circumstances. However, the latest census indicates the presence of 3,810 animals (according to data obtained from the Consejo Nacional de Camélidos Sudamericanos of the Peruvian Ministerio de Agricultura), which is still less than the figure given by Grimwood (1969). The fall in numbers can be attributed almost exclusively to hunting (Tables 2.1 and 2.4).

To this information I can add some data. For example, we know that until the nineteenth century, guanacos were seen in the zone of Chosica,

TABLE 2.1. WILD CAMELIDS IN CENSUS AND CONSERVATION AREAS, 1992

Country	Vicuña	Guanaco
Argentina	15,900[1]	20,887[1]
Bolivia	12,047[2]	
Chile	27,927[3]	19,856[3]
Peru	97,670[4]	1,347[4]

[1] Cajal and Puig 1992:Table 2.38, Table 3.39.
[2] Villalba 1992:Table 5.43.
[3] Glade and Cunazza 1992:Table 7.47, Table 8.49.
[4] Hoces 1992:Table 9.52, Table 10.54.

quite close to Lima (42 km from the capital city), as well as on the heights of Huarochirí, also in the Department of Lima. It also seems that these animals lived in the Santa Valley in the Department of Ancash (Bonavia 1991:113–114). Although Wheeler (1985a:29) reports that Franklin (1975:195) found some guanacos in the Department of La Libertad, I was unable to determine whether Franklin meant the Calipuy Reserve or wild specimens at some other site.

I believe it is worth noting that there is evidence that these animals have long vanished from the Callejón de Huaylas. Lynch (1980a:13) inquired into this matter and says that camelids have not lived there "in the recent past."

The specimens remaining in the Central Andean Area live in rugged zones far from the reach of man, as was noted by Flores Ochoa (1975b:300).

According to the recent data of Hoces (1992:53), some guanacos remain in the Salinas y Aguada Blanca National Reserve, to the south of the Departments of Arequipa and Moquegua. In addition, a group was recently found close to Mount Salcantay, in the province of Anta, Department of Cuzco. Some animals also remain in the Pampa Galeras National Reserve, and its zone of influence, but these are no more than a few specimens (see Table 12.2).

In nineteenth-century Patagonia there were still huge herds of these animals, because Darwin (1921:237; 1969:171) saw "on the banks of the St. Cruz . . . one herd which must have contained at least five hundred" guanacos.

Raedecke (1976) studied the guanaco at the southernmost tip of the continent, where the geomorphology exhibits quite peculiar characteristics, and the main Andean cordillera in the province of Magallanes reaches only 2000 masl and occasionally 3000 masl (Raedecke 1976:20).

The ancient distribution of the guanaco in Magallanes included all of the pampa and the timber line, where they are found today. They could also be found in the forests of coigüe as far as the Beagle Canal and the Navarino and Gable Islands, as well as other minor islands (Raedecke 1976:12, with data from Bridges 1948).

At present, "the populations are isolated and occupy different and their own geographical sites. Besides the guanacos in southern Isla Grande, the remaining populations are limited to small groups of several hundred specimens at most, and are spread over a great expanse of land. These populations are usually found close to escape routes or hiding places where they will be safe from hunters and dogs, such as in the woods, volcanic lava fields, hills, ravines or the cordillera" (Raedecke 1976:34). The highest guanaco density is found in the farthest reaches of the province of Magallanes, east of the cordillera, in Última Esperanza and the Isla Grande of Tierra del Fuego, owing to the ruggedness of the land and its remoteness (Raedecke 1976:35). Thus, we cannot simply say, as Sumar did (1988:26), that the guanaco area is limited to the southernmost part of South America.

The guanaco population has fallen dramatically since the arrival of the Spaniards, but unfortunately there is no study of the Peruvian guanaco that presents an actual overall view (Novoa and Wheeler 1984:121; Wheeler 1984c:78; 1985b:69). The reader can find data on the extinction of these animals in Koford (1957), Raedecke (1979), Franklin (1983), Wilson and Franklin (1985), and Flannery et al. (1989:89). Franklin (1981:63) writes that the number of guanaco must have been "immense" before the coming of the Europeans—tens of millions in Patagonia and Tierra del Fuego, and millions on the dry Andean slopes. Franklin notes that George C. Musters saw herds of 3,000 even in 1871. Wheeler (1991:18) believes the population of these animals fell "to almost 7 million" in the nineteenth century, but she does not give her source.

According to Franklin (1982:469), the guanaco is still the most widely distributed South American camelid, if in small numbers, despite being in danger of extinction. At the turn of the twentieth century the guanaco was so abundant in Patagonia that the hostile sheep breeders of Santa Cruz wanted its complete eradication, on the grounds that it was harmful to sheep and was a "national plague" (Albes 1918; Allen 1942; Dennler de la Tour 1954). The Department of Agriculture of Argentina also began experiments to domesticate the guanaco and use its skin, wool, and meat, but some believed it was silly to replace domestic animals with guanacos or to believe that the wool, skin, and meat of the guanaco could be better than those of sheep (Strook 1937). This made the number of guanacos fall dramatically when competing with livestock, but especially because the young animals were hunted for their highly prized pelt. The fences that interfered with their traditional migratory routes also had a significant role in this process (Franklin 1981). Hundreds of thousands of guanaco pelts were exported from Argentina for years (Allen 1942). Franklin (1981) collected the data for eight years in the 1970s (based on personal communication by Ricardo Ojeda) and concluded that 400,000 guanaco pelts were exported from Buenos Aires (wool is not included), which yielded $3 million in taxes. This figure, of course, does not include the black market. According to this same source, in 1979 the "*chulungueros*" (also known as "*chulungueadores*")—that is, the hunters of *chulengos* (young guanacos or offspring)—had great success in the market because more than 86,000 *chulengo* hides were exported from Argentina, and some 70,000 in 1981 (Cajal 1981).

According to Raedecke (1976:12), the highest density of these animals was in Tierra del Fuego and east of the Andean cordillera. Franklin (1982:475–476) prepared an important review of these conditions.

In Peru the guanaco is a rare, endangered species. The number of animals has fallen steeply, especially due to intense hunting. As already noted, there is a small number of these animals close to Pampa Galeras, which has grown because the animals are protected. A national reserve was established in 1981 at Calipuy, and the population

there was the biggest remaining population left in Peru until the year of Franklin's study (1982), and also the northernmost (Grimwood 1969; Franklin 1975). According to the 1996 census, the national reserve with the most animals at present is Salinas y Aguada Blanca, in the Departments of Arequipa and Moquegua.

In Chile the number of guanacos was rapidly falling and approaching a dangerous level (Miller et al. 1973), until in 1975 the Servicio Nacional de Parques y Forestales began a campaign for their protection in Tierra del Fuego and in the Parque Nacional Torres del Paine, in the far south (Franklin 1981). Until 1982, when the report I am using by Franklin was published, Tierra del Fuego had one of the highest numbers of guanacos in the continent, some 12,000 animals. This population seems to have increased, because 17,224 animals were reported in the aforementioned National Park and in Tierra del Fuego in 1992. In all of Chile there were 19,856 (Glade and Cunazza 1992:Table 8.49; see my Table 2.1). At present there are four reserves for guanacos (Wheeler 1995:277).

In Argentina there was great killing. In the fifties the number of guanacos was falling and the species had been eliminated in most of northern and southern Argentina. Some believed guanacos were in danger of becoming extinct (Dennler de la Tour 1954). According to Franklin (1981), a comparison with ancient times showed that the guanaco was beginning to be scarce in Patagonia, to the point that Howard (1970) claimed it had been practically eliminated from the Patagonian pampa. The current condition of the guanaco in Argentina has not yet been well defined (Franklin 1981), but such massive hunting with no careful management and no preservation program definitely cannot continue for long. It is estimated that in 1992 there were 20,887 guanacos in Argentina (Cajal and Puig 1992:Table 3.39; see my Table 2.1). Wheeler (1995:277) says there now are 14 reserves in this country.

The guanaco probably was never very plentiful in Bolivia, but it has now practically vanished as a breeding population (Cardozo, pers. commun. to Franklin 1982:476). It is estimated that no more than 200 animals survive. Torres (1984) mentions just 54 individuals (Villalba 1992:44; see my Table 2.2). Strangely enough, the species is not protected either in Bolivia or in Paraguay (Wheeler 1995:277).

To understand this phenomenon we must bear in mind other factors. For instance, a mortality of 15% is assumed for the first month of life (Raedecke 1976:129). Raedecke (1976) is further convinced that the pressure from hunting by the Indians did not by itself limit the number of guanacos except in places of permanent occupation. For Raedecke, the constraining factor was the lack of food in wintertime. "With these pastures, the number of guanacos for this period was then at the maximum carrying capacity possible" (Raedecke 1976:12).[5]

Raedecke (1976:136) suggests the causes of and figures for guanaco mortality. The major factor is starvation, which accounts for 74.1%. Accidents come second, at 11%, and hunting is third. In the case of the *chulengos*, hunting accounts for 9% of the mortality, and for 2.7% in adults. Because of its significance, Raedecke emphasizes winter mortality of the *chulengos*, with death from starvation being involved in 45% of the cases. Raedecke also notes that in most animal populations, one of the most important causes of death is predation by man and other animals.

Raedecke (1979) himself later apparently adjusted his figures (in his dissertation, which I have not read but which is cited by Flannery et al. 1989:94–95) and noted that death by starvation reached 81% among the guanacos of Tierra del Fuego. However, accidents, poaching, and sarna (mange) also played a significant role. Sarna is caused by the same mite that attacks the llama and is detectable in 13.3% of the guanacos, although it kills no more than 2.8% of the cases. There is also infant mortality, which, however, is hard to determine. This may be due to the fact that each year, only about half the adult females have offspring that survive.

Raedecke (1976:147) established that a given area can support a limited number of animals; any increase above this figure cannot be sustained for long. In the case of Tierra del Fuego, the population is controlled by the lack of factors necessary for survival, such as food and protection. The decrement in these survival factors can be directly attributed to the numerous and large populations of sheep that remain in the area. For Raedecke (1976:148–149), no increase in the carrying capacity of the guanaco can be anticipated anywhere without a concomitant decrease in the number of domestic livestock.

In terms of the various kinds of guanaco groups, the local population of Tierra del Fuego is estimated at 112 males for every 100 females, still according to Raedecke (1979, cited by Flannery et al. 1989:94). For every 100 adult females there are 37–51 offspring of each sex, 18–26 yearling females, and 11–15 females aged 2–3 years. Most of the guanacos die by the time they are 12, but some old males can live up to 16–18 years. In 1981, Franklin (1981:63) estimated that the number of guanacos for the entire continent was between 50,000 and 150,000 animals.

Torres (1992a:35) recently noted that commercial hunting carried out intensively but without any rules is a threat to the stability of the guanaco population. Argentinian and Chilean sheep ranchers kill the guanaco because it competes with their livestock for food, and also facilitates the transfer of diseases. In other areas the lack of protection favors poaching.

The figures on the export of hides or pelts do not show the real number of guanacos killed. Hunting often produces more kills than the official statistics indicate, and, what is worse, the export of guanaco products is carried out without necessary technical support. Besides, the exact distribution of the species is unknown, and nothing is known about the existing animal population density.

At present, Peruvian law prohibits the export of guanacos and their hybrids, their semen, or other reproductive materials except for research materials previously authorized by the Ministerio de Agricultura (Decreto Supremo No. 007-96-AG, Article 37, 1996). Law No. 26496 (1995) and its Reglamento (Decreto Supremo No. 007-96-AG, 1996) regulate the property and commercialization of these animals and their hybrids. There are also sanctions for hunting them. According to these laws, the guanaco and its hybrids are protected by the state, and ownership of the herds and their products is granted to the peasant communities on whose land the animals live. It is also legislated that the wool and its derivatives must be taken from live animals in authorized shearing.

TABLE 2.2. NUMBER OF CAMELIDS IN SOUTH AMERICAN COUNTRIES, EXCLUDING PERU

Country	Year	Guanaco	Vicuña	Alpaca	Llama
Ecuador	1974				2,000[3]
	1982				2,000[6]
	1984				2,000[7]
	1985				2,000[10]
	1988			"Some"[9]	2,500[9]
	1991		482[16]	100[16]	9,687[16]
Bolivia	1974				1,800,000[3]
	1977			500,000[4]	
	1982	200[6]	4,500[6]	300,000[6]	2,500,000[6]
	1984	200[7]	2,000[7]	300,000[7]	2,500,000[7]
		54[14]			
	1985	54[16]		300,000[10]	2,000,000[10]
	1988	"Some"[9]	4,500[9]	300,000[9]	2,550,000[9]
				362,844[17]	
					2,233,020[17]
	1991		12,000[16]	324,336[16]	2,022,569[16]
	1992	"Unknown"[13]	12,047[13]		
Chile	1974				64,000[3]
	1977			80,000[4]	
	1982	20,000[6]	10,000[6]	500[6]	85,000[6]
	1984	13,000[7]	1,000[7]	500[7]	85,000[7]
	1985			80,000[10]	85,000[10]
	1988	20,000[9]	10,000[9]	10,000[9]	75,000[9]
	1991	25,000[16]	30,000[16]	27,585[16]	70,363[16]
	1992	19,836[13]	27,921[13]		
Argentina	1974				75,000[3]
	1982	550,000[6]	9,000[6]	200[6]	75,000[6]
	1983	578,700[16]			
	1984	109,000[7]	2,000[7]	200[7]	75,000[7]
	1985			200[10]	75,000[10]
	1988	550,000[9]	10,000[9]	"Some"[9]	100,000[9]
	1991		23,000[16]	400[16]	135,000[16]
	1992	550,000[13]	23,000[13]		
Colombia	1982				200[6]
	1984				200[7]
	1985				200[10]
Paraguay	1985	53[16]			

Sources:
[1] Fernández Baca A. 1971.
[2] Franklin 1973:78.
[3] Cardozo 1974a:11.
[4] Flores Ochoa 1977b:Table 3, 42.
[5] Flores Ochoa 1979:231.
[6] Franklin 1982:Table 2, 475.
[7] Novoa and Wheeler 1984:Table 14.1, 117.
[8] Brack Egg 1987:65–73.
[9] Sumar 1988:Table 1, 26; 1992:Table 2, 86.
[10] Flores Ochoa, 1990b:Table 1, 92.
[11] INIPA, 1990, Julio Sumar in a letter on 17 October 1991.
[12] INIPA 1981, in Sumar 1992:Table 1, 85.
[13] Torres 1992b:31.
[14] Torres 1984.
[15] Flores Ochoa 1990b:Table 2, 95.
[16] Wheeler 1991:Table 1.1, 18.
[17] San Román C. 1993:Table 2, 250.

2.8.2 The Vicuña
(*Lama vicugna* or *Vicugna vicugna*)

The vicuña (Figures 2.3, 2.4) is the smallest member of the Camelidae family. On average, an adult animal can weigh 38.5 kg (San Martín and Bryant 1987:5; Sumar 1988). Torres (1992a:32) says the weight ranges between 40 and 50 kg, but Wheeler has more detailed data. She indicates that *V. v. mensalis* has an average height at the shoulder of 86.50 cm in females and 90.43 cm in males. The average overall length is 96.33 cm for females and 110.73 cm for males, with an average weight of 33.24 kg for females and 36.22 kg for males (the data are from Paucar et al. 1984). Wheeler admits that there are some disagreements regarding length because published measurements range from 137 to 181 cm and from 144 to 175 cm (Hofman et al. 1983; Gilmore 1950; Pearson 1951). Wheeler also indicates that some researchers have reported heavier weights ranging between 45 and 55 kg, or 30 to 65 kg (Gilmore 1950; Miller et al. 1973). There are no statistics for *V.v. vicugna*, according to Wheeler (1991:23).

FIGURE 2.4. Drawing of vicuña that illustrated the work of Buffon (1830:Pl. 83, Tome 17). *Courtesy of Danièle Lavallée.*

The taxonomic classification of the vicuña has long been debated. We need only recall what Molina wrote in 1782 (Bk. IV, 313). He noted that Buffon believed that the vicuña was the wild alpaca let free, but pointed out that this clearly was a mistake, because the alpaca and the vicuña belong to the same genus but are different species. The debate continues, as we saw when discussing the taxonomic issues (Chapter 1), and specialists do not agree on whether this is a different genus from all other camelids or whether the vicuña is also *Lama*. I am unable to pass judgment and will respect the position taken by the various scholars mentioned here.

Two vicuña subspecies have been described. The *Lama vicugna vicugna* (or *V. v. vicugna*) predominates in the south (between 18° and 29° S latitude) and is apparently different from the northern form, *Lama vicugna mensalis* (or *V. v. mensalis*, which lives between 9°30′ and 18° S latitude) in its greater size, bigger molars, lighter color, and some other characteristics. However, study of a larger series is needed to validate this subspecies (Novoa and Wheeler 1984:121; Wheeler 1995:277). Wheeler (1995:278) comments that a dubious variety of vicuña was described (*V. v. elfridae*; Krumbiegel 1944), based solely on animals in German zoological gardens.

There is little information on vicuña hybrids, but Koford (1957:215) collected important data.

FIGURE 2.3. Vicuña (*Lama vicugna* or *Vicugna vicugna*). *Hernán Torres, courtesy of the author.*

Thus, we know that in 1845, Father Cabrera obtained a herd of 20 paco-vicuñas after 21 years of work (for more details, see Chapter 1). In 1893 Belón formed another small herd and was able to cross the hybrids (Madueño 1912). Koford notes that "[n]o vicuña hybrids have as yet been obtained, apparently due to the frequent sterility of the hybrids and the lack of genetic knowledge, but the efforts persist." However, we know that the first paco-vicuña was born in Calacala in 1931. Twenty years later Paredes had about 50 hybrids, the result of crosses between vicuñas and alpacas, llamas, and paco-vicuñas. In 1952 Paredes had about "three fourth parts" of individuals obtained crossing a female paco-vicuña with a male vicuña.

According to Wheeler (1985a:29), the Peruvian vicuña *Lama vicugna mensalis* (*V. v. mensalis*) frequents all puna regions between 4000 and 4800 masl, and, unlike the guanaco, its ecology, biology, and behavior have been well studied (Wheeler cites Koford 1957; Dourojeanni 1971; Brack Egg 1979; Otte and Hofman 1979; Franklin 1974, 1980, 1982, 1983, but neither Dourojeanni nor Brack Egg appear in her bibliography).

To judge from Koford (1957:155), Wheeler's statement (1985a:29) is based on studies made after 1957, because prior to that year no systematic study of the vicuña's habits existed. Koford stressed that almost all that had been published on the habits of this animal was "based on repetitions of the statements, many true but some false, of J. J. von Tschudi . . . (1844–1846, in German)." Koford likewise believed that the essential data are in the studies by Cabrera and Yepes (1940: 256–269 [1960:83–85]) and Gilmore (1950:429–454), which Wheeler did not mention. He also notes that the observations made by Pearson (1951:161–168) must be considered.

Franklin (1982:474) notes that if the grazing mammals from the African or North American plains are compared with the camelids that occupy high-altitude pastoral zones of the Andes, the latter must cope with the extreme heat, cold, and aridity. The vicuña has several anatomical and physiological characteristics that enable it to survive on the puna better than any other domestic animal. For Sumar (1988:26), these characteristics include the extraordinarily fine and thick fleece, unusually low energy requirements, the surprisingly high weight of the newborn animals, and the special anatomical features of the toes, which end in broad elastic pads.

Brack Egg (1987:74) has described the adaptive characteristics of this animal. These are, first, the mimetic cinnamon color of its coat, its extremely fine and warm wool, and its blood (14 million red corpuscles per mm^3), which allows more efficient use of oxygen at high altitudes. It also has callous feet better adapted to a stony surface or road, incisors that grow continuously to a certain age and thus prevent premature wear due to chewing hard and dry grass, and finally the ability to reach a speed of 45 km/hour, which enables the vicuña to defend itself from predators like the puma and the fox.

Some specific comments have to be made regarding speed, because Koford (1957) does not agree with Brack Egg (1987:74). According to Koford (1957:173), the moderate speed of a galloping vicuña is 20 km/hour. It can run at 30 km/hour for only a short time (Koford quotes Hall 1937: 472). Koford cites Howell (1944:65), who suggests that the running ability cameloids developed was to cover great distances between food and water rather than to escape from predators. Koford therefore emphasizes that the vicuña has a great ability to jump and climb on rock surfaces, and that this is what protects it from predators and dogs, rather than its speed. These animals in fact have great jumping ability. Koford notes that a fence 1.8 m high is needed to stop a vicuña, because it can jump up to 2.10 m long and usually 60 cm high. He saw one vicuña jump 1.35 m high.

Before moving on, it is worth noting the great fineness of vicuña wool. On average, the diameter of the fiber is only 13.2 microns, while the wool from a fine Merino sheep has an average diameter of 22.8 microns. In a wool-quality grading system in which the quality of the wool is based on the finest count to which the fiber can be spun, fine sheep wool is of grade 62–64, llama and alpaca wool is 56–60, but the vicuña wool grades between 120 and 130 (Koford 1957:214, based on the *American Wool Handbook* 1948).

Vicuña pregnancy lasts 11 months, according to Brack Egg (1987:74), and approximately 11.5 months in Franklin (1982:477), with a variation of about one or two weeks. These data are correct, although Koford reports that some scholars (e.g., Romero 1927:140) noted that gestation lasted 10 months, without indicating the data source for these assertions. Koford's data (1957: 176) indicate that this period actually lasts about 11 months. The infants are born in the rainy season, that is, in February–March, with the highest percentage of births in the morning, between 9 a.m. and 2 p.m., and on sunny days. Afternoons are avoided, as these are stormy. This is very important because the nights are extremely cold above 4000 masl, and the optimal environmental conditions—higher temperatures and lower rainfall—occur precisely in the late morning and early afternoon. The data gathered in 1969–1970 show that 10%–30% of infants die in the first four months, but the cause is unknown. Foxes are believed to be responsible (see Brack Egg 1987: 74; Franklin 1982:477; Sumar 1988:27). However, Koford (1957:164–165) noticed in Huaylarco (on the Arequipa-Puno highway) that about half the young died during the late fetal or infant stage. Neither predators nor humans were the cause, because the same thing happens in protected sites. This is another case that has to be studied.

Some of these animals can live up to 10 years in the wild, but the average is surely less. The longest life span appears to have been recorded in 1904 at the New York Zoological Park, where a vicuña died at age 24 (Koford 1957:165).

I have not found detailed information on the diet of these animals. Based on a study by Malpartida and Flórez (1980), San Martín and Bryant (1987:40) indicate that at Pampa Galeras "the vicuña carries out a great selection of plants, but . . . this depends on the condition of the pastures. Its selectivity is thus most limited in pastures that are in poor condition." We must bear in mind that unlike the guanaco, which grazes and browses, the vicuña just browses (Wheeler 1991:26). In fact, the vicuña is the only wild ungulate that develops favorably in the pastures of the Andean altiplano, with its fluctuations in temperature and frosts and its scant rainfall, which entail short but predictable growing seasons and low plant productivity. These animals were once common in the coastal region, in the highest parts of the western

slopes of the Andes, where they found scant annual vegetation, and in the lomas, with more of their typical vegetation (Franklin 1975:191–192; 1983:619; Grimwood 1968).

As is well-known, in the Andes the altitude of the biotic zones decreases as one moves south. The reason the vicuña does not move down beyond 20° S latitude is that the southern part is relatively arid and the appropriate fodder grows only above 3600 masl (Koford 1975:219).

Sumar (1988:26) says that "[t]he vicuña . . . only lives in wild state." Flannery et al. (1989:89, 91) have a similar position: "Most zoologists believe that the vicuña has never been successfully domesticated; it resisted all attempts at domestication by the Inca and continues to be difficult even to hybridize with the alpaca (Fernández Baca and Novoa 1968)." However, this does not seem to be completely true. There is an interesting datum in Rivero (1828), whom I prefer to cite verbatim: "The warm climate [*Rivero* means the Peruvian coast] does not seem to be an obstacle for them to live in; at present I have one [a vicuña] in the mining house at the Casa de Minería in Lima that has endured two hot summers, and in Huánuco, which is extremely warm, there are two; it is claimed that these animals do not mate when domestic." Brack Egg (1987:62) reports that more than 30 years of experiments in vicuña domestication in Puno were interrupted in 1767 with the expulsion of the Jesuits from Spain and her colonies. The order had herds of up to 600 animals there. The Jesuits had managed to tame the animals after these were born in captivity for several generations (Brack Egg gives León 1932a as his source; see also Madueño 1912:12). Cardozo (1975a:107) remarks that "[t]here have been, and are, attempts to tame and domesticate [the vicuña], but their number is insignificant. For this reason the vicuña is still considered a wild, nondomestic species."

Koford's commentary (1957:173, 215) is most important in this regard. He saw a domestic vicuña in Calacala in the altiplano, and notes that it was a house pet and a docile animal. They live very well in zoological gardens. Koford presents the statement made by Sr. Paredes, a vicuña breeder who has worked on this since 1919. In 1951 Sr. Paredes had a herd of 400 animals.

Shortly after birth, the young vicuñas were taken to the corral, where they were caught with a lasso and fleeced. These animals cannot be treated like cattle or sheep. Even the ones regarded as house pets became vicious on reaching maturity. At Calacala, most of the males had to be castrated to prevent their fighting.

Penned groups are difficult. Males and females do not fight when free, but do fight in captivity. At Calacala the vicuñas have lived in domestic conditions for generations, but they are too nervous and quarrelsome to be managed efficiently. Koford (1957:215) therefore believes that "[i]f vicuñas were amenable to domestication, it is probable that the ancients who domesticated the llama and alpaca would also have domesticated the vicuña." Here Koford makes a serious mistake. The llama and alpaca are the result of the domestication of wild forms, as was seen in Chapter 1 (see Figure 1.1), and the vicuña could be the ancestor of these domestic forms, or of one of them.

Since the time of Bolívar there were pressures, also according to Koford (1957), for the domestication of these animals. Emilio Romero and Luis Maccagno recently tried to domesticate them.

The vicuña has a most interesting social organization based on a specific herd structure and territorial system. It is one of the few ungulates that defend an annual territory for feeding and a separate one for sleeping. This territorial system is the basis on which the vicuña population, and the way it uses its environment, is organized. It should be noted that the spatial unit of the vicuña is not the marital couple but the family band. In its territory a male vicuña defends itself from another male because it is protecting its territory, not its females. The territory of these animals can be subdivided according to its characteristics into permanent territory of family groups, marginal territory for mobile family groups, territory for groups of males, and finally, one just for males that are physically and sexually mature. The troop of males without territory forms a reserve supply of males. The territory is delimited by communal mounds of dung, a characteristic of all South American camelids, but not to the same degree or in the same way. The territory of the vicuña is retained all year long, not just during mating season (see Franklin 1978; 1982:468–480; Koford 1975:205).

It is interesting that Koford (1957:211) saw vicuñas grazing 43 m apart from groups of alpacas and llamas. He counted up to 75 vicuñas in a plain where hundreds of llamas and alpacas were grazing. However, according to what Sr. Paredes told him, in a group of mixed lamoids, each animal tends to remain with its own species.

According to Dourojeanni (1973:12), the main mortality factor among newborn vicuñas is the climate (particularly low temperatures, strong winds, and abundant rains). This leads Dourojeanni to believe that perhaps in earlier times the vicuña lived in zones with more benign climatic conditions. Brack Egg (1987:74) confirms this, because for him the major mortality factor is pneumonia in the newborn offspring. Lightning and the lack of pastures due to droughts are also important factors.

The distribution of the vicuña has varied over time. In the seventeenth century, Cobo (1964a: Bk. 9, Chap. 58, 367) said of the vicuña that "It lives only in the highlands of Peru, on the coldest *páramos* and among the snow-covered cordilleras." Tschudi (1885:94; 1891:96; 1918:205; 1969:124) noted that "[t]he vicuña has a bigger habitational area than the alpaca, it is found both in central and southern Peru, and in parts of Bolivia." It is clear that both writers lacked information, because the distribution of this animal was even greater at the time they wrote. Even more important, this error has persisted. For example, in 1969 Hershkovitz wrote, "The limits of the present geographic range of the vicuña coincide with those of the altiplano. In effect this once wide-ranging species is now endemic to the altiplano" (Hershkovitz 1969:61). Actually, this is not true, because although the vicuña is a species restricted to the arid altiplano, its range also extends to the rugged pampas of the adjacent coastal mountains (Gilmore 1950:453).

Although all scholars agree that the vicuña extends from the northern reaches of central Peru, not all accept the same limit. For Pascual and Odreman (1973:35), the limit lies in the Department of Junín, while for Novoa and Wheeler (1984:121), Grimwood (1969:66), Koford (1957: 157), Hoces (1992:51), Wheeler (1991:21), and Torres (1992a:32) it lies in the Department of Ancash between 9°30′ S latitude and 10° S latitude, then runs south along the Andean cordillera and the coastal mountains, including western Bolivia and as far as northwestern Argentina and northern Chile. The southern limit in Chile is 29° S latitude in the province of Atacama, but it formerly extended up to Coquimbo, while in Argentina the limit lies between the provinces of La Rioja and San Juan. Hoces alone (1992:51) claims that the limits lie on the present frontier between Bolivia and Chile, that is, 18° S latitude. Some scholars mention the presence of the vicuña in southern Ecuador, but this, as Franklin (1982:474) notes, has yet to be confirmed. Cabrera (1961:II, 324) also questioned this. However, Brack Egg believes the vicuña "[i]s extinct in Ecuador." (See Gilmore 1950:451; Franklin 1975:191, 1982; Novoa and Wheeler 1984:121; Pascual and Odreman 1973: 35.) Even so, we should not forget that according to Cieza de León (1984:Pt. I, Chap. LXXX, 236), in pre-Columbian times there were vicuñas in the Huamachuco zone in the Department of La Libertad. Several scholars believe the historical and present base of this animal is southern Peru (Franklin 1982:474; Albes 1918; Koford 1957; Jungius 1971).

As Franklin (1982:474) correctly noted, the Andean puna is to the vicuña as Patagonia is to the guanaco. The altitudinal distribution of the puna lies approximately between 3700 and 4800 masl, though Brack Egg (1987:73) places the upper limit at 5200 masl, while both Hoces (1992: 41) and Torres (1992a:32) place the lower limit at 3000 masl. We know from Koford's studies (1957:157) that in 1950, two-thirds of the vicuña population lived above 4250 masl. This is a harsh environment of sparse vegetation and cold temperatures in semiarid and undulating high-altitude grasslands. The vicuña is ecologically confined to the high puna because it is an animal that grazes strictly on the high puna grassland, as the land at lower elevations quickly turns to scrubland valleys and slopes with intensive cultivation, or the barren foothills of the Andean coastal desert. The two conditions that favored the vicuña's sedentary habits are the uniform rainfall pattern in extensive areas of the altiplano and the abrupt transition from the altiplano's grasslands to the lower elevations of brushland (Franklin 1982:474; 1983:619; see my Figure 2.5).

We know, according to Koford (1957:164), that about 60% of the vicuña population lives in a dry and saline zone, south of the latitude of the Salar Uyuni, in Bolivia (20° S latitude). However, Koford clearly indicates that the limits of this animal's range seem to be determined by the availability of food and freedom from harassment, not by the thin air (Koford 1957:157).

As for the need to obtain moisture in food, Koford lists the alpaca, the vicuña, the llama, and the guanaco in decreasing order. The increasing geographic range of the lamoids' distribution suggests that seasonal access to fresh food and the various degrees of tolerance for dry food are crucial factors in determining the limits of their range (Koford 1957:161–162).

Brack Egg (1987:73) studied the vicuña in Peru. He notes that this animal is found from Ancash to Puno and Tacna, throughout 14 departments. Hoces (1992:51) concurs. This is significant, because 30 years earlier Koford (1957:218) gave the Department of Pasco as the northern limit for the vicuña. The only detailed analysis I have found is Grimwood's (1969:66–69), who in his report, published 18 years before Brack Egg's (1987), managed to find this animal in only 10 departments. This report is worth citing because of the detail and significance of the data.

At present, the vicuña is almost extinct between 9°30′ S latitude and 13°30′ S latitude, with the sole survivors being some nine to ten groups that are widely separated and involve fewer than 150 individuals in all. These seemed doomed, wrote Grimwood (1969:66–69), and it was unlikely that they would be able to survive for more than a few years. However, in 1992 there were 329 vicuñas in two protected zones, one in Ancash and the other in La Libertad (Hoces 1992:Table 9, 52; my Table 2.3). In 1997, 623 animals were counted; 72 vicuñas were counted in the Department of Cajamarca (according to data from the Consejo Nacional de Camélidos Sudamericanos of the Ministerio de Agricultura).

Conditions are slightly better south of 13°30′ S latitude, but they are good in just one site: Pampa Galeras, in the province of Lucanas, Department of Ayacucho (for more details on this reserve, see Franklin 1983:583 et passim). Here there were some 1,200–1,300 animals in 1965, in

Figure 2.5. Current and past distribution of the vicuña. *Prepared by the author and drawn by Osvaldo Saldaña.*

--------- Approximate distribution in pre-Columbian times. The area for which no data are available but which we can assume was occupied by these animals is shown with a dotted line.

———— Approximate distribution during the conquest and Colonial times.

▨ Present distribution.

an area of 600 km². In 1992 there were a total of 61,147, according to Hoces (1992:Table 9, 52; see my Table 2.3). I do not know how accurate this figure is, because according to Lizana Salvatierra (1993:A1), "Pampa Galeras is abandoned [since 1989] and without technicians specialized in the preservation and management of vicuñas, a fact that has spurred more poachers to appear in the area." Although it is true that this is an article in a newspaper and must be taken with a grain of salt, from what I was able to discover it turned out to be true in those years. The Pampa Galeras National Reserve is once again under control, now that the serious problems caused by terrorism are over. Even so, in 1997 there were just 51,668 animals (according to data from the Consejo Nacional de Camélidos Sudamericanos of the Ministerio de Agricultura).

In other places there are populations quite removed from each other, each made up of some groups, but only in one case do we find more than 100 individuals. There also are some 500–600 animals in semidomestic conditions in the Hacienda Calacala[6] in the province of Azángaro (Department of Puno), but about a third of them are hybrids from a cross with a llama or an alpaca. It is worth reviewing the conditions present in each department. I shall use the evidence in Grimwood (1969) and compare those data with the data in Hoces (1992; see my Table 2.3) and from the Consejo Nacional de Camélidos Sudamericanos of the Ministerio de Agricultura (1997), including some data published in newspapers in 1993 (Lizana Salvatierra 1993:A1), as well as other data I collected.

In La Libertad there were no vicuñas in 1969. The presence of 70 vicuñas was reported in 1992, and this number fell to 29 in 1997.

In 1969 there were six or seven small groups in the Department of Ancash with about 35 animals, in the provinces of Yungay and Bolognesi.

TABLE 2.3. CENSUS AND CONSERVATION AREAS FOR THE VICUÑA IN PERU

Name	Department	Hectares	Number of Vicuñas
Los Libertadores-Wari Region (Subproject Pampa Galeras)	Ayacucho[1]	527,223	59,097
	Huancavelica	34,000	882
	Apurímac	55,202	1,168
Inca Region (Subproject Cuzco)	Cuzco	120,000	1,596
Andrés Avelino Cáceres Region (Subproject Huancayo)	Junín	133,440	3,474
	Huánuco	50,000	470
	Pasco	9,560	24
Lima Region	Lima	1,038,800	3,667
Arequipa Region (Subproject Arequipa)	Arequipa[2]	331,288	2,966
José Carlos Mariátegui Region (Subproject Puno)	Puno	1,832,767	21,363
	Tacna	181,190	2,196
	Moquegua	76,530	438
Chavín Region (Subproject Huaraz)	Ancash[3]	28,000	259
La Libertad-San Martín Region	La Libertad	5,568	70
Total		4,423,568	97,670

After Hoces 1992: Table 9.52
[1] Includes the Pampa Galeras National Reserve.
[2] Corresponds to the Salinas y Aguada Blanca National Reserve.
[3] Corresponds to the Parque Nacional Huascarán.

The species is extinct in the rest of the department. However, there still were vicuñas in the Callejón de Huaylas in the 1980s, but these were "restricted to extremely high elevations where [the vicuña] is rarely seen, and then only in precariously small herds" (Lynch 1980a:13). Some 250 animals were reported at the Parque Nacional Huascarán in 1992. The 1997 census counted 594 animals dispersed among the provinces of Recuay, Bolognesi, Ocros, Yungay, and Wari.

In the Department of Junín, a small group was reported in 1969 west of Lake Junín, but the species was extinct in the rest of the department. In the 1970s, Matos and Rick (1978–1980:33) noted that the vicuña "had almost vanished in the northern sierra. Their density is poor, albeit important, south of the Nudo de Pasco." In fact, in 1977–1979 there were vicuñas in what was then the SAIS Ramón Castilla in Junín, between Jauja and Tarma. In 1984 there were some in northern Junín in the Corpacancha and Conocancha area, and their presence in the SAIS Túpac Amaru was confirmed in 1979–1986 (Domingo Martínez Castilla in litt., 9 May 1997). Even so, the 1997 census counted 10,515 vicuñas in the provinces of Junín, Yauli, Tarma, Jauja, and Huancayo.

In 1969 there were only three groups in the Department of Lima, with fewer than 20 individuals each, in Matucana and Canta. In 1992, 3,667 vicuñas were reported. Later, 16,961 animals were counted in the 1997 census, spread over the provinces of Yauyos, Oyón, Cajatambo, Huarochirí, Huaral, and Canta.

A group of fewer than 20 animals has been reported in the Department of Huancavelica, close to the lakes of Castrovirreyna, and two small herds are known in the southern tip of the department, close to Córdova and Huachuas. The 1997 census recorded 6,740 animals in the provinces of Castrovirreyna, Huancavelica, Huaytará, and Angaraes.

Besides the group at Pampa Galeras, which consisted of 51,668 animals in the 1997 census, the vicuña is found in Ayacucho close to Negromayo, where there are perhaps 150–200 animals, even though only 55 have been seen. There may be some animals close to Andamarca.

In 1969 there were very few vicuñas in the Department of Apurímac, in the provinces of Andahuaylas and Abancay. The presence of small

groups has been reported in the province of Antabamba. Between Andahuaylas and Puquio-Chalhuanca 84 animals were seen, 55 of which were concentrated in six groups in Pampa Chuquibamba. Two separate herds were seen in the Pachachaca Valley, one with seven animals, the other with 18. Some 1,168 animals were counted in 1992 in the protected area, and 11,551 in 1997, dispersed throughout the provinces of Grau, Andahuaylas, Aymaraes, Abancay, and Antabamba.

According to 1969 data, vicuñas had long been absent in the Department of Cuzco, in the provinces of Quispicanchi, Paucartambo, Calca, and Urubamba, if indeed they ever lived there. They have vanished from the Ocongate area, where they lived some time ago. Small groups have been sighted in the La Raya area, in the Vilcanota Valley, and some probably lived in parts of the provinces of Canchis, Canas, Chumbivilcas, and Espinar. In 1992 it was believed that 1,596 animals lived in this department. However, according to the 1997 census, there were 2,817 in the provinces of Canchis, Quispicanchi, Chumbivilcas, Espinar, and Paucartambo. Canchis was the province with the biggest population— 51.44% of the department's total.

In 1969 the situation was somewhat obscure in the Department of Arequipa. Local hunters claimed that there were still vicuñas. The presence of 2,966 animals in the Salinas y Aguada Blanca National Reserve was reported in 1992. Here, however, the available information is contradictory. In 1993 it was said that "[i]n Arequipa, some 10,000 vicuñas have no protection against poachers due to the lack of interest shown by the Ministerio de Agricultura in taking the necessary measures that would warrant their preservation" (Lizana Salvatierra 1993:A1). The threat of extinction faced by the vicuñas in this department was recently announced. Mauricio de Romaña, president of Protección de la Naturaleza (Prodena), claimed in an interview that "[t]he vicuña's predicament in Arequipa is alarming, and I dare say that the species is in danger of extinction because there is no control whatsoever" (Anonymous 1997b). However, the latest 1997 census gave a total of 2,898 animals distributed throughout the provinces of Condesuyos, Arequipa, Castilla Alta, and Caylloma.

In 1969 there were small groups of vicuñas in the Department of Puno, south of the provinces of Carabaya and Sandia, but no sure data were available. The vicuña had vanished completely from the province of Azángaro, except in semi-domestic conditions in the Hacienda Calacala. The presence of small groups has been pointed out in the haciendas south of the province of Chuquito, but these groups comprise fewer than 150 individuals. Some 70 vicuñas were known in the southwestern corner of the province of Lampa, and there might have been some in the province of Melgar. However, in 1992, it was said that 21,363 animals lived in the department. The Puno Subproject, which included Tacna and Moquegua, had a total of 23,997 vicuñas, a number that, according to the 1997 census, had fallen to 15,321. However, in 1993 the newspapers claimed that according to data from the Consejo Nacional de Camélidos Sudamericanos, "[c]onditions are alarming in Puno. The vicuña population has fallen in the last four years from 23,000 to 5,000 due to poachers" (Lizana Salvatierra 1993:A1). However, this last census gave a total of 14,307 animals distributed throughout the provinces of Chuquito, Putina, Carabaya, Collao, Azángaro, Melgar, Puno, Lampa, Huancané, Sandia, Yunguyo, and Moho. Chuquito has the highest number of animals (33.33%) and Moho the lowest (0.10%). Some groups are known in the provinces of General Sánchez Cerro and Mariscal Nieto, Moquegua, with a total of 294 animals according to the 1997 census.

The reports of both Grimwood (1969) or Hoces (1992) do not mention the Department of Ica. However, the 1997 census records 100 animals here, all in the province of Chincha. Grimwood concluded in 1969 that the total was fewer than 2,500 animals. He admitted that a complete survey was impossible, so it can be assumed that the total population of the species may be as low as 5,000 individuals. Grimwood believed that to say there were 10,000 animals in 1969 was to make too high a claim.

Here a comment is in order. Twelve years before, Koford (1957:164) had estimated that the total South American vicuña population was 400,000 animals, more than half of which were in Peru. In other words, there should have been more than 200,000 vicuñas in Peru, which is about 190,000 more than Grimwood asserted in 1969 was "too liberal" an estimate. According to Hoces (1992), the total Peruvian population in 1992 was 97,670 vicuñas (see Tables 2.1 and 2.3), while the latest census, carried out in 1997, gives a total of 102,780. If these figures are accurate, then the change in 1969–1992 was truly remarkable for the increase in the number of animals. The decline from 1957 to 1997 is likewise remarkable. Grimwood was convinced that dogs and hunters were partially responsible for the death of young animals.

The picture becomes much bleaker when we bear in mind that the 1982 estimates considered that the Peruvian vicuña population accounted for 75% of the world vicuña population, followed by Chile with 10%, Argentina with 10%, and Bolivia with just 5% (Brack Egg 1987:73).

The distribution of the vicuña outside Peru is as follows, according to Koford (1957:218). To the east in Bolivia, the vicuña is found in the vicinity of Potosí and Cochabamba. In Argentina, its eastern frontier is the Santa Victoria cordillera. Little is known of its distribution in southern Bolivia and Argentina. However, we know that the vicuña is abundant in the zone of Jujuy and in western Salta. There are vicuñas in the provinces of La Rioja and San Juan, and to the west of Catamarca, on the border with Chile. In Chile the vicuñas reach approximately the same latitude as San Juan in Argentina.

The most recent information available dates to 1992 and gives 12,047 vicuñas for Bolivia (Villalba 1992:Table 5.43), 27,927 for Chile (Glade and Cunazza 1992:Table 7.47), and 15,900 for Argentina (Cajal and Puig 1992:Table 2.39). These are data for animals in census and conservation areas (see Table 2.1).

Using data collected between the 1950s and the 1970s (Koford 1957:162–163; Jungius 1971:139), Hesse (1982:209) reported that juvenile individuals (i.e., less than a year old) made up 17%–19% of the whole vicuña population.

Everything seems to indicate that thousands of years ago, the vicuña was not confined to the highlands, as Koford (1957:218–219) correctly notes, because fossil remains have been found in deposits dating to the Pleistocene and Early Holocene in the lowlands of the Argentinian pampa, close to Buenos Aires (López Aranguren 1930b:120–122). In this regard, López Aranguren wrote, "Since this specimen comes from Lake Chichi, it might seem strange that a species now confined to the highest Andean plateaus could have lived in what is now the province of Buenos Aires. However, the Museum has a fragment (9-341) of an incomplete and very fossilized jaw that cannot be separated from that of the present-day vicuña either by its characteristics or size, and the same thing happens with the mandibular symphysis from Luján, presented by Ameghino [1889] in his plate XXXVI. . . . Perhaps the vicuña from the province of Buenos Aires represents a local form" (López Aranguren 1930b:122). To explain this, Cabrera and Yepes (1940:268; 1960:84) suggested that the climatic conditions in the pampa have gradually changed, or that perhaps a local vicuña form adapted to this new way of life. These animals naturally have physiological characteristics that enable them to live at high altitudes. (This was discussed and explained earlier in this chapter.) But besides exhibiting great efficiency for life in a high-altitude environment with low oxygen content, the vicuña has great tolerance for variations in atmospheric pressure. So much so, in fact, that this same animal that lives at high altitudes can also live in a very healthy state in zoological gardens at sea level. Furthermore, here they reproduce quite well. For this reason, Koford (1957: 218–219) posits, and I concur, that the main factor restricting the vicuña to the highlands is almost surely not physiological. Instead, it seems that the species was widespread throughout South America, as shown by the fossil remains found in the lowlands, and began to become extinct there when competition with other ungulates became more severe. However, it must not be forgotten that the coming of man and later the great disturbance wrought by sheep herding are two of the factors that forced the vicuña to leave much of its original habitat (i.e., in the lowlands) and seek refuge in the highlands. So it is quite possible that sheep herding has brought about great changes in the puna grasslands throughout the century. The present and remarkable dominance of coarse bunchgrasses might be due in part to the heavy utilization of the most succulent plants by the sheep (Koford 1957:212). Koford concludes that

"as change in elevation goes hand in hand with change in climate, vegetation, the numbers of humans and livestock, and other factors, critical evaluation of the influence of each factor on the distribution of vicuñas will not be possible until our knowledge of the ecology of the Andean highlands has been greatly advanced" (Koford 1957:219). Cardich (1987b:22) disagrees. He claims that "it is unlikely that [the vicuña] had an ecological behavior different from the present one in a not too distant past, such as the Early and Middle Postglacial." No argument is presented to support this claim. Cardich simply repeats Troll's ideas (1931, 1935, 1958) and then mentions Koford (1957), so he either has not read Koford or he misunderstood him, because we have just seen that Koford's position is exactly the opposite of Cardich's.

The problem is clearly far from being solved, but there are reasons to believe, according to Rick (1980:21), that the vicuña was the first camelid that occupied the puna in the past, and at present it is the only wild camelid surviving in that environment. If the guanaco were present, it must have been eliminated in a most selective way. Rick (1980:21) believes that the predictable nature of the vicuña made it an easy prey for hunters, while its social organization was far more easily disturbed than the pattern known for the guanaco. This agrees with Lynch (1980a:13), who noted that: "From observations in southern Peru and Bolivia, it appears that their natural range is considerably greater (even below 4000 m), and that vicuña populations may have been numerous enough to contribute substantially to early human subsistence. Even in the 1940s Pearson (1951:121) calculated a frequency of about one vicuña per 120 acres in the 5000 m-high altiplano of southern Peru. He found that this compared surprisingly well to a figure of one deer per 25 acres on some of the best deer land in the Unites States."

The fall of the vicuña to the status of an endangered species is clearly a postconquest phenomenon (Novoa and Wheeler 1984:121) that might come to an end in our time. I find the discouraging account given by Flores Ochoa (1967: 25; 1979b:25–26) not just an isolated case but a generalized phenomenon. Flores Ochoa discussed the area of Paratía, the district of the same name in the province of Lampa, Department of Puno.

On the most remote and highest peaks, far from the predatory reach of the misti, or mestizo [N.B.: Flores Ochoa's translator made a mistake here, for the original text said *blanco* (white man), not *mestizo*], one sometimes sees isolated specimens of the beautiful and slender vicuñas.... Numerous herds of magnificent animals once grazed on the hillsides. Now they have become a memory, the subject of nostalgic stories told by men forty or fifty or older. The mestizos, with their greed and their automatic weapons, did not hesitate to destroy an entire herd of vicuñas in a day in order to acquire and sell the extremely fine wool.

Despite this somber picture, Flores Ochoa (1975b:300) still hopes that the vicuña is "undergoing a recovery, even though it has yet to overcome the critical level." Sumar (1985:12) believes that although this camelid was on the verge of extinction in the 1950s, "its numbers have grown thanks to the appropriate measures applied in Peru with international technical assistance." However, Flannery et al. (1989:89) studied the vicuña in the eighties and were not as optimistic; they believed the vicuña was facing extinction. Recent reports in the newspapers seem to confirm this, for the deaths of a great number of animals at the hands of poachers are continually reported (e.g., Anonymous 1999). To this should be added the lethal effect the 1997 cold wave had on camelids in general (e.g., Anonymous 1997a).

Finally, it is worth noting some differences between the guanaco and the vicuña. Although this is not too well documented, as Franklin (1982:474–475) points out, it seems that the geographic and ecological divide of the guanaco and vicuña is related to altitude. Some scholars claim that both live together (Lydekker 1901:1002), with the guanaco living throughout the vicuña zone (Koford 1957:211). Others have noted that they both live in the same mountains, but never together (Osgood 1916:203). Still others claim that some guanacos live on the puna (Link 1949:45), and that this animal lives from sea level to over 4000 masl, including areas that lie below the vicuña's altiplano habitat (Miller et al. 1973:59).

Franklin (1982) believes there are few areas where the guanaco and vicuña live in close proximity, given the sharp decline in both species. Franklin claims to know only two of them: one on the western Andes in southern Peru (14° S latitude), in the Pampa Galeras Vicuña Reserve, the other on the eastern slopes of the Andes in central Argentina (32° S latitude), in the San Guillermo National Reserve. In these two cases both species are adjacent, but separated by the habitat and by altitudinal differences. Guanacos end and vicuñas begin in the transition zone between the mountain communities and the flatlands, at about 4000 masl. The vicuña uses the grasslands in the high puna, while the guanaco occupies the lowland mountains with shrub, as well as desert areas. According to Franklin (1982:481), "[v]icuña and guanaco social behavior and social systems are basically the same, yet there are a number of subtle and important differences."

The social organization of the guanaco is usually more varied and flexible than that of the vicuña, given its wide distribution in South America, its wider distribution at high altitudes, and its occupation of several kinds of habitat (Franklin 1982). On the other hand, the vicuña population is sedentary, while the guanaco population is sedentary and migratory, or migratory. Guanaco migrations are altitudinal or lateral (without changing the altitude) because of snow and droughts. In the 1940s, Allen (1942) noted that the guanaco moves downward from the eastern Andean spurs to lower altitudes when summer begins in temperate Patagonia. On the western slopes of the Peruvian Andes, the guanacos also make changes in altitude movements when they move to and from the coastal lomas.

We thus find that similar environmental conditions explain the similarities between these two species. Most differences between the sedentary vicuña and the migratory guanaco result from having fodder in the vicuña's environment, and the seasonal availability of food for the guanaco due to the snow mantle (Franklin 1982:483). Readers interested in more data on the vicuña and its management should see Hofman et al. (1983).

At present, the export of the vicuña and its hybrids is forbidden in Peru, as well as export of its semen and other reproductive materials (Decreto Supremo No. 007-96-AG, Article 37, 1996). Export is allowed only for scientific research and must be authorized by the Ministerio de Agricultura. Like the guanaco, the vicuña is also subject to Law No. 26496 (1995), which sets down the property and commercialization system and the penalties for hunting. With this law, the vicuña and its hybrids come under the protection of the state, and ownership of the animals is granted to the peasant communities on whose land they live. These communities likewise own the products derived from these animals, namely, the wool and its derivatives obtained from live animals and from authorized shearing.

At present, three kinds of land where vicuñas live (called Apéndices) are recognized in Peru, in accordance with an international convention. The first affords total protection. In the second, shearing for commercial purposes is allowed, and a third is almost a free territory, even allowing commercialization of the animals. In this regard it is worth recalling that Article 17 of the Supreme Decree No. 007-96-AG, the Reglamento of Law No. 26496, regulates the "special permission for the internal commerce" of these animals." The Conference of the Convención Internacional de Comercio con Especies en Peligro de Extinción (CITES), held in Miami in 1994, included all of the Peruvian population within Apéndice 2 and likewise allowed the export of the wool, which had previously been banned, as only the export of cloth was allowed (Rony Garibay Suárez, pers. commun., 10 August 1999).

2.8.3 The Llama (*Lama glama*)

The llama (Figures 2.6, 2.7, 2.8) has been defined as an "[a]dmirable example of the adaptation of an organism to the physical conditions of the environment" (Viault 1895:201), but this is only half true, as already noted. At present, no population of wild llamas is known to be in its homeland (Franklin 1982:465). For Wheeler (1991:19, 27), they seem to descend from *Lama guanicoe cacsilensis*.

The llama is the biggest of the four species, reaching a height ranging between 1 and 1.20 m at the shoulder. Its weight when alive is given by most scholars as around 110 kg, though some give as much as 155 kg (Gade 1977:114; San Martín

and Bryant 1986:2; Franklin 1982:Table 1, 465). Sumar (1988:25) claims that adult males weigh 116 kg, with variations that range between 66 kg and 151 kg; this variation includes the weight indicated by Torres (1992a:32) of 125 kg, and an adult female of 102 kg with the upper limits ranging between 70 and 150 kg.

According to Franklin (1982:482), Rosenmann and Morrison (1963) found that llamas have a more efficient respiratory adaptation to heat stress and dehydration than the domestic ruminants, but are unable to alter the body temperature through hyperthermia (Schmidt-Nielsen et al. 1957), which is advantageous for preserving water in desert conditions. While the guanaco has the "thermal windows" that Morrison (1966, see above) discovered, in conditions where water is scarce the llama is able to feed on more food and expel less urine than goats adapted to dry environments. The llama is able to reduce its metabolic energy far more than sheep and goats even when food is limited (Rübsamen and Engelhardt 1975).

The usefulness of the llama in the Andean world is due to its ability to carry loads, precisely because it is one of the best adapted animals in the American continent (Custred 1977:65).

We have already seen, when discussing the general characteristics of this family at the beginning of this book, that the ability these animals have to go for a long time without water has been greatly exaggerated. This is reflected in Prescott (2000:806; 1955:114), who wrote that "[t]he structure of its stomach, like that of the camel, is such as to enable it to dispense with any supply of water for weeks, nay, months together." I do not find it worth insisting on this point after all that has been said about it. What is worth discussing concerns the food of these animals. In this regard the llama "seems to fall within the group of ruminants classified as dry and fibrous fodder consumers, which is how Van Soest (1982) classifies the Old World camels" (San Martín and Bryant 1987:41).

San Martín and Bryant (1987:37) noted that the selective characteristics of llamas observed by San Martín (1987) resemble those seen in the Old World camelids. The latter actually discard the thick and luscious vegetation and seek the dry pastures usually avoided by other animals (Yagil 1985).

FIGURE 2.6. Llama (*Lama glama*). *Wilder Trejo, courtesy of the author.*

FIGURE 2.7. Llamas illustrating the first edition of Pedro de Cieza de León's *Crónica del Perú, Primera Parte.*

FIGURE 2.8. Drawing of llama that illustrates the work of Buffon (1830:Pl. 83, Tome 17). *Courtesy of Danièle Lavallée.*

According to San Martín et al. (1989:97, 108), the llama seems to be better adapted than the alpaca (and, of course, than an ovine) to subsist eating the low-quality fodder found in the most arid regions of the Andes. These animals "make more extensive use of the coarse clustered fodder [found] in the highest vertical stratum. They also spend more time in each pasture, have a lower bite-rate, and select diets that are lower in quality (San Martín 1987) than those of the alpaca, which indicates that the llama apparently takes more time to eat coarse fodder. Llamas are apparently able to use such coarse fodder due to their physiological adaptations that allow them to compensate for these low-quality diets (San Martín 1987)."

It is worth recalling that the llama grazes and browses, and that this enables it to adapt to the most diverse ecological conditions (Wheeler 1991: 31). However, although the llama is such an important animal, there are no studies on the botanical composition of its diet (San Martín and Bryant 1987:30). It follows from the data in Cardozo (1954) and Franklin (1982), which are based on visual observation only, that the llama has a greater preference than other ruminants for dry, tall, and fibrous fodder, while the alpaca prefers pastures that grow in wet soil. These observations on the llama's selective characteristics made San Martín and Bryant (1987:33) believe that this animal is adapted to dry environments: "The theory is supported by the present distribution of the llama. Seventy percent of the llama world population is found in the Bolivian altiplano . . . where the annual rainfall ranges between 250 and 450 mm. Meanwhile, in the Peruvian altiplano, which only has 25% of the world's population, the annual rainfall ranges between 500 and 900 mm (Tapia 1971). In the Peruvian altiplano the llama population is similarly concentrated mostly in dry areas (dry puna) (Novoa and Wheeler 1984; Tapia and Flores 1984). On the other hand, it has been observed that this species is susceptible to problems of flatulence when it grazes on wet lands, a problem rarely found in alpacas (Sumar, personal communication)." San Martín and Bryant take as a starting point only the present distribution of these animals, and do not consider that it was different in the past.

As regards the gestation period, in llamas it is 10.5 months. Flannery et al. (1989:99) observed in the Yanahuaccra-Toqtoqasa area in Ayacucho that births take place anytime between January and March, but are usually more frequent in February. Shimada and Shimada (1985:6) in turn indicate that the llamas taken to the United States give birth all year long, while in the altiplano they do so between January and March, a point of agreement with Flannery et al. (1989:99). However, Shimada and Shimada (1985) claim that the animals can give birth all year long when they are isolated and are paired each month. As regards birth, we know that among the Aymara, llamas are not helped at all, but the newborn do receive special care (Tschopik 1946:521).

One point on which there apparently is no agreement is the breeds or varieties of llamas that exist. Although Gilmore (1950:436) accepted that llama traits are vague and that these must have been far more marked in prehispanic times than at present (he mentions the five-toed variety), he noted (1950:437) that today there seem to be "several vague breeds." Gilmore mentions the common llama, the large, burden-bearing llama of the altiplano, which he, however, lists with a "?," and the small Riobamba llama (Ecuador), which also is associated with a "?." In truth, there are no definitive data. Gilmore adds, "In pre-Columbian times, in the Highlands and on the Coast of Peru, there also existed several breeds, some as indefinite as those today: (1) Small llama on the Coast (definitely not an alpaca; specimens seen from Pachacamac, south of Lima); (2) normal-sized llama with sunken forehead from the Coast (specimens from Pachacamac); (3) large burden-bearing llama (?) of the highlands, utilized especially by the Inca army; and (4) aberrant llama with five front toes from Chancay, Central Coastal Peru." As shall be seen throughout this study, there are no data that can support the presence of these breeds or varieties. Wheeler (1991:29) was emphatic on this point, noting that "we cannot talk of the presence of Andean llama breeds," even though she later changed her mind in light of new evidence (Wheeler et al. 1992; see below).

Wing (1975b:33) is far more cautious in this regard, for she claims that "at least" two llama breeds are distinguishable, but she likewise insists that "with most vague characteristics." These breeds are the big pack llama and the common small-sized llama. Wheeler et al. (1992:468–469) explain this problem well. They believe that the lack of written prehispanic sources on llama and alpaca breeding and loss of the orally transmitted knowledge the specialists had make it difficult to assess the significance of European conceptions of animal breeding, and how the latter influenced animal management. At present, llamas are raised as pack animals, and three different phenotypes are known, though it is possible that more exist. Almost all Andean llamas are of the *q'ara* (*ccara*) type. This means that they do not grow wool, do not have hairs on the face, and have limited growth of wool. Less common is the *ch'aku* (*chaku*) type, which, as its name indicates, is raised for meat, has heavier fleece, and has hairs growing on its forehead and ears. The hairs characteristic of the third type fall in between the two previous types. Color tends to be patchy and ranges from white to brown, black, or gray. The llama's wool is usually coarse and hairy, and it is generally agreed that llamas are not, nor have they ever been, bred for wool, while the alpaca is specifically kept for it. Two phenotypes can be distinguished based on the characteristics of the wool. According to Wheeler et al., almost 90% of alpacas currently have short and crimped hairs of the *huacaya* variety, while 10% have the long and wavy hairs of the *suri* type (see Wheeler 1991). The *huacaya* type apparently recalls the Corriedale sheep breed, and the *suri* of Lincoln. There are animals with intermediate-type hairs, but these are rare (see also Wheeler 1995:286–287). Recent studies by Frank and Wehbe (1994) identified seven different kinds of wool in the Argentinian llama population, thus giving rise to the possibility that there are more than the three varieties mentioned. A different kind of classification was proposed by Cardozo, who divided llamas into brachymorphic (round short profile, abundant wool) and dolichomorphic (narrow elongated profile, sparse wool) (Cardozo 1954:61).

The genetic factors controlling these traits are unknown, and the *huacaya/suri* 9:1 ratio is due to chance breeding. The colors in the hairs range from white to black and brown, including intermediate hues, but tend to be uniform throughout the body. According to Novoa (1981), all llamas and 80% of alpacas are today under the control of traditional herders, who do not raise animals by selecting for specific phenotypes. The European concept of breeding, with record keeping for the herds, is not part of the native Andean stockbreeding practices. However, studies by Wheeler et al. (1992) seem to show that in prehispanic times, there was some knowledge that allowed for the selection of specific traits in a group of animals, thus giving rise to real breeds. This point is discussed in Chapter 4.

Interestingly, Gilmore (1950:428) believes that religious and other stimuli can help maintain some breeds, which can then be developed by isolating and crossing them. Religious needs would thus be responsible for the black and white llamas, as well as for the white alpacas, and perhaps even for the five-toed llama found in archaeological sites.

The versatility of the llama is well known in the Andean world, and this is something that becomes evident when talking with the herders. I find the case of Q'ero (Cuzco) to be the best example. Here the community makes extensive use of this animal, which plays a major role in its ecology, as it allows produce from different altitudinal levels to be transported and used. The Indians note that the llama is far more tolerant than the alpaca of the various pastures and rugged terrain that characterize the zone, and that the llama needs less care. Llamas usually graze without having to be looked after by the herders (Webster 1971a[b]:177).

One of the llama's most important activities is as a pack animal. Today, however, it obviously does not have the same importance as in prehispanic or Colonial times, despite serving this function in highland communities. It is best to leave this point for later on, so as to weigh its real importance.

The range and distribution of llamas is an important subject. Although several scholars have said that it roughly coincides with the borders of the old Inca Empire (e.g., Franklin 1982:467), there is no assurance that this distribution was due to the Inca. On the contrary, there is some evidence, provided later, that seems to refute this. It is interesting that Prescott (2000:807; 1955:115) thought that the *ichu* (the name the natives give to the hard-leaved and prickly graminaceous plants with hard pointed leaves, such as *Stipa*, *Festuca*, *Calamagrostis* [see Soukop n.d. (1987):218]) was responsible for the area occupied by the llama, and that "the absence of it is the principal reason why they have not penetrated to the northern latitudes of Quito and New Granada." This is very doubtful.

Troll (1935:142) considered that the northern limits of llama raising coincided with northern Peru, or with the line where the puna ends. The wet *páramo*, constrained south of this line by the eastern slopes of the mountains, extends from there westward and shortly thereafter encompasses the whole width of the Andean cordilleras slightly to the north. For Troll, the frontiers of the puna lie on the eighth southern parallel, at about the city of Trujillo. Troll said that "[i]t is understood that it is not fully impossible to raise the llama also in the 'paramo zone.' But the fact that optimum puna conditions are not found there is shown by the acclimatization attempts carried out by the Incas in present-day Ecuador" (Troll 1935: 142).

Troll (1935:142) also claimed that "it is very difficult to take these animals to the humid jungle zone to the east, even for a brief time." This is only half true, for as we shall see later on, there are instances of llamas living in these regions, as indicated by historical sources.

Some scholars, such as Flores Ochoa (1982: 63) and Flannery et al. (1989:89), accept that the llama can live in almost all ecological levels of the central Andes because it tolerates a wide range of habitats, from the highlands to the coast. Others, however, like Wheeler (1985a:29), clearly state that their "favorite habitat lies over 3000 masl, particularly on the vast steppes of the puna," something Gilmore (1950:436), Novoa and Wheeler (1984:117), Sumar (1988:25), and Gade (1977:116) also say in other words. This is not correct, as we shall see. What is true is that, as Flores Ochoa (1990b:92; emphasis added) notes, "*at present* . . . [these animals are found] . . . in lands that lie over 4000 masl," an idea shared by Cardozo (1954). Llamas prefer drier areas and the herders know it, so in the wet season, the animals are taken to lower zones (Browman 1974a:191). For this reason, the llama is also common in the land of the vicuña (Koford 1957:212).

Regarding the geographic range of the llama, some scholars believe it extends from southern Colombia to Chile and Argentina (see Novoa and Wheeler 1984:117 and Fig. 14; Sumar 1988:25; Wheeler 1991:27), while for others the northern limit lies in central Ecuador (in the Riobamba zone) and the southern one in northern Argentina, with Chile left aside (Gilmore 1950:433; Gade 1977:116). Palermo (1986–1987:68), for example, emphatically states that the llama "does not live south of 27° S latitude (in the province of Catamarca, Argentina), save for isolated specimens in zoological gardens and some small groups recent-

ly introduced in the provinces of Tucumán, La Rioja, San Juan, and Córdoba." There are other minority positions, such as Pascual and Odreman (1973:34), who claim, based on a study by Cabrera and Yepes (1940), that the llama "at present lives from the Department of Huánuco to the south," though they later add that according to Troll (1968:28), "there still are llamas in the páramo above Riobamba" in Ecuador. The other position is held by Cardich (1974:35), who claims that "in past times the area between Lake Junín and Lake Poopó, which is given by Browman (ibid. [1973]) as the place to raise llamas, would have to be extended in its northern limits at least up to 8°30′ latitude, the northern limits of the Cordillera Blanca." This point is discussed later in the context of the paleontological, archaeological, and historical data (see Figure 2.9).

What should be pointed out, and the specialists do agree on this, is that the highest concentration of llamas has always been in the altiplano. Tschudi (1885:94; 1891:96; 1918:206; 1969:124) long ago said: "The region where the llama is found in highest density, and this from very ancient times, is the province of Collao, particularly in the area around Lake Titicaca, but the geographic range of the llama has diminished throughout the centuries. The llamas were probably found in a vaster area than today in pre-Inca times and certainly before colonial times, especially in zones to the west and north." Wheeler (1985a:29) notes that at present, the area occupied by the llama tends to be restricted to a zone around Lake Titicaca which has a radius of approximately 400 km. The llama is, in fact, scarcer in areas more to the north, and there are some areas, such as the Callejón de Huaylas, where according to Lynch (1980a:13), it has never lived "in the recent past." However, older people do recall the presence of herds of llamas in the early twentieth century at a nearby site, the puna on the eastern slopes of the Cordillera Blanca and the western slopes of the eastern cordillera, an area that corresponds to the famed site of Chavín de Huántar (Miller and Burger 1995:424).

"The llama . . . Its importance is due to its ability to carry loads and in transportation, where it is irreplaceable for herders and complementary for agriculturalists who use its help, albeit in smaller

FIGURE 2.9. Current and past distribution of the llama. *Prepared by the author and drawn by Osvaldo Saldaña.*

----······   Approximate distribution in prehispanic times. The southern area for which no data were found is shown with a dotted line.

——   Approximate distribution during the conquest and Colonial times.

▨   Present distribution.

amounts because they sometimes have one or two animals which suffice to help them in their short-range commercial movements, either to go to the Sunday markets or to carry their harvest to their dwellings" (Flores Ochoa 1990b:85). For Gade (1977:118), if llamas are still kept, it is only because they are better than donkeys at high altitudes.

No llama census exists, just population esti-mates. The last one was for 1997 and added up to 1,119,777 animals (Ministerio de Agricultura, Oficina de Información Agraria; see Table 2.4). The export of these animals is allowed, but no statistics are available. It is estimated that some 700–800 llamas and alpacas leave Peru every year, most of them alpacas (Rony Garibay Suárez, pers. commun., 10 August 1999).

TABLE 2.4. NUMBER OF CAMELIDS IN PERU

Year	Guanaco	Vicuña	Alpaca	Llama
1964		5,000[8]		
1965			3,304,000[5]	
1967		5,713[8]	3,290,000[16]	
1971			3,200,000[1]	
1973		10–15,000[2]		
1974				954,000[3]
1976			2,444,800[15]	1,361.050[15]
1977			3,865,000[4]	
1978		55,500[8]		
1980		61,822[8]	2,402,305[10]	
1981			2,490,000[12]	1,361,050[12]
1982	5,000[6]	62,000[6]	3,020,000[6]	900,000[6]
1984	5,000[7]	50,000[7]	3,020,000[7]	900,000[7]
1985	1,600[16]		3,020,000[10]	900,000[10]
1986			2,510,912[16]	989,593[16]
1988	5,000[9]	65,000[9]	2,490,000[9]	1,361,050[9]
1990			3,037,000[11]	1,080,000[11]
1991		100,000[16]		
1992	1,347[13]	97,670[13]		
1997			2,675,695[17]	1,119,777[17]

Sources:
[1] Fernández Baca A. 1971.
[2] Franklin 1973:78.
[3] Cardozo 1974a:11.
[4] Flores Ochoa 1977b:Table 3, 42.
[5] Flores Ochoa 1979:231.
[6] Franklin 1982:Table 2, 475.
[7] Novoa and Wheeler 1984:Table 14.1, 117.
[8] Brack Egg 1987:65–73.

[9] Sumar 1988:Table 1, 26; 1992:Table 2, 86.
[10] Flores Ochoa, 1990b:Table 1, 92.
[11] INIPA, 1990, Julio Sumar in a letter on 17 October 1991.
[12] INIPA 1981, in Sumar 1992:Table 1, 85.
[13] Torres 1992b:31.
[14] Torres 1984.
[15] Flores Ochoa 1990b:Table 2, 95.
[16] Wheeler 1991:Table 1.1, 18.
[17] Oficina de Información Agraria, Ministerio de Agricultura.

2.8.4 The Alpaca (*Lama pacos*)

At present, no "wild" or "feral" populations of alpacas are known to be in their homeland (Franklin 1982:465). As we shall see in Chapter 5, the evidence today seems to indicate that this species evolved from the vicuña. Even Wheeler (1991:32) believes the alpaca derives from the subspecies *Lama vicugna mensalis*.[7]

The alpaca is a smaller animal than the llama (Figure 2.10) and measures approximately 1 m at the height of the shoulder, according to Sumar (1988:25). Gade (1977:114) indicates this varies from 0.60 to 0.90 m. Sumar (1988:25) claims that the average weight of a male adult is 64 kg and that of a female 62 kg, while San Martín and Bryant (1987:2) indicate, with data from Condorena (1980), that the mean weight for the *huacaya* variety is 62 kg, and 64 kg for the *suri* variety. Gade (1977:114) mentions a range of 68–102 kg, and Torres (1992a:32) gives the figure of 70 kg. Franklin (1982:Table 1, 465) gives a range of 55–65 kg. As Wheeler (1991:34) notes, these variable figures are due to the fact that the weight of the fleece is not taken into account.

The harelip of these animals allows them to feed on short and tough grass. Their teeth are periodically renewed, and this enables them to chew the tough puna grass. The alpaca is characterized by its ability to cut the grass without harming its roots, whereas the roots are damaged by sheep because they pull them up (Flores Ochoa 1967:76; Flores Ochoa and Palacios Ríos 1978:84).

According to San Martín and Bryant (1987:41), the alpaca is classified as an opportunistic animal, and in the group of intermediary animals in pasture selection. This group is characterized by its use of a wide variety of plants. The alpaca is an animal that specialists characterize as highly adapted to its environment, and it has the capacity to change its selection of plants in native pastures according to their availability (San Martín and Bryant 1987:37–38). The alpaca thus mainly eats tall graminaceous plants during the rainy season and small ones in the dry season (San Martín and Bryant 1987:28, based on the study by Tapia and Lascano 1970).

Flores Ochoa (1967:23) says that the alpaca's favorite food is a variety of *ichu* the natives call *rama pasto* (*Calamagrostis* sp.), but he also notes (1979:228–229) that the alpaca is quite adaptable to tough, green, and juicy grasses such as the *khunkuna* (*Dislechia muscoide* [a mistake; the correct name is *Distichia muscoides*], *Plantago rigida*), *paqo*, and *kuli*, all grasses that need an environment with abundant water. In summer the alpaca easily finds fresh food because the prairie is covered by

FIGURE 2.10. Alpaca (*Lama pacos*). *Jorge Flores Ochoa, courtesy of the author.*

green vegetation, thanks to the rains, but once it vanishes, particularly in September and October, the driest and warmest months, the alpaca can suffer from not having the type of food to which it is adapted. For this reason, herders have developed an irrigation system that enables them to modify natural conditions and provide their animals with the grasses that they prefer. Two kinds of irrigation systems are known. One is called the *oqho*, *waylla*, or *bofedal*, an artificial swamp that requires permanent irrigation; the other is called *qarpay*, a temporal irrigation system.

Webster (1971a[b]:177) gave a good description of this phenomenon for the herds of alpacas raised by the Q'ero (90 km east of Cuzco), herders at altitudes of 4000–4600 masl. The animals here eat a hard, high-altitude grass and apparently suffer occlusions and masticatory deficiencies. If the animals are taken to lower altitudes to graze in harder soils and more humid environments, not only do they contract fatal diseases, the production of wool also decreases. Topic (1987:832) has also emphasized the diseases caused by changing pasture zones, noting pulmonary problems, trouble with gums and teeth, intestinal diseases, parasitosis, problems with hooves, and even malnutrition.

Alpacas' strong dependence on water has often been noted (e.g., Gilmore 1950:442; Flannery et al. 1989:91; Browman 1974a:191; Maccagno 1932:15). Gilmore (1950:442) has even suggested that perhaps the ecological barrier restricted the range of this animal. Flores Ochoa (1975b:301) provides a good description of this aspect in the life of the alpaca:

> Alpacas are very susceptible outside the altitudinal limits indicated, so the herder must always be careful to provide them with the best living conditions, leading them to green moist grasses and abundant water. This is why transhumance is the major characteristic of high-altitude pastoralism. In summertime the animals are taken to the "lower" parts between 4000 and 4300 m, where the grass is green and abundant thanks to the rains. In winter they are moved to the high zones where the puna valleys abound, brooks and rivers run, and there are lakes and springs that allow tough and spongy

high-altitude grasses that are always green, as was already noted.

Alpacas' limitations "place them in weak survival conditions" and make the herders' intervention necessary (Palacios Ríos 1990:66). Besides indicating dependence on fresh pastures and humid localities, Palacios Ríos also noted that alpacas "[c]an survive on dry grass, but the output of meat and wool is of poor quality." The animals can even lose locks of hair, which dry up. The meat also becomes insipid and fibrous, and is no longer good for making charqui or dried meat (see also Flores Ochoa and Palacios Ríos 1978:85).

These animals have another important limitation: they cannot look for food below the snow. Snow can fall at any time of the year. The snowfields can cover great expanses in the dry season or winter (i.e., between May and October) at altitudes that are below 4200 masl. In general (for there are no fixed terms), the snow disappears just a few hours after it falls, and usually does not last for more than two or three days. If it lasts longer, it becomes a real disaster for the herders. The snows fall with more intensity in the rainy season, that is, from December to March, particularly in areas above 4300–4400 masl, and can last for several days without a break. In these cases the snow can reach a depth of several centimeters. In these conditions, alpacas are unable to find food and die of hunger. This is what causes seasonal movement of the herds, with transhumance to the higher parts over 4500–4600 masl in the dry or winter season, where the water from melting glaciers forms springs and ponds with fresh grass, and a descent to lands below 4400 masl in the rainy season, when the rains produce new pastures (Flores Ochoa and Palacios Ríos 1978:85–86).

Wing (1975b:33) has noted the presence of two varieties of alpaca, a long-haired one and a short-haired one. Flores Ochoa (1967:76) also distinguishes two varieties, but he defines them as "major," thus implying that there are others. The ones he mentions are the fine-wool *suri*, whose infants are very delicate, and the coarse-wool *waqayo* (*wakaya*), which is more resistant and stronger.

Alpacas have a pregnancy cycle that lasts 11 months, and offspring are born in the rainy months, when it is not as cold and the pastures are better, so the mothers have good food and can provide milk. In addition, the animals are born during the day, and this prevents the newborn from having to withstand the very low temperatures that characterize night time on the puna (Flores Ochoa and Palacios Ríos 1978:84). However, the mortality of the offspring is very high, while alpaca fertility is low. This is because many males and females are sterile and aborted births are frequent. For this reason, herds grow slowly, which is slowed even more by diseases. Despite being adapted to its environment, the alpaca is a delicate animal. The major cause of mortality among the newborn is diarrhea (hemorrhagic septicemia) (Flores Ochoa 1975b:304; 1967:77; Maccagno 1932; Custred 1977:65–66). Sumar (1988:25) notes that it is precisely due to this high neonatal mortality and low fertility rates that the use of alpacas in most farms comes to only 6%–9%.

It has often been said that the alpaca is the member of the Camelidae family best adapted to high altitudes, but at the same time it is claimed that the alpaca can have trouble adapting to lower altitudes (Gilmore 1950:432, 444). It has even been said that it "suffers malaise" when brought down below 1000 masl (Flannery et al. 1989:89). When discussing the physiology of the alpaca at the beginning of this chapter we saw that it actually is not adapted to high altitude, but instead is an animal that can live better there than others because it has special attributes. In other words, its adaptation is genotypic. Gilmore (1950:444) unwittingly contradicts himself, because after noting this presumed adaptation, he adds: "though they live long and breed in North American zoos."

It is clear that the alpaca became important once its wool was prized and began to be exported, first to neighboring countries and then to Europe (Flores Ochoa 1990b:87). For Andean Indians, on the other hand, the alpaca has always been, and still is, an important source not only of wool but also of protein in their diet (San Martín and Bryant 1987:2), and even as a pack animal, although this practice is not as widespread. The alpaca is a willful animal in comparison with the passive llama, and does not venture far from its pasturelands (Koford 1957:154).

The alpaca is the least widespread species in this family, as Tschudi (1885:94; 1891:96; 1918:

205; 1969:124) had already noted. Scholars agree that its habitat today is the highlands, land that lies above 4000 masl and up to 5200 masl, though it seems that its ideal habitat lies between 4300 and 4800 masl, with temperature variations of ca. ± 15°C and abundant wet areas or *bofedales* (Flores Ochoa 1990b:92; Novoa and Wheeler 1984: 117 and Fig. 14.2; Wheeler 1985a:30; Custred 1977:66; Cardozo 1954:93–94).

Interestingly, this habitat is reflected in Andean ritual because, as Flores Ochoa (1974–1976: 259) explains, "the relation between the *illa*[8] and the *qocha*[9] in the *señalu q'epi*[10] is not just a symbolic representation of the alpaca's favored habitat to which it is best adapted, and where the best results are obtained in the production of meat, wool and 'multiplico' [increase in numbers], which is why they want to keep them there. . . . The preferential position the pacocha's *illa* has in the *haywarisqa*,[11] over and above that of llamas and sheep, also indicates that the alpaca is the animal best adapted to, and the most valuable one in, the puna."

Scholars do not agree on the distribution of the alpaca, but it is not true that there are few studies on this point (Dedenbach Salazar 1990:96). Dedenbach Salazar (1990:96–97) gives the area comprised by the northern and central sierras of Peru (especially Cajamarca, Piura, Huamachuco, and Huarochirí). Flores Ochoa (1990b:85) includes almost all of the high puna in central and southern Peru, while Gilmore (1950:441–442) says the alpaca is limited to southern Peru, northern Bolivia, and the northern tip of Argentina. In turn, Wheeler (1985a:29), Gade (1977:116–117), and Pascual and Odreman (1973:35) favor the altiplano, with an approximate radius of 200–400 km around Lake Titicaca, but Wheeler has accepted that the alpaca extends from Cajamarca to northern Chile and northwestern Argentina (Wheeler 1991:32; 1995: 288). Gade (1977:117) even specifies that the alpaca "is not found north of 11° in the central Peruvian sierra (i.e., about as far as the city of Junín) or south of 21° (i.e., southern Bolivia and northern Chile). There are therefore no llamas or alpacas at 37.50° in the central sierra of Chile, as claimed in a popular textbook on economic geography, or as far south as 32°" (see Figure 2.11).

Flores Ochoa (1982:64) believes that the reason for this distribution of the alpaca is that it is

FIGURE 2.11. Current and past distribution of the alpaca. In the specific case of this animal, delimiting its area of distribution in past times is far more difficult, given the problems for identification mentioned in the text. *Prepared by the author and drawn by Osvaldo Saldaña.*

--------   Approximate distribution in prehispanic times.

------   Approximate distribution during the conquest and in Colonial times.

▨   Present distribution.

coupled with the humid puna, "and does not have a remarkable presence in the northern páramo that begins in the central Peruvian highlands and extends as far as Venezuela." However, Wheeler (1985a:29) recalls that the alpaca once occupied almost all of the Andean lands, from Ecuador to northwest Argentina, and Gilmore (1950:441) specifically indicates that the southern limit is the Argentinian locality of Catamarca.

Peruvian legislation has banned the export of alpacas since 1843 (Novoa and Wheeler 1984: 126), but in 1991 the door was opened to export with the Decreto Legislativo No. 635 and the Decreto Supremo No. 042-91-AG. These, however, were far too general as laws. The Decreto Supremo No. 008-94-AG was therefore issued in 1994, which laid down the rules for export. Thereafter the animals had to be publicly auctioned, an export quota set, a minimum and a maximum width of wool-thread in microns were required of the wool, its colors were specified, and finally, the number of animals in each variety that could be exported was determined (Rony Garibay Suárez, pers. commun., 10 August 1999).

In 1987, Brack Egg estimated that Peru had 90% of the world's alpaca population (Brack Egg 1987:75). Today it is almost impossible to calculate this figure accurately because a significant number of these animals live in several countries throughout the world (see Chapter 12). There are only estimates. Yet, using the data available for other South American countries (see Table 2.2) and for the rest of the world (see Chapter 12), it can be assumed that the number of animals must be more than 408,000. Considering that the last estimate made in Peru in 1997 gave a figure of 2,675,695, we can conclude that at present, 85% of the world's alpaca population lives in Peru. In 1982 Franklin noted that the alpaca was replacing the llama in Peru (Franklin 1982:468). Current figures are proving him right, because the 1997 estimates (Tables 2.4 and 12.1) gave 1,555,918 more alpacas than llamas.

We can end this chapter by noting that there are no data on wild populations of llamas and alpacas, even though a mix of wild and domestic camelids occurs in the wild (Novoa and Wheeler 1984:123). Camelids have at least four very distinctive physiological characteristics: oblong red blood cells, induced ovulation, the recycling of nitrogen during digestion, and gastric pouches in the stomach. The reproduction of these animals is not fully controlled, and the low fertility index just mentioned is due to a poor adjustment in reproduction in captive conditions. Besides, their life at high altitudes obviously exerts an influence. The only characteristic that has been taken advantage of in the reproduction of llamas and alpacas is their ability to reproduce in just a single season (see Cardozo 1975a:88, 106).

NOTES

[*] Monge C. wrote the section on camelid morphology and physiology.

[1] The term "pre-adaptation" is often used in evolutionary biology and is inadequate from a literal perspective. By pre-adaptation we should mean a chance genetic change that enables adaptation to a new habitat (Carlos Monge C., pers. commun., 28 June 1994).

[2] The term "allopatric" is applied to different species or subspecies that do not occur simultaneously and that have different geographic areas of distribution.

[3] The endostyle is a small enamel column or style located on the crown, at the union of the two great cusps of the upper molars (Rusconi 1931).

[4] The term used is "base of tail," which is not clear. The term "TL" usually means from the tip of the nose to the tip of the tail. The term "SL" usually means the tip of the nose to the base of the tail, where the tail joins the body.

[5] The carrying capacity is the maximum number of individuals of any species that a given area can support.

[6] Koford (1957:173, e.g.) writes Calacala, while Grimwood (1969:67) prefers Cala Cala. I follow Koford, who I believe is right. Stiglich (1922:176) also spells it as Calacala. This site is 15 km southwest of San Antonio de Putina and 130 km north of Puno.

[7] Kadwell et al. (2001) say that archaeozoological evidence links alpaca origins to the vicuña at 6000–7000 years B.P.

[8] These are representations of the *paqochas* (alpacas). They are figures carved in hard, fine-grained stone that represent alpacas (Flores Ochoa 1974–1976:252).

[9] The *illas* have a hole on their back called *qocha* (lake, pond, lagoon) where beer, *chicha*, and alcohol are poured during the *haywarisqa* ceremony (Flores Ochoa 1974– 1976:252).

[10] A bundle where the sacred objects used in pastoral rites of fertility and propitiation are kept (Flores Ochoa 1974–1976:248).

[11] A long and complicated ceremony held between late December and the carnival. It is related to the fertility and increase in the number of livestock animals (Flores Ochoa 1974–1976:247).

CHAPTER 3

NOTES ON THE PALEONTOLOGY OF THE SOUTH AMERICAN CAMELIDS

with the collaboration of Christian de Muizon[*]

3.1 ORIGINS

The continent of South America was separated from North America during much of the Tertiary, between 65 and 2 million years ago. It was only 3 million years ago—that is, toward the end of the Tertiary—that the Isthmus of Panama rose and South America became connected to North America (see Marshall et al. 1979; Marshall 1985).

This land bridge opened the door to a great faunal exchange between the two continents, but some animals (the Procyonidae) had migrated earlier (Simpson 1980; Marshall 1985), and the ground sloths moved north (Webb 1985). Several migrations from North America to Asia also took place during the second half of the Tertiary. These migrations are documented in the fossil record, and I refer to them throughout this chapter. Only those connected to the history of the camelids are considered here.

When the land bridge of the isthmus made the Great American Interchange possible, most families of land mammals crossed from North to South America and vice versa around 2.5 million years ago, in the Late Pliocene. At first, animals moved in both directions, producing a similar mix on both continents (Webb 1991). However, the impact of the interchange waned in North America after 1 million years. During the Pleistocene,

groups that had originally entered from North America continued diversifying at an exponential rate, with the result that only 10% of North American genera derive from immigrants from the south, whereas more than half the modern mammals of South America entered as immigrants from the north. Furthermore, six immigrant taxa that entered North America became extinct, whereas only two immigrant families became extinct in South America (Webb 1991).

Webb has posited a two-phase model to explain the asymmetrical outcome of this exchange of land mammals. During the first phase, a humid interglacial phase, the tropics were dominated by rain forests, and the major biotic movements developed from the Amazon to Central America and southern Mexico. During the Ice Age, which was drier, the savanna habitats extended widely along the tropical latitudes. According to Webb, this hypothesis predicts that the immigrants from the north were usually able to reach higher latitudes in South America than the taxa going in the opposite direction to North America. Northern families achieved greater diversification in their phylogenesis at the low latitudes of North America before the interchange. Webb emphasizes that these hypotheses have yet to be proved.

The Camelidae originated in the North American continent (Carroll 1988:514), and a group of them, the Lamini tribe, migrated to

57

South America; another group, the Camelini tribe, migrated to Asia, and then probably dispersed to Africa and Europe. As is the case with humans as well, many hypotheses and conjectures have been proposed to explain the migration of fauna, and particularly to establish the precise time such migrations took place.

Camelids are artiodactyls whose origins lie close to the Paleocene ungulates. They were originally grouped with the Condylarthra, known earlier from North America than from South America. They appeared in the Late Eocene as part of the great radiation of selenodont artiodactyls.[1] It should be borne in mind that artiodactyls are quite varied (they have crescent-shaped teeth with cuspids; see Webb 1965) and that their differentiation began in North America. After the Isthmus of Panama emerged, the Tayassuidae, the Camelidae, and the Cervidae entered South America, attaining maximum diversification in the Late Pliocene and in the Pleistocene and declining thereafter. Although we know that camelids appeared in North America during the Eocene, it has still not been established when and where the first South American camelids differentiated (Hershkovitz 1969:67, 26). In general, the mtDNA-based study of Stanley et al. (1994) on the molecular evolution of the Camelidae family agrees with the data from the fossil record but does not clear up the problem. On this subject they wrote:

> If estimates of the timing of the Camelini-Lamini divergence from the fossil record are correct (11 my), our data suggest an average rate of nucleotide substitution of 1.6–1.8% per million years. Using this estimate, our finding of 10.3% sequence divergence in the genus *Camelus* may have begun in the early Pliocene (5–3 my ago), possibly before migration of the Camelini to Asia, but no fossil material of *Camelus* is available from North America to help resolve this question (Harrison 1985). Similarly, a 6.7% divergence between the wild South American species seems high if these genera appear 2 my ago. (Stanley et al. 1994: 4)

We therefore must try to trace the history of this family with the evidence currently available.

Those camelids that appeared in the late Eocene in northern South America were the first modern families of artiodactyls (i.e., even-toed ungulates), followed by pigs, peccaries, and deer in the Oligocene, and giraffes, pronghorns, and bovids in the Miocene (George 1962). The origins of camels, both South American as well as Asian/African, go back to North American ancestors. In fact, this was an exclusively North American group for more than 40–45 million years of their evolution. The critical dispersal to other continents took place in the Late Miocene for the Old World and in the Late Pliocene for South America (Webb 1965).

Poebrodon, the first Camelidae, comes from the North American Upper Eocene (40–36 mya). This form is poorly known, had low-crowned teeth, and was probably the size of a big rabbit. An ancestral camel (*Poebrotherium wilsoni*) appeared in the Lower Oligocene. This animal was somewhat larger, looked like a guanaco, was the size of a sheep (ca. 60 cm), had a gap between its incisors and its canines, and had two toes on each foot. Its low-crowned premolars and molars were adapted to browsing on leafy vegetation. Perhaps it lived in an environment ranging from the open flatlands to the mountains in the west. It is possible that these ancestral camelids were quite advanced and efficient runners compared with contemporary ruminants. Among the most advanced artiodactyls of that time, the ancestral camel had noticeably long legs, as well as a long and powerful neck (Webb 1972).

Camelids diversified in the Early Miocene (i.e., 24–16 mya), probably as a result of the spread of savannas in the great North American flatlands at that time, which provided this family with an environment and an important catalyst to evolving. The major changes were in morphology, feeding habits, and locomotive adaptations. Forms like *Protolabis*, *Aepycamelus*, and *Procamelus* appeared at this time.

The two major contemporary groups of camelids, camels and South American camelids, appeared in the Late Miocene. The first known Lamini is the genus *Pliauchenia* from the Clarendonian (Middle to Upper Miocene, 18–14 mya).

The genus *Paracamelus* appeared in North America in the Upper Miocene. It has also been found in the Early Pliocene in China and Russia (David Webb, pers. commun., 5 May 1994) and in the Spanish Turolian (7 mya; Morales Soria and Aguirre 1980). This means that the Camelini moved to the Old World before this date with the genus *Paracamelus*. The first known camels in Africa date from the Lower Pliocene (5–3.5 mya). They rapidly spread west along the dry Eurasian belt and reached the southern area around the Mediterranean, in East Africa, and as far as China to the east, through the Gobi Desert. These Old World camel forms eventually developed into the two present-day species, the two-humped Bactrian camel (*Camelus bactrianus*) of the mountains and the Mongolian steppe and the one-humped dromedary or Arabian camel (*Camelus dromedarius*) of Southwest Asia and the deserts of northern Africa (Gauthier-Pilters and Dagg 1981). In northern Africa, the dromedary is known from 2700 BP (Lhote 1987), while the earliest fossil camel remains in Asia are those of Shar-I-Sokhta, in central Iran (Compagnoni and Tosi 1978), which were dated to 4600 BP and are considered to be *Camelus bactrianus*. *Camelus dromedarius*, the dromedary, was found in about the same period, but only in the Arabian Peninsula. Although the Shar-I-Sokhta camel is considered a more domestic form, few data are available that would let us identify when the domestication of camelids began in the Old World (Scossiroli 1984:237).

Meanwhile, long-legged *Hemiauchenia*, which lived in Hemphillian times (Upper Miocene-Lower Pliocene), became diversified in the central and southern latitudes of North America and appears in various North American contexts as *Hemiauchenia vera*. The same genus and series of species continued living in North America throughout the Pleistocene. It was around 3 million years BP that the Panama land bridge gradually closed the Bolívar Channel,[2] separating the two continents. One of the most spectacular and well-documented exchanges took place then (Marshall et al. 1982), including the invasion of *Hemiauchenia*, which appeared in the Andes and the South American plains at the beginning of the Pleistocene (see Franklin 1982:458–462; Marshall 1985).

According to Webb (1974:170), traits similar to those of the llama had already begun to develop in certain progressive species during the Miocene, while the camelids were still an exclusively North American species. Thus, in Clarendonian times (Late Miocene), the genus *Pliauchenia* had many characteristics of the Lamini, which allows it to be classified in this tribe and makes it the oldest known representative (the author cites the studies of J. T. Gregory [1942] and Webb [1965]). The history of the Lamini tribe throughout the Pliocene and Pleistocene is complex. The main obstacle to a clearer understanding of this history is the complex distribution of the Lamini throughout both continents, and their rapid evolution. One of the early groups of North American llamas spread south and settled widely, and then returned in part to North America (see Figure 3.1).

One of the figures included by Webb (1978: Fig. 1, 397) shows the distribution of the hypsodont mammals in South America. The Macraucheniidae appear there almost from the beginning of the Miocene to the end of the Pleistocene. The Macraucheniidae lived in the South American steppes from the Middle Miocene to the Pleistocene as long-necked browsers, mainly along river banks. In the Middle Tertiary, the fauna living in the Andean region was the woodland, savanna, or scrub type (Webb 1978:402).

In the Late Miocene, the native fauna of the savanna in both American continents show a marked decrease in diversity. The savannas turned into steppes in the temperate latitudes of North America, and the vertebrates that lived in this environment were decimated. Something similar happened in South America but even earlier, during the Middle Miocene. On both continents, however, some savanna vertebrates adapted rapidly to the steppe, and then to desert conditions. By the Middle Pleistocene, the faunal exchange between both continents had come to an end precisely because of these changes and the rise of an environment with moderate humidity.

Open-land vertebrates were among the groups arriving from the north that had remarkable success, the llamas among them. A second major climatic change took place in the Late Pliocene and Early Pleistocene, and the resultant cooling increased the range of the savannas, thus

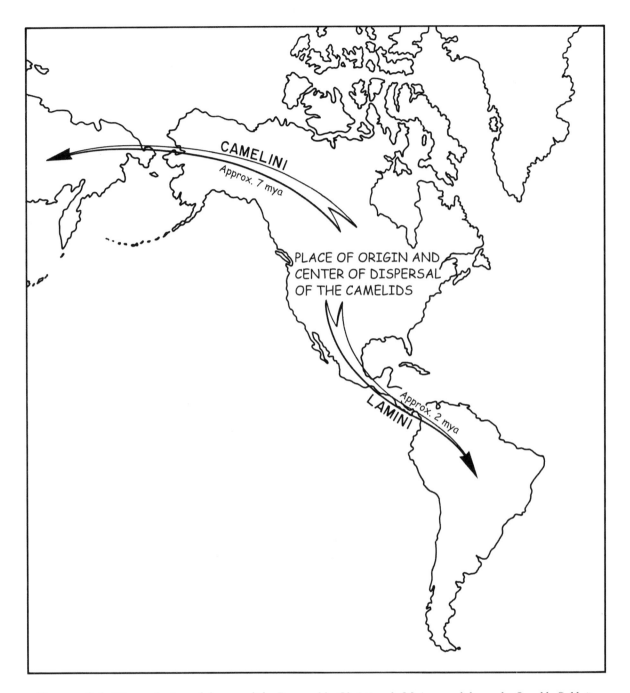

FIGURE 3.1. The evolution of the camelids. *Prepared by Christian de Muizon and drawn by Osvaldo Saldaña.*

facilitating the great faunal exchange associated with this period. Several groups of animals took advantage of the more open vegetation to migrate north to south, and vice versa. The animals native to the savanna had to compete with the newly arrived immigrants. It was at this moment that the litoptern Macrauchenia and the notoungulate Toxodon remained as the most remarkable rep-resentatives of the native ungulates. On both continents, the final decline of the savanna animals occurred by the end of the Pleistocene. It was then that camelids disappeared from North America. Interestingly, several savanna verte-brates belonging to the mixed fauna of the exchange survived at latitudes lower than those they had occupied prior to the Pleistocene extinction.

These taxa with a restricted post-Pleistocene diffusion include the llamas (Webb 1978:416–418). Webb (1978:405, Table 1) shows the mixture that took place between grazing animals throughout the Late Pliocene and the Early Pleistocene. *Hemiauchenia* and *Lama* appear in the north, and *Palaeolama* in the south. These genera were all probably adapted to the savannas.

Webb and Marshall (1982) carefully analyzed the faunal displacements between the two continents and divided these movements into a series of strata based on Simpson's study (1980). What they call Stratum 3 comprises the last 7 million years in the history of South American land mammals, and it was in this period that an extraordinary event took place, which the authors call "the Great American Faunal Interchange." This stratum can be divided into two parts, based on the immigrants, the time of their arrival, and their means of dispersal. Stratum 3a comprises the Procyonidae (and possibly also the Cricetidae rodents), which arrived as passive migrants[3] from Central America about 7 mya. Stratum 3b comprises those taxa that walked to South America from Central America across the Panama land bridge after it appeared around 2.8 mya (Webb and Marshall 1982:40; Marshall 1985).

According to Webb (1978:404), in the last million years of the Pliocene and the first million years of the Pleistocene (i.e., from 3 to 1 mya), two dozen North American genera extended over South America, and a dozen South American genera moved to North America. The llama was among those groups that settled in the south. A superficial analysis of this mixture of inter-American fauna shows that it was widely based on, and predominantly adapted to, the savanna.

During the Middle and Late Tertiary, a separate evolution of the savanna biota took place in the temperate latitudes of each American continent. The Great American Faunal Interchange presents a truly dramatic ending to this story. From a geographic perspective, it is surprising that this mixture of autochthonous temperate fauna extended widely over the tropics. From a temporal perspective, it is to be noted that although the evolution of this distinct fauna took 40 million years, the mixing took place in just 2 million years (Webb 1978:403).

Webb's perspective on the possible routes followed by the savanna vertebrates as they dispersed throughout South America after arriving from the north is extremely interesting. Webb (1978:413–416; see also 1991:274, 278) posits the presence of two major routes, one he calls the "Andean Route" or "Upper Route," the other the "Eastern Route" or "Lower Route." These are two savanna corridors with different characteristics and diverse taxa. The Andean Route probably was a more direct natural route, presenting a more uniform, non-woodland natural path to the vertebrates. Webb deems that at intervals, the vertebrates might have followed the Pacific spurs instead of the altiplano (highland plateau). During the glacial intervals, glaciations and increased precipitation combined with colder and drier conditions in the lowlands, and so we may assume that there was a change in the limits for many savanna vertebrates, centered around midpoint elevations. The Andean Route and the Eastern Route converge in Argentina. Webb and Marshall (1982:43) believe that "[t]he 'Andean Route'. . . was an entirely appropriate dispersal route for most of the North American land mammals involved. The ranges of these immigrants presumably expanded along the lines of least resistance, some concentrating on the altiplano, others on the Pacific or Amazon slopes. In diverse ways the Andes allowed them to get a foothold deep within the continent of South America (Eisenberg 1979)."

According to Marshall (1985:75, Fig. 6), one of the South American animals that dispersed to North America was *Palaeolama*, for which there is evidence in South America from the early Ensenadan (ca. 1.5 mya) and, in North America, from the mid-Irvingtonian (ca. 1.4 mya). However, the recent discoveries of the Leisey Shell Pit change the picture, for they correspond to the early Irvingtonian (ca. 1.6 mya) (Webb and Stehli 1995).

Among the North American animals to disperse to South America was *Hemiauchenia*, known in the southern continent since the Uquian (ca. 2 mya) and in North America from the Hemphillian (over 5 mya). Among the North American pseudo-dispersal animals to South America are *Lama*, for which there is evidence in South America dating to the late Uquian (ca. 2 mya), and

Vicugna, whose remains have been found dating to the early Ensenadan (ca. 1.5 mya).

Marshall (1985:74, 76) explains what he calls "dispersal events" in the following terms. The sudden appearance of an immigrant marks a dispersal event. Since the exchange was potentially reciprocal, then a dispersal event recorded in North America should have a "twin" in South America. Marshall distinguishes seven of these events, of which only three are of interest here.

Type 1 is exemplified by an immigrant taxon in continent B that comes from continent A, with fossil evidence in A that temporally precedes or co-occurs with the evidence obtained in B. These taxa are the ones most securely identified among the types of dispersants. They include *Hemiauchenia*, which moved from North America to South America.

Type 6 represents the pseudo-dispersers and comprises those taxa belonging to a family on continent A that developed in B from an ancestor that dispersed from A. These taxa, like those in Type 7, are dispersants that are really autochthonous to B and developed from allochthonous families in A. These taxa have been confused with true dispersants. The best examples are *Lama* and *Vicugna*, which evolved in South America from immigrant (Type 1 dispersers) *Hemiauchenia* (Webb 1974).

Type 7, which Marshall (1985) defines as pseudo-dispersers, comprises the taxa derived from a continent A family that developed from an ancestor in A that dispersed to B and in B, and later dispersed to A. A possible example would be *Palaeolama*, which was believed to have originated in South America from *Hemiauchenia* (Type 1 dispersed to South America) and then dispersed to North America (Webb 1974). However, with the discoveries made in the Leisey Shell Pit (Webb and Stehli 1995), the position of *Palaeolama* must be reexamined.

Still according to Marshall (1985:69, 71), the first immigrants to South America that used the Panama land bridge were the Camelidae, represented only by *Hemiauchenia*. *Palaeolama* was among the first immigrants to North America over this same route. Thus, it appears that two synchronous and reciprocal events happened in the Late Tertiary: the first between 2.8 and 2.6

mya, the second between 2.0 and 1.9 mya with *Hemiauchenia*, which moved to South America. There was only one dispersal event in the Early Pleistocene at the time of the maximum glaciation 1.4 mya, with *Palaeolama* scattering over North America (Marshall 1985:77). However, this position is now questioned with the evidence from the Leisey Shell Pit (Webb and Stehli 1995).

Interestingly, the members of the families that "walked" from South America to North America gave rise to virtually no divergence, whereas about 60% of the genera of North American "walkers" evolved in situ, which shows that the immigrants were subject to considerable diversification after their arrival (Marshall 1985:387).

Webb and Marshall (1982:40) wrote: "The most dramatic change in the generic composition of the South American land mammal fauna began in the late Miocene and continues into the present. . . . [T]his interval is the most revolutionary chapter in the history of the land mammal fauna of South America. Stratum 3 taxa rose to hegemony, and now comprise at least half of the Recent land mammal genera in South America." Webb and Marshall (1982:45) state that two mechanisms have been posited to explain the great success of the taxonomic diversification of Stratum 3. One ascribes the increase in secondary diversification to a limited number of immigrant stocks in South America (i.e., an autochthonous diversification), while the other posits that much of the increase derived from a continuous and accelerating migration from Central America (i.e., an allochthonous diversification). Currently, however, neither of these positions is regarded as wholly correct, and the evidence suggests that both mechanisms were operative. In the case of the Procyonidae, the diversity of the fossil taxa in North America and of recent ones in Central America argues strongly in favor of the allochthonous hypothesis. On the other hand, the diversification of the Camelidae in South America accords better with the autochthonous hypothesis, since the centers of diversity known in the past and in the present occurred in South America, and specifically in the Andean region.

According to Webb (1974:210–211), *Hemiauchenia* extended its range in South America quite early in the Pleistocene. The oldest discoveries on this continent are in Argentina, in the Uquian or Chapadmalan deposits (Simpson 1940; Kraglievich 1946; Pascual et al. 1966; Patterson and Pascual 1969). These South American species were presumably derived from North American populations of the Blancan age. Still according to Webb (1974), studies do not indicate this correlation or show a special relation between any particular species of Nearctic and Neotropical *Hemiauchenia*. Comparisons will be possible when more material from Uquian strata is available. There was no other kind of llama in Blancan times, when *Hemiauchenia* spread from temperate North America to temperate South America. It was during the Middle Pleistocene, when the Tarija fauna accumulated in Bolivia, that the extinct genus *Palaeolama* and the modern *Lama* developed.

It seems evident that *Palaeolama* differentiated from *Hemiauchenia* after settling in South America. The center of origin of *Palaeolama* and the modern llamas would have been the Andean region. The limb bones became stockier, the metapodials shorter, and the epipodials longer. David Webb (1974:210) noted that these changes in the limbs imply adaptation to a rugged terrain, one of the major characteristics of the Andean region. However, when going over the matter once more with Stehli, they wrote: "This view appears reasonable when applied to the modern llamas of South America, although they are by no means confined to mountainous habitats; but it seems more tenuous in the context of the Gulf Coastal Plain in North America. More plausible perhaps is the suggestion of Graham (1992) that stout limbs were more adaptive to predator escape in scrubby and forested habitats" (Webb and Stehli 1995:640). On the other hand, the mastication characteristics that distinguish *Palaeolama* from *Hemiauchenia* include the shallow jaw, low-crowned cheek teeth, and cervid premolars, which imply a mixed diet with less grass and more shoots and leaves from shrubs, thorn bushes, or trees.

Webb (1974:210–211) wrote that the common ancestor of *Palaeolama* and the present-day llamas is implicit in the shared morphological characteristics, but this is not documented with any precision. When the deposits at Tarija accumulated, the biggest *Palaeolama* and the smallest

Lama were already distinct (even though in Tarija, the latter genus is better represented than the first). Both, however, share some special characteristics, including similarly proportioned limbs and their Andean distribution center. There is no doubt that their divergence took place early in the Pleistocene, perhaps shortly after the Andean stock differentiated from *Hemiauchenia*. The peculiar modification in the premolars that distinguishes *Palaeolama* happened after it diverged from *Lama*. The common ancestor must have resembled *Palaeolama* more than *Lama* in its great size, relatively big premolars, and the weak "buttresses" in the lower teeth. *Vicugna* also differentiated from *Palaeolama* quite early in the Pleistocene, probably as a different lineage of *Lama*. The oldest discoveries made in the Andean region seem to clarify these relationships (Webb 1974).

Figure 9.11 of Webb's study (1974:212) shows the geographic distribution of the Lamini in the Pleistocene and in recent times. It seems clear that in Early Pleistocene times, the distribution of *Hemiauchenia* extended from its native land in North America to South America. Part of this original stock adapted quite rapidly to the rugged terrain and the modified food resources of the Andes, and the radiation to *Palaeolama*, *Vicugna*, and *Lama* took place. The latter rapidly extended its area east and south over most of South America, where it coincided with the area previously established by *Hemiauchenia*. *Vicugna* stayed in the high Andes. *Palaeolama* expanded its area dramatically from what seems to have been its Andean origins. While *Lama* spread east and south, *Palaeolama* spread west and north. It thus expanded to the Pacific lowlands of Peru and Ecuador as *Palaeolama aequatorialis*. Then it must have moved north through Central America, around the Gulf Coast to Texas and Florida, where it began to settle in the Lower Pleistocene (lower Irvingtonian, ca. 1.5 mya [Marshall 1985: Fig. 6.75]). There it widely overlapped with *Hemiauchenia* as *Palaeolama mirifica* in the Middle and Late Pleistocene. The llamas then dispersed in the Pleistocene to South America, radiated rapidly, and returned in part (*Palaeolama*) to North America. *Palaeolama* and *Hemiauchenia* became extinct on both continents by the end of the Pleistocene, and only *Lama* and *Vicugna* survived in

South America (Webb 1974:211, 213; interested readers can find information in Simpson [1950: 383]. Other studies, like Novoa and Wheeler [1984:120–121], just repeat the primary data from Webb [1974]. See my Figures 3.2 and 3.3.

With the finds made at the Leisey Shell Pit, however, the picture changed and must be restated in new terms. Webb and Stehli (1995:630) wrote: "The precise relationship of *Palaeolama* to other lamines remains problematical." The data

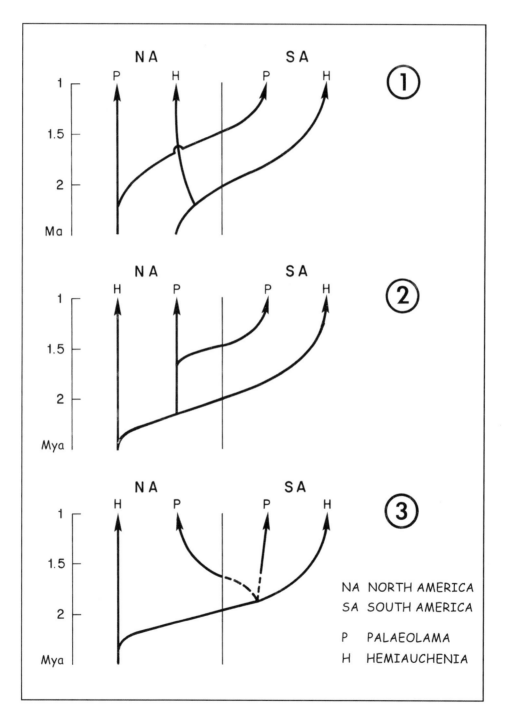

FIGURE 3.2. Possible explanations of the origin of wild South American camelids, from the evidence currently available. *Prepared by Christian de Muizon and drawn by Osvaldo Saldaña.*

from the *P. mirifica* sample from the Leisey Shell Pit supplied important comparative elements. Webb (1974:210) noted that several morphological characteristics indicated that *Palaeolama* and the modern llamas have a common ancestor, but noted that this "is not precisely documented." By the time the Tarija deposits had accumulated, the larger *Palaeolama* and smaller *Lama* were already quite distinct. Webb and Stehli (1995:630) add, "Harrison (1985) followed this proposal and placed *Palaeolama* as the sister group of living Laminae. This is hardly a convincing phylogenetic analysis." With the evidence available, "the relationships of *Palaeolama* and other llamas must be acknowledged as an unresolved polychotomy" (1995:631). For now, the geographic history of *Palaeolama* will remain debatable as long as there is no well-defined sister group (and not just the other lamine genera). The continent of origin remains uncertain even if one follows the dubious practice of taking the site of its first stratigraphic appearance as the likeliest place of origin. In North America, the oldest discovery of *Palaeolama* is the early Irvingtonian of the Leisey Shell Pit. The attribution to this genus of an older llama, a new species proposed by Dalquest and Mooser (1980) belonging to the late Hemphillian of the local fauna from Ocote, close to Guanajuato (Mexico), was corrected by Montellano (1989), who ascribed it to *Hemiauchenia* on the basis of more complete materials. The oldest well-documented discoveries of this genus in South America are of a *P. weddelli*-type species in Tarija, Bolivia, from the Ensenadan (approximately the Middle Pleistocene). These species differ from the Leisey samples of *P. mirifica* mainly in their larger size. On this basis it can be surmised that the samples are closely related, and thus may indicate immigration from North to South America, contrary to what Webb (1974) and Montellano (1989) proposed. Webb and Stehli wrote:

> On the other hand, Mones (1988) proposed a downward extension of the South American range of *Palaeolama* to Uquian (Plio-Pleistocene) based on associated faunal evidence . . . in the upper San José Formation of Uruguay. Marshall et al. (1984) also cited a Uquian range for

Palaeolama in Argentina, although their evidence has not been presented explicitly. The resolution of this crucial biogeographic problem will depend ultimately on resolving the phylogenetic problem, that is, identifying the sister group or *Palaeolama*. We reject the suggestion of Marshall et al. (1984:21) that *Hemiauchenia* (and other proposed South American taxa) "are most likely sub-genera of *Palaeolama*" for . . . many reasons. . . . Presumably the critical evidence regarding the origins of *Palaeolama* resides in Pliocene sediments somewhere in lower latitudes of the two American continents. (Webb and Stehli 1995:631)

Of note, it is likely that some extinct species may have existed by the late Pleistocene, such as *Lama owenii* and *Lama angustimaxila* in Argentina and Bolivia, and even genera such as *Eulamaops parallelus* from the Upper Argentinian Pleistocene (Pascual and Odreman 1973:34). Marshall et al. (1984:60–61, Table 1) have assembled the radiocarbon dates obtained for the extinct fauna we are interested in. For *Lama*, we have a date of 12,600 years BP at Los Toldos (Argentina; Cardich 1977), another of 11,010 years BP at Arroio Touro Passo (RS, Brazil; Bombin 1976), and finally a date of 13,460 years at Huargo (Peru; Cardich 1973 [there are older, undated remains at the site]). Four dates are known for *Palaeolama* but with some doubts, for the datum appears with a question mark: 14,150, 19,620, 14,170, and 16,050 years BP for Pikimachay (Peru; MacNeish et al. 1970). I believe this information should not be taken into account for the moment. The subject is discussed at length later in this chapter. I should add that there is a date of 11,100 years BP for *Lama* sp. at the site of Quereo (Chile), but there are also *Lama* spp. in older strata that have had dating problems (Núñez et al. 1983, 1987).

Adaptation is clearly one of the points that most interest us. Webb (1974:208–210) is one of the researchers who has discussed the subject the most:

> Stratigraphic considerations indicate that within the llama tribe, limb proportions evolved in the opposite direction from

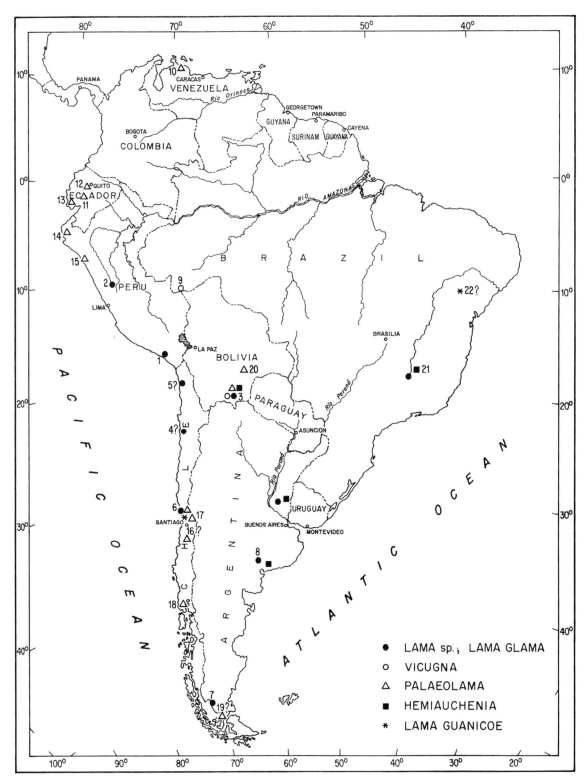

FIGURE 3.3. The distribution of fossil remains from the Lamini tribe genera in South America. Unnumbered sites have been taken from Webb (1974:Fig. 9.11, 212) and their location is only approximate. They do not appear in the text. Question marks indicate either that the location of the site is uncertain or that the identification of the discovery is doubtful. Each case is explained in the text. The site legend for this map is on the facing page. *Prepared by the author and drawn by Osvaldo Saldaña.*

what might ordinarily be expected. *Hemiauchenia*, with "more advanced" cursorial proportions, is geologically more ancient than *Palaeolama* and the modern llamas. This unexpected progression is confirmed by the limb proportions in *Pliauchenia*, a lamine genus of Early Pliocene (Clarendonian) age from North America. Although complete propodial elements of *Pliauchenia* have not been described, the epipodial and metapodials are adequately represented in the sample of *P. magnifontis* Gregory (J. T. Gregory 1942) from Big Spring Canyon, South Dakota. The ratios of metapodials to epipodials in this species agree rather closely with those of *Hemiauchenia* (range, 0.81–0.85). The same is true of other Pliocene Camelinae, such as *Procamelus grandis*. Thus, it would appear that *Palaeolama* and the modern llamas secondarily reduced their 'speed ratios'; and the relative lengths of their metapodials and increased lengths of their epipodials during the Pleistocene. Quite possibly such changes took place as these llamas came to inhabit the Andean slopes and the altiplano. Their stockier limbs would thus represent an adaptation to the rugged terrain of their apparent homeland. Maneuverability and jumping must have become more important than sheer speed. The longer proportioned limbs found in *Hemiauchenia* and *Pliauchenia* were better adapted to the flat open country of the Great Plains of North America and the Pampas of South America. Thus, the unexpected shortening trend in evolution of limb proportions may be explained as an adaptive shift in locomotor style from one appropriate to open flatlands to one geared for irregular terrain.

To solve these problems and trace the evolution of these animals, as well as their adaptation to various environments, more evidence will have to be found and their actual distribution in the South American continent after their arrival studied. As Hoffstetter (1952:314) noted, the Pleistocene terrain has camelid remains "not just in the area currently occupied by the group, but also in the Argentine and Uruguayan pampas, as well as in the intertropical area (eastern Brazil, the Andes, and the Ecuadorian side)."

Between 1 and 3 mya, the camelids began first to adapt to a dry environment, and only recently to high altitudes. Kent (1986:1) believes that this adaptation was caused by the development of locomotive and feeding mechanisms suitable for areas with low plant resources, as in the punas or altiplano, which have rocky land and a seasonal snow cap. These mechanisms were developed due to the relation between the phylogenetic history of the camelids and the requirements of their environment.

Figure 3.3. Site legend.

Lama		*Palaeolama*		*Vicugna*	
1	Tirapata	3	Tarija	3	Tarija
2	Huargo	6	Quereo	9	Río Jurna
3	Tarija	10	Muaco		
4	Antofagasta	11	Punín	*Hemiauchenia*	
5	Tarapacá	12	Quito	3	Tarija
6	Quereo	13	La Carolina	21	Lagoa Santa
7	Cueva de Eberhardt or	14	Talara		
	Cueva del Milodonte	15	Pampa de los Fósiles	*Lama guanicoe*	
8	Flamenco II	16	Tagua Tagua	22	Toca de Boa Vista
		17	Chacabuco		
		18	Monte Verde		
		19	Cueva de Fell		
		20	Ñuapua		

Kent (1987) himself has discussed this subject in another study, and his view concurs with mine (Bonavia 1982a:394), so it is worth citing him:

Most of the South American pre-Pleistocene camelid fossils are found in non-Andean localities, primarily eastern Argentina (Cardozo 1975:74, map 3). Although some sampling and/or preservation bias might partly account for this distribution, *it would appear that the earliest South American camelids were not initially adapted to the high elevations, and possibly that their presence at high elevations is a post-Pleistocene phenomenon. Thus, the earliest high-altitude camelids may have entered their present range contemporaneous with and perhaps because of human hunters.* (Kent 1987:171; emphasis added)[4]

I shall now try to establish the distribution of fossil forms on the basis of the discoveries that have been made. First, however, I would like to insist that the Family Machraucheniidae is autochthonous and not related to the Camelidae, which emigrated from North America. Rather, it belongs to the Litopterna, an order of mammals that originated in the Didolodontidae, an ungulate family that appeared in the South American Paleocene. This point must be emphasized because I have found some confusion among nonspecialists in this regard, owing to a vague similarity in shape between camels and the camelids that invaded South America.

As Patterson and Pascual (1968:428, 430) point out,

The Macraucheniidae, alone in the order [Litopterns], achieved large terminal forms equaling the larger true camels. They were the only South American ungulates known to have evolved a long, graceful neck, which accounts in large part for their rather camel-like appearance. The chief peculiarity displayed by them was a progressive reduction of the nasal bones accompanied by unmistakable evidence of the evolution of a short proboscis. Proterotherids and macauchenids [sic] seem to have thrived in the evolving pampas areas.

They add (1968:443) that the Macraucheniidae were "vaguely camel-like . . . and the invading camels would offhand seem the most likely competitors, yet the record does not bear out this possibility: macrauchenids and large species of *Palaeolama* coexisted."

3.2 *Lama* sp.; *Lama glama*

According to López Aranguren (1930:16), we owe the first descriptions and illustrations of fossil llamas to Paul Gervais (1855).

In Chapter 1 on systematics we indicated that two extinct species of llama are currently accepted, *L. owenii* and *L. angustimaxila*. Let us turn to the evidence for their presence in South America.

I was unable to find much data for Peru. According to Dedenbach Salazar (1990:81), Cabrera (1932) noted the presence of *Lama* species in Pleistocene lands in Peru. The correct date of Cabrera's study is 1931, and he never said so. When mentioning Peru he refers to the "living" species, not the fossil ones (Cabrera 1931:115). Pascual and Odreman (1973:34) simply state, "The first remains of some species of the genus *Lama*, and perhaps *Vicugna*, appear particularly in high Andean caves; such is the case of the discovery of presumably llama and vicuna remains in Casa del Diablo, close to Tirapata at 1819 m, reported by Nordenskiöld (1908)." The site of Tirapata seems to be in the highlands of Mollendo in southern Peru.

Cardich (1973) found llama remains in different strata of Huargo Cave in the Department of Huánuco. He simply mentioned "fossil remains of camelids . . . from 11,510 BC" (Cardich 1974: 35, n. 4). These were all classified by Pascual and Odreman (1973:31–32) as *Lama* sp. The most ancient layers, the ones that interest us now, are the ones numbered from 7 to 10, and in them there is an association with extinct fauna such as *Equus* sp. and *Scelidotherium* sp. Layer 8 has a radiocarbon date of 13,460 years. Pascual and Odreman (1973:37) make an interesting observation: "[Do t]he camelids present in all layers of Huargo Cave belong fully or partially to one of the living species?" They add, "Since *Palaeolama* is relatively frequent in the Upper Pleistocene sediments, both for the Andean and the coastal areas of Peru and Ecuador, is their absence in Huargo Cave due to the heterogeneity of the respective sediments? Or is it due to human preference?" They end by noting in their conclusion the "[p]ossible presence of some of the present-day species of camelids or related extinct species." However, it is worth noting that the association of these bones with human remains has been questioned "owing to the absence of a lithic industry" (Kaulicke 1979:104), and was so noted by Cardich (1973:26) himself: "As for the presence of man, there are few traces in the lower layers (10, 9, 8 and 7)," and the ones he indicates are quite doubtful (see Cardich 1973:26–27).

For Bolivia, Cabrera (1931:115) notes the presence of *Lama glama* in Pleistocene lands. There are no more details. Gilmore (1950:435) discusses the subject. He reports remains belonging to the Late Pleistocene and early Holocene found in the intermontane valleys that were not subject to glaciations, specifically in Tarija (Gilmore [1950:435] mentions *Ulloma* too, but this is a mistake, because this locality is in the altiplano). Gilmore states these are llamas. He does not give much information. Marshall et al. (1984: 32) report having found *Lama* in Ensenadan strata in the locality of Tarija (in the southern Bolivian Andes).

As regards Brazil, Dedenbach Salazar (1990: 81) reports that Cabrera (1932 [1931]) mentions *Lama* remains in lands belonging to the Pleistocene. This does not appear in Cabrera (1932 [1931]). Marshall et al. (1984:41, Fig. 8) indicate that *Lama* is found in strata that might belong to the Holocene and that are older than the Late Pleistocene.

For Uruguay, there is only a vague datum by Marshall et al. (1984:56), who report the presence of *Lama* in Pleistocene strata, also reported by Cabrera (1932 [1931]) according to Dedenbach Salazar (1990:81). This, however, is a mistake.

There is somewhat more information for Chile. Marshall et al. (1984:44–45) say there are *Lama* among the Pleistocene mammals in the provinces of Antofagasta and Tarapacá, in central Chile, and in the caves in the far south. Núñez et al. (1983, 1987) in turn refer to the site of Quereo,

in the ravine of the same name, close to the town of Los Vilos, between 25 and 180 masl. They report (1983:31) "some evidence of camelids . . . they are big-sized, juvenile animals" in Quereo Level II, with an age ranging between 13,000 and 11,000 years BP, and specify that these are *Lama* sp. (1987:172).

The data are not too specific for Quereo Level I, dated between 25,000 and 22,500 years BP. I therefore prefer to quote the authors themselves:

> The increase of camelids in this level is remarkable, and the dispersal and degree of destruction of the remains is compatible with natural factors. They represent several individuals of different age and level of mineralization. In the case of *Paleolama* [sic] sp., juvenile and adult individuals, possibly juvenile males of different sizes, appear within the same species because there are remains of *Lama* sp. The size that can be deduced for the oldest individuals of *Paleolama* [sic] sp. is of great proportions. . . . Most of these materials cannot be separated from those belonging to *Paleolama* [sic] sp. Other remains of smaller size . . . *would instead belong to Lama* sp. (Núñez et al. 1983:31; emphasis added; 1987:172)

Finally, Núñez et al. (1983:45) report that remains of *Lama* sp. have been found far south in the famed Eberhardt Cave, also known as Milodont Cave, in the level dated to 3300 years BC.

The evidence for Argentina is slightly more extensive. Cabrera (1931:110) reports: "*Lama glama* (Linné), whose remains are also found in the Pleistocene, as has been noted by Dr. López Aranguren." This is supported by Marshall (1988), who indicates there is *Lama* in Uquian lands, that is, in the Late Pliocene or Early Pleistocene. Meanwhile, Kraglievich (1946:327) wrote that "the presence of camelids in the Chapadmalan, particularly of the genus *Lama*, is proven," that is, he is referring to the end of the Pleistocene. On the other hand, according to Dedenbach Salazar (1990:81), López Aranguren (1930[b]) has reported the presence of *Lama* in post-Pleistocene levels. I was unable to find this citation, and it is evidently a mistake.

Pascual and Odreman (1973:34) summarized the problem and wrote:

> According to Cabrera (1961:321), "In wild state, the llama has been extinct for centuries, perhaps since precolumbian times." According to several scholars, it was present during the Pleistocene in Bolivia and in the pampas region of Argentina. If we agree with López Aranguren (1930:40; the reference is wrong, the correct one is 1930[a]:24. To it we must add 1930[b]:110, which explains the point better) and Cabrera (1931:115; 1961:320) that *Auchenia ensenadensis* Ameghino, 1889, is a synonym of *Lama glama*, then we must admit the presence of this species since at least the Middle Argentinian Pleistocene (Ensenadan Age; see Pascual et al. 1965). Its presence in the Upper Argentinian Pleistocene was noted by López Aranguren (op. cit.) and Cabrera (1931).

The correlation of the mammal-bearing sediments of Tarija with those of the Argentinian Pleistocene has been discussed and is most doubtful; however, and despite the presence of some noncharacteristic taxa, the total faunal assemblage so far known can be correlated with that of the Upper Argentinian Pleistocene. If this were so, and according to López Aranguren (op. cit.) and Cabrera (1931), *Palaeolama crequi* Boule, *Auquenia intermedia* Gervais, and *Auquenia castelnaudi* Gervais are synonyms of *Lama glama*, this species would also have been present in Bolivia during the Upper Pleistocene.

In his more recent synthesis, Webb (1974:174) only indicates the presence of *Lama owenii* in lands from the Argentinian Lujanian (Late Pleistocene), and *Lama angustimaxila* in the Uquian (Early Pleistocene and Sub-Recent) of Argentina and Bolivia.

In his report on the site of Los Flamencos II, Las Encadenadas, a Buenos Aires lake in the Partido of Saavedra, in the province of Buenos Aires, Austral (1987:94) notes the presence of *Lama* sp. in the structure of his Layer C. This corresponds to the Ensenadan and the Lujanian, "which date from the Meso and Neopleistocene," according to Pascual et al. (1965:190–191) (Austral 1987: 97). (For the location of the sites mentioned, see Figure 3.3.)

3.3 *Vicugna*

Neveu-Lemaire and Grandidier (1911:45) report the presence of vicuña fossil remains in Pleistocene lands from southern Brazil, but I believe that their classification of *Lama minor*, which as far as I have been able to ascertain has not been accepted by specialists, must be taken with care. On the other hand, Simpson and Conto (1981) report the discovery of *Vicugna* sp. when discussing the site of Jurna River, in the state of Acre (on the border between Peru and Bolivia).

Cabrera (1931:116) notes the presence of *Vicugna* in the Bolivian Pleistocene, which was confirmed by Webb (1974:174) and by Marshall et al. (1984:32), who specified that these are remains found in Tarija (south of the Bolivian Andes, at 1950 masl), in Ensenadan terrain that belongs to the Middle Pleistocene and Sub-Recent. This was confirmed by later studies (Wheeler 1995:277, based on MacFadden et al. 1983).

The information regarding Argentina is contradictory. Gilmore (1950:451) wrote that the vicuñas arrived in the humid pampas of Buenos Aires during the Late Pleistocene and Early Holocene (his statements are based on studies by López Aranguren [1930a, 1930b] and Cabrera [1931]). Using the same sources (plus a 1961 study by Cabrera) for *Vicugna vicugna*, Pascual and Odreman (1973:35) wrote: "The fossil remains of this genus assigned to a single living *Vicugna vicugna* species, are known only in the Upper Pleistocene of Argentina and Bolivia."

Marshall et al. (1984:29) have a different position. It is true that while discussing the typical fauna of the Uquian they indicate the presence of Camelidae *sensu lato*, but when discussing *Vicugna vicugna*, they note its association with Ensenadan terrain (or lower Pampean) and state this is the first time these remains have been found. In other words, they mean the Middle Pleistocene and Sub-Recent, or an earlier age than the other scholars. This is supported by Webb (1974:174). However, a recent review of these materials concluded that the vicuña developed from the guanaco in the Holocene. In fact, mtDNA studies support a divergence of at least two million years between both (Wheeler 1995:277, based on the studies of Menegaz et al. 1989 and Stanley et al. 1994).

(For the location of the sites mentioned, see Figure 3.3.)

3.4 *Eulamaops*

This extinct genus is only known in strata corresponding to the Late Pleistocene of Argentina. Specifically, it is the species *E. parallelus* from the Lujanian (see Webb 1974:174–175).

3.5 *Palaeolama*

Palaeolama is the biggest genus (Webb 1985:377), and the earliest finds come from the strata of the early North American Irvingtonian age (ca. 1.6 mya) (Webb and Stehli 1995). "The geographic distribution of *Palaeolama* in North America, thought by Webb (1974) to be restricted to Florida and the Texas Gulf Coast, extends as far north as Edisto Island, South Carolina (Roth and Laerm 1980), and southern Missouri (Graham 1992), and as far west as southern California (Conkling pers. com.)" (Webb and Stehli 1995: 630).

The North American dispersals to South America are recorded first in rocks from the Ensenadan Age. Assuming these dispersals represent "twin" events, the Ensenadan remains could begin in, or be earlier than, 1.4 mya. This estimate is consistent with the ages known for the Ensenadan rocks and fauna (which correspond to the Tarija Formation) and Uquian (Uquia Formation) (Marshall 1985:77).

We know that during the Pleistocene, *Palaeolama* lived over a vast expanse of South America and diversified into different species, given the importance of its stratigraphic depth and its geographic scattering (Hoffstetter 1952:346). Besides, among the genera of South American camelids, it is the only one that managed to adapt to quite different environments, such as the tropical

region, the inter-Andean corridors, and the Argentinian pampa. It was mainly a browser (Webb and Stehli 1995:640). The other genera seem to have been strictly limited to the temperate or cold areas of the southern region and did not reach the lower latitudes except at high elevations. It follows that biologically, *Palaeolama* is the closest representative to the branch that could have migrated from North America through the inter-tropical zone (Hoffstetter 1952:321–322).

Hoffstetter (1952:375) explains that "[t]he genus *Palaeolama*, which recent studies had considered monotypical, actually underwent an important diversification in the enormous dispersal area it occupied. We were forced to identify different species, some of which lived in the tropical area, others in the Andean corridors, and a particular branch adapted to life on the pampas. A specific differentiation in relation to different geological levels is likewise observed in the Andean region."

Specialists disagree on the division of this genus, as was indicated when discussing the classification of these animals. Thus, Hoffstetter (1952: 317–320, 346) accepts six different divisions, all South American, which he considers species. First, we can distinguish a form from the Argentinian pampas (northeast Argentina, comprising the provinces of Buenos Aires and Santa Fe) called *P. paradoxa*. Then there are three species from the highland Andean valleys: *P. weddelli* (found in Tarija, Bolivia), *P. reissi* (from the Ecuadorian Andes, mainly in the area of Punín), and *P. crassa* (also from the Ecuadorian Andes). Finally, there are two species that inhabited the tropical lowlands: *P. major* (from Lagoa Santa and Minas Gerais, in Brazil) and *P. aequatorialis* (from the Ecuadorian coast). Still according to Hoffstetter (1952:317–320, 346), this assemblage seems very homogeneous in terms of the skeletal characteristics.

Webb (1974:175–176), however, does not agree with this position and indicates in a critical review that *Hemiauchenia* was originally included by mistake in the genus *Palaeolama*. Webb accepts that Hoffstetter (1952) recognizes the separation at the species level but has not realized its more significant importance (generic). This division affects not only the status of both species but also the taxonomic arrangement of all Pleistocene llamas, with the exception of the modern genus *P. weddelli* of Tarija (Bolivia), which typifies a widely dispersed group of species with particular characteristics, and *Hemiauchenia paradoxa*, which typifies a more ancient group of species even more widely dispersed, also with particular characteristics. Webb (1974:175–176) accepts *Palaeolama* based on *P. weddelli* of Tarija, wherein he includes the species from the Bolivian plateau, the Andean and coastal areas of Peru and Ecuador, and the Gulf Coast and the plains of Florida and Texas. It is because of all this that Webb (1974:175–196) accepts only *P. weddelli*, *P. aequatorialis*, and *P. mirifica*, which is not considered here because it is a North American species.

I now turn to the distribution of *Palaeolama* remains on the South American continent. In Venezuela, all scholars refer to a single discovery, that of Muaco, in the state of Falcón, close to the Vela del Coro, made by Cruxent (see Rouse and Cruxent 1963; there are two reports on the paleontological remains, by Royo and Gómez 1960a and 1960b, which I was unable to see). When discussing the site, Lynch (1978:476; 1983b:113) mentions only "one camelid" and gives dates of 16,000 and 14,000 years BP (see Rouse and Cruxent 1963), but states there are serious questions about its associations. This find is also mentioned by Marshall et al. (1984:58), but without any comment; however, these same scholars (1984:11) report that the remains of animals introduced by the Europeans were found in Muaco beside the extinct fauna. I believe the find is questionable.

As Hoffstetter (1952:316) has stated, the Ecuadorian fossil fauna, like that of all the intertropical area, does not include any representative of the genera *Eulamaops*, *Lama*, or *Vicugna*, which seem to have been localized in the southern, temperate, or cold part of the continent, at least to judge by the current ecology of the last two genera (see also Hoffstetter 1952:374). Pascual and Odreman (1973:34) offer a very similar statement.

Hoffstetter (1952:371) reports the discovery of *P. reissi*, claiming this "could be a descendant of *P. weddelli*," and assigns it to the Puninian of the third glaciation. Marshall et al. (1984:49–50) are more precise and refer to two fossil zones, both between 2300 and 3100 masl in the Andean re-

gion. One is typified by Punín, south of the Riobamba River, where *Palaeolama* (*Protauchenia*) was found, the other is close to Quito by the Chiche (or Chichi) River, where *P. crassa* was found, but Marshall et al. indicate that this "is not the direct ancestor of *P. reissi* from the Punian" and that it belongs to the second glaciation. The authors also report that there are several problems with dating the remains, and in any case they lie outside the limits of the carbon 14 method. It is not clear whether Hoffstetter (1952) and Marshall et al. (1984) are referring to the same find or not. Webb (1974:177) apparently disagreed with the classification of Marshall et al. (1984), because when discussing the remains found along the Chiche River, close to Quito, he classified them as *P. weddelli*.

On the Ecuadorian coast, all scholars apparently refer to one find, that of La Carolina (in southwestern Ecuador, north of the Santa Elena peninsula). Lemon and Churcher (1961:429) classify it as *Palaeolama* and ascribe it to the Upper Pleistocene. Webb (1974:181) agrees with the date and says it is *P. aequatorialis*, whereas Marshall et al. (1984:50) only say *Palaeolama* (*Astylolama*), but in regard to the date they say it is "apparently Pleistocene."

Several discoveries were made in Peru that must be discussed. The first was in the vicinity of Talara, in the far north (Department of Piura). Marshall et al. (1984:54) note that this is *Palaeolama* (*Astylolama*) and ascribe it to the Lujanian. On the basis of data in Bryan (1973), Marshall et al. give it two radiocarbon dates, of 13,616 and 14,418 years. These are the studies made by Lemon and Churcher (1961:429; see also Richardson III 1978:282). Webb (1974:182) is far more specific about the classification of the remains because he ascribes them to *P. aequatorialis*, and as for their age he gives the Late Pleistocene [?].

The remains of *Palaeolama* from the Pampa de los Fósiles in the Cupisnique area, north of the Chicama River Valley (in the Department of La Libertad), were identified by Hoffstetter (Marshall et al. 1984:54; Chauchat 1987:23; 1988:58). Claude Chauchat found the remains on the surface in 1972, hundreds of meters east of the Piedra Escrita site, among the sand dunes of the great sandy ground of Cupisnique, at the exit from the ravine of the same name. These remains were left in place. Frédéric Engel later passed through the site at an unknown date and proceeded to collect the bones and excavate, but without any scientific methodology. This was verified by Chauchat and Hoffstetter, who visited the site in 1975. Engel later allowed Hoffstetter to prepare casts of the bones, which are in the Musée National d'Histoire Naturelle of Paris. Engel reported that he had ground up the bones and sent them to Japan to be dated, but the laboratories replied that this was impossible, because the bones did not have enough collagen. It was thus believed that the primary evidence for the Cupisnique *Palaeolama* had been destroyed. However, in May 1994, Claude Chauchat and the author verified that the Museo de la Nación in Lima exhibited in its paleontology section a "hind leg of *Palaeolama* sp." from El Zorro in the Cupisnique Ravine. It was also stated that the corresponding date was "Pleistocene–Holocene, 9000 years," and that the bones belonged to the collections of the Museo de Antropología y Agricultura Precolombina of the Universidad Nacional Agraria, La Molina, in Lima (Catalogue No. V. 6178). According to Chauchat (pers. commun., 1 June 1994), the site is the one he discovered. The specimen is actually a hind leg comprising the basipodium, metapodium, and acropodium. Christian de Muizon (pers. commun., 28 June 1994), who prepared the casts that are in Paris, had seen the piece on exhibition, and apparently it was not the same bone shown to Hoffstetter. This would mean that Engel found more bones than he initially reported, or that he later returned to the site and obtained more samples.

In Chauchat's opinion, *Palaeolama* is not too common among the fossils in this area. Ubbelohde-Doering found some bones whose whereabouts are unknown (Claude Chauchat, letter of 24 June 1993). I cannot date these remains, but mastodon bones from the same area have an age of 15,000 years BP, horse bones 25,000 years BP, and Scelidodon 16,000 years BP (Claude Chauchat, letters of 19 September 1992 and 24 February 1993; see also Collina-Girard et al. 1991, who, however, are mistaken in saying it is a *Scelidotherium* [p. 127]). It is most likely that the aforementioned *Palaeolama* remains fall within this range.

A detailed analysis is needed to remove all doubts about the discoveries made in the Peruvian highlands. Marshall et al. (1984:55) thus mention that *Palaeolama* was found at Pikimachay (in the Department of Ayacucho) in association with early man, and quote MacNeish, Berger et al. (1970), MacNeish, Nelken-Terner et al. (1970), and MacNeish (1971). Although it is true that after the citations, the authors "note that the occurrence of *Palaeolama* is not confirmed, and that various generic names are used in a broad sense," I find their critique of the research team that worked in Ayacucho far too generous. I believe that thus far there is no evidence for the remains mentioned. Let us turn to the sources.

MacNeish, Berger et al. (1970:976) discuss the Ayacucho Complex and the extinct fauna and write: "The other identified bone was a small fragment of a large mandible, most assuredly in the Camelidae and probably paleollama [sic], *but again Hoffstetter warned of taxonomic problems*" (emphasis added). The study by MacNeish, Nelken-Terner et al. (1970) describes the finds made in Pikimachay Cave, and they refer first to Zone j (to which they give an age of 14,100 years BC), superficially discussing the bones of the extinct fauna, stating that "there were some the size of a camel or horse" (1970:13). There are two entries for Zone h. In one case they say: "bones of paleo-llama [sic]" and give it an age of 12,200 years BC (1970:14–15). They then refer once more to Zone h and mention "paleo-llama" [sic] (1970:33).

MacNeish (1971) analyzed the subject once more. He mentions Zone j when discussing the discoveries made at Pikimachay, to which he now gives a radiocarbon date of 19,600 years, and says there was "perhaps an ancestral species of horse or camel" (1971:40). When describing Zone h (which for him has an age of 14,150 radiocarbon years) he says there was "[a] kind of ancestral camel" (1971:42). When dealing with the subject of the Ayacucho Complex in Pikimachay Cave, MacNeish et al. (1975) once more mention many remains of extinct fauna, including "six llama bones" (1975:15), while the summary only says "camelids" (1975:75).

It is important to note that these reports are all extremely vague and raise serious doubts about the archaeological context. Even more significantly, there is no mention of *Palaeolama* in two of the three volumes of the final report of the Ayacucho Project thus far published (see Mac-Neish et al. 1983a, 1983b), which are the ones that could have included this information. On the basis of what has so far been presented, I believe the Ayacucho data should not be used without a careful analysis of the sources, as has unfortunately happened (e.g., Lanning 1970:90), until a detailed paleontological report of the findings is published.

Finally, I would like to note that when referring to the Peruvian Andes, Webb (1974:177) states there are *Palaeolama weddelli* corresponding to the Late Pleistocene. I do not know what site he means.

For Chile, Casamiquela and Dillehay (1989: 207) wrote: "The bones of paleo-lama [sic] have been found at several paleontological sites in central Chile (Casamiquela 1969). Primarily a steppe or tundra-grassland adapted animal, the camelid is known to occasionally extend its zoogeographic range into the deciduous forests near open environments, the case in the past and today on the Isla de Tierra del Fuego." It seems, however, that *Palaeolama* is not common in Chilean archaeological sites, even though remains ascribed to *Lama Major* [sic] or giant guanaco were identified in central Chile (in the Chacabuco zone: Fuenzalida 1936; Núñez et al. 1983:44). Núñez et al. (1983) believe the area of Quereo (to which I refer later) was optimal for these animals, to judge from the data available for this site and neighboring lands. Without going into more detail, Marshall et al. (1984:44–45) state there is *Palaeolama* in Chilean Pleistocene strata.

So it is that Lanning (1970:102) discusses the Tagua Tagua site in the Cachapoal River Valley, in the province of O'Higgins, and notes the presence of *Palaeolama* with an age of 11,380 radiocarbon years. However, it should be noted that the only source given by Lanning is Montané (1968), where the date really appears, as well as the discovery of extinct animals, but *Palaeolama* does not figure among them. I do not know whether Lanning found later data.

In the locality of Quereo (close to Los Vilos), Núñez et al. (1983:31) found remains of *Palaeo-*

lama sp. in the strata they call Quereo 1 that range between 25,000 and 22,500 years BP. There are young and adult individuals, possibly young males of different sizes. "The height deducible for the biggest *Paleolama* sp. [sic] individuals is of great size, somewhat bigger than the specimen housed in the Museo de la Plata (Argentina)." Camelids make up 5.63% of the animal remains in layer Quereo 1; "the high frequency of camelids characterizes this level, in agreement with a dispersal area of *Paleolama* sp. [which was] included in this layer to the range of species hunted" (Núñez et al. 1983:36).

Núñez et al. state that *Palaeolama* is "[n]ot common in archaeological contexts, though it was previously identified in the Central Chile basin (e.g. at Chacabuco) as *Lama Major* [sic] or giant South American guanaco (Fuenzalida op. cit. [1936]). It seems that its environment in the region of Quereo was optimal, to judge by its significant record in level I and in the land around Graben (eastern dunes)" (Núñez et al. 1983:44). This was later confirmed, and the authors repeated that "paleolama [sic] and extinct megafauna might have been hunted and butchered *in situ*" (Núñez et al. 1987:167).

To avoid any confusion over the chronology, it is worth adding a comment made by these same scholars (Núñez et al. 1987:170, 172). Regarding Quereo and Tagua Tagua, they wrote:

> The aforementioned sites hold evidence that shows there might have been occupations prior to the eleventh millennium BP. Actually, the remains of Pleistocene fauna with traces of human activity compatible with the hunt of at least a horse and paleolama [sic] . . . were found in the Quereo I level (Núñez et al. 1983). It is a record of broken bone fragments, bone artifacts occasionally percussed, bones with cut marks. . . . Two radiocarbon dates of associated wood from this level gave dates of 11,600 ± 190 and 11,400 ± 145 years BP. These dates are considered wrong because they are synchronous with the upper level (Quereo II), distant 1.30 m from differentiated sediments. They can be taken as dates rejuvenated due to

contamination with subterranean waters. In fact, Level I is obviously earlier because of its stratigraphic position. Because of its correlation with the Tagua Tagua site, it is tentatively posited that this level corresponds to the Laufen interstadial (Middle Würm), which is between 25,000 and 22,000 years BP. Level Quereo I presents remains of big herbivores with their smaller predators . . . *Paleolama* sp. [sic], *Lama* sp.

A date of 11,100 ± 150 years BP was given to level Quereo II, in association with *Lama* sp.

Dillehay found remains of *Palaeolama* at Monteverde, in the province of Llanquihue, 35 km west of Puerto Montt and at only 50 masl (Casamiquela and Dillehay 1989:207). This is just one *Paleo-lama* [sic] bone and is, according to the authors, the first paleocamelid found in the forested areas of south-central Chile. Casamiquela and Dillehay explain that "[a]lthough it is possible that this animal was adapted to forest environments in the late Pleistocene period, today this ungulate can only be found on the eastern side of the Andean mountains in the open, arid environment of the Argentine and Chilean Patagonia, and in the open drier desert and puna grassland of northern Chile, northern Argentina, Peru, and Bolivia."

Lanning (1970:102) reports the presence of *Palaeolama* in the context called Magallanes I in Fell's Cave, with an age of 11,000 years BP. Remarkably enough, however, the report by Bird does not mention *Palaeolama* at all, nor was it found by the other researchers who worked there later (see Bird and Bird 1988:134–187; Markgraf 1988:197–201).

Speaking of Brazil, Marshall et al. (1984:41, Fig. 8) certify the presence of *Palaeolama* in strata which might be "Holocene and earlier than the late Pleistocene," without indicating the location of the finds. Lynch (1990a:19; 1990b:154) mentions the remains in the Toca da Esperança Cave in the state of Bahía, where a camelid of uncertain age was found, allegedly dating to the Middle Pleistocene. This is in fact one of the studies by Lumley et al. (1988), which has been strongly criticized (see Lavallée 1989).

When discussing Brazil, Marshall et al. (1984: 43) wrote:

A fact of geochronological importance in South America, when using land mammals as time indicators, is the differential survival of certain genera in different areas. *Scelidodon*, *Tetrastylus* and *Palaeolama* (?), apparently long gone from temperate Argentina, were living up to the beginning of the Holocene in tropical Brazil, and other subtropical and tropical areas of South America.

For Bolivia we have Webb's (1974:177) data, which mention the presence of *Palaeolama weddelli* in the "probable" Middle Pleistocene of Tarija, in the southern Bolivian Andes, at 1950 masl, although Webb adds that it is often given an age belonging to the Early Pleistocene. Marshall et al. (1984:33) also refer to these remains, but make a comment that is worth repeating. They wrote: "Most recently Webb (1974:176) proposed an Ensenadan Age for the Tarija fauna on the basis of [the] stage of evolution of the *Palaeolama* and *Lama* species. However, there are several species of *Palaeolama* from this fauna (Hoffstetter, unpublished material in the MNHN, Paris), some of which are poorly known" (Marshall et al. 1984: 33). In addition, they note that according to the calibration made by Macfadden et al. (1983), the Tarija formation extends between 1.0 and 0.7 mya, perhaps less. They specifically place the fauna of Tarija in the Ensenadan, but indicate that some fossils from this locality might be more recent.

Marshall et al. (1984:34–35) also note that *Palaeolama* remains were found at Ñuapua (in the Department of Chuquisaca of southeast Bolivia), in contexts called Ñuapua 1 and 2. Ñuapua 1 corresponds to the lowest stratum, which probably is of Lujanian age, but it might include a part of the late Ensenadan, while Ñuapua 2 is a Holocene layer where there are associations with humans.

Marshall et al. (1984:51) report the presence of *Palaeolama* s.l. (which includes *Hemiauchenia*) in Paraguay, and make the following comment (1984:52): "It is unfortunately impossible to know whether the *Palaeolama* from Paraguay . . . is related to *P. major* (Brazil and the Bolivian

Chaco) or to *Hemiauchenia paradoxa* (Argentina); the latter species is characterized by its elongated metapodials, which indicate an adaptation to an open environment (Webb 1974)." These same scholars (Marshall et al. 1984:56) claim that there is *Palaeolama* in the Pleistocene deposits of Uruguay, but do not give more data on the discovery (or discoveries).

Finally, we know from Marshall (1988:383) that *Palaeolama* was also found in Argentina, in rocks from the Late Uquian (the Late Pleistocene–Early Pleistocene), and that it originally and definitely came from the north.

(For the location of the sites mentioned, see Figure 3.3.)

3.6 *Hemiauchenia*

Hemiauchenia (= *Tanupolama*), a long-legged fossil llama, is known in the semi-dry, temperate areas that extend from the North American West (including Mexico) to Argentina (Webb 1974:197). In North America it is the most common genus, with a wide dispersal over the continent, particularly in California (Webb and Stehli 1995:633). It is found in the strata belonging to the Upper Miocene and Pliocene; then it extended to Middle America (Dalquest and Mooser 1980; Montellano 1989) and spread to South America in the Early Pleistocene. In the Andes, the stock became differentiated, so that *Palaeolama* and *Lama* appeared as distinct genera in the Middle Pleistocene land of Tarija (in Bolivia), with quite clear differences. *Palaeolama* spread into the Gulf Coast region and coexisted in Florida with *Hemiauchenia*. These major genera of llama finally became extinct on both continents (Webb 1974:213). Webb and Stehli (1995:637) restated the problem in terms of the discoveries made at the Leisey Shell Pit in Florida. They wrote: "The fact that *Hemiauchenia* and *Palaeolama* were already distinct by the early Irvingtonian in Florida suggests that they emigrated from North America to South America as separate taxa. On the other hand . . . this question remains open until a sister group for *Palaeolama* has been clearly identified." Only the *H. paradoxa* and *H. major* forms are known in South America; the rest are from North America.

H. major has been found in the Late Pleistocene strata in the caves of Lagoa Santa (Minas Gerais), in Brazil, and *H. paradoxa* in the Lujanian and Chapadmalan deposits of Argentina, in the Early and Late Pleistocene (Webb 1974:198–199).

It is worth noting that these animals were mixed feeders (Webb and Stehli 1995:640).

(For the location of the sites mentioned, see Figure 3.3.)

3.7 ALPACA (*Lama pacos*)

Relying on López Aranguren (1930[a and b]) and Cabrera (1931), but with some reservations, Gilmore (1950:441–442) indicates that these animals were in the humid pampas of Argentina, close to Buenos Aires, in the Late Pleistocene and Early Holocene. López Aranguren (1930b:123) specifically notes: "As for *Lama pacos*, we only know of post-Pampean remains, and we are currently unable to state whether it was a more modern immigrant or a form derived from *Auchenia gracilis*." Pascual and Odreman (1973:35) add that "[a]ccording to Cabrera (1961:II, 323), 'its extinction as a wild species must have been prior to that of the llama. . . .' Its fossil remains ("*Auchenia lujanensis*," Ameghino 1889) are found in Argentina in the Upper Pleistocene of the Pampas region; besides this reference, its remains are unknown in a fossil state in other, older sediments of the pampas, or in any other sediments of geologic antiquity in the rest of the continent." These remains were apparently not considered either by Marshall (1985, 1988) or Marshall et al. (1984), or by Webb (1974).

3.8 GUANACO (*Lama guanicoe*)

I was able to find very little information on this animal as a fossil. Gilmore (1950:447) reports remains in Colombia from the Late Pleistocene and Early Holocene, but the only proof presented is the fact that there is a place known as Páramo de Guanaco. Gilmore (1950:447) also notes the presence of guanaco in some parts of Paraguay during the Late Pleistocene and Early Holocene, but without giving much evidence.

Gilmore (1950:447) also reports the presence of this animal in Late Pleistocene and Early Holocene times in northern Argentina, while according to Dedenbach Salazar (1990:81), López Aranguren (1930[a and b]) also notes it, but for post-Pleistocene times. The information in Dedenbach Salazar (1990:81) is in this case wrong. López Aranguren (1930b:106) wrote: "In the collections of the Museo of La Plata I was able to examine a great number of mandibles belonging to this species [she means *Lama guanicoe guanicoe*], which come from the Argentinian Pleistocene." Further on she specifies that *Lama guanicoe* Lönnbergi "is none other than a big-sized guanaco of Santa Cruz" (López Aranguren 1930b:109). Pascual and Odreman (1973:35) wrote in this regard: "So far, its fossil remains have only been recorded for the Argentinian Pleistocene (Cabrera 1931)."

Recent newspaper reports tell of the discovery of fossil guanaco remains in Toca de Boa Vista, north of the state of Bahia, in northern Brazil, which are 10,000 years old (Anonymous 1993: B1). The information should be taken with reservations.

In the case of Chile, Núñez et al. (1983:44) report that

> [i]n Quereo there are scant remains of this species [they mean fossil *Lama guanicoe*], and its forms recall those of living camelids. Its absence at archaeological sites is striking, insofar as the remains are present in paleontological deposits of the Argentinian Pleistocene. On the other hand, its association and contemporaneity with the remains of non-fossil guanacos recovered at early sites such as Los Toldos, Fell's Cave, and Pomsonby is not clear. Their presence in central Chile had already been posited, together with a series of Pleistocene forms: mastodons, horses, deer, edentates, and carnivores at a depth of 4–7 m in a detritus deposit located in the Chacabuco hacienda (Fuenzalida 1936).

I have not found any reference to this point either in Marshall (1985), in Marshall et al. (1984), or in Webb (1974).

(For the location of the sites mentioned, see Figure 3.3.)

3.9 Ecological Changes

To understand this great faunal movement, a clear idea of the ecological changes that took place over time is essential. In this case the Pleistocene and the Holocene are key, as are the changes in the climate and the flora along the valleys and the *páramo*-puna, as well as the part played by the natural barriers created by the last glaciation (and by all glaciations in general), which prevented or at least hindered and affected the movements of fauna. I am not able to carry out an analysis of this kind. Instead, I want to present some useful pointers that could help nonspecialists. Interested readers will find a summary in Simpson (1971).

The formation of the savanna is one of the most important problems that must be understood; its explanation has shed new light not just on faunal but on human movements as well (see Bonavia 1982a:64 and passim). Marshall (1988: 384) provides an excellent synthesis, which I shall try to summarize here.

All "walkers" that have been preserved as fossils represent taxa that were apparently tolerant of, or were specifically adapted to, the savanna ecosystem. This indicates that there must have been a continuous corridor along the American tropics, or at least a mosaic of open habitats, as indicated by Webb's studies (1978, 1985). The history of the savanna habitat has been studied in detail by Raven and Axelrod (1975) and Webb (1977, 1978). We therefore know that lowland areas in tropical latitudes remained cool and dry when the glaciers advanced in temperate regions and highland tropical areas, thus causing a shrinking of the wet tropical forest habitats, giving rise to island-like refuges and to the expansion of the drier savanna (Haffer 1974; Van der Hammen 1974). The opposite happened when the glaciers retreated. Several marine regressions took place in the Caribbean area during times of glacier advance, and this created optimal ecological advantages for the reciprocal dispersion of savanna biota between the Americas (Cronin 1981). This phenomenon has been documented for the period between 3.2 and 1.4 mya, and coincides approximately with the fossil remains. The distribution of the savannas consisted of separate habitats when the glaciers retreated, but in times of glacier advance the savannas united along a corridor that ran along the eastern side of the Andes—what Webb (1978) calls the "High Road" or "Andean Route." This corridor gave rise to a north-south road that enabled the dispersal of the savanna biota in South America. The corridor continued along the Panama land bridge up to the southern part of the United States, and extended eastward into Florida. This happened at a time of glacier advance that caused a fall in the sea level of some 50 m, so that the corridor widened. Thus, when the glaciers advanced, the savanna habitat in the southern United States and the southern part of South America were mutually accessible. The last savanna corridor appeared between 12,000 and 10,000 years BP (Bradbury 1982; Markgraf and Bradbury 1982).

In the specific case of Peru, we know from Campbell's studies (1973:5108; textual citation reproduced by Richardson III 1978:282) that "[t]he coast of northern Peru where the Talara Tar Seeps are located is desert today. . . . The paleo-avifauna indicates that during the Wisconsin glaciation northwestern Peru was a savanna woodland or a savanna with extensive riparian forests. During the summer months, regular heavy monsoon rains occurred, followed by a winter dry season with continuous cloud cover similar to the present season." We can assume that this type of ecology could have exerted its influence as far as the northern part of the Department of La Libertad, and this could explain the aforementioned discoveries of fossils.

Glaciers are known to have played a crucial role in this process, but they represent a dynamic phenomenon. Thanks to studies by Wright (1980: 253; 1983), we know that most of the modern glaciers in the central Peruvian Andes are restricted to altitudes above 5000 masl, or at a slightly lower elevation on the western side. In fact, the glaciers were particularly big in the Western Cordillera because the mountains are higher and gather more humidity. The traces left in this area by the glaciers show that the ice of the Pleistocene climbed up most of the mountains and extended below 4000 masl in some Andean valleys. Wright (1980: 255) believes the snowline came down from 4600 m to 4300 m in the Western Cordillera during the Ice Age. The modern snowline came down from 4900 m to 4800 m, so the depression in the snow-

line increased from 300 m in the west to 500 m in the east. The larger size of the ancient glaciers presupposes a most significant climatic and vegetational change in that interval.

We know that in Peru's central highlands there were two marked cold oscillations around the end of the Pleistocene, one between 18,000 and 16,000 BC, the other around 11,000 BC. The optimum climaticum happened in the Andes at approximately 9500 BC. Meanwhile, in the Vilcanota Cordillera, the glaciers reached their climax in 26,000–12,000 BC, with two subsequent advances, one in 10,000 BC, the other in 9000 BC. It was from this last date that the snowline more or less reached its present level or slightly below (see Bonavia 1991:40).

The abovementioned increase in the size of the glaciers

produced the descent of the glacier tongues some hundreds of meters to an altitude that lies below the present limits; the humid mountains were then covered by an icecap at more than 4000 masl. Locally, the lower limit of the glaciers can descend even more to the level of the great peaks (predominance effect), but just as well in the humid mountains: so it was that recent glaciers (the last Quaternary cold wave) came down to 3100 m NE of the Bogotá savanna, in the lower part of the peaks, and climaxed around 3600–3700 masl. The fainter contours of an older glaciation are found at 2900 masl. The lower limit of Quaternary glaciations rises in the dry Andes. . . . The decrease in the size of the glaciers, and the expansion of cold and humid surfaces, limited the use of the high mountains even for the hunters. Meanwhile, valleys and inter-montane Andean basins and deep valleys which are at present relatively dry and in a sheltered position, and whose height ranges between 1500 and 3400 masl, were warm. The humidity due to rainfall and to springs fed by mountain waters was better distributed both in space and time, and vegetation was far denser than it is at present, with the corresponding fauna. These basins were the favored sectors for hunters and gatherers.

During the phases of glacial regression, warming was accompanied by the reestablishment of the rhythm of rainfall between the seasons. This was characterized by the conquest of recent moraines by herbaceous vegetation; the colonization of proglacial accumulations by a hydrophilous vegetation, and of debris cones by trees (*Polylepis*). Andean grasses and sparse plantations were grazed by herds of guanaco and deer. For hunters, these were favorable sites. The vast expanses of the "puna" allowed for rapid movement. Occupation was instead limited to some privileged biotypes in the dry basins around ponds, and vegetation became ever more xerophytic. (Dollfus and Lavallée 1973:80–81)

A similar picture at a local level has been drawn by Lavallée and Julien (1980–1981:100), who specifically refer to the Junín area. They explain that herds of animals used the high-altitude regions once the Quaternary glaciers retreated. In the terminal times of the Pleistocene, the big, cold, and humid expanses were free of ice and were once again covered with herbaceous vegetation, where animals began to proliferate at an unknown date, perhaps before 9000 BC.

A similar phenomenon took place in the southern part of the continent, with only the altitudinal limits varying. Thus, in the punas of Atacama it seems that the snows came down farther than they do at present, between 700 and 1200 masl; their current limit lies between 5500 and 6000 masl. This must have occurred at the time of the maximum expansion of the glaciers, but no dates are available (Yacobaccio 1986:2). The glaciers in this area reached the present limits around 8000 BC (Markgraf and Bradbury 1982).

Markgraf (1988) discusses the temperature rise that occurred before 8000 BC in the southern part of South America, and how the decrease in grasslands affected the megafauna, that is, the big herbivores. She notes that "[o]nly the guanaco, apparently less specialized in its forage requirements, managed to survive these problemat-

ic times, even though its numbers were greatly reduced at first (Bird 1938)" (Markgraf 1988:201). When discussing the fact that precipitation increased after 6500 BC, Markgraf states that this could have sufficed for the guanaco population to grow, as well as for the hunters (Markgraf 1988:201).

Markgraf (1988:200–201) also analyzed the terminal phase of the last glaciation in Tierra del Fuego and Patagonia, and explains that before 9000 BC, the dominant vegetation in South America was a treeless and herbaceous steppe between 50° and 54° S latitude. It was in general colder than in any part of present-day Patagonia. The xeric steppe appeared in 8000–9000 BC, thus suggesting a rise in temperature and a heightened lack of humidity. This lasted until 6500 BC, when the woodland expanded, thus implying an increase in rainfall and a thermal gradient similar to the present-day one in these areas. These conditions have apparently lasted until the present. There are some signs that in the Early Holocene, between 6500 and 4000 BC, there was more humidity and warmth than in the Recent Holocene (i.e., between 4000 BC and AD 1000). This sequence was interrupted by certain arid periods between 4000 and 3000 BC, and AD 1000.

3.10 GENERAL CONSIDERATIONS

In conclusion, we can note that despite its geographic and ecological diversity, the southern Andean region has a native fauna that is not essentially different from and only slightly more varied than that of the altiplano, even though this is a relatively uniform life zone. The recent habitats of the mammals were established in the last ice age. The centripetal fauna included marsupials and edentates from the oldest strata, caviomorphs from the middle strata, and camelids from the most recent strata. They might have been derived from Patagonia east of the Andes, which, it seems, was never covered by ice during the Pleistocene (Hershkovitz 1969:59).

In 1931, Cabrera prepared a synthesis in which he tried to draw some general conclusions on the problems of the Argentinian camelids. I find it useful to cite some of his ideas because these are still relevant and can be generalized, even with the passage of time. Cabrera (1931:89) stated that López Aranguren had reviewed the fossil camelids of Argentina a year earlier, in 1930, and had drawn two conclusions: "1. That the four forms of this group currently alive are also found in fossil state, which seems to prove their specific difference, and 2, that the number of fossil species is far less than has been so far accepted." Cabrera insists that although in that same year, Rusconi (1930b) used different criteria and tried to separate the forms López Aranguren considered synonyms, "the group systematics, as far as the extinct forms are concerned, must still be corrected in a restrictive sense, i.e. that the number of valid forms is even less than what Dr. López Aranguren admits." This follows from, and is confirmed by reading, Webb's study (1974).

NOTES

* Christian de Muizon assisted the author with the general topic of paleontology, and discussed with him the possible explanations of the origin of wild South American camelids. He prepared Figures 3.1 and 3.2.

1 I must draw attention to the fact that in a recent and widely read publication, Torres (1992a:32) states that camelids originated in the Pliocene, which is wrong.

2 In Franklin (1982:462) we read "Bolivian Channel," but this clearly is a mistake for Bolívar Channel.

3 By passive migrants I mean those animals that did not enter the South American continent by their own means but instead were transported by natural means, such as on tree trunks or the remains of drifting vegetation.

4 I have to point out that Sumar also discussed this subject but actually plagiarized Kent's study, which does not even appear in Sumar's bibliography. To prove this contention, all one need do is compare the paragraph from Kent (1987:171) I have transcribed and that appears in the article by Sumar (1992:82).

CHAPTER 4

ARCHAEOLOGICAL DATA
FROM PERU

This chapter reviews the available archaeological data on camelids, following two guiding principles. The first is a chronological one, beginning with the earliest preceramic remains and ending with the Inca. Then, within each epoch the materials are divided into subcategories, with those from the highlands discussed first, then those from the coast, and finally those from the *ceja de selva*,[1] if any.

4.1 PROBLEMS IN IDENTIFYING DOMESTICATED CAMELIDS

Before beginning, it is worth pointing out something that scholars know but the general public does not—the problems identifying camelid bones. First, there are problems in distinguishing wild camelids from domestic ones; second—and this is a major problem—it is still archaeologically impossible to clearly distinguish the four species on the basis of their bones, particularly llama, alpaca, and guanaco bones (see Pollard and Drew 1975:296; Franklin 1982:467; Stahl 1988: 357). As Novoa and Wheeler (1984:123) have pointed out, there are very few osteological traits that can be used to identify camelid bones at the species level. Osteometric analyses have therefore been limited to the family, and do not reveal the process of domestication. There is a decrease in size from llama to guanaco to alpaca to vicuña, and this characteristic was used "with success" by Wing (1977[a]). (Readers who want to expand their knowledge of the osteological traits of South American camelids can read Pacheco Torres et al. 1986 and FUCASUD 1994, 1995, 1996.)

Wing has tackled this problem and has probably had the most success so far. Regarding camelids, Wing wrote that "[t]hese are highly variable animals, difficult to distinguish on the basis of fragmentary osteological remains, and have therefore been identified only to the familial level" (Wing 1980:150). Using contemporary camelid bones, Wing managed to divide archaeological camelids into different size groups on the basis of multiple variables. One group includes the bigger camelids, the guanaco and the llama, and the other has the smaller ones, the alpaca and the vicuña (Wing 1972, 1977a, 1977b). This grouping by size was adopted by Hesse (1982a), with some simplifications.

Wheeler (1982b) in turn tried to use tooth eruption and tooth wear to establish differences between members of the Camelidae group. In her method, the morphology of the incisors plays a crucial role, "since few, if any, species-specific, post-cranial skeletal characters have been established for the South American camelids (Wing 1972, 1977[a]; Wheeler Pires-Ferreira, Pires-Ferreira, and Kaulicke 1976; Hesse 1982a[b]; Miller

1979)" (Wheeler 1984a:401; see also Wheeler 1995:281).

For obvious reasons, this is not the place to go into detail on the dental traits of the Camelidae. Interested readers will find a description by Wheeler (1982b). Here only the characteristics of the incisors are considered. Wheeler (1982b:12–13) believes her results are valid for all lamoids, because she was unable to find a significant difference among the three animals she studied—llamas, alpacas, and the alpaca-vicuña hybrids. Llamas and guanacos have spatulate (spoonlike) incisors, with enamel covering all the surface of the crown (on the lingual and labial aspects), and have a different root structure. In contrast, vicuña incisors are rootless and not spatulate, almost square in cross section, and have enamel only on the labial surface.

The alpaca, in turn, has incisors with traits that fall between those mentioned. They are not spatulate, the cross-section is more rectangular than square, the enamel is the same as that of the vicuña, and the root structure is the same as that of llamas and guanacos (Wheeler 1982b:12–13). According to Wheeler (1984a:401), "[t]his difference is clear-cut and unmistakable, so there is no possibility of erroneous identification in samples from the earliest occupation levels where there is no evidence for the presence of domestic forms."

Shimada and Shimada (1985:18) noted that Wheeler (1982[b]) believes different camelids can be identified by the shape of their incisors, but Kent (1982[a] and personal communication to the authors, 1984) shows that this is not always possible. After studying 100 alpaca mandibles, Kent in fact stated he found several that did not have the traits of the "alpaca incisors" described by Wheeler. Shimada and Shimada then made the following observations regarding their studies in Lambayeque, which will be mentioned in due time: "All the camelid incisors examined from sites in the Lambayeque region have been of the llama/guanaco type, according to the Wheeler classification system. However, well-preserved incisors that allow confident assignment are infrequent and we cannot eliminate the possibility of alpaca herding and breeding on the coast." Although llamas and guanacos cannot be distinguished osteologically, Shimada and Shimada

seem to suggest that these are llamas (Shimada and Shimada 1985:18). I discuss this subject in more depth when dealing with the discovery of camelids in the Lambayeque area. This qualification is necessary because the method is apparently not completely reliable.

I would like to draw the attention of specialists to a study by Herre (1953) that has gone unnoticed. Herre shows that the mastoid sinuses (tympanic bullae) of wild and domestic South American Tylopoda play an important role in providing support for the lower jaw, and keep the hyoid bone in place. This explains the particular shape of the mastoid sinuses. The domestic forms of the *Lama* sp. exhibit an important change in the mesotympanal area. The anulus tympanicus changes its form and position, and the width of the external auditory meatus becomes irregular. These modifications in the animal indicate a functional diminution of the auditory apparatus. These morphological changes are not considered primary changes in the sense organs, but they must be compared to what Vau (1936) wrote about dogs. A modification in the anulus tympanicus takes place after domestication, as can be seen when a ram is compared with domestic sheep. This anatomical feature, were it detected in archaeological remains, would complement the other evidence and possibly help confirm or reject the domestic condition of the animals.

4.2 THE PRECERAMIC PERIOD (CA. ?–1800/1500 BC)

Before reviewing the discovery of camelids at various archaeological sites, it is worth making a general statement. Franklin (1982:466) stated that "[w]hen early man, a hunter and gatherer, arrived in South America about 10,000 to 20,000 years ago, a number of camelids were there for his use, including *Paleolama*, *Hemiauchenia*, *Lama*, and *Vicugna*."[2] This statement is generally correct, because the extinct animals were already living in the South American continent when the first humans arrived. As proof of this statement, in some sites there are clear associations that show co-occurrence. However, these associations are not common, and it would seem that they are more frequent in some geographic areas than others. Statements of this kind should be used with caution and with due awareness of exactly what is being said. On the other hand, we should not take the opposite position, or forget the presence of extinct animals, and state that "[w]hen humans first entered the Andes, they discovered and hunted two native wild members of the camelid family, the guanaco (*Lama glama guanicoe*) and the vicuña (*Vicugna vicugna*)" (Flannery et al. 1989: 89).

4.2.1 The Highlands

I will first review the high-altitude sites on the puna, and then those in the valleys. (For the location of the sites mentioned, see Figure 4.1.)

In the Callejón de Huaylas, the sites where camelid remains have been documented are located in high-altitude areas north and south of Guitarrero Cave (in the lands of the Shupluy peasant community, on the left bank of the Santa River, at 2580 masl). The faunal variety in these sites is low, with cervids and camelids dominant. Lynch (1971:145) noted the abundance of camelids at sites in this area, and Wing (1977a: 839, Table 1) confirmed their early presence at Guitarrero Cave, at 2580 masl in the province of Yungay, Department of Ancash. This has been widely shown by the studies made by Lynch (1980a) and Wing (1980), who investigated the faunal remains and reported their presence in Complex II. Wing showed that their number increased in Complexes III and IV. I refer to this site below when discussing valley sites.

Camelids constitute 75.5% of faunal remains at PAn 12-58, a site located at a higher elevation than Guitarrero Cave, on the headwaters of the Santa River (Department of Ancash, province of Recuay), which belongs to the late Ice Age, with a date of 9690 BC (Wing 1980:157, Table 8.2). Then there are PAn 8-126 (Department of Ancash, province of Huaylas) and PAn 12-53 (in the same department, province of Recuay), which have both a preceramic and a later occupation, where camelids, respectively, constituted 83.3% and 86.5% of faunal remains. Finally, there is site PAn 12-57, which had only a preceramic occupation, and 95.1% of the faunal remains were camelids (Wing 1980:157, Table 8.2). Unfortu-

nately, no more data are available for these sites, about which Wing (1980:156) noted that "camelid remains constitute 76% of the fauna in the early period sample and 83–96% in the later samples." (These "later samples" belong to later periods, which will be considered later.)

There is another site in the Callejón de Huaylas at 3970 masl called Pampa de Lampas, whose exact location I do not know, and which should belong to the late preceramic. A great number of camelids were present, because figures of 64% and 80% have been reported (Wing 1978: 169, Table 1).

There are few data for Lauricocha Cave (in the northeastern part of the Raura cordillera, inside the former hacienda of the same name, in the province of Dos de Mayo, Department of Huánuco). Regarding the earliest layers, called Lauricocha I, Cardich simply says that "[t]he tarucas or tarugos (*Hippocamelus antisensis*) had a major role in the diet" (1960:108–109) and that these "are proportionately slightly higher than the camelid bones" (Cardich 1980:132). When discussing the Lauricocha II component, Cardich explained that "South American camelids (*Lama glama*, *Lama guanicoe*, *Vicugna vicugna*, etc.) *seem* to predominate among the animal remains that served as food. There also appear fragments of *Hippocamelus*, but in a smaller percentage" (emphasis added). In an earlier study, Cardich (1959:104) simply said of "Horizon II" (which evidently corresponds to what he later called Lauricocha II) that "there appear *much osseous materials* from the food, and in general comprise the great herbivores of the Andean periglacial fauna: llamas (*Lama glama*), vicuña (*Vicugna vicugna*), guanaco (*Lama guanicoe*) and surely other camelids" (emphasis added). Lauricocha III is not mentioned at all. In a much later work, Cardich (1980:132–133) said of Lauricocha II and III as a whole that camelid bones were more abundant (the data were repeated in Cardich 1987b).

Wheeler Pires-Ferreira et al. (1976:486, Table 2) later tried to study these materials, but apparently only managed to reconstruct the data for Levels 20 and 21, which correspond to Lauricocha I (ca. 7000–5500 BC), where camelids constitute 59.1%, and Levels 12–14, which correspond to Lauricocha III (ca. 4200–2500 BC),

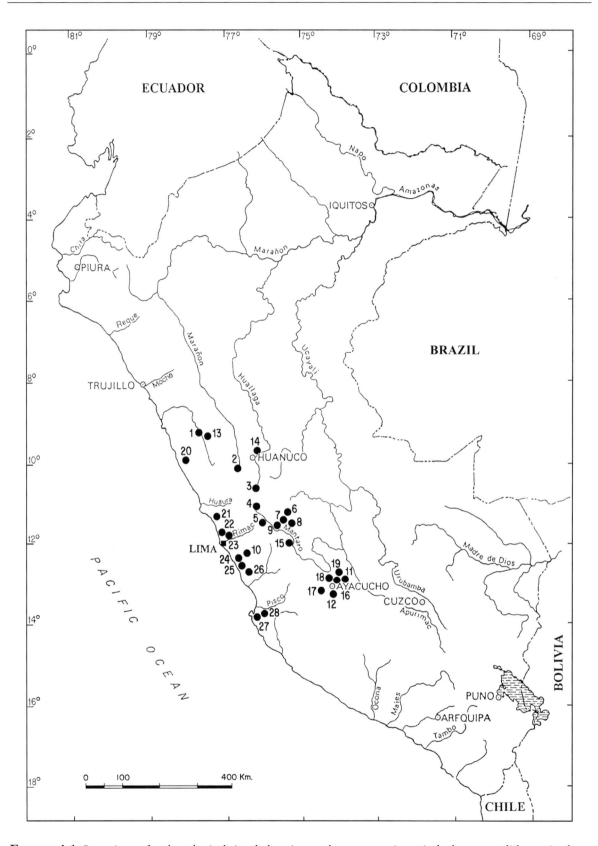

FIGURE 4.1. Locations of archaeological sites belonging to the preceramic period where camelid remains have been found. The site legend is on the facing page. *Prepared by the author and drawn by Osvaldo Saldaña.*

where camelid remains constitute 84.7% (the data were repeated in Wheeler 1977:6, Fig. 2, and Wing 1986:253, Table 10.5).

Cardich (1987b:14) later wrote of his Lauricocha III period:

> There is a great presence of camelid bones as food refuse, as well as bones used. Perhaps this was connected to the practice of at least partial herding, leading to the domestication of the llama and the alpaca. After the first studies, we tentatively wrote that "it would not be impossible for the Lauricocha II high-altitude hunters to have evolved into the pre-Chavinoid hunter-livestock raisers" (Cardich 1958:20). Faced with the presence of the Lauricocha III remains in the first report (Cardich 1960:117), we mentioned a change in regard to previous phases, and simply stated that "we await new evidence that can help us understand these events better." Lately, the sophisticated studies of Danièlle Lavallée and her team have made important contributions on camelid domestication at Telarmachay in the Junín puna and also in the central Peruvian highlands, where the transition from hunters to herders would have taken place between 6500 and 3800 BP. (Lavallée et al. 1982:86)

In this same study, Cardich (1987b:25) adds: "These hunter-gatherers of the Lauricocha tradition probably attempted to domesticate the Andean camelids at the highest altitudes, perhaps with newborns and based on their exceptional contact with them. However, not all species were receptive to domestication. They failed with vicuñas and guanacos, but were successful with the alpaca and the llama."

It is worth pointing out two facts. First, Cardich uses the 1982 study by Lavallée et al. and ignores their final 1985 report. This had serious consequences for his study, because this is precisely where the problem of camelid domestication was set down with a wealth of detail. Even worse, when discussing this subject Cardich states that "[t]hey failed with vicuñas and guanacos, but were successful with the alpaca and the llama" (1987b:25). In other words, Cardich believes all four modern lamoid forms have always existed and that domestication started with them, when it is well known that the domestic forms were derived from some wild forms. Chapter 1 also showed that the phylogenetic problem is quite complex and has yet to be solved. I return to the problem of domestication and the various positions later.

From the same scant data furnished by Cardich, it follows that there were abundant bone remains in the strata excavated. I visited the zone during the excavations and can certify this, but it is clear that these remains were not given due consideration. Many doubts still linger over the work done at Lauricocha as regards the stratigraphy itself, and no studies were made by specialists. Flores Ochoa (1982:67) wrote: "The study of Lauricocha [Cardich, 1964] . . . evinces the scant attention that was paid at the time to the animal bones, and the lack of almost any attempt

---

FIGURE 4.1. Preceramic Period Site Legend

1	Cueva del Guitarrero	11	Ruyru Rumi	21	Río Seco del León
2	Lauricocha	12	Chupas	22	Ancón
3	Piedras Gordas	13	Huaricoto	23	El Paraíso
4	Pachamachay	14	Kotosh	24	Paloma
5	Panaulauca	15	Callavallauri	25	Chilca
6	Telarmachay	16	Pikimachay	26	Asia
7	Acomachay	17	Jaywamachay	27	Cabezas Largas
8	Cuchimachay	18	Ayamachay	28	Santo Domingo de Paracas
9	Uchcumachay or Tilarnioc	19	Puente		
10	Tres Ventanas	20	Los Gavilanes		

to identify them by species, age or gender." Kent (1987:173) criticized the work done by Cardich at Lauricocha, noting that he failed to connect his discoveries to the process of domestication and believed that the camelids of Lauricocha II were hunted. Domestication had apparently already begun. In my opinion, Cardich's attempt (1987b:14, 25) to explain this process almost 30 years after the excavations at Lauricocha, apparently to defend himself in the face of the evidence found in the studies by other scholars, in no way fills the great gap on this subject at Lauricocha, because no concrete evidence is furnished. Kent (1987:173) is actually right, but I doubt that these materials can really be reexamined, given the way in which they were excavated. What should be done, in light of the significance of the sites in this area, is to excavate anew, with a more rigorous methodology. (For the critiques leveled at Cardich's studies, see also Lavallée 1990:33.)

Piedras Gordas in Champamarca, a suburb south of the city of Cerro de Pasco, in the department of the same name, is another site for which the data are inadequate. The research was carried out by Hurtado de Mendoza (1987), who excavated using arbitrary levels and failed to keep a three-dimensional record of the artifacts, because "these appeared in such great numbers that the process of individually recording them considerably slowed down the excavation" (sic) (Hurtado de Mendoza 1987:212). No further comment is needed.

As regards the fauna, it is known that "a considerable number of animal bones" were found. However, a serious contradiction appears, because it is stated that it was not possible to "determine the genus and species in most of the camelid and cervid remains," and that only 29% of the collection had been studied (Hurtado de Mendoza 1987:228). However, the following page of the report literally reads: "The only two peculiarities observed are as follows: 1. The vicuñas (*Vicugna vicugna*) and guanacos (*Lama guanicoe*) seem to be more abundant in the lower levels and surrender their relative popularity to the llama (*Lama glama*) and above all to the alpaca (*Lama pacos*) in the upper levels; and 2. The presence of hybrid auchenid forms is noticeable

from level 6, particularly of guarizos (*Lama glama* × *Lama pacos*)" (Hurtado de Mendoza 1987:229). In other words, it was first stated that the animals were not identified to the genus and species level, and then not only are the four known species named but also a hybrid, something I have not found in any other report.

Thus, camelids predominate at all levels, according to Hurtado de Mendoza (1987:230). Table 9 of Hurtado de Mendoza (1987:230) provides a tabulation that comes close to this assertion, but not quite. The table shows that camelids predominate at all levels except Level 1. In the oldest level they make up 71.9% of the remains found, and 79.7% in all. According to the report, the oldest level has an age of 8000–9000 BC, but there is only one date for Level 11, which gives 7995 radiocarbon years. The first three levels are associated with ceramics.

According to Table 11 (Hurtado de Mendoza 1987:234), adult animals prevail over the subadult, or those under one year of age, at all levels. However, it is striking that the difference is not too large at many levels, whether they are old or recent.

The report mentions a "decline in the puna ecosystem due to excessive fluctuations in the environmental temperature" which made the cervids scarce, so the hunters turned to camelids instead (Hurtado de Mendoza 1987:233). However, an examination of the percentages presented shows that the decline took place from the beginning to the end of the cave's occupation. In other words, the data are not consistent. Besides, we know that the "[f]all in the remains of cervids in the punas of Junín seems to be more a cultural than an environmental phenomenon (Wheeler 1984[a], 1985[b]); this is also seen in the fauna from other sites in the area [this refers to Junín] (Wheeler Pires-Ferreira et al. 1976; Wing 1986; Moore 1988[a in my bibliography] 1989)" (Baied and Wheeler 1993:147).

This study has other problems, which the author himself notes. Hurtado realized that "[w]hen separately quantifying the bones from camelid and cervid hands and feet, and legs and arms, considerably fewer bones from hands and feet were found than was to be expected, given the anatomy of these animals." But "against what could be expected, there are considerably fewer bones of cervid feet than camelids." We can ask whether these are real conditions or were instead altered by sampling problems. Several statements are then made about the possibility that camelids were hunted in localities closer to human habitations, while cervids were hunted in more distant localities, but without any real evidence being presented (Hurtado de Mendoza 1987:234–235). The hypothesis put forward by Hurtado is as follows. He believes that the site was "always" used as a camp by the hunters: "The consistent proportions of different kinds of artifacts at all the levels of the site suggest that it retained the same economic function during the eight or nine millennia of its occupation, including a recent phase in ceramic times." Hurtado de Mendoza then claims that these were the economic activities of a base camp, where the animals were killed, food was cooked, hides were prepared, and so on. He notes that there is some variability, but cultural change correlates with environmental changes. Thus Phase I, which corresponds to Levels 12 and 11 (9000–8000 BC), holds great concentrations of animals, which reflects intense hunting. In Phase II, which corresponds to Level 10 (8000–7500 BC), "faunal remains fall to 48% of the concentration index recorded in Phase I," which Hurtado interprets as an effect of the predation of previous times. This is why the hunters used a less predatory tactic. There now was an "alternating sequence of moments with a relative stability and a series of crises of interaction between human groups and the environment. It was this crisis, the first of which is particularly detectable in Phase II, that killed a way of life that had advanced in the Late Pleistocene, and whose survivals seem to be detectable in the levels ascribed to Phase I." Phase III, which is related to Level 9 (7500–7000 BC), shows "a marked increase in hunting productivity." This would indicate the increase in the concentration of bones as a result of apparent climatic improvement, and the arrival of new human groups with a "qualitatively more refined cultural baggage." Phase IV, defined on the basis of Levels 8 and 7 (7000–5000 BC), was a time of intense cold in the sev-

enth millennium BC when two new kinds of lithic points appeared, which "must have lowered the availability of the cervids . . . [and] led to greater pressure on the camelids." Phase V (Levels 6 and 5, 5000–3000 BC) shows "a serious decrease in the availability of cervids, which placed greater pressure on the camelids." The cervids once again appear in important numbers in Phase VI (Levels 4 and 3, 3000–1500 BC) in the face of insufficient numbers of camelids, "which had suffered a great predatory impact since the beginning of the Piedras Gordas occupation." Finally, hunting activities took place in Phase VII (Level 2, 1500–500 BC), with an increase in cervid bones (Hurtado de Mendoza 1987:235–239).

Finally, Hurtado de Mendoza (1987:204) classifies his camps into four groups: base camps with a "high occupational complexity," secondary base camps with just a few animals killed, base camps for the manufacture of lithic artifacts with few animals killed, and base camps with a greater emphasis on working stone.

All this, of course, clearly contradicts the work of other colleagues, with the difference that the latter presented their supporting evidence while Hurtado does not. I wonder how these assertions can be made without excavating extensively, without seriously studying the remains, and after excavating just one site with an inadequate methodology. The bone remains analyzed by Altamirano and Guerra were not studied properly. There is no basis for the climatic studies. Finally, I believe that such a precise chronology cannot be established by an excavation made with artificial levels, and with just one radiocarbon date. This study does not merit further consideration.

Pachamachay, 7 km west of Lake Junín, in the department and province of the same name and at 4250 masl, is certainly an important site. It is estimated that the remains of more than 12,000 camelid MNI (minimum number of individuals) were found here (Rick 1980:267).

Several studies were made at this site. Ramiro Matos worked here in 1969 and 1970. He excavated the site anew in 1973 with Peter Kaulicke, and John Rick excavated it in 1974–1975 (Rick

1980:57). We thus have a series of reports that detail the work done. This can be misleading if we are not careful to specify which excavations we refer to, since the data are not always clear. Among other things, it is not always specified whether the data in Matos's excavations include those he made with Kaulicke. For this reason I essentially rely on the so-called Matos and Rick excavations.

Let us begin with the excavations made by Ramiro Matos. From the available data it seems that these were made with arbitrary levels, which, as is well known, presents serious problems. In a preliminary report, Wing (1975a:79) wrote: "Most of the fauna is composed of camelids, mainly domestic forms. A great part of the camelids are juvenile specimens, which make up 56% of the sample so far studied. Most of the young animals are 18 months old." This last figure was later corrected, the author noting that these animals were in fact newborns (Wing 1986: 248). This same report presents a table that notes that camelids make up 81.97% of the fauna excavated (Wing 1975a:80). Wing (1977a:839, Table 1) repeats this in a later report but specifies that the sample belongs to the late preceramic (ca. 2000 BC).

Matos and Rick (1978–1980:44) indicate, apparently with more complete data, that camelids accounted for 97.6% of the faunal remains for the early preceramic and 96.2% of remains for the late preceramic.

In his final report on the work done at Pachamachay, Rick (1980) explains that Wing gave him more complete data after the 1975(a) report, and notes that these are the materials Matos excavated in 1969 and 1970. (I understand that Jonathan D. Kent examined the materials from this cave and included them in his dissertation [Kent 1982a], which I was unable to read.) However, in 1988(b), Kent published data which I assume are part of this dissertation. They are discussed later. From Rick's data it follows that camelids constitute 97.30% of the remains in the early preceramic strata and 96.06% in the late preceramic. Rick also explains that in the early preceramic sample, 36.6% are adult animals, 5.6% are subadults, and 57.7% are juveniles, whereas in the late preceramic 52% are adults, 16.5% are subadults, and

31.5% are juveniles. In all, camelids make up 97% of the total fauna at this site (Rick 1980:234, Table 10.1).

In a later study, Rick (1983:146, Table 5) repeated the data. (All that is indicated is that the percentage for the early preceramic was 96.10% instead of 96.06%, perhaps because of a misprint, as the edition was carelessly proofread.) Here he gives for the early preceramic an age that varies between 8000 and 2000 BC, and for the late preceramic an age that varies between 2000 BC and AD 350. Attention must be drawn to a fact that might mislead the reader who has not seen Rick's 1980 report. Fom the excavations made by Matos, which are reproduced in Table 5 (Rick 1983:146), Rick says, "Wing (1975) published a preliminary report on the osseous material, giving us complete data (Table 5)." This means that the data in Table 5 repeat the preliminary data in Wing (1975a), which is not true. These are the full data Wing gave Rick, who published them in 1980 (234, Table 10.1). This is a case of mistranslation.

Rick (1980:326–329) divided the preceramic occupation of Pachamachay into five phases. Phase I (10,000–7000 BC) corresponds to an occupation by nomadic groups, few in number, dedicated to hunting the vicuñas. In its first stages, Phase 2 (7000–5000 BC) shows the first signs of sedentism, with camelid hunting as the main activity. The occupation of the site decreases toward the end of this phase, and the evidence for vicuñas decreases. This phenomenon continues in Phase 3 (5000–3000 BC), and there is a change toward the end of this phase, for there is an increase in population that continues in Phase 4 (3000–2200 BC). Rick interprets this as the use of more conservative hunting methods and the control of herds of animals. Hunting was almost replaced by herding in Phase 5 (2200–1500 BC). In the following phases the site became a temporary camp because its population had moved to the neighboring lake.

When Rick (1980:266) published his final report on Pachamachay, he still did not have the study of the fauna. This was included in a later study. Here he followed the division into phases of the preceramic occupation of the cave that we have just seen. In Phase 1, the oldest, camelids constituted 80% of the faunal remains, 100% in

Table 4.1. Camelid Remains Found at Pachamachay (1975 Excavations)

Phase	Date	Camelids (%)	Lama pacos (%)	Lama glama (%)	Lama guanicoe (%)	Vicugna vicugna (%)
Phase 1	10,000 BC–7000 BC	80[x]				
Phase 2	7000 BC–5000 BC	100[x]				
Phase 3	5000 BC–3000 BC	100[x]				
Phase 4	3000 BC–2200 BC	90[xx]	3.33			
Phase 5	2200 BC–1500 BC	93.45[xx]	1.27	0.09	0.05	
Phase 6	1500 BC–800 BC	93.22[xx]	2.67	0.12	0.24	1.58
Phase 7	800 BC–400 BC	86.32[xx]	0.47	0.47	0.47	
Post-7	400 BC–AD 500	69.34[xx]	1.55	0.52	0.26	0.52

[x] Unidentified camelids
[xx] Includes both unidentified and identified camelids
The table was prepared using Kent 1988b:Table 3, 133–136, and Table 1, 128.

Phase 2, 100% in Phase 3, 90% in Phase 4, and 93.45% in Phase 5 (Rick 1980:144, Table 4) (see my Table 4.1).

Here it is explained that these are the results of the excavations Rick made at the site in 1975, which were included in Kent's dissertation (1982a). The statistics were based on just a few bones, and Phase 5 is considered the "terminal preceramic." Here they note that only the bones in Phase 5 could be identified to the species level, and the results were "11 vicuñas, 2 llamas, and 27 guanacos. Since this is one of the first analyses that dares to identify the camelids by species, it can be subject to a certain percentage of error, but in any case we can see that herding was already established at least during the final decades of the preceramic" (Rick 1983:145).

Wheeler Pires-Ferreira et al. (1976:486, Table 2) repeated the data from Matos and Rick (1978–1980:44), with just a slight variation in the decimals of the percentages. Wing (1986:253, Table 10.5) also repeated the data, introducing just a small variation in the decimals of one of the percentages. Wheeler (1977:6, Fig. 2) also repeated the data, although only partially.

There is only one discordant note: Pires-Ferreira et al. (1977:153) give a percentage of 94.1% for camelids in the late preceramic at Pachamachay. As can be seen, this figure does not appear in any of the reports previously cited, and I do not know its origin. It is surely an error.

The data Kent (1988b) includes on Pachamachay are not just significant but also revealing, for they contradict Rick's position (1983: 145) in several respects, and also expand it. I find it necessary to present them in detail. However, it must be noted that the differences between the analyses made by Kent (1998b) and Wing (1975a) reflect the fact that Kent had access to the complete collections, whereas Wing worked only with the materials from the first excavations.

Kent (1988b:137, 140–141) discusses Rick's position. First he notes that Rick believes that vicuñas were the major resource in the preceramic period, particularly in Phases 3 through 5, so the bones of this animal should predominate. Second, if sedentism was the rule toward the end of Phase 5, as Rick believes, then there should be camelid bones of all age groups. Besides, if the population attained its maximum level in this period, then there should be a greater accumulation in Phases 4 and 5. Third, "If the focus of the settlement had moved to the shores of the lake in Phase 6, and if the cave then became a seasonal camp, as Rick suggests, then the mortality profile for camelids can reflect this fact, showing the season or seasons the site was occupied. Identifying the camelid species can also show if the camp was used by herders, hunters, or both." Fourth and last, if the domestication of camelids took place before the middle of Phase 5 or 4, then the identification at a species level can be evidence for it.

From the analysis of the materials, Kent (1988b:137, 140–141) concludes that vicuñas never predominate in any phase, whereas domestic camelids are more numerous than wild ones by a ratio of two in almost all phases analyzed. The absence of pre-Phase 4 remains makes it impossible to appraise the presumed predominance of the vicuña in the early preceramic. Second, the data support year-round sedentism in Phase 5. There are juvenile camelids and fetuses/newborns. The increase in population is evident in the amount of remains from Phase 5. We do not have enough bones to show the maximum increase in population that presumably took place in Phase 4 (see Table 4.1).

Third, Rick suggests that the cave was a seasonal camp in Phase 6. However, the remains include juvenile animals and fetuses/newborns. Bones from vicuñas, guanacos, and cervids have been identified, thus indicating that the economy was a hunting economy. At the same time there are domestic camelids, which indicates that herding was practiced. Kent tends to believe that both activities were practiced by the same people. In Phase 7 there is a seasonal occupation, no vicuñas, and the cervids increase. Fourth, bones from domestic camelids appear in all layers of Phase 5, so they must have been domesticated earlier. "The length of the process of domestications has not been well defined, but it seems to have begun long before the beginning of Phase 5. Based on osteometric analysis, the first domesticated camelid that appears in Pachamachay is the alpaca" (Kent 1988b:137, 140–141). This agrees with the data from Telarmachay (Lavallée et al. 1982:92; Wheeler 1984d:198).

Kent (1988b:132) notes that the proportion of camelids ranges between 5.85% and 13.68%, which is higher than the 2% suggested by Rick (1980:266, 295), who quoted a study by Wing (1975a) in which Wing analyzed only the materials from the slope where Matos excavated in 1969–1970. Besides, the age of 64% of the remains can be established. Adults then account for between 43% and 59%, and fetuses/newborn animals account for between 9% and 20% (see details in Kent 1988b:Table 4, 138, and Fig. 2, 139). This also differs from Rick's data, which clearly indicate the predominance of juvenile animals.

Kent (1988b:132, 136) insists that "[t]he vicuñas in no time or phase comprise the biggest part of the camelids identified; this fact has some implications for the model of dependence of the people who lived in the site, based essentially on the hunt of vicuñas. The presence of guanacos in this area is also noteworthy, because nowadays they are not found here, nor is there any historical documentation of their presence close to the site."

As regards Phase 4, Kent (1988b:136) comments that only one bone of all the remains can be attributed to the alpaca species, whereas materials from Phase 5 reveal the presence of four camelid species. The most frequent species is the alpaca, followed by the vicuña, the llama, and finally the guanaco. The number of alpacas is twice that of vicuñas, and this proportion is found in all succeeding phases. Kent therefore concludes that "[o]bviously the data are related to the hypothesis of vicuña-dependent, sedentary hunters of the Preceramic."

Kent (1988b:132) likewise notes that the frequency (RF) of camelids falls after Phase 5, changing from more than 90% in Phases 5 and 6 to 86% in Phase 7 and 69% in post-Phase 7 (see Table 4.1). For Kent, this reflects the changes in the way the herds were cared for, and is the opposite of the trends seen during the earliest phases of the preceramic, when the percentage of camelids increased according to the data in Pires-Ferreira et al. (1976).

Kent (1988b:132) likewise observed that most of the bones correspond to the late phases, beginning with Phase 4 (3000 BC), so it is not easy to verify the part of the model that refers to the early and middle preceramic. The conclusions reached refer only to the late preceramic and subsequent periods. It is to be noted that Kent managed to ascertain that 28% of all the bones examined belong to camelids, which make up 90.35% of all identified bones.

Flores Ochoa (1982:68–69) made a comment on Rick's work at Pachamachay that I find interesting and therefore reproduce verbatim, as it can be misleading if attention is not drawn to it. We must bear in mind that it was written both before the publication of Rick's latest book (1983) and before the appearance of Kent's report (1988b). It reads thus:

> The use of the vicuña was supplemented with the use of different kinds of plants, as well as lake and river resources, because the puna was far richer than is usually believed. To state that camelid remains correspond to the vicuña is interesting, even though these were in no condition to enable vicuña bones to be clearly distinguished from possible alpacas or llamas. This would let us suppose that these camelids were dominant and sufficed to support sedentary human populations, at least in the area where Pachamachay lies. Alpacas and llamas could be found at lower-altitude zones, perhaps even at 3,500 meters or less, on the floor of the Mantaro River Valley itself. Were this possibility to be verified with greater precision, we would have more elements to support the conjecture that the alpaca and llama populations had a wider dispersion, and that their present niche at altitudes over 4,000 meters is due to social and cultural factors which allowed the expansion of exotic species brought by the European invaders.

Kent's data (1982a), included in the latest book by Rick (1983:145), seem to support the position held by Flores Ochoa, but we should not forget that the sample does not seem to be significant; besides, it refers to just one of the phases, the last one, of the site in question. Thus, in his last report, Kent (1988b:132) clearly says that "[v]icuñas at no time or in any phase comprise the biggest part of the camelids identified." This would seem to invalidate the comment made by Flores Ochoa. Wheeler (1985a:29), however, does not agree and considers that vicuñas "must have been one of the most numerous big herbivores that frequented

the punas and abounded in the land immediately around Telarmachay." The data from this site support Wheeler's position, but the point is still open to discussion.

Rick's studies have been severely criticized by several scholars. Interested readers can find the data in Wheeler (1984d:196).

Panaulauca is another site in the Department of Junín (in the province of Tarma, at 4150 masl). Wheeler Pires-Ferreira et al. (1976:486, Table 2) report that camelids represent 26% of faunal remains in Level 7, the most ancient one (7000–5500 BC), 87.6% in Levels 4 through 6 (5500–4200 BC), and 85.8% in the upper Levels 1–3 (4200–2500 BC). The data were later repeated by Wheeler (1977:6, Fig. 2) and Wing (1986:253, Table 19.5). The materials from this site were studied by Moore (1984), a study I have not seen. The results are summarized in a later work (Moore 1988a:156). Here Moore notes that "camelid percentages ranged from a low of 71% in Phase 1 to 87% in Phase 4. . . . This proportion then increases only slightly to about 90% at the end of the sequence." The data do not agree with the early phase as presented by Wheeler Pires-Ferreira et al. (1976), who give 26% of camelid remains, while Moore (1988a) gives 71%.

I find the studies Moore (1988a:156, 157; 1993) made of the Panaulauca materials of great interest. According to them, the wear in the cheek teeth of grazing animals is evidence of human management of herds and their pasture resources. The amount of wear varies according to the silica content of the forage and the potential abrasiveness and amount of soil the animals take in while feeding. Traditional herders control the dental wear of their animals as a measure of potential longevity and health, and try to reduce the rate of wear by moving the herds to fields that have not been exposed to intensive herding. The study of teeth from archaeological specimens from Panaulauca has provided a rate of dental wear throughout the transition to herding. The wild animals that were hunted have a higher proportion of tooth wear, whereas the early domestic animals have lower rates, thus suggesting the impact of human management. Present-day animals in a well-managed herd have rates lower than any prehistoric flock, thus suggesting that the early herders had more or less effective control of the health of their animals. This information can be coordinated with other evidence to understand the fragile nature of some early food-production systems.

Danièle Lavallée headed a team that worked in the area of San Pedro de Cajas, in the Department of Junín, province of Tarma. Almost all the sites this team explored lie between 4000 and 4500 masl, "always in the immediate vicinity of the wide steppe of the puna . . . [which] . . . approximately draws an E-W arc whose curve follows the lower limits of the upper puna and goes along the headwaters of the ravines, so that these sites simultaneously control hunting and pasture lands, and the routes to lower-altitude ecological floors" (Lavallée 1979:115). No preceramic or Formative site has been found there below 4000 m, even though the area has several shelters or caves that apparently have not been lived in since before the Early Intermediate period. However, the fact that all of these puna sites are close to the routes to lower zones makes Lavallée believe that there might be early settlements in the neighboring valleys, and that high-altitude and valley sites here might perhaps have exchanged products (Lavallée 1979:115).

The site chosen for study was Telarmachay, a rockshelter in the area of San Pedro de Cajas, 4420 masl (see Lavallé et al. 1985). Telarmachay has yielded abundant faunal materials; 400,000 animal bones are estimated to have been removed, 137,985 of which had been studied by 1985 (Wheeler 1985b:63). Pires-Ferreira et al. (1977:153), Wheeler (1977:6, Fig. 6), and Wing (1986:253, Table 10.5) provided preliminary data, but there is no need to cite them, as a report is ready that, although not definitive (because there still are more osteological materials to be studied), is clearly the most detailed yet published on a highland site. I return to it forthwith.

Interestingly, in all of her preliminary studies Lavallée had noted that camelids "in all cases" accounted for most of the material from the preceramic levels at San Pedro de Cajas, in a proportion varying between 70% and 85%.

I shall now analyze the specific case of Telarmachay. In general, camelids and cervids comprised between 97.85% and 99.15% of all the animals throughout the preceramic occupation.

The cultural materials at the site were divided into seven phases, the first three of which are ceramic (early and late) and the rest preceramic. The report for the first three phases has not yet been published.

Phase VII, the oldest (ca. 7000–5200 BC), gives 64.73% for the camelids, including the vicuña and the guanaco. Of these remains, 47.2% are adult, 15.7% juvenile, and 37.08% fetuses/newborns. Phase VI (ca. 5200–4800 BC) has 77.84% camelids. These remains show an evolution in hunting modes: 49.4% of the remains are from adults, 14.3% are from juvenile animals, and 36.33% are from fetuses/newborn animals. Phase V was divided into Lower and Upper, with Lower Phase V in turn divided into Lower Phases V 2 and V 1. Lower Phase V 2 (ca. 4800–4000 BC) has an exceedingly high proportion of camelids and cervids, which comes to 99.15%, 81.69% of which corresponds to camelids. The vicuña and guanaco are always present. Of these remains, 46.2% were from adult animals, 18.5% from juvenile animals, and 35.28% from fetuses/newborns. The significance of these camelids is that they show the changes in dental morphology that, according to Wheeler, indicate the beginning of domestication. In Lower Phase V 1 (ca. 4000–3000 BC), camelids comprise 85.94%. This, however, poses a problem. In a previous report, Wheeler (1984a:398, Table 1) said that the adult animals made up 43.0%, juvenile animals 11.8%, and fetuses/newborns 45.2%. However, in the 1985 (1985b:66) report she claims that in this phase, fetuses/newborns make up 56.75%, and no mention is made of the proportion of adult and juvenile animals. Perhaps new calculations were made. In Upper Phase V (ca. 3500–3000/2500 BC), the camelids are as high as 85.51%, 25.3% of which are adults, 6.4% juveniles, and 68.21% fetuses/newborn animals. Finally, in Phase IV (ca. 3000/2500–1800 BC) camelids constitute 88.64%, 17.2% of which were adults, 9.8% juvenile, and 72.99% fetuses/newborns (see Wheeler 1982a:5, Fig. 2; 1984a:398, Table 1; 1985b:65–67; 1994: 14–15). Spunticchia (1989– 1990:58, Fig. 4) prepared two interesting graphs that show this quite well. One is on the appearance of different animals at Telarmachay; the other one shows the use the inhabitants of this site made of the camelids.

Wheeler (1984a:401; 1985b:68) made the following comments on these results. A cumulative difference of –24.97% for cervids and +23.91% for camelids is detectable between the earliest occupation of Telarmachay, corresponding to Phase VII and beginning around 7000 BC, and the last preceramic occupation, Phase VI, which ended in ca. 1800 BC. This change did not happen suddenly but was instead gradual and constant throughout the sequence. In other words, we clearly see that the number of camelid remains increases at an inversely proportional rate to that of cervid remains. This probably reflects the continuous process of adaptation of man to the puna.

On the other hand, the proportion of camelid fetuses/newborns rose to 56.75% throughout Lower Phase V 1. This figure is significantly higher than in the earliest levels (37.08% in Phase VII, 36.33% in VI, and 35.28% in Lower Phase V 2) and more than the normal average of 35%–40% for the present populations of guanaco and vicuña (pers. commun. Franklin to Wheeler, and Franklin 1978) (Wheeler 1985b:71).

As regards the remains of guanacos among the fauna of Telarmachay, Wheeler (1985a:29) comments that the available data (she cites Neveu-Lemaire and Grandidier 1911) prove these animals live in the highlands and valleys according to the season. Their presence at Telarmachay indicates that they lived there in the preceramic period. (Wheeler 1983 is given as source but it does not appear in the bibliography. This clearly is a mistake. It must be 1984b in her bibliography, which corresponds to my 1984. The subject is discussed on page 401. This had also been discussed in Wheeler 1982a), but it is not known whether they frequented the puna all year long.

Some test pits were also made at Acomachay, in the ravine of Allacurán, 4520 masl in the area of San Pedro de Cajas. Two sites were chosen, Acomachay A and B (see Lavallée and Julien 1975). At Acomachay A, Test Pit 1, layers 2, 3, and 4 formed one single occupation unit corresponding to the late preceramic (2500–2000 BC). The sample is small, and in it camelids constitute 94.8%, 89.5% of which are adults and 5.3% juvenile animals. "These results must be taken with reservations due to the small size of the sample." Only layer 2 had animal bones in Test Pit 3 of the same site, and it also corresponds to the late preceramic. The camelids are present in a proportion of 93.3%, 86.6% of which are adults, and 6.7% juveniles (Wheeler 1975:122–123, Table 3).

At Acomachay B, the test pit did not allow the site to be dated, but layers 2 and 3 are considered to be preceramic. In any case, camelids predominate in all layers (90.9% for layer 1, 91.2% for layer 2, and 81.9% for layer 3) (Wheeler 1975:123, Table 4). If the materials from both test pits are combined, we have a total of 94.1% for the camelids (Wheeler Pires-Ferreira 1975:126, Table 7). This information was reproduced by Pires-Ferreira et al. (1977:153), Wheeler (1977:6, Fig. 6), and Wing (1986:253, Table 10.5).

Cuchimachay is another site in the area of San Pedro de Cajas (see Lavallée and Julien 1975) where test pits were dug to see their contents. In the first it was found that layers 3 through 8 corresponded to the late preceramic (2500–2000 BC), and camelids generally constituted 82.3% of the identified fauna, with adults predominating at 59.8%, while juvenile animals constituted only 22.5%. Only layer 6 was identified as belonging to the late preceramic in Test Pit 2, and here camelids predominated over other animals by 77.1%. In this case, too, adults made up the majority (56.1%), in comparison with young animals (21%) (Wheeler Pires-Ferreira 1975:120 and 122, Tables 1 and 2). If the results of both test pits are combined, camelids accounted for 80.5% of all faunal remains (Wheeler Pires-Ferreira 1975: 126, Table 7).

Another cave in the punas of Junín has the kind of information needed. This is Uchcumachay, which is also called Tilarnioc. It is located in the province of Yauli, in the Department of Junín, at an altitude of 4050 masl (Wheeler Pires-Ferreira et al. 1976:483). It is always said (e.g., Wheeler Pires-Ferreira et al. 1976) that Ramiro Matos discovered this site. This is not true. It was discovered by George O. Kirkner, who made the first excavations there, the materials from which are kept in the Prehistory Laboratory at the Universidad Peruana Cayetano Heredia in Lima, which were examined by John Rick.[3] It was Kirkner who told Matos of the site. The preceramic levels were numbered 4 to 7. Camelids accounted for 84.8% in Level 4 (dated to around 2500–4200 BC),

82.3% in Level 5 (4200–5500 BC), and 54.9% in Level 6 (5500–7000 BC). In Level 7, the earliest (7000–10,000 BC), there were no camelid remains (Wheeler Pires-Ferreira et al. 1976:484, Table 1, and 486, Table 2). The data have often been repeated (E. Pires-Ferreira et al. 1976; Wheeler 1977:6, Fig. 2; E. Pires-Ferreira et al. 1977:153; Kaulicke 1979:107–108; Wing 1986:253, Table 10.5). There is just one problem. When discussing Uchcumachay, Wing (1980:160, Table 8.4) gives different percentages for the camelids than those given above. She gives 85.9% for Level 4, 82.6% for Level 5, and 56.8% for Level 6. There are three reasons why I assume this is a mistake. First, Wing (1980:160, Table 8.4) gives Wheeler Pires-Ferreira et al. (1976) as her source, which I also used. Second, in Wing's (1980) Table 8.4, the disagreement with the data in Tables 1 and 2 of Wheeler Pires-Ferreira et al. (1976) just involves percentages, as the amount of bones used for the statistics coincide. And finally, in a later study, Wing (1986:253, Table 10.5) herself includes exactly the same percentage figures as are in the original study by Wheeler Pires-Ferreira et al. (1976).

I must point out that Lavallée and Julien (1980–1981:122, n. 3) noted that the dates established for the Uchcumachay levels, published by E. Pires-Ferreira et al. (1977; these are the same dates previously published by Wheeler Pires-Ferreira et al. 1976), were determined on a purely comparative basis with dates from other regions in Peru, and so must be taken with due caution.

We now turn to the Tres Ventanas caves, located at the headwaters of the Chilca River Valley, 3926 masl in the province of Huarochirí, Department of Lima (Engel 1970c:426). There are several studies by Engel of these caves (actually three), but he mentions the camelids in only two of them. In one, Engel (1970a:56) cites Caves 1 and 2, but does not present an analysis of the fauna and simply says it "primarily consists of Auquenidae." In another study from this same year, Engel (1970b:428) wrote: "All levels yielded abundant animal bones, particularly from auchenids." The analysis of these materials was never published. It is worth recalling that the excavations Engel and his team made here present serious methodological problems, and so there is no way of knowing the contexts of these bones. Fur-

thermore, it is not known who made the identifications, so we are not even sure that the bones classified as "auquenids" really are Camelidae (for a review of the problems regarding the stratigraphy of these sites, see Bonavia [1984]).

To finish this section I will discuss two sites in the Ayacucho area studied by the project headed by MacNeish (see Appendices A and B at the end of this book). The first is Ruyru Rumi, in the vicinity of Quinua at 4032 masl, in the province of Huamanga, Department of Ayacucho. The so-called Occupations 1 and 2 correspond to preceramic times. Occupation 1 (6800–6200 BC) had "more deer bones than camelid bones." According to the archaeologists who worked here, Occupation 2 (3400–2700 BC) allows one to surmise that its occupants exchanged camelid meat with people from higher-altitude areas, but no evidence is presented to support this assertion (MacNeish and García Cook 1981b:127).

The other site is Chupas (3496 masl in the province of Huamanga, Department of Ayacucho). Here there were "possibly camelids" in Zone E of Occupation 2, dated between 4710 and 4610 BC (Vierra 1981b:143).

Thus far I have examined high-altitude highland sites. I now turn to lower-altitude sites in the valleys. There is one in the Callejón de Huaylas. Camelid remains seem to be abundant in archaeological sites here, as was mentioned earlier (Lynch 1971:145).

I have already mentioned Guitarrero Cave, which is located at 2580 masl in the province of Yungay, Department of Ancash. This clearly is a valley site, but it should be noted that its deposits include materials from higher altitudes, which is why it was mentioned previously.

At this site there are some problems with the interpretation given for the figures for camelids. It was initially reported that these animals represented 10% in 8600–5600 BC, while the figure rose to 17% around the year 5000 BC (figures calculated in terms of the MNI; Wing 1977a:839, Table 1). Wing gives definitive figures in her final report (1980). In a first table (1980:Table 8.1, 154–155) she presents the percentages in terms of the weight of the bones. Here, the camelids in the so-called Complex II (for the chronology, see Lynch 1980a) (ca. 8500–6000 BC) account for

5.5%, in Complex III (ca. 5000 BC) 22.7%, and in Complex IV (which includes late preceramic and later materials) 24.2%. In a second table (Table 8.4, 160), the percentages were also calculated in terms of the MNI. It turns out that camelids account for 4.9% of remains in Complex II, 17.4% in Complex III, and 40% in Complex IV. What I do not understand is why in a later study Wing (1986:256, Table 10.6), when referring to the MNI, gives 3.2% for Complex II and 12.1% for Complex III.

Either way, camelids were absent in the earliest strata of Guitarrero Cave (Complex I), and then constantly increased (10%, 33%, 35%) (Wing 1980:156). For this reason, Wing (1980: 163) believes that the domestication of camelids took place in the occupations dating to Complexes III and IV.

In the Callejón de Huaylas is Huaricoto, another important site located at 2750 masl in the province of Carhuaz, Department of Ancash. In a preliminary report it was said that there was a clear predominance of deer and large wild camelids, probably the guanaco, in the faunal materials belonging to the Chaukayan phase, in the late preceramic (2200–1800 BC) (Michael Sawyer, pers. commun. [1982] to Burger and Burger [1985:123, 125]). In a later study (R. Burger 1985a:507, Fig. 2, and 532, Table 2), it is reported that camelids in the Chaukayan phase were exactly 25.5%.

The famed site of Kotosh, 5 km west of Huánuco on the Higueras River (in the Department of Huánuco, in the province of the same name), also has some problems with the figures presented. Wing (1972:331, Table 3), who studied the materials, notes that in the Mito phase—which according to the Japanese archaeologists (see Izumi and Terada 1972) corresponds to the late preceramic period (although I believe it should be related to the Initial period; see Bonavia 1994:Fig. 3, 24)—15% of the bones are from camelids. This figure was later reused by Wing (1977a:839, Table 1). However, when this same scholar referred to the camelids from Kotosh Mito in another study (Wing 1980:160, Table 8.4), she said they accounted for 17.8% of the bones and gave as the source her own aforementioned study (Wing 1972), in what evidently is a contradiction. It is even

stranger that years later, Wing (1986:256, Table 10.6) mentioned Kotosh Mito as having 12.3% camelids, this time without giving a source. I cannot understand the discrepancy.

It is interesting that llama bones were found in one of the niches of Building J in the Temple of the Crossed Hands (Izumi and Sono 1963: 153). I believe it would be far more correct to speak of camelids, because Wing (1972:239) is quite emphatic in noting that it is extremely difficult to identify the species in these animals. This discovery was commented on by Lathrap (1970: 105), who wrote: "The occurrence of cameloid bones as a sacrificial offering in one of the wall niches of the Temple of the Crossed Hands is of great interest. It is probable that the bones are those of the llama and thus an indication of animal domestication and a flourishing pastoral economy. The use of the llama as a sacrificial animal is a typical cultural pattern in the Central Andes; even today most ceremonies of propitiation, divination, curing, and black magic must be sanctified by the sacrifice of a llama."

In the Department of Junín (in the province of Huancayo, district of Chupaca) there is a site called Callavallauri (at 3340 masl) that was studied by Hurtado de Mendoza and Chaud (1984). They report that this shelter has a preceramic component in which camelids appear in significant proportion. However, the study is so confusing and has such serious methodological problems that there is no way of knowing the real age of the finds.

In the Ayacucho area, the project headed by Richard MacNeish has not yet published the final report on faunal remains, so we must rely on preliminary reports (for a description of the sites, see MacNeish et al. 1981; see also Appendices in this volume). I should point out that all the reports thus far published on the research done by this project were written in a muddled way, and it is difficult to correlate the data provided by the various authors. I turn first to Pikimachay Cave, located at 3000 masl in the province of Huanta, in the Department of Ayacucho. MacNeish, Nelken-Terner et al. (1970) discussed the subject in a very vague manner in the second preliminary report. When discussing Zone j (to which they assigned a date of 17,650 BC), they noted the presence of "llama," and of "camel or horse" in Zone i1, which has a date of 14,100 BC (MacNeish, Nelken-Terner et al. 1970:13). Then, when discussing Zone h, which should have a date of 12,200 BC, they said there were "some llamas" (besides the "paleo-llama" I mentioned in the chapter on paleontology) (MacNeish, Nelken-Terner et al. 1970:15).

Interestingly, in a later study MacNeish (1971) used a different terminology when discussing these same findings at Pikimachay. Thus, when discussing Zone j (which he now dated to 19,600 BC), what he had called "llama or deer" now became "perhaps an ancestral species of horse or camel" (MacNeish 1971:40). He then discussed Zone h, for which he gave a similar date (14,150 BC), and said here there was "[a] kind of ancestral camel." There is no way of knowing whether by this he meant what he had previously called "paleo-llama" or "some llamas" (MacNeish 1971:42). Finally, in this study he mentioned a discovery made in Zone f1, dating to 8860 BC, that was not mentioned in previous reports, and said "modern deer and llamas" were found here (MacNeish 1971:43).

Regarding this same cave, García Cook (1974: 19) wrote that here there were "some camelid remains [that] probably belong to domestic llama or alpaca (*Lama glama, Lama pacos*)."

When discussing the Chihua phase (6550–5100 BC) and Pikimachay Cave, Wing (1975b: 34–35) says there was "earlier evidence of domestic lamoids," and that in Pikimachay, one-third to one-half were young animals about 18 months old.

Another report by Wing (1977a:839, Table 1) gives far more data, but it is not easy correlating those data with the data of MacNeish and his team. Here Wing presents a complete chronological sequence for Pikimachay, with the respective percentages of camelids. The data are worth repeating: 5800–4600 BC, 3%; 4600–3100 BC, 26%; 3100–1700 BC, 32%; 1000–550 BC, 19%; 500–150 BC, 57%; and AD 500–1550, 44%.

However, I do not understand why when she later mentions Pikimachay, Wing (1986:256, Table 10.6) gives 10% for the camelids in the "early [phase] of the cave," and 38.3% for a "late" phase, which add up to a total of 43.2%.

The other cave in Ayacucho is Jaywamachay, in the province of Huamanga, at 3350 masl. Here, acording to MacNeish, Nelken-Terner et al. (1970:27), there were "bones which seem to be mainly of llama and deer" in Zone I (which is dated to ca. 8000–7000 BC), and "many llama" in Zone H (7030 BC). This could be misleading were it not for the report by García Cook (1981: 71), which specifies that camelids accounted for 28% of bones. García Cook does not use the word "llama," which is something quite different. Zone E is then mentioned (ca. 6500 BC), for which it is simply said that there were "llama" (MacNeish, Nelken-Terner et al. 1970:36). The report by García Cook (1981) gives some additional information that the scholars cited above do not present. He first mentions Zone G (to which he ascribes a period ranging between 7100 and 6800 BC), where camelid remains make up 21%, and specifies that these are the remains of immature animals found in an area of activity within the site (García Cook 1981:67, 72, 78). García Cook then mentions Zone F (6675–6425 BC) with 24% camelids (García Cook 1981:72 78), and finally says deer predominate over camelids by 3 to 1 in Zones C (6435–6165 BC), D (6535–6285 BC), and E (6600–6300 BC) (García Cook 1981:67, 73).

Curiously enough, no mention is made in these reports of any camelids found in Zones J1, 2, and 3 (with an age of about 10,000 years), whereas MacNeish (1971:44) claims here were "extinct species of deer and possibly of llamas."

Wing (1986:253, Table 10.5) indicates that camelids represent 24.5% in this cave between 10,000 and 5500 BC. Referring to approximately this same period, Flannery et al. (1989:91) mention the actual presence of guanacos, and explain that "there was no evidence of domestication at that time" (they give MacNeish et al. 1981 and Flannery n.d. as their sources).

Ayamachay (located at 3000 masl, in the southern part of the province of Huanta, in the Department of Ayacucho) is a site with some problems in the interpretation of the data. MacNeish (1981b:121) specifically refers to Occupation 3, Zone VI, to which he ascribes a date ranging between 3600 and 3000 BC. He says that here "[t]he camelid bones suggest an occupation by herders." However, Wing (1986:256, Table 10.6) gives more precise figures for this same site. In the period 5500–2500 BC, camelids amount to 1%, and 6.3% in the period 10,000–5500 BC. What cannot be known is where this comes from, as the sources are not indicated in the table, and I was unable to identify any study that deals with this site in the bibliography.

The last site in Ayacucho is Puente (2582 masl, in the province of Huamanga, Department of Ayacucho), which also presents some problems. We have two reports on it by Wing that present different figures and for which the sources are unknown. First we have the data published in the 1970s (Wing 1977a:839, Table 1), which give a sequence for the site with the proportion of camelids found. For the period 7100–5800 BC the percentage is 2%; it is also 2% in 5800–5100 BC, 3% in 5100–4700 BC, 10% in 4700–4600 BC, and finally 8% in 4600–4300 BC. However, years later Wing (1986:256, Table 10.6) mentioned only three periods: one in 10,000–5500 BC with 2.7% for the camelids, another in 5500–2500 BC with 3.3%, and one in 2500–1750 BC in which the proportion of camelids reportedly increased to 26.7%. This apparent contradiction cannot be explained with the existing data.

I feel that given the confusion surrounding the Ayacucho data, it is useful to include here some of the conclusions that appeared in different reports. In this case I shall not concern myself with any site but with the phases established around them (see MacNeish et al. 1983).

"[C]amelid bones" are mentioned for the Huanta phase in the last report published (MacNeish et al. 1980:7). No date is assigned when the data on this phase are summarized, but an age of 12,000 BC is given in Table 1-1 (MacNeish et al. 1980). This agrees with a previous datum (MacNeish, Nelken-Terner et al. 1970:35), which said the actual expression used for this same phase was "llama or camel bones."

The Puente phase (9000–7100 BC) is considered one of hunters, "mainly of deer and a few camelids." However, the data on hunting are essentially based on the types of lithic points found, which I find rather conjectural (MacNeish et al. 1980:7). Earlier, in a preliminary report it had been pointed out that for this phase, the faunal remains were "mainly of llama and deer" (MacNeish, Nelken-Terner et al. 1970:36).

As regards the Jaywa phase (7100–5800 BC), Lumbreras (1974:37) wrote: "Llama remains, assigned by MacNeish . . . imply that a similar process [of domestication] was under way simultaneously in the highlands. It has not been possible as yet to prove that . . . the . . . animals were domesticated." However, in the last published report the word "domestication" does not appear at all. What is said about this phase corresponds to a seasonal life pattern because the camelids were hunted at higher elevations. There is an interesting datum, even though no more information is furnished, which says that the "presence of camelids" was detected in the feces (it is not specified whether these were human or not, but it can be assumed) (MacNeish et al. 1980:8). It would be interesting to know what kind of camelid remains are detectable in the feces.

The Piki phase is assigned a date that ranges between 6700 and 5000 BC. It is said that the "limited number of deer or camelid bones suggest hunting was not important," but then it is added, "[h]owever, hunting, both of the ambushing type [that is, specialized hunting of camelids] and the individual stalking (of deer), were mainly dry-season activities in camps at high elevations" (MacNeish et al. 1980:9).

The following phase, called Chihua, which was dated between 4500 and 4330 BC, apparently is a crucial phase, because according to the authors of the report, the evidence indicates "that the people were now penning and pasturing domesticated animals rather than taming them as in the previous phase. Llama bones also suggest that some sort of control, such as selective hunting, killing of male juveniles, breeding of wild and/or tamed and/or semi-domesticated camelids occurred even though herding of fully domesticated llama or alpaca was still not done." Further on they add that "the diet seems based mainly on the products of their hunting of deer and camelids, and other herding activities they may have done with camelids" (MacNeish et al. 1980:10).

Finally, I list the Cachi phase, dated between 3100 and 1750 BC, which for the authors corresponds to a seminomadic life, and in whose sites corrals are observed. "Preliminary analyses suggest that camelids, perhaps semi-domesticated guana-co, were herded in the Puna zone during the dry seasons when some hunting and plant collecting were done. With the coming of the wet season the people moved to the Low Puna or humid Woodlands camps to grow potatoes with a cor-responding de-emphasis on hunting and herding activities." The authors posit an exchange of products between lower and upper zones, and conclude that the "[e]vidence for such an ex-change system is not only the large camelid bones found in low elevation sites, where camelids *could not live*" (emphasis added). It is worth noting that when the authors list the plants cultivated and the domestic animals, they say "camelid (possibly do-mesticated)" (MacNeish et al. 1980:11). I empha-sized the author's belief that camelids were un-able to live in zones below the puna because I disagree. I return to the subject later in the chap-ter. Although when summarizing the research, MacNeish (1981c:222) agrees that these were "perhaps domestic" animals, a point repeated by MacNeish and Vierra (1983a:128), these same authors contradict other chapters in the same re-port. Still discussing the Cachi phase, MacNeish and Vierra (1983b:185) suggest "camelid herd-ing," but doubt that the corrals were for camelids, as they use the word "probably" (MacNeish and Vierra 1983b:235). However, on the problem of domestication, MacNeish (1983:272–273) wrote literally about camelids that these were "herded, tamed or [were] domesticated camelids," and then added that "[t]he other major change is in the use of domesticated, or at least tamed camelids; camelids account for about 20 percent of the diet of the highlander and 5 percent of the consumption of people at lower levels, where they may well have been received in vertical ex-change from the Puna dwellers." MacNeish ends by saying that "[t]he other distinctive aspect of the highland was the importation of the concept and practice of herding camelids; perhaps in Cachi times some sort of wild guanaco or pale-ollama [sic] were herded, but later, by ceramic times, the animals herded were alpaca and llama." Again, this clearly contradicts the data I cited ear-lier from these same authors. Besides, I do not understand how MacNeish dares suggest that

humans bred *Palaeolama* when there is thus far no evidence of it in all of the South American con-tinent, and there is not a single paleontological report (as I have already discussed) that shows for sure whether this fossil animal really exists in the remains excavated in Ayacucho. This kind of con-jecture should not be included in a scientific re-port.

Finally, and to avoid mistakes, I would like to point out that MacNeish et al. (1975:15) first noted the presence of six llama bones in an Aya-cucho Phase(15,500–13,000 BP), but the authors themselves said they were "in doubtful con-texts"—so much so that they were not mentioned later.

4.2.2 The Coast

Thus far I have discussed the evidence for cam-elids in highland preceramic sites. I will now try to analyze the subject for the coast in this same period. In 1982 I tried to state the problem (see Bonavia 1982a:392 and passim), and will now re-state it with more data.

On this point there are differing opinions be-tween scholars who believe there were camelids on the coast in early times and those who deny it. Lanning (1967a:63), for instance, was emphatic: "As yet there is no evidence of the llama . . . in coastal sites of this period" (Lanning means the late preceramic), even though in 1960 he himself had said this was somewhat possible. Cohen (1978a:259) says practically the same thing for the preceramic in general, obviously following Lan-ning, who was his teacher. Lumbreras (1974:37) and Tabío (1977:211) are also of the same opin-ion, and claim that these animals appeared on the coast only during the Initial period. Writing specifically on the central coast, in the Ancón-Chillón zone, Cohen (1978b:122; 1978c:27) later said that camelids only appeared in the Early In-termediate period.

Cardich (1980:117–118) also emphatically denied the presence of camelids on the coast. However, his argument is no more than a reiter-ation of what other scholars had said: the "phys-iological adaptation [of the camelids] to life in the high altitudes, especially the alpaca . . . the vicuña . . . the llama . . . and possibly the guanaco to a lesser degree." Cardich then adds that this is "compounded by the fact that the coastal climate is harmful for the llama." His sources are the above-referenced studies by Maccagno (1932) and Jensen (1974), especially the latter, which is questionable and lacking in scientific support.

However, it is interesting that even in the 1980s, Isbell (1986:142) could write: "Economic interdependence and intensive exchange of food between highland farmers and seashore fishers during preceramic times seems very unlikely. The distance is great, especially *without domesticated llamas as beasts of burden*" (emphasis added). In 1992 Wheeler et al. still held that "[t]o the best of our knowledge, llama herding began on the coast approximately 1400 years ago (Pozorski 1979; Shimada and Shimada 1985), but the first clear evidence of alpaca breeding in this zone comes from the site of El Yaral with an age of 900 to 1000 years. Alpaca may also have been present on the north coast of Peru at this same time (Shimada and Shimada 1985)" (Wheeler et al. 1992:470). In a later study Wheeler et al. (1995:833) take a similar position, only here they state that "the earliest use (although surely not the oldest) of camelid fibre in Peruvian textiles comes from textiles preserved in coastal sites approximately dated to 2500–700 B.P. (Novoa and Wheeler 1984)." This occurred despite the fact that in the 1970s Wing wrote that "[b]y about 2000 BC both animals (camelids and guinea pig) were introduced into the central Peruvian coastal food economy. One millennium later the use of these animals spread both north and south along the highland from Central Peru, and the use of camelids became more intensive on the coast of Peru" (Wing 1977b:17). She later insisted once more that there are camelid remains in the period 2500–1750 BC, in the late preceramic, although emphasizing that these "are abundant in coastal sites only after 450 AD" (Wing 1986:255). There was as well a review of the problem published in 1982 (see Bonavia 1982a:392–395).

Burger (1985b:276; 1993a:31) is one scholar who begins to accept the evidence when he writes that "camelid remains are rarely recovered in refuse from late Preceramic or early Initial Period sites on the coast." Quilter (1991:395–396)

agrees, and believes that camelids were brought to the coast in preceramic times. However, it is significant that when discussing and describing Peru's *kjoekkenmöeddings* in the early twentieth century, Uhle (1906:13) claimed there was a "species of *auchenia*" in them. He added that "[d]eer are still found everywhere, in sites not far removed from the coast, but the auchenia species (huanacos) are missing, especially in the North." Uhle's data have been confirmed, and although much evidence is not available, it does exist, and I will review it shortly. However, it must be pointed out that I am convinced that the lack of data is due not to the absence of remains in preceramic sites but to two other reasons. First, most excavations have been very restricted, often no more than test pits, to the point that Peruvian archaeology does not have studies of sites in their entirety, particularly for the preceramic. In other words, the sources available usually are limited and not too significant. Second, faunal and botanical remains were not considered important until quite recently, or were studied only cursorily. Many valuable data were lost. This is easily verified by noting that specialists in osteological identifications often did not participate in these projects. It is clear that there will be several surprises when more systematic work begins.

The first site with data is on the north-central coast. This is Los Gavilanes, in the province of Huarmey, Department of Ancash (for more data on the site, see Bonavia 1982a). A local chronology subdivided into three preceramic epochs—called Los Gavilanes 1, 2, and 3—was established. The site specialized in maize storage, and the refuse from everyday waste materials is scant.

Part of a Camelidae cranium was found in a context dating to Epoch 3, which has a date of 2200 BC. This is an adult animal (Bonavia 1982b:200), and I assume it is a llama. It represents 0.1% of the faunal remains excavated at the site (Wing and Reitz 1982:Table 19, 192–193). (In a later study Wing gave 0.9% [Wing 1986:258, Table 10.7], but this figure corresponds to the preliminary analyses, which were later corrected.) Alpaca (*Lama pacos*) hair and some textiles with alpaca wool were found in this same context

(Bonavia 1982a:102–103, Table 5, and 297, 302, Photograph 78; 1982b:201). There also were textiles with alpaca wool thread in an Epoch 2 context (2800 BC). Other woolen remains were in uncertain contexts and might belong to Epoch 1, or in any case Epoch 2.

It should also be pointed out that abundant llama (*Lama glama*) excrement was found in contexts from all three epochs (no absolute dates were established for Epoch 1, but it is earlier than 3000 BC) (Bonavia 1982c:225–226). The excrement was analyzed and the diet of the animals was reconstructed (Jones and Bonavia 1992). I will refer to this in due course.

All the evidence at this site suggests that the deposits of Los Gavilanes in Epoch 3 were filled with maize, which was brought to the site from the nearby valley on llamas (Bonavia 1982a:271, 272–273, Drawings 64 and 395). It is clear that llamas also went to the site in the two previous epochs, during the Los Gavilanes occupations 2 and 1, but no more details are available about this. We can deduce that these animals were eaten sparingly, given the few bone remains in the refuse.

In the northern part of the Department of Lima (in the province of Chancay) is the site of Río Seco del León (usually known in the literature simply as Río Seco), which belongs to the late preceramic. The data are quite muddled. In the first report (Engel 1956:134) it was pointed out that "there probably are camelid bones" among the faunal remains. Then the report says that "[s]ome bones from Río Seco (Chancay) might be from llama" (Lanning 1960:41). However, Wendt (1963:237; 1976:19) claims there were "cervids and camelid (guanaco?) remains," and reports that this was a lower mandible.

Despite being rich in refuse from preceramic times, the famed Ancón site (in the department and province of Lima) was never the subject of a detailed study, and we really have scant data. Maldonado (1952a:73) reported: "I collected loose auchenid excrement from the great midden of Miramar, in Ancón, which was dispersed all over the bulk of the midden, and comes from all cultural periods in this zone, from the base of the mounds to the top itself." Maldonado explained that the excrement "preserved its natural shape." These remains obviously cannot be dated. It is

possible, from the location of the sector, that they belong to the Middle Horizon or later times, but there might be an older component. I find the datum interesting, notwithstanding its ambiguity.

Moseley (1972:29) reports that "several cameloid bones" were found in the Ancón-Chillón area, during the excavation of one of the sites belonging to what Lanning called the Encanto Phase, which goes from 4200 to 2500 BC. Rick (1983:32) in turn simply comments, without any real evidence, that in the lomas of Ancón there "probably [were] pastoral animals like camelids." We can assume he means one of the studies Lanning made in the 1960s.

One of the well-known sites for the late preceramic period, located close to the mouth of the Chillón River (also in the department and province of Lima), is El Paraíso. There are several reports on this site, but none has any data on camelids. The only data come from Reitz (1988b: 35), who says that "[i]n the ¼-inch samples . . . Camelidae [and other animals] were identified"; she gives Table 10 as a source (Reitz 1988b:53), which is a summary of MNI and biomass at El Paraíso (but here the subject is touched on in global terms as "terrestrial mammals"), which in turn refers to an unpublished report by Wing (1985).

The only other data we have are from Quilter, who worked at the site. In a paper read at the First Convention of Andean Archaeology (3 June 1988), which I attended, Quilter said there were camelid coprolites but that these were not in a secure context. Quilter et al. (1991:280) later noted that in general, the remains of land mammals were scant in the sector of El Paraíso they excavated. They mentioned two camelid bones, also from an uncertain context, that might come from a later occupation superimposed on the preceramic one and dating to the Early Intermediate period.

What Engel (1966b:62, 65; 1967:265, 267) has reported is the discovery of wool at El Paraíso, although in small amounts, which is "possibly from the guanaco *Lama huanacus* [sic]."

South of the Peruvian capital but still in the province of Lima (in the department of the same name) is the site of Paloma, which has an occupation that goes from ca. 5700 to 3000 BC (Reitz 1988a:316). Benfer (1983:5; it was published in 1984) wrote that the most recent level of the site held guanaco remains. In effect, Reitz (1988a:316) specifies that "Camelidae, probably guanaco (*Lama guanicoe*) were . . . identified in the grab samples. Guanacos are known to frequent the coastal plain during the foggy winter season (Grimwood 1969). Grimwood even describes the animals as having once been plentiful on the coast (1969:71) . . . samples suggest that young animals were among the guanaco frequenting the lomas of Paloma, but do not necessarily indicate that the animals were born there." In another report, Reitz (1988b:40, Table 2) herself notes that cf. Camelidae, possibly guanaco, amount to 1.0% at the site, and Camelidae, guanaco, to 5.7%. It is specified that at least one of the animals was less than 18 months, another was subadult, and another adult. The age of many others could not be established (Reitz 1988b:34).

Just south of Paloma is the Chilca Valley (in the Department of Lima, province of Cañete). Here, a maxillary bone "that seems to belong to an auquenidae" was found at a site called Chilca. The presence of vicuña wool is also mentioned (Engel 1964:149–150, Fig. 11). According to Engel (1966a:80) this would be "vicuña" and dates to 3025 BC. Wing (1977b:16) confirms the discovery but does not specify what species this is. Moreover, the date she gives disagrees with Engel's date (1966a:80). Her comment is as follows: "Period 7 (2500–1750 BC) also marks the time that camelid remains are first recovered from a coastal site, the Chilca site." In a later study Wing (1977b:Table 17) notes that these remains amount to 3% of the faunal remains at the site, although she later gave this as 4.4% (Wing 1986:258, Table 10.7).

Regarding Chilca, there is another piece of information that must be taken with some caution. Fung (1969:64), probably speaking of the occupation prior to the late preceramic, claims that "[t]he dwellers of these houses did not have cotton, and used vicuña wool to make thread instead." Fung attributes this to Donnan (1964) and Engel (1966a). There is a problem, however, because this does not appear in either of the two articles cited. What Engel (1966a:31) says is quite different: "[C]lothes made out of vicuña or guanaco skin" exist in an area that extends from Chilca to Nasca, and at a much earlier date. However, in another report, probably the one Fung had in mind, Engel (1964:149) wrote that vicuña wool "is present in Chilca, where it was used to make string." We can only wonder why Engel never again repeated this in any of his other studies.

An important archaeological site called Asia lies beside the Pan-American Highway in the province of Cañete in the Department of Lima. There Unit 1 was excavated. When discussing long mammal bones, Engel (1963:52) claims these are hard to identify and could be either deer, vicuña, or alpaca. But these are the opinions of a non-specialist, just as in most of Engel's reports; a report by a faunal specialist was never published. In addition, the discovery of a bag and other woolen textile fragments was reported, and it was claimed that "[v]icuñas and alpacas were possibly kept for their wool, but we suspect that fox skins, other felines, and wild animals might have been used" (Engel 1963:25). The cavalier way in which the subject is discussed is clear and needs no further comment.

Engel also excavated the Cabezas Largas site in the Paracas area (Department of Ica, province of Pisco). The site dates to 3000 BC. Among other items, several skeletons were exhumed, wrapped in mantles of "vicuña skin" (Engel 1960:15). He specified that "[t]his is the not yet dressed hide of an auchenid skin, as Peruvians say, quite possibly a vicuña." Only fragments survived of these mantles (Engel 1960:17). Engel (1960:23) commented: "The abundance of skins, which can only belong to auchenids, argues for contacts with the highlands. However this does not preclude that a non-domesticated auchenid managed to live in a shaded valley close to the Pisco River." Once again, no specialist analyzed the material, so the identification made by Engel's workmen is suspect. It would be more reasonable to conclude that these are skins of guanacos, which abounded on the lomas of the south coast. Engel, however, was convinced that this was a vicuña, since he reiterated in another study (Engel 1964:149–150) that vicuña wool had been found in Paracas and in Río Grande (Department of Ica, in the province of the same name), but noted that no bones from this animal were found at Paracas. No details are given for these finds, as usual.

Furthermore, Engel (1964:149) himself claims that "vicuña wool is abundantly used in

Paracas and Río Grande." The Paracas remains would belong to the preceramic V (in Lanning's 1967a terminology), and those from Río Grande to the late preceramic (VI).

Villorrio 514, Santo Domingo de Paracas, is mentioned in one of the last reports published by Engel (1981), and was given a date of 4000–3000 BC, although suggesting there is no accurate date. Mammal bones are mentioned here, and these "apparently [were] from Camelidae" (Engel 1981:34). Then it says that "some camelid hairs" were found in Tent VIII (Engel 1981:36). When describing Tomb 2b, which held two bodies, one an adult, the other a child, we read that the head of the child "was wrapped in a vicuña skin." The remains of a similar skin were apparently draped over the body of the adult. Lastly, a vicuña skin was found in Tomb 3, prepared in the shape of a bag (Engel 1981:36). There is no certainty to these identifications.

Finally, Engel (1981:24) again reports guanaco footprints in Pampa Colorada, south of Ocoña (Department of Arequipa, province of Camaná), in a sediment below a layer of sand that seems to date to the early Holocene. Nothing can be inferred from this datum.

4.3 INITIAL PERIOD (1800/1500 BC–900 BC)

As has already been noted, several authors have asserted that domestic camelids were present in the Initial period, emphasizing that the llama was on the coast (e.g., Lumbreras 1974:37), while others argue it was on the coast and in the highlands (e.g., Tabío 1977:212). However, these general statements contain no specific data. Let us turn now to the available data.

4.3.1 The Highlands

I will begin with the high-altitude areas, following the same sequence used with the preceramic period. (For the location of the sites mentioned, see Figure 4.2.) There are three sites in the Callejón de Huaylas. The first is Tecliomachay, on the western slopes of the Cordillera Negra, in the headwaters of the Sechín River (Lake Canchiscocha), and in the Cercocancha Ravine at 4650

masl. According to Malpass (1983:7), here there is just one occupation from the late Initial period. This corresponds to a group whose components "were engaged in domesticated llama herding." Only 27.5% of the bone remains have been identified, and 91% of these are camelids.

The second site in the Callejón de Huaylas is PAn 12-51, but I do not know its exact location. From its designation, it has to be in the province of Recuay (see Bonavia 1966:13). I know it is in the highlands and apparently has an important occupation from this period, where camelids accounted for 84.8% of the faunal remains (Wing 1980:157, Table 8.2).

Finally, based on a study by Sawyer (1985), Burger and Miller (1995:452, Fig. 15) note that 40% of the faunal remains at Huaricoto are camelids.

For the Junín area we have data for some of the same sites extensively discussed in the section on the preceramic period. Therefore, I will not repeat their geographic locations.

The picture at Pachamachay is confusing in regard to the ceramic occupation, because so far neither Matos nor Rick has published the data showing the excavated material. In the first report (Matos and Rick 1979–1980), the materials were presented as a "tentative overall division," and the only thing that follows from it is that the pottery belongs to the "Regional and Formative Development." The "Ceramic Epoch" is assigned 97.6% of the camelids (Matos and Rick 1979–1980:44). There should theoretically be materials from the Initial period, given the dates assigned to the various levels (Matos and Rick 1979–1980:46, Fig. 6).

Rick repeats exactly the same data in another report (1980:243, Table 10.1) and claims they come from the excavations made by Matos. Finally, Rick (1983:146, Table 5) repeats the same percentage in a final study, insisting once again that these are the materials excavated by Matos. In regard to the ceramics, he says only that these belong to the "Formative Epoch," which is obviously quite vague.

For Telarmachay, the data for ceramic periods have not yet been published.

From the beginning, Lavallée (1979:116) noted that "the formative levels contain an even higher amount of osseous material (up to 70/75% of all remains collected), with an even higher proportion of camelids (90% on average)." Lavallée and Julien (1980–1981:106) later repeat that resources were essentially based on the consumption of camelids during the Formative period, at a rate of 82%–90%. They also add that Stratum III (dated to 1400–1500 BC [for the dating, see Lavallée and Julien 1980–1981:102]) has a proportion of newborn and very young camelids of more than 60% with respect to other animals of the same species (specifically 21% adults, 18% juvenile, and 61% very young). Meanwhile, in Stratum II (dated between 700 and 200 BC), the proportion of very young individuals is 35%, while juvenile animals represent 16% and adults 49%.

The data are also vague and scant in the case of Uchcumachay (Tilarnioc). Kaulicke (1979: 108) simply says that more than 90% of the bones in the Formative levels are camelids.

In the highland valley sites we come first to Huacaloma. This is located on the outskirts of the suburbs of the city of Cajamarca, to the south (in the Department of Cajamarca, in the province of the same name), and at an altitude of 2700 masl. Two phases are distinguishable, one called Early Huacaloma, which corresponds to the early Initial period, and another called Late Huacaloma, which corresponds to the late Initial period, prior to the arrival of the Chavín phenomenon in Cajamarca. Terada and Onuki (1982:253) at first claimed there were abundant deer bones in the faunal remains of the Early Huacaloma period, and "an extreme scarcity of llama bones." This was later confirmed by Melody Shimada (1982: 308–309), who specified 60% for deer and 16% for camelids, although admitting that it was a small sample. She made the following comments: "[T]he materials recovered are likely to be secondary context refuse, and the camelid bones found there may have been processed elsewhere at the site." She then adds that "[m]ost of the camelid bones are from immature (unfused) individuals, and, considering the dependence upon deer hunting, it seems reasonable to suggest that these are wild forms." Years later, when writing on the study of new collections from the Early Huacaloma period, Melody Shimada (1985:291) insisted that the faunal sample was extremely small and that there were no camelids in it. However, she did recall that the samples collected in 1979 did have some camelid bones.

Shimada, Elera et al. (1982:142) also agree that the animals found in this phase were wild, and note that "the abundance of projectile points [present in the Early Huacaloma phase] falls significantly in succeeding periods, when the camelid bones predominate among the faunal remains. The strong dependence on deer hunting during these two periods confirms the hypothesis that camelids were present in their wild state."

When discussing the materials from the Late Huacaloma phase excavated in 1979, Melody Shimada (1982:309) noted that camelids were almost completely absent. She added, however, that it was not clear whether the sample was too small or whether this reflected reality. Shimada was right in her surmise, because the picture changed with later studies. The appearance of camelids and the decline in importance of hunting can be seen in the new materials. This transition seems to have been gradual, to judge from the report, as there was a moment when deer and camelids appeared in almost equal proportions (54% and 46%, respectively, of the artiodactyls identified). Besides, fewer than half the camelid bones belong to adults (M. Shimada 1985:292).

According to Shimada (1985:292), the measurements of the camelids fall within the biggest guanaco/llama form. She has estimated that 45% survived the first year and 37% survived past the first 42 months. Of the remains found (which are not many), most are adults or juveniles, and 22% are fetuses/newborn forms. (N.B.: The percentages do not appear in the report; they are mine, based on the data presented therein.) Besides, Shimada notes that fewer than half the camelids killed in this and subsequent periods seem to have been used primarily for food, because they were adults.

This change in the economy of the Late Huacaloma phase was also discussed by Terada and Onuki (1985:272). They wrote that "[a] significant change occurred at the end of the *Late Huacaloma Period or the beginning of the Layzón Period* when camelid domestication was introduced. The people of the Late Huacaloma Period seem to have been reluctant to accept domestication at first, but

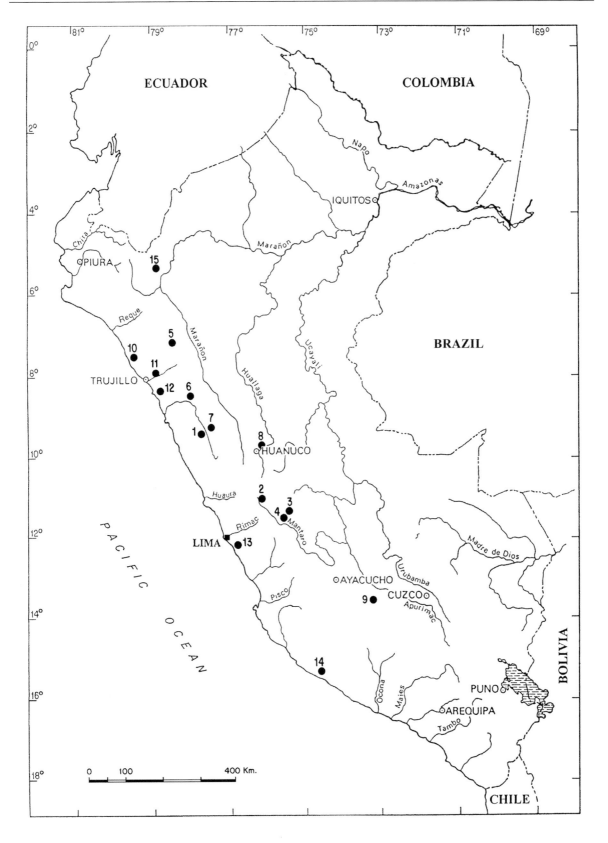

FIGURE 4.2. Locations of archaeological sites dating to the Initial period where camelid remains have been found. The site legend is on the facing page. *Prepared by the author and drawn by Osvaldo Saldaña.*

the growing difficulty of capturing deer may have led them to turn to camelid rearing, which had a long history in a more southern region" (emphasis added). They later insist (Terada and Onuki 1985:273) that "[d]omestication of the camelid was introduced as an important subsistence activity."

There is an important problem with this site that must be stated, which was also pointed out by Burger (1984a). According to Terada and Onuki (1982:261), the Late Huacaloma phase resembles the Urabarriu and Chakinani phases of Chavín de Huántar, so they conclude that Late Huacaloma may be contemporary with the Chakinani phase. On the other hand, Melody Shimada (1982:318) claims that the domestication of these animals had taken place long ago in the Layzón phase (which is later) of Cajamarca and in Chavín de Huántar.

Now, according to Burger (1984a:432), the transition to an ever-increasing use of domestic camelids at Chavín de Huántar occurred in the Urabarriu phase, and the almost exclusive dependence on llamas for meat happened then in Chakinani and Janabarriu. This is why Burger notes that if Terada and Onuki are right, then Shimada is mistaken, because the phenomenon occurs much earlier at Chavín de Huántar and is perhaps contemporary with the Late Huacaloma phase. This has yet to be explained, but I think that Burger is correct.

The site of La Pampa is in the northern part of the Department of Ancash, south of Corongo, in the province of the same name. It is on the Manta River, a tributary of the Santa River, at 1800 masl. There is a report by De Macedo on the osteological component of this site that presents only the identification of the bones, without the slightest comment (Macedo 1979:97–98). From Table 1, which provides the only data, it follows that there are almost no camelids in the Yesopampa and Tornapampa phases (one bone in

each). The Yesopampa phase corresponds to the Initial period and Tornapampa is not too clear, whereas the Caserones phase is within the Early Horizon (see Terada 1979).

At the site of Huaricoto in the Callejón de Huaylas (discussed earlier), there is material belonging to the Initial period. This can be divided into early and late components. The earliest component, called the Toril phase by Burger (1985a: 507, Fig. 2; 532, Fig. 2), must be the final part of this period, because he gives it a date of 1500–1400 BC and believes that it was contemporary with Kotosh Wairajirca. The percentage of camelids is 31.3%. There is another, later component that falls between the Initial period and the Early Horizon, dated between 1400 and 700 BC. It is contemporary with the Kotosh Kotosh phase in Huánuco and with Urabarriu in Chavín de Huántar. Here, camelids represent 26.8% of the total faunal remains.

The Kotosh materials were studied by Wing, who, when referring to the Kotosh Wairajirca phase, initially noted that camelids came to 20% (Wing 1972:331, Table 3). A similar (21%) figure was later cited (Wing 1977a:839, Table 1). However, in a more recent report (Wing 1980:160, Table 8.4) the figure for camelids rose to 28.8%, though in this case both the Kotosh Wairajirca and Kotosh Kotosh phases seem to have been combined. Although there is little material, Wing (1972:336) noted that these remains include the following: "[T]he large camelids are guanaco and the small camelids are vicuña" and "the evidence of increased abundance of camelids at Kotosh . . . indicates [the] spread of greater use of camelids in the highland valleys both north and south of Central Peru" (Wing 1977b:16).

Wing (1972:331, Table 3; 1977a:839, Table 1) notes that during the Kotosh phase camelids comprise 31%.

FIGURE 4.2. Initial Period Site Legend

1	Tecliomachay	6	La Pampa	11	Caballo Muerto
2	Pachamachay	7	Huaricoto	12	Huaca Negra
3	Telarmachay	8	Kotosh	13	Cardal
4	Uchcumachay	9	Waywaka	14	Hacha
5	Huacaloma	10	Puémape	15	Michinal

I was unable to find specific data for the Aya-cucho area. All that is available are the general comments made by MacNeish et al. (1980:12) on the zone, which is between Andamarca (on the heights of Chupas at 3600 masl) and Wichqana (close to the city of Ayacucho), in a period the authors estimate extends from 2213 to 1670 BC. MacNeish et al. (1980:12) mention the presence of storage sites and "corrals, often with camelid bones, [which] may indicate that these centers, at intermediate elevations, were the redistribution centers socially linking not only low and high elevation communities but uniting the lower elevation agriculture economies with those of the seminomadic high elevation herd and root crop subsistence patterns." This will obviously have to be verified with empirical data.

Grossman prepared a report on the site of Waywaka, close to Andahuaylas at 2490 masl in the Department of Apurímac, province of Andahuaylas. He published the results of an early identification made in 1971, and then a later one made in 1983. Grossman (1983:63, Table 3) initially divided the Muyu Moqo style into two phases corresponding to the Initial period, one early and the other late. The percentage of camelids in the Early Muyu Moqo phase amounted to 42.9%, and for the Late Muyu Moqo phase 77.8%. Grossman (1983:73, Table 4) later refined his sequence and split the Muyu Moqo style into A, B, and CD. Muyu Moqo A and B correspond to the early Initial period and have a camelid component of 72.7% and 64.3%, respectively. Muyu Moqo CD was assigned to the late Initial period, with the camelids comprising 68.6%.

Grossman (1983:64-65) notes the problems in defining what a domestic animal is, and says he does "not think that we can reasonably suggest that a group was domesticating and controlling access to large numbers of camelids until or unless we can demonstrate that they utilized large numbers of these animals in big concentrations outside their natural habitat." Basing his conclusions on Wing (1977[c]:124), insofar as the data might indicate low species diversity with the clear predominance of one or two species, and taking into account the low density of all the identifications at Muyu Moqo, as well as the lack of a clear predominance of camelids over cervids and guinea pigs, Gross-

man concluded that it was doubtful that the inhabitants of this site raised the animals. And although it has been pointed out that domestication was probably a long process involving humans, Grossman believes that in the case of the Muyu Moqo phase, this process is not evident at Waywaka until the Qasawirka occupation, which dates to the Early Intermediate period. We should note that this is the only contrary opinion. I discuss it later.

4.3.2 The Coast

I shall now review the data available for the Initial period on the coast.

Víctor Vásquez (in a letter dated 17 July 1992) tells me that only two bones belonging to *Lama* sp. were found at the site of Puémape (on the coast of the Department of La Libertad, in the province of Pacasmayo, district of San Pedro de Lloc), in association with Cupisnique remains.

Although I have no specific data, it seems that the Cupisnique tombs Rafael Larco Hoyle (1941: 164) excavated in the Chicama Valley included llama offerings, because he wrote that he had found "some bones . . . probably belonging to llama." The presence of this animal in the Cupisnique occupation is confirmed by Bird, who was working then in the area and knew of Larco's research. Bird remarked that "sacrificed llamas" were found in one of the structures (Bird 1954:3).

There are somewhat more data for the Caballo Muerto complex, 17 km inland along the middle part of the Moche River, in the Department of La Libertad. The specific reference is to the Huaca Herederos Chica, one of the earliest in the complex, which corresponds to the transition between the Initial period and the Early Horizon.

The presence of camelid remains has been known since the first reports by Shelia Pozorski (1976:101–102, 105, 111), but in low percentages, as only four bones were assigned to *Lama glama*. These bones show cut marks, evidently made while dismembering the skeleton. All subsequent reports, and the comments made by the authors who took up this datum, include the exact percentage of camelids present at this site, which is 14.4% (S. Pozorski 1976:336; 1979a:169, Table 1; 1983:30, Table 5; Pozorski and Pozorski 1979:

427, Table 4; Shimada et al. 1982:141; Novoa and Wheeler 1984:124; Shimada 1985:9–10).

Shelia Pozorski (1976:111) comments that the use of camelids as food was an important component of the subsistence pattern at Caballo Muerto, because animals that large could potentially provide a stable source of protein for the inland people and were as reliable as shellfish. Shelia Pozorski insists that there is no certainty in the identification and that these were "probably" *Lama glama* (1979a:174).

Interestingly, the cervids occur in a slightly higher percentage, 17.6%, and they, along with the camelids, supplied the people of Huaca Herederos Chica with almost all of the meat consumed. Based on the evidence of camelid domestication in the highland, Shelia Pozorski (1983:33) says that "there is little doubt that the camelid identified at Caballo Muerto was introduced into the Moche Valley from the highland in domesticated form. Butchering marks on some of the camelid bones attest to their use as food although they may also have served as beasts of burden, supplied wool, or been used in ceremonial functions. Camelid remains were discovered at several later mounds of the Caballo Muerto complex, a factor which suggests that, unlike deer, camelids persisted as an important meat source." Pozorski and Pozorski (1979:428) present a similar idea but add, "The continuous presence of these animals suggests that, unlike deer, camelids persisted as a meat source throughout the duration of the Caballo Muerto occupation."

Pozorski and Pozorski (1979:430) suspect that deer were rapidly and massively eliminated from the Caballo Muerto zone, and that was why its inhabitants adopted domestic camelids. Thomas Pozorski (1982:232) adds that initially, the llama occupied third position, behind shellfish and deer. The deer later disappeared and the llama replaced this part of the diet, competing with and perhaps surpassing marine products.

Thomas Pozorski (1982:232) assumes that llamas grazed on the banks of rivers and irrigation canals, and cites Cardozo (1954:66-67), pointing out that "llama prefer the wild plants of the altiplano; therefore, in the past they probably preferred wild plants on the coast to cultivated ones. Some foraging may have occurred upvalley. Most

likely there were never so many llamas that special cultivated fields were needed to feed them." We must be careful with citations from Cardozo, as I have already noted. Most of his studies are highly speculative, and in none of them does he show a good use of sources; moreover, he does not know the archaeological literature. Shimada et al. (1982: 141) and Shimada and Shimada (1985:9–10) merely repeated this, as well as Novoa and Wheeler (1984:124). This last study is an overview, and it is striking that for the Initial period only the data from Caballo Muerto are mentioned, with the other information noted here ignored. Also striking is that Novoa and Wheeler state that the "[e]vidence for the first spread of domestic camelids to the coast comes from . . . Caballo Muerto," when there are data for preceramic times.

Slightly more to the south, in the Virú Valley (still in the Department of La Libertad, province of Trujillo, district of Virú), is a site that goes under several names. It is called Huaca Negra, Huaca Prieta of Guañape, and Templo de las Llamas, and even goes by the geographic name of the place, Cerro Prieto of Guañape. Here, Strong and Evans (1952:27–34) found four llama burials. Raymond M. Gilmore, who studied the remains, said these were llamas or alpacas, with the first possibility more likely.

However, there is a problem. Willey (1953), who worked in this zone, mentions Huaca Prieta of Guañape, which corresponds to V71. However, this site does not correspond to the location shown for Huaca Prieta of Guañape on the large map included with the book, for it is inland and more to the north. This clearly is a mistake, because Willey (1953:44) specifies that the site is located "near the beach." And there is another problem. Willey (1953:56) clearly says that "two llama burials" were found here. What is not clear is whether Willey meant the discovery made by Strong and Evans, and made a mistake in saying two instead of four burials, or whether these are two additional burials supplementing those found by Strong and Evans. It seems that the latter is correct because Willey's observations came later than those by Strong and Evans, since he mentions in his book the study made by these two scholars (Willey 1953:57). Either way, Willey's comment (1953:48) on Huaca Prieta of Guañape, which he defines as a "special building," is interesting. He wrote: "The llama burials found within it may be votive offerings. Or could this, and the other community buildings of the period [Willey here means the Guañape Period, the Virú Valley occupation that corresponds to the Initial period], have been llama corrals?"

This discovery has been cited several times (Lumbreras 1974:52; Shimada and Shimada 1985: 10; S. Pozorski 1976:240; Pozorski and Pozorski 1979:430–431). Pozorski and Pozorski (1979:4 31) believe the camelids "may have been furnished from an inland center for ceremonial use in the temple."

For the central coast I have no specific data, just a general statement by Lanning (1967:89), who said that "[l]lama bones are commonly found in Initial Period refuse deposits."

In the Lurín Valley (in the department and province of Lima) there is an important site whose study is still unfinished. This is Cardal, which has been dated between 1170 and 740 BC. Here, camelid bones "appear in small number. The species has not yet been identified, but these camelids seem to be an important source of meat (Miller ms.)" (Burger and Salazar Burger 1992: 124).

To finish the coastal data I must mention the site of Hacha, in the Acarí River Valley (department of Arequipa, province of Caravelí), less than 200 masl. The mean radiocarbon date given is 925 BC. According to the first report, no camelid remains were found here, just a painting that I find interesting. The wall of a building called "Structure 2" is "decorated with a series of marching camelids" (Ridell and Valdez 1987–1988:8, 10, Fig. 10). It actually is unusual to find camelids decorating coastal structures, and to the best of my knowledge this was only repeated at Paramonga at a much later date (the Late Horizon), but this decoration was never adequately described (see Bonavia 1985:169, Fig. 121).

However, in a later study Valdez (1996:99) specified that "[m]ost of the animal bones were recognized as belonging to Andean camelids, but their poor state of preservation did not allow further studies." Valdez (1996:101) believes the bones found at Hacha are from guanacos, to judge by the teeth. (Valdez followed the methods set

down by Wheeler 1984[c in our bibliography].) However, I believe he is wrong, because we know there were llamas on the coast in preceramic times, as has already been seen: "Considering that llamas were already domestic animals and that only guanacos are likely to live on the coast, llamas could be eliminated" (Valdez 1996:101). He adds (1996:101) that another argument for this would be that no evidence of high newborn mortality was found at Hacha (still following Wheeler 1984 [c in my bibliography]). Based on these criteria Valdez assumes that the scene depicted in Hacha Structure 2 depicts guanacos.

4.3.3 The *Ceja de Selva*

There is also information for the *ceja de selva*, specifically for the site of Michinal. This is a lateral valley located on the left bank of the confluence of the Tabaconas and Chinchipe Rivers, 3 km from their confluence. This is in the Department of Cajamarca, in the province of San Ignacio, district of Chirinos. The site is at an altitude of 510 masl (Miasta 1979:45–46). Miasta excavated here and reported that cervids predominated among the faunal remains at the site (63.63%), "followed by camelids (31.81%). . . . The cervids were likewise being consumed from the first period up to strata V, where they were substituted, or better still, supplemented, by camelids (13.17%)." This study is deficient and confusing, and the chronology is not clearly presented. Apparently there are preceramic strata without camelids. These appear in a context that probably dates to the Initial period (see Miasta 1979: 193–197). It is a real shame that better data are not available for a site like this, which is extremely interesting for this subject, owing to the natural environment. At present, these lands belong to the very dry tropical forest (ONERN 1976:63) and are in the semiarid humidity provinces of Holdridge's Bioclimatic Diagram (1967:Fig. 1). From ONERN (1976:64) we know that seasonal pastures formed by native grasses develop during the rainy season and become good standing hay in winter to be consumed by goats and cattle. It may well have fed camelids, and the data available make me believe that the environment during the Initial period must not have been all that different from the present one.

4.4 EARLY HORIZON (900 BC–200 BC)

Some believe the Early Horizon was the time when domestic camelids became crucial for the economy of native groups (e.g., Shimada 1982: 146). This is questionable.

4.4.1 The Highlands

Here I review the evidence for the Early Horizon using the same criteria as were used for earlier periods. I begin with the highland sites. (For the locations of the sites mentioned, see Figure 4.3.) The first evidence, although general, refers to the Huamachuco area, in the mountainous part of the Department of La Libertad. McGreevy and Shaughnessy wrote that "[i]f it is assumed that the *jalca fuerte* (3700–4000 masl) is the preferred setting for camelid herding, the lack of *jalca fuerte* sites is surprising. Either there were very few domesticated camelids until later in time, or the prehistoric herding regime was different from that documented further south in the Central Andes, where remains of settlements of full-time herders are common in the *jalca fuerte*" (McGreevy and Shaughnessy 1983:240). The authors then discuss the presence of domestic camelids at Huacaloma in Cajamarca and assert that "[t]here is no obvious reason why camelids would not be present in Huamachuco by this date as well. A different herding regime is suggested in which settlements are located in the upper *quechua* or lower *jalca*, and the *jalca fuerte* is used for herding on a daily basis on communally held land. Under such a regime, the only herding-related structures that would be expected in the grazing area would be windbreaks for the shelter of herders. Intensive animal care (e.g., culling, breeding, tending, etc.) would be carried out at lower altitudes in the permanent settlements" (McGreevy and Shaughnessy 1983:240).

Site PAn 12-57 is in the Callejón de Huaylas, apparently in the puna area of the province of Recuay. The only thing we know about it is that it has a Chavín occupation. According to Wing (1980:157, Table 8.2; 158, Table 8.3), camelids make up 96.1% of faunal remains in terms of bones recovered, and 97.1% if calculated by weight. In this same study Wing (1980:160, Table 8.4) gives a figure of 96.2%, but it surely is a mis-

print, as the first figure is later repeated in another study by Wing (1986:253, Table 10.5).

In the province of Recuay there is another site that I believe is in the puna, PAn 12-51, which I am certain belongs to this horizon. Camelids make up 88.5% of all faunal remains (Wing 1986: 253, Table 10.5).

Two sites are known in the puna of Junín: Pachamachay and Telarmachay. Both have Early Horizon occupations. At Pachamachay, camelids make up 98.2% of the faunal remains, while in Telarmachay, early and late phases can be distinguished. In the first phase camelids make up 84.3% of all bone remains; they are 86.1% in the late phase (Wing 1986:253, Table 10.5).

In the Department of Ayacucho there is a site called Tukumachay that is in the southern part of the department, on the road to Cuzco, at an altitude of 4350 masl. (MacNeish [1981c:238–239] includes it in Table 8-13, which describes Fig. 8.8, but the site does not appear in this figure.) This site has an occupation with Wichqana ceramics that has been dated between 1200 and 900 BC. Camelids were found here, and these "were almost certainly hunted or herded." From the description we can see that chosen cuts were brought to the site, and there are no head parts. It is presumed that the animals were killed somewhere else (Vierra 1981a:133).

For the site of Chupas, 25 km from the city of Ayacucho at 3600 masl in the province of Huamanga (Department of Ayacucho), Ochatoma (1992:198) wrote: "The presence of camelid bones in association with Chavinoid ceramics would demonstrate . . . the significance that raising camelids had in this period." Ochatoma based this statement on data from Cruzat's dissertation (1967).

Farther south is the famed site of Qaluyu in the Department of Puno (in the province of Azángaro). Here, camelids also played a significant role at this time, because they constitute 72.7% of the bone fragments (Wing 1986:253, Table 10.5).

Pucara is another major site in this same department but in the province of Lampa, at an altitude of 3930 masl. Novoa and Wheeler (1984: 123) wrote: "Faunal remains from the type site of Pucará [sic], excavated by Elías Mujica and analyzed by Jane C. Wheeler, indicate that domestic

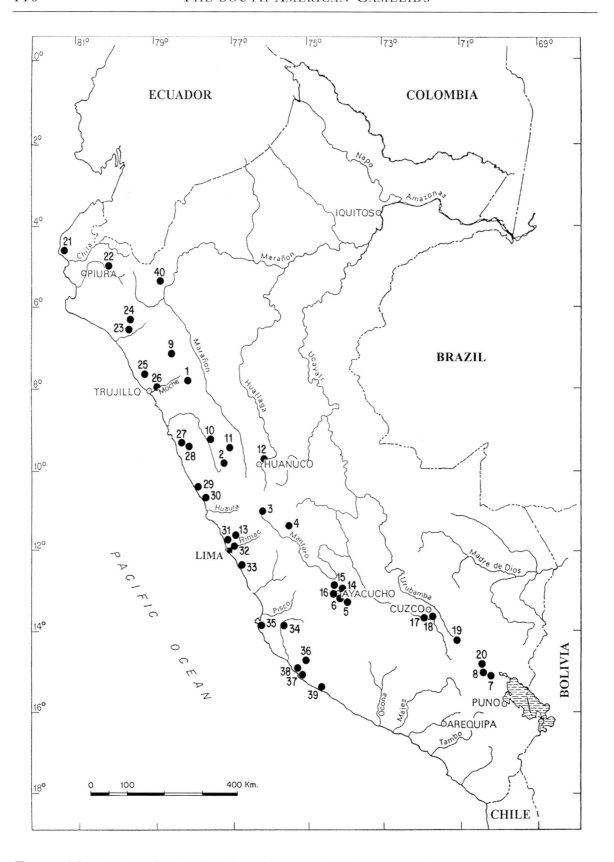

FIGURE 4.3. Locations of archaeological sites dating to the Early Horizon where camelid remains have been found. The site legend is on the facing page. *Prepared by the author and drawn by Osvaldo Saldaña.*

camelids played a predominant role in the economy of this early urban center, but little additional information is available from other sites in the area (Wing, in press) and new evidence of early alpaca domestication in Junín suggests that specialized breeding for wool production may have been the real contribution to the Pukara culture."

I now continue my review of the highland valley sites. At the site of Huacaloma, there is an Early Horizon occupation called Layzón, dated between 500 and 200 BC (Terada and Onuki 1985: 273). According to Terada and Onuki (1982:255), it was at this time that the number of llama bones increased. Melody Shimada (1982:310–311) studied the faunal remains. She asserts that camelids suddenly appeared and there was a complete reliance on them after Late Huacaloma. Shimada even suspects that the majority of the unidentified artiodactyls are in fact camelids. These animals account for about 84% of the bones, while deer account for only 4%. She wonders whether these are guanacos, llamas, or both, but states that there is much evidence that these are domestic animals.

Shimada believes that if her hypothesis is correct (M. Shimada 1982:310–311), then there are two possibilities: they were either brought there or raised locally. Given their abrupt appearance, she assumes they were brought from the central Andes, where—according to the information available—these animals were domesticated (this contradicts Terada and Onuki [1985:272],

who posited a process of local domestication). Shimada assumes that domestic llamas and alpacas appeared first in the Kotosh and Chavín periods. (Shimada presumably meant Kotosh Kotosh and Kotosh Chavín. This has not been confirmed, as we will see later.) Based on data from Wing (1972) and Miller (1981), Shimada posits that the same process of the deer-camelid relationship that took place in Kotosh and Chavín now developed "independently" in the Layzón period of Huacaloma (M. Shimada 1982: 311). In this case the economic emphasis on camelids did not occur prior to 200 BC, although when it did happen, it did so rapidly. This possibility, according to Shimada, has interesting cultural implications because the iconography and ideology associated with Chavín de Huántar during most of the first millennium BC were not connected to any important economic change. The evidence from Huacaloma suggests that this economic change came about later—unless there was a break in the occupation prior to the Layzón period (this has been confirmed, as I will discuss later). Bearing in mind that there is a stylistic similarity between Salinar and Layzón pottery, the contacts between the coast and the highlands were perhaps more important than those between north and south. One reaches the same conclusion, says Shimada, using an alternative possibility, namely, that camelids were domesticated during previous phases. If the samples from Huacaloma are representative, they demonstrate

FIGURE 4.3. Early Horizon Site Legend

1	Huamachuco	14	Pikimachay	28	Moxeque
2	(Puna of the province of Recuay)*	15	Wisqana	29	Bermejo
		16	Jargan Pata	30	Faro de Supe
3	Pachamachay	17	Marcavalle	31	Ancón
4	Telarmachay	18	Minaspata	32	Huachipa
5	Tukumachay	19	Pikicallepata	33	Curayacu
6	Chupas	20	Q'Ellokaka	34	Cerrillos
7	Qaluyu	21	Pariñas	35	Paracas
8	Pucara	22	Ñañañique	36	Cahuachi
9	Huacaloma	23	Huacas Lucía y Cholope	37	San Nicolás
10	Huaricoto	24	Huaca La Merced	38	Huaca del Loro
11	Chavín de Huántar	25	Huaca Cortada	39	Chaviña
12	Kotosh	26	Huaca de los Reyes	40	Cerezal
13	Huancayo Alto	27	Pampa Rosario		* Exact location not known

that these animals were not economically significant until the Layzón period. The introduction of domestic llamas and pastoralism at a regional level seems more likely. Shimada used the study by Espinoza (1974:48), which notes that deer hunts were a "magic rite" in the Cajamarca-Huamachuco area in historical times, to suggest that this might have far more ancient roots.

Melody Shimada (1985:291) says that there would be one possible alpaca if Wheeler's dental traits method (1982b) was used for Cajamarca. All the rest would correspond to the llama/guanaco forms. There would seem to be eight specimens in the EL phase and two in the Layzón period, which, as we saw, must correspond to the transition between the Early Horizon and Early Intermediate period. In regard to this scenario, Shimada (1982:149) comments that there was "a major economic shift from deer hunting in the Late Huacaloma Phase to the abrupt appearance and overwhelming dependence on domesticated camelids (most likely llamas) from the onset of the Layzón Phase." In the second campaign, a team of Japanese archaeologists excavated at the Layzón site, also in the Cajamarca basin. A context belonging to the Layzón period was found here. Although in this case it is a ceremonial context, the faunal materials do not differ much from those of Huacaloma for the same period. Camelids thus outweigh cervids, 73% to 27%. In comparison with Huacaloma, the survival curve shows a gradual diminution in deaths, with greater survival after the first 12 months but about the same percentage after 42 months.

Out of a minimum number of eight individuals, two are adults, four are juveniles, and two are fetuses/newborns (i.e., 25%; my percentage) (Shimada 1985:293).

Several sites were discovered in the Cajamarca area by the Japanese archaeological expedition (see Terada and Onuki 1985:271). These sites have not been described or analyzed. Limited data are available on two of them in Shimada's (1988) report. She mentions the sites of Huacariz and Kolguitin. In the first case we can assume that it is Huacariz Grande, mentioned by Terada and Onuki (1985:271). This site is located 6.5 km southwest of the city of Cajamarca (Shimada 1985:132). It has a Late Layzón phase occupation,

and it has been established that camelids compose 93% of its remains and that these are big llama/guanaco forms (Shimada 1985:134).

In the Early Layzón phase at the site of Kolguitin, 7.5 km northeast of Cajamarca city (Shimada 1985:132), camelids represent 67% of faunal remains, and at least half were adult animals. We can assume there were large and small camelids at the site, to judge from their dental characteristics (Shimada 1985:133).

There are some data for the Early Horizon in the Callejón de Huaylas, but they are vague. For some of these sites there are few references available, and we can only guess their location thanks to their codes (see Bonavia 1966). This is the case of Chopi Jirca (PAn3-3), in the province of Carhuaz, which was studied by Gary Vescelius. Vescelius never published a report, either for this site or for all the others he studied or explored (see Bonavia 1966:59). According to Wing (1986:256, Table 10.6), the site belongs to this period and camelids represented 88.9% of the fauna. PIRC and Pi are two other sites studied by Vescelius; all that is known is that they are in the Callejón de Huaylas. They have an Early Horizon occupation, according to Wing (1986:256, Table 10.6), and the percentage of camelid bones is, respectively, 90.2% and 54.6% of the total remains studied.

Finally, Burger (1985a:507, Fig. 2; 532, Table 2) has reported on the Early Horizon levels of Huaricoto. Burger discusses what he calls Early Capilla (700–400 BC), which is contemporary with the Kotosh/Chavín phase at Kotosh and Urabarriu and Chakinani at Chavín de Huántar. The percentage of camelids from this phase was 68.6%. Then came the Late Capilla phase (400–200 BC), which is contemporary with Kotosh/Chavín at Kotosh and with Janabarriu at Chavín de Huántar, with 55.5% being camelid remains.

Still in this area we find Chavín de Huántar (Department of Ancash, province of Huari, district of Chavín de Huántar). Lumbreras provided both a preliminary and a final report on his work at this site, from which we can glean information on the fauna recovered (Lumbreras 1989 and 1993, respectively). In the preliminary report Lumbreras said that during Chavín times there was "—apparently abundant—camelid raising"

(Lumbreras 1989:38). When discussing the Huarás epoch, Lumbreras states there were "many camelid bones" (Lumbreras1989:118). However, most of the references are for the Gallery of the Offerings, where "thousands of broken camelid bones" were found, according to Lumbreras (1989:130). Then he explains (1989:158) that the broken plates and bowls found there held "camelid ribs." Lumbreras insists on this identification on page 186 (1989) and then specifies, for this same place, that "there are camelids (especially alpacas?)" (1989:205). However, the only actual enumerated data are the camelid bones found in the Gallery of the Offerings, which make up 15.86% of the total (Lumbreras 1989:207); but this figure is inconsistent with the oft-repeated abundance or the "thousands" of bones mentioned. It is true that 55.14% of the bones are reported as unidentified and that these correspond to a "large mammal," "mainly Artiodactyls" (Lumbreras 1989:207), which clearly might include camelids.

In his final report, Lumbreras (1993:291–298) devotes an entire section to camelids, but little is actually added to the data in the preliminary report. An attempt is made to summarize the findings on these animals, both in archaeological and in distribution terms, but the results are quite poor. First of all, I must indicate that the quantity of camelid remains is slightly higher than previously believed. Although no figure is given as a percentage, we can deduce that it was 20.27% (Lumbreras1993:Fig. 34, 292). It is then explained that all body parts from these animals are represented in these remains in at least 30% of the reported specimens, but Lumbreras then adds that "this representativeness is relative because there are few cranial remains (mostly mandibles), while leg and thorax bones are generously represented" (Lumbreras 1993:296). It is worth noting that only one example allows us to assume that a complete animal was placed there, for in all other cases the bones were disjointed (Lumbreras 1993:296). Lumbreras hypothesizes that the camelids were part of "food" offerings, some "roasted and probably parboiled or cooked in water; they were placed as given 'portions' according to the abundance and/or succulent condition of the meat" (Lumbreras 1993:296–297). To support this hypothesis Lumbreras (1993:297) says that "there are camelid bones in all cells, adult, juvenile and infant, save for Cells 5 (which does not have bones of infant animals), 7, and 9 (which does not have bones of juvenile animals). A trait equally common to all cells is the presence of front and hind leg.... All also have 'portions' of arms (scapula or humerus), thigh (femur), 'loin' (the region of the coccyx), ribs and 'chops' (vertebrae-ribs), and some cranial remains." There were also camelid bones in the main corridor, where there was "a constant deposition of adult, juvenile and infant animals" (Lumbreras 1993:298). The lack of precision in a report of this kind is striking. Lumbreras (1993:298) writes: "There are 640 camelid bones, i.e., over 15% of the total recovered, not counting those in the category "large mammals" which number 2018, and with which they come to more than 50%." It so happens that in Figure 34 (Lumbreras 1993:292), the number of camelid remains is 641 and the number of large mammal remains is 2,028. Bearing in mind that the total number of bones found—that is, camelids plus other animals—is 3,161 (including those found in the cells and in the corridor), camelids compose 20.27%, a lot more than 15%, and large mammal remains compose 64.15%. If we add the figures for camelids and large mammals, they come to 84.43%, which is also considerably more than the 50% noted by Lumbreras. Interestingly, several bones had marks left by rodents and cuts made with sharp or serrated instruments, particularly on the long bones. Only a few had ochre stains, which might be due to chance or might be the result of some rite (Lumbreras 1993:298). There thus can be no doubt that the context of these remains, which were jumbled with the remains of carnivores and rodents, was a ceremonial one.

The zoological analysis presented in this report (Cardoza 1993) is very poor. The only thing done here is to identify the material at a taxonomic and anatomic level, trying, in the latter case, to identify what part of the skeleton the bone remains corresponded to and the age of the animal. In the case of the camelids, the taxonomic identification reached only as far as order and family (Cardoza 1993:Table 1, 371). It is striking that the total number of bones mentioned in this study (3,133; Cardoza 1993:371) is not the same as that given in Figure 34 (3,161; Lumbreras 1993:292).

The only part of the conclusions that can be salvaged says that "the animals were laid like portions for a meal. For the most part, only the thighs and ribs of the large animals, that is, camelids and cervids, were placed like portions of food. In the case of younger and/or infant animals there are parts of the body like the cranium, the sternum, etc., while the smaller animals like Canidae, birds and rodents, among them the viscacha [sic] and the guinea pig, were offered whole." Besides, "[m]ost of the material was burned" (Cardoza 1993:393). In his preliminary report, Lumbreras suggested that the camelids were mainly alpacas (Lumbreras 1989:205). The final report says that "we must assume that the camelids represented in the gallery are essentially llama and alpaca" (Lumbreras 1993:295). The only support for this statement is that "the domestication of camelids had been accomplished in the Central Andes" by the time the Gallery of the Offerings was used. Subsistence based on the meat of these animals therefore spread to "lands where these did not yet exist, were scarce in their wild form, or simply did not form part of the consumption habits of the northern peoples" (Lumbreras 1993:295). Lumbreras then insists that these were domestic animals, as we would otherwise have to consider "large herds of guanacos and/or vicuña that could have made such a careful selection by age groups possible" (Lumbreras 1993:298). These arguments are groundless. Domestication had, of course, been accomplished by the Early Horizon. This is not new. However, to say that the custom of using camelids spread to areas where these "did not yet exist, [or] were scarce," meaning to the north, is to ignore the facts. The dispersal of these animals is shown in my Figures 2.2, 2.5, 2.9, and 2.11 (in Chapter 2), which are based on the data available, and it is clear that camelids lived in the Callejón de Huaylas and its vicinity. Nor do we need to think in terms of large herds of wild animals if we consider that the Gallery of the Offerings was used for a long time, as follows from Lumbreras's study. In other words, the animals were not all killed at the same time. Besides, Lumbreras seems to forget that a great number of animals can be trapped in a *chaco*, clearly a very an-

cient Indian custom. Nor is there a "selection by age groups," as he claims. To disprove this, all we need do is study the data presented by Cardoza (1993:372–377). These clearly show there is no constant in the figures that could indicate the intention of choosing animals by their age. Finally, it is absurd to argue, as Lumbreras does, that since domestication had already taken place, the remains excavated at Chavín de Huántar must correspond to domestic forms. Did domestication entail the disappearance of all wild camelids?

The point here is the following: How could Lumbreras have identified the camelid bones and even raised the possibility that these were alpacas and llamas, when Carmen Rosa Cardoza, who studied the animal remains (see Lumbreras 1993: 207; Cardoza 1993:Table 1, 370), was unable to do so?

Besides, identifying the materials found in the Gallery of the Offerings is not easy because remains belonging to various Chavín phases, mainly Urabarriu and Chakinani, were found there (for a more extensive discussion of this subject, see Burger 1984b:173–183, 186–187).

For this same site of Chavín de Huántar we have more accurate data in Miller (1984), who studied the faunal remains excavated by Burger (1984a) in the area occupied by the present town of Chavín de Huántar, on the left bank of the Huachecsa River (see Miller and Burger 1995: 427). It was on the basis of these studies that Burger (1984b) prepared his three-phase sequence for Chavín de Huántar: Urabarriu, Chakinani, and Janabarriu. Miller wrote that "[t]he Chavín fauna, during all three phases [he means the aforementioned phases], is dominated by large herbivores of the families Cervidae and Camelidae" (Miller 1984:283). He then specifies that hunting was important during the Urabarriu phase (900–600 BC according to Burger [1985a:507, Fig. 2]), shown by "the presence of guanacos or vicuñas among the . . . camelid bones. . . . In contrast, during the Chakinani (600–400 BC according to Burger [1985a: 507, Fig. 2]) and Janabarriu (400–200 BC according to Burger [1985a:507, Fig. 2]) phases, the importance of hunting appears to have been supplanted by the herding of llamas and/or alpacas or trading with camelid herding residents of the alturas zone" (Miller 1984:283–284; for more details on the chronological sequence for Chavín de Huántar, see Burger 1984b, 1998).

The percentage of camelids in the Urabarriu phase is 67.33%; it is 95.60% in the Chakinani phase and 95.10% in the Janabarriu phase (Miller 1984:285–287). In 1981 Miller (1981) had already clearly noted that the economy was based on the hunting of cervids, vicuñas, and llamas in the Urabarriu phase, but with camelids as the major source of protein. Then he added that "the Urabarriu faunal assemblage probably represents a transitional phase in the hunting strategy of the inhabitants of the Chavín area at a time when domesticated camelids were beginning to be introduced from highland areas to the south, but before they had achieved the dietary monopoly observed in the later Chakinani and Janabarriu periods" (Miller 1984:284).

Miller (1981) believes that although the llama might not have been present in the earliest phases, a dramatic change did occur in the Janabarriu phase, when the people began to raise llamas, thus suggesting that hunting was abandoned for the raising of domestic llamas. Miller also believes that the economic increase emphasizing domestic camelids did not take place in the northern highlands before 200 BC, but that when it did happen, it did so rapidly. This same scholar (1979; 1981:12) notes that camelids were raised for two years in the Janabarriu phase, after which 40%–50% of them were butchered. However, a large percentage of the population survived for 42 months (I was unable to read the studies by Miller and have taken the data from Melody Shimada [1982:311–312]).

I believe Burger (1989:563; 1993b:71) made a most important comment on this subject:

> Both the hypothetical high-status and low-status areas of Chavin de Huantar depended on llama meat as the principal source of animal protein during the Janabarriu Phase, but a comparison of the ages of the animals being consumed reveals that the upper-status area was consuming younger, more tender animals than the nearby lower-status workers. Moreover, an analysis of selective representation of camelid bones from both sectors of the site indicates that these animals

were not being slaughtered at Chavin de Huantar, but instead were being butchered elsewhere, presumably in contemporary villages like Pojoc and Waman Wain near the high pasture land.

In a later study Miller and Burger (1995) confirm these assertions and explain them in more detail. It is only in this last report that percentages are presented for camelid occurrence in faunal remains that are slightly different from those previously indicated. So, for the Urabarriu phase we now have 68% and not 67.33%; 94.6% for the Chakinani phase and not 95.60%; and 93.7% for Janabarriu instead of 94.19% (Miller and Burger 1995:429, Table 1).

Miller and Burger suggest the meat came from the highlands as charqui, or freeze-dried meat. On the basis of ethnographic studies conducted by Miller in southern Peru (1979), with excavations of modern garbage and surface collections, Miller and Burger believe they have found a pattern that shows the predominance of head and foot bones, while leg bones appear in smaller numbers.

Excavations done in a peasant community show that head and foot bones are five times more frequent than leg bones. After noting several characteristics of charqui-producing practices in the southern puna, Miller and Burger claim that the preparation of this product typically includes the bones with the meat and uses parts of the animals' body except for the head and feet, which are immediately consumed. Potentially, the production and transportation of charqui would yield a sample with an overrepresentation of podial and cranial elements in high-altitude sites where charqui was prepared, and a concomitant overrepresentation of leg and axial elements in the highland or coastal valleys receiving the end product.

These are the criteria Miller and Burger used to interpret the osteological materials from Chavín de Huántar. In fact, the study of the bones, analyzing the differential representation of the various body parts, and bearing in mind the proportion of the five main groups of carcass recovered among the materials in the three phases of Chavín de Huántar (head, axial elements, forelimbs, hind limbs, and feet), yielded the following results. Head and foot bones abound in the Ura-barriu phase, whereas in the Chakinani and Jan-abarriu phases there is a strong reduction in the carcass group and a corresponding increase in leg elements (Miller and Burger 1995:438–440). Miller and Burger believe the Chavín de Huántar river valley is far too narrow to provide enough pasture for flocks of camelids. This would have given rise to an exchange of products between the valley and the puna (Miller and Burger 1995:442).

In this last study Miller and Burger tried to establish to which camelid species the bones found at Chavín de Huántar belong. Despite the problems in making this kind of inference, they concluded that llamas predominated over vicuña remains in the Urabarriu and Chakinani phases, while llama predominated in the Janabarriu phase. This would mean that whereas hunting was still practiced in the first two phases, it had almost disappeared in the last one (Miller and Burger 1995:427–438).

(When discussing the presence of alpacas, Miller and Burger claim that it is significant that "alpaca fiber does not appear on the north coast until the time of the Gallinazo culture" [Miller and Burger 1995:435]. They then add that "[c]amelid wool does not appear in textiles until the Early Horizon" [Miller and Burger 1995: 450]. This is not correct, as we have seen, for al-paca hair, and some textiles with alpaca wool, were found at Los Gavilanes, a late preceramic site on the border between the central coast and the north coast [Bonavia 1982c:102–103, Table 5, 297, 302, Photo 78; 1982b:201].)

Miller and Burger (1995:451) suggest that the llamas of Janabarriu times might represent a di-rect import of a southern camelid stock, rather than animals whose ancestors adapted to the northern environment more than 2,000 years ago. Here they follow the ideas presented by Miller and Gill (1990). (This was later expanded in Miller and Burger 1998.)

However, Lidio Valdez does not fully agree with this position. He specifically disagrees with Miller and Burger (1995) as regards the use of charqui. Valdez argues that "[t]he consumption of charki cannot be archaeologically established. They believe [he means Miller and Burger 1995] that the absence of camelid heads and legs at sites like Chavín de Huántar indicates charki consump-tion, but this is simply wrong. In fact, the same patterns result when fresh meat is transported, something I myself saw in Ayacucho. Besides, Miller and Burger ignore other models that ex-plain this pattern better, and seem to be fully un-aware that the animals' head and legs are used in a different way, thus resulting in a different kind of destruction of these bones, as well as a spatial distribution for them that is different as well. In addition, there are examples of camelid legs used as offerings. These are all effects that Miller and Burger do not consider. Their interpretation is far too simplistic" (Lidio Valdez, pers. commun., let-ter, 30 March 1999).

During the Kotosh Chavín phase at the famed Kotosh site, the percentage of camelids reached 50% of the total faunal remains (Wing 1972:331, Table 3). However, in a later study Wing (1986: 256, Table 10.6) mentioned Early Kotosh, to which she assigned 20%, and Late Kotosh, with 48.6%. In this case it is not clear whether Wing has split the Kotosh Chavín phase or whether "Early" means Kotosh Kotosh and "Late" stands for Kotosh Chavín. In her comment, Wing (1972: 336) mentions both phases and says Kotosh Ko-tosh and Kotosh Chavín are similar and differ from previous phases. Both saw a remarkable in-crease in the number of camelids, and among them the smallest showed a relative increase. They were probably guanacos and vicuñas. There also are re-mains that probably are domestic forms of the llama and alpaca. Wing (1972:336) repeats that "[i]t is in these periods that we have the first clear evidence of use of domestic llama and alpaca al-though hunting wild camelids was still practiced."

A site called Huancayo Alto is in the upper Chillón River Valley (in the Department of Lima, district of Canta), at about 2000 masl. Dillehay (1979:27) studied the area and explains that in the middle Chillón River Valley there is no sign of population movement between the lowlands and the highlands. However, there are "enough camelid bones" and highland ceramics for us to surmise that some relations existed with the Canta or Junín groups. Dillehay is thinking of seasonal movements, because he notes that "[t]here are no corrals to indicate that the inhab-itants of Huancayo Alto housed camelids."

Farther south in Ayacucho, there are some rather vague data on the presence of camelids in the Early Horizon for sites that are already known. Thus in Pikimachay at this time, camelids make up 41.1% of the fauna according to Wing (1986:256, Table 10.6), while in Jaywamachay half of the animal remains correspond to camelids in what García Cook (1981:74) calls Zone B, which is presumably associated with the Rancha and Paracas styles. However, there is no report from specialists.

MacNeish and Vierra (1983a:186) note, while discussing a series of sites belonging to populations living in the riparian woodland zone, in a very general discussion of the "Rancha pattern," dated between 500 and 200 BC, that their economic activities were agriculture and managing "camelid herds and llama caravans," but without explaining the basis for this assertion.

In this same area there is a famous site known as Wisqana, located on the confluence of the Totora and Pongora Rivers (in the province of Huamanga, Department of Ayacucho). Wing (1986: 257, Table 10.6) mentions an Early and a Late Wisqana. For the first period she says that camelids made up 100% of the sample, whereas for the second it was 79.3%.

In 1966 Augusto Cruzat worked in the area of Chupas (also in the province of Huamanga) and prepared a dissertation (1967) that I was unable to obtain. Ochatoma (1992:197–198) refers to it and notes that the sites excavated were Solar Moqo, Osno Era, and Kichka Pata. The dominant ceramic there was a "Chavinoid type, Chanapata and Paracas." Ochatoma comments that "[a]ccording to the author [he means Cruzat], the presence of camelid bones in association with Chavinoid pottery would show the importance camelid raising had at this time."

Ochatoma (1992) himself reports the work he did at Jargan Pata (in the province of Huamanga, district of Ayacucho), which belongs to the Early Horizon. He reports having found there "numerous human and animal bones, including those of camelids" (Ochatoma 1992:196).

In the Department of Cuzco (in the province and district of the same name) there is a famous site, Marcavalle. It is southwest of the city of Cuzco and located at 3314 masl. Camelids are among the most important remains in the faunal assemblage of this site, but the percentages differ in various reports. Wing (1986:257, Table 10.6) notes that the percentage of camelids is 94.2%, while Miller (1979:135; this is his dissertation, which I was unable to obtain; the information comes from another of his studies [Miller 1984: 284]) says it is 79.3%. I do not know whether both scholars worked with the same samples, but probably not. However, Flores Ochoa (1982:70) refers to Miller's dissertation when noting that camelids in Marcavalle constitute 82.5% of the animal remains. Flores Ochoa (1982:70) adds that more than one species of camelid was consumed, but llamas seem to have been the most numerous. Of the bones, 51% were not fused, which shows that 30% of the Marcavalle camelids were killed before they were a year old. Flores Ochoa concludes, still on the basis of Miller's study, that the people in this locality relied more on big camelids, possibly llamas, and perhaps some guanacos that were probably hunted. When referring to the work done at Marcavalle, Wing (1972:338) agrees that its people relied almost exclusively on camelids, and also notes that these were certainly domestic. Even so, the differences in the percentages between these scholars have yet to be explained.

The Minaspata site lies in this same area, 35 km from Cuzco and close to Lake Lucre (in the Department of Cuzco, province of Quispicanchis, district of Lucre), at 3100 masl. Based on the data of Miller (1979), Flores Ochoa (1982:70–71) indicated that camelids here represented 71.6% of the fauna excavated. No period was specified, even though the site was occupied from the Early to the Late Horizon. I assume that the data are for the Early Horizon.

Wing (1986:257, Table 10.6) also refers to this site, but she notes the percentage of camelids in the period that actually corresponds to the Early Horizon is 67.4%. The data do not agree with those presented by Flores Ochoa (1982: 70–71). Again, I do not know whether these are the same samples or not.

South of Cuzco lies Pikicallepata (in the province of Quispicanchis, district of Sicuani) at 3410 masl. According to Wing (1977b:16), here was an "increased abundance of camelids." This site, which has an important Early Horizon occupation, has been subdivided into an early and a late phase. Camelids account for 57.8% of faunal remains in the first phase and 76.7% in the second (Wing 1986:257, Table 10.6).

Farther to the south, we have no data for the famous Pucara site (in the department of Puno, province of Lampa, district of Pucará), just an assertion by Lumbreras (1974:37), who stated that "[t]he earliest alpaca remains known are associated with the Pukara culture . . . and are dated around 200 BC."

There is a site known as Q'ellokaka, 47 km northwest of Pucará at 3930 masl (in the same province of Lampa). It also belongs to this period, according to Wing (1986:257, Table 10.6), and camelids make up 75.9% of the faunal remains.

4.4.2 The Coast

Specialists seem to agree that camelids begin to abound in archaeological sites on the coast during the Early Horizon. Wing (1977b:17) thus wrote that "[t]he coastal procurement pattern is initially based on a dependence on marine resources and marine-dependent resources, such as sea lions and guano birds, augmented by hunting, primarily deer. At about the time of Christ, following the development of methods of farming land along the coast, herd animals [camelids] were also maintained in this region." Wing (1986:255) later repeated: "Camelid remains are abundant in coastal sites only after 450 BC." This was also accepted by Kent (1987:173), but he relied on the aforementioned data by Wing. (To prevent mistakes, it is worth noting that in his study, Kent [1987] mentions Wing 1982. This work was then in press and not published until 1986.)

Lange Topic (1987:832) also wrote that "[t]he data Shimada and Shimada have compiled leave little doubt that llamas were present at least on the north coast in substantial numbers from the Early Horizon on, serving as beasts of burden, ritual offerings, and source of meat." Burger (1989:561; 1993b:69) confirms this: "Analysis of Early Horizon refuse on the South, Central and North Coast point to an increase in the consumption of llamas. In fact, this highland meat source may displace marine resources as the principal source of animal protein at inland sites in some coastal valleys. The latter shift suggests a greater dependency on adjacent highland areas, and a formidable increase in exotics is characteristic of many coastal and highland Early Horizon occupations." I will now present the data I managed to compile.

On the northern part of the north coast in Pariñas (in the Department of Piura, province of Talara) is a site whose designated name is PV7-18, and for which data are scant. The site belongs to the transitional period between the Early Horizon and the Early Intermediate period (450 BC–AD 300/650), and the data on camelids at the site are apparently contradictory. Wing (1977b: Table 10) first wrote that camelid remains at this site were as high as 15%, but in a later report (Wing 1986:259, Table 10.7) she gave 46.2%. I do not know how to explain the difference.

Also in the Department of Piura (in the province of Morropón, district of Chulucanas) there is a most important, well-studied site which lies on the outskirts of the city of Chulucanas: Ñañañique. A "small amount of camelid bones" was found in the Ñañañique phase, which corresponds to the Early Horizon (eleventh to seventh centuries BC; Guffroy 1992:102). According to Guffroy (pers. commun., letter of 15 February 1990), camelids are present in small numbers in the soil and sediments of the Ñañañique phase. No definitive data are yet available, but it is known that the quantities are small. It is interesting that while there is a high percentage of young animals among the bones of deer and large mammals, those of camelids belong mostly to adults. Camelids are present with no major increase during the subsequent phase, known as Panecillo. There are no data for the Encantada phase, but they dominate the very small sample of the Chapica phase (this would coincide with what was observed at Pirincay [Ecuador] and in more southern sectors, such as Cajamarca). The condition and small number of camelid bones do not allow them be classified at species level. Two hypotheses can explain their presence in the earliest levels: either these are domestic camelids from the southern Andes that were perhaps used for long-distance transportation (which would explain the eventual predominance of adult individuals) or, as is less likely, these are wild camelids hunted in high-altitude areas. According to Guffroy (letter of 29 April 1991), the interesting point here is that the final percentages show the same proportions as those of the Early Huacaloma phase in Cajamarca, and greater than those of Late Huacaloma (where they almost vanish). Bearing in mind the geographic location and the climate of Upper Piura (in comparison with the more southerly Andes), this is slightly surprising. Guffroy wonders whether one of the hypotheses might not be the survival of wild herds (previously present in the department, according to James B. Richardson III) up to the first millennium BC in the Upper Piura basin.

In the Department of Lambayeque (province of Ferreñafe, district of Pátapo) lies the Huaca Lucía-Cholope within the Batán Grande complex, which has been dated between 1400/1300 BC and 700/600 BC. Izumi Shimada, Carlos Elera, and Melody Shimada (1982:140–141) report that the midden holds a considerable amount of seashells despite its inland location (about 50 km). Land resources comprise camelids, dogs, guinea pigs, lizards, birds, and rodents. Although the sample is not very big, and bearing in mind that the guinea pig sample is probably not representative, the camelids emerged as the most important source of land protein. Several observations led the authors to believe that the camelids were domestic. First, they are present at all levels except 16 and 17, where few organic remains were found. Second, both adult and young members are represented. Third, most of the body parts were found, including feet and cranial bones, which one would not expect to find if the camelids had been brought in as processed charqui (dried meat), as is known from ethnographic data for the Cuzco area (Miller 1979:97–100) and close to Cotahuasi, in the province of La Unión, in the northern reaches of the Department of Arequipa (Inamura, pers. commun. to the authors, 1980). Fourth and last, camelid dung hills have been found, suggesting that these animals were raised locally (Izumi Shimada 1982:145).

Regarding the third point, Inamura lived and traveled with llama herders, and reports that the large adult male llamas rapidly lose weight during their journeys and that animals meant to be turned into meat are killed at their place of origin and converted into charqui even when the transaction was to take place just a short distance away (the same data are summarized in Izumi Shimada [1982:145]).

Moreover, it is worth noting that no bones of fetal/newborn animals were found among the camelid remains of Huaca Lucía-Cholope, but there are young animals (Shimada and Shimada 1985:8). This made Shimada and Shimada wonder whether conditions in coastal corrals were different from those in the highlands, which, according to Wheeler (1982a), led to the infestation of enterotoxemia. Bearing in mind the dry environment and the evidence showing that animals were raised all year long, Shimada and Shimada concluded that the low index of fetal/newborn animals in the remains of the Lambayeque and La Leche valleys may have been an indication that this disease did not appear on the north coast. An alternative possibility would be that the camelids were not often born in this area but in the grasslands, or in corrals in other parts of the valley (Shimada and Shimada 1985:21). On the other hand, camelids compose 90% of all identifiable bones at a nearby site in the middle La Leche River basin called Huaca La Merced, which also dates to the Early Horizon (Shimada and Shimada 1985:21).

In the Department of La Libertad there are several important monuments dating to the Early Horizon, but there are no data on faunal remains for them. We know that camelids have been found at Huaca Cortada, Huaca de los Reyes, Huaca de la Cruz, and Huaca Guavalito (Pozorski and Pozorski 1979:428).

In the Casma Valley (Department of Ancash, province of Casma) is the site of Pampa Rosario, 16 km inland and at 150 masl. It belongs to the Early Horizon, and "[i]ts middens . . . [contain] remains of camelids, probably llamas (*Lama glama*)—both skeletal evidence and dung" (Pozorski and Pozorski 1987:70). Llama bones were found in this same valley (in the district of Casma) at Pampa de las Llamas around Waka A, on the right bank of the Casma River, south of Cerro San Francisco and in the "Archaeological Complex of Moxeque," as Tello (1956:52) called it. This huaca, according to Tello (1956:52), is associated with Chavín ceramics, but no more data are available.

In this same department but in the province of Huarmey (in the district of the same name) is

Bermejo, for which I was unable to find data. The occupation apparently corresponds to the transitional phase between the Early Horizon and the Early Intermediate period. The only thing known is that camelids account for only 1% of the faunal remains (Wing 1977b:Table 18).

Farther to the south, in the department of Lima (in the province of Chancay, district of Supe Puerto), is the Faro de Supe site, excavated by Willey and Corbett (1954:140–141), who found the hoof and part of a lower leg bone belonging to *Lama glama glama* in a tomb (No. 8). (The details of Skeleton No. 8 are actually on page 19 of this report, but only a "llama" is mentioned there.) Willey and Corbett believed that all the tombs excavated at this site were contemporary and corresponded to what was then called the "Early Ancón-Supe style," equivalent to the Early Horizon.

Ancón, one of the most important sites in all of Peru, is in the Department of Lima (in the province of Lima, district of Ancón). It has a significant Early Horizon occupation, but I have no data on its faunal remains. Rosas Lanoire (1970:266) noted that "[t]he presence of llamas is supported by a layer of guano from this mammal located on stratum III of Test Pit 5 in the Tank Sector, but with no evidence of bone remains." This would mean they were beasts of burden.

The Huachipa site in the Rímac River Valley (in the department and province of Lima, in the district of Ate) apparently dates to the Early Horizon. According to Altamirano (1983b:34), here "South American camelids" represent 83% of the identified fauna. The report says that these are "possibly llama (*Lama glama*)" and that 3% of the remains correspond to adults and 97% to young animals. No infant animal was found. "The absence of corrals as well as of infant animals perhaps indicates that their presence on the coast was due to an economic exchange with societies settled at high-altitude ecological tiers." Altamirano concludes that the main source of meat in the diet of the Huachipa people came from camelids, and insists these were "possibly brought from other ecological tiers" (1983b:34). There is not much support for this assertion.

South of Lima is the site of Curayacu (in the province of the same name, in the district of Pu-

cusana), for which no faunal data are available. Wing (1977b:Table 19) initially said (with some reservations) that camelid bones could be estimated at 11%, but later (Wing 1986:259, Table 10.7) gave only 0.4%.

In the Ica Valley (in the department and province of the same name, district of San José de los Molinos), in the eastern upper part, is the important site of Cerrillos that Wallace (1962:312) excavated. Regarding camelid remains, all we know is that "[o]f the bone material, llama was by far the most common at all levels."

The Paracas area (Department of Ica) is much talked about in Peruvian archaeology, but the lack of data on animal remains is once again regrettable. Novoa and Wheeler (1984:123) say of the Pucara-style textiles found at Paracas that this is "the earliest use of alpaca wool in the Andes," along with Alto Ramírez, in Chile. This is incorrect, as I pointed out earlier, because alpaca wool (see earlier discussion) was used for textiles in pre-ceramic times (see Bonavia 1982b:201).

Geismar and Marshall (1973:3, 5) report camelid remains dating to "Late Paracas" at the sites of Cahuachi, San Nicolás, Chaviña, and Huaca del Loro. However, there are no specific data in this report.

There is a vague claim by Novoa and Wheeler (1984:124) that "[o]n the coast of southern Peru . . . llama feet and woolen textiles are frequent in burial mounds starting about 500 B.C., but it is not known if herds were present in the area or not."

4.4.3 The *Ceja de Selva*

To conclude the discussion of the Early Horizon, I must mention a site located on the northern *ceja de selva*, where the Cunía or Cerezal River joins the Tabaconas River. This is in the Department of Cajamarca, but I do not really know whether it is in the province of San Ignacio or Jaén. This is the Cerezal site, located at an altitude of 510 masl, which in Tosi's terminology is the tropical very dry forest floor (see ONERN 1976).

According to Miasta (1979:82), cervids represent 14.7% of faunal remains at this site and camelids 5.8%. "[T]he presence of camelids is mostly noted in layer II (4.4%) and 1.4% in layer IV, and [is] absent in layers III and V." Miasta (1979:84–85) wrote:

Of the camelidae species [sic], we were
able to identify *Lama glama guanicoe* [sic]
[*guanaco*]. . . . There is no evidence that
the other bone remains of the Camelidae
species [sic] belong to the genus *Lama
glama* [sic] [*llama*]. In this regard, ethnography (Espinoza Soriano) (54) [the
reference is to Espinoza Soriano 1973]
tells us of the presence of domesticated
llamas in the area in the late period.
However, the archaeological context records the presence of llamas since the
Initial period, (55) [the reference is to
Lanning 1967(a)] both on the coast and
in the highlands. In any case, if all specimens belong to the *Lama glama guanicoe* genus [sic], it is interesting to detect
its presence in high temperature climates like that of the selva alta or rupa
rupa. It is more likely that these were
captured in their natural habitat and
were brought back alive to the lowland.
The wall paintings of Faical, San Ignacio, picture camelids, but we were unable
to identify the genus. Their presence in
their natural habitat is anyway clear in
early periods (the altitude of Faical is
over 2150 masl). . . . On the other hand,
a change in climate might take place in
these latitudes, as long as the domestic
llamas were not present.

This study is muddled and has serious mistakes, as the reader has probably noticed. The
chronology is not clearly shown anywhere in the
report. Camelids apparently appear in the strata
belonging to the Early Horizon, but this is not
certain. This is a real shame, for it is an ecological area where we would not expect to find the remains of these animals, and it is also the only
datum available for this period.

4.5 EARLY INTERMEDIATE PERIOD
(200 BC–AD 500)

4.5.1 The Highlands
Following the same criteria, I will begin my review with the highland areas. (For the location

of the sites mentioned, see Figure 4.4) The first
site for which I have some vague information is
Huachanmachay, in the Callejón de Huaylas. The
site is at 4500 masl on the Cordillera Negra, in
the headwaters of the Casma River. Malpass
(1983:5) found a sequence at this site that goes
from the preceramic period to the Early Intermediate period, and notes that the last occupation
corresponds to llama herders, but this association
is not clear. Only 18.4% of the bone remains have
been identified, and 82% of them correspond to
camelids (Malpass 1983:4). In the introduction
(Malpass 1983:2), these animals are said to be domestic llamas. No more details are given on their
identification.

Browman (1974a:188, 195) believes that until
AD 500, the human groups living farther south,
in the area between Jauja and Huancayo, depended on domestic camelids for 50% or more of their
subsistence, but no details are given.

The data I have for Ayacucho in this period
are vague, and without any supporting data. Mac-
Neish (1981c:224) mentions the Huarpa phase
(300/200 BC–AD 200/300) when synthesizing his
research in the area and states that there is "evidence of caravans of camelids." When discussing
their Ocros occupation, Pongora phase (which
they date to AD 310–525), MacNeish et al. (1980:
13) state that "[s]ome of the llama bones are more
robust, suggesting that camelids, besides providing meat and wool, were now a major factor in
long-range transportation." When discussing the
same occupation at the Chupas site, which I have
already mentioned, Vierra (1981b: 144) says there
"possibly [were] camelids." Finally, MacNeish and
García Cook (1981b:127) describe the Ruyru
Rumi site close to Quinua (in the province of
Huamanga) at 4032 masl. They note that Occupation 4 corresponds to Huarpa (350 BC–AD
250) and contains "camelid bones." Until the faunal data for all of these sites are published, there
will be no way to get a clear picture from this information.

There are slightly more data about the highland valleys. For the Department of Cajamarca,
Melody Shimada (1982:311–312; 1988:134) subdivided the Early Intermediate period remains in
the Huacaloma site into the phases Cajamarca A,
Cajamarca B, and Early Cajamarca. She notes

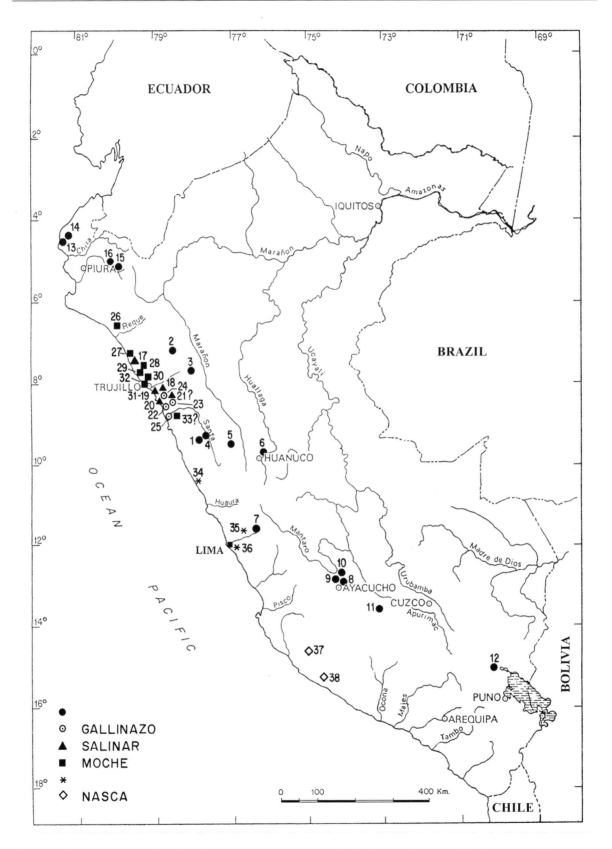

FIGURE 4.4. Locations of archaeological sites dating to the Early Intermediate period where camelid remains have been found. The site legend is on the facing page. *Prepared by the author and drawn by Osvaldo Saldaña.*

that in the Cajamarca A phase, camelids, which were quite likely domestic, were still the most important economic resource, as they comprised 85% of the diet, against just 4% for deer. It is assumed that these were llamas and alpacas. Of these animals, only 56% survived beyond the first year, and 37% lived beyond 30 months. This is an increase in comparison with previous periods and was perhaps due to bringing in more animals and an abrupt killing that took place in animals between 3 and 3.5 years of age. Only 14% survived more than 42 months (i.e., 3.5 years), suggesting their primary use as meat.

Conditions were similar during the Cajamarca B phase because the survival gradient was similar. Only 53% of the animals survived the first year. The presumed killing took place between 2 and 2.5 years, with 33% living beyond 3.5 years. Although camelids continue to be significant during the Early Cajamarca phase, the dependence on them had reached its highest point, as they accounted for 99% of all faunal remains. The other species present are represented by just a few bones. However, the survival gradient differs from that of previous phases, as there is no indication that new animals were introduced at the site, and 76% of them survived the first year. Deaths occurred between 2 and 2.5 years of age.

It is not clear whether the animals died of natural causes or were slaughtered, although the second possibility seems more likely, but those that survived the first year were apparently kept alive until the second or third year and then killed. However, a high percentage of the population was alive after 42 months, or 3.5 years.

Few animal remains were found in the 1985 campaign at Huacaloma in the Early Huacaloma period, and there were no camelids there (M. Shimada 1985:219). Meanwhile, in the Late Huacaloma period there were just a few camelid bones, with cervids being absolutely predominant (M. Shimada 1985:292). A transitional period between Late Huacaloma and Layzón was surmised during the 1979 campaign, which would seem to have a date of ca. 530 BC (Terada and Onuki 1985: 273). This was confirmed with research conducted in 1982.

Shimada's 1982 report (p. 310; see also Shimada 1988:133) noted that the importance of camelids increased abruptly in the Layzón period. With this new, intermediate phase known as EL, the transition appears more gradual. In this context, then, we have deer and camelids present in almost equal proportion, respectively 54% and 46% of the artiodactyls identified. Fewer than half of the camelid bones belonged to adult animals. Of these, all were from large guanaco/llama forms.

Of these animals, 45% survived their first year, and 37% survived more than 42 months. Of a minimum number of nine individuals, four are adults,

---

FIGURE 4.4. Early Intermediate Period Site Legend

1	Huachanmachay		*Salinar*	28	Pampa Río Seco
2	Huacaloma	17	Puémape	29	El Brujo
3	Huamachuco	18	Cerro Arena	30	La Poza
4	Cueva del Guitarrero	19	Moche	31	Huaca de la Luna
5	Chavín de Huántar	20	V-434	32	Huaca Pelada
6	Kotosh	21	V-604	33	(Santa)*
7	Calancancha				
8	Pikimachay		*Gallinazo*	34	Bermejo
9	Ayamachay	22	Huaca Gallinazo	35	Huancayo Alto
10	Wisqana	23	Castillo de Tomaval	36	Lomas de Atocongo
11	Waywaka	24	Huaca de la Cruz		
12	Qaluyu	25	Suchimancillo		*Nasca*
13	(North of Talara)*			37	Cahuachi
14	(North of Talara)*		*Moche*	38	Tambo Viejo
15	Tamarindo	26	Huaca del Pueblo		
16	Loma Valverde	27	Pacatnamú		* Exact location not known

three are juvenile, and two are fetal/newborn animals (i.e., only 22%; this percentage does not appear in the report and is my calculation). On the basis of many analyses, it can be concluded that at least half of the camelids killed in this period and the next were adults, which suggests they were mainly used for meat (Shimada 1982:292).

These studies also show a significant increase in camelids when the remains of the EL phase are compared with those of the Layzón period. In the latter, camelids make up 91% of the artiodactyl bones identified. In this case the measurements also indicate big guanaco/llama forms. On examining the survival gradient we find that 40% got beyond the first 12 months, and 15% reached 42 months. And out of a minimum number of 18 individuals, 6 were adults, 8 were juveniles, and only 4 were fetuses/newborns (22%; this percentage does not appear in the report and is my calculation) (Shimada 1982:293).

There are two other sites in the Cajamarca Valley which have already been mentioned: Huacariz Grande and Kolguitin. Both have an occupation that dates to the Cajamarca phase. In the case of Huacariz Grande, the stage of the Cajamarca phase to which the occupation corresponds is not specified; it is too long and goes from the Early Intermediate period to the beginning of the Late Horizon (see Shimada 1988:Table 1, 139). Nor is it indicated how many camelid remains were found. All that is reported is that these correspond to big and small forms, presumably llamas and alpacas (Shimada 1988:134). The occupation corresponds to the Early Cajamarca phase in the case of Kolguitin, which is the limit between the Early Intermediate period and the Middle Horizon. Although no figures are given, in this case it is stated that the proportion of camelid and deer remains are almost equal, clearly an anomalous case. However, this might be due to the scant number of remains found (Shimada 1988:134).

McGreevy and Shaughnessy (1983:240–241) worked in the nearby zone of Huamachuco, and specifically refer to the Condebamba River basin, part of which is located in the Department of La Libertad (in the province of Huamachuco), part in the Department of Cajamarca (in the province of Cajabamba). Its altitude ranges from 3200 to 4200 masl. These scholars did not find camelids in the archaeological sites, but their comments are important:

> No large corrals are in evidence, but faunal remains from other Huamachuco area sites (Cerro Sazón, Marcahuamachuco) indicate that camelids were in this area from at least the Early Intermediate period on. The relatively low frequencies of camelid bones suggest that the animals were not kept or managed in large enough numbers to require large corrals; instead, animals may have been kept in household compounds. Modern evidence shows that people both farm and herd in this lower area[4] . . . at short distances from their habitations without the use of large corrals.

The authors then comment that Cieza de León (I will discuss this in the following chapter) notes that later there was a large number of camelids in this same area. In fact, they found some corrals dating to this period at altitudes ranging between 3200 and 4200 masl. However, McGreevy and Shaughnessy (1983:240–241) consider that "the evidence still indicates the *jalca fuerte* [that is, between 3700 and 4200 masl] is not a major habitation location for pastoralists. Instead, an extended territory is suggested. This again would point to lower habitations, at the interface of *quichua/jalca* zones [about 3200–3700 masl]."

Lange Topic et al. (1987:833) support the statements made by McGreevy and Shaughnessy (1983:240–241) and comment that camelid bones are not frequent in the excavation in the Huamachuco area. They accept that some llama herding was carried out among the stubble fields or in pasture lands above 3800 masl, but insist that the importance camelids had here was far from what it was in societies farther south. However, Shimada and Shimada (1987:837–838) do not concur. They indicate that the presence of big and small camelids has been recorded in the Cajamarca basin since 200 BC (here they cite the studies by M. Shimada 1982, 1985a [1985 in my bibliography], and 1985b, which is as yet unpublished). There is also ceramic evidence of contacts between coast and highlands, including Huamachuco and the Callejón de Huaylas, from the beginning of the Christian era up to AD 800/900. According to Shimada and Shimada, it is worth bearing in mind that "[i]ntensive herding does not necessarily imply large herds." They believe that McGreevy and Shaughnessy (1983:240–241) are correct in suggesting that instead of using large corrals, the animals might have been kept in the enclosures of family houses. This, Shimada and Shimada (1985) believe, corresponds to what they posited for the north coast.

There is practically no information for the Callejón de Huaylas for this period. All we know is that camelids represent 7.95% of the faunal remains in the Guitarrero Cave, which was mentioned earlier, in the occupation corresponding to Complex IV, dated between 450 BC and AD 650 (Wing 1986:257, Table 10.6).

Burger (1985a:507, Fig. 2; 532, Table 2) initially noted that the amount of camelid remains among the faunal remains at the site of Huaricoto, which I have already mentioned, came to 65% during the Huaraz phase (200 BC–AD 100), which is contemporary with the Sajarapatac phase of Kotosh, and the Huarás phase at Chavín de Huántar. However, Miller and Burger (1995:452, Fig. 15) later stated that the proportion was 95%, based on a study by Sawyer (1985).

According to Lumbreras (1989:118), "many camelid bones" were found at the famous Chavín de Huántar site in the Huarás period, which corresponds to the Early Intermediate period, but this is not documented.

In Huánuco we know for the Kotosh site that camelids accounted for 41% of the remains in the Sajarapatac phase, which corresponds to this period (Wing 1972:331, Table 3; even though in a later study this same scholar [Wing 1980:160, Table 8.4] corrects the figures and gives 49.5% instead). Wing (1972:366) notes that the major difference when compared with previous phases is the relative decrease in the proportion of small camelids. Guanacos and llamas predominate, and some are probably alpacas.

In the last phase at Kotosh, which is called Higueras and corresponds to the Early Intermediate period, there was a relatively large increase in camelids among faunal remains, because they reach 64% of the total. Moreover, for every two small ones there was one large one. The first are clearly

alpacas, the others llamas and guanacos (Wing 1972:336–337, 331, Table 3). (In another study Wing [1986:257, Table 10.6] indicates that at Kotosh, the percentage of camelids was as high as 69.6% between 450 BC and AD 650, but she does not specify what Kotosh phase she is referring to. It should correspond to the Higueras phase, but then this would correct the data presented above.)

Farther south is a site called Calacancha in the Department of Lima (in the province of Huarochirí, district of Huanza). Its occupation corresponds to the Early Intermediate period and the beginning of the Middle Horizon. According to Kaulicke (1974–1975:32), "the frequency of camelids is remarkable."

For the Department of Ayacucho there is limited information from sites already mentioned. Wing (1986:257, Table 10.6) notes that camelid remains in the strata dating to the Early Intermediate period at Pikimachay (Ac 100) come to 66%. In Ayamachay the figure is 25% (Wing 1986), and at Wisqana it is 96.6% (As 18; Wing 1986); see also Appendices A and B in this volume.

The Waywaka site, already mentioned, is in the Department of Apurímac. Grossman (1983: 72–73, Table 4) explains that settlement size increased during the Qasawirka occupation, which falls within the Early Intermediate period, accompanied by a massive increase in either the number or the density of camelid remains in the middens. Camelid remains constitute 91.5% of all animal remains and are llama and alpaca, according to George Miller, who analyzed them. For Grossman, the increase in animal bones recovered both in the primary levels of the Qasawirka phase as well as in the fill of the deposits is consistently several times higher than in the Initial period: "The relative bone count alone clearly shows that between the end of the Initial period and the Early Intermediate period, a significant change in the accessibility of camelids had taken place for the peoples living at the site of Waywaka. Because the Muyu Moqo and Qasawirka peoples lived at the same ridge-top location, these contrasts suggest that a major shift in economic orientation had occurred between the two occupations."

Grossman notes (1983:75–76) that this new trend toward camelids in Waywaka is also suggested by two small llama statuettes found in the

Qasawirka strata. This leads Grossman to believe that the significance of these animals in the Early Intermediate period was not just economic but also religious. Bearing in mind the ecological and nutritional needs of llamas and alpacas, Grossman believes that the animals killed and used at Way-waka were not raised there. He believes the site, which is between 2900 and 3100 masl, is below the altitudinal range needed to raise herds, and therefore assumes that it was controlled from the highlands.

Farther south, in the Department of Cuzco, there was also an occupation at the Marcavalle site that dates to this period. According to Wing (1986:257, Table 10.6), camelid bones here represent 64.9% of all animal remains. Meanwhile, at Minaspata they represent 65.2% for this same period (Wing 1986:257, Table 10.6) and 42.1% at Pikicallepata (Wing 1986:257, Table 10.6).

In the Department of Puno, data are available only for the Qaluyu site (which is in the province of Azángaro, district of José Domingo Choque-huanca, 4 km north of Pucara and at an altitude of 3930 masl). The percentage of camelids in the strata corresponding to the Early Intermediate period is 8.6% (Wing 1986:254, Table 10.5).

4.5.2 The Coast

Far more information is available on coastal fauna for the Early Intermediate period. In this case I will try to separate the remains corresponding to each coastal culture as far as possible, following the same chronological and geographic sequence.

Little information is available for the Department of Piura. Fonseca and Richardson (1978: 306) indicate that there is evidence that herds of camelids were kept in two great sites of the Sechu-ra period (500 BC–AD 500) north of Talara. Richardson himself personally confirmed this to Shimada and Shimada (1985:10), although the latter add that it is not clear whether these animals were used for domestic or ceremonial purposes.

On the other hand, Kaulicke (1991:414) reports the work carried out in "Upper Piura," that is, on the coastal part of the province of Morro-pón, in the sector known as Tamarindo. There is an important occupation here dating to the Early Intermediate period. The bone materials come "from all excavated sectors. . . . Camelids predominate in the bone material (between 60% and 80%), and cervids to a lesser degree. In the case of the camelids these are llamas (*Lama glama* [sic]) according to Altamirano, who is in charge of analyzing the material, and an animal whose size lies in between the llama and the alpaca (chasa-llama)."

Kaulicke claims that the presence of camelids and the fact that they were probably raised in the region leave the possibility open that these animals were used to transport goods from the coast to more distant areas. These animals were also used in sacrifices, and their wool was used to prepare textiles, ropes, and so on (Kaulicke 1991: 415). He adds that "Upper Piura has evidence of relatively big settlements in the Early Intermediate period. Its people depended on an advanced agriculture supplemented by camelid breeding and marine products" (Kaulicke 1991:418–419). There is, however, one problem. From the study it cannot be determined in what period the animal remains appear, Vicús or Moche.

Kaulicke (1993) later reported on his work in the area. He says there is evidence of "camelid raising" at the Loma Valverde site, close to the cemetery of Yécala (1.5 km southeast of the hamlet of Vicús, in the Department of Piura, province of Morropón). At the bottom the site has an occupation that must be contemporary with Salinar, and on top another corresponding to Early Gallinazo (Kaulicke 1993:101). The data are unfortunately too vague and do not offer specific information.

4.5.2.1 The Salinar Culture

Víctor Vásquez S. (pers. commun., letter of 17 July 1992) reports that a significant amount of llama dung was found in the Salinar strata at the site of Puémape (Department of La Libertad, province of Pacasmayo, district of San Pedro de Lloc), thus clearly showing that these animals got at least as far as this site on the coastline.

Shelia Pozorski (1979a:175) mentions the sites of Cerro Arena and Moche, both in the Moche Valley. The first belongs to the Salinar culture. Pozorski wrote: "An examination of animal protein sources . . . reveals an emphasis on the domesticated llama to an extent that most other potential meat sources were of minor importance (Table 1).

This suggests that llama herds were being controlled and maintained by both polities [i.e., Cerro Arena and Moche] for the purpose of supplying meat and possibly labor and wool." In the case of Cerro Arena, camelids make up 92% of animal remains (Shelia Pozorski 1979a:169, Table 1).

The presence of camelids in Cerro Arena is confirmed by Brennan (1978; I was unable to read this study, which is cited by Shimada and Shimada [1985:10]). Shimada and Shimada (1985) indicate that it is not clear whether these animals had domestic or ceremonial uses.

In regard to Cerro Arena we know there was a large accumulation of aeolian sand at the southernmost part of the site. Thirteen human burials were found here in 1995 during work on the Second Stage of the Chavimochic Project. Ten of the burials included camelid remains. There were, in addition, 18 burials of camelids alone, among 28 animal burials in all.

All of the burials were in the sand but under a layer of irregular stones that had been laid without any visible order. The stones were at a depth of 60 cm. The human remains lay extended on their backs, while the camelids lay on their right or left side, but always facing west. It is interesting that seven human corpses showed traces of red pigment (believed to be hematite), while two camelids also had red pigment on several parts of the body. Some of the camelid limbs had been tied with ropes. Establishing the cultural connection of these remains is very difficult because of the lack of cultural elements. However, the form of the human burials, the presence of red pigment, and the closeness to the eponymous site (see above) allow one to assume that they belong to the Salinar culture.

Of the camelid remains, 12 individuals were studied by Víctor Vásquez Sánchez and Teresa Rosales Tham. They concluded that 10 animals, or 83.33%, were buried between birth and three months of age. The other two, or 16.66%, were between three and six months old. These animals were all identified as llamas (*Lama glama*) (José Carcelén Silva, pers. commun., letter of 14 May 1999; see also Anonymous 1995). I have been told that another cemetery solely of llamas was found in the Chao River Valley, but I do not know its cultural affiliation.

West (1971a:52) makes reference to the diet of the people of the Puerto Moorin phase in the Virú Valley (Department of La Libertad, province of Trujillo and district of Virú), which belongs to the Salinar culture. West notes that "[l]and mammals included llama and perhaps deer; both are relatively uncommon." Shimada and Shimada (1985:10) also confirm the presence of camelids in the Virú Valley at this time (but they use data from West [1977]). However, in this case they once again add that it is not clear whether these animals had domestic or ceremonial roles.

In the lower part of the Virú Valley, northeast of the river, south of the Pan-American Highway, and close to the littoral, is a site called V-434 which also belongs to the Puerto Moorin period. West (1971b:29) initially reported that llama remains were found during excavaton of the middens, and noted that these comprised 15% of the faunal remains according to Wing (1977:Tables 4 and 5).

In a preliminary report, Reitz (1976:4, 6) wrote that the Camelidae were a recent introduction to the area, and explained that all occupation levels at site V-434 belong to the Puerto Moorin period.

Reitz (1976:6, 8, 9, 11) mentions the Camelidae in general terms, but does specify that she was able to distinguish young and adult forms. Reitz explains (1976:12–13, 15–16) that some of the camelid bones were burned and had chew marks, even dog bites. All the evidence indicates the human population of V-434 used marine and beach resources almost exclusively, besides camelids. Reitz is emphatic in noting that "[t]he domestic flocks were the most heavily exploited resource in terms of dietary contribution. They probably grazed in the lomas." She also says that domestic camelids were not eaten continuously but "from time to time."

However, Reitz (1976) is far more specific in Figure II, noting that the Camelidae composed 13.2% of the faunal remains (a figure similar to that given by Wing [1977b:Tables 4 and 5]). A large camelid corresponding to a guanaco/llama (1%) was identified, as well as a small one, a vicuña or alpaca (0.3%).

In her final report, Reitz notes that these clearly are llama or alpaca remains, and that in overall terms they represent 45% of the remains studied (Reitz 1979:78–79).

After noting that the guanaco prefers to live at an altitude of more than 5000 masl, Reitz (1979:87) writes:

> They are attracted to lomas; but in an agricultural context outside the lomas area on the coast, the camelid remains are probably domesticated llama (*L. glama*) or alpaca (*L. pacos*). The presence of camelid remains on the coast outside the lomas area may constitute evidence of domestication (Wing 1977[b]). They could be pastured in the lomas or near the houses and fields. They may have been on the coast only seasonally, when conditions permitted sufficient plant growth for their support.

Further on Reitz indicates that the most important resource for the inhabitants of V-434 was the herds of domestic camelids, and that llamas could have been easily obtained. However, she insists that "[i]t should be noted . . . that camelids may not have been a regular meat source, but one reserved for ritual occasions (West 1979: personal communication)" (Reitz 1979:88–89). However, it is worth pointing out that Wing (1986:259, Table 10.7) later noted that camelids in V-434 composed 50% of the faunal remains, a figure higher than that given by Reitz (1979:88–89).

Wing (1977b:Tables 4 and 5) mentions another site in the Virú Valley called V-604 which I cannot find, but it seems to be contemporary with V-434, even though the presence of camelids is very low, a mere 4%.

4.5.2.2 The Gallinazo Culture
The data available on camelids in the Gallinazo culture are restricted to the Virú and Santa River valleys.

When discussing the economy of this culture, Strong and Evans (1952:213) wrote that "[f]rom the relative abundance of llama bones, wool, and droppings it is apparent that animal domestication played a considerable role." It should not be forgotten that little was known of the process of domestication of camelids when this statement was made, and its significance lies in that it indicates the presence of these camelids in the Virú Valley.

The famed Huaca Gallinazo (V-59), the site for which the culture was named, is in this valley, on the northern side of the lower part of the valley (in the province of Trujillo, district of Virú). When Strong and Evans (1952:85, Table 5) worked here they found a significant quantity of mammal bones and suggested that these probably came from llamas. These same scholars also studied the Castillo de Tomaval (V-51), in the middle Virú River valley, and noted having found some worked bones that were probably llama bones, in a context belonging to the Gallinazo occupation (Strong and Evans 1952:121). They also noted that "[l]lama . . . or deer bones occur in nearly every level from bottom to top and are far more abundant than are large sea mammal bones" (Strong and Evans 1952:125). Shelia Pozorski (1976:246) also examined the refuse at this site and concluded the bones probably came from llama.

Strong and Evans (1952:168) also studied Huaca de la Cruz (V-162), in the middle part of the Virú Valley, and found "a clay floor covered by animal (presumably llama) dung" in the strata belonging to the Gallinazo culture. They state that "[l]arge amounts of llama dung and much vegetational material suggested that at one stage in their occupation these structures had housed animals." Interestingly enough, Strong and Evans (1952:197) likewise report that three tombs (12, 13, and 16) held evidence of "llama sacrifices by beheading (two in burial 12)." Shelia Pozorski (1976:247) also noted that the remains of domestic llamas, excrement, and especially bones are quite common in the Gallinazo strata at this site.

Wilson has studied the Santa River valley, which separates the Departments of La Libertad and Ancash. When describing the period he calls Early Suchimancillo/Middle Gallinazo, which corresponds to the initial Early Intermediate period, Wilson notes that it is at this time that one finds the first evidence of structures resembling corrals. Two of them were dated, one in the Quebrada Silencio and the other 7.5 km down the valley. There are no cultural remains in either of them, but they seem to have been built to pen animals. Wilson assumes these were llamas or alpacas because he managed to find figures which depict camelids drawn in the desert resembling those of Nasca. Wilson also notes that contacts with the highlands were strengthened at this time (Wilson 1988:171, Fig. 80).

Wilson has also found corrals in the upper and middle valley that correspond to what he calls Late Suchimancillo, or the late Early Intermediate period, or more specifically the final stages of Gallinazo. These corrals are found on both banks of the river and are located at more or less regular intervals, beside old roads. One of them is associated with an important center, the rest with small hamlets. Wilson believes that these corrals were strategically located beside the routes followed by caravans of llamas (Wilson 1988:189, 193).

4.5.2.3 The Moche Culture

Most of the data on camelids in the Early Intermediate period obviously correspond to the Moche culture. It is a shame that it has not received due consideration, even though it is one of the most studied cultures in ancient Peru. The starting point has always been the incorrect assumption that camelids cannot thrive on the coast. I shall cite but one example. Benson (1972: 86) wrote: "Llamas were certainly brought down from the highlands, and, judging from the pottery representations . . . they were used as pack animals." She then added: "Since llamas are indigenous to high altitudes, it is also questionable how well they would have endured on the coast if they were used as work animals. They may have been used only on trips to and from the coast, and perhaps, since it had a special meaning from its association with the mountains, the llama may have been used only to transport ritually significant material. The appearance of sacrificed llamas in burials reinforces this theory. It is also possible that they were eaten, but they cannot have been plentiful enough to have been a major food source" (Benson 1972:89).

These same ideas have been used to explain why these animals vanished from the coast. Larco Hoyle (1938:92) thus wrote: "All animals pictured in [Moche] pottery are found at present, save for some species which vanished because they were not native to the land, as happened to the llama, an animal of great historical significance which is only found in the high Andean plains . . . from whence it was no doubt brought beforehand to the coast and acclimatized, overcoming the difficulties."

These statements clearly show that the facts have been distorted by the prejudices held, both now and then, against the possibility of camelids living in environments that are not exactly at high altitudes, as will be shown throughout the present study.

It is clear that the highest number of camelid representations made in pre-Columbian times belong to the Moche culture. However, all representations only show llamas, whether painted or modeled in clay. Lavallée (1970:65) made a special study of animal representations in ancient Peru and asserts that the alpaca, guanaco, and vicuña were never depicted by Mochica artisans.

One intriguing problem has yet to be solved — almost all the Mochica depictions show short-necked llamas, which are quite different from the present-day animals. Horkheimer (1958:26–27)[5] was one of the scholars who studied this subject the most, and wrote: "We assume that there was another subgenus of domesticated auchenid because we find hundreds of representations of an auchenid-like pack animal which has a very short neck in the huacos of the Mochica potters, who are usually distinguished by their realistic and precise reproduction of the environment.(5) The existence of this short-necked animal explains why so many chroniclers mention 'carneros' and 'ovejas' ["rams" and "sheep," i.e., male and female in Spanish], terms which are poorly adapted to the long-necked race. The coastal auchenid possibly disappeared due to the great plague (mange) of 1544–45(6), when the herds had already been decimated by the Spanish host, anxious for meat to eat. In any case, 'the abundance of llama skeletons in the cemeteries close to the lomas' (Tello 1942, p. 607 [my reference 1942:19]) proves that llamas were kept at lower altitudes than today in pre-Columbian Peru, at least temporarily" (footnote 5 is Schmidt [1929:176, 1; 179, 1; 180, 1 and 2], and 6 is Garcilaso de la Vega [1609, Bk. VIII, Chap. XVI, which corresponds in my bibliography to the 1959 edition, Bk. 8, Chap. XVI, 148]).

Horkheimer later returned to the subject (1961:31–32) and stated that it is usually believed that this animal, a native of the highlands, gave the people there a monopoly on the transportation of the goods exchanged between the coast and highlands. However, Horkheimer cautioned that when "examining the representations of thousands of Mochica huacos we find that this animal almost always has a short neck, i.e., it is essentially different both from the llama as well as other members of the auchenid family. We therefore must consider that there was a variety that became extinct after the Conquest."

Although the observations made by Horkheimer are correct as far as the Mochica representations are concerned, this presents several problems and raises several questions, which will be taken up later. Nonetheless it is worth mentioning that the idea posited by Horkheimer has not been wholly forgotten, and that some students still posit the existence of a now vanished variety of coastal llama. Based on the work they did in one sector of Chanchan called Loma Roja, or South Unit No. 2, Chayhuac, Vásquez and Vásquez (1986:n.p.) thus assert that "there is now no doubt" that a camelid species acclimatized to the coast from very early times once existed, and that this "clearly is a variety of llama different from the one living in the highlands (pers. commun., Altamirano 1985)." However, Vásquez and Vásquez do admit that more evidence is needed on this subject. I will return to this subject later.

Actually, at present many specialists agree that in Mochica times there was a significant number of camelids on the north coast. Novoa and Wheeler (1984:124) even mention "great herds" in the Trujillo zone.

Shelia Pozorski (1979a:176; 1982:180) has analyzed the bones in several sites of this period and believes that Mochica society supported itself almost exclusively with herds of llamas that were under the centralized control of the state. The llamas provided the Mochica with virtually all the animal protein consumed. Pozorski (1976:123) notes that a review of the resources of animal protein recorded for the Moche culture shows a concentration on domestic llama meat (more than 90%), to the point that all other potential sources of meat were of less importance. The evidence suggests that animal raising was controlled and supported by the Mochica government in order to have an animal for meat, transport, and wool. The evidence for this control is visible from an archaeological perspective and shows in the age of the animals killed for their meat, as well as in

the pattern present in the markings left by the butchering, which clearly exhibit standardization. There is also a huge amount of excrement in archaeological sites. The herds were kept inland, still according to Pozorski, and gave the Moche state a security earlier human groups had been unable to attain.

Shelia Pozorski (1976:113–118) extensively discussed the different aspects of this subject. Pozorski states the problem of camelids adapting to the coastal environment, but notes that much of the evidence available is still contradictory. For example, she mentions that Cardozo (1954:65) argues that the soft plant diet on the coast produces an abnormal growth in dentition, which does not occur when the llama feeds on hard highland grasses. This, however, contradicts Wing (1973:9), who maintains that only the vicuña has incisors that grow continually, implying that the teeth in all other camelid species are not overly affected by the nature of the food eaten. Although it is true that this problem will have to be studied in the future, the archaeological data attest to the presence of a large number of animals in the coastal area that evidently managed to survive. For example, Pozorski studied the fauna excavated from the huacas at Moche (the Huaca del Sol and the Huaca de la Luna) and concluded that its people were almost wholly dependent on llama meat. Besides, a considerable amount of excrement shows this. It follows that herds were kept in the vicinity. Pozorski admits that large herds of llamas were perhaps kept in more distant areas inland and at a higher altitude, in places not cultivable and could therefore stand herding better.

Some cultural factors pertaining to the raising of llamas and the processing of their meat, briefly mentioned above, can be analyzed in terms of the great many bones excavated. This assessment considered the age of the animals, the frequency of the bones burned, the frequency of the type of bone, and the marks left on them by human activities.

Bone fusion was used to assess the age of the llama population, at least partially. It is worth recalling that Wing (1972:330) determined several stages in the age of camelids using a sequence of fused epiphyses. The sixth stage in the sequence corresponds to the eruption of the teeth, which

Cardozo (1954:91) established as a chronological age equal to 36 months. However, Pozorski considers that all other studies are relative because they are not linked to chronological age.

A partial age profile for the llama population was reconstructed once the age of the Moche bones was organized in terms of the progressive fusion stages, and the results, shown in terms of the percentage of unfused bones, were plotted (readers who want more details will find them in S. Pozorski 1976:Appendix II, Fig. 18). It turns out that the high rate of unfused bones is more significant even in the later stages of fusion. The distribution shows that the animals consumed in Moche were usually very young adults. It can be concluded, following the studies of Uerpmann (1973:315–316) and Perkins and Daly (1961:101), that the methodical exploitation of a group of animals of a specific age is to be expected among people who keep domestic herds for specific ends. Although it is true that Collier and White (1976:96–102) criticize the use of data indicating the methodical exploitation by age and sex to argue for early domestication, Pozorski believes that the data on age can be used to determine how the herds were used, because the domestic status of the llama is not in question (see S. Pozorski 1979b:150).

The data for the age of the animals in Moche indicate that those consumed came from herds kept mainly for meat. Once animals have been selected for raising, it is far more profitable to kill them immediately after they reach adult size. At this point they do not gain more weight but still consume fodder. The high proportion of young adult animals in the faunal samples supports this position.

On the other hand, the frequencies for the bones can occasionally be used to discover the methods used in processing the animal, but the best direct evidence is obtained from the analysis of cut marks and blows preserved in the bones. Pozorski (1976) used all this information in an attempt to reconstruct the procedures used by the Moche people from the moment a llama was killed up to when it was used (for more details, see S. Pozorski 1976:Appendix II), bearing in mind that the procedure followed becomes standardized when a group of people primarily depends on an animal for its meat.

From this analysis it is deduced that the animal was killed and skinned, and the carcass was dismembered by cutting through the joints. Although many of the smallest pieces were probably cooked with the bone, the meat was cut from the biggest bones to make it easier to handle. The long bones were then broken and split open to extract the marrow.

The frequency of burned bones in Moche samples was appraised in terms of body part (see S. Pozorski 1976:Appendix II and Table 65). It was verified that less than a third of the bones from a given section were burned. The highest frequency was for the ribs and vertebrae, which might mean that the meat was roasted. On the other hand, the lower frequencies of the other bones might be attributed to their being cooked in some kind of vessel.

Bone parts, especially the long and dense ones, were only occasionally used to make tools. In the Moche samples, the two proximal metapodial fragments show deep grooves where the fragment was cut before making the artifact (S. Pozorski 1976:117–118).

Besides its utilitarian value, the llama had an important role in the beliefs of the Moche people. This is reflected in the presence of llama remains in burials. Larco Hoyle (1938:92) had already observed that llama remains are found in most of the Moche tombs. Unfortunately, there is no systematic data on this point. Llama bones were found only in some tombs in the excavations made in the flat area between the Huaca del Sol and the Huaca de la Luna in Moche, but some important data are available. The bones found were usually skulls, feet, or the lower part of a leg, and these were the parts of the animal that had little or no food value. On the other hand, it has been verified that llama remains were found with the dead of both sexes, and were also present in child and adult burials (Donnan and Mackey 1978:381). We also know that llama sacrifices were common, but this is a phenomenon that appeared before the Moche (Donnan and Foote 1978:406).

The representations of llamas left behind by the Mochica (especially those of Moche II and III [Larco Hoyle 1948]) are excellent and show different aspects of the activities and life of these animals. Llamas appear loaded in different ways and positions, carrying men in different ways, resting, mating, females with offspring, and so forth. Several of these camelids appear with a lateral rope attached to the ears through a perforation (see Donnan 1978:Fig. 174). Shimada and Shimada (1985: 19) suggest that this might reflect a coastal custom, as opposed to a kind of harness sometimes worn by highland llamas. This is possible, but the llamas with a harness made with a rope are also frequently depicted in Moche (see Donnan 1978: Fig. 175), as well as loaded animals that have neither rope nor harness (see Donnan 1978:Fig. 176). A study is needed to ascertain whether Shimada and Shimada are right. Apparently they are not.

The Mochica also used to make cuts in the ears as marks of ownership (see, e.g., Donnan 1978:Fig. 178). This is still done in the altiplano, although it is not a general custom, according to Tschopik (1946:521; interested readers wanting more information on ear marking are referred to the study by Palacios Ríos [1981:222–223]).

Although there is no direct evidence, I can suggest that the llamas formed caravans to transport products, not just between the highlands and the coast but also along the coast. On this point Lavallée (1970:65) and Shimada and Shimada (1985: 20) agree. The latter (Shimada and Shimada 1985) believe that llama caravans and rafts were the two means of transportation used by the Mochica within their domain, that is, between Vicús in the north and Huarmey in the south.

However, we should be aware that llamas have certain limitations, particularly when carrying loads. We should therefore not overstate their importance, as Larco Hoyle did, claiming that llamas had "the same role that the horse had . . . in the eastern civilizations" (1938:92). There is a huge difference in strength between these two animals. Horkheimer (1958:27) says that camelids "could not become either mounts nor draft animals [which is why] Andean societies did not become pastoral peoples. The Andean region therefore did not give rise to the process so often repeated in the Old World: pastoral peoples who dominate agricultural communities." This is both true and false at the same time. It is mistaken insofar as the presence of pastoralism in ancient Peru, a point to which I return later. But it is true because Andean pastoralism was different from

traditional Old World pastoralism, and thus its consequences were clearly different.

Let us now see what the archaeological data are for llama remains in Mochica contexts. One of the most important archaeological sites of Peru is in the Department of Lambayeque (in the province of Ferreñafe, district of Pitipo)—the Huaca del Pueblo. We know that an occupation at this site extended from Moche IV to the Middle Horizon. In the Moche context, the remains of llamas appear in primary association, often with houses. These remains include different body parts of the animals, including some from fetuses or newborn animals, juveniles, and adults. Some of the bones show traces of the killing (Shimada and Shimada 1985:14–15).

Although it is not useful, I must mention a study by Chimoy (1985:167, 169–170). He claims to have identifed a "*Lama* sp. llama (Mamalia Camelidae)" and a "Short-Necked Llama" (Mamalia Camelidae) [sic]" through the analysis of "Moche and Chimu" ceramic specimens. Chimoy relied on the study by Shimada et al. (1981) to conclude that "[t]he evidence suggests that the people domesticated the 'llama' in corrals [? sic] and fed it with 'algarrobo [*Prosopis* sp.] and/or maize'; there are remains of 'llama' dung with 'algarrobo' seeds." However—and this is the worst part—Chimoy identifies goats, bulls, sheep, and "*Homo Sapiens*" [sic] in pre-Columbian times, and states that the material should be re-examined "because the presence of these animals in the South American continent is very debatable. Perhaps it is assumed that mistakes could take place in the taxonomic determination" (Chimoy 1985: 170–172). No comment is needed.

Llama sacrifices were recorded in funerary contexts within the great urban center of Pacatnamú, in the northern part of the Department of La Libertad (in the province of Pacasmayo, district of Guadalupe). Llama crania, bones, and teeth were found in tombs belonging to the Moche culture (Ubbelohde Doering 1959:19, Fig. 21; Donnan and Cock 1986:69, 81, 98).

Shimada and Shimada (1985:10) report that Bankes (1971; a dissertation I have not read) excavated a site dating to Moche IV in the province of Trujillo (district of Rázuri) in this same department. There were llama bones, but it is not known whether these were for domestic or ceremonial use.

An ossuary of camelids almost 1 m thick has been found in the same province but in the district of Magdalena de Cao, almost 500 m southwest of Huaca Blanca (also known as Huaca Cao Viejo), in the El Brujo complex. The place is called Paredones and is a cemetery. In some cases even the hair of the animals can be seen. The site seems to be Late Moche or early Middle Horizon (Segundo Vásquez and César Gálvez Mora, pers. commun., 30 August 1991). After receiving this information I visited the site on 15 April 1993 with Víctor Vásquez Sánchez, and ascertained the existence of what seems to be a burial of some 30–50 llamas. The context is evidently a funerary one. There are corrals for the animals in the vicinity, according to Segundo Vásquez Sánchez (pers. commun., 15 April 1993).

Víctor Vásquez Sánchez (pers. commun., 17 July 1992) told me that camelid excrement was found at the site of La Poza in the Parte Alta sector (in the Moche Valley, district of Huanchaco, 18 masl). The site has Salinar, Gallinazo, and Moche occupations. Although it was excavated by Donnan and Mackey in 1969, Escobedo and Rubio in 1982, Castillo in 1986, and Deza and Segundo Vásquez in 1988–1989, no report has been published so far.

Uhle excavated some Moche tombs close to Trujillo and in front of the famed Huaca de la Luna (in the province and district of Trujillo), where he found bones of sacrificed llamas (Menzel 1977:60). Shelia Pozorski excavated refuse corresponding to the Moche culture in the Huaca del Sol, in the same locality, and established that camelid bones represented 68.1% of the remains in the diet at this site. Shimada and Shimada (1985:10), however, note that this sample should not be taken as representative of the ordinary diet of the Mochica; it might well be a ceremonial banquet because this was a sacred site.

Extensive excavations have been carried out recently in both the Huaca de la Luna and in the flat area between this huaca and the Huaca del Sol (Chapdelaine et al. 1997:Fig. 29, 72; Cárdenas et al. 1997:Table 7, 133, 132). Camelid remains were found here. The study of the bones shows that these were probably adult llamas, originally used as pack animals, that were then killed to be used as food (Cárdenas et al. 1997:144). Some llama remains were found in Plaza 2 of the Huaca de la Luna, but their context is uncertain. However, there were more bones from these animals in two sectors of the urban area known as the Sector de Tuberías and the Taller de Alfarero that correspond to two occupations, Moche III and IV, respectively (Uceda 1997:Fig. 1, 18).

According to reports of the work done later in the South Urban Area (see Chapdelaine 1998: Fig. 85, 88), camelid remains were found in three of the Architectural Components (9, 7, and 5). There were also camelid remains in the tombs found in Architectural Group 16 (see Vásquez Sánchez and Rosales Tham 1998:Table 30, 187). Fifty-one bones in all were measured, and "[t]hree camelid species have been identified: *Lama guanicoe* 'guanaco,' *Lama glama* 'llama,' *Lama pacos* 'alpaca,' and it is possible that the sample includes bones belonging to hybrids (waris)." It is interesting that the alpaca predominates over llama, and there is only one case of guanaco (Vásquez Sánchez and Rosales Tham 1998:189). These remains all belong to late Moche and "even go far beyond phase IV," as there is a group of radiocarbon dates that go from 1280 to 1530 BP (Vásquez Sánchez and Rosales Tham 1998:181). It is worth noting that an "age structure" was established through an osteometric analysis, which showed that the age of most of the animals ranged between three months and four years, with old animals a minority; their age age ranged between 13 (just two cases) and five years (Vásquez Sánchez and Rosales Tham 1998:178–181).

Finally, I must mention a site that no longer exists, Huaca Pelada, which was close to the beach resort of Buenos Aires (in the province of Trujillo, district of Víctor Larco). Julián Castro Burga A. excavated here and submitted a dissertation to the University of Trujillo in 1949. He evidently was not an archaeologist. José Eulogio Garrido, the director of the journal *Chimor*, published some excerpts taken from this dissertation along with his own comments, but in such a way that it is not always clear whether the text is the work of Castro Burga or Garrido.

The discovery of a complete camelid skeleton in one of the rooms in the huaca, obviously an offering, was reported (Director de Chimor 1959–1960–1961:14). However, a later comment mentions an "enormous amount" of bones and even states that it was a vicuña (Director de Chimor 1959–1960–1961:27–28). The chronology is unclear; it could be either Moche or post-Moche.

While working in Huaca de la Cruz (V-162) in the southern part of the middle Virú Valley (mentioned earlier), Strong and Evans (1952) noted llama dung in a room and found a knee from this animal, all in a Moche context (Strong and Evans 1952:135, 137). Close to the monument they also found a "very large amount of llama dung" (1952:168). There are several other sites in the Virú Valley where camelids have been found but whose exact cultural affiliation I do not know. They may date to the Early Intermediate period. I also do not know their exact locations, because the site designations are not the same as the ones used in the Virú Project (see Willey 1953). Virú 632 poses a problem because camelids were initially given the percentage of 14% (Wing 1977b: Table 7), and then 3.5% (Wing 1986:259, Table 10.7). There is also a problem with the Virú 633 site, where camelids constituted 0.7% of the fauna (Wing 1977b:Table 7), and then this percentage was corrected to 2.1% (Wing 1986:259, Table 10.7). Virú sites 368 and 636 were said to have 2.1% camelids, while Virú 604 had 0.8% (Wing 1986:259, Table 10.7).

In regard to the Moche occupation of the Santa River Valley, Donnan (1973:123–124) wrote:

> By far the most common bone material in refuse was camel [sic]. In some instances it was impossible to determine whether the camel [sic] bone was of llama. It is interesting to note that llama bone occurred in abundance in some levels and was absent in others. It is generally fresh looking, seldom showing signs of weathering. Vertebrae and ribs are the most commonly found camel [sic] bones, and they are generally found unbroken. Only one complete long bone was found—all others were extensively broken. The bones have almost no scratches that could be attributed to butchering. Skull bones were generally lacking except for jaws and teeth.

Wilson (1988:220) in turn mentions "three possible corral enclosures" associated with roads. One of them is quite large, as it measures 110 m × 95 m. There is no refuse in them, but Wilson assumes these were corrals. They date to the late Early Intermediate period and the beginning of the Middle Horizon.

4.5.2.4 Other Cultures

The site of Bermejo is in the Department of Ancash (province of Huarmey), north of the mistakenly named Fortress of Paramonga. Camelid bones there are relatively scant, for they compose only 1.4% of faunal remains (Wing 1986:259, Table 10.7). I do not know its exact cultural affiliation, but the context corresponds to the Early Intermediate period.

For the central coast, and specifically for the Ancón-Chillón area, Cohen (1978b:122; 1978c 27) claims that llama bones and coprolites begin to appear only from this period, and at a "remarkably late date." This does not seem to be true, however, or at least it does not follow from the study by Patterson et al. (1982:71, n. 8), who explain that camelid remains begin to predominate in the refuse deposits of the Chillón Valley during Epoch 7 (which corresponds to the Early Intermediate period)—the time when camps with Lima 6 pottery appear in the lomas in this area. This suggests, still according to Patterson et al. (1982:71, n. 8), "that herding was becoming a more important economic activity in the coastal social formations *than it had been earlier*" (emphasis added). This supports information supplied by Dillehay (1979:27–28), who noticed that a larger amount of camelid bones suddenly appears at the site of Huancayo Alto in habitation rooms, which would mean that the inhabitants were a mix of *chaupiyunga* and highland ethnic groups.

In the Lurín Valley, Patterson et al. (1982:69) confirmed that a minor presence of herders from the upper part of the valley is detectable in the Atocongo lomas, as well as an increase in the number of herders in these pastures from the lower part of the valley.

4.5.2.5 The Nasca Culture

Little information is available for Nasca and most of that is vague. Maldonado (1952a:73) states that he found a "thick layer of auchenid dung, easily recognizable and identifiable" in the Cahuachi zone (Department of Ica, province and district of Nasca). He also mentioned having seen "a corral, 20 × 20 m" at Paredones, close to Nasca, "with a deposit of auchenid dung almost 40 cm thick," and that "its characteristic form can be seen; in the deepest part the mass is amorphous because it was deposited in corrals where it endured daily the pressure and deformation caused by the legs of these animals."

For Cahuachi, Strong (1957:31) reports abundant llama remains found in a ceremonial or sacrificial context in the structure he called the Great Temple (Silverman's Unit 2 [1993:301]). For this same site, Silverman (1985:87) notes that 99% of the fauna Strong recovered (she means the 1957 study) is camelid. Silverman actually quoted the report by Geismar and Marshall (1973), which is most inadequate. The material studied covers a very long period that extends from "later Paracas through Huaca del Loro culture phases (Strong 1957:7), ca. AD 200–900 (Strong 1957:46)" (Geismar and Marshall 1973: 5), in other words, from the late Early Horizon to the Middle Horizon. Moreover, the materials came from several sites: Cahuachi, San Nicolás, Chaviña, and Huaca del Loro (Geismar and Marshall 1973:48–51). The report gives a total of approximately 900 bones, of which probably 894, or 99%, are from camelids. The report is very vague, as I have already noted, and the data do not appear in the tables; I deduced them. Nor is there a table of the bones not from camelids.

Geismar and Marshall believed the sample was far too small to be representative (I disagree), and that fragmented camelid bones predominated in it. No pattern was observed in the fractures, but there were bones with marks left by butchering, chewing, and burning (Geismar and Marshall 1973:6). Geismar and Marshall proposed three possible interpretations regarding the distribution of size in these bones: first, that this difference might represent the presence of a domestic form and a wild species; second, that the difference in size might represent a difference in age within the same species; and third, that this difference in size might represent some combination of the first two possibilities, that is, young do-

mesticated animals, young wild ones, old domesticated animals, old wild ones, and so on (Geismar and Marshall 1973:7–8). Using Wing's (1972) tables for epiphyseal fusion, Geismar and Marshall (1973:7) concluded that there are instances of adult and young animals in all size ranges, thus invalidating the second possibility presented in their interpretation.

However, Geismar and Marshall (1973:3) considered that "an inventory of the collection indicates that the sample is relatively homogeneous, composed predominantly of camelids." They believed that in comparison with what corresponded to the camelids, the small number of remains from other animals supported the hypothesis that "a selection was made from a controlled source or, in fact, that animal domestication was practiced" (Geismar and Marshall 1973:7).

Silverman (1988:413) later confirmed the presence of llama excrement at Cahuachi, apparently in an Early Nasca context. She found remains of llamas in burials (Silverman 1988:421) and insists that these animals were used for ritual use (Silverman 1988:424). The data were confirmed and expanded in Silverman's final report (1993). For Unit 18 she notes (see Silverman 1993:Fig. 2.4, 20; Fig. 5.19, 70) that "[a]bundant camelid bones were observed on the surface of this area in 1984, probably corresponding to Orefici's recent discovery." The bones must belong to an Early Nasca occupation, to judge from the ceramic sherds found on the surface (see Silverman 1993:71).

In the East Zone of Cahuachi (see Silverman 1993:Fig. 2.6, 25), "Camelid remains (mostly of young individuals) and human bones are scattered on the surface" of platform WW3 (Silverman 1993:86). Llama bones were also found in a funerary context. Silverman recorded several of these findings. I mention here two that I find interesting. First, there is Tomb 10 from Unit 19 (see Silverman 1993:Fig. 2.4, 20). This is a llama burial where the animal was specially prepared, and most of the bones had been eliminated. Only the cranium and leg bones remained. This seems to have been a desiccated animal. It was extended and laid on its right side, with the legs extended in front. It was oriented to the north. As Silverman comments, "In this particular case, we may speculate that the internal organs of this llama had been used in a divination ritual" (Silverman 1993:199; see also Fig. 14.6, 200). The tomb could not be dated, but it probably corresponds to the Middle Horizon (Silverman 1993:202).

The other case is a tomb found close to what Silverman (1993:Fig. 14.18, 204) called "Burial Area 1," which was excavated by Strong's project. In Tomb 27 the adobe bricks around the body had been burned, and there were incinerated camelid bones (Silverman 1993:209).

When discussing the faunal remains, Silverman points out that the data from Cahuachi confirm the position of Shimada and Shimada (1985). Llama excrement frequently occurs in her excavations. Silverman suggests its presence in the kancha in Unit 16 is directly related to fact that this was a place where pilgrims assembled, and where llamas delivered and carried away goods. Strong (1958:31) found abundant remains of camelids in the upper part of Unit 2 (his Great Temple). But Silverman notes: "We did not recover butchered camelid bones, although Orefici presumably did since Valdez (1988[b]) speaks about camelids as a source of meat protein." Silverman then mentions her Tomb 10 in Unit 19 (see above) and the discovery of a llama leg in Unit 128, which was interpreted as a special ritual associated with the building. The data from Valdez (1988[b]:34) suggest that the animals were killed and eaten at the site. Silverman says, "I agree with Valdez Cárdenas (1988[b]:32) that this means that camelids were not being kept/bred/herded at Cahuachi." She notes that the data on camelids in Orefici and Strong are consistent with hers (Silverman 1993: 304). Silverman then comments that "[l]ike Strong (1957:31), I would explain the significant presence of camelid remains at Cahuachi as evidence of their use in ritual (including ritual feasting)" (Silverman 1993:202). She also suggests "that camelids, notably llamas, were being sacrificed at Cahuachi and consumed in ceremonial episodes" (Silverman 1993:304).

It is to be noted that Silverman's position does not agree with that of Geismar and Marshall (1973), but we saw that their report does not adequately support the data, so doubts still linger. Nor does Silverman agree (as she claims) with Orefici, because he considers that "camelid raising was an alternative to agricultural work" in the Nasca region (Orefici 1992:203). However, as we will see later, Orefici's work was not at all reliable.

Valdez (1988b:32–33) reported the most recent work done on the Nasca culture in Cahuachi, by an Italian mission. He notes that 3,002 out of 4,154 bones (or 72.2%) came from Camelidae and belong mostly to adult animals. Valdez also states that a "minimum and insignificant amount of coprolites . . . [and] a marked absence of fetuses of recently born and tender [animals]" appeared during excavations. He therefore considers that the animals were not raised at the site, and that there was some exchange with herders who lived "above 2000 masl." Valdez identified two species, *Lama glama* and *Lama pacos*, but notes that wool from *Lama vicugna* was also found. He believes (1988b: 34) that the animals were sacrificed at the site, because remains of almost all body parts were found. The animals were probably used essentially for meat, and in some cases the broken bones indicate that the marrow was extracted.

Valdez later sent me the definitive data from the study of the Cahuachi camelid remains. From them it follows that 3,382 bones out of a total of 4,919 belonged to camelids, or 69.7% (Lidio Valdez, pers. commun., letter of 30 March 1999). In a later report, Valdez (1994:677) noted that the association of camelids with ovens and vessels suggests the animals might have been sacrificed. He specifies that all parts of the body are represented in the sample, suggesting that the animals were taken alive to Cahuachi and sacrificed at the site. Valdez also explains that projectile points are rare in Nasca sites, but even so, some were found inside the ovens. We can therefore assume the points were used for the sacrifices. This is confirmed by the fact that most of the bones belong to adults, suggesting that these were domestic animals, as the adult-young ratio in wild animals is thought to be more equal. However, there is a mistake in Valdez's study. We read: "Of all other units at Cahuachi, 3382 belonged to adults and only 822 to young ones (Valdez 1988[a in my bibliography])" (Valdez 1994:677). In fact, according to information supplied by Valdez himself (in a letter of 30 March 1999), 3,382 is the total number of camelid bones identified, 2,560 of which correspond to adult animals (75.6%) and 822 to young animals (24.3%).

The presence of a high number of adults would also show, according to Valdez (1994:677), that domestic animals were controlled, and that adults were preferred for sacrifices, while the young animals were kept in the herd. Valdez (1994: 678) adds that the meat of the sacrificed animals was probably eaten.

Silverman (1993:304) objects that the report by Valdez (1988b) does not indicate the type of screening used in the research at Cahuachi, and thus questions the validity of his assertions regarding the absence of llama coprolites. One interesting piece of data recorded by Silverman, derived from a personal communication given her by the members of the Italian mission, is that "the multiple burial of more than sixty sacrificed camelids was discovered to the southeast of the Unit 19 mound" during the 1991 campaign (Silverman 1993:28).

Orefici (1992:96) discussed the Valdez thesis (1988a), regarding the absence of fetuses or newborn animals at Cahuachi and Tambo Viejo, while the presence of young or adult animals is noteworthy. He suggests two hypotheses. The first is that llamas lived on the coast from May to November in the lomas and then returned to the highlands, and the second is that the rearing areas were far removed from habitational sites and religious centers. Orefici notes that the remains found at Cahuachi are not of a domestic but of a religious type, and believes that the absence of excrement supports the second hypothesis. The problem is that he (1992:225, 227, 234) contradicts himself, because when discussing the Great Pyramid of Cahuachi, he mentions finding not only the bones of camelids that were possibly sacrificed but also dung, which was perhaps used as fuel. Then Orefici (1992:230) reports the bones of sacrificed animals when discussing the Stepped Temple, where camelids prevailed. However, it is almost impossible to know the exact context of these discoveries, and the data provided by Orefici are very ambiguous (see Bonavia 1993).

When discussing the Nasca culture *largo sensu*, Orefici (1992:92, nn. 12, 93) says that "camelids predominate" and that "[t]he most common finds belong to domestic species: llama (*Lama glama*) and alpaca (*Lama pacos*)." He insists (1992:162, n. 11) that "camelid legs and ears" are commonly found in the tombs at Cahuachi, and that "camelids (together with guinea pigs) were the most frequent offering" (1992:164) in Nasca tombs. These claims are not supported.

Valdez, however, continued working with these materials, applying Wheeler's model (1984c); that is, he assessed the samples of incisors from the Cahuachi remains. Valdez noted that the clear absence of vicuña was confirmed, but that nothing could be said about the guanaco. On the basis of an osteometric analysis done according to guidelines laid out by Wing (1972), Miller (1979), and Kent (1982a), Valdez concluded that two groups of camelids are clearly distinguishable, one of large animals (guanaco/llama) and the other of small ones (alpaca/vicuña). Valdez concluded that "it is still difficult to clearly distinguish the two bigger or smaller species based only on their bones." Even so, he believes that osteometric analysis is now the best option when trying to identify the four camelid species, always following the criteria established by Miller (1979) and Kent (1982a). The results obtained for Cahuachi indicate that the bone collections correspond mostly to alpacas and the rest to llamas, but this is not definite.

Valdez believes that since the values taken from the Cahuachi collections fall midway between the guanaco/llama and the alpaca/vicuña group, then the analysis confirms the absence of guanacos and vicuñas. Valdez applied Greenfield's method (1988) and concluded that alpacas were raised mainly for wool, and used as meat only when they were adults (Lidio Valdez, pers. commun., letter of 23 February 1999).

Valdez (1988b:33) also reported the archaeological excavations that Francis A. Riddell carried out in Tambo Viejo (Department of Arequipa, province of Caravelí, district of Acarí). Here there was a "presence of camelids . . . during the Early Intermediate period because despite the lack of remains from fetuses/newborn and infant animals which indicate raising, there is a good sample of coprolites which seem to have been used as fuel." The statement is confusing, as can be seen, and the required evidence is not presented. Based on Troll (1980 [1958]:33), Valdez then comments that "it does not seem relevant to discuss the breeding of these animals on the south coast," where they "must have entered under certain circumstances or for ritual purposes." He then speculates that the animals were taken there in winter to take advantage of the lomas and because there are not enough pastures in the puna; besides, at this time temperatures fall and young animals die. In the rainy season they would have gone back to the highlands. "We thus understand the absence of coprolites at Tambo Viejo, as well as the absence of remains of fetuses/newborn and infant animals." This is all speculation, and is full of traditional biases.

4.6 MIDDLE HORIZON (AD 500–900)

Unfortunately, there are relatively few data for this important period of Andean history, particularly in regard to the highlands. There is no question that this was a time of large-scale, intense movements of people and armies, or that caravans of llamas must have played an important role. This is one of the best examples of how, in the end, the samples we have are not that significant.

4.6.1 The Highlands

Izumi Shimada (1982:182), referring to the Department of Cajamarca, and specifically to the Middle Horizon, shows that faunal remains indicate the continuity of earlier subsistence patterns. The only noteworthy point is a "possible" increase in the use of camelids, which were used as meat —perhaps their primary use— and for other purposes, mainly as pack animals. This follows from the higher proportion of camelids with less than 3.5 years in the samples, in comparison with previous periods.

This was confirmed by Melody Shimada (1982:312–313). Based on the sample excavated at Huacaloma for the Middle Horizon, she claims that camelid remains show these were the most economically important animals, because they constitute 97% of the animal remains. (For the location of the sites mentioned, see Figure 4.5.)

Shimada makes an interesting point: Although there was continuity in these animals' importance, a change in their management is noticeable once we compare these data with data from previous periods. It seems that new individuals were introduced at the site after the first year, when 68% of those recently born survived. There is a possible killing off of these animals between two and three years of age, which corresponds to the pattern found in previous phases, but half the population survived to 3.5 years. Maintenance of these animals after this age might indicate an increase in their use in diverse activities, something that did not happen earlier, as camelids were used essentially as meat. The fact that they were maintained for more than 3.5 years might mean they were used as pack animals. An increase in younger animals is also noticeable, and two different populations of llamas and alpacas are distinguishable. Melody Shimada notes that although exchanges between river basin agriculturalists and highland herders had occurred in previous periods, in this time period the first were gaining land from the latter, and ended up controlling the camelids in the Middle Horizon.

The data were confirmed in another site excavated by this same project: Huacariz, which was already mentioned, where large and small forms, probably corresponding to llamas and alpacas, were found (Shimada 1988:132, 134).

There is almost no information for the central highlands. According to Browman (1974a: 190), the Huari conquest of this region caused a profound crisis in society: herding became a secondary strategy for the exploitation of natural resources, while agriculture became the primary one.

In Ayacucho, we know almost nothing about camelids in the Middle Horizon occupations of Ayamachay and Pikimachay except that they comprised 82.4% and 53.9% of the fauna (Wing 1986:257, Table 10.6). On the other hand, camelid remains were found in the excavations at Conchopata, in layers A, B, and C, which go from the end of the Early Intermediate period to Middle Horizon Epochs 1 and 2 (according to the chronology of Menzel 1968), and comprised 21%. Their analysis showed that their consumption was indiscriminate, since both adults and young animals were included in the faunal remains and were the major source of protein for the people (Pozzi-Escot 1985:120).

Wheeler's research (1986:291–292) on material from sites located in the Colca Valley, east of the town of Coporaque (Department of Arequipa, province of Cailloma, district of Coporaque), shows that the primary source of animal food came from Camelidae, mainly the domestic llama (*Lama glama*), and occasionally from wild vicuña (*Lama vicugna* [*Vicugna vicugna*]) and perhaps the wild guanaco (*Lama guanicoe*). In strata dating to the Middle Horizon, adults constituted 72.7% of camelid remains, young animals 19.7%, and fetuses/newborn animals 7.6%. The predominance of adult animals apparently means that an effort was made to keep the animals alive for as long as possible.

Wheeler notes that the pattern can be taken as an index even though the data are scant, in that the main use of adult llamas was as beasts of burden, which prevailed over their use as meat. This would explain the low frequency of remains from young animals and is supported by ethnographic data. The low frequency of fetuses/newborns (7.6%) perhaps suggests herds were efficiently managed, thus lowering the mortality for animals in this age group, but it must not be forgotten that this figure could be an artifact of the sampling method.

4.6.2 The Coast

The data for the coast in the Middle Horizon are far more abundant. When discussing the Middle Horizon in Lambayeque (mainly with data from the Batán Grande and Pampa Grande group), Izumi Shimada (1982:172) notes that remarkable continuity from previous periods can be perceived, particularly due to the economic significance that domestic llamas had. Remains from young and old animals were found in the excavations.

With regard to this same area, Shimada and Shimada (1985:18) discuss the study by Wheeler (1982[b]) in which she posits that camelid species can be distinguished by the shape of their incisors. Kent (1982[a] and personal communication to Shimada and Shimada) had apparently shown that this method was not always applicable. All teeth in Lambayeque are of the llama/guanaco type when assessed by Wheeler's methodology (1982b). However, the incisors are not always well preserved, and the possibility that alpacas were present cannot be ruled out. Although the llama cannot be distinguished osteologically from the alpaca, the scope of their functions that can be deduced archaeologically suggests these were llamas.

The careful management of the herds of domestic llamas is deduced from the number of animals killed, or that died, from the beginning of the Middle Horizon on. For Shimada and Shimada (1985:18), it is hard to imagine that the hunt of the wild guanaco could have supported the population deducible from the existing urban centers (see Schaedel 1972). We know that at this time, agricultural activities were intense and extensive, and the survival and management of any big herd of animals therefore had to be under human control.

Shimada and Shimada (1985:14–15) specifically report for the Huaca del Pueblo that 14 sacrificed camelid fetuses/newborns were found in association with the Lambayeque style (which they call Sicán) at the Batán Grande complex. There could well be more, for the area has not been fully excavated. With regard to the site of Sapamé, which is also within the Batán Grande area and has an occupation that extends from the Middle to the Late Horizon, Izumi Shimada (1982:173) indicates that during the excavations Melody Shimada observed a stratum with llama dung mixed with algarrobo (*Prosopis* sp.) leaves and fruit. Algarrobo seeds were found in dung pellets. This shows that the llamas were fed with fodder present in the locality.

Shimada and Shimada (1985:20) believe that caravans of llamas carried minerals to Batán Grande from the highlands.

In his study, Izumi Shimada (1982:173) mentions an unpublished report on the work done at Huaca Chotuna (Department and province of Lambayeque, district of San José), which Melody Shimada gave Christopher Donnan in 1980. This report also shows that llamas predominated as the main source of meat.

I do not know the exact location of Huaca Julupe, which is also in the Department of Lambayeque, to the northeast of La Leche River. According to Shimada and Shimada (1985:15), the monument corresponds to the end of the Middle Horizon and the beginning of the Late Intermediate period. There are traces of camelid sacrifices.

In this same department (in the province of Chiclayo, district of Saña) is Pampa Grande, an important site studied by several specialists. The

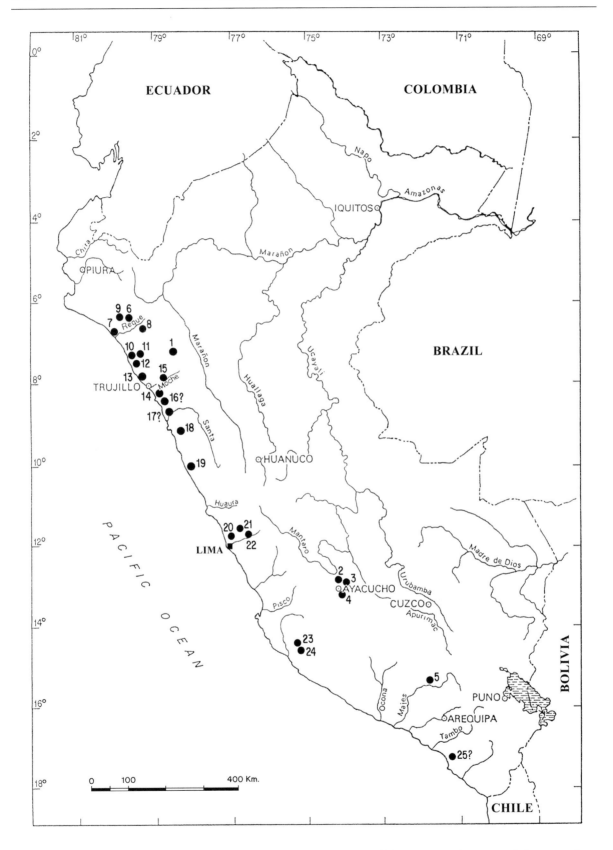

FIGURE 4.5. Locations of archaeological sites belonging to the Middle Horizon where camelid remains have been found. The site legend is on the facing page. *Prepared by the author and drawn by Osvaldo Saldaña.*

occupation of this great urban center dates to the last phase of Moche, or Moche V. Izumi Shimada (1982:161) considers that the presence of llamas at this site had a "critical economic significance."

In fact, all data both from explorations and excavations show that camelids predominated in organic remains in terms of number, volume, and weight. All bones in the skeleton are also well represented, which shows that the animals were butchered at or close to the site. The age ranges from fetal/newborn to adult, and shows that 43% of the animals survived for 42 months, with a diminution between 24 and 30 months, which might "reflect maintenance until adult weight was attained, at which time their function and use were determined." This could mean diversified use, such as for transportation and for food. Study of the remains is not always easy, as it sometimes is difficult to determine the sex of the animals. Furthermore, it is evident that sacrifices of a religious kind also took place, as we shall see later (Shimada and Shimada 1985:13).

The analysis of refuse shows that camelids were the animals most used by the people of Pampa Grande, and that their bones were used to make tools. We know from ethnographic data that a significant percentage of the bones were lost, and it is also possible that this happened in pre-Columbian times, perhaps due to dogs and guinea pigs. This is one factor that should not be overlooked when interpreting archaeological evidence (Shimada and Shimada 1981:38).

There are, however, some problems regarding the significance of the camelids at Pampa Grande. It was initially estimated (Shimada and Shimada 1981:38) that these remains made up approximately 87% of the total volume of bones analyzed. A qualification was added, specifying that "[t]his is a conservative figure [because] Melody Shimada estimates that more than 90% of the unidentified bones belong to camelids." Izumi Shimada (1982:161) later noted that out of 5,007 identified bones, 4,345, or 86%, belonged to camelids. The figures are approximately the same in terms of weight, because out of a total of 12,086 g of bones, 10,678 g, or 88%, is from camelids. However, Wing (1986:259, Table 10.7) estimates 70.2% for the camelids. There clearly is some discrepancy in the figures, but either way, the numbers are significant and show the predominance of camelids.

Now, 50% of these remains were from young individuals and the rest from adults, respectively above or below three years of age. This suggests multiple uses, as was previously mentioned, with their use as food predominating (Shimada and Shimada 1981:38; I. Shimada 1982:161–162).

A contradiction thus arises in the data presented by Shimada and Shimada. They claim (1981:41, Table 8) that only 3% of the llama bones recovered bore marks of butchering, but Izumi Shimada (1982:162) later wrote that these marks are "frequent" on almost all parts of the skeleton. This discrepancy must be cleared up. This slight disagreement over the figures notwithstanding, I fully agree with Shimada (1982:161) that "it is difficult to deny that llamas were the primary protein source for the Pampa Grande population." It is worth emphasizing that camelid remains were found not just in refuse but also directly associated with houses (Izumi Shimada 1982:162).

As mentioned, there is evidence of camelid sacrifices at Pampa Grande, specifically in the

FIGURE 4.5. Middle Horizon Site Legend

1	Huacaloma	10	Pacatnamú	19	PV35-4
2	Ayamachay	11	Cañoncillo	20	Ancón
3	Pikimachay	12	San José de Moro	21	Huancayo Alto
4	Conchopata	13	El Brujo	22	Cajamarquilla
5	(East of Coporaque)*	14	Moche	23	Pacheco
6	Batán Grande	15	Galindo	24	Huaca del Loro
7	Huaca Chotuna	16	Virú 631	25	(Moquegua)*
8	Pampa Grande	17	(Santa)*		
9	Huaca Julupe	18	PV31-32		* Exact location not known

Huaca Fortaleza, the biggest monument in the urban complex. The burial of a boy with a camelid was found there, then a burial with three young camelids associated with two human limbs, and there is evidence of llama sacrifices in the post holes on top of the huaca (Shimada and Shimada 1981:52; 1985:13; I. Shimada 1982:162).

Even more remarkable are the discoveries made by Haas (1985:400–402), who excavated Huaca Grande, in the same complex. On the first platform he found 863 llama bones in one of the rooms. A pile of disjointed llama bones was found in the fill of the second platform. There were also concentrations of llama bones in four of the 14 rooms discovered, all disjointed, and in one case an articulated skeleton was found. Haas considers that the great number of llama bones present in a place where no evidence of domestic activities has been found clearly indicates ceremonial use. Interestingly, Haas (1985:407) determined that the llama bones recovered and studied belong to adult animals, and apparently no particular body part predominates. This differs from the findings at Huaca Fortaleza, where very young animals were sacrificed (Shimada and Shimada 1985:13).

Some large enclosures were found at Pampa Grande. Izumi Shimada (1982:156, 162–163) believes these are related to the llama caravans; he is convinced that much evidence supports the use of llamas as beasts of burden, and of caravans for transportation between sites. First we have the representations left by the Mochica. In addition, in Pampa Grande there are several main streets joining central areas with outlying ones that often end in open areas with no architectural remains or artifacts. Shimada believes these were the places where the caravans arrived. The rectangular enclosures around them with limited access served as formal sites of economic and social transactions related to the materials they carried, either loading or unloading them. A comparison of the wide avenues that led to the terminals shows the streets that converge around them are narrower and more winding. An important argument for this position is the central location of these terminals and their proximity to manufacturing areas (see also Izumi Shimada 1978. In addition, we know from Shimada and Shimada [1981:63] that Izumi Shimada also discussed this

subject in his doctoral dissertation [1977], which I have not read.)

Shimada and Shimada (1981:63) admit that these data must be reappraised in a wider context. We should not forget the location of Pampa Grande, which lies inland and had no intense interaction with the communities close to the shoreline to allow supplementary marine products to be obtained. It is conceivable that camelid raising was the major source of animal protein. Furthermore, the available data seem to indicate that all camelid products were used at Pampa Grande, just as in the highlands.

Izumi Shimada (1982:162) concludes that for all of its economic and religious roles, Pampa Grande evidently needed large herds of llamas that were raised and maintained in the vicinity. He believes this did not pose a problem, owing to the dietary and climatic tolerance of these animals, which allows them to consume a wide variety of fodder and live at different altitudes. In addition, Shimada considers that the evidence of cattle raising in Colonial times, which moved from the lower part of the valley to the upper part, shows this might have been done with the camelids: transhumance to procure fodder and pasture grounds. Here Shimada introduces an important caveat: the extensive deforestation, and its adverse effect on water retention, that has taken place over time up to the present perhaps gives a distorted view of reality, so that the present-day fodder and pasture grounds do not reflect their distribution and extent in pre-Columbian times.

Pacatnamú, a site already mentioned, has an important Moche V occupation. It was estimated that 20.89% of the tombs in the corresponding cemetery had camelid remains in them, between infant and juvenile animals (Donnan and Cock 1985:115–121). In addition, inside the North Courtyard of Huaca 1 there is a place called the "room of the offerings." A great number of offerings made with different materials were found here. There were also remains of human beings and llamas, all burned to a greater or lesser extent. Although the human remains seem to correspond to a secondary burial, the preliminary analyses "seem to show that at least one adult llama (3 years old) was sacrificed in situ" (Donnan and Cock 1985:75). This specimen exhibited polydactyly, which makes one assume that the animals were selectively crossed (Donnan and Cock 1985:117). The last report specifies that "[t]wo features of this camelid are noteworthy: its hind feet are four-toed rather than the normal two-toed, an occurrence related to inbreeding (Altamirano n.d.); and the bones are warped, indicating they were fresh at the time of the fire" (Cordy-Collins 1997:285).

The site of Cañoncillo, on the lower part of the Jequetepeque Valley (Department of La Libertad, province of Pacasmayo), also belongs to the Middle Horizon. Mammals contributed most of the animal protein, approximately half of which came from the llama. It is estimated that about half of these were young, but there also are many adults, which were used for wool and to work. Many bones bear cut marks caused by being skinned (Pozorski 1976:253).

San José de Moro, on the right bank of the Chamán River, in the Department of La Libertad, province of Chepén, district of Pacanga, is an interesting site still under study. I present the data courtesy of Castillo and Donnan. A tomb was found with a small complete camelid, and there were camelid bones in boot-shaped tombs. These are just crania and the lower part of the legs. There were at least two small camelids in another tomb with a child. The remains of these animals were also placed in tombs with subterranean chambers, at the feet of the main individual, some of which seem to have been complete animals. Two complete camelids were found in the tomb called M-U30, one at the feet of the coffin, the other one on top of it. There were also fragments of these animals cut into pieces and specially placed in the niches inside the chamber (Castillo and Donnan 1993 MS.:24–25, 29, 43). Castillo and Donnan comment that

> [t]he remains of at least three camelid specimens were found inside the chamber, as well as parts and fragments. These were found on top of the ceramic offerings and in between the crucibles, which determined their poor preservation. . . . The crucibles therefore appeared both on top and below the camelid bones, so some

were laid down before the camelid bones and others after them, from which we can deduce that during the funerary ritual, these two kinds of offerings were placed simultaneously. A complete adult camelid was found south of the chamber, lying on its left side over the coffin. Another camelid was found north of the chamber, between the north wall and the beams. This animal seems to have been lying on its flexed legs. The parts or fragments of camelids found were carefully cut and placed in groups of approximately the same size. The cuts of meat found in the chambers contrast with those in the boot-shaped tomb, for while legs and crania stand out in the latter, ribs and the upper part of the legs are given prominence in the former. This differential distribution reveals a stratified access to the meat of camelids, and therefore to proteins from this source. Not all the niches were full; some just had camelid bones and crucibles. Others had architectural models and metal miniatures. (Castillo and Donnan 1993:45; interested readers can find more data in a later report, Castillo and Donnan 1994)

A considerable number of burials belonging to the Lambayeque culture were found at Huaca Blanca, in the El Brujo complex, which I have already mentioned. According to Víctor Vásquez Sánchez (pers. commun., letter of 18 December 1992), each burial has at least two complete camelid crania with their front and hind limbs, and in some cases there were three or four crania. One burial had eight of them. All these crania were well preserved, with preserved skin, hair, and ears. They are so well preserved that even the eyelashes of the animals remained. Vásquez reports that the remains of the plants eaten are still on the teeth. This will probably allow the diet of these animals to be reconstructed with great accuracy. Segundo Vásquez (pers. commun., 15 April 1993) estimates that there must be the remains of at least 100 animals there.

Shelia Pozorski (1976:142, Appendix 1, 370, Table 30) confirmed that most of the meat dating to the Middle Horizon "Early Chimu" occupation of the Moche huacas (the Huaca del Sol and the Huaca de la Luna) came from domestic llamas. These in fact compose an estimated 37.6% of all animal remains, which is slightly less than half the animal protein consumed. This means that the total sample of llamas was not as large as that recorded for the Mochica occupation of this same site and did not allow an equally detailed analysis to be made (see also S. Pozorski 1979a: 169, Table 1).

On the other hand, we know that a cranium and the four lower legs from a large llama were found in one tomb in this same area that dated to the same time (Donnan and Mackey 1978:254–255).

An important archaeological site called Galindo, occupied during Moche V, lies on the border between the upper and the lower Moche River Valley (also in the Department of La Libertad, province of Trujillo, district of Laredo). This is one of the sites for which we have more data on camelids.

Bawden (1982:314), who studied the site, reports there was abundant llama excrement in several residential structures, and a corral for these animals was found close by. These structures usually include llamas at the site, and suggest that the functions related to their use were in the hands of individuals who lived there, in dwellings close to the corral. Bawden assumes these were under the control of a secondary branch of the elite.

Regarding another group with a large domestic residence, two additional rooms, and an enclosure, Bawden reports that it had "a large llama corral" at one end, and comments that its presence shows the way the goods stored in the adjacent rooms were transported (1982:316).

Bawden (1982:309) likewise studied a ceramic workshop and found some deposits with remains of cane, wood, and "a considerable amount of llama manure." This evidently was the fuel used to fire the pottery. In addition, close to the workshop is an enclosure that served as corral for the llamas, which were almost certainly used to transport the clay. Interestingly, several fragments of rope were found in the vicinity.

Shelia Pozorski (1976:130) studied the faunal remains and found llama excrement 20 cm thick in one of the enclosures studied by Bawden. Pozorski (1979a:169, Table 1) estimated that camelid bones constituted 69.4% of all animal remains, thus making it possible to state (S. Pozorski 1976: 129) that domestic llamas were the major source of meat for the people at this site. This was confirmed by the fact that a great number of the bones studied were burned. (Pozorski 1976:Appendix 1, Tables 16, 18, 20, 22, 24, 26, 28 [pp. 344, 347, 350, 354, 357, 361, and 366] shows the corresponding percentage of llama bones found in various excavations; in only one case was it below 50%, in another slightly above this figure, and in the remaining cases it was above 80%, with a single case of 95%. See also Pozorski 1976:147–148 and 1982a:181.)

Pozorski (1979a:176) believes that the research conducted at Galindo allows us to say that at this time, herds of llamas were maintained in the Moche Valley, just as had been established for the Early Intermediate period.

Pozorski (1976:129–136) also offered a tentative reconstruction of llama use at this site, based on their remains and focusing mainly on raising, meat processing, and meat distribution system. For this purpose she used data for age, cut marks caused by humans, the frequency of burned bones, and the kinds of bones that remained. According to her study, a significant relationship exists between food remains and specific house structures.

To reconstruct the age distribution and order in the herds, the age series was appraised in terms of epiphyseal fusion, following Wing's (1972:330) method. This work showed that about half the animals were killed for their meat and were adults older than those studied from the Moche site. Even so, a high percentage was butchered at the optimum age for a young animal. For Pozorski, the older group might suggest the use of animals used to carry loads and provide wool (see Pozorski 1979b:150).

The marks detected on the bones that are human in origin are similar to those found on remains from the Early Intermediate period. They suggest that the methods used to skin animals and disjoint their bones were efficient, which is probably why they lasted.

Three of the excavations indicate that within one residential structure there were separate areas to prepare food. A higher proportion of forelegs appeared in Cut 2, along with a moderate quantity of hind limb parts. Hind limbs were prevalent in Cut 3, and to a lesser extent the part close to the vertebral column. Foreleg parts were scarce in this case. The body parts most frequently found in Cut 4 were the skull, the vertebrae, the innominate region, and the hind limbs. This led Pozorski to deduce that the proportion of body parts used in each kitchen was different. On the other hand, the remains used in each unit varied, which might indicate some sort of hierarchy among social groups, perhaps extended families.

Other data show much simpler homes. Cut 1, for example, held only forelimbs, Cut 5 just the neck and shoulder girdle regions, and Cut 6 the upper body, showing a higher frequency of skulls, thoracic vertebrae, and humerus sections. Cut 7 emphasizes the lower part of the body, that is, the limb-lumbar-vertebral-pelvic parts, although some other elements also show the use of other body parts.

Some interesting observations were also made. It was found that the meatless parts of the feet were always missing the third phalanx, and if they appeared, it was only in a small percentage. On the other hand, toe bones were missing, perhaps due to the skinning process, which might have left them attached to the skin. A greater use of metapodials and the first and second phalanges as raw material for tool manufacture was also observed.

Finally, it is worth emphasizing that a large percentage of bones was burned. However, we must admit the possibility that some may have been burned after the meat was eaten and others burned incidentally during cooking.

I have already mentioned that camelid dung was used as fuel to fire the pottery. However, Shelia Pozorski (1976:130) managed to determine that in a large proportion of cases, this was also the fuel used in the homes. Concentrations of camelid manure were found close to several houses and seem to have been intentionally accumulated. There were also traces of dung inside the houses.

The data I have for this horizon in the Virú Valley are minimal. All I know is that there were camelid remains in a site called Virú 631. However, it is not known exactly in what proportion because the same source gives the proportion as both 8% (Wing 1977b:Table 6) and 1.7% (Wing 1986:259, Table 10.7).

At this point is is worth introducing some general considerations presented by Daniel Julien (1981:1) when discussing the area between the Chicama and the Virú River Valleys from the Middle to the Late Horizon. After analyzing the existing evidence, Julien admits that camelids might have been raised; however, he believes the data are still inconclusive. Referring specifically to the Virú Valley, he adds, "[t]he presence of great amounts of camelid excrement as well as bones gives rise to the possibility that raising activities perhaps took place in the abundant areas with salt grass;[6] even if this is still in question, it still is a possible alternative hypothesis, given the presence of this material" (D. Julien 1981:5).

During his fieldwork in the Santa Valley, Wilson (1988:255–256) found at least 16 "possible" corrals for llamas, five or six of which are beside the desert. The others, though farther away, are always connected to the desert and associated with the remains of roads that went up the valley and toward Chao. Wilson remarks that "considering the large number of probable corral enclosures found along the major roads and communication routes of the Early Tanguche [i.e., the early Middle Horizon] settlements system, it seems likely that llamas were widely used as a means of transporting goods from place to place in the valley as well as between separate regional systems along the coast."

I have no data for the nearby Nepeña Valley (Department of Ancash, province of Santa, district of Nepeña) save for a bit of information Proulx (1968:75) supplied for site PV31-32, close to the famous site of Cerro Blanco, which corresponds to the transition between the Middle Horizon and the Late Intermediate period. A tomb was found at the site with llama hair and wool.

I studied a small site close to the beach southwest of Los Gavilanes, in the Huarmey Valley (in the province of the same name). I call it PV35-4 (Bonavia 1982a:417). It was a temporary camp occupied by a small group of people who came from farther up the valley and spent just a few days here. Among other remains we found llama dung, which, to judge by its amount, must have come from a single animal. From this it can be surmised that trips of this kind were made with these animals, which were kept here during the travelers' visit.

I have no data for the sector from the north-central coast to the central coast, specifically between Huarmey and Ancón.

I know that Uhle discovered the foundations of old corrals at Ancón, with llama dung on the most inland part of the site (Menzel 1977:42). Menzel (1977:43) established that these corrals correspond to the first stages of the Middle Horizon. She remarks that this suggests that there was a flow of goods between Ancón and other areas, probably including the highlands too (Menzel 1977:42).

I also found the remains of llamas in a stratigraphic context that corresponds to the Middle Horizon, and verified that a variety of maize from the highlands was used in Ancón at this time, thus supporting Menzel's suggestion (Bonavia 1960: 202–203; 1962:73).

Dillehay (1979:28) studied the Chillón Valley and wrote about the site of Huancayo Alto (Department of Lima, province of Canta, district of Santa Rosa de Quives), remarking that a greater highland participation is perceived for the Middle Horizon. This is reflected in "a quantitative increase in camelid bones," pottery, and highland architecture.

Unfortunately, there are no data for the valley of Lima at this time. We only know of a reference by Maldonado (1952a:73), who reports having found a big corral almost 30 × 30 m in the ruins of Cajamarquilla "where the auchenids . . . left a thick layer of dung more than 40 cm thick." I have also seen camelid dung in the refuse in these ruins. It is known that the city was abandoned in Middle Horizon Epoch 2B, but there is also a late occupation, so it is hard to establish precisely to what period the dung belongs.

Menzel (1968:78) refers to the study Ronald L. Olson conducted at the site of Pacheco in 1930, on the grounds of the former hacienda Soisongo (Department of Ica, province and district

of Nasca). Olson excavated a room adjacent to the sector Tello excavated in 1927. There he found many llama bones in association with Nasca 9 and Chakipampa B ceramics, which belong to Middle Horizon Epoch 2B. The Huaca del Loro, which Strong studied (1957:36), is in this same department, on a tributary of the Nasca River, the Tunga (Department of Ica, province of Nasca, district of Marcona). In his report Strong says he found rooms full of sacrificial items, including the remains of llamas. He does not, however, give more details.

This is all the reliable information available for this area. Orefici (1992:96 and n. 21) claims that the only sites with evidence of penned camelids belong to this period and to the Late Intermediate period, but he does not present any evidence to support his claim.

To conclude the Middle Horizon sites, Goldstein (1990:36) vaguely mentions some Tiahuanacoid sites in the Department of Moquegua. These sites apparently date to Tiahuanaco V, or the late Middle Horizon, but it is not clear from the article. Goldstein notes the presence of camelids, "whose remains we found at all Tiwanaku sites."

4.7 LATE INTERMEDIATE PERIOD (AD 900–1440)

As was the case for the Middle Horizon, for the Late Intermediate period I have more data for the coast than for the highlands, but I will begin with sites in the highlands. (For the location of the sites mentioned, see Figure 4.6.)

4.7.1 The Highlands

Hocquenghem (n.d. [1989?]:115) refers to the highlands of Piura, which she defines as the inhospitable *páramo* above 3500 meters. She explains that herds of vicuñas found refuge in the *jalca*, and herds of alpacas and llamas were raised. The wet pastures below the *páramo* and along the San Pedro River must have been good for alpacas, which need soft, evergreen plants, while the tough grass on the slopes could provide sustenance for llamas, which need a tougher grass (Hocquenghem's references are to Flores Ochoa

1985 [which does not appear in her bibliography. This must be a mistake and is probably my 1975a], 1977, 1988; Palacios Ríos 1977 [my 1977b], 1980; Brougère 1980; Gundermann 1988; María Fernández, pers. commun.).

Hocquenghem (n.d. [1989?]:116) notes that these animals must be penned in corrals at night to protect their young from foxes and condors, and protect all the animals from pumas and thieves. Hocquenghem considers that perhaps the main role corrals had was to collect the dung produced overnight, to be used as fertilizer later. She notes that remains of ancient corrals still exist in the highlands, close to the present-day hamlet of Cajas and at Los Altos, between Frías and Chalaco. On the other hand, by day the herds must have been driven to places where they could graze, the herders knowing well that pastures had to be made to last all year long. Ditches or stone walls were built to separate grazing zones. The remains of these divisions still exist in Los Altos and close to Cajas, where the peasants call them "pucaras." Hocquenghem (n.d. [1989?]:156, 162, Table 3) found historical evidence that "camelids were raised . . . in the lands of the Guayacundos of Caxas." These lived in the Jívaro lands of Cajas and Ayabaca (Department of Piura, provinces of Ayabaca, Morropón, and Huancabamba [Hocquenghem n.d. [1989?]:Map 7, 122]). The occupation probably dates to the Late Intermediate period.

I have already mentioned the Huamachuco zone, and specifically the Condebamba River basin, where McGreevy and Shaughnessy (1983:241) reported they were unable to find camelid remains. Their comment is for "later periods," so this might very well mean the Late Intermediate period or the Late Horizon. McGreevy and Shaughnessy explain that they found only two corrals, one in the *jalca fuerte* (3700–4200 masl) and another in the upper *quichua*/lower *jalca* (3200–3700 masl), where there is evidence of herding on a larger scale. "However, the evidence still indicates the *jalca fuerte* is not a major habitation location for these pastoralists. This again would predict habitation lower, at the interface of the *quichua*/*jalca* zones."

From the studies done by Parsons (1988; I was unable to read this study, which is cited by Dedenbach Salazar [1991:90]), it is known that

farther south, in the Department of Junín, there is a concentration of corrals in the punas of Tarma around Lake Junín that dates to the Late Intermediate period. It can be assumed that this animal resource was heavily exploited at that time. In addition, Bonnier (1986:102, 111, and n. 30) notes that camelid bones predominated in this time period at the site of Rakasmarka (also in the Department of Junín, province of Tarma, district of Palcamayo), at 3900 masl. She remarks that there were wild (vicuñas) and domestic animals.

Cuelap (Department of Amazonas, province of Luya, district of Tingo) can be included among the valley sites. Arturo Ruiz (pers. commun., 19 August 1992) notes that a great quantity of llama remains were found in the strata belonging to this period in the excavations he made.

For Ayacucho we once again have only very general information. MacNeish and García Cook (1981a:124) wrote that "camelid bones" were found at the site of Rosamachay on the Huarpa River, 2650 masl, in the strata with Rancha and Chanca ceramics that date to this period. This would "indicate that the people were perhaps herders or travelers who made a brief stopover in the cave." Wing (1986:256, Table 10.6) reports that camelid remains in the Late Intermediate period strata at the sites of Ayamachay (Ac 102) and Pikimachay (Ac 100), respectively, made up 35.5% and 52.9% of the faunal remains studied.

El Yaral is a particularly important site 50 km from the Pacific Ocean at 1000 masl, on the west bank of the Osmore River, about 20 km south of Moquegua (in the department of the same name, province of Mariscal Nieto, district of Moquegua). It is in one of the driest deserts in the world. According to data provided by Wheeler (1993:19–22, 23–24; 1995:285–286, 288) and Wheeler et al. (1992:470–471; 1995:834), perfectly preserved and naturally desiccated llamas and alpacas were found here from AD 950 to 1350. This presented a unique opportunity to study the physical characteristics of these animals before the conquest. The work was done by Don Rice in 1986, with the collaboration first of Geoffrey Conrad and then of Jane Buikstra (see García Márquez 1988; Rice et al. 1989; Watanabe et al. 1990).

The site of El Yaral belongs to the Chiribaya culture and was first occupied ca. AD 950, with

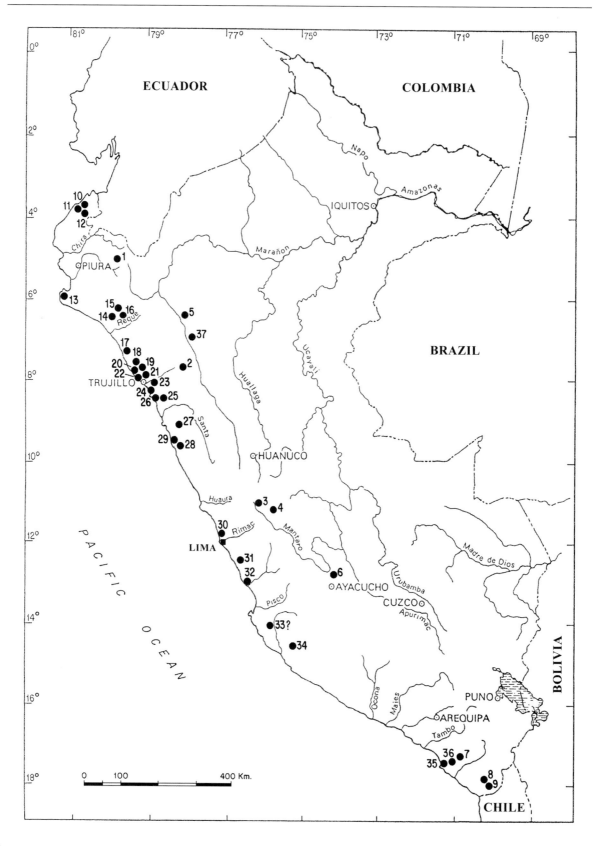

FIGURE 4.6. Locations of archaeological sites dating to the Late Intermediate period where camelid remains have been found. The site legend is on the facing page. *Prepared by the author and drawn by Osvaldo Saldaña.*

an occupation of less than 400 years (the radiocarbon dates obtained were AD 980, 990, and 1160). According to work by Bermann et al. (1989) and Stanish (1989), the Chiribaya culture developed near the present-day port of Ilo and in the Osmore River basin, without any interference from the early Tiahuanaco occupations. However, it was people bearing this culture who carried the herding traditions from the highlands to the coast, and laid the foundations of what would later become the Chiribaya culture. These scholars believe the llamas and alpacas of El Yaral descend from the original Tiahuanaco stock because there is no evidence that Chiribaya was in contact with other contemporary groups from the highlands. "This is particularly important because it means that not only do these animals predate the Inca empire, but also that they came from what both the inca [sic] and spanish [sic] conquerors considered to be the heartland of alpaca and llama production" (Wheeler et al. 1992:471). The site of El Yaral is large and has architectural remains. The excavations made in two of the biggest buildings uncovered burials of llamas and alpacas below the floor. Everything seems to indicate these animals were ritually sacrificed with a blow that fractured their cranium. They were then buried in shallow holes and covered with sand. The bodies turned into natural mummies, associated with the other offerings, because of the extreme dryness of the land. Wheeler et al. (1992: 471–473; 1995:835) explain that their exceptional preservation made possible the first systematic

analysis of pre-Columbian fibers from these animals. Skin and fiber samples taken from 11 standardized sites on the left side (six along the midline from neck to tail, four at mid-rib from forelimb to hind limb, and two midway down the forelimbs and hind limbs) of four alpacas and six llamas were studied. Special attention was paid to the diameter of the fiber to obtain data on the composition of the fleece, but microstructures were also studied with scanning electron microscopy and skin histology. Altogether, 200 fibers were analyzed (a standard measurement for the textile industry). The data permitted an assessment of the fineness and uniformity of the fibers and yielded the first direct information on the characteristics of the fleece in pre-conquest llama and alpaca.

The measurements of the llama and alpaca fibers of El Yaral revealed the presence of four distinct groups of animals. The data showed a natural variation in the diameter of the fiber across the body, with coarser samples coming from the neck and the legs. The finest fiber was found along the back, with the diameter tending to gradually increase from the mid-rib height on. Taken together, the eight samples correspond to the fleece, and represent the portion of the fiber that is shorn for use in textile manufacture. Based on eight complete samples per animal, the average diameter of the fiber for the two groups of alpacas was 17.9 (SD ± 1.0) μm and 23.6 (± 1.9) μm. The llama fleece likewise included a group with a finer fiber of 22.2 (± 1.8) μm, as well as a coarse fiber at 32.7 (± 4.2)

FIGURE 4.6. Late Intermediate Period Site Legend

1	(Provinces of Ayabaca and Huancabamba)*	13	Bayóvar	26	(Lower Virú Valley)*
2	Huamachuco	14	Huaca del Pueblo	27	PV31-29
3	(Near Lago del Junín)*	15	Cerro Sapame	28	Huaca de las Llamas
4	Rakasmarka	16	Cerro de los Cementerios	29	Manchán
5	Cuelap	17	Pacatnamú	30	Ancón
6	Rosamachay	18	Caracoles	31	Chilca
7	El Yaral	19	Cerro de la Virgen	32	Cerro Azul
8	Tocuco	20	Huanchaco	33	(Near Ica)*
9	(Near Tocuco)*	21	Chanchan	34	Huaca del Loro
10	Cabeza de Vaca	22	Loma Roja	35	San Gerónimo
11	Pirámide del Sol	23	Huacas del Sol y de la Luna	36	Estuquiña
12	San Pedro	24	Choroval	37	Patrón Samana
		25	V-313		* Exact location not known

μm. Analyses of variance indicated that the difference between these ancient groups was highly significant (Wheeler et al. 1995:835–837). According to Wheeler et al., the hereditability of the fineness of the fiber has not been established for llamas and alpacas, even though values of 0.22, 0.27, and 0.38 have been given for the weight of the fleece. However, it was noticed that the diameter of the fiber usually increases with age and the number of shearings. There are likewise signs that a rich diet can increase the diameter of the fiber, while a finer fiber can be obtained by keeping the animals in a restricted state of nutrition and during periods of physiological stress (Wheeler et al. 1995:837) (Wheeler et al. relied on studies by Chávez [1991], Bustinza [1991], Carpio [1991], and Frank and Wehbe (1994)].) The problem here for a comparison is that the size of the fiber's diameter in present-day llamas and alpacas varies considerably as a result of differences among the samples studied and the various measurement techniques used. Variations ranging between 9 and 88 μm have been reported for alpaca fibers, and between 8 and 144 μm for llamas. This indicates the presence of coarse, protective hair, which, though variable, is significant. A comparison of hair from the El Yaral animals with hair from contemporary specimens shows that alpaca hair is finer in the archaeological specimens. The extra fine Chiribaya wool, defined on the basis of the remains from El Yaral, measures 10–14 μm smaller in diameter than fiber from *huacaya* alpacas of the same age, while the fine Chiribaya wool is between 4.7 and 8.5 μm smaller in diameter. The diameter of the fiber in the six El Yaral animals showed the existence of animals with fine and coarse fiber. The one called fine Chiribaya on the basis of archaeological remains has fiber on average 22.2 ± 1.8 μm in diameter. The comparative samples from present-day animals "vary from 18.8 μm undercoat/39.8 μm hair in the more heavily fibred *chakus* to 20.1 μm undercoat/73.1 μm hair in *ccara*, increasing to 22.0 μm undercoat/ 42.2 μm hair in *chakus* to 25.2 μm undercoat/77.7 μm hair in *ccaras* at 24 months" (here Wheeler et al. quoted Macquera 1991). Wheeler et al. explain that although these figures cannot be directly compared with those for El Yaral, the ancient fleece without hair is as fine as the modern ones with the same characteristics. It is interesting that abundant coarse hair is found in all contemporary samples, whereas only one of the five ancient llamas exhibited fine, visible hair in the fleece. Two of the animals had very fine hair, and two were single-coated with no perceptible hair in the fleece. The sixth animal showed a contrast, for it had a coarse hairy coat with an almost equally coarse undercoat. The diameter of the fiber in this animal is 32.7 ± 4.2 μm, larger than that seen in contemporary llamas (Wheeler et al. 1995:837–838).

On these grounds Wheeler et al. conclude it is possible there were three varieties of domestic camelids at El Yaral that yielded different fibers. They include the alpaca's fine and extra fine Chiribaya fiber, and the llama's fine Chiribaya fiber. They add that it is usually believed that llamas were used only as pack animals. However, in some communities these animals are selectively bred to produce fiber. (For this, Wheeler et al. accepted the studies by Dransart [1991a, 1991b]. However, Dransart does not touch on this point in the first study, and furthermore he is emphatic in claiming that "at present it is not possible to distinguish the fleeces of different South American camelids" [Dransart 1991a:316]. I was unable to check the second study, which is a dissertation.) The fiber from fine Chiribaya llama suggests that this practice, now nearly gone, has very ancient origins. Wheeler et al. accept that various environmental factors can affect the fiber, so the existing data do not suggest that this kind of variety ever existed (Wheeler et al. 1995:838).

Nevertheless, Wheeler et al. present some interesting arguments in an attempt to prove there is some evidence that the animals from El Yaral were selectively bred. First, there is the presence of uniform characteristics in the fleece that are not found in present-day llamas and alpacas. Moreover, the existence of a single llama coat with a uniformly fine fiber at El Yaral shows the presence of a variety or breed of animals that is now unknown. Second, at El Yaral the fibers have a uniform color. There are only two multicolored specimens. The llama's fine Chiribaya fiber includes a pure white animal, two red-brown, and two gray/beige. We know that in Inca times, pure-colored llamas were chosen for religious rituals. The alpaca remains of El Yaral include animals with brown and white hair, black, brown, and a vicuña-colored one. Finally, the third argument presented by Wheeler et al. is that there are some hints that in past times the fiber might have grown faster than in present-day animals. In this regard they compared the data relating age and fiber length in present-day animals (based on Cardozo [1982], Bustinza [1991], and Chávez [1991]) with the same kind of data obtained for the El Yaral remains. Wheeler et al. admit that the present comparative sample is not fully valid, as the sample comes from animals raised at high altitudes. Wheeler et al. in fact note that there is a generalized belief that some relationship exists between altitude and the fineness of the alpaca fiber (they relied on studies by Carpio [1991] and Bustinza [1991]), because the quality of the pastures decreases with altitude and the stressors induced by the environment become far more severe. However, Wheeler et al. admit that the relationship between these factors and the diameter of the fiber has yet to be shown. The comparison was not carried out with alpacas because these animals were shorn before being sacrificed (Wheeler et al. 1995:837–839).

At Chiribaya Alta, another site contemporary with El Yaral, 140 llamas were excavated in funerary contexts. A preliminary study shows the animals in locally raised herds were selected for certain qualities, and animals not considered useful or fertile were eliminated, so as to obtain an output of fiber of a quality not found nowadays (Wheeler et al. 1995:839).

The data from El Yaral tend to show that present-day animals have coarser fiber.

> In contrast to the pre-conquest alpacas and llamas from El Yaral, today's animals are characterized by a lack of uniformity. Coarsening of the fiber, increased hairiness, and increased variation in fiber diameter across the fleece of huacaya and suri alpacas, as well as the apparent disappearance of fine fiber llamas, can almost certainly be explained by extensive hybridization produced by events of the conquest. Taken together, the four pre-conquest breeds cover the range of fiber diameter measurements in today's animals, and crossing with the hairy Chiribaya coarse fiber llama could have produced this re-

sult. The probable role of hybridization in the evolution of today's llamas and alpacas remains to be studied, as does the possibility that a genetic bottleneck occurred with the massive destruction of native Andean livestock during the 16th century. (Wheeler et al. 1995:839)

As Wheeler et al. (1992:472–473) had previously said, "a breakdown in controlled breeding between the fine and extra fine El Yaral breeds would alone account for the variation observed today." Furthermore, "[t]he most probable cause of coarsening and hairiness in both huacayas and suris would be through hybridization with the coarse fiber llama breed, a not improbable scenario amid the chaos and destruction of the conquest. Clearly, however, such a process would not have affected just the alpaca gene pool. The El Yaral mummies indicate the possibility that extensive crossbreeding between alpacas and llamas may have occurred during the sixteenth century and has played a much more important role in the formation of today's livestock than has been realized" (Wheeler et al. 1992:472–473).

This certainly is one of the most important discoveries made in this field. Some of the questions raised will be answered once their study is complete. Even so it is worth bearing in mind, as Wheeler et al. suggest, that "there is a clear and urgent need to locate and protect possible relict population of pre-conquest alpaca and llama phenotypes" (Wheeler et al. 1995:839).

To finish with the highlands, Trimborn (1975: 23, 25–26) reports that to judge from the amount of bones and dung found on the surface and in excavations, llamas were evidently raised at the site of Tocuco, in the Caplina Valley of the Department of Tacna (in the province of the same name) at 1340 masl (this was confirmed by Kleemann 1975:92–93). Trimborn (1975:26–27) mentions another site in this same valley "beyond Tocuco," where llama remains and dung were found. Although its chronological position is not clear, the site seems to belong to this period.

4.7.2 The Coast

There is much more information for the coast than for the highlands. Altamirano (1987:44) reports

Chimu sites around the city of Tumbes (in the department and province of the same name, in the district of Corrales), such as Cabeza de Vaca, the Pirámide del Sol and San Pedro, which are just 4 km from the sea. A lot of camelid bones were found on the surface of these sites. The sites apparently belong to the Late Intermediate period. The same investigator (Altamirano 1987:40–43) discusses the site of Bayóvar, on the Illescas Peninsula (also in the department and province of Piura, but in the district of Sechura), where, he notes, *Lama glama* constituted 47.53% of the bone remains analyzed from the Chimu occupation. He specifically mentions 37 individuals, 13 of which are "infants," 14 juveniles, and 10 adults. Altamirano suggests that the 13 "infant" animals, constituting 35.13% of the total, indicate "mortality," but this is a normal proportion for captive animals. On the other hand, it indicates the high frequency of animals classified as juvenile and "infant" that are between seven months and four years of age. Altamirano explains that the remains included skulls, mandibles, and phalanges of animals that were certainly killed on the spot, and correspond to at least five individuals. Camelid meat was eaten roasted, boiled, and as charqui. However, the evidence used to reach these conclusions is not presented. Altamirano notes that the fact that adult animals account for the smaller percentage (27.02%) of the remains indicates that llamas were used for transportation. Further, the huge amount of coprolites at the site indicates that llamas were raised in the vicinity. Altamirano ends by stating that the presence of camelids of various ages implies that the occupation was continuous. This conclusion is supported by the abundance of llama dung at all levels. It is a shame that Altamirano did not publish more on the subject, because his data are invaluable.

In an overview of the Lambayeque Valley, Shimada (1979:14) notes that "camelid remains predominate in sites located both on the coast and in the highlands," and points out that there are "different species" in the sierra. When trying to explain the problem posed by maintaining camelids, Shimada and Shimada (1981:63) state that a review of the environmental variations in the Lambayeque Valley, both in time and space, indicates there was once a vast expanse with relatively succulent vegetation. For this reason Shi-

mada (1977) and Mackey (1977) posit intervalley transhumance, according to seasonal variations.

Let us now examine the actual data for the Lambayeque sites. First, we know that a remarkable amount of camelid dung, in places almost 3 m thick and 5 m wide, was found in the Huaca del Pueblo de Túcume (Shimada and Shimada 1985: 15). This monument corresponds to the Late Intermediate period. The presence of llama coprolites was confirmed by Víctor Vásquez S. (pers. commun., letter of 17 July 1992), who also noted that dung from these animals was found at Huaca 1 in this same complex in the excavations undertaken by Hugo Navarro, although the cultural association is not clear.

The Cerro Sapame site is on the confluence of the La Leche and Lambayeque River valleys. The remains of juvenile and adult llamas were found, as well as a thick layer of their dung. There is a layer 50 cm thick, with a lot of excrement mixed with algarrobo seeds (*Prosopis chilensis*) and a little bit of soil. This layer is repeated in several parts of the site. Everything seems to indicate that it is material derived from cleaning the corrals, and that the animals were fed algarrobo. Camelid dung was also found at a lower level but mixed with stalks, leaves, and maize cobs (Shimada and Shimada 1985:15).

Izumi Shimada, Epstein, and Craig (1982) studied Cerro de los Cementerios in the La Leche River valley, close to Batán Grande, which dates to this period but was also occupied during the Late Horizon. Metal was worked here, and the authors believe it was transported by caravans of llama. A considerable quantity of camelid remains was recovered in Sector IIIe. Shimada, Epstein, and Craig believe the big male llamas, capable of carrying loads in excess of 45 kg, were used to transport the minerals. They also found a great rectangular stone enclosure close to the Cerro Blanco mine that was empty and lies close to the road connecting several of the site's sectors. This might have been used as a corral for the llamas (Izumi Shimada, Epstein, and Craig 1982: 959, n. 23).

More than 1,500 animal remains from this metalworking center were analyzed; 81% belong to camelids, which would therefore constitute the primary protein source. Interestingly, the bones were found in all metalworking contexts, but most were in small cavities in the floor, between the workshops, and were mixed with remains of plants and other animals. This apparently is the refuse from meals consumed by the people who inhabited the site.

Although there were camelids of different ages, the survival curve shows that around 60% lived more than 3.5 years, which implies they were used for more than food. The curve does not exhibit an abrupt fall, which could perhaps reflect selection at a given age, but also gradual death or killing. This supports the hypothesis that the animals were used to transport metals. The raw materials were brought from a place 3 km away, but the finished materials could have been carried much farther. There also is a network of roads that connects the sites in the area (Shimada and Shimada 1985:15).

Donnan and Cock (1984:66; 1985:117–121) specifically refer to Huaca 1 in the preliminary reports of the work they carried out at Pacatnamú. The presence of three llama sacrifices was detected in the walled complex adjacent to this huaca on the south side. Most of the animals were infants, light-colored, and carefully wrapped in textiles. The animals were only 2–3 months old according to Altamirano (1984; cited by Donnan and Cock. I was unable to obtain this report). Another shallow burial of a camelid 3–4 months old was later found on the summit of this same huaca, which dates to the end of the Chimu occupation of the site. Only the head and limbs were buried.

A llama burial was found in the West Complex of this same Huaca 1. In this case, the head and the limbs were also buried. Its age was estimated at between 2.5 and 4 years, and it has polydactyly. Finally, the complete skeleton of a llama 2 months old and wrapped in textiles was found south of Huaca 36 in the course of excavating a stratigraphic pit. All these burials had offerings associated with them.

On the northern side of the Moche Valley (Department of La Libertad, province of Trujillo, district of Huanchaco) is a secondary site called Caracoles. The occupation dates to the Late Intermediate period and the Late Horizon. Llama remains were found, but these are not too impor-

tant, as they constitute only 3.9% of all the fauna (S. Pozorski 1979a:169, Table 1; 1982:186).

Also in the Moche Valley is a Chimu occupation on the Cerro de la Virgen, just 5 km northwest of Chanchan (province of Trujillo, district of Huanchaco). Llama bones, coprolites, and wool were found here. Although llama provided the major source of meat from one single species, its animal protein constituted only 35.9% of the total meat consumed (S. Pozorski 1975:224; 1976:178; 1979a:169, Table 1; 1982:188).

Donnan and Foote (1978:399, 403, 405) describe 17 tombs of children found in Huanchaco (in the district of the same name in the province of Trujillo) that date from the Late Intermediate period. "[E]ach . . . was associated with the remains of one or more young llamas." In an area of 600 square meters, almost every unit excavated showed remains of llamas or of llamas and children, indicating that this kind of burial was quite common in the area. Donnan and Foote note that all the llama skeletons showed evidence of antemortem injuries. Although there are some difficulties in identification, Donnan and Foote tend to think these remains are *Lama glama*. The fact that the skeletons are not fully developed and have formative dentition patterns indicates these are young animals. All were approximately the same age at the time of their death, which Donnan and Foote believe was between six and three months. They suggest these were sacrifices.

At Chanchan, the Chimu capital city, there is very clear evidence that camelids were used at the time the city was full of life. We know, for instance, that, to judge by the refuse, llamas were the major source of protein for the inhabitants during a reoccupation of the Rivero compound ("Ciudadela"), for they constituted 55.1% of all animal remains (S. Pozorski 1976:164; 1979a: 169, Table 1). On the other hand, a great number of llama bones were found among the refuse discarded by looters in Unit BB, adjacent to the Laberinto and Tello compounds. In addition, llama burials with their bones still articulated were found on the floor and on the bench of one room. Another burial of a llama was on the upper part of a platform. Topic states that "the significance of the llama to Unit BB was as great as or greater than was the case at Huaca Prieta de Guañape or Aspero" (J. Topic 1982:159–160).

The remains of a great many llamas lay below the floor in an area close to the west side of the platform in the Laberinto compound, in the front courtyard. Llama bones were also found close to the other platforms. Conrad, who worked here, notes that a "considerable number of llamas" were killed for sacrifices, and these remains were then buried close to the aforementioned platform (Conrad 1982:100, 103).

Thomas Pozorski (1979:135–136; 1980:242) describes the burial platform of Las Avispas, in the southeast sector, outside the northeast corner of the Laberinto compound. He refers to the lateral Courtyard B on the west side, which was pillaged by looters, and notes that hundreds of complete or almost complete camelid bones that probably correspond to *Lama glama* lay on the surface. Everything suggested they formed part of funerary contexts and were not food remains. Unfortunately, no intact tomb was found, but from the number of bones it can be surmised that the llamas were sacrificed on the platform when the individuals were interred. Human sacrifices also took place here. Camelid remains are rare outside Courtyard B.

Topic reports a domestic complex in the city, one to which caravans went. He mentions two of them found in the Bandelier compound. He notes that the two complexes were located in the center of the site at the terminus of the transportation network. The one most intensively studied (J. Topic 1990:Fig. 13) had a communal kitchen, large corral-like rooms, a platform filled with llama burials, and rooms with multiple sleeping benches (J. Topic 1990:Fig. 14). The other complex appears to be similar. They were capable of housing about six hundred people. They undoubtedly housed exchange specialists. Some artifacts suggest exchange with the sierra or beyond. "Along with other exotic goods, it is likely that two essential raw materials, alpaca wool and metal ingots, were brought into Chan Chan by caravan" (J. Topic 1990:161; see also p. 164).

Shelia Pozorski and John Topic found a significant amount of llama dung in stratified deposits in the marginal neighborhoods of Chanchan (S. Pozorski 1976:154). It was likewise established that a great part of the area occupied by the marginal neighborhoods was meant to be used as a corral for the llamas, as shown by the remains of dung just mentioned above (J. Topic 1980:273). On the other hand, a section of the floor in the kitchen of each dwelling was found to have been formed by heavily trampled llama dung. This means the animals were also kept in the houses (J. Topic 1982:151).

The area where foreigners lived in these marginal neighborhoods was also identified. These had big, corral-like rooms and a huge number of llama burials (J. Topic 1982:167). Topic (1978:6) also identified remains of a small group of inhabitants of the marginal neighborhoods who probably spent just a short time in the city, whom he believes were merchants. Topic found them in two places, and in each case there was a small associated mound. One of them was excavated and found to have a fill of sacrificed llamas. According to Shimada (1982:165, who read Topic's dissertation, which I have not), a macaw (*Ara* sp.) and *mishpingo* seeds[7] associated with the llama remains were also found here.

In these marginal neighborhoods there was a series of platforms. In one of them John Topic (1980:277) found several bones of young llamas. The tombs of complete llamas were found on the floor and the earthen bench of a room. There were also llama skeletons below the room, on the top part of the platform's fill. Topic believes the platform with the llama tombs was probably an altar. Shelia Pozorski (1980:182 and Table 1) established that most of the meat eaten in the marginal neighborhoods was llama meat, with a range of 55%–80%. (Shelia Pozorski [1979a:169, Table 1] gives 65.7% in another study, which I assume is an average.) This is supported by the high percentage of llama bones recovered in the excavations. The bone samples analyzed come from John Topic's excavations as well as from Pozorski's. Big, stratified accumulations of llama dung were found in them, which had been placed mainly in the artisan sector. The llama bones recovered often made it possible to reconstruct the process of butchering, the supply systems, and even the taming techniques. Not all the units held the same type of remains. The front part of the animal predominates in some units, with the hind limbs in second place. The skull and the spinal column are well represented. In others, all body parts are found, with a slight emphasis on the hind section. There are also foot segments with no flesh, while the lower part of the foot is missing the third phalanx, which probably fell off during skinning. The metatarsus and the first and second phalanges are well represented. The fact that there is a low incidence in the marginal neighborhoods of burnt bones, which originally had had meat, suggests that most of the food was cooked in vessels (see also S. Pozorski 1976:153–156, 375, Table 32; 380, Table 34; 385, Table 36; 390, Table 38; 1980: 191 and Table 1; 1982:182). We also know that the mean age of about half the animals sacrificed was that of old llamas, which suggests they were important in Chanchan as pack animals and fiber producers. At the same time there is also a high percentage of bones from young animals, which evidently were sacrificed at the optimum age for eating (S. Pozorski 1980:182–183, 189, Table 1). It is worth pointing out that the supply pattern for meat in the marginal neighborhoods of Chanchan is closely correlated with that of the earlier urban centers of Moche and Galindo (see above). The slightly older age of the remains at Chanchan perhaps reflects an increase in the use of the herds for wool and as pack animals (S. Pozorski 1979b:150).

Shelia Pozorski (1976:199; 1979a:180) notes that the data obtained from agricultural sites in the area close to Chanchan and in the capital city itself indicate llama butchering was the officially approved and easiest way to get meat in the Late Intermediate period. The major source of animal protein used at Chanchan is identical to that used in the earliest sites. She notes, however, that the same exact resources were not used in the satellite administrative communities, but some llamas were always sent to them.

Finally, personnel from the Universidad Nacional of Trujillo excavated an artificial mound in the Chanchan sector called Chayhuac. The mound, called Loma Roja, is located beside the road (the *vía de evitamiento*) that leads to the beach resorts of Huanchaquito and Buenos Aires. The remains belonged to Middle and Late Chimu (Víctor Vásquez S., pers. comm., letter 17 July 1992). Segundo Vásquez and Víctor Vásquez S.

(1986:n.p.) report that abundant camelid dung was clearly visible in the stratigraphic profile, where it formed thick, brown-colored layers. For Vásquez and Vásquez, this means the camelids had not been taken to the coast temporarily but were being raised close to the littoral instead and fed mainly on salt grass ("grama salada," *Distichlis spicata*).

Besides the coprolites, an upper mandible with its teeth, two metapodials, and phalanges belonging to *Lama* sp. were identified (the animal was identified as a llama through an analysis of the incisors following Wheeler's methods [1984a]). Interestingly, these remains "come from Layers 'L' through 'A,' and reinforce our conclusion on the acclimatization of these camelids to the coast" (Vásquez and Vásquez 1986:n.p.).

Llama bones were also found in Chimu graves at Moche, between the Huaca del Sol and the Huaca de la Luna (Donnan and Mackey 1978:381).

At the site of Choroval, between Las Delicias and Salaverry, south of the Moche River (also in the province of Trujillo, district of Víctor Larco Herrera), there is an occupation that dates to the Late Intermediate period. It was estimated that here camelids constitute 17.8% of the remains from the animals used (S. Pozorski 1979a:169, Table 1).

Farther south, in the Virú Valley, Willey (1953: 319) reports finding llama burials in a cave in the middle part of the valley (V-313) that dates to the La Plata period—that is, to the Chimu occupation.

Daniel Julien (1981:12) reports the presence of camelids between the Chicama and Virú valleys after the Middle Horizon and says that several interpretations can explain this. They might have been highland animals periodically taken to these sites by people from the highlands who moved down to exchange products. It is also possible that they lived on the coast, in which case they might have been brought there in small numbers as pack animals or in bigger herds, and were also used as a source of food and fiber. Julien cites West (1978), who claims that although inland sites along the Virú Valley have camelid remains, these are much more common in the lower part. Julien explains that the salt grass ("grama salada," *Distichlis spicata*) close to the beach could have fed the animals all year long, with quite low maintenance needs.

He also claims that ethnographic data collected by West (1978) from present-day fishermen in the Virú Valley support this model, because they now frequently raise animals such as sheep, goats, pigs, and/or ducks.

Here it is worth recalling that to account for the presence of camelids among the Chimu, Troll (1935:139, quoting a study by Latcham [1923]) claimed they "had land in the highlands to raise llamas." This has not been proved and reflects the bias this scholar had against camelids living in the lowlands.

For the Nepeña Valley we only have the data Proulx (1968:71) gave for site PV31-29, close to Cerro Ceylán (Department of Ancash, province of Santa, district of Moro). The occupation probably dates to the Late Intermediate period. Llama hair was found in the refuse.

With regard to the Casma Valley, it is worth explaining a remark made by Horkheimer because it may be misleading. In regard to the great number of llamas sacrificed in pre-Columbian times, note 10 reads as follows: "Tello, 1940, p. 608 assumes that the huge piles of llama bones in the 'Huaca de la Luna' (almost 13 km east of Casma) are the remains of sacrificed animals" (Horkheimer 1973:59). This footnote did not appear in the original version (Horkheimer 1958:27), while the reference to Tello did appear but was 1942, not 1940. On the other hand, this version does not mention the "Huaca de la Luna." This is a mistake that was probably introduced by the translator, since it is the Huaca de las Llamas. This reference and the correct name were added to the second German edition (Horkheimer 1960:43), which was the basis for the Spanish edition. The location of the site is correct, as it is on the right bank of the Moxeque River, 1 km east of Huaca Moxeque (Department of Ancash, province and district of Casma). Tello did not, however, mention this name. With regard to "the auchenids," Tello wrote: "[T]he until now mysterious accumulation of bones from these animals, most of them burned, in one of the buildings located close to the Temple of Moxeque, in the Casma Valley, would not correspond, as local legends claim, to the droves of llamas loaded with gold to ransom Atahualpa which were buried there by the Indians on hearing the news of the death of the Inka, but to lla-

mas sacrificed perhaps in pre-Columbian ceremonies" (Tello 1942:19–20). Here Tello does not mention what period these remains might belong to. However, in a study published after his death (Tello 1956:291) he indicates the presence of a huaca called "Las Llamas" while discussing the "Ruins of El Purgatorio" in the Casma Valley. Then he explained that the monument was "Sub-Chimu" (what is now Chimu), and notes that the *huaqueros* (looters) made a ditch "which laid bare many llama bones" (Tello 1956:303).

Manchán, an important Chimu site, is in the coastal desert south of the lower part of the Casma Valley (still in the province and district of the same name). Camelid bones were found here. Altamirano (1983a:65) says these are frequently of a reddish brown color, some black and others burnt white. This is why he deduces that all were roasted and some burned. There are, however, inconsistencies in Altamirano's report. He claims to have identified *Lama glama* and *Lama pacos*, and notes that young llamas predominate, but gives two different figures: 65.24% and 64.39% appear on the same page. Altamirano also says that to judge by the bones, many were used to prepare charqui (freeze-dried meat). The total amount of camelid bones at this site would be 11.01%. He adds that "commoner people had a diet based on the consumption of camelid and fish meat" (Altamirano 1983a:70). Altamirano (1983a:71) claims that these remains show "the high consumption and use [of camelids] during the late pre-Hispanic periods." However, he gives the total figure for these animals as 11.09%, whereas 11.01% appeared on the previous page and in the table. Based on his study of the mandible, Altamirano claims that this was *Lama glama*. Nonetheless, when discussing the camelids he later says that "infants" accounted for 6.77% of remains, juveniles (between 1 and 2 years of age) for 81.54%, and adults for 11.68%. "The llama was the major source of meat in the human diet, possibly in relation with the exploitation of the lomas and the use of the settlement to raise domestic herds" (Altamirano 1983a:72–73). Altamirano in fact insists that the llama and the alpaca were raised "at the same site and in the lomas" (Altamirano 1983a:73). Using the data furnished by Altamirano, we can determine that he used the bones of 38 individuals. He insists that young animals were used the most, and that their meat was stored as charqui (Altamirano 1983a:73). It is a shame that there are so many inaccuracies, and that the evidence to support these assertions was not presented.

In a report on the central coast, Cohen (1978c: 27) writes that to judge by the remains found in the refuse, domestic animals became important quite late. Llama bones and coprolites begin to appear in the Late Intermediate period, and it was only then and in the Late Horizon that "we have evidence for a diffused herding of flocks in the lomas region." This assertion is clearly wrong in the light of what has already been seen. On the other hand, in regard to the Ancón area, the evidence of llama remains in the distinct stratigraphic context which was pointed out for the Middle Horizon, continued into the Late Intermediate period (Bonavia 1960:202–203; 1962:73).

There are almost no data for the area immediately south of Lima. Engel (1970c:21) discusses the lomas of Chilca in very vague terms, and says corrals were used during the occupation phase he calls Puerto Viejo, which corresponds to the Late Intermediate period/Late Horizon. These corrals have camelid bones and dung inside them.

The lomas of Chilca are also mentioned by Engel (1970c:27–29) in regard to the time the "Cuculí" occupied them, that is, during the Late Intermediate period. Engel explains that the abundant use of textiles made from animal fibers, characteristic of the dress worn by the "Cuculí," perhaps indicates camelid raising. He claims there is evidence that camelids were extensively used, but only for transportation and wool production, not for meat, because the Cuculí ate fish. The evidence for this assertion was the discovery of "corrals and tambos which might have sheltered the animals during journeys and transhumance; these tambos are full of guano and auchenid bones."

Wing mentions "Chilca" in two of her studies, in connection with the Late Intermediate period, but I do not know what sites she meant. In the first case (1977b:Table 15) Wing indicates a presence of just 0.1% of camelid bones and 0.2% in the second (1986:259, Table 10.7).

Cerro Azul (Department of Lima, province of Cañete, district of Cerro Azul) was clearly an im-

portant site both before and after the Spaniards arrived. Its major occupation corresponds to the period under discussion, the Late Intermediate. There was "llama excrement on the floor" in the large open-air courtyard (*canchón*) of a residential tapia structure in the southwestern part of the site. This excrement suggested that the "llama caravans coming from the *chaupi yunga* or 'mid valley' were received here." Besides the excrement in the canchón itself, there were llama bones among the food remains, albeit not in great number (Marcus 1985:4, 7; 1987:397). Marcus et al. infer that the people of Cerro Azul occasionally consumed llamas from the caravans that arrived at the site, and at the same time received charqui (freeze-dried meat), probably brought to the coast by these same caravans (Marcus et al. 1999: 6565–6566, 6568–6569).

Gilmore (1950:436) refers to the Department of Ica and notes that there is evidence that the "Chimac" (clearly Chimu) and the "Chincha" had pastures for their llamas in the highlands. However, his source is a study by Estruch (1943: 118). This publication is just part of Estruch's dissertation and was published without its bibliography. It is essentially comparative osteology, which I cannot appraise, but the historical sections are poor and full of mistakes. Estruch almost certainly used Troll (1935:139) as his source (who in turn quoted Latcham [1923], who wrote that "the Chimu and Chincha had land in the highlands to raise llamas"), because his dissertation was presented to the Universidad Nacional de San Agustín of Arequipa, and the Spanish version of the article by Troll (1935:139) was also published in the journal of this same university.

According to Gilmore (1950:436), the fact that the pastures are in the high-altitude zones is explained because artificial and natural pastures and fodder plants are not cultivated along the coast. He accepts the presence of llamas on the coast, but always beginning with the hypothesis that these animals moved down from the highlands, following the ideas presented by Maccagno (1932).

On examining the data collected by Uhle, Menzel (1977:13) found data on a burial chamber discovered close to Ica that dated to the Late Intermediate period. The remains of one or more sacrificed llamas were found here. Several llamas were in another tomb, and these had silver adornments, indicating their great importance.

We also know that Strong (1957) found evidence of a sacrificed llama during his excavations at Huaca del Loro in Nasca, in a context dating from the end of the Late Intermediate period, close to a room he called the Circular Temple.

For the Department of Moquegua there are data from San Gerónimo, close to Ilo (in the province of Mariscal Nieto, district of Ilo), just 200 m from the sea. There is an important Chiribaya occupation that dates to this period. Camelid remains are among the refuse. Jessup (1990:161–162) says the amount of bones from these animals caught his attention, and that these remains "do not just represent adults, but young animals too." Jessup also explains that camelid feet are commonly found in these graves; once he found up to nine llama skulls in one grave. On the other hand, all textiles in this area are made of wool, which indicates that "the auchenids were an important part of the overall economy, and that some, if not all, were raised on the coast."

A "great number of camelid bones" were found at the site of Estuquiña Calana, in this same area but in the Moquegua River basin, in the district of the same name and in the period Stanish (1990:129) calls Estuquiña. There also was "a similar abundance of camelids" at another site called P1.

Finally, there are two very vague but important sets of data. Wheeler (1991:27) wrote that "there is evidence that [llamas] were reared 900 years ago on the south coast, in the Moquegua region (Wheeler, unpublished manuscript)," and that "it is possible that [the alpaca] extended to the south coast (Wheeler, unpublished material) 1000 to 900 years ago" (Wheeler 1991:32). This was later confirmed (Wheeler et al. 1992; Wheeler 1995:287).

4.7.3 The *Ceja de Selva*

To complete the data corresponding to the Late Intermediate period I shall mention a site located in the northern *ceja de selva*, in the Department of Amazonas (in the southern part of the province of Chachapoyas, district of Chuquibamba). This is Patrón Samana, located in the mountains that

look east of the Marañón River, at 3750 masl. Llama bones with incisions on them were found in one of the houses here, whose occupation belongs, according to Schjellerup (1992:355, 357–358, 361), to the "thirteenth century."

4.8 LATE HORIZON (AD 1440–1539)

There is no question that the number of camelids was extremely high at the time of the Inca Empire, but there is no way of knowing whether there were more animals than before. As Tschudi correctly noted, "[a]lthough female llamas usually have just one infant, just like all other species of aukenias, these multiplied on a large scale owing to the extreme care taken in dealing with the herds, and despite the fact that a large number of animals were used, either as sacrifices or as human food" (Tschudi 1885:98; 1891:100; 1918: 212–213; 1969:129). Flannery et al. (1989:117) asked why the herds were bigger in Inca times than at present. First, they suggested that the flocks of individual communal villages (*waqcha-llama*) were not significantly larger than the ones found today. They also suggested that some *ayllu* flocks might have been aggregated so that the herds of 50–100 families grazed together.

The biggest herds belonged to the state, the Church, and the curacas (*qhapaqllama*). These herds were managed with a herding strategy completely different from the one they studied in Ayacucho in the 1970s. The state separated the animals into groups, so that males, females, pregnant females, and females suckling their young (*uñas* [but Gonçalez Holguin 1952 gives *yuñalla* for the lambs]) were separated. In addition, the temple animals that were destined for sacrifice were kept segregated by color. The state management of llamas was gradually disrupted, starting with the Spanish conquest.

Flannery et al. (1989) also note that the biggest flocks the Spaniards described were usually found on the vast flat or slightly rolling parts of the puna, such as the altiplano. However, it is worth noting that the llamas reported by the Spaniards did not share their lands with sheep. Today, a sector in the puna with 100 llamas and

100 sheep could almost certainly hold more llamas if the sheep were removed.

Finally, Flannery et al. (1989) state, just to give an example, that the Inca flocks did not face the same morbidity factors the authors documented in Ayacucho. In Inca times, hundreds of pumas, foxes, and wildcats were eliminated with the *chaku* (see Chapter 7 [7.3.1]). It likewise seems that cattle rustling was then quite rare. Finally—and I discuss this more later—these scholars suggest that the mite (*Psoroptes equi*) was unknown before the Spanish conquest and was introduced by sheep. Camelid mortality due to pumas and other animals, rustlers, and mange must have been lower in Inca times. By this, Flannery et al. (1989:114) do not mean that raising camelids had no problems in the Inca Empire. On the contrary, they were many, such as frosts, intestinal parasites, and the stress of altitudinal change. What Flannery et al. want to show is that the obstacles encountered then were not the same as those encountered today, and that the mortality factors pointed out by the herdsmen in the Ayacucho communities they studied, Yanahuaccra and Toqtoqasa, were not as significant in Inca times.

On the other hand, we know that the Inca introduced camelids into all areas where these had been missing, and even gave pastures in the nearby puna to the people of areas newly incorporated into the empire when they had none (Murra 1978:90, 93).

It also seems that in the Inca Empire, llamas had a higher economic and religious significance than alpacas (Franklin 1982:468). As Flores Ochoa (1990b:86–87) explains, the royal insignia of the Sapa Inka was a white llama called *napa* (for its whiteness). The animal walked in front of the ruler when he moved through the streets of Cuzco. In addition, the Inca held in great esteem the officials in charge of managing llamas (*llamakamayuq*), with whom he met during the festivals to participate in their dances, such as the *llama llama*, or dueting with the llamas, as can be seen in the drawing by Guaman Poma de Ayala (1936:Fig. 318; my Figure 4.7). And whenever there was a drought, a (black) llama was tied up and given no food or drink so that its moans would attract the rains (see Guaman Poma de Ayala 1936:Fig. 254; my Figure 4.8).

Flores Ochoa (1990b) explains that depictions of alpaca are few in Inca (and pre-Inca) iconography, in contrast to the abundant depictions of llamas in various media. Even the heavens of the Andean constellations have llamas but no alpacas. "The Milky Way, the immense river of Andean mythology, is crossed by llamas. The eyes of the biggest ones are the stars Alfa and Beta, in the western constellation of Centauri" (Flores Ochoa 1990b:87).

For more effective control, there were hunting grounds belonging to the state and the church in the highlands (Murra 1978:84). The famed *chaku* were also held, where a great number of people participated over wide expanses, and in which the killing of animals was controlled. Later I review the data the Spanish chroniclers left on this point. (The reader may find a review in Franklin [1982:468], but this ecologist only used secondhand sources and did not cite a single chronicle.)

Llamas were protected by the state (Squier 1974:7), and we know that the counting and inspection of the flocks took place within a ceremonial context. The royal initiation rites took place in November, and camelids played an important role in them; on this date, a census of all flocks belonging to the state and church was made. This coincided with a ceremony and sacrifices held throughout the empire that were meant to increase the size of the herds. Special ceremonies were held for this purpose, and the best herders received prizes. The royal mummies were also asked about the well-being of the flocks in coming years (Murra 1978:102).

Murra (1978) notes that at present it is almost impossible to make a quantitative comparison of the flocks belonging to state and church. There were several church flocks, carefully separated by color because these belonged to different cults—those of the Sun, Thunder, and several sanctuaries. Murra notes that according to one source used by Román y Zamora ([1575] 1897), more than one million llamas were dedicated to the sun. Although this estimate is uncertain, it is quite possibly close to the truth, considering the dedication with which the *quipu camayoc* carefully kept their records. (Murra [1978:102] notes that the same estimate reappears in Murúa [1962–1964], but since the wording is almost identical it may

FIGURE 4.7. This drawing by Guaman Poma de Ayala refers to "[t]he Naricza-arani festival of the Inca. In the festival they sing and dance *naricza* with a *puca* [red] llama. They sing with a slow beat [*conpas*] at the pitch of the ram [*al tono del carnero*] for half an hour, saying y-y-y, and the Inca begins to sing like the ram at his pitch [*al tono*], and then they begin singing the couplets [*coplas*]." *After Guaman Poma de Ayala (1936:fol. 318).*

FIGURE 4.8. Drawing by Guaman Poma de Ayala that mentions the month of "October" and shows "*Uma raymiquilla,*" the "procession that asks water of god . . . in this month another hundred white rams were sacrificed to the major huacas, idols, and gods so they would send water from heaven, and they tied other black rams in the public square and did not give them water so they would help plead." *After Guaman Poma de Ayala (1936:fol. 254)*

be presumed that the Mercedarian friar copied the figure from Román y Zamora, whose book was published in 1575, or that both authors had a common, unknown informant.)

Moreover, it is also known that llamas were used on a large scale in Inca times to carry loads, and that great caravans continually traveled across the Andes (Rowe 1946:219; for more details see Dedenbach Salazar 1990:168). The respective data left by the Spanish chroniclers will be reviewed later. Franklin (1982:467) made a statement in this regard that can be misleading: "Hundreds of thousands of llamas were employed in silver and gold mining." This clearly is a mistake because although these animals were no doubt used in mining in Inca times, no data can indicate numbers. Meanwhile, there is much

information for the Colonial period, and it is certain that Franklin was confused.

I only intend to show some aspects of the use of camelids in Inca times. Readers who want more detailed data can see the study by Dedenbach Salazar (1990:225 et passim), where an ethnohistoric and linguistic analysis is given on the role these animals played in the life of the Andean peoples in Inca times.

According to Dedenbach Salazar (1990:174–176), camelids were reared in Inca times more for transportation and wool than for meat. She even states that little camelid meat was eaten. We certainly lack data in this regard, particularly from an archaeological perspective, a field where almost nothing is known, but still, this does not seem true in light of historical data. It is true that the guinea pig (*qowi*, *ccoui*, or *cuy*, *Cavia porcellus*) was one of the most important sources of meat for the Andean peoples in this period (Rowe 1946:219), but I believe that camelids also played a major role.

As regards carrying loads, no one can doubt that llamas were used for this task on a large scale. However, there is one aspect that needs to be studied more, as noted by Hyslop (1984:302–303): loads were also carried by human bearers, who could carry the same or heavier weights than llamas and could be more easily managed, as llamas are bad-tempered. However, Hyslop admits that it is still uncertain whether the backs of men or camelids were more responsible for the movement of goods in the Inca Empire. I return to this point when discussing the caravans.

As Tello (1942:11–12) wrote, those who reared llamas and alpacas in high-altitude areas were at the same time herdsmen and hunters who lived off the resources derived from cattle raising and from exchanging products like wool, meat, skins, and textiles for lowland agricultural products.

Raising these animals did not just mean looking after them, as at present, or searching for pastures and water. It was a far more complex process because it was carried out on a large scale, much as agriculture was in the lowlands. For example, raising alpacas requires special pastures in marshy areas, which are created by diverting waters from the ravines to the plains and mountain slopes to flood extensive areas. Besides, the llama and alpaca herders knew the puna in all of its topographical and climatic aspects. They were thus in a position to choose the best pastures and the most sheltered areas to put up their dwellings and corrals to provide protection from vultures and felines, the two greatest enemies of the herds. They were also acquainted with the periods when epidemics would break out and how to fight them, and the way to efficiently select animals as beasts of burden, for meat, or for wool. This herdsman knew how to determine the most optimal season for breeding. He noticed that llama gestation took nine months, much like that of man, and to ensure reproduction he separated females from males and only let them mate in certain periods, so that their offspring could enjoy the best time of the year, the highland summer,

which is warm and has the most advantages for survival, while avoiding winter because of the intense cold and the lack of soft and adequate pastures (Tello 1942:11–12). Moreover, the use of llamas varied considerably, depending on whether they were state animals or from the local *ayllus* (see Flannery et al. 1989:114).

Camelids had a most significant role in religion, as I have already noted, but this is a subject I do not intend to discuss because it is so vast that it deserves a study by itself. Interested readers will find the essential data in the study by Dedenbach Salazar (1990:181–183 et passim). (Tello [1942:12] also provides a good description of a 1790 fertility ritual.) However, it is important to remember that llama sacrifices were an important part of religious rituals not just during the days of the empire but also before it (Menzel 1977:13). Thousands of llamas were sacrificed in the Capacocha, a ceremony held in the Aucaypata plaza in Cuzco (Rostworowski 1988b:79).

But the Inca Empire mainly used state flocks for military purposes. The military animals carried provisions and other loads, and in case of emergency they themselves could be used as food (Murra 1975:140). As Troll (1958:30) correctly wrote, "[t]he Inca armies were accompanied by great herds of llamas. As beasts of burden and as meat supply, they provided the army with shock troops against other peoples who could not do this, and who could only get meat by hunting." (See also Tschudi 1885:104, 1891:105; 1918:223; 1969:135; Troll 1958:30; Flannery et al. 1989:114. And I must once again include Dedenbach Salazar [1990:177–180], who studied this subject extensively.)

There are conflicting opinions about camelid flocks on the coast, and llamas in particular. Some, like Izumi Shimada (1982:163), believe that "their herding and breeding was possible on the coast and [was] most likely practiced" (Shimada refers in general terms to late periods), even suggesting that llama caravans could have been connected to the trade in guano and algae (*cochayuyo*) from the coast (I. Shimada 1985:XVI). Novoa and Wheeler (1984:124) are also emphatic: "State-owned herds of llamas were maintained on the coast by the Incas until the time of the Spanish conquest." Murra (1975:132) in turn believes the flocks were moved each year from the highlands to the low-lands and vice versa, to exchange highland prod-ucts for coastal ones. However, we should ac-knowledge that Murra primarily refers to the southern part of the central Andes.

I believe there can be no doubt that a signif-icant number of camelids lived on the coast dur-ing the Late Horizon, but there is no real evi-dence of an Inca state policy for establishing flocks on the coast. I found no data on this sub-ject, and Franklin Pease is of the same opinion (pers. commun., 26 May 1992). With the data so far presented, it is clear that there was a signifi-cant number of these animals on the coast before the arrival of the Inca. The Inca must have con-tributed to their expansion, but everything seems to indicate that what led to it was not a conscious and definite policy.

When describing the llama, Gilmore (1951: 435) states that in "Inca times" there were "small-er breeds" on the Peruvian coast and in Ecuador. Gilmore does not give his source for this point, and I have not found any save for some archaeo-logical data related to Ecuador.

Before beginning the study of the archaeo-logical data we should mention that Hyslop (1984:302) believes that to some extent, llamas were responsible for the distance separating the *tampu* (roadside lodgings), because it is less than the distance a person can walk. According to Hys-lop, this shorter distance was because the llama is delayed because he is browsing and grazing along the journey. Corrals that were possibly used for the llamas are associated with the Inca *tampu* from west-central Argentina to Ecuador. This shows that during the Inca Empire, camelids were commonly used for transportation. However, the lack of archaeological data is really astonishing, considering the great importance camelids had in the Inca Empire. The Late Horizon is the least known period, and one to which archaeologists should dedicate their endeavors. Let us review the evidence.

4.8.1 The Highlands

I begin with the highland sites. (For the location of the sites mentioned, see Figure 4.9.) As I have mentioned several times, McGreevy and Shaugh-nessy (1983:240–241) traveled from the province of Huamachuco in the Department of La Liber-tad to Santiago de Chuco, a district to the south-west of Huamachuco, without finding evidence of intensive camelid raising. They have confirmed the presence of two very large, late pre-Colum-bian corrals, one in the *jalca fuerte*, the other in the upper *quechua*/lower *jalca*, which are evidence of large-scale herding. However, they indicate that the *jalca fuerte* was not the most important dwelling place for herders, who probably lived in the lower area, on the border between the *quichua* and *jalca* zones.

In this regard, McGreevy (1984:170–171, 174, 193–194, cited by Theresa Topic et al. 1987: 833; I was unable to read this dissertation) sug-gests that the great flocks of camelids in the Hua-machuco area mentioned by Cieza de León (to be discussed in the following chapter), and which still existed several years after the Spanish inva-sion, were due to the Inca policy of increasing the llama population throughout the empire. Part of this policy was the introduction of intensive herd-ing in areas where this strategy had not previous-ly been part of the traditional pattern of land use.

Theresa Topic et al. (1987:833) insist that camelid bones are not particularly frequent in ex-cavations conducted in the area. They admit that herds of llamas were probably maintained in the fields of stubble and in pastures above 3800 masl, but here camelids seem to have been less impor-tant than in more southerly, contemporaneous societies.

There is a report by Wing for the famed Inca urban center of Huánuco Pampa, located close to Urqumayu, 12 km from La Unión and at more than 3800 masl (Department of Huánuco, province of Dos de Mayo, district of La Unión). It says only that camelid bones constituted 86.6% of the faunal remains analyzed (Wing 1986:254, Table 10.5). However, this same scholar pub-lished an exhaustive report some years later that has valuable data (Wing 1988). Llama and alpaca bones predominate among the remains of Huánuco Pampa, as expected. The cultural asso-ciations of the samples indicate specialized state activities, in terms of the type of architecture where they were found (pers. commun., Craig Morris to Wing). Nonetheless, it is interesting that no evidence of animal rearing, such as corrals, has been found at the site. Wing clearly says that among the remains at Huánuco Pampa, one cannot expect to find more than one segment of the breeding practices carried out by the Incas. The faunal remains studied were excavated in Zones II, III, and V of the site (see the plan in Wing 1988:Fig. 176). On examining the plan, these correspond to the central part of the city.

The sample Wing worked with comprised 33,513 bone fragments and 8,836 identified teeth. The identifiable component represents 26.4% of the total sample for the three zones. The percentage for each zone is 27.3%, 26.3%, and 25.6%, respectively. There is little difference among these zones. This suggests that the bones were subjected to several destructive factors before and after being deposited, as their fragmentation is such that the samples cannot be identified.

Among these remains, camelid bones predominate, constituting 84% of the samples in the three zones. In Zones III and V they represent more than 88%, while in Zone II they make up 51% of all samples identified, and 67.1% of the remains if European domestic animals are excluded (see Wing 1988:Tables 1–3, 172–173) (Wing 1988:167).

The remains of European domestic animals are scarce, less than 3% in Zones III and V, but they are abundant in Zone II, where they constitute 24%. These remains were not considered, given the objectives of Wing's study, even though there was a small period in which the Inca and European occupations overlapped.

Elizabeth Wing wrote: "The remains of herd animals, camelids, are overwhelmingly abundant at Huánuco Pampa leading one to conclude that they were central to that part of the economy concerned with animal use" (Wing 1988:168).

Wing explains that she used two methods to distinguish the four camelid species, but "with limited success" (Wing 1988:168). One of them was based on the differences in size between the two big species, the guanaco and the llama (Wing 1972). The other method took into account the morphological differences in the incisors (Wheeler 1984[a]; see Wing 1988:Fig. 2, 177).

Naturally, the starting point is that most of the camelids at Huánuco are domestic, and this assumption is based on documentary evidence.

Using Browman's study (1974[a:194]) we can expect to find some mixture with wild camelids, given that in ancient Peru, domestic and wild animals were caught in the *chacos*. (On these occasions some domestic animals escaped and returned to the wild.) Deer were then killed to get meat, while the vicuñas were shorn of their fleece and then set free, and some guanacos were captured and included in the flocks of llamas. On the basis of the methods indicated, we get approximately the same number of llamas and alpacas, if we accept that most of the animals at Huánuco Pampa were domestic.

On the basis of body measurements, 55% of the remains fall into the range for llamas and 45% within the range set for alpaca. If we follow the morphology of the incisors, 44% were llamas and 56% alpacas. (These are conflicting results and show that one or both methods are incorrect. I draw the reader's attention to this point, which I discuss later.)

On the other hand, the samples that are identifiable to the species level are too small to give an indication of the specialized use of these animals in the different zones of the city.

Wing (1988) therefore states that one conclusion from the whole assemblage is that approximately the same number of llamas and alpacas were butchered, even though their specialized use differs— for fiber/wool on the one hand, and for transportation on the other. We should also bear in mind that pack animals carry their loads and then leave, not leaving behind any archaeologically recoverable traces (save dung).

An attempt was made to identify the habitat of these animals through the study of the phytoliths stuck to the teeth, which show the vegetation eaten (Armitage 1975). The phytoliths were scraped from the surface of the molars in four specimens and were analyzed by Deborah Pearsall. The four animals from which adequate samples were taken were probably alpacas. The phytoliths included 29% "Fustucoid [sic, for Festucoid] grasses"[8] and 43% dicotyledon types (Pearsall in a letter to Wing, 13 March 1980).

Wing presents these data without any interpretation. A working hypothesis is that the samples correspond to animals from herds that carried loads from the lowlands, and that they would

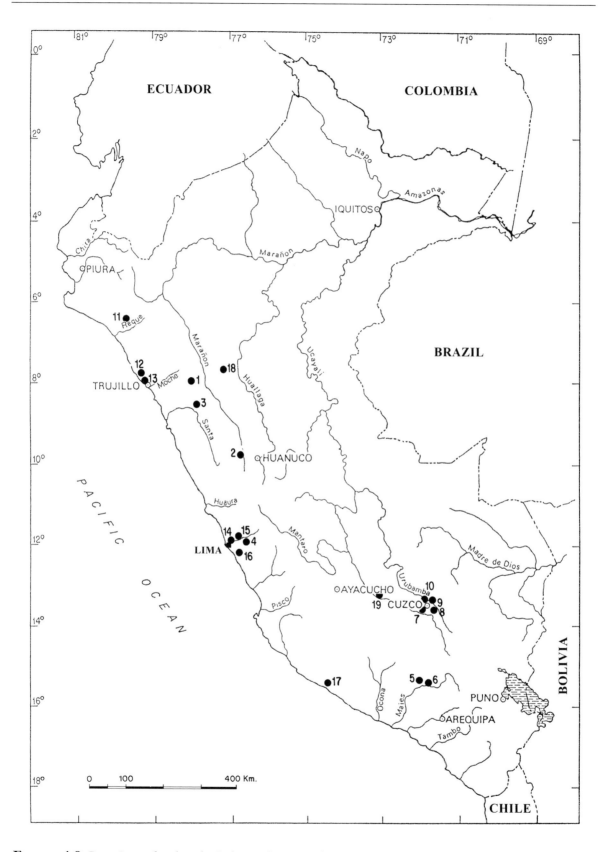

FIGURE 4.9. Locations of archaeological sites dating to the Late Horizon where camelid remains have been found. The site legend is on the facing page. *Prepared by the author and drawn by Osvaldo Saldaña.*

have phytoliths from that food in their teeth. But Wing admits that much more work is needed to reach valid conclusions.

The age distribution of camelids at Huánuco Pampa indicates a mature population. Age determination was based on the pattern of tooth eruption, on tooth wear, and on the substitution of 134 dental specimens in the sample. The biological model used was Wheeler's (1982[b]).

Only 4% of the animals studied were less than nine months old, and 92% of the population was alive at two years of age, 78% at three (see Wing 1988:Fig. 3, 177). Three years is the critical age, because it is then that the animals first give birth, trained as pack animals, and shorn. It is a well-known phenomenon that the young in most populations have a high death rate. This happens in flocks susceptible to bacterial infections in corrals (Wheeler 1984[a]). "The virtual absence of remains of these young animals suggests that the herd animals were not reared at Huánuco Pampa, but only brought to the site as mature, productive individuals. This is supported by the lack of corral features at the site" (Wing 1988:169).

Wing explains that when studying size in animals, there are two bones that often remain intact and can therefore be measured along several planes. These are the calcaneum and the astragalus. The pair of measurements that can be taken more often and clearly distinguish contemporary alpacas and llamas are the greater height (aa in Wing 1972) and the greatest breadth at the sustentacular process (b'b' in Wing 1972) of the calcaneum. These two measurements were taken in 121 specimens (Wing 1988:169, Table 5 and Fig. 4, 175 and 178). Most of the samples from the site, and all contemporary ones that were measured, conform to a single regression line. Some archaeological specimens deviate from what was expected. An element that is narrower than expected due to its height can be called gracile. Two specimens fall outside the confidence interval of the regression by being more gracile. In turn, six elements are robust, as they are broader than would be expected from their height.

Similar examples of deviation for the regression indicator are found in measurements made on the astragalus. These measurements are for the greatest length (c'c' in Wing 1972, or GLI in von den Driesch 1976) and greatest width (aa in Wing 1972; BD in von den Driesch 1976) (Wing 1988:Fig. 5, 179). Just as the measurements of the calcaneum fall close to the regression line, so do most of the measurements of the astragalus taken from the Inca site and from contemporary animals. Once again there are eight specimens that fall outside the confidence intervals. Six can be characterized as gracile and narrower than expected from their length, while two are robust and relatively broader than expected in relation to their length.

Most of the unusual specimens were associated with the walled compounds, but unfortunately, none of these astragali or calcanea can be directly associated with each other, so there is no way of knowing whether any come from the same foot. However, given that these two bones articulate with each other, it is possible that a gracile astragalus would articulate with a gracile calcaneum, and so they would be proportional to the body as a whole.

All gracile astragali come from the walled group, VB5. Five of the deviating calcanea and one robust astragalus come from another walled compound, IIIC4. Five of the specimens were found outside the walled compounds (IIB2, IIB5, IIIC9, IIID1a, and VA3) (see Wing 1988:170, Fig. 1, 176). Wing wonders whether these deviating

FIGURE 4.9. Late Horizon Site Legend

1	Huamachuco	8	Minaspata	15	Cajamarquilla
2	Huánuco Pampa	9	Pachatusan	16	Pachacamac
3	La Pampa	10	Wanakauri	17	Tambo Viejo
4	Sucyahuillca	11	Batán Grande	18	Ruinas del Abiseo
5	(Near Achoma)*	12	Médanos de la Hoyada	19	Pampaconas
6	(Near Coporaque)*	13	Chanchan		
7	Qhataq'asallaqta	14	Complejo Pando		* Exact location not known

specimens correspond to significantly different animals or whether these are aberrant data. Wing concludes:

> I believe that even though the number of deviating specimens is relatively low, the data from the measurements of these two elements support each other in respect to the type and relative frequency of the deviation and in the predominant contexts in which they were found. It must be remembered that these are probably remains of domestic animals which are by nature variable. Since the majority of these deviating specimens were found in association with specialized contexts at the site, it is possible that they may have come from special animals. They may represent animals that resulted from breeding experiments. Similar deviations must be sought in other Inca sites, as well as descriptions of unusually proportioned animals in the documents. (Wing 1988:170)

Wing concludes that although the Huánuco Pampa data conform in several respects to the measurements she would expect to find in faunal remains, there nonetheless are new data that will have to be verified with more studies of faunal remains from Inca sites.

The aspect of this faunal set that does satisfy expectations is the predominance of native domestic animals, particularly herd animals. It could have been predicted that the age structure of the flock would show fully mature animals, primarily including those used for transportation or for wool production. Although these findings were predictable, it is important to have confirmation on the basis of faunal remains.

On the other hand, this study provides information that cannot be predicted on the present state of our knowledge. It leads to a more detailed understanding of animal use in an Inca urban settlement. The problem of identifying camelid species diminishes at Huánuco Pampa because we can surmise that the animals are either llamas or alpacas. The analysis made with two different methods indicates these were used in equal numbers. However, establishing a correlation between any of the sectors in the city and the flocks detected is not feasible.

Among the bone remains there are some—16 in all—that significantly deviate from the rest. These specimens are markedly more robust or more gracile than average, and do appear mostly in walled compounds. Wing suggests they could be experiments in hybridization (Wing 1988: 168–171).

I would like to make two comments on Wing's work. The first refers to her study of phytoliths. Although it is true that the evidence provided by these studies is quite vague, it does not necessarily indicate that these are lowland plants, as Wing believes. *Ichu* (*sensu lato*) is a typical plant of the puna. And there are high-altitude plants among the dicotyledons. Actually, no interpretation is possible if more precise identifications are not made, at least to the genus level.

The second comment concerns the function of one of the areas in the city of Huánuco Pampa that Wing mentioned: IIB. Although it is true that there is not much evidence, according to the studies by Morris and Thompson (1985:79) it can be inferred that this sector was the home of *aklla*, weavers and chicha makers, but it must be admitted that this does not help at all in interpreting the data.

In regard to the Tarma-Junín area in the central highlands, we have already seen that here the use of camelids was quite common from early times. Wing (1972:336–337) has shown that their number increased during Inca times, and that there was an inverse relationship between large and small forms. There was a wide range in size between both forms. The larger ones are probably guanacos and llamas, and the small ones vicuñas. In statistical terms, camelid remains constitute 70% of the fauna analyzed (Wing 1972: 331, Table 3). (In a later study Wing [1986:257, Table 10.6] gave 86.4%.)

We know that in the Department of Puno, the Lupaqa exploited great flocks of llamas and alpacas in the cordilleras behind Chuquito (in the province of Puno), Pomata (in the province of Chuquito), and Zepita (also in this province). There are plentiful data for this area in historical sources.

Let us turn now to the valley sites. The site of La Pampa (Department of Ancash south of the province of Corongo, on the Manta River, a tributary of the Santa River) is at 1800 masl. According to de Macedo (1979:97–98, Table 1), a greater concentration of camelid bones is evident in the Caserones period occupation (Late Horizon) than in previous periods. The number of bones is not too large, for there are only 18 specimens.

I have no data at all for the area between the northern and the central highlands. According to Rostworowski (1978:43), Sucyahuillca (1410 masl), in the Department of Lima, in the province of Huarochirí (district of San Bartolomé), seems to have been a sacred place of the Yungas and to have been connected with the cult of Pachacamac. The llama herds belonging to the famed coastal oracle grazed at this site, close to the present-day town of San Bartolomé. Rostworowski says that it can be surmised that this was a Yunga colony established in the highlands to look after the flocks of animals needed for sacrifices at Pachacamac. This information, found in historical sources, is extremely important, but an archaeological study of this area is needed and has yet to be done.

In a footnote, Engel (1970d:11) mentions the "dry highlands of Chilca" and reports having found numerous camelid bones in corrals "that were used during the last pre-Columbian centuries." Engel does not present much evidence or actual data, as usual.

Farther to the south, in the Department of Arequipa (in the province of Cailloma, district of Achoma), and in particular in the Colca Valley, at Achoma, Shea (n.d. [1985?], n.p.) reports that in pre-Inca, Inca, and Colonial times this was a land of herders, who were also in contact with herders from the nearby puna.

Coporaque is also in the Colca Valley and in the province of Cailloma, in the district of the same name. Archaeological studies have shown that camelids were the primary source of food—mainly llamas and occasionally vicuñas, perhaps guanacos. Although the occupation dates to the Late Horizon, some doubts arise because the site is in disarray and may include remains from the Colonial period. Study of the remains indicates that 72.9% of the bones belong to adult animals, 16.4% to juvenile animals, and 10.7% to fetal/newborn ones. The low frequency of young animals would indicate that these were mainly pack animals, which had priority over meat animals. Further, the low frequency of fetuses/newborns suggests that the herds were efficiently managed, thus reducing the mortality for animals in this age group, even though this figure could also result from the way in which the samples were collected (Wheeler 1986:291–292).

The site of Qhataq'asallaqta is in the Department of Cuzco (in the province of the same name). It is at 3600 masl, on a hill southeast of Cuzco, on the city's outskirts. The site was occupied during the Late Horizon and early Colonial period. Camelid bones constitute 96% of the site's fauna, and their study shows that more than one species was used—at least llama and alpaca (Flores Ochoa 1982:69–70).

Twenty-three percent of all bones analyzed were not fused, indicating that they belonged to young animals, and just 2% had been killed before they were one year old. This shows that the inhabitants of Qhataq'asallaqta selectively killed those animals that were beyond their most productive age, both for fiber production as well as for transportation.

On the other hand, it has been confirmed that at this site the small and medium-sized camelids, such as the alpaca, were used more. It is not very likely that vicuñas were hunted at such a low-altitude site. From this we can deduce that alpacas were the animals most used by the people of Cuzco in Inca times (Flores Ochoa 1982:69–70; the data he uses were taken from Miller's dissertation [1979], which I have not read).

Minaspata is another site already mentioned; it is 35 km from Cuzco on the banks of Lake Lucre, at 3100 masl (in the province of Quispicanchis). It seems to have been continuously occupied from Early Horizon to Inca times (Flores Ochoa 1982:69–70). Camelid remains belonging to the Inca occupation come to 81.5% of the fauna recovered at the site (Wing 1986:257, Table 10.6.).

Flores Ochoa (1982:70–71) reports that near Cuzco there are at least two more sites with optimal conditions for large-scale llama and alpaca herding. "The first one is on the slopes of the sacred guardian mountain of Cuzco, the Pachatusan, where large corral structures, irrigation systems for pastures and abundant pastures are found. The other one is near the city itself, on the slopes of the mythical Wanakawri mountain (personal communication of Dr. Luis Barreda Murillo). We can also mention several large *kanchas* or corrals like those of Qoriqocha, less than ten kilometers from the city of Cuzco, which can house several hundred animals where a single llama is now hardly seen."

4.8.2 The Coast

Let us now review the scant data available for the coast. On the basis of their work at Batán Grande (Department of Lambayeque), Shimada and Shimada (1985:20) are convinced that during the Late Horizon, large caravans of llamas delivered minerals from the highlands and took away finished products along the coastal strip.

A site known as Médanos de la Hoyada (which mistakenly figures in publications as Medaños de la Joyada) is in the Department of La Libertad (province of Trujillo, district of Huanchaco), 14 km northeast of Chanchan. This is a cluster of sunken garden plots (locally known as *puquios*) quite close to the beach that is surrounded by deep midden deposits. The midden belongs to the Late Horizon, even though there are remains from a Colonial occupation, too. It was a residential area, and the presence of a significant amount of llama coprolites and bones has been confirmed (Kautz and Keatinge 1977:90).

We know that once it was no longer used at the end of the Late Intermediate period, the Ciudadela Rivero in Chanchan was reoccupied by small groups of people who settled in its front sector. This must have occurred after the city was conquered by the Incas. At this time, too, llamas were the main source of protein for its inhabitants. Camelid remains made up 55.1% of the animal remains studied in one of the excavations (S. Pozorski 1976:199; 1980:189, Table 1, and 191).

I was unable to find any data whatsoever for the coast between Trujillo and Lima.

For the Lima Valley there are the data in Ramos de Cox et al. (1974–1975:9). The information concerns a huaca called Corpus I in the Pando Complex, close to the campus of the Pontificia Universidad Católica del Perú. The text is reproduced verbatim here because it is quite garbled. It reads: "It is clear that the llamas used to transport goods were brought from higher altitude areas and were domesticated [sic!] for this end. The concentration of excrement in Corpus I made us think . . . it was a possible stop for the llamas used for transportation. We believe this proposal is highly likely, considering that this stop was located precisely in Corpus I and close to the Huaca La Luz. Was this the area dedicated to the more specialized production and its transport? There is also a possibility that saddle-bags (bags made of netting) were made for the llamas as part of the textile production." No other evidence is presented in the report, and there is no way to ascertain what period the huaca and associated midden correspond to. We can assume they are late.

In the Lima district of Lurigancho lies the famed city of Cajamarquilla, whose major occupation dates to the Middle Horizon but which remained partially occupied up to the Late Horizon. In the 1950s I saw enclosures with a significant layer of llama dung, and in some dwellings there were camelid bones. I assume these belonged to the last occupation of the site.

Maldonado (1952a:73) reports having collected a "significant amount of loose and ball-like auchenid dung" in the midden of the Temple of the Sun at Pachacamac (in the Lurín Valley, Department of Lima, district of Lurín). According to Maldonado, these remains came from the excavations by Strong. Although Maldonado does not give his reference, we can assume he means the work of Strong and Corbett (1943).

Finally, there are reports that Riddell and Menzel found the burial of a whole llama while working at Tambo Viejo, in an Inca site in the Acarí Valley (Department of Arequipa, province of Caravelí, district of Acarí; this is a manuscript by Riddell and Menzel dated 1954, and quoted by Donnan and Foote [1978:406]).

4.8.3 The *Ceja de Selva*

The data from the *ceja de selva* are interesting. It has been shown that the refuse in one of the buildings in the ruins of Abiseo (mistakenly called Pajatén or Gran Pajatén, and whose name is perhaps Yaro [see Bonavia 1990]), located at 2850 masl in the Department of San Martín (in the province of Mariscal Cáceres, district of Huicungo), included the remains "of a big auchenid, either guanaco (*Lama guanicoe*) or llama (*Lama glama*), but the remains studied, particularly the skull parts associated with teeth, certainly belong to the *Lama glama* species" (Macedo 1968:57–58).

Thirty camelid bones were later found in other excavations, seven of which belong to young animals, including a fetus. Cornejo and Wheeler (1986:n.p.; I obtained only part of this report) wrote:

> The identification made enables us to deduce that not only were camelids being used, but also that the breeding process was carried out within the Park [i.e., the Abiseo reserve]. The record of fetus and "nelmantos" [sic] shows it. It is therefore necessary to consider the adaptive process of camelids to this ecology. It is possible that some of the quadrangular structures identified as corrals (Deza 1973 [I am not familiar with this study. The corrals must be in the area of the ruins]) were intended to raise camelids, and this enabled camelid reproduction to take place within the adaptive process and crucial for the survival of these species. There is also evidence that some camelid bones had been worked and turned into artifacts.

Wheeler (1991:27) is then emphatic in saying that "there is evidence that [llamas] were bred . . . at the site of Gran Pajatén, located in the upper ceja de selva of San Martín" (Wheeler ms.).

After making the first investigations of the Ruinas del Abiseo in 1966 (Bonavia 1968:75–76) I noted, when discussing the camelid bones, that I believed these animals had not lived in the area and were the result of exchanges with neighboring highland peoples. It is possible that I was mistaken. Church (1991:21) excavated Building No. 1, where I had also worked, and he discovered camelid remains. He wrote:

> Only the collection of the Abiseo phase has been studied, but the record of fetuses and newborns suggests that camelids were raised in the vicinity of Gran Pajatén (Cornejo and Wheeler 1986). Although it was a preliminary analysis, restricted to

the late sample, the implications the control of camelid herds has for long-distance trade are obvious. The possibility that these animals were raised in what is now a tropical forest merits intensive research. Above all, it is worth establishing whether this area was always wooded during the period indicated, or whether some technique was used for preparing pastures within an ecological management system. . . . Camelids apparently formed a significant part of the region's diet.

The data obviously are of great importance, but it is essential that the final report be published, so that we know the exact number of camelid remains found and their context. However, the suggestion of a possible climatic change and possibly different flora when the Ruinas del Abiseo were inhabited should be taken with the utmost caution. There is no evidence of this. The only changes that took place, and those were on a limited scale, were made by man who cut the trees to build edifices and agricultural terraces (*andenes*).

Finally, and still within the *ceja de selva*, we have the data provided by Lyon (1984:6) for the province of La Convención in the Department of Cuzco, in the area of the Pampaconas River. "There it was said the Inca had herds around Pampaconas (Valenzuela 1906:108) which would have supplied both meat and fiber as well as sacrificial animals. Although they are not mentioned, there may have been other herds on the grasslands occupying the high ridges separating, for example, the Paucartambo from the Urubamba and Yanatili rivers."

This information is invaluable, and archaeological work should be undertaken in the zone.

4.9 ICONOGRAPHY

One topic not discussed here are the depictions of camelids left by pre-Columbian cultures, which can certainly provide valuable data. But these are different kinds of data from those dealt with in this book, which is why they are not considered. They require a separate study which I was unable to undertake. Here I limit myself to reviewing the most remarkable depictions I have seen personally or are in books, so that the reader can appreciate the quality of the data available. Ceramic representations were chosen because they show some realistic details. In addition, there are many camelids in rock paintings from various periods (e.g., Rick 1983:Fig. 54, 184–185; Bonavia 1972:Figs. 44–45, 137–138) (see my Figures 4.10–4.12) and in several other media, such as textiles (e.g., Harcourt 1962:Plate 31B; Tello and Mejía 1979:412).

Although I did not carry out specific research in this area, there is no doubt that most of the depictions of interest here correspond to coastal cultures from various periods. There must obviously be some among the materials from highland cultures, but apparently not with the variety and number of coastal specimens.

Perhaps one of the earliest depictions in ceramics was left by the north coast Cupisnique people (ca. 1200–600 BC). It is a modeled vessel in the shape of a llama lying down. It has a short neck (Larco Hoyle 1941:90, Fig. 122).

Among the modeled ceramic figures of the Vicús Vicús culture (ca. 300 BC–beginning of the Christian era) are several that are clearly llamas. These figures are general likenesses, with exaggerated anatomical details (such as the ears), and the short neck of the animal is striking (e.g., Gastiaburu 1979:95). There are several problematic pieces in this culture, making it difficult to know if they are really camelids. This is the case with some pieces in the Museo Nacional de Antropología, Arqueología e Historia of Lima.

The Museo Amano in Lima has a Gallinazo-style (ca. AD 1–200) vessel modeled in the shape of a llama that strongly recalls the Vicús Vicús pieces. It is an animal lying down, but carrying a load. However, the culture with possibly the most depictions of this kind is Moche (ca. 200 BC–AD 500). There are some beautiful ones among the pieces from Vicús that are labeled Vicús Moche, especially a llama lying down with a load of wood. The llama has a bundle tied to each side and a rope tied to a hole in one of its ears. Another example, also of an animal lying down, has some kind of saddlebag in which what appears to be a vessel has been placed, on each side. In this case there is no fastening of any kind. A modeled head of a llama is also remarkable (Gastiaburu 1979:95).

There are several modeled llama heads from the Moche culture in the Museo Nacional de Antropología, Arqueología e Historia in Lima. Some have a harness (see Instituto de Arte Peruano 1938: Lám. 17), others do not (Figure 4.13). There is also a series of representations of llamas carrying loads. Some have different kinds of harness, such as a halter (Figure 4.14); other have none (Figures 4.15–4.16). A few have a rope inserted in a perforated ear. The load is crossed over the animal's back, like a single bundle (see Horkheimer 1973: the first photograph facing p. 80; an identical piece is also in Benson 1972:92, Fig. 4-16 [or Donnan 1978: 112, Fig. 174]) (Figures 4.14–4.16). There are animals lying down and standing up.

There also are figures of llamas with some kind of saddlebag and a vessel on each side (for similar pieces, see Donnan 1978:113, Fig. 176; Bonavia 1994:Photo 216 right, 286).

The Museo Nacional de Antropología, Arqueología e Historia in Lima has a piece that shows a

FIGURE 4.10. Rock paintings of Cuchimachay (Department of Lima, province of Yauyos, district of Tanta). The depiction probably shows two llamas that are pregnant. It is difficult to establish the date of these paintings, because rites to propitiate the fertility of camelids continued throughout Colonial times. The style of these paintings is probably late, and is not connected to the early paintings of the hunter-gatherers (see Bonavia 1972; Bonavia et al. 1984:13). *Photo by Duccio Bonavia.*

man riding a llama. He is lying down lengthwise over the animal, his head is at the tail and he is holding onto the llama's neck with his legs (catalog no. K/6593). There are other cases that are variants on the preceding one. As for the men, they are in the same position but go over the load of the animal. The latter is placed transversely (Figures 4.17, 4.18; see also Benson 1972:Fig. 4.14, 91). There is a whole series of these vessels, and the animals are shown with or without harness. In some cases they are standing up, in other cases lying down.

In the Moche culture llamas were also ridden in other ways. One representation shows the rider in the same position in which the load was carried, namely, lying down on the animal's back transversely. The animal has a rope attached to a hole in the right ear, with the other end is held by the man (Figure 4.19). In another case a warrior rides

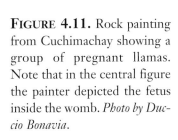

FIGURE 4.11. Rock painting from Cuchimachay showing a group of pregnant llamas. Note that in the central figure the painter depicted the fetus inside the womb. *Photo by Duccio Bonavia*.

FIGURE 4.12. Rock painting in Cave No. 3 of Chaclarragra (Department of Huánuco, province of Dos de Mayo), about 5 km northeast of the caves of Lauricocha, showing a wild camelid hunt. It was painted in dark red. The painting measures 1.40 m wide. It must date to around the time when the hunters of the cave of Lauricocha lived. *After Cardich (1964/66:Fig. 114, 135). By permission of Augusto Cardich, Lima, 10 November 1994.*

FIGURE 4.14. A loaded llama in resting position. The sex of the animal is not indicated. The animal has a simple, halter-like rope harness. In this case the bundles are strung with ropes in a very peculiar way. Moche IV (22.8 cm long and 17 cm high). *Photo by Duccio Bonavia. MNAAH (C-01458 [1/2434]).*

FIGURE 4.13. The head of a llama made with great realism. Moche III (16.5 cm long and 23.2 cm high). *Photo by Duccio Bonavia. MNAAH (C-01454 [1/ 2397]).*[9]

FIGURE 4.15. Loaded male llama resting. Moche I (22.7 cm long and 18.5 cm high). *Photo by Duccio Bonavia. MNAAH (C-01479 [1/2431]).*

FIGURE 4.16. Loaded llama resting. In this case the artisan did not indicate the sex of the animal. Moche IV? (23.7 cm long and 24.6 cm high). *Photo by Duccio Bonavia. MNAAH (C-01461[1/2435]).*

FIGURE 4.17. Llama in resting position. The animal is loaded; on top of it is a man lying face down. He holds onto the neck of the llama with his legs while grasping the area around the animal's tail with his hand. The artisan has not indicated the sex of the llama. Moche IV (23.3 cm long and 22.9 cm high). *Photo by Duccio Bonavia. MNAAH (C-54519 [1/2448]).*

FIGURE 4.18. Loaded male llama. Here the animal is standing up. On top of the bundles a man lying face down grasps the llama's neck with his legs while holding onto the back part of the animal with his hands. The llama has a simple, halter-like rope harness. Moche IV (20.3 cm long and 22.6 cm high). *Photo by Duccio Bonavia. MNAAH (C-01482 [1/2449]).*

FIGURE 4.19. Male llama at rest. On top of it a man lies face down and in this case crosswise. The animal has a rope tied to a perforation in its right ear, which the man holds with both hands. Moche IV (25 cm long and 24.2 cm high). *Photo by Duccio Bonavia. MNAAH (C-01463 [1/2441]).*

FIGURE 4.20. Mochica warrior riding a male llama. His left hand holds a club and the right one holds a rope inserted through the right ear of the animal. The vessel is broken and is missing the stirrup spout that went on its back. Moche III? (19.6 cm long and 19.6 cm high). *Photo by Wilfredo Loayza. MNAAH (C-69209 [104618]).*

a llama in much the same way as he would a horse. The animal's left ear is pierced and the rope is fastened to the hole, while the other end is held in the right hand of the "rider" (Figure 4.20).

Another example shows a llama carrying a man with an amputated leg. The animal has a hole in its left ear to which a rope is fastened, with the end held in the man's left hand (Figure 4.21). In only one case is a llama ridden by a mythical animal.

Figures of llamas with an infant (e.g., Donnan 1978:115, Fig. 179) or mating (Larco Hoyle 1938:91, Fig. 60) are also known. In both cases there is much realism in the representations. Attention should also be drawn to the few pieces in which the animals have an ear notch, like a "marking" ("*señalización*"; see Donnan 1978:115, Fig. 178). Most notable is that all Moche specimens show short-necked llamas.

The art of the Nasca culture (ca. AD 1–500) also has camelid representations. In the Museo Nacional de Antropología, Arqueología e Historia

of Lima there is a series of highly stylized llama figures painted on vessels in different positions. An example is a vessel showing a person pulling two llamas that are tied at the neck with ropes; one of the animals is loaded (Figure 4.22). Some depictions seem to be guanacos. In some cases they are shown wounded by darts (Figures 4.23, 4.24), and even the hunters appear (Figure 4.24). These recall a well-known piece in the Museo Amano (Museo Amano n.d. [1972]:32, No. 0012), in which a guanaco hunt is shown.

However, there are a great many ceramic modeled pieces, both large and small. They are coarse and do not have either the beauty or the finesse of the Mochica pieces. Some of these llamas have a rope around the neck (Figure 4.25).

In the collections of the Leyden Rijksmuseum voor Volkenkunde there is a very important vessel (Figure 4.26). It is a modeled piece showing a woman with a load of wood on her back held by a sash, apparently made of cloth, that goes over her

FIGURE 4.21. Man riding a male llama. Here the left foot of the rider has been amputated. With his left hand he holds a rope that passes through a perforation in the left ear of the animal. Moche IV? (20.1 cm long and 28 cm high). *Photo by Duccio Bonavia. MNAAH (C-55037 [1/3754]).*

FIGURE 4.22. Vase painted with highly stylized motifs. The painting shows an individual leading two llamas with ropes tied at the neck. The upper animal is not loaded, while the second one below is loaded. The sex of the animals is not shown. Nasca 5 (13.3 cm long and 14 cm high). *Photo by Duccio Bonavia. MNAAH (C-11616 [3/5139]).*

forehead; in other words, she holds the weight with her head. On her left side is a llama carrying a load (Sergio Purin, pers. commun., letter of 8 December 1994). The llama has a rope arranged like a halter that goes through the upper part of the head in front of the ears and then comes down both sides to go round the neck. From there it extends and is held by the left hand of the woman. The realism of the llama is remarkable.

I have already mentioned that we know of few examples of camelids in highland culture styles, nor are many illustrated in catalogs of exhibits or private collections. However, a llama accompanied by a man, perhaps a herder, is relatively common in the Recuay style from the Callejón de Huaylas (ca. AD 1–500); in other cases a warrior carrying a shield and wearing a big headpiece is depicted. These figures are usually sculptures of poor quality, with inaccurate proportions. The animals' legs are too long in relation to the neck, which is too short. One of these pieces is in the Museo Amano of Lima (and a very similar one was illustrated by Lavallée and Lumbreras [1986:209,

FIGURE 4.23. Vase with a most stylized representation of a camelid, probably a guanaco, that has been hit by a dart. The plants depicted with the animal are presumably cacti. Nasca 5–6 (14.6 cm wide and 16.5 cm high). *Photo by Duccio Bonavia. MNAAH (C-10368 [35/1199]).*

FIGURE 4.24. A highly stylized scene on a ceramic vase showing a hunter throwing a dart with a spear thrower. Camelids visible to the right and left of this individual are almost certainly guanacos, which scatter in panic from a hail of darts. At the lower left we see that one of the animals is hit by one of these darts. Nasca 7. *Photo by Duccio Bonavia. After Schmidt (1929: 349, right).*

Fig. 191]; see also Schmidt [1929:239, lower left]). There are several of these pieces in the Museo Nacional de Antropología, Arqueología e Historia of Lima (Figures 4.27, 4.28).

The artisans of the Huari culture (ca. AD 500–900) also left us beautiful examples of llamas. Their large votive ceramic llamas from Pacheco (Department of Ica, in the province and district of Nasca), modeled in a most realistic style (Menzel 1968:17), are quite famous. (Some are now housed in the Museo Nacional de Antropología, Arqueología e Historia of Lima; see Figures 4.29, 4.30). There are also smaller pieces from the same site (Figure 4.31). In addition, there is a modeled figure of a llama skull which shows an amazing realism (Figure 4.32; llama skulls, but stylized in this case, are often represented in the Northern Huari style; see Instituto de Arte Peruano 1938:41). There also are other figures of llamas, with or without a load, dating to the transitional period between the Early Intermediate period and the Middle Horizon on the south coast.

A private collection in the United States has a beautiful Huari piece that shows a modeled llama (Katz 1983:272, Fig. 147).

The Museo Nacional de Antropología, Arqueología e Historia of Lima likewise has modeled figures of llamas in the north coast Lambayeque style (ca. AD 700–800). In one example, a man is shown lying face down over a loaded animal and, just as in Moche times, the man's head is at the animal's tail and his feet at the neck. Another

FIGURE 4.25. Male llama with a very long, halter-like rope harness that lies over the animal. Although the head is very well made, the body is disproportionate. Nasca 4 (24.7 cm long and 20.5 cm high). *Photo by Duccio Bonavia. MNAAH (C-54285 [21/91]).*

FIGURE 4.26. Modeled vessel of a woman carrying on her back a bundle of firewood held by a band, presumably cloth, that is supported over her forehead. In other words the weight is borne by the head, which has been artificially deformed. In her left hand the woman holds a rope to which a loaded llama is tied. From what can be seen, it seems to be a bag, presumably made of cloth, with something inside it. Interestingly, the harness the animal wears is different from the one found with Mochica llamas. Nasca 5 (16.5 cm high). *Courtesy of Rijksmuseum voor Volkenkunde, Leyden (Rmv 3277-10; EH 358).*

FIGURE 4.27. Llama-shaped ceramic vessel. The piece is incomplete, because there is small break on the right side. There probably was a warrior like the one in Figure 4.28. The depiction is not as realistic as that of the Moche style. Huaylas (14.9 cm long and 12.2 cm high). *Photo by Duccio Bonavia. MNAAH (C-64013 [c.c.P./732-RA]).*

FIGURE 4.28. Richly attired warrior with a shield in his right hand. A llama is on his left. In this case the animal has no harness. Huaylas (20 cm long and 20.8 high). *Photo by Duccio Bonavia. MNAAH (C-55033 [1/1091]).*

FIGURE 4.29. Vessel representing a most realistic male llama. Robles Moqo, Middle Horizon 1B (55 cm long and 69.5 cm high). *Photo by Duccio Bonavia. MNAAH (8/7121).*

FIGURE 4.30. Large ceramic vessel in the shape of a remarkably realistic male llama. Robles Moqo, Middle Horizon 1B (56 cm long and 75 cm high). *Photo by Duccio Bonavia. MNAAH (C-60592).*

FIGURE 4.31. Vessel representing a male llama. Robles Moqo, Middle Horizon 1B (21.2 cm long and 20.1 cm high). *Photo by Duccio Bonavia. MNAAH (8/7715).*

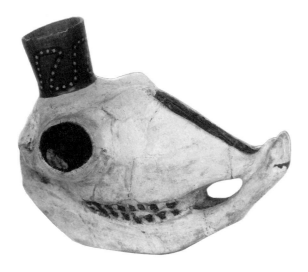

FIGURE 4.32. Vessel representing a camelid skull. Robles Moqo, Middle Horizon 1B (23 cm long and 17.2 cm high). *Photo by Duccio Bonavia. MNAAH (C-5535 [8/7707]).*

interesting piece shows a man mounted on a llama in the normal position with some kind of "bridle" that leaves the animal's neck and is an extension of a halter-like harness (Figure 4.33). The collections of this same museum have two modeled llamas in the Teatino style (ca. AD 700–800).

Camelid figures also abound in the Chimu culture (ca. AD 900–1400). In the Museo Nacional de Antropología, Arqueología e Historia of Lima I found an interesting Early Chimu vessel modeled in the shape of a female llama (Figure 4.34). In addition, there are several modeled llama heads that undoubtedly copy Mochica counterparts (Figure 4.35). There are also a great number of llamas shown with and without a "halter," and others that are loaded (Figure 4.36) in the Mochica manner; some are harnessed, others are not.

Although our sample presents only a brief overview of the subject, I have the impression that in terms of variety, Chimu artisans left behind a much larger series of depictions than the Mochica. On many vessels we find scenes of men handling different-sized llamas (Figures 4.37, 4.38). On other pieces we see men leading animals with ropes (Figure 4.39). Llamas were also depicted loaded with a bundle and with a man on top lying face down, with the legs at the neck of the animal and the head over the tail, just like among the Mochica (Figure 4.40).

A series of vessels—and I believe this is the most important point—shows men riding llamas

in different ways. A "rider" is shown kneeling over the animal and holding the left ear of the llama in his left hand, without using the harness (Figure 4.41). In another case a man is standing or seated on something over the animal (Figure 4.42). There also are men riding llamas face down, as already described, but with pack animals.

There is also a beautiful piece showing a seated man loading a young female llama (Figure 4.43), similar to the Chancay-style vessels with the same motif.

Finally, there are many llamas depicted lying down, with their hind legs tied to their forelegs (Figure 4.44). There are male and female animals. In most examples, the animals appear to have a short neck. A notable exception is the vessel in Figure 4.37, which shows a very long-necked animal.

There are a great number of modeled representations of llamas in the Chancay culture (ca. AD 900–1400). The Museo Nacional de Antropología, Arqueología e Historia of Lima has a remarkable collection. It should be emphasized that these are extremely crude depictions, and the animals are obese, with short legs and a shorter neck than in the Mochica representations. Almost all are small. Few examples have the size of a regular vessel. These llamas are shown with and without harness (see Instituto de Arte Peruano 1938:Lám. 49a [there are two very beautiful pieces in a private collection in the U.S.; see Katz 1983:305, Figs. 179, 180]; my Figures 4.45, 4.46, 4.47).

FIGURE 4.33. Vessel showing a llama with a mounted man. Note that the animal has a halter-like harness like the one worn by the Mochica llamas, which was used by the "rider" with his left hand. Lambayeque (18.5 cm long and 15 cm high). *Photo by Wilfredo Loayza. MNAAH (C-27802 [36/1600]).*

FIGURE 4.34. Vessel shaped like a female llama. Early Chimu (late Middle Horizon, early Late Intermediate period) (38.5 cm long and 22.2 cm high). *Photo by Duccio Bonavia. MNAAH (C-30294 [m/13.92]).*

FIGURE 4.35. Vessel showing a llama head. Chimu. (20.5 cm long and 16.4 cm high). *Photo by Wilfredo Loayza. MNAAH (C-064099 [85633]).*

FIGURE 4.36. Loaded llama in resting position. In this case the sex is not indicated. Chimu (19.5 cm long and 14.7 cm high). *Photo by Wilfredo Loayza. MNAAH (C-27777 [2/1455]).*

FIGURE 4.37. Ceramic vessel with a llama with its legs tied and a man who tries to hold her by clutching her neck with his left hand and the animal's right ear with his right hand. The interesting feature of this piece is the llama's long neck. Chimu (24.4 cm long and 23 cm high). *Photo by Wilfredo Loayza. MNAAH (C-27795 [1-11]).*

FIGURE 4.38. A vessel that has on its upper section a modeled representation of an individual behind a llama that is lying down. He is holding the right front leg of the animal with one hand and with the other is doing something on the tail or hind part of the animal. Chimu. *Photo by Duccio Bonavia. After Schmidt (1929:259).*

FIGURE 4.39. Modeled vessel showing a man holding a llama. The halter-like harness is striking in this piece as it apparently is not a simple rope but an elaborate band, and the rope that holds the animal is very thick. Chimu? *Photo by Duccio Bonavia. After Schmidt (1929:239, bottom right).*

FIGURE 4.40. Vessel representing a llama that is apparently standing up, carries a bundle, and bears a man lying face down on top, who holds onto its neck with his legs and to its tail or hind part with his hands. The llama has a halter-like harness, like those of the Mochica llamas. Chimu (18 cm long and 18.7 cm high. *Photo by Wilfredo Loayza. MNAAH (C-27775 [2/1445]).*

FIGURE 4.41. Vessel showing a llama that seems to be lying down with a man kneeling on top. He is holding the left ear of the llama with his left hand. The animal seems to have a halter-like harness. Chimu (21 cm long and 16.3 cm high). *Photo by Wil-fredo Loayza. MNAAH (C-27803 [2-1443]).*

FIGURE 4.42. Llama, apparently in resting position, mounted by a man. Curiously, the man is standing or seated on something that is on top of the animal, which seems to have a halter, but it is not handled by the individual. Chimu (22 cm long and 25 cm high). *Photo by Duccio Bonavia. MNAAH (C-27788 [Q-145]).*

FIGURE 4.43. Vessel with a seated man who has a female llama on his back. The individual holds the hind legs of the animal with his left hand and the front legs with his right hand. Its small size makes it seem a young animal. Chimu (17.5 cm long and 24.8 cm high). *Photo by Duccio Bonavia. MNAAH (C-54949 [1/3191]).*

FIGURE 4.44. Female llama lying down with its legs tied. The right ear of the piece is broken. Chimu (21 cm long and 19.5 cm high). *Photo by Duccio Bonavia. MNAAH (C-27807 [1/2466]).*

FIGURE 4.45. Llama. It could be surmised from its shape that the intention was to represent a pregnant animal. This is probably not so, for all camelid figures in Chancay style show this characteristic, as is the case of the animal in Figure 4.49, which is male. It should be noted that the front legs of the animal were reconstructed. Chancay (43.2 cm long and 25.5 cm high). *Photo by Wilfredo Loayza. MNAAH (C-67002 [103, 997]).*

FIGURE 4.46. Llama. Chancay (14.6 cm long and 7 cm high). *Photo by Duccio Bonavia. MNAAH (C-39059 [58697]).*

FIGURE 4.47. Ceramic llama carrying textiles. Note that the halter-like harness differs from that in other examples from different cultures. In this case the rope does not go around the neck but is instead fastened to the perforation in the left ear before tying the snout. Chancay. *Photo by Duccio Bonavia. After Schmidt (1929:257, top).*

There are also pieces with a llama carrying a smaller one on its back (Figure 4.48), and a series of vessels depicting a man carrying on his back a llama with its legs tied up.

Two examples of typical Chancay workmanship merit a brief mention. One is a modeled figure of a llama with its ears notched and pierced (Figure 4.49); the other is the head of the animal with its ears notched, just as in the aforementioned Moche case (see Instituto de Arte Peruano 1938: Lámina 49b). They are quite possibly ownership markings like the ones made nowadays, which are known as "*señalización*" (marking; see Palacios Ríos

1981). Llama heads are also common (Figure 4.50). Finally, I would like to mention an unusual piece: a gold llama attributed to the Chancay culture in the Museo de Oro del Perú.

I have not seen many representations of camelids in south coast specimens from the Late Intermediate period (ca. AD 900–1400). However, these will surely be found if a proper search is made. A llama-shaped vessel dates to the end of this period and exhibits some realism despite its coarseness (Figure 4.51). These animals were still being depicted on ceramics in Colonial times (see Menzel 1976:Plate 41-583).

FIGURE 4.48. A very crude representation of a llama with its off-spring on top. This type of figure is relatively common in this culture. Chancay (17.8 cm long and 13.2 cm high). *Photo by Duccio Bonavia. MNAAH (C-64137 [104, 640]).*

FIGURE 4.49. Male llama with a rope around its neck and a textile on its back. It is worth noting that the ears were notched along the edges. These probably are owner-ship markings, similar to the *"señalización"* (marking) currently done by herders. Chancay (27.2 cm long and 16.8 cm high). *Photo by Duccio Bonavia. MNAAH (C-65848 [87606]).*

FIGURE 4.50. Vessel in the shape of a llama head. Note that although camelid representations in the Chancay style are usually very crude, in this case the head is very well made and is most realistic. Chancay (17.3 cm long and 17.4 cm high). *Photo by Duccio Bonavia. MNAAH (C-39076 [54683]).*

Camelids were still pictured in pottery during the Late Horizon. I recall having seen llamas pictured in Inca-style ceramics (ca. AD 1400–1500) but do not recall their characteristics. In an Inca-style vessel a llama is shown lying down as it is attacked by two felines (Figure 4.52). More common are llama figures in silver, in a very naturalistic style with the accurate proportions of the animal (Figure 4.53), but the pieces are quite varied (Figure 4.54). However, the most representative ones are the small, stylized stone sculptures of llamas and alpacas. These figurines have a hole on their back and were originally filled with llama fat or coca, when used in rituals. They are known as *illa* or *conopa*.[10] There are many of them in the Cuzco Museo de Arqueología (see Rowe 1946: 248). These stone *conopas* are distinguished by careful carving, always in stones of a fine texture, grain, and color (see Bonavia 1994:photo 194, 257; my Figure 4.55).

There are other llamas carved in stone in a more realistic style (Figure 4.56).

I have also seen some llama figures in Provincial Inca style, one of which was illustrated by Katz (1983:315, Fig. 190). It depicts a man carrying an animal. The piece has been classified as "Recuay-Inca," but I believe it is a Provincial Inca piece of undefined style. I find no connection with the Recuay style.

FIGURE 4.51. Llama-shaped vessel. Ica (final phases of the Late Intermediate period). *Photo by Duccio Bonavia. After Schmidt (1929:303).*

FIGURE 4.52. Ceramic vessel with a highly stylized modeled camelid being attacked by two felines on its rim. Inca (15.6 cm × 10.2 cm). *Photo by Duccio Bonavia. MNAAH (C-55126 [4/184]).*

FIGURE 4.53. Silver llama. On its back is a red blanket inlaid with incrustations of gold and cinnabar. Inca. *American Museum of Natural History of New York (No. 327113). Adolph Bandelier Collection. Courtesy of Department Library Service, American Museum of Natural History.*

FIGURE 4.54. Male llama, possibly silver. Coastal Inca. *Photo by Duccio Bonavia. After Schmidt (1929:404, top section, second row, center).*

The Museo Nacional de Antropología, Arqueología e Historia of Lima has modeled llama heads quite similar to those in the Chimu style. The Mochica tradition is still distinguishable in some of them, but the Inca influence is also evident. During the Late Horizon, camelids were pictured throughout all of the empire in the local styles, but always with Inca influence. There are also llamas loaded in the Moche fashion, with or without harness, and several modeled llamas with no load, with or without harness.

Two specimens deserve a separate description. One is a modeled representation of a man pulling a female llama with the rope she has at the neck. The animal is resisting and has opened its mouth, baring its teeth in a gesture typical of camelids. This is one of the few cases in which the llama is pictured with a long neck and normal body and limb dimensions (Figure 4.57). The other piece shows a man with some tool in his hand, apparently washing the back of the animal or something similar (Catalog No. C-55123).

In this same museum there is a sculptured head of a llama in coastal provincial Inca style. It is a *pakhcha*[11] (ca. AD 1480–1500; my Figure 4.58).

Finally, I must give an explanation to prevent future mistakes. Nachtigall (1966a:195) wrote: "Notching the ear tips—hence the term *señalada* or mark—which can be particularly observed in northern Argentina, has been adopted from the Spanish custom of marking the sheep in this fashion, because llamas stand out individually for their color and do not need that supplementary

FIGURE 4.55. Stone representations of alpacas known as *illa* or *conopa*. Inca. (The one on the left measures 9.2 cm long and 5.5 cm high; the one on the right is 9.4 cm long and 6.2 cm high.) *Photo by Duccio Bonavia. MNAAH, Left: (L-8807 [15/134]), right: (L-8808 [96628]).*

FIGURE 4.56. Sculpture in green stone of a llama lying down, represented quite naturalistically. Its cultural affinity is not easy to establish, but it probably corresponds to late Inca times. It comes from Carhuacatac, between Tarma and Tarmatambo (province of Tarma, Department of Junín). *Private collection. Drawing by Pablo Carrera.*

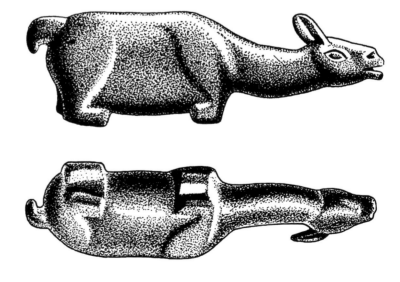

0   5cm

FIGURE 4.57. Ceramic vessel showing an individual pulling a female llama. In this case the width of the rope used is curious. The attitude of the animal is interesting, particularly in regard to its snout, which is most characteristic of a llama when it resists doing something it does not like. The piece is broken, as the animal's ears and the upper part of the aryballos which the man is carrying are missing. Chimu-Inca (25.5 cm long and 16.5 cm high). *Photo by Wilfredo Loayza. MNAAH (C-55034 [2-5138]).*

FIGURE 4.58. Vessel in the shape of a llama head. It is an imitation of an Inca *pakhcha*. Coastal Inca (19.2 cm long and 15.6 cm high). *Photo by Duccio Bonavia. MNAAH (C-63921 [C.C.P. /883-RA]).*

marking. Nor has it been taken up in all of Peru, but particularly in the puna of Atacama, according to what I observed there. The final sacrifice of a llama or alpaca is also done by slitting the throat." This statement is definitely not true. We have seen that there are quite clear illustrations of "markings" on the ears in Moche and Chancay ceramics (e.g., Donnan 1978:Fig. 178, 115; and my Figure 4.49). As regards the Inca period, Murra (1964a:79) says that "[a]ll llamas or alpacas in the herds were ritually and publicly marked, and it was known whether they belonged to a given family or belonged to the 'community.'" (The later English version specified that the notches were made on the ear. See Murra 1965: 191 and Wheeler 1981:41). Flannery et al. (1989: 111) remark that this was done "presumably by cutting a notch in its ear, as is still done in Ayacucho during the August *chupa* or *waytakuy*." The "marking" (*señalización*) process in the Aymara area has been described by Palacios Ríos (1981: 222–223). For these reasons I find Nachtigall's observations wrong both for the past and the present.

NOTES

1.   In Peru, this is a synonym for the *selva alta* (high forest), which is also known by the Quechua term *rupa-rupa*. For more details on this zone, see Peñaherrera del Águila (1969:43–45, 77–78) and Pulgar Vidal (n.d. [1980?]: 143–164). This is "the montaña, a wet, cloud-forested transitional zone between the highlands and the jungle rainforest of the Amazonian lowlands" (Willey 1972:82). It corresponds to the "very humid mountain woods" of Tosi (1960).

2.   Sumar (1985:11) makes the same statement, but it is not his. It was taken from Franklin (1972:466), because his words are a literal translation of the phrase just cited, but without the source given.

3.   At present, these materials are in the Universidad Nacional de Trujillo.

4.   McGreevy and Shaughnessy mean the lower *jalca*/upper *quichua*, i.e., 3200–3700 masl.

5.   There are three versions of this study. One is from 1958, a mimeographed text published by the UNESCO Program for the Study of Peru's Arid Zones. It is in Spanish. A German edition appeared in 1960, and a Spanish translation in 1973. Bibliographical mistakes were introduced when the book was translated into German, but even more serious was that the meaning which several phrases had in the original version was changed. Unfortunately, instead of using the original text, the 1973 Spanish edition was translated from the German edition, incorporating the mistakes the latter had. This is why I prefer to use the first edition. For the reader's convenience, the citation in the original version (1958:26–27) corresponds to (1960:42–43) in the German edition and to (1973:58–59) in the second Spanish edition.

6.   It is also known as alkali grass and spike grass. Salt grass is most commonly used. "There are two varieties, the one that grows in contact with salt water is sometimes called 'seashore salt grass.' The other, which grows on alkaline soils in contact with fresh water, is sometimes called 'interior salt grass'" (Lawrence Kaplan, pers. commun., letter of 18 February 2000). In this case it is the "seashore salt grass" variety, which in Spanish is called *grama salada* (*Distichlis spicata* [Poaceae]).

7.   This is a mistake. The correct name is *ishpingo* or *espingo*, and several scholars have identified the seeds as being of the *Nectandra* genus. The identification is thus far not confirmed, however.

8.   *Festuca* is a plant of the family of the Poaceae of a cosmopolitan gender. In general, peasants give the name of *ichu* to all plants of the grass family with hard and prickly leaves, including *Festuca* (Soukop n.d. [1978]:180–181, 218).

9.   The acronym MNAAH means Museo Nacional de Antropología, Arqueología e Historia. The numbers correspond to the catalog number of the piece.

10.   However, it is worth pointing out that the term *conopa* does not exclusively refer to these representations but was used for sacred objects in general believed to be producers of agricultural produce or livestock, and took the shape of the product itself. The *conopas* were also called *illa*, but it is possible that these labels are not just synonyms (Pease 1992:98).

11.   This name is given to vessels that have a spout on the front and are thought to have been used to worship water.

CHAPTER 5

THE PROCESS OF DOMESTICATING THE SOUTH AMERICAN CAMELIDS

The domestication of camelids certainly took place in the preceramic period. The subject is complex, however, with many facets to it and specialists who do not agree. (I will not get into a discussion of what a domestic animal is, for this goes far beyond the scope of this book. Readers who want to review the essential characteristics of a domesticated animal can find them in Dedenbach Salazar [1990:18], which provides a good summary; see also Thévenin [1961] or Zeuner [1963]). Not only are there disagreements among researchers, but many facts are ignored, giving rise to even more confusion. For example, in one archaeological dictionary (Whitehouse 1983:286; there is also a 1988 edition), the only camelid to appear is the llama, with a brief description that says that "[t]he first clear evidence of its domestication (dating to the Initial Period) comes from ceremonial burials in the Viru Valley and from remains at Kotosh." And Bustinza (1970a:28) naively states that "[c]onsidering that dress is a primary necessity, and that Andean man lives in an environment that is at present cold, [it follows that] the alpaca was domesticated after the llama."

In scientific works there are problems as well. Wheeler (1984a:395) correctly notes that statements on the origins and development of Andean pastoralism have been mostly based on conjecture, supported by indirect evidence. It was only in the 1970s that this problem began to be stud-ied with empirical evidence drawn from archaeological and zooarchaeological data. Several researchers from various disciplines became interested in the problem of domestication. Like Wheeler (1984a), I will mention here the geographer Troll (1931 [1935]); zoologists Gilmore (1950), Hemmer (1975), Herre (1952), Krumbiegel (1944), and Steinbacher (1953); geneticists Capurro et al. (1960); veterinarians León (1939) and Vallenas (1970b); paleontologists López Aranguren (1930a, 1930b) and Cabrera (1932); archaeologists Bird (1954), Lumbreras (1967), and Rick (1980); and ethnohistorians such as Murra (1965). In the case of Lumbreras (1967) and Murra (1965), however, the problem was approached from a purely theoretical point of view.

The most important contributions undoubtedly came from the fieldwork directed by the archaeologists Danièle Lavallée and John Rick, whose different perspectives gave rise to a discussion, which is always healthy for science. Undeniable progress was made with the systematic and detailed research of Elizabeth Wing and her team, who analyzed most of the osteological material from the archaeological sites in the central Andean area, and also of Jane Wheeler, though on a lesser scale. To these contributions we should add those made by William Franklin, George Miller, and Jonathan Kent. Their major contributions appeared in the 1980s. No publications are mentioned here because it was largely thanks to their data that I managed to carry out the study that led to this book, and the various references appear here and there throughout its pages.

The most direct evidence on the origin of domestic animals comes from the bones, dung, fibers, and textiles found at archaeological sites. The most important data used by zooarchaeologists in studying domestication are changes that occur in the bodies of the animals, the high frequency of morphological variations, changes in the survival curve, and an increase in the number of animals of one species relative to others living in the same area.

The changes usually detected are a decrease in body size and an increase in morphological variability. This is evident in the osteometric studies done in the Old World on sheep, goat, cow, and pig remains. There are problems, however, because these changes have not been recorded in South American camelids. This is because in the Old World there is a great difference in size between the ancestral wild forms and their domestic ones, whereas in the South American camelids there is almost no variation between the guanaco and the llama, or between the vicuña and the alpaca. Furthermore, the bones of all four species are similar except for the skull and the mandible, to the extent that it has thus far proved impossible to determine some characteristics that would enable archaeologists to assign camelid remains to species. What is usually done is to separate the large animals from the small ones, but in both groups there are remains of a domestic animal and a wild one. (Scholars do not agree on this point. Although the presence of a size gradient is accepted, for some scholars the guanaco is larger than the llama, which is larger than the alpaca, and the alpaca is larger than the vicuña [Kent 1982a; Wing 1972], while other scholars regard llamas as the largest and guanacos as the smallest [Franklin 1982; Novoa and Wheeler 1984]. These inferences are based on samples of present-day animals, but so far no evidence shows that the current measures hold for the past. In the case of llamas, at least, it seems that modern llamas are not the same. We find a good example of this point in the conclusions reached by Miller and Burger [1998: 270–276]). The available data on the process of do-

mestication of the llama and the alpaca are therefore essentially based on the changes observed in the survival curve and an increase in the amount of camelid remains among archaeological bones (Wheeler 1991:37; see also Kent 1986:1).

Wing (1980:163) studied the materials from Guitarrero Cave in the Callejón de Huaylas, starting from the premise that data on "animal domestication in the Andes are primarily changes in patterns of animal use rather than anatomical changes in the animals themselves." Based on the evidence from nearby sites, such as Chavín and Kotosh, Wing concluded that domestication seems to have taken place in the occupations corresponding to Complex III and IV, that is, ca. 3000 and 4000 BC.

At present, however, the largest amount of data on camelid domestication undoubtedly comes from the central highlands, from the Junín area in particular. Specifically, I refer to the studies conducted at Telarmachay (Lavallée et al. 1985), Pachamachay, and at several nearby sites listed in Chapter 4 (see Section 4.2.1; see Rick 1983, 1990). However, it should be pointed out that Lavallée and her team focused on a single site, while Rick studied several, and there are major differencess between the two studies. First, the Telarmachay study is exhaustive and was carried out with a flawless methodology that did not leave anything to chance. Meanwhile, at Pachamachay the work was only partial, just as at other sites in the area; its methodology has questionable points; and the study of the fauna that has been presented is a summary and remains incomplete. In addition, it is not possible to gain a good understanding of the fauna in context for the various occupations or the role of fauna in each of them. Just as in many studies by U.S. archaeologists, more attention was paid to the development of a theoretical model than to the search for real archaeological data and its follow-up, which are essential in archaeological research.

The transition from hunting to raising camelids has been documented at Telarmachay but not at Pachamachay.

In Telarmachay, an in situ development can be posited on the basis of the pattern of animal use practiced uninterruptedly since preceramic times, a development that goes from a generalized hunting of puna ungulates between 7000 and 5200 BC to specialization in hunting guanaco and vicuña around 5200–4000 BC, and to the emergence of the first domestic animals around 4000–3500 BC. Finally, herding fully domesticated animals appeared after 3500 BC (Wheeler 1984a: 405; 1985b:68; 1995:280).

The presence of domestic animals was established on the basis of an increase in the frequency of camelid remains and changes in survival curves and in dental morphology (Wheeler 1991:38). Hence, Wheeler's main argument for believing that animal domestication had taken place by the upper Phase V is the presence of remains of fetuses and newborn animals. She explains this fact by arguing that not only is an important hunt or massacre of newborn animals uneconomical, it is also historically unreported. Besides not being useful from the perspective of meat consumption, the wool, the hides, the tendons, dung, and the big bones, all of which could have multiple uses, are lost. Although Wheeler does not reject the possibility that the killing of young animals or pregnant females might in some cases have been due to massacres or other unknown factors, other causes seem to have been at work in the case of Telarmachay. On the basis of studies made in wild camelid populations (i.e., Raedecke 1979; Franklin 1978), Wing observed that the mortality in newborns is relatively low, whereas in animals that are domestic—that is, the llama and the alpaca—the mortality in young animals is quite high (she also used the study by Fernández Baca [1971]). It is known that the main cause of this mortality is enterotoxemia, diarrhea caused by the bacteria *Clostridium perfringens* types A and C, which can cause an epidemic (Moro and Guerrero 1971:9–14). The disease is caused by the extreme infestation of these pathogenic germs in the filthy enclosures full of excrement where the animals are kept, and complications arising from failure to transfer the passive immune globulin from the mother to her offspring. The possibility that another species of *Clostridium* intervenes, namely, *C. welchii*, whose epidemiology is connected with filthy corrals, was also mentioned in a later article (Lavallée et al. 1982:90). This, however, is not mentioned in the final report (Wheeler 1985b:71–72). In the Andes, enterotoxemia almost exclusively affects newborns

during the rainy season, between December and late April. Wheeler considers that this disease was brought about by domestication, from the moment the animals were kept in corrals. This disease does not appear among wild camelids (see Wheeler 1984a:403; 1985b:71–72; Novoa and Wheeler 1984:124; Baied and Wheeler 1993:148 [based on Ramírez 1987 and Garmendia et al. 1987]; Wheeler 1995:280 [based on the data in Leguía 1991 and Ramírez 1991]; and Wheeler 1993:15).

Wheeler admits that the identification of camelid species is very difficult. However, she believes, as was already explained, that some identification is possible through dental analysis. She noticed that there are guanaco and vicuña incisors throughout the first three phases (VII, V, and lower V), but those of vicuña are always more abundant. A third kind of incisor appears in lower V 1 that corresponds to the modern alpacas and becomes more common starting with upper Phase V and in Phase IV (Wheeler 1985b:78; 1993:16).

Wheeler (1985b:78) concluded:

If the incisors of this kind really correspond to the first domestic alpacas, and if we take into account the other evidence of domestication in lower [Phase] V 1, this modifies what was thought of the date it appeared and its origins: domestication would no longer have appeared around 2500 BP in the Titicaca Basin (Lumbreras 1967;[1] Bird 1954), but 6000 BP in the puna of Junín, and we have to consider the serious possibility that the vicuña had some role in its rise, for the evolution shows the predominance of small camelids with vicuña-type incisors over small camelids with alpaca-type incisors. And although the vicuña incisors exhibit intermediate characters between those of llamas/guanacos and vicuñas, this possibility is reinforced by the fact that they seem closer to those of the vicuña due to the localization of their enamel and the late forming of the root, and because these characteristics inherent to the vicuña predominate in the well-known cross of llama/vicuña.

Wheeler is, however, most prudent in her final sentence, for she says that although the data seem to indicate that the alpaca is a domestic vicuña, "much still remains to be done to confirm this hypothesis" (see also Wheeler 1991:40–41). Interestingly, in a later study she emphatically admitted that "the alpaca [descends] from the vicuña" (Wheeler 1995:281). Her database has not changed vis-à-vis her previous studies, and all that has been added is "1991, unpublished data."

However, it must be pointed out that Shimada and Shimada (1985:18), Kent (1982a and personal communication to Shimada and Shimada) noted that this method cannot always be applied. The data are in Kent's dissertation, which I was unable to obtain, but an interesting proof of them is in the studies by Wing (1988) and Miller and Gill (1990).

Going over the previous argument, Wheeler (1984a:406) notes that the Telarmachay data may indicate early domestication and raising of the alpaca, and these data seem to correspond to Wing's (1977a:852)[2] observations, insofar as "the predominance of the small form in the northern part of Peru south to Tarma, and the predominance of the large form in the southern part of Peru north to Ayacucho . . . suggests the earlier development of a small breed in the more northerly region of the Peruvian Andes." These conclusions are based on the study of camelid remains from 11 sites.

With regard to the possible presence of llamas at Telarmachay, Wheeler (1991:40) wrote: "It was not possible to establish whether llamas also appear at this moment because their incisors cannot be distinguished from those of the guanaco; however, the presence of big neonate remains suggests the possibility that they were present." Still, this seems to be questioned in a later study (see Baied and Wheeler 1993). In a more recent article Wheeler (1995:281) emphatically stated that on the basis of the work done at Telarmachay, it can be concluded that "the llama descends from the guanaco." She did not present new data save for the "unpublished data" of 1991, just as in the case for the origins of the alpaca. Later, Wheeler once more reviewed this subject, and though still defending her position on the guanaco as the ancestral form of the llama, she was now far more specific and stated that it was the subspecies *Lama guanicoe cacsilensis* (Wheeler

1995:277). The only argument Wheeler presented was the "geographical distribution" (Wheeler 1995:277). Curiously, the bibliographical reference given is "(Wheeler 1984, 1986, 1991)." First, the date 1986 is wrong. This was published in 1985 (1985a in my bibliography), and the pages given by Wheeler, 21–59 (Wheeler 1995:294), are wrong; they are 61–79. Wheeler made the same mistake in a previous study (Wheeler 1993:28). However, a revision of the bibliography shows the following. In a 1984 study (1984a in my bibliography) it is clearly stated that Wheeler does not accept the *L. g. cacsilensis* variety (Wheeler 1984a:401). The 1985 study does not include a single reference to this subspecies (Wheeler 1985b:61–79). Finally, although the presence of this subspecies was accepted in 1991, it was not associated with domestication (Wheeler 1991:40). Wheeler's position is therefore far from clear.

Baied and Wheeler (1993:148) insist that the animal domesticated was the vicuña. At the same time, they note that the evidence for the domestication of the llama is less clear-cut, but it "seems to have happened" at the same time somewhere to the south (they rely on Wing 1974 [this is a mistake, the date is 1977a] and 1986, although to tell the truth, Wing does not say so in the latter study). Baied and Wheeler are quite explicit in accepting that "both alpaca and llama domestication must have occurred many times in different parts of the moist puna. The pollen data from Telarmachay may indicate that overgrazing within the vicinity of the site became a problem shortly after domestication occurred (van der Hammen and Noldus 1985)."

Finally, what Lavallée et al. (1985) propose is that Telarmachay was a seasonal camp throughout its entire occupation. The people who lived there came from base camps that probably lay at lower altitudes. At first they were there only to hunt, and then to manage domestic herds. These scholars believe that the evidence for this is the low density of lithic artifacts and the presence of numerous bones from fetuses or newborn animals.

Unfortunately, Rick's studies (1980) do not present sufficient data to draw inferences about the process of animal domestication. Rick himself admits this when he notes that "[t]he Pachamachay data do not give many direct clues about the process of camelid domestication" (Rick 1980:329). I find the deductions Rick makes on the basis of the relationship between a given type of lithic artifact and the process of domestication overly speculative. He ends by noting that at the very earliest, this could have taken place at the end of his Phase 4 (2200 BC), "but most probably w[as] not well under way until sometime in Phase 5" (Rick 1980:329)—that is, between 2200 and 1500 BC. Rejecting the proposal by Mac-Neish et al. (1975:26) that the remains found at Pachamachay include "some evidence of domesticated or tamed llama and probably alpaca" (but these authors did not justify their assertion, which seems to be mere speculation, as Rick points out), Rick (1980:333) once more states that "[n]ot until the post-preceramic Phase 6 does camelid herding have any major impact on local organization."

In his conclusions, Rick (1980:334) notes that he would not be surprised to find evidence that the actual raising of fully controlled, domestic camelids was a late development in the puna, in comparison with other areas. First of all, there is no actual reason to believe the predomestic form of the modern llama comes from the puna. Here, animals like the vicuña can be managed in their wild state with a high certainty of procuring food without the great investment of energy needed for herding. With the sole exception of Phase 7 (dated between 800 and 400 BC), few data known for the puna suggest the early domesticated camelids were used for transportation. For Rick, this is one of the aspects of the domestic camelids that make them immensely superior to wild, managed species. The data for the puna suggest that in preceramic times, there was little need for interregional exchange, something that could have lowered the value of maintaining domestic camelids. Other, lower zones might, however, have had less secure resources for the camelids, in part because of poorer flora and far more marked seasons. Perhaps the intermontane valleys are similar to the puna as domestication sites. The preceramic puna lifestyle was eventually transformed by the development, or introduction, of domestic camelids. However, complex puna societies were slow in developing, in comparison with other parts of Peru.

Finally, what Rick (1980:265–266) posits is a model based on three proposals that "will be con-

sidered separately, but all three are interrelated, forming the essence of the subsistence-settlement strategy of the society."

- First, because the vicuña is the most productive, stable, accessible, and nonseasonal *contemporary* resource of the puna, then it or its prehistoric relatives had the same role: it was the basis for the sustenance of the population (emphasis added).

- Second, human groups did not need to leave the puna in any season because there was the possibility of having access to vicuñas or similar camelids year round.

- Third and last, the population became sedentary as a result of the behavior and density of the vicuña, lived almost all year long in a base camp, and used the neighboring areas.

I would like to emphasize two of Rick's proposals that are important for the ensuing discussion. First, we saw that for one of his proposals (the first), Rick uses present conditions (Rick 1980:266). The present condition of camelids cannot be assumed to be a good indicator of the past. Far too many changes have taken place. Second, Rick himself states that "[t]he hypothesis of year-round occupation of the Junín puna *cannot be conclusively proven with the evidence at hand*, although a wide range of seasons is indicated in the floral and faunal remains" (Rick 1980:270; emphasis added).

In fact, Rick's thesis contradicts that of Lavallée et al., as was noted by Wheeler (1984a:405; 1984d; 1985b:68), who also criticized Rick's work, with good reason. Wheeler (1984d:196–198) says the puna is not the optimum, benevolent environment Rick claims. Instead, it is an inhospitable and unpredictable environment with very humid and very dry seasons, occasional droughts, and a fluctuating carrying capacity. On the other hand, while it is true that the vicuña is a dominant species, Rick has arbitrarily eliminated the huemal deer and the guanaco. Besides, he relies on contemporary data, which invalidates his model. His population estimates are hypothetical, and he does not present other alternatives. The model posited by Rick is synchronic because he treats 26 preceramic sites as if they were contemporaneous, even though only three can be dated. From this he draws his conclusions on sedentism. Finally, Rick lacks empirical data with which to confirm his model.

Rick's interpretation of the fauna is based on the unpublished 1969–1970 excavations made by Matos that Wing studied but which were then revised, and so are not valid. Wheeler essentially refers to the high percentage of juvenile animals that were thought to have been killed in the dry season to prepare charqui, and which were actually neonates born in the rainy season (data published by Wing in 1986, but communicated to Wheeler in 1984). Wheeler also indicates that at the time she wrote (in 1984: Wheeler 1984d:196–198) there was no way of tying the faunal remains to the stratigraphic ones, for the first had not yet been studied. Wheeler also insists on emphasizing the invalid nature of the archaeological arguments presented by Rick to prove that domestication was a late event. Furthermore, according to Wheeler, the arguments Rick used to reject early domestication are negative. On the basis of the analysis made by Wing, Rick concluded that the specialized hunting of the vicuña took place after 10,000 BC because there is no evidence of a decrease in deer remains. However, reinterpretation of the data by Wing, and the fact that a change from the generalized hunting of camelids and deer to a more specialized camelid hunting took place in the earliest levels of Uchcumachay and Telarmachay, which are nearby, indicate that the date given by Rick is far too early.

Starting with an analysis of variations in lithic density, the source of raw materials, debitage remains, and styles, Rick suggests a progression from not too mobile hunting to hunting from a sedentary base. This would have led to the abandonment of Pachamachay and its hinterland and the establishment of a balanced hunting system based on the rational use of the camelid population, allowing the sedentary occupation of Pachamachay for some 1,500 years. Wheeler (1984d) notes that in preparing his model and his interpretation, Rick assumed that the early hunters replaced all other predators of the vicuña, which regulated the food chain at its highest levels. However, he ignored the 10%–30% mortality from natural predation detected in the first four months of life in the Pampa

Galeras National Vicuña Reserve in Ayacucho, with an estimated annual harvest rate of 20%. If we take this factor into account, then man only increased the predation level, so it is clear that the early hunters could not have retained a vicuña-hunting lifestyle for long, and management of the movements of the animals must have been required quite early to ensure productivity.

The hunting of wild camelids would have remained an important activity even if an attempt were made to increase the number of herds, but several new animals—pumas, mountain cats, foxes, and perhaps condors—were instead added to the list of predators, to judge by the studies at Calacala in the Department of Puno, which show that natural depredation spirals when animals are enclosed. An increase in the frequency and variety of projectile points would thus not be unexpected in the domestication process, because the enclosed animals had to be protected. Of course, the only levels where puma and mountain cat bones are found date to after 4000 BC, which is the date that coincides with the domestication of the alpaca at Telarmachay, thus supporting Wheeler's proposal.

Although Kent (1988b) supports some of Rick's ideas (1980:265–266), on others he disagrees. Kent also questions certain points made by Lavallée et al. (1982). Kent (1988b:137) says that Rick's position on the domestication of camelids prior to mid-Phase 5, that is, between 2200 and 1500 BC, is not correct because it may well have taken place earlier. Kent (1988b:142) notes that domestication took place at the end of Phase 4, and "possibly earlier." Besides, Kent notes, the cave was occupied all year long and was a base camp for the exploitation of several environmental zones. The inhabitants of Pachamachay were not essentially vicuña hunters, insofar as the remains that predominate belong instead to alpacas, "together with other domesticated animals." The guanaco is one of these wild animals, thus implying that it lives at a higher altitude than posited by Cardozo (1954:42), and indicating that guanacos can be found in the same areas occupied by vicuñas; however, they are very scarce in the deposits at the site. Kent also notes that the behavior of the guanaco is not substantially different from that of the vicuña, thus implicitly rejecting Rick's model, which does not consider the guanaco in relation to the settlement at Pachamachay (Wheeler 1984[d]:196). The behavior of the camelids seems to be quite similar among the four species (Kent uses the term "varieties," which I do not believe is correct) (von Pilters 1954; Raedecke 1979), "and does not seem to have much to do with validating the model." Based on one of his studies (Kent 1982[a]), Kent (1988b:131) also indicates that there are variations in the timing of the heat according to the latitude and the availability of grass, and the birth date naturally varies with it. It seems that in the northern latitudes this happens at a later date than in southern latitudes, and that possibility should also be borne in mind for Pachamachay.

Kent (1988b:142–143) then notes that the age profile in Phase 5 indicates a relatively high proportion of fetuses/newborn animals (27% according to Fig. 2 [1988b:142–143]), whereas in Phase 6 the percentage drops to 13% and the percentage of adults increases. Year-round occupation is observed in Phases 4, 5, 6, and post-7, but not in Phase 7. This was a period of occupation during the rainy season. Given the absence of vicuñas, we can detect a change in the way of life, in the sense that these animals were perhaps not available owing to the high population density in previous phases, which gave rise to overhunting and "broke"[3] their land "to the point of interrupting their normal breeding and feeding cycles." Rick (1980:328) discussed the point and believes that competition with domestic animals might also have exerted an influence. The age profile in Phase 5 is different from that in the contemporaneous levels at Telarmachay. Here, according to Lavallée et al. (1982:72), the percentage of fetuses/newborn animals comes to 70%.[4] Kent wonders why there is a difference in this age group between sites. Lavallée et al. (1982:88) claim that the high percentage of fetuses/neonates at Telarmachay suffices by itself to indicate that the bones come from domestic animals, but there also are domestic camelids at Pachamachay. Domestication, Kent says, cannot be the only cause for the high number of fetuses/newborn animals. This perhaps happens with goats and sheep at Old World sites, but not necessarily in Andean sites.

For Kent (1988b:142–143), this problem highlights the danger inherent in using age profiles alone to infer domestication, without consid-

ering other kinds of evidence. The risks in accepting the high rates of fetuses/newborn animals as an index of domestication were detailed by Collier and White (1976), and there are, in fact, several factors that could have given rise to their high frequency. The argument also applies to the fauna at Pachamachay in its Phase 5, and to the decrease in these remains in the following phase.

Lavallé et al. (1982[:88–90]) suggested the possibility that because its movements are restricted, a herd of domesticated camelids might be exposed to some highly infectious diseases like enterotoxemia. According to Fernández Baca (1971, 1978) and Moro et al. (1958), however, other diseases occur under the same conditions. In addition, although Kent does not reject the possibility that a disease may have been responsible for the high percentage of fetuses/newborns at Telarmachay, he points out that no known methods can provide direct evidence of any disease that could have affected the camelids at Telarmachay or Pachamachay. Perhaps the experiments Lavallée and Wheeler (1985) are conducting to detect the parasite *Clostridium perfringens* will provide more concrete evidence.

Kent insists that not all bone evidence from one site necessarily derives from a single cause. The high proportion of bones from fetuses/neonates at Pachamachay could be the result of very complex factors arising from the management of too many herds and the demands of exchange, if that were practiced. We should also take into account nutritional factors, because "[t]he neonatal disease is probably just one factor among several which affected the configuration of a mortality profile. A difficult question remains: Since a high percentage can be attributed to, say, enterotoxaemia, why are the bones of these animals found within the deposits in the cave, and not rotting in the puna?" (Kent 1988b:143).

An interesting and almost unknown position is that of Spunticchia (1989–1990), who introduced an interesting intermediate position into the Rick–Lavallée debate.

After analyzing the study by Lavallée et al. (1985), Spunticchia (1989–1990:161–164) states that he agrees with their proposals. He adds only that "[w]e find that the hypothesis that the base camp was located in the valley is correct, even if it is theoretically difficult to exclude a mobility model with the periodic use of more fields spread over the extent of the 'puna' ecosystem" (Spunticchia 1989–1990:164). He also claims that "leaving aside the methodological considerations" (Spunticchia 1989–1990:165), Rick's data seem to confirm his position (Spunticchia 1989–1990:164–166).

After analyzing the environmental and climatic data, Spunticchia claims that at the time under discussion, the puna appeared as a group of microenvironments, some complementary, others not, of which at least one, the lacustrine one, must have been particularly attractive for human groups, given its high potential for marine and plant resources (Spunticchia 1989–1990:167–168). Bearing in mind that there are approximately 35 km between Pachamachay and Telarmachay, Spunticchia posits the following hypothesis:

1. From a synchronic perspective, the hunter-gatherers or the herder-horticulturists used different settlement models in relation to the following: the geographic position of the site within the ecosystem and the number and type of microenvironments, complementary or not, that are close to the one under consideration that had the potential to ensure a balanced diet.

2. The settlement models might be subject to variations from a diachronic perspective, as a result of better adaptation of the groups to the environment (i.e., with more knowledge and a higher technological level)" (Spunticchia 1989–1990:168).

With regard to the first hypothesis, Spunticchia explains that the sites of Telarmachay, Uchcumachay, Panaulauca, and so forth are all located on the edge of the puna and close to valleys. It can be assumed that the focus and attraction for all the groups living there was the valley, not the puna, because the sites may represent the upper limit of a land controlled from base camps in the valleys. The reasons for this kind of settlement could be manifold. First, there is the need to balance the diet with meat and vegetables. The puna presented advantages for the first resource, the valley for the second one. Second, there is the need to maintain control over several ecological niches so as to have access to these resources all year round,

and perhaps prevent conflicts with groups living at high altitudes that controlled other resources. It is worth emphasizing that Spunticchia considers that camelids were not part of these possible conflicts because they were accessible in several environments. The presence of zones with different resources in a given area, both plant and animal, must have been decisive when choosing the kind of settlement in those sites that lay inside the puna. This must have been the case for Pachamachay. Spunticchia therefore believes that for the puna sites, the sedentism model may well have been the one that best matched the effort/benefit ratio. Any other solution would have entailed an unjustified expenditure of energy for the human groups. Thus, for nuclei established inside the puna, for example, a transhumance model oriented toward the valleys could entail efforts that were physical (due to distance) and physical-biological (changes in altitude and climate), efforts that were cultural (adaptation to new environments), and perhaps conflict due to clashes with other groups.

With regard to the second hypothesis, Spunticchia starts from the premise that human groups reached the Junín puna by climbing up from the highland valleys and needed an adequate period to adapt, both physically and culturally. Once this stage was over, it was likely that some of these groups would have become established in the most attractive areas for subsistence.

According to Spunticchia, from a diachronic point of view, it is conceivable that the valleys were the strong attraction for these groups, for the development of plant cultivation and the spread of species requiring a warmer climate. This led to the establishment of base camps at lower altitudes. From then on, any interest in the puna was perhaps essentially limited to camelid raising, perhaps developed by herders from the valleys, particularly at the time that births or shearing took place. It is possible that some nomadism existed among these herders within the puna microenvironments to meet the pasture needs of the animals (Spunticchia 1989–1990:167–175).

In brief, what Spunticchia posits is that we analyze the customs of the camelids and the conditions present in the Junín puna during the Holocene, when transhumance was not the only possible model in the middle preceramic, when populations practicing specialized hunting and had close relationships with camelids. From this perspective, the hypothesis of sedentism might have a rationale if limited to the site under discussion. However, Spunticchia explains that he does not agree with the biomass estimates made by Rick. In other words, what Spunticchia believes is that site location may have played a major role when choosing an adaptive model. This is why the hypothesis of a model for seasonal occupation (transhumance) is put forward for sites located at the edge of the Junín puna (such as Telarmachay), which were outposts of populations established in the middle to high parts of the valleys, and why a stable occupation could have existed at sites located inside the puna, like Pachamachay, close to microzones that had sufficient and diverse resources (Spunticchia 1989–1990: 184–185).

This proposal, which had not been previously presented, is interesting and seems feasible from a theoretical point of view. However, I must emphasize that whereas much concrete data exist for Telarmachay and in sufficient numbers to consider them valid, the data for Pachamachay are scarce and the results are based more on conjecture than on evidence. Other sites in the puna with characteristics similar to Pachamachay must be analyzed to establish whether Spunticchia's thesis is valid, and whether the data support it.

Hurtado de Mendoza (1987:241) proposed a different hypothesis for the Pasco region, adjacent to Junín. He posited that after the passage of several thousand years, the constant predatory pattern brought about a breakdown in the ecosystem that was no longer sustainable, thus forcing the human population to change its strategy, i.e., abandoning hunting and gathering and turning to animal domestication. However, this process would not have been limited to human activities but would instead have co-occurred with environmental changes that modified the equilibrium of the system. I must note that Hurtado de Mendoza conducted his research without any serious study, his database is poor, and he does not present any evidence to support his position, which, by the way, contradicts the ideas of real archaeologists who have worked in the area.

As was already stated, the final faunal reports of the Ayacucho Project have not yet been published, but summaries by Elizabeth Wing and Kent V. Flannery were specially prepared for this volume and they are presented at the end of this book. Although I mentioned domestication when analyzing all the data for the Ayacucho sites (see Chapter 4), I must point out that dating the various Ayacucho phases follows MacNeish et al. (1983).

For the Jaywa phase (7100–5800 BC) there is one single mention from MacNeish, Nelken-Terner et al. (1970:37), who claim that in this phase "there are hints of the domestication of llama." This evidently was speculation by the authors, as the statement was never repeated.

I found three references to the Piki phase (5800–4400 BC). The first is by MacNeish (1969: 38), who was categorical and spoke of "definitive evidence" of the domestication of llama. He changed his mind the following year, for he then claimed that "some of the llama bones . . . may be those of domesticates" (MacNeish, Nelken-Terner et al. 1970:37). This last claim was corroborated by García Cook (1974:19), who mentioned "some camelid remains [that] probably belong to domestic llama or alpaca."

The data on the Chihua phase (4100–3100 BC) presented by those who worked in the field are confusing. MacNeish et al. (1980:10) mention camelids being herded and penned, but MacNeish and Nelken-Terner (1983:10) suggest some form of control, such as a selective hunt, the killing of young males, or the "breeding of wild and/or tamed and/or semi-domesticated camelids . . . even though herding of fully domesticated llama or alpaca was still not done" in the region. In this regard I have more confidence in Wing's data (see Wing, Appendix A, this volume). She says of the Chihua phase in Pikimachay Cave that "there are earlier signs of domestic lamoids" (Wing 1975b: 34–35). Then Wing states that "[t]he earliest indications of domestic lamoids in valley sites are from the Chihua period . . . at Pikimachay Cave, in the Ayacucho valley, where we found a moderately high dependence on camelid" (Wing 1977a: 848). Wing's data are confirmed by Flannery et al. (1989:91), who, when referring to Pikimachay Cave, say it was here that the first domestication of camelids in Ayacucho took place (see also Flannery, Appendix B, this volume).

But apart from these data, which I believe are valid, MacNeish's team is once again guilty of a contradiction because they cannot agree on whether or not there were domestic animals in the following phase, called Cachi (3100–1750 BC). Let us see what they state in this regard. First they claim that "camelids, perhaps semi-domesticated guanaco, were herded in the Puna zone," but when listing the products from the puna we find "camelid (possibly domesticated)" (MacNeish et al. 1980:11). This statement was later supported by MacNeish (1981c:222) and by MacNeish and Vierra (1983a:128). However, MacNeish and Vierra (1983b:185) contradict themselves, because with regard to the same point they explicitly say "camelid herding." MacNeish (1983: 272–273) repeated it when mentioning animals that were "herded, tamed or [that were] domesticated camelids." He adds that "perhaps in Cachi times some sort of wild guanaco or paleollama [sic] were herded." I have already commented on this point, and will not go over it again. The most prudent position is to accept the data presented by Wing and supported by Flannery, both renowned scholars (see Appendices A and B of this volume).

I find the claims made by Lumbreras too hasty. He wrote (1974:37; emphasis added): "Clear evidence for domestication appears about 5000 BC both in the highlands and *on the coast*. In the highlands, the best documented finds are from Ayacucho, where the 'Cordilleran complex' of highland . . . animals (Lumbreras 1967) had begun to take form about this time. Andean domesticated animals are the llama (*Lama glama*) and alpaca (*Lama pacos*)." I have already discussed the Ayacucho findings, and although there is some evidence it is not that "documented," as Lumbreras claims, and it was even less so in 1974, when his book was published. Besides, the evidence does not have the age he claims. His claim that domestication took place on the coast "about 5000 BC" is gratuitous, and no evidence is presented. In fact, he contradicts himself: just a few lines below the above citation Lumbreras writes, "On the coast, domesticated llamas date between 1500 and 1000 BC" (Lumbreras 1974:37).

There are several other points to discuss about domestication. One of them is the geographic area where it took place, and there are several positions on this point. I review first the group of scholars who support domestication in the highlands. Kent and Wing favor the central Andes. Kent (1987:175) is emphatic in this regard: "[I]t seems that the most likely focus for the earliest camelid domestication was in the central Andean puna. This occurred between 6000 and 5000 BP, as alpaca or alpaca-like animals are identified along with contextual and biological evidence for new herd management practices. These changes occurred rapidly, and I suggest a punctuated equilibrium model for camelid domestication in this area." Wing (1977b:17) also favors the central Andean puna.

Latcham suggested that the altiplano (1922: 82; 1936:611) was the area of camelid domestication, and he started his hypothesis noting that the Titicaca Basin is the center of the geographic distribution of these animals throughout the Andean cordillera. Troll (1931:277; 1935:160, 178–179) also maintained that the Titicaca Basin must have been the center of domestication, arguing that it presents the highest concentration of camelids, and also has the best resources for them. These claims influenced several scholars. For example, Gilmore (1950:437) said that the llama was domesticated in the high central area of southern Peru, Bolivia, northern Chile, or northwest Argentina. Gilmore cited Latcham (1936) to indicate the area around Lake Titicaca. In regard to the alpaca, this same scholar (Gilmore 1950:445) also gives the altiplano as the zone of domestication. Flores Ochoa (1975b:297) likewise leans in this direction, as he writes that "[t]he Lake Titicaca basin (Murra 1964[a]) surely was the zone of camelid domestication, because their highest concentration when the European invaders arrived was here (Cieza de León 1941; Cobo 1956 [1965]), and here 65% of the Andean population of alpacas and llamas still lives." This statement by Flores Ochoa summarizes the arguments used by most of the scholars who hold this position.

In this regard we must offer an explanation. Like Flores Ochoa (1975b), several scholars credit John Murra (1964a:76; 1965; 1975:118) with positing that the altiplano was the center of camelid domestication. Scholars have failed to read carefully what Murra wrote. He never claimed this. What he said was that "[t]he Titicaca area, where the highest concentration of domestic and wild species exists, has been suggested as the place of domestication" (Murra 1964a: 76–77; the statement is repeated in his other studies, e.g., 1975:118, where only a citation by Latcham [1922:82] was added).

The argument presented is clearly unsound because, as Wheeler Pires-Ferreira et al. (1977: 160–161) correctly pointed out, the distribution in question is in part due to the economic disturbances brought about by the Spanish conquest, which practically led to the elimination of camelids in several parts of Peru and Bolivia. The Junín puna is a good example because there is ample evidence there for the abundance of camelids in pre-Columbian times, but now there only are a few small groups of vicuñas and a few llamas and alpacas. What brought about this major change was the introduction of sheep. Flores Ochoa has also emphasized the introduction of sheep (see, e.g., Flores Ochoa 1979a).

In all of her studies mentioned throughout these pages, written alone or co-authored with other scholars, Wheeler obviously comes out as the strongest champion of the Junín area as a center of camelid domestication. Even so, it is significant that she has not closed the door to the possibility that this process might have taken place in other areas. Wheeler is quite emphatic in stating that "[s]imilar evidence documenting the domestication process should be forthcoming as more archaeological research is undertaken at high elevation sites in the area between Junín and the southern boundary of the Lake Titicaca basin" (Novoa and Wheeler 1984:124; Lavallée 1990:38 follows Wheeler). Those who favor the Junín area as a center of domestication are Pearsall (1978: 394) and Matos and Ravines (1980:205); however, they cite in support only a study by Matos (1976) that does not appear in their bibliography.

Lumbreras (1974:37) is also of the opinion that the center of domestication might have been in several zones, but he considers that the "more probable areas are those of Lake Junín or particularly Lake Titicaca," and also claims that the Ayacucho area "was certainly not a primary cen-

ter" in this process. He does not present any concrete evidence to support his claims.

Another area where the early domestication of camelids has been suggested is the Salar de Atacama in northern Chile, at an altitude of 2500 masl. Specifically, this is the Puripica 1 site studied by Hesse (1982a, 1982b), which has an age that ranges between 4815 and 4050 BP (2865–2100 BC). Here, small camelid forms related to the vicuña and large ones related to the guanaco have been identified. Hesse (1982a:296, 210) believes the big ones were domesticates. The data were commented on by Baied and Wheeler (1993: 148–149). Besides the studies by Hesse, they also mention an article by Dransart (1991[a]), whom they credit with "research on the fiber remains which confirm their presence [of camelids] at 3,200 to 2,600 BP." This is striking, because Dransart (1991a:312) ends his comment on Puripica by stating that "[i]t is to be greatly regretted that the deposits are too humid to allow for the preservation of animal fibre remains." Leaving this aside, Baied and Wheeler (1993:148–149) comment, in contrast to Hesse (1982a), who suggested a possible independent domestication center, that "the lack of time depth makes it impossible to determine if the herds were introduced into the area already domesticated, or if they were brought under human control locally. Whatever the case, it has been suggested that the shift to herding occurred at this time in response to oscillating climates (Druss 1978; Hesse 1982a, b), but direct palaeoecological evidence is lacking, as are data from the adjacent dry puna zone which might document the spread of domestic animals from the north." I agree with this statement and feel that there is not enough evidence available to make a definitive inference.

In conclusion, I must note that Baied and Wheeler (1993:147–148) state that the puna ecosystem must have been a primary domestication center for the ungulates, comparable to the Near East (for goats and sheep), Mediterranean Europe (for cattle), and Tibet (for the yak).

Shimada and Shimada (1985:19) are the only scholars who have drawn attention to the possibility of studying the process of camelid domestication in other, non-high-altitude, areas. They note that most of the data available on the diet and climatic flexibility of domestic camelids make it possible to expand the positions thus far held. We should bear in mind that high-altitude caves and sites were sought for such study because of the current distribution of camelids, while the coast has been much studied, particularly in search of early sites, to solve problems concerning the importance of marine resources.

All of this has meant that the intermediate areas have been forgotten. I agree with this position and am convinced that when sites in these intermediate areas are excavated, there will be many surprises, not just in terms of animal domestication but also with respect to plants. I have long suggested this (see Bonavia 1982a:413).

There is also a group of scholars who do not take any position. An example is Mujica (1985: 121–122), who provides a good summary of the state of our knowledge, but in the end he fails to set forth his own position. (I must point out that this study has two references to studies by Lumbreras from 1967 and 1970. He is said to have claimed that the domestication of the llama might have occurred outside the Titicaca Basin, while the alpaca would have been domesticated there. In his 1967 study, Lumbreras definitely did not make these claims. I was unable to read the 1970 study, as it is an internal report of the Museo de Arqueología of the Universidad Nacional Mayor de San Marcos.)

Flannery et al. (1989:116) also do not take a position and just state that "domestication may have taken place over a large area of the guanaco's original range," even though the current data come from Ayacucho and Junín. However, they do caution that more care is needed when using the sixteenth-century distribution of the llama, because this distribution reflects regional economic specialization imposed by the Incas. Neither Stahl and Norton (1987:382) nor Franklin (1982:467) take a definite position.

There can thus be no doubt, as Flannery et al. (1989:91) correctly point out, that the study of camelid domestication is still in its infancy, notwithstanding the research that has been done. Nevertheless, one should realize that the field has advanced more in the last 20 years than ever before. As recently as the sixties, scholars were writing that "the domestication of American camelids

happened immediately after cultivation was discovered, or was introduced, in the area of agricultural groups" (Nachtigall 1966a:196–197), while others gave the Initial period as the earliest date for the domestication of llamas (see Lumbreras 1967:267 or Tabío 1977:211–212). Archaeological data now seem to confirm that llamas and alpacas were domesticated in the Andean mountains around the fourth millennium BC, at an altitude that seems to lie between 4000 and 4900 masl.

But as Novoa and Wheeler (1984:123) correctly point out, "identification of the ancestral form, or forms, from which they were domesticated remains a matter of debate. . . . Much more archaeological, palaeontological, and biological research is needed . . . before this problem can be resolved." Here I am not going to repeat the various positions taken by the specialists, because this was discussed in Chapter 1 on phylogeny and taxonomy.

It is important to recall, as Craig (1985:27–28) notes, that a great guanaco population, the main species for early man, existed after the extinction of the megafauna and before the domestication of the llama and the alpaca. However, the vicuña may have been equally important in the southern reaches of the central Andes. Craig (1985) suggests that an unidentified proto-lamoid species might have developed rapidly during the Late Pleistocene and might have presented the genetic stock for domestic animals. Craig believes that the habitats these ancestral animals preferred were perhaps intermediate between those of the guanaco and the llama/alpaca.

However, Craig (1985:28) goes even further, for he claims that "[t]his *Paleolama* [sic] stock may have been present in the so-called short-necked llama of northern coastal Peru, and in the diminutive *wiquinche* domesticate of forest Indians in central Chile. These two animals have not survived." While the first part of Craig's statement seems to merit attention, the last phrase is highly speculative, so although there is some possibility that a smaller animal did exist, as we shall see later, short-necked llamas have thus far not been identified at an osteological level. Moreover, this is an extremely complex problem, because size variations are very common in camelid fossils. They include forms that have the same size as the present vicuña, guanaco, alpaca, and llama, as well as camelids much bigger than these—there are even bones the size of a dromedary's (*Camelus dromedarius*). This suggests the possibility that there still are other ancestral forms that we do not know about (Wheeler Pires-Ferreira et al. 1977:163).

In fact, the gap between initial domestication, which probably began around 6,000 years ago, and the development of specialized breeding to produce llamas and alpacas similar to present-day animals lasted about 4,000 years, as Baied and Wheeler (1993:149) note. It was at the end of the preceramic, some 3,800 years ago, that domestic llamas and alpacas were common in the highland valleys of Peru and northern Chile (Hesse 1982a; Wing 1986). Llamas were raised on the Peruvian coast 1,400 years ago, according to Pozorski and Shimada and Shimada (Pozorski 1979[b]; Shimada and Shimada 1985 [I have already noted our disagreement as regards these points, because I believe this happened at an earlier date]), and in Ecuador (Wing 1986; Stahl 1988 [this study does not appear in the bibliography cited by Baied and Wheeler]; Miller and Gill 1990). There is evidence 400 years later of alpaca production on the south coast of Peru (Wheeler, Russel, and Stanley 1992). However, Baied and Wheeler end by noting that it was with the Inca Empire that the distribution of llamas and alpacas reached its maximum extent, as a pack animal throughout the Andean lands, while accompanying the imperial armies.

I will just add that Flannery et al. (1989:91) believe that such early domestic camelids probably still looked a lot like the guanaco, since they had presumably not yet undergone selection for the wide range of coat colors and fiber qualities that we see in today's llamas and alpacas.

On the other hand, there seems to be agreement among specialists that the alpaca was the first domestic camelid, which follows not just from the Telarmachay data (Wheeler 1985a; Novoa and Wheeler 1984) but also from the Pachamachay data. Kent (1987:175) claims that his own studies (he relies on his 1982a dissertation) show complete agreement between the data from these two sites, showing that the alpaca was the first identifiable domesticated camelid. According to San Martín et al. (1987:97), Wheeler

gave a lecture at a meeting on veterinary medicine (see Wheeler 1988b) where she reported recent evidence that apparently indicated that alpacas were domesticated in the moist regions of central Peru, while llamas were domesticated in the drier areas of the Peruvian southeast. No more data are available.

The notion that the domestication of the alpaca was a marginal and extremely late phenomenon in Andean history, as was claimed by Maccagno (1912:2) and repeated by Cardozo (1975a: 105–106), must be discarded. It is even doubtful that "interest in wool was the major incentive for the domestication of these animals," as Bennett and Bird (1960:260) believed.

With the data provided by archaeology, it is very difficult to determine what motivated man to begin the process of animal domestication. Lavallée (1978:38) did try to provide an explanation. She believes that "[i]t is highly likely that [the hunters] were forced by the disturbances they themselves had wrought upon the balance and stability of animal populations: when animals are systematically killed, this may bring about a change in the reproductive process within a species, thus making the animal group diminish." This claim by Lavallée is interesting and valid, but I believe that in the specific case of the Andean camelids, two other factors were of major importance in the process of domestication. On the one hand, we have their inborn territorial habit (I obviously refer to the wild species.) A group of hunters that managed to establish its camp close to a herd had the chance to learn its behaviors and understand its social and ecological mechanisms. In other words, they could become fully acquainted with the behavior of these animals. On the other hand, the herds diminished as continuous killing took place. It is a known fact that natural selection does not account for everything in the case of small populations; here, genetic drift has a major role. This is an accidental change in the genetic makeup of small populations that can make them accidentally diverge from the original ones at any given moment. We must bear in mind that the difference between populations and species that originate in genetic drift are not necessarily adaptive and can in fact be neutral (see Futuyma 1979:25–26). This perhaps led to forms that were not useful for hunters, who realized that they needed to foster the breeding of those animals that most suited them, and for this, the animals had to be isolated from the group. If this took place, it must have been the beginning of domestication. However, I cannot ignore an observation made by Stothert and Quilter (1991:34), who suggest that domestication might have happened in different ways in various sites. They give the southern Andes as an example of a place where the control of the animals might not have been just for food but also to ensure greater access to camelids as beasts of burden or for their hair (these scholars rely on second-hand information, a citation Wheeler and Mujica [1981:76] took from Lynch [1983b]). Terada and Onuki (1985:272) suggest there was domestication in Cajamarca during the late Initial period, and the same thing could have happened elsewhere (see also Bonavia 1999:137–138).

The wool argument was used by Latcham (1922) to point to the altiplano as a domestication zone for camelids (according to Wheeler [1991: 41], Lumbreras [1971] holds the same position. I was unable to examine this study.) Strangely enough, Novoa and Wheeler (1984:123) and Wheeler (1991:41) claim that the use of alpaca wool began in Pucara around 500 BC, to judge from some textiles found at Paracas (on the southern coast of Peru) and at Alto Ramírez (on the northern coast of Chile). They claim that these textiles constitute "the earliest use of alpaca wool in the Andes." This gave rise to the hypothesis that alpaca domestication took place in the altiplano at the time of the Pucara culture. This position is groundless if we recall that wool from these animals was found in the preceramic (in 1982 we reported [see Chapter 4 and Bonavia 1982b:201] that specialists had identified wool from *Lama pacos* in the preceramic coastal context of Los Gavilanes. Although it is true that it probably came from a highland center, this apparently was in the Callejón de Huaylas or in the central sierra.)

Two interesting suggestions have been made and should be taken into account in future studies on this subject. The first is by Browman (1974a:188), who noticed that a distinctive characteristic of llama and alpaca herding is that it is integrated into the hunting and gathering ecosystem in which the structure was introduced and is usually maintained. In this case, innovation is a conservation strategy whose initial effect is to preserve the means of life of llamas and alpacas, under the stress of changing environmental conditions. The transition from hunting wild guanaco to the herding of llama and alpaca simply ensured that the animals were close by when they needed to be hunted. Camelid herding is an adaptation to a semiarid prairie ecosystem that can tolerate animal herding but is not especially appropriate for cultivation. It is the most efficient way to exploit the environment and its resources in the highlands of central and southern Peru.

The second suggestion, made by MacNeish (1981a:149 et passim), is that the seasonal movements of humans must be understood in order to understand the process of domestication. Plants are not a good index, given their poor preservation in highland sites. Animals are therefore the most important marker. However, the interpretation of these remains still poses several problems due to the lack of zoological and ecological studies. MacNeish thus suggests that a good marker would be deer and camelid fetuses, as the bones of these animals permit an approximate identification of the season that corresponds to the period of their biological cycle. MacNeish believes that the fusion sequence method proposed by Wing (1972) may be very important.

NOTES

1   To tell the truth, Lumbreras did not say this, nor did he mention this point in his work. Significantly, when discussing the same subject in a later article, Novoa and Wheeler (1984:123) no longer gave this source.

2   Wheeler says page 181, which is certainly a mistake. There are several mistakes of this kind in this article. In the final report on Telarmachay, Wheeler (1985b:78) once again cited this same article by Wing, and once more gave the wrong page, i.e., 181.

3   "*Rajados*" in Spanish [translator].

4   To prevent mistakes, it must be pointed out that Kent (1988b) used the preliminary report by Lavallée et al. (1982), not the final one (1985). Second, the exact figure included in Table 2 of Lavallée et al. (1982:72–73) is 73%, and was later corrected to 72.99% (Lavallée et al. 1985:65, Table 2).

CHAPTER 6

HISTORICAL DATA

This chapter focuses on documents that provide data on Andean camelids. This is an extremely difficult task because far more information is available than I expected, and I was unable to study all of the documents; such a study will have to be done in the future.

The chronicles are cited according to the classification of Porras (1986). Quotations are used when necessary, but in most cases the data are summarized to save space.

6.1 SOURCES ON THE INCA PERIOD

This study analyzes camelid data from the documents left by chroniclers. It would be interesting to have data on the distribution of camelids in the Andean[1] territory during Inca times. The lack of data is disappointing because, as Dedenbach Salazar (1990:83) correctly points out, "this cannot be studied with the help of ethnohistorical sources." What can be done, and what I will do throughout this chapter, is to try to trace the presence of camelids in the central Andes by examining the accounts of eyewitnesses as they traveled across Peru. I am aware that this approach gives an incomplete picture. I will also try to ascertain how reliable the data are in the case of writers who copied information from others.

6.1.1 The Highlands

Let us turn first to the highlands. On the basis of data found in documents, Espinoza Soriano (1973b:30–74) reports that in 1460, the Inca ordered a group of *mitimaes* from the Huanca region, in the central highlands, to move to Chachapoyas, in what is now the Department of Amazonas. Espinoza believes that "[p]erhaps the camelids mentioned in the 'Relación de Chuuimayo' had their origin in this population movement, but it would have had to have taken place a few years earlier."

Only one indirect piece of information has been found for the Department of Cajamarca, mentioned by Espinoza Soriano (1976–1977:141, n. 3). Using sixteenth- and seventeenth-century *probanzas* and *visitas*, Espinoza says in regard to Pariamarca that "[s]ince Inca times" a small group of Guambo or Huampu *mitimaes* was relocated "in a part of this land known as Atunpariamarca, half a league from the city of Cajamarca," where they had "as habitat a sheepfold, a corral, and great pastures, which shows that they were also used by imperial Cuzco to raise State animals."

When describing the "province of Guamachuco" (today's province of Huamachuco in the Department of La Libertad), Cieza de León (1984:Part I, Chap. LXXXI, 236) says, "Before

207

the Spaniards entered this kingdom there was . . . a great number of sheep:[2] and in the highlands and wilderness there lived another great number of wild and sylvan livestock called Guanacos and Vicunias." Then he adds that "[t]he Incas had in this province (so I was told) a royal grove, where none of the natives could enter to kill the wild animals, of which there were many, on pain of death . . . . It is very pleasing to see how the guanacos jump . . . and on entering the stockade many other Indians with their Ayllos and sticks, they kill and take the number their lord wants: for these hunts would take ten thousand or fifteen thousand head of livestock, or the number he wanted: there was so much of it."

Koford (1957:218) discusses vicuñas, but with no real evidence, and finds the data in Cieza de León "inconclusive for they seem to be based on hear-say." Throughout this chapter we shall see the veracity of this chronicler, who clearly says, "so I was told" (Cieza de León 1984:Part I, Chap. LXXXI, 236). Furthermore, Cieza wrote: "I am used to giving the reader an account of the way this narrative was written, to satisfy him that I am neither making it up nor adorning my account with what was not so, nor ever happened" (Cieza de León 1909:291). There is no reason to reject his data. Cantú (1992:134) recently made an excellent point about the difference between the idea of time for Indians and for Europeans: "While Cieza was not always a historian aware of this critical methodological problem, which was far beyond the cultural scope of a sixteenth-century man, his history nonetheless still stands as one of the most invaluable sources for the study of Inca history due to the richness of its contents, the basis of his data, the sensitivity and insights about ethnographic and ecological data, and the intellectual curiosity and honesty of its author." In addition, Koford (1957:157) himself accepts that the present northern limits of the vicuña lie above 10° S latitude. Why wasn't it farther north in pre-Columbian times? As far as we know there are no substantial environmental differences that would have prevented this northern extension.

Camelids clearly lived in the Huánuco area in Inca times, something that is evident from the *visita* made in 1549. I was unable to locate the sites that follow, but they must be within the area

where several ethnic groups once lived, in what are now the provinces of Huánuco and the western part of Pachitea in the Department of Huánuco (see the maps in Ortiz de Zúñiga 1967). Thus, Conapariguana (1972:24–25) mentions the towns of Pillco, Pacha Coto, Xigual, Cochamarca, Guarapa, Quillcay, Curamarca, and Ñauça, and says these were "the mitimaes placed by Tupa Ynga and there are three other towns of the Yarus who are also mitimaes placed by these caciques and the principales called Yacan and Quirucalla and Maraya [missing] who do not have principales who are shepherds from the time of the Inca, and are still so and watch over the animals."

Mori and Malpartida ([1549] 1967:292–293) visited the same area and indicated that they were in the town of Mual. From the statements made by the caciques and people there, Mori and Malpartida found that "in the time of the Inca they herded livestock." That same day the inspectors went to the town of Caracara and found "two Indians and three widows [who] were placed here as shepherds in the time of the Inca." It is extremely interesting that this Indian group paid tribute to Cuzco, which included "two hundred and forty Indians" to "look after the sheep" of the Inca (Mori and Malpartida [1549] 1967:306).

From this point on, there apparently are no more data on the altiplano.

Cieza de León (1985:Part II, Chap. LXIII, 182) says that when the Inca went to Collao during the reign of Guaynacapa, "they brought an account of the many herds of livestock they had." Cieza de León (1985:Part II, Chap. CII, 279) also said of the Colla groups in Collao (particularly of the towns of Horuro, Assillo, and Assángaro) that they "consider themselves very rich both in livestock and provisions. When the Inca ruled this kingdom they had many herds of sheep and rams [*carneros*] in all of these towns." Curiously enough, Betanzos (1987:Chap. XLIV, 189) repeats almost the same information. When writing of the "province of Collasuyo" he says that Guayna Capac "had reports [which said] there was much game and deer and livestock and vicuñas . . . which are mountain sheep." The Peruvian exploits of Betanzos are obscure and romanticized, as Porras (1986:309) noted. There apparently is no sign that he was ever in the alti-plano. However, at one point in his book, Betan-zos (1987:Chap. XLV, 191) seems to claim he was indeed there, and notes that Collasuyo is a land "full of livestock."

Interestingly enough, Murúa (1922:114) also discusses the conquest of the lands of Collao by Huayna Capac, and says that when the Inca real-ized the Colla were subdued, "he brought from there a great many livestock of the land of all kinds, because that province was overflowing with livestock." The animals were distributed through-out the kingdom among "major lords and indi-viduals and the poor, and what remained, which was much, he gave to all the caciques to keep and use, naming people to look after it. Said livestock is counted, cared for and looked after; they kept an account of that from the world ["de los mundo"] and the dry meat of those who died, and of this livestock they took much to sacrifice for the Sun and the other huacas."

These are the only data for the highlands, and they are really not abundant. Even less informa-tion is available for the coast.

6.1.2 The Coast

Lorenzo Huertas transcribed a manuscript (AGI Justicia, Leg. 457, F) in the Archivo de Indias in Seville, and kindly told me about it. From the document we learn that in 1566, don Gregorio Gonzales de Cuenca (1566:MS., F. 1509) investi-gated the deaths of three Indians at the request of don Juan, the cacique of Collique. When men-tioning the execution site, the cacique declared that "the aforesaid Indians were taken in front of some corrals made of earth and stone where they say in ancient times Guayna Cava had rams [carneros] and sheep of the land in said corrals." It is even specified (Gonzales de Cuenca 1566:MS., F. 1062) that the exact place was "in some land called Sifunacoll," in the jurisdiction of Collique, in other words, most probably in the present-day district of Saña, in the province of Chiclayo.

Rostworowski (1972:285) has studied the visi-ta Martínez de Rengifo made in 1571 as ordered by viceroy Toledo, wherein the ethnic group that lived in the Chillón Valley, on the central coast, is mentioned. There it says, among other things, that the Indians "looked after 300 head of live-stock in the lands of the Inca, whose meat was taken to Cuzco as food for the mamaconas."

When discussing the upper part of the neigh-boring Rímac River valley, Espinoza Soriano (1983–1984:173) indicates the presence of llamas and wild vicuñas "in great abundance." It would seem that these data correspond to Inca times, but it is not certain.

Finally, in a document dating to 1558 Ortega Morejón and Castro (1974:101) give an account of the Chincha Valley both before and during the time of the Inca Empire. They state that when a curaca from the valley died, "if he who succeeded him was from his pachaca he inherited all of his estate, land, clothes, women, and livestock."

As I mentioned at the beginning of the chap-ter, the distribution of camelids in Inca times can-not be established with the available ethnohistor-ical sources.

6.1.3 State Control of Camelids

From the chroniclers we learn the way camelids were controlled by the Inca Empire. For example, when describing the province of Collao, Sancho de la Hoz (1968:Chap. XVIII, 331) states that its peo-ple did not use the sheep, "[not] because [in] that Province of Collao there is not a good number of sheep, but rather because the people are so sub-servient to the lord, whom they must obey, that not one is killed without his permission, or of the head-man [principal] or governor who is in the land on his orders, for not even the lords and caciques dare kill any without his permission." This point is im-portant because the chronicle was written in 1534 (Porras 1986:108–109) and, as Sancho himself points out, it is an account given by the "two Chris-tians" the governor sent from Cuzco to reconnoi-ter Collao, and evidently refers to Inca rule. Pedro Pizarro (1978:Chap. 16, 110) is not that detailed when describing this same region, but he clearly means the Inca. And when discussing the Indians who looked after the livestock, Pizarro says they made sure that "none of these animals were killed." This same chronicler notes, when describing "the order the natives had," that "[t]hese lords had a house where they killed the livestock of this land every day, and from there distributed to all lords and great orejones. The livestock of the land breeds little but there are many in this land, because it was all for the lord and nobody took it if he did not will it. This livestock of the land multiplied little even though there was a great number of it in this land, because it was all for the lord and none took it if he [the lord] did not will it so. These animals were used as beasts of burden and as meat when needed" (Pedro Pizarro 1978:Chap. 33, 240).

Zárate (1968:Bk. I, Chap. XI, 140) notes the orders Guaynacaba and his father gave in order to have much livestock, and "how out of these sheep of the land a certain amount was dedicated to the sun and was to be placed on the fields as tithe, and of these they bred many. For if it was not Guaynacaba himself for his army, it was deemed a sacrilege to have anyone touch them, and when the Inca needed them, he could take twenty and thirty thousand of them in one day just by organizing a hunt . . . that is called a *chaco*."

Gutiérrez de Santa Clara (1963:Bk. III, Chap. LXIII, 251) also discusses the herds "of rams [*carneros*] and sheep of the land itself," and how these were distributed all over the land and were dedicated to the cult of the sun. Gutiérrez is quite specific about the care with which the sacred herds were tended, for "these were most faithfully tended as a saintly and sacred thing. None dared take a ram [*carnero*] nor a sheep no matter how much in need they were unless it was the Inca himself, who took them whenever he was in need, mainly when at war against some rebels, or when conquering some lands anew. If by chance some Indian of any state, quality, or condition whatsoever killed a ram [*carnero*] of these . . . he was damned and excommunicated and sacrificed for it, saying the gods bode him ill."

Cieza de León (1985:Part II, Chap. XVII, 46) in turn describes the conquests of the Inca and notes that "if in any of these provinces there was no livestock, then the Inca commanded that so many thousand heads were given them." He immediately adds that "they should not dare kill or eat any offspring for the years and time he indicated."

In 1571, in his *Relación de los fundamentos acerca del notable daño que resulta de no guardar a los indios sus fueros* ("A report on the principles explaining the serious harm that would result if traditional rights of the Indians are not respected"), Polo de Ondegardo (1916:61–63) gave the

best description of the steps taken by the Incas vis-à-vis the herds of camelids when they conquered new lands. Polo specifies that it was commanded that the animals should stay in the place they were, "save for what he [the Inca] counted, and it was ordered that no female should be taken as tribute." It was noted that these same regulations should apply to animals who were dedicated to religion. Then he "[d]ivided the pastures and hunting grounds, not to make them common to the people [*hacerlo consçegil*], but so that no province would take the livestock of another, nor hunt—when he gave them permission—but all in their own established district."

The insistence on not killing females is worth noting, because even the people to whom "gifts were given . . . [for] a small service, were likewise forbidden to kill females." Polo de Ondegardo (1916:65) then repeats that "[t]he Incas had ordered that females from these animals belonging to the community were not to be killed, nor from any others, and so they breed greatly because neither those of the Inca nor of the church were killed . . . and only the males were killed even when the wild livestock was taken in *chacos*, and this was not done without permission to each in their homeland." What measures were to be taken with a sick animal to prevent the disease from spreading were also established. So "[c]onsidering it is a most contagious disease, it was ordered that [a]ny tame head [*res*] who got the mange was not to be fed or healed, and should instead be buried deep, and this was done."

It seems that moving animals to Cuzco was also regulated, and was done for the sacrifices on orders from the Inca. This took place in the "month of February, in the number indicated and rams [*carneros*] were always taken because, as was already stated, neither in sacrifices nor in any other provisions was any female spent, and so forth as regards livestock" (Polo de Ondegardo 1916:95).

Polo de Ondegardo (1916:63) suggests that other rules applied before the Inca. His extremely interesting and important observations on this point show the jurist's desire that these rules be followed to preserve the species. Polo de Ondegardo wrote: "In regard to the livestock, it seems many regulations were prepared at different times. Some are so useful and helpful for their preservation that it would be good to follow them now."

In his 1561 report to Licenciado Briviesca de Muñatones on the perpetuity of the encomiendas in Peru, Polo de Ondegardo (1940:135ff.) had already made a detailed description of the strict, careful, and wise Inca rulings on "livestock."

Garcilaso de la Vega (1966:Bk. 5, Chap. X, 260; 1959:86) left important data on "[t]he caravans used for carrying supplies." He says that "[a]lthough it was a common practice for the Indians themselves to carry loads, the Inca did not allow them to do so in his own service except in case of necessity. He ordered that men were to be exempted from all sorts of labor from which they could be spared, saying that they were to save their efforts for other operations from which they could not be spared and on which they were better occupied, such as building fortresses and royal palaces, bridges and roads, terraces, aqueducts, and other useful works for the benefit of the community on which the Indians were constantly engaged."

The regulations the Inca had for livestock are recorded in other documents, although with some differences. Thus, we have an anonymous document published by Rostworowski (1970:167–169) that was prepared in Chincha and quite possibly dates to the late sixteenth century, in which several ordinances given by Topa Ynga Yupanqui are mentioned. On folio 267v of the document we find that the people who had "been given livestock, which are the sheep of the land," should also receive herders. These had special tribute requirements. Then the document adds that the Inca "commanded that this order should be followed both in the highlands and on the coast" (Rostworowski 1970:167).

Like Polo de Ondegardo, "whose administrative reports are sources of the utmost importance to understand the Andes" (Pease 1992:142), this anonymous document mentions the strict control exerted over these animals. It is thus said that the Inca "[c]ommanded that it be decreed that no man should kill rams [*carnero*] nor sheep [*oveja*] to eat who was not a cacique with a thousand Indians as subjects, and the rest were allowed to bleed the rams [*carneros*] and eat the blood they drew, and so some bled them from a vein they have on

the mandible and the blood they drew they ate" (Rostworowski 1970:169).

When discussing this document, Rostworowski (1970:160) notes that the data cannot be generalized, nor can we know whether these regulations applied to all of the empire because we do not have other, similar data. She believes that perhaps this was ruled for Chincha because "these animals do not thrive in the coastal climate" and had to be preserved to carry loads to the highlands. I disagree, as was previously pointed out. Furthermore, this same document mentions not only the rulings on domestic camelids but also the "livestock which has no owner and breeds in the wilderness, who are known as Guanacos, Vicuñas, and these could not be killed without permission from the Inca, on pain of death" (Rostworowski 1970:168). In a later study, Rostworowski (1981:52) mentioned the care taken to avoid killing livestock for food, or at least that it was done in a quite different way from the Spaniards. However, no sources are given.

In the seventeenth century, Cobo collected the data available on the subject in a chapter entitled "The way tame livestock was distributed . . . and how the hunting grounds and wilderness belonged to the king." Cobo nearly repeats what Polo had reported, for he says that the Inca had commanded two things for the preservation of the camelids. First, that "any animal [*res*] that caught *caracha* (a kind of disease like mange [*sarna*], which these animals have and from which many die) should be buried alive and deep, and no one should try to heal it or kill it to eat, so that others would not catch the disease, which is extremely contagious." The second ruling was that females should not be killed, "not even in sacrifice nor for any other reason" (Cobo 1964b:Bk. 12, Chap. XXIX, 123).

Cobo was more specific about wild camelids.

The hunting grounds of the wild animals like the guanaco, vicuña, and deer were also delimited; there was no ban on these animals, which besides are harmful, save that these hunting grounds were not divided into parts, as was done with the land and the tame livestock. Instead each province had its own, so that the people who lived in one did not go hunt in another jurisdiction. The Inca also established royal hunting grounds and his own, so that none could hunt in them without his permission or that of his governors. This permission was given at certain times for a limited number, depending on what need there was. It was also forbidden to kill the females. According to the order they had in their *chacos* or hunts, they could well carry out both orders. (Cobo 1964b:Bk. 12, Chap. XXIX, 123)

In fact, the *chacos* were one of the control methods, and the Spaniards witnessed some of them, as will be seen below. Pedro Pizarro (1978: Chap. 16, 111) says that "[e]ach year they made human fences in which they took these vicuñas and guanacos and sheared them for the wool, which was used to make clothes for the lords, and the animals which died were dried in thin pieces under the sun . . . and those alive were released. In the wilderness there were great gurdas [sic], as noted, and these hunts were made on orders of the lords. Some of them were sometimes present and rejoiced in them. This order was followed in all the wildernesses of this realm."

Polo de Ondegardo (1916:65) also described the *chaco* and how the Incas forbade that females be killed. However, even more important is his following comment: "[I]t was a great oversight not to preserve this custom, nor banning the chacos [carried out] given the excessive [loss of life], and they [the camelids] are dying out. This could still be remedied if an order was now given."

We find a good description of the *chaco* in Garcilaso de la Vega (1966:Bk. 6, Chap. 6, 325–326; 1959:162–163), who dedicated an entire chapter to the "[s]olemn hunting excursions made by the kings throughout the country." It is worth reproducing part of this text:

> At a certain time of the year, after the breeding season, the Inca went out to the province that took his fancy, provided that the business of peace and war permitted. He sent out twenty or thirty thousand Indians, more or less, according to the size of the area that was to be beaten. These men were divided into two groups, one in

a line to the right and the other to the left, until they had made a great enclosure which might consist of twenty or thirty leagues of land, or more or less according to the area they decided to delimit. They followed the rivers, brooks, and valleys that had been fixed as the limits for the year's hunting and avoided entering the area set aside for the following year. They shouted as they went and observed all the animals they startled. It had already been arranged where the two lines of men were to come together to close the circle and close in on the game they had collected. They also knew from observation where the beasts had stopped, and the country they chose was clear of trees and rocks so as to facilitate the hunt. Having enclosed the game, they tightened the circle forming three or four rows of men, and closed in until they could take the game with their hands.

Garcilaso then explains that the animals included many that were dangerous, such as "lions, bears, many foxes, wild cats . . . and other similar creatures that do harm the game," which were killed. He then says of the vicuñas that these were taken in the *chacos*, and "twenty, thirty, or forty thousand head were taken, a very fine sight which gave rise to much rejoicing." The other animals, like "[f]emale deer of all kinds," were released, just like the guanacos and vicuñas were let loose once they had been sheared. Garcilaso (1959:164–165) specifies that these animals were controlled with the *quipu* and gives several more details.

Cobo (1964b:Bk. 14, Chap. CVI, 269) also discussed this subject, but he evidently took his information from Pedro Pizarro (see above). However, Cobo adds that ten or twenty thousand Indians took part in the *chacos* in Inca times, and "They also used to pen the animals gathered in said way, in a corral made between hills and in narrow areas. This kind of hunt was called *caycu*." Readers who want more information on this point should see Rowe (1946:217).

Here a comment made by Pease is relevant. Pease states that the ethnic lords, the people, and sometimes even the Inca participated in the *chaku*.

In the chronicles, the *chaku* appears as a tribute rendered to the state (Tahuantinsuyo). "It is clear that *chaku* was a term that designated several tasks, because we should distinguish those intended to gather domestic herds and those in which wild animals were gathered. It does not seem likely that the herds of camelids and an indiscriminate hunt were combined in just one activity" (Pease 1991:87–88).

With regard to the ownership of these animals in Inca times, almost all herds belonged in part to the ethnic groups and in part to the state, with several subdivisions, which I will not go into here (see Murra 1975:120). However, according to Rostworowski (1981:52), and to judge from the data from the *visitas* and the *encomenderos*, on the coast llamas belonged only to the lords, not to the common people.

We should also note that in the Inca Empire there were great state warehouses where dry meat was stored. When describing these storerooms, Cobo (1964b:Bk. 12, Chap. XXX, 126) says they held "abundant llama charque or jerky."

6.1.4 Camelids and Religious Beliefs

Camelids certainly were an important element in ancient Peruvian beliefs. There are abundant data on this subject, and I do not intend to dwell on them. Some examples will be given instead. When explaining the division between the sacred and the profane, Polo de Ondegardo (1940:184) comments that "in all provinces there was much livestock belonging to the Sun, the Inca, and some specific guacas . . . but as for the livestock, when the Spaniards arrived and for a long time afterward everything became confused, one with the other, and each became the master of what he kept and joined it to his own, and from it paid his tribute." When describing "the sacrifices and things they sacrificed," Polo de Ondegardo (1917:37) describes the kind of animal chosen for these tasks. Guaman Poma de Ayala (1936:f. 255) also has some information on this point. When describing the month of "October," Guaman Poma gives an account of what items were sacrificed so that "water was sent them from the heavens," and among other items mentions 100 white rams [*carneros*].

This information was repeated by the late chroniclers. Thus we find in Calancha (1975:Bk. II, Chap. XII, 850–851) the huge number of llamas that were sacrificed in the empire. Cobo (1964b:Bk. 12, Chap. XIX, 98) in turn compiled news on the war between Huascar and Atahuallpa. Many sacrifices took place in Cuzco when Quizquiz reported his victory and the seizure of Huascar, and "many children and llamas were killed in various ways."

We should bear in mind that the Indian religious tradition has not disappeared and that several rites are still practiced, some in their original form, others merged with or concealed in Christian ceremonies. As an example I will mention an interesting piece of information generously provided by Frank Salomon (pers. commun., letter of 12 December 1990). The Archivo Arzobispal in Lima holds an untitled file (Idolatrías y Hechicerías, Legajo 14 in the classificatory system established in 1989–1990) prepared in 1741 in San Francisco de Guisa, a subject of San Juan de Quivi. The file has nine folios. In it we find that people used to sacrifice a llama and to spray the walls of the church with its blood in order to strengthen the building. It is also reported that two stones, male and female, were worshipped as huacas. Some llama figurines painted with the blood of these animals were on top of the more rounded stone, which was male. (Readers who want more information on this point should see Flores Ochoa [1988c] and Tomoeda [1988, 1993].)

6.2 THE FIRST REPORTS

Several scholars have tried to establish the first time that a European saw a camelid, and the account that survived of this event. All, as far as I know, mention the diary of Pigafetta, to which we will turn forthwith. However, I would like to draw attention to a map shown to me by Ramiro Castro de la Mata. This map was probably prepared by Piri Reis in 1513 with data from the third voyage of Vespucci (published by Levillier 1948:43). The caption reads, "A Detail of the Turkish Planisphere of Piri Reis (1513). Cape Polonius, Cape Castle?" On this map, an animal with a tail was drawn south of the Río de la Plata. Its characteristics are like a guanaco's except for the tail. The addition of a tail is not surprising, considering the abundance of fantastic animals in contemporary maps—if it really is a guanaco. Furthermore, Piri Reis would have had the animal described to him, because he could not have seen it himself. If my assumption is correct—that is, if the animal drawn is a guanaco—this would be the first Western depiction of this camelid.

Antonio Pigafetta embarked with Magellan on his famous voyage around the world. He sailed as his servant but rose in esteem, and eventually became one of the men in this expedition who stood out the most (Trucco 1937:946). His diary (Pigafetta 1927:56–57) includes a description of the Port of St. Julian in Patagonia (49°30′S latitude), where Magellan's expedition wintered. On May 19, 1520, the writer recorded his first contact with a Patagón.[3] Among other things, Pigafetta says: "His dress, or rather his mantle, was made of a very well-sewn skin of an animal that is abundant in this country. . . . Strange animal: This animal has the head and ears of a mule, the body of a camel, the legs of a deer and the tail of a horse, and neighs like the latter."[4]

Interestingly, there is no entry for guanaco in the small vocabulary of Patagonian terms prepared by Pigafetta (1927). However, when describing the customs of the Patagonians, he wrote as follows: "They brought four of the animals I have mentioned, tied to some kind of halter, but these were small and of the kind they use to catch the big ones. For this they tie up the small ones to a shrub; the big ones come to play with these animals and then the men, who are hiding in the shrubbery, kill them with their arrows" (Pigafetta 1927:59). He continues: "The following morning [the native] brought the captain one of these big animals we have spoken of" (Pigafetta 1927: 60), and then adds that "[t]hese people dress, as I have already said, with the skin of an animal, and with these skins they also cover their huts, which they move to and fro" (Pigafetta 1927:63). From the text it follows that these were guanaco skins.

Arbocco Arce (1974:10) says the same thing but does add an interesting detail. He wrote: "[T]he wanaku was the first auchenid seen by the Spaniards in 1520, on Magellan's voyage. The boatswain Francisco Albo says of them: 'They are like camels without the hump.'" Albo was in fact a mate of Pigafetta and one of the 18 survivors of Magellan's expedition. He was the navigator on

one of the ships of the expedition and left an account entitled "*Derrotero del viaje de Magallanes*" (An Account of the Voyage of Magellan), which has more geographic and navigation details (Trucco 1936:83). I was unable to study this account. (Sumar [1992:83] mentions this phrase without giving his source, but he certainly got it secondhand.)

It should be noted that there is a third account of this voyage by a Ligurian navigator, perhaps Pancaldo (Trucco 1936:83), called the *Derrotero* or *Roteiro*, which might include some references to camelids. I was unable to find a copy of this account.

Attention must be drawn here to a misleading citation from Sumar, who wrote: "It is possible that the first account the Spanish conquistadors had of the Andean camelids was from an Indian chief in the Darién. Pressured by captain Pascual de Andagoya to tell of gold and the new lands, the chief modeled a lump of clay in the shape of a llama and told the conquistador: 'the people who have these rams [*carneros*] are that of the Kingdom of Gold'" (Sumar 1992:82–83). This statement does not in fact appear in Pascual de Andagoya's account (1954). The event described figures in the description (later we will go into more detail) that Vasco Núñez de Balboa received from the cacique Tumaco, and which was reproduced by Las Casas (1981:Bk. III, Chap. XLIX, 600) and Herrera (1945a:I Decade, Bk. X, Chap. III, 241). This was confirmed by Trimborn (1954: 32). By his own account Pascual de Andagoya (1954:224) reached the Darién after Vasco Núñez de Balboa. Finally, Andagoya does not mention rams [*carneros*] but sheep [*ovejas*] (Andagoya 1954: 246, 248). I do not know where Sumar found this citation, if it is correct; Sumar's article is full of mistakes, and sources are missing or wrong when given. But the fact remains that he did not check with the original source.

It is odd that when describing the arrival of Columbus to the island of Isabella, Irving (1854:38, 44) credits him with having said that "there were . . . animals other than lizards, mute dogs, some kind of rabbit the Indians called utias, and guanacos. The Spaniards looked at the latter with horror and disgust under the impression that this was some wild beast . . . but then they saw how tame it was and found the Indians considered it an exquisite dish." Irving (1854:38, 44) then refers to Haiti and says that its inhabitants "caught . . . the guanaco . . . without making the slightest effort." This author clearly got mixed up or used the term "guanaco" without knowing what these are, and credits Columbus with what he never said; the term does not appear in his diary or letters (Colón [1492–1493] 1956, 1958).

Several other scholars discussed this subject, but in regard to the Andes. One of them is Tschudi (1885:94–95; 1891:97; 1918:207; 1969:125), who credits Diego de Ordaz (1531) (this citation does not appear in his bibliography) with having obtained the first news of llamas from the Indians who lived in the area of the Meta River, a tributary of the Orinoco River. According to the Indians, the llamas lived on the Andean plateau of New Granada. I was unable to verify this assertion.

Ridout (1942:401) considers that "[t]he first civilized man who heard of the vicuña was Núñez de Balboa in 1513. They were later seen by Alfonso [sic] de Molina, one of Pizarro's men." This evidently is one mistake after another, due no doubt to careless use of the sources. As we shall now see, Núñez de Balboa is credited with having heard of these animals, but he definitely did not see them. As for Molina, he did see them and they were llamas, not vicuñas (see Cieza 1987: Part III, Chap. XX, 55).

There are several accounts of the case of Núñez de Balboa—for example, Pietro Martire d'Anghiera, "the European popularizer of the voyages of Columbus, who managed to include the expedition that discovered Peru in his last letter of 1526" (Porras 1986:31). In his Third Decade (1944:Bk. V, Chap. III, 205), written in 1516 (Ramiro Castro de la Mata, pers. commun.), Pietro Martire indicates that "Vasco boasts that in his deals with Tumaco he heard amazing secrets of the riches in that land which he does not want to disclose at present. [Vasco] notes that Tumaco said them to his ear." We can glean what these secrets were from the writings of Las Casas, but it is also worth noting that Tumaco was a cacique with land close to the Gulf of San Miguel (see Map II in Trimborn 1954). Las Casas wrote his history between 1527 and 1559, stating: "It was said that this cacique Tumaco told Vasco Núñez, pointing to-

ward Peru, that down the coast there was a great amount of gold and some animals on whom the natives place their loads, and out of clay he made a figure like that of the sheep of that land, with the neck they have which seems that of a camel. The Spaniards were amazed. Some said he lied, some wondered if these were camels, others whether these were deer or tapir, which abound in several parts of Tierra Firme and which are like small calves, but they differ in that their legs are small, almost a palm from the floor, and I believe they lack horns. And this was the second hint Vasco Núñez had of the riches and the condition of Peru" (Las Casas 1981:Bk. III, Chap. XLIX, 600). When Herrera wrote his noted *Décadas* in the early seventeenth century, in the First Decade (1945a:Bk. X, Chap. III, 241) he practically copied Las Casas verbatim when narrating the encounter of Vasco Núñez and the cacique Tumaco. Around that time Montesinos took from an unnamed source an account of the conversation between Tumaco and Basco Núñez (as he spelled it). This is a brief account: "[A]ll that coast long continued without end, and everywhere there was much gold, and the natives had animals that carried it, and they showed this by making the animals out of clay" (Montesinos 1906:32).

Trimborn (1954:32) gives the date when Vasco Núñez discovered the Gulf of San Miguel and heard the cacique Tumaco. It was in 1513. Here Trimborn summarizes the accounts of Las Casas (1981:Bk. III, Chap. XLIX, 600) and Herrera (1945a:Bk. X, Chap. III, 241), but adds that "this suggestion was made by the cacique Bononiamá." Trimborn based this on Rubiano (p. 193), an author of whom I know neither name nor date because he was not included in the bibliography. As far as I am aware, the cacique Bononiamá just mentioned gold and nothing more. No animal was mentioned. Also intriguing is the fact that Trimborn's only primary source is Herrera (1730; 1945a in my bibliography) and not Las Casas (1981:Bk III. Chap. XLIX, 600), whom Herrera copied. The other authors cited by Trimborn are late ones (Winsor 1886; Means 1928; Krickeberg 1922).

I feel obliged to mention the way sources are misused and the serious errors this causes. Cappa (1888:Tomo II, 38–39) describes the discovery of the Mar del Sur by Balboa without giving his source, and says that "[c]acique Cheapes modeled a quadruped in clay." He clearly got his information second-hand. The same happened to Gilmore (1950:433), who, basing his statement on Ignacio de Armas (1888:71), tells of an Indian cacique who described and "drew" a figure for Balboa in Panama that the Spaniards interpreted as a camel, a (female?) elk, and a tapir. This, as can be seen, is a far cry from the original account presented above, and it also led Gilmore (1950:433) to make a statement that needs no further comment: "This may mean, if the story is true, that llamas were diffused culturally from the Inca [sic] in Ecuador to the mountains of northern and central Colombia shortly before the sixteenth century, or that the cacique was a 'traveling' man."

From the data presented it follows that the first person who heard of the camelids probably was Vasco Núñez de Balboa, at least according to the written sources. If the animal that appears on Piri Reis's map is indeed a camelid, then it would be earlier, for the map was published in 1513. Pigafetta's account narrates what seems to be the first encounter Europeans had with this animal before the Spaniards undertook the discovery of Peru—again, so long as the animal pictured on Piri Reis's map is not a camelid.

6.3. TERMINOLOGY

A significant problem in this kind of study is interpreting the terminology used by the chroniclers when they were discussing native animals. More than one scholar has discussed this point. Shimada and Shimada (1985:17) noted this problem, namely, having to interpret what a "*carnero de la tierra*" and an "*oveja de la tierra*" were in sixteenth- and seventeenth-century sources. Quite often they seem to be synonymous. Shimada and Shimada (1985:17) also note that modern writers have repeated this inconsistency. Thus Antúnez de Mayolo (1981:58) says the "*carneros*" are llamas and the "*ovejas*" alpacas, while Rostworowski (1981:51) identifies the "*oveja*" with the llama.

This is clearly a complex problem. Benavides (1986b:391) provides a good example when discussing Coporaque (in the Colca Valley) in the 1591 *visita*. All of the livestock was described as "livestock of the land" (*ganado de la tierra*), that is, llamas and alpacas. From the text it can be deduced that the distinction between sheep (*ovejas*) and rams (*carneros*) perhaps applies to ages or species.

The problem is far more complex, however. We owe the best study on this subject to Dedenbach Salazar, who shows (1990:84) that the chroniclers actually use four terms: *ganados* (livestock), *carneros* (rams), *ovejas* (sheep), and *corderos* (lambs).

Dedenbach points out that the term "livestock" comprises "big" and "small" animals. The *Diccionario de Autoridades* (1976 [1726–1737], cf. 2.3.2.3.3) and the sources agree on this distinction and point out that the first term refers to cows and the second to sheep, goats, merinos, and pigs. Some sources also use Quechua terms, but in most cases these are in Spanish, with "of the land," "of [the] Indies," or "of the Indians" added; "*ganado*" is also used. Garcilaso de la Vega (1959:Vol. III, Bk. 8, Chap. 16, 146) mentions "big livestock" (*ganado mayor*) and defines it as "huanacullama," so these references do not always refer to European livestock.

Most chroniclers use Quechua words for wild camelids, although with Spanish spelling. Thus they say "guanaco" or "vicuña," but also "deer," without specifying whether it is a taruca (*Hippocamelus antisensis*), a lluychu (*Odocoileus virginianus*, white-tailed deer), or a camelid.

Dedenbach Salazar (1990:84) explains that "Antúnez de Mayolo (1981:58) says that 'we have to bear in mind that sheep was the name given to the alpacas, and rams [*carneros*] to the llamas.' I cannot confirm this, however. Some documents seem to use a consistent terminology, but most of the sources use terms arbitrarily." Here I agree with Dedenbach Salazar, but the study by Antúnez de Mayolo (1981) completely disregards the facts, is full of mistakes, and should not be relied on or taken into account.

In brief, it is evident that many terms were applied to the various kinds of domesticated camelids, while only two exist for the wild ones (Dedenbach Salazar 1990:70).

Not only are the data marshaled by Dedenbach Salazar excellent, but her study is unique in its category, so I feel obliged to summarize the most relevant points here. Dedenbach Salazar explains that in the Quechua sources, the term *llama* seems to refer primarily to domesticated camelids in general, but also to "a type of camelid opposed to other domesticated camelids (geographic accounts and descriptions, late chroniclers) like *paqu* and *wakaywa*, and lastly to all camelids (the comprehensive chronicles of the Indies)." Dedenbach Salazar, however, admits that this correlation is not consistent, and an interesting trend appears. *Llama* originally designated the domestic camelids in Quechua itself, as shown by the definitions in the dictionaries, but the way in which the Spanish speakers began to use the term may have supplied it with other meanings, some more restricted, others more extensive, according to each writer, and, quite possibly, according to their command of Quechua (Dedenbach Salazar 1990:71, 97).

Dedenbach Salazar (1990:84–85) correctly indicates that to draw a conclusion on the type of livestock present in terms of geographic areas, would "be to set out from non-verifiable assumptions." For example, it is incorrect to say there were no camelids on the coast, so "livestock" must therefore mean European animals, or that since most of the livestock in Collao were camelids, then "livestock" must mean them. This kind of argument can be misleading, and Dedenbach Salazar quite clearly shows that the sources must be studied one by one.

In a footnote, Dedenbach Salazar (1990:85, n. 6) presents one example that I prefer to quote in full:

> One example is a citation from Rostworowski about camelids (1981:51): "In 1549 the cacique of Atico called Chincha Pulca also declared having had some carneros and ovejas which were seized by Maldonado, his first encomendero (Galdós Rodríguez 1977:77)." Earlier this same document had said that "the Ynga sent them wool and they made him cumbe clothing and took it where he ordered them to," and "that the Ynga had in Parinacocha some livestock of sheep [*ovejas*] and they gave him thirty Indians to watch over them." From this it follows that the people of Atico had no livestock of the

land, for if they had indeed had some, these would have been pastured at a higher altitude. It is therefore very likely that these were ganado ovino.

I do not agree with this example, as I will discuss later.

Dedenbach Salazar (1990:86) also indicates that the later references should be carefully studied because everything seems to indicate that the European sheep were rapidly introduced into South America, something that might be misleading.

The confusion Dedenbach Salazar (1990:86) has in mind becomes clear in, for example, the *Vocabvlario de la lengva general de todo el Perv llamada lengua qqichua o del Inca* by Gonçález Holguin (1952), which dates to the early seventeenth century. Here we find the following meanings for "carnero": "Ram [*carnero*] of the land. Llama," "Hairy ram of the land kid to eat [*chivo de comer*]. Paco and smooth [*raso*] karallama," "Woolly ram [*carnero*] for carrying loads. Huacayhua apakllan a chacnana," "White, spotless ram [*carnero*] the Indians sacrificed, huacarpaña," "Male ram [*carnero*]. Orco llama female China llama" (Gonçález Holguín 1952:449). "Sheep" [*oveja*] in turn are "Sheep of the land. Llama," "Sheep of Castille. Castille llama," "Wild Sheep. Vicuña and huanacu," "Very woolly Indian sheep. Pacco" (Gonçález Holguín 1952:611). In this lexicon, then, the terms are intermingled, with llamas being indiscriminately called "ram" [*carnero*] or "sheep" [*oveja*].

To summarize, in general, the terms used for llamas were: livestock of the land, ram, ram of the land, rams of this land, sheep, and sheep of the land (*ganado de la tierra, carnero, carnero de la tierra, carnero de esta tierra, oveja, oveja de la tierra*). Santo Tomás is the only one who gives "wild ass" (Dedenbach Salazar 1990:53), to which the term "sheep of Peru," used by Zárate (1968:176), should be added. This shows the initial confusion over the terminology, for Domingo de Santo Tomás was one of the first students of Quechua who arrived in Lima with the first group of Dominican friars brought by Friar Vicente Valverde in 1538. He authored the first Quechua grammar and dictionary. The *Vocabulario* was published in Valladolid in 1560 (see Porras 1953:25–26).

Meanwhile, the alpaca was called very wool-ly Indian sheep, a kind of woolly ram [*carneros*] of the land, hairy ram [*carneros*] of the land, hairy rams [*carneros*], and small woolly rams [*carnerillos*] of the land (Dedenbach Salazar 1990:54).

For readers who would like more information on this point, I recommend the chapter in Dedenbach Salazar (1990:57–80) titled "The Classi-fication of Camelids according to Written Sources," where the data of each chronicler in Quechua and Spanish are studied. Here I will give some examples which shall often be used in this study.

Enríquez de Guzmán (1960:139), writing in 1535–1539, mentions both "sheep" and "rams." Fernández de Oviedo wrote in 1550 despite never having been in Peru, but he had good informants, such as Diego de Molina, the navigators of Pizarro, Ordóñez, and others, and so his chroni-cle is of "great historical value" (Porras 1986: 178–179). Oviedo wrote: "On the coast they call this animal *col* and in the highlands *llama*, the male or ram of these is called *urco*, and the lamb *uña*" (Fernández de Oviedo 1959a:Bk. X, Chap. XXX, 53). Las Casas (1948:9–10) also did not come to Peru. He wrote in 1552–1561 that "[t]he native inhabitants of those regions call a species of sheep *llamas* and the rams *urcos*," but in his works "guanacos" and "pacos" also appear.

Gómara (1946:243, 276), who also never went to Peru and wrote in 1552, discusses "sheep" and "pacos," but he explains that there are "two kinds of pacos which the Spaniards call sheep." Mean-while Zárate (1968:Bk. III, Chap. II, 176), an eye-witness (1544–1545), calls the llama "the sheep of Peru."

Cieza de León (1984:Part I, Chap. CXI, 294–295) traveled all over present-day Peru between 1547[5] and 1550. He tells us that "[t]he natives call the sheep llamas, and the rams vrcos," but goes on to add that "[t]here is another lineage of this live-stock that they call guanacos," and "there likewise is another kind of these sheep or llamas that they call vicunias," as well as that "[t]here is another kind of domestic livestock that they call Pacos." When discussing the "sheep" in the third part of his book, the chronicler explains that "[t]hese an-imals, which the Indians call as I said in my First Part, the Spaniards called sheep because they saw

wool on them and were so tame and domestic" (Cieza de León 1987:Part III, Chap. XXII, 62). Later he adds, "they call a certain lineage of rams 'pacos'" (Cieza de León 1987:Part III, Chap. XXXVIII, 110).

Father Acosta (1954:136), who traveled in Peru between 1572 and 1586, in turn tells us that the "livestock of the land [is what] our people call the rams of the Indies, and the Indians in the gen-eral language call *llama*." Meanwhile, in the late sixteenth century Matienzo (1967:89–90) men-tions "a livestock . . . we call sheep and rams. . . . The natives call the sheep *llamas*, and the rams *urcos*," but immediately adds that "[t]here is an-other lineage . . . they call *guanacos*. . . . Another kind of wild mountain livestock . . . they call *vicuñas*. . . . There is another they call *pacos*."

Garcilaso de la Vega (1966:Bk. 8, Chap. XVI, 513, 512; 1959:146–147) is not too explicit either. He wrote that "[d]espite the great difference be-tween them and European animals, the Spaniards call them rams and sheep." At the beginning of the chapter he says: "They are of two sorts, one larg-er than the other. In general, the Indians call them by the name llama, 'cattle.' . . . To distinguish, the larger are called huanacullama because of their similarity to the wild animals called guanaco." We can therefore assume that for Garcilaso, rams were llamas and the sheep were guanacos.

For Rivera and Chávez y de Guevara (1974: 169), who wrote in 1586, these seem to be syn-onyms: "rams of the land . . . and are called gua-nacos."

Between 1613 and 1653, Father Cobo amassed an unbelievable amount of data on the flora and fauna of Peru. In the words of Porras (1986:511), Cobo "had the gift of analyzing and giving definitions, of classifying." The chronicler thus clearly mentions the camelids, calling each by its name. When discussing the llama, Cobo says that "the Spaniards call it *ram of the land* even though it has no horns"; of the guanaco he com-ments that "these the natives of Peru call guana-cos"; and as for the alpaca, he explains that "these are called pacos," immediately adding that "the Spaniards call them woolly rams unlike those for cargo, which they call plain [*rasos*] rams" (Cobo 1964a:Bk. 9, Chap. LVII, 365–366). The only camelid he does not mention clearly is the vicuña, but he does give a good description instead (Cobo 1964a:Bk. 9, Chap. LVII, 367). Interestingly, the renowned Jesuit evidently realized that there was some ambiguity in the use of the terms, because he says "*llamas*, which we call sheep and rams of the land" (Cobo 1964b:Bk. 13, Chap. XXI, 201).

Finally, Herrera (1945c:313), who did not know Peru and who wrote between 1601 and 1615, just says: "the rams they call of the land . . . which the Indians call Llamas." He also mentions the "vicuñas."

This brief review clearly shows great confusion and the difficulties in distinguishing camelid species on the basis of written sources, as Dedenbach Salazar has clearly shown.

Horses must have caused the same amazement and awe in the Indians as camelids caused in the Europeans. On this point, Pedro Pizarro (1978:Chap. 7, 28) recounts what Atahuallpa was told by the messenger he had sent to Poechos to spy on Pizarro and his army. The chronicler credits him with the following words: "[T]hey are bearded thieves who came out of the sea and ride some rams like those of Collao, which are bigger than any in this land." In other words, it would seem that the Indians at first thought the horses were some kind of llama.

6.4. THE FIRST CAMELIDS TAKEN TO EUROPE

Of all the various things the Spaniards found in America that were unlike their own, it was the camelids that captured their attention the most, to the point that these animals were included among the items Pizarro decided to take back to Spain as samples. Porras (1978:126–127) wrote: "Pizarro left with Pedro de Candia carrying some of the Indians he had picked up while passing through the province of Motupe. . . . As samples of the land he also took some llamas, polychrome woolen textiles, and gold and silver objects collected during the voyage. Indians, llamas, ponchos, and silver would be the first to announce the existence of Peru to European eyes."

There are accounts of the departure of these animals. Andagoya (1954:242) wrote between 1541 and 1542, that is, before coming to Peru.

When discussing Tumbes and Pedro de Candia, Andagoya wrote: "From here he brought the sample of sheep and the account with which he came to Spain." Later there is the testimony of Cieza de León, which confirms that on his return from his second voyage, on passing through Tumbes Pizarro "departed after first taking some sheep the Indians gave him, which the captain ordered be healed and kept to take as sample" (Cieza de León 1987:Part III, Chap. XXIV, 69). Cieza even gives a brief account of the effect these animals had on their arrival at Panama: "They were amazed with the sheep on seeing their size, and prized their wool because they made such fine clothes with it" (Cieza de León 1987:Part III, Chap. XXV, 71). So "[w]ith this . . . Pizarro prepared for Spain, taking the sheep he had brought with him as proof of his account" (Cieza de León 1987:Part III, Chap. XXVI, 74). We also know by the description Cieza made (1987:Part III, Chap. XXVI, 75) that some llamas remained in Central America, because after Pizarro's voyage Almagro sent a ship to the Gobernación of Nicaragua, where Hernando de Soto, Hernán Ponce, and other Spaniards were. "They found out from Ribera what Peru and the city of Tunbes were. They saw the sheep and some blankets."

This is also confirmed by Molina "El Chileno" (1968:300),[6] who wrote his account in 1552, according to Porras (1968:35).

In the early seventeenth century, Herrera also recounted the voyage of Pizarro in his Fourth Decade, and mentioned the sheep he took with him, as well as the amazement these caused in Panama first, and later in Spain, but with no original data. Herrera plagiarized Cieza de León almost word for word (see Herrera 1945b:Bk. II, Chap. VIII, 179; Bk. III, Chap. I, 185; Bk. VI, Chap. III, 293). As Araníbar (1963:122) points out, Herrera "began a systematic pillage without stopping to acknowledge his sources."

I do not know whether the trail of the animals taken to Spain has ever been followed, or what their fate was. Neither is there any documentation of the amazement they must surely have caused on their arrival in Europe. According to Cieza de León (1987:Part III, Chap. XXVII, 76), when Pizarro reached Spain "[a]ll looked at the sheep he had taken." I have only found the 1550 by report Fernández de Oviedo (1959a:Bk. X, Chap. XXX, 53) which says: "These sheep have already been seen in Spain because the Marquis himself took them to Castille, where they are now famous. In this city there are some brought from that land." Guaman Poma de Ayala (1936:f. 370) took up the account of Pedro de Candia, which tells that when talking of Peru in Spain, it was "said of the rams [*carneros*] of the land that there were small camels."

Del Busto gives the most fantastic fictional accounts. In one of his studies he recounts how Pedro de Candia traveled with Pizarro to Spain: "The travelers disembarked in Seville and for the first time in the history of Europe, a herd of auchenids rhythmically crossed the city." Del Busto then explains that in Seville, the travelers heard the emperor was outside Castile, and so they left for Toledo, the home of the queen and the Council of the Indies. "This traveling retinue of Indians and llamas, led by a soldierless captain and an artilleryman with no cannon, stopped on the banks of the Tagus. What happened there is not clear, but the chronicle of Cieza has spared us part of this embarrassing situation. For instance, it is reported that the courtiers carefully observed the camels of the Indies, and the Tallanes who wore their dress beautifully" (Busto 1960–1961: 386). This is not true. In this chapter, Cieza (1987: Part III, Chap. XXVII, 76–79) only points out what was mentioned above, that is, "[a]ll looked at the sheep he had taken" (Cieza de León 1987:Part III, Chap. XXVII, 76). The Tallanes are not mentioned.

In another of his books, del Busto (1973:118) again wrote about this same event: "Likewise the Tallan interpreters with their multicolored mantles and gold and silver jewelry. Finally, half a dozen of those 'camels of the Indies' that were by now called the 'sheep of Peru' [5]." Note 5 (Busto 1973:180) says: "*Ibidem* [Herrera y Tordesillas, Antonio de, 1945]. IV Decade, Bk. VI, Chap. III, p. 293 of vol. V." (Exactly the same reference is repeated in del Busto 1978b:44.) The reference is wrong. This does not appear in Herrera.

The same book narrates the reaction of the counselors of the court: "The eyes of the counselors shone with this [the reference is to gold], and then looked at the Indians dressed with beau-

tiful mantles and golden crowns and diadems on their head. The three were good-looking and had nothing in common with the Caribs. The old and myopic counselors must have made them come closer to admire their silver bracelets, their anklets with copper rattles and sandals made of fiber. . . . *But what was entirely new for them—even though they smelled very bad—was this ruminant quadruped, an exotic and haughty animal that gave wool and looked like a hump-less dromedary*" (Busto 1973:119; emphasis added). No direct reference is given, but everything seems to indicate that del Busto is once again using Herrera (1945b:IV Decade, Bk. VI, Chap. III, 293, Vol. V). (This same reference is repeated in del Busto 1978b:45–46).

It should be noted that in this case, del Busto (1973) specifies "half a dozen" camelids, something he did not mention in his 1960–1961 study. As far as I was able to ascertain, it is not known precisely how many animals were taken to Spain. The reference to Herrera (1945 [1945b in my bibliography]) is incorrect. Herrera, I repeat, did not write what del Busto attributes to him. To avoid any doubt I prefer to cite Herrera's Fourth Decade to the letter, even though it might seem redundant. On leaving Tumbes, Francisco Pizarro "embarked *some* sheep to take as samples" (Herrera 1945b:Bk. II, Chap. VIII, 79; emphasis added). When Pizarro reached Panama, "[t]hey were amazed by the size of *the* [*las*] sheep" (Herrera 1945b:Bk. III, Chap. I, 185; emphasis added). On arriving in Spain, Pizarro received permission from the King to head for Toledo, where the court was, and there "all looked at the sheep he had brought" and the emperor "[a]dmired the strange shape of *those* animals" (Herrera 1945b:Bk. VI, Chap. III, 293; emphasis added). Herrera could not have said otherwise because he plagiarized Cieza de León, as noted. It is remarkable that del Busto did not go to the primary source.

6.5. THE SIGNIFICANCE OF CAMELIDS DURING THE CONQUEST

It is clear that camelids played a major role on both sides during the Spanish conquest. This becomes obvious when reading contemporary documentation. I do not want to dwell on the subject. Instead, I have chosen two facts that seem to reflect what I am saying.

During the third voyage of Pizarro, Diego de Almagro sent Nicolás de Ribera to convince the men in Nicaragua to join the conquest of Peru in his role as eyewitness. Some camelids were among the items Ribera took to show off the riches of the new land. This was recorded by Cieza de León (1987:Part III, Chap. XXVI, 75), as was already mentioned, and was later copied with slight modifications by Herrera (1945b) in his Fourth Decade (Bk. VI, Chap. II, 291–292), who wrote that "Nicolás de Ribera showed quite well the riches of that land, he exhibited the blankets, and the sheep, with which he raised the spirits of many."

On the other hand Cristóbal de Mena (1968: 140), who seems to have been an eyewitness,[7] tells that when Pizarro and his army marched to Cajamarca, "two Indians came with ten or twelve sheep by command of Atabalipa, and gave them to the governor." Mena adds that they spent five days in those highlands, and that "one journey" before arriving "to the encampment" the Spaniards received from Atabalipa "many sheep as presents."

We thus see that Europeans immediately realized how valuable these animals were, and used them advantageously in two different ways that were of great import for the success of the conquest of Peru, namely, to convince the men of Nicaragua and to impress the King of Spain and his court. It was not just gold, as is often said, that played a major role. On the other hand—and there is no need to belabor this—we see the value these animals had for the natives. They were the first presents the Inca sent to the invaders.

6.6. EUROPEAN DESCRIPTIONS OF CAMELIDS

A point I find worth emphasizing has to do with the descriptions of camelids provided by the chroniclers and some later writers. In them we see the amazement of the Europeans who were seeing these animals for the first time, but also the problems they had in describing them, and above all in understanding them.

The Spaniards who first saw the camelids mentioned them but did not describe them. The first description, although brief, is perhaps that by Enríquez de Guzmán (1960:139), who wrote between 1535 and 1539 (Porras 1986:155–157). He mentions dress and "linings" when discussing what lies "below the equinox line": "These linings are made with wool from said sheep, which are big and are used as beasts of burden. They tame them as we do there [in Spain]." This paragraph is interesting because it clearly shows the chronicler was not familiar with the animals and probably believed they were tamed like horses, whereas the way men treat llamas and alpacas, which are tame animals, is quite different.

Fernández de Oviedo, I once again repeat, was never in Peru, and the section on Peru in his chronicle was written in 1550 (Porras 1986:179). Although Las Casas said of his writings that these were no more than "empty talk save when he discusses trees and herbs" (Porras 1986:177), they actually "have a great historical value as regards the conquest of Peru. They are a testimony of direct actors. . . . Prescott does not consider it a history, but rather a collection of notes for a great history" (Porras 1986:179). Besides, Fernández de Oviedo personally had the opportunity of seeing a camelid, but we do not know whether this was a llama or an alpaca.

Fernández de Oviedo (1959a:Bk. X, Chap. XXX, 52–53) titles one chapter in his book "The Sheep and Livestock in the Southern Lands, in Tierra Firme, in the Gobernación of New Castile Where Atabalipa Was King." Here Fernández de Oviedo notes that

[i]n New Castile and the Gobernación of the Marquis Don Francisco Pizarro, where Atabalipa was a most rich and powerful lord, the Indians have three kinds of sheep: some small like goats from Guinea, and some slightly bigger, and others bigger than all. The big ones are the size of a small ass but have skinny legs, [and] a long neck similar to that of camels, save that these [sheep] do not have a hump like them. However, in legs and hands and all the rest they are quite similar to camels. They ruminate like sheep, and are such that the Indians use them to carry loads and carry on them

what they will, so long as it is a moderate weight.

Fernández de Oviedo then goes on to say what this animal was called, as was previously mentioned. He then discusses the llama and adds:

And they are nice animals to see, very tame and domestic. The medium-sized ones are of the three kinds I mentioned. These are the ones that have a most fine wool that resembles silk, with which the Indians make very rich cloth. The Adelantado Don Diego de Almagro gave me one of these bigger sheep in the city of Panama, and I embarked in a caravel at Nombre de Dios, and coming by sea it died in this gulf and we ate it. It seems to be one of the best meat in the world. I have not seen the other two kinds of sheep of that land. [However, s]ome citizens of this city of Santo Domingo who have been in that land say that one and the other are good meat. They are the same color as the sheep in Spain, white and black, and both colors combined, and the wool is plain and not merino wool, and for the most part the big ones are rare and their hair short, but the wool is longer on their back. The middle-sized ones are a bright reddish color and white, the two colors combined in these, or each on their own. Among the smaller ones, which are wilder, one can chance on a field upon a wild herd of five hundred and a thousand of them, black and very delicate.

It is almost certain that Fernández de Oviedo had a llama, so that the descriptions of the animals he had not seen would correspond to the alpaca and vicuña.

Further on, when reporting "several matters in the gobernación of Francisco Pizarro," Fernández de Oviedo gives some news that Pedro Corzo told him. According to del Busto (1973: 178), Corzo was the first non-Spanish pilot who sailed along the coasts of Peru. It seems that during the third voyage of Pizarro, he sailed as far as Lima.

According to Fernández de Oviedo (1959c: Bk. XLVI, Chap. XVII, 94), this Pedro Corzo had "a long experience . . . in these matters of the Indies, where he has for several years been sailing and walking by sea and land." It thus seems Corzo traveled in Peru. Lima and its surroundings are mentioned, as well as some animals and plants, but there is no way to ascertain whether these pertain to the coast or the highlands. Fernández de Oviedo (1959c:Bk. XLVI, Chap XVII, 94) credits the following description to Corzo: "There are some animals the size of deer with a cloven hoof, and all in all they are like deer save for the hair, which is coarse and very thick, are not horned nor do the Indians eat them. They are like the animals called *mufros* in Italy, and these walk together in great herds of five or six thousand, more or less." This clearly is a camelid. There is no way to ascertain whether these were llamas, as del Busto (1973:178) assumes. They could well be alpacas. (We should note that the reference given by del Busto [1973:183, n. 113] is mistaken. He gives Book VIII, when it actually is Book XLVI of Fernández de Oviedo [see above].)

In his chronicle, Gómara (1946:276) discusses the "[q]uality of, and the Atmospheric Conditions in, Peru." When discussing the "cold highlands," Gómara says there were "[t]wo kinds of pacos, which the Spaniards call sheep, and some . . . are domestic and others wild." Later on, when discussing Collao, Gómara (1946:243) states that "Fernando and Gonzalo Pizarro caught [there]. . . much sheep, which are more camelid-like [acamalladas] from the withers on, even though they look far more like deer. The ones called pacos grow a very fine wool . . . but walk slowly. This goes against the impatient temperament of the Spaniards. Tired, they turn their head to the knight and spit a noisome water. When they tire too much they fall, and do not rise again until they have no load on even if they are beaten to death." I find Gómara's comments extremely interesting. Although he was never in Peru, Gómara must have had an informant who was well acquainted with these animals. In this instance, it could not have been Zárate, with whom he agrees on other points (Porras 1986:191). Used to driving horses, the Spaniards evidently must have had many difficulties in getting used to the behavior of the camelids, which is completely different from that of horses.

In his chronicle, Las Casas (1948:9–10) gives a good description of the camelids, but it was clearly copied from Cieza de León and so is not worth including here. I also have the impression that Las Casas had access to the work of Gutiérrez de Santa Clara because when describing the urcos, he says that these are "ass-like beasts slightly bigger than those in Sardinia" (Las Casas 1948: 9), while Gutiérrez de Santa Clara (1963:Bk. III, Chap. LVII, 235) says of these animals that they are "the size of Sardinian asses." The comparison both writers draw with Sardinian asses is striking and is too similar for it to be a coincidence.

The chronicle Zárate wrote (1968:Bk. III, Chap. II, 176–177) is a firsthand account, but it pays little attention to camelids. These are compared with camels, "to which they would be very similar in shape, were it not for the fact that the camel's hump is missing." Then he adds, "This animal is most beneficial because it has a very fine wool, specially those called pacos, who have their locks long. They eat little, especially those who work, eat maize, and can go four and five days without water. The meat is delicious and healthy like that of Castilian rams." In this case it is interesting to discover how an eyewitness who certainly did see these animals makes a mistake in their size by comparing them with the camel.

As for Gutiérrez de Santa Clara, there are some discrepancies among scholars. Araníbar (1963:109, 119) questions his presence in Peru and points out that perhaps he was in Collao around 1546. Although he does not state it conclusively, Porras (1986:241) accepts his Peruvian sojourn and attributes much importance to his chronicle. Even so, the descriptions Gutiérrez de Santa Clara gave of camelids lead to some confusion. If he personally did see them, his experience was such that he did not become familiar with the animals.

Gutiérrez de Santa Clara (1963:Bk. III, Chap. LVII, 235) describes "five kinds" of camelids, two of pacos and one each of guanacos, urcos, and llamas. Of the latter he says they "are very big, and loaded like horses or mules which go in droves, and there are huge herds of them which the Indians raise, and they are very tame . . . and walk har-

nessed like ass- or mule-like beasts. . . . The other kind is called urco. These are the rams they eat, and very many are brought to the cities to be sold, and are weighed by the butchers for all, and it is a good meat delicious to eat, and these are the Sardinian ass-like rams, which are very fat." Immediately after, Gutiérrez de Santa Clara adds, "there are other kinds of rams whom some call guanacos. These are like deer [*venados berrendos*] who run a lot and live in the wild in big herds." Last of all Gutiérrez mentions the "two kinds of pacos, which strictly speaking are the rams [*carneros*] and sheep [*ovejas*] of this land. Their wool is very coarse and used to make coarse mantles and blankets, and mattresses and tapestries of the land which are now made here. The other wool is very fine, and from it very rich mantles and dresses are made for the great lords, and of this wool was made the tassel the Yngas had as royal insignia, which no other person of any state and condition whatsoever could bear on his forehead, on pain of committing a crime of *lessae majestatis* [sic], and now after the Ingas died the curacas wear them."

From these descriptions it follows that something is wrong. Gutiérrez de Santa Clara calls the llamas by their name, but when discussing the urcos he makes a mistake because he is clearly talking of llamas too; that is, Gutiérrez de Santa Clara is distinguishing the llamas used as beasts of burden from those that were eaten. There is no problem with the guanacos. Where Gutiérrez de Santa Clara once again goes wrong is when he mentions the "two kinds of pacos." The description shows he is confusing alpacas with vicuñas, and thinks both are "pacos."

Strangely enough, Cieza de León, whose descriptions are so detailed, is somewhat spare when discussing the "Rams [*carneros*], Sheep [*ouejas*], Guanacos and Vicunias which Abound in Most of the Peruvian Highlands." Cieza de León says (1984:Part I, Chap. CXI, 294–295):

> The natives call the sheep llamas, and the rams vrcos. Some are white, others black, and still others brown. Their size is such that some rams and sheep are as big as small asses, have long legs, and are fat at the belly. Their neck and shape resembles that of the camel. The heads are elongated and

look like those of the sheep in Spain. The meat of this livestock is very good when fat, and the lambs [*corderos*] are better and more tasty than those of Spain. It is a very domestic livestock and does not make noise. . . . They eat the grass in the fields. When they complain, they whine and lie down like camels. There is another lineage of this livestock which they call guanacos, of this shape and size: they are very big and big herds of them go wild in the fields, and they run jumping so speedily that dogs who are to catch them must be fleet of foot. Besides these, there is also another kind of these sheep or llamas they call vicuñas. These are faster than the guanacos, but smaller. They live in the wild, eating the grass the Lord grows there. The wool of these vicuñas is excellent and is all so good that it is finer than that of the merino sheep of Spain. I do not know whether cloth can be made with it, but I do know the clothes they made for the lords of this land are well worth seeing. The meat of these vicuñas and guanacos is somewhat like that of wild meat, but it is good. In the city of La Paz I ate dried meat from one of these fat guanacos at the inn of Captain Alonso de Mendoça, and felt it was the best I had ever seen. There is another kind of domestic livestock they call Paco, but it is very ugly and woolly. It is the same shape as the llama or sheep [*ouejas*], save that it is smaller. When young, the lambs look a lot like those of Spain. They give birth to one of these sheep once a year, and no more.

We see that the descriptions given by Cieza de León are brief but correct, and he does not confuse the species. Apparently there were two kinds of llamas for him, the "llama sheep" and the "rams vrcos," which seem to be the llamas used as pack animals and as meat. In this he agrees with Gutiérrez de Santa Clara.

La Gasca, who presided over the Lima Audiencia and was in Peru in 1547–1550 (Markham 1941:125–132), left us his *Descripción del Perú* written in 1551. Of the camelids he said, "They like-

wise had great herds of the sheep of the land, and these are like camels save that they are not that big, nor do they have a hump" (La Gasca 1976:49).

Strangely enough, despite having made a series of very detailed and perceptive observations of other aspects of the Andean landscape, Father Acosta apparently only described the llama and the vicuña. In my opinion, when describing the llama he confused it with the alpaca, and was apparently unaware of the guanaco. Acosta wrote:

> Peru has no greater riches or advantage than the livestock of the land which our people call rams of the Indies, and the Indians in the general language [i.e., Quechua] call *llama*, because all things considered, the animal yields the highest benefits with less expenditure than all known animals. From this livestock they get food and dress, like the sheep in Europe, and take more than needed, for they use it to take and carry their loads. On the other hand, they do not have to spend on horseshoes, saddles, packsaddles, or barley because they serve their masters for nothing, and are satisfied with the grass found on the fields. In this way God provided sheep and donkeys in one same animal. Since these are poor people He did not want to have them incur any expenditure, so the pastures are many in the highlands and this livestock does not ask for more expenditures, nor does it need them. . . . Their meat is good if tough [*recia*], that of the lambs [*corderos*] one of the best and most delicious that are eaten. (Acosta 1954:136–137)

As regards the vicuña, Acosta says that these "graze and live in very high mountains, in the coldest and uninhabited places which are there called punas. The snow and the ice do not bother them, but they seem rather to enjoy it. They go in herds and run speedily. . . . It is understood that they do not breed a lot, so the Inca kings forbade the hunt of the vicuña, unless it were for festivals and on their orders. . . . The meat of the vicuña is not good even though the Indians eat it, and they made *cusharqui* with it" (Acosta 1954: 135).

Although Licenciate Matienzo, a judge in the Audiencia of Charcas in the late sixteenth century (see Porras 1986:548), could have left us an original description of the camelids, he failed to do so. The one he gave (see Matienzo 1967: 89–90) was simply copied from Cieza de León (see above) almost to the letter. This might have been due to the dislike he had for all things Indian, as his descriptions are all a "hostile ethnography," as noted by Porras (1955:154).

Garcilaso de la Vega left a detailed account of the camelids, to which he dedicated two chapters of his chronicle. Here I will recall just the most important parts. In the chapter on "[t]heir tame animals: the flocks they kept" (1966:Bk. 8, Chap. XVI, 512–516; 1959:146–152), Garcilaso reports that these animals "are indeed so tame that a child can drive them anywhere, especially those that are used as beasts of burden." He then explains that these animals "are of two sorts, one larger than the other. In general the Indians call them by the name of *llama*, 'cattle.'. . . To distinguish, the larger are called *huanacullama* because of their similarity to the wild animals called *guanaco* from which they only differ in color, the tame animals being of all colors, as horses are in Spain . . . while the wild guanaco [*huanacu*] has only one color, a polished chestnut with the flanks of a lighter brown." Garcilaso then adds that "[t]hese creatures are as tall as European deer, and resemble no other animal so much as the camel, without its hump, and without a third of its bulk. They have long smooth necks." After specifying the use that could be given to the skin of this animal (which I discuss later), Garcilaso notes that "[t]he llama ["livestock" in the Spanish edition, D.B.] is also used by Indians and Spaniards for the transport of merchandise to all parts, though they travel best between Cuzco and Potosí where the land is flat. The distance is nearly two hundred leagues; and they also go between the mines and many other places." However, the chronicler warns that "they must not be pressed beyond their usual pace or they tire: they then lie down on the ground and there is no getting them up whatever one does, even if the burden is removed. One can then flay them, and there is no way of getting them to move. If one persists in trying to get them to rise and goes up to them to lift them, they defend themselves with the dung they have in their maw which they bring up and spit at the person nearest to them, aiming it as [sic] his face for preference. They have no other weapon of defense, not even horns as deer have. Despite the great difference between them and European animals, the Spaniards call them rams and sheep." Garcilaso continues thus:

Although the flocks are so large and the distances so great, the animals do not put their owners to any expense by way of food, lodging, shoes, packsaddles, trappings, girths, cruppers, or any of the numerous other requirements that carriers need for their beasts. When they reach the place where they are to spend the night, they are unloaded and turned loose in the fields where they graze on whatever grass they find. They keep themselves in this way for the whole journey and require neither grain nor straw. They will eat maize if given it; but the creature is so noble that it can manage without even when working. It does not wear shoes, for it is cloven-footed, and has a pad instead of a hoof. No packsaddle is needed, for their wool is thick enough to take the weight of their load, and the carriers take care to arrange the packs on either side, so that the strap does not touch the spine, which might prove fatal. The packs are not attached with the cord carriers call twine [*lazo*], since as the animal carries no frame or saddle, the weight of the burden would cause it to cut the flesh. The packs are therefore sewn together by the canvas, and though the sewing rests across the backbone, it does no harm provided the strap is kept aside."

The chronicler then says:

Of the smaller stock, the *pacollama*, there is less to say, for they are of no use for transport or any other service, but only to eat, their flesh being slightly inferior to the other, and for their wool, which is long and excellent. It is used for making . . . textiles.

In the following chapter (Garcilaso 1959:Bk. 8, Chap. XVII, 152–153) on "[t]he wild flocks," the chronicler explains as follows:

Before the arrival of the Spaniards, the Indians of Peru had only the two types of domestic cattle we have mentioned, *paco* and *guanaco* [*huanacu*]. They had more wild cattle, but treated it like the tame, as we have said in speaking of the periodic hunts that were held. One of the wild varieties was called *guanaco* [*huanacu*], a word also applied to the larger tame llama which resembled it in size, shape, and the quality of its wool. . . . While the females browse in the valleys, the males keep watch on the high hills and give warning by whinnying like horses if they see people. If they are approached, they take flight, driving the females in front. Their wool is short and coarse, but is also used by the Indians for textiles. . . . There is a wild animal called *vicuña* which resembles the smaller domestic animal, the *paco*. It is a delicate creature with little flesh, but much fine wool. . . . The *vicuña* is taller than the largest goat. The color of its wool is very light chestnut, or what is called *leonado*. It runs so fast that no greyhound can catch it. . . . It grazes in the highest deserts, near the snowline.

Koford (1957:212) wrote: "It seems . . . that vicuñas were not well known at first hand by Garcilaso de la Vega, because for basic information he draws on the writings of Acosta." This is a gratuitous critique. True, Garcilaso had read Acosta, as well as Gómara, Zárate, and Cieza de León (see Porras 1986:397), but in this case he did not use any of them. It is to be doubted that he had ever seen a vicuña in his youth. Nevertheless, his description, although not exhaustive, seems to be essentially correct.

When writing of the camelids in 1586, Cabeza de Vaca (1965:340) specified that these "are tame and [the Indians] need not use grains nor horseshoes, packsaddles, bridles or headstalls; a rope suffices to lead them as they see fit." In this same year Carabajal (1965:206) reported, while describing the corregimiento of Vilcas Huamán, that "[i]n this province there is livestock of the land, which are camel-like rams, albeit smaller, from which the wool the natives dress with is taken, and these rams are so tame they are used to carry loads. . . . There are some other wild animals almost like the rams themselves, which are called guanaco and vicuña."

The Mercedarian friar Murúa (1922:154) also included in his chronicle a description of the camelids, but it is not an original one because it was copied from Gómara (see above).

The account left by Ramírez (1936) is interesting because it was written by a clergyman who lived in Peru for several decades, and was written "solely on the basis of the personal experiences of the author, [who was] . . . an eyewitness of the very century of the conquest" (Trimborn 1936:xii). His account dates to 1597.

When discussing the highlands, Ramírez wrote:

In these punas or paramo lives much livestock of the land. . . . [T]his livestock of the land is all the riches of the Indians of Peru because it is most useful for several purposes, they eat the meat, dress with the wool and load them with their merchandise and other things they see fit. . . . [T]hey walk from sunrise to noon, and at that time unload them and let them loose and they graze until nightfall, when they collect them and tie them with some kind of breast harness, ten animals by ten animals which ruminate all night long, like oxen. Peru could not feed itself if there were no herds of these rams, of which the Spaniards buy very many from the Indians.

Here Ramírez explains how the Spaniards purchased the animals at one price and then sold them at a higher one, immediately adding that "there are many sales of this kind and for a high number of animals. This livestock, which is used to carry loads, is the most plentiful in Peru and in some provinces there is more than in others. There are some differences between them, like in the breeds of horses in Castile (fol. 9r). It is a live-stock with little wool, and what it has is very coarse. It is used to make ropes." Ramírez then adds, "the shape and size of these rams is like that of camels: they have the neck and head like camels and the feet and hands cloven with nails like the rams from Castile; it is a small livestock in comparison with our beasts" (Ramírez 1936:17–18).

The clergyman then describes the "pacos" and says "[t]here is another kind of livestock they call pacos, which is slightly smaller than what has been said but this animal likewise provides wool. . . . [A]mong the Indians there are many of these animals in all the highlands and it is of less value than the others" (Ramírez 1936:18).

Ramírez also discusses the guanacos and vicuñas, which he rates as "wild animals, of the same size and shape as the tame ones. One is called guanaco; these guanacos are swift animals, they are bred in the paramo and punas of a very cold land almost among the tame livestock. The guanaco has little wool of a color that lies between white and brown. The Indians hunt them with ropes and dogs. . . . the other species is called vicuña and is a smaller livestock than all of the above-mentioned ones, and swifter than the rest. It has little wool." Ramírez then describes the fineness of the wool from these animals (Ramírez 1936:18).

Father Ocaña wrote at a later time, in the seventeenth century, but not only is the account he left of the "rams" interesting, it is also original in some respects, as, for example, when describing the sexual organs of these animals and the way they mate, which the chroniclers do not mention. Ocaña was in Potosí on 18 July 1600 and seems to have written his account in Peru in 1605, though perhaps he finished it in Mexico, where he died in 1608 (Álvarez 1987). The description Ocaña gave of these animals came from his stay in Potosí. He wrote of

rams, which are like one-year old calves, long-legged and slimmer in all their limbs than calves.[8] The rams have about five long cuartas [which is equal to 0.2089 m; Llerena Landa 1957:43] from leg to back, which is almost a vara [or 0.8359 m; Llerena Landa 1957:205] and a half; the neck is very high, without horns; the head,

both in males and females, has slightly long ears; and the features of the face are the same as those of a ram; the neck is almost an incomplete vara long, slightly less; the tail is small, so that by the neck and by their lying down and getting up they seem to be some species of camel. The males have their [sex] organ below the belly, like the rams of Castile, but it is not straight and instead turns back like a half crook, and so when they urinate they hit outside the back legs with their urine, as if the penis had strength because it turns a lot, and when mating the female lies on the ground and kneels down. . . . What they eat is a grass they call icho, which is like the . . . esparto grass of Spain. (Ocaña 1987:166–167)

In regard to llamas, Ocaña wrote: "This livestock takes a lot of work, does not eat at night and dies easily because it is very delicate, particularly in warm lands where the warmth hurts it much and then it gets much mange [carache]" (Ocaña 1987:182).

The observations Ocaña (1987:182) made on the behavior of the llama are interesting.

These Indians walk with the rams of the land, which are raised unhurriedly. And they have so much of it that if one of these rams lies down with its load it will not get up even if the Indian beats it a thousand times or kills it, nor do they get up, no matter how much their tail is twisted or their testicles squeezed. And the Indian sits down on the road beside the ram and looks at it for two or three hours, until the ram itself gets up. And the other Indians walk on ahead with the rest. And this is why Spaniards do not walk with them, just Indians, because it has happened that when some Spaniards came with them, the animals lay down because they were tired, which is a common occurrence, and the Spaniards tried to make them get up and beat the animal to death and left them dead on the road. This is why they do not walk with them, just the Indians, who are unhurried.

Oliva (1995:11), who was in Peru between 1593 and 1642 (but he wrote only in 1630), simply noted of the vicuñas and guanacos that these are "animals that do not exist in Europe, as big as wild goats and swift in running."

Although Father Cobo is known to have used the data of at least Pedro Pizarro, Gómara, Molina, Acosta, Garcilaso, and Ramos Gavilán (Araníbar 1963:125) because he wrote in 1653, in this case his description seems to be original. I find it important, like all that Cobo said on the flora and fauna. In his monumental work he described all four camelid species separately. This is important, because we have seen that great confusion exists among almost all sixteenth- and seventeenth-century writers.

First of all, Cobo (1964a:Bk. 9, Chap. LVII, 365) discusses the llama:

It is as big as a small ass a vara and a half tall, slimmer, and with long legs, which are small and have cloven hoofs; fat at the belly, with a small tail a palm long that it always carries erect; a large head similar to that of a sheep; a neck similar to that of a camel because it is long and slender; it ruminates and grows wool like a sheep, for which reason, and because its head and legs are similar, the Spaniards call it *ram of the land*, even though it has no horns. There are brown, black, and white, and combinations of these colors.

Cobo (1964a:Bk. 9, Chap. LVII, 366) then describes the alpacas, noting that "[o]f the tame llamas, some are to carry cargo and others not, and only their wool and meat is used. These are called pacos and are of the same color and shape as the pack animals, save that they are slightly smaller and not as strong, and grow a longer, finer, and uniform wool all over its body, including the neck and head. For this reason the Spaniards call them woolly rams, unlike the pack rams, which they call plain rams."

Of the guanaco, Cobo (1964a:Bk. 9, Chap. LVII, 366) notes that "[t]here are three differences among the llamas, some [are] tame and the others wild. These the natives of Peru call guanacos, which are in all like the tame and domestic llamas, save that all are brown and are never

tamed and domesticated, and their wool is shorter and rough." He then describes the vicuña (1964a.:Bk. 9, Chap. LVIII, 367) as a "wild animal smaller than the llama, and much similar to it in disposition and build. It is the size of a goat, thinner and with longer legs, cloven hooves, a long and slender neck, and the upper part of the body and most of it is a golden color tending to a light leonado [chestnut], with the belly and the lower chest white. . . . These are the fleetest and fastest animals that are known." Later on, Cobo (1964b: Bk. 11, Chap. XIV, 40) adds that these animals "never come down from the high sierras and very cold paramos where they live."

Guaman Poma de Ayala does not give this kind of description, but he does note how the news reached Spain, and how these animals were described: "and he said there were small camels— of the rams of the land" (Guaman Poma de Ayala 1936:f. 37(0); 1969:72).

Later writers also left some descriptions of the camelids. In the early seventeenth century the writer known as the Anonymous Portuguese, now known to have been Pedro de León Portocarrero (see Lohmann 1967), said: "In these mountains walk many guanacos, which are like the rams of Peru, and neither these nor the vicuñas are raised on the coast. These rams are bigger than ours, taller and longer, and the neck is very long up to the head, and their wool well done, it is white and in others dark brown. This is the best livestock and the most beneficial one in the world, because they work." León Portocarrero then notes the way they are loaded and the distance they cover, and finishes his description by stating that "their sustenance is nought but what they eat in the pastures. And if they get tired on the road, nothing can be done but to let them rest, because they do not want to get up until they are rested even if beaten to death, and the one that flees runs faster than a horse" (Anónimo Portugués 1958:80). The writer clearly was not too familiar with these animals and was somewhat confused, because he did not realize that it is the llamas that carry loads, and they differ from the guanacos and the vicuñas.

Herrera was never in Peru, as is well known, and he wrote in the seventeenth century. We should bear in mind that he "proceeded to make a systematic pillage [of the writings of other people] without acknowledging his sources" (Araníbar 1963:122). We also saw that Herrera copied Cieza (see also Pease 1992:144). As regards camelids, Herrera gives quite a good description, but I have been unable to establish his source. In his Fifth Decade (1945c:Bk. IV, Chap. IX, 312–313), Herrera discusses "[d]iverse animals and birds of Peru." He says:

> [T]he vicuñas of Peru and the rams do not exist in New Spain; the rams are called of the land and are tame animals, and most useful. The vicuñas are wild, do not have horns, and do not exist in all the world save for Peru and Chile. They are bigger than goats, smaller than calves, are almost tawny-colored, and live in the highest mountains, in the coldest and uninhabited parts that are called the puna. They go in herds and run swiftly. On seeing travelers they flee and place their young in front. . . . The livestock of the land of Peru are a great treasure, especially the rams which the Indians call llamas. From them the Indians get their dress, food, and transport, because these carry their loads. As for food, they are satisfied with the grass in the pastures. Of these animals some are woolly, others plain, and the latter are better for carrying loads, are bigger than the big rams and smaller than calves. The neck is like that of a camel in many colors. . . . This livestock likes cold climates. The plain rams tend to get scared and run up the mountains. . . . The pacos usually get mad and bored with their load, and they lie down mad and do not get up even if they are killed. In this case the Indians sit down beside them and pamper and tame them, and they get up though sometimes they take two or three hours.

Although the description is generally correct, it is clear that the writer had not seen the animals he was describing and so did not distinguish the various species clearly, as Cobo did.

It is clear, and in this I agree with Dedenbach Salazar (1990:109), that the most detailed and accurate descriptions are evidently the ones given by Garcilaso de la Vega and Father Cobo. However, in general we find that the Spaniards were unable to give original descriptions or to avoid drawing comparisons with the fauna they knew, owing to the great difference between these animals and the ones they were used to. On the other hand, their greatest difficulty clearly was distinguishing the llama from the alpaca, a difference not always grasped, whereas the differences among these animals and the vicuña and guanaco were certainly more obvious.

The comments Flores Ochoa (1990b:85–86) made in this regard are most interesting. Flores Ochoa believes it is clear that llamas were more important than all the other species in the sixteenth century, because the chroniclers mention them far more often. On the other hand, they paid less attention to the alpaca, so their descriptions were confusing or subject to more than one interpretation when they did. Flores Ochoa finds it hard to believe that when the chroniclers mentioned the "livestock of the land" in a general sense, that is, without distinguishing the llama and the alpaca, it was because they could not identify them or distinguish one from the other. To support his position, he indicates that the chroniclers show their familiarity with the Andean fauna when describing wild camelids, clearly distinguishing the vicuña from the guanaco, and the two of them from the domestic species. The descriptions made by Cobo and Acosta, the most famous of all, pay more attention to the llamas and are not so careful or detailed when discussing the alpaca. Flores Ochoa believes that these omissions in such "diligent" observers cannot be taken as carelessness or lack of attention. He concludes that

> [i]t is therefore possible that the herds of alpacas were not that many and were concentrated in some parts of the highlands. These herds were meant to give the fibre needed to weave clothes, possibly the ones known as *qompi*, the finest textiles. Llamas may have supplied the fibres needed for the common dress or *awasqa*, as well as that used for sacks, ropes, slingshots, and other items meant for a rugged use. Only a brief span observing alpacas and llamas is needed to see which parts of

their anatomy distinguish them. The frequent reference to llamas therefore reflects the prevailing conditions of their greater importance and number, for they are found everywhere. This emphasis also shows in religious activity when it comes to llamas. No mention is made of alpaca sacrifices. Llamas are the animal offered in various religious festivals held in the city of Cuzco, in the shrines and in other sacred places. (Flores Ochoa 1990b:86)

The position held by Flores Ochoa is interesting, but I do not fully agree. As a "man of the Andes," he uses concepts that are precisely those the Europeans did not understand. If the descriptions reproduced here are read carefully, we find much confusion and ambiguity. This was not intentional. It was definitely due to the problems presented by having to describe animals never seen before, with whom the Europeans were not in contact long enough to understand and distinguish among them. It should not be forgotten that European observers were often intrigued by the strange behavior of these animals, which they frequently emphasized and found so different from that of horses or asses. This shows both strangeness and lack of understanding. On the other hand, the frequent comparison with the camel is not just due to the need for a contrast with an animal known by Westerners; it also conceals their bafflement at this resemblance, which for them was incomprehensible.

6.7 THE GEOGRAPHIC DISTRIBUTION OF CAMELIDS, BASED ON WRITTEN SOURCES

Much has been written on the spatial distribution of camelids, both before and at the time of the arrival of the Europeans, but as far as I am aware no one has studied the problem using the existing sources except Dedenbach Salazar (1990). Curiously, attention has almost always focused on the llama, with the other camelids ignored. Let us look at examples picked at random. Gilmore (1950:433) wrote that before the eighteenth century, in both historic and prehistoric times, the distribution of

the llama was along the Peruvian coast and part of the Ecuadorian coast (Guayaquil, Puná Island), northern Ecuador, southern Colombia (Pasto?), south as far as Santiago de Chile, and perhaps even to El Chaco in Paraguay. For Gilmore, this distribution coincides in general with the Inca Empire and was perhaps a result of its conquests, even though he admits, based on Murra's work (1946), that the llama existed in Ecuador "several thousand years earlier."

In regard to the llama, Pascual and Odreman (1973:34) state that its distribution "reached as far as Ecuador in the late seventeenth or early eighteenth century." Flannery et al. (1989:116) conclude that "[b]ecause llama trains traveled so widely in the Andes, the animal had been distributed from Ecuador to Chile by the time the Spaniards arrived." I do not intend to question these statements but rather want to point out that we usually begin with the assumption that the distribution of camelids the Europeans found was due to the Inca expansion. We will see that this is not entirely true. Furthermore, to draw a conclusion of this kind, we need the most exhaustive study possible of the sources.

Dedenbach Salazar (1990) is, I insist, the only scholar who has carefully studied the sources, although with a different goal from mine. I intend to expand the database. However, her observations may be taken as the yardstick for any future study of this kind. She has clearly pointed out the difficulties present in the study of data taken from the chronicles. To prevent any distortion, I prefer to quote her exactly. She wrote (1990:84):

The most reliable data on the distribution [of camelids] in Inca times certainly are in the early sources of the conquest. However, these travelers were probably unable to get an accurate idea of the number of livestock because their concerns were not of a statistical nature. In these animals they just saw a potential food resource for their campaigns of conquest. The inspectors instead wanted an exact figure on which to levy the tribute, but here we should bear in mind not only that there already was a large number of European livestock (see Espinoza Soriano 1971

[1558–1598]), but also that they were unable to record the exact number due to the lack of cooperation of the Indians, and that verifying the figures these had given was impossible.

Dedenbach Salazar (1990:96–97) insists there are few accurate references on the differences in habitat of llamas and alpacas. She then adds that there are references to llamas in Canaribamba and Cuenca, and to pack animals, that is, llamas, in Quito and in Porco-Arica to the south. The sources also indicate that alpacas were found in the northern and central highlands of Peru, too. These sites are Cuenca, Cajamarca, Piura, Huamachuco, and Huarochirí. However, Dedenbach Salazar insists that "we should bear in mind that the term *llama* could just as well have referred to domestic animals in general." According to the sources, there were also vicuñas and guanacos in the south, that is, Bolivia (Dedenbach Salazar cites Stouse 1970), Collao and Ayacucho, in central Peru (Ruiz 1952:76 is cited here), and, according to Cieza de León, in the highlands of Ecuador too, even though Novoa and Wheeler (1984:121) give the Department of Ancash as the northernmost point for the vicuña. I shall later expand on this subject.

6.7.1 General Considerations

Let us see what sources are available. I will begin with some writers who do not give exact data for specific zones, only general estimates.

First we have Sancho de la Hoz (1968:Chap. XVI, 327–328), the secretary of Pizarro, who wrote his account in 1534. In regard to the area extending from Tumbes to Jauja, Sancho de la Hoz stated that "[t]here is much livestock of sheep who go in herds with their herders who watch over them, removed from the cultivated fields, and have a part of the province to winter." Porras (1959:192) published a letter to the King by Fray Tomás de Berlanga, dated 3 February 1536, which among other items, recorded "very large herds of sheep," but without specifying where.

In his chapter on camelids, Cieza de León (1984:Part I, Chap. CXI, 293) also describes their distribution: "especially in this kingdom, in the

Gobernación of Chile, and in some provinces of the Río de la Plata."

When discussing the llama and the alpaca, Father Cabello Balboa (1951:Part III, Chap. 5, 232), writing in 1586, noted in turn that "the kinds of livestock like the ones we first mentioned, called llamas, and vicuñas and pacos, have not been seen nor found save far away in what we now call Peru as far as the straits of Magellan, because in the lands below [*tierra de abajo*] (of which I do not speak as it is derived and accesory to this one) these species of livestock were not found, nor have they ever been seen, but deer and fallow deer of different species do live."

Meanwhile, Father Cobo (1964a:Bk. 9, Chap. LVII, 365) says that llamas "are only born in the sierras of Peru, and spread all over the cold lands covered by the empire of the Inca which are, besides Peru, the kingdom of Chile and the provinces of Tucumán and Popayán."

León Pinelo was in Lima in 1621 and mentioned in his study (1943:53) a mistake made by Juan Fabro Linceo, who stated the llama lived from Potosí to Caracas, "because they only live from the Charcas to Popayán, and their biggest number and service is from Potosí to Cuzco and to the coastal valleys, because few reach Lima."

The seventeenth-century documents included in the *Juicio de Límites* (published in 1906) mention llamas, alpacas, guanacos, and vicuñas in the following areas: Cajamarca ("Descripción general del Perú," JdL 1906, Vol. 1, 372), where woolen cloth "from the sheep of Peru" was made; Huancavelica, Cangallo, Chuquito, Pacajes, and Sicasica ("Relación de los pueblos entre Lima y Chuquisaca" [n.d.], JdL 1906, Vol. 3; "Descripción de La Paz" [1651], JdL, 1906, Vol. 11) (taken from Dedenbach Salazar 1990:92).

These general statements do not amount to much when assessing the distribution of camelids in the Andes. For this we need concrete data. I will start with the coast, for which there are general data, and specific data for sectors or localities in it. Let us turn first to the general data.

6.7.2 The Coast

First we have an interesting item in the *Noticia del Perú* whose author is a matter of dispute. Prescott had him as a "Conqueror" (Prescott 2000:929, n.

14; it should be noted that the Spanish edition [1955:257, n. 18] reads "*conquistador anónimo*"), whereas Carlos M. Larrea concluded in 1919 that its author was Miguel de Estete. Porras agreed (see Porras 1986:116–118). Rowe (in Gasparini and Margolies 1977:Chap. II, 344, n. 20) still believes the chronicle must be considered anonymous, and Pease (1989:175, n. 2) believes it must be later. Be that as it may, toward the end of his manuscript the writer discusses the Inca roads. While describing the coastal roads he digresses to discuss their characteristics and describes not just the customs of the people but also its geography, flora, and fauna. It is while discussing the coast that he says "there is abundant livestock" (Estete 1968b:396).

There is a late chronicle written by Andagoya in Spain in 1541–1542, before he went to Peru. However, this man knew all participants in the conquest and, in Porras's words, "his veracity is not to be ignored" (see Porras 1986:70). Also, when writing of the coast, he mentioned, among other things, that "[b]esides the sheep there is a good number of deer, partridges and other fowling birds different from those in Spain" (Andagoya 1954:248). It is clear that the writer meant the coastal valleys, and perhaps the lomas too.

Gómara was never in Peru, and he published his book in 1552. "[He] most likely ... had a well-versed informant who had been in Peru" because "his chronicle is one of the most elegant and documented ones on the discovery of the New World" (Porras 1986:190–191). He describes in some detail the "[q]uality of, and the atmospheric conditions in, Peru," and says of the coast that "[t]here are rabbits, foxes, sheep, deer and other animals" (Gómara 1946:276).

Another account written around this same year (i.e., 1552) was that of Molina "el Chileno," who had been in Peru since 1535, as he participated in Almagro's expedition to Chile. Molina also wrote a general description of the Peruvian coast and is one of the Spaniards who pointed out the damages wrought by the conquest. Of the land between Huarmey and Chincha, he reports that it "abounded ... [with] ... livestock" (Molina 1968:313).

I find the data in Ramírez most interesting. Although later than the previous data, as it was written in 1597, this was a "clergyman who lived in Peru for decades," and had the virtue of dedicating himself to collecting not just historical and geographical data, but also to ethnography, something not too common at the time (Trimborn 1936:XII). As he himself declared, "[m]ost of the news in this description, and the most important and remarkable, I have seen and done, and of others I have ample and sufficient news" (Ramírez 1936:11). When describing the "sandy plains" he said "[t]here is little livestock of the land and it is most degraded" (Ramírez 1936:14). This might be hinting at the killing of these animals that took place in the first years of the conquest, which must have been more marked on the coast than in the highlands. However, according to the testimony of León Pinelo (1943:53), in the early seventeenth century there still were llamas "in the coastal valleys."

It is clear that more evidence will certainly be available once the data scattered in archival documents are studied (see Rostworowski 1981:51). However, when discussing the earliest data the chroniclers left about the coast, Dedenbach Salazar (1990:87) considers that the use of camelids for wool production must have been restricted. She believes the main reason for this was the climate, something I find incorrect. But this makes Dedenbach Salazar believe that coastal camelids must have been used mainly as beasts of burden and as sacrifices. In this regard she wrote, "[b]earing in mind the vertical organization of the economy of several Andean ethnic groups, the data [i.e., those of the chroniclers] might refer to the highlands of the respective coastal cities, which probably were either controlled by the coastal people, or were frequented by them through commercial contacts." Although something like this must have happened, we should not forget that the model of "Andean verticality" has been overgeneralized and exaggerated. True, it did exist, but not in all of the land. In fact, it seems to have been practically nonexistent on the north and central coasts.

Proceeding with her discussion, Dedenbach Salazar (1990:98) reviews the studies done by Rostworowski (1981:51ff.) and Shimada and Shimada (1985:17), which present data confirming the presence of camelids on the coast in Inca times. In the case of Rostworowski, Dedenbach Salazar concludes that not all of the data explicitly refer to camelids. On the other hand, besides the *visitas* (inspections) mentioned by Shimada and Shimada, there also are two documents concerning the south coast (which I mention later) in which the presence of camelids is confirmed, to wit, one by the cacique of Acarí in 1593 (*Visita de Acarí* 1973), the other the testament of Diego Caqui in Tacna (1981). According to Dedenbach Salazar (1990:98), "[w]hat stands out in these documents is that there does not seem to have been as many domesticated camelids on the coast at the time of the conquest (see Ramírez [cited above]) as in the highlands, and even on the coast itself in prehistoric times (Donnan 1973; Pozorski 1979a, 1979b; Moseley and Day, eds., 1982; Shimada 1982; Shimada and Shimada 1985). And although we have to bear in mind that some early chroniclers declare having seen much livestock of the land on the coast . . . one gets the impression that the number of domesticated camelids fell considerably since pre-Inca times." This point will have to be discussed at length at the end of the book, but first we need to review all of the data in the chronicles.

Rostworowski (1981:50) has shown the significance of the lomas. She states that these were used not just by highland populations but by coastal ones too, who had their herds of camelids there. Rostworowski believes the llamas grazed on the lomas in winter because of lack of pastures on the coast. Here she agrees with Dedenbach Salazar (see above) in that the animals were not raised either for the wool or for meat, but mainly for transport. Rostworowski is conclusive in saying that "[t]ransport must have been the major concern of the Yungas, hence the need to adapt the animals to the coastal environment" (Rostworowski 1981:50). This use of the lomas should be taken into account because it is a frail ecosystem and camelids are not as destructive as European-imported animals. Camelids certainly could have used them without causing much harm. I disagree with Rostworowski in that man did not have to adapt the camelids to the coastal environment. Camelids already had the physiological characteristics essential for living on the coast, and were surely there before man arrived.

Murra (1975:119) is another scholar who insists that coastal camelids were mainly used to carry cargo. However, he only mentions "the beasts that remained on the coast to carry the *wanu* from the islands to the irrigated fields."

These are the only general accounts of camelids I have found for the coast. Let us turn now to the data available on specific geographic locations. I begin with those chroniclers who witnessed the events that took place in the early stages of the conquest and who therefore described what they saw, or wrote what they heard straight from the mouth of the Indians. Even so, it is worth bearing in mind that "[t]he route followed by the Spaniards from the present Ecuadorian coast to Cajamarca is full of inaccuracies" (Pease 1992:130).

According to Porras (1986:53), the first report dates to 1528 and is possibly somewhat earlier because it corresponds to the second voyage of Pizarro. This is the famed *Relación Sámano-Xerez*, which mentions "that town of Calangome . . . [where] there are four towns together of a lord, which are the said Calangome and Tusco and Çeracapez and Çalango; there are many sheep there" (Sámano-Xerez 1968:13–14). The data apparently correspond to the account the writer received from the Indians sailing in the aforementioned "raft of the Tumbesinos." He clearly misunderstood what the Indians described, because besides the sheep [*ovejas*] he also mentions "pigs, cats, and dogs" (Sámano-Xerez 1968:14). According to del Busto (1973:114, n. 5), the Spaniards must have met the raft in the first fortnight of December 1526.

This is an extremely important document for, as Pease (1992:159) says, it is "the first account of the contact between the Spaniards and the Andean people." There are, however, some problems regarding this document on which the specialists do not agree. The first problem is the author. The chronicle bears the name of Sámano and Xerez, yet it is an anonymous account that was copied by Juan de Sámano, secretary to Charles V. According to Porras (1986:53), Jiménez Placer "has proved" that the author was Francisco de Xerez, who was later secretary to Pizarro and the author of another important chronicle. Porras accepts him as such. Pease (1992:137), however, has his doubts and believes that it "was possibly written by Bartolomé Ruiz, the pilot" of one of the Spanish ships.

The second problem is the location of the "town of Calangome" whence came the chronicler's informants, for he clearly states "Calangome, where they are" (Sámano-Xerez 1968:13). Porras (1968:13, n. 18) wrote: "The chronicler no doubt referred to Inkan towns. The inaccuracies in reproducing the geographic names make identifying the four towns subject to the lord of Calangome or Calangane impossible, though it can be held that it is Túmbez" (see Porras 1937, 1967). I do not know what data Porras used to reach this conclusion, but Cieza de León (1987: Part III, Chap. X, 31) must, in fact, have received a similar account, because he explains that after being in the Isla del Gallo, the pilot Bartolomé Ruiz reached Coaque and then his ship found a raft whilst journeying "to the sunset," which he specifies belonged to the "natives of Tunbez." He then explains that they "[s]howed spun and unspun wool which was from the sheep, which showed what art they are of, and said there were so many they covered the fields." But there is another piece of evidence that shows Tumbesinos were traveling on the aforementioned raft. These are two passages from Cieza which I had not noticed, and that were pointed out to me by John H. Rowe (pers. commun., 7 September 1993). The men agreed to continue the journey south once the ship that had gone to Panama returned to the Isla de la Gorgona. Cieza de León (1987:Part III, Chap. XIX, 51) clearly states that "[t]he Indians from Túnbez came in because they knew the language and it was convenient to take them and have them as interpreters." The voyage continued "straight to the sunset up the coast," that is, southward, and "when they had sailed for twenty days they saw an island that was off Túnbez and near Puná, which they named Santa Clara." The chronicler adds: "When they saw the islet, the Indians from Tunbez who were aboard the ship *recognized it and cheerfully told the captain how close they were to their land*" (Cieza de León 1987; emphasis added). The evidence is conclusive. There is no question that people from Tumbes were aboard the raft. I do not know if its ten occupants (Cieza de León 1987:Part III, Chap. X, 31) were

all Tumbesinos, insofar as the three women do not seem to have been on board the ship Pizarro was sailing toward Tumbes. But, I repeat, there is no doubt that a group of them were from Tumbes.

Prescott (2000:855; 1955:173) accepts this account. He says, when mentioning the encounter with the raft and its occupants, that "[t]wo of them had come from Tumbez, a Peruvian port, some degrees to the south; and they gave him to understand that in the neighborhood the fields were covered with large flocks of the animal from which the wool was obtained." It should be noted that the source used by Prescott was the chronicle now known as Sámano-Xerez, and he does not mention Cieza de León in this regard (see Prescott 2000:855, n. 14; in the Spanish edition [1955] this is p. 173, n. 13). Del Busto (1963–1970a:53) also agrees because he mentions "the raft of the Tumbesino Tallan" when describing the voyage of the conquerors. Although in his 1973 study del Busto did not mention the place the raft came from when describing the encounter of Ruiz's ship with the raft (Busto 1973:85–87), a footnote reads: "raft of the Tumbesino Tallan" (Busto 1973:114, n. 5)—a statement twice repeated later (Busto 1973:103, 113).

However, not all scholars agree with this identification. Murra (1946:804) says of Calangome that "Rivet and Jijón locate [it] in Manabí." He did not give his sources, and I was unable to locate them. The study by Jijón might be Jijón y Caamaño (1941).

Holm contradicts Porras and states the following: "He goes as far as saying that 'Calangane,' so easily identified etymologically and from the statements made by the men on the raft captured by Bartolomé Ruiz with the present-day Salango, which town 'does not appear in other chroniclers,' and instead leads one to believe the lord of Tumbes was called Cacalame" (Holm 1982; 1985:14). Estrada Icaza (1987:6–7) touches the subject in passing in the introduction to one of his books, but it is clear that he accepts Holm's position because he mentions the "raft from Salango" which, he says, is better known as "the raft of the Tumbesinos."

Stahl (1988:361–362) is one scholar who pretends to discuss the subject but in truth makes no contribution to it. Stahl repeats the position of

Ecuadorian scholars (he cites Jijón y Caamaño [1941:91, 392], Holm [1953:81–83; 1982:14–38] and Estrada [1957:21, 43, 103]). Stahl considers that Porras can be easily contradicted with the argument that the objects in the raft can be identified as Ecuadorian. His argument is not convincing. I do not intend to join the debate, but it is clear enough that there were Tumbesinos aboard the famous vessel. I will return to this point when discussing the Ecuadorian area. However, it is worth noting that the study by Estrada (1957) does not mention the raft, as Stahl pretends. Worse, the page 103 he mentions does not exist. Furthermore, Stahl (1988:361) makes the mistake of placing the encounter between the Europeans and the raft in the first voyage of Pizarro, when it actually occurred during the second voyage.

Estete ([Anonymous?] 1968b:354), who was already mentioned (see above), narrates the events that took place in the second voyage the Spaniards made (late 1527–early 1528), and the account Pedro de Candia gave of Tumbes. Estete wrote: "In this town they began to see the sheep that are in that land, and they took aboard the ship some of the sheep the Indians gave them of their own free will." Estete then confirms that "[f]inally, on taking aboard some Indian boys and said sheep." This happened in early 1528. Del Busto also discussed this point. This has to be pointed out, for there are several inaccuracies that can be misleading. Del Busto wrote: "The Tumbesinos traded heavily in a reddish ware carried on big and woolly rams, which can well be called 'the camels of the Indies' due to their rhythmic gait" (Busto 1973:108). His his sole reference is Busto 1960–61. The same words appear in that study (Busto 1960–61:385), except that the phrase "and woolly" is omitted. Note 28 (Busto 1960–61:400) cites Herrera ("1945, III Decade, bk. x, chap. V, p. 97 of Vol. V") and Guaman Poma de Ayala ("1936, folio 370") as his sources. First, as far as Herrera is concerned, this apears on page 98, not 97 as cited by del Busto, and second, the account does not fit. As for Guaman Poma, we have an incorrect reading because the reference is to the arrival of Pedro de Candia in Spain.

Ruiz de Arce (1968:417) describes the island of La Puná (La Punan) on the third voyage (late 1531–early 1532), and says: "We found ten

sheep." This was confirmed by Pedro Pizarro (1978:Chap. 1, 5), who was on the island and wrote: "There were some sheep which they were given."

Del Busto (1973:146) also described the events that took place in the Island of La Puná, but again colored his account with fantastic flourishes that distort the truth. He wrote: "[W]hat most astonished the raw recruits was the discovery of some 'sheep of Peru,' of the kind called 'camels of the Indies.' These were so fat they could not breed, but the soldiers felt their wool to their heart's content and compared it with Castilian wool (71)." It should be noted that del Busto (1973:182, n. 71) gives as his sources Ruiz de Arce ("p. 88, 89, 1953" [1968:417 in my bibliography), Pedro Pizarro ("p. 27, 1944" [1978:Chap. I, 5 in my bibliography]) and Trujillo ("p. 51, 52. 1948"). We have seen (see above) the first two references, neither of which mentions "camels of the Indies" or the attitude of the soldiers. As for Diego de Trujillo, del Busto used the 1948 edition and I used the 1968 edition, which "has been taken from the edition made by the Escuela de Estudios Hispano-Americanos of Seville in 1948" (Editores 1968:10), that is, the one he used. It includes a description of La Puná (Trujillo 1968:17), but no mention is made of the "sheep" mentioned by Ruiz de Arce (1968:417) and Pedro Pizarro (1978:Chap. I, 5).

When describing Tumbes, Ruiz de Arce (1968:419) indicates "[t]here were many sheep." The host of Pizarro then left Tumbes. Estete (1968b:365) says that "after a three-day journey he reached said Tallana river and a town on it that is called Puechos" (around May 1532). After describing the town, Estete adds that it was a "much bigger land than Túmbez; it abounds in food and livestock of that land." This was confirmed by Xerez (1968:205), who says Pizarro found "livestock of sheep" on arriving at Poechos.

The account of Ruiz de Arce (1968:420) continues with the passage of the Spaniards through "Tangaraya," that is, San Miguel de Tangarará, where "much sheep are raised." When describing the people of this land, Ruiz says that "[i]nland, they use the sheep and load them." San Miguel de Tangarará, in the Chira River Valley, was the first Spanish settlement founded in 1532 (Rostworowski 1989:177).

"From Tangaralá . . . [Pizarro] moved to Piura" (Pedro Pizarro 1978:Chap. 7, 27) and then to "Sarrán," the present-day Serán to the south of the Department of Piura, close to Loma Larga (see del Busto 1963–1970b:57–60; see also Map 3 in Hocquenghem n.d. [1989?]:31). When describing the provinces, Pedro Pizarro (1978: Chap. 29, 220) explains that "Tangaralá, La Chira, Poechos is another; Piura, Sarán, Motupe, Çinto . . . is another." Xerez (1968:210) describes the stay of the Spaniards in "Zaran," where the cacique brought them a "supply of sheep." Moving on, they stopped anew at Motupe where, according to Xerez (1968:214), the Indians "sacrifice sheep."

I believe it is worthwhile summarizing here some comments made by Petersen (1962:360–361) on the climate in 1532. In this regard, Petersen points out that both Lears (1895) and Murphy (1926) believed the conquistadors were able to attain their feat only because their arrival coincided with one of those rare years of abundant rainfall, something both of these scholars believe is essential for crossing the desert between Tumbes and Piura. Petersen adds that "[t]his concept certainly has its roots in the mistaken assumption that the expeditionary force had to cross the semi-desertic coastal strip." In this study and other previous ones (see Petersen 1935, 1941), Petersen carefully studied the climatological conditions in the Peruvian northwest and concluded that Lears and Murphy were wrong. He likewise shows that the march of the Spaniards through the land of the Tumbes took place in a normal year, that is, one with little rainfall. He concedes that the chroniclers point out that in May 1532 the Tumbes River was rising, but Petersen points out that this is usual in most years. The water flow of the Tumbes River essentially depends on rains that fall inland, and can swell even when there is no rainfall on the coast. Petersen also rejects the presumed march along the seashore, and posits instead that it took place along the road in inland Tumbes, which skirts the hills of Amotape (today known as the Cerros de la Brea) and is the main thoroughfare on horseback. This agrees with the route mapped by del Busto (1963–1970a).

It is worth pointing out that Quinn et al. (1986) ignored the aforementioned studies and posited an El Niño at the time of the Spanish arrival. Hocquenghem and Ortlieb (1990:327; 1992) refute them, showing they are mistaken.

Cieza de León visited the northern lands of Peru in the mid-sixteenth century and passed through Piura la Vieja, "which was on the piedmont, close to the present-day city of Chulucanas" (Hocquenghem and Ortlieb 1992:212), so he was well acquainted with the land he was describing. Cieza certainly had good informants, and no one has questioned his assertions. In the third part of his chronicle (Cieza de León 1987:Part III, Chap. XX, 53), Cieza de León recounts the arrival of Pizarro at Tumbes in the second voyage, which must have happened in April 1528, according to del Busto (1963–1970:53). Cieza de León explains that the Indians sent rafts to the ship bearing, among other items, "a lamb the virgins of the temple gave to be taken to them." Cieza de León explains that Pizarro received this gift "with great contentment," emphasizing that it was a "lamb" [un cordero]. This is interesting, because other chroniclers do not describe the episode in this way. Prescott (2000:869; 1955:188) also mentioned this event but gave no source. He certainly did not use Cieza de León, but possibly a later source. So, when recounting the rafts the Tumbesinos sent to the Spanish ship, Prescott says these went "with a number of llama, of which Pizarro had seen the crude drawing belonging to Balboa [we already saw this, see above], but of which till now he had met with no living specimens. He examined this curious animal, the Peruvian sheep—or, as the Spaniards called it, the 'little camel' of the Indians [a term that was actually little used]—with much interest, greatly admiring the mixture of wool and hair that supplied the natives with the materials for their fabrics." Here Prescott is mistaken, because Pizarro and his men had previously seen llamas in the Island of Puná (see above). He makes other mistakes, too, as when he says there were "bananas" on the raft (1955:188).

In this regard, del Busto (1973:104 [the same appears also in 1978b:37]) once again distorts the truth. He wrote: "That same afternoon the curaca or lord of Tumbes sent Pizarro—in ten or twelve rafts—much fruit, vessels with water and fermented maize liquor and a sheep of the land, of the kind that gives wool. The Spaniards received all amidst much rejoicing and were most amazed by the auchenid, a quadruped that re-called the profile of a humpless dromedary." His only source is Herrera (1945. "III Decade, bk. X, chap. IV, p. 95–96. Vol. V") (Busto 1973:115, n. 31). Now in his Third Decade, Herrera (1945b: Bk. X, Chap. IV, 95) wrote: "[T]hen ten or twelve rafts were prepared in which they sent much food, and fruit, and vessels with water and chicha, and a lamb [*cordero*] which the Virgins of the Temple gave for them." This account clearly is a copy of Cieza de León (1987:Part III, Chap. XX, 53), who says the gift was from "the Virgins of the Temple," not "the curaca or lord," as del Busto says, who also adorned the original citation with de-tails of his own invention.

Cieza de León (1987:Part III, Chap. XX, 55) tells how, when Alonso de Molina landed at Tumbes, he saw, among other things, "much cul-tivated land and fruits, and some sheep." The chronicler (1987:Part III, Chap. XXI, 58) also says that Pizarro sent Pedro de Candia to verify what Alonso de Molina had recounted. Candia landed and, after verifying it, requested permission to re-turn from the native lord, who sent several rafts to the ship with produce from the land, and "to the captain [Pizarro] he sent with this same Candia a beautiful ram [*carnero*] and a very fat sheep [*oveja*]."

Herrera (1945b:Bk. 10, Chap. V, 97–98) pla-giarized Cieza de León in his Third Decade, adding just a description of the sheep, for he says these "are small camels." This definition was evi-dently taken by Prescott (see above). The infor-mation was later altered, as in the case of Cappa (1888:Vol. II, 70), who described the amazement of Molina at the "huge herds of llamas" he saw in Tumbes. In other cases this was modified, as in del Busto (1966:36, 38).

Cieza de León (1987:Part III, Chap. XXII, 61–62) continues his account of the second voy-age of Pizarro and relates that when the Spaniards arrived as far as Collique, "which is between Tan-gara[ra] and Chimo," the Indians greeted them and gave them "five sheep," among other things. To be on the safe side, Pizarro landed a sailor called Bocanegra, who was so enthusiastic about what he saw that he wanted to stay behind. An-other sailor was sent to see if what Bocanegra said was true, and Cieza recounts how "Juan de la Torre saw herds of sheep, vast cultivated lands,

many green irrigation canals, so beautiful that the land seemed so propitious there was nothing with which it could be compared."

In this case too, Herrera (1945b:III Decade, Bk. 10, Chap. VI, 99–100) copied Cieza, intro-ducing minor additions of his own invention, while del Busto (1966:39) provided a poor de-scription of the event. He wrote: "Further on they disembarked in the land of the Chimú and Malabrigo. Won over by the exotic life of these Indians, the sailor Bocanegra deserted. Juan de la Torre went down in search of him, and found him 'well and happy, and with no intention of return-ing,' and left him amid the dressed-up Chimú, who were taking him to see the herds of 'small camels.'" It is clear that in this case his source was Herrera (see above) and not the original in Cieza de León. However, it is striking that del Busto (1973) later used this description once more, but added some words that changed its content. When recounting how Juan de la Torre went in search of Bocanegra, del Busto says he returned to the ship "leaving him amid the dressed-up Chimú who were carrying him on a litter to see the *enormous* herds of 'small camels'" (Busto 1973:109; emphasis added to show the words added to the 1966 edition).[9] Del Busto (1973: 115, n. 39) gives as his source Herrera (1945 [1945b in my bibliography]: "III Decade, bk. X, chap. VI, pp. 98 to 100 of vol. V"). Indeed, Herre-ra does not use the word "enormous" at all. I here cite Herrera literally, so the reader can see what Herrera wrote: "They sailed until reaching Co-laque, between Tangara and Chimo, the sites where the cities of Truxillo and San Miguel were later founded. The Indians came out to receive them with much delight, bringing food; they sup-plied water and wood and gave five sheep." It was then that the sailor Bocanegra decided to remain behind. On this occasion Pizarro sent Juan de la Torre "to see whether this was invented by the Indians; he returned saying the sailor was fine and happy, and with no intention of returning. . . . [H]e said he had seen *herds of sheep*. . . . These first Castillians called these animals sheep for the wool they had, and they were so tame and domes-tic being small camels, as was said" (Herrera 1945b:III Decade, Bk. X, Chap. VI, 99–100; em-phasis added). The part about the litter is true, because it also appears in Cieza de León (1987: Part III, Chap. XXII, 62).

There are several sources covering the journey back from the second voyage. First there is Gómara, who must have had good informants even though he was not in Peru. Porras (1986: 191) believed that Gómara had perhaps read Zárate, whose book was published after his (in 1552) even though he wrote it before, but this does not seem to have been the case. Gómara (1946:225) says: "Pizarro sailed to Motupe, which is close to Tangarara; from thence he returned to the Chira River and took many deer-like sheep [*ovejas cervales*] to eat, and some men as interpreters, in the town they called Pohechos." Zárate (1968:Bk. I, Chap. II, 117) meanwhile says that "on the river they call Puechos or La Chira, he took some livestock from the sheep of the land."

Cieza de León (1987:Part III, Chap. XXIII, 63) also mentions another episode that must have taken place around this time. Before arriving at Collique, the ships passed by a cape they called Aguja, and then entered a port they named Santa Cruz. Between this port and Collique, Pizarro sent Alonso de Molina by land in search of firewood. Molina could not return due to the heavy sea, and stayed with the Indians and was with "the Capullana lady." When Pizarro returned for him, she sent many rafts to the ship "with provisions and five sheep said lady sent."

Petersen (1962:360) tried to reconstruct the route followed by the conquistadors. He believes that when Pizarro returned on 3 May 1528, after having sailed as far as the Santa River, he accepted the invitation the cacique of the Tallanes made and landed in a place Romero and Romero del Valle (1943) identified as the Punta Capullana (4° 29′05″), 3 km southwest of Lobitos. Petersen, however, thinks that the landing and the feasts in honor of Pizarro took place 29 miles (47 km) farther to the south, close to the mouth of the Motape River, now the Chira River, in whose valley were several towns ruled by women caciques whom the Spaniards called "Capullanas" because of their characteristic dress. The source given by Petersen is Lizárraga (1938 [1968 in my bibliography]).

Still in connection with the second voyage, Cieza de León (1959b:Part III, Chap. XXXVII, 106) tells that on returning to Tumbes, Pizarro wanted to punish the Indians who had killed two Spaniards but "[f]ound none or a few, so they stole what they could both of sheep and other items."

For the third voyage of Pizarro there is a note by Fernández de Oviedo (1959c:Bk. XLVI, Chap. II, 35), who described the Spaniard's departure from Tumbes; after mentioning a Turicaran River they reached the town of Puecho, that is, Poechos. Here "the banks of a river were found, well peopled and supplied with many provisions from the land, and sheep." This must have happened on 16 May 1532 (Busto 1963–1970b:57). Although it is true that Fernández de Oviedo was never in Peru, he wrote the section on the conquest in 1550, and according to Porras he may have received the aforementioned information from Pizarro's pilots (Porras 1986:178).

Almagro arrived at the coast of Tumbes by late December 1532, after the events of Cajamarca, and "[t]hose who came on the ship were much delighted with these news [i.e., what their comrades told them about the Indians]; as many of them as was deemed convenient landed, and rejoiced when they saw so many herds of sheep in the fields" (Cieza de León 1987:Part III, Chap. XLVII, 140).

In a newspaper article Zevallos Quiñones (1984:A2) discussed the voyage of Pedro de Alvarado from San Miguel to Pachacamac. Zevallos says Alvarado took with him "llamas as pack animals." No source is given, and only a letter sent by Almagro to the King on 1 January 1535 is mentioned, which I have not seen. This is possible because Zárate (1968:Bk. II, Chap. XII, 170) narrates the events that Diego de Almagro and Pedro de Alvarado experienced on their journey from Quito to Pachacamac, from whence it follows that the cacique of the Cañares told them Quizquiz, a "captain of Atabalipa," came with an army, "and he had mustered as many Indians and livestock as he had found from Jauja downwards." Zárate later describes the army of Quizquiz, who "went in the center with the main body and the livestock." Zárate (1968:Bk. II, Chap. XII, 171) mentions that "more than fifteen thousand sheep were left on the field" after the battle.

This is also confirmed by Gómara (1946: 236), who recounted the clash between the soldiers commanded by Alvarado and the rearguard of Quizquiz's army before reaching San Miguel.

Here Gómara says they "[l]eft fifteen thousand sheep . . . they were the sheep of the Sun, because the temples had great herds of them, each in its land. None could kill them on pain of sacrilege, save the King in times of war and when hunting. The kings of Cuzco invented this so they could always have a supply of meat in the continuous wars they waged."

It is likely that the Spaniards used the llamas, for we know that while in "the snowy Pass . . . [a] great number of his [Pedro de Alvarado] horses, too, had perished. . . . Such was the terrible passage of the Puertos Nevados" (Prescott 2000:1002; 1955:333). This must have happened in 1534, because when telling the story of friar Marcos de Niza, Porras (1949–1950:198) says "[h]e embarked with Pedro de Alvarado in the Port of la Posesión in Nicaragua, in January 1534, and disembarked with him in the bay of Caraques, to take possession of Quito before Pizarro's lieutenants arrived" (for more details, see Pease 1992:133).

A piece of information I find interesting appears in the "Relaciones geográfico-estadísticas del Perú, fechas por el orden de las instrucciones y memoriales que mandó despachar Su Majestad en 1577 (por real Cédula de Felipe II, despachada en San Lorenzo el Real en 25 de Mayo de 1577)." Therein figures the "Relación de la Ciudad de San Miguel de Piura" (Anonymous 1925: 84, 96; Salinas Loyola 1965a:38, 44), where the following is specified: "77. Livestock of the land are sheep and of the kind that went to Spain, there are all kinds and are raised in the jurisdiction of said city. The price of each sheep of the land is six pesos, and those of Castile a peso and less; goats at half; pigs at five; cows at six." Farther on, it is added, evidently for Inca times: "178. That in almost all major provinces they had land dedicated to the Sun and livestock of sheep, and guacas and shrines" (Hocquenghem n.d. [1989?]:149 cites this account).

When describing Piura, the Carmelite friar Vázquez de Espinosa (1948:1177/372) does so vaguely: "In the valleys of Piura there are very good offspring and herds of Merino rams. . . . [I]n the highlands there are offspring of the larger cattle [*ganado mayor*] and mules." We do not know exactly what was meant by "ganado mayor." Perhaps this was cattle, but it might also have been

llamas or alpacas. Besides, his account dates to the early seventeenth century and Vázquez de Espinosa apparently did not go to Piura, because he is said to have been in Cajamarca only in 1615 (Clark 1948:X).

There are no data regarding the first stages of the conquest in the Department of Lambayeque. The earliest found refers to the 1540 *visita* of Jayanca (district of Lambayeque; La Gama, 1540 [1975]). We find here specified, according to Espinoza Soriano (1975a:253), that "[i]n Jayanca there also was a hamlet of potters . . . another exclusively of llama herders (or "shepherds") . . . it was formed by three married couples but we do not know how much livestock they had."

From this inspection it follows that the *visitador* demanded to be shown "by sight all towns and ranches and sheepherders." In the list figure "[t]wo other towns with twelve homes with five Indians, three sheepherders" and then "[a]nother town of sheepherders with around eight homes, three Indians" (La Gama 1540 [1975]:262, 265). The cacique of Jayanca then explained what "[t]ributes [were paid] to the Inga and the encomendero" Francisco Lobo. After declaring that he had not know Guaynacaba or whether his father had paid him any tribute, the cacique specified they paid tribute to Francisco Lobo every two moons, and "sometimes [gave him] six sheep," among other items. Meanwhile, they gave Maxo and Pacura, their "prencipal," "seventy loads of maize and five sheep as tribute," besides several products and animals (La Gama 1540 [1975]: 270–271).

Lorenzo Huertas (pers. commun., 6 February 1992) kindly showed me a document from the Archivo General de Indias in Seville (AGI Justicia, Leg. 458), dated 1566. Here reference is made to the clothes the Indians had to carry on their back from Jayanca to Trujillo. Several complaints appear because this was done without the help of herds. Although llamas are not clearly mentioned, everything seems to indicate these were llama herds.

Ramírez-Horton (1982:125) studied the early sixteenth-century north coast occupations, which included that of "keeper of the herds." Her Table 1 lists the artisans identified in the six polities studied with the help of documents, and their respective dates. The following are listed for "Shepherds/Herders": Jayanca 1540, Pacasmayo 1582, and Sinto 1566. Ramírez-Horton (1982: 127) then explains that around 1550–1560, the chiefdom of Cinto, which was part of the Spanish repartimiento of Lambayeque, had "herders," and that the Lambayeque *mitima* also served in that capacity.

Shimada and Shimada (1985:17) also state that during his 1566–1567 inspection, the *oidor real* Don Gonzales de Cuenca gathered several testimonies from the Indians in northern Peru, including the Lambayeque region, on large and small herds of rams [*carneros*] and sheep [*ovejas*] of the land. The *visita* (Archivo General de Indias, Seville/Justicia 461, 856v) mentions eight stone and mud corrals where the rams [*carneros*] and sheep [*ovejas*] of the land belonging to Guaynacapac were kept.

These same scholars (Shimada and Shimada 1985:17) note that Don Diego Mocchumi, lord of Túcume, had nine sheep of the land at his death in 1574 (Archivo Regional de Trujillo/Corregimiento Residencia 30.VI.1576). Another document from the Archivo Regional de Trujillo (Corregimiento Ordinario 118-1584), which dates to 1582, mentions the ruler of Moro, who had rams of the land.

Finally, Rostworowski (1981:51) says that in 1580 the Indians of Lambayeque filed a suit against the vecinos of Trujillo because the torrential rains of 1578 had destroyed not just their homes and land, but also the network of irrigation canals, thus impeding cultivation. This sparked a major famine that hindered them in paying their tribute (BN-A-534). In various depositions the Indians declared they had been forced to sell their "sheep of the land and horses at a low price to buy food" (fol. 210v). In this same source, one witness declared that Don Martín, the cacique of Lambayeque, had lost "a great number of livestock of sheep he had which died, and this because of said rains."

For the Saña region, all we have is the information found in Ramírez-Horton (1982:127). In her Table 2 she notes that in 1564 the señorío of Saña, a part of the Spanish repartimiento of the same name, had "herders" in Chontal.

Cock (1986:177) discusses the properties of García Pilco Guaman, the curaca of Moro-Che-

pén (in the province of Pacasmayo, in the Department of La Libertad), which he reconstructed with the testament drawn in 1582. In this regard Cock wrote: "Pilco Guaman also claimed the property of cows and bulls, and 860 heads of sheep and goats. No llamas or alpacas are mentioned, but it is known that these lived in Moro in 1582 (Ramírez-Horton 1982:134, n. 14). However, in the auction, the sandals of Pilco Guaman were described as being made of llama wool (ADT, CO Le. 154, Exp. 204). The llamas were perhaps not included in the list because they belonged to the chiefdom of Moro-Chepén, or because they belonged to Pilco Guaman but his family deemed them too important to be sold in the auction at his death. The refusal to sell camelids is not unusual, as it can be found in the testament of Don Alonso Caruatingo, lord of the guaranga of Guzmango, in Cajamarca (Rostworowski 1977b [1977a in my bibliography])."

However, it is not entirely correct that llamas were not sold at that auction. There is evidence that at least one was sold. Zevallos Quiñones (1989:97) published information on the auction of the estate of Don García Pilco Huamán, cacique principal of the repartimiento of Moro, held on 7 August 1582 in the village of Nuestra Señora de Guadalupe (i.e., the present Guadalupe). At this auction, "3 pesos were paid to Juan Chacquerref, from Tecapa, for 'a ram of the land.'"

Ramírez-Horton (1982:127, 134, n. 14) reports the presence of 200 *mitimaes* from the Spanish repartimiento of Moro who were in the señorío of Pacasmayo and who remained loyal and subject to their lord. "Some of these probably were herders in charge of llama herds, because an Indian official ("*governador*") of Moro owned llamas and wore sandals made of their wool in 1582 (ART/CoO: 11-VIII-1582)."

I was unable to find much information on the Moche Valley. There is the testimony of Zárate, who despite having been in Peru for just one year, from 1544 to 1545, "conveys to us the conversations and the recollections, and sometimes perhaps the written accounts, of the protagonists of these events" (Porras 1986:218). When discussing the Chimo Valley—that is, today's Moche Valley—Zárate laconically says that "it abounds . . . in cattle" (1968:Bk. I, Chap. VII, 127). However, Jorge

Zevallos Quiñones (pers. commun., letter of 22 June 1993) let me know of a document he found in the archives at Trujillo. It is recorded therein that in the city of Trujillo, Juan Álvarez (f. 4), the man who auctioned the annual management of the slaughter houses owned by the city, agreed on 16 March 1543 on a system for provisioning animals with Melchor Verdugo. An Afro-American "shearer," sheep and yanaconas would be needed to look after the animals. It was specified that these would have to be "good sheep from the Yungas," priced at 4 tomines each. Rams would also be provided. Zevallos Quiñones explains that Melchor Verdugo was the encomendero of Cajamarca and a major rancher, and asks: "Does this mean that camelids were brought to Trujillo to be butchered as a popular staple?"

Zevallos Quiñones (pers. commun., letter of 22 June 1993) has found another document in which one Ortuño de Artazu declared in Trujillo, on 17 May 1561, that López de Córdova (f. 489v) sold to the merchant Baltazar de Zamora "a hundred rams of the land, four more or less which are not sick with the mange [*carache*], nor any other defect whatsoever, and the smallest of them is over a year and a half." López de Córdova would deliver the first 50 animals in Trujillo within ten days, and the others four days later. He offered more rams within five months. Zevallos Quiñones reports that the part with the price is unfortunately missing. He also adds that Artazu had his livestock in the Carabamba area, in the province of Otuzco (Department of La Libertad, district of Salpo). Castañeda Murga tells me in this regard that the testament of the seller Ortuño Artazo records that he was the owner of the ranch and obraje of Carabamba. Castañeda Murga adds that "[t]he *obraje* in this area suggests to me that in pre-Columbian times there were native textile workshops or mitma weavers, and lots of livestock." To reach Carabamba one climbs up the Chao and Virú valleys. From a seventeenth-century document I know that when cattle were brought from the highlands to provision the city of Trujillo, they were taken along Virú, and "when pastures were lacking in the sierra they stopped in the middle valley (in the Cucur area) so the cattle could graze for some days before continuing the journey to Trujillo" (Juan Castañeda Murga,

pers. commun., letter of 10 August 1998). To give an idea of what prices were like, another document records that on 5 February 1573, Juan de Orozco declared that Diego Muñoz Ternero (fol. 68) sold "rams of Castille," gelded, over a year old, to Alonso Morillo at half a peso each, for a total of 300 pesos. From these documents it follows that the sale of rams of the land in the city of Trujillo was an everyday matter.

For the Chao Valley, Zevallos Quiñones (1991:12) gives us an interesting datum that concerns the 1565 testament of Antón Elmo, who ordered that an Indian woman from Chao called Leonor be paid for a llama his (watch?)dog had killed. The relevant section of the testament reads thus: "Item, I declare that a dog of mine killed a ram of the land of Leonor, [an] Indian woman from Chao, I command that six pesos be paid for the ram from my goods" (Elmo 1991 [1565]:20).

There is a gap in the information between the north and central coast, and for the northern part of the Department of Lima there is just one 1607 document cited by Rostworowski (1978:140). It refers to the Huaura River Valley, in the province of the same name. This is a lawsuit between people from the coast and the highlands. The curaca of Végueta rented land to the father of one curaca from Andaxes (in the highlands of the present-day province of Oyón), who offered to send the coastal lord a gift of "rams of the land and green and dried potatoes as a sign of friendship" (BN, B-1937, fol. 6v, año 1607).

There is an interesting piece of information for the zone of Lima in Volume 1 of the Libro de Cabildos de Lima (Torres Saldamando 1888:Vol. I, 250). Here it is explained that some decrees were passed in Los Reyes [Lima] on 10 December 1538 to protect the Indians. The treasurer Alonso Riquelme stated that he had been informed that some Spaniards who came from Cuzco and others living around the city of Lima would steal many sheep from the Indians and take them to Lima to sell them. Riquelme declared that this was harmful not just for the natives, but also for the kingdom. This is most interesting, as it is confirmed by the testimony of the accountant Zárate (1968:Bk. I, Chap. VII, 127), who, when describing the Lima Valley, says that "the land is teeming . . . with livestock." In this case it is firsthand testimony, for Zárate was in Lima between 1544 and 1545. Now, it is true that the term "livestock" can also mean European animals, but in this case it is worth recalling what Cobo said in the seventeenth century, namely, that the "livestock of Castile" only began to abound in the Lima River Valley around 1548. It can therefore be assumed that these were camelids. In fact, when discussing the "abundant provisions" available in the city of Lima, Cobo (1964c: Fundación de Lima, Bk. I, Chap. XIII, 316) noted that "[m]eat is no less abundant than bread, even though at first this or that was lacking for some years, until the seeds and livestock brought from Spain multiplied, and so the vecinos of this city lived with maize and other vegetables, and *llama meat, the animal we call ram of the land*. Castilian livestock began to be killed in 1548, because it had greatly multiplied" (emphasis added).

Ramos de Cox (1971:68–70) in turn reported the archaeological research carried out in a pastureland called "Mantaru" in the Fundo Pando, inside the urban area of Lima, close to the campus of the Pontificia Universidad Católica del Perú. Here it is explained that east of this pasture there is another one called "Corpus." According to Ramos de Cox, the earliest occupation apparently corresponds to "a corral for llamas over the remains of a mud wall building that was used until the sixteenth century." She adds that "[a]lthough [it is] incomplete and cut by a more recent mud wall, this unit shows the presence of llamas." Then she speculates about the name "Mantaru" and concludes that these were highland groups that lived there and dedicated themselves to trade. The study unfortunately does not present any evidence, nor does it have any support for the assertions made; it is not possible to tell whether there is any evidence to support the conclusions. Finally, Ramos de Cox (1971:68–70) wrote: "The presence of llamas in the coastal market in these early years of the Spanish invasion can be explained by these animals being used for transportation, which were perhaps found in the various corrals in Lima: when discussing the abundance of provisions, Bernabé Cobo (Cobo, 1936, p. 316 [this citation corresponds to the one previously mentioned, but in the 1964c edition]) explains that llamas were killed for the Indians,

and llama meat they sell in the market." The account Ramos de Cox gives of Cobo is not entirely correct, but in any event does not serve to prove her claims.

Guillén Guillén (1978:46) has studied the testimony of Don Diego Cayo Inga, the grandson of Pachacuti Inga Yupanqui, an eyewitness of the sack of the shrine of Pachacamac, in the Lurín Valley. According to Don Diego, the priests from the shrine "had sent messengers to all provinces in the territory of Peru." More than 20,000 Indians arrived carrying gold objects and "many other things such as rams [*carneros*], sheep [*ovejas*], deer of all kinds." These were obviously camelids, but there is no report of what happened to them. According to Estete (1968a:247), this must have happened in February 1533, because Hernando Pizarro and his men arrived at Pachacamac on 30 January.

I was unable to find information on the southern valleys as far as Chincha. However, when describing "[t]he town of Cañete," Guaman Poma de Ayala (1936:f. 1038) specifically notes that it "does not have any livestock."

As noted earlier, Gómara is one of the chroniclers who never went Peru but had good informants, according to Porras (1986:191), and who "amassed data and news that do not appear in other chroniclers." His book was published in 1552. In it, when mentioning the herds of camelids, Gómara claims that "in Chincha there were great herds that were traded" (Gómara 1946:276). This is confirmed by Cieza de León (1984:Part I, Chap. LXXIII, 220), who, when writing about the Chincha Valley, said that "[o]f the sheep of this land there are almost none, because the wars which the Christians had with each other consumed the many they had." This is a pathetic account of what once existed and was destroyed!

In one of her studies, Rostworowski (1977a: 269–270) discusses the Ica Valley and mentions the will of the old curaca Don Fernando Anicama and his son. Anicama, who shared the lordship of Lurín Ica with another curaca, died in 1571. In his will he said there were "rams of the land" in Ica. Among other things, his son declared he had 100 llamas, while in general his father left his wife "sheep of the land." (Rostworowski also discusses this in another study published in 1981, p. 52.)

Vázquez de Espinosa (1948:1359/451) likewise describes the Ica Valley and says "there is in it much larger cattle [*ganado maior*], goats, pigs, rams." The "rams" in this case undoubtedly are "of the land," since, when discussing the "town of Ica and its fertile vineyard valleys," Vázquez explains that "a great amount of wine is produced [here], much of it is taken in rams through the province of Chocorvos to Castrovirreyna, Guancavelica, Guamanga and other places" (Vázquez de Espinosa 1948:1354/450). He clearly means llamas. This piece of information is invaluable because, as Vázquez de Espinosa himself says, in 1617 he was in the neighboring Chincha Valley (Vázquez de Espinosa 1948:1343/445); if he did not get the chance to go to Ica, he at least got the information firsthand. This means llamas still lived in this valley in the seventeenth century.

Farther south there are some data on Acarí, in the Department of Arequipa (province of Caravelí). In 1593 Don Alonso Satuni, from Guallaca Ayllo, in the Hurinsaya moiety, declared having "sixty heads of livestock of the land" (*Visita de Acarí* 1973:158). When discussing the second part of this inspection, which deals with the coast and was carried out in the town of Chaviña in the same year of 1593, Rostworowski (1982:242–243, 245) mentions the "Loma Ayllo fishermen" (see *Visita de Acarí* 1973:170). She is indirectly suggesting that this ayllu must have managed the resources in the lomas and on the coast, and in this case there might have been camelids in the lomas, because they were a pastureland for these animals. Rostworowski (1982:242–243, 245) specifically says of the "[a]yllo called Loma" (see *Visita de Acarí* 1973:172) that "in pre-Columbian times it must have dedicated itself to herding camelids." She reaches this conclusion by comparing it with a previous study on the lomas of Caringa (see Rostworowski 1981:42).

In the *Visita de Atico y Caravelí* (Rostworowski 1981:51; Galdós Rodríguez 1977:77) it is said that in 1549 the cacique of Atico (in the district of Caravelí, Department of Arequipa) called Chincha Pulca declared having had some rams [*carneros*] and sheep [*ovejas*] which had been seized by Maldonado, his first encomendero. Dedenbach Salazar (1990:85, n. 6) argues that in this same document it says "the Ynga sent them wool

and they made cumbe clothing and took it where he ordered them to," and "the Ynga had in Parinacocha some livestock of sheep and they gave him thirty Indians to watch over them." On this basis Dedenbach Salazar suggests we can deduce the people of Atico did not have any livestock of the land, and that if they had had some heads, these would perhaps have been able to graze in places at a higher altitude. She concludes that the citation refers to European sheep. This is possible, but I do not find the argument convincing, particularly in light of the evidence of camelids in the area at that time.

To finish with the coastal information, there are some data on the province of Tacna, in the department of the same name. I mean the Sama Valley, because the 1567 *Visita a la provincia de Chuquito* includes the statements of several individuals (Diez de San Miguel 1964:124, 126, 127, 129, 247). First, the witness Juanes de Villamonte declares that "the Indians in this town of Sama . . . have some sheep as property of their community which are about six hundred, more or less, and they had more and these have diminished, and they have goats and some cows." When talking of the Sama Indians too, the witness Juan de Matute claimed that they "have much livestock of the land which they trade, and pastures to maintain them." In turn, we read in the statement the witness Pedro de Bilbao made on these same Indians that "they have . . . rams of the land in good numbers which they trade with the Indians from the province of Chuquito and Pacaxes." The witness Marcos de Silva also said that "they have "rams of the land." Further on in the same inspection we find that "[t]he Indians in the province of Sama are Yunga and Indian mitimaes placed by the Ynga . . . and have some livestock those from the sierra bring in exchange for said food."

A very interesting piece of information appears in the testament of Diego Caqui, the curaca of Tacna, dated in 1588. We read: "I also declare that one Francisco de Cárdenas, a merchant [*mercader de flete*], owes me a hundred rams of the land on which I carried some merchandise of Castilian clothes some seven years ago, or probably less, and on the way he sold forty rams of the land of those carrying the merchandise without my permission" (Caqui 1981:213). Further on we read: "I also declare as my belongings one hundred breeding sheep of the land which are looked after by an Indian named Ticona, a native of Putina; they must have multiplied" (Caqui 1981:214).

6.7.3 The Highlands

Let us now turn to the historical data available for the Andean highlands. First I review the general data, then turn to the specific.

When describing the land between Cajamarca and Cuzco in 1532, Xerez (1968:235) said that "[i]n all this land there is much sheep livestock, much goes wild because they cannot tend as many as are born." This remark by Xerez was commented upon by Dedenbach Salazar (1990:86): "Now, it is not known whether he actually saw 'untended', i.e., untamed, llamas and/or alpacas, or whether those that for him 'go wild' were instead the wild guanacos and vicuñas. Either way, there must have been great herds of camelids if the Spaniards killed 150 heads a day." Zárate (1968: Bk. I, Chap. VIII, 130) made a similar statement a few years later, but unlike Xerez he distinguished the llamas and alpacas from the guanacos and vicuñas, because when describing the highlands he remarked that "[t]he Indians have many wild sheep, and other domestic ones." Meanwhile, in the late sixteenth century and early seventeenth century, Murúa (1964:154) pointed to the highlands as a hunting ground for vicuñas and guanacos.

In 1597 Ramírez prepared a description of the Andes. When discussing "the province of Chinchaysuyu, between Cuzco and Guamanga," he said that it "has much livestock of Castile and of the land" (Ramírez 1936:40). He then mentioned the "Collasuyu," approximately the southeastern part of the Inca Empire, and specified that here "much livestock of the land is raised, which is the best Piru has, and is used for the droves of rams, of which there are many, and this is best and has a better price than all the rest. There are several ranches of sheep of Castile belonging to Indians and Spaniards which grow very well, and there is a great number of them" (Ramírez 1936: 42). Finally Ramírez describes "[t]he province of Condesuyo"—in other words, roughly the southwestern part of the Inca Empire—and notes that

it "abounds in birds and livestock of the land" (Ramírez 1936:41).

As was already pointed out, Herrera, the Cronista Mayor de Indias, whose *Décadas* date to the early seventeenth century but are not firsthand, nevertheless provided an account of the kingdoms of Peru that is interesting and certainly correct. In his Fifth Decade, Herrera (1945c:Bk. I, Chap. V, 172–173) says that "in the sierra great herds of wild goats are raised which they call vicuñas, and the pacos and guanacos, which together are commonly called sheep of the land, and rams, and many donkeys." Of the altiplano he says that there is "a multitude of livestock of Castile and of the land."

In Chapter X of his grand work, Father Cobo, that "great inquirer about life," the "great naturalist . . . [who] has the gift of analyzing, defining, and classifying," in the words of Porras (1986:510–511), studied "[t]he first difference in climate found in the Sierra of Peru" (Cobo 1964a: Bk. 2, Chap. X, 74). Here we find invaluable data which I cannot help but cite here, particularly because they hold information not found in other chroniclers. Cobo wrote:

> Abundant pastures grow in the rest that is not taken or buried under the snow, such as the savanna, [the] slopes and some knolls and hills that are not as steep as the cordilleras; the grass is, however, coarse and dry, for which reason the meat of the livestock that grazes on it is not as nutritious and tasty as that of Spain, Chile, and other temperate lands in America. In the past the Indians of these punas and pasturelands raised a great number of domestic livestock of the land, and vicuñas to hunt for their precious wool. And the Inca kings had pastures distributed among the neighboring peoples. Today they have in all of the sierra almost the same boundaries that were set by the Inca.

He adds: "Besides these tame livestock brought from Spain and those of the land, there is much game of guanaco, vicuña." Cobo finished preparing his work around 1653.

It is somewhat striking that, in the face of these assertions, Murra (1978:90) could write that "[t]he camelids were really abundant only in the Collao, the land of the Aymara-speaking ethnic groups." This certainly is not true.

I will now analyze the specific data available for the various parts of the Andean highlands, starting from the north. Cieza de León left some data on the Department of Piura, specifically on what is now the province of Huancabamba, or places not too far removed from there. The chronicler wrote: "There used to be a great number of the livestock they call sheep of Peru: at this time there are very few left due to the hardships the Spaniards caused them . . . and some guanacos that go through the heights and wilderness." He then adds that "[t]here are breeding grounds for livestock everywhere (Cieza de León 1984: Part I, Chap. LVIII, 185). It is worth emphasizing the chronicler's insistence on the killing of animals the Spaniards made in the first years of the conquest. To really appraise the value of Cieza's data, we must recall that he arrived in Peru just 15 years after the Spaniards had begun to cross what is now Peru.

Dedenbach Salazar (1990:86-87) made a mistake when discussing these same facts, because she remarks that the chronicler meant San Miguel de Piura. However, a proper reading of Cieza (1984:Part I, Chap. LVII, 185) shows that he is evidently describing the highlands of Huancabamba and not of San Miguel, because between the lines we read: "There are several rivers and some small brooks in the valleys and wetlands: their water is good and delicious." This is exactly in the paragraph on camelids, which I cited, and which evidently does not refer to the coastal area. Strangely enough, Rostworowski (1981:51) made the same mistake.

Writing of this same region, Cieza de León tells how in October 1532, while Pizarro was still in the lowlands of Piura, in the town of Pabur, he sent a detachment under Hernando de Soto to explore an area close to Huancabamba, where the news Pizarro had received made him suspect there might be an Inca garrison. The Spaniards "[t]raveled until they reached what they call Caxas, a province in the highlands. They saw great buildings, many herds of sheep [ovejas] and rams [carneros]." The chronicler reports a skirmish between Spaniards and Indians, and tells

how the latter came out "carrying strong ropes" to tie up the Spaniards, "believing they were some pacos who would promptly let themselves be ensnared." He immediately explains that "they call a certain lineage of their rams 'pacos'" (Cieza de León 1987:Part III, Chap. XXXVIII, 110; to see the route followed by Soto, see Map 3 in Hocquenghem n.d. [1989?]:31).

Hocquenghem (n.d. [1989?]:105) writes about the páramo that lies "in the heights of the Andean Cordillera and the Guamaní Cordillera, at an altitude of more than 3,500 meters," in the province of Huancabamba. Hocquenghem notes that "[b]efore the conquest there were vicuñas in these high areas, which according to Cieza de León vanished around 1550." I was unable to locate this citation in Cieza. It might be chapter LVIII of Part I (1984:185), where it says that "[t]heir clothes are from wool of these sheep and of vicunias, which is better and finer." However, vicuña clothes could have been brought from the south, as Cieza never specifies that there were vicuñas here. As we have already seen, when mentioning "some guanacos" in this same paragraph, he quite clearly says those "who live in the heights and in the wilderness" (Cieza de León 1984:Part I, Chap. LVIII, 185). Nor did I find any reference concerning this subject in the third part of the *Crónica del Perú* (Cieza de León 1987).

For the Chachapoyas region there are few, but good, data. Cieza de León (1987:Part III, Chap. LXXXIX, 296) tells how Alonso de Alvarado left Trujillo for Chachapoyas and, after being in the province of Chillao, arrived with his men at a town called Longlos, where they saw "some herds of sheep." It is not easy identifying the place, which is in the present Department of Amazonas, but Stiglich (1922:375) says that this is an old designation given by the inhabitants who belong to a part of the province of Luya. They were the ones to the west of the Utcubamba-Marañon divide. In fact, in the same *Diccionario Geográfico del Perú* by Stiglich (1922), we find a town called Chilla in this same area, and a river with that name that "climbs down to the Marañón almost in front of the ravine of Chalán."

Cieza de León (1987:Part III, Chap. XCIII, 310) then mentions the expedition of Alvarado that set out to conquer the Chachapoyas. It seems that some Indians feigned peace and sent the Spaniards an idol "with some sheep."

The testimony of the Carmelite priest Vázquez de Espinosa is of great value, for he himself tells us that he was in Chachapoyas in 1615 (see Vázquez de Espinosa 1948:1191/378), so his data are firsthand. He wrote a "Description of Caxamarca." We must bear in mind that the corregimiento of Cajamarca included the region of Chachapoyas. It was precisely when discussing "the provinces of the city of the Chachapoyas" that Vázquez de Espinosa noted that "close to Leimebamba, walking to the Chachapoyas, is the province of Caxamarquilla del Collay . . . it is all . . . mountainous and very temperate, where there are some offspring of larger cattle [*ganado mayor*] and livestock of the land" (Vázquez de Espinosa 1948:1189/377). He therefore meant the southern part of the province of Chachapoyas. Vázquez de Espinosa then described the "city of the Chachapoias" (in the province of the same name) and said: "Five leagues from Chachapoias is the town of Luya." When discussing the "livestock" he specified that there are "many offspring of larger cattle [*ganado mayor*], and some of livestock of the land, of mules and the best horses" (Vázquez de Espinosa 1948:1194/379).

I shall now review the Department of Cajamarca. It seems that there already were a great many camelids here when Atahuallpa arrived, besides those in his retinue. Fernández de Oviedo (1959c:Bk. XLVI, Chap. V, 47) says: "And on arriving at Caxamalca as he had commanded it, because he felt it was a good land that abounded with much livestock of sheep and other provisions for his army." It is true that Fernández de Oviedo was never in Peru, but he wrote in 1550 and, according to Porras (1986:178), used the testimony of Diego de Molina to describe the events at Cajamarca, so we can trust the truthfulness of his assertions.

It is known that the Inca brought with him a very large herd of llamas, because Porras (1959: 59) published a letter from licenciate Espinosa to the emperor, dated in Panama on 21 July 1533, giving an account of the events at Cajamarca. Here it is specified that the Inca and his retinue "brought over ten thousand sheep loaded with their provisions."

We know that Atahuallpa sent Pizarro some messengers as he climbed up toward Cajamarca bearing "ten sheep," among other items (Xerez 1968:218). A testimony of major importance is that of Ruiz de Arce, who was an eyewitness to the events that took place at Cajamarca. It is not known when he wrote his memoirs, but it certainly was after 1540 (Porras 1986:130). Porras (1986: 129) is right when he calls him an "essentially objective, and unfortunately pithy, narrator," because all he says when describing Cajamarca is: "there are many sheep" (Ruiz de Arce 1968:425). This was later confirmed by Cieza de León (1987: Part III, Chap. XLIII, 124), who, when discussing the same chapter, said there were "many herds of sheep" on the fields.

After taking the Inca prisoner, the Spaniards headed for his encampment outside the city, at the site of the present Baños del Inca. Here they "found innumerable provisions, both meat and dried beef, as well as those sheep, some for pack animals, others for food" (Estete [Anonymous?] 1968b:378). Xerez (1968:232) says that all these "sheep" were set loose by the Spaniards, but "[t]he Governor ordered . . . that the Christians kill every day as many as they needed."

An anonymous document originally published in 1534 seems to summarize two letters dated in that same year and apparently inspired by one that Gaspar de Espinosa sent to the King in 1533, which possibly used information from a letter Pizarro sent to the Cabildo of San Miguel (see above) (Porras 1967; Soto 1992:11). The document tells us that before setting south, Pizarro was given a present by a cacique: "there were four sheep, the size of the sheep of the land, which is that of a five- or six-month-old pony. And with them two herders of gold, the size of small men" (Anonymous 1992d:XXXIII).[10] This document would date to 1534 acording to Porras (1937, 1967), but John Rowe (pers. commun., 23 August 1992) believes it must be from 1539.

In the 1540 *visita* of Cajamarca (see Espinoza Soriano 1967:26), the Indians declared that the tribute they paid the Inga and the encomendero was of "sheep," among other items. During the inspection, Cristóbal de Barrientos made the following comment: "In this land there seems to be little livestock, and said lords of Cajamarca say it is so." This seems to contradict the later observations of Cieza de León (see below), so the comment Espinoza Soriano (1967:19) made in this regard is important: "In 1540 the curacas declared that there were very few herds of llamas in their province. Barrientos himself records that he also got the same impression during his travels, but we should not forget that he covered quite little of the Cajamarquino lands."

The testimony of Cieza de León (1984:Part I, Chap. LXXVII, 228) slightly more than 15 years after the events of Cajamarca, is of interest because he reported that "a lot of livestock [is still] raised," and that "[t]he Indians . . . are quite clever . . . when it comes to . . . raising livestock."

We know that some of the *mitimaes* who lived in Cajamarca between 1550 and 1560 and who were loyal to the curaca of Saña were "shepherds" (Ramírez Horton 1982:127).

But the killing of camelids in Cajamarca, from the beginning of the conquest to the end of the sixteenth century, must have been tremendous. Lorenzo Huertas (pers. commun., 6 February 1993) found a document in the Archivo General de Indias of Seville (AGI Justicia, Leg. 458) recording a charge levied against Gonzáles Cuenca in 1566 for having authorized some residents of Cajamarca to have the Indians carry goods to Trujillo. No llamas were available for this task.

Today, camelids are being imported from the altiplano to try to develop them here once more, but everyone knows there are no camelids in Cajamarca, and this appears in treatises on geography (e.g., Peñaherrera del Águila 1969:282). However, a datum lost in Kosok's (1965:131) writings is interesting. Kosok was in Peru between 1948 and 1949, and while in Cajamarca he noted that "the llama, faithful servant of the Indians, has completely disappeared, though traditions among a few of the very old inhabitants in Cajamarca indicate that it survived there up to a hundred years ago"—that is, the first half of the nineteenth century.

6.7.4 The Expedition of Hernando Pizarro to Pachacámac

I will now briefly depart from the strictly geographic approach so far used in this analysis of the distribution of camelids according to historical sources because I want to analyze what the "twen-

ty on horseback and some riflemen" (Estete 1968a:242–243) under the command of Hernando Pizarro, Francisco Pizarro's brother, saw. (Xerez [1968:241] confirms they were "twenty on horseback," but he does not say how many of the others went, just "some on foot.") The detachment was headed for Pachacámac to ensure that the ransom promised by the Inca would be honored. As Porras noted, this was the first European expedition in the still unconquered Inca Empire (Porras 1986:116). Miguel de Estete was with these riders as inspector (*veedor*). His travel notes—because that is what they are—are therefore extremely important for my discussion, because these constitute "the first geographic document of Andean Peru" (Porras 1986:116). Let us see what he noted.

The expedition set out from Cajamarca one "Wednesday, day of the Epiphany (commonly known as the feast of the Three Wise Men), on 5 January 1533" (Estete 1968a:242; Xerez 1968: 241), and reached Corongo after passing through Cajabamba, Huamachuco, Andamarca, and Totopampa. At Corongo, Estete (1968a:244) noted that "all along the road is a great number of livestock, with herders who look after it." Corongo is at 3192 masl, in the northern part of the Callejón de Huaylas, in the province of the same name. (Dedenbach Salazar made a mistake when she placed this citation on "the outskirts of Jauja." The chronicler [Estete 1968a] literally says "went to sleep at another small town called Corongo").

The voyage continued and the riders passed lengthwise along the Callejón de Huaylas (2200–3000 masl), where "the Indians of said town lent many sheep. . . . All that land is full of livestock . . . so that going along the road the Christians saw the herds of sheep coming down the road." When pasing through "Guarax," which must be the present-day town of Huaraz (in the province of the same name), Estete again mentions that it was a place "where there is much livestock. . . . They had two hundred heads in one corral just to feed the captain and his men" (Estete 1968a:245).

Estete explains that they went to another town he called "Pachicoto," which must be present-day Pachacoto in the province of Recuay, and then moved to "Marcara," which certainly cannot be the present-day Marcará because it is north of

Huaraz, close to Carhuaz, and must instead be the town of Marca, in the southern part of the province of Recuay. The chronicler quite clearly states that here they began to descend to the coast. Of the lord of this place he wrote, "this belongs to lords of livestock who have their herders in it, and take them there to graze in certain seasons" (Estete 1968a:246). (In this case Fernández de Oviedo [1959c:Bk. XLVI, Chap. XI, 69–70] says he is following Miguel de Estete, but the way he changes some geographic terms is interesting.)

There is no other reference to camelids from this moment on until the group returned. From Pachacámac they headed along the coast to Huaura, and there began to climb up into the highlands. They reached "Aillón," which might be the present town of Aillón. It appears in Stiglich's *Diccionario Geográfico del Perú* (1922:60) and is in the northern part of the province of Huaura (Department of Lima). According to the chronicler, this town depended on a larger one he calls "Aratambo," which I was unable to locate (nor did del Busto find it [1963–1970b:58]). The chronicler notes it had "much livestock" (Estete1968a:250). From here the group moved to another town called "Chincha," which was subject to Cajatambo and which was not found either. However, it is noted that "this town has much livestock." The detachment reached Cajatambo (in the northern part of the Department of Lima), "where there is much livestock, and all along the road there are many sheep corrals" (Estete 1968a:251). (In this case, Fernández de Oviedo [1959c:Bk. XLVI, Chap. XII, 73] claims to follow Estete, which he does without major changes, except instead of saying "Chincha," as Estete does, he says "Chinchi," and he replaces "corrals of sheep" with "corrals and shepherds.")

From this site the Spaniards once again headed south instead of turning north to Cajamarca. When going through Junín (in the present department of the same name), Estete described it as "a plain where there is much medium-sized livestock, like that of Spain, and with a very fine wool" (Estete 1968a:251).

Estete (1968a:252) mentions the many small lakes in the Junín area and notes that "there is much livestock here." He immediately adds: "A river that goes to the town of Pombo leaves this small lake. . . . There is much livestock along this river." Pombo is the name of an old town in the Pampa de Bombón, the present-day Pampa de Junín (Dedenbach Salazar 1990:87).

Estete (1968a:255) then continues describing the Junín area and repeats that "there were many herds of livestock." Then he mentions the town of Tonsucancha and notes that here "there is much small livestock of very good wool." Murra (1978:82) believes this last passage marks the first time the alpaca is mentioned. He is almost certainly right. Tonsucancha is in the southern part of the province of Dos de Mayo, in the Department of Huánuco (see the map prepared by McK. Bird to illustrate the 1562 *Visita de Huánuco*. It is included in Ortiz de Zúñiga 1967, Vol. 1).

Although in this case Fernández de Oviedo (1959c:Bk. XLVI, Chap. XII, 74, 76) also says that he is citing Estete, some variants make me believe he had another source, which is why it is best to cite him. Fernández de Oviedo wrote that on Tuesday the 10th, Hernando Pizarro and his retinue "slept in a hamlet of herders beside a small freshwater lake that goes down three leagues to a plain, where they saw such an amount of medium-sized livestock, like that of Spain and with fine wool, that their numbers were much to be admired." Then he mentioned Friday the 14th, when the group "left Pombo to go to Xauxa; and this day he went to sleep in a town called Chacamarca." Fernández de Oviedo explains that here there are several neighboring towns, "and around them a vast number of livestock, of which the Spaniards thought there was more than a hundred thousand heads." Of Jauja he wrote: "It abounds with provisions and livestock." On Friday the 20th the group began the return journey to Cajamarca and went as far as Pombo. "Wednesday they left said town of Pombo, and went through a plain full of many herds of livestock to sleep in some big dwellings." On Friday the 27th "he went to sleep at another small [town] called Tonsucancha. . . .This town is of much small livestock, of very good and fine wool resembling that of Spain."

The last citation from Estete (1968a:255) concerns the road between Bombon and Cajatambo, and specifies that "[a]nother day they went to sleep in another town called Guaneso. . . . it is a land with much livestock." This undoubtedly is the archaeological site now known as Huánuco Viejo or Huánuco Pampa, in the province of Dos de Mayo, in the district of La Unión, Department of Huánuco (at 3800 masl). Interestingly, though Fernández de Oviedo (1959c:Bk. XLVI, Chap. XII, 76) here cites Estete (1968a:255), he nonetheless writes "Guanaco," and not "Guaneso." This Huánuco Viejo must not be confused with the present-day city of Huánuco. Stiglich (1922:515) said of the ruins of Huánuco Viejo that "[t]hey were discovered by Hernando Pizarro on 26 March 1533. The Spaniards first called it Huaneso and then León de Huánuco, when Gómez de Alvarado founded a town over these ruins, intended as the capital of this corregimiento. The rebellion against the whites headed by Illarupa depopulated it. This is why Don Pedro Barroso founded the present-day Huánuco a few days later."

6.7.5 More Data on the Highlands

Let us return to the description of the land from north to south. We have seen Cajamarca and will now study the conditions in the department of La Libertad. Cieza de León (1984:Part I, Chap. LXXXI, 236) discussed Huamachuco, in the present-day province of the same name. The chronicler wrote that "in the land of this province of Guamachuco [there is] a great number of livestock of sheep: and on the heights and in the wilderness roam an even greater number of wild livestock called guanacos and vicunias." However, I must point out that on this same page Cieza mentions the "royal hunting grounds" the Inca had there, where only wild livestock was kept, and specifies these were guanacos. Hunts were held here, and "of these hunts they took ten thousand or fifteen thousand heads, or the number he wanted: so much there once was."

Porras (1959:216) published a letter from Francisco Pizarro to Pedro de Alvarado, dated Los Reyes [present-day Lima] on 9 July 1536, which says: "I named Señor Alonso de Alvarado captain to people the lands behind Trujillo inland, and he found a land very rich in livestock." It is hard to give an exact location for the site, but it could be Santiago de Chuco.

Vázquez de Espinosa (1948:1180/373) also talks of "[t]he province of Guamachuco," and comments there was "[much livestock] of the land" and "big [and abundant] offspring of large and small livestock."

Here as in Cajamarca there is evidence that a great killing of animals took place, because Espinoza Soriano (1971a:54) reports the conditions in the province of Huamachuco during the eighteenth century, and notes that no camelids are mentioned, just European animals. For example, in San Pedro de Chuquisongo, one of the largest and most important haciendas and estancias in the corregimiento, there were "a hundred thousand heads of lambs [borregos] and rams [carneros]," which were still being raised in the nineteenth century.

Cieza de León (1984:Part I, Chap. LXXX, 234–235) described "the Conchucos, and the great province of Guaylas," that is, present-day Callejón de Huaylas, and noted that here "[i]n past times there was such a big number of livestock and sheep [ovejas] and rams [carneros], that there was no counting them." He immediately explains that "the wars finished with this throng there was [and] little [now] remains." This means there were still camelids here in the mid-sixteeenth century. This is confirmed by the 1558 Visita of Guaraz, because Varón Gabai (1980:57) notes that "[a]s regards livestock, they mention having 'sheep of the land,' goats, pigs, and some 'sheep of Castile . . .' and they are looked after by old Indians and boys." Dedenbach Salazar (1990:100) says that in this case it is not altogether clear whether this is livestock of the land or of Castile, but I believe there can be no doubt about it, to judge by what Varón Gabai reports (1980:57). Besides, when discussing the Visita of Llaguaraz, a repartimiento close to the present-day city of Huaraz, which also took place in 1558, Varón Gabai wrote: "Herding is probably the major activity in the repartimiento, and a slight displacement of native livestock by European livestock can be perceived" (Varón Gabai 1980:65). This shows beyond any doubt that there were camelids in the zone. However, it is worth bearing in mind the rest of Dedenbach Salazar's (1990: 100) comment: "[W]hen examining the visitas for the number of heads we must bear in mind that to avoid paying too much tribute, the Indians remained silent about a great part of their livestock, something that must have been easy to do because the dispersed herding made it difficult and even impossible to control or verify these claims." In other words, it is possible that there were more animals than the Indians declared.

Silva Santisteban (1964:134) has a datum on the corregimiento of Conchucos, on the northern part of the Callejón de Huaylas, which says that there were many sheep "of all kinds of livestock, though the most part is sheep of Castile and cows." Silva Santisteban based his statements on the seventeenth-century documentation of Francisco López de Carabantes (but this title does not appear in the bibliography). It is true that the kind of sheep in question is not clearly defined, but we can assume there were camelids, because it is stated that these were "of all kinds."

The testimony of Vázquez de Espinosa is once more of great importance. He described "the province and valley of Guaylas," specifically the "town of Requay" (i.e., the southern part of the Department of Ancash, in the province of Recuay), and emphatically stated that "in it there are great herds of rams . . . of the land," explaining that there also were "rams of Castile" (Vázquez de Espinosa 1948:1379/459), so no doubts remain. We know by the Carmelite friar's own words that he was in Chavín de Huántar in 1616, that is, in this same area, so his data cannot be questioned (see Vázquez de Espinosa 1948:1372/457).

There are fewer data for the Department of Huánuco, slightly more to the south. Aguilar y Córdova remarks that "[t]he province of Guanuco [is] so called for an animal with this name known in those lands." This clearly is a skimpy reference that must be taken with due caution. (I was unable to find this source and took the citation from Espinoza Soriano [1975b:16], who attributes it to Aguilar y de Córdova [1599:Bk. I, Chap. VI; 1938 edition]. However, this same author is cited by Dedenbach Salazar [1990:93], but she calls him Aguilar y Córdova and gives 1578 as the date of the document. Moreover, although both scholars apparently refer to the same 1938 edition, the page numbers do not agree: for Espinoza Soriano the range is 321–325, and for Dedenbach Salazar it is 321–345.)

A far more precise bit of information refers to the site of "Rondo de los Queros" (to the west of the Department of Huánuco. It probably is the Runtu that appears on the map of "[t]he Highlands of Huánuco" prepared by McK. Bird, appendix to Ortiz de Zúñiga 1967). During the 1562 inspection of the Chupachos, a vecino declared that "he has twenty heads of livestock of this land, males and females, and several have died because of the mange [caracha]."

For this area we once again have the testimony of Vázquez de Espinosa (1948:1363/454), who was here in 1616. He comments "[o]n the province of the Guamalies," "where the tambo of Guanuco el Viejo is," which was already mentioned when recounting the voyage of Pizarro's brother. He explains that this tambo "is on a great plain full of livestock . . . of the land" (Vázquez de Espinosa 1948:1375/458–459).

I was unable to find data on the lands to the south as far as the highlands that lie on the border between the Departments of Lima and Junín. Rostworowski (1978:161) studied this zone and explains that Causso figures among the towns that formed the ayllu of Canta and must have been on the "plains of Bombón," where the former Hacienda Corpacancha lies (i.e., north of the province of Yauli, in the Department of Junín). The headman of this town declared in the 1553 visita that they lived there because they were "shepherds of the sheep of Canta, which in due time come to graze there" (fol. 130v). Rostworowski, however, notes that in this case, as in others, there are contradictions, "which possibly mean they did not declare the truth, and had hidden their livestock in distant places" (Rostworowski 1978:161).

The 1549 inspection of Canta mentions another 16 "seasonal" towns intended for specific tasks. Caruacaya or Caruacayán is mentioned in the midst of the puna, behind the Cordillera de la Viuda, and was "considered a town of shepherds" (Rostworowski 1978:170).

Caraguayllo, in the Esquibamba moiety, located in the chaupi yunga, was also visited at this time (1553). Esquibamba is a hill in a ravine close to the town of Arahuay. The ravine runs parallel to the Chillón River, upriver from Santa Rosa de Quives. The visita says: "Also, we asked them about the livestock, they said they did not have it because it was yunga and was in warm land"; nevertheless each town had to pay a tribute of "twenty sheep," among other items (Rostworowski 1978:165–166,

255). In this same *visita* it is said that Yeso was one of the towns that formed the ayllu of Canta, and was "already in the yungas [and] there were no sheep" (Rostworowski 1978:162).

Rostworowski (1988b:76) recounts a conflict between the Canta and the Chaclla in this same region. "In 1549, during the Peruvian sojourn of licenciate Pedro de la Gasca, the encomenderos Nicolás de Ribera el Mozo—encomendero of the Canta—and Francisco de Ampuero—encomendero of the Chaclla—tried to force a peace between the curacas and decided that the Chaclla would sell their land to the Canta for two hundred 'sheep of the land,' i.e., llamas." It seems that the Chaclla resisted because they did not understand what a "sale" was and refused to receive the sheep of the land. Ampuero then commanded that they be sent to his house in Los Reyes (present-day Lima), where he sold some and sent the rest to the butcher. It is worth noting that with the victory of Túpac Yupanqui, a hundred Chaclla *mitimaes* from Mama, now the site of Ricardo Palma in the Rímac Valley, close to Chosica (province of Lima), settled in Quivi (present-day Quives).

From the data in Cosme Bueno (1951:34) we know that in 1764, "[m]uch livestock [was] raised" in the province of Canta, "as well as the other wild animals of the Sierra, such as vicuñas and rams of the land, which are different from the European ones."

Finally, Dedenbach Salazar (1990:100) comments on Rostworowski's (1988b:76) research in Canta. She mentions her statement that the whole truth was not told to the inspector regarding the animals, with only five sheep mentioned. In other words, the livestock was hidden from the *visitador*. Dedenbach Salazar brings up the archaeological research done by Dillehay (1979), which was mentioned in the previous chapter, and concludes that "[t]his evidence can also make us consider that Canta must have had relatively large camelid resources."

There are interesting data for the province of Huarochirí (Department of Lima) and the neighboring province of Jauja (Department of Junín). Dávila Brizeño inspected Huarochirí in 1571 and 1572 and wrote the well-known *Descripción y relación de la provincia de Yauyos*. In this region is the famous Mount Pariacaca, which dominates the local cordillera. Dávila Brizeño (1965:159) wrote that "[i]n the wild of this Cordillera of Pariacaca, on one slope and the other, there are a great number of wild livestock of the land which they call vicuñas and guanacos." Vicuñas were still seen in the early seventeenth century (1605?), because after passing through the heights of Pariacaca, Father Ocaña wrote that "[i]n the fields there are some vicuñas" (Ocaña 1987:238). I worked in the area in 1960 and 1983 and did not see any.

Some of these animals still lived in the neighboring province of Yauyos in the late eighteenth century, because in 1764 Cosme Bueno (1951:32) wrote that "[i]n the hills there are many vicuñas and huanacos." The Yauyos were a warlike and fierce people who lived in the Tupe region, at the headwaters of the Cañete River, in the present province of Yauyos, in the Department of Lima (see Rostworowski 1978:32).

Dedenbach Salazar (1990:85) discusses the significance camelids have in the rites and traditions of Huarochirí collected by Ávila (1608? [1966]), which indicates that this area must have had a great number of them. Dedenbach Salazar notes that Spalding (1984:44, 193) found huts and corrals in the puna of Huarochirí with the help of eighteenth-century documents. I saw some of these corrals in the area of Mount Pariacaca in 1983, and found some small herds of llamas and alpacas.

We now turn to the Department of Junín. Ruiz de Arce (1968:426), who wrote after 1540 but who was in Peru between 1532 and 1534 (see Porras 1986:127–130), left an account of his voyage to Cuzco. When crossing the Meseta de Bombón he found that "there are many wild sheep." When decribing the "land of Bombón," Cieza de León (1984:Part I, Chap. LXXXIII, 240) said: "These natives of Bombón had much livestock, and although consumed and wasted by the wars, as can be surmised, still they have some, and great herds of wild ones are seen along the heights and wilderness of their land." Did Cieza mean vicuñas? It is possible, because when discussing Junín in 1614, less than half a century after Cieza's travels, Ayala (1974–1976:276) wrote: "vicuñas, of which there are many in this land."

Finally, there is the testimony of Vázquez de Espinosa (1948:1367/456), who described the "province of Chinchacocha," that is, Junín, and reported that "[m]uch livestock of the land grows in this province."

There are some specific data for the Tarma zone. Cieza de León (1984:Part I, Chap. LXXX, 234) mentions "Tamara and Bombón." This of course is a mistake by the chronicler, who possibly wanted to say Tarama, a name other chroniclers use for Tarma. Cieza said that "[i]n past times there was such a large number of livestock and sheep [*ovejas*] and rams [*carneros*] that there was no counting them."

According to Arellano (1988:78–79), in the eighteenth century most of the ayllus of Tarma had lost access to, or had less, pasture than they had had since the resettlement program (the *reducciones*) of Toledo. The factors that unleashed this phenomenon were, on the one hand, the increase in Indian *forasteros* and Spaniards working in the mines, and, on the other, the lands rented to other herders by the curaca. In 1571 the ayllus on the left bank had almost no lands, or had completely lost them. Besides, several haciendas and estancias on the right bank passed to non-Indian hands. Arellano (1988:80) explains:

> The presence or maintenance of camelids never appears in the titles to the haciendas in Tarma, even the haciendas belonging to Indians. All haciendas record [European] sheep as the main product, save for Huancal, which only raised cattle, mules, and asses, the kind of livestock that was also raised in other estancias, as well as horses. It is interesting that no camelids are mentioned, even though we know there were abundant llamas, vicuñas and huanacos in the punas thanks to the observations made by a Spanish botanist in the second half of the eighteenth century, who spent eleven months in Tarma. He adds that vicuñas and huanacos are found from 5° S. Lat. (26). It is especially interesting that he does not mention the alpaca, and when it does appear it is in regard to Cuzco. From this we can assume that the alpaca was not known as the

major wool producer, which would explain, on the one hand, the intensive use of wool from sheep, and the remark that vicuña and huanaco wool was used for accessories such as handkerchiefs, socks, caps, gloves, etc., in combination with cotton and/or silk. The ponchos were instead made of carnero wool combined with cotton. This raises the question of whether or not the alpaca was known and raised in the Central Sierra, and was exterminated due to the intensive raising of sheep. We could likewise assume that alpacas never extended (even in pre-Columbian times) beyond the southern Peruvian altiplano.

Note 26 cites Ruiz (1952, T 1:76, 82–83, 93). Ruiz does not appear in the bibliography, but he must certainly be Hipólito Ruiz López, who went to Peru in the eighteenth century with José Antonio Pavón y Jiménez, another botanist.[11]

For the Jauja area (Department of Junín), we once again have the testimony of Ruiz de Arce (1968:427), for whom "[i]n this land only sheep and maize grow." Although he did not specify what sheep these were, we can assume they were of the land because of the early date he was in the area. Espinoza Soriano (1972:86–87), who wrote that in 1533 the Jauja Valley was still peopled and full of livestock and green chacras, concurs. He adds that "Pizarro realized that the only truly plentiful products in the valley were maize and auchenid livestock."

When discussing the Huancas who lived in the area, Espinoza Soriano (1972:35, 41) then comments that only a minority lived in the puna, looking after "hundreds of herds of llamas." However, he notes that the livestock was not abundant because the llamas were few in comparison with those in the kingdom of Chinchaycocha. These were used as pack animals.

In 1597 Ramírez (1936:35) also wrote that Xauxa "is a valley that abounds in all kind of provisions and fruits and much livestock both of the land and of Castile."

However, in the late sixteenth century there were still vicuñas and guanacos, because in his *Descripción de la provincia de Xauxa*, dated 1582, Vega (1965:171) reports that "the game available in the heights of this valley are . . . vicuñas and guana-cos." In the early seventeenth century, Vázquez de Espinosa (1948:1339/443–444) gathered infor-mation on the "valley and province of Xauxa," and wrote that "on the road to Tarama . . . many rams of the land . . . around here there is much feral livestock of the land, or wild, which are guana-cos." He immediately adds there also was "much [livestock] . . . of our merinos."

In 1586 Ribera and Chavéz y de Guevara (1985:187) wrote of Huancavelica: "And clothes they make out of the wool of the livestock native to this land, of which they also breed some that serve to eat and carry [loads]." Rivera and Chavéz y de Guevara (1985:192–193) then describe the "fierce and wild" animals in the area. They explain that "there are rams of the land smaller than camels, and they are called guanacos, and other smaller ones they call vicuñas, they are very swift . . . and all this livestock is for food."

That same year, 1586, Cantos de Andrada et al. (1965:307) wrote an account of the "Villa Rica de Oropesa," the present-day city of Huancaveli-ca (in the department and province of the same name). They made the following comments: "[A]nd [in] all the sierras of this kingdom . . . live wild livestock, vicuñas and *guanacos* and deer; there was a great number of the vicuñas and *gua-nacos* in the time of the Inga, because the natives were not allowed to kill any without his permis-sion, and since then they have belittled it and taken the liberty both in this and in amusing themselves with bad customs. . . . Around this villa there is little livestock of this wild kind due to the business the people do and to the domestic live-stock, both of the land and of Castile, which lives and breeds a lot."

Slightly later, in 1597, Ramírez (1936:37) noted for this same area that "the Indians in said land have much livestock." When describing the "Province of Castrovirreyna" in the eighteenth century, which corresponds to the present-day province of the same name in the Department of Huancavelica, Cosme Bueno (1951:74) said that "[w]hat abounds is the livestock due to the abun-dance of pastures. There also are llamas or rams of the land, vicuñas . . . and huanacos."

I will now review the data for the Department of Ayacucho. Ruiz de Arce (1968:427), who, as we recall, was in Peru between 1532 and 1534, men-tions Vilcas. This is the present-day town of Vil-cas Huamán, in the province of Cangallo. He said that "[i]t is a land of much sheep; there are many wild [ones]."

In 1557, when describing the province of Huamanga, Damián de la Bandera (1974:140) said: "[I]t has wilderness and some livestock, and they hunt the wild ones with which they support themselves and dress, and prepare jerky[12] which they call *charque*."

There is important information for 1586. First, there is a *Descripción del corregimiento de Vil-cas Guaman* (the present-day Vilcas Huamán) by Carabajal (1965:206). Here it is says that "[i]n this province there is livestock of the land. . . . There are other wild animals which are almost like the rams themselves, called guanaco and vicuña."

This same *Descripción* includes references to a series of "parishes," which I was unable to locate, but this is in any case irrelevant. In "The account of the parish of Guancaraylla and its dependen-cies [*anexos*]," Carabajal says: "[A]nd of the wild animals they have guanacos and vicuñas." He then mentions the parish of Quilla and Colca, and notes that "[t]he natives of these three towns have rams of the land in small numbers . . . and they have the wild animals, vicuñas and guanacos." Meanwhile, in the parish of Pabres (Papres), "[t]hese natives have rams of the land for their use and sustenance . . . and of the wild ones guanacos [and] vicuñas." Exactly the same was noted for the parish of Chuiqui, while "some rams of the land in small numbers" are mentioned when describ-ing the parish of Guanpalpa. Finally, Carabajal says of the parish of Pacomarca that it had "rams in small numbers . . . there are guanacos, vicuñas" (Carabajal 1965:212–218).

Around this time Monzón, Gonzales, and Arbe (1965:228) wrote on Atunrucana, in the ju-risdiction of the city of Huamanga (in the province of the same name). With regard to the Indians, they said that "most of them have livestock of the land in the heights that are called *puna*, which are big rams and small pacos, and in the same way all go to look after it and heal it of careche, which is what in Castile they call scabies [*roña*], so they are not ashamed to look after their livestock." Monzón again, now with Quesada, Sánchez de Haedo,

Gutiérrez de Benavides, and Taipemarca (1965: 246), described, also in 1586, the province and repartimiento of Rucanas Antamarcas, in the jurisdiction of the city of Guamanga. This is the present-day province of Lucanas. Here it says that "in this province there are pack rams of the land called guacaes (5); there are pacos which are small livestock, with much wool they use to dress and good meat to eat, and all are quite tame livestock. There is another fierce [*bravo*], wild, undomesticated livestock they call guanacos, slightly smaller than the rams of the land, good meat. There are vicuñas that are small, though in build they resemble rams." Note 5 explains that the term *guacaes* comes from "*hucayhua* or *huacahuya*, the name of the biggest pack llamas."

This same Monzón, now with Saravia, Frías, and Taypimarca (1965:223–224), described San Bartolomé de Atunsora, in the province of Rucanas and Soras, in the jurisdiction of the city of Guamanga. In this case the description also dates to 1586. Here "there is domestic livestock, which are rams of the land used as pack animals, and other smaller ones they call pacos, who have wool . . . they are almost of the same build and kind and good to eat . . . and there is a reasonable number of this livestock; there are other wild livestocks called guanacos and vicuñas, of the same kind, temperament and build as the other rams of the land."

In 1586 too, Rivera and Chávez de Guevara (1974:151) described the "city of Guamanga and its district [*términos*]"—that is, the present-day city of Ayacucho (province of Huamanga)—and explained that here "there are cold *páramos* the Indians call punas. They are uninhabited and serve as pastures for the livestock of the Indians, which requires this kind of land; it gives no wood other than a somewhat dull and short hay said livestock grazes on." Further on (1974:158) it is explained that "clothes they make out of the wool of the livestock native to this land, of which they also use to breed some who serve to eat and carry." Then (1974:168–169) they say: "there are rams of the land smaller than camels, and are called guanacos, and other smaller ones they call vicuñas; they are swift and have a very soft and delicate skin clothes are made of."

When describing the environs "[o]f the city of San Joan de la Vitoria de Guamanga," that is, the city of Ayacucho in the early seventeenth century, Vázquez de Espinosa (1948:1432/487) notes: "and in the midst of the frigid puna [there are] many offspring of livestock of the land." While describing the bishopric of "San Joan de la Vitoria de Guamanga," Vázquez de Espinosa discussed "Vilcas" (i.e., Vilcas Huamán), explaining that "all the province is fertile in provisions, there are in it offspring of large and small livestock, and of livestock of the land" (Vázquez de Espinosa 1948: 1436/488–489). Then he says that after going through "Abcara" one crosses "the uninhabitable puna," and "in those cold plains there are only miches, which are like herders' huts where they keep a great number of livestock of the land. In these punas [there is] an endless number of wild livestock they call simarron [feral], which are vicuñas and guanacos" (Vázquez de Espinosa 1948:1438/489). This description is for the highlands of Nasca.

Finally, we have two (eighteenth-century) passages in Cosme Bueno, one (1951:79) on the province of Vilcas Huamán that says "[t]here are all kinds of large and small livestock"; the other (1951:77) on the province of Parinacochas, says "[t]here is a lot of livestock due to the abundance of pastures. . . . Many huanacos breed in the hills."

Fornee, Luque, Placencia, Gallegos, and Cevereche (1965:19) describe the Department of Apurímac in 1586. They specifically mention the corregimiento of Abancay, Zurite, Guarocondor, Anta, and Puquibra, and note that "around these towns there are . . . guanacos." Around this same date Fornee, Murcia, Salazar, Quipquin, and Cevereche (1965:23) mention the localities of San Antón de Chinchaypuquio, La Visitación de Nuestra Señora de Zumaro, La Encarnación Pantipata and Santiago Pivil, and note that "there are . . . guanacos." Fornee, Vaez, Velázquez, and Cevereche (1965:26) report on the localities of San Sebastián Pampaconga, Sant Juan de Patallata, Santa Ana Chinta, and Santiago Mollepata, where "there [also] are . . . guanacos." Finally, Fornee (1965:29) mentions the towns of Santiago Hamancay, Santa Catalina de Curavacu, and Sant

Pedro de Sayvita, but in this case he notes that "there are ... vicuñas."

Vázquez de Espinosa (1948:1483/511) described the provinces of Abancay and Andahuaylas in the same department, and said that "between Amancay and the province of Andaguaylas lie the provinces of the Cotabambas, Cotaneras and others of the Quichua nation, rich in livestock of the land. . . . [C]lose to this province is the province of Guamampalla, from whence to go to the coast toward the west, more than 30 leagues of cold and uninhabitable puna are covered, where only vicuñas and guanacos live."

This same author (Vázquez de Espinosa 1948:1480/510) described the district of "Andaguaylas," which "has fertile fields, and in them big offspring of livestock of the land . . . and large and small livestock from Spain." Vázquez de Espinosa (1948:1488/513) notes that the province of Chumbivilcas is "southeast" of the province of the Aymaraes, and between it and the Apurímac River, "[w]hich is big and inhabited by many people and livestock of the land." He ends with the province of Aymaraes (between the Abancay and Apurímac Rivers), explaining that in the puna "there only are some huts called miches, where the Indian herders are looking after their livestock which is of the land. . . . [T]here are guanacos, vicuñas," but "at present there also is a great number of livestock of Castile, merino rams." Vázquez de Espinosa adds that the province has "[l]ivestock of all kinds." Finally he explains that the neighboring province of "Vmasuyus" is "rich in livestock of the land, and at present has many offspring of ours of Castile." He concludes by saying that "on the east it borders with the wilderness and deserts of the sierra and cold puna, inhabited only by vicuñas [and] guanacos" (Vázquez de Espinosa 1948:1485–1486/512, 1487/513).

In her study, Dedenbach Salazar (1990:93) notes that Lizárraga (1968:Bk. I, Chap. LXXIX, 60) mentioned "livestock of the land" in Andahuaylas. This is a mistake, because Lizárraga never said so.

One would expect to find lots of data for the Department of Cuzco, but that is not the case. However, the scant data available are interesting. First I will review what was said regarding the environs of the city that was the capital of the Inca Empire. There is a description in the chronicle of

Ruiz de Arce, but these evidently are references he selected from other people. It is known that he left Jauja and returned to Spain, and did not continue with Pizarro's army, which was going to Cuzco (see Porras 1986:127). Ruiz wrote: "This land is most rugged, with many sierras, it rains a lot and snows, there are many sheep. Nothing is raised save for maize, and this is harvested once a year" (Ruiz de Arce 1968:433).

The testimony of Pedro Pizarro, cousin of and page to the Marquis, is important. because he was one of the men who entered Cuzco with Pizarro in 1533. Porras (1986:137) says that "[t]he testimony of Pedro Pizarro is one of the most truthful and direct on the actors and events of the conquest. . . . He always says the truth." The events in question happened after 1536, because the chronicler says that "food was lacking, especially meat," after the uprising in Cuzco and the Spanish seizure of the fortress. Hernando Pizarro then sent men "to Pomacanche, provinces that are toward the Collao, thirteen or fourteen leagues from Cuzco." He immediately adds that "thus we were and went some thirty or twenty-five days, and gathered up to two thousand heads, and returned to Cuzco with them without any hindrance" (Pedro Pizarro 1978:Chap. 20, 144). The site of Pomacanche, which the chronicler mentions, is in the district of Acomayo, in the province of the same name.

Cieza de León (1984:Part I, Chap. XCIII, 260) also wrote on the Cuzco region. His testimony dates to the mid-sixteenth century, because he joined the army of La Gasca and witnessed the end of the uprising of Gonzalo Pizarro on 9 April 1548 in Jaquijaguana (Araníbar 1967:n. 2, li). Cieza said: "[I]n the valleys there are herds of cows and goats and other livestock, both Spanish and native livestock." In other words, the impact of the conquest is perceivable just over 16 years after the European arrival.

In 1597 Ramírez (1936:36) wrote that Cuzco was "a most fertile land, overflowing with food and livestock and fruit, both of Castile and of the land."

In the seventeenth century, Vázquez de Espinosa (1948:1603/556) noted in his description of Cuzco that "the Spaniards have . . . their haciendas and outbuildings . . . ranches of large and small livestock, pigs, and livestock of the land, with many mules and horses for their service." In other words, the Spaniards were raising camelids.

Acuña, Prado, Quispihaqueua, Yllanes, and Vilcacuri (1965:324–325) left a description of the town of Libitaca, which was "close to the city of Cuzco." Livitaca actually is in the southern part of the Department of Cuzco, in the district of Chumbivilcas, in the province of the same name. They wrote: "[T]hese towns are surrounded by hills and in them are . . . guanacos and vicuñas." The description was prepared in 1586. This same year Acuña, with Sánchez de la Cueva, Quispe, Sánchez Goliardo, Rigon, Nina Chaguayo, Medina, and Supanta (1965:323), described the town of Belille, also "close to the city of Cuzco," which I was unable to locate. They said that "high in the hills . . . there are . . . vicuñas."

Vázquez de Espinosa (1948:1609/558) later mentioned the corregimiento of Quispi Canche (i.e., the present province of Quispicanchis), which adjoined "the province of the Canches," and noted that here the Indians "breed much livestock of the land."

There are several references for the southern part of the Department of Cuzco and part of the Department of Arequipa. Thus, Cieza de León (1984:Part I, Chap. XCVIII, 270) writes of the "land of the Canas" and "the Canchis," and specifically of Ayaviri that "it is well supplied with provisions and livestock." Murúa (1964:243–244) regarded this region as the richest in livestock of the land. Although it is true that the claims made by Friar Murúa, who wrote between 1590 and 1600, must be taken with due caution (see Araníbar 1963:106, n. 2), in this case they must be correct because we know that he was in Cuzco, in the parish of Capachica (Porras 1986:477).

Oré (1992 [1598]:VIII, 32v; Dedenbach Salazar [1990:93] gives the folio as 32) also said that the province from Cuzco to Arequipa, which belonged to the Canas and Canchis, as well as that of the Collahuas and Cauanas, were "rich in livestock of the land." Finally, Bueno (1951:103), who began his description of the provinces of Peru in 1764 (Porras 1955:224), says of the "Province of Canas and Canchis or Tinta" that "[l]ivestock is raised in the highlands due to the abundance of pastures, and also very many vicuñas, huanacos."

As for the Department of Arequipa, we have the testimony of Ramírez (1936:45), who lived in Peru until 1580 (Porras 1955:222). In 1597 he wrote of Arequipa that there were "some punas of very cold land where much livestock of the land and of Castile is raised."

Benavides (1988:44) in turn claims that notarial documentation from sixteenth-century Arequipa indicates the presence of great herds of camelids in the Colca zone (in the province of Cailloma, Department of Arequipa). These herds were of llamas and alpacas. Benavides (1986a:387) reports that at the time of the Spanish invasion, the Collaguas were known for their large population and large herds of camelids. (She cites Pease [1977], Málaga [1977], and Cook [1982].) We know that the Collaguas occupied part of the Departments of Puno and Arequipa, but the Colca Valley was the heart of the colonial province of Collaguas (Pease 1981:195). Ulloa Mogollón et al. (1965:331)[13] wrote of this province as follows: "In this province, in its highlands and pastures, there are some animals they call guanacos which resemble the livestock of the land, and vicuñas. . . . They have domestic livestock of the land, which abounds in this province. There are sheep of Castile and goats, albeit few." This was so to the point that Bueno (1951:86) claimed, even in the eighteenth century, when describing the "[p]rovince of Collahuas and the site of Cailloma," that "there are abundant large and small livestock, rams of the land, vicuñas."

The evidence for this is the llama trade that developed between the "caciques of the Collaguas" and the Spaniards in this area in the late sixteenth century. This is recorded in the documents from the Archivo Departamental (now Regional) of Arequipa. I include as evidence some parts of these documents, which were kindly given me by María A. Benavides. The transcriptions of these documents were made by hand in a notebook by a group of Alejandro Málaga's students at San Agustín University in Arequipa in 1984. They were typed by Laura Gutiérrez Arbulú. This was carried out as part of the project William Denevan directed in the Colca Valley (María A. Benavides, pers. commun., 14 June 1996).

One document records that "Don García Chuquiango and Don Alonso Caquia and Don

Felipe Chacha and Don Joan Vica, caciques principales of the repartimiento of the Collaguas, in the hanansaya moiety of the encomienda of Francisco de Noguerol Ulloa, vecino of this city of Arequipa," "sell and give in sale to you Hernando de la Fuente . . . three hundred and four pack rams of this land chosen big and fat and none can have the mange [*carache*] and they cannot be old and two and half years or more and cannnot have the fang worn [*rroçado*] to your satisfaction" (Escribano: Diego de Aguilar, Protocolo No. 2, Registro No. 8, 19 de octubre de 1568. Sección Notarial, Año 1568, fol. 383v–384v).

Another of these documents records that "Don Gregorio Congori and Don Alonso Cama caciques of the repartimiento and Collaguas Indians . . . in this said city of Arequipa . . . we sell and give in sale to you Diego de Frías . . . two hundred male native sheep [i.e., llama and alpaca], thirty of them long-fleeced and the rest plain males, chosen from among three hundred and fifty healthy males and clean of the mange [*carache*] to your satisfaction, free of mortgage or pawn, for a price of four silver pesos de plata corriente for each ram, and the gold pesos they come to at said price" (Carta de venta, escribano: Diego de Aguilar. Protocolo No. 3, 24 de marzo de 1569. Sección Notarial, Año 1569, fol. 136v).

This same year, "Don Gregorio Concoro and don Joan Carbacava Indian headmen of the Collaguas . . . bestow and acknowledge by this present letter that we sell and give in sale to you Hernando de la Fuente . . . ninety and two for each ram of this land good and clean of the mange [*carache*] and the rest plain [*rrasos*] to your satisfaction, and they have to be from three to three and a half years and must not have the incisor worn [*rroçado*], [be] free of mortgage or pawn for a price and amount of three and a half pesos corrientes each" (Escribano Diego de Aguilar, Protocolo No. 3, 5 de noviembre de 1569, Carta de venta. Sección Notarial, fol. 496v–497v).

It is also recorded that "[i]n the city of Arequipa . . . before the most magnificent Martín López de Caravajal alcalde ordinario in this city and its jurisdiction . . . Don Pedro Chunvivilcas and Don Francisco Yanaguara from the encomienda of Licenciate Gomes Hernandes, and Don Pedro Llamquicha de Cavana from the en-

comienda of Juan de la Torre, and Don Luis Poma of Luis Cornejo appeared . . . they will give forty-two Indians, each the part that befalls them to Rodrigo Carmona . . . to go with him to Chule pass of this city and take under their responsibility, and are all the rams of the land he has and shall look after them and graze them. . . . In said pass they shall load all the merchandise and bulk goods and wine jugs said Rodrigo Carmona has to carry and bring on said livestock" (Escribano Diego de Aguilar, Protocolo No. 5, 5 de julio de 1571. Sección Notarial, Cuadernillo de Registro No. 6, fol. 285v–286r).

Finally, there is a document dating to 1584 which records that "I, Francisco Retamosso, vezino of this city of Arequipa . . . give . . . you Gonçalo Zegri . . . ninety rams of the land are given me as payment [sic] and have to give and pay [me] as their encomendero for the tribute assessment of Saint John's Day tax term on June first of this present year good and clean of the mange [*caracha*], according to what said caciques and Indians are bound to give me according to said tribute assessment . . . each ram at four pesos corrientes of eight reales" (Escribano Diego de Aguilar, Registro No. 1, 4 de enero de 1584, Sección Notarial, Protocolo No. 14, fol. 21v–22r).

Cieza de León (1984:Part I, Chap. LXXVI, 223–224) mentions the border area in the mountains between the Departments of Arequipa and Moquegua, that is, the "province of Condesuyo," where some Spaniards had their encomiendas. Cieza wrote: "The Hubinas and the Chiquiguanita, and the Quimistaca and the Collaguas, are people subject to this city. Of old they were much peopled, and had much livestock of their sheep."

Then there is a series of data amassed in 1586 on different parts of the Department of Arequipa that I was unable to locate. Acuña, Flores, Serra de Leguizamo, and Quevedo (1965:312, 315) describe the "partido of the Condesuyos and Chunbibilcas" (which perhaps corresponds to the present province of Condesuyos), and report that here there were "vicuñas and guanacos." They then mentioned the town of Hontiveros de Alca, which also had "vicuñas . . . and guanacos." Acuña, Cabreras, Maybire, and Ossorio (1965: 317) then mentioned the town of San Sebastián de Llusco and noted the presence of "vicuñas . . .

and guanacos." When describing the town of Capamarca, Acuña, Chanaca, Quispi, and Anues (1965:319) in turn noted that there were "guanacos and vicuñas . . . [in] the ravines and heights of this town." Finally, Acuña, Aytara, Chuqui Taipi, and Chununco (1965:321) wrote of the town of Colquemarca and noted that "high in the mountains these towns are surrounded by . . . there are . . . vicuñas . . . and guanacos, like rams, and these roam in the punas."

To finish with this department, I once again have to cite Vázquez de Espinosa (1948:1391/467), who noted in his description of the city of Arequipa that "[c]lose to the city, to the northeast, is the province of the Condesuios . . . where there are . . . many offspring of the livestock of the land."

But there is no question that most of the data pertain to the altiplano. Thus Estete ([Anonymous?] 1968b:397), one of the conquistadors who apparently did not go there but was in Cuzco (Porras 1986:116), and so had at his disposal information from people he was acquainted with, described the Collao region and said "there is a great abundance of livestock, and those from the warm region provision themselves here."

Pedro Pizarro (1978:Chap. 16, 110), an eyewitness whom I have already mentioned, wrote: "[B]ecause these people from the Collao looked after the herds of the Sun and of the one who ruled on earth, in large numbers because they had extensive pastures in their land and a vast wilderness. A large number of wild livestock was raised in this wilderness, which they call guanacos and vicuñas [and are] similar to the domestic stock [*ganado manso*]. Guanacos are a big and plain livestock, with little but mostly fine wool from which clothes were made for the lords. This wild livestock is so swift that few dogs can catch them, no matter how fast they are."

Zárate spent just one year in Peru, between 1544 and 1545, as was already mentioned, and he aparently did not travel to the southern highlands. However, in his book (1968:Bk. III, Chap. XII, 197) he mentions the Collao, and says that "in it there was much livestock of the sheep we mentioned."

Polo de Ondegardo prepared his reports between 1554 and 1559, and his works were used by Acosta, Calancha, and Cobo (see Araníbar 1963:

123). Polo de Ondegardo deals briefly with the Collao, and when he reports that "the livestock" needs "extremely cold" lands to live in, he adds, "just as in all of Collao, and to the sides as far as the city of Arequipa, toward the coast, just as in the Caranges, Avllagas and Quillacas and Collaguas, and all those lands, almost all of them part of the province of Cuzco; all of these lands, although they have been so considered by those who have seen them, could be deemed uninhabitable were it not for the livestock" (Polo de Ondegardo 1916:63).

In the late sixteenth century Murúa (1964) also said that "[t]he Collao is a cold land . . . and is teeming with sheep." Father Ocaña (1987:197) also was in the altiplano in the early seventeenth century, and said that it was "[a] most flat land, where there are extensive pastures for the livestock of the land, which is much raised, and that of Castile too." Around this time Vázquez de Espinosa (1984:1609/558) described "the flat land of Collao . . . [the] province of the Canas," and noted that this was "where they breed livestock offspring in great numbers, both of the land . . . and the Merino from Spain." He then added (Vázquez de Espinosa 1984:1615/562) that "[t]hey raise in their district a great amount of livestock of the land, Merino rams from Castile and cattle." In 1649 Contreras y Valverde (1965:5) had already described the region "they call the province of Collao," noting that it had "many pastures, countless livestock of the land and Castile, both large and small, vicuñas, [and] guanacos."

In the eighteenth century, Cosme Bueno described the province of Paucarcolla (between Carabaya, Larecaja, Lampa, Omasuyos, the coastal cordillera, and Chuquito), where "[t]hey also raise rams of the land or llamas. There are few vicuñas" (Bueno 1951:123). Cosme Bueno then described the province of Lampa (Department of Puno) and mentioned "its climate [which] is very cold," noting the presence "of large and small livestock, which much abound due to the abundance of good pastures. There also are rams of the land, vicuñas" (Bueno 1951:113).

For the area north of Puno, where the Collas live, we have first of all the data from Cieza de León (1984:Part I, Chap. XCIX, 271), which correspond to the first half of the sixteenth century.

Cieza said that this was a land with an "extensive wilderness that is full of wild livestock." In the early seventeenth century Guaman Poma de Ayala (1936:fol. 77–78) noted that the Collas have "much livestock of the land." However, Dedenbach Salazar (1990:85) cautions that Guaman Poma did not know all of the places he described, so his data must be taken with due caution. Interestingly, the Indian chronicler often does not specify what "livestock" he is discussing, whereas in this particular case he specifies it was "livestock of the land."

Cieza de León (1984:Part I, Chap. CIIII, 282) described the Indians of "Xuli, Chilane, Acos, Pomata and Cepita," and claimed they were "rich in livestock of their sheep." In 1567 the headmen of the Urinsaya moiety of Juli made a statement claiming "that some Indians have livestock and other do not, and all the estate [haciendas] the Indians have is livestock of the land" (Diez de San Miguel 1964:120).

In the 1567 *Visita of Chuquito* there are several depositions on the chiefdom of Pomata (in the province of Chuquito). First, the witness Damián de Salazar declared that "there is little livestock from Castile," but said he had "heard say they have much livestock" (Diez de San Miguel 1964:44). Bernardino Gallego claimed that "the others, who are the Aymaraes he mentioned, have most of them livestock of the land . . . and these Aymaraes have some transactions and advantages [grangerías] freighting their livestock to carry loads to Potosí which they carry on their rams . . . and these said Aymaraes likewise hire themselves with the Spanish to load their livestock and go to various places" (Diez de San Miguel 1964:45). And then this same Gallego, "[a]sked whether he knows, or has heard say how much livestock of the land there is in this province, said he does not know this beyond the fact that most of the Aymaraes Indians have livestock, twenty and thirty and a hundred of them, and more or less, and others have three and four heads, and there are Indians who have a thousand heads despite not being a cacique." Asked "how much community livestock they have, this witness said that from what he has heard from the many caciques he believes they have much community livestock, but he does not know how many" (Diez de San Miguel 1964:46).

In this same *visita* there are abundant data on the province of Chuquito. Thus, Martín Cari, cacique of the parcialidad of Anansaya, declared that "some Indians—the minority—have sheep [ovejas] and rams [carneros] of the land, though they used to have more" (Diez de San Miguel 1964:17). And Pedro Cutimbo, "governor who was" of the same province, mentioned the lands of Martín Cari and Martín Cusi, and stated "that in all of said province they have much livestock of the land . . . that there is little livestock of Castile, which are sheep and goats, and there are no cows, and there likewise are many pigs in this province." He then added that "in times past in the time of the Ynga there was a vast number of community livestock, so much so that there was no pasture for so much livestock as there then was even though there is so much land in this province, and now there is much community livestock, as will be seen in each town in the province, and he does not know whether there are other community goods other than livestock" (Diez de San Miguel 1964:39–40).

The Indians from the Anansaya moiety of Chuquito then declared as witnesses, and "said they have livestock." As regards the livestock of Castile, "they said there is some livestock of sheep and pigs of Castile, but it is not much" (Diez de San Miguel 1964:85).

Another witness in this *visita* was Bernardino Gallego, a "resident in this province of Chuquito" whom we have already mentioned. His statement records "that this witness has visited the puna of this province and most of it, and he has seen much livestock and he knows it belongs to the Indians of this province, particularly to the towns of Acora and Juli, towns this witness reported as being very wealthy in livestock, in such numbers that the Indians who watch over the animals and their owners know not how many they have, and he likewise knows there is livestock belonging to the community because as corregidor, this witness commanded the people of Chuquito to bring a hundred heads belonging to the community to make up for the lack of food so provisions for the poor could be bought, and they easily did so, and brought half the pacos and half the rams of the land." The same witness "said that all the main towns in this province have livestock of Castile to give the friars a ration of meat . . . and he does not know how many each may have, other that he understands it is not much, and besides this he knows that the town of Pomata has over one thousand sheep of Castile, and the town of Juli has just as many." When the abovementioned Bernardino Gallego was "[a]sked how much livestock of the land is in this province, both of individuals and communities, he said that this witness knows neither amount, but he knows the livestock is much, and that all usually have livestock, little or many, and he has heard say of an Indian who is not a cacique but a headman, who is one Don Juan Alanoca from Chuquito, [that he] has over fifty thousand heads, and he has heard say that the other Indian headmen have over a thousand heads" (Diez de San Miguel 1964:49–50).

Alonso de Buitrago was another "resident in said town of Chuquito" who testified in the *visita*. He said: "[O]f this livestock of Castile the Indians have up to two thousand heads . . . he cannot know the number of livestock because there is many, and it is said that some caciques have ten thousand sheep, and other Indian headmen a thousand, and other Indians a hundred more or less, and the poor three and four, and ten and twenty, and in this way some a few and others a lot." But he added that "in all the towns there is community livestock and he does not know in what amount" (Diez de San Miguel 1964:55).

Pedro de Entrena had also lived in Chuquito, and he declared that "the Indians in this province have much livestock of the land, and likewise have goats and sheep from Castile . . . they hire themselves for Potosí, to go with loads on their rams, and they go to the coast to exchange their livestock and wool for maize and other items with the Yungas." He adds, "there is much livestock, and the Aymarae Indians usually have some no matter how poor they are . . . there is much community livestock" (Diez de San Miguel 1964:57, 59).

Bernardino Fasato, also a resident in this province, "has seen in it that the Indians have much livestock of the land, because when this witness was travelling to the coast through the puna, which is where they graze, he saw huge herds of livestock belonging to the Indians of Juli, and so do the Spanish merchants who live and travel in this province, because they also have many

pacos." Fasato also declared that "he had not seen . . . community livestock . . . but he has heard say and it is a most public and notorious fact that in all towns in this province they have them by themselves and set aside" (Diez de San Miguel 1964: 60–61).

Another *visita* that concerns the province of Chuquito was made in 1570–1575. One of the items in the Cuaderno III is on "[l]ivestock belonging to individuals." Here we read: "We found that the Indians in the towns of Jule, Pomata and Çepita and Yunguyo privately own ninety nine thousand, three hundred and fifty six heads (99,356)." Then it reads under "Community": "Also sixty thousand, three hundred and forty one heads of the community were found in seven major towns in this province, which derive from what belonged to the Sun and the Inga (60,341). They have some livestock of Castile, privately and in common, albeit a little, and some pigs and goats privately" (Gutiérrez Flores and Ramírez Segarra 1970:38).

Flores Ochoa (1977a:23–24) made an important comment about this *visita*. He said: "The inspection Gutiérrez Flores (1572) [sic] made of the province of Chuquito is illuminating because it shows the rapid destruction of the livestock. Gutiérrez Flores reports just 159,697 head, 99,356 of which belonged to 'individual Indians' and the remaining 60,341 to the community. This figure is in stark contrast with the 50,000 animals in the hands of some individuals just five years before."

In Cuaderno IV there is an additional comment about this same *visita* (Gutiérrez Flores and Ramírez Segarra 1970:41). It reads: "For the most part all Indians in said province have much livestock of the land in particular, which is very useful and advantageous for them, both for the wool they have to make their clothes, and because the pack rams of this province have value and price in it and everywhere else[, and] they are rented to carry coca and other things to the sites of Potosi and Arequipa and other places, and to carry loads on their rams for travellers who pass through said province every now and then."

For Chuquito we have the 1597 data in Ramírez (1936:47). He noted that "[m]uch livestock of the land is raised, and it is the best and biggest of Peru, they are taken away a lot for the caravans and to carry loads [and] is a great business many Spaniards take part in." Ramírez then described the value of the exchange and concluded that "each year more than six thousand rams are removed with this business."

Castro y del Castillo (1906:204–205) also tells of the vast number of camelids in the "Gobernación y provincia de Chuquito," a document slightly later than the one by Ramírez, as it is dated in 1651. Here it is explained that

> [a]ll around the lake there also are livestock ranches belonging to the aforementioned jurisdictions, much cattle and sheep that came from Spain, and most of it livestock of the land, which is a kind of ram, of which some are slightly smaller than an ass, and are used in all journeys or movements in these provinces, both inside it and to communicate with the distant and remote ones, with the mountain of Potosí in particular being much relieved, as they carry on their back all that is needed as supply, and take down the metals to the ingenios. So just the rams dedicated to these journeys are said to be innumerable. There is another species of ram which they call alpaca, smaller in build and very woolly, whose wool, almost silk-like, is used by the Indians for their everyday dress and that of special occasions, and by the Spaniards for exquisite blankets and coverlets and other expensive goods. All animals of the land of this kind in these provinces of the lake, which have puna, produce the bezoar stones, an antidote against poison, but the most prized ones are those produced by the vicuña, a kind of wild animal that lives in the harshest corners of the puna, where the Indians snare many both to eat their meat as well as to sell the wool, which is so prized in other parts and particularly in Spain.

Castro y del Castillo concludes by remarking that "[f]rom here the expanse of the lake can be seen, which acts as a gallery wherein the Ynga, the king of old of this realm, could see the multitude of livestock grazing from the mountain slopes [down] to its shores."

This animal wealth persisted to the eighteenth century, because the description of the province of Chuquito left by Cosme Bueno (1951:125) reads thus: "There is abundant livestock, cows, sheep and pigs, and also llamas or rams of the land, with whom they trade instead of asses. . . . There also are alpacas, huanacos and vicuñas."

Finally, and to finish with the highlands, in the seventeenth century Barraza (1937:197) left a datum for an island in Lake Titicaca which he did not name, just noting that it was the biggest one. Here the Indians "hoard abundant livestock of all kinds, which they leave in the fields cultivated the year before so that they graze, and no one needs to watch over them because they are locked in well, and there is no danger of their being stolen."

6.7.6 The *Ceja de Selva*

The *ceja de selva* obviously cannot furnish information of the kind under discussion. However, it is interesting to find historical data for this region, much like the archaeological data presented on it.

First of all there is the account left by Palomino (1965:187), which concerns the northern part of the Department of Cajamarca, specifically the Jaén area. The account dates to 1549, and the provinces are not well located, but even so some have been identified. Palomino's campaign must have been an "entrada"—an expedition for the conquest of new lands in Chuquimayo. The description says that Palomino went from the province of Perico "to the province of Cherino. This is a province with many people and well inhabited; a strong river flows through it, and the population is on one and the other side. . . . It is a rugged land. . . . There is a forest [*montaña*] on top of it all, outside the settlement." Its inhabitants "[w]ear the dress of the Perico; shirts to the navel; they are woolen ones because they have sheep." The landscape is then described, and all in it corresponds to the *ceja de selva*. "From Cherinos I came to this province of Silla and Chacainga, where I have populated the city of Jaén. These are highlands and very high ones at that, albeit not too rugged, and of a good climate." From thence Palomino moved to "[t]he province of Copallen [which is] a high and rugged sierra. . . . It is a temperate land. . . . They go to war with many feathers in many colours . . . and many middle-sized sheep like the pacos; they are very fat." Palomino then arrived at Loma de Viento, which "is a high land. . . . There are sheep. . . ." Finally, Palomino reached the Vagua River valley.

Taylor and Descola (1981:12) analyzed this information and concluded that the expedition went to the Chinchipe River basin. Dedenbach Salazar (1990:90) managed to identify the locality of Copallen, which corresponds to Copallín Nuevo and is east of Bagua (in the province of the same name), in other words, in the Department of Amazonas.

Espinoza Soriano (1973a) studied the ethnic groups in the Chuquimayo basin in the sixteenth and seventeenth centuries and located the sites mentioned in the Palomino document (1965). He notes that the province of Chuquimayo actually corresponds to the Chinchipe River and says that "[i]t was a fertile land, with pastures that maintained huge herds of livestock" (Espinoza Soriano 1973a:32). Espinoza based his statements on Palomino (the 1549 copy; I used the 1965 edition, see above) and an anonymous source (1582:82). However, there is a problem with this anonymous source. In the footnote only 1582 is mentioned, yet the bibliography includes two anonymous sources with the same date, which are distinguished by the letters a and b. It can be assumed that it is a, because this one discusses Jaén, but page 82 is mentioned in the text, and this does not correspond to any of the two entries in the bibliography, neither a nor b, because pages 28–33 are given for a and pages 21–27 for b. The text by Palomino does not mention "huge herds of livestock," but I do not know whether these appear in the anonymous source I was unable to read.

Espinoza Soriano (1973a:37) placed the province of Perico in the highlands of Jaén, three miles upriver from the Chinchipe River. In effect, there is a place that bears the same name south of the province of San Ignacio, almost on the border with the province of Jaén. Using once more the two abovementioned sources, Espinoza Soriano says that "[t]hey also raised auchenid livestock." This coincides with the data in Palomino (1965). The same thing seems to appear in the 1582 anonymous source (p. 28). Once again it is not indicated whether this is a or b, but from the page number it can be deduced that it is a.

The province of Chirinos, also acording to Espinoza Soriano (1973a:36), is "ten to sixteen leagues from the city of Jaén." It actually is on the river of the same name in the province of San Ignacio. In this case Espinoza Soriano based his statements on the data found in Palomino (1965), when the latter says the Indians "[h]ad pastures and auchenid livestock." Espinoza Soriano (1973a:42) placed the province of Copallín close to Jaén, and the data on the camelids are based exclusively on Palomino (1965).

Last of all, Espinoza Soriano (1973a:42) mentions the Canas of Cacahuari, who lived in Lomas de Viento. The Spaniards apparently gave the place this name because it was a plateau, "most famous and large," which the Indians called Cacahuari. According to Espinoza Soriano, "[a] sixteenth century document states that they lived in the 'district [*términos*] and jurisdiction of the city of Jaén de Bracamoros, in the province of Chuquimayo.' This actually means the site lay within the limits of the Jaén government, but not that it really was within the ethnic 'province' of Chuquimayo." It is strange that although Espinoza Soriano mentions "a document," the reference always mentions two: Martínez (1567) and the Memorial of Alejos de Medina (1561–1562; AGI Justicia 1082). There is no way of knowing which of the two holds this information or whether it appears in both. Either way, Lomas de Viento should be on the Chinchipe River, not far from Jaén. Espinoza Soriano (1973a:20) says that the pastures in the region enabled raising livestock, based in this case on the data in Palomino (1965) and the Anonymous source (1582b:30).

According to Espinoza Soriano (1973a:20), the ethnic groups that lived in the provinces of San Ignacio and Jaén were preferably concentrated on the basins of the Chuquimayo or Chinchipe, Chirinos, and Tabaconas Rivers, and west of the Marañón River. "They encompassed lands with frigid cordilleras and others of a warm selva alta, which is permanently humid and has an abundant flora."

There also is the 1545 account by Alvarado (1965:162) on the conquest of the Chachapoyas (possibly the southern part of the Department of Amazonas). This account mentions the province of Chilio or Chillao, which I was unable to locate with any precision. There the conquistadores "walked through the sierra alta full of woodlands. ...The Christians reached one of these places belonging to the principal lord called Conglos, where they found many provisions and some herds of sheep."

One of the most important expeditions was led by Pérez de Guevara (1965:168–169) in 1545. This expedition is recounted in a letter Pérez de Guevara sent to Gonzalo Pizarro "on his expedition to Rupa-Rupa [another name for the *ceja de selva*]." The Spaniards reached the Motilones zone after traveling between two cordilleras "without being able to cross the cordillera anywhere due to the horses, which did me no good, but in all this land I covered there always was the news that on crossing the cordillera there were savannas, and much people and sheep." Pérez de Guevara then says they reached "the savanna" and "on entering the first villages we saw some sheep and I tried to collect as many as possible." From here the Spaniards went to the town of Mocomoco, where Quechua was spoken, "and from them found out how close I was to Goánuco."

Marcos Jiménez de la Espada (1965c:170–174) wrote a note on this expedition and located the Motilones in the vicinity of the Moyobamba River. Jiménez de la Espada indicates that Pérez de Guevara reached as far as the Huallaga River, crossed it, and went upriver instead of returning to San Juan de la Frontera de los Chachapoyas, skirting the eastern cordillera along its eastern slopes, and arrived at the moors and punas close to Huánuco. He was unable to locate Mocomoco. This is correct, because Pérez de Guevara must have entered the present-day province of Moyobamba, in the northern part of the Department of San Martín, reached the Huallaga River (possibly in the province of San Martín), followed it upriver through the province of Mariscal Cáceres, and from there entered the northern reaches of the Department of Huánuco.

On the other hand, the Motilón tribe should not be confused with the Motilones of eastern Colombia. The Motilón live on the Moyobamba River (San Miguel) near Moyobamba, Lamas, and Tarapoto, and spread along the Huallaga River to Chasutino (Steward and Metraux 1948:598). So the data are consistent.

Zárate (1968:Bk. IV, Chap. XXII, 237) recounts this expedition because he was in Peru the same year that Pérez de Guevara sent his letter to Gonzalo Pizarro, and he undoubtedly got his information there. Zárate, however, distorted the data, because he states that "Mullobamba" is the source of the Marañón River, and that the Spaniards heard there of lands where "camels and hens like those of New Spain are raised, and sheep which are somewhat smaller than those of Peru." Gómara (1946:249), who was never in Peru and who wrote his chronicle in 1552, clearly copied his description from Zárate (1968:Bk. IV, Chap. XXII, 237) but changed his data, for he mentioned "camels, Mexican turkeys [*gallipavos*], and sheep smaller than those from Peru." It should be noted that Romero (1936:367) did not look up the original source and based his text on Gómara, from whom he literally took the word "camels" without realizing these were camelids. As Tschudi (1885:95; 1891:97–98; 1918:207–208; 1969:125) wrote: "Camelids evidently were llamas." However, he added the following: "I cannot imagine what the writer meant when he mentioned sheep smaller than those from Peru." What Tschudi did not realize was that the last phrase was invented by Zárate, for Pérez de Guevara did not write it.

It is worth emphasizing the preconceptions held against camelids, how some scholars cannot accept that these animals are able to live outside their "high-altitude environment." We can mention Tschudi (1885:95; 1891:97; 1918:208; 1969:126), who wrote: "According to these sources [he means the chroniclers and contemporary documentation] . . . it seems that in Inka times there were also llamas in some regions of the warm eastern lands of South America. In his *Voyage to the Equinoctial Regions of the New Continent*, Hauff edition, p. 275, Humboldt believes that this tradition was derived from the fact that the domestic animals of Quito and Peru must have begun climbing down the cordilleras and spread throughout the eastern part of South America. But I find this explanation of the naturalist scholar groundless, because it is clear that the llama cannot have spread in great herds through a climate which is wholly deadly for it. *The high altitude animals* with long and thick wool cannot spread through the humid and warm lands of the woodlands. . . . There, living conditions for the aukenia species are as unfavourable as can be without an adaptation. The local extent of llamas in pre-Hispanic times should be taken with the utmost caution, as shown by the previous notes given by some conquistadors and gold prospectors" (emphasis added). This subject will be covered later, but I find that the account given by Pérez de Guevara and others I mention later must be correct. In any case it is true that these were not big herds.

In this regard, Dedenbach Salazar (1990:96) comments that data for camelids in the tropical forest region, or in the jungle itself, are sparse but do exist. So it is said that sheep from "Piru" are found in Zamora, a humid forest [*montaña*] region; in Luya, close to Chachapoyas; in the "cold climate" ["*temple frío*"]; and fat, middle-sized sheep like the "pacos" are found in Copallín. This scholar concludes: "Since there are no archaeological data that prove the opposite (Lathrap 1970; Harris 1985), we can assume these must have arrived through commercial exchange (Lathrap 1973), and were perhaps raised in small numbers." I agree with her in general but differ when she asks the following question: "Was this a special breed adapted to the tropical climate?" (Dedenbach Salazar 1990:90). I believe the answer is always the same: no. Camelids were able to live in other climates and conditions without much trouble. The term "adaptation" is inappropriate.

Finally, to end the chapter, I am forced to correct a mistake that might in the future prove misleading. Del Busto (1975:522–523) cites friar Martín de Murúa (1964) and writes: "When discussing the highlands and the *ceja de selva*, the mercedarian friar Murúa states," and then includes the citation, which mentions that a great number of camelids were raised. The reference del Busto gives for Murúa in his note 32 is as follows: Bk. I, Chap. III, vol. [T] 1, p. 153. First, on carefully reading the passage from Murúa it turns out that the *ceja de selva* is mentioned at the beginning of the page, but this has nothing to do with the animals. When mentioning these, Murúa clearly states: "Because as I said at the beginning, in the punas a countless number of livestock are raised." Second, the reference given by del Busto is wrong. The correct one is Book III, Chapter III, volume

[T] 2, p. 153. (I do not include the full passage from Murúa here because it appears in Chapter 7 of this book, in a discussion of the number of camelids present in Peru at the time the Europeans arrived, according to the written sources.)

NOTES

1 The term "Andean" includes what has traditionally been known as coast, highlands, *ceja de selva*, and part of the tropical forest, because these areas are physically and culturally inseparable from the Andes. I clarify this because the term is often used to designate only the highlands.

2 To avoid any confusion over the terms used by the Europeans, I suggest that the reader turn to section 6.3 first.

3 A study of the diary shows that this note was written after October 1524. The diary was published in Venice in 1536, under the title of *Relazione del primo viaggio intorno al mondo* (Trucco 1936:83).

4 In a note to the 1812 Italian edition of the *Diary* of Pigafetta, Carlo Amoretti, a renowned Italian naturalist, wrote: "This animal is the guanaco (the *Camelus huanacus* of Linnaeus), similar to the one naturalists call llama and vicuña, a species of camel or sheep renowned for its precious wool. The description given by this writer fits the guanaco perfectly, and all sailors report that the Patagonians dress with their skin." This note appears in the 1927 edition (see Pigafetta 1927:n. 1, 57) without any other explanation.

5 "[B]ased on a mistaken statement by Cieza himself," Porras (1962 [see 1986]:223) gave 1548 as the date when the chronicler arrived in Peru. This can actually "be set in mid-1547" (Araníbar 1967:li, n. 2). This mistake was not pointed out in the second edition of Porras (1986:281).

6 An anonymous account was attributed to Molina—who Porras (1986:315) says is incorrectly called "the Almagrista"—which is the one used here (Porras 1968:201). Porras (1986:317) raises the possibility that the author may have been Bartolomé de Segovia. For more details, see Porras (1986:315–319).

7 According to Pease (1989:174, footnote), the text was published anonymously in Seville in 1534. It was Porras who attributed it to Mena in 1935, even though it was published only in 1948. Some believe Xerez could have written it. Rowe (pers. commun. to Pease) believes Mena cannot have been its author.

8 Footnote 64 by Álvarez (1987) is here inserted into Ocaña's text to explain that the rams here described are llamas, but several mistakes are made in the footnote.

9 In Spanish, del Busto first said "*llevaban*," "taking him." He then added "*en andas*," so it now reads "carrying him on a litter." Translator's note.

10 The version of this document that we know was written in French. I translated it into Spanish. The original account reads thus in modern French: "*Parmi les pièces qui y figuraient, il y avait quatre brebis, de la grandeur des bre-*

bis du pays, qui sont de la grandeur d'un poulain de cinq ou six mois. Et, avec elles, deux pasteurs d'or, de la grandeur de petits hommes" (see Soto 1992).

11 The scientific expedition "of Ruiz and Pavón" is always mentioned (e.g., Porras 1955:57, 354, 400, 459), but their first names never appear. However, it is to be noted that both the "Índice onomástico general" and the "Índice onomástico de cartografía" in Porras's above-mentioned book on sources (1955:579, 581, 591, 592) make a mistake, because they include "Pavón, Hipólito" and "Ruiz, Manuel." I do not know whether this mistake was made by Porras or by Carlos Araníbar Zerpa, who prepared the indexes, as noted in the colophon of the book (Porras 1955; Steele 1964; Tauro 1988:Vol. 4, 1573; Vol. 5, 1838).

12 This term drives from the Spanish *charqui*, which in turn comes from the Quechua word *chharqui*.

13 To avoid mistakes, it must be pointed out that Denevan (1988:79–86) reproduced Chapter 9 of the "Relación de la Provincia de los Collaguas," but attributed it to Juan de Ulloa Mogollón, whereas it was authored by several individuals, as recorded on page 86 of Denevan's reproduction. Our citation appears on page 84 of Denevan's work.

CHAPTER 7

CAMELID HERDS AND CARAVANS

Numbers, Killings, Abuses,
Regulations, Rituals, Fertility, and Diseases,
According to Documentary Sources

7.1 CAMELID HERDS IN COLONIAL TIMES

This chapter evaluates the data on camelid herds in written sources. There is not much general information; most refers to specific areas. All the information dates to the post-conquest period.

In 1561, Polo de Ondegardo (1940:171) described how the Indians brought food from far away by "livestock" ("*ganado*"). Garcilaso de la Vega (1966:Bk. 8, Chap. XVI, 514–515; 1959:149) left an interesting description of these caravans. He wrote:

> Although the flocks are so large and distances so great, the animals do not put their owners to any expense for food, lodging, shoes, packsaddles, trappings, girths, cruppers, or any of the numerous other requirements that carriers need for their beasts. When they reach the place where they are to spend the night, they are unloaded and turned loose in the fields, where they graze on whatever grass they find. They maintain themselves this way for the whole journey and require neither grain nor straw. They will eat maize if given it; but the creature is so noble that it can manage without [food] even when

working. It does not wear shoes, for it is cloven-footed, and has a pad instead of a hoof. No packsaddle is needed, for their wool is thick enough to take the weight of their load, and the caravan drivers take care to arrange the packs on either side, so that the strap does not touch the spine, which might prove fatal. The packs are not attached with the cord they call twine [*lazo*], since the lack of a frame or saddle means that the weight of the burden would cause [the twine] to cut the flesh. The packs are therefore sewn together by the canvas, and though the sewing rests across the backbone, it does no harm, provided the strap is kept to one side.

In another passage, Garcilaso de la Vega (1966: Bk. 8, Chap. XVI, 513; 1959:147) explains that "[t]o avoid tiring them, each flock has forty or fifty animals not carrying a load, and whenever any beast is found to be flagging, its burden is at once removed and transferred to another, before it lies down; for once it does this, there is no solution but to have it killed."

Another excellent description of camelid caravans was provided by Father Cobo (1964a:Bk. 9, Chap. LVIII, 367). It reads thus:

> The major trade the Spaniards currently conduct in this kingdom of Peru [is] in

267

caravans of these rams with Indian pack drivers. On them coca, wine, maize, flour, and other provisions are carried, both to the mines of Potosí and other places. These animals and beasts of burden are the least expensive in the world because they need no iron shoes, harness or headstall, [for] they carry their load without these things. Nor do their owners have to pay for their sustenance, as they eat nothing but the grass they find in the pastures where they stop each day before noon, [and] throughout all the highlands of Peru they do not lack pasture all year long. Those who ply their trade with this livestock usually pass through deserts and uninhabited areas, for greater access to pastures. On the way there is no bad road for them; since they are agile and bold they enter mires with no danger at all, cross steep slopes and pass through bad roads jumping across them with their load, where mules would be unable to walk. The herds are always big, because each usually has five hundred or more rams, and a thousand or two thousand, with eight Indians for every hundred animals who lead, load or unload them.

Cobo finishes thus: "They are so calm and tame that unlike other animals, they do not need to be tamed, save for the first time a load is placed on them, they receive and carry it as compliantly as if they had been born tame."

Herrera (1945c:V Decade, Bk. IV, Chap. IX, 313) is quite concise, saying only "five hundred and a thousand go in one of these herds." According to Dedenbach Salazar (1990:168), the *Descripción del Virreinato del Perú* (1965:80) mentions droves or caravans of 400 animals or more. This description dates to about 1615.

When the Spaniards began their trading activities in the early Colonial period, the native population was forced to hire its labor and livestock for transportation (Pease 1992:191).

The first reference available for a specific region was given to me by Lorenzo Huertas (pers. commun., 16 January 1991) and concerns Cajamarca. In the documentation he examined, which

dates to about 1566, Huertas found complaints filed by Indians who argued they had to carry goods to the coast on their shoulders. Huertas was unable to find any evidence of transportation using llamas in this period, and shows that it was widespread in the Inca period. This is very important.

Lorenzo Huertas (pers. commun., 16 January 1991) likewise claims that although there is evidence that camelids lived in the northern part of the Department of Lima in the sixteenth century, the data available on transportation to the coast with these animals are scant. Pack camelids are also mentioned for Cajatambo right up to the eighteenth century, but they were also used for ritual purposes, food, and wool.

Cantos de Andrada et al. (1965:306) described the mines of Huancavelica around 1586. They explained that "trade between Guancavelica to Chincha, [a distance of] thirty and six leagues . . . is carried out by rams of the land." The distance comes to slightly more than 150 km if the unit is given in Castilian leagues, and to slightly more than 200 km if it is a common league.

Ramírez (1936:36–37) described Huancavelica in 1597, and how "quicksilver" was taken "to the sea" and shipped to "the port of Arica, and from thence [it] is taken in caravans of rams to Potosí." In his "description of the city of Castrovirreyna and its district," Vázquez de Espinosa also mentions the mines and notes that "the metal is carried from the mines to the mills on rams of the land."

This same Vázquez de Espinosa (1948:1432/ 487) says, when describing the "city of San Joan de la Vitoria of Guamanga"—that is, present-day Ayacucho—"[i]n their district . . . much wine is made in its valleys, and brought on rams [*carneros*] from the valleys of Ica, Ingenio, and Nasca [that are] toward the west."

Garcilaso de la Vega (1966:Bk. 8, Chap. XVI, 513; 1959:147) mentions the case of Cuzco, and in this case his testimony is important because it is firsthand. He says: "In my time there were in Cuzco flocks of six hundred, eight hundred, and a thousand and more head to carry/transport these goods. Flocks of less than five hundred were thought nothing of."

In the late sixteenth century, in 1555–1560, Lizárraga (1968:Bk. I, Chap. LXXXI, 63) discussed the coca-growing fields "a three- or four-day march from Cuzco," in "a land called the Andes, where these coca fields are." When describing how coca was carried in baskets, he says they were "taken away on rams of the land." Ramírez (1936:39) touched on the same point slightly later, in 1597, but his account is different. Ramírez wrote: "[T]hese baskets are taken from the warm lands to the sierra, which is a cold land, carried by Indians on their back or by mule caravans, because the rams of the land cannot stand such a warm and wooded land, [and so] many die. In the highlands they place it [the coca] in their storerooms and from there load it on caravans of rams to [take to] Potosí." Lizárraga later specifies that "a ram carries four baskets for many leagues." So in one case we find that coca was taken from the ceja de selva on llamas, while in another it was "carried by Indians on their back," or on mules. I do not know whether this discrepancy is due to the poor data collected by one of the writers or whether such a change actually took place in slightly over 35 years. Curiously, however, Vázquez de Espinosa (1948:1606/557) explained in 1628, when describing the province of "the Andes of Paucartambo," where "the best coca" grows, that this and other fruits were removed on llamas. He literally says, "which they load on rams of the land to take to Potosi, Horuro and other places." In other words, he agrees with the account given by Lizárraga.

There certainly must be more data for this area, because Cuzco was the midpoint between Lima and Buenos Aires, and this was the place where mules and llamas came and went (Denegri Luna 1980:X).

The inspection by Diez de San Miguel in the province of Chucuito took place 36 years after the conquest. It includes important testimonies on the point under discussion. Thus, Martín Cusi, cacique principal of the Urinsaya moiety, "said that the Indians of this province have rams of the land and pacos [alpacas], and go to the yungas to trade," and that "before now the caciques leased rams to the Spaniards to take cargo to Potosí" (Diez de San Miguel 1964:29).

Martín Cari, cacique principal of the Anansaya moiety, in turn declared that 40–50 Indians from Chuquito were given to him each year to go to Moquegua, Sama, and other places with "the rams" to bring maize. This trip lasted 2–3 months (Diez de San Miguel 1964:21).

The witness Alonso de Buitrago, "resident in this said town of Chuquito," declared that "there is much livestock of the land . . . and they go with their livestock to the yungas to load maize and chili pepper and other items, and they rent livestock to carry wine and coca, and other merchandise" (Diez de San Miguel 1964:53–54). Another witness, Martín de Leguiña, declared in the town of Ilave that the Indians "have livestock of the land . . . and hire themselves and their livestock out [to go to] the coast and Cuzco and Potosí, and they go with their livestock to trade maize and chili pepper in the yungas." Martín Bello and Francisco de Santander made almost identical statements (Diez de San Miguel 1964:134–136).

The "secret" inspection Gutiérrez Flores and Ramírez Segarra (1970) made of the province of Chuquito in 1572 provides an important deposition concerning the use of caravans of llamas—and the uses of these animals—in the late sixteenth century, not just to develop trade on a previously existing basis, but also of the abusive methods churchmen used to make a profit. I will cite some examples.

The first one concerns "a chacara [field] and the community livestock taken" by Friar Agustín de Formicedo. It explains how this friar seized the produce of this community field and sent "two thousand hanegas[1] on rams" (Gutiérrez Flores and Ramírez Segarra 1970:Cuaderno II, 4r/4v, 18). This same document lists "Frai Agustín de Formicedo: Indians and rams he took without paying [for them]." The following is specified: "They also gave five Indians fifty rams, which went to the coast of Arequipa, to the Çama Valley; they brought forty loads of maize without paying freightage" (Gutiérrez Flores and Ramírez Segarra 1970:Cuaderno II, 10v/11r, 25). The same friar sent another "five rams and two Indians" to the "Camata Valley" for maize and coca "and paid them nothing for the round trip" (Gutiérrez Flores and Ramírez Segarra 1970: Cuaderno II, 11v/ 12r, 26).

The inspection also mentions "Fray Domingo de Mesa. He used the work of many Indians and rams, and other things, and did not pay them." In fact, Mesa sent several items to be sold in the Camata Valley "with eight Indians and fifty rams of the land" (Gutiérrez Flores and Ramírez Segarra 1970:Cuaderno II, 13v/14r, 27).

The document then lists "Frai Estevan de Tordesillas. Rams and Indians they gave him and he did not pay [for them]." Tordesillas used a caravan of "forty rams with sacks for the Moquegua Valley and four Indians who went with them" (Gutiérrez Flores and Ramírez Segarra 1970: Cuaderno II, 15r/15v, 29).

Finally, there is an interesting comment in this inspection. "Most of the Indians in said province have much livestock of the land . . . which is very useful and advantageous . . . because the pack rams of said province have value and price both there and elsewhere. They are hired to take coca and other items to Potosí and Arequipa and other parts, and to carry loads on their rams for the travelers" (Gutiérrez Flores and Ramírez Segarra 1970:Cuaderno IV, 1r/1v, 41). Readers interested in more details should see the *visita* itself or the comments made by Flores Ochoa (1970:esp. p. 69). (It should be noted that he mentions the "inspection [made by] Gutiérrez Flores," when it actually is the "inspection [made by] Gutiérrez Flores and Ramírez Segarra.")

The records of the notary Diego Dávila, held in the notary of Dr. Víctor Cutipe, in Moquegua, hold interesting data for Juli, Chuquito, and Zepita (close to Desaguadero). These records are for 1587–1588, 1590, and 1593–1595. Although not explicitly indicated in all cases that these are pack animals, this can be deduced from the context.

I turn first to the records for Juli. It is worth noting that this was an important road junction where flocks of llamas from Cuzco, the coastal valleys, and the eastern lowlands between Cuzco and the tropical forest (which extend up to the Bolivian *yungas*) gathered and were replaced (Pease 1992:249). The aforementioned notarial records include a power of attorney that Gabriel de Montalvo gave Alonso de Medina in 1588 to purchase "livestock of the land" (Dávila 1984: 192). In 1590, Juan de Peñafiel and Andrés de Espinosa bound themselves to pay Pedro Serrano "for 165 rams of the land" they had purchased from him (Dávila 1984:199). In this same year

Jerónimo de Huzeda, "a resident in the Moquegua Valley," bound himself to pay Pedro de Olazárraga for having taken on the debt of his creditor Jerónimo de Vargas "for the remainder of ninety rams of the land" (Dávila 1984:199–200). In 1591 Juan Pizarro de Sosa, a clergyman, gave power of attorney to his brother to hire livestock of the land (Dávila 1984:212). For 1594 we find a promissory note for 194 "carneros rasos of the land" (Dávila 1984:288), and a sale of 120 "rams of the land" (Dávila 1984:289). Finally, there is the will of Pedro de Olazárraga, dated 1594. His estate included six "rams of the land" (Dávila 1984:302).

For Chuquito, there is a 1593 contract between Gonzalo de Mazuelo, Gabriel de Porres Rivas, and Hernando Caballero Páez to "place in said Moquegua Valley the required livestock of the land, provisioned by the Yçangas and Guascas Indians" (Dávila 1984:256). For Zepita, there is a 1594 sale of "300 rams of the land," and a promissory note incurred by Martín Alonso the same year to pay 4,250 pesos he owed Francisco de Portillo for the purchase of 300 "rams of the land" (Dávila 1984:285). This same year Martín Alonso of Zepita bound himself again to pay Jerónimo de Huzeda 4,250 pesos he owed him for the purchase of 300 "rams of the land" (Dávila 1984: 286). Another sale of 125 "rams of the land" in Zepita is also recorded for 1594 (Dávila 1984: 186), and a promissory note for 125 "rams of the land" (Dávila 1984:287).

Data are likewise available for Arequipa. In the 1586 *Relación de la provincia de los Collagua* by Ulloa Mogollón et al. (1965:332) we read:

> The deals and profit they make are in livestock of the land, wool, and meat, which Indians from the highlands bring from where the livestock is raised to the valley Indians, and they trade them for maize and quinoa, and this is the main deal between them. The Spaniards come to this province to buy and trade . . . rams of the land for the transportation and trade in wine they take from the city of Arequipa and its valleys to the cities of Cuzco, Chuquito, and Potosí, and for the harvest of coca in the Andes.

Pease (1981:194) discusses his studies in Collaguas and notes that it was clear the people had a far-flung "trade network" throughout the sixteenth and seventeenth centuries. This follows from data that mention a "trade" in Andean hands, as well as another between natives and Spaniards, who led "auchenid caravans between the coast, Cuzco, and Potosí, loaded with wine produced in the coastal valleys of Arequipa, Moquegua or Tacna, and other merchandise." In 1597 Ramírez (1936:45) mentioned the wine manufactured in Arequipa, explaining that it was "taken out in earthen jugs and loaded on rams; one carries two of them quite safely anywhere."

With regard to the great movements of llama caravans between the altiplano and the coast, in the late sixteenth century Acosta (1954:136) said as follows:

[T]hey take herds of these rams loaded as in a caravan, and three hundred, five hundred or even a thousand go in one of these caravans. They carry wine, coca, maize, chuño and quicksilver and any other merchandise, [as well as] the best of them, which is silver, because they take the bars of silver seventy leagues from Potosí to Arica, and to Arequipa in another time, which came to a hundred and fifty. And it was something I often marveled at, seeing these herds of rams with a thousand and two thousand bars, and much more, which come to over three hundred ducats, with no guard or consideration other than just a few Indians to guide the . . . rams and load them up, and at most some Spaniards; they slept all night long in the middle of the countryside, with no care other than the abovementioned. And in such a long road and with such a small guard, no silver was ever missed: so great is the safety in which one travels in Peru.

There are some interesting data on Potosí, in Bolivia. For instance, Father Ocaña (1987:181) stresses the great amount of firewood needed at the time for smelting in Potosí, and notes that this was brought in caravans of llamas. Ocaña (1987:167) then describes Potosí and the use llamas had there:

Most of the provisions, like wine, flour and fruits are brought on these rams. . . . The wine they bring to Potosí usually comes from Arequipa, [a distance of] one hundred and eighty leagues. These same rams carry the silver bars belonging to the King and individuals coming down from Potosí. They carry them all of the one hundred and eighty leagues there are to the port of Arica in eight days, because there are fresh rams every four leagues. And they walk day and night without the silver ever stopping. The silver bars for Our Lady of Guadalupe came down on these rams on 17 March 1601; I had the royal officials ship them on these rams because they are swifter than the mule caravans.

Dedenbach Salazar (1990:169) disagrees with the length of the voyage pointed out by Ocaña; I return to this point at the end of the chapter.

Much wine from the coast was sent to Potosí. The testimony of Friar Benito de Peñalosa, who according to Herrero (1940:112) was in Peru "around 1622, as he himself notes," is interesting in this regard. This is a little-known tract titled *Libro de las Cinco Excelencias del Español, que Despueblan a España*, published in Pamplona in 1629. Herrero discussed this work and even reproduced some passages. I am interested in one of them, which reads thus: "There are Spaniards who go trade with four and six thousand rams of the land to Potosí and elsewhere. Each ram carries two *botijas* [earthen jugs] of wine; only Indians know how to do this because of the time and patience needed to load and unload said rams, and each Indian has five and twenty rams under his care, which they call *ayllo*" (Peñalosa 1629:124, cited in Herrero 1940:113–114).

When describing the province of "the Pacages" (Sucre-Potosí, Bolivia) in the Audiencia of Charcas, Vázquez de Espinosa (1948:1413/478) explains that the wine taken to Oruro and La Paz came from Tarata and Putina, "where they have livestock offspring." However, he notes that the wine was carried on mules from Arica up to San

Pedro de Tacana, and "from here they are carried by the rams."

With regard to the coast, I was able to find data on caravans only for Ica and the far south. Pedro León Portocarrero, also known as the Anonymous Portuguese ("Anónimo Portugués") mentioned in Chapter 6, described the voyage from Pisco to Chincha in the seventeenth century: "From Pisco it's six leagues to Chincha, an Indian site with a good port. Here they bring quicksilver from Guancavelica on rams, and here it is shipped to Arica, whence it is taken to Potosí and other mines. The guanacos who carry the quicksilver to Chincha [then] go to Pisco and Arica [to] pick up wine, and return with it to the mountain" (Anónimo Portugués 1958:111). The writer certainly made a mistake in saying that the caravans were of "guanacos." They were clearly llamas.

In this case the testimony of Vázquez de Espinosa (1948:1333/441) is also invaluable, as he was in Chincha in 1617, as already noted. He describes the wine produced there and says that part of it was taken by sea to other places, "for most of it is taken to the sierra on rams to Guamanga, Cusco and other places." He also mentions (1948: 1354/450) the neighboring "villa of Ica, and its fertile valleys with vineyards," where "á great amount of the abovementioned wine is picked up; much of it is taken on rams to the province of Chocoruos, Castrovirreyna, Guancauelica, Guamanga, and other places."

However, most of the data on the movements of llama caravans on the coast refer to the far south. Lorenzo Huertas (pers. commun., 16 January 1991) has reviewed an enormous amount of sixteenth-century documentation for the regions of Tacna, Moquegua, and Arequipa. He tells us there are many contracts for llamateers who went to the coast to carry wine to Potosí, Cuzco, or Puno. This was also confirmed by Masuda (1981: 175), who noted, when writing of the Indians in Camaná, Tambo, Vítor, Siguas, Locumba, Moquegua, and other towns, that they "hired themselves to drive the llamas the Spaniards had, 'which are many and in great numbers, with which they trade and carry the merchandise that goes from one end of this kingdom to the other' (Jiménez de la Espada ed., 1965 [1881–97]: I, 350 [1965a in my bibliography])."

Catherine Julien (1985:192) studied the transport of guano by caravans of llamas in the area between the coast and the Puquina zone in the eighteenth century. She wrote that the beasts of burden were a great advantage for transporting the guano extracted on the coast. The supervisor of the 1792 inspection reported that half a fanega of guano did not suffice for a Puquina agriculturalist. It is explained that this amount corresponds to about 15.5 kg, or between a third and half of the load carried by a llama (in a note, Julien [1985:223, n. 11] explains that the calculation was based on the weight of the excrement of domestic poultry, i.e., 55 cubic feet equals 1 ton). Julien adds: "People from the highland were excellently equipped to transport the guano back to the cultivation site. Llamas, and the ropes and sacks made from llama fiber, were available to them."

There are data on the Moquegua zone in the notarial records of Diego Dávila, to which I have already referred (see above). Here it is recorded that Alonso de Estrada gave power of attorney to Andrés de Villegas in the Moquegua Valley in 1591 to "hire livestock of the land to carry wine, coca, [and] food to the villa of Potosí" (Dávila 1984:217). In this same year and place, Sebastián Durán bound himself to use his livestock of the land to carry 1,000 botijas of wine to Potosí (Dávila 1984:217). Power of attorney was likewise given to Juan Vallejos "for the purchase of 250 rams of the land," and to hire them from the port of Arica or anywhere else as far as the villa of Potosí (Dávila 1984:231).

There also is a mortgage made in 1593 in the town of Moquegua, of 300 earthen jugs of wine that had to be taken on 150 "rams of the land" from Moquegua to the Collao region (Dávila 1984:272). That same year in Moquegua, power of attorney was given to Hernando Caballero Páez to "hire and freight the necessary livestock of the land, provisioned by Ysanga and Guasca Indians," to carry 2,000 botijas of wine to Potosí (Dávila 1984:278). Power of attorney was given to Cristóbal de Cortázar, Diego Lozano, and Miguel Díaz Zarco in the "valley of Cochuna and Moquegua," again in 1593, to purchase 700 "*carneros rasos* of the land" (Dávila 1984:283).

Lorenzo Huertas (pers. commun., 16 January 1991) tells me that the notary records in the

Archivo de Moquegua hold a contract dated 26 August 1593, between don Gonzalo Mazuelo and the "lord of his caravan," Andrés Díaz de Villegas, who had 500 pack llamas, to carry botijas of wine. Six contracts were signed this same year to carry produce; they required about 2,000 llamas, according to estimates by Huertas. The same notarial records indicate that transporting wine to Moquegua and Potosí "on livestock of the land" took three months, and that the Indians specializing in these tasks were few.

This trade was still active in the seventeenth century, because when Vázquez de Espinosa (1948: 1409/476–477) wrote of Moquegua, he described the many natural resources available there, emphasizing fruits and sugar. He explains that these were taken "on rams to the province of Chuquito, and all of the highlands."

There are some data on the Sama Valley in the 1567 inspection of the province of Chuquito, which has been cited several times. Here there is a statement by Juan Matute on the Indians from Sama who had much livestock of the land; to carry guano from the islands, "they go for it with rams eight or nine leagues" (Diez de San Miguel 1964: 126). Pedro Bilbao, another witness living in the Sama Valley, said of the "rams of the land" that "others bring them to this valley for maize and wheat and chili pepper, and for the cotton harvested, . . . and sometimes . . . they have some work going to the sea for guano, which is eight leagues [away], because without it there is no maize, and they bring it on their rams" (Diez de San Miguel 1964:127).

Vázquez de Espinosa (1948:1412/478) also mentions the Sama Valley, but to talk about its great output of "agi" (chili pepper). He specifies that between the valleys of Locumba and Sama, "almost two hundred thousand baskets are collected in the two of them, which rams take to Potosí, Horuro and all of the sierra."

Pease (1982:108) in turn discussed a curaca in Tacna in the late sixteenth century, who had a trading firm "using droves of auchenids [caravans of camelids]."

From Cantos de Andrada et al. (1965:306) we know that in 1586, "there are 90 leagues from the port of Arica to Potosí, and the quicksilver is carried with rams." I do not know which leagues he used. If they were Castilian ones, the distance is slightly over 377 km, but if they are the common ones then it is slightly over 501 km. Ramírez (1936:46) tells how the quicksilver from Huancavelica reached Arica by sea in 1597, and "from there the quicksilver and the other clothes are removed on rams of the land and taken to Potosí."

In the late sixteenth century, Lizárraga (1968: Bk. I, Chap. CXI, 94–95) made a most interesting description of the trip from Porco to Arica which I believe is worth quoting at length. It reads thus:

The royal road from Potosí to Arica passes half a league away from Porco on the right hand, which are 100 plain journeys, very cold; [there are] some sand flats which are not too rough for horses but [are so] for rams of the land—when loaded —and for mule caravans, which is why the caravans of rams taking quicksilver and merchandise . . . from Arica to Potosí . . . must have completed their [day's] journey by nine in the morning, that is three leagues, having begun the day at three before dawn and even before. This is because although most of the sierra cannot be inhabited because of the extreme cold—and most of this road is so—from nine in the morning to four in the afternoon the heat from the sun is so strong that it is worse than on the plains and in the warm valleys. This road is very tiresome due to the unsteadiness of the weather and [to the fact that] for three or four journeys there are no *tambos* [inns] where one can lodge, save for some poorly built walls. And once the descent to Arica begins [the road] is most exhausting, because there is not a drop of water in the 20 leagues that extend from the point one begins climbing down the ravine known as Contreras. This is where the pack rams of the land are at risk, and many drop dead. Once the ram lies down there is nothing to do but to unload it and leave it there. Here it dies of hunger and thirst. If they ate sand and did not take water for eight days they would come out very fat. Seeing so many bones of rams in this ravine is a crying shame for what the lords of the rams lose (and this is the best road). For this reason they take for the cargo half more than needed. Once in the highlands they do not have this risk because neither pastures nor water are wanting, and once the ram completes its [day's] journey nothing will make him move forward. The mule caravans cover these 15 leagues in half a day and one night.

This must have been a major route because Porco, which is quite close to Potosí, was an ancient mine that had been previously worked by the Indians and was abandoned by the boom in Potosí (Pease 1992:238).

Del Busto (1975:523–524) succinctly discussed the route followed by the caravans of llamas that carried the minerals from the mines in Huancavelica to the port of Chincha, and then from Arica to Potosí, as well as the journey back to Arica, now with metal bars. (Del Busto cites the above-cited passage from Lizárraga but makes a mistake when he gives the chapter as XCI; the correct number is CXI.)

When describing the "City of San Marcos of Arica," Vázquez de Espinosa (1948:1415/480) says that "the *vecinos* [residents] have more than a thousand mules with great caravans to transport merchandise to Potosí, Charcas and all the lands above, as well as to carry the silver to the port, and there also are great caravans of rams." Later, Vázquez de Espinosa (1948:1417/482) mentions the great amount of totora (a sedge) close to Arica, explaining that "with it they load [*estriuan*] the ships with their cargo of wine and [all] the rest, and all the caravans are prepared to carry their loads to Potosí, [and there] they make textiles [*seroncillos*] to load the rams with wine and quicksilver."

Dedenbach Salazar (1990:168–169) prepared a good summary of the distance covered each day by the caravans. She notes that Murúa (1964:II, 154) indicates they covered two or three leagues every day (which according to her is ca. 10–15 km),[2] and three or four leagues according to Ramírez (1936:17). The Anansaya cacique Martín Cari, in Chuquito, listed some distances covered by the caravans and the days needed. To go from Cuzco to Potosí and back took four months (Dedenbach Salazar notes that the same point appears in GdlV [1. VIII, Chap. XVI, II:316], which is correct; GDIV is an abbreviation of Garcilaso de la Vega [in my bibliography: 1959:Bk. VIII, Chap. XVI, 150]), two from Ilo on the coast to Cuzco, one month from Chuquito to Arequipa, and 25–30 days from Chuquito to La Paz (Diez de San Miguel 1964:17). Dedenbach Salazar notes that if we take the distance between the corresponding sites on modern roads, we get an average of about 12 km per day. This coincides with the data left by the chroniclers, but not entirely with contemporary data. West (1981a:70) describes a voyage with a caravan of llamas in which he participated. Here 289 km were covered in 22 days, including a seven-day rest. In other words, the daily distance covered was 19 km. (To be accurate, in his article West gives a distance of 180 miles. If we assume he used the imperial mile, then the distance covered was as noted. If, however, he used the Spanish mile, which does not seem feasible, then the distance covered would have been 250 km, in which case a day's march would have been about 17 km.) Custred (1974), however, studied the interzonal trading voyages made by the community of Alccavitoria, in the province of Chumbivilcas, Department of Cuzco, and gave a greater distance, although for unloaded animals. Custred specified that llama caravans began their journey around eight or nine in the morning and finished their walk at about four in the afternoon, which meant a daily eight-hour march, including rests taken along the way. The journey from Alccavitoria to Ocapata, a distance of 150 km, took them six days. On average, llamas covered 25 km a day. Fifteen to twenty llamas, and one driver, were needed to travel the same distance from Ocopata to anywhere else (Custred 1974:276). Dedenbach Salazar (1990: 168–169) ends her discussion by stating that the time a journey took naturally depended on the type of road that had to be taken, and the terrain to be covered.

Dedenbach Salazar (1990:169, n. 58) has an important footnote which refers to the above-cited passage from Ocaña (see above and Ocaña 1987:167), where the chronicler notes that the distance from Potosí to Arica—180 leagues—was covered in eight days. Dedenbach Salazar does not believe this is possible, as it would mean covering more than 100 km per day. (I once again disagree with the calculations made for the distance; if these are common leagues, then the daily march would be 125 km, and if Castilian leagues then it would be 94 km per day; however, this in no way affects her argument.) As Dedenbach Salazar correctly points out, it is hard to believe that such a journey could be made under such conditions, even allowing for the presence of fresh animals every four leagues, as stated by Ocaña, and even if they walked both day and night.

It is hard to reach any conclusion regarding the number of animals the caravans had. We saw they were estimated as 300–1,000 animals. There is some agreement on these figures. Cobo (1964a) is the only author who mentions caravans of 2,000 animals.

There is one last comment I would like to make, about a term used by del Busto (1975:517–518). He wrote: "Just as mule drivers were muleros or muleteers, and camel drivers were camelleros, those who drove the great llamerías from the mines to the ports were called *carnereros*, because their trade was to take the caravans of 'carneros de la tierra'.(28)" I do not intend to question the statement made by del Busto, but I was unable to find the term "carnerero" in the chronicles or in historical sources. Pease (pers. commun., 30 May 1993) does not recall having seen it either. Strangely enough, del Busto's footnote 28 cites Vázquez de Espinosa. No edition or page number is given; it just reads, "book III, chap. VI." This work does not include a bibliography, just footnotes. On examining them, I found that del Busto (1975:420, n. 11) used the 1948 edition of Vázquez de Espinosa, so the reference corresponds to page 340 (1096), where that term does not appear.

7.2 HOW MANY CAMELIDS LIVED AT THE TIME OF THE SPANISH CONQUEST AND COLONIAL PERIOD?

Establishing the number of camelids in the central Andean area at the time of the Spanish conquest, even an estimate, would be most interesting. This is impossible, however. There is no way to establish it. Besides, as was noted in Chapter 6,

there was never any early Spanish law or administrative regulation that sought to collect statistics of this kind, particularly in the early years of the conquest. The inspectors (*visitadores*) did get them later, but then it was more a matter of finding out how much food these animals represented. I do not intend to present any estimate but instead give an overall idea of the number of camelids in terms of the data available.

First of all, there is some general but nonetheless significant information. The account written by Andagoya before coming to Peru, as was already noted, is a late but important chronicle because "it was written by a captain acquainted with all the participants in the conquest" (Porras 1986:70), and so his story is clearly truthful. Andagoya wrote: "The number of sheep living and raised in that land was so extraordinary that it was remarkable" (Andagoya 1954:246–247). This awe is also reflected in the writings of Bartolomé de las Casas (1948:9) when he writes that "[i]n those kingdoms there was an immense amount of sheep and in such numbers that it cannot be believed. The herds and flocks usually had twelve, fifteen, and twenty thousand heads."

Zárate (1968:Bk. I, Chap. XI, 140), as was already noted, briefly mentioned the hunts made by the Incas. He noted that "in a day he [the Inca] could take twenty and thirty thousand of them." Zárate means the "sheep of the land."

As usual, Cieza de León (1985:Part II, Chap. XVI, 43) made a clear appraisal of the state of affairs: "[I]n this kingdom of Peru there was a great number of domestic and wild livestock, *urcos*, rams and pacos, vicuñas and sheep, *llamas* [some words illegible in the original manuscript] in such fashion that both the lands inhabited and uninhabited were full of great herds, as there were, and are, excellent pastures everywhere so they could be well raised. . . . And they multiplied so much that it is not to be believed how many there were in the kingdom when the Spaniards entered it."[3]

Garcilaso de la Vega (1966:Bk. 6, Chap. VI, 326; 1959:163)[4] is the chronicler who provides the best description of a *chaco* (hunt). He clearly based his writing on his own recollections and on the data he received, because he talks about the age of "[t]he Inka Kings of Peru." After explaining how it was that harmful animals were killed, he ends by saying, "The number of deer of various kinds [*corzos y gamos*] and the large sheep they call huanacu with coarse wool, and of the smaller vicuña, with very fine wool was very considerable. Naturally in some areas they were more plentiful than others, but often more than twenty, thirty, or forty thousand head were taken, a very fine sight which gave rise to much rejoicing." Thus, when discussing the present—which for him was the late sixteenth century—he concluded, somewhat despondently, "That was in former times. Those in Peru can say how few have escaped the destruction and waste caused by the arquebus [crossbow] for guanacos and vicuñas are now scarcely to be found except in places where firearms have not reached."

However, we should bear in mind that Garcilaso mainly referred to wild camelids; llamas and alpacas apparently still survived in significant numbers despite the slaughter. About this same time Murúa (1964:153) noted that "in the punas [they] raise an infinite number of cattle and livestock—sheep and of the land." And when Pedro de León Portocarrero (Anónimo Portugués 1958:80) traveled across the Peruvian highlands in the early seventeenth century, he saw Indians "taking four hundred [animals] per herd, others five hundred, and others even more." Oliva (1895: 11), who was in Peru about the same time, also mentioned the "infinite [number] of vicuñas and guanacos." Finally, in 1597 Ramírez (1936:21, 25) described the people of the Andean highlands and mentioned that they "raise much livestock," and as regards the "riches of the Indians," he said that it lay among the "highlanders," "because of the livestock they raise."

7.2.1 The Coast

The information for the coast is scant. However, María Rostworowski (pers. commun., 12 March 1991) kindly provided me with the data in a 1573 document on the parish of Pachacamac and Caringa, held in the Archivo Nacional of Santiago de Chile. It belongs to the inquest made by Rodrigo de Andrada (Varios, Vol. 64). This document is part of the lawsuit filed by his heirs. The statement made by 55-year-old Don Juan Cucho Gualle in the town of Pachacamac, on 19 September 1573, is on folio 22r. The witness declared that the first encomendero, a man named Hor-

doñez (Ordóñez) "had himself given 2,000 rams of the land [by the Indians], because at the time many were raised in this repartimiento." On folio 25v is the statement made by another witness, Don Alonso Choque Guamani, curaca of Caringa, who said they "gave the encomendero much livestock of the land, of which there were many then." According to Rostworowski (pers. commun., 12 March 1991), the abovementioned encomendero was one of the first, a long time before Vaca de Castro, the encomendero in 1544.

For the south coast there are some important but scattered data. The *visita* made at Acarí in 1593 (1973:158) lists "seventy heads of livestock of the land" when recording the estate of the "caçique prinçipal," Don Alonso Satuni. The notarial records in the Archivo de Moquegua for 1593 hold six contracts for the transport of goods. Lorenzo Huertas did some calculations and concluded that 2,000 llamas were needed for this. According to Huertas, the trade in "livestock of the land" at that time was a significant activity. This livestock was usually traded in "the provincias de arriba," that is, in Zepita, Potosí, and Oruro. One of the contracts specifies that a mule was 13 times more expensive than a llama, but the latter had the advantage of being better adapted to carry the kind of botija used to transport wine, and the llama also needed less fodder (Lorenzo Huertas, pers. commun., 16 January 1991).

The same archive records the establishment of a company to transport wine (fol. 586n). The contract was signed by two partners in Moquegua on 29 March 1615. It is stated therein that each had to contribute "700 carneros rasos of the land" (Lorenzo Huertas, pers. commun., 16 January 1991).

Finally, the deeds to the Hacienda Para, in the Archivo Departamental de Tacna, include the 1588 testament of Diego Caqui (1981:213–214), which has already been mentioned but is worth repeating here. Among other points is the following: "I likewise declare that one Francisco Cárdenas, merchant, owes me the freight fare [for] a hundred rams of the land on which he carried some merchandise of Castilian clothes; it is now about seven years—perhaps fewer years than more—and on the way he sold forty of the rams of the land who carried said merchandise without my permission."

Then we read: "I likewise declare as my property one hundred ovejas de vientre of the land that are tended by an Indian called Ticona, native of Putina; they must have multiplied."

7.2.2 The Highlands

For the highlands I have data on a zone that must now be part of southern Ecuador. As we saw, Zárate (1968:Bk. II, Chap. XII, 170–171) tells how Diego de Almagro and Pedro de Alvarado, who were heading from Quito to Pachacamac, ran into Quizquiz. The latter had a great army and the "livestock he had found in Jauja and lower down [*abajo*]." Indians and Spaniards clashed before reaching San Miguel, in the province of Chaparra. After the battle, "all the clothing that the Indians were unable to take to the sierra they burned that night, [and] fifty thousand sheep remained on the field."[5]

The studies made by Silva Santisteban (1964) show there is no documentation on camelids for the province and corregimiento of Cajamarca in the late seventeenth and early eighteenth centuries. This is important, because we saw that there were camelids in Inca times.

Cieza de León (1984:Part I, Chap. LXXXI, 236) described the neighboring province of Huamachuco, where the chronicler was told that here "[t]he Incas had . . . hunting grounds" where many chacos [hunts] were held. Cieza finished thus: "because in these hunts they would take ten or fifteen thousand heads of livestock, or the number wanted: so great was the number of them." The fact that he does not say more may mean there were no camelids by the time he traveled in the area.

There are not many data for the Callejón de Huaylas, but even so they are highly significant, as this is an area where camelids have practically disappeared. There is a 1558 inspection of the repartimiento of Guaraz made by Diego Álvarez (1968–1969:18), corregidor of Huánuco. Here we read: "All the caciques, headmen and Indians subject to them declared that they have the following livestock: one thousand eight hundred and three sheep of the land, four hundred and fifty-three goats, and a hundred and fourteen pigs, and nine sheep of Castile, [and] a mare. They bring [the livestock] from the land of their repartimiento, [where it is] in a part where it does no harm,

and it is tended to by old Indians or children, and they have good pastures for said livestock."

Varón Gabai (1980:57) discussed this *visita*. He notes that interestingly enough, European animals had acquired great importance in the local lifestyle despite the higher number of native livestock. He adds: "The curacas declared owning 1,203 'sheep of the land' (that is to say llamas and alpacas), and as regards Spanish livestock they listed 452 goats, 114 pigs, nine sheep of Castile and one mare" (Varón Gabai 1980:57). These figures differ from those previously given. This might be a mistake in the transcription, but both accounts could be true, as there are two copies of the *visita* (Franklin Pease, pers. commun., 16 October 1993).

Varón Gabai (1980:65) then discusses the inspection of Llaguaraz—a repartimiento close to present-day Huaraz—also in 1558. Livestock raising probably already was the main activity at that time. A slight displacement of native livestock by European livestock is perceptible, for one of the informants says: "I declare that all have six hundred and thirty-one sheep of the land and two hundred and twenty-eight goats, and a hundred and eighty-three pigs and one mare . . . and they have them on their own lands because of the good pastures they have for this."

Finally, Silva Santisteban (1964:134) reports that one hacienda in Huaraz had 20,000 heads of sheep and another more than 12,000 in the late sixteenth and early seventeenth centuries. The source does not specify what kind of sheep these were, but they were presumably European animals.

For Huánuco we have the following data from the *visita* made by Ortiz de Zúñiga (1967) in 1562. Although tedious, it is worth presenting here the figures for camelids given by the informants, but a few points must be made first. The inspector used the terms "livestock of this land," "carnero [ram] of the [or of this] land," or "sheep of the [or of this] land." Second, although in some cases only "carnero of the land" or "sheep of the land" are used, in many others, both are distinguished, saying for instance "two sheep and one carnero" (see, e.g., Ortiz de Zúñiga 1967:175), thus showing that the distinction was intended. However, there is no way to establish what the distinction referred to. Were they llamas and alpacas? Varieties of llamas? Llamas for meat and pack llamas?

There are also cases—although a minority—where only "carnero" or "sheep" is used. I believe these are camelids in both cases, because the statements say "sheep of Castile" and "carneros of Castile" when European animals are meant.

Third, it is significant that exact figures are always listed. Only in one case does the *visita* read "more than two sheep of the land" (Ortiz de Zúñiga 1967:261). However, in some instances the number belonging to "all the Indians in this town" (e.g., Ortiz de Zúñiga 1967:193) is given at the beginning of the inquest, and then comes the house-by-house inspection. When the individual figures are added up, the total never squares with the total given. In the case mentioned above, 36 animals are listed at the beginning, but the sum of the statements gives a total of 26. I used the total figure for my sums.

Finally, we should not forget that there probably were many more animals, because one witness declared that "many [of his animals] died of mange [*carache*]" (Ortiz de Zúñiga 1967:172).

We now turn to the data available, which I have summarized. They come from the towns of Canchapara, Ambi, Quinoa, Rumar, Chanlla, Guancayo, Rondo, Pecta, Achinga, Atcor, Queros, Guayan Queros, Guaoya, Manchaguachi, and Mantacocha Quira Quilcay (Ortiz de Zúñiga 1967:99, 107, 116, 150, 160, 163, 165, 171–178, 182, 184, 188, 193–199, 202, 203, 206, 210, 231, 261; for the location of these towns, see the maps included in Ortiz de Zúñiga 1967:Vol. I).

The sum of the statements for the "livestock of this land," "rams [*carneros*] of the land," and "sheep of the land" is a total of 233 animals. Then come 9 "rams" and 15 "sheep," for 257 animals in all.

In the repartimiento of Juan Sánchez, of the Cuzco "*mitimaes*," there were "two sheep of this land." The towns of San Francisco had 108 "sheep of this land"; the cacique of the town of Caure declared 40 or 50 "sheep of the land"; eight "sheep of the land" and a "ram of the land" were listed in the town of Nauça"; and Yacha had 10 animals between "rams [*carneros*] and sheep of the land." The Ananquichuas *mitimaes* had a single "sheep of the land" (Ortiz de Zúñiga 1972:32, 39, 61, 65, 67, 86, 87, 98).

The town of Paucar-Yachas is another one where the total does not square with the individual figures. First, it is reported that the town had a total of 21 "sheep of this land," but later the individual statements concerning "sheep," "rams," and "livestock of this land" amount to 11 (Ortiz de Zúñiga 1972:104, 106–108).

Fifteen animals, between "sheep of the land" and "rams of the land," were declared in the town of Ocrumarca-Yachas (Ortiz de Zúñiga 1972:111–112). Coquín is another town for which we are first told that it totals 20 animals. However, between "rams," "lambs," and "rams and lambs of this land," the house-by-house inventory comes to 9 (Ortiz de Zúñiga 1972:134, 138, 139, 141, 143). The same thing happens in the town of Chuchuco, where a total of 15 "sheep of this land" is given, but between "sheep," "rams," and "sheep and rams of this land," the house-by-house accounting comes to 14 (Ortiz de Zúñiga 1972:149–154).

Something similar happens in Caure, a Yachas town. We are told that the residents among them had a total of 59 animals, "livestock of the land." But between "sheep," "rams," and "sheep and rams of this land," the individual accounting yields 46 animals (Ortiz de Zúñiga 1972:160–169).

In Curamarca (where *mitimae* Indians resided) there were just two "rams," and there were eight "sheep of this land" in Quilcay. The "common livestock of Quilcay, Curamarca, and Nauça" are then listed, and only European animals are mentioned, but at the end of the statement it is pointed out there were also "three head of livestock of this land besides the ones declared by the headman" (Ortiz de Zúñiga 1972:179, 180, 192, 196).

In the town of Guarapa there were only three "livestock of this land." However, when the inspectors list the Yacha, the document reads: "In the moiety of the Yachas eighty-six head of livestock of this land were found," whereas the *mitimaes* had "[t]wenty-six head of livestock of this land" (Ortiz de Zúñiga 1972:199, 251).

It is worth noting that almost all witnesses had a large number of European animals, which I have not analyzed (in some cases 30 goats, 50 sheep of Castile, or 101 pigs are listed), while in other cases the Indian witness only had European animals.

Mellafe (1967) and Dedenbach Salazar (1990) made interesting comments on this *visita*. I will summarize them.

Mellafe (1967:337–338) believes that as far as livestock is concerned, much information was kept hidden in the statements. Among the Chupachu there were, according to the statement made by their curacas, 152 head of "livestock of the land," 100 sheep of Castile, 37 pigs, and 180 goats. There apparently were no cows, and yet it must be noted that some communities in the region already had some in small numbers. Mellafe believes that much can be inferred from the figures, even though these "are not reliable." Perhaps the most striking fact is the low number of camelids compared with the relatively high number of "sheep of Castile." Mellafe notes that these and the goats were the cheapest in trade with the Spaniards, and that the encomenderos usually gave them as compensation for special tasks or excess tribute payment. On the other hand, the abundance of sheep in towns with a textile tradition is understandable, as it promised a good woolen textile activity at the individual, communal, or entrepreneurial level.

Mellafe believes there are several reasons for the low number of llamas, besides the fact that the truth was concealed and false statements were made, even though in this specific case these must have been the most significant reasons. These animals were in great demand during the Spanish conquest and the civil wars, both as pack animals and as meat. Moreover, the encomenderos demanded them as part of the tribute, and the Chupachu Indians themselves, who declared having just 152 head in all, used to give their encomendero 104 head a year until 1549. Shortly thereafter, llamas—together with the Indians— became a major aspect of transportation, and it was only in the early seventeenth century that demand for them fell as the mule appeared.

The "livestock of the land" had begun recovering from the wear and/or population decline brought about by the conquest and the civil wars at the time the *visita* was held, but at the same time transportation was passing into the hands of Spanish and mestizo entrepreneurs, so that the animals that were now increasing were used in mines and cities, and the Indian communities

were therefore unable to use them. Finally, the fact that Huánuco lay at the center of major trade routes certainly sped up the phenomenon (Mellafe 1967:337–338).

Dedenbach Salazar (1990:103–105) makes other points. She notes that estimates made in the last century show that camelids can decrease remarkably fast. She gives the alpacas in the Department of Puno as an example (using the data in Flores Ochoa [1984:34]). These fell by almost 50% in 1970–1980, so a similar phenomenon in the sixteenth century is not unlikely. Although the Huánuco Indians mention abundant pastures, they also state that mange (*caracha*; see above) was the reason for their decrease. So there are several possibilities: either the "livestock of this land" decreased a great deal since 1532 because the Spaniards seized it; due to diseases; or because the Indians only declared to the inspectors the number of animals present in the towns, not those grazing in the highest sierras (as assumed by Morris and Thompson [1985:154]). However, this entailed the risk of the inspector seeing the animals during his journey. It is for this reason that Dedenbach Salazar asks "if all individuals questioned in the *visita* declared having little livestock and having exchanged these products with other groups, and yet at the same time stated having been or given sheepherders to the Inca and having good pastures, [then] should there not be more camelids in the puna in the vicinity of Huánuco, and is it not likely that they [the natives] omitted them in their statements? What type of herding practices did they use? Did they (1) use the punas in their own zone, and (2) get camelid products from other groups?" Dedenbach Salazar does not accept Murra's (1972: 431–432) argument that the Chupachus managed the herds on a regular and immediate basis. She proposes "direct access" with "double residence" and barter. Another possible explanation would be the model proposed by Fuji and Tomoeda (1981:57), namely, that the *punaruna* (men of the puna) living in the puna looked after their own animals and those of the *llaqtaruna* (men of the towns).

Dedenbach Salazar concludes that despite the low number of animals listed in the *visita*, the reports of Cieza de León, Estete, and Vázquez de

Espinosa (examined in the preceding chapter) concerning abundant livestock in the Huánuco region are believable.

Turning now to the Junín zone, the reader will recall that on Hernando Pizarro's journey back from Pachacamac, the Spaniards spent the night in the town of Chacamarca while en route from Pombo to Xauxa. There were many towns around Chacamarca, and "around these was a very great number of livestock, and to the Spaniards it seemed that there were over a hundred thousand head" (Fernández de Oviedo 1959c:Bk. XLVI, Chap. XII, 74; he says this information came from Estete; however, it is worth noting that the passage does not appear in this chronicler [Estete 1968a]).

For this same area Flores Ochoa (1977b:23) wrote: "The Huanca were paying their tribute with camelids. Those of *Hatun-Saya* gave 58,673 llamas and alpacas to the Spaniards in 1533–1544. The *Saya Urin-Huanca* gave 514,656 animals to Pizarro on October 1533, and 27,958 in 1534–1544 (Espinoza Soriano 1972, cited by Browman 1973:41). There are no data on how much was given by the third *saya* of *Hanan-Huanca*, but it is possible that it gave a similar number as the other sayas (Browman 1973:41)." This is interesting because it gives a total of 601,287 animals, which is no mean number, but it does not appear in the article by Espinoza Soriano (1972), which Flores Ochoa did not cross-check; he took the data from Browman (1973), which I was unable to read.

On the other hand, Guacrapáucar (Guacora-pacora; 1972:201–215) stated in Lima in 1558 that the "three moieties of Luringuanca, Ananguanca, and Xauxa" gave 5,184 llamas, 2,534 sheep, 95 alpacas, and 691 yearling lambs from the time Francisco Pizarro left Cajamarca to the defeat of Hernández Girón. This comes to 8,504 animals in all. Although the source does not specify whether the sheep and the yearling lambs were "of the land," we can assume so, even though some European animals may have been present.

This same Guacrapáucar (1972:216–259) then requested the Lima Audiencia to bear witness to "the services [rendered] by his moiety of Lurinhuanca and himself since the arrival of Francisco Pizarro." Ten witnesses did not answer and two said yes when asked, in 1560, whether their livestock had been seized or they had given some.

The *Probanza de servicios* dated 1561, Lima, includes a "Memoir of the curacas of Atunxauxa." Among other points, the "Capítulos del memorial" asked "the witnesses presented on behalf of the caciques, headmen and the Indians of Atunxauxa given in encomienda to Gómez de Caravante" if they knew whether the caciques sent Francisco Pizarro "sheep and rams," if on arriving in the valley Captain Soto was given a "great number of sheep," if "they rendered a great number of sheep" on the arrival of Alonso de Alvarado, and whether, "besides this, said soldiers stole[6] a great number of sheep" (Cusichaca et al. 1972:260–278).

The "Memoir" in the aforementioned document lists the number of animals rendered or "rancheados." I find this interesting, for it not only indicates the number of animals present, it also shows how the local fauna was destroyed.

For the "Provisions [sent] to Cajamarca," the Indians delivered 20 rams of the land. The conquistadors received in Jatunxauxa 1,275 sheep and 12,045 rams of the land, but the Spaniards seized 29,231 animals among sheep, rams, and pacos. At the "[r]eturn of Francisco Pizarro to Jauja" he was given 296 sheep of the land and 61 rams, and 120 sheep and 6 lambs for the "[e]xpedition of Pizarro to the valleys of Pachacamac and Lima." For the "[e]xpedition of Quizo Yupanqui," 60 sheep and 10 lambs of the land were given; the "[e]xpedition of Alonso de Alvarado" received 846 sheep of the land and 50 lambs; 61 sheep of the land in the "Battle of Páucarbamba and expenditures"; 10 sheep for the "[e]xpedition of Pizarro to Cuzco"; 54 sheep of the land for the "[e]xpedition of Captain Mercadillo"; 2 sheep of the land for the "[p]assage of Captain Rodrigo de Salazar"; 10 sheep of the land and 1 lamb for Rodrigo de Mazuelos; 60 sheep of the land as "[p]rovisions for Captain Mosquera"; and 54 sheep of the land as "[p]rovisions for Captain Gómez de Alvarado."

The "[p]illage made by Diego Hernández" cost the Indians 28 sheep of the land and 10 lambs of the land, and the "[p]rovisions for General Pedro Alvarez Holguín" 118 sheep of the land and 17 lambs of the land. In the "[p]illage made by Almagro el Mozo," the Spaniards "stole"

12,902 sheep of the land and 13,045 lambs. Five sheep and 23 lambs were delivered as "[s]upplies for the army of Vaca de Castro," and 81 sheep for the "[e]xpedition of Captains Pedro de Puelles and Pedro Vergara," the same number given at the "[r]eturn of Vaca de Castro." Twenty-five lambs were given as "[a]rmament for Viceroy Núñez de Vela," and in the "[p]illage by Gonzalo Pizarro" the Spaniards "stole" 19 sheep. In the "[p]illage by Francisco de Carvajal" 25 sheep were stolen, and 192 sheep of the land were stolen in the "[p]illage by Juan de Acosta." Seven sheep of the land were given as "[p]rovisions for Gabriel de Rojas," and 173 sheep of the land and 3 lambs as "[a]rmament, provisions and soldiers for La Gasca." Fifty-two sheep of the land and 7 lambs of the land were given as "[s]upplies for Pablo de Meneses and the Royal Army," while the Indians gave 46 sheep of the land, 4 ovejas partidas, and 6 lambs of the land for the "[a]rmament and provisions for Captains Juan Tello and Miguel de La Serna." Last of all, 21 sheep of the land and 3 lambs of the land were supplied at the "[r]eturn of Pablo Meneses."

At the end of this document it is recorded that "[o]f all said things . . . thus far nothing has been given them or been paid" (Cusichaca et al. 1972: 279–320). From this it follows that the Indians supplied or had stolen from them 12,065 rams of the land, 16,652 sheep of the land, 13,250 lambs of the land, 29,231 sheep, rams and pacos, and 4 ovejas partidas. This gives a total of 71,202 animals, which is no small number.

The statements made by the witnesses include some telling phrases. Thus we find the following: "[T]hey had stolen it all, together with another great number of sheep and rams of the land," "great number of . . . sheep," "and a great number of sheep and rams," "and a great supply of . . . sheep flesh," "which [he] took with him a great number of rams and sheep," "[a]nd this witness saw that on some occasions some of the soldiers of said Alonso de Alvarado went outside the valley in search of sheep, and [they] brought [back] some livestock. And as regards this lawsuit, this witness knows that said soldiers could not refrain from taking a great number of livestock [belonging to] said Indians of Xauxa, and this witness does not know the number of things said caciques gave to said Alonso de Alvarado, or were taken from them, and he refers to the quipus the Indians will have about this." Almost all witnesses repeat the phrase "[a] great number of sheep." I have listed the most significant statements (Cusichaca et al. 1972:346, 348, 355, 359, 373, 384– 385).

Dedenbach Salazar (1990:105–106) also discussed the data in these documents. (To avoid confusing the reader, it must be noted that in her study, Dedenbach Salazar lists these sources under Espinoza Soriano, who published them. She gives 1971 as their date of publication, whereas I give 1972; both dates are correct, for the journal has the latter date on the cover and the former one inside.)

Dedenbach Salazar considers that the general terms used in this source—that is, sheep and lambs—seem to refer to camelids, even though it is not specified whether these are "of the land" or "of this land." The argument she used to support this is as follows. In the document we read the following: "[T]hey gave said marshal five hens to eat which were the first in the Kingdom of the Highlands" (Cusichaca et al. 287). "From this it follows," explains Dedenbach Salazar, "that the great numbers of livestock must be camelids because, if the first hens still lived at that time, then it is not likely that the European livestock would already have multiplied to give thousands of head; pigs were also given, however, but in lesser numbers. These animals are always listed as food resources (e.g., the 'Probanza de Servicios':284, 290)." It is for this reason, and the fact that "pacos" are listed in small numbers, that we can deduce that the "sheep" and "rams" in this document mean that the llamas were used as pack animals and meat, whereas European sheep could not have been used as beasts of burden. The fact that the Spaniards were given many textiles also allows us to assume they did not want to take away the source of the raw material, be it sheep or alpaca, preferring instead finished textiles.

Espinoza Soriano (1972:123) recounts the battle between the Cuzqueños and the Spanish-Huanca army in 1536. He explains that during the pillage, the Cuzqueños took 451 llamas and 70 lambs.

Ruiz (1952:Chap. XII, 76; cited by Dedenbach Salazar 1990:96) confirms that in the mid-seventeenth century, "[an] infinite [number of] vicuñas [were] raised in the punas of Bombon and in all other punas in said province."

In the Archivo General de Indias (Patronato 95-B Ramo Tasaciones de los Yndios y visitas para que conste que son pocos y dan poco), Rostworowski (1978:228, 230, 238–241, 247–248, 250, 252–254) found a *visita* to the moiety of Canta. Here it is recorded that the site of Rococha depended on Canta and "has sheep," that the "Indians" from Canta "have a great number of sheep," and that 39 head "of ovine livestock of the land" belonging to the cacique, 29 from the headmen, and 86 from the people were brought "from said town of Canta." On the other hand, the census made in the moiety of Canta recorded one sheep; at Carcas two were found that belonged to the headman, and 10 to the "Indians." So 167 animals were listed in all.

Then, in the *visita* of the moiety of Locha we find that 9 head of "sheep livestock of the land" were declared in the town of Ayas, 10 in Hurco belonging to the headman, and 18 to the "remaining natives." This gives a total of 37 animals.

The *visita* of the moiety of Lachaque records 10 animals belonging to the headman, 8 to another headman, and 12 belonging to the natives, for a total of 30 animals in all.

In the *visita* of the moiety of Copa, the headman had 10 heads, all other headmen 5 in all, and the natives 45. This gives a total of 60 animals. In the *visita* of the moiety of Ysquibamba, the headman declared 27 heads and the natives 54, or a total of 81 animals. These documents record that an annual tribute of "one hundred" sheep was given to the Inca in the town of Rococha, and that Nicolás de Ribera El Mozo, "who at present holds them in encomienda," had received "in the last ten months . . . four hundred sheep," among other things. Another witness declared that "eight hundred sheep and rams" were given to the same individuals (Rostworowski 1978:230–231).

And the *visita* Dávila Brizeño (1965:161) made in the highlands of Lima in 1571–1572—the famed *Description of the Yauyos*—records that while acting as magistrate, he punished some caciques from Anan Yauyos "and took from them four hundred head of livestock of this land."

For "Castrovirreyna and its district" we have the testimony of Vázquez de Espinosa (1948: 1446/492), which shows the dramatic increase in European husbandry, and how native livestock was no longer even mentioned for the area. He wrote: "In the year 1610 there were four cattle ranches and five sheep [ranches] and five goat [ranches], and one of mules . . . and in these ranches there were 1,600 cows, 5,000 sheep, 12,000 goats, and four hundred *yeguas de vientre* [brood mares]. At present there are many more because they grow well, and they are increasing." Strangely enough, I was unable to find such figures for Cuzco. All I can do is repeat the passage from Pedro Pizarro cited earlier (1978:Chap. 20, 144), which tells how when meat was scarce in the city after the Indian uprising of 1536, Hernando Pizarro sent men to the district of Acomayo in search of livestock. From Pomacancha they brought "up to a thousand head of cattle."

The zone of Puno in the highlands is certainly the area for which more data are available. The major source is the famed *visita* made by Diez de San Miguel in 1567. It is recorded that the inspector's statement was based on the Indian *quipus*. It is noted that in the province there were 45,000 "and as many head of livestock of the land," and 25,000 "and as many head of livestock from Spain." If we add "the livestock each town, moiety, and ayllu have . . . it is clear that there are more than eighty thousand head" (Diez de San Miguel 1964:211).

The "Provision for the management of community livestock in the province of Chuquito" says that in the inspection made by Garci Diez de San Miguel, and the data he submitted to licenciate Lope García de Castro, member of His Majesty's Council, President of the Audiencia and Chancellery of this City of Kings, it is recorded that "the Indians in said province of Chuquito have over forty thousand head of livestock of the land [belonging to] the community, and another number of livestock from Castille" (Diez de San Miguel 1964:273).

Then we have the statement given by the witness Martín Cari, cacique principal of the Anansaya moiety. "Asked what number of livestock said Indians keep for him, he said they keep three hundred head for him." In the Anansaya moiety he declared "up to four hundred heads of livestock of the land" (Diez de San Miguel 1964:22–23). The same witness was then "[a]sked what number of livestock they [the Indians] have, of the land or from Castile; he said as noted that said Indians have livestock, but he does not know how many" (Diez de San Miguel 1964:18).

The same *visita* records that Martín Cari requested protection for having declared 30,000 "head of community livestock." Martín Cusi did the same for 2,000 "head of livestock of this land of all kinds." These caciques charged each other with not having disclosed all the animals they had. Don Pedro Cutimbo, governor of the province of Chuquito, stated that livestock had been concealed, as the Hurinsaya moiety had "more than" 8,000 head, and there were 2,000 "more or less" in Anansaya. It was likewise noted that what was declared in Acora and Juli was correct (Diez de San Miguel 1964:167–169, 171).

The statement made by the headmen of the Anansaya moiety in Acora was as follows: "[W]hat livestock does each Indian have? They said that some have two hundred head, others a hundred, others fifty, others ten and four and three, and more or less than these numbers they have declared, and some Indians have none . . . and have no livestock from Castile, save the caciques who have some goats, and these are few." Of the community livestock in the town of Acora it is said "from the *quipus* of the Indians [it follows] there are in said partido of Acora thirteen thousand and five hundred and thirty heads of livestock of the land, rams and pacos belonging to the community of the moiety of Anansaya, and to Hurinsaya six thousand and nine hundred and ninety five head of said livestock, as recorded in said *quipus*" (Diez de San Miguel 1964:92, 88–89).

Then comes the statement made by Martín Cusi, cacique principal of the Lurinsaya moiety. "Asked what number of community livestock there is in the Lurinsaya moiety . . . of this town of Chuquito [he replied that] the community has four hundred and fifty head of livestock of the land, and a headman who calls himself Don Luis Cutipa is in charge, and he will later on give an account" (Diez de San Miguel 1964:30). Don Luis Cutipa, "headman of the Lurinsaya moiety," declared "the community livestock the Lurinsaya moiety in this town of Chuquito has . . . , [he] declares five hundred head of sheep of the land, adult females [*hembras grandes*] and female pacos, a hundred and seventy three are of adult sheep [*ovejas grandes*] and the rest pacos, and he has not tallied them this year, nor does he know how many there are." It is then said that Lurinsaya had more animals than Anansaya, and that "from the deposition . . . it seems to have two thousand head of livestock"; the witness explained himself by saying that "said Lurinsaya moiety said it has no more livestock than it has declared because much livestock died with the snowstorms and blizzards that occurred a year ago, and another part was used up with the wine earthen jugs which the community has spent in providing food and meat for the Indians who go to Potosí to work and pay His Majesty's tribute" (Diez de San Miguel 1964:78–79).

The witness presented by the headmen of the Urinsaya moiety in turn declared "that some Indians have sheep by threes and sixes and more or less, and some years where food is scarce they take them to the yungas to sell [them]." When asked about the number of livestock of the land, they declared "they do not know it because some Indians have three sheep and others six and ten, and a hundred and two hundred, and others more or less, and the number they have cannot be known." As for Castilian livestock, "they said there is little . . . and that some Indians—albeit a few—have two or three sheep, and others have hogs" (Diez de San Miguel 1964:80).

Martín Churi from the Lurinsaya moiety "said that in Cotahuasi—which is some four leagues from this town [he meant Chuquito]—he has six hundred and fifty seven heads of livestock of the land . . . and he does not know the livestock they have in other parts, save that he understands they have more livestock" (Diez de San Miguel 1964:88).

The community livestock of the town of Chuquito is then discussed. "The *quipus* [recording] the community livestock were to be brought and counted. It seemed that the Anansaya moiety of this town of Chuquito has a thousand and nine hundred and fifty and one head of community livestock of this land, pacos and rams, and the Lurinsaya moiety two thousand and thirty head" (Diez de San Miguel 1964:88).

Then there is the testimony given by Bernardino Gallego, a "resident in this province of Chuquito," who, when "[a]sked how much livestock of the land is in this province, both [belonging to] individuals and the community, said that this witness does not know the number of either, but he does know it is a lot of livestock, and that all usually have a lot or a little livestock, and he has heard it said by Don Juan Alanoca of Chuquito, an Indian who is not a cacique but a headman, [that] he has over fifty thousand head of livestock, and that he has heard the remaining headmen say they have over a thousand head." He then added that "there is much community livestock" (Diez de San Miguel 1964:50).

Alonso de Buitrago, another resident of "said town of Chuquito," declared that "of this livestock from Castile the Indians probably have up to two thousand head." However, he then said of the sheep that "he cannot know how many livestock there are because it is a lot, and it is well known there are caciques with ten thousand sheep, and other headmen with a thousand, and other Indians with a hundred and more or less; other, poor ones, have three and four and ten and twenty, and in this order some a little and others a lot." Later he noted that "in all the towns there is community livestock and he does not know in what number" (Diez de San Miguel 1964:55).

Melchior de Alarcón also supplied information about Chuquito. He said, "there are many of them [Indians] who are rich in livestock . . . and other individual Indians have some livestock," noting that "the damage [wrought by their going to the coast] is grievous because their livestock is squandered" (Diez de San Miguel 1964:139).

For the Anansaya moiety of Ilave we have the statement made by the cacique Don Francisco, who said that "the Indians of both his moiety and those of Urinsaya have sheep, some a little and others many; some [have] two hundred, some three hundred and others fifty and more or less, up to ten and six and four." When he made his statement concerning the town of Ilave, "[i]t was ordered that the *quipus* of the community livestock be brought . . . and they were seen and counted, and it was found that the Anansaya moiety had six hundred and ninety and three heads of livestock of the land, and in Urinsaya a thou-

sand and four hundred and twenty eight head" (Diez de San Miguel 1964:108, 113–114).

The Indians in Juli stated that "up to half the Indians have livestock of the land and the other half do not have them. Those who have them [have] three hundred head, and two hundred, and a hundred, and eighty, and fifty, and twenty and up to three . . . and they do not have livestock from Castile." The *quipu* were counted to declare the community livestock; all three moieties in Juli, i.e., Urinsaya, Anansaya, and Ayanca had 16,846 among "rams and male pacos, and females." The two moieties of Pomata Anansaya and Pomata Urinsaya had 216 head of "livestock of the land" and 2,135 rams and sheep from Castile. In Zepita, 2,347 head of "livestock of the land" and 90 head of "livestock from Spain" were counted (Diez de San Miguel 1964:116, 122–123).

Bernardino Gallego likewise made a statement concerning the chieftainship of Pomata. "Asked whether he knows or has heard say what number of livestock of the land there is in this province he said that he does not know, save that most of the Aymaraes Indians have livestock, [some] twenty, and thirty, and a hundred, and more or less, and others [have] three or four head, and there are Indians who have more than a thousand head who are not caciques." Asked how many community livestock they have, this witness said that he believes they have much community livestock because he has heard many caciques say so, but he does not know how many" (Diez de San Miguel 1964:46).

For the province of Chuquito there is another important source. This is a "secret lay inspection made by . . . Friar Pedro Gutiérrez Flores . . . and Juan Ramírez Segarra" in 1572 (1970), or five years after the inspection made by Garci Diez de San Miguel.

Here we find a reference to an abusive sale of livestock belonging to the Anansaya and Urinsaya moieties of the town of Batalla, made by friar Agustín de Formicedo. This town had "over a thousand and two hundred head of said livestock, of which he chose the best and sold them to a vicar in the city of La Paz." It turns out that the total number of animals sold was 275 head of "sheep of the land" and rams, 631 head of "pacos sheep" and "pacos," and 29 "big pack rams"

(Gutiérrez Flores and Ramírez Segarra 1970: Cuaderno II, 5r/5v, 5v/6r, 6r/6v, 19–21). This same document includes an "Account . . . of community livestock." Here are recorded the "[l]ivestock [belonging] to individuals. We found that [in] the towns of Jule, Pomata, Çepita and Yunguyo, individual Indians have ninety and nine thousand, three hundred and fifty six head of livestock. 99,356." And "Community. Item in the seven *cabeceras* [major towns] in this province were found seventy thousand three hundred and forty one heads of community livestock which come from [those belonging to] the Sun and the Inca. 60,341. They have some livestock from Castile, individually and as a community, although they are not many, and some pigs and goats in particular" (Gutiérrez Flores and Ramírez Segarra 1970:Cuaderno III, 38).

Further on we read: "Item. Since in this province [of Chuquito] there were found at present seventy thousand or so community head of livestock of the land, big rams and male and female pacos, besides those in possession of individual Indians, which is a lot, as will be seen in the account of the tribute-paying Indians and their means, which said livestock is and comes from that which the Inca and the Sun had and [were] sacrificed to their guacas. It is understood that there is much more than has been discovered because of the many there were in the time of the Inca, and it seems almost impossible that what is missing was spent and consumed, particularly since it is an established fact among the Indians that two parts out of three of this livestock multiply each year." Then we read: "Item each year four thousand big rams and many more, more or less, can be taken each year to be sold according to the offspring there are" (Gutiérrez Flores and Ramírez Segarra 1970:2v/3r, 43 and 3r/3v, 44).

Finally, this document records the great number of abuses, to say the least, committed by the friars of the order of St. Dominic "who live in the parishes of said province" of Chuquito. Here the great number of livestock taken from the Indians appears. I discuss later, but it is worth repeating now a complaint made against the damage wrought by the systematic theft of livestock by the friars. It is stated that had this not happened, the Indians would "now [have] five or six thousand [more] head with which they could comfortably pay their tribute at present" (Gutiérrez Flores and Ramírez Segarra 1970:Cuaderno II, 8r/8v, 23).

A "Padrón de los mil Indios Ricos de la Provincia de Chuquito" (Register of the thousand rich Indians of the Province of Chuquito) by Gutiérrez Flores, dated 1574 (1964), is also known and was part of a larger document that has not been located. Here the Indians listed come to 985 only, and a total of 131,521 head are mentioned; the Indian who owned 920 heads had the most, and the one owning 50 had the least. It turns out that each Indian had 133 head on average (Gutiérrez Flores 1964:306–348).

Then there is an analysis of "[t]he rich Indians in this Province besides those who are in the register, the moieties and towns." In all, 379 Indians were registered, with a total of 23,268 heads, and a per capita average of 61 animals. Most had 108 heads, and those with less had 30 animals (Gutiérrez Flores 1964:349–362). This inquest was held to levy tribute on the Indians.

There are many commentaries on the Chuquito document, some of which are interesting and worth including here. Murra (1970:53) discussed Garci Diez de San Miguel's *visita* and explained that only part of the thousands of pages from this document were selected in Lima and sent to the Council of the Indies in 1568. What was published in 1964 is an incomplete version. Garci Diez de San Miguel opposed the members of the Audiencia who wanted to increase the tribute paid by the Lupaqa because these Indians were officially considered rich. "This impression was due," explains Murra, "to the fact that the herds of llamas and alpacas were convertible to European currency, and besides seemed inexhaustible." Garci Diez de San Miguel estimated that the community animals came to 80,000 head, and suggested to Governor García de Castro that the beasts the caciques and headmen had should have a manager to prevent waste and to centralize responsibility. According to Murra, this suggestion, which was accepted by the governor, "has an Andean flavor to it."

Dedenbach Salazar (1990:106–107), in turn, discusses all three sources from Collao—the 1567 *visita* by Garci Diez de San Miguel, the inspection carried out by Gutiérrez Flores and Ramírez

Segarra in 1572 (which she mistakenly attributes solely to the former), and the "Register of the thousand rich Indians" prepared by Gutiérrez Flores in 1574. Dedenbach Salazar notes that about 30,500 head are declared in the 1567 *visita*, but Inspector Diez de San Miguel (1964:211) states there were more than 80,000. In the 1572 inspection five years later, 60,341 heads were declared. Dedenbach Salazar asserts that "[c]onsidering the inaccuracy in the figures of the 1567 *visita*, the total number of community livestock must have remained stable."

Dedenbach Salazar also notes that it was only in the second *visita* (1572) that the privately owned livestock was counted, which came to 99,456 head belonging to 12,271 Aymara tribute payers. She believes that an average of 8.1 head of livestock per family is too low a figure if one intends to essentially subsist on these animals. Dedenbach Salazar believes that the result must have confused the inspector, and it was perhaps because of this that the "Register of the thousand rich Indians" was prepared, which gives completely different figures seven years after the first inspection. Thus, 1,278 rich Indians had 156,274 head of privately owned livestock, which gives an average of 122.3 head per family. All Indians with more than 40 to 1,700 head were considered "rich," but of these, only three had more than 1,000 animals each. The confusion wrought by the data was recorded by the inspector himself at the end of the list: "[W]hat is recorded in the registers is far less than what said Indians have individually, and thus it was declared by some Indians before said lords that caciques and Indians had agreed not to declare more than a tenth part of their estate in the padrones and record of their estate, and it was verified that an Indian who declared forty head had one thousand, one hundred and seventy, from which can be inferred the great number of livestock in said province, and what little transparency and surety can be had in such a brief span" (Gutiérrez Flores 1964:363). So it is in no way surprising, says Dedenbach Salazar, that different statements made by different Indians in the 1567 *visita* give such varied figures. For instance, it was said that the headman Juan Alanoca of Chuquito had over 50,000 head of livestock (Diez de San Miguel 1964:50), whereas in 1574 Halanoca, from Ayllu Guarico in Chuquito Anan-saya, declared no more than 100 head (Gutiérrez Flores 1964:80v, 308). If it is the same individual, then this is too great a discrepancy.

Dedenbach Salazar (1990:106–107) likewise notes that the ratio of livestock to herders found in several depositions does not seem plausible. Martín Cari declared having 300 head of livestock that were tended to by more than 80 Indians; the *visitador* doubted the truthfulness of this statement (Diez de San Miguel 1964:21–22, 32–33). In this same *visita* a witness declared that an Indian usually kept over 250 "carneros" (Diez de San Miguel 1964:164). The figures given by Flores Ochoa (pers. commun. to Dedenbach Salazar; Flores Ochoa 1977a [1977d in my bibliography]:136) of 200 alpacas and 60 llamas on average for a modern nuclear family show first, that an individual can herd up to 65 animals, and second, that it is quite likely the Lupaca chiefs had more livestock than a nuclear family.

Flores Ochoa (1970:67–68) also made some comments on the "secret inspection" made by Gutiérrez Flores and Ramírez Segarra (which he mistakenly attributed only to the former). He points out that we must take into account the figures in this *visita* regarding the size of some of the herds. He indicates that in the province of Chuquito there were in all 159,697 head, 99,356 of which belonged to "individual Indians" and the remaining 60,341 to the community. (Here Flores Ochoa cites Gutiérrez Flores and Ramírez Segarra 1970, "Cuaderno I, 1r." The citation is mistaken; it should be Cuaderno III, 38). Flores Ochoa considers that this number is small, particularly since the list of tribute-paying Indians gives a figure of 12,271. Bearing in mind the number of tribute-paying Indians, each would have about eight animals if we consider only the herds belonging to "individual Indians," and 13 if we include those belonging to the community. Flores Ochoa believes that today, a person with such a number of head would be considered a "*wahcha*" [sic], that is to say poor and very unlikely to maintain a household; he would thus be forced to have a bigger herd. In his estimate Flores Ochoa does not include the Uru Indians (of whom there were 3,198 in 1572) among the tribute payers, because Diez de San Miguel (1964:112) indicated they did not have livestock.

The herds (Flores Ochoa 1970:67–68) dwindled rapidly—1,137 heads were sold on just one occasion, ranging from "big sheep of the land" to "lambs of the land," that is, reproductive animals and very young offspring, "thus depriving the herds of the possibility of growing, or even of retaining a constant level to replace those [animals] sold, killed, or the ones who died from diseases."

Flores Ochoa also wrote in another article (1977b:23–24): "The visita of Gutiérrez Flores (1572) [Ramírez Segarra, the second author, is not cited] . . . is informative as it shows the rapid destruction of herding. It merely considers 159,697 head, of which 99,356 belonged to "individual Indians" and the remaining 60,341 to the community. This figure is in stark contrast to the herds of 50,000 animals held just five years before by some individuals."

Dedenbach Salazar (1990:107, n. 31) does not fully agree with this comment. For her, the decline in camelid number interpreted by Flores Ochoa (1977b:23 [the same entry in my bibliography]) cannot have been that stark if we bear in mind the "Register of rich Indians" (Gutiérrez Flores 1964), which shows the presence of great herds even two years after the inspection by Gutiérrez Flores (and Ramírez Segarra; the second inspector is once again not given), even though none had 50,000 head, as they had in the 1567 visita (Diez de San Miguel 1964). According to Dedenbach Salazar, this does not prove that these great herds did not actually exist.

Nor should we forget, as Pease (1991:86) correctly notes, that the data recorded by the Spanish officials might be distorted, since the herds belonging to the Inca were immediately turned over to the Crown or distributed among the Spaniards at the time of the conquest. This understandably caused some response among the indigenous people, who began to hide animals in various ways. In other words, they took them to remote areas or included them in the communal or individual herds.

Only a few references were found for the Department of Arequipa that refer to the Colca Valley. These are documents held by the Archivo Departamental de Arequipa[7] and are signed by the scribe Diego de Aguilar (Benavides 1988). There is a bill of sale issued in Yanque in 1568, stipulating a sale was made to a Spaniard of 304 llamas 2.5 years old or more. A similar document signed in Yanque Collaguas in 1569 stipulates the sale of 200 "rams of the land" to a Spaniard. Two other bills of sale from Spaniard to Spaniard were made in 1575. Both were drafted in Arequipa. One established the sale of 200 alpacas, the second one the sale of 100 "rams of this land" (part from Collao and part from Collaguas). Finally, there is a 1584 document, a part of a power of attorney given by one Spaniard to another to pick up 90 "rams of the land" at Laricollagua. Interestingly, in all cases it is specified that the animals should not have the mange (carache; Benavides 1988:40–42). Benavides (1988:44) notes that for the encomenderos, camelids were an important element in trade.

The 1591 visita of Yanque Collaguas Hanansaya lists 17 tribute payers who were registered as yanaconas (retainers) of Don Juan Halanoca, cacique principal; it is specified that each of them had between 10 and 50 llamas or alpacas (Benavides 1995:21). Based on documents from this same archive and from the Archivo Arzobispal of Arequipa, Benavides concluded, also for the Colca Valley, that there were 1,035 camelids in Coporaque in 1604–1615, and 600 in Cabanaconde in 1569–1645 (Benavides 1995:Table 4,37). However, Benavides notes that "for the Colonial period [the data] come from the visitas and must be taken with caution as they reflect the statements made by the caciques and their subjects, who had a vested interest in feigning the lowest possible number . . . of livestock" (Benavides 1995:18).

Other data worth noting appear in the tribute roll enforced in 1549 on orders of Pedro de La Gasca, the president of the Lima Audiencia; the document lists the amount of goods per repartimiento, including the number of "sheep," "lambs," and "pack rams." It is recorded that the "repartimientos of Guamanga" (including Xauxa, Soras, Lucanas, and Cacayacure) had 951 sheep, 64 lambs, and 100 pack rams in all (La Gasca 1983–1984:53, 65–66, 69). The "repartimientos of the city of Cuzco" (comprising the repartimientos of Aymaraes and Asangaro) had 216 sheep, 6 lambs, and 50 pack rams (La Gasca 1983–1984:74–75). "Two hundred and fifty-three sheep,

30 lambs, and 170 pack rams were found in the "repartimientos of the City of Arequipa" (comprising the repartimientos of Chuquibamba and Collaguas) (La Gasca 1983–1984:78–79). In the "repartimientos of the City of Kings" (which comprised the repartimientos of Caxamalca [province of La Nasca], Andax[es] and Guamachuco), there were 303 sheep and 9 lambs (La Gasca 1983–1984:85, 87, 92, 94), and 324 sheep and 4 lambs in the "repartimientos of the City of León" (province of Guanuco)" (which comprised the repartimientos of Chupachos, Luringuaylas, and Yaros) (La Gasca 1983–1984:98, 100–101). These figures add up to a total of 2,047 sheep, 113 lambs, and 320 pack rams.

Years later Viceroy Francisco de Toledo ordered the "general inspection of Peru" in 1570–1575, which is known as the tribute roll ("*Tasa*") of Toledo. It contains more in-depth data on the subject. The tribute roll lists the number of "rams of the land" and "pacos." It begins with the "City of Cuzco and [the] Province of Collao," including the "Royal Audiencia of La Plata," which shall not be considered here. The sites listed are Arapa, Caman, Asangaro of Antonio Quiñones, Asangaro of Captain Martín de Alarcón, Taraco, Nunoa, Cauanilla and Oliberos, Cavana, and Lampa. The sum of the rams of the land in all these groups is 1,128 head (Toledo 1975:87–88, 90–97).

Then we have the "province of Cuzco," with the following sites listed: Andaguaylas the Great, Collana, Aymara, Taype, Ayllo, Cotabambas and Omasuyos, Parinacochas, Guainacota, Alca, Cayo Aymaraes, Challuanca, Mudca and Pairaca, Llusco Aymara, Haquira Yanaguaras, Achambi and Cotagucia, Pichagua, Yaure, Curatopa, Colquemarcas, Chilata, Yanaoca, Tinta, Cotahuaras, Piti Yanaguaras and Mara Yanaguaras, Vililli, Cacha, Capac Marca, Omacha, Sangarara, Yanaguaras Malmayas, and Hancoba. A total of 2,980 rams of the land and 100 pacos were recorded here (Toledo 1975:115–118, 121, 123–131, 133–136, 142, 150, 154–155, 158, 163, 171).

In "Arequipa and its jurisdiction"—which includes Arones of His Majesty, Laricollagua, Collaguas, Cavana, Arones and Ocoña, Chachas and Ucuchachas, Chuquibamba, Andaguaychicha, and Viraco—1,053 rams of the land were record-

ed. In Guamanga—which comprised the Lucanas Andamarcas, Guachos Chocorbos, Quichuas and Aymaraes, Totos, Hanan Chilques, Hurin Chilques, and Tanquiguas—1,775 rams of the land were recorded (Toledo 1975:261–262, 264, 268, 271, 276–278). If we add up all of these figures, we get a total of 6,936 rams of the land and 100 pacos.

7.3 SLAUGHTER, ABUSES, AND ILLICIT APPROPRIATION

One of the factors that must have contributed greatly to the decline in camelid numbers in the central Andes is the indiscriminate killing that took place, particularly in the early years of the Spanish conquest. Then came the theft of livestock from the Indians, and the poor use made of them. This is a point that needs to be documented in order to appreciate its magnitude, even though it will never be possible to fully reconstruct the actual numbers.

Xerez (1968:232) narrated the events of Cajamarca. I have already mentioned that once the Inca was seized, Pizarro ordered that all the llamas found around the camp of Inca Atahuallpa should be set free. At the same time he "ordered . . . that the Christians kill every day as many as they needed." Xerez (1968:235) also says that while the Inca was held prisoner in Cajamarca, "[a]mong the Spaniards who are with the governor they kill[ed] each day a hundred and fifty, and it looks as if they are not missed in this valley, nor would they be, even though they [the Spaniards] were a year in it. And the Indians usually eat them in all of this land." The chronicler means sheep.

The choirmaster Cristóbal de Molina "El Chileno" also mentions the events at Cajamarca, but his manuscript probably dates to 1552 (Porras 1986:317). He states that the Spaniards "stole" many things there, among which he lists "sheep." He then adds:

> Besides, since each Spaniard took over such a great number of retainers, for them to feed it was necessary to follow no order in the livestock, and this they did to such an extent that it often happened that some Spaniards killed 10 or 12 sheep just for the

marrow. I will tell what I saw in Cuzco three years after this: one night a Spaniard entered a corral belonging to another one and stole 50 or 60 sheep, and I even believe they were more, and that night beheaded them all. The following day, when the other [Spaniard] found that his herd had decreased, he sent many men in search of them, and they spied the corral and house of the Spaniard, and found all the sheep dead, and each was as big as a calf.

Further on Molina adds: "At this time and for the next 12 years there wasn't a Spaniard, no matter how poor he was, who passed or walked through a town that he was not given a sheep or lamb to eat, by him and his hunting dogs, and if the cacique or lord did not do so, they beat them" (Molina 1968:302–303).

Santillán also narrates the events at Cajamarca. Although the Licentiate was in Peru, for he arrived around 1548 and left in 1562, he wrote his chronicle in Spain in 1563 (Porras 1986:324–326). His testimony is interesting because he collected his data twenty years or more after the events, yet the memory of how the Spaniards squandered the native livestock at Cajamarca was still strong. Santillán (1968:409) wrote: "It is said that [the Spaniards] killed a great number of sheep just to eat the brains, and the rest they abandoned, and to find a fat sheep they killed ten or twelve. Others supplied the slaughter houses, [while] others took great herds of livestock in the expeditions, and in this way used up almost all the livestock there was in this land, with as much diligence as if God had commanded them to do so with the Amalekites. And so, there being in this land more livestock than grass, they left it almost with none." That final phrase is heart-breaking.

Now is the time to clarify a point. When describing what happened after the Inca was seized, Fernández de Oviedo (1959c:Bk. XLVI, Chap. VIII, 59) wrote: "Since there was such a great number of them and they cluttered the camp, the governor ordered that all the sheep should be set loose on the fields, and that the Spaniards should kill every day as many as they needed." Porras (1986:178) notes in this regard that Fernández de Oviedo "usually collects direct oral testimonies

from participants in these events, like Diego de Molina for the episodes at Cajamarca." However, in this case it is clear that Fernández de Oviedo took his data directly from Xerez (1968; see above), because it is copied almost word for word.

Pedro Pizarro (1978:Chap. 34, 244) recorded an important point when he described a hunt (chaco) held in Jauja in honor of Francisco Pizarro. He says that "eleven thousand and so heads of wild animals were killed . . . most of them were wild livestock of the land." Were these vicuñas or guanacos? Probably both, but it scarcely matters. What is significant here is the number.

Romero (1936:364-365) cites a passage from a book I was unable to obain that I find revealing and therefore I reproduce it here.

In the acrid critique he makes of the deeds of the conquistadors in his book *Coloquios de la verdad*, written around 1555 or 1560, Pedro de Quiroga criticizes the wanton destruction of the llama by the Spaniards, which they called "rams of the land" to distinguish it from the "rams from Castile," imported into Peru after the conquest. In the dialogue between the Indian Tito and the Spaniard Barchilón, the former says: "You know—for you saw it at the time you won this land (it would be better to say lost and destroyed it)—what number you found of sheep of this land, which is such a useful livestock as no better has ever been seen throughout the world, nor with such long or useful products, or such precious meat and wool—[yet] you destroyed it without qualms; you went after it with spears and swords as if they were not your animals or were harmful and wild ones; you inflicted such great death on them that the stench in the towns and fields you crossed was unbearable. What people committed such monstrosities? Or if you understood what the Indians said of you and your blunder—"Our gods have placed us in the hands of witless people [and] we have fallen[8] under the power of madmen"—you would have realized your mistake because you diminished and put an end to such an essential thing for you and for us, so that [although] there [once] were so many sheep of this kind that they covered the fields with their multitude, they are now so wanting that you kill us to make us give you what you killed."

Espinoza Soriano (1972:94–96, 260) studied and discussed a Probanza prepared in late 1561 at the request of the three caciques of the repartimiento of Atunxauxa. The inquest was held to establish "what they have served His Majesty with in the time of the disturbances caused in these kingdoms, and conquests and discoveries."

Espinoza Soriano explains that during all of his stay in Jatunsausa, Pizarro received from the Huanca all they could give him, particularly male and female llamas. What the Spaniards most consumed in the herds was the meat from males. For example, it is recorded that the Indians gave Pizarro 12,045 male llamas and just 1,275 females during his first stay in Jatunsausa (this gives a total of 13,320 animals; however, the source does not say llamas but "sheep," nor does it mention male animals but "rams of the land" instead) (Cusichaca et al. 1972:279–280). However, the important thing here is that, even so, thefts still took place. Espinoza writes: "For the theft of llamas in October 1533 . . . was huge. The second item in the *capítulos* of the Memorial states that *said soldiers stole a great number of sheep and [clothing and] goods* (128)." (Espinoza omitted the words between square brackets; the correct reference is Cusichaca et al. 1972:265.) All of these items were carefully recorded in the *quipus*. Still according to Espinoza Soriano, whereas the Huanca zealously looked after the female llamas in their herds to favor their increase, the Spaniards instead hunted and quartered them, because their meat was far tastier than that of males. "Thus it was that they stole entire herds, which reached the astounding number of twenty-nine thousand, two hundred and thirty-one female llamas, including some males and pacos too." (The source actually says "Sheep, rams, and pacos"; Cusichaca et al. 1972:280). In most cases this was done just for the brains or the neck, which were believed to be the most delicious part of these animals. According to Espinoza Soriano, "[t]he most incontrovertible

evidence of the thefts or *rancheamientos* made by the Spaniards at Jatunsausa on October 1533, as well as in 1534, is provided by an eyewitness of these events: Pedro de Alconchel. He said: '[. . .] He knows that besides what said Indians gave, as was told, the Spaniards who entered the Xauxa Valley with said Marquis took and stole a great number of sheep and rams from said Indians of said valley'" [Cusichaca et al. 1972:363]. Ribera el Viejo tried to save the Spaniards' honor by claiming that the native helpers brought from Cajamarca were the thieves: "[. . .] He knows that the yanaconas and Indian helpers whom said Francisco Pizarro and his men brought with them in their service stole and took a great number of rams and sheep and maize and potatoes and clothes and other things from said caciques of the Xauxa [Atunxauxa] Valley [. . .] (131)." [Espinoza Soriano mistakenly places Xauxa in the citation; the correct name appears in square brackets. Cusichaca et al. 1972:373–374.] (Notes 128 and 131 in Espinoza Soriano refer to Cusichaca 1561:fol. 4r and 67r. This actually is an incomplete reference because it should read Cusichaca et al.; as for the inquest, it was held at the request of Cusichaca, Eneupari, and Canchaya, the three caciques of Atunxauxa. It should be noted that the citation by Espinoza Soriano refers to the folios in the original document, which is hard to follow in the transcription published in 1972; in each case I indicated this with ". . ." in square brackets, above.)

Espinoza Soriano (1972:104) then explains that the Spaniards made a tremendous waste of provisions from 27 October 1533 to 20 April 1534, to the point where the deposits were emptied. The boy Canchaya, who returned from Cuzco to Jatunsausa with Pizarro, verified "how the deposits of maize and other provisions which the caciques of said valley of Xauxa had, which he [Canchaya] had left behind full, were [now] empty; he found out from the Indians in said valley and the soldiers of said Marquis that they had all been pillaged, [together] with a great number of sheep and rams of the land (152)." (The source given by Espinoza is Cusichaca 1561 fol. 13v, which corresponds in the transcription to Cusichaca et al. 1972:346.) This is interesting testimony, as it was given by Baltazar Canchaya, "headman of said repartimiento of Ananxauxa, given in encomienda to said Gómez de Cara-vantes," who was between eight and ten years old at the time of the events discussed (Canchaya et al. 1972:345).

Espinoza Soriano (1972:137–138) empha-sizes the theft committed by the soldiers of Al-varado and how the Indians provided them with many llamas, but even so the soldiers "stole" a great amount of livestock. On this there is the tes-timony of Francisco de Illescas, who declared that "this witness saw that sometimes some of the sol-diers of said Alonso de Alvarado went out of said valley in search of sheep [and brought back some cattle (*ganados*) with them]; of this case this wit-ness knows said soldiers could not refrain from bringing much livestock from said Indians of Xauxa . . ." (212). (The words in square brackets are my additions; note 212 reads: Cusichaca 1561, f. 72v. In the transcription it corresponds to Cu-sichaca et al. 1972:384–385).

I will not reproduce here the immense amount of data on the "rancheamiento" of ani-mals and the abuses committed by the Spaniards found in the above-cited document, which Es-pinoza Soriano (1972) discussed. The way An-dean camelids were decimated without any con-trol at all is shocking. On the other hand, the records of the Lima town council show that pro-visions favoring the natives were passed in late 1538; the treasurer Alonso Riquelme stated that he was informed that some Spaniards who came from Cuzco and others living in the vicinity of Lima were seizing and stealing much livestock from the Indians to sell in the city (Torres Sal-damando 1888:I, 250; see also Rostworowski 1981:51; this last text erroneously gives 1900 as the date, but the bibliography includes the cor-rect one, 1888).

A letter Pascual de Andagoya sent to the em-peror in 1539 lists the abuses committed by the Spaniards. We read: "[T]hus all soldiers and veci-nos brought all of the Indians clothes and food, and sold them in the market at such low prices that they gave a sheep for half a peso, and killed all they wanted just to make candles, and although this damage was light, more is to be expected in future [for] the Indians are left with nothing to sow and not having livestock nor ever had it, they cannot fail to die of hunger, because maize does not grow in that land [Cuzco] more than once a year" (Porras 1959:371).

The accountant Zárate (1968:176–177) also discusses the killing of the animals: "And all over the land there are these public slaughterhouses, because at first these were not needed, for each Spaniard had his own livestock, and on killing a sheep the neighbors sent to his home for whatev-er they needed, and so they sometimes got their provisions."

Cieza de León showed deep concern about this phenomenon and warned that "if the Span-iards had not hurried to diminish it with war, then no count or sum [would be possible] because of the many there were. But as I have said, the great pestilence of war, which the Spaniards had be-tween them, came upon the Indians and [the] live-stock" (Cieza de León 1984:Part I, Chap. CXI, 294). And when he describes "the Conchucos and the great Provinces of Guaylas, Tamara and Bom-bón," he clearly states the huge number of live-stock there was, but bitterly comments that "the wars however killed the great number of livestock there was, in such a way that of this crowd little remains, because if the natives do not keep them to make their clothes and wool, they will be in trouble" (Cieza de León 1984:Part I, Chap. LXXX, 234–235). He says almost the same when describ-ing the Chincha Valley, of which he said that there almost remained no "sheep of this land," "because the wars the Christians waged between them-selves put an end to the many they had" (Cieza de León 1984:Part I, Chap. LXXIII, 220). We should not forget that this happened toward the mid-six-teenth century.

Around the same time, Licenciate Santillán (1968:441) wrote that "[t]he Spaniards seized and destroyed all of the livestock which was presented to the Inca and the Sun when they entered the land, and afterward." Then he lists the tribute the Indians had to pay, and remarks how it was in some sense similar to what was given to the Inca. He ex-plains that in some places the tribute had to be ren-dered in cash as it could not be tendered in wool, "because the Spaniards have impaired the live-stock from which they took [their tribute] in the time of the Inca." Besides, his words show the im-pact of the new culture that was asserting itself. He actually notes that in some places the wool had to be given to the encomendero, but it was not from camelids, "because there already is a great number of sheep from Castile" (Santillán 1968:426).

The position of Polo de Ondegardo is very important because he was councilor to viceroys and governors, and "[n]ever was history his direct goal, he just resorted to it by accident, in support of administrative measures regarding the Indians" (Porras 1986:335). Polo discussed the abuses inflicted by the Spaniards, noting in this regard that

> in this assumption some also have tried to take their animals claiming that they belonged to the Sun or the Inca, and had their will before there was justice like there now is, taking a great number [of livestock]; and it certainly was very clear that when president Gasca rated or had the land rated in His Majesty's name, when he ordered that tribute be tendered with livestock, it was not his will that the Indians should give what they had and used as their own, but only what belonged to what they gave to the Inca and his religion. After I understood this point I strictly condemned some who almost had it as their trade, and in such standing had taken a great number from the Provinces of Ayncaraes and Chunvivilcas. This was then reported to the Audiencia, and then to the viceroys, and it has never again been allowed. (Polo de Ondegardo 1916:62)

In the late sixteenth century, Licenciate Matienzo (1967:90), a judge in the Audiencia of Charcas, described the camelids in general, but his comments concur with what has thus far been stated, except that he ascribes part of the responsibility to the Indians—and here he clearly was biased, because he does not say that the Spaniards also ate camelids. The chacos were already under Spanish control by the time he mentioned them. He states that "[a]ll of this livestock is most useful. There used to be an infinite number, both of the tame and wild ones, but the wars that took place in this kingdom between the Spaniards diminished them, and the greatest harm to the tame [camelids] was that the Indians eat them, [as well as] a disease called *caracha*, which is like the mange and which they catch, and has killed many. The

mountain animals have been destroyed in the *chacos* they hold, as they often kill five hundred and more in one *chaco*."

Garcilaso de la Vega (1966:Bk. 6, Chap. VI, 326; 1959:163) does not often mention the ravages wrought by the Spaniards. However, when describing the *chacu* he says there was a great number of vicuñas, immediately adding this: "That was in former times. Those in Peru can say how few have escaped the destruction and waste caused by the arquebus [crossbow], for guanacos and vicuñas are now hardly to be found except in places where firearms have not reached."

Koford (1957:214) claims that the vicuña was freely hunted during the 300 years of Colonial rule, and it was only in 1825 that Simón Bolívar prohibited their killing. This is not true. As we shall later on see, there were laws ("*ordenanzas*") for the protection of this animal; the fact that they were not enforced is another matter.

Espinoza Soriano (1976–1977:145) studied documents on Pariamarca, which "is next to the Cajamarca-Chilete highway, and beside Carambayoc," which was pillaged in 1544. The tribute-paying population of Pariamarca was subject to the Pomamarca huaranca. The aggrieved curacas could not complain because the pillage of men and livestock directed against Cuismancu had been ordered and carried out by higher-ranking curacas. A lawsuit was therefore filed in 1565, and Martín Muqui declared that "he understood that it had been done because they had many black sheep." According to Espinoza Soriano (1976–1977:n. 5, 145), the deposition of Cristóbal Calvajoque during the trial held in 1565 to clear up the pillage of Pariamarca records that "said Indians of Pariamarca had many sheep and they should go to take them and bring the [Indians] tied, because such rich Indians should not be in Pomamarca and pay tribute and serve Hernando de Alvarado. And so said Don Diego and Don Pedro went to said town of Pariamarca and brought many Indians tied up and took their sheep and brought them to this town of Caxamarca, and gave them to said Baltasar Colquicuzma so he could use them." From this Espinoza Soriano deduces that these Indians were herders who were strongly persecuted because they raised black llamas.

The lawsuit dragged on into 1566 in Caja-marca, and "instead of being scared, Don Antonio Condorpoma demanded that the curacas of Chu-cuimancu and Caxamarca be punished for the usurpation they made." He added that "they went to said town of Pariamarca and took them by force, inflicting much ill treatment on said Indi-ans. The worst thing was that not satisfied with the above, Don Baltasar Colquicuzma and said Indians [behaved] cruelly with the aforesaid [and] took two hundred and fifty black sheep from this town, which we estimate come, with the offspring they had in twenty years, to more than twenty thousand pesos" (Espinoza Soriano 1976–1977: 151). We should bear in mind that the color of the llamas is very important in magical and religious rites, black ones having a special role. In addition, black wool was much used in folk medicine (Es-pinoza Soriano 1976–1977:142–143).

The Huánuco *visita* likewise documents the destruction of camelids by the Spaniards. Ortiz de Zúñiga (1972:59–60) left a sad comment: "[A]fter the Spaniards entered this land they asked for more tribute than they used to give to the Inca and took it by force, whipping them and taking what they had with other ill treatment . . . and so deprived them of the livestock of the land."

The *visita* Diez de San Miguel (1964) made to the altiplano holds ample data on the tribute the Indians had to pay the friars in the area. The inspector's conclusion was that "in past times" each friar was given two rams of the land every month. Then the Indians gave them 200 sheep and 100 rams, but in this case it is not specified that these animals were from Castile, so they must be camelids (see below; Diez de San Miguel 1964: 228).

This same *visita* records the statement made by Don Francisco, cacique of the Anansaya moi-ety of Ilave, which is most interesting because it shows how the Indians managed to convince the friars to give them tribute in European livestock instead of using native animals. Here we read: "[B]oth moieties used to give each friar two big rams every month, and they agreed with them to give them Castilian livestock so that they could support themselves with the offspring, and so they gave them three hundred sheep from Castile and seventy rams" (Diez de San Miguel 1964:

108). Don García, cacique of the Urinsaya moi-ety, likewise declared that before they "took away" two rams of the land each month for every friar, but then they gave them 300 sheep and 70 rams of Castile so as to give them no more rams of the land (Diez de San Miguel 1964:112).

Something similar appears in the donation the Indians of Ilave made to the friars, because they were bound to give "four rams of Castile [to the monastery] every month." This is why they donated "seventy rams of Castile and two hun-dred breeding sheep [*ovejas de vientre*] so that they could support themselves with them and their offspring." However, the document likewise records that the Indians did not have these ani-mals and had to buy them to make the donation (Diez de San Miguel 1964:186–187, 196).

This same *visita* records that the friars penal-ized the Indians from Juli, and the punishment had to be paid in livestock. We read: "[I]t seems that some friars gave the Indians one thousand five hundred and ninety three pesos in rams as penalty." Then we read: "It should likewise be or-dained that said friars should not charge mone-tary fines on the Indians, because His Majesty has established that these should not be charged on laymen because they charge too many livestock and money as fine, as Your Worship can ascertain with the inspections made at Ilave, Juli and Po-mata" (Diez de San Miguel 1964:232, 235).

The witness for the headmen from the Urin-saya moiety of Chuquito made a similar state-ment: "[B]efore now they used to give two rams of the land every month to each friar, and to ex-empt themselves from this, both moieties of Anansaya and Lurinsaya gave one hundred and fifty sheep and a hundred rams of Castile at one go, so that they could maintain themselves with meat from the offspring." The witnesses for the headmen of Anansaya made the same statement, with just a few changes in the wording (Diez de San Miguel 1964:83, 87).

The headmen of the Anansaya moiety of Acora declared that they "used to give each friar a big ram of the land each month, and when they did not do so they gave two pacos which were worth as much as the ram, and it must be about five or six years ago that they gave them two hun-dred and seventy sheep of Castile and a hundred rams to maintain themselves with the increase, and from then on they have not given them rams of the land and pacos" (Diez de San Miguel 1964:90). A similar but vaguer statement appears in the statement made by the Indians of the Urinsaya moiety of this same place. They claimed that "before they used to give each friar two rams every month because they bought some sheep and rams of Castile and gave them to the friars." Later on they did not give them anything "because they [the friars] maintain themselves with the offspring of those they gave them of Castile, and they do not remember the [number of] sheep and rams they gave them" (Diez de San Miguel 1964:97).

Finally, the *visita* clearly states the inconvenience that arose when the corregidor appointed the lieutenant of Sama and Moquegua, because to have him happy and to preserve his position the latter gave the corregidor gifts that included "rams of their kind." The abuses the Spaniards living in Sama inflicted on the Indians are also described: "[T]hey take their livestock against their will to bring it loaded with fish and other coastal goods, and do not pay them their salary" (Diez de San Miguel 1964:249–250).

In the secret 1572 inspection of the province of Chuquito, made on orders of viceroy Toledo to look into the abuses committed by the clergymen, the Dominican friars also record several grievances of this kind. They include some connected with the livestock. Cuaderno II records that a friar sold community livestock. He took "one thousand two hundred heads of said livestock, of which he picked the best and sold them to a vicar in the city of La Paz." The document gives some details of the sale. It turns out that 1,384 animals were sold, including "sheep of the land," "paco sheep," "male pacos," and "lambs of said sheep" (Gutiérrez Flores and Ramírez Segarra 1970:19–21). The same Cuaderno includes an important number of animals the friars received or forced the Indians to give them. The Indians complained, stating that "for them it was grievous that said friar Agustín sold the livestock of the land, because [had it not been so] now they would have five or six thousand heads, with which they could now comfortably pay their tribute" (Gutiérrez Flores and Ramírez Segarra 1970:23). Further on, the "Indians and rams" that "Friar Agustín de Formicedo . . . took

without paying for them" from the valleys of Çama, Moquegua, and Yuta are recorded (Gutiérrez Flores and Ramírez Segarra 1970:25). Then it is recorded that in Camata, "Friar Domingo de Mesa took advantage of the service of many Indians, rams and other things and did not pay for them." This friar seized 50 rams and had them sold (Gutiérrez Flores and Ramírez Segarra 1970:27–28).

Wheeler et al. (1992:468) cite the data in the *visitas* to show the fall in the number of animals in the altiplano. However, they wrote thus: "Native llama and alpaca herds also virtually disappeared within little more than a century following the conquest of Cuzco in 1532 (Flores Ochoa 1977 and 1982 [1977b and 1982 in my bibliography])." This article makes several mistakes, which must be corrected. First, it says that the Spaniards entered the Inca capital city in 1532, and since the only source given is Flores Ochoa (1977[b], 1982), the reader might believe that it was the latter's mistake. This is not so. Flores Ochoa never said this, and the mistake is due to Wheeler et al. (1992). Francisco Pizarro and his men entered Cuzco in November 1533. The chroniclers do not agree on the day. According to Prescott (2000: 991; 1955:321), Markham (1941:83), and Porras (1992:3), it was 15 November, whereas for del Busto (1988:135) it was on the 14th. There is another inaccuracy in discussions of the secret *visita* of 1572: some attribute it to Gutiérrez Flores, but, as we have seen, it was carried out by Gutiérrez Flores and Ramírez Segarra. The first two to make this mistake were probably Murra (1970:54) and Flores Ochoa (1977:23), whose error was later inadvertently copied by several scholars. Wheeler et al. (1992) clearly took it from Flores Ochoa but do not cite him despite his inclusion in the bibliography. Also, although both this *visita* and that by Garci Diez de San Miguel are mentioned, they do not appear in Wheeler et al.'s bibliography. Furthermore, citing Flores Ochoa, Wheeler et al. say the herds "virtually disappeared." The truth is that Flores Ochoa did not say this. His (1977b:23) study mentioned "rapid destruction," and the 1982 (p. 72) study "quite diminished," which is something quite different.

Espinoza Soriano (1987–1989:274) studied the Colla kingdom north of Puno, and published the document titled "The Caciques and Indians from Millerea, on the [intention] of concentrating them in the town of Guancane. 1583," which includes the "Petition [made] by the Indians of Guancané (Guancané, 4 December 1583)," signed by Don Felipe Caquia and Don Pedro Hilapay, "caciques principales of the repartimiento of Guancané." They noted, among other points, a situation "which is against justice and a great harm to ourselves and our Indians, because we do not have enough lands to cultivate or graze our livestock due to the fact that all the Indians of this repartimiento are resettled in these towns."

Murra (1970:54) discussed and explained how all herds that had belonged to the Inca and the community were immediately considered to belong to the Spanish king (see also Pease 1991:86). This idea already existed before Toledo and took Phillip II to be the heir of the "Inca and the Sun." According to Murra, this excuse to seize Lupaqa resources had been used "before Gutiérrez's report" (the reference is wrong, because it was Gutiérrez Flores and Ramírez Segarra, as already noted) by Agustín de Formizedo, a Dominican friar who lived in Pomata (who has already been mentioned).

Murra wrote: "[Formicedo] had gathered the lords both of Alasaa and Maasaa[9] in his region and told them that 'it was risky to have said livestock which on knowing that it belonged to the Inca and the Sun, any judge would seize for His Majesty because they belonged to him and not to the Indians, because it was of the Sun and the Ynga . . .' (24)." The point is interesting but there are some inaccuracies in Murra's citation. Footnote 24 reads: Cuaderno II, ff. 5r 9r [sic]. The passage quoted appears on page 19 of the *visita* by Gutiérrez Flores and Ramírez Segarra (1970), and the correct folio is 4v/5r, where not a word is said of "the lords both of Alasaa and Maasaa"; instead we literally read "being in this parish and town," whereas the previous paragraph says "in Machaca . . . in said town."

Murra (1970:54) continues thus: "To prevent this disaster, Formicedo suggested that many head of prized livestock should be sent to Cuzco to be sold, buying instead inoffensive sheep 'from Castile' in Huánuco. When Don Felipe Ticona, 'warden and *quipocamayo* [record keeper]' of the

community's herds, opposed the sale and threatened to 'go and complain before the corregidor,' Formicedo 'imprisoned him in a hovel for three days and threatened to whip him . . . and furthermore threw many stones at him' (25)." Footnote 25 reads: Cuaderno II, f. 5r, 7r. However, the events were not exactly as Murra recounted them. It is true that "said Felipe Ticona, warden and quipocamayo of said livestock, told him that if he did so he would go and complain before the corregidor, and for this reason threw many stones at him, warning him that he would punish him if he complained" (Gutiérrez Flores and Ramírez Segarra 1970:6v/7r, 21), but this was the livestock of Hurinsaya. Meanwhile, when arguing over the livestock of Anansaya, "Don Felipe Ticona, headman of an ayllo, wanted to defend [the caciques] and was the one who most answered back [, so] he seized him and imprisoned him in a hovel for three days and threatened to whip him if he did not have the livestock brought to him" (Gutiérrez Flores and Ramírez Segarra 1970:4v/5r, 19), which is completely different.

Murra ends thus: "Although he collected such gossip that was unfavorable for the Dominic friars, the rationale of Gutiérrez Flores [and Ramírez Segarra] turned out to be that which had been forecasted by Formicedo: according to the inspector, "each year four thousand big rams and as many pacos, more or less according to the number of offspring," could be taken from the 60,000 community animals [see Gutiérrez Flores and Ramírez Segarra 1970:43] "to be taken to Potosí . . . at His Majesty's expense" (Murra 1970:54).

This state of affairs was discussed by Pease (1992:262). He explains that according to the sources, the camelids figure in a "list of rich Indians" shortly after the European invasion, whose status was established by the number of head they had. It is possible that this was due to the fact that the livestock listed as belonging to "the Inka or the Sun" in the reports the Spaniards received was taken to be something that could be distributed among the Spaniards. The data available correspond to Chuquito, but the same must have happened in other places more to the north, such as Cuzco and its environs. The Dominican friars seem to have understood the role of the "community" herds and pressed for its sale to help pay the tribute, but this was forbidden by the Spanish administration, which instead had the aforementioned census of "rich Indians" prepared.

To judge from the data from the Chuquito *visita*, it seems that the Indians did not understand these good intentions and opposed them, believing them to be a ruse, owing to the constant abuses committed by the Dominican friars.

Murra (1978:85) believes that the emphasis placed on the herds as a source of wealth dates to Colonial times. However, Dedenbach Salazar (1990) made one observation that I find most important, namely, that besides being a source of wealth, camelids still played a significant role in the native religious system—something the Christian parish priests realized. Thus, when discussing Dávila Brizeño's account (1586 [1965]) on Huarochirí, Dedenbach Salazar (1990:100) notes that "not only does he give some idea of the vast number of livestock dedicated to the religious cult, he also suggests for the Colonial period that the parish priests took away the animals from the Indian population because they regarded them as 'idolatrous' devices."

In general, scholars agree on the great decline in the number of camelids during the Spanish conquest and the viceroyalty, due to the reasons given above (e.g., Brack Egg 1987:62; Franklin 1982:468, though he only uses Flores Ochoa 1977 as source; however, it is true that page 236 of that study lists some more data on this subject, but it is not the most complete source). But no one, to the best of my knowledge, has made a specific study of this point.

7.3.1 The Hunt

The *chaco* (hunt) was certainly one of the most significant causes of the destruction of the native Andean fauna. This Indian custom persisted in Colonial times, although without the necessary order it had in Inca times, to the point that it was banned. The descriptions some eyewitnesses left of these hunts are excellent. I will cite two examples.

First there is the account by Pedro Pizarro (1978:Chap. 34, 243–244). It reads thus: "[A] hunt of wild livestock was held in Jauja, which the Indians call *chaco*."

A fence was made in Jauja before Don Francisco Pizarro, Don Diego de Almagro and their captains. It was made by Manco Inca, then lord of this kingdom, in a valley where he had many Indians close the fields and drive the livestock into the valley. These Indians walked toward each other driving the livestock to where they wanted it to go until they held each other's hands and formed a circle, thus gathering all of the wild livestock. This they call chaco. It was found that they had killed about eleven thousand calves [*reses*] of wild livestock—which are as I have already noted—so this number of livestock, deer, foxes, partridges, and mountain lion was found [to be inside]: the highest number was of the wild livestock of this land. I wanted to give an account of this because it is something remarkable.

The second account I chose comes from Zárate (1968:130), who wrote as follows:

The Indians make the hunt of these animals a great festival, which they called the chaco, and is as follows: four or five thousand Indians at least, as much as the land can stand, get together and stand apart from each other forming a circle, thus taking up two or three leagues of land [about 11–17 km if this is a common league ("*legua común*"), and about 8–12 km if a Castilian league]. They then come together to the music of some songs they have for this end, and draw closer until they join hands and even link their arms. Thus they gather much game of all kinds of animals, as in a corral, and there they take and kill what they please. The noise they [the men] make is so great that they not only scare the animals, but make many partridges and neblís and other birds fall stunned among them, who confused [as they are] with the number of people and the sound let themselves be taken by hand, and some with netting.

Interested readers should see Gutiérrez de Santa Clara [1963:Bk. III, Chap. LVII, 235–236], who besides giving an excellent account of the chaco also gives a detailed explanation of the bolas the Indians used in it.)

Acosta (1954:135) took note of some of the complaints made against the chacos: "Some complain that after the Spaniards arrived, there was far too little restraint with the chacos of vicuñas, and these are now less." When making a good account of the chaco which they saw in the "city of Huamanga and its jurisdiction," Rivera and Chávez y de Guevara (1974:169) noted that "this is forbidden to prevent said livestock from dying out." The chacos were prohibited beginning in September 1557, as we shall see later on.

Even so, this kind of hunt persisted, with all its tragic consequences. The account given by Juan and Ulloa in the early eighteenth century is interesting in this regard, not so much for their description of the chaco but rather for their critique and the dangers the hunt entailed. Juan and Ulloa specifically refer to a vicuña hunt:

> They catch them with lassos and kill them to get the skin, leaving the meat behind, which is quite good, because their only goal is to use the skin. This is a method that one cannot avoid censuring as an abuse, because an animal that does no harm has to be killed just to remove its wool. In this way these have diminished so much that now they are only found with great difficulty, and before long the species will have died out due to the extreme neglect we have in preserving that which we use. No one dared kill one of these animals when the Incas ruled Peru, and they held a rodeo every year to gather them and fleece them, and then left them to spread all over the fields, and in this way they always increased. This species of livestock would not be so impaired had this been done after the Spaniards entered these countries, but caring only for today's convenience they have never looked after their [the vicuñas'] subsistence, and so they have rapidly extinguished them, sometimes killing the animals with firearms in the hunt, others with cutting weapons once imprisoned, without realiz-

ing that the mistake will not be easily repaired once the species is lost. (Juan and Ulloa 1988:592–593)

7.3.2 Dogs and the Slaughter

It seems that the dog had a major role in the hunting of wild camelids. Koford (1957:212) drew attention to this point, and everything seems to indicate that it was a phenomenon that began "at an early date," even though his only source is Garcilaso (1964:383). However, Koford notes that firearms were also used.

It is true that Garcilaso de la Vega discussed this point, for he wrote that many Spaniards "would take with them a pair of hawks, retrievers, greyhounds, and their arquebus [crossbow]... and hunt, so that by the time they stopped for the night, they would have caught a dozen partridges, or a guanaco or vicuña or deer" (Garcilaso de la Vega 1966:Bk. 8, Chap. XVI, 516; 1959:147). However, we also know from Garcilaso himself that hunting with hounds must not have been easy, at least with the vicuña, because it "runs so fast that no greyhound can catch it, and it must be killed with arquebuses [crossbows]and by trapping, as in Inca times" (Garcilaso de la Vega 1966: Bk. 8, Chap. XVII, 517; 1959:152–153).

There are earlier sources, such as Ramírez, whose work dates to 1597. Ramírez says of the guanaco that "the Indians hunt it with ropes and hounds, which the dogs kill because they follow them for a long distance, not because of their swiftness." Ramírez then specifies that the vicuña is "faster than the res" (Ramírez 1936:18). This simply verifies what Garcilaso noted, in that hounds were used, even though they were not too effective. Father Cobo confirms this in his chapter on the vicuña. He wrote: "They are the lightest and swiftest animals known, and although hunted with hounds they are not caught in the first run; instead, the dogs observe where the vicuña is heading to and run to catch them, or run after them until they tire out, and it happens that the vicuña escapes when the hounds tire as much as the vicuña" (Cobo 1964a:Bk. 9, Chap. LVIII, 367).

However, the custom of having hunting hounds was not only widespread at the time; it was also taken up by the Indians. Lizárraga's account is invaluable in this regard. He wrote the following when describing the road to Potosí in the late sixteenth century: "[S]ome Indians and Spaniards take greyhounds with them in case any guanaco or vicuña appears, so they can hunt it" (Lizárraga 1968:Bk. I, Chap. C, 87).

Lizárraga said much the same when recounting the journey from Atacama to Copiapó. This is an important account because Lizárraga was in Chile (Hernández Sánchez-Barba 1968:X). The Dominican friar describes the "indios de Guía," and explains that these "take greyhounds with their slippers so that their feet are not hurt. They hunt deer and guanaco with them, and the dogs are so skilled in this that if they sight one they will surely catch it. They feed from this meat, which is good" (Lizárraga 1968:Bk. I, Chap. LXVIII, 50).

When describing the Chilean guanaco in the seventeenth century, Molina left a vivid account that confirms the use of hounds by the Indians, a passage that once again confirms the veracity of what Garcilaso de la Vega wrote (Garcilaso de la Vega 1966:Bk. 8, Chap. XVI, 516). Molina wrote thus: "The natives hunt them with dogs, though they usually do not catch any save for the youngest, which fall behind because they do not have legs strong enough to flee. The big ones begin to gallop, or trot away so swiftly that a horse running free-reined would be unable to catch up with them. They stop every now and then to look at the hunters pursuing them, and after giving a big neigh much like that of the horse, soon vanish at an amazing speed. Even so the Indians riding swift horses manage to take them alive, throwing from afar a lasso to the legs." Molina then vividly describes how the Indians use this "lasso," which actually is a bola, explaining that it can be used to both kill the animal or take it alive (Molina 1782:Bk. IV, 319–320).

We should bear in mind that although the dog was not always the direct cause of the killing of the guanaco and vicuña, it nonetheless was an excellent help in the hunt despite its small height, as was noted by Matos and Ravines (1980:202). These scholars have also noted (1980:202) that dogs can run faster than camelids under certain conditions, and divert them from their route. In this regard it is worth recalling that the Tehuelche in Patagonia used the dog a lot when hunting (Cooper 1946a:

143). Koford (1957:206) believes that dogs are at present the major carnivorous enemy of the vicuña. He has collected accounts that note that using dogs to hunt young vicuñas is a common practice, but he admits never having seen this.

Curiously enough, in 1567 Licenciate Matienzo (1967:91) made several proposals to protect the camelids, yet proposed that the Spaniards be allowed to hunt the guanaco and vicuña with dogs, crossbows, and arquebus (see below).

Dogs sometimes pester the herds of llamas and alpacas, but these defend themselves. The account Bayer made in 1751 is interesting in this regard: "They defend themselves from the dogs with their forelegs, because they hit with them with such force that on receiving a blow even the fiercest [hounds] howl, lose their bravery and run" (Bayer 1969:38).

7.4 LEGISLATION

Several legal aspects regarding livestock are significant. Since this subject is beyond my training, I will just include some interesting data I found.

As was already noted, in Colonial times the payment of taxes with camelid products, the vicuña in particular, was accepted (Brack Egg 1987:67).

When discussing this point, Pease (1992:262) notes that the livestock was part of the tax, much like any other good used to pay taxes to the Spanish crown. Its price reached even higher levels after the discovery of Potosí, particularly around this mining center, because of the llama's importance in transporting the mineral. This gave rise to much fraud when the livestock population fell because of the late sixteenth-century epidemics that ravaged it.

Cases like that of the Collagua were frequent. Here, Pease says, the Colonial state exempted the tributaries from paying in livestock because of the decreasing number of animals due to epidemics. However, the tax was always paid in the way stipulated by the corregidor, who deposited the corresponding funds in the royal treasury of Arequipa, but later on used the livestock thus collected to take various products to Potosí, then selling the animals in the abovementioned town at a price several times higher than that ascribed to it in the tribute rolls. "Although the studies made of the significance livestock had in Andean life during the Colonial period have established some important points, particularly its presence and continuity, more research is needed on the way Andean peoples used the auquenids, and whether their traditional criteria changed outside the commerce-related sphere the Spaniards established in the Andes" (Pease 1992:262).

Friar Vicente de Valverde discussed the tithe, the right the Church had to receive 10% of all agricultural produce, in a letter he sent to the emperor from Cuzco on 20 March 1539. Valverde listed the "goods of this land" that could be tithed. These include: "Of animals lambs from the sheep found here," and "[w]ool from the sheep from here, milk and cheese selling it" (Porras 1959:314; for a later edition see Valverde 1969b). It is striking that Valverde lists cheese and milk, camelid products that were traditionally not used. Although it is not stated, the reference probably refers to the milk and cheese from European animals.

It is a matter of record that in 1557 the king gave a royal decree ("*real ordenanza*") forbidding the hunting of *ganado mayor*—camelids, in this case—so as to preserve them. In 1768 the Marquis of Rocafuerte sent the viceroy Manuel de Amat y Juniet a petition made in Yucay, requesting that vicuñas be only sheared but not killed, "so that they do not become extinct" (Brack Egg 1987:62).

In 1561 Polo de Ondegardo (1940:191–192) discussed the cultural differences between Indians and Spaniards cleverly, almost like an anthropologist. Polo de Ondegardo said that "it would be like curing all diseases with one same medicine, [which] would be a greater mistake than leaving them as they now are. I believe that for now the tithe on the livestock of the land should not be discussed, and they should be exempted [from it] for several reasons, particularly because they would have the opportunity to hide and conceal it, and would not use the livestock as would be reasonable. This would make them go looking for it, and since it has remote areas and punas it would be too harsh on them, and with time everything is remedied."

We know from the account given by Licentiate Montesinos, that

[a]round this time (i.e., 1556) there was a great scarcity of meat, and it was so that in Guamanga, a meeting [held by the municipality] on 10 September of this year prohibited killing vicuñas and deer, with the argument that there were many natives who had almost done away with the wild livestock. Due to this, another meeting held [by the municipality] on 22 September prepared regulations for butchering that had been written in Guamanga in the year 1553, but no regulations were then made, save that the arrelde [four pounds] of a cow was set at two tomines and a half, that of male pig at a tomín and a half, and Juan Barbudo bound himself to supply the city at these rates. (Montesinos 1906:243)

It was probably because of events like the preceding one that in 1556, Hurtado de Mendoza Avendaño, Marquis of Cañete ([1557] 1975: 65–66), issued a "Decree from the supreme government . . . banning chacos of guanacos and vicuñas for five years in" the repartimientos of Andamarca, Yauyos, Çaxamalca, Sacari, Parinacochas, the Guancas, Lucanas, the inhabitants of the city of Los Reyes, "and any other individuals to whom it applies and should apply." The decree said as follows:

[T]hou knowest that a report was given to me of how there has been and is a considerable diminution in [the number] of wild livestock of the land, which are sheep, rams, guanacos and vicuñas who lived in peopled and inhospitable areas, due to the disorder in hunting and the chacos, which were and are held every day, and to their death [so as] to use their wool and meat. From this follows much harm to the natives because in past times they used the wool to dress and to trade thanks to the care had in this and their reproduction, [but] with the disorder brought about by the Spaniards and other people the natives have also taken advantage of part of it, and did and do the same in the areas and places where [the animals] are to be hunted. The natives, both caciques and Indians, had and have many arguments which arise every day, for which reason it is most important that these hunts and chacos and the killing of said livestock ceases, and good care is had so that it multiplies and the natives in time use what they used to and it does not die out. Therefore, all this being seen by me and understanding the matter, and what should be done to prevent these inconveniences and remedy future ones, I decree and command that for a period of five years no person of any quality and condition whatsoever can in any way hunt and hold a chaco of any wild livestock, that is guanaco or vicuña, sheep or rams raised in desert lands, to kill and shear the wool.

(Lohmann Villena also discussed this decree [Matienzo 1967:90, n. 1]) and noted that it had been published in the *Revista del Archivo Histórico del Cuzco*, Vol. IV:61–63, Cuzco 1953. Dedenbach Salazar [1990:91] likewise discussed this document.)

However, this provision was still not enforced in the late sixteenth century. For this reason, in 1567 Licenciate Matienzo (1967:90–91) wrote as follows: "[S]aid chacos have been forbidden by decrees [given by] this Audiencia, but these are not enforced as they should be. To preserve this useful and essential livestock, His Majesty commanded in a decree sent to the Audiencias in this kingdom that they send him their opinion of what should be done to preserve this livestock, and in the meantime we should do as we see fit. This has not been done, though there was a great need for it to be done."

Matienzo (1967:90–91) then made an important proposal:

This could be solved with the following laws:

I   That all towns in the highlands have community livestock where there is food to pasture it. The Spanish corregidor of the town should keep an account of it in a book, where the number of animals and their offspring should be recorded, and how many are sold or die, to render accounts whenever they are so required to do so.

II Likewise, that said corregidor and any other judges should not let the cacique nor any other Indian kill a ram, lamb, or sheep of the land to eat, unless it is exhausted, has caracha, or is so old that it is useless as a pack animal, under penalty of a hundred lashes, and the meat thus killed will be sold half for the Indian community, the other for the accuser and the judge.

III Likewise, in Potosí and Porco the justicias ordinarias must have great care in visiting the ranches of the Indians to see if they have any dead sheep of the land or rams, save when it has the mange, and must sentence them with said penalty. The rams of the land must not be forbidden to be killed in the towns, because this would go against the greatest business there is, where the Spaniards sell the Indians so many rams of those which are exhausted from carrying things. These business deals give six thousand pesos and more each year, which all are taken from the Indians who purchase these rams, some to carry food and other items, and others to eat, and all the money taken from these rams comes in fact from the coca, because the Spaniards acquired the rams from the Indians in the lowlands for coca, and the Indians from Potosí and Porco buy them again and pay the Spaniards for them.

IV Item, that town corregidores and tucuiricos must be particularly careful with the livestock of the land, both that belonging to the Indian community or to any individual, and must be counted and registered once or twice a year. Those who report animals that are old, tired, or too sick with the mange should be given permission to kill, sell, eat, or make charqui or dried meat with them.

V Item, neither the caciques nor the Indians can hold a chaco, under penalty of two hundred lashes and having their hair shorn for any cacique or headman on whose orders one is held, or lets them take place, and the person for whom said chaco is held will pay a fine of two hundred pesos, half for the community and half for the Chamber and the hospital for Indians, and if he were a judge he will be fined with twice the number of years he has lived there. If he is a churchman, his prelate will be informed so he punishes him. However, chacos are allowed to take the guanacos and vicuñas and shear them to use the wool, letting them loose after being shorn save for those needed by the Indians to eat that same day the chaco is held, which they can hold with permission from the judge and in no other way.

VI Item, that they can hunt said guanacos and vicuñas with hounds, and the Spaniards with said hounds, crossbows and harquebus.

Matienzo also discussed the distribution of the *tambos* (inns for lodging) throughout the highlands, and he made another proposal besides the order that roads be built. Each tambo should have 60 or 100 rams of the land belonging to the Indians who served in the tambo, because each animal could carry the same as two Indians and walk at a fast pace, which would be most useful. The Indians would be paid for each ram what an Indian was paid (i.e., a tomín), the advantage being that one Indian could lead six rams. Half of the proceeds would be for the Indian and half to care for the livestock (Zavala 1978b:54).

It is worth recalling that some years later, on 6 November 1575, viceroy Toledo issued some decrees. No. 28 read thus: "[M]ake sure some tambos suitable for traveling in the deserted lands of the sierra are in good condition, and have rams in them to carry loads, and Indians who go with them," thus partially fulfilling what Matienzo had requested (Toledo 1978:144–145).

There is no denying the excellence of these proposals, which confirm the dramatic situation of the camelids at that time, because otherwise there would have been no response like this one. But Matienzo was not alone. Falcón, another licentiate, also raised the alert. Falcón was a Spanish jurist who lived in Lima in the late sixteenth century and defended the Indians. He was proxy for several Indian communities. It is not known when he arrived in Peru. In his "Account . . . of the damages and harm wrought against the Indians," Falcón (1918:141) wrote thus:

> [T]he Spaniards cannot be given the waters and pastures the Indians have, and have these in common [share the same resources] with them. . . . For which reason and according to the laws of Castile, the Spaniards cannot graze in the jurisdiction of an Indian town, and the sentence H.M. gave that pastures should be held in common must be understood to hold for lands which the Spaniards lawfully have . . . not to those belonging to Indians or in the jurisdiction of their towns, because it would otherwise be unfair. This is not to be believed was H.M.'s intention, particularly since the Indians have the pastures divided between themselves and one could not graze in somebody else's land, and thus it was commanded by the Inca, so with his authority everything was made a *dehesa* [pasture ground].

Falcón (1918:163) then adds: "All usually pay excessive tribute and far more than what they can pay or their estate is worth. . . . And they never consented, explicitly or tacitly to said tributes, nor was it asked of them or discussed with them, except that at first the encomenderos took everything they could squeeze from them, which was much at the time because they had stayed with the estate, gold, silver, and livestock of the Inca."

A letter the Marquis of Cañete sent to the King from Los Reyes on 17 January 1593 notes that the royal order forbidding the killing of European and native livestock for twelve years was untimely, because the former had multiplied so much that they were wreaking havoc in the fields of the Indians. Strangely enough, the "female camelids," which were apparently not involved, did participate. The letter reads thus: "The cattle and the livestock of the land increase every day, so there is much trouble in making sure they do not harm the native's chácaras, and it is necessary that some heads of cattle be killed and weighed every now and then at the butcher's. Because of this and since a law [*ordenanza*] was already made, we find untimely the royal decree [*cédula real*] of 22 August 1592 establishing that the females of the livestock of the land, or at least the females of cattle, not be killed for twelve years" (Hurtado de Mendoza 1978:183–184).

The damage wrought by the European livestock is likewise clear in a complaint made by Don Felipe Guacrapáucar: "[T]hey do not harvest all that they sow and must buy the food they are forced to pay by the tribute roll, because their encomendero has many *ganados mayores y menores* in said valley among their fields, from which they receive much harm." Due to this complaint, Phillip II issued a "Royal decree for Don Felipe Guacrapáucar over the livestock in the Jauja valley" dated "Monserrat, on 31 January 1564. I the King." It reads thus: ". . . because it is not fair that since they pay him tribute, he eats their bread and pastures with said livestock" (Felipe II 1977:393).

7.5 RITUALS AND SACRIFICES

Despite all the harm inflicted on the native animals, the Indians never stopped holding their rituals and sacrifices, something that is clearly reflected in historical sources. Thus, Cieza de León (1984:Part I, Chap. C, 275), for instance, gave a good account of how the dead were buried in Collao "with some sheep." In 1790 Oricaín (1906: Discurzo V, 332–333) described the rituals the Indians of the bishopric of Cuzco "held in the ranches of livestock of the land." This is a very interesting document with a high anthropological value that is worth citing. Besides, Oricaín exhibits the European lack of understanding of the native culture. He says thus:

> [They] pasture them separately by gender on different hills, and on a given day which is once a year, [they] go to the most isolated areas or most rugged mountain summits, where they have three successive corrals. They take abundant brews [beverages] of the land, brandy, and other sup-

plies, and after a thousand superstitious ceremonies kneeling, kissing the air, opening and closing their arms, they take out a figure of a ram or llama, which is how it is called in their language, made of stone or copper. They place it on a high place and offer it coca leaves, grains of maize of all colors and much chicha, begging it to help those animals breed and stimulate their fertility. Then the rams [males] enter the corral at one end and the females at the other, and in the center of the corral in between they bury coca, maize, and spill chicha. Then they introduce ten or twelve females in it and as many males. The first ones squat and the latter squat over them, which the Indians help with their hand while the Indian women sing songs from their gentility [traditional past] and play their drums. Once the rams [males] have finished mating and are separated as they were, that idol which they call *illa* comes forth again. They repeat their entreaties and pleas and then begin drinking the brews mixed with [camelid] dung, this being an unbreakable commandment, as is spilling part of it on the floor so that the earth can taste of it first, and they get drunk all night long, from which follow much incest, adultery and rape, the necessary outcome of such disorderly conduct. I ask: if the intervention of man is essential for the reproduction of this caste, then how come the guanaco and its kin, the vicuña, give birth every year without the help of a third party? And how [did they do so] from the creation of the world til the time when the Indians domesticated them? This ceremony then is nothing but a signal of an idolatrous gentility [tradition] which they still keep.

Francisco de Barrionuevo reported to the Council of the Indies from Panama on 15 July 1535, on his return from his voyage to the New Castile, that while in Tumbes he helped "a headman" who was dying, and cured him. On being healed, the headman "wanted to sacrifice four sheep," but then settled for just two (Porras 1959:

162; Barrionuevo 1988:4). The Augustinian friars recorded the custom of making llama sacrifices in the province of Huamachuco around 1560 (Dedenbach Salazar 1990:94). In the early seventeenth century the Augustinian friar Calancha (1976:Vol. III, Bk. II, Chap. XIX, 939) noted in his description of Pachacamac that "here the sacrifice of sheep of the land was begun." Meanwhile, late sixteenth and early seventeenth-century Christian religious sources include several references to llama sacrifices, particularly in southern Peru (Dedenbach Salazar 1990:94).

When discussing the Andean organization of Recuay in the seventeenth century, Mansferrer Kan (1984:53, 60) notes there is evidence that in 1622, "the huacas still had their livestock, and this custom did not entirely die out," because the amount of livestock given to some huacas was cut down. For instance, reference is made to "150 rams to huaca Carachuco, and to Huaca Carhua two hundred head no more." A thousand head are mentioned in the inspection Santo Toribio made in 1593, which fifty years later had grown to 6,783, and fallen to 4,749 in 1774. No reference is made as to what kind of "*carneros*" these were. They could be camelids, but it should not be forgotten that a 1643 lawsuit held in Guaraz mentions "ovine livestock," so there is some doubt.

Finally, the fact that strong acculturation took place in this sphere should not be ignored. The term *herranza* (to brand and/or shoe animals) used for some domestic camelid festivities is Spanish, and has no meaning, strictly speaking. As well, Catholicism and the participation of European animals were introduced into the traditional ritual (see Flannery et al. 1989:144). We should now recall the 1741 document mentioned in Chapter 6, which records that in the highlands of Lima (in San Francisco de Guisa), people used to sacrifice a llama and spray the church's walls with its blood so as to strengthen it (Frank Salomon, pers. commun. letter of 12 December 1990).

7.6. THE IMPACT OF ALTITUDE ON CAMELID FERTILITY

When discussing the biological aspects of camelids, we saw that these animals have trouble breeding at high altitudes where they now live. Some relevant historical data were found that are worth presenting.

An inquest held in the *visita* of Chuquito was carried out "at the behest of the caciques of Chuquito." This petition includes two questions that bear on the subject under discussion. I will discuss only one of them, number X, because it concerns native livestock, while the other question asks the same but in regard to European livestock. Question X is

> whether they know etc. that the livestock of this province breeds and multiplies little and with much effort and labor, both because an individual cannot keep and advantageously manage more than thirty or at most forty [animals], that their management is troublesome, both because of the sterility of the pastures and the lack of water in this province. (Diez de San Miguel 1964:148)

The answers given to question X were as follows. In the "Probanza de los indios," Martín de Leguiña says "that he has seen and knows that the livestock of this land multiplies little because the females give birth from year to year and they die a lot, and are raised and maintained with much work" (Diez de San Miguel 1964:151). The witness friar Agustín de Formicedo in turn said "that he sees that the livestock of the land is raised and multiplies very well . . . and that much livestock dies from a disease called *carache* [the mange]" (Diez de San Miguel 1964:154). Friar Tomás de Castillo, another witness, answered "that he had heard say that said livestock multiplies little" (Diez de San Miguel 1964:157), while still another witness, friar Domingo de Loyola, noted "that he knows—and so it is—that much work and care is had for the benefit of said livestock, because when they give birth the herders are forced to feed the offspring and look after it, and also in making [the animals] pregnant, and other things that are useful for their benefit, in which they spend much time and they multiply little because they give birth from year to year and not all do so" (Diez de San Miguel 1964:160–161). Melchior de Alarcón said "he has always understood that said livestock breeds and multiplies poorly and with much work" (Diez de San Miguel 1964: 163).

The answers are extremely interesting because all, save friar Formicedo, concur in noting problems in breeding. Two other depositions agree on this. First there is Martín Cusi, cacique principal of the Urinsaya moiety, who, when "[a]sked how much livestock the four hundred and fifty head held in common by this town of his moiety will give each year, [he] said that they will grow by a hundred and twenty or a hundred and thirty head a year because some miscarry and others die from carache [the mange], while others do not give birth" (Diez de San Miguel 1964:30–31). Luis Cutipa, a principal in the Lurinsaya moiety, was "[a]sked how many head the five hundred female head he says they have in common will give in one year, he said that at most each year they grow by a hundred and sixty head, and at least by seventy or eighty, because the rest die and miscarry" (Diez de San Miguel 1964:79).

The figures are significant in this case, because according to the first witness, only 27%–29% of the community's animals would bear offspring. According to the second witness, at most 32% would give birth, with a minimum of 14%–16%. The figures are extremely low.

7.7. CAMELID DISEASES ACCORDING TO HISTORICAL DATA

The disease most often noted was caracha or mange. Not many data were found on this point, but those few are interesting.

Once again we turn to the Chuquito *visita*. Question 9 of the aforementioned inquest reads as follows: "Item: do they know, etc., that a great number of livestock dies in several years—in most of them—, both due to drought and lack of pastures, as to the sterility of this province" (Diez de San Miguel 1964:148). The "Probanza de los indios" has the statements made by Martín de Leguiña, and includes as witnesses friars Agustín de Formicedo, Tomás del Castillo, Domingo de Loyola, and Melchior de Alarcón. The answers are interesting. The first "said that a great many livestock dies from a mange it catches called carache," while others answered "that he knows

and has seen that in some years much livestock dies from the snow"; "he has heard say that many animals die when it snows a lot," "that he knows that some years are so bereft of water that the livestock suffers a lot," and "he knows that a great number of livestock die every year both from lack of water and of pastures" (Diez de San Miguel 1964:151, 154, 157, 160, 163). We find that Indians did not answer the question, and they apparently did not ascribe as much importance to natural phenomena as to the mange. This is a clear indication that Indians were so used to natural phenomena, which they had lived with since they were born, that they did not pay much attention to them, whereas the variety of mange imported by the Europeans must have decimated the livestock in a way the native Americans had never seen, and this clearly impressed them.

This same inquest asked the following: "XII. Item do they know etc., that the livestock of this province, which is its major estate, has a disease which is carache, which it is very difficult to cure. Many animals die from it, and most of said livestock are sick with said disease" (Diez de San Miguel 1964:148). The aforementioned witnesses answered the following in the same order (see above): "that said disease of the livestock is called carache and is very hard to cure . . . and a big part of the livestock catches this"; "that curing it of carache which is a disease of which many die"; "that he knows and sees that the livestock of this province gets a disease that gives them carache, and much is spent healing it . . . and that much livestock gets sick and dies with said disease"; "it is a most public and evident matter that the livestock of this land gets a disease that they call carache, and many [animals] die and it is very hard to cure." Finally, the last witness declared that "said disease of carache is [found] wherever there is livestock of the land, which is a mange [*sarna*] that forms a big scabie, which if it is not cured when the scab begins to peel off is then very difficult to heal, and the ram or sheep can be in such a condition when he is healed that nothing can be done, and said disease diminishes these animals; if care is not taken, most die with said carache, and it is difficult to cure" (Diez de San Miguel 1964:152, 154, 157, 161, 164).

Garcilaso de la Vega (1966:Bk. 8, Chap. XVI, 513–514; 1959:148) left a description of this disease, which shows it in all of its awful truth. It reads thus:

In the time of the Viceroy Blasco Núñez Vela, in 1544 and 1545, there appeared various plagues in Peru including what is known to the Indians as carache or llama-mange. It was a very dire disease, hitherto unknown. It afflicted the flank and belly and then spread over the whole body, producing scabs two or three fingers high. Especially, on the belly, which was the part most seriously affected and which came out in cracks two or three fingers deep, such being the depth of the scabs down in the flesh. Blood and matter issued from the sores, and in a few days the animal withered and was consumed. The disease was highly contagious, and to the horror of both Indians and Spaniards it accounted for two-thirds of all the animals, both paco and guanaco. From them it was transmitted to the wild varieties, the guanaco and vicuña, but it was not so severe for them because of the colder zone they frequent and because they do not gather in such numbers as the tame animals.

NOTES

1   *Fanega*, a grain measure equal to 6439.48 m².

2   I do not agree with Dedenbach Salazar in the conversion of distances, because if we use common leagues, then a distance of 2 leagues comes to about 11 km, 3 leagues to about 16 km, and 4 leagues to about 22 km. However, using Castilian leagues, a distance of 2 leagues comes to about 8 km, 3 leagues to about 12 km, and 4 leagues to about 16 km.

3   To avoid confusion, I must point out that Sumar (1992: 83) cites this paragraph from Cieza de León literally and ascribes it to Zárate.

4   In the 1959 edition the description of the huanacu has been left out.

5   Sumar (1992:83) wrote: "[I]t is for the occupation of Xauxa [Jauja] by the Europeans that the chronicler Pedro Sancho says that in its hasty retreat, the army of Quizquiz left behind 15,000 camelids and 4,000 prisoners." First of all, the event narrated by Sumar took place in the far north, in "a Province called Chaparra," and not in Jauja. Second, the chronicler who narrated this event was Agustín de Zárate, not Pedro Sancho. The citation by Sumar is exactly the same as the one here reproduced (Zárate1968:Bk. II, Chap. XII, 171).

6   "*Ranchear*" in the original Spanish text. This word must be explained. It is best to turn to someone who was well acquainted with it. When he mentions Diego de Almagro and the "men he brought from Guatemala," Pedro Pizarro (1978:Chap. 19, 121) wrote: "These were the first inventors of *ranchear* (which in our common parlance is to steal), for among those who came to the conquest with the Marquis, not one man dared take a corn cob without permission."

7   Now the Archivo Regional of Arequipa.

8   In Romero in this passage we read "*vendino nos ha*," which does not make sense. It probably is a misprint.

9   The spelling used by Murra is Aymara, as it appears in Bertonio (1956:Part II, fol. 9a).

CHAPTER 8

DATA FROM THE NINETEENTH
AND TWENTIETH CENTURIES

I do not intend to make as exhaustive a study of these centuries as I did for pre-Columbian and Colonial times, because this is a vast field that requires a study unto itself. Here I include only some data that are useful in one way or another to supplement the information presented throughout this book.

It is interesting that large herds of llamas still came to the coast in the early nineteenth century. Tschudi (1966:243, 244), for example, was in Peru in 1838–1842 and wrote: "The Indians often come down with big herds of llamas in search of salt." Tschudi emphasized that "[d]espite all the care and precautions taken, a great number of llamas die in each journey to the coast, and particularly to the jungle, because they cannot tolerate the warm climate." This is partly true, but it includes the assumption that camelids are high-altitude animals. This bias is evident in Forbes, who in 1870 said of the altiplano Indians who came down to the coast with their caravans of llamas, that "the Aymará Indians find themselves completely outside their natural environment in climates that are completely inadequate for their constitution, just like their llamas and alpacas, and die in great numbers if they do not return promptly" (Forbes 1870:219–220; cited and translated by Larrain Barros 1974:134 [retranslated into English.—Trans.]).

There is some scattered information on these animals' distribution that can supplement the data

presented in other chapters. Thus, Tschudi (1885: 94; 1891:97; 1918:206; 1969:125), after citing some chroniclers and travelers (specifically Diego de Ordaz, Orellana, Zárate, and von Hutten) who indicated camelids were present north of today's Peruvian border, said, "Even more interesting and intriguing is the disappearance of llamas from their northern frontier. Although we do not have any positive data on how far to the north they extended, we do find some notes by a few scholars." Interestingly, a man like Tschudi, who made a significant study of the llama for his time, merely raised the question and did not try to answer it.

On the other hand, it is worth recalling the testimony of Raimondi (1943:147), who verified in the late nineteenth century that there still were vicuñas in the Callejón de Huaylas. This caught his attention, for he wrote: "In the heights of Recuay one notices some vicuñas, which are almost unknown in northern Peru." When crossing the heights of the Department of Lima between Caruapampa and San Lorenzo, in the vicinity of Huarochirí, at slightly above 3900 masl, Raimondi noted there were "some llamas" (Raimondi 1945:86). In other words, there no longer were many of these animals.

Another invaluable datum left by Raimondi (1948:34) concerns the Hacienda Chocavento, in the Acarí ravine (province of Caravelí, Department of Arequipa). Raimondi wrote: "The huanaco is quite common in these hills; in the few hours the excursion lasted we managed to see 25 animals in small troops, some of them so tame they let us approach up to a distance of 30 paces." According to Stiglich (1922:390–391), the hacienda is at an altitude of 345 masl. Raimondi described the guanacos when climbing to the mine of San Pedro de Pampa Colorada. He gave no altitude but reached the mine in four hours after covering a distance of 22.5 km, so it cannot have been much higher.

There no longer are any camelids in the northern Peruvian highlands. However, Kosok (1965:131) managed to get some information from old Cajamarquinos, who said that llamas had been seen in the locality up to a century ago. In the 1930s or 1940s these herds had already vanished, and the few left in the haciendas had been brought from Cuzco (Fernando Silva Santisteban, pers. commun., 22 October 1992).

However, slightly farther to the south, in the highlands of Trujillo, Cristóbal Campana recalls there were many llamas when he was a child, around 1936–1938. These were large herds kept by the Northern Smelting Co. to carry the mineral (Cristóbal Campana, pers. commun., 19 August 1992). Around that time Gustavo Padrós was an "arriero" who went from the sierra of La Libertad to the coast carrying minerals and other products with a herd of about 100 llamas. He used the Virú Valley as his route down, and then went along the coast to the port of Salaverry (Cristóbal Campana, pers. commun., 29 April 1994). Today no camelids are left on this sierra.

Masuda (1986:253–254) in turn collected invaluable information for the southern coast that goes back for just a short span, to 1956, "up to thirty years ago. According to the old men of Chala and Atiquipa, all of the land was filled with livestock in loma season, and even highland herders came with their cattle and llamas." Additionally, a witness who was 37 years old in 1981, recalled "that he traveled with the drove of llamas taken to the coast by his father. . . . They took fifteen or twenty llamas on each trip." Leaving Pisquicocha, they went as far as Chala in eight days. The witness recalled that the first task on arriving was "building a hut and leaving the llamas in the pasture." Another informant noted that they left their town, at 4000 masl, "[i]n the win-

ter months . . . with a drove of 8 to 10 llamas headed for the coast . . . they arrived at Chala." The trip took 10 days and was made to collect *cochayuyo*.[1] The informant recalled that "a pack llamo [sic] carried 2–3 arrobas" (i.e., 23–34.5 kg).

There is also some information, if brief, provided by Steward, who discussed the tribes in the montaña and said: "The only domesticated animals were the llamas and alpacas on the upper Marañón River, llamas and guinea pigs kept by the Jívaro" (Steward 1948:519). Later on Steward added, "The aboriginal Jívaro were unusual among Montaña tribes in keeping a few llama and guinea pigs" (Steward 1948:621). Steward unfortunately does not specify which Jívaro community had this custom or whether all of them had it, for the Jívaro are widely distributed. The Antipa live from the Santiago and the Zamora Rivers to the Upper Marañón, while the Aguaruna occupy the right bank of the Marañón, between the Nieves and Apaga Rivers. The Huambiza in turn live on the right bank of the Morona and Mangosia Rivers and on the left bank of the Santiago River, from the Cuticu cordillera to the Marañón River, while the Achuale inhabit the area between the Pastaza and Morona Rivers. In addition, there "probably" are Jívaro groups in the heights of Loja (Palta and Malacata). In general terms, the Jívaro occupy part of the eastern lands of northern Peru and southern Ecuador (see Steward 1948:618).

I also managed to find some reports on llama caravans, which obviously are very interesting. Now, I have no intention of presenting data on those areas where the caravans are still in use,[2] such as Junín, for example (see Hurtado de Mendoza 1987:201), or Cuzco (see Custred 1974) and the altiplano (see Koford 1957:212; West 1981a), but instead to note some data on areas where these have vanished, or are about to vanish.

In the trip he made throughout Peru in 1863–1877, Squier (1877:245; 1974:132) described his descent from the Andes to the altiplano and discussed a site called La Portada at 3800 masl. Squier wrote: "The merchants of Tacna have built here a rude enclosure for the droves of llamas that come from the interior with products for the coast."

Cristóbal Campana (pers. commun., 19 August 1992) provided me with data on the life of

Don Víctor Huamanchumo, who still lives in Chimos Beach close to Samanco, south of Chimbote (district of Samanco, province of Santa, Department of Ancash), where he became mayor. Huamanchumo's grandfather, who lived there in the 1920s, told him that "they divided the sea among themselves from Huanchaco onward. Then he came with his grandmother, who was pregnant, and found a place to the south, because in the north everything was full." He returned to Huanchaco to request his degree. Among other things, Huamanchumo said they found a salt pan and a "salitronera." Campana explained that a salitronera is a reddish salt that is not harmful and can salt the fish without hardening it, leaving the fish in a flexible state. The interesting thing is that his grandfather left for Recuay with herds of llama and took potatoes, chile pepper (*ají*), and ollucos as far as the Nepeña Valley, returning to the highlands with salted fish.

Cristóbal Campana (pers. commun., 19 August 1992) himself told me that in 1936–1938, his father, Manuel María Campana Castillo, worked with Miguel Deza. He traveled from Chacra Camigorco, at 3300–3400 masl in Santiago de Chuco (in the province of the same name in the Department of La Libertad), to Tamboras, a distance of about 60 km. Manuel Campana took with him a caravan of 50–60 llamas, each of which carried 6–7 arrobas, that is, between 69 and 80.5 kg. Campana rode a llama and carried arrobera saddlebags, that is, half an arroba on each side (amounting to 11.5 kg). We should bear in mind that Mr. Campana was 1.75 m tall and weighed 75 kg, which means that the llama carried a weight of 86.5 kg in all (Figure 8.1).

Campana also recalls that in 1940–1943 Gustavo Padrós, whom I have already mentioned, had a herd of about 100 llamas and went with his animals from Ancash to the heights of Trujillo.

Matsuzawa (1978:670) in turn was told by Antonio Trujillo that the people of San Marcos, 8 km downriver from Chavín de Huántar (in the province of Huari, Department of Ancash), took their loaded llamas across the Cordilleras Blanca and Negra to Casma on the coast to buy sea products and other items not found in the highlands. The distance is 80 km by air, and the journey took six days.

There are some data from the higher parts of the central coast that go back to the 1940s. These data were collected by Dillehay (1987:438) from local informants and tell of the movement of goods and people before the modern highway to Lima was built. There were networks of indirect trade between groups in the puna of Junín and Canta. Small caravans of llamas from the Junín puna took textiles, dried llama meat, cow meat, and salted trout to Canta in return for fruit, peanuts, and maize from the middle valley, as well as ropes, radios, and plastic items that came from Lima.

More to the south, Valderrama and Escalante Gutiérrez (1983:76) narrate the journeys llameros made to the coast, from Huancavelica to Ica. "They stayed between three and four days on the places on the coast where they went because the heat affected the llamas and gave them the plague, which is why they hired corrals from the owner of the house they reached, and paid children (sons or relatives of the corrals' owners) to see that the llamas did not lie down on the ground during the day, because if they did so their stomach swelled and they did not get up again. [The children also] prevented them from walking, and it was only in the afternoon that they drove them to the brooks to have water." This is interesting because it shows how the abrupt change in environment affected the animals, who needed some time to get used to it.

Masuda (1981:190) has information that was given him by Shigueyuki Kumai, who was working in Soras, in the Department of Ayacucho. The llameros carrying *cochayuyo* from Camaná got as far as this place. From what the old men said, it seems that the herders in this area used to exploit this marine resource far more actively. Today this activity is losing importance. Even so, Masuda managed to obtain data on herders' activities on the south coast during the survey he made in 1978. Two informants in Chala reported that during the lomas season, the herders got very close to the coast, where they left their llamas to some keepers. The herders climbed down the banks to "fish algae." An informant at La Bodega, which is between Atico and Ocoña, said that in 1976 the herders came and left the herd on the upper part, where the pastures are, while they went down to "fish algae." Another interesting

FIGURE 8.1. Photograph taken after 1936 in the highlands of Trujillo. The man seated on the llama at right is Manuel María Campana Castillo, father of Cristóbal Campana. The other individual seated on the left is Miguel Deza. The boy holding Campana's llama is Augusto Reyna. The group had just finished a long day's march. *Courtesy of Cristóbal Campana.*

testimony was obtained at Agua Salada, where herders with llamas come in wet years, but they did not go in either 1978 or 1977 because those were dry years.

Masuda notes that there are many archaeological sites in the area where this town of *cochayuyeros* is located, including the famed Inca settlement of Quebrada de la Vaca, and there are corrals too. The informants indicated that the herders used these corrals to keep their herds while they collected the *cochayuyo*.

Concha Contreras (1975:73) discusses this same area and specifies that the journey to the coastal valleys was undertaken not just to acquire products, for about 30 years ago (i.e., until 1945) several herders traveled as *fleteros* (freighters). In other words, the merchants in the area hired herders to take or bring cargoes to and from the

coast on their llamas. Some llameros report that they traveled as far as the port of Chala to take sugar and rice, but with the highway this activity has now disappeared.

When describing the site of Tocuco, in the Department of Tacna, Trimborn (1975:18) in turn wrote: "On 4 September 1970 we saw a caravan of llamas with cargo going up valley, close to Pachía, and it can be taken as an example of barter or 'cambalache.'"

Pease (1981:197) mentioned a study carried out in 1977. He wrote: "Besides verifying the presence of men from Sibayo on the coast, it was found that part of the movements were made with auquenids, and part by road." This is interesting, because this observation was made at Punta Coloca, north of Mollendo (province of Islay), on the coast of the Department of Arequipa, whereas

Sibayo is in the upper part of the Colca Valley at 3810 masl, in the province of Cailloma of this same department.

Browman (1990:42) believes that at present, the caravans survive essentially as a way of diminishing risks, as they provide the peasants with an option that ensures a minimum economic sustenance in times of crisis, such as the oil crisis in the 1970s or the drought of 1982–1983.

As I have noted elsewhere in this book, ancestral traditions survive among the Andean people far more strongly than is usually thought. Bayer left a beautiful description of the *chaco* in the mid-eighteenth century. He wrote:

> The Indians hunt the vicuñas in the following way. They drive them towards the hills of a valley which they close off with a big rope, to which they tie several white locks of wool and feathers. The vicuñas are scared when the wind moves these objects and so dare not jump over the rope. The Indians then diminish the enclosure formed with this rope until the vicuñas are very close together, and then enter the circle, throwing their *liwis* at their legs to entangle them and make the animals fall to the ground. The *liwis* are formed by a lasso that ends in three ropes, each of which ends in a lead ball. Then they cut their throat. The Indians are so skilled in this hunt, and at the same time so favored by luck that in one day they catch more than 40 or 50. (Bayer 1969:42–43)

Tschudi also described and participated in the early nineteenth-century *chaco*, so his testimony is invaluable. Tschudi wrote:

> The Indians rarely use firearms to hunt the vicuña. The Indians go out to trap vicuñas in the so-called chacu in April or May, when the horses return to the winter pastures. Every family in the towns in the puna must send at least one man to participate in this hunt. The widows go with them as cooks. All the group, which is formed of seventy, eighty, or more people, climb to the heights of the puna, where the herds are. They take sticks and enormous bundles of string. The sticks are dug into the ground on any appropriate plain some fifteen paces one from the other, and joined with the string at one or two and a half feet high. In this way they form a circular enclosure, and leave aside an opening about one hundred feet wide. The women hang colored rags from the string, which are moved by the wind. The men, some of them on horseback, disperse as soon as the chacu is ready and scare the herds of vicuñas from several miles around to make them go into the circle. When enough are inside they close the fence. The frightened animals do not dare jump over the string with colored rags that the wind moves, so they are easily killed with the bolas. These weapons consist of three lead or stone balls, two heavy ones and a light one, secured to long ropes made of vicuña tendons joined at the ends. To use them, the Indian places the lightest ball in his hand and makes it whirl the others over his head in big circles. At an appropriate distance, fifteen to twenty paces from the target, he lets the balls in his hand go and the three of them fly whirling round to their destination point and wrap around the first solid object they find. They usually aim at the hind legs of the animals, which become bound so tightly that they impede all movement and make the animal fall down to the ground.

> The vicuñas thus hunted are killed and their meat is distributed in equal parts among those who participated in the hunt. The skins belong to the church: these were given to the parish priest or sold to carry out the necessary repairs in the temple. In 1827 don Simón Bolívar issued a decree whereby the vicuñas caught in the chacu were no longer killed, just fleeced. The law was in force for just a year because these animals were almost impossible to fleece. It is said that a similar law existed in Inca times, but that adult males could be killed.

If guanacos come in the chacu, they break the string or jump over it because they are not as easily scared as the vicuñas, in which case the latter follow them. Much care is therefore taken not to include guanacos in the dragnet. The stick and the ropes are picked up once the vicuñas are dead, something the women do very fast, and the chacu is raised once again a few miles farther on. The hunt lasts for a week. The number of animals killed on these occasions varies. I took part in a chacu in Altos de Huahuay for five days. 122 vicuñas were caught. (Tschudi 1966:246–247)

Koford (1957:218, using the 1847 English edition of Tschudi, a translation of the original 1846

edition) gives the site just as Huayhuay, thus making it seem that this was the way it appeared in the original text. I was unable to compare this with the original version, but the 1966 Spanish translation, which was translated straight from Tschudi's original text (1846), as noted, says Altos de Huahuay, so I assume this is the correct name. Koford suggests that Huayhuay probably corresponds to Huayllay, which is to the west of Lake Junín. I believe instead that it is Huaihuai, in the province of Yauli, district of La Oroya, where there was an estancia with this same name (see Stiglich 1922:483).

I cannot end this chapter without noting that camelids have never lost their importance in the popular arts, and that these animals have been depicted with different materials (e.g., Figures 8.2 and 8.3). This subject merits further study.

FIGURE 8.2. Ceramic vessel representing a llama with two individuals playing panpipes. The piece was made in 1964. Quinua (Department of Ayacucho, province of Huamanga, district of Quinua) (23.5 cm long and 26 cm high). *Duccio Bonavia. Author's collection.*

FIGURE 8.3. Llama-shaped ceramic vessels. The pieces were made in 1964. Quinua (Department of Ayacucho, province of Huamanga, district of Quinua). The vessel on the left measures 17.5 cm long and 6 cm high, and that on the right measures 11.5 cm long and 10.5 cm high. *Duccio Bonavia. Author's collection.*

NOTES

1 Several marine algae are known in Peru by the name *cochayuyo*: *Nostoc commune, Ulva fasciata, Grateloupia doryphora, Gigartina chamissoi, G. glomerata, G. paitensis, Rhodoglossum denticulatum, Porphyra columbina*, and *P. leucostica*. It must be *Porphyra columbina*, since this passage refers to the Department of Arequipa, where it is used to prepare soups and *picantes* (see Acosta Polo 1977).

2 I am aware that the term caravan is inaccurate in the strict sense. The *Diccionario de la Real Academia Española* (1992: 402) defines a caravan as a "group of people in Asia and Africa who got together to make a trip in safety," which clearly is not the case in the Andes. However, this term has been much used in the scientific literature as a synonym for *recua* (drove), which comes from the Arab *rakuba*, caravan. I therefore took this poetic license.

CHAPTER 9

ARCHAEOLOGICAL AND
HISTORICAL DATA FROM OTHER
LATIN AMERICAN COUNTRIES

This chapter presents all the data found on camelids in other American countries. In this case the search for data was not exhaustive because I had not planned to move outside the central Andes. The data presented here are what I found, and the research was expanded in just a few cases. I do not intend to present a complete picture of this subject. Instead, these data will help other scholars who want to prepare a study like this one.

This chapter begins in the north with Nicaragua.

9.1 NICARAGUA

The information I present on Nicaragua is brief, and above all quaint. Porras (1967:19) mentions a letter to the king sent by Pedrarias Dávila (whose real name was Peter Arias or Pedro Arias de Ávila), then governor of Panama, in April 1525. This letter was set in rhyme in Italian and was included by Henry Harrisse in his *Biblioteca Americana Vetustissima* (1866). It is odd that the landscape is mentioned while describing Nicaragua, and we read: "*e pecore ussono di fine vello*" ("there are rams of fine wool"). Needless to say, there were no sheep in Nicaragua. On the other hand, Pizarro was still on his first voyage when Pedrarias wrote to the king, so the "rams" of Peru had not yet been seen and were still unknown. However, it is true that more than one line of evidence, some of which has al-

ready been mentioned, shows that news of the Peruvian riches had spread far to the north. Was this just a fantasy of Pedrarias? It is possible, but then we have to ask why he mentioned rams. I cannot imagine what animal gave rise to this confusion. Or had the news of the Andean "rams" been passed on to the Spaniards by the Indians, and this is what Pedrarias meant? I do not know. The truth is that this would be of no importance were it not for another enigmatic bit of information from nearby Costa Rica, to which I now turn.

9.2 COSTA RICA

In the late seventies, Snarskis (1976) published a report on his archaeological research in the eastern lowlands of Costa Rica. (I am aware that Snarskis published a previous study in 1975 on this same subject, but I have not seen it.) Among other points, Snarskis (1976:348, 350) wrote:

A final, provocative bit of evidence for South America-Costa Rica contacts is a ceramic effigy vessel from Línea Vieja which appears to portray a member of the camelids, a faunal family native to an Andean habitat. Although the vessel lacks precise data on provenience and associations, it is definitely in the El Bosque style, placing it around the beginning of the

Christian era.[1] Zooarchaeologists Dexter Perkins and Patricia Daly of Columbia University (personal communication) state that the effigy is much more like a llama or guanaco in total morphology than a deer. The important role played by the domesticated llama in pre-Columbian [sic] cultures is best documented in the Peru-Bolivia area, but the family Camelidae ranged as far north as Ecuador and Colombia, and the sixteenth-century Panamanian chief, Tumaco, amazed the Spaniard Vasco Núñez and his men by modeling in clay a long-necked beast which they immediately recognized as a "camello." Pointing toward the south, Tumaco went on to tell the Spaniards that much gold was to be found there, and that the people of that region used such long-necked creatures as beasts of burden (de las Casas 1961:291 [A detailed account of this event appears in Chapter 6, with the citation from las Casas, which in the 1981 edition used here corresponds to Bk. III, Chap. XLIX, 600.]). The author has seen other Costa Rican vessels depicting similar animals with bound eyes blindfolded and tied-down cargo. It is conceivable that these represent captive deer such as are sometimes portrayed on Maya polychrome ceramics (Gordon Ekholm, personal communication), but the possibility remains that some aboriginal Costa Rican potters had occasion to see either an actual llama or a representation detailed enough to allow them to reproduce faithfully the spraddling padded toes and deeply cleft, pendulous upper lip characteristic of the American camelids.

The photograph Snarskis presents (1976:Fig. 6, 350; my Figure 9.1) is captioned "El Bosque zoomorphic effigy vessel, probably representing a member of the family Camelidae." In effect, there can be little doubt that this is a camelid, specifically a llama. I cannot imagine what other animal it might be. The only other possibility would be a deer, but I do not think it is, particularly because of the shape of the feet and the cleft

lip, as Snarskis correctly noted. It is striking that the animal was depicted with a short neck, exactly as the Mochica artisans figured it, and perhaps like the small prehispanic llamas from Ecuador. Is this just a coincidence?

On the other hand, this is a piece that in temporal terms corresponds to what was called the Early Intermediate period in the central Andean area. Was there already some kind of contact or relation between the central Andes and the lands far to the north at that time? Is this in any way related to the Nicaraguan "rams" mentioned by Pedrarias? These are just questions, and as such they remain. Michael J. Snarskis wrote to me in regard to this subject, noting among other points that he has "another photograph of a ceramic piece from SW Costa Rica that depicts a camelid even more realistically. It can be dated between AD 1000–1500, and is of the type called Tarrago Galleter (Linares 1968)" (pers. commun., letter of 12 August 1994). The vessel concerned is later than all the other pieces mentioned, in temporal terms corresponding to the Late Intermediate period/Late Horizon of the central Andes. At this time there

FiGURE **9.1.** Ceramic vessel whose base represents a camelid, almost certainly a llama. El Bosque (Costa Rica). *Duccio Bonavia. After Snarskis (1976:350). By permission of the author and the Society for American Archaeology.*

undoubtedly were contacts between Central and South America, so in this case it is far easier to explain the representation. This does not, however, answer the question raised above, which indicates that a study of this subject is in order.

Fonseca and Richardson (1978) present some additional data that are of great interest. They refer to the Las Huacas site, in the Nicoya Peninsula of Costa Rica, on the Pacific Ocean. Here they describe two fragments of sculptures representing animal heads found in the Velasco 1 Collection, which come from the cemetery of Las Huacas. They were given a date of AD 180–525. These pieces were described by Hartman (1907: Lám. 39–14, 22). The first is made of steatite, the other of serpentine. In both cases the materials are found in Costa Rica. The pieces were shown to Otto M. Epping of the Carnegie Museum of Natural History, who identified them as camelids. In effect, the illustrations in the study by Fonseca and Richardson (1978:Figs. 2, 3, 4, 5, and 6, pp. 303–307, respectively) are definitely camelids, and may well be llama or alpaca. I reject the possibility of their being either guanaco or vicuña (Figures 9.2–9.3, and 9.4–9.5).

Again according to Fonseca and Richardson (1978:300 ff.), materials identified as Maya were identified at the cemetery of Las Huacas. On this basis they suggest the possibility of maritime relations between the Maya and South America by raft, but they do not dwell on this point.

The interesting thing is that Fonseca and Richardson (1978:307) add one more piece of information concerning these two camelid heads, the piece reproduced and described by Snarskis and the other pieces Snarskis mentioned (1976:350), which show camelids with their bundles. Fonseca and Richardson recall that Fischer (1882:187, Lam. 8) illustrated a camelid head in stone from northwestern Costa Rica. They comment: "The modeling in clay of a camelid by a Panamanian chief for the Spanish and the fact that he related to the Spanish that they were used as beasts of burden, reflects knowledge of llamas in Central America during the sixteenth century (de las Casas 1961:291 [1981:Bk. III, Chap. XLIX, 600, in my bibliography])."

The data are intriguing, but with the information available, any comment would be no more than unwarranted speculation.

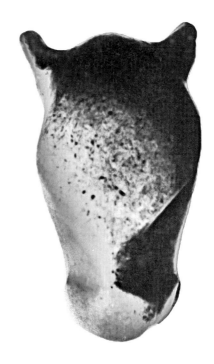

FIGURE 9.2. Stone sculpture (steatite) in the shape of a camelid head. Las Huacas (Costa Rica). *Duccio Bonavia. After Fonseca and Richardson (1978:Fig. 3b, 304). By permission of the Carnegie Museum of Natural History, Pittsburgh.*

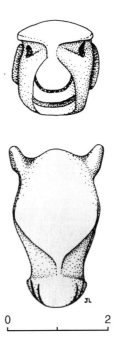

FIGURE 9.3. Drawings of the sculpture in Figure 9.2. *After Fonseca and Richardson (1978:Fig. 5, 306), by permission of the Carnegie Museum of Natural History, Pittsburgh.*

FIGURE 9.4. Stone sculpture (serpentine) in the shape of a camelid head. Las Huacas (Costa Rica). *Duccio Bonavia. After Fonseca and Richardson (1978:Fig. 2b, 303). By permission of the Carnegie Museum of Natural History, Pittsburgh.*

FIGURE 9.5. Frontal view of Figure 9.4. *Duccio Bonavia. After Fonseca and Richardson (1978:Fig. 4b, 305). By permission of the Carnegie Museum of Natural History, Pittsburgh.*

9.3 VENEZUELA

In the case of Venezuela, the facts are easier to explain. Gil (1989) drew my attention to this information. In his study, Gil recounts some aspects of the history of Orinoco and explains that in 1531, the Crown signed the agreement with the Welser Company. "The government of Venezuela thus became a monopoly of the German bankers" (Gil 1989:42).

The Germans ruled from Coro. Ambrosius Alfinger and Nicholas Federman "had eagerly sought the desired pass to the Mar del Sur in their expeditions (1529, 1531, and 1530, respectively). In August 1530 Micer Ambrosio, preparing a report in Coro, boasted not just of having conquered innumerable towns but also of having begun exploring the Lake of Our Lady of Maracaibo." In addition, he had heard rumors of "people dressed with varvas [?], having stone houses and sheep" (Gil 1989:43). For this reason, on 12 May 1535 Jorge de Spira "took the road to Maracapana, what we call the plains of Mata, because there was news of much riches toward that land" (138) (note 138 reads: A.G.I. Patron. 27, 18 fol. 3r; A.G.I., Justicia 65, no. 4 fol. 162). Spira thus went inland along the Sierra Nevada in search of gold, but he failed. The discoveries therefore moved west, "so the mythical region evidently acquires a marked Peruvian flavor: the cacique Guaiguri told Spira that on the other side of the sierra 'the Christians would find much gold and silver and tame sheep, like the ones found in Peru' (140), and the captain-general garnered similar information that always placed the gold toward the setting sun at Zubiairi, on the Pampanene and the Bermejo Rivers (141)" (Gil 1989: 44–45; note 140 is G. Fernández de Oviedo, *Historia*, XXV 11 [BAE 119, p. 37b]; note 141 is G. Fernández de Oviedo, *Historia*, XXV 13 [p. 41b;

p. 42a]; XXX 14 [p. 43a]; see Hutten, *Diario*, pp. 363–364).

According to the probanzas prepared at Coro in 1538, all agreed that the riches lay beyond the Pampamene River. Gil notes that

> [t]he Guaipiés of that river, who seem to be the present Guayupés . . . made many wry faces and gestures and gave heavy blows to big jars as a sign that an infinite amount of gold and silver in vessels, jars and other containers lay ahead, because the natives in this region do not use other metals in their tableware. And to indicate they had humped [*corcobadas*] sheep called emas, as was noted by Juan de Villegas (fol. 22v), Francisco de Villegas (fol. 47r), Jerónimo de Cataño (fol. 40r), and Damián del Barrio (fol. 37v) [A.G.I. Justicia 990, no. 2, 10], "they lay on the floor on hands and feet and bleated," as specified by Frutor de Tudela (fol. 14r), who also knew that the cacique lord of Toroiva had been with his father in that land as a child, from whence he had brought a sheep loaded with gold, an Indian copy of the golden fleece. (Gil 1989:45)

This is the region of the Sierra Nevada where the Pampamene and Bermejo Rivers run.

Gil then discusses the search for El Dorado and the "low hills and the lake of Caranaca, deep inside the Llanos." He explains that "the lake of Manoa became the lake of Caranaca." Here it is explained that the tradition that spread throughout the second half of the sixteenth century had the Ordás expedition going upriver, when an Indian tried in vain to make the captain turn down the Caranaca River. Gil notes that this river appears seven leagues from Cabrute on a map of the Orinoco that is contemporary with Ortal, with the following caption: "Carranaca River, through which Alonso de Herrera entered and went forward, and from this river to the inlet of Meta there are XXXVII leagues, along the big river of Huyapari." "Thus it was that Caranaca received unto itself all the Eldoradista pomp and finery. In effect, the presbyter Bernardo del Vallejo Velasco knew well that these people, dressed and of much reason, 'were of the lineage that had come down from

Piru in the time of the Inca, because it was said they had rams [*carneros*] like those of Peru' (656)" (Gil 1989:163; note 656 is A.G.I., S Dom 15, 1 no. 8 fol. 8v–9r). This happened in the late 1500s. The Meta River is on the border between Venezuela and Colombia, and flows into the Orinoco River.

In fact, Fernández de Oviedo recounts the voyage of Governor Jorge Espira in 1536, when he entered through "the nation of the Caquitio," and then came "to a strong river called Apuri" and another "they call Darari," and then "up to another river called Cazavari," where they found out that "in this same sierra there was a cacique called Guaigueri, who would give the Christians a complete account of everything." In fact,

> [t]his cacique made them understand that on the other side of the sierra the Christians would find much gold and silver, and tame sheep like those found in Peru, and they keep them at night in their corrals; and it is a land of savannas and lacks wood, and all the vessels in the Indians' tableware is of gold and silver. And in two moons' journey they would meet a cacique or king, an important lord whom they call Caciriguey, where those riches are; and this great lord, who is very powerful and rules a big population, has big houses of prayer or mesquites where certain ceremonies take place on certain days of the week. And finally, he gave many details of those riches, and that the sierra was rugged but would be crossed over without danger, and that he wanted to go with the governor to show the Christians what he said and the way [there]. (Fernández de Oviedo 1959b:Bk. XXV, Chap. XI, 37)

So the Spaniards visited the Mazopide and Guaipié Indians, and, "having news that the headwaters of the Meta were in this sierra," they crossed the Oppia River, which lies 190–200 leagues from Coro (Fernández de Oviedo 1959b:Bk. XXV, Chap. XI, 38). The group thus reached the headwaters of the Meta River and had serious clashes with the Guaipié. Continuing their voyage the Spaniards reached the Papomane (or Papomene) River, where the Guaipié also lived, while the Chogues were slightly farther on. The rich nation was "eight marches . . . from the Bermejo River." "These Indians added more: that pots and large jars [*tinajas*] and all other vessels in the Indians' tableware in that rich land were of gold and silver; and they named the fine gold by its name, and the low [quality metal] and the silver accordingly. And they said what the sheep were like, and named them the way they are named in Peru, llama, and said how they brought them tame and placed them inside their corrals." And "[o]ne of those Indian headmen claimed that he had been in the land he talked about, and said he had seen with his own eyes those great riches he was describing" (Fernández de Oviedo 1959b: Bk. XXV, Chap. XIII, 42).

It follows that the news of Peru's riches had spread as far as the northernmost corner of South America. However, this happened several years after the feats of Pizarro and his men, which no doubt impressed the natives, and this almost surely accelerated the diffusion of the events. Besides, in this case we are dealing with a zone that lies close to the northern marches of the Inca Empire, which had expanded there shortly before the conquest.

9.4 COLOMBIA

Steward (1949:718) wrote that the llamas "never reached Colombia." This is not entirely correct.

9.4.1 Archaeological Data

There are only archaeological data for the San Luis site (Ipiales, in the Department of Nariño). Uribe (1977–1978:116) reports that the camelid remains found here belong to animals of all ages, from newborns with their milk teeth to old adults of both sexes. The remains were classified as belonging to the "Order Artiodactyla, Family Camelidae, Genus *Lama*."

From the report it follows that this is a midden, but no data are presented on the amount of bone materials found. Only two artifacts made with llama bone are presented, and Plate 2-4 shows three camelid mandibles. As for the date, Uribe (1977–1978:116) says: "It was not possible to obtain carbon to date the midden. The cultural materials from San Luis are identical to those

obtained by the archaeologists from FINARCO at the Piluán site, for which they obtained a date of AD 1450 (Groot, Correa & Hooikas 1976)" (this was repeated by Stahl 1988:Table 1, 358).

If the date given by Uribe is correct, and if it is taken with the caution with which radiocarbon dates must be taken, then one might think that these animals arrived in Colombian lands as a result of the Inca expansion. If we accept the chronology proposed by Rowe (1946), Topa Inca reached Quito in 1463–1471. Huayna Capac later expanded his domain as far as the Ancasmayo River, that is, as far as the border between Ecuador and Colombia. However, it should not be forgotten that these animals lived in Ecuador before the Late Horizon, so the Colombian camelids might have derived from this stock.

9.4.2 Historical Data

Novoa and Wheeler (1984:124) state that "[e]thnohistorical sources record that llamas accompanied Inca armies as far as southern Colombia . . . during the period from AD 1430–1532." Unfortunately, they do not give their source, which I was unable to trace, nor does Franklin Pease (pers. commun., 26 May 1992) recall any data on this.

I was unable to make a thorough study of the historical sources, but I do have some data. There is a reference in this regard in Cieza de León (1984:Part I, Chap. XXVI, 89), who described his travels from Popayán to Cali in the sixteenth century. While describing the "province of Buga," Cieza notes that when Chritóual de Ayala died, all of his possessions were sold, including "one sheep of those from Peru," for which 280 pesos were paid.

In 1537 the Audiencia of Santo Domingo reported to the king the arrival of Captain Francisco César after crossing the sierra of Urabá. In this letter we read that César "reached a land where he found people dressed with the same clothes of Peru, and sheep, and other many things from that place" (letter dated 30 May 1537, A.G.I., S. Dom. 49, Vol. II, No. 57; Audiencia de Santo Domingo, 1989).

When describing Popayán in the early seventeenth century, Guaman Poma de Ayala (1936:

fol. 990) noted that there was "little livestock." It is true that Guaman Poma did not know all of the places he gave an account of, and this evidently is one of them, but in this case his data are correct.

Tschudi (1885:95–96; 1891:98; 1918:208–210; 1969:126–127) wrote the following in this regard:

> I have not found clear data on the llama in the works of the old chroniclers of Peru, [or] of its northwards dispersal, and the references that say there were many in New Granada are vague. I find it hard to believe that this is accurate, because the truth is that besides the lack of reliable news on the presence of aukenias [sic] in the high plateaus of New Granada, the people in these localities only used cotton to dress but not wool, because they had no domestic animals to provide them with it. Besides, so far no objects have been found among the many antiquities removed from thence that depict a llama, whereas in Peru these objects abound (127).

Tschudi is actually right, because the only references I was able to find in the chroniclers are the ones included here (see above).

Footnote 127 (in the 1918 edition, 1 in the 1885 and 1891 editions, 10 in the 1969 edition) in Tschudi is a letter Alfons Stübel sent him. We read the following: "In Colombia the llama has not become naturalized anywhere, nor has it been imported as a beast of burden. The reason must lie in the climate, particularly in the strong rains that fall on the Cordillera almost all year long, and which turn the roads into bogs where llamas can hardly walk. In Pasto I saw two of these animals which were kept as curios. More to the south the llama is only found as a domestic animal, in the vicinity of Quito, . . . so the Equatorial line can well be taken to be the northern frontier [of the land] inhabited by this species." Further on he adds: "As for the vikuña [sic], it is not found . . . in Colombia" (Stübel 1885:96; 1891:98; 1918:209; 1969:126). In this case the 1969 translation is very poor, as it even distorts the original meaning. Furthermore, instead of "Pasto" we read "Pasco." This footnote by Stübel is interesting, but the idea of rainfall as a constraint for the llama is sheer nonsense.

Interestingly, when discussing the Inca period in Colombia, Bennett (1946a:54) mentions the environmental constraints present there for both man and culture, and which brought Inca expansion to an end. Here Bennett says that the llama and the alpaca were not raised in Colombia. He attributed this in great measure to the fact that the high *páramo* is not suitable for these animals, and indicated that not even the wild forms, like the vicuña and the guanaco, were found there.

Pulgar Vidal (1950:19/85) in turn made a general comment, but without any specific data. All he said was, "No chronicler specifically notes whether auchenids were found among the native peoples of New Granada, but they do recall that the host of Belalcázar took llamas as pack animals, which became an item of trade and their meat the tidbit in the feasts the Spaniards gave to one another. Pedro Cieza de León says that 280 pesos of good gold [*buen oro*] were paid in Cali for a sheep from Peru." I was unable to find the reference to the llamas taken by the Spaniards.

9.4.3 Additional Data

The presence of some llamas in Nariño in modern times is noted here merely in passing, because this datum has to be taken with due caution, as it comes from Cardozo (1974a:11), who is not a reliable source.

As for the guanaco, I found only one, quite specific reference that is most interesting, even though it needs checking. It comes from Rivero (1828), who laconically wrote the following: "I have found this animal in the cordillera of Bogotá." Curiously, Gilmore (1950:433, 447) supports the possibility that guanacos lived in southern Colombia based solely on the toponym "Páramo de Guanaco," which I was unable to find.

We see that although it is true that there were few camelids in what is now Colombia, their presence cannot be denied. It cannot be determined when they might have started living there. Although it is possible that their arrival was a result of the Inca expansion, everything seems to indicate that it was through an indirect route, not with the Incas. The literature must be examined to establish whether the Spaniards actually took llamas with them, as Pulgar Vidal claims. At present we do not know how many camelids there are. In 1982 there were 200 llamas, according to Franklin (1982:Table 2, 475; see my Table 2.2).

Troll (1958:29) mentions the study by Hellmich (1940:89 [which I have not seen]), where an account of an (apparently failed) attempt to introduce llamas and alpacas in Colombia is given. It reads thus: "Nothing shows with more clarity the fact that the Andean *páramo* and the Andean puna are two regions of a completely different biogeographic nature, than the recent state-sponsored attempt to provide agriculturalists with a source of resources and better living conditions through the import of llamas and alpacas. All of the animals distributed for reproduction in the upper levels of the Colombian Andes died because the wet climate of the *páramo* steppe was unfavorable for them."

9.5 ECUADOR

There is far more information on Ecuador, both historical and archaeological. I begin with the former.

9.5.1 Archaeological Data

First, there is a general statement for the coast. This is by Wing (1986:252), who notes that camelid remains have been found in burials dating to 500 BC–AD 1155. However, she based her comments on a study by Hesse (1980), which I discuss later.

To avoid confusion, I must point out that Lanning (1967b:5) mentions the presence of camelid remains in the Santa Elena peninsula, north of the Gulf of Guayaquil. However, these are not archaeological but paleontological remains, for Lanning based his statement on Hoffstetter's data (1952), which were discussed in the section on paleontology in Chapter 3.

There are some interesting data for the Ecuadorian Formative. First, Bruhns (1989:67) mentions the presence of camelid depictions in Chorrera-style pottery. This was confirmed by Olaf Holm (letter of 18 November 1991), who told me that "[i]n the ceramics of the Chorrera

culture, approximately 1200–200 BC, I have seen a ceramic vessel modeled in the shape of a llama. It is an ungulate and I do not believe it can be any animal but the llama. It comes from the North Coast of Ecuador."

Pirincay, in the highlands, is an important Formative site. It is worth noting what Miller and Gill (1990:51) thought of this site before beginning their research: they were convinced they would not find camelids in Pirincay because there was no mention of them either in the 1582 *Relaciones Geográficas de Indias* ("Relación geográfica de San Luis de Paute") (Jiménez de la Espada 1965:273 [1965a in my bibliography]), or in the data from Chobshi Cave, close to Sigsig (Lynch and Pollock 1981), or in Tequendama, in the Bogotá savanna, Colombia (Correal and van der Hammen 1971), or in contemporary zoology (Patzelt 1979). This, however, was not so.

The site is in the southern sierra of Ecuador, on the lower Paute River, 2 km south of the town of the same name and 47 km northeast of the city of Cuenca, in the province of Azuay. The Cuenca Valley is the southernmost and driest of the three major depressions in the highlands separating the snow-capped peaks and the cold *páramo* of the eastern and western cordilleras of Ecuador. The Paute River valley forms an important corridor between the large Cuenca Valley, with its easy access to the coastal plains of southern Ecuador and northern Peru, and the tropical lowlands to the east. Here an hacienda of the same name controls the flow of traffic from the highland valleys of Cañar and Cuenca, and the tropical lowlands to the east (Miller and Gill 1990:51; Bruhns 1988–1989:71; 1991b:1; Bruhns et al. 1990:221).

Miller and Gill (1990:51) note that although the Cuenca Valley is considered dry by the standards of the Ecuadorian altiplano, the average rainfall there is higher than in comparable highland valleys in Peru. Even though the vegetation around Pirincay has been considerably modified, there are still ecozones that extend from the *páramo* to lush grasslands and dense woodlands. Miller and Gill likewise note that Bennett (1946 [1946a in my bibliography]) and other scholars believe that this area was isolated both in present and in past times. However, it seems clear that since the late Pleistocene it has been a great transversal corridor of the southern Ecuadorian Andes. The routes that started in this basin to the northwest led across the Cañar Valley to the Gulf of Guayaquil, or southwest to the Jubones Valley, to the southern coast, and from there to Peru. The Paute Valley, on the eastern divide, forms a natural corridor to the Amazonian lowlands, as was already noted.

The site of Pirincay has a chronology that extends from the Late Formative to the First Regional Developments period. Radiocarbon dates give a series of dates that go from 1220 ± 120 BC to AD 190 ± 45. In Ecuador it is contemporary with Cerro Narrío and the coastal Chorrera culture. The lowest levels at Pirincay are contemporary with Peru's Early Horizon, while the upper layers correspond to the Early Intermediate period (Miller and Gill 1990:52). Three major periods have been established for this site, called Early, Transition, and Late. The first is given a date of ca. 1000 BC, the second a date of ca. 300 BC, and the third a date of ca. AD 100 (Miller and Gill 1990:53).

At first only the Pirincay camelid remains dating to the Formative were reported (Wing 1986:250). Today we know there are no camelids in the layers representing the early and transitional phases. Around 400 or 500 BC (but in later reports 300 BC is also given), we find a major change in the economy of Pirincay in the form of the introduction of camelids, which soon acquired a great importance (Bruhns 1988–1989:76; 1991b:1–2; Miller and Gill 1990:Tables 1, 2, 3, 52, 53).

Only two shallow pits with a great number of bones were mentioned in the preliminary work published on Pirincay, most of which were identified by Miller as being of the llama/alpaca type. It was also noted that the camelid remains were articulated. The presence of sewing artifacts made from cervid and camelid bones is reported (Hammond and Bruhns 1987:53; Bruhns 1988–1989:72).

It was later reported that the llama bones were found in the refuse belonging to the layers that lie, in absolute terms, between 400 BC and the first century AD, and were abundant around the pits. Although the first analyses showed that the llamas found in the pits were young (from one to two years old), suggestions of a greater age range have appeared. Everything seems to indicate that these pits are the remains of sacrifices or ceremonies (Bruhns 1989:70; 1991b:3–5). Bruhns (1991b:6) wrote: "[T]he camelids . . . indicate that we are dealing with ritual rather than ordinary consumption of camelids."

In 1989, Bruhns stated that the types and species of animals at Pirincay are the same as those found in the early Chobshi site. However, in the fifth century BC, hunting was abandoned for camelid raising. Based on the data of Miller and Gill, llama bones, and perhaps those from alpacas, constitute almost 95% of the fauna in late layers (Bruhns 1989:66). It was later explained that when camelids arrived at Pirincay, they did so in a peculiar way. The fauna recovered in the earliest layers of Pirincay shows a wide range of resources, with a preference for deer that later changes to more than 80% of the bone weight for some species of camelid, which Bruhns defines as a "small llama." It is explained that several of these llamas were placed in special pits along with very large hunks of jars, ollas, cups, and bowls (Bruhns 1991b:7).

Miller and Gill (1990:55–56) presented more data in a later report. They specified that the "fauna shift[s] dramatically" in the Late Pirincay phase. Instead of the range of a large and widely varied fauna found in the components called Early and Transition, we now find one dominated by camelids. Their total weight comes to 83.4%. Miller and Gill wrote, "To our knowledge such an abrupt transition from the hunting of wild game, principally deer, to almost complete reliance on camelids, is unparalleled at any other site in the Andes."

As examples, the gradual shift from hunting to animal management in Junín, Kotosh, Huacaloma, Huaricoto, Chavín de Huántar, and Waywaka is mentioned (these sites are all mentioned and discussed in Chapter 4 of this book). And after noting that in the central Andes, the transition took place approximately from 5500 to 4200 BC in the south and in 850 to 200 BC in the northernmost sites (the literature cited is Wheeler-Pires Ferreira, Pires Ferreira, and Kaulicke 1976; Wing 1972; Shimada 1982, 1985; Sawyer 1985), Miller and Gill write:

But in all cases the shift away from the hunting of cervids appears to have occurred in a slow incremental fashion and was always preceded by a long period of minor, secondary exploitation of camelids. The transition at Pirincay is conspicuously distinct from the Peruvian pattern and has all the hallmarks of a faunal dietary revolution. Both the Early and Transition levels at Pirincay as well as the faunal assemblage of the Cueva Negra de Chobshi, some 80 km to the south (Lynch and Pollock 1981), suggest that camelids did not even exist within the Cuenca region prior to the beginnings of the first millennium BC. This interpretation is corroborated by their absolute absence at all other reported sites in the northern Andes of Ecuador and Colombia dated prior to 650 BC (Correal and van der Hammen 1971; Wing 1986). On the coast they appear to have been absent until the time of Inca expansion during the late 15th century (Byrd 1976). (Miller and Gill 1990:56)

Bruhns et al. (1990:232) stated something similar and insisted on the total absence of camelids in the Ecuadorian highlands prior to the introduction of domestic forms from the south. Then they said:

> The camelids of Pirincay, substantially earlier than previous discoveries of this animal, indicate an introduction of pastoralism towards the end of the Formative, rather than with the Inca invasions as has been previously suggested. . . . Based on osteometric evidence (presented in detail in Miller and Gill 1990), the Pirincay camelids represent a previously undocumented form of llama, intermediate in size between the contemporary llama and alpaca. This undersized llama, reported in historic times as "the small llama of Riobamba, Ecuador" (Gilmore 1948 [a mistake; the date is 1950]:437) has not been identified archaeologically before.

Bruhns (1989:67) discussed these finds and noted that it is very difficult to indicate the origins of these animals, and furthermore, there is to date no other evidence of camelid herding in Ecuador during the Formative. Moreover, the research carried out by Guffroy and his colleagues in Loja did not yield any evidence of camelids south of Azuay, "although it is said that Dr. Mathilde Temme discovered some camelids, also in Loja, in a context some centuries later than Pirincay." Bruhns ends by insisting once more that there is a complete lack of concrete evidence with which to explain the origins of the Pirincay camelids, as well as to understand the reasons behind this "revolutionary shift" in the economy of the Paute Valley, with respect to both food and transportation.

In another study, Bruhns (1991a:4) posits the possibility of contacts with the south, but she notes that there are no data to know how these were carried out. Bruhns likewise notes that whereas Peruvian sites have evidence of llama and alpaca, "Pirincay only shows one camelid species: a small llama." However, she believes that some "Peruvian customs" appear together with these animals, such as "the sacrifice of animals with the burning of the skeleton and the burial of the burned bones with offerings in pots (probably with chicha inside), the festivals where camelid meat and chicha were consumed, and the disposal of the partially articulated bones, cups and jars in special pits."

Here I shall try to summarize the conclusions drawn from the Pirincay findings by Bruhns, on the one hand, and Miller and Gill on the other.

I begin with Bruhns (1991b:2). She believes that since the site is just half a day's walk from the high-altitude *pajonal*, it can reasonably be assumed that most of the llamas were not kept close to it. The study William Middleton made of the phytoliths found in the teeth of the animals showed that these were herded in a great variety of ecological zones, including the high-altitude *pajonal* and all the zones around the site. It likewise seems that the food eaten by the llamas included maize, beans, and the grass in ravines and plains, rivers, and the tropical forest that lies between the *pajonal* and the valley. This would suggest the llamas were brought from their pastures when necessary, and that the animals grazed close to the fields where crops were grown, or were fed domestic plants. However, it is possible that most of the animals were raised in the high-altitude part of the region and that only some of them grazed close to the site. The problem is that the Pirincay sample is small, probably representing just 11 animals (Karen Olsen Bruhns, letter of 22 January 1992; she based her conclusion on Middleton's report [1992]). Bruhns recalls that today sheep and cattle are fed after the harvest with maize stalks and leaves, along with the remains of the bean vines that grow wrapped around the maize. There are no ethnohistorical data on camelid husbandry in Ecuador, nor are there any studies on the small herds in the provinces of Imbabura and Chimborazo, so nothing is known of the way these animals are raised by the natives of the northern Andes. According to Bruhns, the modern llama and alpaca raisers who use the stock imported from Chile in 1980 move their herds according to the seasons from the *pajonal* to the pastures in the lower-altitude forested and cleared-forest grasslands, because the remains of high-altitude *pajonal* are poor in nutrients. Furthermore, a liver fluke has been introduced and is the cause of a considerably higher mortality in those herds that have been pastured exclusively on high-altitude grasslands. The animals kept in other places, such as archaeological parks, are essentially decorative and are practically not looked after at all. Now we know, Bruhns says, that the inhabitants of Pirincay had a small variety of llamas, and that those killed at the site had grazed in different vegetation areas. The most interesting point, as Bruhns emphasizes, is that some very distinctive customs appeared together with the llamas that give us some information about this animal, such as data on ritual or perhaps husbandry customs, that might have arrived with the animals when these were taken there from Peru.

Miller and Gill (1990) studied the camelid bones and present an in-depth discussion that I find extremely important for the topic under discussion. First, Miller and Gill note that to judge by the values V (the variation coefficient) has, the Pirincay camelids seem to have constituted a homogeneous breeding population and a single

species. And according to the strictly statistical data, the Pirincay camelids do not derive from the same alpaca or llama population at La Raya in Peru (Miller and Gill 1990:57).

Miller and Gill then discuss different possible ways to interpret the remains (Miller and Gill 1990:57–63). The first possibility is that they belong to a vicuña. This possibility is rejected, for three reasons: first, the morphological characteristics of the incisors; second, the differences in osteometry; and finally the present distribution of these animals. Miller and Gill believe that its northern limits lie at 9°30′ S latitude in the Department of Ancash, that is, 800 km away from Pirincay. They admit they take this position even though both Cieza de León (1973:110–155 [1984:236]) and Vázquez de Espinosa (1942:279–290 [1948:373]) noted the presence of vicuñas farther north in the sixteenth [and seventeenth] century, but according to Miller and Gill other writers do not do so. (Later in the chapter I review some historical sources.)

The second possibility is that they are guanacos. Miller and Gill believe that this identification also must be rejected if we use the osteometric data. However, they admit that most of the available measurements are derived from Tierra del Fuego animals, which seem to have reached a larger size than those living farther north. They also reject as uncertain the possibility of a relationship with *Lama guanicoe cacsilensis* (a doubtful form already discussed in Chapter 1). For Miller and Gill, another argument against the remains being those of guanacos is biogeography. Although Cieza de León (1973:121 [1984:143]) notes the presence of guanacos in Ecuador, the northernmost distribution of this species is 8° S latitude, in the Department of La Libertad; Miller and Gill believe this must have been the farthest extent of its range in antiquity (they base their conclusion on the studies by Novoa and Wheeler 1984:121, and Franklin 1982:473). Miller and Gill insist that other early writers likewise do not mention the guanaco or the vicuña (Vázquez de Espinosa 1942:361–385 [1948:338–360] and Jiménez de la Espada 1965[a]:271–281 are given as examples), but if these animals were rapidly becoming extinct in 1550, as Cieza de León noted, then they could have completely vanished 30–80 years later. If wild camelids were endemic in the Ecuadorian highlands, they would have been hunted by the men who lived in the Cueva Negra de Chobshi, or in the earliest phases of Pirincay, but this did not happen. Miller and Gill believe that Wing's (1986) analysis of the fauna in other early sites of Ecuador supports their position. Everything seems to indicate that the camelids found at Pirincay were domesticated and introduced from the south. For Miller and Gill, the presence of *Cavia* would be one more proof.

The third possibility is that the Pirincay animals were alpacas. If this was proved, it would go against several lines of evidence. A first consideration is distribution. Alpacas do not go beyond 10° S latitude in the Junín puna (they cite Gilmore 1947 [a mistake; the date is 1950]:433, Franklin 1982:473, and Novoa and Wheeler 1984: 119), and the species might have been more restricted prior to the Inca conquest. Miller and Gill note that currently these animals are concentrated in the altiplano (Bustinza 1970); they cite the hypothesis that claims their center of domestication was there (Lynch 1983[a]), as well as the fact that these are animals that live above 4000 m (Cardozo 1954:94). They also note the alpaca's preference for the bofedales (here based on Cardozo 1954:94; Gilmore 1947 [1950]:442; Webster 1973:120; Palacios Ríos 1977[a]:37–39) and the fact that it is considered the most ecologically specialized of the four species (Franklin 1982: 474). However, Miller and Gill admit that although the alpaca seems to be restricted to the puna, it has also been noted (Flores Ochoa 1982: 69) that this is an artificial distribution. This is why it is believed that the alpaca occupies an ecological refugium, which they admit has been "incorrectly interpreted as environmentally inflexible." They then mention the importation of alpacas to Australia (Vietmeyer 1978), and recently to Ecuador (Meisch 1987) and coastal California; however, Miller and Gill believe that this is an acclimatization through strange means, and in no way a diffusion mechanism.

Miller and Gill emphasize that archaeological studies have shown that the alpacas concentrate in the puna. They mention the studies made at Kotosh, in the Callejón de Huaylas, Tarma,

Huacaloma, and Manchán (Wing 1972, 1977[a]: 849–850; Shimada 1985:296; Altamirano 1983[a]: 65). Miller and Gill then write:

> In all of these cases, however, the alleged alpacas occur in the faunal sample in tandem with larger camelids interpreted as llamas. It is as if camelid herding was introduced into the north-central Andes during the Initial Period and Early Horizon as a single pastoral package that inextricably linked the two domesticated species. The Pirincay camelids of alpaca size are unique in this regard. They occur alone, neither preceded nor accompanied by camelids of llama size. If the Pirincay camelids were alpacas, they were living in a unique situation unparalleled in the archaeological record and for which we have no good ethnographic analogues. Even today, when motorized vehicles have contributed to the drastic reduction of the llama population of southern Peru, most alpaca herders maintain a few llamas for occasional pack work (Flores Ochoa 1968; Palacios Ríos 1977[a]). Thus, a population solely of alpacas at Pirincay would imply that their function at the site was not linked to long distance transport. (Miller and Gill 1990:60)

Miller and Gill then note that at present, the role alpacas had in prehispanic times in separate areas is under discussion, with the primary use limited to use of the meat or fiber. So, if the alpaca was used at Pirincay to produce wool, this would have necessitated llamas for transportation. For Miller and Gill, the presence of alpacas without llamas would imply that the wool was only used locally, and that regional exchanges were carried out by human means. This, however, contradicts the evidence, for it has been verified that at Pirincay, contacts were made that ranged from the Pacific coast to the Amazon. On the other hand, the advantages the llama had for transportation were obviously clear to those who imported the Pirincay alpaca from some southerly place where both species were available.

If the Pirincay animals were alpacas used for their meat, then pack llamas were not needed.

However, it is hard to imagine, as Miller and Gill note, why alpacas and not llamas were chosen for this purpose.

The incisors are what casts doubts on their being alpacas. With a single exception, these correspond to the guanaco/llama type. The study was made using the method of Wheeler (1982[b]), as well as by comparisons with living animals. This was independently confirmed by Kent (personal communication to the authors). However, Miller and Gill admit that the method is not secure, so the results remain doubtful.

Miller and Gill also ponder the possibility that it is a llama-alpaca hybrid, what is known as a *wari*. However, they do not believe the possibility that the Pirincay population could have been made up of just these animals. The *wari* are the first-generation offspring of a cross between llama and alpaca. F_2 generation offspring tend to revert back to the parental stock, whether llama or alpaca. In this way a self-sustaining population in Pirincay was impossible.

Finally, Miller and Gill (1990:60) discuss the last possibility, namely, that the Pirincay animals were llamas. The osteometric analyses show that these animals are smaller than llamas. But since all other possibilities have been rejected, the case must be reexamined. The materials used for the comparisons came from La Raya and collections in U.S. museums of an uncertain pedigree. It was noted that the use of pack llamas has been dropped at La Raya with the advent of modern transportation, so the animals must be considered to be of average size. Pack llamas are traditionally believed to be the biggest and come from Collao, a point supported by the chroniclers (Miller and Gill base their conclusions on Gilmore 1947 [1950]:437, Pedro Pizarro 1978: 28, and Vázquez de Espinosa 1942:629 [1984: 294]). Miller and Gill strongly emphasize that in Inca times the llamas from Collasuyo were believed to be the biggest, and they consider, based on data in West (1981[b]), that those from the Departments of Puno and Cuzco are still the largest.

Citing Gilmore (1947 [1950]:437), Miller and Gill indicate the presence of a small llama at Riobamba (Ecuador). They criticize the fact that Gilmore does not point out how much smaller these are, or indicate his source. However, they find it significant that Riobamba is less than 100 km from Pirincay. Unfortunately, there is little information with which to compare the Ecuadorian animals with those of northern Peru, "but the few reports available that provide osteometric data comparable to those recorded at Pirincay do suggest that a camelid smaller than the contemporary llamas of La Raya and slightly larger than the contemporary alpaca, existed in the northern Andes from the Early Horizon of the Peruvian sequence and the late Regional Development period of the Ecuadorian sequence." Miller and Gill conclude that "[t]his Huacaloma population matches almost exactly the measurements of the Pirincay population" (Miller and Gill 1990:61– 62). In the case of Huaricoto, the bones belonging to the Early Horizon are slightly bigger than those of Pirincay but smaller than those of a present-day llama. Conditions do not change in the Huaraz phase, but the average falls in the Middle Horizon (Miller and Gill based this statement on a personal communication from Sawyer).

On the other hand, the studies by Miller (1979) at Marcavalle on Early Horizon materials show a bimodal distribution, with a small group closely resembling contemporary alpacas and a large group resembling present-day llamas. Wing's study (1977[a]) using materials from southern and central Peru, all from highland sites, shows a trend toward big camelids that falls slightly from south to north, thus exhibiting the pattern found from Marcavalle to Huaricoto and Pirincay.

Miller and Gill then explain that the osteometric analyses show a pattern that is

that of a classic cline[2] in which members of a single species, having diffused across geographical space from their historical point of origin, adapt progressively to slightly different environmental niches and become ecotypes in the process of incipient speciation. Following such a clinal model, llamas would be seen as having achieved their greatest size in the southern puna Andes near the center of their diversity and domestication and to have incrementally decreased in size as they spread into the páramo Andes to the north. Such a size gradient conforms to Bergman's Rule (recall that the area is south of the equator) and the general tendency of mammalian populations within the same species to have less body bulk near the equator, and those with greater bulk to be found farther from it. Differences in vegetation and other environmental factors between the puna and the páramo also may have contributed to diminution in body size. The cultural context in which these bones and their measurements occur, however, precludes a simple evolutionary explanation in which barriers to inter-population gene flow are limited to landforms, climatic regimes, animal behavior, etc. For the biogeography of a domesticated animal to express itself as a latitudinal cline, we suspect that cultural barriers, such as discontinuous trade relationships, must also have influenced the free flow of genes. At this point in time, however, we are prepared only to point out the existence of the camelid cline phenomenon; definition of the mechanism by which it occurred must await further research. (Miller and Gill 1990:63)

The final conclusions drawn by Miller and Gill (1990:63–65) are lengthy but worth summarizing. They raise three major points. First, in contrast to the highlands to the south, where a camelid-based economy flourished between the second and third millennia BC, the inhabitants of southern Ecuador depended on wild animals, mainly cervids, just before the beginning of the Christian era. Everything seems to indicate that camelids were completely absent from these latitudes before their introduction from the south in a domestic form during the Regional Development period (ca. AD 100). (This obviously is a contradiction, for earlier dates corresponding to the Late Formative had previously been given; see above.) This perhaps agrees, Miller and Gill say, with a single "interaction sphere" connected with the expansion of the Chavín cult during the Early Horizon (500–200 BC). In the northern Andes, the simultaneous association of camelid herding and Chavín cultural materials is found at the sites of Kotosh, Pacopampa, Huaricoto, Guitarrero Cave (Complex III/IV), and Huacaloma.

The second point is that if the absence of wild camelids in Ecuador is confirmed, then Cieza de León's claims of the existence of vicuñas and guanacos close to Tomebamba and Loja will have to be discarded. It is unlikely, say Miller and Gill, that these wild animals could have been completely absent in 100 BC, present around AD 1500, and then absent once more in the present day without a major climate change. Despite the statement by Cieza de León, Miller and Gill believe that in this case he perhaps used southern patterns he was more acquainted with to emphasize the decline wrought by the Spaniards.

Third, and this is the most suggestive point drawn from the Pirincay data, is the existence of a previously unidentified camelid form whose size lies in between that of contemporary llamas and alpacas. Miller and Gill believe they have a smaller llama, and hypothesize that this was the form that prevailed in the north from about 10° S latitude, from the Early Horizon to the Middle Horizon. If this were true, Miller and Gill believe that it would prove three things.

First, it would prove there is an unexpected degree of polymorphism in the llama, which obviously puts into question any method of osteometric discrimination exclusively based on statistical evaluations, such as the discriminant analysis decision rules (Wing 1972), stepwise discriminant analysis (Kent 1982[a]), or bivariate scatterplots (Miller 1979). In other words, all methods of analysis would have to be redesigned.

Second, it would cast doubt upon several zoo-archaeological identifications of alpacas, especially in areas far from the normal distribution of *Lama pacos* in the southern sierra. Instead of the alpaca, the small llama would prove Wing (1977[a]:852) right in that there was a "predominance of the small form in the northern part of Peru south to Tarma, and the predominance of the large form in the southern part of Peru north to Ayacucho."

Third, the admonition by Shimada and Shimada regarding the inappropriate application of southern contemporary camelid management models should be given due consideration. Miller and Gill agree with Shimada and Shimada (1987:

837) that camelid raising cannot have been "static or homogeneous, in time and space." If these smaller animals were raised in the north, it might have been practiced in a way that does not necessarily parallel that of the south.

Miller and Gill then cite Stübel (Tschudi 1965: 126–127 [two mistakes are made here: first, the date is 1969, and second, the correct thing is to cite Stübel as the author of this letter, even if this is noted in the text. For the contents of this letter see below]), and based on this letter wonder whether the animals might have been used as beasts of burden in a limited way due to their small size and because they did not do their job properly. Miller and Gill wonder what their primary function could have been if this were true.

It has been frequently argued that some relationship existed between llama domestication and verticality (Murra 1965; Custred 1979). However, others believe that the horizontal development of civilizations was based on a strong regional exchange where llamas had a major role. Such is the case of Lynch (1983[a]:10) for the Tiahuanaco culture (but I believe that his position falls within the first group). The article ends quite forthrightly with the following question: "[I]s the undersized llama a prehistoric reality, or an osteometric mirage?" (Miller and Gill 1990: 65).

Bruhns later had the kindness of telling me that the study done by Miller and Gill (1990) did not include work on the materials from the 1985–1988 field season. "In these seasons we found very big numbers of camelid remains, particularly in the 'ceremonial' pits cut in the pavement c. 300 BC" (letter of 22 January 1992). The paper presented at the Society of American Archaeology meeting in 1991 discusses these pits (see Bruhns 1991b).

Guffroy made some comments on the work done at Pirincay that are of interest here. He wrote me as follows: "The significance of the camelids is clear in the upper levels of Pirincay. However, I believe that we must be careful with their chronological positions (see the C^{14} dates). As with the rest of late Cerro Narrío, these phases could be more related to the Early Intermediate Period than to the late phases of the Early Horizon. We have no specific data for Catamayo

(1,200 m), save for its absence in the Formative levels" (letter of 15 February 1990).

(It should be noted that Stahl [1988:Table 1, 358] assigns a date of 1000 BC to the finds made at Pirincay. However, he based his estimate on an unpublished manuscript by Bruhns and Hammond, which I do not know and which must obviously be a preliminary report.)

To finish with the Pirincay finds, we must consider that Bruhns et al. (1990:232–233) believe that metallurgy and camelid herding are two imports, the result of the hypothetical incorporation of southern Ecuador into an "interaction sphere" connected to the Chavín phenomenon. This would have taken place between the sixth and the second century BC. It is worth noting that Bruhns et al. refer to cultures connected with Chavín, rather than to a direct Chavín influence. They note that the remains of sacrificed llamas were found in association with maize, and believe that this is likewise an imported phenomenon.

Bruhns et al. (1990:232–233) say in closing that the Pirincay excavations demonstrated that interregional exchanges were important for the early cultures of the northern and central Andes, even though this connection with the south was unexpected, because it had previously been thought that camelids had been introduced into Ecuador only by the Incas. On the other hand, there was no evidence of exchanges with Peru, or else it came from nondocumented finds: "Pirincay's deep, undisturbed archaeological deposits have provided new and provocative information concerning these long-distance contacts and the changes in focus of these over a thousand years of prehistory."

North of Pirincay, apparently in a higher-altitude area, there is a site called La Florida where camelid remains have been found. No more data are available because this site was only mentioned in the article by Miller and Burger (1995:452, Fig. 15), based on a personal communication from León Doyon.

There are also data on two sites in the upper Chanduy, in southwestern Ecuador, 25 km from the Chanduy estuary and a short distance from the Pacific Ocean. According to Reitz (1990: 28–29), these sites correspond to the Guangala phase of the Regional Development period, between 100 BC and 800 BC. It is noted that in terms of biomass, camelids (llamas and guanacos) in the El Azúcar 30 site account for 0.136%, and 0.263% of the remains in El Azúcar 47B (Reitz 1990:Tables 2 and 18). Reitz's Table 22 shows that the camelids have the Andes as habitat or are domestic forms.

I shall summarize the most important data in Reitz (1990:28–29). She notes that camelids can belong to one of three species, the wild guanaco (*Lama guanicoe*), the domestic llama (*Lama glama*), or the domestic alpaca (*Lama pacos*). The wild guanaco is unknown in the southwestern coast of Ecuador, while the domestic llama is found on the Ecuadorian coast (Franklin 1982: 473). According to Reitz, the alpaca does not seem likely at such an early date. Besides, only two camelid bones have been identified at El Azúcar, an unmodified one and a modified and burned one.

Reitz notes that camelid remains had been previously found in tombs on the Ecuadorian coast, at the Ayalán and Peñón del Río site, but these contexts are more recent than those from El Azúcar (Hesse 1980, 1981; Stahl 1988; Wing 1986). There also are reports of camelids at Pirincay (see above), which must be contemporary with those from El Azúcar. (N.B.: Reitz agrees with Guffroy [see above] as far as the chronology is concerned.) When discussing Miller and Gill's study (1990) and the small llamas they identified, Reitz comments that it is hard to establish whether their size measurements agree with those recommended by Angela von den Driesch, data which were used for the study of the El Azúcar remains. Assuming they were, "[i]t is interesting to note that the range for the proximal first phalanx reported by them was 17.4–20.00 mm, with a mean of 18.4 mm. The phalanx recovered from El Azúcar 30 is at the upper end of their range and hence may be from a small llama. Differences among measurements based on phalanges must be used with caution since there can be several millimeters of difference between phalanges of the forefoot and hind foot. Nonetheless, these comparisons are interesting, although more study is needed. The metapodial could not be measured" (Reitz 1990:28–29).

Reitz later told me (pers. commun., letter of 12 November 1991) that "I admit that this is very early for such a find in Ecuador, but these two bones are really easily identifiable as camelid, so I feel sure that the identification is correct."

We also know from Elizabeth Reitz that these remains correspond to the materials excavated by María Masucci. She wrote me: "I made no effort to identify the bones beyond the Camelidae since they were a highly modified distal metapodial and an unmodified fused first phalanx. These were from widely separate areas of El Azúcar. Given the early data, I assume they were more likely to be guanaco, but they could have been trade items nonetheless" (Reitz, pers. commun., letter of 3 April 1992). This comment is interesting insofar as Reitz disagrees with her previously stated opinion (see Reitz 1990:28–29) that the remains from El Azúcar might belong to small llamas like those from Pirincay.

The last piece of information I was able to find for this period belongs to Cerro Narrío, which is north of Cuenca. They are vague data, mentioned in passing by Bruhns (1991b:7). All she says is: "Cerro Narrío's camelids . . . associated with the intrusive Integration Period burials." Stahl (1988:Table 1, 358) likewise refers to these findings (at least apparently so) without any explanation, but gives them a date that is evidently wrong, for he says "1000–1500 BC." (Stahl took the data from Collier and Murra [1943:68] and from Braun [1982], which I was unable to obtain.)

All the other data I have correspond to the late period between AD 500 and 1550.

Thus, Holm (1985:9), in a study on the Manteño-Huancavilca culture, wrote: "The first Spaniards said they had seen many 'lambs [*borregos*] of the land' on the coast, which at the time meant llamas (*Lama glama*), the American camelid that was unknown to the Spaniards. We found bone remains of llamas in excavations, but do not have any proof that this animal from the altiplano was domesticated on the Ecuadorian coast."

When discussing Holm's work (1985; Stahl uses the first edition [1982], which is identical to the 1985 edition I used), Stahl (1988:358) comments that he knows only two contexts on the western side of the lowlands where camelids have been found: the cemetery of Ayalán (Hesse 1981), and the one he himself identified at Peñón del Río.

The site called Peñón del Río lies east of Guayaquil, on the left bank of the Guayas River, close to the town of Durán. According to Stahl (1988:358–360), the occupation corresponds to the Milagro phase (ca. AD 1350), and there were camelid bones among the animal remains. Stahl wrote that according to Kent's method (1982[a]), and notwithstanding the small size of the sample, "the few measurements fall clearly into the larger guanaco/llama category," and everything seems to indicate that these are domestic llamas.

Stahl (1988:358–360) notes that the distribution of this Milagro-Quevedo complex includes most of the Gulf of Guayas, from the Andean spurs to the coastal mountains on the west, the province of Esmeraldas in the north, and the hybrid Milagro-Quevedo/Manteño-Huancavilca complex in the southern part of the coast of El Oro. However, Stahl believes that the Milagro-Quevedo phase is poorly documented and that the geographic limits are vague.

Stahl says that we must bear in mind that there were strong exchanges in the area connected with Mexico and Peru. He believes that the inferences to be drawn from this cultural scenario regarding the northerly distribution of camelids in Ecuador becomes clearer if we bear in mind that its westernmost and lower areas were never incorporated into the Inca domain, which ended on the coasts of Tumbes (he cites Murra 1946: 809). Stahl (1988:358–360) notes that save for the island of La Plata, some 23 km off the coast from Manabí, and the occasional and isolated find, there is no archaeological evidence of any Inca occupation in the western lowlands (according to the studies by Dorsey 1901; Estrada 1954:90; 1957a[b]:23; 1957b[a]:40; 1962:8; and Marcos and Norton 1981).

With these data, Stahl (1988:358–360) wonders how the few unequivocally identified camelid remains found in contexts in the western Ecuadorian lowlands dating to pre-Inca times can be explained. Stahl believes that there were mechanisms for the introduction of camelids prior to the Inca expansion. These might have been exchanges made with the highlands overland, or sea trade using balsa craft. The bones themselves might have been introduced with other camelid products exchanged, or as desired objects too. Stahl also suggests that a live animal might have been introduced as an "exotic luxury," or as a pack animal that was not returned to the highlands, or as a source of food or for the wool. He ends as follows: "More important, coastal herds may have existed in Ecuador at least by the time the Spanish arrived. Some supporting clues are to be found in the earliest European accounts of the area" (Stahl 1988:360).

Several scholars mention the Ayalán site. This is in the Santa Elena peninsula, in Punta Anllulla, close to Potrero de los Ceibos, on the estuary of the Guayas River.

Hesse's preliminary report (1980:139) is actually quite vague and discusses only camelids cursorily. We read that 68 tombs, 37 of which contained associations of animal bones, were excavated during the work carried out by Ubelaker in a great cemetery. From among these 37 tombs, there is a series of radiocarbon dates for 25 of them that give dates of AD 730–1730. For the other tombs, a series of radiocarbon dates that ranges from 500 BC to AD 1150 was established. Hesse comments that "[w]hile there is some overlap in the date ranges, the two burial types probably represent chronologically distinct occupation phases." It is then noted that there were llama (*Lama* sp.) bones among the remains associated with the graves. However, he does not indicate which of the two burial contexts is associated with these remains.

I have not read the report by Ubelaker (1981), just the appendix by Hesse (1981) which studies the fauna, but the chronology is not discussed here. All it says is that in the cemetery of Ayalán there were "small quantities of animal bone in association with the human skeletal material." The more common herbivores include "some form (or forms) of camelid, *Lama* sp." A more precise identification was not possible (Hesse 1981:134).

In the discussion it is noted that the camelids are outside their wild range, that the volume of animals associated with human burials "do not seem to represent" a deposit of food refuse, and that the remains are of skull and mandible fragments. Table 68 indicates that of the camelid bones found in urns, 17% came from heads and 12% from postcranial bones, while those outside the urns comprised 15% of heads and 12% of postcranial bones. Hesse notes that a group of bones that could not be determined were classified as *Lama/Odocoileus*, and were not included in the percentages (Hesse 1981:138).

When discussing these finds, Dedenbach Salazar (1990:89) points out that they belong to the Manteño and Milagro phases of the Later Integration period. Stahl (1988:358) also discusses them and cites the studies by Ubelaker (1981) and Hesse (1981), but breaks no new ground. Hesse ascribes them to the Milagro phase (ca. AD 500–1500). Novoa and Wheeler (1984:124) mention the finds without making any comment, but claim they are "[d]omestic camelid remains" and cite Hesse (1981). Frankly, Hesse never mentions the word "domestic."

Finally, it is to be noted that Wing (1986:Table 10-6, 259) also mentions these bones but without specifying amounts. Furthermore, she based her work on Hesse's preliminary study (1980).

The site of Sacopampa is in the northern highlands of Ecuador, in the province of Imbabura. It lies in an intermontane valley. According to Wing (1986:Table 10-6, 257), camelid remains here account for 26.9% of the fauna and correspond, in temporal terms, to AD 650–850. Wing says she obtained the data from Alan J. Osborn. Wing (1977b:16) first said: "The first vestiges of camelids in northern Ecuador are seen in the Sacopamba Site (Period 12, AD 650–840), and they are abundant in the following period at Cochasqui (Oberem, pers. commun.)." (Although the 1986 reports says Sacopampa, this one reads Sacopamba. Novoa and Wheeler [1984] also call it Sacopampa, while Stahl [1988] writes Sacopamba. I was unable to determine which is the correct spelling of this name, but assume it is Sacopampa.) Table 3 then shows that camelids constituted 20% of the faunal remains. In their discussion of these finds, Novoa and Wheeler (1984:124) mention llamas and alpacas, something which Wing did not mention, choosing instead to speak about camelids, and they claim that these "are common in the valley sites of Sacompampa." Stahl (1988:Table 1, 358) also mentions this site but gives it a later date, for he

says "AD 1250–1525," basing this statement on a study by Athens (1980:271–272) that I was unable to find.

Stahl (1988:Table 1, 358) mentions the presence of camelids in Sequambo (La Concepción, in the province of Imbabura) without much data, and dates them to AD 1250–1525. Athens (1980: 273) is once again his source.

Cochasquí, a site in the hacienda of the same name in the province of Pichincha, to the northwest of Quito, is important, but we do not have much data on the fauna. The site is in a longitudinal, intermontane valley at 2900–3000 masl.

Novoa and Wheeler (1984:124) discuss Cochasquí (which they date to AD 850–1430) and mention llamas and alpacas, specifying that these "are common." However, they are based on a then unpublished study by Wing, which appeared in 1986. In this article Wing (1986:Table 10-6, 257) actually mentions the presence of camelids in Cochasquí but without giving the number or specifying the species. Stahl (1988:Table 1,358) mentions the presence of camelids without giving additional information. His source is a study by Fritz and Schönfelder (1987:Table 4) that I was unable to obtain.

In this case it is Kaulicke (1989:244–245) who provides more data. He discusses an Early phase and a Recent phase. Both belong to the Late period (see Willey 1971:Fig. 5-4, 259). The Early phase is dated between AD 950 and AD 1250, and the Late phase to AD 1250–1550, the latter comprising a pre-Inca (AD 1250–1525) and an Inca period in AD 1525–1534 (Meyers 1989:196–197; Schönfelder 1981:259). Camelids represent 80%, 59.3%, and 84% in the Early phase, and 84.1% and 78.8% in the Recent phase.

Kaulicke says that most of the remains belong to llama. He repeats the hypothesis that in Ecuador there were neither guanacos nor vicuñas, and also follows the size of the bones. Kaulicke admits that there might be a small number of alpacas. He notes that young animals account for 6%, 6.7%, and 6.25% in the Early phase and 13.46% and 12.2% in the Late phase.

Also according to Kaulicke (1989:246), "camelids supplied most of the meat, apparently through the sacrifice of animals, especially adults. For Wing, this probably indicates other uses as suppliers of wool or of transportation, something that would not be surprising if the llama actually does predominate. In this regard it is also worth insisting that the small-sized camelid remains found in the settlement do not exclude the possibility that alpacas were raised as wool suppliers in the second occupation phase."

Horacio Larrain Barros (1980a:171) also discusses northern Ecuador. He notes that camelid herding—llamas and perhaps alpacas—was "certainly" in use in Pichincha and Imbabura, "and their remains have been found in graves." Not many data are presented. Bruhns (letter of 22 January 1992) in turn tells me that there really are not many references to Ecuadorian camelids. "The tombs on the slopes of Mt Pichincha excavated by León Doyon held evidence of the use of camelids in the form of one single woolen textile with copper decorations. The other textiles in the tombs are all of cotton. These tombs were dated to AD 500."

A discovery of camelids for which there are no data is Quinche (on the river of the same name, in Pichincha), which was mentioned by Stahl (1988:Table 1, 358). Stahl's source was a study by Jijón y Caamaño (1912:66, n. 1 [this must be a mistake, the date should be 1914]) that I was unable to find. This find apparently belongs to a late date.

Pucara de Rumicucho is a site in the Quito area. Horacio Larrain Barros (1980a:171) wrote: "The recent excavations the Banco Central made in the pucará of Rumicucho (1976–77), close to San Antonio de Pichincha, have yielded abundant remains of llama (*Lama glama*) bones together with Inca pottery (Almeida, personal communication 8.VI.77)." (There actually is a study by Almeida Reyes [1982] that I was unable to find, and from its title it follows that camelids were indeed discovered at the site mentioned.)

Stahl (1988:Table 1, 358) mentions the discovery of camelid remains at the Cerrito de Macají site, in the Riobamba Valley, province of Chimborazo, and gives it an age of AD 500–1500. He does not give any details. His sources are Jijón y Caamaño (1927:15) and Meggers (1966:148), which I have not read. Stahl (1988:Table 1, 358) himself notes the presence of camelid remains at Challan (Zula, Chambo River, in the province of

Chimborazo), which must have the same age as the Cerrito de Macají remains. No details are given either; the information is just attributed to Collier and Murra (1943:19–22), which I was also unable to obtain.

For the famous site of Tomebamba, all I have are the data sent to me by Olaf Holm (pers. commun., letter of 18 November 1991). "I saw some corrals in the paramo of Cajas west of Cuenca, on the way to the coast, walls of gigantic stones which for me may have been corrals for the palaces of Tomebamba (present-day Cuenca), but no one has studied them so far. However, the site of ancient and Inca Cuenca inevitably must have had [sic] enormous llama droves [flocks] nearby."

Cardozo (1974b:142), whose data must always be taken with the utmost caution, cites Jijón y Caamaño without giving any other bibliographical reference and says, "but he also mentions from his own sources the shaft tombs of Imbabura and their respective culture. Of them Jijón y Caamaño says: 'they had llamas.'" This is far too vague. Cardozo probably meant one of the studies by Jijón y Caamaño (1914, 1920), where he established an Inca period and three pre-Inca periods. There is no way to determine which of them Cardozo meant.

I was unable to locate Oña. Bruhns (letter of 2 January 1992) tells me that Presley Norton had told her that Matilde Temme found camelid remains in the excavations at Oña, but nobody knows whether these actually are camelids, and if so what species and period they belong to.

Finally, there are some very general data. Guffroy (1987:874) notes the absence of great camelid herds in the province of Loja in prehispanic times. On the other hand, Bruhns (1991a:8) mentions relations between Peru and Ecuador in the Middle Horizon. She discusses the presence of lime in Azuay and Cañar and notes that "[t]he possibility that lime was exchanged for coca cannot be rejected due to the existence of pack llamas in this area." When discussing the Inca conquest of Ecuador, Bruhns (1991a:9) comments that "[w]ith their permanent villages, agricultural peoples with harvests of maize, potatoes and quinoa, and the grasslands [*pajonales*] full of llamas, the verdant valleys of Azuay and Cañar were very attractive for the imperialist Incas."

9.5.2 Historical Data

Thus far I have analyzed the archaeological data I managed to assemble. The historical evidence follows. However, before reviewing the data for specific areas, let us take a look at some general data on the Inca conquest of present-day Ecuador.

First, it is worth emphasizing that accumlated mistakes and unsubstantiated facts have been combined and converted into axioms. I specifically mean the practice of repeating that it was the Incas who took the llamas to the north. I will give just one example because I come back to this matter later. The example is from Gade (1977: 116), who wrote: "It is possible that it was the Incas who spread the llama to the farthest reaches of their empire, from northern Ecuador . . . although nowadays they are no longer found in those zones."

Bennett (1946a:49) had a better grasp of the problem, because when discussing Inca times he noted that "although they [the llamas] may have existed before, they certainly increased in numbers and importance." Bennett then notes that the llama and the alpaca did not adapt well to the Ecuadorian *páramo*, so although they are still found there, it is only in small numbers because it "is obvious that Ecuador is a poor country for breeding these animals."

In other words, it cannot be denied that the Incas took domestic camelids with them in their expansionist campaigns, but this does not mean the animals did not already live in the area, as the archaeological data show.

As for the Incas, they were apparently interested in making these animals develop in the Ecuadorian area, as follows from Cieza de León (1985:Part II, Chap. LVIII, 167), where we read that Topa Ynga commanded that he be continually informed of what he had left behind in Quito. One of the things that most interested him was "the growth of the livestock."

Cieza de León (1985:Part II, Chap. LXVIII, 198) himself narrates the Quito campaign and mentions the Pastos. Cieza says that Guaynacapa "commanded the headmen to pay tribute and they said they had nothing to give him, and to levy it he commanded that each house of the land was

bound to give him in tribute, every so many moons, a somewhat big cane tube full of lice. At first they laughed at this order, but later, seeing that despite the many they had they were unable to fill so many cane tubes, they raised the livestock that the Inca had ordered be left there, and they gave in tribute what multiplied and the food and roots there are in their land." Horacio Larrain Barros (1980a:325) discussed this passage, and it is best to cite him exactly:

> Here Cieza indicates that the Inca left auchenid livestock (llama, and perhaps alpaca?) among these northern groups. Does Cieza mean the Pastos themselves? In any case, it seems that the Quillacinga are not included here. If we assume that the Incas imposed the woolly auchenid livestock to groups located farther north than the Pastos, then they a fortiori must have imposed the livestock on the latter. In any case, it seems obvious that the Pastos paid tribute in two major ways: (a) with the offspring of the woolly llama livestock, and (b) with the various tubers that thrived in their high-altitude environment.

In 1582 Gaviria (1965:286) in turn wrote as follows in his account of the "town of Santo Domingo de Chunchi": "They were taught all the idolatries after the Inca came . . . and were made to sacrifice sheep of the land in these lands." When discussing Gaviria's account, Dedenbach Salazar (1990:89) notes that everything seems to indicate that camelids were long known in Ecuador, because the animal surely cannot have been a recent import if it formed part of a religious rite, for then the whole rite would have had to be new.

While discussing Inca influence in Ecuador, Salomon (1980:280) notes that although the introduction of the llama must have had a great impact on the economy, there is no trace of a more direct intervention so as to reorganize the economy of the chiefdoms.

Some scholars accept the presence of camelids in Ecuador, both domestic and wild ones, but do not present any actual data. Thus Flores Ochoa (1979a:226) points out that there were herds of llamas in this area until at least the eighteenth century, but his source is of uncertain reliability because he cites Cardozo (1954:65). The same happens with Brack Egg (1987:62), who accepts the presence of the vicuña in Ecuador and claims that it became extinct during the Spanish conquest and in Colonial times, but without citing his source.

Let us now turn to the actual data found in the written sources, starting with the coast. First we must note Estrada Ycaza's position regarding Cieza de León. Estrada Ycaza (1987:122) believes that Cieza's major informants on the Ecuadorian coast were first, his comrades, then the Indians in the Ecuadorian sierra, the Indians in the Peruvian highlands and coastlands, and last of all the boatswains. Besides, Estrada Ycaza (1987:132) believes that Cieza did not know the Ecuadorian coast and that "he obtained very little information during his hasty passage through the Ecuadorian sierra, and almost all the descriptive material he collected on the coast came from the 'old Indian captains of Guaynacapa' (Crónica, XLVII [1984:Part I, Chap. XLVII, 156]) he interviewed in Cuzco, whose testimony was not too objective or truthful." Estrada Ycaza insists on this point in his study (1987:141 and 153).

It is possible that Estrada Ycaza is partly right, but Cieza de León was quite straightforward in his writings, and, as he himself noted in the last pages of his *Guerra de Quito*, "I always give the reader an account of the way in which I write this narrative, to satisfy him so that he knows I am not making it up nor adorning it with what did not happen, nor wasn't" (Cieza de León 1909: Chap. CCXXXIV, 291). Thus, when discussing the town of Manta, he clearly says, "[s]o some Spaniards of the first who discovered this kingdom told me" (Cieza de León 1984:Part I, Chap. L, 162). When discussing the coast of Manta "[i]nland," Cieza says that "[t]hey found some livestock of the sheep they call of Peru, though not as many as in Quito, or in the provinces of Cuzco" (Cieza de León 1984:Part I, Chap. L, 162). Camelids apparently vanished in this area because in Lizárraga's writings (1968:Bk. I, Chap. III, 4), which date to the early seventeenth century and discuss the land from Manta or Puerto Viejo "inland," we read that the area "abounds in . . . cows and sheep, and is abundant with many horses and no mules." Although Lizárraga quite

honestly says "I have not seen it," we know that his data are correct. Although he does not specify what kind of sheep these are, we can assume they are from Castile, given the context of European animals where they appear. In this regard Stahl (1988:362) notes that perhaps the camelid herds vanished in this area slightly before the Spanish conquest, as happened in other places.

Several early observers wrote about the Island of La Puná, because the Spaniards arrived here in the third voyage of Pizarro. Pedro Pizarro (1978: Chap. 5, 18) discusses the Island of La Puná and notes that "[i]n this island they found five sheep of the land so fat that they could not breed, and when they killed them not even two arreldes[3] of meager meat were found on them." Pizarro then describes the Ecuadorian coast: "[T]hey have no sheep, nor did they find more than I have recounted as far as Tumbes," that is, the five sheep mentioned above (Pedro Pizarro 1978:Chap. 5, 19). This testimony is most important, because Pizarro was there and describes what he saw.

Gómara (1946:226) distorts the truth, because he said the following of these same events: "Puna goes down 12 leagues, and is from Túmbez as many leagues. It was full of people, of deer-like sheep [*ovejas cervales*], and of deer." Cieza de León (1987:Part III, Chap. XXXV, 103) is more reliable, but he mistakes the number of sheep. When describing the deeds of Juan Piçarro, Sevastián de Velalcáçar, and other envoys of Pizarro, Cieza de León says that "[t]hese who went found seven sheep; they killed them and quartered them to eat." Interestingly enough, Herrera (1945c:IV Decade, Bk. Nine, Chap. I, 51) also gives the same number of animals as Cieza de León but mentions their fatness, which Cieza de León omitted: "[O]n this occasion they found seven sheep, so fat they had no meager meat."

Dedenbach Salazar (1990:86) discussed the animals found in the Island of La Puná: "We cannot say with due certainty whether the livestock found in the Island of La Puná . . . were camelids that had arrived there through some exchange, or sheep which arrived before the Spaniards themselves also through an inter-American exchange." The point is that the reference that gave rise to this comment is Cieza de León (1987:Part III, Chap XXXV, 103; see above), but Dedenbach

Salazar forgets that Pedro Pizarro (1978:Chap. 5, 18) literally says "sheep of the land," so there is no question that these were camelids, quite possibly llamas.

Szaszdi (1988:7) reports that Salazar de Villasante listed the products of La Puná in the late sixteenth century: "There is in the island much livestock of sheep and goats of the Indians." However, he does not say whether these are "sheep of the land." To judge by the coat of arms granted to the cacique of this island in 1560, wherein European rams [*carneros*] appear, we can deduce that Salazar de Villasante does not mean camelids. This makes me assume that by this time there were no more camelids on the island.

Before reviewing the data on the Ecuadorian highlands it is worth citing a general observation Horacio Larrain Barros (1980a:171) made about Chapter XXXIV of the first part of Cieza de León's chronicle (1984:113–114), titled "Wherein we end the account of what there is in this land as far as the jurisdiction of the town of Pasto." Larrain Barros comments that the data therein presented on food are invaluable, insofar as it is pointed out that deer and bird hunting supplied the diet with the proteins essential as a "complete" [sic: I assume this is a misprint for "complement"] for a diet mainly based on vegetables. Unlike the more southerly tribes that lived in the "Callejón interandino," the significance hunting had for the groups living there suggests, according to Larrain Barros, that these groups depended on a mixed agricultural and hunting economy. The ethnic groups who lived near the Guayllabamba River to the south (Caranqui, Cayambi, Quitu, Puruháe, Cañari, Palta) instead must have had a mixed, agropastoral type of economy. According to Larrain Barros, north of the Guayllabamba River (i.e., north of the Pasto and Quillacinga), there is no reference whatsoever on the use of camelids, "even though their biome [sic] was particularly adequate to raise them." Larrain Barros remarks that the Inca conquest of this land was under way around 1526–1527, when Huayna Capac and his captains were crushing the Pasto. He ends by noting that "[t]here thus apparently was not enough time to settle either agricultural *mitimaes*, nor wool-producing auchenid livestock (llamas and alpacas) under their super-

vision. Auchenid raising (llamas, possibly alpacas) was certainly in use in Pichincha and Imbabura, and their remains have been found in burials."

Let us turn now to some specific facts. Cieza de León (1984:Part I, Chap. XXXIX, 126) gives an account of the struggle between the Otabalo and the Carangue for the treasure of the Inca. This happened in the province of Imbabura, in northern Ecuador. Otabalo "called most of its Indians and headmen and chose those he believed were the most disposed and swift, and these he commanded to dress with their shirts and long blankets. And taking thin and perfectly finished staffs they should get on the biggest of their rams [*carneros*], and place themselves on the heights and hillocks, so that they could be seen by the people of Carangue. And he, with more Indians and some women, feigned great fear, and showing they were going in fear arrived in the town of Carangue, saying how they were fleeing from the fury of the Spaniards, who had come upon their towns riding their horses." Cieza de León adds: "This news caused alarm and was held to be true because the Indians on the rams [*carneros*] appeared on the heights and slopes. And since they were far away, they thought what Otabalo claimed was true, and began to flee heedlessly."

The presence of llamas in the towns of Otavalo was confirmed by Paz Ponce de León (1965: 237) in 1582, after the introduction of European livestock. Paz Ponce de León said that "[t]he provisions they used and had, they now use; they eat meat from the rams of the land and of Castile; in earlier times, only the caciques and lords ate them."

Cardozo wrote about the Carangue or Caranqui, who lived close to the frontier with Colombia. He says that among the Caranqui, "the meat from the llama was a dish fit for the caciques alone" before the Spaniards arrived. Cardozo attributes this to Jijón y Caamaño, but without giving the corresponding bibliographic data (Cardozo 1974a:10). In another article published that same year, Cardozo wrote: "[W]e can cite Jijón y Caamaño, who noted that for the Caranqui, 'the meat from the llama was a dish fit for the caciques alone.'" Cardozo says that Jijón y Caamaño gave Sancho de Paz de León as his source (Cardozo 1974b:142). Although Cardozo does not say so, the passage from Jijón y Caamaño apparently comes from his 1951 study, but Sancho de Paz y León does not appear in the bibliography. This has to be taken with due caution, given the carelessness with which Cardozo uses the data. I was unable to verify his claim, but I wonder whether this is not Paz Ponce de León (1965), whom I have cited and who claimed that among the people of Otavalo "only the caciques and lords ate them [the llamas]."

There are more data for the Quito region. It seems that there were camelids here before the arrival of the Incas. Salomon (1980:196–197) studied the ethnic lords of Quito and found that the house of a ruler ideally was not just a center of political activity, but also the center of a cosmic order. Salomon cites Atienza ([1575] 1931: 167) to indicate the way the house "of a lord, [a] powerful cacique" was built. Among other ceremonies, Atienza indicates that they "sacrificed the animals they prized the most, such as live deer, rams [*carneros*] of the land, guinea pigs."

Salomon (1980:230) analyzes the impact the Inca conquest had in the zone of Quito. When discussing transportation, he notes that "[a] third type of traffic was the transportation of bulk goods: the mindaláes and their carriers, armies, and the herds of llamas bearing State supplies. Although the 1573 Anonymous (Salomon used 1965 [b in my bibliography]:213; this corresponds to 1938:28, which I used; see below) confirms the use of llamas to carry loads around Quito . . . these were not many and most of the cargo was moved on human shoulders."

It seems that there were many camelids in the zone of Quito when the Spaniards arrived. Tschudi (1885:98; 1891:100; 1918:213) recalls the killing of llamas by the Spanish soldiers "just to eat the brains," but he immediately adds in a footnote: "It seems they also did so to eat the heart, which they much liked, and so it is stated by an eyewitness, Alfonso Palomino, an officer of Belalcázar, the conqueror of Quito, in his true account of what happened in the ancient kingdom of Quito, page 2; and so they consumed more than a hundred thousand llamas in a not too big district, and in just a few months. (See Velasco "*Historia del R. de Quito,*" I, pág. 133 [Velasco 1946 in my bibliography])" (Tschudi 1891:footnote 2, 100; 1918:footnote 132, 213. This footnote was added to the 1891 edition; it does not appear in

the 1885 one. It also appears in the 1918 edition but was omitted in the 1969 one.)

Fernández de Oviedo (1959c:Bk. XLIX, Chap. III, 241) also collected data on the province of Quito and described its animals. "Of the animals they say there are many . . . which are common in Tierra Firme [the mainland], domestic and wild (like those big sheep from Peru, and the smaller ones)." This is most important, because Fernández de Oviedo clearly means the llama and the alpaca. What cannot be established is what "wild" meant for the chronicler. Were they vicuñas?

Cieza de León wrote twice about the area of Quito. The first is a general description that, among other points, reads thus: "In the jurisdiction of this city of Quito there was a great number of this livestock that we call sheep, which looks more like camels."[4] Further on he specifies that he means the "sheep [*ovejas*] and rams [*carneros*] that we call of Peru" (Cieza de León 1984:Part I, Chap. XL, 130). Then he recounts the founding of the city of Quito by Velalcácar, and observes that "there was such a huge number of a livestock so beautiful and lovely, which we have all seen in some places covering all of the fields, [and of which] there is now so little that almost none are left, but to stuff themselves with brains some killed five or six sheep, others killed as many to have marrow pie prepared" (Cieza de León 1987:Part III, Chap. LXXVIII, 256). Cieza's permanent concern for the indiscriminant killing of the animals, which reappeared when describing Peruvian land, is remarkable. This agrees with what Tschudi noted (see above), but I do not know whether Cieza de León was the source of Velasco (1946).

When discussing this passage, Dedenbach Salazar (1990:86) says it is "quite likely that [Cieza de León] also means camelids." I do not believe there can be any doubt about it.

Benzoni (1985:117–118) likewise described the Quito region in 1547–1550. He said "[t]hey had a very great number of sheep as big as asses and similar to camels, of which it is said that not long ago they had a disease like leper from which many died, but an even worse leper were the Spaniards, who have almost completely destroyed them." This testimony is invaluable because Benzoni was in the province of Quito and in the city of the same name, and was acquainted with the Flemish Franciscan friar Jodoco, from whom he found out about the early days of the conquest (see Radicati di Primeglio 1967:XXIV–XXV; 1985:20–21).

In the mid-sixteenth century Zárate recounted the voyage of Gonzalo Pizarro, who set out from Quito in search of the land of the "Canela." Zárate said: "[H]e set out from Quito, taking with him 500 well-provisioned Spaniards, hundreds of horses with their replacements, and more than 4,000 Indian allies, and 3,000 heads of sheep and pigs" (Zárate 1968:Bk. IV, Chap. II, 200). When Pizarro returned famished to Quito after being betrayed by Orellana, the Indians sent him "help [that] reached him more than 50 leagues from Quito" because "the vecinos of Quito had supplied a great provision of pigs and sheep" (Zárate 1968:Bk. IV, Chap. V, 205). It is not easy to establish whether these were sheep of the land or of Castile. It can be assumed, from the number and the early date, that these were sheep of the land.

Gutiérrez de Santa Clara (1963:Bk. III, Chap. LVII, 238) recounted the voyage Francisco de Orellana made with Gonzalo Pizarro in 1543. He says that "[w]hen Gonzalo Pizarro went by land from Quito to this conquest" he reached a town called Zumadoco, which I was unable to locate. "Gonzalo Pizzaro moved forward with 220 men, 150 horses, 4,000 Indian allies, and 3,000 sheep and pigs, and reached the town of Quixos." As regards the sheep, the same comment made regarding Zárate applies in this case.

Thanks to an account of "[t]he city of San Francisco del Quito," we know that in 1573 there were still camelids in the vicinity of the city of Quito. In a note, Jiménez de la Espada (1938:9) indicates that the original document is in the Ecuadorian Academy of History. The account was prepared in Spain and "seems to be a sister to the accounts of Piura, Loja, and Zamora prepared in Madrid by Juan de Salinas Loyola, governor of Yaguarzongo and Pacamurus, on command of licenciate Juan de Obando and answering the inquiry's 200 questions."

Question 72 was answered in the following way. "The other livestock is like that of Spain save for the sheep of the land, which are the size of an ass, [and] have the feet and hands, neck and head

of a camel" (Anonymous 1938:26). The answer given to question 78 was that "[i]n the land there only are cows, sheep, goats, pigs, mares, sheep of the land." In question 79 we read that "[o]nly the sheep of the land is held to be native, of which there are few and the natives use them, because they usually load them" (Anonymous 1938:28; see also Anonymous 1965b:213).

Herrera certainly used the data just cited (Anonymous 1938), because he says, when discussing "[w]hat can be said of the province of San Francisco del Quito," that "there were many sheep [ovejas] of the land, rams [carneros]" (Herrera 1945c:V Decade, Bk. VI, Chap. VI, 375). Herrera then takes the description of the city of Quito from Anonymous (1938) and attributes it to the "District of the City of Quito," saying that "there are not many sheep of the land, because the Indians ordinarily load them" (Herrera 1945d:V Decade, Bk. X, Chap. X, 144).

In the seventeenth century Vázquez de Espinosa (1948:1096/340) noted something very interesting when discussing "San Francisco del Quito." He said "from here we begin to find the rams [carneros] of the land of Peru." His testimony is crucial because the Carmelite priest was in Ecuador in 1614 (Clark 1948:VII), and this is clearly an account of a personal experience. Vázquez de Espinosa was more ambiguous when describing the "Bishopric of Quito." He mentioned the town of Hambato and the "province of the Puruae, which is most cold and [is] peopled by Indians and very many offspring of livestock, especially of rams [carneros], for in these districts there must be over 600,000 rams [carneros]." It is not clear whether these were rams [carneros] of the land or of Castile.

In the inspection made in 1559 in the town of Yuga, to the southwest of Tumbaco (at 2250 masl) and east of Quito, we find that "they served the Yngas in looking after their livestock of the land which were sheep" (San Martín and Mosquera 1990:166).

Cieza de León (1987:Part III, Chap. LXXXVII, 284–285) tells "how Velalcáçar moved the city of Riobamba to the Quito," from where "[t]hey set out several times in expeditions to fight with the lords who had rebelled, and fought in crags and won several skirmishes [alvarradas].

The [Spaniards] came in such disorderly fashion over the livestock of sheep that they wholly diminished the great number there was of them." Velalcácar then sent one Juan de Anpundia in search of Copeçopagua. Cieza de León tells that the latter sent three or four men, and then "Quingalinbo and other captains of the Inca came in peace, so he returned to Quito with much livestock as provisions."

In another passage, Cieza de León (1987:Part III, Chap. LXX, 231) details the adventures of Captain Sevastián de Velalcácar, his arrival to Quito and his departure for Cayanbe. On the way the Spaniards arrived at a town called Quinche, which was beside Puritaco. Cieza de León then says that "Velalcácar then returned for his men and they all walked to Cayanbe, where they saw the fields full of herds of very big and beautiful sheep [*ovejas*] and rams [*carneros*]."

Cieza de León (1984:Part I, Chap. XLI, 135) then gives an account of the expedition the Spaniards made from Quito to Tomebamba, where lay the "great habitations of Tacunga," and where the Incas had left many *mitimaes*. He says the Indians had very many rabbits, pigs, hens, and "therefore of sheep, lambs and rams [*carneros*]."

As Dedenbach Salazar (1990:86) correctly noted, in these cases the chronicler mentions sheep, lambs, and rams [*carneros*] but in a context wherein several European animals appear, so it not clear whether Cieza de León actually means camelids.

However, when describing an area more to the south, that of the Ambato River, in the midst of the Ecuadorian cordillera where the Sincho and Pillares Indians lived, Cieza de León (1984:Part I, Chap. XLII, 138) himself noted that there were herds of deer "and some sheep and rams [*carneros*] of those called of Peru."

Cieza de León is likewise not too specific when discussing another southern area in the vicinity of Riobamba, where Pedro de Alvarado went during his march to Quito. He said: "They saw in the fields some herds of sheep, which caused much rejoicing." In this place the Spaniards rested, "eating good lambs of those they found, which are peculiar and taste better than the best in Spain." When sending a detachment in search of the men who had stayed behind in the

cordillera, Alvarado ordered them to "take 25 sheep." Cieza de León then explains how the rescue party arrived with the sheep, and how much this helped them (Cieza de León 1987:Part III, Chap. LXVII, 218–219). Further on the chronicler tells that Diego de Alvarado, brother to Pedro, discovered a town called Ajo in the plain. Cieza de León once again says that "they found many sheep" when Pedro de Alvarado arrived (Cieza de León 1987:Part III, Chap. LXXII, 236).

Although he is not too specific, I believe that Cieza de León in this instance meant camelids. In this regard there is an interesting note by Stübel (1885:96; 1891:98; 1918:209, fn. 127) in a letter addressed to Tschudi in 1885. We read:

Further to the south [of Colombia] the llama is only found as a domestic animal in the vicinity of Quito, but it is not too abundant here either, so we can well consider that the Equatorial line is the northern limit [of the area] inhabited by this species. It is only around Riobamba, [and] certainly due to the sandy soil, that the llama has the same importance as a domestic animal for the Indians that it has in Bolivia, but here it is not used for long voyages, for example to the coast; it is only used as a pack animal for short distances.[5] I cannot say anything for sure about the presence of the llama south of Riobamba, towards the Peruvian frontier. I likewise want to state that there are a small number of llamas in wild state on the northern side of Chimborazo, as has been said, in some pastures planted at 4500–4800 masl. This was repeatedly assured to me, but I never had the chance to see them for myself. As for the vikuña, it is not found either in Colombia or in Ecuador. As far as I know, this animal is limited to the Bolivian altiplano, including a part of Peru.

The Puruhá groups lived slightly more to the southeast (see Murra 1946:Map 7, 787). There is a document called the *Relación de Xunxi* which describes the "province of the Puruaes." It is undated, but must be from the late sixteenth century (1582?). Here we read: "It is a temperate land, it is below the Chimborazo volcano." The account mentions the sacrifices made of "sheep of the land," and adds that "today there are many at the snow line, which the Indians do not kill, nor go to them to do them evil." However, when Diego Ortegón arrived, "to remove this vice from them, [he] ordered many Spaniards to go kill these sheep." The account then says of the natives that "[t]hey have sheep, sheep of Castile, goats, cows, many oxen and sheep of the land" (Paz Maldonado 1965:262).

Salomon (1980:286) wrote on the Puruháes, who, as the document says (M and R 1557), live in the Riobamba Basin in what is now the province of Chimborazo. These Indians fell under Inca control shortly before those in Quito. Salomon notes that the data on camelids are scant and confusing, perhaps because of the destruction of the herds which was in general due to the Pizarrista wars. Salomon explains that there were 100 llama heads in Guayllabamba, which according to Tiqui had been "dadas" (presented) by Padilla (perhaps as a restitution, or as investment capital for their reproduction; fol. 244r). The llamas Padilla held as his were probably the remnants of State herds, because the encomenderos usually claimed as theirs the possessions of the Inca State that were under the care of their grantees. On the other hand, State herds are known in the valley of the Chillos (close to Quito, to the south), but unlike the Puruháe, the lords there do not seem to have had their own herds (fol. 252r). Salomon finishes by pointing out that llama herding was organized in a pyramidal mita, just like maize agriculture in distant areas (fol. 252r).

There are several accounts of the zone of Cuenca. First, when describing Tomebamba, the present-day Cuenca, Cieza de León (1984:Part I, Chap. XLIII, 143) says: "In past times, before the Spaniards won this kingdom, there was a great number of sheep of those of the land all over these sierras and fields, and a greater number of guanacos and vicunias. But with the haste the Spaniards had in killing them, so few remain that almost none are left." This same chronicler narrates the adventures of Velalcácar, who had remained in San Miguel and from there went to Quito in search of treasure. On reaching Tomebamba, "[t]hey saw great herds of sheep [*ovejas*] and rams [*carneros*]"

(Cieza de León 1987:Part III, Chap. LVIII, 183). It can be assumed that these were animals of the land, because we know that in 1582 sheep of the land were being used in the offerings in Alusi, "in the jurisdiction of the city of Cuenca" according to Italiano (1965:288). That same year Gómez et al. (1965:284) refer to Canaribamba, also "in the jurisdiction of the city of Cuenca." They wrote: "There likewise are some rams [*carneros*] which these natives call llamas, animals the size of a year-old calf, and the neck is the length of more than a vara and a third, neither more nor less." And although Guaman Poma de Ayala was never in this region and does not specify what livestock ["*ganados*"] he means (his data must therefore be used with care, as Dedenbach Salazar [1990:85] correctly notes), when describing the city of Cuenca he says there were "few livestock" (Guaman Poma de Ayala 1936:fol. 998). Dedenbach Salazar 1990:97) believes there were llamas and alpacas at this site. Although she does not give her source here, I believe she is using Cieza de León (1987:Part III, Chap. LVIII, 183; see above).

When describing the journey between Tomebamba and Loja, which is more to the south, Cieza de León (1984:Part I, Chap. LVII, 182) says that "[t]here are many guanacos and bicunias which are like their sheep."

The presence of camelids in the region is confirmed by Salinas Loyola (1965b:296), who wrote on the "city of Loxa" in 1571 or 1572 and noted that there were "[l]ivestock of the land itself, sheep." He then added: "What they most prize is the clothes they wear and [the] sheep of the land themselves, due to the service and wools they have from them" (Salinas Loyola 1965b:303). Because of the wool, it is possible that in this case these are alpacas.

Herrera (1946b:VIII Decade, Bk. II, Chap. XVI, 384) wrote "[o]n the founding of the city of Loxa" in 1546, and noted that "[a]s for the animals, there are . . . sheep of the land and of Castile." Although Herrera was not in Ecuador, I find that his information is correct. For this city there is also a questionable reference (as was pointed out above) by Guaman Poma de Ayala (1936:fol. 1010), who says there was "some livestock."

In 1561, Salinas Loyola (1965c:197) narrated the voyage he made in 1557–1558 starting from

Loxa. Salinas Loyola says the Indians "[h]ad livestock of sheep of those from Piru." Then he tells how he founded a town called Valladolid. Of it he says: "The city of Valladolid . . . is . . . up to twenty leagues away from Loxa, part of which were gone over in north-south direction. . . . They had livestock of sheep of those from Peru" (Salinas Loyola 1965c:204–205). (Valladolid is on the river of the same name, a tributary of the Mayo River, at less than 2000 masl.) Leaving the settlement, Salinas Loyola "crossed a sierra of mountains that lasted leagues" until he reached Cumbinama, where "there was a great number of sheep from Piru" (Salinas Loyola 1965c:198). From there he moved to Coraguana, which "is a hunting ground," until he reached the present Santiago River and then the Marañón River (Salinas Loyola seems to be describing the region of Jaén) and a province called Cungarapa, where "[t]here are sierras and plains, and the land has a very good climate. . . . There were sheep, albeit not many" (Salinas Loyola 1965c:200).

I must point out that the account just cited actually includes several untitled accounts, all by Salinas Loyola. I have used them all as one.

Aldrete, one of Salinas's captains, left an account of the Gobernación of Yahuarzongo and Pacamurus dated in 1580. Aldrete mentions the city of Loyola, which is 24 km from the city of Loxa (on the Vergel River, a tributary of the Chinchipe River). He describes it thus: "It is a land of knolls and savannas, and the natives are all dressed in wool because they have many sheep of the land, which grow the bezoar stones in their stomach" (Aldrete 1965:150). Aldrete described the city of Valladolid in 1557, the year of Salinas Loyola's first expedition. He says that it was a land of "knolls and rugged land," and notes that fights broke out among the Indians to "steal the sheep." At the same time Aldrete explains that these Indians "went dressed with wool from these sheep" (Aldrete 1965:151–152).

Several modern scholars have discussed these points. One of them is Caillavet (1987), who studied the Bracamoros who lived between Loja and Valladolid, that is, in the Malacatos Valley, in the Sabanilla Cordillera. The Bracamoros stubbornly resisted the Incas, but were finally defeated. Caillavet says: "According to the sources them-

selves, the description of their way of life lets us see features that are no doubt due to Inca rule: the possession of llamas and guinea pigs" (Caillavet 1987:303). Caillavet based her study on the 1571 *Relación* of Salinas Loyola.

Hocquenghem (n.d. [1989?]:125) also studied the Bracamoro ethnic group, which she places "[n]orth of the land of the Tabaconas and east of the land of the Guayacundos, on the Atlantic slopes of the Andean Cordillera, in the upper Chinchipe River valley. It is a region of paramo and selva alta." (Hocquenghem cites the aforementioned *Relación* of Aldrete [1965], but does so incorrectly: Jiménez de la Espada, ed., 1965, Vol. III, pp. 151–152; she also cites the account of Salinas Loyola [1965c; see above] and in this case the reference is also incorrect: Jiménez de la Espada, ed. 1965, Vol. III, pp. 197–198) (Hocquenghem n.d. [1989?]:129–130). Hocquenghem says that the exchange of products between the Bracamoros and the highland peoples is confirmed by the llamas and the guinea pigs these people of the Amazon Basin had (Hocquenghem n.d. [1989?]:131).

Readers who want to expand these points should read Taylor and Descola (1981), who discussed the aforementioned documentation and locate the human groups in the "Upper Chinchipe region, north of the fifth southern parallel." This zone is not far from the frontier between Peru and Ecuador (see Taylor and Descola 1981: Map no. II, 11).

In another of his *Relaciones*, Salinas Loyola describes the city of Zamora, some 20 leagues "more or less" from Loxa, or more precisely "to the east of that of Loxa." It was peopled in 1549 and formed part of the Audiencia of Quito (but "before its government was one with that of the cities of Quito, Loxa, Jaén and the rest around it") (Salinas Loyola 1965d:125). Zamora is below 1200 masl. Salinas Loyola himself (1965d:125) explains that "[t]he valley and site where it is peopled is called Camora in the language of the natives themselves. . . . Besides said name, the land where it was peopled was called all together Poro-auca, which means 'Indians of war' who had not given obeisance and domain to the Incas, the natural rulers of Piru."

There is not much information in this *Relación*, but what there is is extremely interesting and important. Point 77 reads: "The natives themselves had and raised sheep of Piru, and of [the animals] from Spain the ones that best live are cows, pigs and goats" (Salinas Loyola 1965d:129). Point 164 specifies that "[w]hat they [the natives] prized the most were the sheep from Piru because they raised and had them, and others they call *cuys* [guinea pigs]" (Salinas Loyola 1965d:133).

Taylor and Descola (1981:15) said in this regard as follows: "The hinterland of the city of Zamora came to be known under the name of 'Province of Nambija,' and we have several documents on this region that were published by Espada [it should be Jiménez de la Espada]: the 'Relación de la Ciudad de Zamora de los Alcaides' by Salinas de Loyola, dated in 1582 (RGI 3:125 and following), and two documents prepared by Álvaro Núñez, an encomendero of Zamora, that probably belonged to a series of visitas made in 1582 (RGI 3:136 and following, and 139 and following)." A mistake was clearly made here. According to Jiménez de la Espada himself (1965b:116), the date of the "Relación de la Ciudad de Zamora de los Alcaides" is 1571. The account dated in 1582 is the "Relación de Zamora de los Alcaides, dirigida a la Audiencia de Quito," which was written in effect, as Taylor and Descola (1981) correctly point out, by Álvaro Núñez for Francisco de Auncibay, a "magistrate in the royal chancery of Quito." Curiously enough, only plants, not animals, are mentioned in it.

The description of the weather in both accounts just mentioned is interesting, and they concur. Salinas Loyola (1965d:125) wrote: "The jurisdiction of said city and the site where it is peopled is warm and wet. . . . It is a land of much vapors produced by the wetness of the land and because it rains a lot." Núñez (1965:137–138) in turn says: "The land is all warm and wet and very sick, both for its natives and for the Spaniards." This clearly is the ecology of the Amazonian slopes of the Andes.

Taylor and Descola (1981:52) do not agree with Salinas Loyola and question his statements. In their study they analyze the Jívaro groups of the highlands and the *ceja de montaña*, and comment that "[t]he raising of llamas and guinea pigs, common in the Chinchipe River basin, seems to have been almost absent from the Zamora Valley." They

add that "Salinas mentions the presence of llamas (something that is to be doubted), guinea pigs and copper axes, which bear witness to relations of exchange with the groups in the montaña."

Caillavet (1978:49) also comments on the Salinas Loyola account (1965d), and when discussing the Indians says that "[p]otential contacts and exchanges with the Palta ethnic groups conquered by the Incas would thus explain the raising of llamas and guinea pigs in the sixteenth century by the groups of the Amazonian lower slopes of the Andes." This goes beyond the subject of this study; I can only note that there are two Palta (or Xiroa) people, one living on the upper *ceja de montaña*, the other Palta Serrano, and both are closely connected (see Taylor and Descola 1981:16).

(To prevent mistakes it must be noted that Herrera [1947:VIII Decade, Bk. V, Chap. XIII, 115] took the data on the presence of "Sheep of the Land" in the city of Çamora [he does write it with a "Ç"] from the Relación of Salinas Loyola [1965d], as was noted by Jiménez de la Espada in an endnote to Herrera's chronicle [1947: VIII Decade, Bk. V, Chap. XIII, 135].)

I left a pasage from Guaman Poma de Ayala for the end because of its vagueness. However, it is worth citing because Larrain Barros (1980b) discussed it. This passage, from the Indian chronicler (Guaman Poma de Ayala 1936:fol. 168), implies that there was "livestock" in the region when the Incas arrived. The passage reads thus: "The thirteenth captain Capac Aponinarua Andesuyo. The other captains who went with Guayna Capac Inca to the conquest of Tomibamba [and] Quito . . . and these other Indians remain infidels in their towns in the montaña and are still to be conquered, and there are many other Indians on the other bank—this is a sierra toward the Mar del Norte to Margarita [where] there are very many Indians, gold and silver and livestock and infidels—the land has yet to be explored."

Larrain Barros (1980b:207) believes that around the time that Guaman Poma de Ayala wrote, that is, before 1613, the Yunga populations had not been incorporated de facto into the Spanish Crown, even though the information provided by this chronicler indicates that these had somehow served the Inca. The habitat indicated—the "tierra de la cierra hacia la mar del norte" (Larrain Barros uses the 1956–66 edition of Guaman Poma)—refers to the lowlands, not the highlands, that is, to areas on the banks of the rivers that form the Amazon (such as the Ucayali, the Urubamba, and perhaps the Tigre, the Santiago, and the Napo Rivers), the zone where Guaman Poma places (although in a vague and imprecise way) his Antisuyo and Chuncho Indians who were subject to the Inca. Since the Indian chronicler did not leave his homeland (Larrain Barros [1980b:140, fn. 7] and bases his statement on the study of Ludeña de la Vega [1975]), Larrain Barros believes that he perhaps had access to groups of *mitimaes* and their Quechua-speaking caciques, who may have served as informants.

To complete this review of the data on camelids in the Ecuadorian area, I want to mention two observations Salomon made. The first (Salomon 1980:141–142) concerns the chronicle of Cieza de León. Salomon draws attention to the tense of the verbs Cieza used. Thus he wrote: "*They have* much cotton, with which they make their clothes to dress and pay their tribute. In the jurisdiction of this city of Quito *there was* a great number of this livestock that we call sheep, and which more precisely resembles a camel." (Salomon cites p. 130 of the 1962 Cieza de León edition, which corresponds in my bibliography to Cieza de León 1984:Part I, Chap. XL, 130; emphasis is Salomon's.) Here we see that when the chronicler mentions the livestock he does so in the past tense, while when discussing cotton he does so in the present tense. For this reason Salomon believes that "these are herds destroyed by the Spaniards, or lost because they were no longer tended by the Incas."

Salomon (1980:141–142) then comments that since Cieza de León was in Quito before 1550, he must have seen clothes manufactured in Tahuantinsuyo. "The association of wool with the imperial economy is also reflected in the 1559 visita, where the statements about camelid-wool textiles are invariably in the past tense. They only refer to the output for the State, never in benefit of a cacique or private individuals (M and SM 1559:fol. 838v, 856v). By 1573 this output was no more, few llamas were left around Quito and were usually used as beasts of burden (Anonymous [1573] 1965[b in my bibliography]:213)." Salomon like-wise notes that the native textile industry underwent several abrupt changes in 1400–1600. A local tradition of using plant fibers survived throughout the area partly as a local tradition, and partly as an Inca tradition of cotton weaving.

Salomon (1980:141–142) ends by saying that

[w]e must also bear in mind the fact that there was little to choose in the interval between the decline of the Inca herds of llamas and the growth of the herds of sheep. In fact, the period 1534–1548, when native society had to fall back upon its own resources, can be seen in more than one respect as a revival of some pre-Inca institutions. It had been a period of growth for the cotton industry. As far as Quito is concerned, the introduction of the wool industry by the Incas was mainly a device of the State sector. Its goal was the increase of the tribute revenue and the accumulation of strategic reserves. Unlike most of the southern provinces, Quito did not have herds cared for at the chiefdom level or below. . . . The communities in the Chillos valley only reported an output for the Inca State (M and MS 1559:fol. 817v, 838v).

Salomon's second observation (1980:279) explains that llama raising seems to have moved forward with the Inca military frontier in terms of both distance and speed, because according to the sources the llama was known "of old" (Salomon cites Paz Ponce [sic: de León] [1582] 1965:237) in the locality of Otavalo and was consumed by the local chiefs, even though, Salomon notes, these animals are not mentioned as a source of fiber for clothes. To the south, in Quito, llamas are only mentioned as beasts of burden (and Salomon cites Anonymous [1573] 1965[b in my bibliography]:213). Farther south the llama was associated with the textile industry and ceremonial sacrifices (Salomon cites Paz Maldonado [1582] 1965:262; Gómez et al. [1582] 1965:284; Salinas [sic, Loyola] [1571?] 1965[b in my bibliography]: 303). Finally, Salomon ends by noting that in Xunxi, the "ceremonial and devotional use (of the llama) was specifically described as an Inca innovation."

Finally, I cannot avoid discussing several points for which I found some data, or to which attention must be drawn for one reason or another.

One subject that warrants careful study in the future by specialists is, in Gilmore's words (1950: 435), the possibility that "smaller [llama] breeds" lived in Ecuador. This point apparently had no support at the time Gilmore wrote his study, but it is once more under discussion, owing to the abovementioned archaeological evidence.

Geography is one aspect that has always been considered relevant for the dispersal of the camelids, that is, the variations that take place north of the Peruvian border. Troll (1935, 1958) was one of the many convinced that the binomial warmth/humidity had a major role in this, but as a constraint. Troll wrote:

> It goes without saying that it is not wholly impossible to also raise llamas in the "paramo zone." But the fact that the optimal puna conditions do not exist there is proven by the attempts at acclimatization the Incas made in present-day Ecuador. The llama has survived there until the present day but only in small numbers, and not without reason. Most of the Ecuadorian llamas are found in the central provinces (Chimborazo, Tungaragua and León), where the paramo is especially dry due to the fact that the soil is formed by very permeable "volcanic tuffs." On the other hand, there the llama is only for local use, and is not used for long voyages. From all that has been published on llamas in Ecuador, it clearly follows that successful breeding places cannot be established there. (Troll 1935:142)

Miller and Gill (1990:51, footnote) also discuss this point. They note that the *páramo* of Ecuador, Colombia, and Venezuela can be compared only approximately, but are significantly different from the punas of Peru and Bolivia. The *páramo* zone in the Equatorial Andes is characterized by irregular terrain that lies approximately between 3800 and 4700 masl, and has a permanently cold and wet climate, with precipitation all year long. Meanwhile, the puna lies between 4100 and 4800 masl, is considerably drier than the *páramo*, and has a seasonal rainfall. (Miller and Gill base their claims on the studies by Cuatrecasas 1968; Troll 1968; Pulgar Vidal 1987 [n.d. (1980?) in my bibliography].) Miller and Gill (1990:51) then note that despite the lack of studies on Ecuadorian zooarchaeology, it is clear that the endemic fauna of the region was significantly different from that found in the analogous intermontane valleys of the Peruvian area. They say that Wing (1973:3 [I do not know this study]) has estimated that up to 80% of the animals were different in both areas.

These same scholars (Miller and Gill 1990: 49–50) are quite emphatic in noting that the geographic distribution of the domestic camelids, as well as the habitat of the wild species, has "become synonymous with the highlands of southern Peru and northern Bolivia." Actually, the possibility that camelids might have been previously abundant in northern Peru and Ecuador was considered in the nineteenth century by only a few scholars. Miller and Gill mention Cieza de León's statements (1973:110–155 [1984:143–182 in my bibliography]) about the presence of vicuñas and guanacos in the Tomebamba region and in some other parts of Ecuador (we saw that there actually are more data than that in Cieza; see above), and that the llama and possibly the alpaca may have developed there under Inca control. Miller and Gill end with the following question: "This commentary, even if accepted without reservation, does not, however, address a more fundamental question: were the domesticated camelids, as has long been assumed, introduced only with the Inca conquest of Ecuador in 1463 under Thupa Inca or does their Ecuadorian tenure have greater antiquity? It also leaves unresolved the related question: which of the domesticated camelids (llamas or alpacas) might have been present in these northern climes?" They do not give an answer. However, it is somewhat striking that, having studied the Pirincay remains, Miller and Gill now have some reservations about the presence of camelids in Ecuador in pre-Inca times. I return to this point later.

Stahl (1988:357) is also interested in this issue. He begins with the premise that at least the domestic llama was distributed in high-altitude and western lowland contexts of Ecuador. He recalls that the number of these animals in Ecuador is very low at present, and are found mainly in the high-altitude basins of Cotopaxi, Tungurahua, Cañar, and Chimborazo (his data come from Franklin 1982:475; Sick 1963:133; Acosta-Solís 1965:362).

Stahl notes that after observing the concentration of llamas in the Chimborazo Valley in the late nineteenth century, Gonzáles Suárez (1969: 208, n. 8; the original edition dates to 1892) made a statement that was somewhat insistently repeated throughout the following century, i.e., that the introduction of the llama was a recent event, one ascribable to the Inca expansion. Gonzáles Suárez suggested that the llama was taken by Aymara colonizers sent as *mitmajkuna*. (I am not well acquainted with the history of Ecuador, but to judge by the dates it seems that Gonzáles Suárez was influenced by Tschudi, who, as we shall see, published his ideas in 1885.) It is interesting, says Stahl, that the idea that the llama was taken to Ecuador by the Incas has persisted despite the fact that Jijón y Caamaño (1927:15) identified remains from these animals in pre-Inca archaeological contexts in Cerrito de Macají, and that Murra (1946:792; see below) stated that there might be llamas in association with Ecuadorian pre-Inca remains. (Stahl lists Troll 1931:264; Murra 1946: 804; Gilmore 1950:433; Sick 1963:133; Wilbert 1974:24; Salomon 1986:82.) Stahl (1988:357) concludes that "[d]espite their relative paucity in highland Ecuador, there is definitive evidence of a pre-Inkaic presence of camelids in the area, with some suggestion of a great antiquity."

Gilmore (1950) wrote something in this regard, but without any real support. He stated that "the guanaco may have ranged to Ecuador and southern Colombia ('Páramo de Guanaco') at that early time [he means pre-Inca times], and the midden bones which have been identified as those of llamas may actually be those of guanacos" (Gilmore 1950:433). Gilmore then insists on this point, namely, that the "alleged" llama bones in the early sites of Ecuador could be guanaco (Gilmore 1950:447). Gilmore based his conclusion on the fact that the guanaco ranged as far as Colombia in the late Pleistocene and early Holocene, in *páramo* lands.

I will now summarize some of the arguments presented by the scholars who defend one or the other position, i.e., those who claim the Incas were responsible for introducing camelids in Ecuador, and those who posit that these were already there in pre-Inca times.

Tschudi (1885:96; 1891:98; 1918:209–210; 1969:126–127) discussed the point with scientific honesty, for he wrote: "It is not known whether there were aukenias [*Auquenien* in the original; I have retained the spelling of the 1918 translation] in the present-day republic of Ecuador at the time of the kingdom of Quito or of the Skiris, but it is believed that after the Incas conquered Quito they took many llamas to the land, especially with Wayna-Kapax [*Khapax* in the original]. . . . We have no precise data so far that can let us set the northern limit reached by the various species of aukenias in the very highlands of Peru."

Troll is far more categorical than Tschudi, but his arguments are weak. On this subject Troll wrote: "Before Inca rule there was no kind of 'Auchenia' in Ecuador, not even the two who live in a wild state, the vicuña and the guanaco. The llama has survived in that country to the present, but without any particular significance. Actually, both the llama and the alpaca are really animals characteristic of the puna, and in Ecuador the puna is replaced by the paramo" (Troll [1931] 1935:140). It is interesting that the only source cited by Troll is Gonzáles Suárez (1890–1892:195 [although I was unable to find a copy, I have data that say the correct date is 1890–1903], which according to Stahl has a 1969 edition), and that he strangely enough ignores Tschudi. In a later study Troll (1958:28) would repeat once again that in Ecuador the llamas "were belatedly introduced by the Incas."

Steward (1949:718) also discussed this point and said that "[l]lamas became important in Ecuador only after the Inca conquest." When discussing hunting in the Quito region, Salomon (1980:135) commented that "this, augmented by the camelids of Inca origin, provided a supplement of meat."

As can be seen, all the arguments—whenever these are presented, for they often are nothing but statements—are very weak and mainly of a geographic nature. They begin with the assumption, fallaciously converted into an axiom, that camelids are puna animals. Let us now see what the scholars who accept the presence of camelids in Ecuador in pre-Inca times have said.

Murra has often been cited when arguing that camelids were introduced into Ecuador by the Incas (e.g., Miller and Gill 1990:51), but this is incorrect. When noting that Cieza de León mentioned the presence of llamas in Ecuador, Murra (1946:804) said that these were *probably* of Inca origin" (emphasis added). In later studies Murra (1964a:118; 1965; 1975:118) cited Troll (1931: 266; 1958:29) in regard to the llama and alpaca being high-altitude animals which were therefore artificially introduced in Ecuador by the Inca state. However, Murra did not discuss this position, nor did he support it. His position was, however, clearly stated in a previous study. There he wrote: "Gonzáles Suárez [no date is given, but it obviously is 1890–1903] suggests that the llama was introduced by the Inca, and this is probably correct insofar as extensive use is concerned. *Nevertheless, archaeological evidence points to pre-Inca presence of llamas in both Imbabura and Cañar*" (Murra 1946:792; emphasis added). His position is therefore clear. (It is worth recalling that Imbabura lies north of the Equator and at more than 4000 masl, while Cañar is south of it at about 2,000 masl.) While discussing the Inca contributions to the Ecuadorian area in this same study, Murra (1946:810) notes that "[l]lamas were brought north in large numbers and provided Ecuadoreans with a beast of burden and a new and reliable protein supply."

The data Cardozo (1974b:143) presents are certainly interesting, but the problem with him, as I have repeatedly noted, is that he is an unreliable source. I was unable to check his sources, so he is cited with due caution. According to him, although Gonzáles Suárez (Cardozo cites a 1973 edition) "believes that llamas were massively imported by the Quichua and Aymara, he admits the pre-Inca presence of camelids. . . . In Otavalo there would have been, according to this Ecuadorian priest, 'one or perhaps two [pre-Hispanic domestic species], to wit, the llama and the dog.' This historian is perhaps the most skeptical as regards presenting the broad geographic distribution, as was done by Juan de Velasco (1946) and the Spanish chroniclers." Were this passage and the comment correct they would be most signif-

icant because they would show that several scholars, including Stahl (1988) and Murra (1946; see above), have misread Gonzáles Suárez. This point is open to revision.

Cardozo (1974b:143) also says that a "late nineteenth century Ecuadorian, Padre Wolf (1892), . . . mentioned the llamas of 'Tungurahua, Chimborazo and León'." There is no way of finding out whether this means pre- or post-Incaic camelids. (Besides the scholars mentioned here and in the section on archaeology, Cardozo [1974b:143] mentioned the following, who, according to him, defend the presence of camelids in pre-Inca times. The data could not be checked and should be taken with due caution. They are Jijón y Caamaño [1951], Larrea [1971, 1972], Andrade Marín [1954]; cited by Salvador Lara [1971] and Noriega [1947].)

Wing (1977b:17; 1986:260) accepts the presence of these animals in Ecuador both on the coast and in the sierra by at least AD 700. Stahl (1988:355) is also definite: "[C]urrent archaeological evidence definitely documents the presence of pre-Inkaic camelids. A limited amount of historic evidence may even suggest the indigenous maintenance of herds in lowland environments prior to the cataclysmic arrival of the Spanish."

According to Pulgar Vidal (1950:19 [85]), camelids were also imported to Ecuador in Colonial times, but the only source he mentions is W. H. G. Kingston, and he does not give any more information because his article lacks a bibliography. According to Pulgar Vidal, Kingston gives an account of his voyages along the upper basin of the Cauca River (close to the Vinagre tributary) and reports that there were several llamas in one park that had been brought from Upper Peru, animals that were also occasionally used as pack animals. Pulgar Vidal concludes thus: "Finally, we know that some hacendados in the savanna of Cundinamarca have brought auchenids, who for several reasons could not acclimatize and died."

I have only found two references in Cieza de León to the killing of camelids in Ecuador during the Spanish conquest. One concerns Velalcácar while he moved the city of Riobamba to Quito, for which reason the Spaniards made several 'entradas.' Cieza de León wrote in this regard: "The

[Spaniards] fell upon the livestock of sheep in such disorder that with their poor administration they wholly diminished the great number there was of them" (Cieza de León 1987:Part III, Chap. LXXXVII, 284). The other reference (Cieza de León 1987:Part I, Chap. XLIII, 143) appears when Cieza de León discusses Tomebamba (Cuenca) and indicates the great number of sheep "of those of the land," and the "greater number of guanacos and vicunias" there was before the Spaniards arrived. He immediately adds that "with the haste the Spaniards had in killing them so few remain, that almost none are left."

Wing (1975:36; 1977a:852 and Table 1, 839) has drawn attention to the fact that there are no camelid remains in the early samples from Sigsig (in Chobshi Cave, 2400 masl), which date to a period that extends from 10,000 to 8,000 BP. It is interesting that until then Wing still believed that "[l]ater sites in southern Ecuador may include introduced llamas."

I have Bruhns Olsen's comment (pers. commun., letter of 22 January 1992) on the latest discoveries made, which I find interesting and prefer to quote. "So far no camelids have been reported in coastal Formative or Regional Development Period sites. I believe that camelids were not too dispersed in Ecuador until the Integration Period. Donald Collier says they did not exist at Cerro Narrío; that the bones found there were all from deer (I have also noticed on several visits to the site that the bones found besides looters' shafts, both old and new shafts, were all from deer or animals such as dogs, etc.). There are almost no data on excavations to the north. I do not recall whether camelids were found in Athens' excavations close to Otavalo, but I do not think so. Surely the problem is not the absence of camelids in Late Formative highland sites (and later), but the lack of excavations in the sierra." Bruhns Olsen adds:

> So far we have abundant remains of a small llama in the levels from c. 3–4000 BC to the abandonment of the site in the first century of the Christian Era. There are no camelids at Chaullabamba, a site belonging to the early phases of Pirincay. At Cerro Narrío there are no camelids,

neither in the early (contemporary with Chaullabamba and Pirincay) nor the late layers. Guffroy did not find camelids during his research in the Catamarca valley, in Loja, but it is possible that these animals exist to the north in Oña, a more verdant area with pajonales. The province of Loja is semidesertic, but there are pajonales and some valleys with water, and I feel that the absence of camelids reflects the lack of investigations. Guffroy's research was not too extensive and we know nothing of the results of Temme's work.

Stahl (1988:361) made a comment on the data in the chroniclers which I find interesting and suggestive. He points out that wool-work is constantly mentioned among the coastal people of Ecuador, yet the tribute list for Quito studied by Salomon (1986 [1980 in my bibliography]) indicates the predominance of cotton clothes. Stahl also notes that the source of these textiles does not appear in the documentation. This is a point that historians should study.

I found only a few data on camelids in Ecuador in the nineteenth and twentieth centuries. I include them just as a reference, because there certainly are far more data.

After his visit in 1802, Von Humboldt ([1849, 1876] 1971:145) mentioned the llamas "that wander over the western slopes of Chimborazo." He explained that according to what the Indians told him, the llamas "went wild . . . when Licán, the ancient home of the lords of Quito, was reduced to ashes." Von Humboldt then comments that "[t]he herds of llamas are just as many as what I saw in the plateaus between Quito and Riobamba" (Humboldt 1971:146).

In 1885, after discussing some data on camelids in Colombia which have already been mentioned (see above), Stübel (1885:96, fn. 1; 1891:98, n. 1; 1918:209, n. 127; 1969:126–127, n. 10) made the following comment: "Further to the south the llama is only found as a domestic animal in the vicinity of Quito, but neither is it too abundant here, so we can well consider that the Equatorial line is the northern limit [of the area] inhabited by this species. It is only around Riobamba, [and] certainly due to the sandy soil, that the llama has the same importance as a domestic animal for the Indians that it has in Bolivia, but there it is not used for long voyages, for example to the coast; it is only used as a pack animal for short distances." (The 1918 translation mistakenly says "long" instead of "short.") Interestingly, in the 1940s Troll (1958:28) noted that "there are still llamas in the paramo above Riobamba, they are used to carry cargo locally." Nachtigall (1966a: 193) recently pointed out a small llama center in central Ecuador, close to Riobamba.

Although in this case Cardozo (1974b:144) seems to be reporting a personal experience, he is not really to be believed because he is not specific and his statement is conditional. Cardozo explains that while traveling in the provinces "from the bridge at Rumichaca, the llamingos were probably only to be found in some districts like Chimborazo or Bolívar. The gracile vicuñas have disappeared."

In another article Cardozo (1974a:10–11) mentions the extermination of the vicuña in Ecuador. He adds that "[t]oday the land of the llama is limited to Chimborazo, Bolívar and Cotopaxi."

Cardozo (1974a:11) then notes that the llama population of Ecuador has fallen dramatically: "The latest data collected by a student preparing his dissertation on this subject indicate that the llama population in Ecuador has reached the critical figure of some 2,000 animals." It is hard to tell whether the data are correct.

Salomon (1980:56) discussed domestic animal raising in the Ecuadorian area, but he takes Troll's position (1931, 1958) as his starting point and as an axiom, wherein the llama is a "member of the puna biotype." Salomon wrote in this regard that "[s]ave for the high and dry lands in the modern-day province of Chimborazo, nowhere in the northern Andes has camelid raising successfully persisted in historical times." This observation is certainly interesting, but it would be worth researching the subject and determining the real reasons that curtailed this development, as I have shown that Troll's position is wrong.

Franklin (1982:Table 2, 475) indicates that in the late eighties there were 2,000 llamas in Ecuador, which according to Stahl (1988:357) were mainly distributed in the high-altitude basins of Cotopaxi, Tungurahua, Cañar, and Chimborazo. The source from which Franklin (1982:Table 2, 475) took the number of animals he gives is not clear, because the sources for Table 2 are listed en masse. They include Novoa (1980), Cardozo (1980), Guzmán (1980), Cajal (1981), Brack Egg (1981), and Rottman (1981). I assume that the data on Ecuador were taken from Cardozo (1980), but I do not know this study. I believe it is Cardozo (1974a:11) because as we saw, he had already published this information, crediting it to "a student preparing his dissertation on this subject." The information is therefore of *incertae sedis*.

From Bruhns (1991b:2) we know that Chilean alpacas began to be imported into Ecuador in 1980, and the caption in one of the figures in another study by Bruhns (1989:Fig. 16, 70) shows that camelid herding began anew in 1985, for the first time since the Spaniards arrived. (Apparently this did not happen in 1987, as Wheeler [1991:32] said, based on an article by Meisch [1987].) These are alpacas raised close to the town of Zhoray, at half a day's walk from Pirincay. The news was confirmed by Bruhns (letter of 22 January 1992), who explained that Stuart White imported alpacas from Chile and was raising them in the province of Cañar, in the pajonal above Pirincay. "Stuart told me that due to the parasites and also to the low concentration of nutrients in the high-altitude straw, the animals had to be taken to fields at lower altitudes for part of the year, particularly at birth time."

I recently found out through a newspaper article (Anonymous 1992b:B6) that the repopulation of Ecuador with vicuñas began in 1988 through an international agreement. Thanks to a Peruvian donation of 100 vicuñas and another 80 animals from Chile, the number of specimens had multiplied by 1992. The article said that everything seemed to indicate that in the next 15 years the 100 Peruvian vicuñas would multiply until they reached 7,950 individuals, which, added to the 6,300 specimens with Chilean blood, would "restore the presence of this almost extinct camelid to the Ecuadorian Andes." According to the article, the Peruvian vicuñas were adapting quite well to their new habitat, while 20 animals of the Chilean group died due to the journey and to stress. These animals are apparently living at more than 4000 masl, in the *páramo* of Mechahusca and El Sinche, in the province of Tungurahua and Bolívar. (For additional data see Table 2.2)

There is one linguistic matter I would like to point out: the use of the term *llamingo* in Ecuador for llamas. Cardozo (1974b:143) claims that Gonçalez Holguin (1952; it is cited in the text but does not appear in the bibliography), Juan de Velasco (1946), and Cordero (1955) "include the words llama and llamingo as authentic Quechua words." For obvious reasons I will not discuss the term llama. I have not read Velasco or Cordero, but the term *llamingo* definitely does not appear in Gonçales Holguín, nor is it a Quechua word. Cardozo himself (1974b:139) says that according to Salvador Lara, this is an Ecuadorian term. Romero (n.d.:Vol. I, 223) gives an absurd explanation: "[I]n Ecuador they are called *llamingos* because they lost height on adapting to the Equatorial sierra."

I consulted Cerrón Palomino, who explained that in Ecuador the ram [*carnero*] is called llama, while the Ecuadorian term *llamingo* was created for the llama. Etymologically, the suffix *-ingo* in Quechua means similar to or derived from, and it was from thence that it was mistakenly taken. This clearly is a modern creation because the term does not appear in the chronicles and is used in the Callejón Interandino, from Cañar to Otavalo, that is, in the area known as intermontane Quichua (Rodolfo Cerrón Palomino, pers. commun., 18 April 1994).

Finally, I am morally bound to comment on two studies Cardozo did on the camelids in Ecuador. These articles include such mistakes that they must be taken with due caution, as they can be misleading. I find it strange that no one, to the best of my knowledge, has so far pointed this out, and that these and other studies by Cardozo are often cited in the literature without any critique whatsoever. In one of these studies Cardozo (1974a:10) said literally: "Carlos Manuel Larrea noted the presence of mastodons close to Quito, *which could well be the ancestors of the camelids*" (emphasis added). No comment is needed. Another article (1974b) has several serious mistakes. For example, Cardozo claims that "the camelids emigrated from North America to the south *at the end of the Pleistocene and the beginning*

of the Quaternary" (Cardozo 1974b:140, emphasis added). In addition, the references are incorrect, and the authors who are cited in the text do not appear in the bibliography. The content is poetic, not scientific. Furthermore, Cardozo cites Ecuador chroniclers who are discussing Peru, with really astounding ignorance.

9.6 BOLIVIA

I did not find many data on the area comprising present-day Bolivia. Particularly missing are the archaeological data, about which I know almost nothing.

9.6.1 Historical Data

Although not abundant, the historical data suffice for an overview. As for the other regions, I present the data from north to south for practical reasons.

I only have one reference for Inca times, from Murra (1975:140), who explains that the Inca had herds of llamas in their frontier strongholds. The Inca garrison could use the herds assigned to the Sun close to Cochabamba, an area where skirmishes often took place. Murra based his comments on Murúa (1946:Bk. IV, Chap. XIV, 410). When discussing Cochabamba, Murúa (1925:52) says, "There used to be a *puquial* (spring) with livestock of the Sun, which the Incas invented to have meat in time of war, and also when on the frontier."

For Colonial times I begin with the province of Omasuyo, one of the most famous in Collao, northeast of Lake Titicaca. An undated document by Mercado de Peñalosa et al. (1965:337), which must date to 1586, mentions Viacha, in the province of Omasuyos and "on the royal road to Potosí," where "[t]hey have some livestock of the land."

A later document by Castro y del Castillo (1906 [1651]:219) on this same province reads "And in the paramo called puna, bleak because of the rigid instability of the weather, one sees herds of wild animals which are guanacos, similar to the rams [*carneros*] of the land, and these behave like the camels. The vicuñas, who are given this name because of their swiftness, seem small does by their shape, the whiteness of their bright reddish-colored wool—white on the chest—is delightful, they have a long neck and a small head."

Cosme Bueno, who wrote in the second half of the eighteenth century, also mentioned the province of Omasuyo, in the bishopric of La Paz. He says: "In it all kinds of livestock and some wild animals breed, such as . . . Huanacos, Vicuñas" (Bueno 1951:120).

For La Paz there are some interesting data. For example, the famed 1573 tribute roll of viceroy Toledo (1975:43–45, 47, 48, 50, 54, 55, 58, 60, 61, 63) indicates that there was a total of 2,366 rams [*carneros*] of the land in the repartimientos of Callapa, Hayo, Machaca, Coquiabiri, San Pedro and Santiago, Calamarca, Viacha, Caquingora, Guaqui, Puno, Tiaguanaco, Paucarcolla, Lacxa and Siqui Sica, all part of the district of the city of La Paz. Curiously enough, no paco appears.

Mercado de Peñalosa et al. (1965:334) describe "the province of the Pacaxes, in the jurisdiction of the city of La Paz," in an undated document that must be from around 1586. Here it says that "[t]he livestock of the land is abundant." Further on we find "guanacos and bicuñas" when the animals of the region are described (Mercado de Peñalosa et al. 1965:340), but it is noted that some repartimientos, such as Guaqui, beside the cordillera of the same name, "[d]o not have much livestock of the land" (Mercado de Peñalosa et al. 1965:337).

A document by Castro y del Castillo (1906: 206–207) dated 1651 also gives an account of the province of the Pacaxes, in the bishopric of La Paz. It reads thus:

[M]uch livestock of the land is grown, and also from Castile, sheep and cows, because it abounds in pastures and water, all good. There are many of the animals called vicuña, which is quite wild [*son campestres*], and although the people raise some in their homes which they catch when young, on neglecting them these follow their inclination and go looking for the wild ones. Although the meat is eaten [it is] bad because it is thin [*flaca*] and insipid, they are prized for the wool and the bezoar stones . . . ; this animal does not

have bile, they move in herds and the males do not mix with the females but live apart, and only one male is found in each herd of the females, whom he heads, and the natives say that when another male mingles with them the two said males fight and the winner stays with them while the loser returns to the herd of his own sex. When they run to catch them, the male places the females in front of him and he stays behind, as if waiting for the hunter to strike, so he is the first to die. When one of these animals gets sick with the disease they call *caracha* [the mange], which is like the mange, the rest bite and push it away because it is a contagious disease, and the latter animals go alone without ever mingling [with the rest] until they die, because they do not heal without treatment, like the rams [*carneros*] of the land, who never heal unless they are treated with fat and sulfur, because the mange spreads until the animal is killed. This animal is very swift and a greyhound barely catches up with it, [and] they flee from where people are. So they live in the emptiest and coldest lands, and their wool is very cold [sic], and in the warm lands it is much prized by the Indian women for its color and softness, and the Spaniards who treat with them [i.e., the Indian women] make weighed bundles [of fleece and skins?: *tratan de regalo hacen pesadas della*]. . . . There are other animals almost of the same species and likewise wild, and wilder and bigger which are called guanacos, although there are not many of these.

Castro y del Castillo (1906:216–217) comments: "This province of Pacaxes was one of the richest in Peru because of the many rams [*carneros*] of the land it grew, because of old the quicksilver was carried from Arica to Potosi on them, and the volume of iron for all the provinces from Lima upward. And as the Indians diminish out of fear of the Potosi mita, so have these rams [*carneros*] dwindled. On missing one and the other, these two kinds of trade were taken over by the mules which come from Tucumán, but most of the wine and the food are carried on these rams [*carneros*], as is the coca, the amount of coca consumed in the kingdom is high because it is the most essential and desirable sustenance for the Indians."

In 1586 Cabeza de Vaca et al. (1965:340) also described the area of the city of La Paz, and noted that "in the smooth and flat places, which are cold lands, there is a large number of vicuñas and guanacos. . . . There also are rams [*carneros*] of the land, which in their language they call llamas."

Eleven years later, in 1597, Ramírez (1936: 50–51) described "the town of Chuquiabo or city of La Paz, which is commonly called Chuquiabo." Ramírez noted that "in the highlands there are several estancias with livestock from Castile with much pastures and water, so there is much livestock of Castile and of the land, and they are very good." He then described "the province called Pacaxas," whose "people are very rich in livestock." Farther on Ramírez mentioned "livestock of the land, which they barter and ship in large amounts and take to Potosí."

In his *Décadas*, Herrera collected news on "Alonso de Mendoça [who] peopled the town of La Paz in Chuquiabo, and other matters regarding this province . . . called the province of the Pacasas." Herrera noted that the province "had many pastures and waterholes for the livestock of the land and the sheep and rams [*carneros*] of Castile, because a large number of them grow here" (Herrera 1947:VIII Decade, Bk. V, Chap. III, 94). Of the "great snow-capped cordillera," Herrera says, "and in the cold land guanacos and vicuñas [live]" (Herrera 1947:VIII Decade, Bk. V, Chap. III, 94).

Vázquez de Espinosa also has several entries for the region. I do not know if he was actually there, but it is possible that he did go while in Arequipa in 1618. In his book, Vázquez de Espinosa discusses "[t]he city of Nuestra Señora de La Paz, and other provinces in its district." He states that "all of these provinces are inhabited by a lot of people, and in them there are many offspring of livestock of the land" (Vázquez de Espinosa 1948:1628/569). When describing "[t]he province and corregimiento of Caracollo and Sicasica, [which] is bordered on the north by the province of Omasuio," Vázquez de Espinosa indicates that "there are livestock, pastures [?], and

in them offspring of livestock" (Vázquez de Espinosa 1948:1631/570). It is true that in this case he did not specify whether this was livestock of the land or of Castile, so the doubt lingers on. However, when discussing once more the "city of Nuestra Señora de La Paz" and the "[p]rovince of the Pacages," Vázquez de Espinosa mentions "plenty offspring of livestock of the land, of the best [kind] there is in that kingdom." Of the "Cochabamba Valley," Vázquez de Espinosa says there are "great herds in the cold deserts of guanacos, vicuñas" (Vázquez de Espinosa 1948:1632/570).

Vázquez de Espinosa (1948:1633/571) also described "the province of Paria, in the district of the archbishopric of the Charcas." He explains that "[t]he province of Paria limits with the abovementioned of the Pacages and Calacato," immediately adding that "there is in it a great number of offspring of livestock of the land."

Many data are available for the famed zone of Potosí, and there must certainly be far more than I was able to examine. In 1573 Benino (1965:363) described the hill and the mines of Potosí. When listing the problems the mines had, Benino says these would be far greater had "God not provided a solution, which is transportation with the livestock of the land, since there is so much of it, the livestock is supplied in great numbers and with great ease."

Colque Guarache (1981:239–240) presented an interrogatory in La Plata in 1574 (but dated to 1575) for the witnesses "on his succession and that of his predecessors, and the service that said don Juan and his father made to His Majesty and his conquerors in subduing and pacifying this kingdom." The two questions that concern us are nos. 23 and 26. They read thus: "Item. Do they know that Don Juan Guarache, my father, spent more than four thousand rams [*carneros*] of the land as pack animals in this city, [in the] tambo [inn] of the Aullagas, and [in the] town of Potosí, and in all the other parts he was serving the side of His Majesty and his defenders." The other question reads: "Item they are asked whether he knows that Don Juan Colque was named captain by viceroy Toledo in the campaign against the Chiriguano [Indians], and that in it he took with him 'five hundred rams [*carneros*] of the land,' among other items." This shows the number of animals an individual had at that time.

Ramírez (1936:55) was quite laconic when describing the "town of Potosi" in 1597. He said that it "has some pastures where the livestock of Indians and Spaniards feeds." Ramírez then described the mining of metal and explained that once the Indians have brought it to the surface "on their back," it is "loaded on rams [*carneros*] and taken to the mills where it is ground and refined until it turns into silver" (Ramírez 1936:57). Ramírez (1936:57) also mentions the "repartimiento [allocation] of Indians" and comments that "the ones and the others are in Potosi to fulfill their obligation and increase the number allotted to them, as well as to carry out their deals and business, and both are in Potosi until the same number comes from the corresponding town who take their place, they bring their wife and children with them and their animals loaded with clothes and food, all of which is eaten and sold in Potosi."

Lozano Machuca (1965:60) discussed the province of the Lipes, close to Potosí, in 1581. He mentioned the Uros, who "live on hunting the guanaco and vicuña." Of the Aymaraes, Lozano said that these were "a people rich in livestock of the land."

The data in Lizárraga are very important, not just because of their reliability but also because he was there. Lizárraga said: "I have said this because I will not speak from hearsay, save for a little, and then I will say I heard it from reliable people; the rest I saw with mine own eyes and, as they say, touched with my hands, so that what was seen is true, and what was heard no less" (Lizárraga 1969:Bk. I, Chap. II, 4).

This Dominican friar described the city of Potosí in the province of Charcas. He said that "the Indians [are] . . . very rich . . . in livestock, although in livestock those of Collao have the advantage" (Lizárraga 1968:Bk. I, Chap. C, 85). When describing the marketplace of Potosí, Lizárraga presents an interesting datum that does not appear in other contemporary chroniclers. He noted that "here the dung of the rams [*carneros*] of the land is sold, and they certified to me that each year they sell an amount worth 10,000 pesos and more" (Lizárraga 1968:Bk. I, Chap. CIII, 89).

Finally, when describing the road to Potosí, Lizárraga noted that "some Indians and Spaniards take greyhounds with them to hunt any guanaco or vicuña that might appear" (Lizárraga 1968:Bk. I, Chap. C, 87). The presence of these animals around the town of Potosí and its mines was vouched for in 1603 by one Anonymous writer (1965:373), who said that "[i]n this mountain there likewise was a great number of game like vicuña, guanacos."

The account left by Ocaña (1987:166–167) is another important testimony because he was in Potosí on 18 July 1600. When describing the famed mines, Ocaña noted, among other points, that "[t]he metals in this hill are brought down in rams [*carneros*] . . . each of these rams [*carneros*] is worth eighty reales in Potosí, and in other lands from where they are brought they are worth less. The ordinary mita[6] . . . is twelve thousand of these rams [*carneros*]. . . . All supplies are brought on these rams [*carneros*], such as wine, flour and fruit."

Vázquez de Espinosa also wrote about the "[r]ich and famous mountain of Potosi." I do not know whether he actually went there, but it is possible that he did so in 1618, when he visited Arequipa. Vázquez de Espinosa claimed that Potosí was "uninhabitable by man because of its coldness," so that "in that desert there were only ganacos [sic], vicuñas" (Vázquez de Espinosa 1948: 1647/576). The Carmelite priest then described the way the metal was refined, noting among other points that "others are dedicated to taking it down from the mountain to the mills in droves of rams [*carneros*], and each day more than 8,000 of these muleteers drive the rams [*carneros*] up [the mountain] for this task" (Vázquez de Espinosa 1948: 1653/580). Vázquez de Espinosa then describes the food in the "imperial town of Potosi." "In this town," he writes, "they consume rams [*carneros*] of Castile, merinos . . . cows which are brought in great numbers from the provinces of Tucumán, Paraguay and Buenos Aires, and pigs," but immediately adds "and much from the land, which is the most common food of the Indians, more than 800 thousand ducats are spent each year" (Vázquez de Espinosa 1948:1665/587).

By Vázquez de Espinosa (1948:1676/593) we likewise know that in the province of Chayanta, between Potosí and La Plata, there was a zone "peopled with Indians with livestock of the land," on one side of the road to La Paz.

For the "[r]epartimiento of the Villa of Plata" we have the figures in the 1549 tribute list of La Gasca (1983–1984:80–83). The repartimientos of Macha and Chuquicota are listed here with a total of 422 sheep and 3 lambs. The 1573 tribute roll of Toledo (1975:87, 88, 90–110, 112, 114) lists the "[c]ities of Cuzco, La Plata and the Province of Collao" and includes the repartimientos of Arapa, Caman, Asangaro of Antonio Quiñones, Asangaro of Captain Martín de Alarcón, Taraco, Nunoa, Cauanilla and Oliberes, Cavena, Lampa, Manoso, and subject towns, Atuncolla, Ayavire and Cupi, Oruro, Xullaca, Caracoto, Caquixane, Nicasio, Achaya, Pucara and Quipa, Llalli, Macari, Caminaca, Sangarara, Ullacache and Omachiri. A total of 2,161 rams [*carneros*] of the land is listed, but curiously, no paco appears.

In 1597 Ramírez (1936:53) described "the territory of the city of La Plata [which] is of the Indians called Charcas." Ramírez specifies that "the Indians raise much livestock of the land [which is] not as good as that previously listed," that is, the livestock found in La Paz and Cuzco.

We once again have to resort to the account of Vázquez de Espinosa. When discussing the "[t]erritory of the city of Chuquisaca or La Plata, province of Charcas," Vázquez notes that there were "many offspring of livestock, both of the land and of Spain" (Vázquez de Espinosa 1948:1677/594), and that "[t]he general occupation of the vecinos of this city is tilling the land and raising livestock" (Vázquez de Espinosa 1948:1711/608). When specifically describing "[t]he city of La Plata," the Carmelite priest mentioned "the animals in the district of this city" and noted that "in the sierras of Caracara, which are barren and cold, there are vicuñas and guanacos" (Vázquez de Espinosa 1948:1740/615), as well as "in the Mojotorio valley and in others, because everything abounds and it is well-supplied with [and] it has offspring of livestock of the land which are the rams [*carneros*] that carry the wine, maize, wheat, flour, timber and whatever is needed to provision the city" (Vázquez de Espinosa 1948:1741/615). He then explained that the city of La Plata had

"great and very rich provinces," specifically noting that "the town of Paspaya" had "much offspring of livestock" (Vázquez de Espinosa 1948:1744/616). Though in this case Vázquez de Espinosa did not specify what kind of livestock this was, we can assume that these were camelids.

In the early seventeenth century, friar Calancha (1975:Bk. II, Chap. XL, 1173) also discussed the province of Charcas and mentioned the city of La Plata, which was also known as Chuquisaca. He then described the various kinds of dog breeds used to hunt different animals and noted that "dogs without race or law" were used to hunt "*gamos*," which for Calancha is a synonym for vicuña.

In the first half of the sixteenth century, Zárate recounted how Alonso de Toro left Cuzco to face Diego Centeno, and explains that on the way he and his men "entered a despoblado [uninhabited land] more than forty leagues long until they reached a place called Casabindo, [which is] where Diego de Rojas entered the Río de la Plata." It is hard to determine exactly where the *despoblado* mentioned by Zárate was, but it must be close to "the town of Plata," that is, Chuquisaca, because that is as far as Alonso de Toro went. What is interesting here is that Zárate (1968:Bk. V, Chap. XXIV, 296) explains that to enter the *despoblado* they "took great pains to take with them a great number of rams [*carneros*] loaded with food."

Lizárraga (1968:Bk. I, Chap. CVII, 91) described the provinces of Chichas and Lipes, which according to him extended "[f]rom this town of Potosí, going slightly downwards to the east" for 50 leagues, and then to the "right hand." In Lipes there were "abundant supplies and livestock, both of the land and of ours." When describing these same provinces, Vázquez de Espinosa (1948:1760/621) says that "[t]he province of the Lipes is 50 leagues to the west-southwest of the city of La Plata, and borders to the east with said [province of] Atacama." He expands what Lizárraga had pointed out (see above), noting that it was "full, and is covered with wild livestock such as the guanaco, vicuña."

Vázquez de Espinosa left some data on the southern "[p]rovince of Carangas," and describes it thus: "[T]his province has several other towns which are very rich . . . due to the very many off-spring of livestock of the land they have, because these are the biggest in Peru and the ones who work the most, and are thus worth more than those in the other provinces. Through its midst runs the royal road from the port of Arica to Potosi" (Vázquez de Espinosa 1948:1643/575). In this same chapter the Carmelite friar continues his description of the province, which has on its borders to the north Pacajes, to the east Porco and Potosí, to the south Atacama, and to the west, Arica and Tarapacá. Vázquez de Espinosa insists that here "there is a great number of wild runaway livestock, which are guanacos and vicuñas . . . besides the tame livestock of rams [*carneros*] of the land, and of Castile" (Vázquez de Espinosa 1948: 1644/575).

Two sources have data on the province of Paria. Ramírez (1936:52) mentions "the Lake of Paria" and claims that those who live there "are very rich people due to the many livestock they raise, both of the land and of Castile." In his description of this "province of Paria, in the district of the archbishopric of the Charcas," Vázquez de Espinosa (1948:1635/571) indicates that "in the sierra and cold land . . . there is a great number of wild [livestock] like guanacos, vicuñas," but adds that this is "beside the tame livestock." He then says "in Sorasora, and on the banks of Paria, where they take the metal on rams [*carneros*], of which there are many" (Vázquez de Espinosa 1948:1636/573).

There are also some interesting data on the 1564 expedition ("*entrada*") to the Mojo (or Mussus). Aleman (1965:276) says that "[l]eaving the Cochabamba valley one goes over twenty leagues of puna where there are many huanacos, vicuñas to hunt." He then says that on the way to "Los Llanos" he reached a province called Machari, "and saw much spoor of livestock; according to what the Indians say, there is in this province a livestock that is almost like the cows from Castile, domestic, which is raised by the Indians and [they] eat its meat. And the night that the Spaniards spent in the town they heard the bellowing of this livestock, which they apparently had in corrals close to the town" (Aleman 1965: 276). Further on Aleman mentions Mahari, which I wonder whether it is the same Machari

he mentioned before (see above), or whether it is another town. Aleman says that here the Indians "had a livestock similar to the cows of Castile, domestic, which they load and eat the meat." Then he describes the Camaniguani site and says that "[a]ll the land here is a cold land; they raise much livestock like that of Peru and wear woolen clothes." Besides, "[i]n Lipira there also is much livestock like that of Peru and the Indians wear woolen clothes." Aleman then describes the Pacaxas, who "[h]ave temperate and cold lands, and many valleys . . . and raise much livestock like that of Peru in punas, and have woolen clothes" (Aleman 1965:277).

From there the Spaniards went to Paitite, which "is a very big province and a very big river runs through it that the Indians call Patite, and it runs through the lake where rises the temple of the Sun where the Indians go to *mochar* [worship]. This is ruled by a cacique principal who is as powerful as were the Incas in Peru." He then explains that "[t]here is in this province much gold and silver. They also raise sheep of Pirú and have much food of maize and other things in such good valleys" (Aleman 1965:277–278). It is almost impossible to establish whether llamas or alpacas were mistaken for the livestock "like the cows from Castile" that Aleman mentions, but these are likely the same because there are other sites with "livestock like that of Peru."

Finally, I found some data on the kingdom of the Quillaca-Asanaque in the southern part of the altiplano, on the present-day border between Bolivia and Argentina. Quillaca, according to Espinoza Soriano (1981:185), lies in the Quillacas and Sevaruyo River basins. It extends from the southern banks of Lake Aullagas to slightly south of the present canton of Soraya de Los Quillacas Asanaques, that is, on the Cordillera Real (also called Cordillera de Condo), to the east of said lake.

According to Espinoza Soriano (1981:177, 179), its inhabitants "were good livestock raisers" and had "herds of strong and fat llamas," but "there likewise was a great number of guanacos and vicuñas in wild condition."

I did not find much documentation on the killing of these animals, and all that I have is for Potosí. At his death, the Jesuit priest Ayanz left a long memorial of uncertain date on the Indians and the Peruvian *mita* which must date to 1596. Of the Indians of Potosí he said that "[o]f all the livestock they take, which is more than 30,000 heads, 1,000 do not return, and sometimes not even 500" (Ayanz 1978:218).

In 1603 an anonymous writer (1965a:380) likewise mentioned Potosí: "This place consumes an enormous amount of livestock of the land, and it seems that the number of livestock that comes loaded with wine and coca and other comestible items, are 40 thousand rams [*carneros*]; besides this, each year 60 thousand rams [*carneros*] arrive that the Indians coming for the mita in the mountain bring, where they carry 40 thousand fanegas of food for their sustenance, and this livestock does not leave for it is all consumed in this town, and the ones and the others come to 100 thousand rams [*carneros*], which at 4 pesos ensayados come to 400 thousand." (This is the information to which Tschudi refers [1885:98 and fn. 3; 1891:100 and fn. 3; 1918:fn. 132, 213].)

After describing the amount of "livestock of the land" used in the mines, this same anonymous writer adds: "And though it was decreed that the Indians should not kill sheep [*ovejas*] of the land it cannot be remedied, and thus we find that in one year 40 thousand heads and more of sheep [*ovejas*] and pacos are killed in the Ranchería [see Chapter 7, note 6], which at 3 pesos ensayados [each] comes to 120 thousand pesos" (Anonymous 1965a:380).

Thanks to the testimony of Simoens da Silva (1980:65) we know that "thousands of llamas" were still used in Potosí in 1930 to carry the minerals.

9.6.2 Additional Data

Brack Egg (1987:62) claims, without citing a source, that the vicuña population of Bolivia was decimated during the conquest and in the viceroyalty.

Bolsi presents data on the presence of camelids in the Bolivian altiplano in 1968 (Bolsi 1968: unnumbered table). At Corregidores, at 4170 masl and south of the Salar de Uyuni, it is mostly llamas that are found. There are llamas and alpacas in Tiahuanaco, at 3863 masl and 50 km east of Lake Titicaca, only in the high part. In Achacani, at 3821 masl and 6 km east of Lake Titica-

ca, there are llamas and alpacas in the highest parts. At Copacabana, at 3815 masl and on the far south of Lake Titicaca, there are llamas and alpacas in the high part. At Kala Kala, at 3800 masl and east of Uncia, there are llamas only in the high part. There are llamas at Chayanta and Aymaya, at 3700 masl and 90 km to the southeast of Oruro. At El Rodeo, at 3600 masl and 30 km north of Potosí, there are llamas and alpacas. In Betanzos, at 3400 masl and 40 km northeast of Potosí, there are llamas and alpacas solely in the high part. Finally, there are llamas and alpacas in the piedmont at Ankotanga and Belén, at 2800 masl and 35 km to the northeast of Oruro.

Villalba recently wrote about the guanaco in Bolivia. She explains that its present distribution is unknown. Its presence has been reported in the southern Andean region, in the Departments of Potosí and Chuquisaca, and in the Chaqueño region to the south and southwest, in the Departments of Tarija and Santa Cruz. Some data are available for the Mochara region and the Estancia Perforación Chaco. Some animals were seen in 1981 in the southern part of Tarija, and perhaps at Pampa de Guanacos, to the southwest of the Department of Santa Cruz. They were also recorded in 1990 west of the Sama Cordillera, southeast of the Department of Tarija. With these data Villalba claims that the area occupied by the Bolivian guanacos can be defined as running between 19°–22° S latitude and 62°–65° W longitude, from 300–400 masl in the Chaco-Beni plains to 3500–3800 masl in the Andean cordillera. There are no data on the number of animals. In 1982 Franklin gave a figure of 200 individuals, and Torres gave a figure of 54 in 1984 (Villalba 1992:44).

Villalba indicates that in Bolivia, the vicuña lives in the altiplano between 14°40′ and 22°50′ S latitude, and between 3600 and 4800 masl. The presence of 12,047 individuals was recorded in 1989. Villalba specifies that two subspecies are distinguishable, which might correspond to the "northern geographic breed" and the "southern geographic breed," both described by Hofman et al. (1983) (Villalba 1992:42; see my Table 2.2).

Lecoq (1987) made an interesting study of the llama caravans that are still used in Potosí. Lecoq concludes that the techniques, the ideology, and the organization inherited from prehispanic times still endure. Second, the old roads are still used, and prehispanic ruins are found beside them. Third, the rituals regarding the camelids do not correspond to a Spanish *Weltanschauung* or beliefs. Fourth, the changes that are taking place in this regard are recent and postdate 1957 (Lecoq 1987:32).

Lecoq managed to verify that in a community in Potosí, a great number of llama caravans go from the altiplano to the middle and lowland eastern valleys, warm and wet, of the Amazon piedmont (Lecoq 1987:2).

Lecoq described the route followed by these caravans. They go to faraway valleys, at distances that range between 150 and 200 km from Lake Titicaca, to places that are east of the Pilcomayo River. These trips take between 2½ to 3 months for a round trip. At other times they head to the middle valleys, to an environment that lies midway between 3000 and 2500 masl. In this case the trips take from 8 to 10 days, that is, a round trip of three weeks to a month (Lecoq 1987:9).

West in turn studied the caravans of llamas that carry salt in Pampa Aullagas, on the southern coast of Lake Poopó, at about 3937 masl. West recovered data saying that in the past they made weeklong journeys to Uyuni (37 km) for salt, and then three-month-long trips carrying the salt along the neighboring valleys (West 1981a:64).

These data are supported by Browman (1990: 42–43), who claims that caravans still are an important means of transportation in some parts of the Andes. For instance, in the Callahuaya region east of Lake Titicaca, in Bolivia, Schoop (1984:41; I do not know this study because the paper by Browman is a summary that does not include a bibliography) recorded weekly traffic of more than 185 llama caravans (with an average of eight to ten llamas per caravan) and 75 trains of mules (with four to six mules on average) carrying 80–100 tons of merchandise, more than that arriving by truck or bus.

9.7 CHILE

The literature on camelids for the Chilean area, both archaeological and historical, must be extensive but I was able to review only a small part of it.

9.7.1 Archaeological Data

I will try to present the data from north to south as much as possible, and will begin with the archaeological data. Atacama is one area for which important data are available.

First, it is interesting that northern Chile shows traces of a pluvial maximum around 12,000 BC, and a probable retreat of the glaciers around 10,500 and 9000 BC. Dillehay et al. (1992:155) wrote:

> Evidence of fossilized animal excrement (Phillipi [sic] 1893) and coastal sand sheets suggest that desert-like conditions prevailed at lower altitudes in both coastal Peru and northern Chile for most of the late Pleistocene. Independent of land-based factors are the effects of El Niño, which may have brought rainfall to the Peruvian coast and marine bioturbation to northern Chile (e.g. Radtke 1987; Rollins et al. 1986). At these times, vegetation might have flourished and herbivorous animals (especially wild camelids) probably appeared (Craig 1984).

Guanaco bones (*Lama guanicoe*) were found in the Quebrada de las Conchas in a slightly later period than that indicated, on the slopes of the coastal mountains, close to Antofagasta. This is an early site, with radiocarbon dates ranging from 7450 to 7730 BC (Llagostera 1977; 1979:314; Lynch 1983b:115). We know that camelid bones dating around to 4200 BC were found in the Salar de Atacama at about 2500–3000 masl at the Laguna Hedionda site (Lynch 1967:15–16).

The llama apparently appears in association with agriculture on the north coast of Chile, in the phase that Bird (1946a:591 and the table in Figure 49) defines as Pichalo I phase (ca. 4000 BC). According to Pollard (1975:296), these appear "inland in northern Chile only around 200 BC."

Dransart (1991a:311) reports that according to Zlatar (1983:22), fibers, "presumably from camelid," were found at the coastal sites of Caleta Huelén 42, belonging to the late preceramic, and Caleta de Abtao, from the middle preceramic. I understand that the dates for Caleta Huelén range between 2830 and 1830 BC, while Abtao has dates that go from 2850 to 2940 BC.

Of great interest to me are the data by Jensen and Kautz (1974:45–46) on Tarapacá 2A, located in a low ravine (1400 masl). They discuss their Period I (pre–food production) and mention the study by True and Crew (1972), which I was unable to obtain.

The authors explain that charcoal remains were found in Tarapacá 2A, in association with abundant guanaco remains and big stemmed points whose context had two radiocarbon dates of 4020 and 3340 BC. The interesting point is that this site was occupied 1500–1200 years later (i.e., ca. 1930 BC) "by people who may have been herding llama, rather than hunting guanaco."

The authors then described an area that was identified as a urinating zone where remains of llama wool were found, which were radiocarbon dated to 1500–2000 BC. In this regard they wrote: "This date coincides with our estimate for the appearance of the domesticated llama, an estimate which we had originally obtained by utilizing the data from Peru. . . . In short, the wool specimens, in combination with the presence of the compacted floors which in turn contained phosphate from urine, suggest that a later population of herders may have corraled their animals at this site."

From the data in Novoa and Wheeler (1984:124), we know that llama feet and woolen textiles are common in the burials of the Chilean north coast from 500 BC, but Novoa and Wheeler point out that "it is not known if herds were present in the area or not."

Núñez discussed the late preceramic on the north coast of Chile in a 1969 study. He explains that since the time of Capdeville (1921) and Uhle (1919) it has been known that the base of the shell mounds of Taltal had a stone industry of a "Paleolithic morphology." On the basis of studies by Berdichewsky (1963), it is now accepted that this context belongs to guanaco hunters. Núñez then comments that "we certainly are ever more positive that in the coastal cordillera there was a period with hunters who practiced a land hunt, especially with guanacos, and who might have come into contact with the immediate coastal region." Núñez, however, says that Bird's studies have

shown that the artifacts in the oldest strata at Taltal, and in general on all the north coast, survived until later times and coexisted with agricultural and ceramic levels. Núñez ends by noting that "[s]tated this way, there were some assumptions that Andean hunting traditions might have reached the coastal region, drawn perhaps by Andean ecological changes" (Núñez 1969:203–204). This agrees with the data available for the Arica region, where we know that several burials were found in the Quiani site, in the Quiani II phase to be precise. In one burial, that of a child, the corpse was found wrapped in leather and tied up with woolen ropes, but it is not known whether these are of vicuña or guanaco. Other bodies were wrapped in guanaco skins (Willey 1971:203, 205).

Núñez then reviewed the data and did not mention hunting at all in the information presented, indicating that previous assumptions were somewhat exaggerated. However, the studies by Niemeyer and Schiappacasse (1963) on the Conanoxa culture, 45 km inland from the mouth of the Camarones River (close to the border with Peru, in a low ravine at 500 masl in the province of Tarapacá), show evidence of guanaco hunting in the late preceramic, because there is a C^{14} date of 2000 BC (Núñez 1969:206).

In fact, we know that two periods were distinguished at Conanoxa. One is a pre-agricultural period that can be given an approximate date, according to Niemeyer and Schiappacasse (1963:142–147), of 3000–1200 BC, the other a period with an incipient agriculture and undecorated pottery (Niemeyer and Schiappacasse 1963:147–150). The findings of guanaco leather, textiles, and string made from the fiber of this animal belong to the first period, while string, thread, and guanaco hairs and wool appear within the context of the second period (Santisteban Manrique 1963:160–161).

Pascual identified the remains as *Lama* sp. and noted that "[this] is quite likely *Lama guanicoe*, but they might belong to *L. glama*" (Pascual 1963:165). Although the title of this study says "Levels 2 and 3," the association of the faunal remains is not clear in the text. From the study by Niemeyer and Schiappacasse (1963:109–111) they would seem instead to belong to Level 1.

According to data in Pollard (1976:18), not only were wild camelid species exploited in the Atacama Desert by 200 BC, but domestic ones were, too. Núñez (1968:208) accepts the presence of livestock by this time in what he defines as the Early Agro-Ceramic period, that is, from the beginning of the Christian era to AD 700 (which corresponds to what Bird [1943] defined as Phase Punta Pichalo III).

Novoa and Wheeler (1984:123) believe that the alpaca wool textiles found in Alto Ramírez, in the Azapa Valley (which should correspond to the early ceramic phase, i.e., around 500 BC), must be considered among the earliest in the Andes. However, we have seen that this wool was used on the Peruvian coast at a much earlier date.

Bennett (1946b:603) discussed the very late times of the Atacameño and defined them as a "basically agricultural and herding" group that occupied the area from the Middle Horizon onward in what is now Argentina and Chile. Here Bennett (1946b:607) discussed the caravans of llamas and added an interesting datum. "One of the characteristic Atacameño artifacts is a wooden toggle, of V-shape, with knobs at each end, which served as a cinch buckle for fastening the packs on the llamas. Wooden cowbells with wooden clappers are also common and were probably used for the lead llama in a train."

Téllez and Silva Galdames (1989:51) discuss Inca times for Atacama la Alta, or what is now the San Pedro de Atacama, but do so vaguely. According to Téllez and Silva Galdames, this site "had a special attraction for the personal interests of the Cuzqueño rulers. Wide expanses for growing maize, innumerable heads of camelids were grazing in the Atacameño oasis and wetlands."

In the higher parts of this same zone are three important sites: Tulán-52, Puripica 1, and Tambillo. Tulán-52 dates to ca. 2390–2320 BC and Puripica 1 to ca. 2856–2100 BC; there are no radiocarbon dates for Tambillo, but (according to Núñez 1981) in typological terms it should date to ca. 3000 BC. The three sites are located along the eastern border of the Salar de Atacama, that is, 200 km east of Antofagasta, at an altitude of 2500 masl (Hesse 1982a:202–203).

In Tulán-52, the most abundant group of animal bones belongs to camelids (Hesse 1982a: 203; Dransart 1991a:311), which according to Wing (1986:Table 10-5, 253) make up 84.8% of the total. Of these, about 32% are of small form, while morbidity is split into different age groups (Hesse 1982a:206). Camelid remains were also found in other later Tulán sites whose antiquity ranges between 900 and 1000 BC. There were thus bones from these animals in Tulán-54 (Hesse 1984:60), and their fiber in Tulán-85 and Tulán-82 (Dransart 1991a:312–314).

Hesse (1982a:206, 210) says in regard to Puripica 1 that 58% of the camelid remains belong to a small form, and that most of the animals died young. He also indicates that the data from this site suggest that the large camelids were domesticated. Novoa and Wheeler (1984:124) and Wing (1986:Table 10-5, 253) indicate that in Puripica 1 these animals account for 76.3% of all the faunal remains. This was corroborated by Dransart (1991a:311–312), who states that camelid remains actually predominate, according to Hesse's data (1982a:203; 1982b:11), and specifies that there is a large camelid form related to the guanaco and a smaller one related to the vicuña.

As for the remains found at Tambillo, Wing (1986:Table 10-5, 253) says they make up 47.4% of the faunal remains at the site. According to Hesse (1982a:206), about 32% of them belong to a small form. He also adds that most of the bones belong to specimens that had matured before dying. In his study, Hesse (1982a:210) used Rick's method (1980:328) and concluded that the animals were successfully hunted in open areas around the site.

Some scholars have tried to summarize the data on the Atacama Desert (readers who want to read more on this point should see Pollard 1971). Troll (1958:29) had already noted that "[t]he culture of the Atacameño in the desert puna based its economic foundations first of all on raising llamas and trading their wool. The concern the people of the warm valleys had for also partaking of the Tierra Fría and the herds of llama follows from some accounts." But to return to the above-mentioned sites (see above), Hesse (1982a:203) clearly notes that the most important category of faunal remains belongs to camelids. He concludes: "Most of the camelids they encountered at the lower elevations of their nomadic round were large animals, perhaps guanacos, and they butchered these individuals at their habitation sites. Small camelids were hunted at higher elevations and these were butchered at the kill site . . . after reoccupation of the region . . . the large camelids were brought under domestication at Purupica-1" (Hesse 1982a:210).

Hesse (1982a:201) then explains that the analysis of the bones from Tulan-52, Puripica 1, and Tambillo "tentatively" suggests that at least two groups, a large and a small one, can be distinguished in the samples, and that different management techniques were used for each kind of camelid. Hesse (1982a:209) insists that the large and small forms in the aforementioned sites were managed differently.

Dransart (1991a:308) in turn is convinced that there is evidence of a change in camelid management between 3050 BC and AD 450. Dransart claims that the switch from hunting and gathering to hunting and gathering in combination with camelid herding is perceivable at both Puripica and Tulán. However, there is no detailed explanation for this change.

Based on the Loa River Valley sequence, from the Vega Alta phase (ca. 800–200 BC) to the Lasana phase (ca. AD 400–1535), Pollard and Drew (1975:296) conclude that the transition from small hunting-and-gathering camps to fortified villages that based their sustenance on raising llamas and irrigation agriculture is evident.

Yacobaccio (1984–1985:168) summarized the conditions in two sites found in the higher parts of the Atacama puna between 2500 and 3650 masl. These are early sites with ages ranging between 855 and 550 BC, and where camelid remains are important. They are San Lorenzo, where camelids account for about 19% of remains, and Tuina, with 40.5%. (See also Yacobaccio 1986:Table 4, 13. For the Tuina site there are additional data in Rivera 1991:11 and Santoro and Núñez 1987:67.) It is worth recalling that camelids are among the few large mammals at high altitudes with which man has established a reciprocal relationship in this part of the world,

over and above the upper limits of agriculture and in an arid environment like the Atacama Desert (Dransart 1991a:304).

Before discussing sites that lie farther south, it is worth noting that Kent (1987:176) argues, based on data for El Niño-related climatic oscillations in the southern Andes, one of which took place in 5000–3000 BC and another in 700 BC–AD 350, that these oscillations might have influenced the adoption of camelid herding in the first millennium BC in sites close to the Salar de Atacama. In this regard it is interesting to recall that the bofedales are considered the best grazing lands for camelids in the Chilean altiplano, followed by the tussock-like grasslands (Troncoso 1982; Baied and Wheeler 1993:150).

Outside the Atacama zone we have the well-known Tagua Tagua site at 1000 masl in the Cachapoal River Valley, in the province of O'Higgins. This is an early site that has been given radiocarbon dates of 9480 BC and 9050 BC. Dillehay et al. (1992:173) note the presence of camelids and give the studies by Montané (1968) and Núñez et al. (1987) as their source. Montané (1968) makes no reference at all to camelids, and I was unable to read the study by Núñez et al. (1987).

Another important site is Quereo, on the Pacific coast south of Los Vilos, in central Chile. Here there are camelid remains in levels Quereo I and II, but these are fossil remains (analyzed in Chapter 3 on paleontology). I mention them here simply to avoid mistakes, because Lynch (1983b: 116; 1990a:26; 1990b:166) simply listed them as "camelid."

Stehberg (1980:18-19) discusses the Andes of Santiago in general terms and explains that between 8600 BC and 6500 BC the Pleistocene fauna became extinct and the native economy was essentially based on camelids and the huemul deer. It seems that by the late preceramic, the diet was based predominantly on camelids, and the first pottery makers were hunting and/or herding these animals as their major activity.

Stehberg and Dillehay (1988:142) studied the Chacabuco and Colina region, in the lowlands (400–1500 masl) of central Chile. This region lies about 20–60 km north of the metropolitan zone of Santiago. Here were collected the remains of camelids (*Lama guanicoe*) that no longer live in the area, thus showing they were important for the economy of prehispanic society (see also Stehberg 1976 [the text says 1976, but in the bibliography it is 1967; I assume the second date is the correct one], Stehberg 1981). Although no period is specified, it follows from the study that it must be a very late one.

Citing Latcham (1922:94), Benavente Aninat (1985:47–48) claims that the guanaco was confused with the llama, and believes that this mistake has been repeated by historians and archaeologists. She thus criticizes Stehberg (1980) when he claims that there were llamas in the Santiago Valley. However, Alonso de Ovalle (1969:72 [or 73?]), whom Benavente Aninat quotes (1985:45), says on mentioning the "sheep of the land" that he recalled "having seen them some thirty years ago in Santiago." Ovalle wrote in 1646. The other references made to archaeologists are very confusing. Finally, Benavente Aninat (1985:47) says: "According to the ethological characteristics of the animal, it was not the 'llama' which populated the area but another camelid species." Benavente Aninat adds that, according to Barros Arana (1872), it is true that the Peruvians brought llamas to the zone, but this attempt failed. However, she immediately adds a doubtful and ambiguous phrase. "Instead they domesticated a similar species, which probably rendered a service similar to that rendered by the 'llama' or 'sheep of the land.'" Then she claims that what the Spaniards saw were guanacos, and that "this animal served as food, *beast of burden*, and its wool was obviously used" (Benavente Aninat 1985:48; emphasis added). All of the arguments put forward by Benavente Aninat to deny the presence of llamas in the zone of Santiago are extremely weak and lack support. A more detailed study of the subject was clearly in order. Besides, to claim so emphatically that the guanaco was used as a pack animal without presenting any argument at all is to go against all of the existing evidence without any support whatsoever.

Dedenbach Salazar (1990:82) mentions a study by Mengoni Goñalons (1983) that I have not seen. It mentioned the discovery in Patagonia of guanacos and another small camelid, now

extinct, dating to 10,600 BC. However, Dedenbach Salazar does not give the exact location of this find.

The renowned Fell's Cave is also in Patagonia. Here Bird (Bird and Bird 1988:187) recorded 20 cultural levels. The guanaco appears in Level 18, with a date of 8130 BC (see also Lynch 1983b:117). Markgraf (1988:196) discussed the context of Fell's Cave and the association of man with the extinct and living fauna, specifically the guanaco. She believes that the abundance of guanaco bones in the early levels and the evidence of a significant decrease in the size of the grasslands that took place shortly before the fauna became extinct suggest that we can assume that environmental changes were a much greater threat to pasture-dependent populations than the pressure exerted by man's hunting. For Markgraf, the palynological data obtained in Fell's Cave support this position.

Camelids, and specifically guanaco, also appear before man in Palli Aike, another famed cave in Chilean Patagonia. There are guanaco bones above the sterile layer, dated to 6689 BC (Bird and Bird 1988:107, 115).

The people of Eberhardt Cave (also known as Mylodont Cave) in the province of Magallanes were also guanaco hunters. According to Lynch, the earliest date is 5800 BC (Lynch 1978:479; 1983b:117). Lynch based his claims on a study by Saxon (1978) which I have not seen. Núñez et al. (1983:45) report that remains of *Lama* sp. were found in the second level (of the evaluation studies made there) which date to ca. 3300 BC (they cite Borrero [1976] and a manuscript by Saxon that Borrero quotes).

Finally, Laming-Emperaire (1968:81) reports that guanacos predominate among the faunal remains found at the site of Ponsonby, Riesco Island, in the Skyring Sea of the Chilean Patagonia, in strata dating to 3720–1750 BC.

Franklin (1982:468) says that early man developed an ethnozoologic culture in southernmost America (i.e., in Patagonia) that did not reach domestication and was centered on wildlife in the plains, specifically the guanaco and rhea. Thus the Tehuelche, hunters in the dry zones of Patagonia, as well as the Onas in the frigid Tierra del Fuego, created guanaco-dependent cultures.

This animal provided them with food, clothing, and ornaments of a religious kind (Franklin cites Bridges 1949 and Cooper 1946a, 1946b). Other tribes, such as the Puelche, the Araucano in the northwest, the Huarpe, and the Querandí, also depended on the guanaco. The warm clothes made with the skin of these animals were particularly important for these southern tribes (Franklin cites Lothrop 1929). Cunazza (1976b:166) also emphasizes this same point and shows how greatly dependent these tribes were on the guanaco, especially the Ona and the Tehuelche.

Finally, I would like to mention an interesting statement by Dillehay et al. (1992:171) concerning the fauna and the weather conditions present at the beginning of the Holocene. They note that the Pleistocene conditions persisted in several areas in the far south during the ninth and tenth millennia BC, and perhaps later too, as deglaciation took place. In the late ninth millennium BC there are more elaborate lithic industries associated with a varied fauna, both extinct and modern, which suggest that the climate improved in Patagonia and allowed some animals to adapt to the environment. This is different from the disappearance of the grasslands on the western Andean slopes and plains of Peru and northern Chile, where the megafauna became extinct perhaps 1,000 or 500 years before, and where the mylodont, and perhaps the horse, abounded up to the eighth and seventh millennia BC. In this period the weather conditions developed and were characterized by an increase in temperature and rainfall (according to the data of Markgraf 1983), as well as by the emergence of an environment that was probably more convenient for the guanaco, which during the Holocene became the primary food source in the grasslands to the east and south.

I was unable to review the literature for the late ceramic periods in Chile, but I will mention two interesting cases. Castillo (1983:6) mentions the Las Ánimas de Coquimbo Complex, on the northern coast of Chile, which dates to AD 900–1100. Twenty-six burials were excavated, 23 of which belonged to adult men or children who were in each case accompanied by one, two, three, or five sacrificed camelids. The animals were placed in such a way that they surrounded the individual, in a protective position, in four different arrangements. In four of the graves the evidence seems to indicate that the sacrificed animals were pregnant. It is also interesting that the dung of these animals was found in the tombs.

Castillo (1983:7) made the following comment:

> The specific behavior of sacrificing camelids, something which goes far beyond the accustomed behavior for food preparation, makes me assume that big herds were managed and that a percentage was set aside for funerary ceremonies, for which the coast of Coquimbo is ideally suited because it has the climate of a steppe with abundant fogs (camanchacas) that enable the presence of permanent vegetation on coastal hills and ravines, which is ideal for herding. If these are domestic animals, it is striking that some were sacrificed pregnant, which I believe is an unusual pastoral practice. On the other hand, the evidence contradicts the assumption that both the llama and the alpaca are animals belonging to altitudes over and above 3000 m, or that their introduction to this region and to central Chile took place at a late date, when the Inca conquered these lands. The cemetery described opens a different, alternate analysis in the face of the general belief that holds the non-adaptability of camelids to coastal areas far removed from the centers with the highest concentration, and at the same time documents a herding group par excellence with a well-adapted population that enjoyed a good diet and managed marine resources, as is reflected by the presence of fishhooks and harpoon barbs.

This statement by Castillo is timely. I will return to this point later.

Rivera (1991:37) noted, while discussing the Regional Development period (which corresponds approximately to the Late Intermediate period of the central Andes, or AD 900–1400), that several strategies based on raising and taming great numbers of llamas and alpacas were applied at this time, giving the people the opportunity to carry out great interregional movements. Caravans of llamas crossed the desert using the signals (now known as geoglyphs) that marked the route (see Núñez 1976).

When discussing the "Late Agro-Ceramic period," that is, AD 1000–1500, based on the research by Niemeyer (1963), Núñez (1969:215) noted that the population of what is known in Chile as the High-Andean Inca settled mainly in the intermontane basins because of the trade routes and hunting and pastoral activities.

9.7.2 Historical Data

Let us now review some historical data. Betanzos (1996:Chap. XXXVI, 151; 1987:164) presents some very interesting information on the Inca epoch. The chronicler recounts that when Topa Ynga Yupangue reached Chile, "he came to a province called Llipi." Here he found that the Indians "had a certain amount of livestock," and so "[t]he Inca ordered these people to send him tribute of . . . livestock." It is a shame that Betanzos did not specify what kinds of livestock, whether llamas (more likely) or alpacas, because these do not seem to be wild animals.

Garcilaso de la Vega (1966:Bk. 7, Chap. XVIII, 446; 1959:67) also describes the Inca conquest of Chile, but his information says nothing of the local fauna. Garcilaso specified that on reaching Copayapu, the Inca "ordered them to take sufficient supplies on beasts of burden, which also were to serve as rations since their meat is very good to eat."

Gómara (1946:237) described the 1535 Chilean expedition of Diego de Almagro, and notes that "[t]here are many sheep, just as in Cuzco." Zárate (1968:Bk. III, Chap. II, 176) also recounts this trip to Chile, but he details the lack of water endured by Almagro's army, "which was solved by carrying sheepskins full of water so that each live sheep carried the skin of another dead one with water." This means that Almagro used a llama caravan in his expedition, just as in Inca times.

Benavente Aninat (1985:44, 47) cites the "Cartas de Relación de la Conquista de Chile" ["Accounts of the Conquest of Chile"] written by Valdivia (1970:171–172), which date to 1551 and refer to the Santiago zone. Here it is noted that this zone "thrived in livestock like that of Peru."

However, conditions were different shortly afterward. In 1558 Vivar (1979:253) discussed the "livestock in this province of Chile," and while describing the European livestock claimed that "this multiplies so well that there are now so many that if the Spaniards had to limit themselves to the livestock of the land, no meat would be eaten."

In his *Décadas*, Herrera (1946b:VII Decade, Bk. IX, Chap II, 220) left a description of the founding of the city of La Serena, Coquimbo, "in the Kingdom of Chile." It reads:

[I]n all the kingdom of Chile there is a kind of tame sheep and a wild [*monteses*] kind, the size of camels and bigger than those of Castile. Their body is usually one vara long, the neck three-quarters of a vara, and they are taller than those of Castile. The upper lip is cleft, with which they throw their spit against whoever annoys them. They do not have a hump like camels, and their meat is slightly drier than that of a Castilian ram [*carnero*]. Their pasture is grass; they are usually white or black in color, and some are ashen. The wild sheep are bright red, a light tawny color. Their wool is long, soft, smooth, and shiny, and more valuable than the wool of Castilian [sheep]. A lock of wool is worth a ducat, and the sheep four and five, and the Castilian sheep twelve reales, and each lock of wool a real. Blankets are made with the wool of the sheep of the land, which resemble those of chamelote [a strong, waterproof textile], [which are] shiny [and] are worn by the rich. They are harnessed by the ears, where they drill a hole and push a thin string through, like grass ropes [*tomica*], which when pulled makes them go wherever they want to lead them, and when loose, run a lot, especially the wild ones, more than a horse.

This description is excellent, and everything seems to indicate that Herrera was describing a llama. The reference to "wild sheep" must mean the guanaco. The source he used is not known, but everything seems to indicate that it was someone who was well acquainted with what he described. There are many similarities with the description Ovalle (1969:72–73; see below) made of these animals, which was published in 1646 according to Benavente Aninat (1985), but the date when it was written is unknown. Herrera perhaps knew of this manuscript, because from Ovalle's own statement it follows that he drew on his own experience when writing.

The passage from Ovalle (1969:72–73) which Benavente Aninat (1985:45–46) reproduced, and which I mention above, dates to 1646. I was unable to examine it, but it is interesting, because the text suggests that he was an eyewitness. Ovalle discusses the animals native to the "Kingdom of Chile." "Among the animals native to that country we can mention first those they call sheep of the land, and are in the shape of a camel, not so coarse or big, and without the hump these have. Some are white, others black and brown, others are ashen." Ovalle then discusses the possible use of these animals to draw the plow, a point that will be discussed later, and then continues with his description:

They are also used even now in some places for trade, to carry wheat, wine, maize and other loads from one place to the other, and I recall having seen them in Santiago [which lies at no more than 520 masl] some thirty years ago, serving as water carriers that brought the water from the river for domestic use, but now they no longer work on any of these tasks because there are so many mules and donkeys that they are used in these and other chores. These sheep have a cleft upper lip through which they spit at those that annoy them, and the children, who are the ones who usually annoy them, on seeing that they want to spit at them flee because they believe, and all usually believe the same thing, that whoever is touched by the saliva is covered by mange wherever it touches them, and since they have such a long neck, which must be about three palmos long, they use these defensive weapons with greater ease. Their wool is highly esteemed, from

which some blankets are woven that seem to be chameloto [a strong, waterproof textile], very shiny. They are harnessed by the ears, where a hole is drilled through which a string is pushed in, which is pulled by the person leading them to take them wherever he will. They kneel to receive their load, and once it is well received and packed get up and take it at an easy pace. . . . The sheep of the land are quite similar to the guanacos, both in the shape of their body and their swiftness, but are completely different in their color because that of the latter is red, a light tawny color and are never tamed, but go forever over the fields from one place to the other in their journeys.

In her study, Benavente Aninat (1985:46) likewise cited Rosales (1877:324, Vol. I), who wrote his *Historia General del Reino de Chile* in 1670 and discussed the camelids. Rosales gives a perfect description of "[t]he rams [*carneros*] we call of the land [, which] are a most domestic livestock." I will give some details. For example, Rosales explains that "[t]hey pierce an ear and push a string through, with which they are led from one place to another, leading them like a horse with the harness." There is another interesting observation. Rosales says that "[i]n Peru they call them Llamo or Paco, and here chillingueque. It is believed that they came here from those provinces because there are so many of them there, while few are found here, and it is great fortune for an Indian to have two or three. And in Peru they have them by the thousands and use them to transport silver, wine, and other merchandise. But on these they do not carry anything and the Indians only raise them with great care for the wool, and carefully look after them, keeping them inside their homes because they are the best property they have to buy women to marry, and are a highly esteemed payment." There obviously are some contradictions between what Rosales said and what the other previously mentioned historians said, but these will be discussed later.

Ocaña (1987:116) makes a brief reference to Chile regarding the point that interests me, and the passage seems to be about southern Chile, but he does not clearly state so. Ocaña writes as follows. "All the land is clean and has plenty of game, particularly guanacos, which are like long-necked rams [*carneros*]." When describing the guanaco and the vicuña, Vázquez de Espinosa (1948:37/15) specified that "[t]hese animals are only found in the Kingdom of Chile, and in the cold [areas] of Peru."

In the eighteenth century, Molina (who must not be confused with either Molina "el Chileno" or Molina "el Cuzqueño" [see Araníbar 1963: 131]) described the alpaca and the llama as beasts of burden and noted that "[t]he Chileans also used the Chilihueque in the same way as beast of burden, but now they have so many mules which have happily reproduced in this climate that they no longer do so" (Molina 1782:Bk. IV, 311).

I also managed to collect some interesting data on specific localities. For instance, Horacio Larrain Barros (1978–1979:68) presents very interesting data on the desert coastal strip between 17° and 30° S latitude, where the Chango lived. Larrain Barros explains that the guanaco (*Lama guanicoe*) came to this area in small herds, climbing down the valley up to the lomas that extended from 500 to 900 masl. They fed on the flowers of the Bromeliaceae, Cactaceae, Poaceae (Gramineae), and other herbaceous annuals that lived there owing to the "camanchaca" [fog] (Larrain Barros gives Weischet 1966 as his source). Larrain Barros notes that Philippi (1860:27, 34) has some references on the discovery of fresh guanaco tracks on the coast close to Miguel Díaz, at "Agua del Panul" north of Taltal, and also in Morro de Mejillones (23° 07′ S latitude), at an altitude of 650 m. On the other hand, Mann (1953) notes the habitual presence of small guanaco colonies on the desert coast that apparently do not undertake any migratory trips between the cordillera and the coast. According to Larrain Barros, Weischet (1966:4) has pointed out the presence of guanacos at the mouth of the Loa River (21° 26′ S latitude). In October 1964 Larrain Barros observed four animals on the slopes of Cerro Moreno (in the province of Antofagasta, at approximately 23° 30′ S latitude), one of which was a young one, in an area with cactaceous plants and lichens (350–600 masl). In May 1972 Larrain Barros also found a lot

of guanaco excrement below some very old tamarugos (*Prosopis tamarugo*) inside the small Pazos ravine in the Pampa del Tamarugal (20° 21′ S latitude and 69° 48′ W longitude).

Larrain Barros (1978–1979:70–71) described the rafts the Chango, using the description Vivar provided in 1558, and notes that the "sinews of rams [*carneros*] (llamas) [were used] to push them between the spines to fix the seam." (Readers interested in more data on the Chango may consult Bird:1964b.)

When Herrera (1945d:VI Decade, Bk. II, Chap. I, 202–203) narrates how "the adelantado Don Diego de Almagro abandoned the conquest of Chile," he explains that "[t]he Atacama Desert separates Peru from the Kingdom of Chile, and now they go to this kingdom along two routes. The first goes along the highlands and the other along the desert." When describing the highlands, Herrera says that "in this desert [*despoblado*] there are a few wild sheep [*ovejas monteses*] which they call guanaco, and they do not reproduce due to the little grass and water there is."

Téllez Lúgaro and Silva Galdames (1989:53) studied the Atacama region in the sixteenth century. They discuss Almagro's expedition, which, on reaching the Atacama Valley, met up with Noguerol and Orgóñez, whom Almagro had sent in advance. They had managed to collect "some maize and livestock."

Vázquez de Espinosa (1948:1752/618) also describes the "[p]rovince of Atacama" and "[t]he Indians of this coast," as well as the great number of fish they catch and "salt." Vázquez then says that "big caravans of rams [*carneros*] loaded with it are taken to Potosi, Chuquisaca, Lipes and all the provinces in the lands above."

There are several references to the Copiapó region. Cieza de León (1987:Part III, Chap. XCV, 319) recounts Almagro's expedition to Chile, which has already been mentioned several times, and says that his army reached Copiapó after the dreadful passage across the cordillera. Here the Spaniards demanded food and help from the Indians, who "came many of them with their sheep, lambs, maize." Cieza then says that the Spaniards who stayed behind tried to cross the cordillera but had to endure great hardships because "there was no wood with which to make a fire other than the dung of the sheep" (Cieza de León 1987:Part III, Chap. XCV, 320). It is also known that when Juan de Herrada left Cuzco with provisions for Almagro, he reached Ortopisa on the way to Copiapó, and his retinue was unable to find food, but after a few more marches they found "a herd of sheep" (Cieza de León 1987:Part III, Chap. XCVII, 326).

Molina "El Chileno" (also known as "El Almagrista") recounted Almagro's expedition too. We must not forget that Porras (1986:317) raised the possibility that this chronicle does not belong to the choirmaster Cristóbal de Molina but perhaps belongs instead to Bartolomé de Segovia, Rodrigo Pérez "or any other . . . but Molina always has the best possibilities [of being its author]." Either way, the chronicler tells how Almagro reached Copiapó after overcoming the hardships of crossing the cordillera, and "this valley had . . . very fat sheep of the land" (Molina 1968:347). This therefore agrees with Cieza de León (1987:Part III, Chap. XCVII, 326).

In their study, Téllez Lúgaro and Silva Galdames (1989:53) describe Almagro's retreat to Cuzco and note that before leaving Copiapó he had carefully prepared his trip. Almagro found out about the Diaguita "lords" and the routes and waterholes in the "Despoblado," and before beginning his trip he collected all the necessary maize and livestock. He also stored as much water as could be carried in the sheepskins they had.

In 1558 Vivar discussed the Copiapó Valley and recounted the clashes between Spaniards and Indians. "Many young Indians died, valiant men who fought manly. Indian men and women were seized, and more than three hundred boys. And gold and clothes were had, albeit not in great quantity. They took sheep and food, for it was a month that we were without meat until the sheep came to the encampment" (Vivar 1979:36). When describing the "Copiapó Valley, and the things in it and the customs of the Indians," this chronicler says that "[t]he Indians are well dressed in cotton and in the wool of the sheep they have" (Vivar 1979:37). On their return to Peru, a group of Spaniards mentioned this same valley and explained that "as they did that, the other Spaniard who was walking in the fields ran into two rams [*carneros*] the Indians were bringing with supplies"

(Vivar 1979:84; see also the 1966 edition under Bibar).

In his *Décadas*, Herrera describes the Kingdom of Chile and mentioned the Guasco Valley south of Copiapó. He says that "[i]n this valley and others there are many partridges and wild sheep" (Herrera 1946a:VII Decade, Bk. I, Chap. IV, 240). Somewhat later he tells how "[a]t the time that Monroy and Miranda left the [Copiapó] valley God deigned to help them with an Indian woman who was going from one town to another, she led a ram [*carnero*] loaded with some sacks of toasted maize, and taking the sacks in the saddlebags they crossed the desert [*despoblado*], which has ninety or more leagues" (Herrera 1946a:VII Decade, Bk. I, Chap. VI, 244).

Fernández de Oviedo (1959c:Bk. XLVII, Chap. IV, 138) also wrote about the trip of Diego de Almagro to Chile, and when mentioning Coquimbo told how Almagro punished the Indians: "And the maize and sheep to pass into the province of Chile and to the Picones was sent for." Later, when describing the vicuña, Molina (1782: Bk. IV, 314) noted that it "abounds in the cordillera of the provinces of Coquimbo and Copiapó, but is usually not found except in the most rugged heights of the mountains: neither the snows nor the ice harms it; on the contrary, it seems it takes advantage of them because if taken to the plains it rapidly loses weight, catches a kind of impetigo and dies: this is the reason why the vicuña has thus far not been taken to Europe." Koford (1957:218) questioned this statement of Molina's (1782:Bk. IV, 314), but without having checked the source, which does not even appear in his bibliography. He just cites Osgood (1943:233), whom I was unable to read.

In Vivar we find news of a zone close to Coquimbo, in the Limari Valley. The chronicler details the departure of Pedro de Valdivia through this valley, and how he reached "[t]he Valley of Cocanbala, which he found empty of people, and for this reason moved on to the Chuapa Valley," where "[g]oing across the valley they seized some Indians who told them where there was much maize which they had hidden in holes, and some sheep" (Vivar 1979:46–47). Vivar continues his account "from the Cocambanbala Valley to the Aconcagua, and of the Indians and things in it," and notes that "[t]his Aconcagua Valley is better and more abundant than all the past ones . . . the sierra is XX leagues from the sea; it has sheep and much maize" (Bibar 1966:37).

In his chronicle, Vivar (1979:74) discusses "how the general Pedro de Baldivia had the city of Santiago reformed, and how they began to raise and plant." Then he notes how "the pigs and sows" multiplied, and how they practiced hunting. He adds, "because it is a warm land and has partridges to hunt, of which there are many, and with wild rams [*carneros*] called guanacos, which have as much meat as a calf" (Sáez-Godoy [1979: nn. 500 and 501] makes some comments explaining the terms carnero and guanaco, but these are irrelevant.)

At this point a passage in Vázquez de Espinosa (1948:1937/681–682) is interesting even if it is not about the point under consideration. It also concerns the "Bishopric of Santiago de Chile." Here is described what "the vecinos of Santiago" had, and 39,250 cows are mentioned "that grow each year by 13,500," 4,270 "mares which each year grow by 1,200," 323,956 goats "which each year grow by 94,764," and 623,825 "sheep, which grow each year by 223,944." It is not specified what kind of sheep [*ouejas*] these are, but everything seems to indicate that they are Castilian sheep. But this is beside the point. The interesting thing is that no European livestock was mentioned in 1558, when Vivar wrote, and only 70 years later (Vázquez de Espinosa wrote his work in 1628 and revised it in 1629) we find not just a huge number of European animals that had already been introduced in the zone of Santiago but also the excellent way in which these animals had prospered.

Palermo (1986:166) has a reference for the Araucano (in middle Chile) which says these had antecedents raising llamas. His source is a study by Montoya (1984:98–99) that I do not know. In another article (Palermo 1986–1987:74) he says that the Indians of the Araucania led their camelids by pulling on a string that went through a hole pierced in one of the ears. This is based on the 1726–1727 edition of Herrera and the works of Molina (1878b) and Rosales (1877–1878). Drawing on the data in Olivares (1864), Palermo (1986–1987:74) notes that the Indians in the Araucania helped the males mate, as was done in

other parts with the llama, whose males tend to have trouble penetrating the vulva of the female due to the shape of their penis. When discussing the *hueque*, Olivares (1864) said, according to Palermo, that "it is a very cold animal and so it multiplies little or, as they say, only with the care of its owners, [something] indecent to do, or unworthy of being mentioned."

Vivar (1979:163) gives us another kind of information on these Indians. He mentions the "Indians of the Mapocho," and says that "[i]n this cordillera and twenty leagues away there are some valleys where live a people called 'Puelche,' who are few." Vivar then explains that these Indians do not dedicate themselves to agriculture but hunt, and observes that "[t]here are many guanacos." He then adds that "[t]he clothes they have are of the fur and skin of the lambs." The Mapuche are Araucano Indians who lived in the southern part of central Chile (Cooper 1946b:687), while the Puelche (a synonym for "Pampeano," of the "pampa"—plains) lived in the Argentinian pampa (Imbelloni 1959:685–686). I wonder whether Vivar made a mistake and meant instead the Pehuence, the neighbors of the Mapuche, who were probably Araucano Indians too.

Vivar (1979:183) mentions another, more southerly province, that of "Conceçion" (Concepción). When describing the customs of the Indians, Vivar explains that they had a kind of armor made out of the skin of sea lions, "[a]nd these capes are lined with the skin of lambs."

Benavente Aninat (1985:45) reproduced a passage on the States of Arauco and Tucapel by Mariño de Lovera (1865 [1594]), a source I was unable to find. There is a problem, however, because although Benavente Aninat included Mariño de Lovera in her bibliography, below the passage we read Bartolomé de Escobar (1865:44 [1]), from whom she apparently took the passage. Bartolomé de Escobar does not, however, appear in the bibliography, and there is no note 1. This clearly is a secondhand quotation that must be taken with due caution.

According to this passage,[7] in these States there were "rams [*carneros*] of the land which are very big and of a different species than in Europe, both . . . for hunting and transportation, and so in . . . parts of Peru there are very big droves of them

. . . they have very big profits. . . . *In some cases* [in the passage cited it is not clear whether the emphasis was in the original text, or was added by Benavente Aninat] . . . it went loaded with presents, something that . . . to the Spaniards by this most tame animal the shape of a camel, although the neck is very thin and raised, and the head small without horns, and the eyes [are] so . . . in this look, they seem to be people . . . they are . . . useful because of the wool rather than for transportation, because . . . more . . . blonde or black serves . . . the size of these animals . . . up to that of a pony [*un cuartago*] . . . tall, but are slightly smaller, and have very thin legs and the nails . . . like the cow."

Vázquez de Espinosa (1948:1960/689) in turn describes the founding of "[t]he Ciudad Imperial" (which is close to Temuco), some "three leagues inland from the sea." He states that "its fields and wetlands were fertile. . . . There were and are at present many offspring of livestock of the land, because it has multiplied much because the land is fertile and has big pastures," but also adds that in those fields there are "cattle and sheep, goats and pigs." This is a most important source.

Vivar (1979:189) has some news on the founding of the city of Valdivia (which happened in late 1551 according to Vázquez de Espinosa [1948:1970/692]). He says that here "[t]here are tame sheep." Later on Vivar describes the customs of the local Indians and explains that in a wedding the bride's father asked for "a certain number of sheep (fifteen or twenty according to their possibilities)" in exchange for the bride (Vivar 1979:191). Vivar later tells how Don Pedro de Valdivia set out to discover new lands on 7 February 1552 and reached a lake he called Valdivia, a "[l]and of much livestock, but it is not loose" (Vivar 1979:198). Given the date, this clearly is livestock of the land (Sáez-Godoy [1979] notes that according to Vázquez de Espinosa, the lake must be some 15 days' journey from Osorno. He gives the page of the passage cited as 694. This is certainly a mistake, because what Vázquez de Espinosa [1948:1975/694] says is very different. He literally says that "two leagues [away from the city of Osorno is] a beautiful lake called Gaeta.")

Vivar (1979:249) continues his description and says that between Valdivia and Osorno there is a place from where the Spaniards head for the sea and go to an island that the chronicler calls "Anquecuy," which had "much livestock."

Vázquez de Espinosa (1948:1975/694) describes the city of Osorno and specifies that "there are 16 leagues from Valdivia to Osorno," and that "[t]here was much . . . livestock . . . wild . . . guanacos, vicuñas. . . ."

Benavente Aninat (1985:45) has an interesting citation from Guerrero Vergara (1880), who in turn copied Juan de Ladrillero (dated as 1558 in Benavente Aninat's text, and as 1557 in the bibliography), and is about the zone of the Strait of Magellan. This is the same voyage Vidal Gormaz (1879) made. The passage reads thus: "Throughout this expanse, from the northern sea up to the cordillera, which are forty-three leagues inside the Strait. . . . There are sheep and guanacos and deer, but with the cold of winter they go inside the mountains where they cannot be found until summertime, when with the heat they come to the river banks out in the open" (Guerrero Vergara 1880:501). Juan de Ladrillero himself (also in Guerrero Vergara 1880:498) describes the clothes of the people: "The clothes of the women are a dress [made with] the skin of the guanacos and sheep . . . and they wear them like the Indians of Cuzco."

We know from data in Benavente Aninat (1985:45) that Vidal Gormaz (1879:516) wrote that Don García Hurtado de Mendoza, Captain General of the Province of Chile, sent several ships to explore the southern territories in 1557–1558. Captain Francisco de Cortés Hojea described the customs of the Indians in the Ancud region (i.e., the Island of Chiloé): "and land at six and four and eight sheep per Indian, and the casiques at 12 and 15, and they only tie one sheep while the rest go loose after them. They only take the woolly ones inside their house, the rest remain out in the pasture with the one tied to a stick driven into the ground and which each one has marked."

9.7.2.1 Chilihueque, Hueque, and Rehueque

There is one controversy over which scholars have long disagreed: the definition of the term *chilihueque* or *hueque* that appears in the early sources, where it is used to define a Chilean camelid.

Palermo (1986–1987:68) states that the most important domestic animal in Chile, "now vanished from the region," was that locally known as *hueque, rehueque* or *chilihueque*, "on whose zoological identification there usually are different positions." Palermo (1986–1987:71) assumes that the term *hueque* was first used by Pedro de Valdivia (1861) in 1550. Benavente Aninat (1985:46) says that the *chilihueque* and the guanaco were still mentioned in 1735 by Luis Tribaldos de Toledo (1864) while giving an account of the animals and livestock in Chile, and he specified that the vicuñas were raised toward the tropics.

In his book, Molina (1782:Bk. IV, 309) describes the vicuña, the *chilihueque*, and the guanaco, noting that it is a "subaltern species of the camel genus" to which the Peruvian alpaca (or paco) and llama also belong to. Molina (1782:Bk. IV, 309–313) himself made a general description of these animals that it would useless to include here, wherein he compared them with the Asian and African camels and established their essential differences.

The definition Molina (1782:Bk. IV, 316–317) left of the *chilihueque*, which he called *Camelus Araucanus* (*Camelus corpore lanato, rostro superne curvo, cauda pendula*), is as follows:

This animal, properly speaking, is called *Hueque*, but the Araucanians, among whom it is found domesticated, began to call it *Chilihueque* or *Rehueque* after the Spaniards arrived, that is to say Chilean *Hueque* or pure *Hueque*, to distinguish it from the European ram [*carnero*], which they call the same because of the resemblance between one and the other. In fact, the *Chilihueque* is shaped in all like the ram [*carnero*], leaving aside the length of the neck and the height of the legs. It has the head shaped like this, the ears oval-shaped and loosely hanging, the eyes equally big and black, the muzzle long and humped, the lips no less thick and hanging, the tail formed in a similar shape but shorter, and all the body covered by a wool this thick but much softer. Their length, measured from the lips to the beginning of the tail, is of about six

feet [1.67 m if these are Spanish feet and about 1.82 m if English feet], but the neck takes up a third of this length. Their height, measured at the hind legs, is of slightly over four feet [1.11 m if these are Spanish feet and about 1.21 m if English feet]. Their color varies and there are white, black, brown and ash-colored. The ancient Chileans . . . used these animals as beasts of burden, leading them down the road by a rope that goes through a hole they made in the cartilage of the ears. This gave rise to the mistake made by those geographers who said that the rams [*carneros*] had grown so much in Chile that they were loaded like mules, and were used to carry merchandise.

Molina then discusses the claim some made that the *hueques* were used to work the land, which is discussed below. Molina concludes: "The Chilihueque is highly prized by the Araucanians, who almost do not kill them even though they like their meat, unless it is to prepare food for an important stranger, or on the occasion of a solemn sacrifice. They dressed with its wool before the discovery of America, but do not use it now that they have European rams [*carneros*] in great numbers, save to make their finest textiles, which are so nice and glossy that they almost seem to be of silk."

It is worth pointing out something not noted by Molina (1782) but mentioned by Benavente Aninat (1985:46) based on Rosales's 1670 work (1877:324), to wit, that the Indians also had a name for the guanaco, which they called *luan*. Rosales apparently explains that the word guanaco "belongs to the language of Peru called Quichua, and that of the Chileans is luan." This was also said by Cabrera and Yepes (1960:77).

As already noted, I was unable to review the literature needed for an in-depth discussion of this point, but I did ascertain that the scholars definitely have different positions.

First, there is the group that believes these animals are llamas. When discussing both alternatives, namely, that it is a selected form of Patagonian guanaco or a llama, Troll (1935:143–144) took the second position based on a study by Lenz (1905). He justified his positions as follows:

It is true that in Chile there were not many specimens of this animal, but even so it was used as beast of burden, its wool was prized, and its meat was even occasionally traded. Its diffusion spread as far as the limits of the Inca colonization of the Araucania, and in the first years of the Spanish conquest it was still found in the Island of Chiloé. Llama raising could not flourish either in the wet southlands of Chile, whose climate is not fit for any "auquenia," not even the guanaco. By the early seventeenth century the llama had completely vanished (in 1614 llamas were still used in Santiago de Chile to carry water [his source is Lenz 1905]). Lenz rightly assumes that these animals were not originally introduced into Chile by the Incas, but were instead brought from the north by the Araucanians themselves, for the original seat of this people extended to the present puna region of northern Chile, where they were in contact with the Atacameño and the Diaguita, both llama-raising groups.

Cooper (1946b:703) takes the same position but does not justify it. He just mentions "[t]he llama (chilihueque, rehueque, *Lama glama*)." Although more cautious, Gilmore (1950:437–438) also does not support his statement. He wrote: "The 'chilihueque' of North Chile was probably the common breed of llama, and was employed as a water carrier (Cabrera and Yepes 1940, p. 263)."

According to Palermo (1986–1987:69), the following scholars have also accepted that the *hueques* are llamas: Cabrera and Yepes (1960 [:77]), Medina (1882), Latcham (1922), Steward and Faron (1959), Mariño de Lovera (1865), and Febres (1882).

Scholars who believe the *chilihueque* was an alpaca include Walton (1811:20), who categorically stated that "the Chileans and the Araucanians [sic] of course have what they called the Chilihueque, or sheep of Chile, which I am sure is the same animal as the Peruvian alpaca in a more degenerate state, and according to Molina they used it as beast of burden in a way not used by the Pe-

ruvians, which was pushing a leather strip through the edge of the ear, to be used as a bridle." This last observation is wrong.

According to Gilmore (1950:438), Marelli (1931:54) also believed the "chilihueque" to be an alpaca. So too did Santa Cruz (1942:6). Of the Mapuche, the latter said that these did not lack meat because they had the *chilihueque*, "which the Spaniards called sheep of the land or by other, similar, names, which Molina called *Camellus araucanus* and is different from the guanaco (the Lúan of the Indians), and which González Nájera, who saw them, said were of two colors, white and black, alone or mixed, which seems to refer instead to the alpaca (*Auchenia paca*)."

Bullock (1958:152–153) also believes that the most important animal the Mapuche had was the *Hueque*, "which the Spaniards later named Chilihueque." For him, "[i]t was a variety of alpaca from Peru which no doubt had the same origin as the llama and the alpaca." Then he says that no llama, alpaca, or *hueque* is known in the wild state. Wilhelm (1978:190) was of the same opinion, because he noted that "[b]efore the Spanish conquest, the most important animal for the ancient Mapuche was the *hueque*, an Auquenid quite similar to the alpaca of Peru (probably a variety of it) which the Spaniards named *Chili-hueque*."

Others believe that the *chilihueque* was a guanaco. According to Benavente Aninat (1985:48), this position was taken by Molina (1901 [1782]), Gay (1847), Philippi (1872), Barros Arana (1889 [dated in her bibliography as 1872]) and Latcham (1922). I was unable to check all of these data, but Molina definitely does not hold this position, for we saw that he clearly distinguished the *chilihueque* from the guanaco (see above, Molina 1782:Bk. IV, 309).

Novoa and Wheeler (1984:121) have an ambiguous position. When discussing the problem of the guanaco, they state without any explanation that "the Araucanians distinguished between the tamed *chilihueque* and the wild *luan*." It follows that Novoa and Wheeler believe these are in both cases guanacos. However, according to Palermo (1986–1987:68) there is no certainty that these terms had this meaning in Araucanian.

I do not believe I can take a position in this matter with the scant data available. I therefore refer to the study by Palermo (1986-1987), which to my knowledge is the best on this subject.

Palermo (1986–1987:68) begins explaining that some scholars, like Latcham (1922), held that the *hueque* was a tamed guanaco, surely based on the area of distribution of these animals, and also on the fact that the llama is not found south of 27° S latitude in the Argentinian province of Catamarca, except for isolated specimens in zoological gardens and some small groups recently introduced into the provinces of Tucumán, La Rioja, San Juan, and Córdoba. Meanwhile, the distribution of the alpaca is at present restricted to Bolivia and Peru (according to data in Cabrera and Yepes [1960], and Davids [1987]). However, says Palermo, in the data of Cabrera and Yepes (1960) there is evidence that in the past, the llama was more widely distributed in Chilean territory, because in the early eighteenth century it extended as far as the city of Santiago.

Palermo (1986–1987:68) notes that in the case of the *hueque*, its presence can be ascertained farther to the south, that is, as far as Ancud (approximately 42° S latitude), in wet coastal zones. We thus see that this distribution coincides with the latitudes and kinds of habitat that guanacos usually frequent, and lies outside the area of distribution of the other camelids. Besides, Palermo says that we should not forget that in certain conditions the guanaco is an animal that can be tamed (he mentions the data in Molina [1878a] and Oporto [1977]. The edition of Molina he cites is a compendium published in Italy in 1776 that I have not seen. I used the 1782 Italian edition, which is unabridged.)

Palermo (1986–1987:68–69) presents two arguments to reject the possibility that *hueque* was a synonym of guanaco. First we have a linguistic argument. In the Araucanian language the guanaco is called *luan*, not *hueque*. It is possible that there might have been two terms, one for the wild animal and another for the domestic one. The second argument concerns the chroniclers' descriptions of the *hueque*'s fleece. Palermo says: "All recall their family resemblance with the camels (leaving aside their smaller size and lack of hump), a common feature to all American camelids, but they usually specifically distinguish between the various kinds of 'sheep of the land'

and describe the guanaco with its reddish color, which is quite different from that listed for the hueques." Mariño de Lovera (1595) says "blonde and black"; Herrera (1601) "white or black, some are ash-coloured"; González de Nájera (1614) says "of two colors, white and black and some all black, others all white"; Ovalle (1666) indicates that "some are white, others black and brown, others ash-colored"; Rosales (1674) states that "the color in some is brown, in others white, and black in some, and in a few these three colors are mixed"; Córdoba y Figueroa (ca. 1740 [1862]) mentions a black specimen; Olivares (1758) says "white, or variegated in black and white"; Molina (1878) that "there are white, black, brown and ash-colored" (this description corresponds to that of 1782 [Bk. IV, 316]). Palermo concludes as follows: "As is well known, these kinds of fleece never appear in the guanacos but do so in llamas and alpacas."

Palermo (1986–1987:69) believes that "the term *hueque*" actually "comprises both species," the llama and the alpaca. Besides the arguments presented, he also adds several lines of evidence that are worth summarizing here.

Palermo notes that Rosales (1877–1878:I, 324) was confused because he claims that in Peru "they call it llama or paco, and here *Chilihueque*." Meanwhile, Wilhelm (1978) states, as was already noted, that the *hueque* probably was a variety of alpaca. Marelli (1931) has a similar position.

Although Molina (1878b:479 [1782:Bk. IV, 314]) rejects the presence of alpacas in Chile, Palermo believes there is evidence to the contrary. Although the wool from the llama has several qualities, it is coarse to the touch, less shiny than that of either the sheep or the alpaca, and not very elastic. Alpaca wool, by contrast, is much longer and has a typical gloss and softness. The chroniclers have interesting data in this regard. Thus, Valdivia in 1551 (1861:55) informed the emperor that in Chile there abounded livestock "like that of Peru, with wool growing to the ground." Herrera (1726–1730:VII Decade, Bk. IX, Chap. II, 191 [p. 220 of the 1946b edition]) said that the blankets made out of this wool "seem to be of chamalote [a strong waterproof textile] [and are] shiny," while Ovalle (1888:I, 91) repeats him. (But it would seem that Herrera copied

Ovalle. This point has yet to be established; see above.) Córdoba y Figueroa (1862:21 [note 18 in Palermo says 1860, but the bibliography reads 1862. I assume the first date is wrong]) likewise discusses the wool of the *chilihueque* and says that it is "shiny and soft," just like Olivares (1864:30), who noted "its very soft and long wool, [which is] therefore most adequate for curious textiles." Molina (1878b:481 [1782:Bk. IV, 317]) explains that this wool was no longer used with the introduction of sheep, and was only used to make fine textiles "which are so beautiful, and so shiny, that they almost seem like silk." It is for all this that Palermo believes that "[t]his insistence on the length, and especially on the gloss and softness of the wool from the hueque, recalls the alpaca (we should remember that the finest *cumbi* textiles were made with this fiber in Peru, from whence camelid raising may have spread to Chile), while the brief description by Valdivia brings to mind the *suri* variety of this species, whose wool sometimes actually reaches to the floor." However, Palermo admits that there simultaneously is another kind of source—like González de Nájera (1889:30), who said that the animal in question had "very long wool, though not as fine as that of our rams [*carneros*]," or Rosales (1877–1878:I, 324), who claimed that "[it] grows a long wool, more than the rams [*carneros*] of Castile, and not as soft"—whose descriptions recall the llama instead.

Palermo (1986–1987:70) later makes the following comment: "On the other hand, we must also bear in mind that llama wool is in general characterized in terms of the present-day specimens from Peru, Bolivia, or northwest Argentina, whereas there might have been an improvement or higher-quality variety of wool in the herds of central Chile; in this regard we must recall that there are differences in the wool of the Argentinian llamas, whose fiber improves from north to south, and come close precisely to the kind of the alpaca in the province of Catamarca (Davids 1978:22). Ultimately, if the Indians east of the Cordillera managed to develop in a brief span a type of sheep superior to the creole sheep as regards the size of the animals and the quality of the fiber, then the 'Araucanians' may well have obtained similar results with the llamas."

Palermo (1986–1987:70) then notes that the llama can live at sea level but suffers on wet soils because it gets mycosis on its feet. On the contrary, the alpaca seeks wet areas. In this regard, the "apparent" raising of the *hueques* in marshy areas of the Chilean Araucania would be explained if at least some populations of *hueques* were actually alpacas, as in the case of Purén (Rosales 1877–1878:II, 240–241) and other southern areas.

Palermo also discusses the climatic factors and what he calls "the adaptability" of South American camelids. He says that despite being adapted to the climate of the puna, llamas and alpacas tolerate conditions of humidity and temperature that differ considerably from it. This is shown by the captive specimens, so Palermo believes there is no need to assume that the climates of central Chile (relatively dry temperate Mediterranean, with a mean temperature of 14°C between the Aconcagua and the Bío-Bío valleys, and rainy temperate, with abundant rainfall and a mean annual temperature of about 12°C from the Bío-Bío River to Chiloé) presented an insurmountable barrier that could have prevented the reproduction of both species introduced by man, and if this had been the case, the identification of these animals with the *hueque* would have to be abandoned. However, Palermo notes that "[t]his does not imply . . . that it was an optimal region to raise them: perhaps their survival in it, especially in the central sector, took a toll in unfavorable birth and morbidity rates that limited reproduction and the growth of populations of this livestock."

Nor did Palermo (1986–1987:71) neglect the study of their diet. He believes that up to the Bío-Bío River, the vegetation has some similarities with what is considered ideal for the camelids under consideration. Woodlands appear ever more south of this river, where there is a transitional area containing alternating patches of prairie and woodland vegetation until they form the Valdivian forest, composed of trees of great size (beeches and others) and an abundance of ferns and epiphytes. Palermo admits that the landscape is different from that more to the north, but notes "that the hueques lived in this zone already transformed by man," and likewise notes "that here we should not forget the grains and the stubble of cultivated plants, in the absence of a more adequate flora."

The conclusions reached by Palermo (1986–1987:71) are as follows:

1) [T]here are no insurmountable ecological barriers for llama and alpaca raising in the Araucania; 2) the chroniclers gave contradictory descriptions of the hueque's wool, some of which recall the fiber of the alpaca, others that of the llama; and 3) there are records of hueques in areas with relatively dry and also wet or marshy land, respectively, preferred by llamas and alpacas, so I believe it is reasonable to assume that the domestic camelids of the Mapuche belonged to both species, and that there eventually were hybrids of them (the so-called "huarizo") which had a better wool than the llama but exhibited a higher morbidity (Cabrera and Yepes 1960:II, 80, 83 [nothing is actually stated here about an increase in morbidity]; Davids 1987:15). A common label for both kinds of "hueque" would be due to their general similarity in external appearance and their possible uses, and the feasibility of their cross. The Dictionary of Erize explains the term as being related in general to the camelids. As adjective, "hueque" means "woolly," which is a generality that goes beyond the specific (Erize 1960).

Finally, Palermo (1986–1987:72) established the distribution of the *hueque* in Chile, stating that in the sixteenth and seventeenth centuries these camelids appeared in all Chilean regions inhabited by man as far as 42° S latitude. In central Chile they went beyond the "Araucanian" lands to the north, because they were used in northerly zones beyond the Copiapó Valley (Córdoba y Figueroa 1862:41; Góngora Marmolejo 1862:11; Herrera 1726–1730:III, V Decade, Bk. X, 228 [1945d:V Decade, Bk. X, Chap. II, 125–126] and III, VIII Decade, Bk. I, 9 [this must be a mistake because the passage corresponds exactly to 1946b:VIII Decade, Bk. I, Chap. VI, 329–330, in the edition I used, and this point is not discussed here]; Rosales

1877–1878:I, 401). To the south they reached at least as far as the Ancud zone. It seems that their distribution between these two points was quite even and comprised both Mediterranean as well as coastal, and even insular, localities.

Palermo supports his conclusions on the distribution of the *hueques* with historical data that are worth including here. Palermo (1986–1987:72) wrote:

Thus for example we find a record of Pedro de Valdivia in Bío-Bío (1549); presents of these animals given to the Spaniards by the confederates of the coastal region of Concepción, Arauco and Tucapel (1550); the sacrifice of hueques on occasion of a treaty in Arauco (1604); an animal presented to Spanish explorers in Puerto Carnero (which owes its present name to this episode, which happened in 1544); the seizure of hueques and European livestock belonging to the Indians in the zone between Arauco and Lebu (1609); the concentration of a great number of specimens by the confederates of Arauco and the Cordillera when summoning other groups to war (1630); the sacrifice of these animals in Tucapel (1604), among the Purene (1586) and the besiegers of Angol (Purene and others) that same year; the purchase of animals in the Island of Mocha by Spielberg's Dutch pirates (early eighteenth century); the sacrifices held during a treaty in the Quellén Valley by the caciques of the Imperial region (1640) and in groups from the tributaries of the Cautín (1639); various uses observed by Núñez de Pineda y Bascuñán in the zone of the Imperial-Cautín rivers (1629); the animals presented to Valdivia in Toltén (1550) and on the southern banks of the Valdivia River (1551); the sacrifices made in the Valdivia region (1646); the gift of one specimen to captain Pastene somewhere on the coast, at approximately 40° S latitude (1544); the sacrifices made during the peace agreed upon in the Osorno zone (1648); the presence of this livestock in Ancud, according to the observations made by the men of Ladrillero (28). [Note 28 says: Mariño de Lovera 1865:44, 122, 133, 138, 230; Medina 1882:181; Núñez de Pineda y Bascuñán 1863:41, 50, 120, 124–125, 140, 200, 278; Ovalle 1888:I, 91; "Poder que dió," 1862:223; Rosales 1877–1878, II:236–237, 240–241, 419, 422, III:81, 163, 324, 335; Tribaldos de Toledo 1864:51; Valdivia 1861:43.]

After the discussion by Palermo it is worth reading carefully Molina's description once again (1782:Bk. IV, 316–317, see above). Although he denies the existence of the alpaca in Chile, Molina provided a description that, if studied beginning with the characteristics of the living species, actually exhibits some features that could belong to a llama, and others to the alpaca.

I will now review some of the archaeological and historical data on camelids in Chile. The first issue is the use of llamas in caravans as beasts of burden.

9.7.2.2 Caravans and Geoglyphs

Núñez studied the geoglyphs of the Chilean desert, which have, among other motifs, complex scenes with caravans of llamas and men leading them (Núñez 1976:150). Most of these geoglyphs lie between the mouth of the ravines of Pampa de Tamarugal and the nearby coast, in sterile sites and in a great desert zone that had to be crossed to establish contact between the coast and the highlands. According to Núñez, these geoglyphs marked the traditional routes for the caravans of llamas. They extended from the middle valleys to the coast (Núñez 1976:177). From the descriptions Núñez (1976:176) gives, it follows that these activities took place from the Middle Horizon onward.

Núñez (1976:179) wrote:

The geographic conditions north of the Camiña River tend to present valleys that end in the Pacific Ocean. However, the valleys and oases are isolated by a desert, with remarkable "despoblados" [uninhabited areas]. With this consideration in mind, the traffic system must have resorted to complex markings in sites of transitory stays between the coastal-agrarian valleys via the altiplano. This thesis is reinforced by the decrease in the number of geoglyphs in the valleys of Arica and southern Peru, due to the fact that the increase in the number of valleys beside the Pacific Ocean does not require such complex traffic markings.

Núñez (1976:180) explains that current ethnographic data show that the caravans of men and llamas hold traditional ceremonies to improve relations between the groups in the altiplano and the coastal oases. These ritual ceremonies are held in the *apachetas*. Núñez therefore believes that these geoglyphs were shrines where the caravans stopped for the night (*paskanas*). He notes (1976:184) that this exchange of products was carried out "along ancient routes of contact that even overcame one of the harshest deserts in the world." Núñez adds that the dynamic potential of the so-called "Atacameño" peoples has been recognized since the time of Uhle (1919) and Latcham (1938). The features that let us identify the use of caravans of llamas have been found both in the middle Loa River (Calama) and in Pisagua (Núñez 1976:187, 189).

The archaeological data provide evidence of distant contacts with the tropical forest and northwest Argentina. For this reason Núñez posits the existence of a most complex network that connected early specialized productions between the coast, the lowland oases, the altiplano and "puna oases," and the eastern lowlands (Núñez 1976: 189). Núñez says that the caravans tended to return to the low-altitude floors after the summer rainy season, and it seems that the routes were regulated by given optimal harvests of maize, algarroba tree, or exceptional "varazones" [a school of fish so big that it reaches the beach] of fish, like that mentioned by Vázquez de Espinosa (1948: 1420/483). Núñez likewise states that the caravans also participated in coastal labors (Núñez 1976:191).

Núñez (1976:192) reports that the archaeological village of Quillagua, which is quite close to the Pacific Ocean and on the Loa River, shows evidence that in late periods, its inhabitants had their living quarters beside great courtyards with extensive guano floors, which seem to indicate that the caravans of llamas used to trade maize, algarroba, and marine products with the highlands were penned here (Núñez based his statements on a personal communication from Cervellino-Téllez).

According to Núñez (1976:193), Meighan (1970) was able to date the village of Guatacondo, which shows evidence of "transportation with caravans," to AD 60. Lanning also managed to date organic samples from a llama path between the sites of Chiu-Chiu and Talabre, in the middle Loa River Valley, which gave a date of AD 250. Núñez (1976:194) also recalls that "there is much evidence of ropes, hooks from llama harnesses and bells in the collections of San Pedro de Atacama." He believes that "[l]lamas and men competed in transportation capacity throughout the Inca land like never before" (Núñez 1976:195).

Núñez (1976:195–196) claims that the caravan traffic reached an amazing level of development in the last century prior to the European arrival. The first Europeans who reached the area saw "exotic caravans of small animals able to transport, as a group, tons of cargo across inhospitable sites." This experience was then adapted to the mining concerns of the colony.

Núñez claims that "[w]ell into the seventeenth century there were 8,000 Andean muleteers in charge of the flow of caravans between the mines and the silver ingenios (V. de Espinosa, 1628–29)" (Núñez 1976:195). In this case the reference is incorrect. What Vázquez de Espinosa (1948:1653/580) actually says (see above) is that there were 8,000 muleteers in Potosí. There obviously must have been more in all of the Andes.

Still using Vázquez de Espinosa (1948) as his source, Núñez then discusses the movements of the caravans between the coastal and the inland oases across the desert, and notes that "[t]he survival of the Andean livestock in this late period shows a remarkable efficiency despite the European campaign for the establishment of breeding grounds for mules. This persistence shows in the transport of guano from the islands of Erica and Iquique in 'big droves of rams [*carneros*]' (1628–29:480-482 [the correct citation is Vázquez de Espinosa 1948:1415/480, 482]), precisely where the most prominent desert geoglyphs are

concentrated." Vázquez de Espinosa's data (1948: 1665/587; 1752/618) also indicate, Núñez says, that dried fish was taken to Potosí along this same route from a big part of the littoral between Arica, Tarapacá, and Atacama.

Núñez (1976:196) says that "[t]he specialization of the movement of pack llamas to the Pacific Ocean and their temporal adaptation in the warm coastal stations would seem to limit the efficiency of the transportation capacity as one of the development mechanisms of highland populations. In one way or another, the loads of guano, bundles of fish, etc., necessitate some permanence which could have been extended if the coast was uninhabited and demanded a temporal exploitation by the tradesmen themselves. Long seasons outside the Andean land, with insufficient water and inadequate fodder, would seem to be an insurmountable problem. However, the early chroniclers (e.g., Bibar 1540 [the reference is wrong: it should be 1558, 1966, as it appears in the bibliography]) show that maize and waterholes or gentle slopes were the key to the passage of these hosts through the same deserts that now concern us. For this reason too it would be hard to determine the ideal size of a caravan or llama train that was just passing through the littoral. Diez de San Miguel (1967:124) counted concentrations of 600 llamas in some coastal oases of southern Peru. These figures seem to be the upper limit. Both present-day caravans as well as those recorded in other early visitas show the lower range: '. . . Both moieties gave forty rams [carneros] with sacks for the Moquegua Valley, and four Indians who went with them to said valley . . .' (Flores Ochoa, 1970:69)."

Núñez (1976:197–198) ends with a most interesting personal testimony. "Though made more than a decade ago, our observations," he wrote, "on the current traffic between the oases and the altiplano (e.g., Pica) show that just a few individuals are needed to drive big caravans of llamas on journeys of more than 15 days. Longer pre-European displacements seem to have taken up to three months (Diez de San Miguel 1967 [1574]). . . . But leaving aside the time used, a humble caravan of 50 llamas with a mean load of 30 kilos per animal carries an approximate volume of 1,500 kilos from the altiplano to the oasis of Pica through a landscape that even today lacks motorways. . . . Caravans of llamas periodically came from Bolivia to the oases of the Atacama puna until just a few years ago."

9.7.2.3 The Caravans

I did not find many references to caravans of llamas in archaeological and historical sources, but those I found are interesting. Téllez Lúgaro and Silva Galdames (1989:48–49) studied the region of San Pedro de Atacama and the Atacameño chiefdoms, which had contacts with the Argentinian Northwest and the altiplano. These connections, they say, were maintained with an active traffic of llama caravans that came to the coast and that possibly carried algae and dried and smoked fish to the people inland. When discussing post-Tiahuanaco times, which is when the chiefdoms were born, Téllez Lúgaro and Silva Galdames note that the vertical and latitudinal complementarity did not fall apart with the collapse of the Tiahuanaco Empire. The caravans of llamas loaded with agricultural, marine, or metal products continued flowing along the ancient roads leading eastward or to the coast.

When describing the town of Arica, Murúa (1925:46) describes the silver that was shipped there and how it came from Potosi "in caravans of mules and rams [*carneros*] of the land."

Téllez Lúgaro and Silva Galdames (1989:66–67) studied the introduction of European plants and animals into the Atacama area in the seventeenth century and the success they had, especially in the exchange and trade of products. One of the most lucrative and persistent activities of the Spaniards and mestizos in Colonial times was the commercialization of dry fish in the highlands. There was also a trade in European goods. These arrived in Chilean ships which unloaded their cargo in Cobija, and had Potosí as their destination. All of these products were carried by llamas or mules, something that was already common in the sixteenth century. "So much so," note the authors, "that in the early seventeenth century the Spanish troops sent from Tucumán verified that in the Atacameño puna, in the words of the official report, 'There . . . is a clear road to Atacama and they come and go with rams [carneros] loaded with wine and other things they bring to

the mines of Cochinoco, which are twenty miles away from Jujui . . .' (128)." (Note 128 is on ANS, FMV, Vol. 50, Ms. "Chile," fol. 4) (Téllez Lúgaro and Silva Galdames 1989:66–67).

9.7.2.4 The Chilihueque or Hueque as a Draft Animal or Beast of Burden

We saw that when describing the Araucanian *chilihueque*, Molina (1782:Bk. IV, 316) said that it was used as beast of burden and was led with a rope pushed through a hole pierced in the cartilage of its ear. Palermo (1986–1987:73) discussed this point and noted that there are some controversies about it. Mariño de Lovera (1865) cites a case that took place on the coast in 1540, in Puerto Carnero, where the Indians gave García de Alvarado a *hueque* loaded with presents. Núñez de Pineda and Bascuñán (1863) claim that in 1629 the *hueque* was used in the area of the Cautín and Imperial Rivers to carry beverages to celebrations. However, in 1674 Rosales (1877–1878) claimed that the *hueques* were not used as pack animals because they were animals carefully looked after. Palermo believes that by that time, the abundance of horses, mules, and donkeys surely made using the *hueque* irrelevant.

A curious fact connected to the *hueques* and the sheep of the land is that these animals are credited with having been used to pull the plow. Benavente Aninat (1985:45–46) cites Ovalle ([1646] 1969:72), who said the following of the Chilean sheep of the land: "The authors cited say that of old in some parts they were used to plow the land before there were oxen in it . . . that when they passed along the Island of Mocha the Indians used these sheep for this purpose." I have not read Ovalle and so do not know what the authors mean.

Tschudi (1885:n. 2, 108; 1891:n. 1, 108) also mentions this author, and notes that "although an Indian is depicted plowing with two llamas in the map accompanying D'Ovalle's (145) account of the voyages in Magellanic lands, this must be taken as no more than a fantastic rendering [made] by European artists." In footnote 145 Tschudi specifies that this appears in Chapter XXI of the aforementioned book (Tschudi 1918:228, n. 145). In the original text, Tschudi wrote the name of the author in Italian because he used the 1646 edition, which was published in Rome. Here appears "Alonso d'Ovaglie." As we saw, the 1918 translation places D'Ovalle in the text and Alonso de Ovalle in footnote 145. I point this out to prevent any confusion. (Several pages on the camelids were removed from the 1969 translation, which is why this part does not figure there.)

Benavente Aninat (1985:46) herself cites Rosales, another scholar ([1670] 1877:324) who wrote as follows: "Gotardo Artusio says that the Indians of La Mocha plowed the land with these rams [*carneros*] of the land pulling the plow. Thus did father Alonso de Ovalle print them in his map, following what he saw in the foreigners, who in this were wrong and led him astray . . . but it is true that in no part of this Kingdom of Chile were they taught how to plow, nor have they been used in this task, for I have been there and I neither saw nor heard that they plow with rams [*carneros*] of the land. Nor do those islanders use a plow to dig the land, but a *luma* [a kind of wood found in Chile] stick."

Molina (1782:Bk. IV, 317) wrote the following in this regard, when describing the *chilihueque*: "Others claim that before the Spanish conquests these Indians used these quadrupeds to work the land, tying them to the plow which they call *Quethahue*, and to tell the truth Admiral Spielberg found that the people of Mocha Island used them for this." Tschudi comments on this point that Ovalle "likewise says that while in Mochica Island [sic, a *lapsus calámi* by this diligent scholar; it should be Mocha], on the southeastern coast of Chile, the Dutch admiral Spielberg saw that the Indians plowed the land with welkes (*huelques* [these obviously are the *hueques*]). Molina, drawing perhaps on Ovalle's statement, also says that the people of Chile plowed with huelques before the land was conquered by the Spaniards. However, these claims have no actual basis" (Tschudi 1885:n. 2, 108; 1891:n. 1, 108; 1918:n.145, 228).

Benavente Aninat (1985:41) mentions Medina (1882, 1952), who in turn says, based on Gay (1947), that the llama was used by the Indians to pull a kind of plow called *queñelvolque*.

This subject was also analyzed by Palermo (1986–1987:73). When discussing the problem posed by the *hueque*, Palermo said: "Their possible use in agricultural tasks has also been dis-

cussed. Both Ovalle (in his map), Molina and Gay mention an ancient use of the *hueque* to draw a kind of plough, a custom the Dutchman Spielberg presumably saw in Mocha Island but which Rosales rejects, stating on the one hand the non-existence of such a tool among the Indians in his time, and on the other only certifying the use of cattle as draught animals in the case of ploughs introduced under Spanish influence (Medina 1882; Molina 1878b; Rosales 1877–78)." This is obviously a typical case of a mistake due to ignorance (see also Figure 9.6), which is then repeated without any study. I find that Rosales's arguments are correct.

9.7.2.5 The Slaughter of Camelids

I have found few data on the killing and disappearance of camelids in Chile.

Téllez Lúgaro and Silva Galdames (1989:59) discuss the events after 1557, when Atacama was still considered a "frontier of warring Indians [*indios de guerra*]." They believe that the conflict "cost the Atacameño society dearly. The columns passing through emptied the barns [*graneros*] and stole the livestock." Cooper (1946b:703) in turn believes that the introduction of mules and sheep diminished the massive raising of the llama.

Stehberg (1980:10) says that in Colonial times, the excessive exploitation of the resources of the cordillera practically led to the exhaustion of the woodland resources of Santiago and the extinction of native herbivores like the camelids and the huemul. (We must bear in mind that the altitude in the foothills of the Santiago region is around 600 masl, but at its highest it reaches 4000 masl, and there are heights of more than 6000 masl.) Stehberg (1980:21) believes that with the coming of the Spanish conquistadors to the zone, and especially when they established their capital city in Santiago, "the pressure on the resources of the cordillera rose to levels never before seen. The natural forests of the pre-cordillera zone and the autochtonous fauna of human consumption (camelid, huemul) were practically wiped out." Brack Egg (1987:62) states categorically, although without citing any source, that the vicuña population was decimated in Chile in the conquest and the viceroyalty. Franklin (1975:191) says the same, and he also does not give his source.

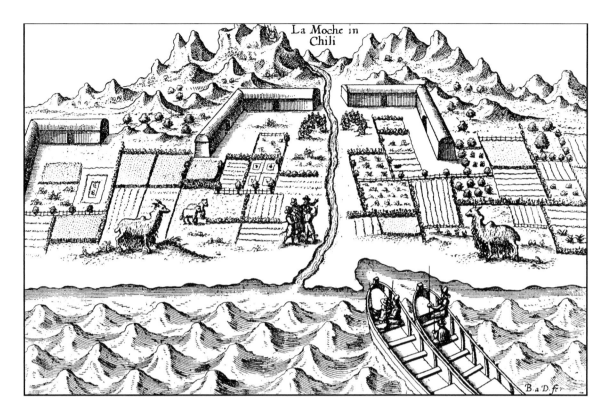

FIGURE 9.6. An engraving showing Mocha Island in Chile. It was published in Rotterdam in 1601 (Cooper 1917:114) to illustrate an account of the voyages of Olivier van Noort (1926). The engraver, Baptist á Doetinchem, seems never to have been in America, and his illustrations were probably based on verbal descriptions and sketches (Omar Ortiz Troncoso, letter of 23 May 1996). This is reflected in the figures of the camelids, possibly llamas, which were depicted with a hump. *After van Noort (1926, Plate 9).*

9.7.3 Distribution of the Guanaco, the Vicuña, and the Llama

I shall now include some modern data only on the distribution of the guanaco and the vicuña in Chile. In the early nineteenth century, Walton (1811:21) wrote that both animals reached Chile from Peru "in wild state naturally," but were only hunted for food and for their skin. Rivero (1828) noted that to judge from what the travelers said, there were guanacos "on all the coast of Chile, which is not the case with the vicuña."

Bittmann (1986:315–316) recently stated, while discussing the north Chilean coast, that the guanaco is one species that until recently climbed down from the highlands to the coast. Then he notes in a historical overview that in prehispanic times the guanaco must have been an important seasonal resource for the fishermen, as was noted

by Philippi (1966:292) for later periods. Remains of these animals have been found in coastal archaeological sites, "which show their skin was used, as well as their bones to prepare tools. Besides, the meat must have likewise been used as food." There are scenes in rock paintings at the El Médano site, illustrated by Mostny and Niemeyer (1953:Fig. 58), which show the guanaco was killed using bows and arrows.

According to Bittmann, several scholars have verified the presence of the guanaco on the north Chilean coast in the present day, "albeit in very small numbers because the vegetation has gradually disappeared." Bittmann gives as an example Bauver (in Pernoud 1942:36), who described the way the fishermen in "the mountains of Cobija" hunted these animals with the help of dogs in the early eighteenth century. Bittmann (1986:315–316) also says that Philippi (1966) described the guanaco hunts of the "Chango" as an alternative way to obtain resources to sustain life. As is well known, this ethnic group lived around the mid-nineteenth century in the southern part of what is today the Second Antofagasta Region. Philippi's description (1966:292), which Bittmann cites and which dates to the mid-nineteenth century, is as follows: "At certain times the life of these people is jeopardized. When the sea is stormy in wintertime for a long spell, so that they dare not fish, they have no other solution but to hunt the guanacos, which at this time of the year are expelled from the altiplano by the cold and the snow and climb down to the coast, where the winter rains produce abundant vegetation, albeit one which rapidly dies out." This same Philippi (1860:34, 67) claims to have found fresh guanaco tracks on the coast north of Taltal (25° 25′ S latitude) and in Morro de Mejillones. On the other hand, Bittmann continues, in our century Mann et al. (1953) indicated the presence of these animals in Morro Moreno, and Weischet (1966) has noted their presence at the mouth of the Loa River (21° 26′ S latitude). Finally, in the 1980s guanaco tracks were seen in the ravines that open on the Taltal zone (pers. commun., Leonel Lazo S. from Taltal to Bittmann 1985), and the excrement of these animals was found in the vicinity of the site called "Los Nidos," which is about 24° 00′ S latitude, 6 km away from the littoral and at 900 masl. Bittmann notes that today, a remarkable amount of fresh water can be found there.

Bittmann (1986:276) then mentions the lomas north of Chile, and explains that the vegetation in this formation can serve as food for the guanaco.

Niemeyer (1985–1986:172–173) reports on a region farther south, in the Copiapó Basin, and notes that here the guanaco has a wider habitat and is often found in coastal areas, while the vicuña has its habitat in the highlands over 4000 masl.

Casamiquela and Dillehay (1989:207) note that the wild guanaco was found in the grasslands of central Chile until the mid-nineteenth century. Lynch (1983b:118) cites Saxon (1978) and notes that he is convinced that the sloth and the guanaco competed for the same habitat in the southern part of the American continent, where the latter's development benefited from human predation, while the giant sloth became extinct.

Glade and Cunazza recently wrote on the guanaco in Chile and report that its highest density is recorded in the Magellan region, specifically on the island of Tierra del Fuego. There are other significant populations in Tarapacá and Aysín. Glade and Cunazza indicate that no scientific studies have been done to distinguish the subspecies, even though the northern breed is traditionally distinguished from the southern one. The present population of guanacos in Chile is 19,856 individuals (Glade and Cunazza 1992:48).

There are almost no data on the llama. Cooper (1946b) alone mentions it while discussing the Araucanian Indians, an ethnic group that, at the time of the European arrival, occupied all the area west of the Andean chain, from the island of Chiloé to at least the Choapa River. Their southern frontier was the Gulf of Corcovado, but the northern one is not that clear; however, it lay somewhere between Choapa and Coquimbo. Cooper mentions the llama while discussing the domestic animals of this group, and states that the tamed guanacos were quite common in the southern part of the area occupied by the Araucanians.

However, a qualification is in order. For Cooper, *Lama glama* is a synonym of *chilihueque* or *rehueque* (this point was already discussed in-depth earlier), but as we saw, this label is not clear, and it most probably refers to both the llama and the alpaca. Cooper's review (1946b:703–704) is actually quite poor. Above all it is a historical review that repeats the data mentioned here while discussing the problem of the *chilihueque* or *rehueque*, just adding references to two authors I did not include, Goiçueta (1852) and Ascasubi (1846). According to the latter, the llama still survived among the Huilliche until the late eighteenth century (Ascasubi 1846:350).

(Readers interested in recent figures for the number of camelids in Chile are directed to Table 2.2.)

9.7.4 The Problem of Domestication

We saw that domestication is one of the most important problems that has yet to be solved, but so far everything seems to indicate that the Chilean area was not one of its centers.

According to Hesse (1982a:203), Druss (1981) suspects that the faunal remains he recovered in his excavations in the Loa River area show evidence of camelid domestication, but that study had not yet been completed at the time Hesse wrote. Prior to this, the earliest evidence of domestication was that found by Pollard and Drew (1975), which indicated the presence of herding. Based on the data from the Tulán-52, Puripica, and Tambillo sites in the Atacama River basin, dated to around 2500 BC, Hesse then states that there was herding with domestic animals.

Pollard and Drew (1975:296) have reported on their work at Vega Alta and Loa. For the first site, the available dates range between ca. 800 and 200 BC, and for the second, between ca. 200 BC and AD 400. This sequence shows a development from small camps of hunter-gatherers to fortified settlements that had agriculture and llama husbandry as their economic foundations.

According to the analyses done by Pollard and Drew (1975:303), they found the remains of wild and domestic animals. "In both Vega Alta II and Loa II settlements, the domesticated camelids were of a different species from the wild analogs, and in each case the domesticated animals were undoubtedly llamas. The identity of the wild camelids was not as easy to establish." The date given for Vega Alta II is 200 BC, and AD 105 for Loa II (Pollard and Drew 1975:298–299). Pollard and Drew show that the domestic *Lama* sp. was not used along the Loa River until long after it was used in Peru (Pollard and Drew 1975:304). This is supported by Pollard (1976:18). (To avoid mistakes it must be pointed out that Dedenbach Salazar [1990:82] took these data and summarized them. However, one mistake slipped into her text. When citing Hesse [1982a], she ascribes "an incipient domestication by 3550 BC" to him, which is incorrect, since Hesse [1982a:203] said "4500 BP," in other words, ca. 2550 BC.)

Rivera (1991:10) writes that "[w]e still do not know how the domestication of the camelids took place, but Hesse (1982[a]) has evidence of domestic llamas and alpacas from 4885 before the present in Puripica, in the Atacama desert." The citation by Rivera is clearly wrong, since the dates Hesse (1982a:203) gives for Puripica-1 are 4815 BP and 4050 BP.

The conclusions reached by Palermo (1986-1987:72) are important. He wrote: "Although a local process of domestication cannot be discarded (as apparently happened with some cultigens), the possibility that these camelids arrived due to Peruvian influence, together with other elements, is very strong. . . . Although these groups, at least those located south of the Maule River Valley, halted the Inca advance, a Peruvian jurisdiction extended northward as far as Copiapó since the fifteenth century, with the presence of troops and *mitimaes* from that source (Córdoba y Figueroa 1862; Fernández de Oviedo 1851–1855, 4 [1959]; Mariño de Lovera 1865; León Solís 1983), so it is quite feasible that the introduction of the hueques can be explained in this way (as well as other previous influences with an Andean provenance in general)."

Stothert and Quilter recently stated that "[t]he evidence for camelid domestication in Chile is late, around 2000 BC at Puripica-1 (Lynch 1983[a]:5; Núñez 1981:155), but it is probable that camelid pastoralism began in the south-central Andes: as early as it did in the central Andes—by 5000 B.P. (Lynch 1983[a]:4; Núñez 1983[a]:61–74 [1983])" (Stothert and Quilter 1991:34). However, their database is not the most complete because it does not include Hesse (1982a). Their date agrees with Baied and Wheeler (1993), even though the latter's bibliography is more up to date. Baied and Wheeler thus state that a second "possible" center of domestication lay in the salty puna of the Salar de Atacama, at a date ranging between 4800 and 4000 BP. Their data come from Hesse (1982a, 1986), Santoro and Núñez (1987), and Núñez and Santoro (1989) [in the bibliography it is listed as 1988, and I included it with this date]. However, Baied and Wheeler (1993:147–148) admit that the data for the dry puna of northern Chile are missing.

Although there is potentially no reason not to accept the possibility of a center of domestication in the Chilean area, the evidence, at least what I know of it, is not clear enough to show it.

As for the guanaco, Gilmore (1950:450) wrote as follows: "For countless centuries, young individuals have been tamed by aborigines for pets, perhaps to be killed later for food and perhaps consciously tamed for this purpose." Gilmore then notes that in the 1950s there were semidomestic herds in Argentina, when he prepared his report, "but they are somewhat refractory to domestication and it is doubtful that the present members ever will be fully domesticated so that they can be raised easily in semiconfinement and utilized as is the llama. . . . Hybridization with the llama confuses this situation" (Franklin [1982:468] repeats Gilmore without further comment).

Raedecke (1976:11) is skeptical on this point, although it follows from his study that some animals might have been domesticated. He wrote: "Barros (1963) explains that the Araucanians tamed the male guanaco and used it as beast of burden, while the females and chulengos were used as food. This is the only reference available on the domestication of the guanaco *but it has not been verified* because most of the Indians did not keep them tamed in great numbers, but hunted them with bolas, arrows, ambushes and occasionally with dogs" (emphasis added).

Scholars in general do agree that the llama was introduced into Chile by the Incas. Thus, Troll (1958:28) wrote that "[t]he llamas, taken to central Chile by the Incas, were in no way able to remain there; they had vanished once again by the early seventeenth century (Lenz 1905)." Bennett (1946a:43) discussed the Araucanian zone and categorically stated that the llamas were introduced by the Incas. Gilmore (1950:438) notes that the southern frontier of this animal marks the limits of Inca expansion. Murra (1975:118) simply accepts what Troll said (1931:266; 1958:29 [the reference is wrong; it should be 1958:28]). Gade (1977:116) concurs but is not that emphatic; he says that "it is possible" that the diffussion of the llama in Chile was carried out by the Inca. Wilhelm (1978:190) and Novoa and Wheeler (1984:124) likewise believe that the diffusion of the llama to central Chile was carried out by the Incas.

9.7.5 Feeding Habits

At the archaeological level, all that is available on the feeding habits of the camelids in Chile is the study by Belmonte et al. (1988). They studied camelid coprolites in a shell mound at Camarones dating to the preceramic period. Unfortunately, the study does not specify from what site the remains come, just that it is at the mouth of the Camarones River and is "Preceramic." (For the Camarones 14 site the dates range between 5470 and 4665 BC, and for Camarones Alto there is a date of 4320 BC. I do not know whether the samples come from any of these sites, or from another I am not aware of.) From the report it follows that Belmonte et al. were unable to identify what camelid species the coprolites belong to. In the studies made, only two of the 12 recorded plants species were identified, *Distichlis spicata* (Gramineae [Poaceae]) and *Scirpus americanus* (Cyperaceae). The unidentified species are monocotyledons, which are absent in the present flora; "the diet of these animals was 100% composed of plants of the Monocotyledoneae class. Of this total, 27.3% belong to the Cyperaceae family, and 18.2% to the Gramineae family. . . . No taxa could be assigned to the following 54.5% of the remaining samples" (Belmonte et al. 1988:51).

9.7.6 Conclusion

From the data presented it follows that camelids played an important role in the life of the native groups in Chile. Palermo (1986–1987:73–74) has shown that these animals were invaluable, at least from the sixteenth to the eighteenth centuries. Many scholars (some of whom were cited here; see above) concur "in that it [a camelid] was one of the most highly prized possessions and the best sign of riches and prestige" (Palermo 1986–1987:73). The products obtained from these animals were varied and ranged from wool to make clothes, to meat; they also acted as pack animals and in ceremonially important roles.

I cannot discuss each of these points, owing both to lack of space and lack of data. I will just add a few more comments on some of them. Palermo (1986–1987:73) states that the chroniclers concur in noting that although highly prized as food, the meat from the *hueques* was not eaten every day, just during certain ceremonies. He cites González de Nájera (1889), who claimed that the Indians only used the blood of these animals in times of dearth, bleeding the heads of living males. However, González de Nájera also claims that the milk of the females was used, despite which Palermo indicates that it is usually noted that these animals cannot be milked. According to González de Nájera, the Indians used the bones to make arrow points.

Palermo (1986–1987:73–74) noted the significant role the *hueque* had in various ceremonies, sacrifices, and as offering in burials, as well as food in special banquets, or in the minga labors in seedtime, and as a gift. He also says that these animals had an important role in trade, and mentions the eighteenth-century example of the Island of Mocha, where the *hueque* was raised to sell it to the moieties on the mainland.

9.8 BRAZIL

9.8.1 Historical Data

It might seem out of place for a book on camelids to discuss an area that lies inside the present territory of Brazil. However, there are interesting data available for the first half of the sixteenth century, which are certainly striking. They concern the trip Orellana made in 1541–1542 to discover the Amazon River. We have the invaluable account of Friar Gaspar de Carvajal (1894, 1986), who was a member of the expedition.

Carvajal wrote: "We walked along this land of Omagua for more than a hundred leagues, and then came to another land of another lord called Paguana, who has many and quite domestic people. . . . Many roads go inland from this town because this lord does not live on the river, and the Indians told us that we should go there, that he would be very pleased with us. In that land this lord has many sheep of those from Peru, and he is very rich in silver" (Carvajal 1894:45–46; 1986:70–71). According to Díaz Maderuelo (1986:n. 91, 70), the chiefdom of Paguana must lie on the left bank of the Amazon River, between the confluence of the Catúa and the Negro Rivers.

The chronicler continues thus:

> [T]he Captain ordered that we were to go in much order until we left this province of Sant Juan. . . . And another day in June [twenty-five] we passed along some islands . . . which is why they did

not stop following us until they had expelled us from their towns. . . . That night we went to sleep outside all inhabited area. . . . and the Indian replied that he was called Couynco (Quenyuc) and was a great lord who ruled as far as where we were. . . . The Captain asked him what women were those [that] had come to help them and wage war on us: the Indian said they were some women who lived inland seven [four or five according to another manuscript in Muñoz (1955)] journeys away from the coast. . . . He said more, that among these women there is a lady who rules and has all the others under her rule and jurisdiction, a lady called Coñon. . . . He said that in the capital and major city where the lady lives there are five very big houses that are temples and houses dedicated to the Sun, which are called by them caranaín . . . and they are dressed with clothes of a very fine wool, because in this land there are many sheep of the kind from Peru. . . . He said more: that in this land, according to what we understood, there are camels which are loaded, and he says there are other animals which we could not understand, the size of a horse and with the hair a hand-span long [geme] and a cloven hoof, and they have them tied up, and of these there are just a few. (Carvajal 1894: 64–68; 1986:83–87) [The 1955 edition (p. 106), which corresponds to the manuscript of Juan Bautista Muñoz and is the one del Busto [1975:510] used, reads: "We also understood that there are camels and other animals which are very big and have a trunk."]

The whole story of the women who attacked them is fantastic. Several details (the Sun cult, "houses . . . with paintings of several colours," "gold and silver idols in the shape of women, and much stonework [sic: *cantería*] in gold and silver for the cult of the Sun" [Carvajal 1986:86]) seem to resemble accounts of the Inca Empire that had reached the area, or came from someone who had visited it. The search for El Dorado must also have had some influence on this (see Díaz de Maderuelo 1986:14–15).

According to this story, these women lived on a part of the lower Amazon, close to Tapajos but a short distance away from the coast, because the interpreter said that "they were some women who lived inland, seven marches away from the coast" (Carvajal 1986:85). There are, however, some inconsistencies because it is also said that these women lived in a "land [that] he says is cold and with little wood, and that abounds in all kinds of food" (Carvajal 1986:87).

In the late sixteenth century (between 1581 and 1586), Ortiguera (1968) discussed the "Journey [*Jornada*] of the Marañón River." Here Ortiguera described Francisco de Orellana's voyage through the Marañón River "down to the North Sea." First he narrates the voyage of Gonzalo Pizarro through the province of Quijos and then on to Zumaco, where the Spaniards founded the city of Ávila. Pizarro stayed there, and Orellana continued the voyage. He moved on to the Irimaraezes, then to the Cararies Islands, and came to Machifaro through the river of the same name. Orellana went forward and passed through several villages that are not to the point. Ortiguera (1968:Chap. XV, 248) says that among other things, "[t]hey found painted on the canoes of some of these Indians some things like those of the Ingas of Cuzco, and on others sheep [*ovejas*] and rams [*carneros*] from Peru, and asking through the interpreters they had with them what those paintings meant, they said that there was this kind of people and animals inland, pointing to some high cordilleras that are visible from the river."

Carvajal's account was known by Gonzalo Fernández de Oviedo, Gómara, Zárate, Garcilaso, Herrera, and Cieza de León (Díaz de Maderuelo 1986:16). Thus, Herrera (1946a:VI Decade, Bk. IX, Chap. IV, 156), when narrating the voyage of Orellana, says that on leaving the town of Machiparo they reached another town, where they found "another lord called Paguana, where the Indians were domestic and gave what they had, and there were sheep of those from Peru."

Díaz de Maderuelo (1986:n. 57, 59) notes that "[t]he first of the settlements subject to the cacique Machiparo can be located close to the confluence of the Teffé and the Amazon Rivers."

He may be right, but conditions were far more complex. These have been carefully reviewed by Chaumeil and Fraysse-Chaumeil (1981:83–84), whom I now summarize. (To follow the narrative better, see "Carte VI" of Chaumeil and Fraysse-Chaumeil.)

The Machifaro are only mentioned in sixteenth-century documents, and there are problems in identifying them. Métraux (Chaumeil and Fraysse-Chaumeil do not give the year, but it might be 1948b) tends to include them in the non-Tupi group of the Middle Amazon, while Grohs (1974) tends to relate them to the Tupi populations. The statement made by Jiménez de la Espada (no date is given, but it must be 1965b and 1965c) relating the Machifalo to the seventeenth- and eighteenth-century Iquitos makes no sense in its historical context.

According to Chaumeil and Fraysse-Chaumeil (1981:83–84), the first to mention the Machifaro (Machifalo) was the Portuguese Diego Núñez (1538, but he does not appear in the bibliography), but it is doubtful that he ever established contact with them. Besides, we must bear in mind that his description was published in 1553–1554, so he may very well have drawn on the account by Carvajal.

"Despite the controversy raging over the exact location of their land, one fact seems to be established: the Machifaro occupied the banks and the islands of the Amazon River east of the Putumayo River, somewhere between the Jutaí and Teffé Rivers. On the other hand, the different positions between the chroniclers are not necessarily contradictory if we bear in mind the dynamics and the migratory possibilities of these groups." Chaumeil and Fraysse-Chaumeil end by stating that "[t]he Machifaro had significant food reserves, especially of maize stored in a kind of barn, and thousands of river turtles held in artificial lakes. They might even have practiced some kind of pisciculture. This naturally is in stark contrast with what we know of the present Amerindians and their scant storage capacity due to the inadequate methods used" (Chaumeil and Fraysse-Chaumeil 1981:83–84). Interestingly enough, Chaumeil and Fraysse-Chaumeil do not mention the accounts of the sheep cited above. Given the sources they used, they seem to have voluntarily ignored them. It is possible that they consider them absurd and unworthy of credit, but it would have been interesting to have their opinion.

As far as I am aware, Tschudi is the only scholar who has discussed the subject. In this regard he wrote: "Nothing can be concluded about the truthfulness of the account that says Orellana saw some llamas belonging to a cacique on the Amazon River beyond the confluence of the Negro River, but this is most doubtful. Were this true, these cannot have been anything but a few animals kept there free, perhaps as a novelty. It is well-known that South American Indians have a penchant for keeping animals. The news given by Philipp von Hutten [not included in the bibliography of any of the editions] that the Priest-King Kwareka of the Amagua nation [the 1918 translation mistakenly says Amagua, but the first translations correctly said Omaguas] had big herds of llamas, simply belongs to the realm of the fantastic" (Tschudi 1885:94–95; 1891:97; 1918:207; 1969:125).

Tschudi, then, does not accept the presence of llamas in the Amazon. The fact that these are the accounts of Carvajal and the Spaniards who were with him does not mean much, because they received the news from the Indians but did not personally see them.

Another piece of evidence that goes against the presence of llamas is a document that seems to belong to Rojas (? [1889] 1986). Herein is recounted the voyage of Captain Pedro de Texeira upriver from the mouth of the Amazon, in 1638–1639. There is no certainty as regards its author, but it most likely is Rojas (Díaz Maderuelo 1986: 22). This does not matter; what does is that this document mentions the "Indian women commonly called the Amazons" (Rojas? 1986:244) and agrees with Carvajal (1894, 1986), with just some variants. However, there is no reference to the sheep of the land. The description goes as far as the Marañón River and is quite detailed regarding the animals and plants (Rojas? 1986:247–248), but at no moment are the sheep, or anything like them, mentioned.

In my opinion, the interesting point is that these Indians knew of these animals. I do not find this striking, because I am convinced that relations between the tropical forest and the highlands have always been intense, and even more so during the Inca Empire (see Bonavia and Ravines 1967).

9.9. PARAGUAY

9.9.1 Historical Data

My data on Paraguay are limited to a single sixteenth-century source by Schmidel (1749, 1962, 1986) which tells of the discovery of the Río de la Plata and Paraguay. Although it is true that the narrative is about lands that now belong in part to Argentina and Bolivia, most of the events took place in present-day Paraguay. "Schmidel's book is a classic that has been republished and copied into modern German several times. It is an intriguing case of a soldier with literary leanings. The really surprising fact about this chronicler is his birthplace. He was a native of Landschut, the ancient capital city of Bavaria (now lower Bavaria), a region whose people are not precisely renowned for being open and cosmopolitan" (Peter Dressendörfer, letter of 16 October 1993). Schmidel wrote his account on his return to Germany, between 1562 and 1565 (Wagner 1986:13; Schmidel 1986:23, 111), but he lived the events he recounts in 1534–1554, so his testimony is of primary importance.

After taking part in the foundation of Buenos Aires in 1536 (see Wagner 1986:n. 4, 115), the "captain general Juan de Ayolas" and his officers "resolved to make an alarde [a military parade]": with four hundred men they "[c]ontinued sailing up the Paraná River to the Curendas" in eight brigs, "in search of another current that from what they say is called Paraguay" (Schmidel 1986:Chap. 16, 39–40). It was during this trip that the expeditionaries met the Cario, and it is then, when describing the animals the natives had, that the chronicler for the first time mentions "sheep as big as a mule" (Schmidel 1986:Chap. 16, 40). He immediately adds: "We reached the Quiloaza. . . ." (Schmidel 1986:Chap. 17, 40–41).

Now, we know that the Quiloaza lived on the islands of the Paraná River, beyond ("arriba", the chronicler says) the Timbú (see Lothrop 1946: 190) who lived on the islands of the Paraná River

which lie upriver from the mouth of the Car-
carañá River, and probably on the eastern banks
of the Paraná River (33° S latitude, 60° W longitude; Lothrop 1946:187).

Schmidel notes that the Cario "travel more
often and farther than any other people in all of
the Río de la Plata. They are excellent warriors
on land, and their towns and cities are on elevat-
ed sites close to the Paraguay River" (Schmidel
1986:Chap. 20, 45). He then tells how the Cario
"had to build for us a big house of stone, earth
and wood . . . [and] we took possession of this
place on the day of Our Lady of Assumption of
the year 1539, for which we gave it that name,
and it still called so" (Schmidel 1986:Chap. 22,
47). Wagner (1986:n. 7, 116) explains that the
year given by Schmidel is when the fortress was
handed over before a notary, but Asunción was
founded in 1537. Its site is to the east of the Central Chaco, and according to the chronicler the
"Cario live in a big and sprawling land almost
three hundred leagues long and wide" (Schmidel
1986:Chap. 20, 44–45), because their last town,
called Guayviano, was "eighty leagues [away]
from the city of Asunción," or about 440 km
(Schmidel 1986:Chap. 23, 49). When describing
the Cario towns, the German adventurer insisted that these "have abundant . . . Indian sheep,
big as mules" (Schmidel 1986:Chap. 20, 44). In
the illustration dedicated to this chapter we see a
llama, among other animals (Schmidel 1962:
Chap. 20, 24; my Figure 9.7). The animal is well
depicted despite the European influence in the
drawing. This shows once more that Schmidel
was a good observer, because according to Plischke (1962:XVIII), the original engravings were
made according to drawings or sketches made by
the chronicler himself.

Now, the Cario are none other than the Indians known today as the Guaraní. In the sixteenth century these were called Carijó or Cario,
while the name Guaraní was established in the
seventeenth century (Métraux 1948d:69). This
was a general term for a very large human group,
but the sixteenth- and seventeenth-century Spaniards distinguished the local tribes with special
names (Métraux 1948d:70). Métraux described
the location of the Guaraní. He explains that
"[f]rom the junction of the Paraná and Paraguay

Rivers, Guaraní villages were distributed continuously up the eastern side of the Paraguay River
and up both sides of the Paraná River" (Métraux
1948d:70). Métraux also noted that "[t]he Cario
(Guaraní), who understood the aim of the Spaniards and who hoped to make them allies in their
raids, were extremely friendly to the Spaniards,
and provided them with food and women. Hence-
forth, the Guaraní served as auxiliaries and
porters in all Spanish expeditions, whether to the
Chaco or to the Andes" (Métraux 1948d:76).

The group continued its trip and reached
"Cerro San Fernando" where the Payaguá lived
(Schmidel 1986:Chap. 24, 50), and which for
Wagner (1986:116, n. 8) is the Pan de Azúcar. Actually, there were two Payaguá groups, a southern
one living at 21° 05′ S latitude and a northern one
at a 25° 17′ S latitude. In sixteenth-century documents, the northern group is called Payaguá and
the southern group Agaz (i.e., the Agace). These
groups are fully extinct (see Métraux 1946:
224–225, Map 4). Schmidel lists the foods these
Indians gave them and among them mentions
"meat of Indian sheep" (Schmidel 1986:Chap. 24,
50).

Later on "[t]hey reached the Jaraye, where
they were received and treated generously." The
chronicler describes their customs and says that
their women "make big cotton blankets as subtle
as those of Arras, weaving on them various figures
in the shape of deer, ostriches and Indian sheep,
according to the skill each has" (Schmidel 1986:
Chap. 36, 66, 68). The Xaraye lived close to the
modern Lake Maniore (18° S latitude, 58° W longitude) and had contact with tribes living close to
the Andes, from which they obtained gold and silver (Métraux 1948a:394; 1946:Map 4). The land
in question is close to the frontier between Bolivia and Brazil, but the Paraguay River they followed is inside Brazil. These are marshlands
about 200 masl or less. They actually are the
headwaters of the Paraguay River.

From thence the group returned to Asunción
to attack the Cario, and it is here that the chronicler describes the camelids. Schmidel wrote:
"We made big bucklers with the skin of the sheep
called 'amidas,' which are big beasts, almost like
mules, of gray skin and legs like cows. There are
many of these in this land. The leather is thick,

FIGURE 9.7. This drawing, which illustrates the work of Schmidel, shows a Cario settlement. In the middle of it is a llama. *After Schmidel (1962:24).*

of almost half a finger. We made almost four hundred of these bucklers" (Schmidel 1986: Chap. 42, 79–80). Wagner discussed this passage and notes that "[u]nder the name 'amida' or 'amate' Schmidel meant the llama, without paying too much attention to the different races [sic] there are of this animal. Its strange figure had to stir the interest of our writer, who made a delicious description of these 'Indian sheep'" (Wagner 1986:17; something similar is in his note 17, 116). As we shall see, I believe that Wagner is wrong.

"After returning to Asunción" in 1548, "they went inland in search of gold" and traveled up the Paraguay River "in seven brigs and two hundred canoes. Those who did not find space on the ships went by land with a hundred and thirty horses." They came to "a nation whose towns are called Naperus, who have nothing to eat but fish and meat." After a long march "we reached a nation whose towns are called maipais" (Schmidel 1986: Chap. 44, 84–85). Now, the Mepene, who are also known as the Mapenuss, Mapeni, and Mepone, were a "tribe of river pirates described by

Schmidel (1903, 164 [1986:85]) . . . [they] lived somewhat to the south of the mouth of the Bermejo River" (Métraux 1946:218–219). The Bermejo River runs parallel to the Pilcomayo River, which is the frontier between Argentina and Paraguay. To certify this location for the Maipais, in Chapter 45 Schmidel says, "Of the Maipais, Chanés people . . . " (Schmidel 1986: Chap. 44, 87) and specifies that the following nation after the Maipais were the Chanés, and adds: "There are fifty leagues to the nation of the Maipais from the hill of San Fernando" (Schmidel 1986:Chap. 45, 88). Fifty leagues is about 289 km, and this is approximately in the central-northern Chaco. According to Métraux (1946:238–239), the Chané are a "western subtribe [who live] along the Andes," that is, on the eastern foothills of the Cordillera Real and the Sierra de Misiones.

According to Schmidel the Maipais "have . . . Indian sheep," and he is quite specific: "The sheep they call 'amida,' of which there are two kinds, the domestic and the wild ones, is used as a beast of burden and as mount. In this journey ["*jornada*"] I myself went more than forty leagues on the back of one of these sheep because I was hurt in one leg. In Peru they carry merchandise on them, like we do with the beasts of burden" (Schmidel 1986: Chap. 44, 86). Schmidel (1962:Chap. 44, 72) left us a picture showing the animals mounted and loaded (Figure 9.8; see discussion in Chapter 10). It should be noted that the term "jornada" does not mean a day. A careful reading of the chronicler's text shows that he meant the journey made. What is more, the second meaning in the *Diccionario de la lengua Española* (Real Academia Española 1992:1206) defines journey as "[a]ll of the road or voyage, even if it is more than one day." As we shall see, this is relevant when interpreting the facts. Besides, it is also significant that Schmidel himself specifies that they covered "sixteen leagues . . . that we walked in four days" (Schmidel 1986: Chap. 46, 90). This means they covered about 22 km per day. We can therefore deduce, also approximately, that the "journey" on which the German rode a llama took more than 10 days.

Continuing with their journey, the group reached another nation called Peionas, where they also "found enough food and meat, like . . . sheep" (Schmidel 1986:Chap. 45, 89). They con-

tinued the voyage to the Barconos, from whom "[w]e requested that they give us food, and they volunteered to bring us . . . sheep" (Schmidel 1986:Chap. 46, 90).

Finally, the chronicler recounts their clash with the Corcoquís. However, these "brought us meat of . . . sheep." Besides, "[t]heir weapons are darts, bows and arrows, and also bucklers made with 'amida' skin" (Schmidel 1986:Chap. 47, 95).

Schmidel specifies that this group lived three days' journey from the Macasíes River (Schmidel 1986:Chap. 48, 95), which for Wagner (1986:n. 20, 116) is the Guapay or Río Grande, whose headwaters are in the Bolivian sierra north of Sucre, that is, the Sierra of Misiones. In effect, the title of Chapter 48 reads thus: "Of the river and the site [called] Macasíes that is close to Peru," and besides this chapter tells "how they send two to Potosi and as far as Lima."

Gilmore (1950:433) discussed Schmidel's account[8] but through the references made by Cabrera and Yepes (1940:258 [1960:78]), and without having read it. Gilmore claims that the presence of llamas in sixteenth-century Paraguay is questionable, but does accept that "the

FIGURE 9.8. Drawing left by Ulrich Schmidel of the "Indian sheep." *After Schmidel (1962:72).*

Spaniards sometimes rode the llama (as Schmidel stated he did) in emergencies, or for sport." However, Gilmore does not present any argument for discussion.

It is striking that when discussing the domestic animals the Guaraní had, Métraux says that "[t]he only domesticated animal in pre-Columbian times was the Muscovy duck" (Métraux 1948d:81). In other words, he ignores the llama, even though Schmidel (1903) is one of his sources (Métraux 1948d:72). It is hard to believe that such a good scholar as Métraux had not read Schmidel's account carefully and did not see the drawings of the llamas he left. I assume instead that like most scholars, Métraux assumed the drawings were a figment of Schmidel's fantasy, or of those who reproduced or elaborated on his sketches, always starting with the assumption that camelids are high-altitude animals that could not have been present among the Guaraní.

I do not agree with Gilmore. The presence of domestic camelids in the areas described is fully feasible when considered within the historical data for the frontier zones here analyzed. Furthermore, Wagner (1986:17) is wrong in saying Schmidel "did not pay much attention to the different races there are of this animal." On the contrary, he was perfectly aware of them. If we read him carefully we find that when describing the camelids in the Asunción region he only mentions "the sheep called 'amidas'" (Schmidel 1986:Chap. 42, 79), whereas when he discusses the foothills of the Andes Schmidel says "the sheep they call 'amida,' *of which there are two kinds, the domestic and the wild ones*" (Schmidel 1986:Chap. 44, 86; emphasis added). In other words, things are quite clear. There were llamas in the Asunción region and farther to the southwest, while in the foothills of the Andes the expedition found llamas and a wild breed that almost surely was the guanaco, which still lives on the Bolivian frontier with Argentina and Paraguay (Villalba 1992:Figs. 4, 45), but could also have been the vicuña. Besides, it is clear that all of these Indian groups had a great exchange of products with the relatively close Andean region—which Schmidel also reached—and that the Europeans had quite precise news of Peru.

On the other hand, I do not find it striking that Schmidel was carried by a llama when he was sick. As we saw, not only is this possible, it is documented both archaeologically and with sources from recent times. From what I was able to deduce, the distance Schmidel covered daily was about 22 km, and it was noted when discussing the caravans (see Chapter 7) that scholars agree that the distance covered by a loaded animal ranges from 15 to 20 km. Besides, we should also bear in mind that Schmidel had to be carried for 10 days, which falls within the normal transportation range of an average pack llama.

Schmidel's account is on all points quite real, and it clearly indicates that these animals could perfectly well live, and in fact did live, in lands that are not precisely high-altitude ones, as is claimed. We should not forget that the Chaco zone is quite low, reaching an altitude of 550 masl in its far north. The southern section of the Chaco is a great depression with salty marshes. In general, this is a land of loess that is covered with layers of salt in several places, particularly in dry lakes and swamps. Of the rivers born in the Andes that come down to the Chaco, only the Pilcomayo, Bermejo, and Salado reach the Paraguay or Paraná Rivers. The rest die out among the sands. Swamps, puddles, and lagoons form during the rainy season. In general, it is a dry land, with lagoons, water holes, and cañadas. However, the lack of water predominates. The climate varies from east to west. The rains are stronger on the eastern side and lessen on the west. There are great variations in temperature, which can fall below freezing point at night in June and August and is very high in the daytime. Temperatures of up to 46° have been recorded. The most common vegetation is of the kind known as "*monte ralo*," or xerophytic forest (Métraux 1946:198–199). The weather in the Asunción region, which is at an altitude of 101 masl, might be of interest. Here the temperature ranges between 27.5°C and 17.6°C, and the mean humidity is of 69%. Rainfall is considerable, with an annual average of 1,039 mm. To compare with conditions present on the Peruvian coast, we have Lima at almost the same altitude, 111 masl, with a temperature ranging between 20.8° and 15.6°C and a mean humidity of 80%, but with a rainfall that comes to only 23 mm a year (Instituto Geográfico de Agostini 1993:645). I discuss this point later on.

Before ending this section there is one point that has to be noted. The context in which sheep appear includes several European animals, something that could make one believe that the chronicler was indulging his fancy, but it was not so. Schmidel narrated what he experienced between 1534 and 1554, and we should not forget that several travelers reached the Brazilian coast and even farther south at quite an early date. Just to give some examples, Vicente Yáñez Pinzón reached the Brazilian coast in 1500, and Cabral landed there a few months later. When Magellan's expedition arrived in 1519, the Indians gave them some hens in exchange for some articles (Dunin-Borkowski 1990:21).

9.10 ARGENTINA

The information on Argentina is considerable, and I have used just a small part of it, perhaps not even the most relevant. In any case, it can serve to illustrate the points under discussion.

9.10.1 Archaeological Data

The data for Argentina are also presented from north to south. Today, the most important fauna in the Jujuy puna include the llama, the vicuña, and the guanaco (Casanova 1946:619). Conditions must have been similar in the past, but obviously with more animals. Fernández (1983–1985:40) reports that a most significant change is visible between the Pleistocene and the Holocene in the Barro Negro site, at 3850 masl in the puna of Jujuy, where camelids replace horses in the faunal remains. Meanwhile, in the strata belonging to the Lower Holocene (ca. 7000 BC) there are abundant bones of indeterminate camelids (*Lama* sp.) (Fernández 1983–1985:36). *Lama* sp. are quite common by the Upper Holocene (ca. 1500 BC–AD 1400), particularly in food refuse, and these are usually young and old animals (Fernández 1983–1985:37).

In the Huachichocana region are several caves that were studied by Fernández Distel (1974, 1986). They lie at the headwaters of the Purmamarca Ravine (or valley) of the river of the same name, a tributary of the Río Grande of Jujuy, in the Department of Tumbaya, province of Jujuy,

at an altitude of about 3000 masl (Fernández Distel 1986:356–357).

I will discuss the two reports published for the cave called CH III, because there are some differences between them relating to "layers" and "levels." The 1974 article noted the presence of camelid remains in mainly three layers. Layer C (which corresponds to Level 1) contained "camelid bone and hoof remains, possibly llama (*Lama* sp.)." This level corresponds to the Humahuaca Inca occupation (Fernández Distel 1974:113). Layer D (which corresponds to Level 2) "consists of a layer about 20 cm thick, formed essentially by camelid guano (llama?). This evidently represents a moment when cave CH III was used as a corral for these animals." In chronological terms, the layer is related to the Late Agro-Ceramic period of Classic Humahuaca (Fernández Distel 1974:106). Layer E (the limit between Levels 2 and 3) contained "camelid bone and hair remains (*Lama* sp.). In Section E1, top, these are the big bones of adult animals, quite possibly *Lama glama*. Section E3, bottom, consists almost exclusively of bone remains of young, newborn or stillborn animals." In chronological terms, this is the limit between the preceramic and the ceramic epoch (Fernández Distel 1974:118). Layer E was initially given a tentative date of 3000 BC (Fernández Distel 1974:118, 122), but three dates were then given whose age ranges between 7670 and 6720 BC (Fernández Distel 1975:13).

The last report claims *Lama* sp. is in Layers C, D, E1, E2, and E3 (Fernández Distel 1986: Fig. 24, 421). These layers respectively correspond to Levels II (AD 1475–1530), III (AD 1000–1400), IV (500 BC–AD 500), V (1000–2500 BC), and VI (6500–7500 BC) (Fernández Distel 1986:378).

Only two comments were made about these remains. First, "[i]n general, the determination of the degree of domestication reached by the genus *Lama* through bone remains has not yet yielded conclusive results in the Andean area" (Fernández Distel 1986:420, n. 128), and "[m]ost of the bones in sub-layer E3 come from stillborn or newborn specimens of the *Lama* genus with incomplete ossification. This, like other subjects, related with the bone materials, has yet to be studied" (Fernández Distel 1986:420, n. 129).

The last report expands the data on Layer D (Level 3), as it is explained that there actually was a layer 20 cm thick formed "essentially by camelid guano (llama?)." This was a time when the cave was used as a corral. Hearths were also found, which show the dung was also used as fuel (Fernández Distel 1986:366, 368). Besides, the last report includes a description of some burials associated with Layer E2 (Level 5), dated between 1000 BC and 2500 BC, one of which held abundant camelid remains (Fernández Distel 1986: 378, 381).

It was also reported that remains of wool were found in Layers B, C, D, and E2 (Fernández Distel 1986:Fig. 15, 389). In this regard it was noted that "it can be anticipated that all are *Lama* sp., but it cannot be established whether they are *L. guanicoe* or *L. glama*. Fewer elements belong to '*Vicugna vicugna*.' In this regard, the piece that most shows the use of this species is a piece of hide with thick hair found in association with the burial of a ferret in layer C. Curiously enough, the possibility that the domestication of *Lama* was well under way at this moment is rejected: this piece indicates that the seizure of wild camelid specimens persisted at all times" (Fernández Distel 1986:390).

The report is truly confusing and the data lack specificity. The bone remains were not identified as to species, understandable given the problems this poses, but the excrement could have been identified. However, it is said of the hairs that the guanaco wool cannot be distinguished from that of the llama but can be distinguished from vicuña wool, and this apparently can be done without any technical analysis. Then the possibility that the domestication of the llama was well under way in the period corresponding to Layer C, that is, between AD 1475 and 1530 (see Fernández Distel 1986:378) is rejected, and the only argument presented is that wild camelids were still being hunted. This does not make sense.

Yacobaccio (1984–1985:168) mentions the cave of Huachichocana (although it is not specified, we can assume that it is the one known as CH III) and indicates that the presence of camelids here is 50%, but without specifying what strata he means. To avoid mistakes it should be noted that Dillehay et al. (1992:179) mention

Huachichocana III and the presence of camelid remains, but attribute the dates to Fernández Distel 1988, whereas the date is 1986. Besides, it is worth recalling that Dillehay et al. only meant the oldest layer in the cave.

Bahn (1994) recently reported, albeit without much detail, the studies Carmen Reigadas is undertaking with microscopic analyses of camelid fibers and adhered follicles. Reigadas analyzed the remains of wool and ropes from Huachichocana III, and reached the following conclusions. Samples belonging to what Fernández Distel (1974, 1986) calls Layer E3 are from guanacos and vicuñas. They are given a date ranging between 10,200 and 8670 BP (i.e., between 8250 BC and 6720 BC). There were also guanacos and vicuñas in Layer E2, dated to 3400 BP (1450 BC). However, from the osteological analysis it follows that some bones correspond to llama, which might indicate a process of domestication. In Layer E1, which has an age of 1400 years BP (AD 550), the fibers are from vicuña and llama, and this is quite possibly a transition period between hunting and raising animals. In Layers D–C the remains indicate a mixture of domestic and wild animals, with the former predominating. The presence of camelid excrement in these layers confirms they were being raised. As well, there are some signs that fibers were selected. According to this article, the most likely date for the domestication of these animals would be 1420 BP, or AD 530 (Bahn 1994:511–512).

This is extremely interesting, but we have to wait for the final report of Reigadas's data. So late a process of domestication is striking. Besides, the dates given by Bahn (1994) do not correspond to those published by Fernández Distel (1974, 1986). These problems have yet to be resolved.

Remains of *Lama* sp., as well as hairs and coprolites of this animal, were found in Layer B of another of the Huachichocana caves, CH IV. However, these were associated with ceramic sherds that are "not representative enough to make an identification" (Fernández Distel 1986:425).

There are only very vague data on Inca Cave 4, in the high puna of Jujuy, close to the Quebrada de Humahuaca, at more than 3000 masl. Lynch (1983b:119) reports the presence of camelids in Level 2, which he dates to 7280 BC, based on data by Mengoni (1982). Dillehay et al. (1992:179) report the same but give an older date, between 8670 and 7280 BC. Their data come from Yacobaccio (1989). Yacobaccio (1986:Table 4, 13) in turn indicates that here the presence of Artiodactyla is 4.7%.

Baraza de Fonts (1986:I) writes that around 7000 BC, the Quebrada of Humahuaca and the eastern fringes of the puna housed human groups "which began the first practices of . . . llama domestication." However, no data are given in support of this claim, and furthermore, this is a most general and superficial publication that does not present any precise data.

Tarragó (1978:490, 497, 502) discussed the Calchaquí Valley, in the province of Salta. She believes that camelids are among the species present in the subdistrict of Jujuy-Tucumán, which were significant for the natives. These animals were intensively hunted in the period 6000–500 BC; there is some evidence of the use of guanacos in the Formative period, 200 BC–AD 1000; and in the Regional Development period, AD 1000–1480, we find through indirect evidence such as textiles that the llama was an important animal in the economy. In this same study Tarragó (1978:493) observes there are still vicuñas and guanacos in the region, indiscriminate hunting notwithstanding, and that herds of guanacos can still be found in Tintin Pampa. The guanaco survives because there are several water puddles and a natural algarroba vegetation.

Farther south we know, from data in Pascual (1960:299), that the most abundant faunal remains in Intihuasi Cave, in the Department of Coronel Pringles (Gónzales 1960), correspond to camelids, at 71%, with cervids coming in second place. Camelids appear in all levels of the cave, as well as in other caverns and "sites" ["paraderos"] in the region. The bone fragments do not let one establish with any certainty the presence of any species other than *Lama guanicoe*. Most of the remains belong to adult animals alone, although there are also some senile ones and, more often, some juvenile ones, which in several cases are newborn individuals. According to Pascual, all the evidence seems to indicate that the aborigines had some preference for young animals. It is worth recalling that the most ancient layer at Intihuasi

dates to ca. 6000 BC and the most recent one to AD 500–1500.

Dillehay et al. (1992:179) report that guanaco remains were found in Agua de la Cueva, close to Mendoza, in strata dating to 8400–7890 BC. These same scholars (Dillehay et al. 1992:168) note that in the northern and southern Argentinian pampa the guanaco is always associated with the remains of early man (they based their statement on the data in Mengoni 1986, Miotti et al. 1989, and Politis and Salemme 1989).

Lynch (1983b:118; 1990a:23; 1990b:161) reported the association of human industry with an "extinct camelid" in Arroyo Seco, Tres Arroyos, in the province of Buenos Aires, but no more data are available. The first study gave a date of 6440 BC, the third 6610–3300 BC. (Lynch took the data from Borrero 1977 and Tonni et al. 1980.)

Los Toldos, in Cañadón de las Cuevas, Estancia de Los Toldos, in the province of Santa Cruz, in the Argentinian Patagonia, is an important site. Here there were abundant guanaco (*Lama guanicoe*) remains in the first three layers according to Cardich et al. (1973:98–121) and Cardich (1987a:102–103), but it was not exclusively used. (Stratum 3 was initially assigned an approximate date of 2500 BC [1973], and later a date of ca. 3050 BC [1987a].) The animal remains in Layers 6 and 7 belong almost exclusively to guanacos, which were almost the only animal hunted in significant numbers. These layers are associated with the Casapedrense industry. At first (1973) they were given a date of 3500 BC, but were then radiocarbon dated to 5310–2900 BC. Layers 9 and 10 represent the Toldense occupation, which was initially (1973) given a date of 6800 BC, but was later changed through radiocarbon dating to 6800–9050 BC. At this time the guanaco appears in a higher proportion but is associated with the extinct fauna and other animals. Its presence is likewise reported in Layers 11 and 12, which were given a date of 10,650 BC, and were only identified as "Level 11 Industry." It should be noted that the 1987(a) report no longer mentions Layer 12.

Later studies discuss stratum 11, the oldest, and say that in the fauna "there is a higher presence of guanaco (*Lama glama guanicoe* [sic]) remains," as well as an association with extinct fauna (Cardich 1984:28). Cardich also says (1984:30) that he found the "remains of a camelid that is not a guanaco and coincides with the bones of *Lama gracilis*, a species identified by F. Ameghino (1889: 581) in the upper Pleistocene of the Buenos Aires pampa, but it could likewise be vicuña, a species that is known much further north but not for the Patagonia" (see also Cardich 1977:153–154). We see that at first Cardich (1977:153–154) left open the possibility that the aforementioned remains could belong to either *Lama gracilis* or vicuña, although clearly leaning toward the first possibility. However, he then claimed emphatically (Cardich 1987a:104) that these were "some bones of a more gracile camelid than the guanaco, which coincide with the bones of a species identified by F. Ameghino (1889:581) for the upper Pleistocene of the Buenos Aires pampa as *Lama gracilis*."

It is worth noting that Cardich (1977, 1984, 1987a) ignores the paleontological literature and made a mistake that can be misleading and must be explained. *Lama gracilis* is actually nothing more than a vicuña. Cabrera (1931:110–111) wrote the following in this regard: "Another species to be dropped is *Auquenia gracilis* H. Gervais and Ameghino. Doctor López Aranguren accepts it [López Aranguren 1930a:26] and assumes it could be the agriotype of *Lama pacos*, but the remains described and which were established as *gracilis* by Ameghino, are completely unrelated with the alpaca *and are no more than vicuña remains*. I have compared them with present-day vicuña specimens from San Juan, from Jujuy and Bolivia, and *there is absolutely no difference. . . .* Ameghino insisted on the validity of the species in 1889, albeit acknowledging its 'great affinity' with the vicuña" (emphasis added).

On the other hand, there are some doubts over the associations of the materials used to date the archaeological strata. This was noted by Lynch (1990a:22; 1990b:160), who likewise noted that Tambussi and Tonni (1985:69) concluded that the faunal remains were the result of both natural and human agents. Lynch is equally categorical in this regard:

Of importance equal to the source of the material dated are the associations of the dates. The sediments and stratigraphy at Los Toldos have been described inade-

quately, but it is clear that the strata were differentiated, at least partly, on the basis of their cultural content. Each stratigraphic unit was "singularizada por la presencia o ausencia, o el diferente porcentaje de elementos tipológicos, particularmente de la industria lítica" (Cardich et al. 1973:98) (Differentiated by the presence, absence or percentage of typological elements, especially of the lithic industry). I maintain that this sort of ex post facto stratification is poor archaeological practice. There is [a] natural-stratigraphic justification for the separation of the topmost cultural layers (1–3) from the Casapedrense units (6 and 7) in the form of a layer (4) of ash and pumice, and the Casapedrense industry is said to be separated from the Toldense layers (9 and 10) by a "semi-sterile" bed 8, *but level 11 has no natural stratigraphic separation from levels 9 and 10, which contain the rest of the Toldense industry*. (Lynch 1990a:22, emphasis in original passage; see also Lynch 1990b: 160)

The data for Los Toldos must thus be taken with due reserve.

Cardich has likewise reported on the site of El Ceibo, in the same province of Santa Cruz. He notes that Layer 12 of Cave 7 corresponds to Level 11 of Los Toldos, and insists that he has also found *Lama gracilis* here (Cardich 1987a:101), a point that was already discussed.

As regards the southern tip of the continent, Laming-Emperaire (1968:87–88) said that the most ancient hunters reached the Strait of Magellan 10,000 years ago, exactly at the beginning of the Holocene. Their economy essentially consisted of the hunt of big herbivores, including the guanaco. Borrero et al. (1985:273) confirm that the main source of food in the Patagonian area in 10,500 BC–AD 1450 was the guanaco.

The guanaco is thus the most important animal from the point of view of the natives. These usually move in small herds formed by an adult male and four to ten females, and sometimes in smaller herds of younger males, with herds of up to 100 heads or just one or two animals being less common (Cooper 1946a:127–128). Based on a study by Mengoni-Goñalons (1978:5), Lynch (1983a:3) explains that the herds of guanacos regularly move between their summer and winter habitats, that is, from the high plateaus to the low-altitude canyons and valleys, or inside the coastal woodlands. These patterns are repeated every year.

For later periods, Raffino, Tonni, and Cione (1977) mention a series of sites in the highlands of northwest Argentina where camelids have been found. Llama and vicuña remains were identified at Las Cuevas, Potrero Grande, and Cerro El Dique, sites with an occupation dating to AD 400–600. Likewise at Tres Cruces and Cerro La Aguada, two later sites that have occupations ranging from AD 400 to 1000.

The data from the Quebrada del Toro, in the province of Salta, are interesting. Llamas constitute 70% of the faunal remains and vicuñas 7% in a context corresponding to the Lower Formative, ca. 200 BC–AD 200. "[T]he camelids of the *Lama* Complex" were the major source of protein for its people in the three sites studied, that is, Cerro El Dique, Potrero Grande, and Las Cuevas. Other faunal remains are of little significance. "[J]uvenile (45%) and adult (55%) specimens are perceivable only in specimens from the Lama Complex. This fact can be interpreted as the result of livestock management, something also supported by other kinds of evidence" (Raffino et al. 1977:16).

Camelids were still the major protein source in both the sites of Tastil and Morohuasi during the Period of Regional Developments, in AD 1000–1470. "The *Lama* Complex" ranges from 57% of the total number of specimens in Tastil to 68% in Morohuasi. Meanwhile, *Lama vicugna* accounts for 28% and 11% at the respective sites. Other faunal elements were not especially representative. The evidence seems to indicate that at this time there was an increase in the hunt of *Lama vicugna*, but llama and/or alpaca raising was the major component of the economy, beside agriculture (Raffino et al. 1977:15, 17–18).

Later, Raffino et al. comment that domestic and wild camelids were the major source of food of animal origin in the Regional Development period. Other animals were used in smaller pro-

portions, but these were obtained solely through hunting (Raffino et al. 1977:23).

As for Inca times, Raffino et al. (1977:21–22) do not present any statistical data because the sample is not too representative, but the *Lama* "complex" is present both with juvenile and adult animals, while only adult *Lama vicugna* individuals appear. "It is evident, given the characteristics of the Inca occupations in the southern Andean Area, that not only do domestic camelids reprise their essential role as a crucial nutritional requirement, they also show their importance as an element functionally connected with transportation and communications." Raffino et al. explain that a big central enclosure used as corral is always found in these Inca sites. This is quite clear, owing to the presence of excrement from these animals.

Finally, Raffino et al. (1977:22) state that two conclusions may be drawn from the three periods discussed. First, llama-alpaca husbandry was the major nonagricultural subsistence activity throughout all of the cultural sequence, that is, the species of the "*Lama* Complex" were the major source of animal protein. Second, *Lama vicugna* was the major species hunted in all periods except the Lower Formative, when *Chinchilla* sp. was the major species hunted. Moreover, the percentage recorded for *L. vicugna* in sites dating to the Regional Development period was the highest for all epochs.

As for the Tucumán region, Troll (1935:142–143) stated, based on the archaeological data, that it was known that llamas were used throughout the land of the "Diaguita civilization" in the Argentinian Northwest as far as the foothills of the Andes, close to Tucumán. Troll likewise indicates that llamas were also raised in isolated fashion in the highlands of Córdoba and assumes that it was a smaller variety, but does not support his claim. He also claims that according to the account of Pedro Sotelo de Narváez (apparently cited by Boman 1908), the Comechigón Indians who lived in this region raised llamas and used clothes made with the wool of these animals.

Nachtigall (1966a:193–194) described a camelid-related ceremony that was depicted in a "ceramic group" published by Stig Rydén in 1932, which belongs to the La Candelaria culture. (As is well known, this culture is pre-Inca and developed between 500 BC and AD 700). The group represents llamas with the male standing out due to the tassels in its ears, which today is still a typical signal in the ceremony known as the Señalada, or the marking of the animals. Nachtigall believes that this find is important because it proves that the corresponding ceremony was not introduced for the first time into the Argentinian Northwest by the Inca, and is instead much earlier. Besides, this is also proof of the antiquity raising llamas has in Argentinian territory.

Márquez Miranda (1946:657) discusses Santiago del Estero and some parts of the Chaco. He indicates that the archaeological research mostly took place in Santiago del Estero (however, we should bear in mind that the data date to the forties). According to Márquez Miranda, subsistence essentially depended on raising llamas among the Indians of the Chaco-Santiagueño culture. The chronology of this culture is not clear, and in temporal terms must correspond to the fifth century of the Christian era onward.

According to Gilmore (1950:435), Rusconi (1930a) reported llama remains found in tombs in the Argentinian Northwest and in the neighboring Chaco.

Camelidae were identified *largo sensu* in some late sites of Santiago del Estero (AD 1220– 1360). It has been suggested that these could be llama, alpaca, and guanaco (Lorandi de Gieco and Lovera 1972:179).

Olivera (1991:64–67) studied the Casa Chávez mounds south of Antofagasta de la Sierra, on the left bank of the Antofagasta River, in the province of Catamarca (in the southern Argentinian puna). The site belongs to the Early and/or Middle Agro-Ceramic period of the Argentinian Northwest. It is reported that *Lama glama/Lama vicugna* account for 90% of the animal remains. Olivera notes that although it is not easy separating the wild species from the domestic ones, some of the mandibles found correspond to *Lama vicugna*, and there is no question about the presence of *Lama glama*, even though the proportion of both species has not been established. However, *Lama glama* seems to be the dominant species.

About 75%–76% of remains of the *Lama* genus belong to newborn/newborn and juvenile animals. Olivera insists in noting that "this high proportion of small-age individuals is repeated in other sites in the central-south Andean Area." The possible presence of alpaca (*Lama pacos*) and/or guanaco (*Lama guanicoe*) is not discarded, but Olivera finds it somewhat unlikely. "As for the alpaca, studies seem to indicate its poor adaptation below 4000 masl, and there is no precise data on the moment when the guanaco retreated from our region" (Olivera 1991:67).

Lorandi (1983:40) mentions an Inca center in Catamarca below the Aconquija, which was studied by Márquez Miranda and Cigliano. They reported that a great number of animal bones, "almost all of them from auchenids," were found below the burned and collapsed roofs of the houses.

A sequence, classified into horizons, was found at Ongamira Cave in Córdoba, at 1150 masl. The dates are not too clear. It is estimated that the transition from Horizon III to IV must date to ca. 1000 BC, while Horizons I and II must have developed in 100–200 BC (Menghin and Gonzales 1954:261).

Camelid remains have been found in almost all horizons. Guanaco remains were in the four horizons, while llama remains appeared only in Horizon III, and possible vicuña remains in I (Pascual 1954:271). Pascual (1954:269–270) wrote: "*Lama guanicoe* comprises the biggest percentage. The examination of the *Lama* remains reveals the presence of a few specimens of bigger size than the common guanaco. . . . All of these remains are most fragmented, so the diagnostic characteristics that distinguish the various species in the genus *Lama* cannot be observed. . . . There also are a few specimens, fragments of maxillary bones and mandibles with some implanted molar teeth of smaller size, that represent specimens of *Lama guanicoe*. They could well be *Vicugna vicugna*." It is known that the guanaco inhabited all of the province of Buenos Aires in the late Pleistocene and part of the Holocene. When the Spaniards reached the Río de la Plata in the sixteenth century, the guanaco was inhabiting from the southern sierra southward and to the west, in the sandy area of Buenos Aires. Its retreat was due to ecological factors related to changes in the climate. The guanaco occupied the intermontane region and the undulating pampa for some time during the Late Pleistocene and the Holocene, when environmental conditions were drier than they are at present (Tonni and Politis 1980).

9.10.2 Historical Data

Let us now review some historical data on Argentina. First there is Espinoza Soriano's (1985–1986) study of the encomienda grant Francisco Pizarro gave to Martín Monge, a *vecino* (resident) of Charcas (Pizarro 1540:561). The "*mitimaes* of Choromatas and Chuyes, which are toward Omaguaca," are listed in this document among the valleys and villages given out in encomienda (Espinoza Soriano 1985–1986:244). According to Espinoza Soriano, the Churumata can be located to the east of what is now the province of Jujuy and north of Salta, between the sites now known as Libertador General San Martín and Nueva Orán (1985–1986:246–247).

It is possible that several Churumata groups were removed by command of the Inca to be resettled somewhere else. What we do know is that a Churumata moiety was relocated in the land of the Omaguaca.

The 1540 document reveals that some families of *mitimaes*, "among whom the Churumatas are listed," lived among the Omaguaca. A document on an encomienda given by Francisco Pizarro to Martín Monge mentions a town "called Chilma of Churumata and Chuis *mitimaes* that lies towards Omaguaca" (Pizarro 1540:561; cf. Canals Frau 1953:507). The encomienda lies between El Tambo del Inga, a town in Mireta (in the province of the Chichas) and the Tambo de los Jagüeyes. It seems that Monge was the encomendero of the sector called Casabindo. Espinoza Soriano concludes that there were Chichas-Orejones among the Churumata, and some of the latter in Omaguaca (1985–1986:252). It is said that these "[r]aised rams of the land [*carneros de la tierra*] or in other words llamas (*Lama glama*), which once domesticated served their owners to trade like droves, for they can carry up to four arrobas and walk for up to three leagues (Lozano 1733:52)" (Espinoza Soriano 1985–1986:253).

The Indians retreated and dispersed in the face of the Spanish advance, and the Churumata as well as the Chichas-Orejones took refuge in the Llanos de Manso (Espinoza Soriano 1985–1986: 255).

The Incas settled a colony of Chicha *mitimaes* in the land of the Churumata. The Chicha had the honorary rank of *Incas by privilege* due to the dangerous role they fulfilled, and so were allowed to pierce their ears and wear the characteristic earspools of their high position. It was for this reason that the Spaniards called them Chichas-Orejones. This migration of Chichas-Orejones was headed eastward, to the Llanos de Manso region, between the Bermejo and Pilcomayo (i.e., in the Gran Chaco region) (Espinoza Soriano 1985–1986:255).

The "Orechones" appear south of the Pilcomayo River on the map by Antonio Machoni (1700), but much farther inland than its mouth in Paraguay, that is, east of the Llanos de Manso. "Pedro Lozano therefore wrote (in 1721) in an account by Father Gabriel Patiño: 'The Churumata Indians, agriculturalists from Peru of whom there are more than five hundred and who use the rams [*carneros*] of the land, are in a valley forty leagues from the Pilcomayo, skirting the cordillera to the south along its foothills, to the part of Los Llanos [de Manso]' (Lozano 1733: 61)" (Espinoza Soriano 1985–1986:256).

Presta and Del Río later published a study "on the Churumata of southern Bolivia." They claim that Espinoza Soriano "overvalued" the encomienda grant given to Monge, which they find imprecise (Presta and Del Río 1993:223–224, n. 1 [the reference to Espinoza Soriano erroneously says 1986, when it should be 1985–1986]). According to this study, the Churumata occupied a bigger expanse of land in the sixteenth and seventeenth centuries than that given by Espinoza Soriano (1985–1986:256), which corresponds to "the eastern Inca frontier region that is delimited by the mesothermal valleys, along the Tomina-Chuquisaca-Tarija axis" (Presta and Del Río 1993:224). Besides the study mentioned, interested readers can find more data in Doucet (1993).

There is interesting information on the Tucumán region. In 1558 Vivar (1979:192) "recounted the departure of Francisco de Villagrán for the kingdoms of Peru." He passed through Potosí and from there went to Omaguaca, crossed the cordillera once more, and reached the city of Nuestra Señora de Talabera de Madrid, or Esteco in the Indian language. Talabera de Madrid was in the bishopric of Tucumán, and Ciudad de Esteco is a synonym of Las Juntas. This place is said to be a "[f]ertile land, and they have some sheep with which they dress." (It seems that Vázquez de Espinosa used this source, even though his citation is different, as we shall see later).

Francisco de Villagrán then went to Xuries. Its location is not clear, but according to Sáez Godoy (1979:193) it must be in the Tucumán zone. Here it is once again noted that "[t]hey have sheep, and in the fields guanacos" (Vivar 1979:193).

Cieza de León (1881:Chap. XCI, 318) also discussed the Tucumán region toward the mid-sixteenth century. He wrote about "[t]he things that most happened and befell Captain Diego de Rojas." When describing this place Cieza says that there were "many ducks, and guanacos there are not a little in those provinces." He then adds that "[t]heir food is maize and meat from the guanacos and sheep they have."

When an expedition was prepared to discover Tucumán under orders from Vaca de Castro, Uncinga, cacique of Chuquicota, gave the Spaniards animals for the trip. In 1575 Uncinga appeared as witness and answered a questionnaire presented by Juan Colque Guarache. To question 11, which asked whether Uncinga helped Vaca de Castro, Uncinga replied that he had "faithfully and loyally served Felipe Gutiérrez and Sotomayor, who were going to the provinces of Tucumán, to whom and to his soldiers he gave much livestock of the land" (Uncinga 1981:246). Dedenbach Salazar (1990:108) comments: "This is an interesting datum because it indicates how the camelids were advantageously used in the warm lands."

Sotelo Narvaez (1965:391) also described the province of Tucumán. Its date is not precisely known, but it must be 1583. He wrote: "This land abounds with pastures and thus has much game ...guanacos...." When discussing the Calchaquí valley, Sotelo Narvaez adds, "and they have the puña, which is the nearby paramo, where they have much game of guanacos, vicuñas" (Sotelo Narvaez 1965:393). In 1594 Barzana (1965:80)

said of this same province that "[t]hey have much game of deer ... there are vicuñas and huanacos as in Peru."

In the early seventeenth century, around 1605, Ocaña (1987:136) described the Tucumán region and noted that there was "[m]uch game in those pampas of ostriches, of quirquinchos, which are like small pigs, and some guanacos...." A little later Vázquez de Espinosa (1948:1764/622) also discussed "the provinces of Tucumán" in his monumental account. Vázquez de Espinosa said that "all the road described, from Potosi onwards, is all rugged land of many sierras, [with] the fields full of vicuñas, guanacos," and he mentioned the area close to the "Omaguaca valley."

While discussing the province of Tucumán in one of her studies, Lorandi (1980:55) indicated that one must not be misled by the fact that the Indians in this region spoke Quechua, dressed like the Peruvian natives, and raised rams [carneros] of the land. She believes that these Andean habits could be the result of productive and administrative functions regulated by the Inca state. It is far too much of a coincidence for them to have been characteristic of peoples living in warm valleys. In any case, it would be far more logical for these populations to speak Aymara instead of Quechua, which reached these southern regions as a result of the Cuzqueño conquest.

In another study, Lorandi (1983:20 [it is the same text published in 1980:161, with some small changes]) noted that the presence of "sheep of the land" in the highlands of Santiago was documented by several sources. Since the route followed by the expedition of Diego de Rojas is not known, some scholars believe that the llamas were in the Santiagueño plains. Lorandi believes this must be rejected, given the heat and warmth of the zone. Most of the land "is taken up by marshlands and does not seem to be a favorable habitat for this animal." In any case, the llamas were able to live in the highlands, where there are pastures that favor raising this animal. However, Lorandi believes that it still is too warm an environment for the natives to have raised large herds of llamas.

Lorandi suggests that llamas were introduced in Tucumán by the Incas, because the provisioning of the state settlements and the transfer of tribute to Cuzco can justify their presence in these lands.

Vivar discusses the "Comechigones" who occupied the area where the city of Córdoba now is (Sáez Godoy 1979:194), and notes that in those lands there were "tame sheep" (Vivar 1979:194). When describing this same city in the province of Tucumán, Vázquez de Espinosa (1948:1785/630) says there were "guanacos and deer in such numbers that they cover the land." Vázquez de Espinosa then describes the land between the cities of Córdoba and Trinidad, and insists there were "uncountable deer, guanaco" (Vázquez de Espinosa 1948:1831/644).

Aparicio (1946:677) says that the Comechigón, a name that takes in all the ethnic groups living in the sierra of Córdoba and San Luis, herded llamas and had large herds. Aparicio even says there is some data about this in Cieza de León, but I was unable to find this, and Aparicio unfortunately does not give the reference.

In 1586, Ramírez de Velasco (1965:76, 77) gave an account of the province of Telan and Zuraca, which lay "70 or 80 leagues from the city of Córdoba." Here he says that "they had many rams [carneros] of the land, of the kind used in Peru to carry loads." He then adds that "while in the city of Mendoza he likewise heard say as something public and notorious that many people came to the kingdom of Chile in search of Christians, and that they brought their loads and bundles on some beasts like asses, and that the Indians said that said beasts had long ears like asses, and came on foot." These must be llamas.

There are interesting sources for the Río de la Plata. Diego Núñez is one of the men who were in one of the expeditions sent by Mercadillo from Mainas (Jiménez de la Espada 1895:214), who years later wrote to King John III of Portugal. Núñez must have reached Machifalo (Machiparo, Machifaro, or Machoparo), "and its natives are mentioned in the famed voyage of Orellana [see above], who rested in one of the towns for some days in May 1542 after seizing it by force" (Jiménez de la Espada 1895:215; see Jiménez de la Espada 1892–1894). Núñez wrote: "I had news that there are sheep like those of Peru in this same land as far as the Río de la Plata, which is the best sign there can possibly be in this land be-

cause where there are sheep the rest abounds." According to Franklin Pease (pers. commun., 2 April 1993), Núñez must have written this shortly after 1542.

Fernández de Oviedo (1959a:Bk. X, Chap. XXX, 53) also collected data on this zone. When discussing the big rams [carneros] he wrote that "[i]n the Río de la Plata there also are those that I said are big from the . . . [embocamiento] inland in said land." The reference does not appear where Fernández de Oviedo says, but when he mentions the testimony of Alonso de Sancta Cruz on the Río de la Plata he notes that "[h]e likewise says that there are those sheep from Peru, and with the wool very long" (Fernández de Oviedo 1959a:Bk. XXIII, Chap. XXX, 367).

When describing the people of "[t]he Río de la Plata" in 1552, Gómara (1946:212) noted that "[t]hey eat fish, of which there are many and fat, and are the main staple of the Indians, but they hunt deer, pigs, boars, sheep like those of Peru, and other animals."

Mandrini (1989–1990:73) made a study of southwestern Buenos Aires. He explains that in the mid-eighteenth century the Tehuelhets supported themselves exclusively with the guanaco (Falkner 1974 is his source). Mandrini explains that in the late eighteenth and early nineteenth centuries the Indians raised only European livestock in this area. However, Mandrini notes there were abundant guanacos, which were hunted (1989–1990:71); his source was García (1836), who journeyed in 1822 to the grasslands south of Buenos Aires, from Morón to Sierra de la Ventana.

The great nineteenth-century guanaco hunts in Patagonia were well illustrated by Musters (1871; see my Figures 9.9 and 9.10), and it is clear that the adoption of the horse by the Indians around 1725 (Cooper 1946a:142) was a factor that greatly increased the killing of guanacos.

9.10.3 The Slaughter of Camelids and Other Recent Data

I do not have any data on the killing of camelids in Argentina. All I have is some information on the guanaco and the vicuña. Raedecke (1976:13) discussed the results of the colonization of Patagonia and its pampa. He says that "the guanaco populations fell from millions of specimens to tens of thousands, and this decline has not been reversed." Franklin (1982:476) reported that 8,000–9,000 guanacos, especially chulengos, were exterminated in 20 years in the Tehuelche Indian

FIGURE 9.9. A hunt of guanaco and ñandu in the Chico River Valley, Patagonia, by a Tehuelche community. It should be recalled that this ethnic group probably began using the horse around 1725. Note the way the second hunter on the left uses his boleadora. *After Cooper (1946a:Plate 40, top), by permission of the Smithsonian Institution Press. The original belongs to Musters (1871).*

FIGURE 9.10. Tehuelche Indians hunt guanacos. The second hunter on the left has thrown his boleadora, which is about to reach one of the guanacos, while another hunter in the foreground is using his weapon to kill a puma. *After Cooper (1946a:Plate 40, bottom), by permission of the Smithsonian Institution Press. The original belongs to Musters (1871).*

Reservation in Chubut. In the early 1950s, the number of guanacos continued falling, and the species has been completely exterminated in most of northern and southern Argentina. Franklin notes that some believe the species is in danger of extinction. (His sources are Gilmore [1950] and Dennler de la Tour [1954].)

As for the vicuña, its population was decimated in Argentina in colonial times, according to Brack Egg (1987:62); however, this scholar does not support his claims.

According to the data of Cajal and Puig (1992), at present the vicuña is found in Argentina in the northwest from 22° S latitude to 29°10′S latitude, at an altitude of 3200-4600 masl. This comprises the provinces of Jujuy, Salta, Catamarca, La Rioja, and San Juan. Cajal and Puig note that whereas in the nineteenth century the vicuña occupied an estimated area of 12,000,000 hectares, it now occupies 9,200,000, that is, a loss of 24%. It is estimated that at present there are 18,400 individuals in Argentina (Cajal and Puig 1992:37), but Torres (1992b:Table 1,31) estimates that there might be 23,000 animals (see Table 2.2). (The figure presented in Table 2.1 corresponds only to the conservation areas of the species.)

Cajal and Puig (1992) report concerning the guanaco in Argentina that its area of distribution used to be three times as large as it now is. It corresponds to a central region with a high concentration of individuals that encompasses the southern part of the provinces of Mendoza, Neuquén, Río Negro, and Chubut, and the Argentinian part of Tierra del Fuego. A second region with a moderate presence of these animals encompasses the province of Santa Cruz, the northern part of the the province of Mendoza, San Juan, La Rioja, and Tucumán, with a corridor formed by the southeastern sector of the province of Catamarca,

northwestern Córdoba, and southwestern Santiago del Estero. There are small populations in the provinces of La Pampa, Buenos Aires, San Luis, and Córdoba. No up-to-date information is available for the provinces of Catamarca, Salta, and Jujuy, where it is assumed that there are very small populations. All in all it is estimated that 550,000 individuals live in Argentina (Cajal and Puig 1992: 39). (The figure given in Table 2.1 is for the conservation areas of the species, but it also is "fragmented data" [Cajal and Puig 1992:Table 3, 39]. See also Table 2.2.)

I have few data on the llama and alpaca in Argentina. They are summarized in Table 2.2.

NOTES

1 Fonseca and Richardson (1978:307) give it a date of AD 300.
2 A cline is a continuous gradation of morphological variations in the population of one species, correlated with its ecological or geographic distribution.
3 The *arrelde* is an ancient Spanish measure equal to 1.842 kg.
4 When discussing this passage, Horacio Larrain Barros (1980a:242, 243 and 245) says in note 269 that there are good descriptions of these animals in Coreal (1722:Vol. I, 352–353), among other authors. I was unable to get this book and do not know what kind of data it has.
5 The 1918 edition mistakenly says "long" instead of "short."
6 The ordinary *mita*, i.e., the standard mining *mita*, which was recorded in the *tasa*.
7 The passage has been copied as it appears in Benavente Aninat, but it is obviously incomplete and lacks a clear context.
8 Gilmore (1950:433) spelled the name as Schmidl. This should come as no surprise because it is also spelled Schmidt and Schmidts. These variants of the name Schmidel are not too important because each time an editor changes it to his or her liking. It is an exact correspondence of Herrera, its Spanish counterpart, which appears as Herrero, Ferrero, Ferreiro, Ferreira, and so on. It is the same family name, taken from the same trade (Peter Dressendörfer, letter of 17 May 1994).

CHAPTER 10

THE FEEDING AND USE
OF CAMELIDS:

Advantages and Disadvantages of
Native and Imported Animals

As historical sources show, European livestock, mainly sheep, goats, cattle, and pigs, rapidly replaced camelids. This replacement happened throughout almost all the Andes, on the coast and in the highlands, the puna included (see Wheeler et al. 1992:468; Baied and Wheeler 1993: 146). But it obviously was a change imposed by the new, dominant Western culture, with the subsequent economic changes it wrought.

Although there might be one, I have not found a thorough study that definitely establishes whether European stock are actually more profitable from an economic point of view than native animals. I therefore do not pass judgment on this issue. But it is a fact that profitability will be low unless camelid herding is protected and developed, and a corresponding and adequate educational campaign is undertaken so that the nonnative world learns how to use and value its products—and little has been done in this regard. What is indisputable is that camelids have several adaptive advantages in the Andean milieu that European animals do not have and have been unable to develop.

Sheep quite early became the animal that most rivaled Andean camelids. However, grazing sheep are far more destructive than llamas and alpacas. The prehensile, split upper lip of the camelids enables them to be highly selective in their diet, so they can pick the most desirable plants or plant part. In marked contrast to sheep, llamas and alpacas do not uproot vegetation with their feet, and never destroy stems and roots. We have also seen that camelids differ from the ruminants of the Suborder Pecora in the morphology of their stomach and in the lack of horns and antlers, and they have soft, nail-covered digital pads instead of hooves. This type of foot has minimal impact on the fragile puna groundcover, whereas sheep hooves sink into the soil (Baied and Wheeler 1993:149). More than one student has sounded the alarm over the environmental damage wrought by European animals, specifically damage to the soil's surface and its vegetation (e.g., Seibert 1983:272). It thus turns out that camelids are ideal animals for an ecosystem like that of the Andes. This issue is of great relevance and deserves its own study. This is beyond my expertise. However, two points are worth expanding. The first concerns the food eaten by these animals, the second the use of camelids and the products derived from them. This discussion will allow readers to understand the advantages and disadvantages to which I alluded.

10.1 FOOD

The food that camelids eat is one of the most important topics not just for understanding the advantages they have over other animals but also for understanding their original distribution in certain apparently hostile environments, such as on the coast. One of the objections made to their diffusion to the coast is the presumed lack of appropriate food. Joyce Marcus (pers. commun., letter of 24 April 1986) raised this objection, and it is also implied by Murra (1975:119) when he states that to understand the ancient coastal agriculture and its relation to that of the highland, "it would be worthwhile finding out how these animals, used to the natural pastures of the highlands, were fed." This point is made even more explicitly by Flannery et al. (1989:115; emphasis added). When referring to the caravans of llamas on the coast, they say these "were *probably* fed kernels, leaves, and stalks of maize *while they waited to return to the sierra, since their normal forage does not grow in the coastal desert*." This clearly shows two things: first, that it is believed that caravans went temporarily to the coast, and second, that the coast has no forage for these animals.

The truth is, there are scant data on this point, in either archaeological or historical sources. What we do have, however, may shed some light on this problem, so although no definitive solution is possible, some guidelines can be set and some quite clear explanations may be given in this regard.

My discussion focuses on the coast because no one doubts the existence of possible foods these animals ate in the highlands.

Several scholars have discussed this point, although not directly, and with more data in some cases than in others. In regard to llamas on the coast, Rostworowski (1981:61) notes that it is possible that these were also fed fruit from the algarrobo (carob tree, *Prosopis* L.) and huarango (*Acacia* W.), just as the other animals were. True, she does not support her assertion, but we shall see that she was right. It is most likely that Rostworowski was influenced by historical sources, which she has used widely and which present some clues in this regard.

According to Shimada (1982:163), Netherly made a similar claim in a study I was unable to obtain (Netherly 1977), where she noted that the coastal valleys provided enough forage to support herds of llamas with the salt grass (variety of a seashore salt grass, *Distichlis spicata*), the vegeta-

tion that grows on the banks of rivers and canals, as well as pods of algarrobo (*Prosopis* sp.).

10.1.1 Archaeological Data

Shimada and Shimada (1985:20) have discussed this point in some depth, but focus on the north coast. They say that in the sierra, camelids graze on the unimproved land around agricultural fields. On the north coast, the complementary ecological zones are quite distant in horizontal terms, and at the same time frequently form a mosaic-like pattern.

Shimada and Shimada repeat the data in Netherly (citing Netherly 1978:36, her dissertation, which apparently has the same data she presented in her 1977 study) on the abundance of food in the coastal zone. They specifically mention algarrobo pods, the vegetation close to water canals, the large tracts of salt grass, as well as less halophytic varieties that do not compete with agricultural fields. Shimada and Shimada add that we should likewise bear in mind the large and lush ravines, like Quebrada de la Montería and the Pampa de Chapparí, in the upper part of Lambayeque—the latter has over 150 km of open pastures that remained uncultivated until the Late Intermediate period. Now these ravines have ephemeral rivers and lush vegetation only in March and May, when it occasionally rains.

Shimada and Shimada also believe that perhaps a transhumance of camelid herds, from the upper to the lower areas, took place, much as happened with the herds of cows in historical times and at present.

Another possible grazing area is the large thorny forest in the central part of the La Leche River valley, which was apparently never cultivated and which at present has about 100 km². The main vegetation there is the algarrobo; today, herds of goats, sheep, pigs, and cows are found there.

In the Lambayeque region in general, the Andean piedmont is farther inland than on the central and south coast, and there is less distance between the valleys in the upper parts, with more open pastures. This is why Shimada and Shimada (1985) assume the maintenance of herds was a result of hard work, and that these were fed in relatively small areas.

A limited variety of cultivated and wild plants used as food are visible among the archaeological excrement found and in its content.

Shimada and Shimada give Pampa Grande (occupied in late Moche times) as an example. Its animals may have been kept in Quebrada Montería, and some food—such as algarrobo pods and maize—may have been taken to the corrals. Llamas were raised in the woods and pasture areas of the Batán Grande region, such as Pampa de Chapparí, with the addition of cultivated and collected plants. They finish by noting that several forages could have been used throughout most of the year, considering the extent of the cultivated areas and the diversity of plants used.

Segundo Vásquez and Víctor Vásquez S. (1986:n.p.) note that llamas tolerate a wide variety of pastures. Thanks to the coriaceous consistency of its leaves, the salt grass produces the wear required by these animals' teeth, and is also an excellent grass that sheep and goats eat today. From the observations made by Vásquez and Vásquez, it seems that in earlier times the consumption of salt grass was more frequent, and toward Chimu times "that excellent forage that is the 'algarrobo' *Prosopis chilensis*" was used. The authors likewise stress that the salt grass (*Distichlis spicata*) growing along the beach could have provided pasture for the llamas, without much care or maintenance.

It is worth noting that *Distichlis spicata*, a member of the Family Poaceae, is native to the Americas and has a wide distribution in areas close to the sea. It thrives in salty soils and is the prototype of the plant communities known as *gramadales*. They cover more or less extensive areas and sometimes extend toward the sandy ground, giving rise to coastal dunes. This naturally is a weed for crops on the low soils of the Peruvian coast, close to the sea (Sagástegui 1973:26; see also Soukop n.d. [1987]:166). It is a drought-resistant plant, too, and is also found close to wastelands. It is quite common and is considered an "important weed" (Anonymous 1973:30–31).

Julien (1981:5) also favors the use of salt grass to feed llamas, and specifically refers to the Virú Valley.

Maldonado (1952a:73–74) examined a great amount of camelid dung, "either whole, slightly ground and diluted in water," from "Nasca, Pachacámac, Ancón and Cajamarquilla" (there is no way of ascertaining to what period the samples could have belonged), but he categorically claims that the animals did not eat algarrobo fruits. He wrote: "If they had eaten algarrobo pods, they would have contributed to the formation of the extensive and old algarrobo forests in the Department of Ica, spreading this plant throughout the Province of Lima. Algarrobo seeds are hard to grind with the teeth of ruminants, and are not digestible. They are easily recognizable in the excrement of the livestock that ate it." Although Maldonado's statement deserves attention, it certainly comes from an unsystematic study, and apparently no major analysis of the coprolites was made. Besides, this contradicts the archaeological data available, although it is true that these are for the north coast.

It is worth recalling some general data on the properties of the algarrobo, which are usually unknown. But first it is worth noting that in folk usage, the term algarrobo refers not just to *Prosopis* but also to *Pithecellobium*, *Piptadenia*, and *Mimosa*. Algarrobo is used to indicate *Prosopis*, as well as *Caesalpinia* and *Ceratonia* (Soukup n.d. [1987]:49). On the other hand, *Prosopis* likewise goes under the common name of huarango, particularly in the south, but the same name is given to the *Acacia* (Soukup n.d. [1987]:211), so the term must be used with care.

The taxonomy of *Prosopis* is quite confusing (National Academy of Sciences 1979:153). Ferreyra (1987:18–21) gives only one species, *Prosopis pallida*, noting that it extends from the north to the south coast (Ica). It is a plant of which 44 species are currently known (National Academy of Sciences 1979:153); in Peru two are distinguished, according to Soukup (n.d. [1987]:336), but there seem to be more.

Prosopis pods are liked by cattle, sheep, horses, mules, donkeys, goats, and a wide range of wild animals, all of which eat it without hesitation. Every year, forests of *Prosopis* in Hawaii, Peru, Argentina, and Chile support animals for a month or more with no other food. In addition, the pods can be collected and stored for later use. It has a feed value roughly comparable to that of barley and maize. Its pods, with a sweet pulp and seeds rich in protein (34%–39% protein, 7%–8% oil), are nutritional and a great benefit for livestock in the dry seasons, when there is no other food. In some species the seeds are digestible only if ground up (National Academy of Sciences 1979: 155). These plants are widely spread, for *Prosopis chilensis* is found in the dry regions of Peru, Bolivia, central Chile, and northwestern Argentina, and can thrive from sea level up to 2900 masl. *Prosopis pallida* is native to the driest parts of Peru, Colombia, and Ecuador, but it is mainly found along the coast (National Academy of Sciences 1979:157–158). Thus, we must not forget this plant when discussing the food of camelids on the coast—and not just there.

DeNiro published the results of an interesting study on the carbon and nitrogen isotope ratio analysis of bone collagen (DeNiro 1988: 119). He worked with bones from two sites, Chilca and Paloma. Unfortunately, no bibliography is given for Chilca; all we are told is that the bones date "from 2100 to 260 years B.P." In other words, this means they date to the period between the Late Intermediate period and the Spanish invasion. There are several sites in the ravine at Chilca, and there is no way to establish which ones DeNiro meant; these presumably are coastal sites, owing to the presence of otariid bones, among others.

DeNiro explains that the isotope ratios displayed in deer and otariid remains, respectively, exhibit a fully terrestrial- and a fully marine-based diet. He then adds: "One of the camelids also has a fully terrestrial isotopic signal, whereas three of the remaining four camelids fall on the mixing line between marine and terrestrial feeding. These three points lie far enough away from the terrestrial box to indicate that at least half of the animals' food came from non-terrestrial sources" (DeNiro 1988:120).

In the case of Paloma, we are well acquainted with the history of the site, which was discussed in Chapter 4. Here again the author points out that "the deer and the otariid give fully terrestrial and fully marine diet isotopic signals"; he then adds: *"whereas the isotope value for two of the three camelids reflect a significant proportion (again, probably greater than 50 percent) of non-terrestrial food sources. The third camelid gives a fully*

terrestrial signal" (DeNiro 1988:120; emphasis added).

In his conclusions, DeNiro notes that the most probable source of nonterrestrial food for the Chilca and Paloma camelids was marine plants "whose collagen isotope ratios fall along the mixing line between marine and terrestrial feeding"—but he likewise points out that these could also be rushes from saltwater lakes. However, DeNiro states that we cannot rule out the possibility that the animals ate other coastal food, such as plants on riverbanks. For DeNiro, in fact, the high content of $\delta^{13}C$ and $\delta^{15}N$ in bone collagen from some camelid remains at Chilca and Paloma "could only be caused by their having fed, for most of their lifetimes, on aquatic resources." He adds that the absence of camelids with a high collagen $\delta^{13}C$ and $\delta^{15}N$ content in bones from prehistoric sites in the highlands indicates the animals either did not eat aquatic food or that none of the lacustrine plants they ate had collagen isotope ratios different from those of terrestrial plants (DeNiro 1988:121).

Curiously, DeNiro does not mention the highland bones he refers to or what sites they belong to, either in the text or in his bibliography—unless he examined the remains from one of the caves found on the upper part of the Chilca ravine. However, it seems that this was not the case, because DeNiro clearly states that his analysis is based on "coastal camelid bones" (DeNiro 1988:121). This part of his report is certainly not clear.

DeNiro admits that his sample was too small, and that more analyses of larger samples are needed to reach stronger conclusions (DeNiro 1988:121).

It is worth noting that DeNiro claims the marine plants were "presumably harvested and provided as forage" (DeNiro 1988:121). But we know from the studies by Reitz (1988a) that the camelid bones from Paloma come from guanacos, that is, wild animals, which humans did not feed. It would be interesting if it was conclusively shown that these animals fed on marine products. I believe it is likelier that their food came from riverbanks or coastal lagoons, common along the Peruvian coast in the Early Holocene. Even so, this result is most interesting and agrees with the data available for the site of Puémape (see below). However, it must be pointed out that Altamirano (1987:44) incorrectly ascribes to DeNiro (misspelled as Deniro) the claim that camelids were fed "yuyos and marine algae." DeNiro never specified what plants he meant, saying only "marine food sources" or "marine plants."

The coastal lomas are another phenomenon that must not be forgotten. At certain times of the year the lomas can provide food, particularly to nondestructive animals such as camelids. Rostworowski has noted that in Colonial times, mules and horses were the main means of transportation, and that transhumance from the coastal forests to the lomas was established, according to the season of the year. "This movement from the littoral to the lomas was no doubt a road also used in pre-Columbian times for the herds of llamas under the control of the lords." Rostworowski likewise evinces that the food of these animals was guaranteed by the fruits and pods of the trees, and the pastures at the lomas (Rostworowski 1981: 52–53).

Although the Peruvian coast presents conditions for the preservation of organic remains that exist in very few parts of the world, the actual archaeological data are unfortunately quite scant; only five cases are known of camelid coprolites from Peru having been analyzed, and one from a Chilean site.

For the preceramic period, Jones and Bonavia (1992) report the analysis of the llama coprolites found at Los Gavilanes, on the north-central Peruvian coast, close to the Huarmey Valley (see Bonavia 1982a; 1982b:200–201). The remains belong to the late preceramic, and the samples are associated with two contexts covering a temporal span ranging from 3200 to 2000 BC. At least 23 taxa were identified in the four samples studied (Jones and Bonavia 1992:Table 1, 839). It should be noted that the analysis was based on samples of pollen grains.

It was assumed that the plants whose pollen appeared at a higher rate were those used to feed the animals. It turns out that maize (*Zea mays*) was one of the most important ones, for its presence ranges between 50% and 6%. *Vicia* is another plant that seems to have been important; although present in just one sample, it reached

51%. The third plant, or plants, appearing in significant percentages (22% and 14%) belong to the Poaceae family, but they definitely are not maize. The genus is hard to identify, but it has been suggested that it could be either *Gynerium sagittatum* (i.e., caña brava) or *Distichlis spicata* (i.e., salt grass).

In the pollen spectra, the low-spine Asteraceae likewise exhibit significant percentages, ranging between 20% and 5%, but the authors tend to believe that in this case it was essentially a pollen rain.

An intriguing fact is the pollen that could not be identified and was labeled "Unknown A" reached 53% in one of the samples. It is possible that this was an important plant for camelid feeding.

The Chenopodiaceae/*Amaranthus*, which reaches 10%, is another pollen that must not be discarded even though it may well correspond to a pollen rain. Jones and Bonavia suggest that llamas perhaps ate *Salicornia fruticosa* while on the road. The remaining families represented in the samples appear in extremely low percentages, and the authors tend to believe they are essentially an aerial environmental contamination, or plants that the caravan of llamas eventually found on their way during their journey; in some cases these could be human food or industrial plants cultivated to give to the animals (Jones and Bonavia 1992:846–849).

A Chenopodiacea, *Salicornia fruticosa*, is one plant that should be studied more in the future. This is a common shrub on maritime beaches, for it is halophytic and establishes pure associations at some sites and is found in significant amounts throughout the Peruvian coastal areas. This is a succulent and cosmopolitan plant, for it is found in both hemispheres (MacBride 1937:469–470), and is used in the Mediterranean area as food for camels (Usher 1974:519). There thus is the possibility that it was also used as food for the llama on the coast. (See Jones and Bonavia 1992:841–842.)

Víctor S. Vásquez (pers. commun., letter of 17 July 1992) has analyzed Camelidae coprolites (it is not specified whether they are from llama or alpaca, although it must be the former) from the site of Puémape, in the province of Pacasmayo, district of San Pedro de Lloc, Department of La Libertad. These studies were made by observing macro- and microremains, without taking pollen grains into account. The excrement was associated with a Salinar occupation, during the transition from the Early Horizon to the Early Intermediate period (ca. 400 BC and AD 100).

The analysis showed the presence of "[a]eriating parenchyma and grains of starch from *Typha* sp., enea rhizomes. Marine diatoms associated with algae tissue, possibly *Gigartina* 'mococho.'" There also were "*Grateloupia* (another type of "mococho"), and a lesser percentage of stoma and trichomes of a nonidentified gramineous plant." Vásquez notes that he believes that in this case, the occlusal wear of the camelid teeth was due to algae with silica in their histological structure.

The presence of *Typha* on the Peruvian coast is normal; what I believe is significant are the remains of algae. In Peru, both genera identified—*Grateloupia* (Cryptonemiaceae) and *Gigartina* (Gigartinaceae)—are algae with some economic significance (Acleto 1971:36, 58). It is true that their use as animal food has not been reported, but others are being used as forage for domestic animals (see Acleto 1971:25–26). Besides, interestingly, both algae are found throughout the Peruvian coastal areas, from Piura in the north to Ica in the south (Acleto 1971:38, 61), and are known under different names. According to Vásquez (pers. commun., 1992), *Grateloupia* goes under the name of *mococho* in the north, whereas on the central coast it is called cochayuyo. *Gigartina* has the same names all along the coast, or it is simply called *yuyo* (Acosta Polo 1977:4).

Jones (1990) analyzed llama excrement from a small Middle Horizon site called PV35-4, which Bonavia excavated on a beach zone in the lower Huarmey Valley (see Bonavia 1982a:417). In this case, the overwhelming presence of pollen belongs to the high- and low-spine Asteraceae. It should not be forgotten that no sector of the Peruvian flora is free of some representative of this family, as there are 204 genera in Peru (Soukup n.d. [1987]:71).

In second place is a type of pollen we were unable to identify; we do not know whether this appeared due to a pollen rain or to plants ingested by animals. Finally, there are significant amounts of pollen belonging to Poaceae that do not correspond to maize. We should bear in mind that except for maize, virtually all pollen grains in this family cannot be distinguished, and the macroremains are far too small and fragmented for them to be identified. We should also bear in mind that the plants in this family are pollinated by the wind, and their pollen is usually taken as a prime representative of pollen rains in natural vegetation. This is a vast family, which makes it hard to suggest which genera this pollen could have belonged to. However, it is possible that the llamas ate *Cenchrus* sp., a plant common on the coast; *Distichlis spicata*, a salt grass, commonly known as grama salada, that is found on the graminales near the seashore that exist in Huarmey; and *Gynerium sagittatum* or caña brava, a common plant in coastal river woods and a common one in Huarmey.

Finally, the fourth plant for which pollen was identified in a significant amount was maize (*Zea mays*). This seems logical because the evidence found at the site showed that the people who camped there ate this plant and almost certainly fed their animals its leaves and stalks, which brought the pollen. Besides, this has been verified for other sites from different periods. But it is worth noting that *Prosopis pallida*, or algarrobo, was identified among the macroremains; this may well have been used by man, but it could likewise be animal food.

Shimada (1982:173) reports that camelid excrement was found at Sapamé, a coastal site within the Batán Grande area, in the Department of Lambayeque, belonging to an occupation that extended from the Middle Horizon to the Late Horizon (i.e., a period ranging between AD 500 and 1500). This is a stratum where the llama excrement is mixed with maize leaves, cobs, and stalks; there is another stratum that has excrement mixed with leaves and algarrobo fruit. Algarrobo seeds were found in the llama excrement. However, no analyses were made in this case.

Víctor S. Vásquez (pers. commun., letter of 17 July 1992) reports that in the excavations Hugo Navarro carried out in Huaca 1, Túcume (Department and province of Lambayeque, district of Túcume)—whose associations have not been noted but which I assume must date to the Late

Intermediate period (AD 900–1440)—he found achira (*Canna edulis*) stomas and foliaceous epidermis in the llama coprolites he studied; maize (*Zea mays*) stomata, elicoid vessels, and parenchyma in a 50:50 ratio; and a similar prevalence of algarrobo (*Prospopis pallida*) endocarps and seeds. Vásquez concludes that "it can be inferred that chala[1] maize, achira and algarrobo were good pastures for the camelidae." What Vásquez is interested in is establishing what vegetables may have contributed to the occlusal wear in these animals' dentition, something essential for their dental development, for no evidence was found of, for instance, gramineous plants, which cause wear because they are coriaceous.

Segundo Vásquez and Víctor Vásquez (1986: n.p.) report on the excavations they undertook at Loma Roja. This is an artificial mound that is part of the urban area of Chanchan (Department of La Libertad, Province of Trujillo, District of Huanchaco). It is close to Vía de Evitamiento connecting Huanchaquito with Buenos Aires. This mound is also known as Unidad Sur No. 2-Chayhuac. Llama coprolites were found here. Two types of analyses, dry and wet, were carried out to reconstruct these animals' food. Fragments of algarrobo (*Prosopis chilensis*) seeds were identified in the dry analysis, covered with their endocarp, small fragments of leaves from a gramineous plant, small blades of salt grass (*Distichlis spicata*), and fragments of an inflorescence of this same species, as well as structures from another gramineous plant. The wet analysis identified algarrobo (*Prosopis chilensis*) stomas and endocarp tissue; gramineous stomata (ca. 20%); stomata, elicoid vessels, trichomes, and a great number of pollen grains from salt grass (*Distichlis spicata*).

For comparative purposes, I include here an analysis of camelid coprolites carried out in Chile (see Chapter 9). Belmonte et al. report the "[o]rganic materials from herbivore coprolites obtained in excavations carried out at the mouth of the Camarones River [in the Province of Tarapacá]. The materials come from a shell mound at the site of Camarones; two squares 1 by 1.20 m were dug; 11 levels were identified, each 10 cm thick, with level 1 being the latest and level 11 the earliest. Coprolites were recovered only in levels 4, 6, 7, 8 and 9, i.e. those belonging to the Prece-

ramic Period" (Belmonte et al. 1988:49). This is all that the article says, so no cultural inference can be drawn. Moreover, the excavation was made using arbitrary levels, but their interpretation is not explained, and the text does not allow us to draw any stratigraphic inference.

The authors note that "[b]ased on the shape and size of the coprolites recovered, several students concur in identifying them as belonging to camelids, but there are not enough elements to distinguish which of the actual species they belong to (C. Santoro, E. Núñez, J. Contreras, and A. Vilaxa, personal communication)" (Belmonte et al. 1988:49).

The analyses consisted in the identification of ordinary cells, stomas, and trichomes, and the description of elements in the foliaceous epidermal tissue (Belmonte et al. 1988:50).

The result was the identification of 12 different species of Monocotyledoneae, two of which could be *Distichlis spicata* and *Scirpus americanus*. (It should be noted that from the report it is unclear whether this is a reliable or tentative identification; on the same page we read: "were identified as," and "are probably represented . . . similar to"). "Although it is clear that the remaining 10 species are not monocotyledons, as these are absent in the present flora, they could not be identified at a species level" (Belmonte et al. 1988: 51). This study actually raises many doubts, and much more could have been gleaned from the materials examined.

10.1.2 Historical Data

Historical sources hold some data on the food eaten by camelids, but they are few and usually vague. Thus, Zárate (1968:Bk. II, Chap. II, 176) says of the pacos: "and [they] eat maize." Gutiérrez de Santa Clara (1963:Bk. III, Chap. LVII, 235) says of the llamas that they "eat maize elegantly."

Garcilaso de la Vega (1966:Bk. VIII, Chap. XVI, 514; 1959:149) in turn devotes a whole chapter to the "tame animals," or llamas. As regards their food, he wrote that they "will eat maize if given it; but the creature is so noble that it can manage without even when working." Just before this he noted the following: "When they reach the place where they are to spend the night, they are unloaded and turned loose in the fields where they graze on whatever grass they find. They keep themselves in this way for the whole journey and require neither grain nor straw."

Dedenbach Salazar (1990:157) discussed the three passages just cited (she only makes a mistake in the case of Zárate, for she gives the reference as "l[ibro]. II, cap. XIII" when it actually is, as we have already seen, Bk. II, Chap. II) as regards maize-based food. Dedenbach Salazar believes this should not be generalized because it can be assumed camelids ate maize in very temperate areas without sufficient pastures. She wonders whether they fed on this plant on the coast—perhaps the herds used for religious purposes. Interestingly, Murra (1978:87) made a similar comment: "Zárate was told that maize was used as fodder, but this may have happened after the invasion, or perhaps on the desert coast." Murra seems not to have read Gutiérrez de Santa Clara or Garcilaso de la Vega in this regard.

I do not concur. Whenever there were crops nearby, maize leaves must always have been the food eaten by these animals. Besides, we should bear in mind that in ancient Peru, the whole plant was moved from the fields to the storage rooms or the place where it was used. This is shown clearly in one of the pictures made by Guaman Poma himself (1936:fol. 1144) that shows the harvest in May. The writer found that this practice is most ancient, for it was already used in preceramic times (see Bonavia 1982a, particularly drawing 64 [272–273] and 376). After the maize is husked, the rest of the plant is an excellent food for the animals. Squier (1877:247; 1974:133), a modern traveler, left an invaluable report in this regard. He tells of passing through La Portada (3800 masl) while moving down from the altiplano to the coast, where he stayed for the night. (This is not exactly a warm place, as stated by Dedenbach Salazar 1990; see above). Squier said, "Before going to bed I went out to the corral. The llamas had been fed each with a handful of maize, and were . . . chewing their cuds."

As far as I have been able to ascertain, no data have survived on the use of the algarrobo as camelid food, except for one case in Argentina, to which we turn later. However, there is invaluable documentation on the use made of this plant with

European livestock, which I find interesting. Thus Ocaña (1987:60) explains that in Piura there were "few people. . . . What there is feeds on livestock from Quito, on rams which are fattened with the algarroba which grows over there." When describing the European rams in Piura, Vázquez de Espinosa (1948:1177/372) likewise says that "they feed on the Guaranga or algarroba, and it is the best and most luxuriously bred ram in the world." The same chronicler mentions the Ica Valley and the European livestock and "rams" in it—it is unclear whether these are also from Castile or of the land—and notes that they "feed on the algarroba that falls from the trees, [and] their meat is good and delicious" (Vázquez de Espinosa 1948:1359/451). Fray Antonio de la Calancha (1976:Bk. III, Chap. I, 1230) in turn describes Guadalupe (in the present-day province of Pacasmayo) and specifically describes the abundance of algarrobales: "[I]t costs the herder or cattle farmer no more than shaking the tree because . . . the treetops in the big algarrobos . . . leave a sweet and good pasture [in the] algarrobo in its pod . . . and with its abundance the livestock grows fat." Pedro de León Portocarrero (Anónimo Portugués 1958:29 [see Lohmann Villena 1967]) left a description of Huarmey (on the coastlands of the Department of Ancash) which says that here "brave horses and mules are grown, which grow strong on the algarroba."

There are some data to show the algarrobo was used by man in other regions. Barzana (1965: 80) mentioned the Calchaquí—a subgroup of the Diaguita (Argentina)—around 1594, noting that "they also feed on a great amount of algarroba, which they collect in the fields every year at the time it ripens, and make huge storerooms with it; when the rains to harvest maize do not come [and] the river does not leave its bed to irrigate the land, [they] overcome their need with this algarroba, which is not just food for they also make a drink of it, so strong that there are never more deaths or wars among them than as long as the algarroba season lasts." Tarragó (1978:493) reports the guanaco feeds on algarrobos in Tintin Pampa, in the Calchaquí Valley (province of Salta, Argentina).

I have gone over this point in some detail because I believe the algarrobo was one of the food staples present in a significant amount on the Peruvian coast; it must have had a major role in the sustenance of the camelid herds kept there in pre-Columbian times. There is no reason why such an invaluable source of food, which the European animals used, cannot have been so for native animals. This point requires a more in-depth study at an archaeological level.

10.1.3. Additional Data

We should not forget that the camelid is an animal that browses on the road and is above all non-selective in its food. Pedro de León Portocarrero (Anónimo Portugués 1958:80 [see Lohmann Villena 1967]) noticed this quite well. When describing the pack rams, he said: "[T]heir sustenance is naught but what they eat in the fields." Troll (1958:30) insisted that llamas need no fodder, for they find their food on the road during their march.

Flores Ochoa (1975b:301) gives a good explanation of the feeding habits of domestic camelids. He notes that llamas eat almost any kind of grass, even the hard and dry ones, and are therefore more dispersed than the alpaca, for the latter looks for green and hard grasses and does not eat dry ones. This is why the alpaca is forced to live in certain special zones with good conditions for hard and green grasses, in humid places with a water supply that can only be found from 4100 or 4200 masl. The grass has a different quality at lower, drier altitudes, and therefore the alpacas tends to suffer from several lethal diseases. The upper limit of the adequate zone lies on average above 4800 or 4900 masl, but in some places it is higher, reaching up to 5000 masl, at least in pasture areas.

Sumar (1992:94) also noted that camelids are adapted to the scant, fibrous and woody high-altitude resources, and under conditions of scant water. Gade (1977:113) noted that both the llama and the alpaca do not depend on man for their sustenance, even in the dry season.

Núñez del Prado (1968:251) studied the community of Q'ero in northeastern Cuzco, a good example of what has just been pointed out. The herding areas used by this community "are spread out among the Q'ero and the Qhollopak'uchu levels," that is, between 3300–4000 and 4000–4800 masl. The former is dedicated to raising bovines, cattle, horses, and some pigs, which feed on the soft pastures found in this environment, and the latter is dedicated to herding llamas and alpacas, which eat the hard grasses growing close to the snow line.

Another good example, described by Shimada and Shimada (1985:4), is that of the llameros [llama breeders] of the Abancay Valley (in the Department of Apurímac), who make two-week-long journeys between sites ranging from 5000 to 1000 masl. Llamas eat almost any underbrush they find, and do the same on reaching their destination. If the caravan arrives after maize has been harvested, they eat the leaves and stalks of this plant. Shimada and Shimada indicate that in some cases, caravans go from the upper parts of the Department of Arequipa to the southern coast to take olives to the highlands, and then the llama feed on olive leaves, without any problem arising. (Shimada and Shimada give as their sources Inamura [1981a, 1981b], and a personal communication]. I was unable to read the latter article, but the data referred to certainly do not appear in the 1981a article.)

One factor that must be borne in mind is that camelids are highly efficient in transforming harsh and dry vegetation into carbohydrates to get energy and metabolic water (Browman 1974a: 191; Fernández Baca 1971:38; Gilmore 1950:436; Flores Ochoa 1968:109).

Flores Ochoa (1979a:226–227, 227 n. 1) likewise mentioned camelids feeding on the harsh puna grass with a high cellulose content that reduces digestibility—precisely those plants that cannot be used by European animals.[2] Based on studies done by Chauca et al. (1970) and Vallenas (1970a), Flores Ochoa writes that the comparative studies of digestibility made in vivo between alpaca and ovines by Fernández Baca, Novoa, and Bardales showed that the former have a higher capacity to digest the raw fiber (cellulose and hemicellulose). This could be due to several reasons, such as ruminoreticular motility, the characteristics of the alpaca's digestive fluids, the activity of the glandular secretions in the cranial and caudal sacs in the alpaca's first compartment, and the qualitative and quantitative characteristics of the ruminal microflora. Of these factors, it seems that the microorganisms present in the rumen play a major role, and could well be responsible for the higher efficiency of the alpaca in converting raw fiber. It is an established fact that digestion in South American camelids is very different from that of ovines, cattle, and possibly other ruminants too. These differences, in combination with the ones already found in motility (i.e., more frequent rumination cycles), more efficient maceration, better mixing and absorption of the ingesta, and possibly a more efficient system for absorbing volatile fatty acids, could explain the improved digestion described for the four species of camelids.

We saw in Chapter 2, which discussed the biological aspects of camelids, that these animals are more efficient than bovines in digesting middle- and low-quality food (see San Martín and Bryant 1987). We also saw that both the llama and the alpaca retain food in the digestive tract for a longer time than other advanced ruminants, thus clearly making them more efficient in the use of low-protein-content vegetation (see San Martín and Bryant 1987). (For more details on the digestion coefficient and a comparison of the various grasses, see San Martín and Bryant 1987)

It has even been said that plants in the high puna have far too much cellulose for them to be used as human food. The optimal converters of this plant energy into animal energy are precisely camelids. Their biological and physiological conditions thus make them essential in the production of energy for a human population (Flores Ochoa and Palacios Ríos 1978:84).

To finish with this point, it is worth recalling the high potential camelids have for living at high altitudes in full productive capacity. This is due, I repeat, to a digestive ability higher than any other species inhabiting the same ecology, where grasses have a higher level of lignification. In other words, the higher rate of converting nutrients into energy translates into a more efficient use of high-altitude pastures by camelids, which in turn means an optimal adaptation to higher altitudes (Matos 1980:96–97).

"On the one hand," writes Matos (1980:97), "all four camelid species (llama, alpaca, guanaco, and vicuña) have a specific behavior, but in general they are similar in their territoriality and their ability to use zonal resources. They are not predatory animals, and do not even cause an imbalance in the environment. Far from it, they represent a conduct translated into an ecological interrelation that as a whole reflects the potential of the biomass."

As has already been noted, feeding camelids in high-altitude zones poses no problem. The major doubt expressed by many students—actually, the majority—refers to the coast. The evidence shows quite the opposite. That is to say, it is a hard fact that herds of camelids could be maintained on the coast on natural vegetation alone, under natural conditions and without the changes wrought by the European invasion. The potential is even greater if we add cultivated plants like maize, of which man uses only the fruit while the rest can be used as fodder.

We should likewise not forget another most important fact: had the practice of camelid raising continued on the coast, the coastal ecosystem would have been preserved in a less destructive way than that brought about by the animals introduced from Europe, particularly on the lomas.

It is worth ending this section citing Terry West, a man who lived with the *llamichos*, or llama herders. He wrote: "More sure-footed on the rugged terrain than the burros (and horses) . . . the native llamas are also cheaper to maintain. They can forage along the route for their food, while burros, the usual beasts of burden in non-pastoral areas, require fodder that has to be grown or purchased" (West 1981a:63).

10.2 USE

Many writers have discussed the many ways to use camelids and their different products. Not all agree on the list, as can be seen in Table 10.1. But in brief, the major ones are wool (fiber), meat, skin (pelt, leather, fleece), fat, bones, dung (*taquia* or *taqya* in Quechua), blood, viscera, and tendons. A specific use is made of aborted animals, or of fetuses with special characteristics. The animals are also used for transportation, ritual sacrifice, and religious ceremonies, and for folk medicine; only a few students list them as a mount or a source of wealth.

TABLE 10.1. USES OF CAMELIDS ACCORDING TO DIFFERENT AUTHORS

USE[a]	1	2	3	4	5	6	7
Wool (fiber)	X	X	X	X	X		X
Meat	X	X*	X		X	X	
Skin (pelt, leather, fleece)	X	X	X	X	X	X	X
Fat	X	X		X	X		
Bones	X	X***		X	X	X	
Dung	X	X****	X	X	X		X
Animal A[b]	X	X		X			
Blood		X**	X				
Viscera		X			X		
Tendons	X			X	X	X	
Medicine (bezoar stones)			X	X			
Source of wealth		X					
Transportation	X	X	X	X	X		X
Mount	X*****						
Ritual sacrifice, religious ceremonies		X	X	X	X		X

[a] Sources: *1*, Browman 1974a:193; *2*, Flores Ochoa 1975a:10, 1975b:306–307; *3*, Gilmore 1950:438; *4*, Novoa and Wheeler 1984:126; *5*, Kent 1987:169; *6*, Dransart 1991a:309; *7*, Sumar 1992:84.
[b] Aborted fetuses or adult animals with special characteristics.
* Fresh or dehydrated; **fresh or as sausage; ***food and other uses; ****fuel and fertilizer; *****only after the Spanish conquest, except in Mochica society.

As an illustration, I mention the case of the community of Paratía, in the Department of Puno, where the llama is mostly used for its wool and transportation and, as a supplement, its meat, blood, leather, and dung (Flores Ochoa 1967:98).

I shall now review some of the uses attributed to camelid products, but not necessarily the most important ones. We should also bear in mind that their significance is only relative, and can be higher or lower according to the community using these goods.

10.2.1 Wool

According to Novoa and Wheeler (1984:126), Peruvian wool began to be exported to the world market around 1830, when the British began exporting it from Arequipa (they base their statement on data in Orlove [1977{a}]). In the 1980s, when Novoa and Wheeler wrote their study, Peru had a mean annual output of 3,400 tons of alpaca wool, which represented 80% of the world's output. About 2,380 tons were exported, and the rest went to local handicrafts.

In 1998 the output of alpaca fiber was 3,505 metric tons (Ministerio de Agricultura del Perú, Oficina de Información Agraria).

Alpaca wool has a high commercial value because it has little kemp,[3] has a low-felting quality, is very fine, and can be woven into lightweight, soft and lustrous fabrics (Novoa and Wheeler 1984:126). There is no need to emphasize the wide use this wool has had since prehispanic times, starting in the preceramic period.

Llama wool is of a lesser quality than alpaca wool (Lanning 1967a:17), but it was nonetheless widely used in the pre-Columbian period and after—so much so that in 1998, Peru had an output of 654 metric tons of fiber from this animal (Ministerio de Agricultura del Perú, Oficina de Información Agraria).

Shimada and Shimada (1987:836) discussed the primary role of alpaca herds, as to whether the fiber was more important than meat. They begin with the hypothesis (as was previously noted) that the alpaca was essentially domesticated for its wool. Shimada and Shimada claim that it is not easy to distinguish the bones of the various camelid species in archaeological remains. Based on a study by Ann Rowe (1977:10), they note that the degrees of fineness in the hairs overlap, with the finest in the llama identical to the ordinary alpaca hair and the finest in the latter identical to the ordinary one in the vicuña, to the point that the analysis of archaeological textile fibers cannot distinguish them. However, it is worth noting that Alberto Pumayala, one of the great specialists in camelid fibers, did not agree and held that it was perfectly possible to distinguish the fibers of the various camelid species (Alberto Pumayala, pers. commun., 30 January 1989). In fact, he and his team in the Ovine and South American Camelid Program of the Universidad Nacional Agraria (La Molina) in Lima did so (see, e.g., Bonavia 1982b: 200–201).

The use of llama wool was not developed at the market level in the modern period, partly due to its coarseness and irregular coloring (Novoa and Wheeler 1984:126).

Interestingly, I was unable to find any data on vicuña wool and its use in archaeological and historical sources, even though its hair is the finest known and was highly valued by the natives (Ko-

ford 1957:155). However, it is known that in 1751 there was a community in the province of Chucuito (Department of Puno) that not only made its clothing with vicuña wool, it even "loaded whole ships [with it] to send to Europe" (Bayer 1969:42). The testimony given by Bayer seems to indicate that this wool was highly valued and used in Europe. According to what Juan and Ulloa reported in 1827, "[t]he Spaniards make no use of it there other than in hats and handkerchiefs, when its delicacy makes it recommendable for other, more important, textiles" (Juan and Ulloa [1826] 1988:587). These same authors report a curious fact. They indicate that in 1737 an Englishman began making hats with vicuña wool in Lima, and these were "of such high quality that they were not second to those of regular beavers" (Juan and Ulloa 1988:587). Today, vicuñas are fleeced in the Pampa Galeras National Reserve and their wool is sold.

Data on guanaco wool are likewise unusual, but we do know that the Patagones used it (Franklin 1981:63; see Torres 1992a:34).

10.2.2 Meat

When studying the great slaughter of camelids that took place during the early years of the conquest we saw that it was mostly for the meat. However, we have verified how little is known archaeologically about the use of camelid meat for food, particularly in the highlands. Everything seems to indicate that its use was widespread. Pedro Pizarro (1968:578) reports something that would seem to contradict this, at least for Inca times. He wrote: "Meat is raised; few ate it save the lords to whom they commanded it be given." Troll had a similar position, because in his frequently cited 1935 study he claimed that "the consumption of llama and alpaca meat . . . had a most insignificant role as only old animals could be killed; we can say that animals were not considered when it came to human food" (Troll 1935: 138–139). In this he agrees with Lanning (1967a: 17), who wrote that llamas were "occasionally [used] as sources of meat" in pre-Columbian times. Romero (n.d.:I, 222–223) even wrote that "[t]he llama was domesticated by the primitive Peruvians as an animal exclusively dedicated to work and not for food." I do not concur, and believe that use of the meat was widespread. In fact, not all scholars agree with Troll and Lanning, and instead support my position. Browman (1974a: 193) thus states that although ethnographic and ethnohistorical sources agree that the consumption of meat from domestic animals was low, the archaeological data indicate that meat consumption was higher when the population density was lower. And Wheeler (1985a:29–30) specifically says of the llama that if its meat is today not valued, and that only old animals are killed because they are useless, "it clearly was not so in the past, when llama meat was the most important source for feeding with meat."

At this point the accounts given by the chroniclers are interesting. Gutiérrez de Santa Clara (1963:Bk. III, Chap. LVII, 235)—who, as we recall, was confused and separated the species of camelids, adding one more—stated that "[t]hese five genres of camelids are very good to eat."

Garcilaso de la Vega is far more specific. First he refers to the llama and the alpaca and says that "[t]heir meat is the best in the world; it is tender, tasty, and wholesome. The meat of the 4- or 5-month-old lamb is recommended by doctors for invalids in preference to chicken" (Garcilaso de la Vega 1966:Bk. 8, Chap. XVI, 513; 1959:147–148). Later on he describes the huanacu, whose "flesh . . . is good, though inferior to that of the tame kind," and the vicuña, "a delicate creature with little flesh," but which "is edible, though less so than that of the guanaco. The Indians appreciated it because they had little meat" (Garcilaso de la Vega 1966:Bk. 8, Chap. XVII, 517; 1959:152–153).

When Monzón et al. (1965:246) described "the province and repartimiento of Rucanas Antamarca, [in the] jurisdiction of the city of Huamanga" in 1586, they said: "There are vicuñas . . . it is good meat." This same year Rivera and Chávez y de Guevara (1974:168–169) described the "City of Guamanga and its jurisdiction," indicating the presence of guanacos and vicuñas and adding that "all of this livestock can be eaten." In 1586 Carabajal (1965:206) likewise left a description of the corregimiento of Vilcas Guamán. Of the llamas he said that "the Indians eat their meat and it has a good taste." He then mentions the guanacos and vicuñas, and notes, "whose meat is

eaten." Finally, when Ribera and Chávez y de Guevara (1965:187, 192–193) described Huancavelica, also in 1586, they mentioned the domestic livestock and said they "are useful as food," and when mentioning "savage and wild" animals they stated that "all this livestock is to eat."

In the description he gave of llamas, Ramírez (1936:18) wrote that "the meat is tough and very sweet; the Indians eat it very well and the tender lambs of this livestock are a delicious food." And when talking about the "pacos" he says they "are usually eaten." When discussing the llama, Cobo (1964a:Bk. 9, Chap. LVII, 365–366) likewise says that "[i]ts meat is like that of cows, though somewhat insipid, but that of their lambs is a dainty," while all he says of that of the alpaca is that they "use [its] meat." Ocaña (1987:166), who likewise saw a great number of llamas when he was at Potosí and almost certainly ate its meat, claims that "the Indians eat the meat from them, and the Spaniards too; and it is like that from cows."

When discussing the "Gobernación and Province of Chuquito," Castro y del Castillo (1906:205) mentioned the vicuñas and specified that "the Indians catch many . . . to eat the meat."

Mejía Xesspe is clearly one of the scholars most concerned with the description and reporting of native foods, and was well acquainted with them. Of the llama he wrote: "The meat from these animals is most agreeable in its first three years, then it takes on the taste of the kind of plants they feed on. When raised on chosen grass, the meat is more agreeable. It is used [both] fresh and dry." Then he presents a list of the dishes prepared with it (Mejía Xesspe 1931:19). Sumar (1992:89) agrees that llama meat has an agreeable taste, but he adds that its smell is peculiar. He indicates that its consistency is firm and elastic; besides it has little fat (just 3.7%) and so is good for human health, whereas the protein content is similar to that of other meats. The dressed carcass of male llamas about a year old yields about 58.13%.

Alpaca meat in turn tastes somewhat like mutton and has a high nutritional value similar to the meat from other domestic animals, but with the advantage of having a very low fat content (1.33%), even lower than that of the llama (Sumar 1988:25). Torres (1992a:33) likewise believes that "[a]lpaca meat is preferred to that of the llama, even though both are quite similar."

However, it must be noted that Flores Ochoa (1990b:96) does not concur with the data in Sumar (1988:25), particularly in regard to the protein content, and claims that this is higher in llamas and alpacas than in other kinds of meat for human consumption, save for that of the horse. He agrees with the low-fat index and points out in addition that the cholesterol content is lower than in ovines. In color, taste, smell, and tenderness it is similar to other red meats; however, the latter are more expensive. According to Flores Ochoa, the variation in the quality of these different kinds of meat depends on the age of the animals, the activity they carried out in life, and the techniques used to kill, process, and store them.

Mejía Xesspe (1931:21) also studied the guanaco and the vicuña; he notes that their meat is rarely used in similar ways and conditions as that of the llama. He adds, however, that guanaco meat is quite agreeable, whereas the vicuña's has an unpleasant taste, particularly once they are two years old, due to the kind of grasses they feed on.

We know from the data in Bayer (1969:42) that around the mid-eighteenth century, "in the province of Juli (now Chuquito, in the Department of Puno) . . . a community called Choquelas . . . fed almost exclusively hunting these vicuñas; they eat their meat." And around the mid-eighteenth century, Juan and Ulloa (1988: 600) noted that vicuña meat "is of good taste and very healthy."

Interestingly enough, there are similar examples for our time. Koford (1957:212) not only describes vicuña meat as "delicious," he also believes that this is precisely one of the reasons for the extinction of this species (Koford 1957:155). He likewise tells of a mining camp he lodged in where vicuña meat was a common staple in the menu. He was also told that a great deal of the charqui (dried meat) sold in southern Peru comes from the vicuña. In Antofagasta (Chile), vicuñas are often killed for their meat in the sulphur mines. According to Koford, it is a favorite of the natives. "The flesh, similar to that of alpaca, is considered 'muy rico' as compared with huemul (deer) venison, which is 'muy seco'" (Koford 1957:213).

There are fewer data for the guanaco, but we do know that in this case this was also one of the reasons for the extinction of this animal (Torres 1992a:34), and that Indian groups like the Patagones ate its flesh (Franklin 1981:63).

Readers interested in the different ways in which llama and alpaca flesh are eaten are directed to Flores Ochoa (1967:40, 80).

Today llamas and alpacas are sacrificed for their flesh once they are no longer useful. When the animals are too old, the meat is processed as charqui and is used by the family or sold (Flores Ochoa 1990b:85). In 1984 it was estimated that each year, Peru consumes 8,000 tons of llama meat and 10,000 tons of alpaca meat. The flesh from these animals is of major importance in the people's diet both in Peru and Bolivia; specialists consider that its consumption could be higher if the conditions in which the animals are raised were improved (Novoa and Wheeler 1984:126).

It is worth citing a specific case, that of the Cotahuasi community in the province of La Unión, on the northern fringes of the Department of Arequipa, at an altitude ranging between 4000 and 5000 masl. In this community up to 50% of the alpacas may die in their first year and are eaten. Here the male alpaca is butchered in its second or third year, while llamas are killed when they are 16, even though most of the animals are killed when they are 10. However, it should be recalled that this can in no way be generalized as there are many regional variants (Melody Shimada 1982:310–311; Shimada based her work on a personal communication from Inamura, 1981, which in turn was based on his experience in the aforementioned community).

I must draw attention to a statement made by Browman (1974a:193): "The meat of a llama is considered equivalent to the meat of two alpacas, historically (Diez de San Miguel 1964:90) as well as today." Browman clearly did not read the source he cites, because Diez de San Miguel said something completely different: "[T]hey used to give each friar a big ram of the land each year, and when they did not do so they gave him two pacos *which were worth as much as the ram*" (Diez de San Miguel 1964:90; emphasis mine). Careless statements like the preceding one, if repeated, can lead to serious mistakes.

10.2.3 Camelid Dung

The use of camelid dung as fuel in the highlands is well known, and its use is widespread, particularly in the puna where no firewood is to be found (see, e.g., Troll 1935:138). It is a fuel that burns slowly and has a high caloric value, so much so that in pre-Columbian times it was used to fire pottery and for metal smelting, a practice that persisted into Colonial times. In saying that "[t]he use of llama dung for fuel may be a relatively recent custom and may be correlated with the deforestation and general vegetational depletion of the Highland area in the past several millennia," Gilmore (1950:440) shows he is completely unacquainted with the Andean world.

An anonymous footnote inserted in Rivera and Chávez y Guevara's *Relación de la Ciudad de Guamanga y sus términos* (1974:154–162, n. 15) was probably written by Efraín Morote Best. The passage on pages 160–161 is interesting insofar as it refers to the mines of Huancavelica and the fuel problem. It explains the problem the members of the Council of the Indies found themselves in owing to the disappearance of *ichu*,[4] which vanished because of its unbridled use as fuel. Laws were therefore passed for the preservation of the plant and its seeds. It was ordered that "plantations of branches, pips and seeds [be made], and the grass and weeds be preserved" because their use was essential for mining. This was all done because of a memoir written by the major miners of the Villa Rica of Oropesa in 1589, and expanded in 1607. One of its most important paragraphs reads as follows: "The idea of Torres de Navarra certainly was worth 4,000 ducats of rent, because thanks to it the mountain of Potosí could be worked, but the Indians had actually already shown the excellence of the fuel through its use in many of their tasks; it was not the only one used due to the lack of firewood, as llama taquia or dung satisfied the need well."

This is confirmed by an observation Ocaña (1987:181) made in Potosí itself. After noting the great amount of firewood needed to smelt the metals, Ocaña added: "The dung from the rams itself is worth money in Potosí; some Indian women go about the fields collecting the livestock's dung and the owners of the mills give them a peso for each sack of it."

Tschudi (1885:105; 1891:105; 1918:224; 1969:135) saw the same thing and wrote that the the custom camelids have of dropping their excrement (*takia*, *otsa*) in a single place was used not just by the ancient Peruvians, but also "by contemporary ones." "These piles of excrement similar to that of goats are carefully collected for use as fuel, particularly for metal smelters, as is still done now. According to the account of Potosí, in 1603 . . . 800,000 loads of otsa were used in these mines, in the smelters."

This datum by Tschudi is of great interest, and its date is consistent with the time Ocaña was at Potosí. (For more details see Dedenbach Salazar 1990:173–174).

Dung clearly was also very important as a fertilizer, and may even be considered essential for intensive cultivation and to attain the highest output possible in the Andes with the technology available in prehispanic times (Grossman 1983:82; see also Dedenbach Salazar 1990:173–174).

10.2.4 Milk

The use of llama milk is a point worth clearing up, as this may lead to serious mistakes by those who are not well versed on this subject. Gade (1969:341, n. 13) prepared a list of publications where erroneous statements were made concerning this point (see also Gade 1993:12, n. 1).

From the data in Tschudi (1885:105, n. 1; 1891:105, n. 2; 1918:224, n. 142; the footnote should have appeared on p. 135 of the 1969 edition but was left out), we know that in *The Naturalist on the Amazon River*, the English naturalist Henry Walter Bates (Tschudi unfortunately omitted the year of publication, merely noting that it was p. 164 of the German translation) stated that the ancient Peruvians used llama milk as food, in the form of cheese. Troll (1958:29, n. 60) gives a similar reference, noting that F. Klute listed the llama as a milk animal in the introduction to the volume on South America of the *Handbuch d. Geogr. Wissenchaft*. (The year of publication is unfortunately omitted, and only p. 41 is mentioned.) When discussing the "rams" described by Ocaña (1987:166), Álvarez (1987:166, n. 64) notes that "their meat and milk" are also used.

Even more serious is that such a noted Peruvian geographer as Emilio Romero said in one of the last editions of his *Perú: Una Nueva Geografía* (n.d.:Vol. I, 222), that the llama cooperated with man, "providing milk."

This idea of the use of llama milk has no basis, is completely mistaken, and must be definitely discarded. The only passage that could raise some doubts because of its ambiguity is by Friar Vicente de Valverde, the bishop of Cuzco. When describing this city and listing items that could be tithed, Valverde wrote: "Wool from the sheep from here [this land], cheese and milk on selling it" (Valverde 1969b:25). It is clear that only the wool was from "the sheep from here [Peru]," whereas the cheese and milk were from Spanish animals. Garcilaso de la Vega (1966:Bk. 8, Chap. XVI, 516; 1959:151), whose acquaintance with the Andean world is unquestionable, categorically said: "The Indians do not use the milk of either kind of animal either to drink or for cheese-making. They do in fact produce very little milk, only enough to bring up their young."[5]

Tschudi (1885:105; 1891:105; 1918:224; 1969: 135; the 1969 edition misrepresents the original text) is likewise categorical in this regard. He wrote: "Although the llama has udders with four nipples and gives abundant milk, they have never been milked either by the Inca Peruvians or their successors up to the present day. The reason for this is the shy and untamable nature of these animals." However, it is to be noted that Tschudi makes two mistakes. First of all, he believes llamas yield abundant milk, which as we saw is not true, and second, he says that the reason why it is not used is the shy nature of these animals. Both statements are wrong.

Troll (1958:29) is likewise emphatic in noting that the idea of the llama as a milk animal "cannot be taken into account," as is Gilmore (1950: 438). Flores Ochoa, one of the great students of these animals, explains that "[t]he milk is not used, and this use was not developed because they [the Indians] claim that it must go to the offspring and because the nipples are not big enough to milk them" (Flores Ochoa 1975a:10; see also Flores Ochoa 1975b:307). We should add that not only are the nipples small, so also are the udders. Franklin (1982:467) and Novoa and Wheeler (1984:126) have a similar position.

While presenting the same position and insisting on the small size of the nipples, Gade (1977:117), however, notes that the main reason why the milk is not used is that this "simply was not an Andean cultural pattern," a conclusion further supported by a study by this same author, which entailed a long analysis of this point (see Gade 1993). Wheeler (1985a:30) is far more cautious; when she states that camelid milk was never used, she notes that "we do not know the reason for this disdain."

We should note that when discussing camelids in Chile, Palermo (1986–1987:73) cites González de Nájera (1889), who wrote that the Indians used the milk, and points out that "it is usually held that these animals cannot be milked." Camelid milk definitely was not, nor is it, used.

Due to lack of space I cannot go into much detail on the use of other camelid products, particularly those from the llama and the alpaca. I will simply cite Garcilaso de la Vega (1966:Bk. 8, Chap. XVI, 512; 1959:146–147), who provides a good description of the uses given to llama skins. He wrote:

> [Llamas] have long smooth necks, the skin of which the Indians flayed and greased with tallow till it was soft and appeared to have been tanned, when it was used for the soles of their shoes. But as it was not really tanned, they used to take their shoes off to cross streams or when there had been heavy rains, otherwise they became like tripe when they got wet. The Spaniards used the skin for making fine reins for their horses, similar to those brought from Barbary. It serves likewise for bands and cruppers for saddles, and for whips and thongs, and straps for girths and light riding saddles.

10.2.5 Camelids as Beasts of Burden

It is clear that one of the most important uses, if not the most important, for the llama was as a beast of burden, because it was the only means available in the Andes for this role. This topic has been discussed in depth in all periods, yet no systematic study has ever been made, and so the authors do not agree on the size of the load or the distance these animals can cover. This is one of the best examples of how little we know about camelids.

General statements abound in this regard, and for obvious reasons I will only mention the ones I find most significant. Moreover, I will provide specific, detailed data on weight, distance, and a few other important variables.

When discussing the Inca roads, Andagoya (1954:246–247) wrote: "[A]long these walked the caravans of sheep loaded with merchandise from one place to another." And when discussing the great number of sheep present, he notes that "it was an admirable thing, and so great the traffic from the sea inland, and from one province to another, quite loaded with merchandise." Garcilaso de la Vega (1966:Bk. 8, Chap. XVI, 513; 1959:147) is quite explicit in this regard. He wrote: "The llama is also used by Indians and Spaniards for the transport of merchandise to all parts, though they travel best between Cuzco and Potosí where the land is flat. The distance is of nearly two hundred leagues; and they also go between the mines and many other places with all sorts of supplies, Indian garments, Spanish wares, wine, oil, preserves, and everything else that is used in the mines. The chief article they bring to Cuzco is the herb called coca." I find this passage by Garcilaso most interesting, because it shows the llama was still an important animal in the viceroyalty, an aspect that has perhaps not been studied by historians.

It is true that it is not always possible to know whether the chroniclers mean llamas or alpacas when they mention beasts of burden. Murúa (1964:154) is quite emphatic in reporting of the pacos, "they are useless for any kind of load." However, Diez de San Miguel (1964:29) recorded the account given by Martín Cusi, the cacique principal of a moiety in the altiplano, who said, "the Indians in this province have rams of the land and pacos, and go trade *with them* in the yungas" (emphasis added). In fact, as we will see later in the chapter, although it is not much used as a beast of burden, the alpaca does fulfill this role.

Ocaña (1987:198), who traveled across Peru, left an account of the Indians in Collao in which he explains that these men carried the Spaniards' baggage on llamas, and noted that "since they set out an hour earlier, they reach their journey's destination at the same time as the passenger riding a good mule." León Pinelo (1943:53) wrote as follows of the llama: "No animal equals them in climbing and going down a slope loaded, and it looks as if God created them in proportion and suited to the Indians' condition, for they treat and understand them as if they were of the same species," and all this "with no more harness than a rope."

Rowe (1946:219) comments that llamas were used in big caravans to carry cargo, and that "the Indians calculated that eight drivers were needed for every 100 animals."

There is one significant point I want to draw attention to. From the sixteenth century to the present day, practically all authors have agreed on one fact: wild camelids were never used as beasts of burden. Except for some modern scholars, whom I cite later, León Portocarrero (Anónimo Portugués 1958:80 [see Lohmann Villena 1967]) for the seventeenth century and Cosme Bueno (1951:77) and Juan and Ulloa (1988:601) for the eighteenth century are the only ones who note the opposite. In discussing the cargo carried by animals and the distance they covered, León Portocarrero mentioned only the "guanaco and vicuña" rams, while Cosme Bueno wrote: "Many huanacos breed on the mountains that are used as donkeys to carry not too heavy cargo, like said textiles and others" (Cosme Bueno 1951:77). Juan and Ulloa in turn said: "These guanacos are a great service for Peru as well as the llamas, because the minerals are carried from where they are extracted to the mills where they are worked, and this cannot be done with other species of animals due to the craggy and rugged condition of the mountains where such burdensome and hard transportation is carried out, so that only guanacos and llamas may walk on them safely, jumping like deer or goats from one boulder to the other, and neither they nor their cargo are in danger" (1988:601). All three cases are certainly wrong. However, some comments are in order regarding the last. Although the first edition of the *Noticias secretas de América* was published in 1826, it was written around 1747. Both authors, as is well known, were members of the expedition headed by Charles de La Condamine to establish the exact shape of the earth (Gómez-Tabanera 1988: VI, X–XI). It is almost a certainty that on this occasion the authors had neither the opportunity nor the occasion to become acquainted with the Andean world, and they probably used second-hand data. It was only later, in 1758–1763, that one of them, Ulloa, returned to Peru as governor of Huancavelica (Pereyra [1940] 1988:712).

I will try to analyze the available data on the load llamas can carry. As a base, I use the arroba, an old Spanish unit that measured capacity and weight and was used in Peru and Bolivia; most of the chroniclers used it. For comparison with modern writers I convert arrobas into kilograms. For practical reasons the authors are grouped according to the number of arrobas they list. It is worth noting that all conversions of weights and distances are based on Llerena Landa (1957).

The only person who claims that "sheep . . . [carry] a weight of up to two arrobas" is Ruiz de Arce (1968:420). Zárate (1968:176) and Cieza de León (1984:Part I, Chap. CXI, 294) give 2 and 3 arrobas. Cieza implies that this was the ideal measure by noting that "the rams take a weight of two and three arrobas *very well*" (emphasis added).

The writers who claim that llamas carry three and four arrobas are Gómara (1946:243), Las Casas (1948:9), Matienzo (1967:89), Garcilaso de la Vega (1959:Bk., 8, Chap. XVI, 147) and Murúa (1922:154). The relevant passage in Murúa is worth discussing. When he mentions cargo he does not say rams but "pacos." There is no way to establish whether he meant alpacas with this term, but there is an interesting datum in Gómara (1946:243) that no other author confirms. Gómara clearly states that beasts of burden were "known as pacos." This would presumably mean that Murúa meant llamas. However, there is one intriguing source. Schmidel (1986), who is frequently cited in Chapter 9, does not use the word paco at all in his description. However, the caption to one of the illustrations in his account, showing a llama on the foreground and three others in the background, three of which are loaded, reads: "Pacos oder Amida: Ein Indianisch schaff" (see Schmidel 1962:Chap. 44, 72, Fig. 30), or "Paco or Amida. An Indian sheep." This may support the hypothesis that Murúa and Gómara were perhaps right. This topic should be studied.

Lizárraga (1968:Bk. I, Chap. XXXI, 63) is the only writer who gives 3–5 arrobas (he wrote "baskets" but explained that each basket weighed 20–25 pounds, and that each ram carried "four and five, and most five"). Those who agree on a weight of 4 arrobas are Enríquez de Guzmán (1960:139) and Cantos de Andrada et al. (1965: 306; the latter, however, mentions "a quintal," a Spanish quintal being equivalent to 4 arrobas). However, the passage in Enríquez de Guzmán is somewhat unique, insofar as he wrote: "The beasts carry four arrobas in weight. . . . Rams take a bit more." Did he mean both llamas and alpacas? There is no way to find out.

Ramírez (1936:17), Herrera (1945c:313), and Cosme Bueno (1951:125) mention four or five arrobas (however, Herrera indicates something that goes against reason, namely, that the longer the journey, the more the animals carried: "and when it is more than one day's journey they carry eight arrobas"). The only writer who gives a figure ranging between 4 and 6 arrobas is Acosta (1954: 136–137). Ocaña (1987:167) gives 6 arrobas, but adds that in Collao the Indians "[l]oad the passengers' cargo of clothes and chests on rams—which they call apires—because a ram takes between seven and eight arrobas" (Ocaña 1987:198). Cobo (1964:Bk. 9, Chap. LVII, 366) also gives 6–8 arrobas. León Portocarrero gave 7 and 8 (Anónimo Portugués 1958:80–81 [see Lohmann Villena 1967]).

In the nineteenth century Tschudi (1966:243) mentioned about 5 arrobas. I say "about" because he used pounds ("125 pounds," actually) and I do not know whether these were Spanish, English, or German pounds (changes in weight are small either way). I assume he was using the English pound, because he ends by noting the animals are rarely "loaded more than a quintal." A Spanish quintal is equal to 4 arrobas, and 125 English pounds is equal to 4.9 arrobas. This would give 5 Spanish arrobas and 5.4 German ones.

Squier (1974:132; he gives the weight in kilos) and Prescott (2000:806; 1955:114) give a hundred pounds. (In fact, Prescott says "little more than a hundred pounds.")[6]

I now turn to what modern scholars have said about the loads carried by camelids. Almost all of them are given in kilograms, so I will standardize measurements. Troll (1935:138) is almost the only one to defend a small load ranging between 11 and 23 kg. (His figures are in pounds, a unit that can change from country to country. I assume he used Spanish pounds, but the difference would not be too great even it were not so.)

A great number of scholars give loads ranging between 20 and 40 kg. Flores Ochoa (1967: 81) noted that an animal can stand up to 40 kg (although in later studies [1975a:10–11; 1975b: 307]) he said "over 50 kg" and "can carry up to 50 and sometimes 60 kg," but only 25–30 kg on lengthy trips). Valderrama and Escalante (1983: 88, n. 17) and Custred (1974:266, 276) claim that 34.5 kg is the load carried by a llama (the weight was given in arrobas). Novoa and Wheeler (1984: 125) give 25–30 kg (but they based this figure on Flores Ochoa [1977b]). On the basis of his field-work, Masuda (1986:254) writes that one of his informants claimed that a "pack-llama carries 2–3 arrobas," that is, 23–34 kg. Lecoq (1987:4) studied the llameros in Potosí, Bolivia, and mentioned a load ranging between 24 and 25 kg; however, when describing the trip he says that "during the journey each must stand a weight of two or three arrobas (25–38 kg)." Two arrobas actually are 23 kg, and three arrobas are 34.4 kg.

Flannery et al. (1989:75) in turn estimate that a llama's cargo weighs 24–25 kg, and note that it is split into two sacks in Ayacucho, with one sack on each side of the animal. Each sack weighs an arroba—12–13 kg according to the authors (Flannery et al. 1989:106). The arroba actually weighs 11.5 kg (its weight with only minor variation is the same for most countries; it has a higher weight in Brazil [14.6 kg], Colombia [12.5 kg], and Portugal [14.6 kg]). Torres (1992a:32) and Sumar (1988: 25; 1992:89) also give a weight of 25–30 kg (but Sumar based his figure on Flores Ochoa [1977b]).

West (1981a:66) lived with Bolivian llameros and believes that the mean load of an adult male llama is 34.5 kg (West gives the weight in pounds—75—and I assume he means Spanish pounds. If they are instead U.S. pounds the weight is slightly less, 34 kg). He notes that for short distances, an animal can carry up to 46 kg. However, llameros load the animals with a weight ranging between about 23 and 28 kg to avoid wearing them out.

Concha Contreras (1975:70) reports that the loads have 34–46 kg in the Apurímac area (he gives the weight in arrobas), but notes the presence of a weight unit in Aymaraes (also in Apurímac) known as "topo" or "llama carga" that is equal to 46 kg.

Three scholars agree on the weight. Browman (1974a:188) says a llama can carry up to 45 kg and more; Inamura (1981:69) gives 46 kg (he gives the figure in quintales, which I assume is the Spanish unit); and Rowe (1946:239) gives the same weight (but he gives it in pounds, which I assume are Spanish pounds). However, in the same above-cited study Browman (1974a:194) states that the weight can be 25–60 kg on longer distances.

In the *Enciclopedia de los Animales* (1970:305) we read that a llama can carry up to 50 kg. Gilmore (1950:438) in turn merely repeats the data in Maccagno (1932), who gives weights ranging from 10–50 kg to 25–60 kg. Franklin (1982:458) also gives 25–60 kg, but he took this from the above-cited study by Browman (1974a).

Simoens da Silva (1980:66) wrote that a "strong and well-trained llama" can carry a single load of 60 kg, whereas a "middle-sized one barely stands 50 kg." He must have made a mistake, however, because he claims that 60 kg equals 4 arrobas or 1 quintal. The Spanish quintal equals 4 arrobas, but these are 46 kg and not 60. Simoens da Silva clearly made a mistake in the conversions. Whitehouse (1983:286) also believes that a llama can carry up to 60 kg over rugged terrain.

We saw that Cristóbal Campana reports (pers. commun., 19 August 1992; see Chapter 8) that llamas carried 69–80.5 kg in the highlands of Trujillo, and that his father's llama carried around 86.5 kg. The claim may not be in doubt, but this could be an exceptional case. However, the animals that appear in Figure 8.1 to the right of the llama mounted by Campana are of the same size.

Before discussing the data on weights, I will review the data available on the distance llamas can cover with their respective loads. Some mention journeys without giving distances. For instance, Matienzo (1967:89) says the animals "walk slowly, though they walk as much as the mule [because] the distance is small." Enríquez de Guzmán (1960:139) simply notes that "they walk unceremoniously [*muy llano*]," and Ocaña (1987:167) says that rams cover "the leagues that are required."

Other writers give specific figures and, just as with weights, I have given them using the figures given. For Colonial sources I use leagues; for modern ones I use kilometers.

Some have noted that loaded llamas can walk 2–3 leagues. So it is with Murúa (1964:154), Acosta (1954:136), and León Portocarrero (Anónimo Portugués 1958:80 [see Lohmann Villena 1967]). However it is worth citing Acosta: "[A]nd if it is a long journey they do not walk more than two or three leagues in one journey ["*a lo largo*"]. The rams (which is what they call these caravans) know their stops where there are pastures and water; here they unload and build their tents and make fire and prepare food, and do not have a bad time, even though this is a most phlegmatic way of walking. A ram of these easily carries eight arrobas or more when it is no more than just one journey, and walks with its load a full day's journey of eight or ten leagues, as the needy soldiers who wander over Peru used to do" (Acosta 1954: 136–137). León Portocarrero points out that pack animals "can take any route, no matter how bad and rugged it is, and cross rivers no matter how heavy their load is . . . and there are some of these rams who walk ten leagues a day over flat land with [a load of] eight arrobas" (Anónimo Portugués 1958:80–81).

Other chroniclers quote a distance of 3–4 leagues, among them Ramírez (1936:17), who adds further that the animals do so "without any packsaddles nor iron shoes [and] every day walk . . . [along] long and rugged roads." Father Cobo (1964a:Bk. 9, Chap. LVII, 366) wrote: "These pack-rams are so tame that a ten- or twelve-year-old boy can easily lead a drove of them . . . and in a long journey covering several days they walk three or four leagues each day; however, if they walk no more than one day they cover their entire journey like a mule." Prescott (2000:806; 1955:114) also gives the same distance but sets it as a maximum: "cannot travel above three or four leagues in a day."

Garcilaso de la Vega (1966:Bk. 8, Chap. XVI, 513; 1959:147) explains that "they can cover three leagues in a day, for they are not capable of heavy work." Herrera (1945c:V Decade, Bk. 4, Chap. IX, 313) in turn wrote that "they do not walk more than four leagues every day, and carry eight arrobas and walk eight leagues when it is more than a day's journey."

The data given by contemporary scholars are in greater agreement. The *Enciclopedia de los Animales* (1970:305) says that a loaded llama walks 10–20 km a day. However, most authors agree in pointing out that the distance is 15–20 km a day. This is the opinion held by Rowe (1946:219), but he considers that it is "a long journey, but over shorter distances they can keep up with a mule."

Flores Ochoa (1967:81) likewise gives a daily distance of 15–20 km, but specifies that it is covered "in journeys that last about ten hours, for periods of up to thirty consecutive days." Then he adds something very important: "It is said that they can go up to five days without food." Browman (1974a:194) states that a distance of 15–20 km is a long journey, but notes that a llama can cover more than 30 km a day in journeys lasting a week or less. He makes a statement similar to the one by Flores Ochoa, in that llamas can go up to five days without food and up to three without water in special circumstances.

Novoa and Wheeler (1984:125) quote the same distance, but they took their data from Flores Ochoa (1977b). Sumar did the same (1988:25).

On the basis of their experience in Ayacucho, Flannery et al. (1989:115) state that covering 15–20 km a day, llamas and their drivers can take up to two or three months to make a complete journey between the highlands and the coast. Franklin (1982:458) claims that a loaded llama can cover a daily distance of 15–30 km, and adds that this is so in rugged terrain, and that castrated animals 3–4 years old or more are usually used. Concha Contreras (1975:87) gives a distance of 25 km and explains that the long journeys made by herders are divided into many days' march which are delimited in terms of distance. In the terminology used by the llameros, these "journeys"—also known as *jornara*—correspond to the distance covered every day. The mean daily march lasts 8–9 hours, and little is eaten during that time, sometimes eating only once a day. In Inamura's experience (1981:69) in the Department of Arequipa, the llameros cover 20–30 km every day. A distance of 25–30 km is likewise noted by Torres (1992a:32).

Gilmore (1950:439) gives a distance of 10–30 km a day, but specifies that Maccagno (1932) wrote that it is 30–35 km a day. Like other scholars I have already mentioned (see above), Gilmore also points out that the animals can walk up to 20 consecutive days and stand up to five days without food, and wonders whether they can go without water too. He believes that the tolerance to thirst is unknown but high. Gilmore cites a study by Romero (1927) I was unable to find, where it is claimed that it is an established fact that a llama can spend three or four days without any water. There apparently is no indication of whether this is while working or not. Gilmore (1950:439) adds that "water probably is derived, during forced abstinence, from carbohydrates and subcutaneous fat, not from a supply in the so-called 'water-cells' of the rumen."

Simoens da Silva (1980:65) is the only author who gives a longer distance. Although he says 21 km is a normal distance, he notes that an unloaded animal may well cover up to 60 km. However, we should note that there is a problem with his conversions, as we pointed out for the weight of loads (see above). Simoens da Silva writes that 21 km equals 3 leagues, which is incorrect: it equals 3.7 common leagues and 5 Castilian leagues. And 60 km is not 10 leagues, as he claims, for 10 common leagues equals 55.7 km, and 10 Castilian leagues equals 41.9 km.

Although a distance of 60 km may seem an exaggeration, we should not forget the indisputable testimony of Cristóbal Campana (pers. commun., 19 August 1992) cited in Chapter 8, which gives the same distance for the caravans that walked across the highlands of La Libertad.

The following section attempts to summarize all of the data in regard to the load these animals can carry and the distance they cover. (Dedenbach Salazar [1990:168] also prepared a summary, but she based her work only on seven chroniclers; furthermore, her citation of Ocaña is wrong.)

First of all it should be noted that although it is true that not many data are available in the chronicles, there are some. So we cannot say, as Murra (1978:88) did, that the chroniclers "[d]o not give us much data either on the distances traveled, the size of the droves, or the nature of the load." Nor can we generalize and say that "[t]he chroniclers estimate the load as 3 or 4 arrobas, and in three leagues the mean distance" (Murra 1978:87). We saw it is not so.

For comparative purposes the units used must clearly be standardized, so all weights in arrobas will be converted into kilograms, and the distances given in leagues converted into kilometers. Since I do not know whether the writers used Castilian or common leagues, I will give both figures. Readers should bear in mind that 1 arroba is equal to 11.5 kg. One common league, an old Spanish itinerary unit that may have been used by the chroniclers, equals 5,572.699 m, while the Castilian league equals 4,190 m.

On studying the data in sources from the sixteenth to the nineteenth centuries, we find that most (five) sources point out that the load a llama carried was 34–46 kg (3–4 in arrobas); second (four) come those sources that give 46 kg (4 arrobas); then (three) those that give 46–58 kg (4–5 arrobas); and fourth (two) the writers who say the load ranged between 23 and 34 kg (2–3 arrobas). Then we have individual statements giving the load as 23 kg (2 arrobas), 46 kg (4 arrobas), 58 kg (5 arrobas), 46–69 kg (4–6 arrobas), 69 kg (6 arrobas), 69–92 kg (6-8 arrobas), 80 kg (7 arrobas), and 92 kg (8 arrobas).

If to those who give an estimate ranging between two figures we add those who only gave one, which fell between the first two (e.g., 2–3 + 2), we find that most (nine sources) claim the load carried by a llama ranged between 34 and 46 kg, followed (three in each case) by those who give weights of 23–34 kg, 46–58 kg, and 69–92 kg. Then come the other two groups, who give the weight as 34–58 kg and 46–69 kg.

There is more agreement among twentieth-century writers. Most (seven) indicate that the weight ranged between 24 and 34 kg, then come two groups (of three each) who claim that it ranged between 45 and 46 kg, and 25 and 60 kg. Others (two sources) hold that the weight varied between 50 and 60 kg, and finally we have individual claims that it ranged between 11 and 23 kg, 34 and 46 kg, 50 kg, and finally 69 and 80.5 kg.

We see there is no agreement between contemporary scholars and those from the nineteenth and previous centuries. Today it is claimed that a llama can carry less than previously noted. Contemporary data come from scholars such as Flores Ochoa who are certainly well acquainted with this topic and have firsthand experience, so their assertions should not be in doubt. On the other hand, it is somewhat difficult to believe that all the chroniclers and later travelers made mistaken appraisals and did not consult people in the know. It is almost impossible to make a definitive appraisal, insofar as there is a relative factor. It is a fact that an animal can carry more in a short trip than in a long one. We are not always told in what situations the data were collected. Second, we should bear in mind that llamas are relatively small animals on which it is easier to load a small load than a large one. So the constraint in low-volume materials or goods is the bulk itself rather than the weight. Finally, we should not rule out a possibility that is ever more clear, namely, that there were stronger animals before and immediately after the Spanish conquest than there are at present, and that the weakening of the species (and its various breeds) was caused by the indiscriminate killing, the loss of control over the herds with the attendant deterioration of the genetic pool, and the absence of any protection for these animals throughout the Colonial and Republican periods.

I would like to point out two examples that cannot be questioned: the big llamas Cristóbal Campana mentions (pers. commun., 19 August 1992; see Chapter 8), which can also be seen in Figure 8.1, and the description given by Schmidel (1986:Chap. 42, 79), which clearly says that llamas "are big beasts, almost like mules." However, it is worth noting that although the figures for the weights do not coincide, it is generally agreed that it is a relatively low one, ranging between 23 and 46 kg. Unfortunately, nothing can be said on this point for pre-Columbian times. It should be pointed out that the Mochica and the Chimu had the custom of making the llama carry an individual besides the load (Figure 4.17, 4.18, 4.19, and 4.40)—a weight that is not irrelevant and that certainly was more than 50 kg. This detail has not been considered, nor has it been properly studied, but it would prove that the chroniclers were not mistaken.

The problem with regard to distances is similar to that of weight. The data given by the chroniclers fall into two major groups (each with three writers)—one that claims that the animals could cover 2–3 leagues loaded, and the other 3–4 leagues. If these are common leagues, then the distance in the first case would be 11–16.7 km and 8.3–12.5 km if Castilian leagues. In the second case we have 17.7–22.2 km if common leagues and 12.5–16.7 km if Castilian. All the other data come from individual claims: 3 leagues (i.e., 16.7 km in common leagues and 12.5 km in Castilian leagues), 4 leagues (22.2 km and 16.7 km), 10 leagues (55.7 km in common leagues and 41.9 km in Castilian ones), and 8–10 leagues (i.e., 44.5–55.7 km in common league, and 33.5–41.9 km in Castilian leagues). If similar individual figures are added to those authors giving a range of two, then the first two groups in the list come at the top. In this case modern scholars tend to agree (seven cases) in giving a distance of 15–20 km; all the rest are individual claims that respectively give distances of 10–20 km, 15–30 km, 21 km, 25 km, 20–30 km, 10–30 km, and 30–35 km.

Once again, modern scholars and the sixteenth- and seventeenth-century chroniclers do not agree. However, the disagreement is not as great as in the previous case. We saw that the chroniclers who head the list fall into two groups. There is a marked difference between the distance given by the first group and modern scholars, but this does not happen with the second group, where there is some agreement, give or take a few kilometers. The chroniclers in the first group give far shorter distances than modern scholars, but there is some agreement with the second group. What must be pointed out is that all distances mentioned by modern scholars do not deviate significantly from the ones most commonly given, for 30 km is the longest distance cited. However, in the chroniclers it is not so; they give 50 km as the maximum distance.

It is hard to tell what accounts for this difference. If there actually was a variety, or varieties, of llamas bigger than the contemporary ones, as several lines of evidence seem to indicate, then it would be logical for their resistance to fatigue to be higher too—i.e., they could cover larger distances. Another point in the chroniclers' favor is that they had to estimate distances in a most rugged and broken terrain which was unknown to them, with irregular landforms where it is not easy estimating distances. I speak from my own experience. Besides, they were unable to fully comprehend the concept of space and distance the Indians had, quite different values from Western ones.

Tschudi, however, made a meticulous study of this problem—save for a few mistakes, which I will point out—knowing full well what he said and with quite accurate observations. It is worth including here the paragraph in question, because as far as I am aware, the only modern scholars who pointed out something similar were Flannery et al. (1989). Tschudi wrote:

> The capacity of the llama as a pack animal has been estimated in various ways by some more or less accurate observers, both in regard to the weight of its load, as well as the distance covered in a day. The load is thus estimated at two to eight arrobas (25–100 kg) and the distance at two to ten leagues (11–55 km). Acosta (l. c bk. iv, chap. 41) gives an exorbitant figure, saying that a llama with a load of two quintales (100 kg [Note: These are centner or zentner, i.e., a German quintal, which equals 50 kg]) covered a distance of ten leagues (55 km) in one day when the journey only took one day. This statement is not credible, because once a llama receives 100 kg in weight—something surely no Indian ever tried, not even as an experiment—it lies down and there is no way to make it get up before the load is removed from it. There is no animal I am more acquainted with in terms of its work capacity than the llama. The claim Acosta makes of ten leagues (55 km) per day [sic] is equally mistaken. Mossbach himself exaggerates when he says (South American Highlands, in Ausland, No. 13, p. 299 [*Südamerikanische Stufenländer.* Ausland 1871, No. 13, p. 299]) that a llama covers about four German miles per day.

During my stay in Peru I dedicated myself to this point with great interest, inquiring among the herders, both Peruvian and Aymarás, obtaining truthful data and a conclusive statement that they as a set rule never loaded their llamas save, at most, with four arrobas (50 kilos) [when] making journeys of three to four leagues a day at most (17–22 km), and that many of the animals did not withstand the trip even with this load if it was a long one. (Tschudi 1885:106; 1891:106; 1918:225–226; 1969:136) [Note: The 1969 edition has edited the original text and omits a large part of this description; the 1918 edition was used here; in comparing the 1918 edition with the original one, the former was found to be complete.]

Tschudi made a mistake in his citation of Acosta. This chronicler says "eight arrobas" and "eight or ten leagues" (Acosta 1954:137). Eight arrobas equals 92 kg, and so is not the two (German) quintales noted in the text by the Swiss scholar. And 10 leagues equals 55.7 km if we take common leagues, and here Tschudi is right. However, he makes a mistake when he contradicts Mossbach, for a German league is equal to 1.852 km, so 4 German leagues is 7.408 km, a distance that a llama easily covers, as Tschudi himself notes. I must likewise point out that 4 arrobas is 46 (not 50 kg), and that 3 common leagues (apparently the ones used by Tschudi) is 12.5 and not 17.5 km; finally, 4 common leagues is 16.7 km, not the 22 km given by Tschudi. Even so, it is worth noting that his data are closer to the chroniclers than to modern scholars regarding the weight. Likewise with the distance he gives. Tschudi wrote in the late nineteenth century. One may well wonder whether an actual change took place since then in these animals, or whether Tschudi made better inquiries than later scholars.

In this regard—and after reading Tschudi—it is worth going over what Flannery et al. (1989) wrote based on their fieldwork in Ayacucho. They begin by pointing out that some writers (they cite Tschopik 1946:533 and Flores Ochoa 1979[b]:95 [also 1967:81]) claim the maximum weight a llama can carry is 40–45 km. They then say:

"Based on our experience in Ayacucho, we regard this as a prodigious burden; in fact, Flores Ochoa says that on 'long trips' of thirty days or so llamas carry only 25–30 kg. Although we do not rule out the possibility that a big, strong capon llama could carry 45 kg, we suspect that 25–30 kg was a more common load, especially on steep slopes. Drivers simply could not risk overloading their llamas" (Flannery et al. 1989:115). In support of this they cite Garcilaso de la Vega (1966:513 [1959:Bk. 8, Chap. XVI, 147]), who said, "To avoid tiring them, each flock has forty or fifty unladen animals, and whenever any beast is found to be flagging its burden is at once removed and transferred to another, before it lies down; for once it does this, there is no solution but to have it killed."

Flannery et al. (1989:115–116) claim they find many sources for the confusion present in this regard in the accounts given by sixteenth-century travelers. First, many of the sources refer to trips taken on relatively flat land, like the plains around Lake Titicaca. They once again cite Garcilaso (1966:513 [1959:Bk. 8, Chap. XVI, 147]): "[Llamas] travel best between Cuzco and Potosí where the land is flat." They then add that the load a llama can carry on a plain is higher than what it can carry in the rugged expanse between Ayacucho and Pisco.

Second, Flannery et al. point out that there is a problem with the popular Spanish weight of the arroba. They wonder what size the arroba mentioned by Garcilaso de la Vega (1966:513 [1959: Bk. 8, Chap. XVI, 147]) was. They believe his informants used the 12.5 kg arroba, but add that this is not certain.

I concur with Flannery et al., but am still convinced that it is possible that stronger animals existed in past times. We should not forget the opinion of Cristóbal Campana, cited in Chapter 8. As regards the arroba, this is an ancient Spanish unit of volume and weight that was, and still is, used in Peru and in Bolivia. It equals 11.5 kg and is a fourth part of the Spanish quintal. Although it is true that this unit varies in South America from one country to the other, save for Brazil and Colombia, the differences are minimal (see Llerena Landa 1957:20–21). This ancient unit has survived with minor alterations, because the 12.5 kg

of the arroba Flannery et al. (1989:115–116) found among the natives of Ayacucho is just 1 kg less than the 11.5 one just mentioned. Strangely enough, the 12.5 kg equivalence is also used in Colombia.

Aside from the aforementioned problems regarding the weight and distance the animals can cover, the observations were carefully made and are correct except in a few cases. I would like to point out the most important ones. First, these animals "walk slowly . . . (Matienzo 1967:89)"; they do so "with no more harness than a rope" (León Pinelo 1943:53) and "without any packsaddles nor iron shoes" (Ramírez 1936:17). Second, they can take "any route, no matter how bad or rugged it is" (León Portocarrero [Anónimo Portugués 1958:80]), because "no animal equals them and it looks as if God created them in proportion and suited to the Indians' condition, for they treat and understand them as if they were of the same species" (León Pinelo 1943:53). Finally, that when these animals " get angry and mad [*abirran*] with their cargo, they lie down with it and there is no way of making them get up; they will rather let themselves be killed than move when they get mad" (Acosta 1954:137). However, "[t]hese pack rams are so tame that a ten- or twelve-year-old boy can easily lead a drove of them" (Cobo 1964a: Bk. 9, Chap. LVII, 366).

Perhaps the two biggest mistakes are those made by León Portocarrero (Anónimo Portugués 1958:80), who claimed it was the "guanacos and vicuñas" who carried loads, and Ocaña (1987:167) when he claimed that "each ram" can walk with its load "as many leagues as are needed."

More data on pack camelids are in order. First, castrated males, as was already noted, become *capon llamas*, or animals essentially used to carry cargo. These *capones* are usually stockier than the *aynos*, or noncastrated adult males, but they fear them because the latter are more aggressive. In Ayacucho a herd of 25–40 llamas rarely has more than one or two *aynos*, which are usually of different ages so that the younger one does not annoy the eldest until the latter is no longer the leader. Herders attain two goals with this strategy. They manage to create a single herd of males and females, and they eliminate the competition typical of wild guanacos, thus ensuring that males that breed the offspring are chosen by the herder and not by natural selection. This has the advantage of letting the herders choose the least aggressive males, which will not kick out their own yearlings (Flannery et al. 1989:97).

So, not all llamas are used for transportation. Females are excluded, not just because of their very nature but because they must be looked after for reproduction. Two-year-old llamas are also left aside, and the *aynos* usually are too. However, these can be included when animals are lacking. Caravans of llamas are mainly formed by gelded males or *capon* llamas, accompanied by male *maltas* 2–4 years of age, which follow the caravan to be used on the road (Flannery et al. 1989:106). Inamura (1981:69) also verified that males begin their training when they are two years old, and are castrated once they are three, so they can begin traveling with the cargo. According to Custred (1974:276), the weight loaded on them must be gradually increased once their training begins. This starts when the animals are 2–3.5 years old with loads weighing two arrobas (23 kg), and when they are four or five years these are increased to three arrobas (34.5 kg).

According to experts, the llamas used in the caravans have most specific characteristics. A llama is tame, docile, and compliant once it gets used to travel. The llameros usually calls it *orqo-llama*, or male llama. As noted in previous paragraphs, pack llamas must be castrated males so that they are strong and disciplined. Were it not so, they would stay close to the females and would not be strong enough for the trip. The animals are castrated when they are two years old, the age at which they can likewise be tamed. It seems that the technique used for this is quite simple, as it consists only in taking the animals on an extended journey along with others that are already used to it, so they gradually get used to and familiar with the group of old llamas. Some cargo may be loaded on them on the way—or at least a poncho—so they get accustomed to carrying something on their back. Pack llamas have a mode of organizing and discipline that is not found in other beasts of burden, like horses or donkeys (Concha Contreras 1975:84–85). (Readers interested in more details on the training of pack-llamas should read Flores Ochoa 1967:81.)

Tschudi (1885:98; 1891:100; 1918:212; 1969:129) makes an interesting observation. He points out that we do not know whether male llamas were castrated before the conquest, as the chroniclers say nothing about this. We do know it was done afterward. As far as I am aware, there is no sign of castration in prehispanic depictions.

According to the data I was able to collect, the animals are considered useful for transportation once they are three years old, as was already noted (see above), but there is no agreement on the age limit. For some it goes up to 7 years, for others up to 10–12. Then they are killed for their meat and leather (Rowe 1946:219; Flores Ochoa 1975b:308; Dedenbach Salazar 1990:167).

One important characteristic of llamas is that these animals browse, finding their food on the way, and are not selective, eating the grass they find (Rowe 1946:239; Browman 1974a:188).

The llama was the sole means of transportation in prehispanic times, but it also had a major role—as shown in Chapter 7—in Colonial times. It was clearly displaced by the mule, which carries three times as much weight (Usselmann 1987:131, based on Manrique 1985). Moreover, the llama's personality does not agree with that of Europeans. However, it still has a significant role in transporting goods, agricultural produce in particular, in several regions of Peru. They are of course losing importance at an ever-faster rate with the development of highways and modern means of communication (Flores Ochoa 1975a:11; 1975b:307–308).

The llama is almost the only animal mentioned when Andean pack animals are listed. Gilmore (1950:446) writes that the alpaca is not used as a pack animal, and Gade (1969:342; 1977:118–119; 1993:3) is of the same opinion. This is wrong, and Flores Ochoa (1977b:16) has criticized Gade for it. According to Flores Ochoa, alpacas actually can also be used as beasts of burden, but on a small scale and for short distances. For this they must have been previously trained, just like the llamas (Flores Ochoa 1975a:11). In another study, Flores Ochoa (1975b:307–308) notes that in this case males are also preferred, but females can be used too. Training begins at two years of age, and an alpaca is fully mature when it is four. Like the llama it can be used until it is

10–12 years old, then it is killed and its flesh and skin used. Torres (1992a:32) also agrees as regards the use of the alpaca as a best of burden, but admits that this is "[r]arely done."

One of the major differences between camelids and other domestic animals is that they are led completely loose, without any leather straps, rein, or whip. Besides, a caravan is usually easily led by a single herder. According to some scholars, a llamero can take 15–20 animals, others say 20–30, and some even claim there are cases where larger groups of 40 or more are driven. Things are different with herds, where a single herdsman can look after 500 animals. If the number is greater than this figure, the herd can be split into complementary herds (Simoens da Silva 1980:65; Franklin 1982:468; Flores Ochoa 1967: 81; 1975b:302. Readers interested in a good description of the management of a herd of alpacas and llamas should read Flores Ochoa 1967).

There is one error to correct: the use of wild camelids as beasts of burden. In this same chapter we have seen that there are four writers who make this claim: León Portocarrero (Anónimo Portugués 1958:80 [see Lohmann Villena 1967]) in the seventeenth century, and Cosme Bueno (1951:77) and Juan and Ulloa (1988:601) in the eighteenth century. Some contemporary writers make the same claim. Herre and Röhrs (1977:253) write: "As an exception, the guanaco was possibly used as pack animal in ancient times, and was probably the first as such." They do not present any evidence or proof to sustain their claim. But Novoa and Wheeler (1984:121) also claim "[t]he Incas utilized guanacos as pack animals." Here the origin of this claim, or the reasons that led the authors to it, are likewise not indicated. I believe it is a mistake and that there is nothing that can substantiate it. I was unable to find any evidence of it. Since I am no specialist in historical sources I discussed the point with Franklin Pease, who pointed out that he had never found any reference to this, either in the chronicles or in any other source (Franklin Pease, pers. commun., 26 May 1992).

10.2.6. Were Llamas Used for Riding?

One subject that has to be cleared up is whether or not llamas may have served as riding animals.

There are contradictory positions about this. However, we cannot take a categorically negative position, as Gade (1969:342; 1977:117; 1993:3) did without having studied this point.

We saw that the only evidence of such use in pre-Columbian times is for the north coast; however, I was unable to make an exhaustive study in this regard, and so the possibility remains open that this custom was more widespread. On the north coast this was a tradition that began with the Mochica as far as I was able to establish, continued in the Lambayeque culture, and persisted at least into the Chimu period (Figures 4.20, 4.33, and 4.41). It must be pointed out that among the Mochica it was not just the invalids (Figure 4.21) who rode these animals, as has been repeatedly claimed (e.g., Larco Hoyle 1945:13; 1946:166). This was more the exception than the rule, at least to judge by ceramic depictions (see Chapter 4). The subject deserves more research. But let us see what the historical sources say in this regard.

When describing the pack rams, Gómara (1946:243) clearly says, "and they even take men on them." It is true that this clergyman was never in Peru, as was already noted, but it is also worth recalling that Porras (1986:191) pointed out that Gómara "collected data and news which do not appear in other chroniclers." Porras was furthermore convinced he "had good and direct informants." Gómara corroborates the evidence left by the Mochica.

Cieza de León narrates the clash between the Ecuadorian Otabalo and Carangue over the treasure of the Inca, which was mentioned in Chapter 9 while reviewing the data on Ecuador. The chronicler tells how Otabalo ordered those of his people who "were the most disposed and swift . . . should get on the biggest of their rams" to fool the Carangue, making them believe they were Spaniards (Cieza de León 1984:Part I, Chap. XXXIX, 216).

Las Casas (1948:9) also said of the llamas that "[on] other occasions men walk on top of them." And when recounting Almagro's expedition to Chile, Zárate (1968:Bk. III, Chap. II, 176) discusses the sheep of the land and says: "[T]he Spaniards have likewise imposed on them that they take a man riding four and five leagues in a day, and when they are tired and lie down there is no way to get them up even if they are hurt or helped, save if their load is removed; and when they take a rider, if they get tired and are compelled to walk they turn their head round and spray [the rider] with a foul-smelling substance which seems to be what they have in the belly." Again, we do not know what leagues the chronicler used. If these were common leagues, the distance the Spaniards covered riding on llama-back was between 22 and 27–28 km; if Zárate used Castilian leagues it was 17–21 km. It should be recalled that this was one of the distances given by the majority of the chroniclers (see above). The objectivity and impartiality of the data provided by the accountant—which I believe are significant—were praised by Porras (1986:215), who even claimed that "his method of documentation, which is at times far too faithful, makes him seem, not a chronicler . . . but a professional historian."

Araníbar (1963:106, n. 2) points out that care must be had with the work of Morúa (or Murúa). However, this chronicler was well acquainted with the Andes and even was in Arequipa in 1600 while writing the last part of his work (Porras 1986:477). Interestingly enough he mentions "pacos" while talking of Collao, and explains these could carry a load—I have already discussed this—and then adds: "[T]hey even accept men on them, but they go very slowly and if they get tired they turn their face towards whoever is riding them and spit a noisome water; and if they get very tired they let themselves down until their weight is removed, even if they kill them" (Murúa 1922:154). This is significant, because the description given by Murúa regarding the behavior of the animals agrees with that of Zárate. The only one point that seems striking is that there seems to be a contradiction in the work of the Mercedarian friar. When describing the paco he says: "[I]t is useless for any kind of load" (Murúa 1964:154), and yet now he claims they "even accept men on them" (Múrua 1922:154). We may also wonder whether Gómara (1946:243) was right when he said that pack-rams were "called pacos." Is it that in one case he meant alpacas, and in the other pack animals? This point remains open.

It is revealing that although he denies it, Father Cobo (1964a:Bk. 9, Chap. LVII, 366) does not rule out the possibility that llamas did have that function. He wrote: "Although they can carry a person, they were never used as mount by the Indians, and even less so now that there are many horses, donkeys and mules." However, it should not be forgotten that Cobo wrote in the early seventeenth century, when there were many European animals in Peru and the Spaniards no longer needed the llamas for their business. Even so, Cobo (1964a) himself left a description of how the Spaniards made the Indians ride llamas when "the corregidores and justices" wanted to flog and shame them. Interestingly enough, this observation made by Cobo was illustrated by Guaman Poma de Ayala in the early part of this same century. In one of his drawings we see the "Corregidor de Minas" ordering that an Indian be punished. The Indian, naked and with his hands tied, rides a llama while another Indian whips him. In his description, when describing the punishments in his "Chapter on Miners," the Indian chronicler noted: "he whips the rest atop a ram" (Guaman Poma de Ayala 1936:fols. 525, 526; my Figure 10.1).

FIGURE 10.1. In this drawing Guaman Poma shows a "Spanish mining magistrate. He cruelly punishes the headmen." The text explains, among other things, that [the magistrate] hangs the headman from the feet, and whips the rest while they ride a ram." *After Guaman Poma de Ayala 1936:fol. 525).*

Tschudi (1885:109; 1891:109; 1918:229–231; 1969:138) discussed these passages from the chronicles. (Note: Besides being a poor translation, the 1969 edition omits part of the original text.) He claimed that "much nonsense" had been said regarding the use of the llama as a riding animal. "The Indians never dreamed something like this." He then says the data Philipp von Hutten gives about the Alto Orinoco expedition regarding "cavalrymen seen in the Omaguas, riding llamas" are "pure fantasy." It seems, still according to Tschudi, that the natives of Río Meta told Diego de Ordaz there was a powerful, one-eyed prince who ruled in the plateau of Nueva Granada, who apparently used animals smaller than deer as mounts (in a footnote on his page 147, Tschudi indicates that it is hard to tell what kind of deer was used for the comparison, whether real deer, the gamo or the South American deer).

Tschudi then discussed the chroniclers: "Agustín de Zárate (l.c., Bk. iii, Chap. ii) tells that in the expedition of Diego de Almagro the Spaniards rode llamas—which actually had no other purpose than carrying the water needed by the soldiers—and that they thus advanced four–five leagues a day (148). López de Gómara says something similar in his *History of the Inka* [sic], Chapter 142." He then notes that it may well be that "a few Spaniards, or perhaps several of them" rode some exceptionally strong animals after losing their horses, and were thus able to walk "25 kilometers" in one day. However, he immediately adds that "it is simply unacceptable that they were able to do this for several consecutive days because armed as they were with weapons and armor, the least they could weigh on average was six arrobas (the two original German editions indicate '75 kg' inside a parenthesis which does not appear in the 1918 translation used here)." Tschudi adds that it is claimed that Spanish common soldiers rode llamas, but that there are no specific data on this.

He then says: "Llamas are absolutely useless as mounts due to their frailness. The brutal Spanish soldiery distinguished itself in its barbarity to man and animals, and destroyed thousands of llamas from overwork." He then cites the account Cieza gave of the clash between the Otabalo and the Carangue which I have already cited (see

above), and notes that "[t]he responsibility for this account falls on the chronicler cited, who accepts it. Even so, it is striking that the Indians only realized they could use the llamas as mount after seeing the Spaniards riding horses, which proves it had never happened before." Tschudi then notes that llamas were not used as mounts later, "it being nothing but an amusement for a young Indian boy every now and then. Finally, I must add that llamas usually are quite averse to these tests."

In another study, Tschudi (1966:244) insists on this point. "The accounts of earlier travelers as regards the use of these animals as mounts . . . are false. Sometimes an Indian boy who has to cross a river and does not want to get wet may sit on a llama, but he gets down as soon as he reaches the other bank."

In this case some comments are in order. The accounts of the Upper Orinoco are clearly an invention (but I admit I do not know the original source of Tschudi's data). As for Zárate (1968:Bk. III, Chap. II, 176), his account is quite clear, for he actually says that on Almagro's expedition to Chile—as we already saw while discussing camelids in Chile (see Chapter 9)—"each live sheep carried the skin of another dead one with water," and then adds, "the Spaniards have likewise imposed on them that they take a man riding." In the case of Gómara (1946:243) we saw that he is quite brief (see above) but precise: "[T]hey even take men on them." Besides, Tschudi makes a mistake when he says that Gómara wrote a "History of the Inka"; he actually wrote the *General History of the Indies*.

I think that Tschudi, just like many other scholars, exaggerates when he mentions Spanish soldiers with "weapons and armor," and simply ignores what actually happened. His vision is distorted, for he is thinking of regular European armies of later times, thus forgetting some essential facts. First, the use of full body armor was decreasing when America was discovered, owing to the introduction of firearms. Second, there were no regular armies in America, which were something new even in Europe. Every man who joined the American adventure did so with the arms at his disposal. This is why it is difficult and even impossible to describe the weapons of the conquistadors in general. Armor and even the corselet had several inconveniences, particularly in a climate like that of South America, and moreover were not that necessary as defense against the types of weapons the Indians had (see Salas 1950: 238 ff.). "Steel armor was useless and even excessive as defense against native weapons due to the inconveniences it caused and the care it required, since all that was needed were the escaupil [a padded garment used against arrows], the coat of mail or the deerskin jerkin, cheaper, easier to put on, more long-lasting and efficient against arrows" (Salas 1950:242–243). It can thus be assumed the soldiers on an expedition like that of Almagro to Chile were not exactly wearing the "armor" Tschudi had in mind. Even so, we shall see later on that it is not strictly true that the llama—or at least some of them—cannot carry the weight given by Tschudi, the "six arrobas"— which, by the way, is 69 kg, not 75 kg as claimed.

As for Cieza de León and his account of the Otabalo and the Carangue, it is not altogether honest to conclusively claim that "the responsibility for this account falls on the chronicler cited, who accepts it." In my opinion Cieza de León is one of the most reliable chroniclers. It was not for nothing that so renowned a scholar as Porras (1986:282) qualified his work as "reliable and truthful." In the case under discussion, Cieza thrice indicates on just one page (1984:Part I, Chap. XXXIX, 126) "they say," "as the people say," and "as is said"; in other words, he was just noting what his informants had reported. I cannot see where he is exaggerating or why what he recounted was implausible, all the more so in light of the other data presented here.

There is one last interesting point mentioned by Tschudi in a footnote (Tschudi 1885:109, n. 2; 1891:108, n. 6; 1918:230, n. 148; this footnote was omitted in the 1969 edition; I cite the 1918 edition). The footnote reads thus: "The league has had several measurements in different periods. First it measured 4000 Castilian varas, i.e. 3345 meters; then it was 5000 varas—4975 meters [both German editions read 4175], and as of 1801 6666 2/3 varas—5572 meters. The official distances recorded in 1845 under President D. Ramón Castilla, whose immediate goal was to standardize the postal service, were calculated at twenty thousand (20,000) feet per league, i.e. 5572 meters." This is the unit used here for the common league (and I noted above that Tschudi used it).

Schmidel (1986), who was widely cited in Chapter 9, is a most important source that no one has cited or discussed. At the risk of being repetitive, I want to recall that Schmidel (1986:Chap. 44, 86) explicitly noted that "[t]he sheep . . . of which there are two kinds, the domestic and the wild ones, *are used* as a beast of burden and *as mount*" (emphasis added). Even more interesting is the invaluable testimony he left of his own experience: "I myself went more than forty leagues on the back of one of these sheep because I was hurt in one leg" (Schmidel 1986:Chap. 44, 86).

A careful reading of the text lets me deduce (see Chapter 9) that the journey was about 22 km per day, and the German chronicler rode a llama for more than 10 days. As we saw in this chapter and in Chapter 7, both the daily journey and the distance agree with the data reviewed. They are fully plausible figures.

But there is more. In his account Schmidel (1962:Chap. 44, 72) included a drawing showing a llama in the foreground led by what seems to be a warrior, for he has a bow in his hand (see Figure 9.8). Behind this animal we see a llama mounted by one individual, and two more llamas behind them. One of the latter has a load tied to what seems to be netting, and the other beside it is mounted by a man. The animals have no kind of harness and are ridden bareback. It is worth noting that the animals have a collar on the neck with which they are led and even handled when mounted. It could be objected that the collars seem to be of metal, which would clearly be a mistake. But on reading Schmidel we saw that the Indians used camelid leather to make several artifacts, and that the group of Spaniards the chronicler was with even used it to make shields with which to protect themselves from the spears and arrows of the Indians. So the collars may well have been leather.

It might be strange that the animals are depicted with scant hair. But if this drawing is compared with the one in Cieza de León (1984:Part I, between pp. 264 and 265, which corresponds to Chapter CXI; my Figure 2.7), we find an amazing resemblance. In this regard it is worth noting that the *Vocabulario de la Lengua General de todo el*

Peru llamada Lengua Qquichua o del Inca, by Gonçález Holguin (1952:270–271), specifies two terms for pack llamas: "*Llama huacauya*, or *huacauia*. The big, wooly pack ram" and "*Llama chunca*. The big, plain pack-ram with no wool." Clearly, it is the latter that Schmidel rode and drew, and that appears in the chronicle of Cieza de León. It should be recalled that Schmidel himself made the original drawings or sketches illustrating his work, which were then copied by some artist (Plischke 1962:XVIII) (see Figure 10.2). And as

Peter Dressendörfer, who is well acquainted with his work, tells me, Schmidel is not a chronicler given to fantasy (pers. commun., letter of 8 December 1993). I would like to insist on this point because although the drawing shows contemporary European influence, particularly in the way humans are depicted and in the bows they carry, this does not show in the animals. The proportions of the body are in general fairly accurate and we are before a stocky animal, so it is clearly a llama. The only striking thing is the size ratio of

FIGURE 10.2. Front page of Schmidel's tract, published in Nuremberg in 1602 (1962). We see the author riding a llama, just as he describes in his narrative (see this volume, p. 386). *Taken from the cover of the 1962 edition of Schmidel, a facsimile of the 1602 edition.*

human to animal. Judging by their proportion, these would have been animals far bigger than the present ones. This would support the hypothesis that in past times there lived varieties of llama that have now vanished.

Rowe (1946:219, 239) accepts that the llama was perhaps occasionally mounted in ancient times, because "it tires so easily under even a small man". He apparently based himself on Cobo alone (1964a, see above). Gade (1977:117) is of similar opinion, but he does not indicate his source.

Larrain Barros (1980a:227, n. 216) discussed the passage in Cieza de León to which I have often referred, when the Otabalo tried to fool the Carangue. He says that an adult llama does not often carry more than 40 kg, and it is therefore impossible that adult men could ride them. He adds, "If this were so, the llama would have been a riding animal for the Inca as well as a beast of burden, like the donkey or the mule. But it certainly is feasible that young boys, who weighed less than the figure given, formed a small mounted caravan."

The only categorical position is that of Pulgar Vidal (1950:18 [84]), who wrote of the "auquénidos" that "according to the testimony of archaeology and folklore, we know they had been so perfectly selected they attained developed specimens they used to ride." However, this is just a mere statement without any value at all, for no sources are indicated.

Sumar (1992:89) writes that the llama "[e]ven served for the transportation of people with some kind of physical defect (cripples, hunchbacks, etc.), as well as sick people who were laid lengthwise along the body of the animal to distribute the weight properly and facilitate their transportation." Sumar, of course, does not present any evidence to support his claims. He is certainly generalizing from the existing examples—listed in Chapter 4—of Mochica representations, in which some disabled individuals are shown riding llamas (Figure 4.21). We also find individuals lying over the animals lengthwise and across (Figures 4.17–4.19), but in this case in all of the examples I have seen, not one shows a crippled individual; all are healthy. It is, of course, possible that Sumar saw a specimen with "sick people," in which case it is a shame that he does not give his source. Even so, Mochica customs cannot and must not be gener-

alized. Sumar himself (1988:Fig. 1, 24) published an illustration showing a hunchback riding a llama, but no details are given. To judge by the photograph, this probably is a small modeled piece (no scale or measurements are given), apparently of metal, perhaps silver. Sumar says it is "from Cuzco." It seems to be a modern forgery or copy. In any case the hunchback does recall the Moche style. It is striking that this person is seated on the llama, which is depicted with a neck that is somewhat too long, and that the individual is holding onto it. If the piece is authentic, it is one of the few cases in which long-necked specimens have been depicted, and as far as I know the only one with a man in that position. Interestingly, Flores Ochoa et al (1994:115) published a sculpture similar to that illustrated by Sumar. It is reportedly silver, is Inca in style, and is in the collections of the Smithsonian Institution (Washington, DC). Provenance is unfortunately not indicated, nor is the scale or its measurements. It seems to be small. A careful comparison of both statues shows that certain details are different, particularly in the legs. No definitive conclusion can be drawn from the poor illustration in Sumar (1988:Fig. 1, 24), with its lack of data, but it probably is an imitation of the original piece in the Smithsonian Institution. The hunchback riding the llama has Mochica-like features, and this could be due to two reasons. Either the piece comes from the north coast—and if it comes from Cuzco, as Sumar claims, then it could be part of the booty seized by the Inca after their conquest of Chimu—or it was made in Cuzco by the smiths we know the Inca took with them.

Again, we should not forget the testimony of Cristóbal Campana (pers. commun., 19 August 1992) often cited here: in the 1930s his father, who weighed 75 kg and carried an 11.5 kg saddlebag ["*alforja arrobera*"]—a total weight of 86.5 kg—would ride his llama for 60 km at one go.

The Llama Taxi program has recently opened in Cuzco, to which I will refer later on. Here an Indian rides a llama before tourists, but covers only 20–30 m. Jorge Flores Ochoa, my informant, notes that the two most important reasons why this activity does not increase is because it is not a traditional activity, so therefore no one has proposed this (Jorge Flores Ochoa, pers. commun., 8 April 2003). With this I in no way mean that the llama was a riding animal. The available evidence seems to indicate that this custom was not widespread in prehispanic times. However, it cannot be denied that these animals were used in this way in some zones or on certain occasions. This was specifically the case on the north coast of Peru (ca. between the Early and Late Intermediate periods) and among the people of the border zone between Argentina and Paraguay, concerning whom Schmidel (1962, 1986) left invaluable testimony. This is supported by the data in the chronicles (reviewed above), which likewise show indisputably that the Spaniards also rode llamas whenever it was necessary. With this evidence I wonder whether culture had some role in this, and not just the size limitation of some llama breeds.

10.2.7 Were Camelids Draft Animals?

There is another controversial point worth elucidating—namely, a mistake made by Humboldt and Bonpland (1815–1826), and then by Humboldt (1849 [1971]), which was popularized by other writers. I was unable to obtain Humboldt and Bompland (1815–1826), but thanks to Tschudi, it is clear that the same passage was repeated by Humboldt (1971).

Humboldt (1971:145) wrote: "Pedro de Cieza de León (1) seems to want to cite an exception that is certainly rare when he says that llamas were yoked to plows in the Peruvian altiplano of Collao." Note 1 reads: "Crónica del Perú, Seville, 1553, chap. 110, p. 264." The reference is wrong because the correct chapter is 111; second, Cieza does not say what Humboldt claims—although it is true that he says "seems to want to cite." I assume that Humboldt seems to have misunderstood Cieza, which reads thus: "Verily in this land of Collao it is a great pleasure to see the Indians go out with their plows on these rams, and seeing them returning home in the afternoon loaded with firewood" (Cieza de León 1984:Part I, Chap. CXI, 294).

Tschudi (1885:107–108; 1891:107–108; 1918: 228–229; 1969:137 [I repeat that the 1969 edition must not be used; it not only misrepresents the original text, it omits part of it, and besides, the title of the chapter in Cieza—which Tschudi also misquotes—is changed into another title that is

likewise wrong] mentions Humboldt's claim and repeats the passage from Cieza (1984:Part I, Chap. CXI, 294). It is strange that a scholar so careful and knowledgeable of the chroniclers copied Humboldt's statement without noticing that the chapter number was wrong. His comment on this passage is quite precise: "As we can see, here there is not the slightest allusion that can let us assume that llamas were used in agricultural tasks, for all it says is that they carried the plows to the field and returned with a load of firewood; they are thus only mentioned in their capacity as beasts of burden. While the Indians worked the land according to their system the llamas grazed, no doubt close by." Then Tschudi insists that no chronicler mentions the llamas as draft animals, and "even though the letter accompanying Ovalle's account of his travels in Magellanic lands [Tschudi refers to the 1646 Italian version, and therefore calls him Ovaglie; there is a Spanish edition published in 1969] shows an Indian plowing with two llamas, this should not be taken as anything but a fantastic drawing made by European artists, works that abound in the travelers' accounts to distant lands published in the sixteenth and seventeenth centuries, and part of the eighteenth century." Tschudi then categorically insists that the llama was not used anywhere or at any time as a draft animal. To support his claim even further, he notes that Garcilaso de la Vega (see 1959:Bk. 5, Chap. II, 67–68) gives a good description of the native "plow" (a foot-plow called *taclla* or *chaquitaclla*), in whose operation "there was absolutely no use for the help of any animal's traction." (Tschudi insisted on this in another study [1966:244].)

Tschudi (1885:108, n. 4; 1891:108, n. 3; 1918: 229, n. 146; this note was omitted in the 1969 edition) indicates that in a paper on the llama, Brandt (1841) tried to deduce from a text by Ulloa that llamas were used in Riobamba as draft animals, but also notes that Wagner showed that he was wrong, using the same text correctly.

Troll (1958:29) also wrote that the use of llamas as draft animals should not be considered. Gilmore (1950:438) says "[t]here are hints" the llama was used to pull the plow, but that this was "doubtful." In other words, his position is not clear. He does note, however, that Romero (1927:

63, which I have not read) refuted a claim by "Theodore de Bry" concerning the plow. I believe I have identified this latter work (Bry 1592), but have not read it. I assume it refers to Chile.

But Gilmore (1950:438) also points out that the llama was never used to pull carriages, and notes that the "toggles for llama harnesses" mentioned by Bennett (1946:607) may instead have been a kind of harness to load cargo on them. In this case his statement is unwarranted, for what Bennett clearly said was that those "toggles for llama harnesses" actually were used "as a cinch buckle for fastening the packs on the llamas."

Franklin (1982:467) made a surprising claim: "Llamas were not employed for pulling plows or carts, and, *though the wheel was a part of Inca toys*, it was not applied to wagons or other such draft vehicles" (emphasis added). The toys mentioned here do not exist, and it is possible that Franklin was confused with Mesoamerica (see Canals Frau 1955:Fig. 4, 22, and 569, n. 11; Stone 1972:190, top illustration). Gade (1977:117–118) likewise denies this use, but curiously enough, claims that

the llama was not associated with cultivation, and concludes that this was probably "because the llama was not a source of food." This does not make sense at all. The llama was somehow connected with cultivation (see Guaman Poma de Ayala 1936:fol. 11050), as produce and work tools were carried on them, and it is a fact that its meat was, and still is, used as food.

Seibert (1983:267) emphatically states there were no draft animals in the Andes, nor the kind of plow animals pull.

When discussing camelids in the Chilean area we saw (see Chapter 9) that the presumed existence of a plow pulled by the *chilihueques* has no basis.[7]

There is also no question that camelids and the goods derived from them had, and still have, many other uses besides those reviewed. One of the most important uses in the prehispanic world was related to religion and sacrifice (see Guaman Poma de Ayala 1936:fols. 240, 254, 270, 880; my Figures 4.8, 10.3–10.5). There are many data on this point, but space does not allow me to discuss

FIGURE 10.3. In this drawing Guaman Poma de Ayala discusses March and explains that "in this month they sacrificed black rams to their idols and gods" to ensure abundance (*abundancia*). *After Guaman Poma de Ayala (1936:fol. 240).*

FIGURE 10.4. Guaman Poma de Ayala describes and illustrates "idols and uacas of the Collasuyo" and explains that "throughout the provice of Collasuyo-Collas...[they] sacrificed black rams and basketfuls of coca leaves . . . [the] Puquinacolla Urocolla sacrificed white rams." *After Guaman Poma de Ayala (1936:fol. 270).*

FIGURE 10.5. Guaman Poma explains that in colonial times the sacrifices made in pre-Hispanic times were forbidden, and the drawing shows "how in the time of the idolatry the ram was sacrificed by pulling its heart out. They should not be killed like this in Christian times, and instead the throat should be slit. He who sacrifices in the old way is a sorcerer and adolater, and these Indians should be punished in this kingdom." *After Guaman Poma de Ayala (1936:fol. 880).*

this topic, which besides goes beyond the scope of this book (see, e.g., Flores Ochoa 1977c; Paz Flores 1988; Tomoeda 1993).

10.2.8 Camelids and Introduced Animals: Advantages and Disadvantages

One point that has not received due attention, particularly from the Peruvian state, despite having been analyzed by a groups of specialists, concerns the advantages camelids have in all areas vis-à-vis imported animals. This is a large topic and includes economic aspects I cannot pass judgment on, and for which I lack data. However, it is worth pointing out the major positions.

Matos and Rick (1978–1980) and Brack Egg (1987) specifically mention the vicuña. The former (1978–1980:33) indicate that unlike imported animals, especially sheep and pigs, the vicuña has the advantage of maintaining the biotic land-

scape in ever-optimal conditions. Brack Egg (1987:75) pointed out four major advantages. First, the Andean area already has the genetic material. Second, the vicuña makes better use of native pastures, and no improvement programs are needed to feed it. Besides, it does not destroy the pastures by trampling on them because it has cushioned and padded feet. Third, it is an animal that thrives in very marginal, semidesert parts of the puna where sheep or even alpaca breeding is impossible or unprofitable. And fourth, it is a wild species that can be managed in its wild state and at a low cost. We should not forget that these are animals that need little attention and are resistant to disease.

Usselmann (1987:133) cites Millones (1982 [which I have not read]), where the latter emphasizes that plantigrade camelids graze without damaging the soil, whereas imported animals, sheep and bovines in particular, besides needing extensive pastures, have caused much erosion with their hooves. In this regard it is worth recalling what Garcilaso de la Vega (1966:Bk. 8, Chap. XVI, 514; 1959:149) said of camelid feet: "It does not wear shoes, for it has a cloven foot, and has a pad instead of a hoof."

It is also worth recalling the characteristics of locomotion in camelids, which are quite unusual and a great advantage in the Andean milieu vis-à-vis imported animals:

> All the camels, including the South American camelids, are natural walkers. In contrast to the trot, which is characterized by moving the legs in diagonal pairs, the pace is marked by simultaneously moving the legs forward as lateral pairs. This movement produces a longer stride, thus allowing the walker to cover a greater distance with a smaller expenditure of energy. Although some trotters use this paso occasionally or are trained to use it, only the camels and South American camelids have a specialized morphologic conformation for the habitual use of the paso. (Wheeler 1991:12–13)

Another important factor is that ungulates use different feeding strategies to exploit their environment. It likewise seems that owing to the larger size of their body, South American camelids are less selective in grazing than ovines. San Martín quantified and compared the diet of camelids and ovines and showed that the latter are more selective than the former, and that llamas prefer a clustered and fibrous fodder. Alpacas in turn are opportunistic animals with a greater capacity than llamas to adapt to different vegetation conditions. San Martín therefore indicates that llamas have a competitive advantage over the alpaca in the driest regions in the Andes (San Martín et al. 1989:98; besides their own data, these scholars used those in Jarman and Sinclair [1979] and San Martín [1987]).

Another important feature of Andean camelids is their higher productivity with regard to other species. According to Novoa and Wheeler (1984:127), the results reached over the last 20 years in biological research have clearly shown the productive superiority of llamas and alpacas at high altitudes. It has been shown that introduced species have a poor adaptation to the puna environment and therefore have low, poor-quality productivity. Moreover, fertility and reproductive rates are low both for sheep and bovines, and altitude-related diseases can cause a high mortality even now, five centuries after they were introduced to this new environment. When compared with native animals, imported ones show a low energetic efficiency regarding fodder and in metabolizing plants in native pastures; furthermore, they have significantly contributed to the degradation of the environment. On the other hand, raising sheep and bovines is a more difficult and more intensive enterprise.

Novoa and Wheeler insist that llamas and alpacas can fulfill a more significant role in the economy of the Andean countries thanks to their proven superiority over European animals: *"They have survived in the marginal high-elevation lands to which recent history has relegated them, but could undoubtedly be more profitably raised on better-quality pastures, and in the lower-elevation zones which they formerly occupied"* (Novoa and Wheeler 1984: 127; emphasis added). These scholars believe the commercial output of the llama and the alpaca—successful in Australia and the United States—shows these animals can be raised outside the Andean area, and that their capacity to use marginal

pastures makes them a major potential resource on a worldwide scale, which thus far has only minimally been exploited. (Here it must be pointed out that Sumar [1988:28] partially plagiarized Novoa and Wheeler [1984:127] without citing them, even though these scholars are mentioned at the beginning of his article.)

In addition, it is a fact that the herds of llamas and alpacas are the safest food resource of the people living in the puna. Camelids have a higher resistance to aridity than sheep, and their mobility furthermore gives security to the herds despite the instability of the climate—one of the most frequent problems for agriculture. As was already noted, camelids are able to turn the plants in pastures with a high cellulose content, and which are useless for human consumption, into an important and useful source of stored protein. Productivity is thus extended to those areas where agriculture is not feasible. And we should not forget that besides the flesh, which is immediately used, camelids provide other goods and services already mentioned (Novoa and Wheeler 1984: 125; they also used data in Thomas [1973], which I have not seen).

An interesting point is that camelids have shown better adaptation to conditions than other species of domestic animals, in the face of unstable weather conditions like the droughts that frequently occur in the Andean area. In the severe drought that struck the Department of Puno in 1956–1957, about 80% of the cattle and horses died; and whereas 40% of the sheep died, only 25% of camelids did so. If these figures are representative enough to be applicable to other parts of the Andean land, then this would suggest that llamas and alpacas are the safest nutritional and economic resource for the peasant population living in those zones (Sumar 1988:24–25).

Flores Ochoa (1979a:227–228) has analyzed the problems raised by alterations and modifications in the landscape due to overgrazing and admits that the high puna landscape has perhaps been affected, and that the lack of trees may be due to it. He is, however, emphatic in stating that there is no real evidence that this damage took place, as Brush (1975), Crawford et al. (1970), and Ellenberg (1958) suggest. (Brush does not appear in the bibliography presented by Flores Ochoa and was taken from Crawford et al.; Ellenberg is misspelled as "Ellensberg" and is also missing from the bibliography, but I assume it is the 1958 paper in my bibliography.)

Flores Ochoa (1979a:227–228) notes that Browman (1974a) claims that overgrazing is a prehispanic phenomenon; "however, the greatest evidence of overgrazing and a substantial modification in pastures indicates that it is due to the introduction of ovines, which have managed to change the plant cover and initiate a process of erosion in many areas, particularly in those over 4000 meters of altitude." Flores Ochoa cites Koford (1957, which also does not appear in his bibliography, as it was taken from Brush [1975], which was likewise not read by him; see above) to indicate that the abundance of inedible pastures could be due to the consumption of the best grasses by ovines. Flores Ochoa ends thus: "It is possible that natural pastures have deteriorated, but it is precisely for this reason that it is worth knowing and understanding the modifications herders make to give the animals grass of a higher quality, and which as far as I am aware of do not cause erosion."

Flores Ochoa is one of the most outspoken critics of the introduction of European animals into the Andean area. He claims that to date, ovines have not attained a full adaptation. But even though "their acclimatization raises problems" they have managed to displace camelids, particularly around Lake Titicaca (Flores Ochoa 1982: 71). We must also bear in mind that a high-altitude habitat is not the best for the development and efficient reproduction of sheep, as they require different pastures not just qualitatively but quantitatively too. The absence of cows and horses in high-altitude zones shows the scant economic significance these animals have for puna herders and lead some to conclude that these animals are not the most adequate to survive in said environment, nor are they profitable in places where llamas and alpacas thrive (Flores Ochoa 1974–1976:259).

Besides, sheep are born at any time of the day and in any season of the year. This is a drawback vis-à-vis camelids because the strong frosts of June—the time most sheep are born—is not favorable, and is also the time with the least grass. In general, bovines are almost unable to stand altitude and do not have any of the necessary char-

acteristics to make good use of the hard and short grass blades characteristic of high altitudes (Flores Ochoa and Palacios Ríos 1978:85).

Flores Ochoa and Palacios Ríos (1978:85) wrote: "The offspring (of ovines) in particular suffer a lot and when a herder raises cattle—more as a sign of ostentation—it is not strange for him/her to bring the offspring to his room to ensure survival." Flores Ochoa explains that it is for this reason that the effort and care required by a big herd of cattle are not within the scope of the energy a herder can invest, and are not an advantage in comparison with camelids. Flores Ochoa and Palacios Ríos (1978) concludes thus: "[I]n the case of ovines we must bear in mind that they deteriorate the plant cover of the high puna and the subsequent erosion can leave large tracts of land naked, making them unproductive. Added to the slow development of vegetation at high altitudes, this factor gives alpacas and llamas an advantage over other domestic animals, [and leaves them] in condition to stand the development of a herding economy. However, at present the herds of puna pastoralists also comprise sheep in numbers that vary according to the zone, altitude, the availability of pastures or the force of external pressures."

One aspect I have not touched on but that is no less important is contemporary transportation. In areas with highways, motor vehicles obviously have definitely displaced camelids. We should not forget that a great part of the Andean highlands still does not have modern transportation and that transportation is still conducted using beasts of burden; where llamas exist, these are the favorite of the natives. This point deserves a detailed study. As an example, I will cite the study Custred (1974) made of the communities living in the province of Chumbivilcas, Department of Cuzco. While discussing the interzonal trading trips made by the comuneros of Alccavitoria, Custred explains that llamas are most used, horses only "sometimes." "Although these can carry twice as much as the llamas, we found that peasants prefer the latter for extended trips, even those who have a great number of both animals. The reason given for this preference is that llamas require less equipment than horses. Only a piece of string and a sack are needed. Furthermore, llamas are quite docile and easy to handle in long trips, thus making the journey less complicated or wearisome. They eat during the day as they walk along a trail, and in the two hours between the end of the day's march and dusk. At night they are placed together in a corral, where they sleep. On the contrary, horses must spend all night grazing. For the traveler it is preferable to have all of the animals in one group because this way they are safer. Finally, llamas adapt better to conditions in the puna. First, they have no problem in finding pastures; second, they are more resistant to the changes in temperature in the puna. Horses instead cannot eat the same kind of pasture, may begin to sweat and run the risk of catching cold from the cold air of the puna—something that never threatens the llamas" (Custred 1974:275–276).

The arguments for native animals are conclusive.

NOTES

1 *Chala* is a misleading term. *The Dictionary of the Royal Spanish Academy* (1992:653) defines it as "[e]spata del maíz," i.e., it means the bracts that wrap the inflorescences of maize. It likewise says that the term comes from Quechua. In this language *challa* however means "[c]aña y hoja de maíz seca" (Gonçalez de Holguín 1952:92), which is quite different. However, it should be noted that on the Peruvian coast herders use the term "chala" for the fully-grown maize plant that is harvested before it matures and is used as fodder. This is what Vásquez meant.

2 It is not beside the point to recall that the vegetation in the puna consists mostly of prickly gramineous plants with hard and pointed leaves, commonly known as *ichu*. This includes *Stipa*, *Festuca*, and *Calamagrostis*. Among these clumps of herbs are other, more delicate, ones which camelids particularly enjoy. (See Soukop n.d. [1987]:97–98, 180–181, 218, 388; Seibert 1983:262 ff.)

3 This is the coarsest part of the wool.

4 See note 2.

5 When they have just given birth, female llamas may produce more than 500 cc of milk in a 12-hour period (Flores Ochoa 1968:109), clearly a minimum amount in comparison with, say, the goat. Average goats produce 1.5–2 liters of milk a day, while select individuals can give produce up to 4 liters a day (Miguel Mori T., pers. commun., 26 November 1993). We should not forget that the goat is far smaller than the llama.

6 The Spanish translation (1955:114) says "*cuatro arrobas*," which is about the same weight, for 100 pounds is 45.35 kg and 4 arrobas is 46 kg.

7 However, in Cuzco there is a program called Llama Taxi, which opened on August 2002. It was the idea of Enrique Ampuero Aquino, an expert on camelid raising and herder training. He designed a cart drawn by llamas which he is now perfecting. Llamas are now used in ecological tourism in the Manu. Tourists are taken to the summit of Aqhanaqo and are brought down in llama-drawn carts up to the cloud forest. They cover a distance of up to 10 km. The tourist then returns through another route and using other means. The carts usually carry two individuals, but they can take up to four people. If the ride is on flat land, the distance covered decreases to 2 km. No special llama breed is used for this activity, the biggest animals simply being chosen. Flores Ochoa believes that if this activity is developed, bigger animals would have to be sought in the herds in Carabaya (Department of Puno) and Bolivia, which have very big llamas (Jorge Flores Ochoa, pers. commun, 8 January and 8 April 2003).

CHAPTER 11

DID PASTORALISM EXIST

IN THE ANDES?

Some researchers have wondered whether pastoralism really existed in the Andes. This is a theoretical subject outside my specialty, and its study must start with a very clear definition of pastoralism, for the term has different meanings, and Old World patterns are inadequate to define or understand the New World phenomenon. The truth is that America has not been considered a geographic area where pastoralism developed. To give just a few general examples, Beals and Hoijer (1958:323) wrote that "[a]ll pastoral peoples are found in the Old World," while Hoebel (1961: 220) practically repeated this when he claimed that "[p]astoralism is a predominantly Asiatic and African economic complex." Kroeber (1963:85) also made the same claim. Hoebel, however, also claimed that "[i]n the New World, only the Navaho were really herders, and only in modern times with sheep acquired from the Spaniards. In the southern plains, tribes like the Comanche were almost stockbreeders but not herders, in the mid-nineteenth century" (Hoebel 1961:220). This clearly is not true. These discussions ultimately hinge on one point. Traditionally, a human group is considered to be pastoral if it depends almost exclusively on animals for food. In this sense, herders do not carry out any agricultural activities, or do so on a very small scale, and acquire vegetable products either by gathering them or through exchange with neighboring groups (see

Beals and Hoijer 1958:323 et passim. For a more detailed discussion of this topic, see Kroeber 1963:85–86).

I must emphasize that the whole problem hinges on the definition of pastoralism used. Although Flores Ochoa (1967, 1968; Webster [1973] takes the same position, according to Flores Ochoa [1975b:297], but I do not know this study) believes it is possible for pastoralism to be an activity wholly independent of agriculture, and that it might have existed at high altitudes in pre-Columbian times (Flores Ochoa 1975b:297), I believe this goes against the archaeological evidence. It is true that it was once thought that pastoralism preceded primitive agriculture (see, e.g., Dittmer 1960:242 et passim; Biasutti 1959), but most researchers today are convinced that it was the other way around. The archaeological data from the Old World prove it (see Jacobs and Stern 1962:98–99; Darling 1956:778). As Jacobs and Stern (1962:99) noted, "[i]n the New World the priority of agriculture in relation to pastoralism is clear."

Archaeological data show that the domestication of plants in the Andean area is far older than the domestication of animals (leaving aside the dog, which reached the continent with man already domesticated). On the other hand, although Andean populations were acquainted with sedentism early on, at the same time they were highly mobile, contradictory though that may sound, and controlled different environmental zones. Everything thus seems to indicate that a mixed economy was practiced from very early times in which there was a continuous exchange of products between groups that tended camelid herds and those that engaged in horticulture or agriculture.

This mixed system was certainly advantageous for pastoral groups because, as Kroeber pointed out, all nomad pastoral cultures that tend to be self-sufficient in isolation face an uncertain future. This is what Kroeber called a "half-culture" or "part culture," that is, cultures that must be supplemented with other cultures with a more varied economy in order to attain a higher level of development (Kroeber 1963:86).

Moreover, a fully nomadic pastoralism, which was and is common in other dry environments, never existed in the Andes, possibly because the domestic camelids were not the only essential source of food (Franklin 1982:466; Gade 1969: 342). What was common was the seasonal movement of the herds, or transhumance; this would be somewhat like the "definite migratory pastoralism" of Darling (1956:779). Such is the case of, for instance, Junín in the central sierra (Wheeler-Pires Ferreira et al. 1976; Browman 1974a:189 et passim), which is discussed in some detail later. I understand transhumance as a form of sedentism because the movements of the herders and their animals occur between two known areas that are always the same. I personally verified this in the 1950s with the goat herders of the lomas of Lachay, on the central Peruvian coast. Not only did the group place its tents on the same site each year, using the stones initially chosen to hold up their tarpaulins, but a great part of their kitchenware and other items were buried to avoid carrying them, in the certainty that they would be found again the following year. My informants stated that the same thing was done in the camps in the highlands.

It is true that in the altiplano there are herds of llamas that graze in conditions that come close to freedom (Koford 1957:212), but these still are animals under the care of their herders. Besides, we should not forget that the domestic camelid does not tend to escape. In all of my long experience traveling across the bleak puna, I have never seen runaway llamas or alpacas, whereas more than once I have met runaway donkeys and horses.

Murra (1964a, 1975) studied the phenomenon in Inca times and has shown that local herding was considered a part-time job to be carried out by the young. Just as today, the herders (*awatiri* in Aymara and *michiq* in Quechua) were children of both sexes. Sometimes they looked after animals belonging to their closest relatives, but more broadly they also fulfilled the reciprocal obligations of their kinship group. When the herds were too large or the pastures too distant, the children were replaced by other adults, who permanently took over the task, which was considered low status (Murra 1975:119–120). Murra asks, "How and when did the transition take place? Who volunteered for this task or was forced into it? What opportunities did they have to abandon herding? These questions are still unanswered. All we know is that after the European invasion and before the Inka expansion there already were herders in the Andes, who did nothing but live in the puna, far from the their centers of origin." It is interesting that Murra specifies that these herders continued belonging to their kinship groups and that, their isolation notwithstanding, they still retained their claims over the fields, which were worked by their relatives (Murra 1975:120).

One of the factors that must be taken into account is that Andean pastoralism is one of the most specialized ecosystems anywhere, because it represents the adaptation of man to an environment that cannot be used in any other way with agriculture or livestock and where the resources used by man are most limited. It is for this reason that Andean pastoralism is highly specialized, in relation both to the animals it uses and the habitat where it is established, the puna.

Besides, in the Andes, herders cannot always use the environment just the way they find it. Instead, they must change it. This is done with certain irrigation methods that enable them to have green pastures for the alpacas in places that lack the necessary humidity for pastures to grow naturally (Flores Ochoa 1975b:301, 306). Flores Ochoa (1975b:306) has given some examples. In the cordilleras of Canchis and Carabaya, in the Department of Puno, there are canal networks to flood lands that are suitable for the growth of puna pastures. This is an adaptive technique because in these regions there is not much land available, and pastures are scarce. For this reason the animals must be moved to the pastures far more often than in other regions, and even taken from the lower to the high areas, and vice versa. In this case the animals are not moved in relation to the seasonal rhythm of summer and winter but when pastures become exhausted.

In the Andes today there is a marked differentiation in herding areas as a result of the introduction of European animals. Thus, herding communities based on camelids and ovines live between 3800 and 4000 masl. Between 4000 and 4200–4300 masl, the communities herd camelids and ovines too, but from this altitude upward one begins to find communities exclusively dedicated to herding camelids. Meanwhile, cattle are restricted to the dry valley bottoms, although in the wettest areas they are found above 4000 masl. Mules and asses are everywhere but are numerically more common in dry areas. Goats are especially common in the dry valleys (Seibert 1983: 269; Flores Ochoa 1974–1976:246). However, it must be borne in mind that herding in the high puna at 4000–4200 masl is a socioculturally adaptive tactic that presents the highest economic advantages for the use of natural, high-altitude pastures. And the animals that yield the highest benefits are the llama and the alpaca, even though sheep, some cows, and a few horses are also found there (Flores Ochoa 1974–1976:246).

It is clear that there was in fact some kind of pastoralism in the Andean area, and that this activity was most sui generis. As Lavallée (1988: 266–267) correctly noted, when drawing comparisons, the similarity with facts from other areas must not hide the marked originality of the American transition toward plant and animal domestication (which the French call *néolithisation*), either in its tempo or in the order of appearance and the significance of its components.

A study of the Old World Neolithic phenomenon shows that favorable local conditions, both climatic and biological, were taken advantage of. In other words, a situation in which animal species, much like several edible cereals and legumes, were potentially capable of being tamed was exploited. A new way of life that brought about the rise of agriculture and pastoralism thus appeared between the seventh and the eighth millennium BC.

In America the process was different, as Lavallée (1988:266–267) correctly notes, and had two independent development centers, one in Middle America and the other in the central Andean area around the seventh millennium BC— perhaps slightly earlier in the Andes. But whereas agriculture played a major role in Mesoamerican agriculture, there were almost no animals to tame (except for the turkey) in the Andean area, where the process was different. Here in the Andes, the practice of complementary ecological strategies played a major role from the very beginning and was the *trait d'union* of agriculture and herding from quite early on. The process began with the wild forms of the vicuña and the guanaco, and then moved on to the domestic forms that developed from the earlier ones in areas that lay above the upper limit of agriculture.

As Lavallée (1988:267) wrote, "This domestication was the origin of a form of pastoral culture unparalleled in the whole world, that developed in an apparently forbidding environment with a harsh climate: the Andean altiplano and the puna."

As Chapter 4 noted, the research undertaken by the French team headed by Lavallée and Julien showed the existence of a displacement system whereby human groups moved among a series of shelters and caves in the puna, in accordance with seasonal changes in hunting grounds first, then changes in herding. This activity changed from a nonspecialized hunting to a more intensive and more specialized hunting activity, one aimed essentially at camelids, until these animals were gradually controlled. The inhabitants of Telarmachay thus had a permanent occupation between the fourth and the third millennia BC, perhaps in the valley lowlands, close to the cultivable lands. This occupation coincided with the rise of the alpaca as a dominant species and the consolidation of agriculture (see Lavallée et al. 1985; Lavallée 1988:272, 278–279, 284).

For this reason, Lavallée (1988) suggests that what she defines as a "proto-raising" of animal species was born at this time. She emphasizes that the animals' natural behavior favored an intimacy and maintenance within a stable and restricted territory. In other words, Lavallée points to the territoriality camelids have, in which she concurs with my position, as stated in Chapter 5.

We therefore arrive at "a veritable pastoral economy," as Lavallée (1988) calls it, based on the exploitation of domestic herds, thus leaving the herders little time for hunting, which lost importance. The only point on which I disagree with Lavallée is her position that in the Andes, the process of domestication was carried out by nonsedentary hunters, who would remain mobile even after becoming herders (Lavallée 1988:286). Transhumance does not entail nomadism but is instead essentially a kind of sedentism.

To strengthen her position, Lavallée (1988) explains that a seasonal occupation system like this can still be used even after animals have been domesticated within an agricultural context. The domestic herds are then moved around a "center" formed essentially by a habitat that is not yet completely permanent, that lies below 4000 masl, and where agriculture is practiced. This is a typical Andean way of life even now. For this reason Lavallée wonders whether domestication was really an innovation. It is clear that control of animal reproduction and the development of new species were major innovations at the biological level. However, the lifestyle of the first Andean herders was essentially untransformed. Proof of this claim lies in the amazing continuity of the lithic tools, the unchanged rate of permanence in high-altitude camps, and the persistence of the occupation of habitats by herders that had previously been used by hunters. These facts are clearly seen in the evidence from Telarmachay (Lavallée 1988:286–287). If analyzed carefully they show, for me, the existence of some kind of sedentism.

It is for all of these reasons that Lavallée says she is tempted to talk of a "conservative innovation," because the way of life implied by nascent camelid domestication was so tightly integrated that it merged with the ancestral behavior of the hunters. "To ascertain this undeniable continuity," notes Lavallée, "we must finally ask ourselves why, when a secure system based on the rational use of the wild fauna existed, humans abandoned it gradually due to the problems inherent to a pas-

toral life. The environmental pressure hypothesis can be discarded, at least as a pressing condition" (Lavallée 1988:286–287).

We must acknowledge that little is known of this phenomenon and more study is required. Lavallée undeniably made a major contribution here.

Flores Ochoa (1979a:226) believes—and I concur—that this pastoral economy is characteristic of the central Andes between Peru and Bolivia, with offshoots in northern Chile and northwest Argentina. Its absence in some parts of Peru, such as the northern highlands, is due to the arrival of the Europeans and the introduction of Western animals.

"Ecological complementariness" has been much mentioned here, and it played a crucial role in the central Andean area. Scholars constantly mention it (e.g., Masuda et al. 1985), and it is a point worth emphasizing. It was seen that raising llamas and alpacas in areas where there can be no agriculture, even of high-altitude crops such as the potato, means that the maximum possible use of the Andean environment is achieved (Webster 1971a, 1971b; Yamamoto 1985:87). These upper limits of environmental use are among the highest in the world, and, according to Yamamoto (1985:87; he based his conclusions on Dollfus 1982), are comparable only to the northern part of the Himalayas.

Yamamoto (1985:92) gives a specific example that is most interesting and illustrative. He notes that maize cultivation in the *quechua* zone has no directly perceivable complementariness in llama and alpaca raising in the puna. However, the cultivation of potatoes is tightly intertwined with camelid raising because their cultivation in lower adjacent areas decreases markedly without camelid dung for fertilizer (Yamamoto based his remarks on Orlove 1977b). The soil would be irreparably damaged without the continuous use of this fertilizer. Yamamoto claims that camelid dung has been little studied as a fertilizer, yet research by Winterhalder et al. (1974) at Nuñoa, a site in the altiplano above 4000 masl, proves it. Here soils have several defects. Today sheep dung is used to improve the soil, and the physical improvement in the soil is evident. Camelid dung clearly has the same effect.

This shows the self-sufficiency of agropastoralism in the southern altiplano of the central Andes, precisely the area where the potato was very important and where camelid herds were numerous. In the words of Yamamoto, "The complementarity that exists between potato cultivation and camelid husbandry, then, can be seen as the best adaptive strategy in minimizing adverse effects, such as soil erosion and deterioration, while maximizing potential productivity within the limitations set by the Andean highlands" (Yamamoto 1985:92).

Readers interested in more data on this subject should read the study by Harris (1985:314 et passim) on northern Potosí, which shows quite well how the ecological complementarity of the ayllus in this area works.

Shimada (1985:XVI–XVII) has stressed the significance llama caravans had for the various kinds of ecological complementarity, a point also noted by several other scholars (see, e.g., Browman 1974a; Inamura 1981a; Lynch 1983a; Masuda 1985; Mujica 1985). Shimada draws attention to a phenomenon that has not received adequate attention, namely, the role of the llama caravans in north coast polities. Part of this tradition survived until not too long ago. I would simply like to add that these caravans occurred not just on the north coast but throughout the length of coast, though probably not in the same way or intensity. For instance, Matsuzawa (1978) described the caravans of llamas that carried goods from the Cordillera Blanca to Casma on the north-central coast.

CHAPTER 12

THE CURRENT DISTRIBUTION OF THE SOUTH AMERICAN CAMELIDS: CAUSES AND CONSEQUENCES

Can a Historically Accurate Statistic Be Determined?

One of the major issues in the subject under discussion is the cause or causes that influenced the current distribution of camelids compared with their original distribution. It is not true that camelid distribution "has not suffered significant alterations," as Gade claims (1977: 119). Nor is what Herre (1969:116) says entirely accurate: "Under the influence of the Conquistadors, the native herds were neglected and changed place, with foreign domestic animals beginning to spread." The issue is far more complex. First, Herre ignores the indiscriminate killings that took place, particularly in the early days of the conquest. Although it is true that some groups of animals changed location, probably the most important cause was the caravans of llamas the Spaniards used in great numbers until well into the Colonial period. But the main point is that the control the Inca state had over the camelids prior to the arrival of the Europeans fell apart. The introduction of new animals was one of the most negative aspects for the survival and distribution of camelids. And the diseases brought by these foreign animals should not be forgotten either.

12.1 THE IMPACT OF THE SPANISH CONQUEST ON THE DISTRIBUTION OF CAMELIDS THROUGHOUT THE ANDES

Dedenbach Salazar (1990:95–96) has studied the distribution of these animals in the early days of the Spanish conquest. She concluded that domestic and wilds camelids still ranged throughout the Andean highlands in the first decades after the Europeans' arrival, from Ecuador to Bolivia, Argentina, and Chile. Dedenbach Salazar notes that on comparing this geographic distribution with the ecological maps of Ecuador (Wolf 1975; Cañadas 1983) and Peru (Weberbauer 1922; *Atlas histórico geográfico* 1969 [the latter reference is incorrect; it should read Peñaherrera del Águila 1963–1970]; Tosi 1960), all the "provinces" listed in historical sources are seen to have had part of their land in the puna or *páramo*.

However, it is clear that Flores Ochoa is one of the scholars who has most studied the problem of camelid distribution and the causes that influenced its change over time. Flores Ochoa managed to establish that the pressures that distort the real conditions of alpaca herding have been

443

exerted since the sixteenth century. The descriptions given by the chroniclers (e.g., Cobo, Cieza de León) clearly confirm that native pastoralism was not limited to the upper ranges of the highlands, as is now the case—Flores Ochoa defines these as "shelter regions"—but was evidently spread all over the vast expanse of pastures in the high puna. The introduction of sheep was part of the Spanish policy for the control of the new lands. But this pressure to introduce sheep did not cease with the end of the Colonial period, persisting instead throughout the Republican period right up to the present day (Flores Ochoa 1979a: 231).

On comparing the areas of camelid distribution in pre-Columbian or early Colonial times with the current one, we find that it has shrunk. All areas now occupied by sheep were originally occupied by camelids. Flores Ochoa considers that the environmental limits of the alpaca were greater and, most important, that their high-altitude specialization (an inaccurate term, as is discussed below) is partly due to a substantial change in the longitudinal and mainly altitudinal distribution of camelids (Flores Ochoa 1982:81; 1988b: 278 [once again, the first study cited has the date 1982 on its cover, but inside we read 1980; I use the first date, whereas others, like Flores Ochoa himself, use the second one]).

To understand what has happened, it is useful to analyze at least some of the factors behind this transformation. One of the major factors was the introduction of European animals. It is worthwhile expanding this point.

Some scholars have discussed the Ecuadorian area. There is an anonymous account from 1573 that specifically refers to San Francisco de Quito. It emphatically says that "[t]he greatest harm the natives receive comes from the Spaniard's cattle, many of whom have their ranches [and so cause] grievous harm to the Indians" (Anonymous 1965b:212). Tschudi (1885:96; 1891: 98; 1918:209; 1969:127) notes, after stating that the Incas took the llama to Ecuador, that "after the Spanish conquest their number diminished considerably, which was due to the llamas being replaced with European pack and wool animals brought to the land by the Conquistadors." Miller and Gill (1990:49) likewise state that the huge fall

in wild and domestic camelids in southern Ecuador noted by Cieza (they cite the 1973 [121] edition, which is the same as the 1984 one, Part I [Chap. XLIII, 143]) is just a small part of a wider pattern of zoological change "that radically altered the biogeographical complexion of the Andes within a mere century after Cieza's writing." They note that just as smallpox ravaged the native population, so the diffusion of horses, sheep, pigs, and dogs (actually, these scholars do not note that these were new breeds of dogs, for the latter already existed in America) "alter[ed] the endemic Andean biota." As a result, some animal communities managed to survive by retreating before European competition to marginal zones, while others became extinct. "The end result was an ecological catastrophe perhaps unparalleled since the end of the Pleistocene (Crosby 1986)" (Miller and Gill 1990:49).

As regards the highlands of what is now Peru, Father Cobo (1964a:Bk. 2, Chap. X, 74) wrote: "At present there is not as much livestock of the land as there was before the coming of the Spaniards, because the Indians have diminished and great expanses of this puna and paramo are occupied by Castilian livestock which thrive in them, such as cows, sheep, pigs and goats, mares, asses and hens." Interestingly enough, Cobo (1964a: Bk. 2, Chap. X, 78) is one writer who clearly noted the problems European animals had living in the Andean highlands. He specifically indicates that once one came down from the upper highlands and reached the highland valleys "mares, asses and mules are born, for outside this third zone upwards they are not raised." And Cobo wrote in the seventeenth century. There is no time for details, but in the writings of Cosme Bueno (1951), published in the eighteenth century, one can find abundant and usually reliable data on the aggressive way in which the European animals entered the Andes.

Fernández Baca (1971:27) concurs in noting that the location of camelids in the high altitudes was brought about in Colonial times by the introduction of European animals. Lynch (1983a:2) takes almost the same position, and likewise notes that camelids were forced to endure the same consequences faced by groups of hunters and gatherers, who found themselves restricted to marginal areas (which in the Andes are usually high-altitude ones), which correspond precisely to the land that is not suitable for raising sheep and other European animals. Lynch is quite emphatic when discussing the south-central Andes, noting that "the politics of Western cultural domination have continued to favor the expansion of sheep into range lands formerly grazed by camelids."

Sumar (1988:24) is of the same opinion, but adds that this has not changed to date. He insists that the competition between native and foreign animals resulted in the disappearance of alpacas and llamas from great expanses of land, even though sheep breeding had its limits, both due to poor pastures and to problems in adapting to altitude. Thus it is that the herds of llamas and alpacas are at present cornered in land lying at or above the upper limit for cultivation, which is marginal for sheep raising.

From the available data it seems that the rate of camelid extinction was not uniform throughout the Andes. According to Novoa and Wheeler (1984:125), the first herds to vanish were those on the coast and in the highland valleys, because the imported animals had no trouble getting established there. The process was slower in the puna, owing both to an unwelcoming climate and to the higher number of camelids present. The introduction of European animals, the Spanish civil wars, and the tribute the natives had to pay also played a major role.

Novoa and Wheeler (1984:125) recall that in 1567, Garci Diez de San Miguel (1964; see Chapter 6) noted the presence of an individual herd of 50,000 llamas and alpacas in Chucuito; five years later, in 1572, Gutiérrez Flores and Ramírez Segarra (1970; see Chapter 6) could count only 159,697 of them throughout the province (Novoa and Wheeler do not give the dates of publication of these sources, which were taken from Flores Ochoa. They make two mistakes, however. The correct reference is Flores Ochoa 1977[b]:23–24. Also, the page numbers printed in the bibliography, 15–22, are inaccurate: they are 15–49. They did not realize that Flores Ochoa also cited Gutiérrez Flores inaccurately, and followed him. The correct reference is Gutiérrez Flores and Ramírez Segarra).

But it is worth noting that the depopulation persisted. For example, there are data for Huancané in 1807, where 81.6% of all animals were sheep and just 1.9% alpacas (Flores Ochoa 1982: 75, citing Macera et al. 1968). Novoa and Wheeler (1984:125) admit the figures cannot be accurate but that the trend is clear. The truth is that by the late sixteenth century the herds were diminishing, while tribute rose and both llamas and alpacas even vanished in areas with great pre-Columbian concentrations.

Novoa and Wheeler (1984:125) explain that under Colonial rule, the demand for Merino and Churro sheep and their wool far exceeded that of llamas and alpacas, so native herders were almost forced to make the change. It was for this reason that camelids were mostly meant for the butchers, to be used as meat for mine laborers, particularly in Potosí, or to carry metal. This phenomenon led to the near-total extinction of llamas and alpacas in a great part of their original area of distribution, whereas sheep herding was eventually limited only by poor pastures and adaptation to high altitudes. This forced the lamoids to find shelter in marginal areas above the upper agricultural limit, where they had no problem surviving.

There is thus no doubt that the European invasion not only depopulated the highlands, it essentially modified the habitat, and not just with the introduction of exotic animals but plants as well (Flores Ochoa 1970; 1979a:225). In this competition with camelids, sheep and cows clearly had the advantage, because there were cultural and economic factors at work with their breeders that complemented the colonization policy then being established. Redistribution and reciprocity, the essential practices in the Andean world, were replaced by European mercantilism, and this certainly altered the Andean agricultural and pastoral world. It also led to the destabilization and modification of the aforementioned economic complementariness, thus entailing a substantial change in the Andean image of the world and its resources (Flores Ochoa 1982:65). The destruction of the herds belonging to State and Church of the Inca period, which the Spaniards appropriated, and their replacement with the aforementioned animals was clearly one of the most important factors at work in this process. It is worth emphasizing that this brought about the depopulation of vast Andean regions, certainly a violent process that resorted to coercion and even imprisonment (Flores Ochoa 1983: 181–182). This is evident in the sixteenth century *visitas*, particularly in the one conducted by Gutiérrez Flores and Ramírez Segarra (1970), where the interest the friars had in introducing sheep is clearly shown.

Flores Ochoa (1982:71–75) insists and shows conclusively that the southern altiplano was one of the major zones of change. His point of departure is a passage in Cobo (1956:365–366 [a mistake, for the correct reference is 1964[a]:Bk. 9, Chap. LVI 365–366]), where only llamas are mentioned, but not the "sheep of the land." Although Cobo only traveled along the western shore of Lake Titicaca and did not visit the puna, it is striking that great herds of alpacas or llamas still existed in the seventeenth century. Today the former live above 4200 masl. According to the data in Bustinza (1970b:29), 67% of the Peruvian alpaca population is concentrated in the Department of Puno, 30% of which is in the province of Carabaya and 11% in the district of Macusani. We should bear in mind, as Flores Ochoa notes (1982:71–75), that these are both cordillera regions, among the highest in the Department of Puno. Still according to Bustinza (1970b:31), if the alpaca populations in the provinces of Carabaya, Melgar, Lampa, and Chucuito are added, they amount to 78% of the alpacas in the Department of Puno. Again, these are high cordilleras. Meanwhile, sheep prevail in the area between Lake Titicaca and the high puna, while cattle prevail close to the lake.

Flores Ochoa (1982:71–75) explains that there are small herds of 30–50 animals on the lake shores, between the plain of Ilave and close to the Desaguadero River. These are alpacas that come down from higher altitudes. There are small, permanent herds near the natural bofedales (marshy areas), close to the Desaguadero River and the lake.

Hocquenghem (n.d. [1989?]:116–117) has studied the cold lands of Piura in the fifteenth and sixteenth centuries and established that bulls and cows entered the pastures with the arrival of the encomenderos and ranchers for meat, cheese, and milk. There also came wool-producing sheep and horses and mules for transportation, both of people and merchandise. The Castilian animals had to compete for the same pastures used by the native livestock, but the Guayacundo people were forced to neglect the native animals because they had to pay tribute in produce prepared with Castilian livestock.

But Hocquenghem (n.d. [1989?]:116–117) explains that in the northern Andes, there is no area above 4000 masl that could have sheltered camelids where cows have not adapted. Camelids thus simply vanished from the highlands of Piura and were replaced by cows, sheep, horses, and mules. This had serious consequences, for these animals have destroyed the pastures with their hooves which then led to erosion. At home, the Indians began to raise pigs, poultry, and dogs to help them look after the herds.

A similar phenomenon took place in the puna of Junín, where the introduction of sheep in Colonial times virtually eliminated camelids (Wheeler Pires-Ferreira et al. 1976:483). Cunazza (1976b:166) also is definite when discussing the use of native resources in Chile, for he states that what did not take place in thousands of years of rational use of the environment happened in just a brief period with the introduction of domestic livestock. This caused a rapid decrease in the guanaco populations owing to the lack of food and the indiscriminate hunting they were subjected to, both because of competition with cattle and to feed the dogs, obtain invaluable pelts from the "chulengos" (guanaco offspring), and make charqui (dried meat) too.

A statement made by Flores Ochoa (1982:65) summarizes this situation. He wrote: "[T]he present picture of the dispersion of South American camelids does not reflect the situation present in the Andes in the mid-sixteenth century, when Pizarro and his men arrived, and after them farmers and herders of plants and herds they were used to husband, exploit and eat in their homeland."

We thus find that in the end, the current location of camelids, and of course more so of the domestic than the wild ones, is more a result of cultural factors *largo sensu* than biological ones. This was what Gilmore (1950:436) believed; he also claimed that this should be ascribed to the fact that camelids do not thrive in other countries where they have been introduced, whereas they live quite well in zoos—a point only partially correct. Franklin (1982:473–474) has also studied this point and admits that the observations Maccagno (1932) and Link (1949) made regarding the fact that llamas prefer to graze in dry lands and not in humid and marshy soil, whereas the alpacas prefer them, cannot be rejected and are correct. Besides, Franklin (1978) himself proved it. We have seen that Koford (1957) noticed that lamoids have a decreasing need of succulent plants, in the sequence of alpaca, vicuña, llama, and guanaco. This makes Franklin (1982:473–474) note that there is a correspondence with the increasing geographic range. Perhaps one of the factors that sets the limits for the distribution of these animals is the availability of seasonal green food and the various ranges of tolerance of dry food. Franklin even states that if the llama derives from the guanaco (something not proved, as was seen), it is paradoxical that it has a more restricted area than its ancestors had. Despite all this, Franklin nevertheless agrees with those scholars who believe that in the end, the current distribution of camelids is due more to cultural than to biological factors.

The groups of Ayacucho herders studied by Flannery et al. (1989) are among the best available examples of the survival of an ancient mode of life. They also show how the process begun in the sixteenth century still continues today. This is a human group that is being pushed toward the margins of society by widespread political and economic changes. Today they live in a region no one wants, but, as Flannery et al. point out, "they live in it on their own terms, still immersed in an Andean cultural tradition whose antiquity can be glimpsed in the sixteenth- and seventeenth-century Spanish documents. There is tremendous resilience in these herders" (Flannery et al. 1989:2).

In this case the cultural pressure shows clearly in the Ayacucho zone of Quinua-Huamanguilla, both on llamas and on sheep, where the animals are forced to use land that would not otherwise be used, simultaneously avoiding any invasion of the cultivated fields in the *quechua* zone (Flannery et al. 1989:39).

The Pampa de Anta (known of old as Xaquixaguana), in the province of the same name close to Cuzco, is another example. Its altitude ranges between 3105 and 3331 masl. Today it is essentially an agricultural zone, but camelid herds grazed here in pre-Columbian times. There are archaeological data from the Early to the Late Horizon that prove it. There are also documents that prove that the lands of Anta were pastures in Inca times, where herds of llamas and alpacas almost certainly were kept. The loss of these animals has resulted from different causes, such as the civil war between Huascar and Atahualpa and, in Colonial times, the demand for pack and meat animals for the city of Cuzco; the Spanish civil wars, which required provisions for the soldiers; the introduction of cows and sheep; and there are even reports of epizootic diseases that apparently had a lot to do with the changes under discussion (Flores Ochoa 1982:76–81).

Another major factor related to these changes was taxation. It is true that a sizable number of domestic camelids still remained between the Inca period and the late sixteenth century, when the inspections (*visitas*) were carried out. However, Murra (1964b:423) notes regarding livestock that "[t]heir number had fallen after thirty-five years of European domination, but enough still remained to be easily sold or bartered when the annual tribute could not be collected with the work of the *mitani* in the mines." Since pre-Inca times, llamas and alpacas had been used as a sort of "bank" or reserve for times of drought, frost, and other calamities. This is confirmed by the statements the Lupaqa made in the aforementioned *visita* conducted by Garci Diez de San Miguel (1964:24): "[T]he offspring of the community cattle are spent as he has declared, so they do not have to take out or put anything in the community chest." As Murra notes, by 1567 this practice had extended to enable the natives to fulfill the European exactions, given that the meat, skin, wool, and energy of the camelids could be turned into commodities, whereas the same was not true of other resources available to the Lupaqa, either agricultural resources or those from the lake. Murra states that when mentioning livestock, the Indians said "hacienda" [asset, estate]. This term was explained by Bertonio ([1612] 1956:Vol. I, 261) years after the inspection made by Garci Diez, in what certainly is one of the best Andean dictionaries: "[T]he Indians call the rams (of the land) alone their *hazienda* [asset]." By this, Bertonio meant that only the llamas were easily converted into cash or other goods, both in the Andean and the European economies (Murra 1964b:423).

Another factor that influenced the decline of camelids, particularly that of the herds of llamas, was the European cuisine. Curiously, only Rostworowski (1981:52) and Tschudi (see below) have drawn attention to this point. We have already seen how the chroniclers themselves emphasized, for example, the fact that animals were killed just to eat their brains.

Some scholars have suggested that changes in the climate could have affected the camelid populations. Thus Izumi Shimada (1985:XVI) hints that the El Niño phenomenon, which influences the contractions of the lomas, might have had a long-term effect on the movements of llamas coming down from the highlands to the coast. Dedenbach Salazar (1990:106) mentions "possible changes in the climate" that could have affected the herds, but her only source is Cardich (1977). There is no denying that small changes in the weather have indeed taken place (though in no way large-scale ones), but I am skeptical regarding the impact this could have had on the large mammals. This issue is often mentioned but only in passing, and no detailed study has ever been made. The archaeological data raise doubts. In any case, it is a topic that should be handled with great care and only after due investigation.

Once again, Tschudi (1885:98–100; 1891: 100–101; 1918:213–215; 1969:129–131) is a scholar who made a serious and documented study of this issue. (I am once again forced to indicate that the 1969 translation is not just flawed; it omits data present in the original and adds others that were not. This is but one example. When mentioning the killing of animals, the original text reads *hunderttausende*, or "hundreds of thousands"; the 1969 translation reads "thousands," which is quite different.) Tschudi believed there were three major reasons behind the decline of these animals. "[F]irst, the excessive work the pack animals were subjected to and their ill-treatment under the churlish Spanish soldiery, to which hundreds of thousands of llamas died in just the early years of the conquest." Then there was the "reproachable

arrogance of these wanton men, who according to one of their own chroniclers (Fernando de Santillán . . . [1986:409] . . .) killed multitudes of llamas just to eat the brains (132) and beheaded ten or twelve to choose the fattest to their taste, leaving the rest unused, plus the consumption of meat by the Conquistadors, which was great and much higher than it had ever been in the time of the Inkas." In footnote 132, Tschudi recalls the statement made by Alfonso Palomino, an eyewitness, who recounted how during the conquest of present-day Ecuador, the Spanish soldiers killed a huge number of llamas just to eat their heart (for more details, see Chapter 9).

Second, Tschudi lists the mange "called koratsa [*Karatssa* in the original] by the Khetsua Indians and Uma by the Aymara." Here it is worth noting one detail. Tschudi says that this disease was present in ancient Peru. "The proof of this can be deduced from the fact that the Indians had special deities that were called upon to free their livestock from the plague. Due [sic] to the change in circumstances brought about by the conquest, the disease took on a most pernicious nature, both intensively and expansively, as it probably did not have in previous centuries." Now, we know that there are actually two different mites, one of which was probably introduced by the European livestock, as was already discussed elsewhere (see Chapter 2). Tschudi then refers to the chroniclers and animal morbidity in the sixteenth century, which have already been discussed, and concludes thus: "I do not know whether it came anew as a plague in the seventeenth and eighteenth centuries, something that is not doubtful, but I do know what happened in 1826 to 1828 and 1839 to 1840, a time when I myself saw hundreds of sick animals."

The introduction of European animals is the third cause listed by Tschudi, one that has already been extensively discussed (see above). Although impassioned, the words used by this remarkable Swiss scientist are accurate. He wrote: "We clearly see that with the number of llamas diminished by two-thirds, and the remaining third exposed to the brutal fury of the Spaniards, their breeding was entirely ruinous; finally a reaction was hindered by the introduction of solidungula and ruminants. Asses and mules, with their higher resistance for work, have gradually lowered enormously the number of pack llamas, despite the greater amount of merchandise traded between the coast and inland." Tschudi also adds that the railroad was another decisive factor in the disappearance of these animals from southern Peru.

Some have tried to ascribe the hunting and decrease of wild camelids to dogs, but actually not enough data have been presented. What I found on this topic is given in Chapter 7. Dedenbach Salazar (1990:163) studied this point and concluded that the use of these animals in hunting was apparently sporadic. This is not true. Although from the available evidence one cannot say that it was a widespread custom, it was indeed used with some frequency. What is more, the natives rapidly adopted it.

Melody Shimada (1982:305) wrote: "Garcilaso de la Vega (1966 [1609]:517) claims that the Spanish killed off guanacos in early Colonial times by hunting with greyhounds, but that vicuñas survived because they were faster than dogs at their altitude." A textual source should not be distorted thus. Garcilaso did not say this. He has two passages concerning this matter, and Shimada cites only one of them. The first one tells how many Spaniards began trading trips to "avoi[d] idleness." "Many of them delighted in accompanying their possessions, and because of the slow movement of the flock, they would take with them a pair of hawks, *retrievers, greyhounds, and their arquebus, and while the flock pursued its unhastening march, they would make off in one direction or another and hunt, so that by the time they stopped for the night, they would have caught a dozen partridges, or a guanaco or vicuña or deer*" (Garcilaso de la Vega 1966:Bk. 8, Chap. XVI, 515, 516; 1959:151; emphasis added). The other passage, the one mentioned by Shimada, discusses the "wild flocks"; when describing the vicuña it reads: "It runs so fast that no greyhound can catch it, and *it must be killed with arquebusses and by trapping, as in Inca times. It grazes in the highest desert, near the snow line*" (Garcilaso de la Vega 1966:Bk. 8, Chap. XVII, 517; 1959:152–153; emphasis added). Shimada is mistaken and attributes to Garcilaso what he did not write.

Clearly, then, the new breeds of dogs the Spaniards brought with them, trained for hunting, somehow influenced the killing of wild camelids. What diminished their effectiveness was the speed of the latter. This clearly follows from the writings of the chroniclers. The actual impact this type of hunting had on the destruction of the native animals is hard to establish. It was probably less critical compared with the other factors under study. Still, it cannot be ignored.

Shimada and Shimada (1985:21) and Sumar (1985:11) have also discussed the problem posed by the disappearance of the camelids and its causes, but do not add anything new. (To tell the truth, Shimada and Shimada [1985:21] credit Rostworowski [1981:60–67] with the statement that European domestic stock displaced the camelids. This is not true; not only does Rostworowski not say this, but the point is not even mentioned in the pages cited. And the terms when she does discuss them [see below] are different.)

Few scholars have specifically discussed the disappearance of camelids from the coast. Lanning (1967a:17) exclusively credits it to the competition by the animals the Europeans introduced: "The donkey carries bigger loads than the llama and coastal Peruvians prefer beef on their tables and sheep's wool in their clothing." Rostworowski (1981:52) ascribes their vanishing from "the plains [the coast]" to several factors; she says one of them "was the desire the Spaniards had for a meat-based diet, and they had a marked taste for tender offspring which they called lambs." Then, Rostworowski continues, the llamas were needed for transportation in the Spanish civil wars. She supports this with Cieza de León, who says of the chiefdom of Chincha that "almost no sheep of the land are left because the wars the Christians had with one another consumed the many they had" (Cieza de León 1941:230 [1984:Part I, Chap. LXXIII, 220]).

Dedenbach Salazar (1990:99) in turn posits several factors to explain the drop in the number of camelids on the coast. One of them is the climate, but she does not provide convincing support for this claim. I believe that climate should not be considered a factor, because once again, there is no archaeological evidence of climate change throughout the Holocene (see Bonavia 1991: 45–47). The second cause listed by Dedenbach Salazar is that the animals converged on the lomas;

overgrazing and the indiscriminate felling of trees led to their loss. This is likely, particularly with the introduction of animals such as the goat, which is of a destructive nature. But we should not forget that the lomas were used since time immemorial, and that camelids definitely did not graze there alone. The third cause listed by Dedenbach Salazar is overgrazing in general, and disease.

Today, the llama's distribution in the Andean cordillera lies approximately in the area between 11° S latitude and 21° S latitude, comprising land in Peru, Bolivia, Chile, and Argentina. The alpaca lives at or over 4000 masl, mainly in southern Peru and Bolivia. I emphasize that this distribution is due to the cultural perturbations that began with the Spanish invasion and culminated in the disappearance of llamas and alpacas, as well as guanacos and vicuñas, throughout much of their original land, essentially in Peru and Bolivia (Novoa and Wheeler 1984:117; Sumar 1992:82; Wheeler Pires-Ferreira et al. 1976:489). Thus, to give no more than a few examples, there are no camelids in the Cajamarca area (Melody Shimada 1982:304), and some vicuñas still live in the Callejón de Huaylas (Tovar 1973:14), while llamas and alpacas have disappeared from Junín (Wheeler 1985a:29) and Ayacucho, where they are found only in the highest areas where the other species introduced by the Europeans cannot survive (Pozzi-Escot and Cardoza 1986:19).

It is surprising that a scholar like Troll, who long studied the Andes and made major contributions, never understood this phenomenon, to the point that in one of his most important studies he wrote: "[W]hat E. Hahn (1896) has emphasized is most interesting, to wit, that the use of llamas and alpacas as domestic and useful animals extends spatially over an area far smaller than the natural area of distribution of the wild forms. Instead, the region where the llama is raised coincides almost exactly with the Peruvian cultural sphere. The less important [sic] alpaca has an even more restricted distribution, and is found only in southern Peru and in some parts of Bolivia. The northern limits of the continuous area of distribution of llamas now lies in central Peru, where the puna ends. It has no role worth noting in the latitude of the Cordillera Blanca (personal communication of H. Kinzl)" (Troll 1958:28). It follows from this passage that Troll did not realize that he was facing an artificial distribution of camelids, and this partly disproves his proposal of a cultural geography which has gained much renown.

Flannery et al. (1989) made a good study of these phenomena and offered some original contributions. They noted that from the available evidence it follows that since the destruction of the Inca state, the environment where camelids grazed changed at a remarkable rate. This is because before 1532 there were several mechanisms to combine small herds with large ones. Under the Inca, the state used its human resources to support the state herds, which apparently were relatively immune to the imbalance in the ratio between males and females, something that did not hold true for the small herds. It was for this reason that these state herds formed an almost inexhaustible reserve. The mechanism whereby a llama moved from one herd to another was an annual ritual called *suñay*, which seems to have existed in Inca times and was used by the herders who looked after the *waqchallama*, individual animals, or those belonging to the *ayllu*. In this way, the survival of the herds was not as challenging as it now is. Flannery et al. therefore believe that the pre-Columbian system was less vulnerable to the demographic problems posed by small herds, there being more potential sources where llamas could be found. A new mechanism that exerted a greater pressure for *suñay* to take place was probably established as soon as the pre-Columbian state collapsed; this is one of the few remaining mechanisms from that age for attaining new breeding stock when needed. But it is likely that the *suñay* had to increase when the transition from a pre-Columbian culture to a new, mestizo culture took place. Before the Spanish conquest, the reserves of llamas were immense, and donations were probably made between blood relatives. Later only small herds, managed at an individual level, existed, so the *suñay* spread. This is what Boyd and Richerson (1985) would explain as a change in the cultural transitions from one generation to another (Flannery et al. 1989:206; readers interested in the models proposed by Boyd and Richerson should read pp. 206–207).

Clearly, no mechanism could succeed in the face of the new Spanish economic policy that imposed its own animals, and even gave them to the natives to make this imposition even more effective. Such was the case among the communities in Condebamba, which received great herds of sheep in the sixteenth century (Espinoza Soriano 1974:94). Flores Ochoa (1970:67) thus believes that the control of livestock, in this new way based on number and quality, became one of the mechanisms the Spaniards used to control the native peoples in the altiplano. At first the Indians were able to handle the European pressure because they still had their communal herds, but their dependence worsened once these were destroyed, and the Colonial government's control became much easier.

The case of the guanaco is even more serious. In Peru this animal is practically extinct, either through excessive hunting or through competition with domestic animals. Small populations survive in Calipuy, in the Upper Santa River and in the Pampa Galeras Reserve, in southeastern Ayacucho. There are larger populations in southern Chile, but here one of the factors, among others, that led to their decline was the expansion of agriculture (Franklin 1975; 1982:476; 1983; Raedecke 1976, 1979; Flannery et al. 1989:91).

Raedecke made an in-depth study of the guanaco populations in Chile. He concluded that the most lasting and probably the major cause of their decline was the change that has taken place in the prairies. This was due to the introduction of sheep and bovines. The guanaco was the characteristic herbivore of the southern pampas, but the introduction of sheep reduced the normal capacity of guanacos here. The populations that survived in the face of pressure from sheep are those that occupy marginal lands that have no economic significance for stock-breeding. However, the morbidity rate must have been high due to the poor conditions of the prairies; moreover, the guanaco was unable to find a safe haven even there, as the growth of the sheep population continued driving them away and forced them to live in forests burned to claim more land. The result is that the current distribution of this species is far more restricted than it once was (Raedecke 1976:14–16).

The guanaco is the only loser in its competition with sheep. Its mortality is mostly due to the lack of food in winter, a result of the summer

expansion of sheep and the overgrazing they cause.

According to Raedecke, several calculations have been made on the competition between the guanaco and sheep. Howard (1969) reports that Argentinian farmers consider that one guanaco eats as much as five or ten sheep. However, others claim it is less and comes to between three and five sheep. Raedecke notes that the correct equivalence is much lower. He based his calculation on the comparative weight of these animals. One guanaco equals 2.5–3 sheep, but the feeding habits of both animals must be taken into consideration, as well as the availability of different plants. By lowering the consumption of plants not eaten by sheep, the figure Raedecke suggests is that of one guanaco per 1.5–1.8 sheep. He concludes that "[t]he competition between both species affects sheep negatively in just 0.03%, whereas the guanacos are greatly affected, for 78% of their morbidity is due to the lack of food in winter" (Raedecke 1976:95–97).

12.2 SOUTH AMERICAN CAMELIDS IN OTHER PARTS OF THE WORLD

Thus far we have presented the South American camelids in their own environment. However, we have seen that these animals were taken to the Old World almost from the first European contact. It is true that at first it was just out of curiosity and to show the riches of the land newly conquered. However, this process has continued, if at a slow rate. I was unable to find the data that must surely exist on this topic; however, and though it might seem useless, this study is needed and should be methodical, that is, it should monitor the life of these animals in other milieus, their behavior, physiology, output, and so on. For the record, this does not mean animals living in zoos but those taken to other countries for specific purposes that are not mere exotic fancy. I am convinced that once this is done, if someone ever undertakes this task, we will find that these animals have no problem living in other environments quite different from the Andean ones that are not precisely high-altitude environments, as generalized in the literature. Instead it will be found that these are among the animals most suited to live in different environments. (Let us take the European case just as an example. At present there are about 7,000 llamas, 2,000 alpacas, and 1,000 vicuñas [pers. commun. of Walter Egen to Javier Delgado, who in turn reported this to the author, letter of 16 April 1999].) Some relevant data are provided here so that the reader can see what I am trying to say.

Camelids, llamas in particular, have been taken to different countries on different occasions after the Spanish conquest. However, I know of no accurate data on the causes of what usually turned out to be failures to adapt. For Franklin (1982: 468), these failures were due to humid conditions and diseases because camelids prefer a temperate climate. Gilmore (1950:432) instead believes that cultural factors played a key role in their failure to adapt. I concur, but this does not mean that Franklin is mistaken—far from it. (Sumar [1988: 25] discusses this point in just a few lines.)

12.2.1 Mexico

Only vague data regarding Mexico were found. In a footnote, Álvarez (1987:n. 64, 166) discusses the "carneros" described by Ocaña (1987), explaining that these were camelids. Among other things he writes: "as were unsuccessful in Mexico the six llamas Pizarro sent as a present to his cousin Hernán Cortés (cfr. Cárcer and Disdier ibid., pp. 423–424)." The passage from Cárcer and Disdier unfortunately does not appear in the book, where the bibliography only figures in the footnotes.

Romero (1936:371) wrote of "the Peruvian llama, transported to distant countries where its reproduction was unsuccessful. The Spaniards took it to Mexico . . . but as with the camel in Peru, this quadruped did not acclimatize anywhere, but Hernández (*Quatro Libros de la Naturaleza, y Virtudes de las Plantas, y Animales que están Receuidos en Uso en Medicina en la Nueva España*, Mexico City, 1615) includes it in his book, gives its picture and gives it the Mexican name of *Pelonichcatl*, noting that it was a species taken from Peru." I was unable to verify these data.

Pease (1992:246) points out one interesting bit of information. He writes that "Cortés was . . . an active trader in early [Spanish] Peru and his ships made many voyages. . . . The operations of

the conqueror of Mexico and his heirs in Peru lasted until the 1570s–1580s; in addition to the export of Peruvian silver in his ships, there are some indications of trade in 'sheep from Peru,' clearly llamas."

I believe Pease is wrong because, as discussed later (see below), the operations of the heirs of Cortés in Peru ended in 1559 and did not continue "[u]ntil the 1570s–1580s" (Pease 1992:246). In fact, the source used by Pease (pers. commun., 16 October 1993) was Borah (1975 [1954]). In his book, the latter explains that the estate of the Marquisate del Valle was established in Mexico at the death of Cortés in 1547. The decision to begin trading with Peru was made then (Borah 1954:37 ff.).

The trips to Peru, four altogether, began in 1554 and ended in 1559 (Borah 1954:35–59). Martín Cortés, son of Hernán, "hired Diego López de Toledo to go to Peru with a power of attorney and the promissory notes signed by Ladrón de Guevara, by Royal Treasury officials of Lima, and others," and collect some outstanding debts that had gone unpaid (Borah 1954:61). This Ladrón de Guevara was captain of one of Cortés's ships who lived in Cuzco since 1559. Diego López arrived in Lima in the summer of 1556 (Borah 1954:40–61; 1975:85–123).

Borah then writes: "He apparently was able to secure payment from the Royal Treasury and all others but Ladrón de Guevara, for their notes disappear from the later list of outstanding debts owing the state." "The debts of Ladrón de Guevara probably never were paid, for they appear as still outstanding in an inventory of the marquisate dated January 10, 1570. In the inventory they follow hard upon a listing of debts owed by Pedro de Ahumada Sámano, deceased (53). The last reminder of the marquisate's Peruvian ventures disappeared during the 1570's when the herd of *ovejas del Perú*—llamas or related species—brought to the Peñol de Xico for domestication in New Spain reached a maximum of sixty head and then died out (54)" (Borah 1954:61–62; 1975:123–124). (Note 53 reads: "Account of the income of the Marquis del Valle in the years 1568 and 1569, prepared by Juan de Cigorondo, accountant of that estate," Mexico City, January 10, 1570, in Paso y Troncoso [1939–1942], Epistolario de Nueva España XI,

57–60. Note 54 refers to the "Inventories and rental agreements on the marquisate properties of the Peñol de Xico 1570–1574," MSS in AGN, Jesús, leg. 273, exp. 4: 8.)

Peñol de Xico is in the Municipio of Chalco, in the state of Mexico, at an altitude of 2300 masl. The mountain of Xico is at the southeastern part of the Valley of Mexico. "It rises like an island in the middle of the onetime Lake Chalco; on its northern and northeastern slopes is the town of San Miguel de Xico." (Data given by Raúl García to José Ochatoma, who in turn passed them to the author in letters of 7 December 1993 and 9 March 1994.)

Regarding what Borah wrote (cited above), all I need add is that these were surely llamas or perhaps alpacas, so the term "domestication" is misused. He probably meant acclimatization or something like that.

I was unable to find out whether there are any other contemporary news accounts of these animals besides those listed.

12.2.2 United States of America

Somewhat more data are available for the United States, and there certainly must be more that I am not aware of. According to Novoa and Wheeler (1984:126), the introduction of camelids to this country has been attempted since 1930. Sumar (1988:25; 1992:86) likewise notes that a small number of animals were taken there in that year, but he does not give more details, and I assume that his source is Novoa and Wheeler (1984:126).

In this case there are some figures available. Sumar (1988:25 and Table 1, 26) notes the presence of 500 alpacas, 5,000 llamas and "few" guanacos. However, in a later study he mentions 8,000 llamas (Sumar 1992:86). Knowing how he handles his data, the information given by Sumar must be taken with all due precautions, all the more so because they do not agree with those of a group of most dedicated scholars whom I will now cite. Based on the data in Tillman (1981), Franklin (1982:468) mentions a "healthy growing" population of llamas ranging between 1,600 and 3,200 head. Novoa and Wheeler (1984:Table 14.1, 117) give the figure of 3,000 in a table, but in the text they say it ranges between 3,000 and 5,000 head. Flores Ochoa (1990b:Table 1, 92) mentions 3,000

llamas, but a figure that is apparently more accurate is given by Shimada and Shimada (1985:5), who give a minimum number of 3,418 head of llamas. They base this figure on the *International Llama Association Breeders and Owners Directory*.

These numbers are interesting because they show how the breeding of these animals has increased in the United States of America. In fact, according to the latest statistics available to me, for January 1999, the Alpaca Registry notes the presence of 20,465 alpacas and 139,734 llamas in the U.S. (Dar Wassink, administrator of the Alpaca Registry, letter of 15 June 1999). These are the registered animals, but I am told that the population is greater because there are unregistered animals. According to another source there would thus be about 180,000 llamas and 20,000 alpacas (pers. commun. from Ingrid Wood to Javier Delgado Sanibañez, who in turn reported it to Elmo León, who sent it to me in a letter of 16 April 1999.)

According to Shimada and Shimada (1985:5), the highest concentrations of llamas are in California, Oregon, and Washington, but there also are some in Texas and Florida. The highest number of animals is in Oregon, at about 1000 masl. However, Novoa and Wheeler (1984:126) report that these animals are well adapted to various environments and altitudes.

Shimada and Shimada (1985:5) draw attention to the great ease with which these animals are obtained, as well as how easy it is to feed them. Izumi Shimada (1982:162) reports that a large herd of llamas has been successfully raised at Patterson Ranch, in eastern Oregon, for more than a decade, with no health problems. At the time of the report it had 425 head. Shimada quotes the owner: "[Llama herds are] easy to maintain having more diet tolerance than horses *accepting any good hay grass or grain and [are] able even to consume moldy hay without harm*" (emphasis added). He then lists the presence of smaller herds in several parts of the United States, in upstate New York and in Florida.

Melody Shimada (1982:306) has chronicled the herds of llamas in Sisters, also in Oregon, at 1000 masl. I do not know whether there is any connection with the Patterson Ranch, but I assume there is none, because the number of animals is different from that given by Izumi Shimada. Melody

Shimada notes the presence of around 400 head, with about 5% infant morbidity.

Flores Ochoa (1990b:91) made an interesting comment on these animals. He notes that with their practice, the breeders in the United States are proving what some Peruvian scientists have claimed, and even showed that there are different breeds of llamas that give high-quality fibers, comparable with those of the alpaca and which can even compete with it. Their quality enables the manufacture of thread that can be used to make textiles with an excellent finish. Still according to Flores Ochoa, the American "llameros" are raising animals specialized in fiber production. They have verified that llamas raised with care and a good diet and that are not used for the harsh task of transportation can yield high-quality fibers. In fact, Novoa and Wheeler (1984:126) report the use of these animals for wool. A newspaper item confirms that "several craftsmen have discovered the quality of its wool for textiles" (Anonymous 1992e:A1).

According to some American friends, llamas are also being used for tourism, to carry cargo by people leading a country life, or as pets. The newspapers recently reported (Anonymous 1992e: A1) that llamas are being used in the United States to carry golf clubs; specifically, they are used as caddies in Talamore, North Carolina. It is specified that each llama can carry 40 kg (this is correct), which is equal to two bags with golf clubs and other accessories. It is also noted that "[t]hey are sociable animals." This news report also confirmed that there are several llama ranches in the eastern United States, where animals are raised for sale and for export. These animals were imported from Chile.

North American ranchers have also discovered that for the herds of sheep, llamas are an excellent protection against coyotes. According to Eliot (1994:133), "[d]uring the 1980s ranchers began using them to guard their sheep, including these in western Montana."

12.2.3 Canada

As of May 1999, 6,028 alpacas and 1,012 llamas had been recorded, according to the Canadian Llama and Alpaca Registry (Bruna Bonavia Fisher, pers. commun., July 1999).

12.2.4 Australia

Alpacas were introduced into Australia in the nineteenth century, but the available data are contradictory. Neveu-Lemaire and Grandidier (1911) write that in 1854, Ledgers [sic] bought 800 animals in the Laguna Blanca valley, east of Antofagasta (between 2600 and 3000 masl), which were to be taken to Australia. Three hundred of them were lost, but the rest reached their destination, adapted, and reproduced well. However, based on data in Vietmeyer (1978), Novoa and Wheeler (1984:126) indicate that it was in 1858, on a commission from the Australian government, that Charles Ledger took 274 alpacas on a four-year-long trip through the Andes and five months at sea. Forty-nine animals were born six months later, and by 1861 there were 417. Novoa and Wheeler note that boyd size and wool output were much increased, despite the animals being raised at low altitudes. The animals adapted quite well to the new environment, but their development was truncated by the introduction of sheep in 1864.

Sumar (1988:26; 1992:86) presents similar if less detailed data. Apparently Vietmeyer (1979) was also his source, but in a different account from that used by Novoa and Wheeler. Sumar (1988: 26) includes a statement ascribed to Charles Ledger which he claims to have taken from Vietmeyer (1979), but without giving a page number. According to this author, Ledger was impressed by the adaptability of the alpaca, and presumably wrote that "[n]o animal in the creation, it is my firm conviction, is less affected by the changes of climate or food." Sumar (1992:86) repeats the same data in another study, but here he does not even cite his source; however, Vietmeyer (1979) appears in his bibliography. Given the carelessness with which Sumar uses data from other scholars, there is no way of ascertaining whether this is correct or not.

According to information received from Chris Tickwell (letter of 9 May 1999), some 26,300 alpacas and 1,000 llamas are estimated to exist in Australia.

Curiously, the newspapers recently published a photograph showing a five-year-old child riding a saddled llama. The photograph was taken in downtown Sydney. Among other things, the caption reads: "Llamas were taken to Australia from the frigid mountains of New South Wales, as part of a tourist campaign" (Anonymous 1992a:B4).

It is worth noting that Australia is the only place in the world where introduced camels turned feral in large numbers and live in natural populations. According to recent estimates, there are 43,000 animals in Australia and 31,570 in the Northern Territory (Heucke et al. 1992:313).

12.2.5 The Philippines

According to Romero (1936:371), llamas were also taken to the Philippines, but they did "not acclimatize." I have no more data on this point.

12.2.6 Spain

In his comments on the description Ocaña (1987) gave of the "carneros," Álvarez (1987:n. 64, 166) explains that Phillip II "wanted to acclimatize the llama in Spain, but failed." In other words, this happened in the sixteenth century. Later, according to Brack Egg (1987:62), llamas and vicuñas were imported into Spain from Peru in the early eighteenth century in order to adapt them, but the venture failed "due to the climate." Álvarez does not give his source, while Brack Egg based his statement on León (1932b).

Today camelids live in Spain only in zoos—an estimated population of 300 animals (Ángel María Martínez Churiaque, letter of 11 May 1999).

12.2.7 France

These animals were also introduced into France (Neveu-Lemaire and Grandidier 1911:60), but no details are available. Sumar (1992:86) claims that this happened in the eighteenth century, but he does not support his claim. Brack Egg (1987: 62) says they died in 1772 in the School of Alfot: a llama and a vicuña that presumably were "the first camelids known in France."

Harcourt (1950:241, 262) reproduced a letter written by Camacho in 1803 in which he offers llamas and alpacas to Mme. Bonaparte. But Harcourt was not acquainted with what actually happened. This was established by Hamann Carrillo (1955–1956). Madame de Beauharnais collected animals beginning in 1792, and Bonaparte allowed her to establish a "menagerie" in the Castle of Malmaison. In 1804 the Spanish king gave

a "royal order" to the viceroy of Peru, stating that acceding to the request made by the Queen of Spain, Madame Bonaparte was granted the vicuñas and alpacas she had requested, commanding the viceroy to get them. (Helmer [1956] reproduced this "royal order" with which "a dozen vicuñas and another dozen alpacas" were granted to Mme. Bonaparte.) Camacho does not appear in the document at all.

In a document dated 1805, the Marquis of Avilés recounts the problems he had in finding and shipping the animals. The shipment was sent by sea to Valparaiso, by land to Buenos Aires, and from there to France by sea. In case the animals died, the Marquis suggested that others should be found in Buenos Aires itself, for it would be "easier to gather them in places close to the Cordillera." Julián García traveled with the animals "with no salary at all," "and was accompanied by three Indians who understood their management and the treatment of diseases that usually assail them" (Hamann Carrillo 1955–1956).

Thirty-seven animals in all were shipped: 10 vicuñas (4 males and 6 females), 3 guanacos (1 male and 2 females), 19 alpacas (9 males and 10 females), and 5 llamas (3 castrated, 1 male, and 1 female). The full cost of the shipment was 5,345.6 pesos reales. Nothing is known of what became of them (Hamann Carrillo 1955–1956).

Curiously enough, lately several news items regarding this subject have appeared in the dailies. It was reported that a group of llamas were introduced close to Avignon. According to the papers, whereas sheep eat only the better-quality, nonweed vegetation such as grasses and soft, leafy plants, and goats tear the bark off trees and devour plantations, llamas do not cause any harm and even protect the soil with their "hooves as soft as cushions." For this reason they were called "ecologic lawnmowers." The news item added that these animals are useful to prevent fires because they keep the areas around forests free of shrubbery. The same report noted that tours with llamas were already offered in the French Alps, and stated that Chile would supply France with 100 alpacas for wool that same year (Anonymous 1990:B5). Another news item published in this same year reported the importation of some 300 camelids, divided between llamas and alpacas, to

France from Chile, "aimed at beginning the official breeding of camelids in Europe for the first time." These were 245 alpacas and 45 llamas, apparently purchased by Patricia Clermont Tonerre, a renowned French breeder who had long dedicated herself to raising sheep. According to the news report, she "opted for Andean camelids because she found that besides giving excellent wool and meat, they can be used to weed the fields as they usually only eat weedy plants which are usually responsible for the forest fires that break out in southern France in the summertime." It was also reported that over half the 250 females in the herd were already pregnant. It was likewise specified that none of the animals would be killed because all had been specifically selected for reproduction (Rosas 1990:B2). According to the latest data available for France, there are now 2,000–3,000 animals, between llamas and alpacas (Jean-Jacques Lauvergne, letter of 4 May 1999). However, only 340 alpacas and 563 llamas had been registered in the Ministère de l'Agriculture et de la Pêche (Etablissement Public National de Rambouillet) of France as of July 1999 (Martine Lavelatte, pers. commun., letter of 28 June 1999).

12.2.8 Other Countries

Neveu-Lemaire and Grandidier (1911:60) likewise report that a not-too-successful attempt was made to introduce camelids into Holland. The same was done in England, but the data available are scant. All that is known is that one Thompson raised a great herd of alpacas for the duke of Derby, who wanted to have them in Scotland (Neveu-Lemaire and Grandidier 1911:60), but when this was done is unknown. Sumar (1992:86) reports the introduction of camelids into England by Walton in 1818, and notes that nothing is known of what became of these animals, for they vanished. Sumar assumes that the mange, which becomes more deadly under a strong humidity and with the summer heat, may have been the cause. But Sumar once again does not give his source.

However, I have been told that today there is a significant number of these animals in England, but the exact number is unknown. It is estimated that there are 2,000–2,500 llamas, and probably the same number of alpacas or slightly more (Paul Rose, letter of 20 April 1999).

As for Germany, in 1996 the Bundestag passed a law stating that not only were llamas allowed in Germany as pets, they were also allowed as stock animals, essentially for their fiber. With this law, a significant increase in the number of camelids present in this country is expected (Javier Delgado, letter of 1 April 1999). At present it is estimated that there are 1,700 llamas, 800 alpacas, and 200 guanacos in Germany (pers. commun. of Walter Egen to Javier Delgado, who in turn reported it to the author, letter of 16 April 1999).

From an interview with Joseph-Anton Kuonen held in Switzerland, it appears that 50 camelids were imported there in 1996 and 150 alpacas in 1998. These animals are in Termen, a small town located at 950 masl (Salas and Cruz 1998; see also Anonymous 1998d).

In Italy there must be some 60 alpacas, 6 llamas, and 6 guanacos in Umbria, not counting those in zoos and circuses. In the Roman countryside there are 60 llamas, and 30 more on the island of Sardinia (Carlo Renieri, letter of 19 May 1999).

It is reported that in 1987 there were 800 alpacas in Russia and also in New Zealand. There were also llamas in both countries (Nuevo Freire 1994:6).

In the case of Israel, it is known that the Ranch Mitzpe Ramon, in the Negev Desert, purchased 175 alpacas from Chile in 1990, which number had grown to 300 by 1994 (Cohen 1994: 116–117). According to more recent data, at present there are about 400 alpacas and 70 llamas (pers. commun. of John Eliasov to Boris Krasnov, reported to Alexander Grobman in letter of 12 April 1999, who kindly reported it to the author).

I have also been told that camelids have been exported from Bolivia to Lebanon (Rony Garibay Suárez, pers. commun., 10 August 1999.)

12.3. IS IT POSSIBLE TO OFFER STATISTICS ON THE NUMBER OF CAMELIDS THROUGH TIME?

One point that is difficult to discuss is the number of camelids that lived in the Andes through time. We often find mere estimates, and in other cases the figures are contradictory. But for our goals here exact figures are not that important. What matters here are overall data showing what trend can be determined for the future of these animals.

Clearly, the hardest time to deal with is the Inca period, because the statistical data that must have existed were lost when the *quipu* tradition, and the *quipus* themselves, were destroyed as a result of the Spanish conquest. As Dedenbach Salazar (1990:84) notes, the chroniclers were unable to present an accurate picture of the number of animals because they had other goals in mind, far removed from statistical considerations. They regarded these animals as a potential food resource for their campaigns of conquest. The point of view of the *visitadores* (inspectors) was different, because they needed the most accurate data possible, as their goal was to levy tributes, but they acted 40 years after the conquest, once two significant events had taken place: the annihilation of the native animals, on the one hand, and the introduction of European animals on the other. It is not always easy to establish in the documentation whether the animals in question are Castilian or "of the land." Besides, there is a third factor that has already been mentioned, namely, the natives did not always give accurate data, and the inspectors had no way of cross-checking it.

Sumar has discussed this point in several of his studies, but he contradicts himself and does not give his sources, as usual. At one point he wrote that there must have been "some ten million animals [camelids]" in Inca times, with llamas as the predominant animal (Sumar 1985:11). Then later on he states that "[w]hen the Spanish Conquistadors reached the Andes in 1532, they found ten million alpacas and at least as many llamas—together with a similar number of wild species—distributed" throughout the Inca Empire (Sumar 1988:24). However, in a third article he claims that the Bolivian historian Jesús Lara (no reference is given, as usual) tried to estimate the number of camelids present in Tahuantinsuyu: "[S]tarting from the number of camelids sacrificed in the religious festivals, [he] reached an estimate of 23 million llamas and some seven million alpacas" (Sumar 1992:83–84).

Franklin gives an estimate for llamas alone. He says that on their arrival, the Spaniards found "[h]undreds of thousands" (Franklin 1982:467), and notes that before the conquest there were

over a million vicuñas (Franklin 1973:78). However, Torres (1992a:34) estimates "two million." Regarding the guanacos, Franklin (1982:476) notes that "[t]here were millions" and accepts the figures given by Raedecke (1979), who considers that 30–50 million is a reasonable estimate for the native guanaco population.

Dedenbach Salazar (1990:97) tried to estimate the numerical distribution of camelids in different ecological zones, but admits that the data are scant. I discussed this point in Chapter 7, and although it is true that not much information is available, we can glean some rough numbers. Dedenbach Salazar suggests that this can be studied at a regional level with the help of different documents, the *visitas* in particular; however, I insist that the various constraints already mentioned should not be forgotten (see above).

When discussing the herds that went with the Inca army, Flannery et al. (1989:114) state: "Some idea of the size of these military llama trains can be gained from Zárate's ([1555] 1853) estimate of 15,000 animals associated with one Indian general." I find this claim incorrect because in this case Zárate meant the army of Quizquiz, one of the captains of Atahualpa who was heading north, but the chronicler clearly specifies that "[he] came with an army of over twelve thousand Indians of war, and *he had gathered as many Indians and livestock as he had found from Jauja downward*" (Zárate 1968:170; emphasis added). The chronicler explains that the army of Alvarado was in poor condition and could not defeat the Indians, who had made a stronghold in a high sierra. On climbing uphill they carried what they could, and the remaining clothes were burned, "[there] *remaining on the field over fifty thousand sheep* and over four thousand Indian women and men who came to the Spaniards" (Zárate 1968:171; emphasis added). This therefore was not a normal army train at all. No conclusion can be drawn from this.

As far as I know, there are no overall estimates for Colonial times. This was confirmed by Franklin Pease (pers. commun., 16 October 1993). Figures like those given by Sumar (1992:85), who claims that by late Colonial times and in the early Republic only 440,000 alpacas and slightly over a million llamas remained, are useless because they are unsupported. There are some estimates that can give us an idea of how many head there were, but only for specific places. Thus, Garci Diez de San Miguel (1964:229) notes the livestock the friars had: "six thousand and nine hundred and four heads of males and females in the seven cabeceras, not counting Sama nor Moquegua, for these do not have livestock." It is specified that this figure was taken from the native *quipus*. However, there is a slight error in this claim. Garci Diez de San Miguel (1964:229–230) himself lists the sheep, rams, male and female lambs, the goats and kids each cabecera had (i.e., Chuquito, Acora, Ilave, Juli, Pomata, Yunguyo, and Zepita). If partial sums are made, we get 4,484 sheep, 812 rams, 1,026 male and female lambs, 271 goats, and 7 kids. The overall sum is 6,600 animals and not 6,594, as claimed by the *visitador*. If we substract the kids we get 6,593. But we must be careful with these figures, because in this case all animals apparently were from Castile and not "of the land."

An anonymous writer (1965a:380) who left a description of Potosí in the early seventeenth century explains that the livestock that went to the mines loaded with wine, coca, and other provisions "are 340 thousand rams," but adds that "besides these, 60,000 rams arrive each year" with the Indians who fulfilled their mita. Ocaña (1987:166) likewise discussed Potosí, where he was in the mid-seventeenth century. He specified that the usual *mita* for the hill of Potosí had 12,000 rams of the land. And the Jesuit Antonio de Ayanz (1978:218) left at his death a lengthy memorandum on "the Indians and *mita* of Peru"; it is not dated but must be around 1596. Ayanz mentions Potosí and notes that the Indians took there "over 30,000 head" of livestock. It is not easy establishing whether the difference in the number of animals mentioned by the anonymous writer and Ocaña was due to a miscalculation or whether a decline in their number actually took place between early and mid-seventeenth century.

Flannery et al. (1989:109–111) have studied the size of herds. They cite sixteenth-century figures (they use as their benchmark the data for the altiplano, to which I have often referred) for the herds of *waqchallama*, that is, the animals belonging to an individual or an *ayllu* at the community level, that indicate they are the same as those available for the herds of Yanahuaccra, in Ayacu-

cho—that is to say, between 15 and 35 animals. Basing their estimates on the data in Murra [1965: 192] and Diez de San Miguel [1964] for 1567–1568), Flannery et al. list Lupaqa families with no animals, others with 2, 3, 10, or 20, and some with 50–100 head. The biggest herds had 500 to 1,700 animals and were managed by several families (relatives of the curacas). Flannery et al. (1989: 109–111) indicate that according to several writers, the Spaniards were interested in controlling and assessing the big community herds and those of the lords and governmental institutions, while much less attention was paid to the small herds belonging to the common people. They cite the *visita* (inspection) made by Ortiz de Zúñiga ([1562] 1957:326), which recorded 100 individuals living in two hamlets with just 33 llamas, while the ethnic chief had only 6 of them.

Flannery et al. (1989:109–111) comment that the biggest herds in the sixteenth century seem to have been those of the communities, or the *qapaq-llama* of church and state. The best references are for southern Peru and Bolivia, specifically around the Lake Titicaca area. Based on a study by Murra (1965), these scholars analyze the *visita* made by Garci Diez de San Miguel. They conclude that in Xuli there was a herd of 16,846 camelids; 3,242 families lived there. In other words, each family had just 5 animals (see Garci Diez de San Miguel 1964:122–123). At Hilaui the herd had 2,122 animals for 1,470 families, or less than two per family (1.4, to be precise). Flannery et al. insist that great herds are mentioned but not the number of animals per family, save in the case of curacas (see the original data in Garci Diez de San Miguel 1964:113–114). It is known that the animals belonging to families (which were marked) grazed together. It is possible that some "ayllu herds" of more than 1,000 animals that the Spaniards saw were actually 50 herds of 20 animals that grazed together, and belonged to 50 families.

Flannery et al. note this problem because from their own experience in 1970–1972, they found it hard to imagine that 16,846 camelids could have formed a single herd. The number of *aynos* (breeding animals) for such a herd would have reached a number hard to control. Such a large number of camelids would have required management quite different from that which these scholars saw at Yanahuaccra and Toqtoqasa, and the sixteenth-century documentation describes precisely this strategy. According to Murra (1965), the animals were separated by sex and age. Pregnant females were separated from pack animals and from females suckling their *uñas* (their offspring) too. This means that the "family group" and the "male group" of the wild guanaco were partially recreated in the sixteenth century, with the addition of some other groups. The great herds of church and state were almost certainly managed in this way because of the labor available to the Inca, who had full-time herders (Flannery et al. 1989:109–111).

This extensive comment by Flannery et al. (1989:109–111) is extremely important. I will return to it because it posits the need for a reexamination of the *visitas* and part of their data, such as that concerning the great herds that have drawn so much attention and about which much has been written. For instance, although Flores Ochoa, unlike Flannery et al. (1989:109–111), does not question the size of the herds, in his studies it is clear that their size caught his attention. For example, when discussing the herd of Juan Alanoca, a rich Indian with 50,000 head (see Murra 1964b:422), he writes: "The number is *surprisingly high* considering that few modern haciendas had that number, and the biggest, for instance on the eastern Lake Titicaca cordillera, had no more than 20 to 30,000 animals. It is striking that a single individual could have so many after several years of European impositions and rape, and the understandable attempts to cover up their real riches out of fear of the Spanish Crown's levies" (Flores Ochoa 1977b:22; emphasis added).

Current data actually confirm what Flannery et al. (1989:109–111) state. It is worth recalling that in the late nineteenth century, when recounting his way down from the altiplano to the coast, Squier (1877:245, 246; 1974:132) noted that "a herd of more than a thousand llamas . . . filed past us." Citing Albes (1918), Franklin (1982:468) says that in the early twentieth century, the herds of alpacas were small. In 1930, Simoens da Silva (1980:60–61) saw herds of "600, 1,000 and even 1500 llamas and alpacas"; he reported that there was an hacienda close to Juliaca with about six thousand llamas and alpacas. Flores Ochoa

(1975b:302) notes that the size of the herds varies widely and that their average size is related to the area and the pastures. Areas with tough grasses, like Phinaya in the Canchis Cordillera, in Cuzco, have an average of 300 alpacas, 100–200 llamas, 50–100 sheep, and 3–5 horses. In Paratía, close to Puno, the average is 300 alpacas, 100–200 llamas, very few sheep, and almost no cows. In Tucsa, at 4200 masl, in the Canchis cordillera, there are on average 80–100 alpacas, 80–90 llamas, and 90–100 sheep, 2–3 cows, and 2–3 horses. According to Flores Ochoa, the richest people have up to 190 or 200 alpacas. At Phinaya and Paratía, the rich have herds of 500 animals and more, whereas the poor can only have 10–15 alpacas, 20 llamas, and 20 sheep; in the worst case they have none at all.

In the Cuzco area, a family on average has over 300 alpacas and 100–200 llamas. Herds rarely number more than 500 animals (Flores Ochoa 1975a:8). All llamas and 80% of the alpacas are controlled by traditional herders; herds of 30–1,000 animals are kept in communal pastures (Novoa and Wheeler 1984:126; they base their numbers on Novoa 1981).

On the other hand, the data in the *visitas* must be taken with some reservation, but they do reflect a rich animal biomass that is now lost. Dedenbach Salazar (1990:198) made a detailed study of this point and concluded that the figures included in the *visitas* are too low. On the other hand, we lack data for the puna. This is probably due to the scant interest the Spaniards had for this region, given that lower ecological regions had agricultural produce and pastures for the animals they had introduced. This must have been the case with Huánuco, for instance.

Moreover, Dedenbach Salazar (1990:198) emphasizes the difference between tribute-related documents and the "*probanzas de méritos*" [a deposition with witnesses]. Lower figures were probably quoted in the tribute-related documentation to avoid the levies, whereas the *probanzas* try to show the number of camelids given to the conquistadors, so a higher figure was obviously stated. We should likewise bear in mind that after the European arrival, disease tends to obscure the actual distribution or number of camelids.

I was unable to obtain statistics on the number of animals per species in each of the Andean countries. The data included in Table 2.2 have huge gaps, but this is what is available for those who are not specifically dedicated to this subject, and it shows the major trends, precisely what interests me here. Besides, in studying Table 2.2 there clearly is some disagreement in the figures given by the various sources used; this could be due to several factors of which I am not aware, and so I cannot pass judgment on these disagreements.

In Peru, things seem to be as follows. In the case of the guanaco, if the data recorded in various sources are correct, then the species is in serious danger of becoming extinct: from an estimated 5,000 animals in 1982, a level apparently maintained for some years, the population dramatically declined to 1,347 in 1992 (see Table 2.4). This is not the case for the vicuña, where we find a significant recovery of the species: from 5,000 head in 1964, its numbers rose to 97,670 in 1992 (see Table 2.4). The alpaca population also fell, for a population of 3,304,000 head was estimated in 1965, and one of only 2,675,695 for 1997. The figures in the various sources disagree more regarding llamas, and the differences are relatively important. Thus, according to one source there were 954,000 animals in 1974, while another records 1,361,050 two years later, in 1976; this increase does not seem real. Even so, no major overall change took place between 1974 and 1997, for there was an increase, from 954,000 animals to 1,119,777 (see Tables 2.4 and 12.1).

For Colombia there are scant data only for 1982–1985 (see Table 2.2), and for llamas alone. The sources indicate that the population of 200 animals remained constant throughout the period.

There are also some figures for Ecuador (see Table 2.2). They indicate that in 1974 there were 2,000 llamas, which apparently grew to 2,500 in 1988 and 9,687 by 1991. Regarding the alpaca, all we know is that in 1988 there were "some," and in 1991, 100. All that is noted of the vicuña is that there were 482 animals in 1991. In the case of both the vicuña and the alpaca, these were animals recently imported from Chile.

According to the sources, conditions in Bolivia are as follows. There is much confusion regarding the guanaco. Two hundred animals were recorded in 1982. But two sources give quite different figures for 1984. Whereas one mentions

200 head, the other one gives 54. "[S]ome" are listed for 1988, while their number was simply unknown in 1992 (see Table 2.2). In the case of the vicuña, the species apparently recovered because it grew from the 4,500 animals recorded in 1982 to 12,047 in 1992 (see Table 2.2). However, if the data are correct, the alpaca population of Bolivia is instead falling, for of the 500,000 animals present in 1977 there were only 324,336 in 1991 (see Table 2.2). We can ask whether the figure for 1977 is correct, because all other sources agree in pointing out that the population remained stable in 1982–1988. The llama increased instead in Bolivia: the 1,800,000 animals recorded in 1974 grew to 2,022,569 in 1991 (see Table 2.2).

The guanaco population apparently remained constant in Chile in 1982–1992, because the variation in these two years ranges between 20,000 and 19,836 head (see Table 2.2). The vicuña population is instead clearly growing, because instead of the 10,000 animals recorded in 1982 there were 27,921 in 1992 (see Table 2.2). To judge from the sources, the condition of the alpaca in Chile is chaotic. The difference in the figures are too high, for a population of 80,000 animals in 1977 apparently gave way to 500 in 1984, grew back to 80,000 in 1985, fell to 10,000 by 1988, and climbed to 27,585 in 1991 (see Table 2.2). This certainly makes no sense at all. As for Chilean llamas, according to the 1974 figures there were 64,000 head, which number increased to 85,000 in 1982, where they remained until 1985, falling to 70,363 in 1991 (see Table 2.2).

In Argentina (see Table 2.2) it seems that the guanaco population remained stable in 1982–1992, with 550,000 animals. The figure given for 1984, which indicates a fall to 109,000 head, is anomalous and apparently without explanation. However, it should be noted that the 550,000 head is an estimated minimum figure, because the census and conservation areas held only 20,887 animals in 1992 (see Table 2.1). In this case the vicuña population also seems to be growing, because from 9,000 animals recorded in 1982 the figure rose to 23,000 in 1992. The figure for 1984 is once again anomalous, only 2,000 animals (see Table 2.2). In this case the figure of 23,000 animals is also an estimate, because the census and conservation areas held only 15,900 vicuñas in 1992 (see Table 2.1). The condition of the alpaca in Argentina is not clear. Various sources indicate that the population remained stable in 1982–1985 at 200 head, but only "some" are mentioned in 1988, and 400 in 1991. The llama population apparently grew in 1985–1991. It apparently remained stable in 1974–1985, at 75,000 head, but in 1988 it rose to 100,000 animals, and to 135,000 in 1991 (see Table 2.2).

For Paraguay (see Table 2.2) all I found was one datum regarding the guanaco. It was published in 1991 but dates to 1985, and mentions 53 animals.

All of these figures have one common factor that must be taken into account. Some authors copy others, and not all use the original source, hence the differences in figures, their repetition, and so forth. In general, conditions are chaotic; it is most likely that we are dealing with estimates, and that the real figures for the camelid populations of South America are unknown.

To show the discrepancies mentioned, I will cite three examples. Franklin (1982:469), apparently based on Grimwood (1969), indicates that in 1960 there was a worldwide population of vicuñas of fewer than 10,000 animals. First I checked the study by Grimwood (1969:66–69), and this figure does not appear. Second, Franklin (1982:Table 2, 475) notes the presence of 85,500 animals for 1980–1981, but it turns out that Sumar (1988:Table 1, 26) gives 89,500 animals for 1988. However, for a date in between—1984—Novoa and Wheeler (1984:Table 14.1, 117) list 55,000 head. Which figure is correct?

The second example is as follows. Flores Ochoa (1990b:92) notes that the Departments of Puno, Cuzco, Arequipa, Ayacucho, Huancavelica, and Apurímac have 2,303,005 alpacas, or 95.92% of the Peruvian alpaca population. He notes that "in 1980 [it was] estimated at 2,402,305 animals." This can make us believe he worked his estimates from this figure, but this was not so. It so happens that 95.92% of 2,402,305 is 2,304,291, and not the figure given by Flores Ochoa. We must therefore assume that the total corresponding to the number of alpacas in the abovementioned departments is based on data that we do not know but that correspond to the year of their publication or slight-

ly before it, when Flores Ochoa prepared his manuscript, or around 1990. Now, if my assumption is correct, it turns out that if we consider the data provided by the INIPA, in 1990 (Sumar, letter of 17 October 1991; see my Table 12.1), the aggregate number of animals in the departments listed by Flores Ochoa came to 2,918,600, which is 96.1% of the total alpaca population. We thus have figures that differ from those of Flores Ochoa.

Later, Flores Ochoa indicates that the remaining animals are found in the Departments of Tacna, Moquegua, Junín, Pasco, Ancash, Ica, and Lima. However, it turns out that the INIPA (1990; see Table 12.1) does not consider the Department of Ica. Both sources agree in that the Department of Puno holds first place in the alpaca population, but whereas for Flores Ochoa it constitutes 46.95%, for the INIPA it is 56.06%. They concur on second place, Cuzco, for the percentages are quite similar: 11.85% and 11.37%, respectively. The difference is slightly higher for third place, Arequipa, for we have 10.91% in one case and 9.69% in the other. The same thing happens with Huancavelica, with 9.57% versus 8.04%. For Apurímac the difference is higher: 8.61% versus 6.40%, and it is even higher for Ayacucho: 7.90% versus 4.53%. The remainder is split between the other departments, but we do not know which are these for Flores Ochoa; for the INIPA they are Ancash, Pasco, Junín, Lima, Moquegua, and Tacna. But it turns out that for Flores Ochoa, this remainder is 4.17%, and 3.9% for the INIPA. The question again is, who is right?

The third example refers to 1988, specifically to the number of llamas and alpacas in the Andes. For this year Sumar (1988:Table 1, 26; repeated in 1992:Table 2, 86) gave 2,800,500 alpacas. However, for this same year Torres (1992a:33) gives a figure by Wheeler (1988[c]; I was unable to find a copy) that comes to 3,320,700 head. In other words, the difference comes to 520,200 animals, which is by no means irrelevant. Which figure is the correct one? Something similar happens with the llamas. Sumar (1992:Table 2, 86) gives 3,593,550 animals for 1988, whereas for Wheeler (1988[c], cited by Torres (1992a:33) there were 3,562,200 head. In this case the difference comes

TABLE 12.1. ESTIMATED LLAMA AND ALPACA POPULATIONS IN PERU

DEPT	LLAMAS						ALPACAS					
	NUMBER			% OF PERUVIAN POPULATION			NUMBER			% OF PERUVIAN POPULATION		
	1981	1990	1997	1981	1990	1997	1981	1990	1997	1981	1990	1997
Piura	—	—	110	—	—	0.009	—	—	290	—	—	0.01
Lambayeque	—	—	195	—	—	0.01	—	—	235	—	—	0.008
Cajamarca	—	—	550	—	—	0.04	—	—	8,232	—	—	0.30
La Libertad	—	—	409	—	—	0.03	—	—	6,228	—	—	0.23
Ancash	1,000	1,500	1,945	0.07	0.14	0.17	5,770	1,000	8,250	0.23	0.03	0.30
Huánuco	2,000	3,600	6,580	0.14	0.33	0.58	—	—	3,235	—	—	0.12
Pasco	22,000	21,500	41,449	1.61	1.99	3.70	800	2,100	27,632	0.03	0.07	1.03
Junín	60,000	68,900	39,200	4.40	6.38	3.50	6,000	9,600	18,350	0.24	0.32	0.68
Lima	31,500	32,000	23,280	2.31	2.96	2.07	30,000	32,500	22,810	1.21	1.07	0.85
Huancavelica	145,000	112,000	140,472	10.65	10.37	12.54	220,000	244,200	211,776	8.84	8.04	7.91
Ayacucho	165,000	89,000	122,815	2.12* (12.12)	8.26	10.96	200,000	137,700	145,000	8.03	4.53	5.41
Ica	850	—	—	0.06	—	—	2,000	—	—	0.08	—	—
Apurímac	150,000	120,000	55,537	11.02	11.11	4.95	210,000	194,100	87,037	8.43	6.40	3.25
Cuzco	153,000	150,000	175,000	11.24	13.88	15.6	286,000	345,600	291,000	11.49	11.37	10.87
Arequipa	195,000	122,000	95,050	14.32	11.30	8.48	290,000	294,300	235,100	11.64	9.69	8.78
Puno	390,000	314,000	368,190	28.65	29.07	32.8	1,207,230	1,702700	1,531,860	48.48	56.06	5.70
Moquegua	24,700	32,000	32,420	1.81	2.96	2.89	15,300	38,800	46,500	0.61	1.28	1.73
Tacna	21,000	13,500	16,575	1.54	1.25	1.48	16,900	34,400	32,160	0.68	1.13	1.20
Total	1,361,050	1,080,000	1,119,777	100.00	100.00	100.00	2,490,000	3,037,000	2,675,695	100.00	100.00	100.00

Sources: The estimate for 1981 was taken from Sumar (1992: Table 1, 85), who used as his source the Programa Nacional de Ovinos y Camélidos, INIPA (1981). The 1990 estimate was provided by Julio Sumar Kalinovski (letter of 17 October 1991), who took it from INIPA (1990). The figures for 1997 are estimates by by the Oficina de Información Agraria of the Ministerio de Agricultura.
*The source reads 2.12. It must be a typo. It should read 12.12.

to 31,350 head. We once again have to ask, which figures are correct?

In Table 2.1, the reader will find the figures for wild animals in the census and conservation areas in Peru, Bolivia, Chile, and Argentina. Tables 2.3 and 12.2, respectively, show the areas of census and conservation for the vicuña and the areas of preservation for the guanaco, both in Peru, with the corresponding hectares they occupy and the number of animals present. Table 12.1 shows the

TABLE 12.2. AREAS OF GUANACO PRESERVATION IN PERU

NAME	DEPARTMENT	HECTARES	NO. OF GUANACOS
Calipuy National Reserve	La Libertad	64,000	1,000
Salinas and Aguada Blanca National Reserve	Arequipa and Moquegua	365,936	148
Partly in Pampa Galeras National Reserve and its area of influence	Ayacucho	25,000	20
Negromayo (the area of influence of the Pampa Galeras National Reserve)	Ayacucho	49,000	15
Huajuma (the area of influence of the Pampa Galeras National Reserve)	Ayacucho	42,000	134
Puno	Puno	262,000	30
Total		807,936	1,347

Source: Hoces 1992:Table 10, 54.

estimated population of alpacas and llamas in Peru, detailing the population by department. These figures are for 1981, 1990, and 1997.

Franklin (1982:Table 2, 475) noted that in 1980–1981 there were 7,543,600 camelids in South America, 53% of which lived in Peru, 3% in Bolivia, 8% in Argentina, and only 2% in Chile. Of this total, the domestic llamas and alpacas, which then comprised 91%, vastly outnumbered the wild species of guanacos and vicuñas, which comprised the remaining 9%. Among llamas and alpacas, the former were more numerous (47%) than the latter (44%), and the guanacos more than the vicuña, even though no figures can be given for this. The respective figures and percentages given by Franklin (1982:Table 2, 475) correspond, with slight variations, to the ones presented in my Tables 2.2 and 2.4. (In this case the available data for Peru in 1996 and 1997 have not been considered, as I was unable to find the corresponding figures for other South American countries.)

Still according to Franklin (1982:Table 2, 475), the alpaca is the species with the highest density, with 3,300,000 animals concentrated in southern Peru and western Bolivia, whereas the highest concentration of guanacos comes to 500,000 animals in an area extending from northern Peru to Tierra del Fuego. The highest percentage of alpacas (i.e., 91%) and vicuñas (i.e., 72%) is in Peru, while the majority of llamas in the continent (i.e., 70%) are in Bolivia, and most of the guanacos (i.e., 96%) are in Argentina. Franklin concludes that "[i]n general, numbers of alpaca and vicuña are increasing, whereas llama and guanaco are declining." I have no intention of questioning the conclusion reached by Franklin, one of the most serious students dedicated to camelids, but the published statistics and those gathered in my tables do not clearly indicate this trend.

Flores Ochoa is undoubtedly the anthropologist most concerned with studying the factors behind the disappearance of camelids, and his ideas have been mentioned in several parts of this book. But he is at the same time one of the scholars most concerned with the future of these animals. He notes that although in urban areas the meat from cows and sheep is worth more than that of camelids, the clandestine and indiscriminate killing in the last ten years [N.B. Flores Ochoa wrote in 1975] has been depopulating the pasture zones, and in urban centers camelid meat is sold as if it were mutton, "or is euphemistically advertised as 'meat of the land,'" as in the markets of Arequipa, or freely sold, with no label at all, in the markets of Sicuani and Puno. (Today one can find alpaca charqui [dried meat] in Lima, which I purchased.) But the killing is not just for food, it also takes place to get skins for the fur industry; in this case young animals are sacrificed which have not grown, had offspring, nor given fiber. According to Flores Ochoa, this activity has passed all tolerable limits and is endangering the survival of high puna pastoralism, as the animals killed cannot be replaced by young ones, thus curtailing the growth of the herds. The picture is complete once we add the measures taken by some herders and the undue takeover by sheep of the high puna (Flores Ochoa 1975a:20; 1977b:46; 1990b:85). Flores Ochoa is quite pessimistic in this regard, warning that "if the curve keeps falling, the alpaca and the llama can vanish before the century is over" (Flores Ochoa 1977b:46). Although this is somewhat overly pessimistic, the danger is clearly real.

It is estimated that 750,000 alpacas were slaughtered in 1970–1974 alone, and the figure might even be higher. These slaughters were usually clandestine because in this same period, the legal permits for normal *sacas* [to remove or kill a given number of animals] came to no more than 180,000 head. This means that 76% (not 77.3%, as noted by Flores Ochoa), or 570,000 animals, were poached just in the Department of Puno alone (Flores Ochoa 1979a:231; he based his figures on data in a manuscript by Canto Tapia).

As regards the vicuña, Brack Egg outlined its history and concluded that it has four major stages. The first stage corresponds to pre-Columbian times, when there was an efficient management of the animal. Brack Egg dramatically calls the second stage the "Age of Extinction," from 1533 to 1964. The third stage, called of recovery, in 1964–1980, is when the population grew from 5,000 to 61,822 animals. This period of recovery was deactivated from 1980 to 1987 due to the "ignorance, the overreacting and the influence of a pseudo-conservationist" (Brack Egg 1987:73). Brack Egg undoubtedly means Felipe Benavides, who had a negative influence on camelids. (If the comments made by Brack Egg [1987:73] are correct, then there certainly are statistical errors in the data shown in Table 2.4.)

Bustinza (1990:17) tried to estimate the mean carrying capacity per hectare in the Puno altiplano. He concluded that it is 0.96 for the alpaca and 0.92 for the llama. Based on these statistics, he assumes that the region could tolerate 3,956,000 alpacas and 4,232,000 llamas—bearing in mind that the riverbank area is at present subject to excessive grazing.

According to Benavente Aninat (1985:48), camelids do not abound in Chile. To prove her claim she refers to several students. She cites Ladrillero (1558 [1880; the bibliography reads 1557]), who declared that at Ancud, "[t]he Indians had four series [sic] and even eight, and the caciques used to have up to twenty." (The citation was apparently taken from Medina [1882] 1952: 186). Valdivia ([1551] 1970) wrote that on his first expedition, the soldiers gathered up to "a thousand sheep" (this was taken from Guerrero Vergara 1880:13). Benavente Aninat indicates that Bascuñán (1646:125; however, it does not appear in the bibliography) wrote that "only important and powerful men had them [camelids]," while González de Nájera ([1646] 1889:68) noted that "there were few who did not have them in herds."

However, Palermo (1986–1987:74–75) does not fully agree with Benavente Aninat, because he believes that the number of domestic camelids in Chile at their peak is hard to determine. Valdivia thus wrote in 1549 that the Spaniards moved toward the Bío-Bío and took "over a thousand head of sheep" from the Indians (Benavente Aninat likewise cites this number); a year later, the area where Concepción was to be founded was "abundant in people, livestock and provisions." The same source reports that in 1551 Chile was a land that "abounded in livestock like that of Peru" (Valdivia 1861).

On the other hand, still according to Palermo, in 1558 the scouts sent to Ancud reported that commoners had between four and five animals each, while the caciques could have up to 20 (Mariño de Lovera 1865; Medina 1882). Palermo notes that there must have been a high number of livestock, given that the population was then numerous. We also know that the Island of Mocha had a significant number of herds for sale in the first half of the seventeenth century. Spielberg provisioned himself there with a "great abundance of rams" (Ovalle 1888).

Palermo (1986–1987:74–75) notes that Medina (1882) claimed that save for the references mentioned, most sources concur that there was scant livestock, a debatable claim at least in regard to the sixteenth and seventeenth centuries. Although it is true that González de Nájera (1889) said in 1614 that "[t]here are few of these rams because they do not have them in herds," there is no evidence that this held true for all of the land, for at this time the native groups living farther south were little known. Considering that according to various estimates, the "Araucano" population had grown by mid-sixteenth century to no less than 500,000–1,500,000 (Cooper 1946[b]; Ribeiro 1985; Rosenblat 1954; Steward and Faron 1959; Hidalgo 1981); that they dressed with woolen clothes; and even bearing in mind the possible use of guanaco fibers obtained through hunting or by bartering with the hunters from northern Patagonia; and even considering the presence of some woolly dogs, we musts conclude per force that the *hueques* must have been numerous. Unfortunately, as Palermo notes, in the best cases most of the historical sources simply ascertain the presence of these animals that were almost worthless for the Spaniards, or simply give vague or ambiguous references, stating that "many sheep of the land," or some thousand "sheep of the land and of Castile" were seized. However, Palermo indicates that in those rare cases when seventeenth-century chroniclers give more or less accurate counts of the animals, their number is remarkably disproportionate vis-à-vis the animals introduced by the Europeans. This is, for example, the case of a Spanish raid in the zone of Elicura, south of Arauco, in 1626, where a booty of 80 horses, some cows, 200 sheep, and just four *hueques* were taken; about this time the raids seized 40 cows, 3,000 sheep, and 30 *hueques* in the Coypu Valley, south of Purén. These animals had disappeared by the late eighteenth century.

Stehberg (1980:35–36) studied the zone of Santiago and tried to quantify the potential primary productivity of the pastures. He reached the conclusion that 42,140 camelids could have lived there in pre-Columbian times. Stehberg indicates that this estimate is quite close to that found in a report by Rafael Herrera dating to 1895, which suggests the management of an optimum herd of 16,000–17,200 head of small livestock, that is, sheep and goats, in the Hacienda Las Condes, and a similar number of cows and horses. Stehberg explains that on extrapolating this figure to all of the area under consideration, and using the corresponding conversion factor to camelid carrying capacity, gives a figure quite close to that estimated using geographical methods.

Stehberg (1980:25) states that in the eighteenth century there were 300,000 guanacos in Tierra del Fuego (he based this on a personal communication from Rottman).

Raedecke is certainly one of the scholars most acquainted with the guanaco problem. He believes that this animal was, and probably still is, the most important wild species in Chile. It was originally dispersed throughout all of the country, and the native cultures depended almost exclusively on it, particularly on the southern plains. Raedecke estimates that in Chile alone there must have been millions of guanacos, and many still live (his report was written in the 1970s) in southern Chile, in the provinces of Aysén and Magallanes. After reviewing the studies by Darwin (1845), Prichard (1902), and Musters (1871), Raedecke categorically states that "[f]rom these references there can be no doubt that in the absence of ovines, bovines and modern man with his weapons, the guanaco must have numbered millions in the temperate plains of South America" (Raedecke 1976:1, 9). Darwin (1845) in fact recalls having seen herds of up to 500 guanacos along the coast of Santa Cruz,[1] and Prichard (1902) writes that during his trip rarely a day passed by without their seeing guanacos, except while in the woods. And he mentions herds of 200–500 animals. Raedecke reproduces the passage: "thousands appeared on the summit of the nearby mountains." Musters (1871) likewise reports having seen "herds of three to four thousand [head]" (Raedecke 1976:9).

Raedecke concludes that at present some 3,000,000 sheep are maintained in Magallanes alone, Tierra del Fuego included, and taking their body weight as basis, an estimate of one guanaco per 2–2.5 sheep can be reached (according to Albrittom 1954). At this rate we can estimate a maximum number of 1,000,000–1,500,000 guanacos in Magallanes before the European arrival. A more realistic estimate would be 500,000–750,000 head, because the natural density of the guanaco is probably not the same as that of sheep, which are managed carefully and continuously. The colonization of Magallanes marked the rapid decline in the guanaco population. From about 750,000 head it fell to less than 20,000. This period was also distinguished by the rapid decline in the condition of the prairies and the poor management of sheep by the herders (Raedecke 1976: 9–10, 15).

Finally, Stehberg (1980:25) reports that it is estimated that in the Norte Chico of Chile, an Aymara family of herders has about 200 camelids (he based his assessment on a personal communication from V. Schiappacasse).

From the above one major conclusion can be drawn. Although the decline of the four camelid species from pre-Columbian times to the present cannot be established quantitatively, it was doubtless catastrophic. I believe that the threat of extinction is quite high for these animals unless serious and permanent measures are taken. The threat is even greater if we start from the premise that we do not even have an accurate picture of this subject. In other words, we do not know reality. The disagreements and gaps in the figures above noted are a good example and prove it.

NOTE

1   In his famous diary, Darwin (1921:237; 1969:170–171) wrote on 23 December: "It generally lives in small herds of from half a dozen to thirty in each; but on the banks of the St. Cruz we saw one herd which must have contained at least five hundred."

CHAPTER 13

CAMELS IN PERU

This book would not be complete without discussing the subject of camels in Peru. It is true that the data are not abundant and that more probably exist, but these suffice for the present. This subject merits more study, especially to find out whether any discussion of the way these animals behaved in the Andes has survived.

Del Busto (1975:510) wrote: "The first news of camels in Peru are due to friar Gaspar de Carbajal, the chronicler of the discovery of the Amazon River, who wrote while sailing down the great Amazon River and before coming to the land of Caripuna, that: [see the original passage below]. This statement is conditional, but clearly a reflection of the Andean auchenid and the peccary of the tropical forest."

First, to avoid any confusion it should be explained that this information was not picked up in Peru but in the territory of what is now Brazil, as was noted in Chapter 9, to which the reader should refer. The passage in Carvajal that refers to the voyage of Orellana in the first half of the sixteenth century tells how the Spanish host passed through the "land and chiefdom of Omagua" and reached "another land of another lord called Paguana." From there the Spaniards continued their journey until they reached the place where another "great lord" lived called Couynco (or Quenyuc), and it was there that they

had news that in another, not so well identified, site there was "a lady" called Coñor, who had much power over a society of women. It was then that the native informant "[s]aid more: that in this land, according to what we understood, there are camels that are loaded, and he says there are other animals that we could not understand, the size of a horse and with the hair of a geme[1] and a cloven hoof, and they have them tied up, and of these there are just a few" (Carvajal 1986:87). Del Busto made a mistake, because he should have said that "[t]he first news of camelids in Peru," not camels. Del Busto is right in noting that this probably was the interpretation the Indian was able to give of the llama. However, I suspect that the Spaniards only heard descriptions of the animals, which is why they just said "camels." Curiously enough, reference is made several times to "sheep like those from Peru" in the preceding pages of Carvajal's chronicle. It is not clear why "camels" appear in the passage cited, but the Indians clearly had some news of the existence of these animals, which they almost surely had not seen.

So del Busto's claim (1975:510)) that "[t]he first news of camels in Peru are due to friar Gaspar de Carbajal" is incorrect.

Garcilaso de la Vega (1966:Bk. 9, Chap. XVIII, 584; 1959:234) gives the required information. He wrote:

467

Before, there were no camels in Peru, but now there are some, though only a few. The first who introduced them—and I do not think that anyone has taken any since—was Juan de Reinaga, a man of noble birth from Bilbao, whom I knew. He was a captain of infantry in the war against Francisco Hernández Girón and his followers and served his majesty well in that campaign. Don Pedro Puertocarrero,[2] a native of Trujillo, paid him 7,000 pesos, or 8,400 ducats, for seven she-camels and one male he had imported; but the camels bred little or not at all.

It should be noted that Garcilaso traveled to Spain in 1560, so he met Juan de Reinaga before that date.

The data in Garcilaso concur with that in Lizárraga (1968:Bk. II, Chap. XI, 118), who in the late sixteenth century recounted the voyage viceroy Andrés Hurtado de Mendoza, Marquis of Cañete, made from Trujillo to Lima by land. Lizárraga wrote that "Don Pedro Portocarrero, vecino of Cuzco and field commander in the war against Francisco Hernández, went to kiss the hands [of the viceroy] in the Guarmey Valley, which is midway, and went supplying the Marquis abundantly, bringing what was needed in his camels and mules up to the city of Los Reyes."

Incidentally, Markham (1941:140) was wrong when he said that "[t]he Marquis disembarked in Paita, sailed along the coast, reached Callao and was received as viceroy in Lima on 29 June 1556." The voyage was made by land. The testimony of Fernández, El Palentino, is quite clear in this regard. In Chapter II of his chronicle, Fernández tells "[h]ow Don Hurtado de Mendoza, Marquis of Cañete, was named viceroy of Peru, and how he began ruling those kingdoms." Fernández says thus: "After taking up his residence in Trujillo and gone over various matters he left for Lima. *And since he was going by land*, every day new matters appeared that had to be attended. The viceroy had a solemn and sumptuous reception as soon as he reached the city of Los Reyes" (Fernández 1963b: Bk. III, Chap. II, 71; emphasis added). (This was also noted by Bromley 1955–1956:140, but he did not disclose his sources.)

Acosta was also in Peru in the late sixteenth century, and his is also a personal recollection. He says that "I saw some camels taken from the Canary Islands in Peru, though not many, and they multiplied there, but not much" (Acosta 1954: 128).

However, it is known that other camels were brought to Peru before these, because Father Francisco Mateos noted the following in a footnote (1964:421, n. 12) to the work of Father Cobo:

> Licenciate Don Fernando de Montesinos, a native of Osuna, writes in his *Anales del Perú* (an unpublished manuscript) for 1552: "A cédula arrived from the king in which he granted Cebrían de Caritate permission to take camels to Piru for ten years, and no other person could introduce them in said time. And among other clauses, we read: 'since they are so necessary for the service of the land, because there no longer are any personal services in it, nor will there be.'" This cédula was publicly read on 23 June of this year. It caused a scandal, etc.

Fernández, also known as El Palentino, left an important testimony in this regard (1963a). He arrived in Peru in 1553 and was city scribe in the city of Los Reyes (Pérez de Tudela Bueso 1963: LXXVIII). His account is important, even though in 1559 the Crown gave a cédula that deprived him of the "oficios de correo mayor y cronista" (Pérez de Tudela Bueso 1963.:LXXXI). Pérez de Tudela Bueso, who studied his chronicle, claims it is based on writings that were sent to the council by President Gasca, and that it is "the most thorough and detailed account available on this subject." He concludes that it is a "solid and reliable testimony" (Pérez de Tudela Bueso 1963: LXXXIV–LXXXV).

It is worth reproducing what El Palentino wrote:

> [A]t the time that the president Gasca left the city of Los Reyes to embark on the Mar del Sur for Tierra Firme, he received a cédula from His Majesty ordering that personal services be dropped, and he sus-

pended the execution of this cédula until His Majesty commanded. For it is to be known that after this, His Majesty granted one Cebrían de Caritate to bring camels to this land for ten years, and in that time no one else could bring them. Among the reasons included in the provisión was one which said these were most needed for the service of the land because there no longer was any personal service, nor would there be. For this reason viceroy Don Antonio de Mendoza and the oidores discussed (the viceroy being sick) whether to eliminate the personal services, the decree of Licentiate Gasca notwithstanding. And so it was publicly read aloud in the city of Lima, on the eve of St John, on three and twenty days of June in the year of fifty-two, with many citizens of the kingdom present. Many were scandalized by it, and so were the soldiers because they were hard-pressed by need. (Fernández 1963a:Bk. II, Chap. II, 288)

To explain the rebellion of Francisco Hernández, Fernández said: "In the second chapter of this second book it was told how on the eve of St John, when the viceroy Don Antonio de Mendoza was close to the end of his days, the provisión on personal services was made public by command of the Audiencia. This was due to some words contained and referred to in a cédula of His Majesty's on the introduction of camels in the kingdom. It was likewise told of the sour reply the judges gave to the petition made by the city" (Fernández 1963a:Bk. II, Chap. XXIV, 327).

Romero (1936:369) discussed the passage in Montesinos, but he used a later edition. Romero said as follows: "Montesinos tells (1642) that a *real cédula* was made public in Lima on 23 June 1552 'granting the Cretan Cebrían de Caritate the privilege to bring camels for ten years, which no other person could do in said time . . . since they are so necessary for the service of the land, because there no longer are any personal services (by the Indians) in it, nor will there be.' The latter statement sparked some disturbances in the viceroyalty." The note by Montesinos is repro-

duced by Zavala (1978a:n. 63, 245), alongside a summary of Cobo's text, and Zavala likewise reports that the royal *cédula* "caused a scandal." (It is to be noted that Montesinos does not appear in the bibliography).

Herrera wrote his famed *Decades* in the early seventeenth century, and there recounted "the dealings of those Retraídos from Peru, and what the viceroy and the Audiencia of Los Reyes did concerning the personal service of the Indians." In the chapter on the rebellion of Egas Guzmán in Cuzco, the historian wrote thus:

While this was happening in Cuzco, one Baltasar de Cariate [sic] presented a royal cédula to the viceroy that granted to him alone permission to take camels to Peru for ten years, in attendance to the fact that there no longer were any personal services, nor would there be, because it seemed that to carry cargo camels was a great relief. Some say that because of the great zeal Don Antonio de Mendoça had to free those natives from all labor, and knowing fully well what the King and the royal and supreme Council of the Indies pressed in this and wanted it, he discussed [the matter] with the royal Audiencia of Los Reyes, heeding the words of the cédula and of the one that arrived at the time of the departure of President Gasca, which absolutely prohibited all personal services despite the Auto, approved when the president departed. Some say that the viceroy and the Audiencia agreed that the cédula was to be executed and they had it read out loud, but it was not so; instead the Audiencia awaited the King's reply concerning the suspension of the execution of personal services since the time president Gasca left Peru, though they had had it [the King's reply] on other matters. (Herrera 1947: VIII Decade, Bk. VII, Chap. III, 179)

As is well known, Cobo presented a lengthy section on fauna in his book, and included a discussion of the camels. Cobo's book dates to the first half of the seventeenth century. His sources are unknown, but they are reliable and more extensive than the ones so far cited. He used both

Acosta and Garcilaso de la Vega (see Aranibar 1963:125), but in this case his data are far more detailed. Cobo (1964a:Bk. 10, Chap. XLIII, 420–421) wrote:

> two species of animals have been brought from Africa to this land, which are: camels and some kind of hens [which are] native to Guinea. Captain Juan de la Reinaga, one of the first settlers of this land, had the camels brought to this kingdom of Peru from the Canaries, which are islands adjacent to Africa, shortly after it was pacified and peopled. Although the first camels that here arrived formed a lineage [*hicieron casta*] and multiplied much, even so they did not spread over the land or go beyond the jurisdiction of this archbishopric of Lima. The owners tamed some to use them, but most went wild and untamed in the sierras that extend from this city to the valley of Ica and are vulgarly known as Las Lomas, because the camels were not highly valued once there were many horses and mules for transportation. They lasted for many years and multiplied a lot until, on the one hand because they lacked the protection and industry of man, as nobody looked after them, and on the other because the maroon blacks killed them to live off of them, they diminished so much that no more than two were found when a citizen of this city wanted to gather all that were left, so they would not die out and their lineage be lost, and both [were] females which were brought to this city, where they lived for some years. The last remaining one died in 1615, and so the camels died out in this kingdom, having lasted in it for over sixty years.

It is in this passage from Cobo that Mateos (1964: 421, n. 12; see above) inserted the passage from Montesinos in a footnote.

The account the Jesuit gives is most interesting. First, he twice says that these animals "formed a lineage [*hicieron casta*] and multiplied much," and that "[t]hey lasted for many years and multiplied

a lot." This indicates that camels did not have any problems surviving in this new environment. Second, it clearly follows that the major reason why their being raised did not continue was essentially cultural. Third, it is interesting that camels found on the coastal lomas was a habitat where they could easily develop, and if they became "wild and untame," as Cobo notes, it means that at the time the lomas were still luxuriant and apparently not much visited. The reason why Cobo presents a more extended account than the writers who preceded him is that, if the last animal died in 1615, as he says, then this happened while he was preparing his *History of the New World* (begun in 1613, finished in 1653 [Porras 1986: 510]), and so the Jesuit certainly knew people who had seen the camels, were acquainted with their history, and passed it on to him directly—if he did not see them himself.

The historian Carlos Romero studied this subject and explained the reasons why it was decided to bring camels to Peru. Romero says that in the early years of the conquest the number of horses in Peru was high and their price was about 600 pesos. Afterward, however, horses became markedly scarce due to the civil wars between the conquistadors, and the price rose, as is recorded in, for example, the *Decades* of Herrera. It was then that the idea appeared of finding a substitute, or at least an auxiliary animal, and so it was decided to introduce the camel in Peru (Romero 1936).

"It seems," writes Romero, "that the first of these imported ruminants were brought by captain don Juan de la Rinaga Salazar, or he had them brought, and the drove was formed by a male and six females. This must have happened around 1550. The following year a man from Crete whom Herrera calls Baltasar de Cariate, and Montesinos Cebrián de Caritate (the latter name is the correct one because the Osunense priest had the cédula before him) was granted the privilege of taking the double-humped ones to Peru for ten years." Romero then discusses the disturbances the royal *cédula* caused in Peru (see above). According to Romero, "the life of the camel in Peru was ephemeral; it hardly lasted more than sixty years. According to some historians they multiplied a lot, according to others a little, and there even are those who deny that they reproduced at all" (Romero 1936:368–369). The data in Romero come from Garcilaso de la Vega and Cobo, whom I have already cited (see above).

It is interesting to review what was said by those whom Romero cites who denied that camels reproduced in Peru: "de Paw (*Recherches Philosophiques sur les Américaines ou Memoires pour servir à l'Histoire de l'Espèce humaine*, Berlín 1768) says: 'Of all the quadrupeds taken to America, those who increased the least are the camels. In the early sixteenth century [sic] some were taken from Africa to Peru, where the cold confused the organs destined for reproduction, and they left no descendants.' Clavigero (*Historia Antigua de Mégico*, London, 1826, vol. II) refutes de Paw and blames the failure on those who wanted to acclimatize these animals in areas that are not too similar to their homeland, and not to the harshness of the American climate, like that author did" (Romero 1936:371). This shows a complete lack of knowledge on the subject, and the arguments are simply absurd. These scholars would have changed their mind had they read Cobo.

Bromley (1955–1956) also studied this subject and sketched the biography of Don Pedro de Portocarrero, to whom, according to Garcilaso de la Vega, Juan de Reinaga sold seven camels. It turns out that Portocarrero was field commander and a favorite of Gonzalo Pizarro, a brother to the conquistador. His homonymous relative was a conquistador in Mexico. Portocarrero paraded the royal standard of Castile during the first entrance of Gonzalo Pizarro into the city of Lima, was a founder and vecino of Guatemala in 1524, and was its mayor in 1527. He married Doña María de Escobar, who was known as "La Romana" and was marrying for the third time. She was the wife of Captain Martín de Estete, who settled in the Trujillo area in 1534 and found the famed treasure of the Huaca del Sol in 1535.

Bromley recounted the arrival to Peru of viceroy Hurtado de Mendoza, Marquis of Cañete, who set out for Lima across the coastal sandy desert. "In Trujillo the ostentatious Don Pedro de Portocarrero caught up with the viceroy and provided him with a party of camels ridden by blacks dressed in the Moorish way to carry his goods and fardels. These camels, the first to be brought to Peru to be used as pack an-

imals in our deserts, were purchased shortly after their arrival by Don Pedro de Portocarrero and his wife Doña María de Escobar in the then enormous sum of seven thousand pesos." The viceroy stopped at Huarmey, as was already said, and complained of the heat, which was when Portocarrero promised that he would provide him with refreshments further on. So it was that on arriving to the Chancay valley and before crossing Pasamayo, which the Spaniards called the "sierra of sand," the party found an artificial oasis with shrubs and plants placed close to big tarpaulins with tables, on which were "jars with refreshments, Castilian wine bags, Peninsular fruits, and many and most fragrant victuals. The snow had been carried on the back of the Indians in just ten hours from the distant mountains of Huarochirí" (Bromley 1955–1956:140; Bromley unfortunately does not give his sources, as was already noted).

Vargas Ugarte (1966:19) also mentions the royal *cédula*, which I have mentioned several times. Besides Montesinos, he also cites El Palentino (i.e., Diego Fernández 1963a; see above) to indicate that rumors of its contents had spread before its publication, because the privilege granted to Sebrián de Caritate, which was already mentioned (see above), was already known in June 1552. However, Vargas Ugarte adds: "[E]ven though Odriozola says that the one favored with the grant was captain de la Reynaga 14." (Note 14 reads thus: "D.H. Vol. 4, p. 100.") Following what Garcilaso de la Vega wrote (1966:Bk. 9, Chap. XVIII, 584; 1959:234), Odriozola is wrong. Juan de Reinaga was the first to bring camels to Peru, and Sebrián de Caritate the second, and he received some kind of monopoly on it.

However, Vargas Ugarte (1966:19, n. 15) cites Garcilaso and credits him with noting that "La Reynaga [sic] transferred his camels to D. Pedro Portocarrero, who used them to carry merchandise on the coast, between Lima and Trujillo." This is not true. Garcilaso de la Vega never says that the animals were dedicated to this task (see Garcilaso de la Vega 1959:Bk. 9, Chap. XVIII, 234; 1966:584). The statement that Vargas Ugarte (1966:19, n. 15) takes from Lizárraga is correct.

Del Busto (1975:511, n. 10) also mentions the license Cebrían de Caritate had to bring camels, but it is somewhat striking that he only uses Vargas Ugarte (1949 [I used the 1966 edition]) and del Busto (1953) as sources, without citing primary sources.

Romero (1936:369) made a serious mistake, because he wrote that the camels brought to Peru were "doubly humped." The camels that arrived in Peru were the one-humped *Camelus dromedarius* that live in northern Africa and had been taken to Andalusia in the Moorish invasions, not the double-humped *Camelus bactrianus*. Gilmore (1950:424) wrote in this regard that "[t]rue camels (undoubtedly the Arabian, or one-humped species *Camelus dromedarius*) were introduced at an early date into the deserts of coastal Peru, but were not integrated culturally and became entirely extinct." A newspaper article on camels in Peru was recently published with very accurate data (Anonymous 1992c:D1).

Whereas camels had no problem living and reproducing in Colonial times, the ones brought to Lima in the 1930s for the zoological garden of the Jardines de la Exposición had a very short life (Romero 1936:372), just like the failed introduction of camels in the southwestern United States. In contrast, their introduction into Australia was successful (Gilmore 1950:432), so much so that this is the only place where camels have gone back to a wild state and now live in natural populations (Heucke et al. 1992:313).

NOTES

1   "*Jeme*" in modern Spanish. This measures the distance between the tip of the thumb and the tip of the forefinger (both extended). It was used until recently in Piura and Cajamarca (Fernando Silva Santisteban, pers. commun., 6 May 2003), and is still in use in Huancavelica (Ramiro Matos M., pers. commun., 6 May 2003).

2   This is the spelling used in the 1966 English edition; the Spanish edition I used reads "Portocarrero."

CHAPTER 14

CONCLUDING REMARKS

Despite all that has been presented, some additional comments are in order on some of the most important points.

Throughout the book we have seen that the interest in camelids began quite early, as in the case of Father Cobo, but Tschudi was probably one of, if not the, first to study the subject scientifically in the early nineteenth century. Besides, despite its clearly being incomplete, even a cursory look at my bibliography shows that much has been written on this subject. Even so, the knowledge we have of this family is most superficial, and in some cases almost nonexistent. Even the phylogeny, one of the most important points, is a matter of discussion, and specialists have so far been unable to agree on whether the four South American camelids belong to one genus or two.

The biological topics were covered only superficially. For a more in-depth analysis, a group of specialists should address the subject in toto. I have mainly emphasized those points that most interest archaeologists and about which most of us know very little.

I have already noted, and will emphasize again later, that there are serious problems in classifying to the species level the camelid remains found in excavations, a point of major importance in resolving issues such as domestication. As long as new indices are not discovered or existing ones developed, certain issues cannot be resolved that are essential not just for the study of evolution and the changes that take place in the animals themselves throughout time, but also for understanding the role these animals have had in the history of mankind.

In this context there is one crucial subject that has been neglected by archaeologists, as I have pointed out (see Bonavia 1991:71 et passim): altitude. This has been my concern since the early seventies, when I began to carry out research and discussions with José Whittembury and Carlos Monge Cassinelli. The hematological characteristics of camelids made physiologists realize that they are not animals adapted to high altitudes but instead have certain acquired and genetically fixed characteristics that allow them to live well at, and tolerate, high altitudes, unlike other animals. To stimulate my colleagues I presented some ideas on this topic in one of my books (see Bonavia 1982a: 394–395), but these ideas seem to have gone unnoticed. I discussed this point with Ramiro Matos more than once, and he noted it in one of his studies, but in a qualified manner and as a mere statement, without raising new questions or presenting concrete data. In presenting the hypothesis that the puna was not always the habitat of camelids, Matos wrote: "It would instead have been a shelter area where they arrived displaced by the hunters. Later on they found a favorable biomass, perhaps as part of the process of domestication, and stayed there, thus forming the puna biotype. At present, high-altitude Andean areas are almost wholly identified with camelids" (Matos and Ravines 1980:196). The idea is correct, but as long as the belief persists that camelids are high-altitude animals, it will curtail research and present a distorted view of reality. We must recognize that camelids have been pushed into shelter areas by human pressure, which imposed cultural patterns alien to Andean reality. We must acknowledge that camelids are a "tropical species that has turned into animals of high-altitude, cold climates," as Cardozo (1975a:89) noted in one of his few insights.

At present, all the evidence indicates that this is an indisputable truth. We have seen, and I emphasize this point later on, that the paleontological specimens that are somehow related to the present-day camelids originally lived in the lowlands, and their displacement to high altitudes is a late phenomenon. Moreover, we must not forget that until the late Pleistocene, the lower limit of the snowline lay below the present level, according to specialists ranging between 4300 and 4600 masl. It was only between 12,000 and 10,000 years ago that the snow cap reached the present level, that is, 4800–4900 masl. There are cases like northern Peru, where the snowline came down as far as 3700 masl (see Hastenrath 1967; Wright 1980; Bonavia 1982a:394; 1991:39 et passim, 74). This means that when man arrived, the snow cap was lower than at present, so not only were the glaciers themselves impenetrable, but probably a sizable area around them too. So, prior to the arrival of man, camelids would have been unable to move to the upper highlands even if they had wanted to do so. One piece of evidence for this comes from Junín, where it is known that the camelids first appeared around 7000 BC, replacing the Pleistocene fauna (Wheeler Pires-Ferreira et al. 1976:487; Wheeler 1977:5; Kent 1987:176). Incidentally, it is worth noting that in the central Andean area there is thus far not much evidence of remains from extinct animals, which are found more to the south. One might wonder whether altitude had a major role in this.

The rise in the intensity of camelid use appears in the puna at an earlier date than in the valleys, to judge from the zooarchaeological remains. At first these were wild animals, because it was only gradually and from the fourth millennium BC, according to current data, that the process of domestication began. Early faunal samples in archaeological valley sites do not include many camelids, at least until 4000 BC. From this date on and up to 1000 BC, we find in these sites an increase in the use of these animals, which rises increasingly as we draw nearer the time of the European invasion. As a matter of fact, there is evidence available for the intermontane valleys from Colombia in the north to the Atacama Desert in the south (Wing 1975b:35, 42–43; 1977a:851, 857; 1977c:124–125; 1978:185; 1986: 250, 255, 261–262).

I posit that man entered South America through the middle-altitude, longitudinal intermontane valleys (see Bonavia 1991:66 et passim) and descended along the continent through the Andes longitudinally, with transverse descents to the coast and probably to the eastern areas.

The picture is clear. On coming south after entering the South American continent, groups of hunters gradually faced the camelids, cervids, and other smaller animals that formed the basis of their diet, even though the presence of the latter is not clear in archaeological sites. Of these animals, only the camelids could easily move beyond the altitudinal limits in the face of the pressure exerted by the hunters, who were hunting them with their ancestral skill, and so sought refuge there. It is clear that at first humans had trouble following the camelids, and when they finally succeeded, humans were unable to follow them to the same extent as at lower altitudes. For example, the archaeological data from Telarmachay show the gradual and constant fall of the cervids and the increase in camelids, and how the process of mankind's adaptation to the puna unfolded (Wheeler 1984a:401; 1985b:68). The price humans have paid since then for living at high altitudes is very high indeed. They were clearly unable to kill a great number of animals as long as the herds were pursued in the intermontane valleys, where these animals had greater freedom of movement. However, the killing was much easier once man found the animals cornered in the puna. This is exactly the picture that the archaeological data reflect. In this regard it is worth mentioning a study, apparently unpublished, by Hoffman (1975), which I was unable to find but which Wheeler Pires-Ferreira et al. (1976:488; 1977:157) cite. Hoffman shows that to judge from the experience with the vicuñas of Pampa Galeras, the critical factor leading to the abandonment of familiar territories and displacement to more distant and inaccessible areas is not man but excessive hunting. One more datum supports the archaeological information.

In pre-Columbian times, the pressure humans exerted on camelids persisted from the moment domestication took place up to the arrival of the Europeans, even though in Inca times even the camelids were controlled through the *chaco*, while herds of domestic camelids were maintained throughout the Andean lands, as evinced by the archaeological data. With the European invasion, the control the Inca state had of the animals was lost, and the pressure that in pre-Columbian times was limited to the wild animals was now turned on all four species, that is, against the domestic animals too. In this case the pressure was far more violent insofar as it was due to an imposed culture where camelids had no defined role. The human-animal relationship, so ingrained in the indigenous inhabitants and which Europeans could not understand, no longer existed. Europeans introduced their own animals, and their enforced raising was a tacit sentence—if not a death warrant or at least a hard life for Andean camelids, whose importance fell after having been crucial for life in the Andes for thousands of years. These animals could do nothing but retreat to high-altitude areas which were inhospitable not just for the Europeans but for their animals too. In addition to all of the aforementioned abuses, the uncontrolled killing during the early conquest and the introduction of hunting with hounds, which decimated the camelid population, were the antecedents that explain the artificial distribution found today.

The current distribution seems to indicate that camelids, both wild and domestic, live at high altitudes only. This is not true. They have lived in almost all habitats and not always, as some say, with migratory cycles from highlands to lowlands. We saw, when discussing the Chilean data, that there is evidence of guanaco colonies living on the desert coast with no migratory cycle to the cordillera. And with regard to the raising of domestic camelids, the archaeological evidence clearly shows this took place not just in high-altitude areas but also in valleys and on the coast. Moreover, a cross section of the South American continent using the historical data available shows that the Spanish chroniclers confirm the presence of camelids from the Peruvian coast to the zone of Chaco and even farther south, across the mountains. Today this is a fact that is slowly being accepted by specialists (e.g., Wheeler 1991:35).

Thus, Wheeler, who in 1985 (1985a:28) wrote of the camelids that they "belong par excellence to the Andean puna," said in 1995 that "[c]learly the coastal desert was not such a bad place to raise llamas and alpacas" (Wheeler et al. 1995:839). A project that is being carried out in the Argentinian Patagonia is very interesting in this regard. Sixty-two llamas from the Jujuy puna were selected and moved to the province of Córdoba to complete some analyses, and from thence to the province of Río Negro, some 3,000 km away from the puna. "The animals underwent an initial period of adaptation, and after a harsh winter were distributed among breeders in areas which had different ecological characteristics" (Hick and Frank 1996:204). To be more precise, the introduction of llamas was carried out in said provinces, between Río Colorado and the 42nd parallel, at Aguada Guzmán, Department of El Cuy. This site is 180 km to the south-southwest of the city of General Roca, in the upper Neuquén River Valley. The llamas were introduced in May 1992 and remained for eight months in a field some 30 km to the west of Aguada Guzmán. They were then split into three groups and given to the breeders in the area, 30 km and 60 km away from Aguada Guzmán, respectively. In 1999 there were seven llama breeders within a 70-km radius.

Other animals were introduced two years after the first introduction in a nearby area in the same province, at Pichileufu, Department of Pilcaniyeu, 60 km east of Bariloche city and 300 km to the southwest of Aguada Guzmán. At present there are five llama breeders, and between them both regions have more than 300 animals (Eduardo Frank and Michel Hick, letter of 9 April 1999). It is worth recalling that the llamas come from a livestock producer in the basin of Lake Puzuelos, 15 km south of Paraje Cieneguillas, which lies at about 3500 masl. The Aguada Guzmán region lies between ca. 600 and 1000 masl, while Pichileufu lies between ca. 700 and 1200 masl. So, except for one flock, the rest of the animals are below 1000 masl, with 150–200 mm of rainfall at Aguada Guzmán and 250–300 mm at Pichileufu (Eduardo Frank and Michel Hick, letter of 27 April 1999). Now, "[t]he animal's adaptation is given by the results of the following parameters: the reproductive rate doubled with regard to that in the puna; the morbidity rate the first year was as expected, and that in the following years below it; the weight of the fleece increased considerably in regard to the national average, together with an increase in the mean diameter of the fiber in regard to the original estimates of the troop, but not in regard to the national mean; the animals completed their growth just like in the puna" (Hick and Frank 1996:204). These results are significant and speak for themselves.

However, there is resistance to accepting that there was an important number of camelids on the Peruvian coast in pre-Columbian times, something evident not just in archaeological but also in historical data. While wild camelids must have existed not just on the coastal lomas but also in the woods of the vegetated river bottoms since ancient times, the domestic animals were introduced on the coast around the third millennium BC (Bonavia 1982a:276; Wing 1986:261–262). It was due to the European invasion that camelids vanished from the coast (Wheeler 1991:27, 32), and there are several reasons for this phenomenon: the indiscriminant killing in the first years of the conquest and then the need for meat, first by the soldiers and later by settlers. We saw that it was only in 1548 that the "livestock of Castile" was killed in the Lima region (see Cobo 1964c: Bk. I of his *Fundación de Lima*, Chap. XIII, 316). To this we should add the introduction of exotic animals by Europeans, the civil wars of the conquistadors, and the need the Spaniards later had for pack animals (see Lanning 1967a; Rostworow-

ski 1981). Dedenbach Salazar (1990) listed other reasons, but I disagree with most of them. She believes that the destruction of the lomas must have been one reason. This could not have been rapid, because the number of European animals initially brought was not very large, and some time was necessary for them to reproduce. In any case this is an interesting subject for study. I do not know whether there is any study of this kind. Nor do I find the overgrazing Dedenbach Salazar mentions particularly important, for the same reason given above. The diseases introduced by the European stock could have affected the camelid population, and we can exclude climate as having had a significant role. There is no evidence of a change in the climate that could have been problematic for these animals. It is worth noting that a herdsman has taken a small number of alpacas from the highlands of Cajamarca to the peasant community of San José (in the Department of Lambayeque, province of Chiclayo, at 5 masl), which are developing normally; indeed, one of them has even had offspring (Sandoval 2001).

But the fact that camelids are altitude-tolerant animals does not mean that they did not have to pay a high price too. It was noted that one of their characteristics is low fertility, which can be assumed to be a physiological limitation acquired due to high-altitude stress. We saw that studies show a differential fertility according to altitude, evidently caused by human intervention. The historical data analyzed here in detail support this. The natives were quite aware of this fact, because in the *visita* of Chuquito they clearly said that the livestock "multiplied little," and one witness declared that only 27%–29% of the animals could breed at high altitudes, while for another a maximum of 32% could do so, with a minimum of 14%–16%. On the other hand, there is high mortality among young camelids in which several factors play a role, and altitude must be one of the most important ones. We have seen that camelids, have developed certain characteristics to lessen the impact of this, such as giving birth during the day. In fact, cultural, climatic, and altitude factors intervene in these phenomena, because it was found that animals in high-altitude habitats only give birth in January to March, whereas those living in the United States do so all year long.

It is worth recalling that despite the time that has passed, imported European animals have been unable to adapt to high altitudes. There they exhibit low fertility, though these imports are affected differently in their fertility. Sheep suffer less than cattle, while horses suffer the most. The latter have developed an intolerance to any physical exertion in this environment, to the point that the natives have coined the term *apunado* to describe this phenomenon. On the other hand, irreversible changes take place in the testicles of rabbits, cats, and rats. The reproductive abilities of poultry are also affected. Monge M. and Monge C. wrote in this regard that "[f]arm animals acclimatized to the rigorous conditions of high-altitude environments are in general smaller in size, reach maturity at a slower rate, and are less productive than their sea-level relatives. Nevertheless, their wool and milk are of better quality" (Monge M. and Monge C. 1968a:200; 1968b:289). Various methods of selective breeding have been applied to raise the output, especially in sheep, because of their greater economic significance and better adaptability to high altitudes. A high-altitude breed called the "carnero criollo" was thus developed. "Introduced by the Spaniards in the sixteenth century, improved through selective breeding, four centuries of inbreeding resulted in sheep with special characteristics [criollo breed]: resistance to a rigorous environment, light body weight (25–30 kg), and poor wool production of nonuniform-color (1 kg per year)" (Monge M. and Monge C. 1968a:200; 1968b:289). Another breed called the Junín breed yielded better results, because animals with a larger body weight and higher wool output were obtained, but always below the optimum conditions.

This shows that almost 400 years of human selection of ovine livestock at high altitudes has yielded poor industrial results. Even the Junín breed required many years of selection, a great part of which was lost with the agrarian reform. It is clear that the results would have been quite different if the same effort had been spent on the alpaca.

South American camelids have long been animals that developed all of their faculties to live in a dry environment, both at high altitudes and on the plains. This caused their great efficiency in digesting mid- and low-quality food to which ovines,

for instance, were unable to adapt. In this regard it is interesting that the Bactrian camel was also closely associated with the open lands and harsh continental climate characteristic of the Eurasian steppes in postglacial times. Today the Bactrian camel still lives in a dry continental climate environment, at altitudes ranging between 500 and 1000 masl, in other words, in large expanses of steppe areas that are covered by a bush-grass vegetation with a true pabular covering between February and May (Compagnoni and Tosi 1978:96, 98). The feral Bactrian camels (*Camelus bactrianus ferus*) that lived in the Gobi Desert at least until 1956, in a land where altitude ranges between 1500 and 2000 masl, also lived in dry and flat valleys with desert vegetation and exhibited a great resistance to thirst (Bannikov 1958:156–158).

But it is not just a matter of the great reduction and cornering of the remaining herds of camelids but also of a serious loss in the genetic pool, which as we saw was maintained in Inca times through the *suñay* (see Flannery et al. 1989). A review of the available data, both historical and archaeological, plus the new contributions made by genetics, shows that in the past there must have been breeds, especially of llamas and alpacas, that are now lost, and it is not known whether they can be recovered or not.

Of all the camelids, the guanaco probably has the greatest adaptability. It is a typical animal of the steppes which can be found in the transition zone of the pampa and lives at altitudes ranging between sea level to above 4000 masl. The available evidence shows that it can be domesticated, though the data from the Chilean area indicate that it is more tamed than truly domesticated.

At present the northern distribution of the guanaco reaches as far as 8° S latitude., in the Peruvian Department of La Libertad, but it should not be forgotten that Cieza de León recorded its presence in Ecuador, both in the Cuenca region and between Tomebamba and Loja. In addition, there are archaeological data from the El Azúcar site, so there is no reason to doubt the information presented by Cieza de León; what is missing is more research. The data on guanacos in the Bogotá zone and southern Colombia in the nineteenth century are not clear. At present the guanaco lives in Peru, Bolivia, Chile, and Argentina.

From historical sources we know that it occupied almost all of Bolivia, and the same can be said of Chile, although it seems that it reached greatest density in Tierra del Fuego. In the case of Argentina, the guanaco lived in the province of Buenos Aires in the late Pleistocene and early Holocene, and extended south and west in the Buenos Aires region. At present it is confined to the southern part of Mendoza, Neuquén, Río Negro, Chubut, and Tierra del Fuego. There are a few in Santa Cruz, in northern Mendoza, in San Juan, La Rioja, Tucumán, Catamarca, in northwestern Córdoba, and in southwestern Santiago del Estero. A few specimens survive in La Pampa, in the province of Buenos Aires, in San Luis, and in Córdoba. Hunting and the introduction of domestic livestock into its territory are the two great threats that can lead the guanaco into extinction unless serious measures are taken. This is because its mortality in the first month of life is relatively high (15%), and the lack of food in wintertime in the far south, where important populations of these animals still survive, reaches 81%. The real density of the existing populations is unknown.

With regard to the vicuña, in pre-Columbian times this animal seems to have been the most abundant of all herbivores living in the puna, and it is likely that the territory of the guanaco and that of the vicuña overlapped (Wheeler 1985a: 29).

The vicuña occupies a habitat that extends from the lomas on the coast to an altitude of 4800 masl, and it can survive in the puna far better than any other domestic animal because it has great adaptive characteristics. Specialists disagree on whether its northern limit lies in the Peruvian Departments of Junín or Ancash but agree that it lies to the south in southern Bolivia and northern Chile. However, there are some data that say it reaches as far as the puna in the provinces of Jujuy, in Salta, Catamarca, La Rioja, and San Juan in Argentina, that is, in northwest Argentina, between 22° and 29°10′ S latitude, and at altitudes ranging between 3200 and 4600 masl. It is worth recalling some historical data. Some data suggest that the vicuña lived in Ecuador at the time of the conquest. Fernández de Oviedo mentions "wild" animals, and Cieza de León specifically reports their presence both in the Cuenca zone as well as in Tomebamba and Loja. We saw there were vicuñas in the highlands of Piura. At present there are some populations of animals imported from Peru and Chile that apparently are developing successfully. The distribution of the vicuña in Bolivia was more extensive in the past, for it has been verified to the northwest of Lake Titicaca, around La Paz, in the Potosí region, and between Potosí and Atacama. There likewise were vicuñas east of Cochabamba, toward the *ceja de selva* and the border with Argentina. In sixteenth-century Chile these animals lived in the cordillera of Coquimbo and in Osorno.

Even so, the climate is still the main reason for the death of newborns. However, it must be conceded that there are scant data on premature birth and the death of offspring. Similarly, little is known about the diet of this species.

This animal is hard to domesticate, although specialists believe it is possible. Although only estimates of the population are available, it seems to be slightly increasing.

It is clear that the vicuña did not live in high-altitude areas for thousands of years. This shows in the paleontological data quite clearly, and its ascent to these areas was evidently possible because of the lack of physiological limitations in the face of the pressure exerted by humans and sheep.

The llama is an animal that developed physiological characteristics essentially suited for life in an arid environment, in a far more efficient way than either goats or sheep. This, plus the aforementioned physiological characteristics, enable it to live well in both the dry environments of the puna and on the Peruvian coast. These animals consume dry and fibrous fodder and reject thick and succulent vegetation. However, its feeding habits are the least-known aspect of the llama.

The size of the llama is an important point that should in be studied in more depth. It is evident that the present standard size is not representative of the species. The historical and oral data assembled clearly show that in past times there was at least one llama breed with a far bigger body than the present one. I will give only two examples. One is the testimony given by Schmidel (1986:79), who, when describing the llamas he saw, said that these "are big beasts, almost like mules"; the other is the testimony of Cristóbal

Campana, which I have mentioned several times (the photograph he gave me is included in this book; see Figure 8.1). We should also not forget the small Ecuadorian llama, known only archaeologically, and which raises several problems. It was seen that the genetic studies in progress on the wool from these animals that proceed from archaeological remains are showing the existence of breeds now lost. Finally, there is one more point that argues for this hypothesis: the evidence that llamas today carry less weight than before. At present, the lack of systematic studies and of significant archaeological populations of these animals not only prevents a comparative analysis; it stands in the way of knowing the real situation itself.

The current distribution of the llama, which some scholars believe has southern Colombia as its northern limit while others give Ecuador, extends across Peru (with a higher concentration in central and southern Peru) and Bolivia to northern Argentina and Chile, but always in the highlands, and is clearly artificial when compared with the paleontological, archaeological, and historical evidence. We saw that the paleontological finds in Peru are scant and not too clear. The rest corresponds to the intermontane valleys of Bolivia, the Brazilian and Uruguayan areas, and Chile, always in the lowlands. Archaeology and history indicate the presence of llamas in Ecuador, and it has been said that these were taken there by the Incas during their conquests. In other words, all camelid remains in this country would be late. Archaeology disproved this, for remains from these animals were found in contexts dating to different periods prior to the Inca conquest. Bruhns (1991a:9) researched and excavated in Ecuador, and takes it for granted that there were herds of llamas when the Incas arrived. The presence of the llama has been verified throughout Peru, with an evident occupation of the coastal strip. Its presence in Bolivia is evident, but there also were significant numbers of llamas in the Gran Chaco region, as far as the frontier between Argentina and Paraguay, and reaching the Río de la Plata. It is not clear how far the llama extended throughout Argentina, but to judge from the historical data it apparently reached as far as the sierra of Córdoba. In Chile it is clear that the llama was herded on the north coast before the Incas arrived, and

in the Regional Development period there were great movements of llama caravans across the desert. Historical data from the sixteenth century seem to indicate their presence in the Magallanes region and on Chiloé Island. However, Palermo (1986-1987) gives 27° S latitude as the southern limit of these animals.

The immense damage wreaked on the Andean lands by Western civilization becomes evident when the current distribution of the llama is compared with the one just presented. This damage is irreversible and can only be partially repaired, however much effort we make.

The alpaca is an animal that has adapted to hard and short pastures. It is an opportunistic animal, as San Martín and Bryant (1987) correctly noted, and can feed on a wide and varied range of vegetation types. It has an ecological barrier other species do not have: water. The alpaca needs wet lands to live in optimum conditions.

Alpacas also suffer stress induced by high altitudes, for they have a very high infant mortality and low fertility. Alpacas today live at high altitudes, and specialists believe that of all the camelids it is the one that best withstands this environment. The distribution of the alpaca is restricted, and there is no agreement on its extent; for some its northern limit lies in Peru's northern highlands, and for others it is in the central Peruvian sierra. Some even believe it is in the altiplano. Nor is there any agreement about its southern limit, because in some cases this is given as the southern part of the altiplano, in others as far as northern Argentina and Chile. Specialists do agree that the distribution of these animals is not continuous. However, we should not forget that in past times the alpaca reached as far as Ecuador.

This is the only case in which the current and the historical distribution have apparently not changed much, but we should note that establishing the distribution of these animals in past times is problematic, owing to difficulties in distinguishing llama from alpaca skeletons among archaeological remains. On the other hand, we saw that the descriptions and labels provided by the Spanish chroniclers do not allow secure differentiation, so this uncertainty keeps this point obscure. However, everything seems to indicate that this species was likewise living at much lower zones than at present. I will repeat here only the findings of El Yaral, located just 50 km away from the sea, at 1000 masl.

The study of these animals shows a whole series of developmental possibilities for the future (if only the State clearly realized this), and also a way to help improve the living conditions of that part of Peru's population that lives in high-altitude areas. I was unable to find statistics, even though these probably exist, on the number of people who at present exploit camelids. According to Flores Ochoa (1975a:19), in 1975 these numbered about 200,000. As Flores Ochoa himself noted, we have the idea that the puna is a poor region, whereas it is a highly specialized ecosystem with quite specific characteristics that are not well known; nor have they been studied enough. More attention should be paid to the puna because this is the region where llamas and alpacas are now mostly pastured, a place that presents an opportunity for the development of human populations living in extremely harsh conditions (Flores Ochoa 1990b:97).

The studies currently in progress of the llama and alpaca remains of El Yaral, which were extensively discussed in Chapter 4, clearly indicate, as Wheeler et al. (1992:473) stated, how much has been lost and—this is the most important point—how much is still to be learned. These scholars clearly indicate that "[m]any of the confused ideas and contradictory statements about these animals coming from both native Andean herders and North American breeders, to say nothing of camelid specialists, are more a heritage of the Spanish conquest than we have realized." And as Wheeler et al. (1992) correctly note, the root of the matter is the lack of written sources in pre-Columbian societies and the loss of the oral traditions on animal breeding and raising transmitted by the specialists, which took place during the first century after the European invasion. What survived was probably quite distorted by the notions and traditions of the Europeans, who were unable to understand the Andean world, much like the animals, who were modified through hybridization. "We have known for a long time," Wheeler et al. add, "that the primary reason why llamas and alpacas are said to have a low reproductive success rate is because native Andean herders tend to raise them as if they were sheep, not because they are inherently infertile or

maladapted to the high Andean environment" (Wheeler et al. 1992). Now, with the data recovered at El Yaral, we are beginning to find out the real potential of these species as fiber producers. What we know of llama and alpaca variations in the Andes is unfortunately too little to even determine whether breeds from before the Spanish conquest still exist in the general population and can be rescued through a general selection. It is now urgent that we identify and preserve these surviving populations before further genetic loss takes place. The genetic markers must be identified, as they can be used to distinguish breeds and their hybrids. The ongoing research on both mitochondrial and nuclear DNA for all four South American camelids is yielding promising results (Wheeler et al. cite the study by Stanley and Wheeler [1992], which I have not seen). The inclusion of the ancient DNA from the llamas and alpacas of El Yaral will be of crucial importance to document the historical development of their present-day counterparts. Since the mummified remains apparently contain the extinct breeds, they can supply the base data that will enable the measurement of the possible effects of hybridization and reduced genetic diversity that the Spanish conquest presumably caused (Wheeler et al. 1992).

There is another aspect that I believe has been neglected and would like to emphasize. That is the possibility of making comparative studies of all members of the Camelidae family, because it seems they have retained important common characteristics despite the long time that has elapsed since the ancestral camelids separated in the northern part of our continent, some heading for Asia, others for South America. Diseases are revealing in this regard. I discussed this point with regard to Andean camelids in Chapter 7, where it was noted that Flannery et al. (1989:102–104) are convinced, and all the evidence undoubtedly supports them, that there was some kind of mange prior to the European arrival, caused by *Sarcoptes scabiei*. Another kind of mange arrived with the introduction of European animals, caused by *Sarcoptes scabiei* var. *ovis*, as well as another type, the *Psoroptes equi* var. *ovis*, which decimated the native camelid populations. Today, one variant of the same mange that existed before the Spanish conquest affects both dromedaries and Bactrian camels, that caused by *Sarcoptes scabiei* var. *cameli* (Kumar et al. 1992; Pegram and Higgins 1992: 72–73).

On the other hand, we saw that one of the lethal diseases that attacks the Andean camelids is enterotoxemia caused by *Clostridium perfringens*,[1] which Wheeler (1985b:72), following the studies by Moro and Guerrero (1971:9–14), says corresponds to Types A and C. However, Wheeler (1991:39) later only mentioned Type A, based on a study by Ramírez (1987). It so happens that camels from Dubai also suffer the effects of enterotoxemia caused by Type A *Clostridium perfringens* (Seifert et al. 1992).

Besides the importance animal parasitism has in pathology, the parasites of the Camelidae family were used for the taxonomic definition of these animals. Many parasites can infect one kind of animal. However, almost all are more restricted and only infect a group of species that are closely related. Some parasites are even more restricted and can only infect one single guest species, or perhaps some species of the same genus.

When the specific characteristics of a guest are extreme, it is possible that the parasites may have been associated with their guests for a considerable span of geological history, so that as the guest evolved, the parasites evolved with it. In these cases the taxonomic order of the guest frequently shows similar or identical patterns. This phenomenon was used to clear up certain taxonomic problems. For example, in the nineteenth century it was noticed that the South American llamas were similar to the camels from North Africa and Central Asia, but the geographic distance between both groups was held to be far too great to place them in one family. When their lice were studied, it was found that these were also similar to each other but different from other lice. The Camelini and the Lamini were grouped in the Camelidae family based on this agreement, and the lice were grouped in the genus *Microthoracius*.

The degree to which the taxonomic outlines of parasites and their guests agree can be used as an index of the age of the association between parasites and hosts. The fact that the ancestral camelids were infested by the ancestors of *Microthoracius*, together with the old age of fossil camelids, indicate that this association existed at least 30 mya (Villée et al. 1968:784).

Four species of the *Microthoracius* genus are known, *M. cameli*, *M. mazzai*, *M. minor*, and *M. praelongiceps*. The first one infests the dromedary camels of Africa, and the remaining three the four species of South American camelids. The only one that has not been found in the vicuña is *M. mazzai* (Dale and Venero 1977:93–96; Escalante 1981:124). I was unable to obtain information on the lice of Bactrian camels.

The fact that the Lamini and the Camelini have similar parasites despite their genetic and geological separation of many millions of years indicates a common biological tolerance that emphasizes the biological constant in the face of a great morphological variability. These two tribes have similar physiological characteristics, as we saw, and also live with the same parasites.

I will try to summarize the paleontological data emphasizing some points, as these are of crucial importance to understanding the evolution of the South American camelids and the changes that took place in them.

Two important groups of present-day camelids appeared in the late Miocene, the camels and the llamas (*sensu lato*). The Lamini Camelidae diversified slightly more (five genera) in the South American Pleistocene than their cousins, the Camelini Camelidae did, in the Eurasian Plio-Pleistocene (three genera). The Lamini Camelidae formed a different tribe (or subfamily) in the late Miocene in North America, where the family had a long Cenozoic history (Webb 1985:369). Their ancestors reached South America toward the late Pliocene, and so far it is not known where and when the first South American camelids became differentiated.

The current evidence says that *Palaeolama*, *Lama*, and *Vicugna* originated in a common ancestor, which could be *Hemiauchenia* or one of its descendants. Several other possibilities also exist (see Figure 3.2).

Worth emphasizing is the fact that the camelids began to adapt to a dry environment between 1 and 3 mya, and only recently to high-altitude environments, that is, in post-Pleistocene times. There is some evidence that leads us to believe that the llamas (*sensu lato*) replaced the hegetotheres in South America (Webb 1991: 271).

We saw that Webb (1978, 1991) pointed out two migratory routes to the south that allowed the big vertebrates to spread throughout South America: on the one hand, the Andean or upper route; on the other, the eastern or lower one. The Andean route was essentially the route of the mid-altitude intermontane valleys. I posited this same route for the entry of man into South America (Bonavia 1991:65 ff., Figure 9). While writing my book I had not yet read Webb's study (1978, 1991), so this is just a co-occurrence of ideas, if used for different ends. Nor had I noticed that in their classic textbook, Bennett and Bird (1960: 21–22) had indicated the Andean route for the entry of man to South America and excluded the possibility of an eastern one. At the time this position was logical, because the environmental changes that took place in the eastern part of the continent, and the presence of a savanna corridor that allowed the movement first of animals and then of man, were not then known (see Marshall 1985:72).

Now, one question that is still pending and needs much more study is when camelids acquired the genotypic adaptation to altitude, namely, the high affinity of hemoglobin for oxygen. At present, the route followed by the ancestors of these animals on their way south is very difficult to determine and will remain so until remains that can act as markers are found, particularly since the shape of the continent was then quite different from what it now is. It is true that Central America has land above 3000 masl, but there are other ways of moving without having to climb up this land. The problem is different in the South American continent. Although in the highland valleys it is possible to stay at a middle altitude, there are times when one cannot avoid crossing great, high-altitude passes. But in fact, if these animals acquired this genotypic adaptation, so far the distribution of fossil remains indicates that it was not used, because they preferred to live in the lowlands. This adaptation became a necessity once human pressure appeared and posed the biggest threat for the camelids through hunting. It was then that the camelids were forced not just to cross high-altitude areas, but to stay at high al-titude permanently. From the perspective of the South American camelids, this story seems logical. However, this interpretation has problems when we try to explain the phenomenon for all of the Camelidae family. It so happens that camels and dromedaries have the same genetic characteristic of the South American camelids of a high hemoglobin affinity for oxygen. It is difficult to decide whether this was acquired in North America or after the ancestors of these animals left the American continent. "It is quite remarkable that the world's richest assemblage of fossil llamas comes from an estuarine deposit in the southeastern United States, far from the center of diversity of living lamines in the Andes mountains of South America" (Webb and Stehli 1995:622). Besides, the groups of animals exchanged between North and South America were not selected by their potential to live in a different continent. The first requisite was their chance presence in the area adjacent to the land bridge (Webb 1991:269). It is possible that this characteristic was acquired by the Camelini after leaving the American continent, during their march west, where this characteristic would have given them a greater chance of surviving in the high-altitude areas of Central Asia and thus favored their westward migration, as I pointed out in a conference given at the Universidad Peruana Cayetano Heredia in July 1988 (see also León Velarde et al. 1991:404). But the presence of a pre-adaptation (see Chapter 2) in these animals from ancestral times is also possible, as Monge C. (1989) posited. This is supported by the study by Poyart et al. (1992), who, based on a study of the molecular structure of hemoglobin, suggest that the ancestor of all four camelid species had a high-oxygen-affinity hemoglobin. In terms of our present knowledge, these evolutionary changes might have taken place in a brief span and not over long periods, as was previously believed. As has been noted, the high hemoglobin affinity for oxygen transmitted to descendants is a high-altitude genetic marker and is accompanied by the absence of polycythemia in the Andean heights. This mark can be acquired in very short spans, as is shown by groups of Andean chickens (León Velarde et al. 1991). It can remain for longer evolutionary periods, as in the case of the guinea pig (*Cavia* sp.) (Carlos Monge C., pers.

commun., 28 June 1994). This change may have been due to chance, and it did not cause problems because it does not hinder life in the lowlands. However, it became an advantage when the ancestral camelids had to invade the high-altitude habitat. This would be a typical case of what Monod (1971) defined as "chance and necessity."

Hemiauchenia arrived in South America in the late Tertiary. It has been found in the Andes and in the South American pampas since the early Pleistocene, and became extinct at the end of the Pleistocene. The oldest remains are in Argentina.

Palaeolama and *Lama* developed in the middle South American Pleistocene. Everything seems to indicate that the Andean region was the center of origin, but there are problems in the case of *Palaeolama*.

We find *Lama* from the end of the Uquian, around 2 mya. It rapidly spread east and south, occupying most of the South American continent, that is, almost the same expanse as *Hemiauchenia*. The most remarkable remains of *Lama* are from Tirapata (Mollendo), Peru, at an altitude of 1819 masl. The finds at Huargo Cave, in the mountains of Huánuco, are of a doubtful nature, as we saw. Moreover, of all the ancestral forms they are the only ones found lower than 4000 masl. In Bolivia there are the Tarija remains, at 1950 masl. Then there are the Brazilian and Uruguayan finds, which obviously are at low altitudes. Quereo, in Chile, is almost at sea level, and the remains found in Argentina are all in the pampa region.

It is worth emphasizing that one of the most striking characteristics of this genus is that it has developed specialized legs for life in a rugged land. This also helped it when it had to restrict its habitat to the high altitudes.

Not many *Vicugna* remains are known, and these come from the Pleistocene surfaces of Brazil, Tarija in Bolivia, and the Buenos Aires region in Argentina. Once again these are low-altitudes sites. In this regard it is worth recalling a comment made by Pascual and Odreman (1973: 34) while studying the remains excavated in the Huánuco region. "The absence of *Lama* and *Vicugna* until geologically recent times in these Andean regions is striking, particularly bearing in mind that their presence has been determined in all the Pleistocene of the southern tip of South America, apparently in quite different biotypes. This problem deserves a more in-depth study due to the implications that its explanation might have in matters of environmental and cultural evolution." *Vicugna* clearly occupied the high altitude habitat only when the Pleistocene glaciers retreated, in the early Holocene (see Wheeler 1995:277).

The origins of *Palaeolama* have become even more problematic with the discovery of the Leisey Shell Pit remains (Webb and Stehli 1995). Christian de Muizon (pers. commun., 27 June 1994) offers three possible interpretations. The first would be that *Hemiauchenia* and *Palaeolama* originated independently in North America and moved separately to South America. The second possibility is that *Palaeolama* appeared in North America from *Hemiauchenia* and moved to South America after 1.6 mya, after *Hemiauchenia* (2 mya). The third scenario is that *Hemiauchenia* moved to South America from North America and originated *Palaeolama*, which migrated from South to North America during the Ensenadan (ca. 1.5 mya). We should bear in mind that the temporal separation between the early Irvingtonian (ca. 1.6 mya), that is, the earliest date so far available for *Palaeolama* in North America, and the Ensenadan (ca. 1.5 mya), the oldest date for this genus in South America, is very small (0.1 my). The fact that no remains with intermediate dates have so far been found is no proof that there are none. The solution to this problem is therefore pending.

We find *Palaeolama* in South America after the beginning of the Ensenadan, that is, ca. 1.5 mya. It extended westward and occupied the Pacific lowlands of Peru and Ecuador. It became extinct toward the end of the Pleistocene.

The distribution of *Palaeolama* is clearer because many remains are known. First we have the Argentinian pampas, the highland valleys of the Andes such as Tarija, in Bolivia, where remains were found, as well as at Ñuapua, in the Department of Chuquisaca, at about 2700 masl. There are remains in the lower tropical zones of Brazil (Lagoa Santa and Minas Gerais), as well as in Paraguay and on the Ecuadorian coast. However, in Ecuador there also are remains in higher areas such as Punín, south of Riobamba, at 2500 masl, or close to Quito, at around 2800 masl. In Peru, the remains in Talara and Pampa de los Fósiles are famous, and in Chile so too are those of Quereo and Monte Verde, sites all found almost at sea level.

Palaeolama adapted to quite different environments, as we shall see, because it occupied tropical zones, corridors between the Andean mountain chains, and a very specific ecology, such as the Argentinian pampas.

Few fossil remains of *Lama guanicoe* are known, and from what I was able to find, some doubts about these exist. Only the discoveries from the Argentinian Pleistocene and the scant remains from Quereo, in Chile, are secure finds.

The Argentinian remains of *Lama pacos* are most doubtful. Everything seems to indicate that the alpaca is a very recent form, and the result of human intervention through domestication.

Clearly, one of the greatest limitations archaeology faces in studying and understanding the problems related to Andean camelids is the difficulty in distinguishing the bones of different species. We saw that specialists still have not found secure criteria with which to separate them. This is a problem that requires more work.

We must also bear in mind that in pre-Columbian times, hybridization was almost surely far more common that it is now, which evidently hinders identification. Browman (1974a:194) made a comment in this regard that I find interesting and timely. Browman specifically referred to the *chaco* and its consequences. Thanks to this custom, the puna herders were able to exert important control over wild camelids; it let them select the males in terms of size and fleece. At the same time the *chaco* was an opportunity to eliminate other males that did not have the desired traits or that had other unwanted or unnecessary characteristics from the group chosen for control and raising. Old animals and others harmful to the species could also be eliminated. A series of mixtures was created by the application of these selective factors to the herds of wild animals, similar to the ones applied to domestic animals, as well as by the introduction of wild guanacos into a potential domestic breeding pool for pack animals, and the flight of some domestic animals that turned feral. These mixtures cause the many problems zooarchaeologists face in distinguishing the various species.

Several other factors have slowed the detailed study of these animals. A review of the archaeological research undertaken in Peru shows that most of it involves monumental sites and cemeteries. Few studies deal with habitation sites, so we have a restricted picture that is mainly focused on ritual. In most cases this was not taken into account and has caused some distortions. For instance, on the coast I observed camelid remains in refuse dating to all periods, and yet, as we saw in Chapter 4, there are only a few studies on this point. Any attempt at quantification, to establish the number of camelids managed on the coast with a margin of error, is thus impossible. Shimada and Shimada (1985:18–19) drew attention to this point gave a good example of it. Richard Schaedel had told them (pers. commun. to Shimada and Shimada) that camelids were rare animals on the coast during the Early Horizon and were imported for specific tasks, as was the case with *Spondylus*. This is obviously not true, as we have clearly seen from the archaeological record.

On the other hand, on the coast there are several big enclosures from different periods, inside most of which are accumulations of camelid dung, almost surely from llamas, but these usually were not studied and have often been simply labeled as "ceremonial." More attention should be paid to them. In this regard I find that the ethnographic data on the subject can be suggestive for the archaeologist (e.g., Flores Ochoa 1987).

When interpreting the faunal data, the archaeologist cannot ignore climate. However, this does not often appear in reports, particularly those on the late periods. The climate played an important role in the life of the camelid populations, especially droughts. For example, we have the case of Pacaraos, in the province of Canta, Department of Lima. Here a great drought took place, first in 1936 and then in 1970, which practically wiped out the camelid population (Lorenzo Huertas, pers. commun., 1991).

There is also another constraining factor that must be emphasized. Reports of archaeological excavations often have paid scant attention to the faunal remains they cursorily describe (when they do describe them). Lynch (1978:476) drew attention to this fact, pointing out some important sites for which the faunal data are missing: El Abra in

Colombia, El Inga in Ecuador, and in Peru, Quiqche, Toquepala, and all sites of the Chivateros Complex on the central coast, as well as Marazzi in Tierra del Fuego, Chile. These are just a few examples. The Peruvian case, with which I am well acquainted, is crucial. In my opinion, a task for the future will be to pay more attention to this point when excavating new sites; and in the case of important sites for which these data were neglected or simply never published, those sites should be excavated all over again. Such is the case of the preceramic site of Culebras, of crucial importance, of which we know almost nothing about its fauna. Here I have tried to make a critical review of the published sites, so that we can see the data we have and how reliable these are.

We saw that although much work has been done on the process of domestication of South American camelids, we still do not have enough knowledge on this subject. However, it is worth noting that the same can be said of camels, as we shall soon see. Although I was unable to read much on this point due to the problems in getting specialized literature, I believe that a comparative study is in order. It is clear that the Lamini and the Camelini have maintained behavior that is in many respects similar, despite the time that has elapsed since they separated. The data on the Camelini can be useful for a better understanding of the Lamini.

Compagnoni and Tosi (1978:100) claim that "[i]n the case of the camel, every inquiry into the process of its domestication comes up against a truly insurmountable barrier—our ignorance of the behavior, habitat, and osteometry of the wild species." Compagnoni and Tosi note that the studies made by Montagu seem to demonstrate that the Khautagai, that is, the *Camelus bactrianus* L. found in the south-central region of Outer Mongolia (Nomin Gobi), is truly wild (for this, Compagnoni and Tosi base their opinion on the work of Bannikov 1958:156–159; Montagu 1965; Namnandorj 1970:15). The problem is that no other detailed studies are known, so the habitat and behavior of *C. bactrianus* can only be reconstructed deductively.

Compagnoni and Tosi (1978:100) explain that these animals probably lived in small herds mostly formed of females and young animals and led by an adult male—in other words, something quite similar to the South American camelids.

Further, Compagnoni and Tosi note that the camels' present habits suggest that it must have been relatively easy to corral small herds once the adult males were eliminated. The young and the females could easily have been kept close to the settlements and their skin, meat, milk, dung, urine, and hair used without any physiological transformation in the animal or any socioeconomic change in the activities of the population using them (Compagnoni and Tosi 1978:100). I find this most suggestive. In my opinion, something similar must have taken place with the South American camelids, all the more so in light of their territoriality. This agrees somewhat with what was presented in Chapter 5.

The comments Compagnoni and Tosi (1978: 100) made on domestication are so pertinent that I reproduce them here:

The particular connotation of the term "domestication" must vary species by species, to the extent that it implies a form of socio-behavioral contract between man and individual members of a gregarious species. Domestication can be said to have taken place only when the process of animal keeping has, to some extent, resulted in changing morphological and behavioral characteristics of the animal and also in altering the socioeconomic structure of the human community. The selection of particular characteristics will be determined by the nature of the interactions between species. Once domestication is conceived of as a selective diachronic process, the domestication of the camel can be said to have begun as soon as man was in a position to continuously exploit the physiological processes of the species, but to have been achieved only when the animal came to be used as a beast of burden and for riding, thus opening up new territories for human occupation in the subtropical desert strip and forcing the human groups who exploited the animal to become specialized in the socioeconomic dimension of their culture.

Such is the logic underlying the development of Bedouin society, which Walter Dostal deduced from the few elements available as early as 1958 (p. 13).

Compagnoni and Tosi (1978:102) specify that the camel's domestication was a slow process of assimilation that reached its maximum specialization only on the Eurasian steppes and in the subtropical deserts, "*under conditions not unlike those of the natural habitats of each species in the wild state*" (emphasis added). This makes me believe that the process must also have been long and slow in the case of the South American camelids, and it must have had several successive stages. The last stage of domestication probably began when man realized that he could use the llama and the alpaca as beasts of burden. This must have been an early process, because we saw that by the late preceramic there is some indirect evidence of this practice on the coast. However, it will be most difficult to archaeologically determine the stages in this process. The specialists who studied the camel agree with this, i.e., that it will be difficult to find actual archaeological evidence of its domestication (see Whitehouse 1983:83).

The most ancient *Camelus* remains, for the species cannot be determined, belong to the Shar-I Sokhta site in eastern Iran and date to 2600–2500 BC. There is dung in the level dating to 2700–2600 BC. The animals are assumed to be domestic (Compagnoni and Tosi 1978:91–92). On the other hand, the Danish Archaeological Mission that worked at Umm an-Nar, on the coast of Abu Dhabi, in Oman, found depictions in relief and osteological remains. They show a convergence between Iran and Oman. These are quite developed agricultural communities that lived around 2500 BC. Apparently, the process of domestication of *Camelus bactrianus* and *Camelus dromedarius* is visible there. In other words, the archaeological evidence indicates that in the third millennium BC the dromedary occupied an area extending from Oman and Sind to the east to North Africa and Palestine to the west (Compagnoni and Tosi 1978:99–100, based on the data in Thorvildsen 1964; Tosi 1974; Free 1944; Mikesell 1955; Zeuner 1963; Epstein 1971; Ripinsky 1975). A comparison of these data with those on

the South American camelids shows that the process apparently began earlier in the Andes, because there was specialization by 6000–4000 BC, and the alpaca, and perhaps the llama too, was domesticated by 4000–3500 BC. And as noted, it is possible that these dates will move back in time, because although not many data are available, they exhibit some consistency in the dates.

A summary of the data presented in Chapter 4 shows that domestication must have taken place in Guitarrero Cave between the third and fourth millennia BC. In general, this date coincides with the data from Telarmachay and apparently Ayacucho. A more recent date has been given for the central sierra (Pachamachay), ca. the second millennium BC. I do not believe that this date will pose serious problems if correct, because the same process was surely repeated in different places, and not necessarily simultaneously. The problem is that the only well-documented data on this process are from the central sierra. We saw that more than one scholar points to another center in the altiplano, which is potentially of crucial importance but for which no concrete evidence based on excavations is available. The Salar of Atacama has also been suggested as another possibility, but the data are very weak.

In any case, at present all scholars believe, and the facts seem to support them, that domestication took place in the puna ecosystem (e.g., Baied and Wheeler 1993).

In some sites domestic animals appear late in the refuse, as at Chavín de Huántar and Kotosh, to give but two examples, but this does not necessarily mean that their domestication was a late event. Rather, it means that herding was a late arrival to the region, even though cultural factors might also have influenced this. In any case, this subject must be carefully studied.

From the few data I collected it seems that Argentina was a marginal zone in this regard and that domestication was a late occurrence, apparently in the first millennium BC or at most in the early second millennium BC (Bahn 1994). However, I am aware that I lack the full information required for a definitive appraisal.

As for what animal was domesticated first, there is as yet no definite evidence. Wheeler (1985b:78) herself very cautiously wrote: "Although the Telarmachay data *seem* to indicate that the alpaca is a domestic vicuña, there is still much to be done to fully confirm this hypothesis" (emphasis added). Then she added that "[i]t was not possible to determine whether the llama also appeared at this moment due to the fact that their incisors cannot be distinguished from those of the guanaco; however, the presence of remains of big newborn animals suggests the possibility that they were present" (Wheeler 1991:40–41).

However, I find it worth noting a point that should be borne in mind in all future research. I mean by this that the zooarchaeologists who studied domestication did not take into account either the paleontological evidence or the suggestions made by paleontologists. It so happens that since the 1930s, López Aranguren and Cabrera pointed out the presence of *Lama glama* fossil remains, that is, llama.

When discussing the names of the species proposed for fossil llamas, López Aranguren discussed *Fossilis* Ameghino (1889) *Auchenia lama* as follows: "With this name Ameghino designates the fossil remains of the current species which 'abound in the Quaternary fields of the province of Buenos Aires.'" She immediately explains that "[a]ccording to the principles of nomenclature, *fossilis* is a synonym for *glama*" (López Aranguren 1930a:25). She emphatically states that "*Lama glama*, extinct at present as a wild animal, appears from the first layers of the Pampean Formation" (López Aranguren 1930b:110). This was accepted by Cabrera (1931:110–115), Cabrera and Yepes (1940), and Pascual and Odreman (1973:34), whom I have already discussed. This is still valid, for Webb (1974:173) recently wrote: "The llama and the alpaca survive as domesticated species, although they may have existed as equally distinctive types before domestication. Such distinctive occurrence is suggested by the fact that Late Pleistocene fossils from Argentina have been referred to each species." I do not know why Webb (1974:173) only refers to the Late Pleistocene, because it follows from both the study by López Aranguren (1930b) and from Pascual and Odreman (1973) that these findings can also be placed at least in the Middle Pleistocene, and not just in Argentina but in Bolivia too. So the possibility that the llama descends from an extinct animal,

similar to the present type of llama, remains open. In my opinion, the problem with the alpaca is different because the paleontological remains mentioned by López Aranguren (1930b:118–119) and accepted by Cabrera (1931:116) have not been considered by either Marshall (1985, 1988), Marshall et al. (1984), or Webb (1974), as was already noted. In this case there are further doubts, all the more so since Wheeler's proposal (1985a) is still far from proven. Be this as it may, the paleontological evidence is very important both for the alpaca and the llama and should not be overlooked, even though the archaeological evidence will have the final word.

In the case of the vicuña there is no problem, as its antecedents can be identified up to the Middle Pleistocene of Argentina and Bolivia (López Aranguren 1930b:120–122; Cabrera 1931:116; Webb 1974:174).

As for the guanaco, we saw in Chapter 3 that although there is fossil evidence, it raises some doubts. To judge by Pascual and Odreman (1973: 35), the Pleistocenic fossil remains of Argentina and Bolivia (noted by López Aranguren [1930b: 109–110] and Cabrera [1931:116]) are reliable. However, these were not taken into account by Marshall (1985), Marshall et al. (1984), or Webb (1974). Given the significance of this point, it is of great importance that these disagreements be resolved.

Chapter 5 presented the arguments and the different models used by various scholars to discover and analyze the process of domestication of Andean camelids. There is no need to go back over this subject, so only a brief summary is in order.

Wheeler (1982a, 1982b, 1984a) proposed that the dental characteristics of the camelids can be used to distinguish the species. However, the studies by Kent (1982a), Wing (1988), and Miller and Gill (1990) questioned the results, particularly Wing (1988), who seems to have quite a solid case. Wheeler's study therefore needs to be reexamined to check the validity of the method.

As for the theoretical models, we saw that Rick's (1980) has serious flaws that invalidate it. The model was presented before the study of the faunal remains was finished, and it was precisely these results which showed serious contradictions, as was noted by Kent (1988b). The model proposed by Wheeler (1985b; see also Lavallée et al. 1985) is far more logical and consistent, but cannot be taken as definitive. It was also seriously criticized by Kent (1988b).

In my opinion, the alternative model proposed by Spunticchia (1989–1990), which tried to reconcile Rick's and Wheeler's models, is most suggestive and should in the future be carefully considered by specialists. The study should be published, so that it is accessible and can engender further discussion and debate. It is apparent, in any event, that despite the progress made in this regard, we are still far from a solution that, even while not definitive, can at least fall within acceptable limits that are not so vulnerable to critique. The road ahead is still very long, and one of the pending tasks is to undertake more systematic research. The Telarmachay materials can be a milestone in this regard, but it is imperative that Jane Wheeler, who is responsible for the study of its fauna, finishes her work so that the results for all the remains are known. I will not stop repeating that one of the most serious problems in Peruvian archaeology is the lack of complete studies of the materials excavated, whose data become irreparably lost in time.

The analysis of the early sites shows that, as Lynch (1978:480; 1983b:118) pointed out, at first the cervids, camelids, and other smaller animals were hunted, simultaneously with the Pleistocene megafauna. However, when the latter began to disappear, the hunter-gatherers were forced to change their economy. I emphasize that the extinction of the megafauna is also not clear and should be carefully analyzed. As was already noted, the scant evidence available for the Andes seems to suggest that the megafauna was not as abundant as it was in other parts of South America, such as in its southern section. Second, the causes of their extinction were various, with the change in the climate and the actions of man clearly being two of the most important. Under the present circumstances it is very difficult to impossible to establish which was stronger. I assume it was the first one.

When discussing the megafauna, Lynch (1991:257) said that "the exploitation of resources by South American Paleoindians was so thorough and ill-adapted to the natural ecology that there was a devastating effect on the species that were

sought." I am not too sure that this reflects reality. The efficiency of the early hunters has been much exaggerated with respect to the megafauna, and the real potential of their hunting equipment has not been sufficiently analyzed. The hunter's challenge was not just the size of the megafauna but also the possibility of mortally wounding animals that no doubt had very thick skin and superior strength. Indeed, it is quite possible that man was more a carrion-seeker than a hunter. In any case, and as far as I am aware, the actual evidence of hunters' weapons left on the remains of bones from megafauna is quite scarce. Much speculation exists about this. The climate, I insist, must have played a significant role, and I wonder whether altitude is not a factor that explains the lesser presence of the megafauna in the central Andes, in comparison with other parts of South America. Human pressure might also have pushed the megafauna to high-altitude areas, just as with the camelids; and as it was unable to tolerate these conditions the megafauna had to move to other, lower areas where it was still possible to live. It is true that this is no more than speculation, but I believe it is somewhat supported by the archaeological data. In any case, this subject still awaits study.

A review of the existing archaeological data shows that the sites available are indeed few, and only a few of these have been studied properly. In most cases the comments made are bold, and not just regarding the fauna.

I shall try to present a summary of this situation. Sites only barely studied but from which data can still be gleaned will be considered here as well studied to avoid going into too much detail.

In the preceramic period there are in the greater highland area a total of ten sites located at high altitudes, and just ten in the valleys. Only three of the high-altitude sites have been well studied (Pachamachay, Panaulauca, Telarmachay), while of the rest some have been cursorily studied as far as the fauna is concerned (e.g., Lauricocha), and others simply do not provide any valid data whatsoever (e.g., Piedras Gordas). It follows from the materials at these sites that camelids are absent in the lower levels or appear in minimal proportions, and that their numbers gradually increase in the upper layers. But in only four sites do we have enough details to distinguish the age of the animals.

Of the three valley sites, only two have been well studied (Guitarrero Cave and Huaricoto, in its Chaukayán phase). In one, camelids are absent at first, then appear and increase in number. In the other there is evidence of guanacos alone.

Before moving on to the coastal sites, I would like to clear up one point, namely, the presence of some form of early sedentism. In this regard Wheeler Pires-Ferreira (1977:18) says: "The possibility of sedentarism in preceramic times based on specialized hunting of the Camelidae has been brought forward. However, the research carried out by Thomas (1973) at Nuñoa in Puno indicates that the energetic efficiency of a pure pastoral adaptation in the puna is marginal at best." Not only do I reject this position, I believe that generalizations of this kind are dangerous. First, more care is needed when using present-day data to study the past. Conditions are different, and although I agree that ethnological data are interesting indexes, these cannot be used in absolute terms. Second, Wheeler forgets that Nuñoa is a pastoral community, while for the early preceramic we have societies that were just beginning the process of domesticating the animals, which was still incomplete. This implies that these groups cannot be classified as herders, but as hunter-gatherers who were in the process of becoming herders. Their economy therefore did not depend solely and exclusively on tamed animals. Finally, I find that in the Andes, the term "sedentism" must be taken with some latitude. I have explained that even transhumance, properly considered, is a form of sedentism. Perhaps an extreme case, but it is sedentism. On the other hand, it is clear that conditions were not the same in all sites. The situation in Lauricocha seems to have been quite different from that at Telarmachay. In the first case there apparently was a continuous occupation, while the second included transhumance to lower-altitude sites. However, both are cases of populations that stayed in the same area. I insist that a determining factor in this regard was the territoriality of the camelids.

As for the coast, most archaeologists have refused to accept the continuous presence of camelids in this area—that is, that they were raised there. Their presence, mostly in late periods, was explained away as a transient introduction from the highlands. This was the case with Lanning (1967a:63), Cohen (1978a:259; 1978b:122; 1978c: 27), Lumbreras (1974:37), Tabío (1977:211), Cardich (1980:117–118), and Isbell (1986).

However, there is a group of scholars who accept that camelids might have been raised on the coast, and admit their presence since preceramic times. Uhle was the first (1906:13), but it was Wing (1977b:17; 1986:255) who showed it with actual data. Others who have not rejected this possibility are Burger (1985b:276), Quilter (1991: 395–396), and Bonavia (1982a:392–395).

Two facts have to be accepted in this regard. First, that coastal samples are not representative except in rare cases. Second, few zooarchaeological studies have been made with materials from coastal sites, and the few existing ones are relatively recent. Thus, much of what has been said was based on speculation, not on concrete evidence.

In view of the existing debate, I find it necessary to present a slightly more in-depth summary of the situation. Preceramic sites with camelid remains are as follows. At Los Gavilanes a single bone, possibly from llama, abundant excrement from this animal, hair, and woven alpaca wool were found (Bonavia 1982a; Jones and Bonavia 1992). The case of Río Seco del León is not clear. Camelid remains, possibly guanaco, have been mentioned (Wendt 1963:237; 1976:19). At Ancón, Moseley (1972:29) reported the presence of camelid bones, which have also been reported for El Paraíso, where guanaco wool has likewise been reported (Engel 1966b:62, 65; 1967:265–267; Reitz 1988b:35). At Paloma, bones that seem to be from the guanaco have been identified in a context dating to the late preceramic (Reitz 1988a:311, 316; 1988b:34, 40). In Chilca there were camelid bones (Wing 1977b:Table 17; 1986: 258), and vicuña skins at Cabezas Largas, but they are more likely guanaco skins (Engel 1960:15, 17, 23). Vicuña skins are also reported at Santo Domingo (Engel 1981:34–36). I have to point out that it was only at Los Gavilanes, El Paraíso, Paloma, and Chilca that a specialist identified the remains. There certainly is not much evidence, but there is one more important datum that must not be forgotten. An important amount of llama dung was found at Los Gavilanes, thus clearly indicating that herds of these animals were used to carry products between the different parts of the valley.

The data are really scant for the Initial period. As we saw, there is only some brief information on some sites in the Callejón de Huaylas and Junín, where camelid remains are found to predominate. The data for highland valley sites are slightly more precise. In Cajamarca there were few camelids at first, and these were probably wild animals, that is, game. An increase takes place around the end of the Initial period, and these seem to be domestic animals. There are few camelid remains in the low-altitude sites in the Callejón de Huaylas; and in Kotosh (Huánuco) the bones found seem to be from guanaco and vicuña. Camelid bones predominate in Andahuaylas, but these are still wild forms.

There are somewhat more data for the coast. At Puémape just two camelid bones were found, apparently llama bones. The llama was an important element in the diet at Caballo Muerto. In the Virú Valley we only have evidence for the use of camelids as offerings. However, on the central coast, llama bones are frequently found in middens. Very few remains were found at Cardal in the Lurín Valley.

An interesting datum comes from the *ceja de selva* in Cajamarca (in the province of San Ignacio or Jaén), where the presence of an important percentage of camelid remains in the refuse has been reported. The report is unfortunately poor.

The data are more specific for the Early Horizon. I know of three sites located in high-altitude zones, in the punas of the Callejón de Huaylas, Junín, and Azángaro. Camelid remains overwhelmingly predominate in all of them, constituting between 72% and 98% of the fauna consumed.

The picture is even more clear for the highland valleys. In Cajamarca we find an almost absolute predominance of domestic camelid remains. The percentages vary at the sites in the Callejón de Huaylas but range between 55% and 92%. An increase in the use of domestic camelids is perceivable in Chavín de Huántar from the Urabarriu to the Janabarriu phases, from 67% to 94%. The case of Kotosh is different. The use of wild camelids continued, and the domestic ones

apparently made their first appearance just then. There is some evidence of the use of domestic camelids in the upper Chillón Valley, while in Ayacucho their remains overwhelmingly predominate at the various sites, just as in the zones of Cuzco and the altiplano. However, the sites known in the latter region are really very few.

For the coast, archaeologists agree that there is a remarkable increase in camelid remains found in the refuse dating to this period. An important amount (46%) was found in the Piura region, and it has been suggested that there were herds of these animals in the Piura River basin. The use of domestic camelids in Lambayeque was of primary importance as a source of proteins and for transportation. Up to 90% of bones of these animals have been detected in the refuse, and there also are big layers of dung. There is some evidence for this time in the Department of La Libertad, but without any concrete data. Bone and dung have been found in the Casma Valley. There is some evidence for the Huarmey and Supe Valleys, but no concrete data.

At Ancón, a thick layer of camelid dung was found, and in the valley the remains, possibly llama, are abundant (83%). In the Ica Valley there is evidence of abundant bones in the refuse, and also as offerings.

For the *ceja de selva* we once again lack clear data from the Department of Cajamarca (in the province of San Ignacio or Jaén), where the presence of camelid remains has been noted. It has been suggested that these were guanacos.

Let's look at some of the commentaries that explain these data. Burger clearly says (1985a: 532), when discussing the site of Huaricoto in the Callejón de Huaylas: "The most striking pattern to emerge is a gradual reorientation from the hunting strategy dominated by large ungulates to a strategy of camelid husbandry, with hunting reduced to playing only a minor role in the economy." He notes that the camelid bones probably belonged to llamas. Then, in a comparative analysis with contemporary sites, Burger (1985a:532) said: "In summary, the Huaricoto faunal analysis, like the studies of the Kotosh and Chavín de Huantár [sic] faunal remains (Miller 1984, Wing 1972), indicates that major changes in the diet of intermontane settlements occurred between

2200 BC and 200 BC. This implies modifications in the social relations of production involved in the hunting or herding of these animals and reinforces the view that a profound cultural transformation occurred in the Callejón de Hauylas [sic, it should read Huaylas] while the Huaricoto shrine flourished." Wing's study (1972:340) confirms this and shows how at first cervids predominate locally, while the camelid remains are probably from wild guanaco and some vicuñas. Then we see a decrease in the hunting of cervids and guanacos while the alpaca and llama begin appearing and gradually prevail.

As for the north coast, Shimada (1982:184) believes that the economic significance of the llama became evident between 1300 and 600 BC, and the llama became the major source of meat for the people of Lambayeque in AD 500–600. Marine resources were still an important part of the diet on the coast, "but even at sites close to the Pacific . . . [they] appear secondary to llama as a protein source."

There is far more information available for the Early Intermediate period. However, there are only records for some high altitude sites in the heights of the Callejón de Huaylas and Junín, where camelids make up important percentages of the animal remains identified in the refuse. More data are available for sites located at lower altitudes, in the valleys. Camelid remains overwhelmingly predominate in the case of Cajamarca. For Huamachuco we can only make some inferences, and it seems that in this period there was no significant number of camelids. We do not know what was going on in the Callejón de Huaylas, but the use of llamas was obviously quite common, to judge by representations in the Huaylas or Recuay style. At Kotosh there is some evidence of the use of both wild and domestic camelids, possibly guanacos. In the Department of Lima the use made of these animals was remarkable, to judge by archaeological reports, even though there are no specific figures. Camelid remains in the refuse in sites in Ayacucho, Apurímac, and Cuzco range between 64% and 97%. Curiously enough, the data available for Puno indicate a low percentage of camelids in the refuse (of no more than 8.6%), but this is just one site, which means that we know next to nothing about the altiplano region.

The data available for the coast for this period are revealing. The presence of great herds of llamas has been mentioned north of Talara. Meanwhile, the remains in Upper Piura reach very high percentages, accounting for 60%–80% of the remains identified.

There is much dung in sites belonging to the Salinar culture, in the Department of La Libertad. In the Moche Valley, the proportion of animal bones in the refuse comes to 92%, and herds of llamas have been mentioned in this area.

The discovery of the Cerro Arena cemetery, discussed in Chapter 4 (Section 4.5.2.1), must also be emphasized. A significant number of newborns and young animals were found here, thus proving that they were raised on the coast. This point will be expanded later in the chapter when I discuss the Moche culture.

The remains of domesticated camelids are also significant in the Virú Valley (40%–50%). Sites belonging to the Gallinazo culture in this valley show a relative abundance of llama remains and much dung.

Regarding the Moche culture, in Chapter 4 we saw in great detail that considerable data are available, and that there can be no question that camelids were much used for different tasks. The study by Shelia Pozorski (1976) established that over 90% of the animal protein used by the Mochica as food came from the llama. In addition, the abundance of dung at Mochica sites is quite common. We also find that camelids played an important role in the beliefs of these people, because there is evidence of their use not just in ceremonial contexts; llama bones, and sometimes complete skeletons, are also frequently found in burials. There is evidence of these activities from the Lambayeque Valley to the Santa Valley.

The study by Vásquez Sánchez and Rosales Tham (1988) is quite revealing in this regard. I noted in Chapter 4 that the osteometric analysis of the Moche bones shows the predominance of young animals. One of the observations made was that young animals and females had to be found to prove that camelids were raised on the coast (Kent Flannery and Joyce Marcus, letter of 30 May 1997). Although it is true that at least one case is known—Cerro Azul, on the south coast, which dates to the Late Intermediate—where the animals studied were adults and no females were present, thus suggesting that these were animals from caravans that went from the highlands to the coast (Kent Flannery and Joyce Marcus, letter 30 May 1997), the truth is that not enough studies of this kind have been made, so I emphasize that the sample is not significant. The case of the Moche animals is important even though their sex was not identified. In this regard I would like to reproduce the comment made by Vásquez Sánchez and Rosales Tham (1998:188): "As for the camelids, to date there are enough data in coastal archaeological sites that support the thesis of herds of camelids raised in this ecosystem. This is reinforced by the various age structures we find . . . [among the animals from the Moche urban area] . . . and by the identification of some corrals around the site, and lately in the middle Chao valley, where great expanses of land set aside for corrals were found in association with a megalithic [sic] citadel, with occupations that go from the Formative [ca. the Early Horizon] to Inca [Late Horizon] (Jaime Deza, pers. commun. 1996)."

As for the central coast, there are data only for the Chillón River Valley, where there is evidence of the use of camelids at several sites, just as in the Lurín River Valley.

The data for the Nasca culture are unfortunately scant and vague. They are mainly focused on data from three sites: Cahuachi, Paredones, and Tambo Viejo. In the first case, dung was found, as well as bones in the refuse, and it seems that there were llamas and alpacas. The report by Geismar and Marshall (1973) mentions figures up to 99% for camelid remains, but the study is superficial and must be taken with due caution. The evidence of a significant amount of dung inside a corral at Paredones is significant, as it shows that llamas were raised at the site. The discovery of excrement at Tambo Viejo supports this interpretation.

The faunal data from the highlands during the Middle Horizon, an important period in the history of pre-Columbian Peru, are quite scarce, and all come from sites located in the valleys. All we know is that the use of camelids increased in Cajamarca, to judge from the evidence from Huacaloma, where it seems that llamas and alpacas were used both for their meat and for trans-portation. They account for 97% of the remains. The percentage is also high in Ayacucho, where the data are somewhat confusing. Llamas used in the Colca Valley were more for transportation than for meat, but wild camelids never disappeared completely.

The data for the coastal strip are more plentiful. The amount of llama remains in the Lambayeque area is quite significant. At Batán Grande there is quite clear evidence that these animals were used not just in sacrifices but also reached the site in caravans of llamas. In addition, the dung was used as fuel. These animals were also used sacrificially at other sites. At Huaca Chotuna it is evident that they were used as meat. However, the most significant site is certainly Pampa Grande, where corrals were found that show the use of great herds of camelids. The highest percentage of the faunal remains found among refuse and in the houses clearly corresponds to camelids (between 70% and 88%). These were also very important as sacrifices.

Shimada and Shimada (1985:16) believe that the survival curve shows a different kind of management at different sites. They agree that the evidence for more breeding and management of the herds on this part of the coast corresponds precisely to the Middle Horizon and later times.

Llamas are common as burial offerings, as has been verified at several sites like Pacatnamú. However, what must be emphasized is their significance as food. A significant example is Galindo, where corrals with a great amount of dung were found, just as at Pampa Grande, which was used as fuel to fire the pottery. Here too the percentage of bones in the refuse is significant (69%). Corrals have been found in several Middle Horizon sites on the north-central and central coasts. Ancón was one important site. Furthermore, at Huarmey it was found that families moved with their llamas from one place to another. The samples are not too clear for the south coast, but there are camelid bones in the refuse and evidence of their use as sacrifice.

Onuki (1985:349) made the following comment: "Economic self-sufficiency was more easily accomplished within the coastal region with less dependency upon interchange with the highlands. If llamas were reared on the coast as argued by the Shimadas (I. Shimada, 1982; Shimada and Shimada, in press [published in 1985]), there were only a few things, exotic materials and mineral ores . . . to be sought in the highlands."

The Late Intermediate period is little known in the highlands as far as the fauna is concerned. However, there is one important piece of evidence that shows there were herds of camelids in the highland areas of Piura. There were also many llamas and alpacas in Junín. Curiously, it seems these animals were rather scarce in the high-altitude areas of Huamachuco. Another interesting datum is the presence of a significant amount of camelid bones in the refuse at Cuelap, in Chachapoyas. Their presence has also been certified in the Ayacucho Valley and in the higher parts of the Department of Tacna.

Once again it is on the coast where we find most of the data available for this period. The archaeological data, both bones and dung, show there were camelids in Tumbes.

The Lambayeque zone is once again the best studied, and the evidence available is incontrovertible. The case of Túcume is remarkable, because layers of llama dung 5 m thick were found here, something that was also verified elsewhere. Shimada and Shimada (1985:16) say of these dung deposits that "some [are] quite immense." There are as well bones from these animals in the refuse that in some places represent more than 80% of the remains. Shimada et al. (1982) proposed—and I agree—that caravans of llamas carried the minerals from the highlands to the coast, and the manufactured objects were then redistributed with this same system all over the coast.

In the case of Chanchan, the Chimu capital city, the percentage of bones in the refuse is very high, particularly in commoner precincts (55%–80%) where we find that almost all houses kept llamas, and everything seems to indicate that these were used both as food and as pack animals. The same holds true in all other neighboring urban centers. In addition, at this time the custom of using camelids as mortuary offerings persisted.

The case is repeated once more in the lower Virú Valley, because the camelid remains are more common in the refuse.

There are no data for the sites on the north-central coast. On the central and southern coasts

there are not many sites with data, but the presence of dung is common there, as are bones in the refuse. In Ica the remains of these animals are frequently found in burials as offerings, just as they are on the north coast.

There actually are scant data for the Late Horizon, but this should come as no surprise, because the archaeology of Inca sites is still to be done.

We saw that in the case of Huamachuco, a problem clearly exists. The studies made in the area actually do not show the number of camelids mentioned by Cieza de León. We may wonder whether the surveys made there have really been exhaustive.

Camelid remains at Huánuco Pampa, one of the most important Inca cities, come to 87%, and similar figures were recorded for high-altitude Inca sites in Junín. Although there are no specific data, camelid remains are common in the altiplano.

Of the sites located in the valleys we know only that there were camelids in the highlands of Lima, as well as in the Colca Valley. Sites studied in the Cuzco region indicate a very high percentage of remains from these animals, which make up to 96% of remains. In addition, corrals frequently appear.

The evidence for the use of llama caravans persists in the Lambayeque region, and north of Chanchan, at a site that lies practically on the beach, there are abundant bones and dung from these animals. Camelid remains come to 55% in the refuse from Chanchan dating to this period.

Camelid dung is common in the late sites of the central coast, and has also been found on the coast of the Department of Arequipa.

We must not forget that there is evidence of camelids at two sites in the *ceja de selva*. In the north, they were found in the ruins of Abiseo (i.e., the mistakenly called Gran Pajatén) and neighboring areas, where the specialist who studied the materials is convinced that the animals were raised there. In the south they were found in the Cuzco region.

The archaeological data allow three major conclusions to be drawn, which I will summarize here, even though they have been presented separately more than once in this book. First, it is evident that far too much has been generalized with too few data. A systematic research agenda that studies the faunal remains of the central Andes in-depth is urgently indeed. Above all, large-scale excavations are needed for the data to be significant. The facts have been far too distorted by small excavations whose results do not give an exact picture of the real significance of the sites. Area studies must be emphasized.

Second, although studies of this kind are few for the coast, it is undeniable that the camelid population was important, and definitely not all the animals eaten or used there for transportation or rituals came from the highlands, but were raised locally. With the available data, it is clear that this was the case at least since the late preceramic. It seems that camelids were more intensively used on the north coast than on other parts of the coast. However, I would not dare to say that this really reflects the facts. This subject has been studied, in more systematic fashion, for the north coast than for other sectors. I am convinced that we shall have big surprises once further studies are made.

Third, it is clear that the use of camelids was concentrated in high-altitude areas during the early preceramic. Gradually, over time, the remains of animals already having a domestic form became more common in the valleys, and this practice became generalized in the Early Horizon. From this moment on and up to the Europeans' arrival, there was no place where camelids were not eaten or used as beasts of burden or in other tasks. We saw that they were even used in some parts of the *ceja de selva*.

We must add one more fact that has not been noted by most specialists. I mean the relationship between El Niño and the greater development of herding that takes place on the coast in this same period. It is true that these are temporal phenomena, but this does not make them less important. The conclusions reached by Makowski after studying the phenomenon on the north coast are interesting. He noted that "the abundance of camelid bone remains in middens adjacent to ceremonial and manufacturing areas suggest that significant-sized herds were perhaps kept all year long on this part of the coast. This brings to mind the rise of livestock production in Piura in Republican times, when the Sechura Desert turns into an African-type savanna, rich in pastures and nutritional algarrobo pods, during the time particularly strong El Niños occur" (Mogrovejo and Makowski 1999:55). The observations made by some specialists describing the effects of the 1925 El Niño are revealing and suggestive, and should be noted. Johnson (1930) says that there was a remarkable increase in the vegetation of the coastal lomas, which in some areas persisted for several years. Murphy (1926:46) noted that on the central coast, and specifically in the Rímac Valley, at Chosica (860 masl), "where ordinarily not six, perhaps not two, llamas might be seen during the course of a year, now became the rendezvous of as many as eight hundred in a single day."

This fully supports the thesis here posited, insofar as these animals were cornered by man at high altitudes in their wild state and returned to the lowlands once domesticated, and taken there by man. What has limited and conditioned the dispersal of these animals is culture, and not just in pre-Columbian times, as has already been seen and shall be discussed later. Besides, this overall picture proves what the physiologists have long said, namely, that the camelids are not specialized, high-altitude animals. If a cross section of the Andes is made and a line drawn across the places where these animals have been found, it traces an arc that begins on the Pacific coastline, encompasses all of the Andes, and ends in the *ceja de selva*. That is to say, it extends from sea level to altitudes over 5000 masl, from the coastal deserts and fertile intermontane valleys to the high punas and the exuberant wet region of the *ceja de selva*, with all of the intermediate life zones that are far more varied than is usually thought.

There is one subject that I must raise here because it is one of the great problems posed by the archaeological data and one whose solution is pending: the enigma posed by the Mochica representations of short-necked llamas. We saw that Horkheimer (1958) believed this was a llama breed that was lost due to the abuses that took place with the arrival of the Europeans. However, his ideas were just that and were never supported with any evidence. As long as insufficient materials are available for the systematic measurement of skeletons, his hypothesis remains unproved. However, what Horkheimer intuitively noted in the fifties might now become possible to confirm, based on analyzing the different size of the animals based on skeletal remains. Even so, it should not be forgotten that this is being deduced only on the basis of ceramic representations. True, given the realistic style of the Mochica artisans, it is hard to imagine that it could be a generalized mistake. It would have to have been an imposed cultural norm, but this is very hard to accept.

Even so, some observations must be made that follow from the study of ceramic collections of the various pre-Columbian cultures, and which no one has noticed. It has been repeated, perhaps following Horkheimer, that the Mochica representations are of short-necked llamas. Once again, there is no exhaustive study of this subject. The problem will definitely have to be restated once this is done.

The main question is: how are llamas depicted in the other styles of the central Andean cultures? I was unable to establish this, so the few observations presented here are essentially based on collections of coastal cultures. They are just suggestions, and should be taken as such.

The Mochica definitely were not the only ones to depict the llama with a short neck. It appears like this in vessels from the Cupisnique, Vicús, Moche, Lambayeque, Chimu, and Chancay cultures. The long-necked llama is more an exception than a rule in the art of these cultures. Christopher Donnan reports having seen a modeled Mochica vessel of a long-necked llama (pers. commun., 15 April 1993). I know of only one Mochica painted scene, but there surely must be others. It is a presentation of offerings in which a loaded llama has been painted. Shown is a long-necked animal of similar proportions to modern llamas. However, in the same scene there is a small llama that could be a young animal, is not loaded, and has a short neck. There are unfortunately no data on the provenance of the drawing (see Hocquenghem 1987:Fig. 34). I have seen several pieces from the Chimu culture depicting long-necked llamas, one of which is published here (Figure 4.37), but there definitely are not many. Both variants exist in the Recuay culture, that is, figures of long-necked (Figure 4.28) and

short-necked llamas (Figure 4.27). I cannot determine which predominates.

The famed Robles Moqo vessels from Pacheco, both big and small (Figures 4.29–4.31), are unmistakably long-necked animals. Here I point out one single Middle Horizon piece from the highlands—a beautiful llama-shaped vessel with a long neck (Katz 1983:272 [Fig. 147]).

The representations of llamas in the Nasca style are not always easy to assess. When depicted realistically they appear with a long neck (Figure 4.26), but when stylized (Figure 4.22) they have a short neck. However, it is interesting that the stylized depictions of the guanaco almost always have a long neck (Figure 4.24).

In the Inca style, the depictions of llamas I recall having seen on pottery have a long neck. This is also true of the pieces in relief (Figure 4.52). The same is true of the many silver statuettes in the shape of a llama (Figures 4.53 and 4.54). We do not know what happened when Inca influence spread throughout the Andes, but the piece here illustrated in the Chimu-Inca style (Figure 4.57) is of a llama with a long neck. An interesting vessel in the Sackler Collection of Recuay-Inca style shows a man carrying a long-necked llama (Katz 1983:315 [Fig. 190]).

We thus find that it was not just the Mochica who left pictures of short-necked llamas. Generalizations are always dangerous, and all the more so when they are unsupported. On the other hand, does the fact that the same culture has depictions of short- and long-necked llamas mean that there were two camelid varieties at that time? Or is it just a stylistic difference? In my opinion, no answer is possible with the data currently available. This will only be possible once a systematic study is made. And when this happens it should not be forgotten that the pre-Columbian artisans, and especially the Mochica, were not only well acquainted with the world they illustrated, they were also excellent observers.

Finally, I would like to emphasize a fact that should be food for thought. Contemporary potters are well acquainted with the llama, and today there are only long-necked llamas. Why then do they depict them with a short neck? (See Figures 8.2 and 8.3.)

To conclude, I am convinced that in ancient Peru there not only existed llamas of different sizes, there were also breeds that are now lost. What we do not know yet is whether art reflects this reality or not.

I should point out that it was clearly Wing who discovered the existence of different-sized camelids, with her careful and unquestionable osteological data. In her writings Wing has mentioned "large" and "small" camelids at least since 1972 (see Wing 1972:335). She then posited that there actually was a small animal form in the north, while another, larger form predominated in the south (see Wing 1977a). The discussion was renewed after the work done at Pirincay in Ecuador, which was discussed in depth in Chapter 9. We should recall that when analyzing the remains of skeletons from this small form, Miller and Gill (1990) rejected the possibility that it was a wild species, also rejecting at the same time that it was an alpaca or wari, that is, a llama and alpaca hybrid. Miller and Gill favored instead a smaller llama breed. They furthermore stated that this could not be derived from or related to the animals currently being raised at La Raya. The chronological discussion raised by the specialists concerning whether these remains date to the Early Horizon or the Early Intermediate period in no way changes the situation.

In fact, no discussion can go beyond Miller and Gill (1990) with the available data, but this definitely proves Elizabeth Wing was right, and perhaps Hans Horkheimer, too. If we add to this the suggestion by Wheeler et al. (1992) that the database of camelid hair lets one argue that an unknown series of llama and alpaca breeds was lost in the chaos of the Spanish conquest, we evidently have enough grounds to accept that the breeds we now know are but a small sample of a great diversity we know nothing of. More work in this line of research is urgently needed to clear up these problems, which are crucial for an understanding of Andean realities.

Historical sources are of great value for the point now under discussion, just as they are important in general for an understanding of the Andean past. We must, however, be very careful when interpreting the data in these sources because, in the words of Araníbar (1963:113), the chroniclers saw "the new things through Christian and romanticizing eyes. Through this way of facile associations, where the things of the new land are measured with the interpretive grid of a foreign cultural habitus."

In Chapter 6 we reviewed and discussed in depth the problem posed by the various terms the Europeans used to label the camelids they encountered. It was found that while in some respects it is possible to close in on the truth of what species the chroniclers meant in each case, absolute clarity on this question eludes us. I have repeatedly noted this throughout the book.

Care is needed on other points as well. One of them is the secondhand use of sources that are not always correct. We also found more than one case in which the facts were distorted by misreading a given chronicler or misunderstanding him through ignorance of the language. In some cases, the repetition of these mistakes has become an established truth. Some facts have likewise been generalized after reading only some of the most well-known sources, without undertaking a systematic review of the sources. It was found that mistakes were made even by specialists who certainly know their subjects and have dealt widely with historical sources. Such is the case of Murra (1975:117), who wrote as follows: "Since there were almost no auquenids in the northern part of the Andean cordillera, the herds of these animals had a limited importance in the economy of this region." This is not true, as is clearly shown by both archaeological and historical sources.

The distribution of camelids in Inca times is one topic for which historical sources are not too helpful. It was seen that for the highlands we have some data only for the zones of Chachapoyas, Cajamarca, Huamachuco, and Huánuco. There are, however, some very interesting and specific data. Such is the case, for example, of Cieza de León (1984:Part I, Chap. LXXXI, 236), who claims that in Inca times the vicuña lived in Huamachuco. We saw that Koford (1957:218) rejected this for no reason whatsoever.

With regard to the coast in Inca times, historic sources mention only the regions of Lambayeque, the Chillón, the upper Rímac Valley, and the Chincha Valley.

Two aspects of the historical data are quite clear, and elicited the admiration of the European chroniclers. The first has to do with the order that reigned in the Inca Empire in relation to the management of camelids, both domestic and wild, and the strict enforcement of the relevant regulations. The second is the great importance of these animals both to the Inca economic system and to their beliefs. The first point has been studied in some detail, although much has still to be explained; the second one has not received the attention it deserves.

It is worth summarizing and discussing the historical data that were presented in depth in Chapter 6. To avoid boring the reader, no citations are given except in specific cases; otherwise, only the authors are identified.

Despite some limitations that I will not dwell on further, the historical data are of crucial importance to show the actual distribution of camelids in the Andes at the moment of the Europeans' arrival. This distribution definitively disproves the mistaken assumption that camelids are high-altitude animals.

As we saw, several writers described what they saw on the Peruvian coast or repeat the data they managed to collect. Many of them specifically mentioned camelids. Such is the case with Estete, Andagoya, Gómara, Molina "El Chileno," and León Pinelo. In the case of Estete we find there was an abundance of these animals, which, according to León Pinelo, were still common in the seventeenth century.

So, when the Spaniards reached the Peruvian coast, they found camelids in Tumbes, Poechos, San Miguel de Tangarará, Piura, Zarán, Motupe, and Collique. The *Relaciones Geográficas de Indias* clearly says that not only were there camelids at San Miguel de Piura, they were also raised there. Cieza de León clearly says that when he passed through Huancabamba there were many camelids, and that these vanished owing to the destruction wrought by the Spaniards.

For the Lambayeque zone there are data only from 1540 on, but there were camelids in this year, and herdsmen in Jayanca until the mid-sixteenth century. We likewise know that there were "sheep of the land" in Túcume in 1574, herdsmen in Saña, and animals at Moro and Guadalupe. We

saw that in 1543 a butcher was sought in Trujillo to sacrifice the llamas needed to feed the local population. There were also llamas in the Chao Valley.

It is likewise documented that "sheep" were sold in Lima in 1538, and Zárate mentions the presence of abundant livestock on this part of the coast in 1544–1545. In 1533 llamas were brought to Pachacamac from the neighboring Lurín Valley. In the Chincha Valley, Cieza de León was told of the existence of great herds of camelids, and he lamented that these had vanished thanks to the Spaniards. For Ica we know that in 1571 the son of the cacique Anicama had a herd of 100 llamas. Vázquez de Espinosa confirms that these animals still existed in the area in 1617. In Atico and Caraveli there were camelids in 1549, while 600 "sheep" and "much livestock of the land" are mentioned for Sama in 1567. The testament of Diego Caqui listed 200 animals in this same area.

In the face of all of this unquestionable evidence, can it still be said that camelids were not raised on the Peruvian coast? The answer is clear. I agree that it cannot be established whether these animals were llamas or alpacas. The llamas were probably more abundant, but we should not reject the presence of alpacas too.

The data are revealing for the highlands, but they do not disprove that camelids did live in the northern part of the land. The data in the chronicles mention "many herds" in the highlands of Piura, and these animals still lived in the vicinity of Chachapoyas in 1615. In Cajamarca, where camelids are now being introduced from the altiplano, there were still many "sheep" when Cieza de León passed through the area. These had vanished by 1566, but in the early nineteenth century the memory of small herds of camelids still survived.

I separated the journey of Hernando Pizarro and his men from Cajamarca to Pachacamac and back from the general context of this book's narrative. I did so because these men were eyewitnesses who were then establishing the first contact with the Andean world. This is the first account of pre-Columbian Peru from the perspective of Western civilization, save for the little-known feat of a group of four shipwrecked Europeans—of whom we know only of the

Portuguese Alejo García and Aleixo Ledesma—who with some Guaraní Indians were in contact with the Andean culture around 1521–1524/1526, and who most probably were the first Europeans who knew of the Inca Empire (see Rowe 1946:208; Métraux 1948c:465–466; 1948d:76; Busto 1988: 22–24), but unfortunately, they did not leave any descriptions.

During the journey the Spaniards saw "sheep" in Cajabamba, and much livestock was said to live from Huamachuco to Corongo. The Callejón de Huaylas, where camelids have almost disappeared, was said to be "most abundant in livestock." Interestingly, there is no reference to any animals from the moment they made contact with the coast until they reached the coast as far as Pachacamac and returned to the highlands through the Huaura Valley. It is hard to believe that at the time there actually were no llamas on this coastal strip, since later chroniclers mention them. We can assume that their attention was drawn by their proximity to the place where the prized gold they had been told of was, and by the fact that they were passing through areas far more peopled than those they had covered in the highlands, as the valleys of the central coast must have been at the time, which even had major settlements.

On their return voyage, great numbers of camelids are mentioned in the ascent up the Huaura Valley, but the exact location of these places is difficult to determine. However, we saw that it is definitely possible that llamas and alpacas were mentioned for Junín. This was probably the first time these two species were clearly distinguished. Great numbers of them are mentioned.

The various sources then insist on the presence of camelids in the Huamachuco zone, in the Callejón de Huaylas and in Huánuco. In the highlands of Lima there are references to the presence of llamas up to 1764. In Junín there was a significant number of animals when Cieza de León passed through there. In Tarma there was "such a great number . . . that they are uncountable" (Cieza de León 1984:Part I, Chap. LXXX, 234). There still were many beasts in Jauja in 1597. There are data for Huancavelica, Ayacucho, Apurímac, Abancay, and Andahuaylas. For Cuzco we should recall that herds of 2,000 head are

mentioned after 1536. Then there are data for Arequipa and the altiplano, where the number of camelids must have been enormous, to judge from the data in the *visitas*.

Although the picture is impressive, its significance lies in that it shows not just the great number of animals extant in the Andes—even if they cannot be quantified—but also the several zones where these abounded, such as the northern highlands or the Callejón de Huaylas, which today are not considered camelid-raising centers.

There also are early sixteenth-century sources on Jaén, Chinchipe, and Moyobamba, in the *ceja de selva*, which mention camelids. This supports the archaeological data and shows the great potential these animals had to live in such varied environments. It should not be forgotten that the Jívaro Indians recently had llamas in the upper Marañón River. True, these were not many, but they obviously could live well there.

A comparison of the distribution of the wild camelids based on historical sources and their present distribution is interesting. For the guanaco we recall that Cieza de León claimed it lived in the highlands of Piura and Huamachuco. We know thanks to the testimony of Dávila Brizeño that these animals lived in the mountains close to the famous Mt. Pariacaca, and according to Cosme Bueno they still lived in the mountains of Yauyos in the late eighteenth century. It is also known that in the mid-eighteenth century there were guanacos in the Tarma zone, and for Jauja there are some sixteenth-century data. For Huancavelica there are data on the guanaco from 1586 to the eighteenth century, and for Ayacucho from 1557 to the eighteenth century. These animals were living in the heights of Nasca in 1600, and around this same time there are data for Apurímac, Abancay, and Andahuaylas. For Arequipa there are data until the eighteenth century, and for the altiplano we have data from the late sixteenth century to the eighteenth century.

There are no problems concerning the southern border of guanaco distribution. There is an essential disagreement about its northernmost extent, because at present this is located in the Department of La Libertad, at 8° S latitude, that is, at about the city of Trujillo, whereas the historical data tell us that in what is now Peru, the guanaco reached as far as the highlands of Piura, and there also were great numbers in Huamachuco. Moreover, we should not forget that Cieza de León said there were guanacos in what is now Ecuador. In other words, it is clear that these animals' territory has diminished tremendously (see Figure 2.2).

The situation is different for the vicuña. Available data on its presence in the highlands of Piura are not clear, but they are clear for the Huamachuco zone.

We know that the vicuña was common in the highlands of Lima in the eighteenth century and that it lived in the vicinity of Mt. Pariacaca in the late sixteenth and even early seventeenth century. The same can be said for the Yauyos zone. Cieza de León recorded its presence in Junín, and there is evidence of it until the early seventeenth century. This species is mentioned in Tarma in eighteenth-century documents, and its presence in Jauja was certified for the late sixteenth century.

There are data for the Huancavelica region from 1586 to the eighteenth century, and almost the same can be said of Ayacucho. Vicuñas lived in the highlands of Nasca until 1600, and there are some data from around this time for Apurímac, Abancay, and Andahuaylas. For Arequipa and the altiplano there are data until the eighteenth century.

The distribution of the vicuña as established using historical sources also differs from the present one in its northern part, while the southern one shows no difference. Today it is not clear whether its limit in the north lies in Junín or in Ancash, but the data in the chronicles show there were vicuñas in the Department of La Libertad, specifically in Huamachuco. In other words, in this case the difference in territory is not as marked as for the guanaco, but even so, the area of distribution of these animals was more extensive in the past than it is now (see Figure 2.5).

Dedenbach Salazar (1990:96) gives a good summary of the historical data on domestic camelids. She says that "[t]he northernmost place where domestic camelids are mentioned is Caranqui, north of Quito, in the province of Quito, and moving southwards they are mentioned in Latacunga, Ambato, Cuenca, Cañar, and Loja. It seems that there was much livestock of the land in the sierras of Cajamarca and Huaraz,

and the richness of the Indians in camelids is especially noted for central Peru: Jauja, Huánuco, Huarochirí. Southward from there almost all of the highlands are mentioned as far as La Paz and Potosí in Bolivia, and Río Estero/Córdoba in Argentina" (see Figures 2.9 and 2.11).

This is correct in general terms. We saw that the data can be expanded for the southern section using historical data (I discuss this a bit more in the following pages). What is actually a hard fact, supported by archaeological data, is the presence of domestic camelids in what is now Ecuador and the northern Peruvian highlands, where they vanished, and attempts are now being made to reintroduce them.

The significance the caravans of llamas had in pre-Columbian times cannot be questioned, since this was the only existing means of transportation besides man. These were very big caravans. Garcilaso (1966:Bk. 8, Chap. XVI, 513; 1959:147) says that those with fewer than 500 animals were not taken into account. The alpaca was also used for transport. This follows from several lines of evidence, including the *visita* of the province of Chuquito. These caravans did not lose their primary importance in the conquest and throughout the viceroyalty, despite the introduction of the mule and the horse. Although the llama cannot carry a great load, it had the advantage that a very large number of llamas could be used, certainly far more than mules or horses. Moreover, they were cheaper to use, because camelids do not need a harness or special foods, either during the trip or when resting. They also had the advantage of not fearing altitude. This has always been a constraint on the use of imported animals, the horse in particular, and must have been a particularly serious problem in the first years of the European invasion, for it took them time to adapt, as is well reflected in the writings of the chroniclers. It should also be noted that wagons were obviously not of much use in the harsh geographic conditions of the Andes, and their use was restricted.

So, in Colonial times, the caravans consisted of 1,000 or 2,000 animals that moved both on the coast and in the highlands. Almost all the wine that reached Moquegua for Potosí was transported by caravan until well into the seventeenth century, as were all the minerals taken from Potosí to the coast to be embarked. Llamas were still used to carry minerals (Figure 14.1) and even guano in the late eighteenth century.

Today the llama is no longer important, given the construction of roads and the use of motorized vehicles. However, caravans are still used, especially in the highlands, in places not yet accessible by road, and there are many such places. According to Masuda (1986), in 1981 products were still moved from the highlands to the coast in Atiquipa. We should not forget that in the early twentieth century caravans went from the Callejón de Huaylas to Chimbote, and that according to the testimony of Cristóbal Campana, in the 1930s and 1940s there were herds of up to 100 llamas in the mountains of Trujillo. The same is true for the mountains of Ica. For the south coast we know that these herds were still seen moving down to exchange products until 1985. This was also quite common on the eastern Andes. Even now, caravans of llamas move down to the Amazonian piedmont in Bolivia.

N. Guérard le fils fecit

A Llamas ou moutons du Berou.   E Plan de la daraggadera.
B Trapiche ou moulin a minerai   F Profil de la daraggadera.
C Buteron ou cour ou lon petri le minerai G La pigne
D Bassins a laver   H Kourneau atirer le visargent

FIGURE 14.1. Engraving showing the mining technology in use in Peru and Bolivia in the early eighteenth century. In the foreground we see two "Llamas or rams of Peru." *After Frézier (1976:pl. XXII,138).*

There is no way to establish how many camelids there were at the time of the conquest and the viceroyalty, not even an estimate, but there is no question that it was an impressive number. It is something we probably find difficult to imagine. The figures in the documents analyzed here show this situation. However, it must not be forgotten that in many official statements, the real number of animals was concealed to avoid theft and taxes.

Even so, the figures are significant. Chapter 7 presented as complete a picture as possible. Here I will just repeat some figures. Two thousand rams are mentioned at Pachacamac in 1573, and around the same date similar-sized caravans came down from the sierra. Quizquiz lost over 15,000 head in southern Ecuador in 1558, on his way back from Jauja. Herds of over 100,000 head are mentioned for Junín, and there were still many in the eighteenth century. Figures of more than 80,000 animals have been given for the altiplano, and 60,000 is a number often repeated. We should bear in mind that this almost always means domestic animals. Wild animals must also have existed in very great numbers. In other words, the figures testify to a truly impressive number and clearly show the loss that has systematically taken place since then, and still continues.

The slaughter of animals that took place throughout Colonial times, especially in the first years of the European invasion, was truly dreadful. This point bears repeating, because historians have thus far neglected it or have treated it only lightly. It will be recalled that 150 animals were killed each day at Cajamarca, and 10 or 12 were sacrificed just for the marrow. Meanwhile, in the "ancient kingdom of Quito" more than 60,000 animals were sacrificed just for their brains and heart. When Quito was founded, five or six animals were killed for their brains and to make "marrow pie."

The case of Cajamarca is heartbreaking. There is evidence that many camelids were here in pre-Columbian times and when the Spaniards arrived, yet these had disappeared by the late sixteenth century. Huamachuco is a similar case, for we know that no camelids were here by the eighteenth century. And in the Callejón de Huaylas, where the Spaniards saw these animals in the early sixteenth century, there were none by the late sixteenth and early seventeenth centuries.

On the other hand, the abuse perpetrated by the Spaniards is quite clear in the Junín zone, where the Indians, as we saw, were forced to hand over animals, and also because the amount pillaged ("rancheado") must have been very high.

It is known that no camelids were left in Castrovirreyna by the early seventeenth century, while the abuses inflicted by the friars in the altiplano are evident in the secret inspection (*visita secreta*) of Gutiérrez Flores and Ramírez Segarra (1572/1970). Flores Ochoa (1977b) is one of the few scholars, if not the only one, who has discussed this destruction.

By 1561 there were few guanacos around La Paz, while in Potosí some 100,000 animals were killed every year and some 40,000 were pillaged.

The *chacos* are another reason for the fall in the number of camelids. While in Inca times this was a way to control the population of these animals and even of making a genetic selection, from the moment the Europeans arrived it became a pastime and an easy way to kill them. The figures given for the animals killed in these *chacos* are striking. Just to give two examples, in one case there were 11,000 dead animals, in another more than 29,000. In most of these cases the animals, I repeat, were killed just to eat the brains and the neck. *Chacos* were still held in the eighteenth and nineteenth centuries, but in a much less destructive way.

There evidently is another reason for the fall in the camelids' numbers that I have not studied much but for which considerable data are available. This is the tribute set and paid in animals. Guaman Poma de Ayala (1936:fol. 877; my Figure 14.2) has one drawing showing "How the *mandón* [low-ranking official] takes the ram [*carnero*] from the old man as tribute."

The abuses committed in this regard are well reflected not just in the *visitas* and the lawsuits; they also appear in the writings of Polo de Ondegardo (1916). To note once again the permanent protest of Cieza de León, who can be considered the precursor in defense of the camelids, is a matter of duty.

True, we should not forget that—as far as I am aware—from at least the late sixteenth century on

there were laws protecting the camelids, but these were not enforced.

Interestingly, despite the time that has elapsed, the memory of the abuses the Spaniards inflicted on the native animals has persisted. The beliefs held in the Departments of Cuzco and Puno are most significant in this regard. These were collected by Flores Ochoa (1974–1976: 258–259). "It is said that there were far more alpacas in ancient times, 'perhaps in the time of the Inca too, which we do not believe will come back.' The alpacas were lost in part because the *ispañulkuna*, 'the Spaniards,' took them, and they diminished in part due to the neglect of some men, as is recounted in a narrative from Paratía, in the Department of Puno."

Dogs also played a role in the killing of camelids, if not on the scale claimed. Everything seems to indicate that this custom was introduced late in the Colonial period with imported hunting dogs, and not in the early years of the Spanish conquest. The truth is that there is no evidence to show that this custom existed in pre-Columbian times. The only evidence I am aware of is the

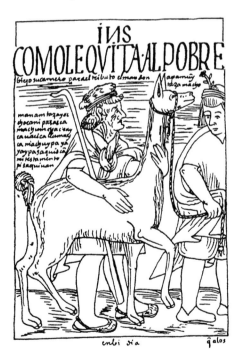

FIGURE 14.2. Guaman Poma de Ayala shows "[h]ow the headman seizes the ram from the poor, destitute old man as tribute." *After Guaman Poma de Ayala (1936:fol. 877).*

Mochica, who used dogs when hunting deer. However, this clearly was a ritual hunt (see, e.g., Donnan 1978:Fig. 262, 179), and there is nothing to indicate that hunting with dogs was practiced by the Mochica on a wide scale.

It is clear that the commercialization of the wool from wild camelids was likewise an important factor for their destruction. I am not in a position to state when this activity began, but it flourished in the eighteenth century (see Juan and Ulloa 1988:592–593) and continued at least until the first half of the nineteenth century (see Tschudi 1966:246–247). Carlos Peñaherrera del Águila tells me in this regard (pers. commun., 16 December 1994) that a careful review of the sources would give a far more accurate picture, including quantifiable data that might be most significant. Peñaherrera tells me, for instance, that the 1794 *Mercurio Peruano* (Vol. II, facsimile edition, Biblioteca Nacional del Perú, Lima 1966, pp. 191, 193–194) records that "100 sacas" of vicuña wool were shipped from Lima between January and late June 1793, and that another "100 sacas" more of vicuña wool plus 1,381 pounds of guanaco wool were sent from Montevideo to Europe.

I have tried to establish the weight of a saca, but was unable to find an exact definition. The 1611 *Tesoro de la Lengua Castellana o Española* defines a saca as "[a] big and ample sack like the wool sacks" (Covarrubias 1943:918), but no weight is given. The 1900 *Enciclopedia Espasa* notes under "wool" that the wool from sheep and castrated rams comes packed in "sacas of some 70 kg [each.]" This corresponds to the wool traded in Aragón and Navarre (Espasa Calpe S.A. n.d.:Vol. XXIX, 527). I assume this is the approximate weight of a saca, because in Colombia there still is a measure called "saco" that equals 5 arrobas, or about 62.5 kg (Llerena Landa 1957:187).

If we accept that a saca equals about 70 kg, then the 200 sacas mentioned in the *Mercurio Peruano* give a total amount of 14,000 kg of vicuña wool. An adult vicuña yields between one-third and one pound of wool (Grimwood 1969:69). This means that each animal yields between 0.153 kg and 0.230 kg of wool, so it follows that between 91,503 and 60,869 vicuñas were killed (it should not be forgotten that these are wild animals) to get this volume.

In the case of the guanaco, it is known that one animal yields about 250 g of wool (Franklin 1982:469). Now, the 1,381 pounds mentioned in the *Mercurio Peruano* are equal to 635,260 g, so 2,541 guanacos are needed to get this amount of wool.

We see that both cases entail a great number of animals. It would be most interesting if someone began to study this topic, for we would thus have approximate figures for the killing of animals throughout time.

The dramatic effects on the native population of Old World diseases introduced into America by the conquistadors are well known, and much has been written about it (see, e.g., Crosby 1972). The animals were not exempted. Flannery et al. (1989) claim that one type of mite responsible for the mange (sarna), different from the local mites, was imported by the Europeans. This is confirmed by Garcilaso de la Vega (1958:Bk. 8, Chap. XVI, 148; emphasis added), who said "it was a very dire disease, *hitherto unknown.*" The mortality from this disease must have been very high, to judge from what the chroniclers said, and another factor contributing to the decline of the Andean camelids.

The data available on lands outside the central Andean area are most interesting, especially those concerning the northern part, even though not much was found. It would be interesting if someone made an exhaustive collection of data, as the results would yield more than one surprise.

The most intriguing data involve Nicaragua and Costa Rica. Clearly, the mere mention of "fine-haired rams [*carneros*]" in a Nicaraguan document should not hold our attention if seen by itself. However, things change if we add the modeled Costa Rican representations. I was unable to examine all the representations mentioned in the archaeological literature, but there can be no doubt that the El Bosque vessel and the stone sculptures of Las Huacas are camelids—llamas or alpacas, in my opinion. Interestingly, the figure in the El Bosque piece is a short-necked animal (see Figures 9.1–9.5).

These representations are perfect, and as far as I am aware, they cannot have been made from descriptions or from oral accounts. The artists who made these pieces certainly saw the animals. The question is, where? No answer is possible.

However, there are only two possibilities. Either some of these animals reached Costa Rica, or the artists—it might have been just one—traveled south as far as present-day Ecuador, where the llama lived—according to archaeological data—at a much earlier date than that of the Central American representations. The journey may well have been across the sea or by land. This posed no problem, since long-distance movements are not the exception but the rule in America. Two possible scenarios took place in these voyages. Either some animals were taken north, or an adventuresome artist returned home and illustrated what he saw. If this actually happened in Costa Rica, then there can be no question that the comment made in the Nicaraguan document is true. Were this the case, it really could be a piece of information received from the Indians. The short neck is another problem. All we know of the Ecuadorian llamas is that these were small. Judging from their representations, the short-necked llama came from the central Andes, including the Mochica ones, which were contemporary with those of El Bosque in Costa Rica. To explain this similarity with the evidence available is practically impossible.

These clearly are speculations, but there is a point of departure that is factual and cannot be denied. The El Bosque and Las Huacas representations are camelids and cannot be any other animals. Only future studies will resolve this fascinating enigma.

The news of these animals collected by the Europeans who entered what is now Venezuela pose no problem. They date to the late sixteenth century, when Pizarro's army had finished its task and once the news of the riches of the Inca Empire had spread. In addition, there were llamas in Ecuador, and the Incas had introduced the big pack llamas, which impressed the natives, and the news spread easily. It seems that at this time, camelids also entered Colombian territory, which adjoins the Venezuelan area under discussion.

The archaeological data on camelids in Colombia come from a single site and are poor. They seem to be late remains that would date to what in Peru is known as the Late Horizon. These remains probably are of animals taken there by the Incas. The Spaniards apparently found camelids from Popayán to Cali and in the highlands of Urabá. These animals, I emphasize, must have been taken there by the Incas.

The archaeological data are far more specific and abundant for Ecuador, but are still inadequate. There undoubtedly is clear evidence of camelids since the period that in Ecuador is called the Late Formative, which corresponds in temporal terms to what is called the Early Horizon in Peru. These are small animals in the case of Pirincay, and there are no osteological data for other sites, but I do not want to dwell on this point because it has already been discussed in depth. There are also data for later periods, almost up to the time of the Inca arrival in present-day Ecuador. We saw that the data are for both coastal and highland sites.

So far, everything seems to indicate that the animals arrived there from the south in a domestic state. Systematic studies are needed in the Ecuadorian area, which clearly has much potential data. Big surprises will be had once these are studied.

The historical data support the archaeological claims, from which it follows that the camelids were in Ecuador before the Incas arrived. It seems that when the Spaniards arrived there were no camelids on the Ecuadorian coastlands, or at least they did not see or describe them. However, in the highlands the Spaniards found them from what is now the border between Ecuador and Colombia until almost the border with Peru—and even, I must emphasize, in the *ceja de selva*. In the nineteenth century there were still llamas in the western Chimborazo area, between Quito and Riobamba. It is interesting that Benzoni (1985:117–118) notes that here the mange (sarna) also ravaged the camelid population.

From the archaeological evidence it seems that what predominated in Ecuador was the llama, with the alpaca apparently arriving with the Incas. However, the historical data clearly indicate the presence of wild species (guanaco, vicuña). This is all subject to reexamination and restudy. Even so, the existence of camelids is undeniable.

The generalized idea, repeated in books, says that camelids cannot survive in the Ecuadorian area owing to the climate, and were introduced by the Inca. However, none of the scholars who hold this position have ever presented a solid argument in support of it. Everything indicates the opposite. What has yet to be established is whether camelids were taken to the Ecuadorian area in the Early Horizon or before, as domestic animals or not, and what happened to the wild animals.

No extensive discussion of what is now Bolivia is forthcoming because I do not have enough data on it, least of all archaeological data. This area raises no problem, insofar as it is one of the areas—among the major ones—where camelids developed. Besides, it still is one of the areas where these animals continue to play a major role in the regional economy. The only striking point is that in the historical documentation examined, no mention is made that may be ascribed to the alpaca. This will have to be studied in the future.

The guanaco was in Chile before man arrived. The archaeological data reveal its presence from Atacama to the southern Patagonia. The only data available for the vicuña are from Atacama and date to about the third millennium BC. The llama apparently reached northern Chile with agriculture and probably spread as far as Santiago. I dare not say whether or not this diffusion was due to the Inca. The presence of the alpaca has also been reported, but the data are not too precise.

However, the historical data specifically indicate that in the early sixteenth century there were many "sheep" in northern Chile, and these still were in the Santiago zone in the seventeenth century. From the information provided by the chroniclers it would seem that there were llamas and alpacas as far as Valdivia and the Island of Chiloé. This is a point that requires more research. And as regards the Magallanes region, there is no question about the immense number of guanacos there.

In Chile, the *hueque*, also known as *chilihueque* or *rehueque*, constitutes a problem. In Chapter 9 we saw that there is some disagreement over the identification of this animal. There are three likely positions. Two minority ones suggest that the *hueque* was the result of guanaco selection or was an alpaca; the third, majority position claims that these were llamas. As noted, I based my review on Palermo's study (1986–1987), which analyzed this point in depth and speaks with authority. Paler-

mo concludes that these could have been both llamas and alpacas. Interestingly, he notes that in the sixteenth and seventeenth centuries the area of distribution of the *hueque* was beyond Copiapó to the north and as far as Ancud to the south. In other words, this corresponds in general terms to the dispersal area of the llama as established with historical data.

Of all the data collected on Brazil, only those regarding the "chiefdom of Paguana"—which, we recall, must have been somewhere on the left bank of the Amazon, between the Catúa and the Negro Rivers—could conceivably be true. We owe this information to friar Gaspar de Carvajal, who was with Francisco de Orellana in the discovery of the Amazon. There are two versions of the chronicle, but "both accounts give a direct feeling of environmental fidelity and narrative sincerity that rule out the hypothesis of any intentional adulteration" (Porras 1986:167). However, Porras says of Carvajal's account that it "always has the vagueness and distance produced by the stealthy passage of the expedition in its river journey. Many of the towns were only seen from the ship without ever stopping in them, others among the confusion and uproar of combat . . . but the ethnographic data are always elusive" (Porras 1986:168). Besides, Porras (1986:169) quite clearly notes that of all the chroniclers, this friar is the one that most "lets his imagination get the best of him." However, one fact is striking. On reading friar Carvajal's account carefully, one can always deduce whether the descriptions are of something he saw himself or of what the Indians told him. For example, one can see that the Amazons are a fantasy because Carvajal himself says that an interpreter gave him the information. When describing Machifaro and other towns, as well as the "sheep of those from Peru," he always does so with what the Indians told him and never claims having seen them. It is therefore strange that when discussing the "lord called Paguana," Carvajal says that he "has many people," explains many details, and then claims that "[i]n this land this lord has many sheep of those from Peru" (Carvajal 1894:45–46; 1986:70–71). In other words, one feels that in this case he is saying what he saw. True, it is hard to accept the presence of llamas in the midst of the Amazon forest, and it is very likely that Father Carvajal let his fancy get the best of him. However, it is not altogether impossible for some of these animals to have reached there from the Andean area, because the relations between these regions have always been intense, and all the more so in the time of the Inca. I admit that it will be most difficult to clear this up, if not altogether impossible.

Although Paraguay appears as a subheading in Chapter 9, it actually is a huge area comprising the banks of the Paraná River, the eastern part of the central Chaco, the border between Bolivia and Brazil, and the border of Argentina and Paraguay (what is known as the central boreal Chaco): in other words, the eastern foothills of the Cordillera Real and the Sierra de Misiones. I have dwelt on Paraguay because Schmidel (1986) is the source we have, and although he discusses all the lands mentioned, the bulk of the events take place in Paraguayan land. Now, there is evidence that says camelids lived in this zone in the sixteenth century. We can deduce that these were llamas in the Asunción zone and in the southwest, while the Andean piedmont had llamas and a wild breed that was probably the guanaco. Although it is true that in general, this is a land with lakes, springs, and ravines, Schmidel's description shows that several areas are wet and have thick vegetation. This is a habitat where camelids are no longer found, but we are once again seeing an extinction resulting from cultural pressure. There are archaeological data showing the presence of llamas in the Chaco (Gilmore 1950:433, 435), and these lived in the Gran Chaco until at least the early eighteenth century.

It is possible that the Guaraní played an important role in the introduction of camelids into these lands. It is known that mining was never practiced in Paraguay, yet metal objects were found in the sixteenth century during the conquest of all of the Río de la Plata basin. The gold and silver obtained by the members of the Solís expedition, from the Guaraní and other Indian tribes in the region, came from the Inca Empire (Métraux 1948d:75–76). In fact, the Guaraní undertook a series of westward migrations from the late fourteenth century on. The third migration, which must have taken place in 1519–1523, entailed the movement of a Guaraní group from the

Itati region to the province of Santa Cruz, now in Bolivia. A fourth migration took place in 1521–1526. On this occasion some white men, shipwrecked members of Solís's armada, were with the Guaraní, as was already noted. One of them was Alejo García, who seems to have played a major role in this migratory movement. This group crossed the Chaco and invaded the frontiers of the Inca Empire, reaching as far as Presto to the northeast of Sucre and Tarabuco to the southeast. The success of this adventure made other Guaraní groups move west (Métraux 1948c:465–466).

A letter Martín Gonzales wrote in Asunción in 1556, which Métraux (1948c:466) cites and took from Gandía (1935:37), has some paragraphs that I find significant. It reads: "These Indians [the Guaraní] go and come back from the land of Peru. As they have no roads and avoid their enemies, they reconnoiter the land ahead of them, settle long enough to sow crops and harvest their food, and go on. So went those who for long had been settled in the Peruvian sierras and those who go today to meet the Christians." The letter adds: "Some Cario [Guaraní] have gone to the mountains along another river, 42 leagues down the Paraguay River, which is called Ypití [Bermejo River]. Cario are established in the mountains near the Ypití River, which also leads, according to what those who came from Perú say, to the city of La Plata."

It is known that some thousands of Guaraní Indians helped Domingo de Irala and Nuflo de Chaves (in this case 2,000–3,000 are mentioned) in their expedition from Paraguay to the Andean piedmont. In effect, the Guaraní not only invaded and conquered the Andean piedmont from the Bermejo River to Santa Cruz de la Sierra, other groups did so from the north and settled beside the Mojo and Bauré groups. On the other hand, the Pitaquiri were also a Guaraní group settled in the Sierra de Chiquitos, north of the first Santa Cruz, while another one lived close to the Jesuit mission of San Xavier, in the province of Chiquitos.

The driving force behind all these movements was no doubt the great desire to get gold and silver adornments and the copper tools they had become acquainted with through the Caracara and Chané Indians, who lived on the frontiers of the Inca Empire. The Guaraní thus became a threat for the Inca, so it was probably during the reign of Huayna Capac that fortified strongholds had to be built as a defense from the Guaraní. The chroniclers reported the war Huayna Cápac waged against the Chiriguano, the descendants of the Guaraní, and how these were defeated by the Inca armies (Métraux 1948c: 465–467).

There is thus no question of the close contacts between the Guaraní and the Andean peoples, so the presence of llamas among the "Cario" that Schmidel (1986) mentions should not prove striking.

The same thing noted for Chile holds true for Argentina: my data are scant and do not let me draw valid conclusions. The potential data here are vast, so a complete picture will be available once someone decides to collect them. It should not be forgotten that in all of South America, Argentina is the one country where not only are the natural sciences greatly developed, this is also a tradition that began early, so the faunal information in the archaeological data is more common than in other places.

Most of the literature I was able to read is on the northern Argentinian highlands. Here, guanacos and vicuñas are present in archaeological contexts since around 8000 BC. However, it is striking that the llama appears much later, with dates that fall within the first millennium BC. The year 530 BC has even been given as the date of domestication. This is obviously possible, but it is somewhat hard to accept, because if domestication took place at a much earlier date in the central Andean area, then it would be surprising that domestic animals did not arrive earlier as well in such a nearby area as northern Argentina, which had continuous contacts with the altiplano and from which it received strong influence. It is true that a date of 2000 BC has been given for domestication in northern Chile, which is also far removed from the date of domestication in the central Andes. Even so, there is a great gap between this date and the one for Argentina. If the data available are correct, this raises an interesting problem. In the future, specialists should dedicate themselves to checking whether the picture is correct, in which case a better explanation will be in order. Or, if it is wrong, they will have to correct it.

As far as I know, in Patagonia only guanaco remains have been associated with early man.

Troll (1935) claimed that llama husbandry reached as far as Tucumán at a late date. This is quite possible, and according to historical data it reached even farther south. What has to be noted is that here, too, the domestic camelids spread a long time before the coming of the Incas.

The scant historical sources reviewed show that in the first half of the sixteenth century there were pack llamas in Omaguaca, and llamas living in the Gran Chaco up to the early sixteenth century. In the highlands of Tucumán there were guanacos and vicuñas, as well as in the sierra of Santiago. In Córdoba there were llamas and big herds of guanacos. In the mid-sixteenth century there were still llamas in the Río de la Plata zone, and guanacos in the province of Buenos Aires in the late eighteenth and early nineteenth centuries.

One of the advantages domestic camelids have is their adaptation to certain kinds of pastures. The llama prefers hard and dry pastures, while the alpaca prefers the hard and succulent ones. This is believed to be due to conditions these animals already had, which have become more marked through time. In this regard it is worth recalling that the ruminants developed more recently than the camelids (Engelhardt et al. 1992:263), and that there are several differences in the digestive physiology of the forestomach of camelids and ruminants, even though rumination and the anaerobic microbial fermentation in a compartmentalized stomach are processes that apply to both cases. Comparative studies show that the morphology and histology of the forestomach are markedly different in both groups. Besides, the basic pattern of motility of the forestomach is different in camelids and ruminants. All biochemical patterns of microbial digestion in the fermentation chambers in herbivores are similar. However, camelids are better able to adapt to extreme diet conditions (Engelhardt et al. 1992:269–270).

To understand these phenomena and the advantages they present in an environment like the Andes, it is worth comparing the digestive processes of camelids and ruminants. First, the anatomy of their stomachs is different, as is the ultrastructure of forestomach mucosa. There is also a great difference in motility, while the biochemistry of digestion is similar, as well as the composition of the saliva and the rates of secretion. However, the utilization of endogenous urea in the camelids is higher than in ruminants when they feed on low-protein diets. Finally, the adaptation to food in arid regions and at high altitudes is clearly superior in camelids as compared with ruminants (Engelhardt et al. 1992:Table 7, 269). This explains some of the problems animals imported to the Andes had and still have.

Now, one of the objections several specialists have raised against raising llamas on the coast precisely concerns food. These specialists have not considered the great adaptive capabilities these animals have, which, moreover, are not selective, so they can eat almost any plant.

We saw, through an analysis of coprolites, which plants were eaten by the camelids that lived on the north-central coast in the preceramic period. True, the plants are not all the same throughout the coast, but I believe these animals had no food problems in the coastal valleys. The possibilities were greater on the north coast, where the coastal strip is wider, especially in the ravines with thorny forests, which are now becoming extinct but which in pre-Columbian times must have been of much greater size.

I will neither analyze nor discuss the wide range of plants the camelids could browse on in the coastal environment. Part of this task was done in Chapter 10. Here I just want to emphasize a group of plants that exist on the coast, which are of crucial importance for these animals' diet. First there is the algarrobo. *Prosopis* occurs in two major zones, Nasca-Ica and Tumbes-Lambayeque, and is found up to 1600 masl (Weberbauer 1945:169). The chroniclers repeated that not only did the imported animals eat this plant avidly, but it was most nourishing for them. The archaeological data show that it was part of the food of coastal llamas. A study specifically made to see what plants should be more developed in poor countries includes the algarrobo because "[l]ivestock relish the pods; cattle, sheep, horses, mules, donkeys, goats and wildlife all eat them avidly" (National Academy of Science 1979:155). Interestingly, in the United Arab Emirates, among the desert plants used as food for race camels is an algarrobo species, *Prosopis cinerea*, which is classified as "Class 1," that

is equal to "very good" (Wensvoort 1992:Table 1, 324).

Besides the algarrobo there are several plants that grow wild on the coast, which were good foods for the camelids: saltgrass (*Distichlis spicata*) common in the coastal gramadales,[2] which is considered an "important weed"; caña brava (*Gynerium sagittatum*), most widespread among the riverine scrub of the coastal valleys; and I also ascribe much importance to the *Salicornia fruticosa*. This is a succulent halophyte plant that is found in great quantities throughout the coast. In the Mediterranean zone it is used as food for camels, as noted. I do not have as yet any concrete evidence that it was eaten by camelids on the coast, but we should not forget that analyses of the llama's coprolites, or of any other camelid's, are few; thus there really is not a significant sample. It has been proved that camelids ate some plants from the Chenopodiaceae family, but the genus has not been identified. This point has yet to be studied.

The evidence for the use of algae to feed the coastal camelids is most suggestive, both by its amount and its easy collection on the coast. This food has been used by man since time immemorial. It is known that some algae are used to feed cattle and sheep in some countries on the European north coast. These are generally used fresh or desiccated, as flour, and mixed with other products. Today in different countries algae flour is mixed with the food of various domestic animals. Experiments show that algae are a high-quality and low-cost food because they contain essential minerals such as iodine and boron, vitamins A and E, and a series of easily digested protein substances (Acleto 1971:25–26).

Besides this, llamas undoubtedly ate the leaves and part of the domestic plants man did not use. The list here could be very large. As an example we can mention the achira (*Canna edulis*), of which man only ate the rhizomes, so a plant with big leaves was left that could provide a significant amount of food for the animals.

A plant I would like to emphasize is maize (*Zea mays*), because for the camelids it was of major importance as fodder. Maize has been cultivated in great amounts since preceramic times and was one of the pillars of the native economy. With maize, man only uses the grains; all the rest of the plant is left to the animals. So the amount available in fresh condition at harvest time, and dry the rest of the year, was very great. The archaeological evidence is quite clear in this regard and clearly shows that the animals ate great amounts of maize. Historical evidence confirms this. It is significant that the text on one drawing by Guaman Poma de Ayala (1936:fol. I1041 [sic]; my Figure 14.3) illustrating the month of April, when "maize matures," says that "the food has to be stored away from thieves . . . and the beasts." Guaman Poma mentions several of them, but drew a llama eating maize.

The ethnographic data Cristóbal Campana (pers. commun., 29 April 1994) gave me is interesting here. The old North Coast mocheros[3] believe that to grow well, all animals must eat maize cobs. They say that "to make them grow into a good hechor you must start giving them maize beginning with the father." *Hechor* is a synonym for stud, and this must be the first male animal.

FIGURE 14.3. The chronicler Guaman Poma de Ayala explains that in April "maize ripens," and that in this time "food should be stored away from thieves, vermin, and rams." The drawing actually shows a thief stealing corn cobs and a llama eating the plants. *After Guaman Poma de Ayala (1936:fol. I1041).*

I am convinced that camelids had absolutely no food problems on the coast. There was enough food to feed big herds.

For a better understanding of this point we need to expand our knowledge with a systematic program of coprolite analysis from these animals, which abound on the coast. We should not forget that one of the least known aspects of Andean camelids is their food. Curiously, few published data exist on the plants camels use as food (Wensvoort 1992:323).

The quality of alpaca and vicuña wool is well known, and it is better than that of the merino from which the famed cashmere is made. We should bear in mind that the vicuña has the finest fiber in the world (5–7 microns; ONERN 1985 [1988]:182). However, it is a generalized belief that llama wool is coarse and hard, and therefore little esteemed. What is not known is that the llama can grow good wool if it is well tended and is not used as a pack animal. This is being done in the Unites States. Recent archaeological studies done at the genetic level have indicated that in ancient times there were several llama breeds in Peru that were lost with the European invasion, and one of them was certainly dedicated to wool. Although this is not explicitly stated in the chronicles, several signs seem to indicate it. A state policy that preserves and improves what remains of this genetic pool is therefore essential.

The use of camelid products and of the camelids themselves in pre-Columbian times was of major importance, not just for the economy of the Andean peoples but for their culture itself. The Andean culture as such cannot be conceived without these animals.

It has been said that little camelid meat was eaten in past times (e.g., Dedenbach Salazar 1990: 174–176). I disagree because this is a completely gratuitous claim. Once again it must be noted that we lack specific indices for many periods, but this does not mean they do not exist. I have shown that there is concrete evidence that camelid flesh was eaten since the preceramic period. This is systematically repeated in all periods, even though no percentages are available for several sites. However, it is significant that camelid bones comprise more than 90% of the remains found in Mochica middens. And at Huánuco Pampa, one of the most important cities in the Inca Empire, this percentage in some sectors comes to 95%, with 67% as the minimum figure. The data from historical sources clearly indicate with a wealth of detail, as we saw, that the Spaniards ate great amounts of it, and in most cases it is stated that it was good meat. I am absolutely convinced that the more studies we have, the more the importance camelid meat had in the diet of Andean people comes through. Let me be clear that this in no way implies an essentially carnivorous diet. In the Andes man always has made a wide-scale use of vegetables, with a balanced diet in which animal flesh of both land and marine origin had an important role. Besides, it should not be forgotten that according to the specialists, camelid meat is quite healthy, so its use should be promoted. It would be a way to develop this kind of husbandry.

Similarly, while little value is attached to camelid dung, this matter is most important for the people living in the high and inhospitable Andean land, especially in the puna, where dry dung is the only fuel available. Moreover, it has a high caloric index. Dung is also the cheapest fertilizer for the fields, and from what I was able to find out, it is of good quality.

The pack llama, a crucial element in pre-Columbian and Colonial times in various aspects of Andean life, has certainly and gradually lost relevance in modern times, with the development of highways and the increase in the number of motor vehicles. Even so, it has not lost all of its importance due to the small scope of the Peruvian road network, in comparison with the vast size of the country. Indian communities therefore continue using llamas to carry their produce. What seems to be an important fact, which for me is definite, is that in past times there were pack animals of a larger size than the present ones. Chapter 10 has some concrete examples that prove this. Besides, there is proof that these big-sized animals—for their species—existed until a relatively short time ago. What must be established is whether this breed (or breeds) is lost or is recoverable.

It should be added that the phenomenon of the "beast of burden" will be better understood once pottery is studied more exhaustively. Perhaps then it will be possible to establish differences between the pre-Columbian custom as regards pack animals, and those introduced by the Europeans. Here I just want to present some ideas. From what I was able to see, the pre-Columbian artisan either clearly indicated the sex of the animals, or made no indication at all. I have not seen a single depiction of a female pack llama. The animals carrying loads are males, or their sex is not shown. Today, it was noted, castrated animals are used, and everything seems to indicate that this practice was introduced by the Europeans. However, a detailed examination of ceramic depictions is needed to ascertain whether there is any sign that the animals were castrated or not. We should not forget that the Mochica left a very clear depiction of a man with the testicles and the penis amputated (see Larco Hoyle 1945:16, lower right). I see no reason why this might not have been depicted if it was a widespread practice with camelids. On the other hand, this kind of analysis lets us see the continuity of certain traditions—for instance, the great resemblance between some Mochica depictions of llamas loaded with vessels (e.g., Donnan 1978:Fig. 176, 113) and the drawing left by Guaman Poma de Ayala showing the transportation of wine in Colonial times (1936:fol. 524; my Figure 14.4).

There are some coincidences in this regard when South American camelids and the camel are compared. According to Compagnoni and Tosi (1978:100, 102), the earliest documentation of camels as beasts of burden dates to the second millennium BC. It is a representation that appears on a plaque from Tell Asmar in Mesopotamia, as well as some figurines of *Camelus bactrianus* from Altyn depe and Ulug depe in South Turkmenia [Turkmenistan] (see also Epstein 1971:567; Calkin 1970:Fig. 4). In Peru the earliest evidence so far available of pack llamas, if indirect evidence, comes from Los Gavilanes, on the north-central coast, and dates to ca. 2200 BC, or about the same time as the camel. It is likewise significant that, as far as I was able to check, pack camels cover only 30 km a day (Beals and Hoijer 1958:325). We should recall there is no agreement among scholars on this point, either ancient or modern, but according to the latter, the maximum distance a llama can cover loaded is 30 km, and slightly more according to the chroniclers.

FIGURE 14.4. With this drawing, Guaman Poma de Ayala refers to the abuses wrought by the "managers, miners and wine merchants." What is of interest here is the way the vessels were carried by the llamas, which is exactly the same as that used in prehispanic times. For comparisons, see illustration 176 published by Donnan (1978:113). *After Guaman Poma de Ayala (1936:fol. 524).*

There is no question that in general, the llama was not a riding animal, but I find that culture had a major role in this. Even so, one cannot say emphatically, like Benson (1972:4–14, 91), that "[t]here are occasional representations of a man on the back of a llama. However, a llama cannot carry a man very far, especially when it is also burdened with saddlebags. The meaning of these pots remains mysterious." This statement includes several mistakes. First, depictions of llamas with men seated on them are not as scant as Benson claims. They cannot be said to be abundant, but they are quite common in all collections with Moche pottery. Possibly the same thing is true in Chimu pottery. Second, Benson did not notice that there are several ways to ride a llama, as depicted in the pottery from Moche and other cultures. I shall list them below. Nor is it true that a llama cannot carry a man very far. The testimonies of Schmidel (1986)

and Cristóbal Campana (Figure 8.1), discussed in Chapters 9 and 10 in great detail, suffice to disprove this. Finally, save in the case of big vessels, the load carried by the llamas is not carried in saddlebags but in great bags, exactly as in the photograph in Donnan's book (1976:Plate 10b). This is another point that requires systematic study.

There are different ways to ride a llama Mochica style. First, it was done with both loaded and unloaded animals. A man always lies face down and lengthwise on animals loaded with a cargo placed crosswise on the animal's back, but facing in an opposite direction to where the animal is going. He grabs the llama with his legs by the neck to avoid falling off and its tail or hind part with his hands (Figures 4.17, 4.18). When unloaded, the llama is ridden just like a horse without a saddle (Figure 4.20), or face down lying crosswise on the animal's back (Figure 4.19). The beasts, both as pack animals and for riding, are driven in different ways, the most common method being the use of a halter-like harness, or a rope pushed through one of the animal's ears. In the Lambayeque culture the man goes astride the animal, exactly as if riding a horse. Besides being seated over the load, face down and lengthwise in regard to the animal, there are two other variants in the Chimu culture with an unloaded animal. In one case the "rider" kneels over the back of the animal and leads it by the ear (Figure 4.41), while in the other it is not altogether clear whether the man is standing on the animal's back or is seated on something (Figure 4.42).

Finally, it should be noted that of the illustrations in this book, eight Moche vessels show loaded animals—in this case it does not matter whether the load is human or not; five are male and three have no sex. None of the Lambayeque and Chimu culture specimens offer any indication of their sex. This clearly is not a representative sample—far from it—but it is an index of what was mentioned above. It is worth noting that when the proposed study is made, this indication of sex will have to be borne in mind, not just for riding and pack animals, but also when studying the ritual role: as it happens, the great, llama-shaped ritual vessels found at Pacheco (Robles Moqo style) are male animals (Figures 4.29–4.31).

I believe that riding llamas was not an occasional practice on the north coast, as has been claimed. This point will have to be studied in other pre-Columbian cultures of the central Andes and adjacent areas where these animals were used.

Jorge Herrera Hidalgo of the Consejo Nacional de Camélidos Sudamericanos del Perú recently told me that there still are some riding llamas in the peasant community of Mayobamba (Department of Lima, province of Oyón, district of Pachangará). Some of these animals were taken by Herrera Hidalgo to Arequipa for an international festival that was held there and were a major tourist attraction (Jorge Herrera Hidalgo, pers. commun., 9 April 2001).

What should definitely be discarded is the belief that camelids were used as draft animals. This was the result either of a wrong reading of the chroniclers or mere fantasy. Besides, in the Andean world there was no tool or device that could have fulfilled this role.

Enough has been said, here and elsewhere in this book, about the advantages the camelids have in the Andean environment in comparison with the animals introduced by the Europeans. Other scholars have emphasized this, and perhaps no one more than Flores Ochoa in several of his studies. Clearly in this, as in other cultural aspects, one must be very careful because all extremes can be harmful. In other words, one cannot seek a return to the past, just as we should not destroy—as is now being done—the legacy of the past for a poorly conceived progress that is solely based on the Western cultural canon. Indigenous and Western cultures must be brought together if Peru is to progress. In this case it would be a crime to lose the knowledge of the Andean people regarding camelid raising. These animals can still contribute much in our time. The advantages they have—being able to live in marginal areas where other domestic animals have trouble surviving, and where they cause no ecological harm—should be used. Camelids can perhaps be used to try to save the coastal lomas, where animals like the goat wreak damage beyond repair; camelids can live there without destroying them. There is enough accumulated experience in this regard that can be useful and cannot simply be allowed to vanish.

Herre (1968) claims that a comparison of the Old and New World cultures leads to the conclusion that in America, animal husbandry was less significant than plant cultivation. Herre furthermore claims that the native animals were only of general importance. I do not concur.

Clearly, a comparison of the number of domestic animals used in the Old World versus in the New World shows them to be fewer in America. Actually, in the specific case of the Andean world, the only other domestic animals were the dog, the guinea pig, and one kind of duck. What should be borne in mind is that in America there are no other animals that could possibly be subjected to a process of domestication, and these few animals were used to the utmost. I am convinced that once we have detailed studies on the use of the guinea pig (*Cavia* sp.), we will find that it was far greater than what is usually believed, and that it had an important position in the indigenous diet. This is another animal that is no longer used for cultural reasons, particularly on the coast.

Both wild and domestic camelids were crucial in the pre-Columbian economy, and while it is true that agriculture was the driving force behind it, it would have been far more limited without the camelids. This was so not just because these animals were the major source of animal protein, but also because they presented several advantages to man. Without their wool to dress in and dung to keep warm, life at high altitudes would have been nearly impossible. Besides, the excrement was the main fertilizer for the fields, and the produce was moved on llamas and sometimes alpacas. In other words, animal husbandry and agriculture were two interactive activities in the vast cultural world of the Andes—two activities, moreover, that I believe are inseparable.

One of the major goals of this book right from the beginning was to establish the actual distribution of the camelids in terms of the existing data. Much has been written, and far too much has been speculated, on the basis of general data or even of assumptions. The maps herein presented include all the sites analyzed, period by period. These sites have in addition been studied in the various chapters of this book. I now present a summary, specifying all doubts and the evidence based on the sites that have usable information, so that the reader can get a clear idea of the state of the art on the matter, and so that specialists will be prodded into further studying the subject.

I begin with the preceramic period (see Figure 4.1). In the highlands there only are six sites with available data, from which it follows that almost all of the information comes from the north-central and central highlands. In the south there are some data only for Ayacucho. However, the only sites with actual identifications are Pachamachay, Telarmachay, and Jaywamachay. In the first case, all four camelid species were identified in the last phase, two wild and one domestic species (probably llama) in the second case, and in the third site the guanaco alone. All that can be said for the remaining sites is that camelids lived there.

On the coast there are nine preceramic sites, of which reliable identifications were made in two cases: at Los Gavilanes on the north-central coast, where the presence of llamas and alpacas has been verified, and at Paloma, on the southern coast, where there is evidence of guanacos. There are four sites with tentative identifications: Río Seco del León and El Paraíso, on the central coast, where it seems that guanaco remains were found, and Cabezas Largas and Santo Domingo de Paracas, on the southern coast, with possible vicuña remains. For the remaining sites all that can be said is that there were camelids.

For the Initial period (see Figure 4.2) there are eight sites in the highlands that extend from Cajamarca to the southern highlands, in the Department of Apurímac. In these sites, the presence of llama has been claimed only for Tecliomachay, in the Callejón de Huaylas, and in Kotosh, where the presence of vicuñas and guanacos has been ascertained. Huacaloma is the only site where the remains are said to belong to domestic animals, even though they have not been identified at the species level, while Waywaka is the only site for which it is claimed that the camelids found were not domestic ones.

For the Initial period, the data are even poorer for the coast. There are data available just for five sites, four of which are on the north coast and one on the central coast. Of these, we know that llamas were used at Puémape, in the Chicama Valley, and at Huaca Negra, in the Virú Valley. The remains from Huaca Herederos Chica perhaps belonged to llama. For Cardal, on the central coast, all we know is that camelid remains were found there.

There are more data for the Early Horizon (see Figure 4.3). In the highlands there are 21 sites that range from the Department of Cajamarca to the Department of Puno. But of these sites, the four species have been identified only at Chavín de Huántar; at Marcavalle, both llamas and guanacos were identified. There are unreliable identifications for Huacaloma (llama, alpaca, and guanaco) and Kotosh (all four species). All that can be said for the other sites is that camelids were used.

On the coast there are data for 19 sites ranging from the northern to the south coast (but on its northern part). However, all that can be said with some certainty is that the remains found at Pampa Rosario, Moxeque, Faro de Supe, Ancón, and Cerrillos were from llama, and perhaps those found at Huachipa too. For the rest, all that can be said is that these were camelid bones.

However, the amount of these remains is significant, to the point where camelids are believed to have been of crucial importance for the economy (e.g., Shimada 1982:146). On the north coast, the percentages of camelid bones in middens reach as high as 90%; in addition, camelids were important in mortuary offerings.

In general, for the highlands in the Early Intermediate period (see Figure 4.4) we can use the data from 12 sites scattered between the Departments of Cajamarca and Puno. Of these it is known that there were llama and alpaca at Huacaloma, llama at Huamachuco, llama and guanacos (and perhaps alpacas) at Kotosh, and llama and alpaca at Waywaka. For all remaining sites, all that can be said is that they had camelids.

On the coast, camelid remains were found at one site in Piura. Three of the remaining sites belonging to the Salinar culture have data of this type. At Puémape the identification went only to the family level, but in Cerro Arena there was llama, and both llama and alpaca were identified at Virú.

All that is known of the three sites belonging to the Gallinazo culture whose faunal remains were studied is that camelids were present. The Mochica sites certainly are the best studied. It turns out that llama bones were identified in six of the eight whose middens were studied, while in the rest all the remains have been identified only to family. Sites with identifications are

Huaca del Pueblo, Pacatnamú, Pampa Río Seco, Paredones, Moche, and Huaca de la Cruz.

On the coast there are only two sites whose faunal remains have been studied, between the Virú and Ica Valleys, but the identification went only to the family level. For the Nasca culture there are data for only two sites, but we know that there were llamas at Cahuachi.

The case of the Middle Horizon (see Figure 4.5) is significant because it is a time of crucial importance for the development of Andean culture. Besides, the central culture, that is, the Huari, was born and had its main developments in the highlands. And yet the faunal data are limited to three sites, one in Cajamarca, another in Ayacucho, and a third one in the Department of Arequipa. Of these, there is evidence for llama and alpaca at the first one only, and of llama, vicuña, and guanaco at the last site. All that is known for Ayacucho, where the capital of the Huari Empire rose, is that it had camelids.

The data for the coast are not only far more abundant, they are also more specific. There are 17 sites between the Lambayeque area and Ica, and it was concluded that there were llamas in 10 of them. These are Batán Grande, Huaca Chotuna, Pampa Grande, Pacatnamú, Cañoncillo, Moche, Galindo, a site in the Nepeña Valley, another in Huarmey, and Ancón. The presence of alpacas is surmised only at Batán Grande. For the others, all that is known is that camelids were present.

The Late Intermediate period sites (see Figure 4.6) in the highlands that have zooarchaeological data are scarce. These are just six, and extend from the Piura highlands to the Department of Tacna. Five of them present specific data. Alpacas and vicuñas have been identified for the upper part of Piura, vicuña at Junín, llama at Cuelap, llama and alpaca at El Yaral, and llama at Tocuco. All that is known for Ayacucho is that it had camelids.

Once more it is the coastal area that has the best and most extensive data for this period. Here there are 18 sites, 10 of which have concrete data. The presence of llama was ascertained in all of them. These are Tumbes, Caracoles, Cerro de la Virgen, Huanchaco, Chanchan, a site in Virú, another in Nepeña, and one each in Casma, Cerro Azul, and Ica. The presence of alpacas is surmised only for Casma. For the remaining eight sites, all

that can be said is that camelids were there. In this case the sites are located all along the coast, from Tumbes to Moquegua.

One would believe that there is much information for the Late Horizon (see Figure 4.9), which corresponds to the Inca Empire. Yet this is not so. There are just five sites with faunal data, four of them with specific data. They belong to the central and southern sierra. Thus we can ascertain the use of llamas and alpacas at Huánuco Pampa, of llamas, vicuñas, and guanacos in the Tarma-Junín zone and the Colca Valley, and of llamas and alpacas in Cuzco. All that can be said for the La Pampa site is that camelids were used.

If we use the ethnohistoric data for Inca times we conclude there were llamas in Chachapoyas, Cajamarca, Huamachuco, Huánuco, and the altiplano. The presence of guanacos and vicuñas is mentioned for Huamachuco.

Only six sites have been studied on all the coast, four of which have evidence proving the presence of llamas: Batán Grande, Médanos de la Hoyada, Chanchan, and Cajamarquilla. All that is known in the cases of Pachacamac and Tambo Viejo is that camelids were there. These sites range from Lambayeque to Moquegua.

The references available from ethnohistoric sources indicate the presence of llamas in Collique (in the district of Saña), in the Chillón River Valley, and in the upper part of the Rímac River Valley, as well as in the Chincha River Valley. The presence of vicuñas is likewise noted for the upper part of the Rímac Valley.

The data for the *ceja de selva* are quite scant, for they are limited to just five sites, each of which belongs to a different period. Michinal is located on the northern *ceja de selva* (in the Department of Cajamarca) and is believed to date to the Initial period. Cerezal, in the same department, has been ascribed to the Early Horizon. The presence of camelid bones has been reported at both sites, and guanaco is also said to be present at Cerezal. Both the chronology and the identification of the bones are unreliable.

Llama bones were found at the site of Patrón Samana, in the Department of Amazonas, though in small number. The site dates to the Late Intermediate period. However, the important data are for the Late Horizon and come from two sites in the Department of San Martín: Ruinas del Abiseo and La Convención. In both cases the identifications were made by specialists, but llama was specified only for El Abiseo. The species could not be identified in the other case. It is worth recalling that in the case of the Abiseo zone, the llamas were raised locally (Cornejo and Wheeler 1986). This is a most important piece of information because raising them in this habitat had previously been ruled out. This is an issue that will require more work in the future.

Some very general conclusions can be drawn from this presentation. What is evident at first glance is the poor quality of the data, as is true for all periods of Andean pre-Columbian history.

If the subject is approached period by period, data are lacking for preceramic times in two important parts of the highlands, the far north and the altiplano. The second is a particularly crucial zone for the subject under consideration. As for the coast, more studies are urgently needed, because the data are exceptionally scant, and there are none for the northern part.

For the Initial period we once again find no data for the altiplano, and for the coast, nothing is known about what happened in the far north and the far south.

In general, there are a fair number of Early Horizon sites, but more precise identifications are required, as well as more specific studies. Furthermore, there are no data whatsoever for the highlands in the far north and for the far southern coast. An important fact that stands out is the high percentage of camelid remains, which indicates a remarkable increase in their use. These obviously are domestic animals.

Early Intermediate period sites are known throughout Peru both in the highlands and the lowlands. Yet to be known is what happened in the highlands of the far north and on the far south coast. But it is worth noting that the poor quality of the data for highland sites is quite prevalent, and that as far as the coast is concerned, there are good reports only for sites belonging to the Moche culture, which really is an exception.

In the last few years much has improved in our knowledge of the Middle Horizon, but strangely enough, there has been no interest in certain aspects of it, such as diet. Practically noth-

ing is known about this point in the vast Andean highlands, which is precisely where the major sites of the Huari culture are and where most of the data that explain the imperial characteristics of this culture were obtained. There are, however, good data for the coastal strip. They are good only in comparison with the whole database under discussion here; in themselves they actually are poor. Even so, there is once again a gap in our data regarding the far north and the far south, where what happened is simply unknown.

For the Late Intermediate period there are several sites that practically cover all the highlands, but the faunal data are scant. The opposite holds true for the coast, where there are data from Tumbes to Moquegua, and of very good quality.

For the Late Horizon, the period in which Inca expansion took place, the data are inadequate: not only are the data for the highlands scant, but they are exclusively concentrated on the central highlands and the Cuzco zone. There are some vague ethnohistorical data for several sites that extend from the southern part of the Department of Amazonas to the altiplano. The coastal data are likewise scant, even though several sites from Lambayeque to Moquegua have been studied.

We saw that studies are scant for the *ceja de selva*. Just four sites are known, two of which have reliable data. However, it is interesting that all four sites are located on the northern sector, but it is clear that there will be many surprises once more sites in the *ceja de selva* are excavated, both in the center and the south. The importance of these finds must be emphasized, because they open new environmental possibilities for life as far as the camelids are concerned.

Peruvianist archaeologists should take two lessons from this analysis. First, there is a pressing need for more interdisciplinary work. The time when an archaeologist worked on his or her own is over. Second, far too much emphasis has been given to the study of large sites with monumental architecture, so studies have tended to gain a better understanding of aspects relating to the political, economic, and social organization, but the perspective of the daily life of humankind has often been lost. In other words, I suggest that

Peruvian archaeology should in the future turn to the excavation of rural habitation sites to understand how Andean people truly lived, in the largest sense of the word. Here the diet is obviously of crucial importance. The significance camelids had in the Andean world will really be understood only when this is done. Everything else is just mere speculation.

Finally, one more point follows from the general analysis here made. The possibility of camelids developing on the Peruvian coast has always been rejected, yet most of the data do not come from the sites in the highlands but from coastal ones. This shows that conclusions have often been drawn without an appropriate study.

As we saw in Chapter 6, the historical data are very rich, though not always specific, and give us a far clearer idea of the problems that concern us here regarding camelids. I shall try to present a summary, even though this is no easy task.

First, it should once more be acknowledged that Dedenbach Salazar's study (1990) is a landmark in this field, and she is certainly one of the scholars who has most honestly and seriously studied the matter. Dedenbach Salazar sketched the location of wild camelids with the data found in the chronicles. She thus concludes that there were vicuñas and guanacos in what is now Bolivia, in Collao, Ayacucho, in all of central Peru, and even in the highlands of Ecuador. She likewise lists the alpaca in Cuenca (Ecuador), Piura and Cajamarca, Huamachuco, and Huarochirí. This does not mean there were no alpacas in other places; what Dedenbach Salazar means is that these are the areas where the chroniclers indicate their presence.

However, when Dedenbach Salazar (1990:98) discusses the coast, she does not deny that camelids were found there when the Spaniards arrived but is skeptical about the number present. Although she does not directly say so, it would seem that she doubts these animals were raised on the coast. If so, I disagree because there are enough historical data to the contrary, which are supported by archaeological data. But let us recall what the chroniclers say. I reiterate that the bibliographic citations are omitted here to make the narrative more fluid, for the reader can easily find them in detail in Chapter 6.

Estete—whom master historian Porras (1986:118) defined as "the first ethnographic chronicler" and who was "chronologically the first of the travelers to write about Peru"—clearly says, when discussing the coast, that there was "abundant livestock." In 1541 it was specified that there were "sheep" on the coast, and in 1552 Molina "El Chileno" also stated that the coast "abounded . . . [with] . . . livestock" when he described the area between Huarmey and Chincha. To judge by the chronicles, the number of camelids had fallen by 1597, clearly as a result of the European invasion. However, in the early seventeenth century there were still llamas on the coast.

But, as we recall, there are more specific data. The encounter with the Tumbesino raft took place during the second voyage of Pizarro. The document is dated 1528, but this event must have taken place earlier, perhaps in 1526. Here the Spaniards got news of the "sheep." Pedro de Candia verified the presence of these "sheep" in Tumbes in early 1528, and in 1531–1532 Ruiz de Arce says that there were "many sheep" in Tumbes. When the latter reached Poechos in May 1532, he ascertained that here the livestock was "abundant." The presence of livestock is then certified at San Miguel de Tangarará, in the Chira Valley, in Serán and Motupe. Cieza de León reaffirmed this for Tumbes when narrating the episodes of Alonso de Molina and Pedro de Candia. And when recounting the story of Bocanegra, who wanted to stay in Collique, and Juan de la Torre who disembarked to inquire into this, Cieza writes that here there were "herds of sheep." There are several references to "sheep" during the return journey of Pizarro's second expedition.

When Pizarro and his army made their third trip they reached Poechos in May 1532 and also found "sheep." Almagro likewise saw "herds of sheep" when he passed through Tumbes in late 1532.

It seems that Pedro de Alvarado took pack llamas with him in 1534, when he traveled from San Miguel to Pachacamac along the coast.

Some evidence shows that llamas still lived in Piura in 1577, while in 1540 we not only find llama herders in Lambayeque; tribute was also paid with these animals. There were still llama herders in the chiefdom of Cinto in 1550–1560. Herds of camelids were also mentioned when the area was inspected in 1566–1567. These animals still lived in Túcume in 1574, though by then they were scant, and herds still lived in Moro in 1582; it is a matter of record that many "sheep" died in 1580 due to the heavy rains that fell on the Lambayeque zone.

Farther to the south there were llama herders in Saña in 1564, while Moche "abounded with livestock" in 1544–1545. In 1561 not only were "carneros" sold in the city of Trujillo (and these were not a few because 100 animals were mentioned), they were taken there from Huamachuco and Cajamarca as food for the lower class. In 1538 there were still domestic camelids around Lima.

Their presence was certified in Chincha in 1552 and in Ica in 1571, where some still lived in the early seventeenth century. Their presence was likewise noted in Acarí, Sama, and Tacna in the late sixteenth century.

In other words, there can be no question that there was a significant number of camelids on the Peruvian coast, most of them possibly llamas, at the time that the Spanish conquest took place and in ensuing years.

No one has questioned the presence of camelids in the highlands at the time of the Spanish arrival, yet some aspects described by the chroniclers are revealing.

The testimony of a group of Spaniards who traveled from Cajamarca to Pachacamac under the command of Hernando Pizarro, the brother of the conquistador, proves revealing in this regard. The group saw much livestock from the moment of their departure to the moment when they reached Corongo, and the same thing happened in the Callejón de Huaylas. On their return to the highlands they also found lots of camelid herds in what is now the province of Huaura, just as in Junín. It should be recalled that small livestock was noted for Huánuco, possibly alpaca. Figures of 100,000 are given.

But besides all this, there are several other references in the chronicles, both general and specific observations, describing a world that to our eyes is unbelievable.

First, there is a consensus that in the period spanning the mid-sixteenth century to the mid-seventeenth century, there was a great number of wild and domestic "sheep" throughout all the Andean highlands.

Thus Huancabamba is mentioned, where no camelids are now left, and yet at the time that Cieza de León was traveling, that is, fifteen years after the European invasion, there was still domestic livestock, as well as guanacos.

When Pizarro reached Cajas he found a great number of "sheep and rams." These also lived in the Department of Amazonas, in the province of Luya, and in the southern part of the province of Chachapoyas in the early seventeenth century.

From the historical sources it follows that camelids lived in the Department of Cajamarca before the coming of the Incas, and it is known that there was a great number of them at the time the Inca (ruler) was captured. It can be assumed that part had arrived with the entourage of the Inca, carrying his possessions, but the rest were local. When Cieza de León made his journey and described the zone, he let us know that the animals still survived there.

From the written sources it also follows that not only were there many llamas in Huamachuco, there were also guanacos and vicuñas. "Sheep" also lived in Santiago de Chuco, Conchucos, and the Callejón de Huaylas, not just in the sixteenth century but also well into the seventeenth century. Their presence is likewise recorded for the Huánuco zone until the early seventeenth century. And not only were there "sheep" in Junín, the presence of vicuñas was likewise noted, and it is known that these lived in the heights of Canta, in the Chillón Valley, until well into the eighteenth century.

Another important zone with camelids extended along Huarochirí, Jauja, and Yauyos—the references are not just to domestic animals but also to wild ones. On Mt. Pariacaca there were vicuñas and guanacos at least until the seventeenth century; besides, camelids were still found in Huarochirí until at least the mid-eighteenth century, and trains of llamas still wound their way across the mountains until the 1940s (Javier Flores Espinoza, pers. commun., November 2002), while there are data for Yauyos—that is, the upper part of the Cañete Valley—on vicuñas and guanacos until the late eighteenth century. In Jauja they are mentioned until at least the seventeenth century.

The Tarma region also often appears as a place with many "sheep," and the same can be said for Huancavelica and Ayacucho. Meanwhile, the presence of guanacos and vicuñas was noted for Ayacucho and the highlands of Nasca until the late sixteenth century and early seventeenth century.

Other zones that according to the chroniclers had many domestic camelids were Apurímac, Abancay, and Andahuaylas, where the presence of vicuñas and guanacos was noted until 1600.

It should be recalled that when Pedro Pizarro was sent with other Spaniards in search of provisions for Cuzco, they went toward the altiplano and returned with 2,000 head. Besides, the presence of guanacos and vicuñas was recorded in the zone. Conditions were similar in Arequipa.

The altiplano appears in historical sources as the place where the number of camelids, both wild and domestic, was simply amazing. There is no point in emphasizing this. The documentation goes as far as the eighteenth century, and I have reviewed it. However, there undoubtedly is more for later times.

There is a remarkable coincidence between archaeological and historical sources as regards the *ceja de selva*, in that the data the chroniclers left only refers to the northern part. The presence of camelids is indicated for the Jaén zone, more specifically in the Chinchipe River basin, east of Bagua, as well as in the southern part of the Department of Amazonas and the northern marches of the Department of San Martín, in the Moyobamba River basin.

From all of the existing historical data it clearly follows that there were camelids all over the Andes when the Europeans arrived, without any altitudinal or other kinds of constraint. They even lived in the *ceja de selva*.

Some traces of this persisted until quite recently, but I was unable to study this in depth. However, it is not beside the point to recall that in the late nineteenth century Raimondi saw vicuñas in the Callejón de Huaylas and guanacos in the lower parts of Caravelí. Or that Kosok collected the testimony of old inhabitants in Cajamarca, who claimed that llamas had lived in the zone until the nineteenth century. I did not see any while I was working there in the 1960s. At present there is a program that intends to devel-

op them once more. The same thing happened in the mountains of Trujillo until the 1930s, according to the testimony of Cristóbal Campana. And curiously enough, the Jívaro Indians had llamas in the Amazon forest around the 1940s.

There is no question that Flores Ochoa is the scholar who has most discussed the current distribution of camelids in several of his studies. He clearly noted that this is wholly artificial and is a direct result of the Spanish conquest. In this regard I used his invaluable teachings and have only tried to expand the database so as to give his position a better basis. I am not sure who was the first to present these ideas, but I get the impression that it was Flores Ochoa, with other scholars simply following him. What is beyond question is that no one presented the data the way he did. However, it should be noted that several scholars fully accept these ideas. Such is the case of Fernández Baca (1971), Lynch (1983a), Sumar (1988), Novoa and Wheeler (1984), and Wheeler et al. (1992), but it is true that not all have followed the same argument. Others have simply made statements without adequate support.

I would like to discuss one point made by Flores Ochoa. In two of his studies (Flores Ochoa 1982, 1988b) he used the concept of altitude specialization to explain the life of camelids that found themselves cornered in refuge zones as a result of the cultural pressures exerted by Western civilization. I believe there was no specialization at all. This point was discussed in Chapter 2. Camelids acquired this genetic mark at some point in their life and simply took refuge in high-altitude areas because they had physiological characteristics that enabled them to withstand altitude better than other animals. This does not mean they did not pay a price for this: they are still paying a heavy toll. It should not be forgotten that in evolution, specialization can lead to failure. This has been shown in more than one case. But it is not so with the camelids. If they were actually specialized to high altitudes, they would simply be unable to survive when taken to lower altitudes or other habitats, as is now being done. And yet they live quite well.

But this small disagreement aside, the fact is that Flores Ochoa was right because the paleontological, archaeological, and historical data support his position. This position must be spread among the Peruvian population if we want to develop an awareness of the need to save our fauna and return it to its original condition as far as possible.

With the data available it is not easy to accurately determine the rate of extinction of these animals. The Incas had clearly found means to prevent it. Extinction began from the moment the Europeans set foot in the Andes. Unlike in other parts of the world, what led to this killing was not the mere pleasure of the hunt or its use for commercial ends. I believe that in this case the two major causes were first, the illiteracy of the majority of the adventurers who participated in that great feat that was the Spanish conquest, and second, a cultural problem. It was the clash of two distinct outlooks, where the stronger one dictated the rules. Some enlightened spirits like Cieza de León realized this, but obviously there was nothing they could do. When later efforts were made to apply regulations that would stop these abuses, first, the greatest damage had already taken place, and second, the law was not enforced. Exactly the same thing is happening today with the killing of vicuñas by poachers.

What is clear is that the places where these animals were eliminated are on the coast and the intermontane valleys. The process was far more gradual in high-altitude zones, and the very harsh conditions present in the puna must have saved the species from total extinction. So much so that, as Hocquenghem (n.d. [1989?]:116–117) notes, there are no camelids in northern Peru, which has no high-altitude zones of refuge.

One of the most complex problems regarding camelids, and which up to a point cannot be solved, concerns numbers. Definitely no figures can be given either for the pre-Columbian period, the conquest, or the Colonial period. In Inca times censuses must have been kept in the *quipu*, but these data have been lost for good. The *visitas* might give at least some general ideas about the number of animals in given areas but, as Flannery et al. (1989; see Chapter 12) note, there are problems in assessing which herds are meant, so these sources must be restudied using another approach. At any rate, this is a subject that must be raised by historians.

My intention in this study was not to review all existing contemporary statistics for camelids in the Andean area. The exact numbers are not too relevant for my objectives, and it is possible that the relevant institutions have far more accurate figures than those I obtained. However, it is an undisputed fact that when the published figures available to the public are compared, one reaches the conclusion that there is a great difference between scholars, even for the same year. This shows that we actually have estimates and not actual censuses, and that in truth, we do not really know how many camelids there are in the Andes (see Tables 2.2 and 2.4). This is certainly a great gap in our knowledge, and one that in the long run will constrain future measures and perspectives. Let me give one example to confirm this. According to statistics that should theoretically be correct (Hoces 1992:Table 9, 52) because they were published by the IUCN/SSC South American Camelid Specialist Group (Torres 1992c), in 1992 there were 97,670 vicuñas (see Table 2.3). Now, according to the statement made by the Instituto Nacional de Recursos Naturales del Perú (Anonymous 1994a:A1), in 1994 there was an approximate population of 57,000 vicuñas. This means that either one of the figures is wrong or the vicuña population fell by more than 40,600 specimens in two years. Either way the situation is chaotic. In 1994 the press reported that the Convention on International Trade in Endangered Species of Wild Fauna and Flora (CITES), held in Miami in November 1994, determined that vicuña wool could be freely commercialized because the danger of extinction for the species was over, even though its situation would have to be carefully supervised. Here it was specified that "[i]n the convention Peru presented the results of a census carried out in that year, which indicated that the vicuña population that is free in the mountains comes to 66,559 [animals]" (Anonymous 1994b:A1). This figure reveals a discrepancy of more than 9,500 animals with regard to that publicly presented that same year by the Instituto Nacional de Recursos Naturales del Perú (see above). This proves the aforementioned chaos.

Although not all scholars agree, everything seems to indicate that if corrective measures are not taken, the llama and the alpaca are on the road to extinction in Peru. Meanwhile, other countries that have understood their importance are trying to develop them. I have no information on what is happening in Peru's neighboring countries.

As regards wild camelids, I base my claims on a recent report by Torres (1992c), who assembled the respective data. Here it is stated that "[s]ustainable utilization of the vicuña is hampered by poaching and illegal trade of its products as well as by the cost, efforts, and difficulties involved in dealing with those unlawful activities. In addition, national regulations have not been updated and adjusted to the needs for protection, management, and utilization of the vicuña. These limitations can be overcome, however, with an adequate policy" (Hoces 1992:21). From the report by Hoces (1992) itself it follows that the predicament of the guanaco in Peru is dramatic. Slightly over 1,300 specimens are left (see Table 12.2). We should bear in mind that conditions have worsened in these years due to the danger posed by terrorism.[4] And although it is true that the vicuña population was recovering, it seems that this effort has waned in recent years due to several factors, and not just terrorism. Recent newspaper articles tell of the killing of "thousands" of vicuñas by poachers in the Pampa Galeras Reserve in 1989 and March 1994 (Lizana Salvatierra 1994:A1).

Conditions are no better in Bolivia. For the guanaco, even "the current distribution . . . is not known for certain," and "[it] is nearly extinct in its original area of distribution" (Villalba 1992:10, 12). As for the vicuña, its population "remains insecure due to the lack of continuity of conservation actions undertaken a few years ago" (Villalba 1992:10; see my Tables 2.1 and 2.2).

In Argentina, "[d]espite the difficulties facing its conservation," the vicuña population "is recovering." Meanwhile, the guanaco "is the largest [population] in South America," according to the report by Cajal and Puig (1992:5; see my Tables 2.1 and 2.2). However, if the report is read carefully, it is clear that protection of the species has huge hurdles to overcome.

Chile is the only country where conditions are encouraging. "The vicuña population . . . has shown a conspicuous recovery, which greatly reduces the risk of extinction that was very high until a few years ago. . . . The guanaco population is also experiencing a satisfactory recovery, despite that

fact that illegal hunting and habitat disturbance persist in some areas of the country" (Glade and Cunazza 1992:14; see my Tables 2.1 and 2.2).

At present there are haciendas on the central Chilean coast, south of Santiago, that are raising llamas practically at sea level (Carlos Monge C., pers. commun., 11 July 1994).

In other words, except in Chile, and to a lesser extent in Argentina, the condition of the wild camelids is threatened.

Chapter 13 mentioned the two attempts made to introduce camels in Peru. Here I want to emphasize a prejudice that must be dispelled. Just as it has been claimed that the llama cannot live anywhere but in a high-altitude environment, so it has been claimed that camels could not thrive in Peru because they were outside their natural habitat. Nothing could be further from the truth. First, the specimens brought must not have been many, and second, they prospered on the coast and turned feral after being abandoned on the lomas of the central and south coast. The camels would certainly have survived had they not been killed by man. Father Cobo (1964a:Bk.10, Chap. XLIII, 421) clearly states that camels "formed a lineage and multiplied aplenty." The major reason why the camels were abandoned was a cultural one. As Kubler (1946:359) correctly noted, they "could not compete with mules and horses." But not, however, because mules and horses were more efficient; rather, simply because these formed part of Western civilization, within whose parameters the Spaniards acted. There is no question that camels could have carried goods across the coastal deserts, at least until the introduction of modern vehicles, in a far more efficient way than either mules or horses. Besides, at high altitudes camels would not have had the problems the latter faced. It is known that on the Chinese side of the Himalayas, i.e., in Tibet, mountaineering expeditions have replaced the human carriers—the role traditionally performed by the Nepalese Sherpa—with Bactrian camels. These climb fully loaded and without any difficulty over 5000 masl (Carlos Monge C., pers. commun., 28 June 1994).

The Australian case should not be forgotten either: today there is a significant population of feral camels, which cannot be precisely said to be in their environment.

To conclude, it is essential to emphasize once again the pressing need of making people conscious of the latent dangers that exist should the transformation of the Andean environment continue without any clear policies based on scientific studies. Seibert (1983:275) was quite clear in this regard when he stated that for thousands of years man had dramatically changed the vegetation and the landscape of the central Andean highlands, with various uses of the land. But while these changes took place in the environment, agricultural techniques and social structures were being simultaneously developed that were well adapted to the ecological requirements of the area. The changes in the social structures have brought about individual-based systems that, through ignorance or selfishness, apply inadequate techniques that are seriously damaging the lands. The elimination or prevention of such damage will not be attained through the importation of exotic techniques; it will only come with the future development of traditional land use techniques. One prerequisite for this development is promoting the renewal of traditional native communities and their organization.

Within this context there can be no question that Andean camelids played a crucial role in protecting the environment against soil erosion, in fertilizing the soil, and in providing the Andean people with a major part of the economic foundations of their sustenance.

It cannot be denied that one of the negative aspects of the Spanish colonization of America—which unleashed one of the most serious crises ever in the history of mankind, as regards the destruction of the native fauna—concerns the camelids. As Wheeler et al. (1992:468) note, "[t]he consequences of this catastrophe on contemporary camelid production are rarely considered and poorly understood."

If the husbandry of native animals still persists, even in deteriorated condition, it is mainly due to the anonymous Indians and the peasant communities that have stubbornly resisted their destruction, whether consciously or unconsciously, but always guided by tradition. "What enabled the survival of the Andean livestock was the re-sistance of the paqocheros against all attempts to deliberately change the constitution of their ecosystem" (Flores Ochoa 1979a:231).

The work that has yet to be done to understand all different aspects of the camelids is still immense. This is a task that can be done in time; all I intended was to indicate some lines of research. In the future this will have to be a collective effort where various specialists intervene. But it will be far more difficult to develop the consciousness of the need to save these animals, and furthermore, to make politicians understand that this is a priority, or that all efforts will be useless unless it is given some continuity. This is an old and painful experience in Peru, one that has not changed over time, as we see today.

Some scholars, like Franklin (1982:485), are optimistic about the future of the camelids. I would like to share this belief but thus far am unable to do so because of all I have seen in the time I spent studying these animals. This is why I would like to conclude with a statement by Flannery et al. (1989: 117), which holds a tragic and harsh reality: "Deprived of the protection of the Inca state, and even of the calamity reserves of the *sapsi*[5] herds, they can appeal today only to their *wamanis*."[6]

NOTES

[1] *Clostridium perferingens* appears in Wheeler (1985b:72) possibly due to a misprint, as it was written correctly in an earlier study (Lavallée et al. 1982:90). Kent (1988b: 143) also misspelled it, however.

[2] A halophilous herbaceous community, comprising plants that tolerate a strong salinity in the environment.

[3] Natives from the community of Moche who live close to the city of Trujillo, in the Department of La Libertad.

[4] This book was first written in Peru in the early 1990s. Terrorism has at present been almost eliminated, but the wild camelids are still threatened by rustlers and poachers.

[5] The communal herds (see Flannery et al. 1989:109).

[6] The spirit of the community. The highest mountain which presides over the life of the town and forms part of the geography of the zone is believed to be the home of the guardian spirit of the community. He is called the *Apu* and is usually a snow-capped mountain. There are other, lower-ranking guardian spirits that are smaller mountains and form the town's defensive garrison against misfortunes like epidemics, etc. These are called *Auquis* or *Wamanis* (*Huamanis*) (see Valcárcel 1967:155).

BIBLIOGRAPHY

"Bibliography is not usually an attractive subject, nor does it provide easy victories, but when it is directed by perspicacity and tenacity it sets a happy course for those who really pursue knowledge."

Luis Jaime Cisneros (1994:1)

15.1 Sources Cited by Bonavia

A
1973   La vida de las palabras. Auquénido. *El Comercio*, No. 73,816, 26 November, p. 2. Lima, Peru.

Acleto Osorio, César
1971   *Algas marinas del Perú de importancia económica*. Serie de Divulgación, No. 5. Universidad Nacional Mayor de San Marcos, Museo de Historia Natural "Javier Prado," Lima, Peru.

Acosta, Padre José de
1954   *Historia Natural y Moral de las Indias*. Biblioteca de Autores Españoles desde la formación del lenguaje hasta nuestros días. Tomo LXXIII, pp. 1–247. Ediciones Atlas, Madrid, Spain.

Acosta Polo, Juan
1977   *Nombres vulgares y usos de las algas en el Perú*. Serie de Divulgación, No. 7. Universidad Nacional Mayor de San Marcos, Museo de Historia Natural "Javier Prado," Lima, Peru.

Acuña, Francisco de, Gaspar Aytara, Juan Chuqui Taipi, and Alonso Chununco
1965 (1881–1897) Colquemarca. Relación fecha por el Corregidor de los Chumbibilcas don Francisco de Acuña, por mandado de Su Ex. a del señor Fernando de Torres y Portugal, visorrey destos reynos, para la discrepción de las Yndias que Su Majestad manda hacer. In *Relaciones Geográficas de Indias. Perú. II*. Marcos Jiménez de la Espada. Biblioteca de Autores Españoles desde la formación del lenguaje hasta nuestros días. Vol. I, pp. 320–321. Tomo CLXXXIII, Ediciones Atlas, Madrid, Spain.

Acuña, Francisco de, Jhoan Cabrera, Pedro Maybire, and Antonio Ossorio
1965 (1881–1897) Llusco y Quinota. Relación fecha por el Corregidor de los Chumbibilcas don Francisco de Acuña, por mandado de Su Ex. a del señor Fernando de Torres y Portugal, visorrey destos reynos, para la discrepción de las Yndias que Su Majestad manda hacer. In *Relaciones Geográficas de Indias. Perú. II*. Marcos Jiménez de la Espada. Biblioteca de Autores Españoles desde la formación del lenguaje hasta nuestros días. Vol. I, pp. 315–317. Tomo CLXXXIII, Ediciones Atlas, Madrid, Spain.

Acuña, Francisco de, Francisco Chanaca, Carlos Quispi, and Miguel de Anues
1965 (1881–1897) Pueblo de Capamarca. Relación fecha por el corregidor de los Chumbibilcas don Francisco de Acuña, por mandado de Su Ex. a del señor Fernando de Torres y Portugal, visorrey destos reynos, para la discrepción de las Yndias que Su Majestad manda hacer. In *Relaciones Geográficas de Indias. Perú. II*. Marcos Jiménez de la Espada. Biblioteca de Autores Españoles desde la formación del lenguaje hasta nuestros días. Vol. I, pp.

317–319. Tomo CLXXXIII, Ediciones Atlas, Madrid, Spain.

Acuña, Francisco de, Andrés Flores, Francisco Serra de Leguizamo, and Juan de Quevedo
1965 (1881–1897) Relación fecha por el Corregidor de los Chumbibilcas don Francisco de Acuña, por mandado de Su Ex.a del señor don Fernando de Torres y Portugal, visorrey destos reynos, para la discrepción de las Yndias que Su Majestad manda hacer. In *Relaciones Geográficas de Indias, Perú. I.* Marcos Jiménez de la Espada. Biblioteca de Autores Españoles desde la formación del lenguaje hasta nuestros días. Vol. I, pp. 310–315. Tomo CLXXXIII, Ediciones Atlas, Madrid, Spain.

Acuña, Francisco de, Alvaro de Prado, Felipe Quispihaqueua, Juan de Yllanes, and Francisco Vilcacuri
1965 (1881–1897) Libitaca. Relación fecha por el corregidor de los Chumbibilcas don Francisco de Acuña, por mandado de Su Ex. a del señor Fernando de Torres y Portugal, visorrey destos reynos, para la discrepción de las Yndias que Su Majestad manda hacer.In *Relaciones Geográficas de Indias. Perú. II.* Marcos Jiménez de la Espada. Biblioteca de Autores Españoles desde la formación del lenguaje hasta nuestros días. Vol. I, pp. 323–325. Tomo CLXXXIII, Ediciones Atlas, Madrid, Spain.

Acuña, Francisco de, Lope Sánchez de la Cueva, Carlos Quispe, Francisco Sánchez Goliardo, Matheo Rigon, Diego Nina Chaguayo, Luis de Medina, and Santiago Supanta
1965 (1881–1897) Belille y Chamaca. Relación fecha por el Corregidor de los Chumbibilcas don Francisco de Acuña, por mandado de Su Ex. a del señor Fernando de Torres y Portugal, visorrey destos reynos, para la discrepción de las Yndias que Su Majestad manda hacer. In *Relaciones Geográficas de Indias, Perú. II.* Marcos Jiménez de la Espada. Biblioteca de Autores Españoles desde la formación del lenguaje hasta nuestros días. Vol. I, pp. 321–323. Tomo CLXXXIII, Ediciones Atlas, Madrid, Spain.

Alarco, Eugenio
1983 *El hombre peruano en su historia. El encuentro de dos poderes. Españoles contra Incas.* Tomos V and VI. Lima, Peru.

Aldrete, Juan
1965 (1881–1897) Relación de la Gobernación de Yahuarzongo y Pacamurus. In *Relaciones Geográficas de Indias. Perú. IV.* Marcos Jiménez

de la Espada. Biblioteca de Autores Españoles desde la formación del lenguaje hasta nuestros días. Vol. III, pp. 147–153. Tomo CLXXXV, Ediciones Atlas, Madrid, Spain.

Aleman, Diego
1965 (1881–1897) Entrada de Diego Aleman a los Mojos o Mussus. 1564. Ultimo Apéndice a las Relaciones Geográficas de Indias. In *Relaciones Geográficas de Indias. Perú. IV.* Marcos Jiménez de la Espada. Biblioteca de Autores Españoles desde la formación del lenguaje hasta nuestros días. Vol. III, pp. 276–278. Tomo CLXXXV, Ediciones Atlas, Madrid, Spain.

Allen, W. R., A. J. Higgins, I. G. Mayhew, D. H. Snow, and J. F. Wade (editors)
1992 *Proceedings of the First International Camel Conference 2nd–6th February 1992.* R. & W. Publications (Newmarket) Ltd., Amersham, Buckinghamshire, UK.

Altamirano Enciso, Alfredo
1983a Pesca y utilización de Camélido en Manchán. *Boletín de Lima*, Año 5, 30:62–74.
1983b La fauna de Huachipa, valle del Rímac. *Boletín*, No. 8, p. 34. Museo Nacional de Antropología y Arqueología, Lima, Peru.
1987 Restos de camélidos prehispánicos en la bahía de Bayovar, Piura. *Boletín de Lima*, Año 9, 52: 37–46.

Alvarado, Alonso de
1965 (1881–1897) Primeros descubrimientos y conquistas de los Chachapoyas por el capitán Alonso de Alvarado. Ultimo Apéndice a las Relaciones Geográficas de Indias. In *Relaciones Geográficas de Indias. Perú. IV.* Marcos Jiménez de la Espada. Biblioteca de Autores Españoles desde la formación del lenguaje hasta nuestros días. Vol. III, pp. 158–164. Tomo CLXXXV, Ediciones Atlas, Madrid, Spain.

Álvarez, Arturo
1987 Introducción. *A través de la América del Sur de Diego de Ocaña.* Crónicas de América 33. Historia 16, pp. 5–30. Madrid, Spain.

Álvarez, Diego
1968–1969 Visita del repartimiento de Guaraz encomendado en Hernando de Torres. La Visita de Guaraz en 1558, Elena Aibar Ozejo. In *Cuadernos del Seminario de Historia*, 9, enero 1968–diciembre de 1969, pp. 5–21. Seminario de Historia, Instituto Riva-Agüero, Pontificia Universidad Católica del Perú, Lima, Peru.

Andagoya, Pascual de
1954   *Relación de los sucesos de Pedrarias Dávila en las provincias de Tierra firme o Castilla del oro, y de lo ocurrido en el descubrimiento de la mar del Sur y costas del Perú y Nicaragua, escrita por el Adelantado Pascual de Andagoya*. Pascual de Andagoya: Ein Mensch erlebt die Conquista. Universität Hamburg. Abhandlung aus dem Gebiet der Auslandskunde. Band 59, Reihe B. Völkerkunde, Kulturgeschichte und Sprachen. Band 33, pp. 224– 261. Gram, de Gruyter & Co., Hamburg, Germany.

Angleria, Pedro Mártir de
1944 (1530)   *Décadas del Nuevo Mundo*. Colección de Fuentes para la Historia de América. Editorial Bajel, Buenos Aires, Argentina.

Anónimo Portugués
1958   *Descripción del Virreinato del Perú*. Crónica inédita de comienzos del siglo XVII. Serie B, No. l. Universidad Nacional del Litoral, Facultad de Filosofía, Letras y Ciencias de la Educación, Rosario, Argentina.

Anonymous
1925   *See* Salinas Loyola, Juan de, 1965a.

Anonymous
1938 (1573)   La Cibdad de Sant Francisco del Quito. In *Quito a través de los siglos*. Recopilación y Notas bio-bibliográficas de Eliecer Enríquez B., pp. 7–64. Imprenta Municipal, Quito, Ecuador. [Author's Note*: See* Anonymous 1965b.]

Anonymous
1965a   Descripción de la villa y minas de Potosí. Año de 1603. In *Relaciones Geográficas de Indias. Perú. II*. Marcos Jiménez de la Espada. Biblioteca de Autores Españoles desde la formación del lenguaje hasta nuestros días. Vol. I, pp. 372–385. Tomo CLXXXIII, Ediciones Atlas, Madrid, Spain.

Anonymous
1965b (1881–1897)   La cibdad de San Francisco del Quito. 1573. In *Relaciones Geográficas de Indias. Perú. III*. Marcos Jiménez de la Espada. Biblioteca de Autores Españoles desde la formación del lenguaje hasta nuestros días. Vol. II, pp. 205–230. Tomo CLXXXIV, Ediciones Atlas, Madrid, Spain. [Author's Note: *See* Anonymous 1938.]

Anonymous
1970   (End of 16th century?) Aviso de el modo que havia en el goviemo de los indios en tiempo del inga y como se repartian las tierras y tributos. In Mercaderes del valle de Chincha en la época prehispánica: Un documento y unos comentarios, María Rostworowski de Diez Canseco. *Revista Española de Antropología Americana* 5:163–173.

Anonymous
1973   *Manual de Malezas en el Perú, comunes en Caña de Azúcar*. Cooperativa Agraria de Producción Casagrande Ltda No. 32. Hortus S.A, May & Baker Ltd., Lima, Peru.

Anonymous
1990   Galos usan auquénidos para prevenir incendios forestales. *El Comercio*, Año 151, No. 79,823. 10 May, p. B5. Lima, Peru.

Anonymous
1991   Matanza indiscriminada redujo el número de alpacas en el país. *El Comercio*, Año 152, No. 80,352. 21 October, p. A16. Lima, Peru.

Anonymous
1992a   También hay llamas en Australia. (Caption of an illustration). *El Comercio*, Año 153, No. 80,595. 20 June, p. B4. Lima, Peru.

Anonymous
1992b   Vicuñas peruanas se reproducen con éxito en Ecuador. *El Comercio*, Año 153, No. 80,615. 10 July, p. B6. Lima, Peru.

Anonymous
1992c   En Pisco galoparon los camellos. *El Comercio*, Año 153, No. 80,721. 24 October, p. Dl. Lima, Peru.

Anonymous
1992d   *Nouvelles certaines des Isles du Peru*. (Facsimilé de l'edition originale), suivi de *Nouvelles du Pérou* de Miguel de Estete, pp. XII–XLII. Amiot. Lenganey. Thaon, France.

Anonymous
1992e   En EE. UU. usan llamas para llevar palos de golf. *El Comercio*, Año 132, No. 80,844. 24 February, p. B1. Lima, Peru.

Anonymous
1993   El guanaco vivió en norte brasileño hace 10,000 años. *El Comercio*, Año 154, No. 80,945. 5 June, p. B1. Lima, Peru.

Anonymous
1994a   En el país hay alrededor de 57 mil vicuñas. *El Comercio*, Año 155, No. 81,347. 12 July, p. Al. Lima, Peru.

Anonymous
1994b   Libremente será comercializada la lana de vicuña. *El Comercio*, Año 155, N° 81,475. 17 November, p. A1. Lima, Peru.

Anonymous
1995   Hallan un cementerio de llamas cerca de las obras del proyecto Chavimochic. *El Comercio*, Año 156, No. 81,718. 18 July, p. B7. Lima, Peru.

Anonymous
1997a Tres mil camélidos habrían muerto durante la ola de frío. *El Comercio*, Año 158, No. 82,484. 21 August, p. B6. Lima, Peru.

Anonymous
1997b Población de vicuñas está a punto de desaparecer en región sur andina. *El Comercio*, Año 158, No. 82,502. 8 September, p. A1. Lima, Peru.

Anonymous
1998a En emirato árabe cruzan camello con llama. *El Comercio*, Año 158, No. 82,637. 21 January, p. A1. Lima, Peru.

Anonymous
1998b Primera cría del camello y la llama debe ser fértil. *El Comercio*, Año 158, No. 82,639. 23 January, p. A12. Lima, Peru.

Anonymous
1998c "Cama" ya tiene tres semanas. *El Comercio*, Año 158, No. 82,657. 10 February, p. B5. Lima, Peru.

Anonymous
1998d Una pasión que nació con un escupitajo. *El Comercio*, Año 159, No. 82,860. 1 September, p. B5. Lima, Peru.

Anonymous
1999 Cazadores diezman población de vicuñas en las alturas de Yauyos. *El Comercio*, Año 159, No. 83,053. 13 March, p. A11. Lima, Peru.

Anonymous Portuguese. *See* Anónimo Portugués

Antonius, Otto
1922 *Grundzuege einer Stammesgeschichte der Haustiere*, pp. XVI + 336. G. Fischer, Jena, Germany.

Antunez de Mayolo R., Santiago E.
1981 *La nutrición en el Antiguo Perú.* Banco Central de Reserva, Lima, Peru.

Aparicio, Francisco de
1946 The Comechigón and their neighbors of the Sierra de Córdoba. In *Handbook of South American Indians*, Vol. 2, pp. 673–685. Julian H. Steward, editor. Bulletin 143, Smithsonian Institution, Bureau of American Ethnology, Washington, DC.

Araníbar, Carlos
1963 Algunos problemas heurísticos en las Crónicas de los siglos XVI–XVII. *Nueua coronica*. No. 1, pp. 102–135. Facultad de Letras, Universidad Nacional Mayor de San Marcos, Lima, Peru.

1967 Introducción. *El Señorío de los Incas*, pp. VII–XCVI. Pedro Cieza de León. Instituto de Estudios Peruanos, Lima, Peru.

Arbocco Arce, Aldo
1974 La verdad sobre el guanaco. *La Prensa*, Año LXXI, No. 31,002. 27 May, pp. 10–11. Lima, Peru.

Arellano Hoffmann, Carmen
1988 Anotaciones del clima, ganado y tenencia de pastos en la puna de Tarma, siglo XVIII. In *Llamichos y paqocheros. Pastores de llamas y alpacas*, pp. 77–84. Compiled by Jorge A. Flores Ochoa. CONCYTEC, CEAC y Editorial UNSAAC, Cuzco, Peru.

Aste Salazar, Humberto
1964 Diferenciación de hemoglobinas en carneros, llamas y alpacas en las grandes alturas, y en vicuñas a nivel del mar. In *Anales del Segundo Congreso Nacional de Veterinaria y Zootecnia*, pp. 332–339. Lima, Peru.

Audiencia de Santo Domingo
1989 Carta del 30 de mayo de 1537 al Rey. In *Mitos y utopías del descubrimiento. III. El Dorado*, p. 56. Juan Gil. Alianza Editorial, Madrid, Spain.

Austral, Antonio
1987 Hallazgo en capa de un artefacto lítico y fauna extinta en la laguna Las Encadenadas. Provincia de Buenos Aires, República Argentina. In *Estudios Atacameños*. Investigaciones Paleoindias al Sur de la Línea Ecuatorial, No. 8 (special issue), pp. 94–97. Lautaro Núñez y Betty Meggers, editors. Instituto de Investigaciones Arqueológicas, Universidad del Norte, San Pedro de Atacama, Chile.

Avila, Francisco de
1966 Dioses y Hombres de Huarochirí. Narración quechua recogida por. . . (¿1598?). Edición bilingüe. Traducción castellana de José María Arguedas. Estudio Biobibliográfico de Pierre Duviols. Museo Nacional de Historia e Instituto de Estudios Peruanos, Lima, Peru.

Ayala, Fabian de
1974–1976 (1614) Une petite chronique retrouvé: Errores, ritos, supersticiones y ceremonias de los yndios de la prouincia de Chinchaycocha y otras del Piru. Edition et commentaires par Pierre Duviols. *Journal de la Société des Américanistes* 63:275–286.

Ayanz, Antonio de
1978 (1596?) Los sucesores de Toledo en las dos últimas décadas del siglo XVI. In *El servicio personal de los indios en el Perú (extractos del siglo XVI)*. Silvio Zavala. Tomo I, pp. 218–220. El Colegio de México, México.

Bahn, Paul G.
1994 Time for a change. *Nature* 367:511–512.

Baied, Carlos A., and Jane C. Wheeler
1993 Evolution of high Andean puna ecosystem: Environment, climate, and culture change over the last 12,000 years in the central Andes. *Mountain Research and Development* 13(2):145–156.

Banchero, Natalio
1973 Mecanismos fisiológicos de adaptación a la hipoxia en la llama (*Lama glama*). *Archivos del Instituto de Biología Andina* 6 (1–4):1–36.

Banchero, Natalio, Robert F. Grover, and James A. Will
1971a High-altitude-induced pulmonary arterial hypertension in the llama (*Lama glama*). *American Journal of Physiology* 220:422–427.
1971b Oxygen transport in the llama (*Lama glama*). *Respiration Physiology* 13:102–115.

Bandera, Damián de la
1974 Relación general de la disposición y calidad de la provincia de Guamanga, llamada San Joan de la Frontera, y de la vivienda y costumbres de los naturales della. Año de 1557. In *Huamanga: Una larga historia*, pp. 137–146. Consejo Nacional de la Universidad Peruana, Lima, Peru. [Author's Note: This was taken from Bandera 1881.)

Bannikov, A. G.
1958 Distribution géographique et biologie du cheval sauvage et du chameau de Mongolie (*Equus przewalkii* et *Camelus bactrianus*). *Mammalia* 22(1):152–160.

Baraza de Fonts, Nélida
1986 De la recolección indiferenciada a la horticultura incipiente. In *The Pleistocene Perspective*, Vol. 2. The World Archaeological Congress, pp. I–II, 1–13. Allen & Unwin, London, UK.

Barraza, Jacinto
1937 Misión especial a la gran laguna de Chuquito sus riveras e islas pobladas de gente bárbara: Apuntes históricos, Carlos A. Romero. *Revista Histórica* 11:196–201. [Author's Note: This is part of the unpublished manuscript of F. Jacinto Barraza, "Historia de las fundaciones de los Colegios y Casa de la Provincia del Perú, de la Compañía de Jesús, con la noticia de las vidas y virtudes religiosas de algunos varones ilustres que en ella trabajaron."]

Barrionuevo, Francisco de
1988 (1535) Carta dirigida al Consejo de Indias. In *D. Diego Tomala, cacique de la isla de la Puná: Un caso de aculturación socioeconómica*, pp. 3–4.

Adam Szaszdi. Museo Antropológico, Banco Central del Ecuador, Guayaquil, Ecuador.

Bartels, Heinz, Peter Hilpert, Klaus Barbey, Klaus Betke, Klaus Riegel, Ernst Michael Lang, and James Metcalfe
1963 Respiratory functions of bloods of the yak, llama, camel, Dybowski deer, and African elephant. *American Journal of Physiology* 205:331–336.

Barzana, Alonso de
1965 (1881–1897) Carta de P. Alonso de Barzana, de la Compañía de Jesús, al P. Juan Sebastián, su provincial. Fecha en la Asunción del Paraguay a 8 de setiembre de 1594. In *Relaciones Geográficas de Indias. Perú. II.* Apéndice III. Marcos Jiménez de la Espada. Biblioteca de Autores Españoles desde la formación del lenguaje hasta nuestros días. Vol. II, pp. 78– 86. Tomo CLXXXIV, Ediciones Atlas, Madrid, Spain.

Bauer, Christian, Harry S. Rollema, Heinz Till, and Gerhard Braunitzer
1980 Phosphate binding by llama and camel hemoglobin. *Journal of Comparative Physiology* [*B*] 136:67–70.

Bawden, Garth
1982 Galindo: A study in cultural transition during the Middle Horizon. In *Chan Chan: Andean Desert City*, pp. 285–320. Michael E. Moseley and Kent C. Day, editors. A School of American Research Book. University of New Mexico Press, Albuquerque, NM.

Bayer, P. Wolfgang
1969 Viaje por el Perú en 1751. In *4 Viajeros Alemanes al Perú*, pp. 29–44. Estuardo Núñez. Relaciones de: W. Bayer, K. Scherzer, F. Gerstaecker, H. Zoeller. Comentarios del Perú 10. Universidad Nacional Mayor de San Marcos, Lima, Peru.

Beals, Ralph L., and Harry Hoijer
1958 *Introducción a la Antropología*. Aguilar, Madrid, Spain.

Becker, E. L., John A. Schilling, and Rodney B. Harvey
1955 Renal function and electrolytes in the llama. *American Journal of Physiology* 183:307–308.

Belmonte Sch., Eliana, Eugenia Rosello, and Nelly Rojas R.
1988 Análisis de restos vegetales de coprolitos de camélidos de la desembocadura del río Camarones. *Chungará* 20:47–61.

Benavente Aninat, María Antonia
1985 Reflexiones en torno al proceso de domesticación de camélidos en los valles del Centro

y Sur de Chile. *Boletín del Museo Regional. Araucanía* 2:37–52.

Benavides, María A.

1986a   Introduction: Ethnohistorical research for the terrace abandonment project, Colca Valley. In *The Cultural Ecology, Archaeology, and History of Terracing and Terrace Abandonment in the Colca Valley of Southern Peru*, pp. 386–389. Technical Report to the National Science Foundation and the National Geographic Society, William M. Denevan, editor. Madison, WI.

1986b   Coporaque in the 1591 *Visita* of Yanquecollaguas urinsaya.In *The Cultural Ecology, Archaeology, and History of Terracing and Terrace Abandonment in the Colca Valley of Southern Peru*, pp. 390–405. Technical Report to the National Science Foundation and the National Geographic Society, William M. Denevan, editor. Madison, WI.

1988   Notarial records for 16th century Collaguas in the Archive of Arequipa. In *The Cultural Ecology, Archaeology, and History of Terracing and Terrace Abandonment in the Colca Valley of Southern Peru*, Vol. II, pp. 38–45. Technical Report to the National Science Foundation and the National Geographic Society, William M. Denevan, editor. Madison, WI.

1995   Cambios en el paisaje agrológico de la Provincia de Collaguas: Un análisis de documentos en los Archivos de Arequipa, Perú. *Revista del Archivo Arzobispal de Arequipa* 2: 15–46.

Benavides, Claudio E., Rubén Pérez, Mauricio Espinoza, Gertrudis Cabello, Raquel Riquelme, Julian T. Parer, and Aníbal J. Llanos

1989   Cardiorespiratory functions in the fetal llama. *Respiration Physiology* 75:327–334.

Benfer, Robert A.

1983   The challenges and rewards of sedentism: The Preceramic village of Paloma, Peru. Mimeograph. [Author's Note: This was later published. *See* Benfer 1984.]

1984   The challenges and rewards of sedentism: The Preceramic village of Paloma, Peru. In *Paleopathology and the Origins of Agriculture*, pp. 531–558. Mark N. Cohen and George J. Armelagos, editors. Academic Press, Orlando, FL.

Benino, Nicolás del

1965 (1881–1897) Relación muy particular del cerro y minas de Potosí y de su calidad y labores, por Nicolás del Benino, dirigida a Don Francisco de Toledo, Virrey del Perú, en

1573. In *Relaciones Geográficas de Indias. Perú. II.* Marcos Jiménez de la Espada. Biblioteca de Autores Españoles desde la formación del lenguaje hasta nuestros días. Vol. I, pp. 362–371. Tomo LXXXIII, Ediciones Atlas, Madrid, Spain.

Bennett, Wendell C.

1946a   The Andean highlands: An introduction. In *Handbook of South American Indians*, Vol. 2, *The Andean Civilizations*, pp. 1–60. Julian H. Steward, editor. Bulletin 143, Smithsonian Institution, Bureau of American Ethnology, Washington, DC.

1946b   The Atacameño. In *Handbook of South American Indians*, Vol. 2, *The Andean Civilizations*, pp. 599–618. Julian H. Steward, editor. Bulletin 143, Smithsonian Institution, Bureau of American Ethnology, Washington, DC.

Bennett, Wendell C., and Junius B.Bird

1960   *Andean Culture History*. Handbook Series No. 15, American Museum of Natural History. New York.

Benson, Elizabeth P.

1972   *The Mochica: A Culture of Peru*. Praeger, New York.

Benzoni, Girolamo

1985 (1565)   *La Historia del Nuevo Mundo*. Museo Antropológico y Pinacoteca del Banco Central del Ecuador, Guayaquil, Ecuador.

Bertonio, Ludovico

1956 (1612)   *Vocabulario de la lengua Aymara*. Facsimile edition. La Paz, Bolivia.

Betanzos, Juan de

1987 (1551)   *Suma y narración de los Incas*. Ediciones Atlas, Madrid, Spain.

1996 (1551)   *Narrative of the Incas*. Translated and edited by Roland Hamilton and Dana Buchanan from the Palma de Mallorca manuscript, Madrid, Spain. University of Texas Press, Austin, TX.

Biasutti, Renato

1959   La classificazione delle Culture.In Renato Biasutti, *Le Razze e i Popoli della Terra*, Vol. I, *Razze, Popoli e Culture*, pp. 674–713. Unione Tipografico-Editrice Torinese, Turin, Italy.

Bibar, Gerónimo de

1966   *Crónica y relación copiosa y Verdadera de los Reynos de Chile, hecha por Gerónimo de Bibar natural de Burgos*, pp. 1–215. José Toribio Medina. Edición facsimilar y a Plana del Fondo Histórico y Bibliográfico, Santiago de Chile, Chile. [Author's Note: *See* Vivar 1979.]

Bird, Junius B.

1946a The cultural sequence of the north Chilean coast. In *Handbook of South American Indians*, Vol. 2, *The Andean Civilizations*, pp. 587–594. Julian H. Steward, editor. Bulletin 143, Smithsonian Institution, Bureau of American Ethnology, Washington, DC.

1946b The historic inhabitants of the north Chilean coast. In *Handbook of South American Indians*, Vol. 2, *The Andean Civilizations*, pp. 595–597. Julian H. Steward, editor. Bulletin 143, Smithsonian Institution, Bureau of American Ethnology, Washington, DC.

1954 Paracas fabrics and Nazca needleworks. In *Textile Museum Catalogue Raisonné*. The Textile Museum, Washington, DC.

1960 Techniques. In Wendell C. Bennett and Junius B. Bird, *Andean Culture History*, pp. 245–299. Handbook Series No. 15, American Museum of Natural History, New York.

Bird, Junius B.,and Margaret Bird

1988 *Travels and Archaeology in South Chile*. John Hyslop, editor. University of Iowa Press, Iowa City, IA.

Bittmann, Bente

1986 Recursos naturales renovables de la costa del norte de Chile: Modos de obtención y uso. In *Etnografía e Historia del Mundo Andino: Continuidad y Cambio*, pp. 269–334. Shozo Masuda, editor. Tokyo University, Tokyo, Japan.

Bligh, John, Isabel Baumann, Julio Sumar, and Félix Pocco

1975 Studies of body temperature patterns in South American Camelidae. *Comparative Biochemistry and Physiology, A Comparative Physiology* 50A:701–708.

Bligh, John, and K. G. Johnson (editors)

1973 International Union of Physiological Sciences: Thermal Physiology Commission. Glossary of terms for thermal physiology. *Journal of Applied Physiology* 35:941–961.

Bolsi, Alfredo S.

1968 Ecología humana del Altiplano. In *Actas y Memorias*, XXXVII Congreso Internacional de Americanistas. República Argentina—1966. Vol. II, pp. 11–74. Buenos Aires, Argentina.

Bonavia, Duccio

1960 Sobre el estilo Teatino. Tesis para optar el grado de Bachiller, Universidad Nacional Mayor de San Marcos, Lima, Peru.

1962 Sobre el estilo Teatino. *Revista del Museo Nacional* 31:43–94.

(Arranger/compiler)

1966 Sitios Arqueológicos del Perú (Primera parte). *Arqueológicas* 9. Publicaciones del Instituto de Investigaciones Antropológicas, Museo Nacional de Antropología y Arqueología, Lima, Peru.

1968 *Las Ruinas del Abiseo*. Universidad Peruana de Ciencias y Tecnología, Lima, Peru.

1972 El arte rupestre de Cuchimachay. In D. Bonavia, W. Espinoza S., W. Isbell, R. Matos, G. Petersen, J. Pulgar Vidal, R. Ravines, and D. Thompson, *Pueblos y Culturas de la Sierra Central del Perú*, pp. 134–139. Lima, Peru.

1982a *Los Gavilanes: Mar, desierto y oasis en la historia del hombre*. Corporación Financiera de Desarrollo S. A. COFIDE and Instituto Arqueológico Alemán, Lima, Peru.

1982b Canidae, Camelidae, Cervidae, Chinchillidae, Caviidae. In *Los Gavilanes*, pp. 200–201.

1982c Camelidae. In *Los Gavilanes*, pp. 225–226.

1984 La importancia de los restos de papas y camotes de época precerámica hallados en el valle de Casma. *Journal de la Société des Américanistes* 70:7–20.

1985 *Mural Painting in Ancient Peru*. Indiana University Press, Bloomington, IN.

1990 Les Ruines de l'Abiseo. In *Inca-Peru: 3000 Ans d'Histoire*, pp. 248–261. Imschoot, uitgevers, Ghent, Belgium.

1991 *Perú: Hombre e Historia. Desde los orígenes al siglo XV*, Vol. I. Fundación del Banco Continental para el Fomento de la Educación y la Cultura. Ediciones Edubanco, Lima, Peru.

1993 Critical review of Orefici Giuseppe, *Nasca: Archeologia per una ricostruzione storica*. Jaca Book, Milano, 1992, 264 pp., 46 fig. *Journal de la Société des Américanistes* 79:302–305.

1994 *Arte e Historia del Perú Antiguo: Colección Enrico Poli Bianchi*. Banco del Sur, Arequipa, Peru.

1995 Los Camélidos y la altura. *Ultramarine Newsletter* 2(2):5–6.

1999 The domestication of Andean camelids. In *Archaeology in Latin America*, pp. 130–147. G. G. Politis and B. Alberti, editors. Routledge, London and New York.

Bonavia, Duccio, Fabiola León Velarde, Carlos Monge C., María Inés Sánchez Griñán, and José Whittembury

1984 Tras las huellas de Acosta 400 años después, consideraciones sobre su descripción del "mal de altura." *Histórica* 8(1):1–31.

Bonavia, Duccio, and Carlos Monge C.
1996   Are the South American camelids high-altitude animals? *Newsletter of the International Society for Mountain Medicine* 6(1):13–15.

1997   La altura: Un reto incomprendido. In *Arqueología, Antropología e Historia en los Andes: Homenaje a María Rostworowski*, pp. 259–274. Rafael Varón Gabai and Javier Flores Espinoza, editors. Instituto de Estudios Peruanos, Banco Central de Reserva del Perú, Lima, Peru.

Bonavia, Duccio, and Rogger Ravines
1967   Las fronteras ecológicas de la civilización andina. *Amaru* 2:61–69.

Bonnier, Elisabeth
1986   Utilisation du sol a l'époque préhispanique: Le cas archéologique du Shaka-Palcamayo (Andes centrales). *Cahiers des Sciences Humaines* 22(1):97–113.

Borah, Woodrow
1954   *Early Colonial Trade and Navigation Between Mexico and Peru*. Ibero-Americana 38. University of California Press, Berkeley and Los Angeles.

1975   *Comercio y navegación entre México y Perú en el siglo XVI*. Instituto Mexicano de Comercio Exterior, Mexico City, Mexico.

Borrero, Luis Alberto, Marcela Casiraghi, and Hugo Daniel Yacobaccio
1985   First guanaco-processing site in southern South America. *Current Anthropology* 26(2):273–276.

Brack Egg, Antonio J.
1987   Historia del manejo de las vicuñas en el Perú. *Boletín de Lima*, 9(50):61–76.

Braunitzer, Gerhard
1979   Phosphate-hemoglobin interaction and the regulation of the hypoxic respiration in the human fetus and the llama. *Clinical Enzymology Symposium* 2:69–73.

1980   Perinatale und Hohenamnung: Die Kontrolle Sauerstoffaffinitat der Hämoglobine ser Sauger [Perinatal and high altitude respiration: Control of oxygen affinity of hemoglobin in mammals]. *Verhandlungen der Deutschen Zoologischen Gesellschaft*, pp. 202–213.Gustav Fischer Verlag, Stuttgart, Germany.

Braunitzer, Gerhard, Barbara Schrank, and Anton Stangl
1977a   Die Sequenz der α-Ketten der Hämoglobine des Schweines und des Lamas (Aspekte zur Atmung im Hochland) [The sequence of α-chains from pig and llama hemoglobins (Aspects on the respiration in highlands)].

Hoppe-Seyler's Zeitschrift fuer Physiologische Chemie 358:409–412.

Braunitzer, Gerhard, Barbara Schrank, Anton Stangl, and Christian Bauer
1977b   Die Regulation der Hohenatmung und ihre molekuläre Deutung: Die Sequenz der β-Ketten der Hämoglobine des Schweines und des Lamas [Regulation of respiration at high altitudes and its molecular interpretation: The sequence of β-chains of hemoglobin from pig and llama]. *Hoppe-Seyler's Zeitschrift fuer Physiologische Chemie* 358:921–925.

Bromley, Juan
1955–1956 El Capitán Martín de Estete y Doña María de Escobar "La Romana," fundadores de la Villa de Trujillo del Perú. *Revista Histórica* (Lima) 22:122–141.

Brooks, J. G. III, and Marsh Tenney S.
1968   Ventilatory response of llama to hypoxia at sea level and at high altitude. *Respiration Physiology* (Amsterdam) 5:269–278.

Browman, David L.
1974a   Pastoral nomadism in the Andes. *Current Anthropology* 15(2):188–196.

1990   Llama caravan exchange rates. (Resumen). In *Trabajos presentados al simposio "RUR 6. El pastoreo altoandino: Origen, desarrollo y situación actua,"*pp. 42–43. J. F. Ochoa, coordinator. 46 Congreso Internacional de Americanistas, 4–8 de junio de 1988, Amsterdam. Publicación auspiciada por la Comisión Ejecutiva del 46 Congreso Internacional de Americanistas, en colaboración con el Centro de Estudios Andinos, Cuzco. CEAC, Cuzco, Peru.

Bruhns, Karen Olsen
1988–1989 Prehispanic weaving and spinning implements from southern Ecuador. *The Textile Museum Journal*, pp. 71–77.Washington, DC.

1989   Intercambio entre la costa y la sierra en el Formativo Tardío: Nuevas evidencias del Azuay. In *Proceedings, 46 International Congress of Americanists*, Amsterdam, 1988. "Relaciones interculturales en el área ecuatorial del Pacífico durante la época precolombina," pp. 57–74. J. F. Bouchard and M. Guinea, editors. BAR International Series 503. Oxford, UK.

1991a   Las Culturas Peruanas y el Desarrollo Cultural en los Andes Septentrionales. Paper read at the symposium "Arqueologia y Etnohistoria del Sur de Colombia v Norte del Ecuador," p. 13. 47 International Congress of Americanists, New Orleans. Mimeograph.

1991b Refreshments will be served. . . . Paper read at the April meeting of the Society for American Archaeology, p. 7. Mimeograph.

Bruhns, Karen Olsen, James H. Burton, and George R. Miller
1990 Excavation at Pirincay in the Paute Valley of southern Ecuador, 1985–1988. *Antiquity* 64 (243):221–233.

Bueno, Cosme
1951 *Geografía del Perú Virreinal (Siglos XVIII–XIX)*, pp. 13–140. Lima, Peru.

Buffon, George-Louis de
1830 *Oeuvres complètes, augmentées par M. F. Cuvier.* Didot F. D. Pillot, éditeur. Paris, France.

Bullock, Dillman S.
1958 La agricultura de los Mapuches en tiempos pre-hispánicos. *Boletín de 1a Sociedad de Biología de Concepción* 33:141–154.

Burger, Richard L.
1984a Review of *Excavations at Huacaloma in the Cajamarca Valley, Peru, 1979: Report 2 of the Japanese Scientific Expedition to Nuclear America*, by Kazuo Terada and Yoshio Onuki (University of Tokyo Press, 1982). *American Antiquity* 49(2):430–432.
1984b *The Prehistoric Occupation of Chavín de Huántar, Perú*. Anthropology, Vol. 14. University of California Press, Berkeley and Los Angeles.
1985a Prehistoric stylistic change and cultural development at Huaricoto, Peru. *National Geographic Research* 1(4):505–534.
1985b Concluding remarks: Early Peruvian civilization and its relation to the Chavín Horizon. In *Early Ceremonial Architecture in the Andes*, pp. 269–289. Christopher B. Donnan, editor. Dumbarton Oaks Research Library and Collection, Washington, DC.
1989 El horizonte Chavín: ¿Quimera estilística o metamorfosis socioeconómica? *Revista Andina* 7(2):543–574.
1993a *Emergencia de la civilización en los Andes: Ensayos de interpretación*. Universidad Nacional Mayor de San Marcos, Lima, Peru.
1993b The Chavín Horizon: Stylistic chimera or socioeconomic metamorphosis? In *Latin American Horizons*, pp. 41–82. Don Stephen Rice, editor. Dumbarton Oaks Research Library and Collection, Washington, DC.
1998 *Excavaciones en Chavín de Huántar*. Pontificia Universidad Católica del Perú, Fondo Editorial, Lima, Peru.

Burger, Richard L., and Lucy Salazar-Burger
1980 Ritual and religion at Huaricoto. *Archaeology* 33(6):26–32.

1985 The early ceremonial center of Huaricoto. In *Early Ceremonial Architecture in the Andes*, pp. 111–138. Christopher B. Donnan, editor. Dumbarton Oaks Research Library and Collection, Washington, DC.
1992 La segunda temporada de investigaciones en Cardal, valle de Lurín (1987). In *Estudios de Arqueología Peruana*, pp. 123–147. Duccio Bonavia, editor. Fomciencias, Lima, Peru.

Bushnell, G. H. S.
1963 *Peru*. Thames and Hudson, London, UK.

Bustinza Menendez, Julio A.
1970a Contribución a la diferenciación específica de los Camélidos Sudamericanos. In *Anales de la Primera Convención sobre Camélidos Sudamericanos (Auquénidos)*, pp. 26–28. Programa Académico de Medicina Veterinaria y Zootecnia, Universidad Nacional Técnica del Altiplano, Puno, Peru.
1970b Distribución ecológica de las alpacas en el departamento de Puno. In *Anales de la Primera Convención sobre Camélidos Sudamericanos (Auquénidos)*, pp. 29–32. Programa Académico de Medicina Veterinaria y Zootecnia, Universidad Nacional Técnica del Altiplano, Puno, Peru.
1990 Carga animal en el Altiplano puneño. (Resumen). In *Trabajos presentados al simposio "RUR 6. El pastoreo altoandino: Origen, desarrollo y situación actual,"* pp. 16–17. J. F. Ochoa, coordinator. 46 Congreso Internacional de Americanistas, 4–8 de junio de 1988. Amsterdam. Publicación auspiciada por la Comisión Ejecutiva del 46 Congreso Internacional de Americanistas, en colaboración con el Centro de Estudios Andinos, Cuzco. CEAC, Cuzco, Peru.

Bustios Galvez, Luis (editor)
1956–1966 *La Nueva Crónica y Buen Gobierno descrito por don Felipe Guamán Poma de Ayala*. 3 vols. Vol. I: Editorial Cultura, Arqueología e Historia del Ministerio de Educación Pública (1956); Vols. II and III: Talleres Imprenta Gráfica Industrial (1966), Lima, Peru.

Busto Duthurburu, José Antonio del
1960–1961 Pedro de Candia, Artillero Mayor del Perú. *Revista Histórica* 25:379–405.
1962–1963 La marcha de Francisco Pizarro desde Cajamarca al Cuzco. *Revista Histórica* 26: 146–174.
1963–1970a Viajes del Descubrimiento del Perú. In *Atlas Histórico Geográfico y de Paisajes Peruanos*, p. 53 Carlos Peñaherrera del Aguila, director.

Presidencia de la República. Instituto Nacional de Planificación, Lima, Peru.

1963–1970b Ruta de los Conquistadores. In *Atlas Histórico Geográfico y de Paisajes Peruanos*. Carlos Peñaherrera del Aguila, director. Presidencia de la República. Instituto Nacional de Planificación, Lima, Peru.

1966   *Francisco Pizarro el Marqués Gobernador*. Ediciones Rialp, Madrid, Spain.

1973   Siglo XVI—Historia Externa. *Historia Marítima del Perú*, Tomo III, Vol. 2. Instituto de Estudios Históricos-Marítimos del Perú, Editorial Ausonia, Lima, Peru.

1975   Siglo XVI—Historia Interna. *Historia Marítima del Perú*, Tomo III, Vol. 1. Instituto de Estudios Históricos-Marítimos del Perú, Editorial Ausonia, Lima, Peru.

1978a   *Historia general del Perú. Descubrimiento y Conquista*. Studium. Lima, Peru.

1978b   *Francisco Pizarro: El Marqués Gobernador*. Librería Studium Editor, Lima, Peru.

1988   *La Conquista del Perú*. Librería Studium Editores, Lima, Peru.

Butzer, Karl W.
1991   An Old World perspective on potential mid-Wisconsinan settlement of the Americas. In *The First Americans: Search and Research*, pp. 137–156. T. D. Dillehay and D. J. Meltzer, editors. CRC Press, Boca Raton, FL.

Cabello Valboa, Miguel
1951 (1840)   M*iscelánea Antártica: Una historia del Perú Antiguo*. Universidad Nacional Mayor de San Marcos, Facultad de Letras, Instituto de Etnología, Lima, Peru.

Cabeza de Vaca, Diego, Juan Gutiérrez de Escobar, and Juan Vizcaino
1965 (1881–1897)   Descripción y relación de la ciudad de La Paz. In *Relaciones Geográficas de Indias. Perú. II*. Marcos Jiménez de la Espada. Biblioteca de Autores Españoles desde la formación del lenguaje hasta nuestros días. Vol. I, pp. 342–351. Tomo CLXXXIII. Ediciones Atlas, Madrid, Spain.

Cabrera, Angel
1931   Sobre los cámelidos fósiles y actuales de la América Austral. *Revista del Museo de la Plata* 33:89–117.

Cabrera, Angel, and José Yepes
1960   *Mamíferos Sud Americanos*. Historia Natural Ediar (2 vols.). Ediar S. A. Editores, Buenos Aires, Argentina.

Caillavet, Chantal
1987   Les groupes ethniques préhispaniques selon les sources ethnohistoriques. In *Loja préhis-panique: Recherches archéologiques dans les Andes méridionales de l'Equateur*, pp. 289–307. J. Guffroy, series editor. Syntèse No. 27, Editions Recherche sur les Civilizations, Paris, France.

Cajal, Jorge L., and Silvia Puig
1992   Argentina. *South American Camelids: An Action Plan for Their Conservation / Camélidos Silvestres Sudamericanos: Un plan de acción para su conservación*, pp. 5–9; 37–41. Hernán Torres, compiler and editor. IUCN, Gland, Switzerland.

Calancha, Antonio de la
1975–1976 (1638)   *Coronica Moralizada del Orden de San Agustín en el Perú, con sucesos egenplares en esta Monarquía*. Crónicas del Perú, Vols. II and III. Edición de Ignacio Prado Pastor, Lima, Peru.

Campbell, Kenneth E.
1973   The Pleistocene avi-fauna of Talara tar-seeps, northwestern Peru. Ph.D. dissertation, University of Florida, Gainesville, FL.

Canals Frau, Salvador
1955   *Las Civilizaciones Prehispánicas de América*. Editorial Sudamericana, Buenos Aires, Argentina.

Cantos de Andrada, Rodrigo
1573   Doctrina de Pachacamac. ANSCH, 1573, Varios, Vol. 64. Archivo Nacional de Santiago de Chile, Chile.

Cantos de Andrada, Rodrigo, Garci Núñez Vela, Gaspar de Contreras, and Francisco Caballero
1965 (1881–1897)   Relación de la Villa Rica de Oropesa y minas de Guancavelica. In *Relaciones Geográficas de Indias. Perú. II*. Marcos Jiménez de la Espada. Biblioteca de Autores Españoles desde la formación del lenguaje hasta nuestros días. Vol. I, pp. 303–309. Tomo CLXXXIII, Ediciones Atlas, Madrid, Spain.

Cantu, Francesca
1992   *Coscienza d'America: Cronache di una memoria impossibile*. Edizioni Associate, Rome, Italy.

Cappa, Ricardo
1888 (1850–1887)   *Estudios Críticos acerca de la Dominación Española en América*, Tomos II, III, and IV. Imprenta de Pérez Dubnill, Madrid, Spain.

Caqui, Diego
1981 (1588)   Testamento de Diego Caqui, curaca de Tacna. Archivo Departamental de Tacna, Títulos de la Hacienda Para, Vol. I. Las relaciones entre las tierras altas y la costa del Sur del Perú: Fuentes documentales. Franklin Pease G. Y. Apéndice. In *Estudios*

Etnográficos del Perú Meridional, pp. 209–221. Shozo Ma-suda, editor. University of Tokyo, Tokyo, Japan.

Carabajal, Pedro de

1965 (1881–1897) Descripción fecha de la provincia de Vilcas Guaman el ilustre señor don Pedro de Carabajal, Corregidor y justicia mayor della, ante Xistobal de Camboa, escribano de su juzgado, en el año de 1586. In *Relaciones Geográficas de Indias. Perú. I.* Marcos Jiménez de la Espada. Biblioteca de Autores Españoles desde la formación del lenguaje hasta nuestros días. Vol. I, pp. 205–219. Tomo CLXXXIII, Ediciones Atlas, Madrid, Spain.

Cárdenas, J., J. Rodríguez, and L. Aguirre

1997 El material orgánico en Huaca de la Luna. In *Investigaciones en la Huaca de la Luna 1995*, pp. 129–149. S. Uceda, E. Mujica, and R. Morales, editors. Proyecto Arqueológico Huacas del Sol y de la Luna. Facultad de Ciencias Sociales, Universidad Nacional de La Libertad [sic], Trujillo, Peru.

Cardich, Augusto

1958 Los yacimientos de Lauricocha: Nuevas interpretaciones de la prehistoria Peruana. *Studia Praehistorica* 1:1–65.

1959 Los yacimientos de Lauricocha y la nueva interpretación de la prehistoria Peruana. In *Actas y Trabajos del II Congreso Nacional de Historia del Perú. Epoca Prehispánica*, Vol. 1, pp. 93–106. August 4–9, 1958. Centro de Estudios Histórico Militares del Perú, Lima, Peru.

1960 Investigaciones prehistóricas en los Andes Peruanos. In *Antiguo Perú Espacio y Tiempo*, pp. 89–118. Librería-Editorial Juan Mejía Baca, Lima, Peru.

1964–1966 Lauricocha: Fundamentos para una Prehistoria de los Andes Centrales. *Acta Praehistorica* (Buenos Aires) 8/10 (Part I).

1973 Excavaciones en la caverna de Huargo, Perú. *Revista del Museo Nacional* (Lima) 34: 11–47.

1974 Los yacimientos de la etapa agrícola de Lauricocha, Perú, y los límites superiores del cultivo. *Relaciones* (Buenos Aires) 8 (Nueva Serie):27–48.

1976 Vegetales y recolecta en Lauricocha: Algunas inferencias sobre asentamientos y subsistencias preagrícolas en los Andes Centrales. *Relaciones de la Sociedad Argentina de Antropología* (Buenos Aires) 10 (Nueva Serie):27–41.

1977 Las culturas Pleistocénicas y post-Pleistocénicas de Los Toldos y un bosquejo de la Prehistoria de Sudamérica. *Obra del Cente-* *nario del Museo de la Plata*, Tomo 2, pp. 149–172. La Plata, Argentina.

1980 Origen del hombre y de la cultura andinos. In *Historia del Perú*, Vol. I, *Perú Antiguo*, pp. 29–156. Editorial Juan Mejía Baca, Lima, Peru.

1984 Paleoambientes y la más antigua presencia del hombre. In *Las Culturas de América en la Epoca del descubrimiento*, pp. 13–36. Biblioteca del V Centenario. Ediciones Cultura Hispánica, Madrid, Spain.

1987a Arqueología de Los Toldos y El Ceibo (Provincia de Santa Cruz, Argentina). In *Estudios Atacameños: Investigaciones Paleoindias al Sur de la Línea Ecuatorial*, No. 8 (special issue), pp. 98–117. Lautaro Núñez and Betty Meggers, editors. Instituto de Investigaciones Arqueológicas, Universidad del Norte, San Pedro de Atacama, Chile.

1987b Lauricocha, asentamientos preagrícolas, recolección vegetal e inicios del cultivo altoandino. *Diálogo Andino*, No. 6, pp. 11–28. Departamento de Antropología, Geografía e Historia, Facultad de Estudios Andinos, Universidad de Tarapacá, Arica, Chile.

Cardich, Augusto, Lucio Adolfo Cardich, and Adam Hajduk

1973 Secuencia arqueológica y cronología radiocarbónica de la Cueva 3 de Los Toldos (Santa Cruz, Argentina). *Relaciones de la Sociedad Argentina de Antropología* 7 (Nueva Serie):85–123.

Cardoza, Carmen Rosa

1993 Identificación de los restos animales depositados en la Galería de las Ofrendas de Chavín de Huántar. In *Chavín de Huántar: Excavaciones en la Galería de las Ofrendas*, pp. 371–393. Luis Guillermo Lumbreras. Materialen zur Allgemeinen und Vergleichenden Archäologie, Band 51, KAVA (Kommision für Algemeine und Vergleichende Archäologie). Verlag Philipp von Zabern, Mainz am Rhein, Germany.

Cardozo Gonzales, Armando

1954 *Los Auquénidos*. Editorial Centenario, La Paz, Bolivia.

1974a Los llamingos en el Ecuador. *Desde el Surco* 1(7):10–12.

1974b Preliminares sobre la Presencia y Trascendencia de los Camélidos en el Ecuador. *Boletín de la Academia Nacional de Historia* (Quito) 123(Enero–Julio):134–146.

1975a *Origen y Filogenia de los Camélidos Sudamericanos*. Academia Nacional de Ciencias de Bolivia, La Paz, Bolivia.

Caroll, Robert L.
1988   *Vertebrate Paleontology and Evolution*. W. H.
       Freeman, New York.
Carvajal, Fray Gaspar de
1894 (1504–1584)   *Descubrimiento del Río de las
       Amazonas, según la relación hasta ahora inédita
       de Fr. Gaspar de Carvajal, con otros documentos
       referentes a Francisco de Orellana y sus com-
       pañeros*, pp. 3–278. Imprenta de E. Rasco,
       Bustos Tavera, No. I. Seville, Spain.
1986 (1894)   Relación que escribió Fr. Gaspar de
       Carvajal, fraile de la orden de Santo Domin-
       go de Guzmán, del nuevo descubrimiento
       del famoso río Grande que descubrió por
       muy gran ventura el capitán Francisco de
       Orellana desde su nacimiento hasta salir a la
       mar, con cincuenta y siete hombres que trajo
       consigo y se echo a su ventura por el dicho
       río, y por el nombre del capitán que le des-
       cubrió se llamó el río de Orellana. In G. de
       Carvajal, P. de Almesto y Alonso de Rojas,
       La Aventura del Amazonas, pp. 37–98. Edited
       by Rafael Díaz. Crónicas de América 19,
       Historia 16. Madrid, Spain.
Casamiquela, Rodolfo, and Tom D. Dillehay
1989   Vertebrate and invertebrate faunal analysis.
       In *Monte Verde: A Late Pleistocene Settlement in
       Chile*, Vol. I, *Paleoenvironment and Site
       Context*, pp. 205–210. Tom D. Dillehay, edi-
       tor. Smithsonian Institution Press, Washing-
       ton, DC.
Casanova, Eduardo
1946   The cultures of the puna and the quebrada
       of Humahuaca. In *Handbook of South Ameri-
       can Indians*, Vol. 2, *The Andean Civilizations*,
       pp. 619–631. Julian H. Steward, editor. Bul-
       letin 143, Smithsonian Institution, Bureau of
       American Ethnology, Washington, DC.
Casas, Fray Bartolomé de Las
1948   *De las antiguas gentes del Perú*. Capítulos de
       la "Apologética Historia Sumaria" antes del
       año de 1555. Los Pequeños Grandes Libros
       de Historia Americana, Serie I, Tomo XVI.
       Lima, Peru.
1981 (1875*)   *Historia de las Indias. II*. Biblioteca
       Americana. Fondo de Cultura Económica,
       Mexico City, Mexico. [*According to Porras
       (1986:202) it should be 1876.]
Castillo, Gastón
1983   Sacrificios de Camélidos en la costa de Co-
       quimbo, Chile. *Gaceta Arqueológica Andina*,
       2(7):6–7.
Castillo, Luis Jaime, and Christopher B. Donnan
1993   *La ocupación moche de San José de Moro*. Uni-
       versity of California, Los Angeles. Manuscript.

1994   La ocupación Moche de San José de Moro,
       Jequetepeque. In *Moche: Propuestas y Perspec-
       tivas*, pp. 93–146. Santiago Uceda and Elias
       Mujica, editors. Universidad Nacional de La
       Libertad [sic], Trujillo; Instituto Francés de
       Estudios Andinos; Asociación Peruana para
       el Fomento de las Ciencias Sociales, Lima,
       Peru.
Castro y del Castillo, Antonio de
1906 (1651)   Descripción del Obispado de La Paz,
       hecha de orden de S. M., por el Ilmo. Sr. D.
       Antonio de Castro y del Castillo para la
       obra de D. Gil Gonzáles Dávila, titulada
       "Teatro Eclesiástico de las Iglesias del Perú
       y Nueva España." Juicio de Límites entre el
       Perú y Bolivia. In *Prueba Peruana presentada
       al Gobierno de la República Argentina*, Tomo
       Undécimo, pp. 184–234. Víctor Maurtua.
       Obispado y Audiencia del Cuzco. Imprenta
       de Heinrich y Co., Madrid, Spain.
Chapdelaine, Claude
1998   Excavaciones en la zona urbana de Moche
       durante 1996. In *Investigaciones en la Huaca
       de la Luna 1996*, pp. 85–115. S. Uceda, E.
       Mujica, and R. Morales, editors. Proyecto
       Arqueológico Huacas del Sol y de la Luna.
       Facultad de Ciencias Sociales, Universidad
       Nacional de La Libertad [sic], Trujillo, Peru.
Chapdelaine, Claude, Santiago Uceda, M. Moya, C.
       Jauregui, and Ch. Uceda
1997   Los complejos arquitectónicos urbanos de
       Moche. In *Investigaciones en la Huaca de la
       Luna 1995*, pp. 71–92. S. Uceda, E. Mujica,
       and R. Morales, editors. Proyecto Arque-
       ológico Huacas del Sol y de la Luna. Facul-
       tad de Ciencias Sociales, Universidad Na-
       cional de La Libertad [sic], Trujillo, Peru.
Chauchat, Claude
1987   Niveau marin, écologie et climat sur la côte
       nord du Pérou à la transition Pléistocene-
       Holocène. *Bulletin de l'Institute Français
       d'Études Andines* 16(1–2):21–27.
1988   Early hunter-gatherers on the Peruvian
       coast. In *Peruvian Prehistory*, pp. 41–66.
       R. W. Keatinge, editor. Cambridge Univer-
       sity Press, Cambridge, UK.
Chaumeil, Jean-Pierre, and Josette Fraysse-Chaumeil
1981   "La Canela y el Dorado": Les indigènes du
       Napo et du haut-Amazone. Au XVIe siecle.
       Bulletin de l'Institut Français d'Études Andines
       10(3–4):55–86.
Chimoy Effio, Pedro J.
1985   Aspectos etnobiológicos de los antiguos lam-
       bayecanos. In Eric Mendoza Samillan, Izumi
       Shimada, Walter Alva, James Vreeland, Her-

mes Suárez, Susana de Alva, Santos Llatas, José Maeda, Pedro Chimoy Effio, Cecilia Kámiche, José Arana Cuadra, Enrique Odar, and Alfonso Samamé, *Presencia histórica de Lambayeque*, pp. 161–174. Ediciones y Representaciones H. Falconí e. i. r. l. Editors, Lima, Peru.

Chiodi, Hugo

1962 Oxygen affinity of the hemoglobin of high altitude mammals (letter). *Acta Physiologica Latinoamericana* 12:208–209.

1971 Comparative study of the blood gas transport in high altitude and sea level Camelidae and goats. *Respiration Physiology* 11:84–93.

Church, Warren B.

1991 La ocupación temprana del Gran Pajatén. *Revista del Museo de Arqueología* 2:7–38.

Cieza de León, Pedro de

1881 *Guerras civiles del Perú*, Tomo II, *Guerra de Chupas*. Librería de la Viuda de Rico, Madrid, Spain.

1909 *Guerra de Quito, de. . . .* Edited by M. Serrano y Sanza. Biblioteca de Autores Españoles, Vol. 15. Madrid, Spain.

1941 (1553) *La Crónica del Perú*. Espasa Calpe S.A., Madrid, Spain.

1973 *La Crónica del Perú*. Colección Austral, No. 507. Espasa Calpe S.A., Madrid, Spain.

1984 (1553) *Crónica del Perú. Primera Parte*. Pontificia Universidad Católica del Perú, Academia Nacional de la Historia. Lima, Peru.

1985 (1880) *Crónica del Perú. Segunda Parte*. Pontificia Universidad Católica del Perú, Academia Nacional de la Historia, Lima, Peru.

1987 *Crónica del Perú. Tercera Parte*. Pontificia Universidad Católica del Perú. Academia Nacional de la Historia. Lima, Peru.

Cisneros, Luis Jaime

1994 Homenaje a Alberto Tauro. Discurso del Dr. Luis Jaime Cisneros en el Homenaje que le rindió la Universidad Mayor de San Marcos. *Boletín Informativo del Centenario de José Carlos Mariátegui* (Lima) 2(7):1–2.

Clark, Charles Upson

1948 Prólogo. In *Compendio y descripción de las Indias Occidentales*. Antonio Vázquez de Espinosa. Smithsonian Miscellaneous Collection, Vol. 108, pp. iii–xii. Smithsonian Institution, Washington, DC.

Cobo, P. Bernabé

1964a (1890–1893) *Historia del Nuevo Mundo*. Obras del P. Bernabé Cobo I. Biblioteca de Autores Españoles desde la formación del lenguaje hasta nuestros días. Tomo XCI, Ediciones Atlas, Madrid, Spain.

1964b (1890–1893) *Historia del Nuevo Mundo*. Obras del P. Bernabé Cobo II, pp. 1–275. Biblioteca de Autores Españoles desde la formación del lenguaje hasta nuestros días. Tomo XCII, Ediciones Atlas, Madrid, Spain.

1964c (1890–1893) *Fundación de Lima, escrita por el Padre Bernabé Cobo de la Compañía de Jesús. Año de 1639*. Obras del P. Bernabé Cobo II, pp. 277–460. Biblioteca de Autores Españoles desde la formación del lenguaje hasta nuestros días. Tomo XCII, Ediciones Atlas, Madrid, Spain.

Cock, Guillermo A.

1986 Power and wealth in the Jequetepeque Valley during the sixteenth century. In *The Pacatnamu Papers*, Vol. 1, pp. 171–180. Christopher B. Donnan and Guillermo A. Cock, editors. Museum of Cultural History, University of California, Los Angeles.

Cohen, David (director-editor)

1994 *A Day in the Life of Israel*. Collins Publishers, San Francisco, CA.

Cohen, Mark Nathan

1978a *The Food Crisis in Prehistory: Overpopulation and the Origins of Agriculture*. Yale University Press, New Haven, CT.

1978b Population pressure and the origins of agriculture: An archaeological example from the coast of Peru. In *Advances in Andean Archaeology*, pp. 91–132. David L. Browman, editor. Mouton, The Hague.

1978c Archaeological plant remains from the Central Coast of Peru. *Ñawpa Pacha* 16:23–50.

Colon, Cristóbal

1492–1493 *Diario de Colón (o Diario de Abordo)*. Fols. 1– 65. Gráficas Yagües, Madrid, Spain. [Author's Note: The original manuscript has not been preserved, but we have the extract made by F. Bartolomé de las Casas, who frequently copied the text literally.]

1956 (15 de Febrero–14 de marzo, 1493) *La Carta de Colón*, pp. 15–22. Reproducción del texto original español, Pedro Posa, 1493, con notas críticas. Impreso en Barcelona, Madrid, Spain.

1958 (1493) *La Carta de Colón, anunciando la llegada a las Indias y a la provincia de Catayo (China). (Descubrimiento de América)*, pp. 1–8. Sacada de la edición de Valladolid: Caracteres de Pedro Giraldi y Miguel de Planes, 1497. Publicada y Comentada por Carlos Sanz, Madrid, Spain.

Collier, Donald

1946 The archaeology of Ecuador. In *Handbook of South American Indians*, Vol. 2, *The Andean Civilizations*, pp. 767–784. Julian H. Steward,

editor. Bulletin 143, Smithsonian Institution, Bureau of American Ethnology, Washington, DC.

Collier, Stephen, and J. Peter White
1976   Get them young? Age and sex inferences on
animal domestication in archaeology. *American Antiquity* 41(1):96–102.

Collina-Girard, Jacques, Jean-Luc Guadelli, and
Pierre Usselmann
1991   Mammifère disparus et premières occupations humaines: L'exemple nord péruvien du
désert de Cupisnique. In *Actes du 116e Congrès
National des Sociétés Savantes*, pp. 111–132.
Chambery, Déserts, PICG 252. Paris, France.

Colque Guarache, Juan
1981 (1575) Primera información hecha por Don
Juan Colque Guarache, cerca de sus predecesores y subcesión en el cacicazgo mayor
de los Quillacas, Asanaques, Sivaroyos,
Uruquillas y Haracapis, y de sus seruicios a
fauor de Su Majestad en la conquista, allanamiento y pacificación deste reino del Piru.
Año 1575. El Reyno Aymara de Quillaca-
Asanaque, siglos XV y XVI, Waldemar Espinoza Soriano. *Revista del Museo Nacional*
(Lima) 45:237–268.

Compagnoni, Bruno, and Maurizio Tosi
1978   The camel: Its distribution and state of do-
mestication in the Middle East during the
third millennium B.C. in light of finds from
Shahr-I Sokhta. In *Approaches to Faunal
Analysis in the Middle East*. Richard H.
Meadow and Melinda A. Zeder, editors.
Peabody Museum Bulletin 2:91–103. Peabody
Museum of Archaeology and Ethnology,
Harvard University, Cambridge, MA.

Conapariguana, Francisco
1972   Visita del repartimiento de Juan Sánchez.
*Visita de la provincia de León de Huánuco en
1562*. Iñigo Ortiz de Zúñiga, visitador.
Tomo II, pp. 24–33. Universidad Nacional
Hermilio Valdizán, Huánuco, Peru.

Concha Contreras, Juan de Dios
1975   Relación entre pastores y agricultores. *Allpanchis Phuturinga* 8:67–101.

Conrad, Geoffrey W.
1982   The burial platforms of Chan Chan: Some
social and political implications. In *Chan
Chan: Andean Desert City*, pp. 87–117. Michael
E. Moseley and Kent C. Day, editors. A
School of American Research Book. University of New Mexico Press, Albuquerque, NM.

Contreras y Valverde, Vasco
1965 (1881–1897) Relación de la ciudad del Cuzco,
de su fundación, descripción, vidas de los

obispos, religiones, y de todo lo demás
perteneciente a eclesiastico desde el descubrimiento de este reyno hasta el tiempo
presente. Fecha por orden de Su Magestad,
por el doctor Don Vasco de Contreras y
Valverde, Dean de la cathedral, consultor del
Santo Oficio, Comisario Apostólico, Subdel-
egado de la Santa Cruzada, Gobernador,
Provisor y Vicario General de su Obispado.
Al Rey Nuestro Señor: en su real Consejo de
Indias. In *Relaciones Geográficas de Indias, Perú.
II.* Marcos Jiménez de la Espada. Biblioteca
de Autores Españoles desde la formación del
lenguaje hasta nuestros días. Vol. II, pp. 1–15.
Tomo CLXXXIV, Ediciones Atlas, Madrid,
Spain.

Cook, O. F.
1925   Peru as a center of domestication. *Journal of
Heredity* 16:32–46, 94–110. [Author's Note:
There is a Spanish translation: El Perú
como centro de domesticación de plantas y
animales. Servicio de traducciones del
Museo Nacional, No. 1, 1937. Imprenta del
Museo Nacional, Lima, Peru.]

Cooper, John M.
1917   *Analytical and Critical Bibliography of the Tribes
of Tierra del Fuego and Adjacent Territory*. Bulletin 63, Bureau of American Ethnology,
Smithsonian Institution, Washington, DC.
1946a   The Patagonian and pampean hunters. In
Handbook of South American Indians, Vol. 1,
The Marginal Tribes, pp. 127–168. Julian H.
Steward, editor. Bulletin 143, Bureau of
American Ethnology, Smithsonian Institution, Washington, DC.
1946b   The Araucanians. In *Handbook of South
American Indians*, Vol. 2, *The Andean Civilizations*, pp. 687–760. Julian H. Steward, editor. Bulletin 143, Bureau of American Eth-
nology, Smithsonian Institution,
Washington, DC.

Cordy-Collins, Alana
1997   The Offering Room group. In *The Pacatnamú Papers*, Vol. 2, *The Moche Occupation*,
pp. 283–292. Christopher B. Donnan and
Guillermo A. Cook, editors. Fowler Museum of Cultural History, University of California, Los Angeles.

Cornejo, Miguel, and Jane C. Wheeler
1986   Identificación del material óseo del sitio Gran
Pajatén. Informe final 1985. Investigaciones
sobre los recursos culturales en el Parque Nacional del Río Abiseo. Apéndice E. (Unpagi-
nated; p. 3.) Center for Andean Studies, University of Colorado, Boulder, CO.

Covarrubias, Sebastián de
1943 *Tesoro de la Lengua Castellana o Española*. Según la Impresión de 1611 en las ediciones de Benito Noytens, publicada 1674. Edición preparada por Martín de Riquer. Barcelona, Spain.

Craig, Alan K.
1985 Cis-Andean environmental transects: Late Quaternary ecology of northern and southern Peru. In *Andean Ecology and Civilizations: An Interdisciplinary Perspective in Andean Ecological Complementarity*, pp. 23–44. Shozo Masuda, Izumi Shimada, and Craig Morris, editors. University of Tokyo Press, Tokyo, Japan.

Crosby Jr., Alfred W.
1972 *The Columbian Exchange: Biological and Cultural Consequences of 1492*. Greenwood Press, Westport, CT.

Cuatrecasas, Jose
1968 Paramo vegetation and its life forms. In *Geoecology of the Mountainous Regions of the Tropical Americas/Geo-Ecología de las Regiones Montañosas de las Américas Tropicales*, pp. 163–186. Carl Troll, editor. Colloquium Geographicum, Geographischen Institut der Universität Bonn. Ferd Dümmlers Verlag, Bonn, Germany.

Cunazza P., Claudio
1976a Enfermedades y parásitos del guanaco (Informe preliminar). In *El guanaco de Magallanes, Chile: Distribución y biología*, pp. 151–165. Kenneth J. Raedeke M. Publication No. 4, Corporación Nacional Forestal de Chile, Santiago, Chile.
1976b Rendimiento de carne en el guanaco. In *El guanaco de Magallanes, Chile: Distribución y biología*, pp. 166–174. Kenneth J. Raedeke M. Publication No. 4, Corporación Nacional Forestal de Chile, Santiago, Chile.

Cusichaca, Francisco, Diego Eneupari, and Cristóbal Canchaya
1972 (1561) Probanza de servicios fecha en la Real Audiencia que por mandado de Su Majestad reside en esta Ciudad de los Reyes destos reinos e prouincia del Pirú, a pedimento de Don Francisco Cusichaca e Don Diego Eneupari e Don Cristóbal Canchaya, Caciques del repartimiento de Atunxauxa, de lo que a Su Majestad han servido en el tiempo de las alteraciones causadas en estos reinos y conquistas y descubrimientos dellos. Lima, 5 de septiembre–13 de octubre de 1561. Los Huancas, aliados de la conquista. Tres informaciones inéditas sobre la participación indígena en la conquista del Perú. 1558–1560–1561. Waldemar Espinoza Soriano. *Anales Científicos de la Universidad del Centro del Perú* (Huancayo) l:260–387.

Custred, Glynn
1969 J. J. Von Tschudi y la etnohistoria peruana. Prólogo a la traducción de "La Llama" de Von Tschudi. In *Mesa Redonda de Ciencias Prehistóricas y Antropológicas*, Tomo I, pp. 120–123. Pontificia Universidad Católica del Perú. Instituto Riva-Agüero, Seminario de Antropología, Lima, Peru.
1974 Llameros y comercio interregional. In *Reciprocidad e intercambio en los Andes Peruanos*, pp. 252–289. Giorgio Alberti and Enrique Mayer, compilers. Perú Problema 12. Instituto de Estudios Peruanos, Lima, Peru.
1977 Las punas de los Andes Centrales. In *Pastores de puna: Uywamichiqpunarunakuna*, pp. 55–85. Jorge A. Flores Ochoa, compiler. Instituto de Estudios Peruanos, Lima, Peru.
1979 Hunting technologies in Andean culture. *Journal de la Société des Américanistes* (Paris) 66:7–19.

Dale, William E., and José Luis Venero
1977 Insectos y ácaros ectoparásitos de la vicuña en Pampa Galeras, Ayacucho. *Revista Peruana de Entomología* (Homenaje a la Universidad Nacional Agraria) (Lima) 20(1):93–99.

Darling, F. Fraser
1956 Man's ecological dominance through domesticated animals on wild lands. In *Man's Role in Changing the Face of the Earth*, pp. 778–787. William L. Thomas, Jr., editor, with the collaboration of Carl O. Sauer, Marston Bates, and Lewis Mumford. University of Chicago Press, Chicago, IL.

Darwin, Charles
1921 *Diario del viaje de un naturalista alrededor del mundo en el navio de S. M. "Beagle."* Tomo I. Viajes Clásicos. Espasa-Calpe S.A., Madrid, Spain.
1969 *The Voyage of the Beagle*. The Harvard Classics. Charles W. Elliot, series editor. P. F. Collier & Son, New York.

Dávila, Diego
1984 Indice del protocolo del escribano Diego Dávila (Siglo XVI). In *Contribuciones a los estudios de los Andes Centrales*, pp. 177–383. University of Tokyo, Tokyo, Japan.

Dávila Brizeño, Diego
1965 (1881–1897) Descripción y relación de la provincia de Yauyos toda, Anan Yauyos y Lorin Yauyos, hecha por Diego Davila Brizeño, Corregidor de Guarocheri. In

Relaciones Geográficas de Indias. Peru. I. Marcos Jiménez de la Espada. Biblioteca de Autores Españoles desde la formación del lenguaje hasta nuestros días. Vol. I, pp. 155–165. Tomo CLXXXIII, Ediciones Atlas, Madrid, Spain.

Dedenbach Salazar Saenz, Sabine
1990 *Inka pachaq llamanpa willaynin. Uso y crianza de los Camélidos en la Época Incaica.* Bonner Amerikanistische Studien, BAS 16. Estudios Americanistas de Bonn, Bonn, Germany.
1993 The Andean classification of camelids according to 16th and 17th century sources. In *European Symposium on South American Camelids*, pp. 67–73. Martina Gerken and C. Renieri, editors. Rheinische Friedrich-Wilhelms-Universität Bonn, Università degli Studi di Camerino, Facoltà di Medicina Veterinaria, Matelica, Italy.

Denegri Luna, Félix
1980 Prólogo. In *Noticias Cronológicas de la gran Ciudad del Cuzco*. Diego de Esquivel y Navia. Tomo I, pp. IX–LX. Fundación Augusto N. Wiese, Banco Wiese Ltdo., Lima, Peru.

Denevan, William M. (editor)
1988 The Cultural Ecology, Archaelogy, and History of Terracing and Terrace Abandonment in the Colca Valley of Southern Peru, Vol. II. Technical Report to the National Science Foundation and the National Geographic Society. Madison, WI.

De Niro, Michael J.
1988 Marine food sources for prehistoric coastal Peruvian camelids: Isotopic evidence and implications. In *Economic Prehistory of the Central Andes*, pp. 119–128. Elizabeth S. Wing and Jane C. Wheeler, editors. BAR International Series 427. Oxford, UK.

Diaz Maderuelo, Rafael
1986 Introducción. In *La aventura del Amazonas*, pp. 7–36. G. de Carvajal, P. de Almesto, and Alonso de Rojas, editors. Edición de Rafael Díaz. Crónicas de América 19, Historia 16. Madrid, Spain. [Author's Note: All footnotes that appear in the text are also his.]

Diez de San Miguel, Garci
1964 (1567) *Visita hecha a la provincia de Chuquito por . . . en el año 1567.* Tomo I, pp. 1–287. Documentos Regionales para la Etnología y Etnohistoria Andinas. Ediciones de la Casa de la Cultura del Perú, Lima, Peru.

Dillehay, Tom D.
1979 Pre-hispanic resource sharing in the Central Andes. *Science* 204 (4388):24–31.
1987 Estrategias políticas y económicas de las etnias locales del Valle del Chillón durante el

período prehispánico. *Revista Andina* (Cuzco) 5(2):407–456.

Dillehay, Tom D., Gerardo Ardila Calderón, Gustavo Politis, and María da Conceicao Morales Coutinho Beltrao
1992 Earliest hunters and gatherers of South America. *Journal of World Prehistory* 6(2): 145–204.

Director de Chimor (Castro Burga A., Julián)
1959, 1960, 1961 Sondeos y desescombro de la "Huaca Pelada." *Chimor*, Años VII, VIII, and IX, Número único, pp. 10–31. Trujillo, Peru.

Dittmer, Kunz
1960 *Etnología General: Formas y evolución de la cultura*. Fondo de Cultura Económica, Mexico City, Mexico.

Dollfus, Olivier, and Danièle Lavallée
1973 Ecología y ocupación del espacio en los Andes tropicales durante los últimos veinte milenios. *Bulletin de l'Institut Français d'Études Andines* (Lima) 2(3):75–92.

Donnan, Christopher B.
1964 An early house from Chilca, Peru. *American Antiquity* 30(2, Pt. l):137–144.
1973 *Moche Occupation of the Santa Valley, Peru.* University of California Publications in Anthropology 8. University of California Press, Berkeley and Los Angeles.
1976 *Moche Art and Iconography.* UCLA Latin American Publications. University of California, Los Angeles.
1978 (1976, 1979) *Moche Art of Peru: Pre-Columbian Symbolic Communication.* Museum of Cultural History, University of California, Los Angeles. [Author's Note: The first edition was published in 1976. A second, revised edition was prepared in 1978 and was reprinted in 1979.]
1986 An elaborate textile fragment from the major quadrangle. In *The Pacatnamu Papers*, Vol. 1, pp. 109–116. C. B. Donnan and G. A. Cock, editors. Museum of Cultural History, University of California, Los Angeles.

Donnan, Christopher B., and Guillermo A. Cock
1984 Primer informe parcial. Proyecto Pacatnamú.
1985 Segundo informe parcial. Proyecto Pacatnamú.
1986 Tercer informe parcial. Proyecto Pacatnamú.

Donnan, Christopher B., and Leonard J. Foote
1978 Child and llama burials from Huanchaco. In Christopher B. Donnan and Carol J. Mackey, *Ancient Burial Patterns of the Moche Valley, Peru*, Appendix 2, pp. 399–408. University of Texas Press, Austin, TX.

Donnan, Christopher B., and Carol. J. Mackey
1978 *Ancient Burial Patterns of the Moche Valley, Peru.* University of Texas Press, Austin, TX.

Doucet, Gastón Gabriel
1993 Acerca de los Churumatas, con particular referencia al antiguo Tucumán. *Histórica* (Lima) 17(1):21–91.

Dourojeanni, Marc
1973 La vicuña. *El Serrano* (Lima) XXII (278): 9–15.

Dransart, Penny Z.
1991a Llamas, herders and the exploitation of raw materials in the Atacama Desert. *World Archaeology* 22(3):304–319.

Dunin-Borkowski A., Cristina
1990 *Gallina araucana prehispánica: ¿Mito o realidad?* Serie, Publicación de Tesis. Asociación Peruana para el Fomento de las Ciencias Sociales, Consejo Nacional de Ciencia y Tecnología, Lima, Peru.

Editores, Los
1968 *Biblioteca Peruana. Primera Serie. Tomo II.* Editores Técnicos Asociados, Lima, Peru.

Eliot, John L.
1994 With llamas on guard sheep may safely graze. *National Geographic* 186(3):133.

Elmo, Anton
1991 Testamento de Anton Elmo, yanacona natural del valle de Chao: 1565. In *Etnohistoria del Area Virú-Santa: Un avance documental (Siglos XVI–XIX). Documentos para la Etnohistoria de la Zona Norte 1,* pp. 19–20. Instituto Departamental de Cultura, La Libertad, Dirección de Patrimonio Cultural Monumental de la Nación. Proyecto Especial de Irrigación Chavimochic. Trujillo, Peru.

Enciclopedia de los Animales
1970 *Los Camélidos.* No. 56, pp. 289–308. Editorial Noguer, Barcelona, Spain.

Engel, Frédéric André
1957 Sites et établissements sans céramique de la côte péruvienne. *Journal de la Société des Américanistes,* n. s. 46:67–155.

1960 Un group humaine datant de 5000 ans à Paracas, Pérou. *Journal de la Sociéte des Américanistes,* n.s. 46:7–35.

1963 *A Preceramic Settlement on the Central Coast of Peru. Asia, Unit 1.* The American Philosophical Society, Philadelphia, PA.

1964 El Precerámico sin algodón en la costa del Perú. In *Actas y Memorias del XXXV Congreso Internacional de Americanistas.* Mexico, 1962. Tomo 3, pp. 141–152. Mexico City, Mexico.

1966a *Geografía humana prehistórica y agricultura precolombina de la Quebrada de Chilca. I.* Universidad Agraria [sic], Oficina de Promoción y Desarrollo, Departamento de Publicaciones, Lima, Peru.

1966b Le complexe précéramique d'El Paraiso (Pérou). *Journal de la Société des Américanistes* 55–1:43–95.

1967 El Complejo El Paraíso en el valle del Chillón, habitado hace 3,500 años: Nuevos aspectos de la civilización de los agricultores del pallar. *Anales Científicos de la Universidad Agraria* [sic] 5(3–4):241–280.

1970a Explorations of the Chilca Canyon, Peru. *Current Anthropology* 11(1):55–58.

1970b La Grotte du Mégathérium à Chilca et les écologies du haut-holocène péruvien. In *Echange et Communications: Mélanges offerts à Claude Lévi-Strauss,* pp. 413–436. Mouton, Paris, France.

1970c Sobre los complejos agrícolas prehispánicos encontrados en las lomas y en los arenales costeños de la cuenca de Chilca. Perú. Paper presented at the XXXIX Congreso Internacional de Americanistas Lima, Peru (p. 37). Mimeograph.

1970d *Las lomas de Iguanil y el Complejo de Haldas.* Departamento de Publicaciones, Universidad Nacional Agraria, La Molina, Lima, Peru.

(Editor)
1981 *Prehistoric Andean Ecology: Man, Settlement and Environment in the Andes. The Deep South.* Recopilación de los Archivos del Centro de Investigaciones de Zonas Aridas "CIZA" de la Universidad Nacional Agraria del Perú. Distributed by Humanities Press for the Department of Anthropology, Hunter College. City University of New York, New York.

Engelhardt, W. V., A. M. Abbas, H. M. Mousa, and M. Lechner-Doll
1992 Comparative digestive physiology of the forestomach in camelids. In *Proceedings of the First International Camel Conference, 2nd–6th February 1992,* pp. 263–270. W. R. Allen, A. J. Higgins, I. G. Mayhew, D. H. Snow, and F. Wade, editors. R.&W. Publications (Newmarket) Ltd., Amersham, Buckinghamshire, UK.

Enríquez de Guzmán, Alonso
1960 *Libro de la vida y costumbres de Don Alonso Enríquez de Guzmán,* pp. 8–361. Biblioteca de Autores Españoles desde los orígenes del lenguaje hasta nuestros días. Tomo CXXVI, Ediciones Atlas, Madrid, Spain.

Escalante, José A.
1981 Ectoparásitos de animales domésticos en el Cusco. *Revista Peruana de Entomología* 24(1): 123–125.

Espasa Calpe S.A.
n.d. *Enciclopedia Universal Ilustrada Europea Americana,* Tomo XXIX. Madrid, Spain.

Espinoza Soriano, Waldemar
 1967   *El primer informe etnológico sobre Cajamarca. Año de 1540*. Separata de la *Revista Peruana de Cultura*, Nos. 11–12. Lima, Peru.
 1971   Geografía histórica de Huamachuco: Creación del Corregimiento, su demarcación política, eclesiástica y económica, 1759–1821. *Historia y Cultura* (Lima) 5:5–96.
 1972   Los Huancas, aliados de la conquista: Tres informaciones inéditas sobre la participación indígena en la conquista del Perú. 1558–1560–1561. *Anales Científicos de la Universidad del Centro del Perú* (Huancayo) 1:5–407. [Author's Note: The date is 1972 on the front cover and 1971 on the back cover.]
 1973a   Los grupos étnicos en la cuenca del Chuquimayo, siglos XV y XVI. *Bulletin de l'Institut Français d'Études Andines* (Lima) 2(3):19–73.
 1973b   Historia del departamento de Junín. In *Enciclopedia Departamental de Junín*, Tomo I. Huancayo, Peru.
 1974   Los señoríos étnicos del valle de Condebamba y provincia de Cajabamba: Historia de las huarancas de Llucho y Mitmas. Siglos XV–XX. *Anales Científicos de la Universidad del Centro del Perú* (Huancayo) 3:5–371.
 1975a   El Valle de Jayanca y el Reino de los Mochica. Siglos XV y XVI. *Bulletin de l'Institut Français d'Études Andines* (Lima) 4(3–4): 243–274. [Author's Note: The *visita* was published without any comment under the title "Visita hecha en el valle de Jayanca (Trujillo) por Sebastián de la Gama" in *Historia y Cultura* 8:215–228.]
 1975b   Ichoc-Huánuco el señorío del curaca Huanca en el reino de Huánuco. Siglos XV y XVI. Una visita inédita de 1549 para la etnohistoria andina. *Anales Científicos de la Universidad de1 Centro del Perú* (Huancayo) 4:5–70.
 1976–1977 La pachaca de Pariamarca en el reino de Caxamarca, siglos XV–XVIII. *Historia y Cultura* (Lima) 10:135–180.
 1981   El Reino Aymara de Quillaca-Asanaque, siglos XV y XVI. *Revista del Museo Nacional* (Lima) 45:175–274.
 1983–1984 Los señoríos de Yaucha y Icoy en el abra del medio y alto Rímac. *Revista Histórica* (Lima) 34:157–279.
 1985–1986 Los Churumatas y los mitmas chichas orejones en los lindes del Collasuyo siglos XV–XX. *Revista Histórica* (Lima) 35:243–298.
 1987–1989 Migraciones internas en el reino Colla: Tejedores, plumereros y alfareros del Estado Inca. *Revista Histórica* (Lima) 36:209–305.

Estete, Miguel de
 1968a (1534)   La relación del viaje que hizo el señor capitán Hernando Pizarro por mandado del señor Gobernador, su hermano, desde el pueblo de Caxamalca a Parcama y de allí a Jauja. Verdadera relación de la conquista del Perú y provincia del Cuzco llamada la Nueva Castilla. Francisco de Xerez. In *Biblioteca Peruana*, Tomo I, pp. 242–257. Editores Técnicos Asociados, Lima, Peru.
Estete, Miguel de (Anonymous?)
 1968b (1918) Noticia del Perú. In *Biblioteca Peruana*, Tomo I, pp. 345–402. Editores Técnicos Asociados, Lima, Peru.
Estrada, Emilio
 1957a   *Los Huancavilcas: Últimas civilizaciones Pre-Históricas de la Costa del Guayas*. Publicación del Museo Víctor Emilio Estrada, No. 3. Guayaquil, Ecuador.
Estrada Ycaza, Julio
 1987   *Andanzas de Cieza por tierras americanas*. Banco Central del Ecuador, Archivo Histórico del Guayas, Guayaquil, Ecuador.
Estruch, Jaime
 1943   Osteología comparada de los auchenidos peruanos. *Revista de la Universidad de Arequipa* 15(18):113–157.
Falcón, Francisco (Licenciado)
 1918   Representación hecha por el Licenciado Falcón en Concilio Provincial, sobre los daños y molestias que se hacen a los indios. In *Informaciones acerca de la Religión y Gobierno de los Incas*. I. Relación de idolatrías en Huamachuco por los primeros Agustinos. II. Relacion de idolatrías en Huarochirí por el R. P. Francisco Dávila. III. Relación sobre el Gobierno de los Incas por el Licenciado Falcón. Colección de Libros y Documentos referentes a la Historia del Perú. Tomo XI, pp. 133–176. Imprenta y Librería Sanmarti y Cía, Lima, Peru.
Faust, Ernest Carroll, Paul Farr Russell, and Rodney Clifton Jung
 1979   *Craig y Faust: Parasitología clínica*. Salvat Mexicana de ediciones, de C. V. Mexico.
Felipe II
 1977 (1564) Cédula Real a don Felipe Guacrapáucar, sobre los ganados del Valle de Jauja. Los trece privilegios, 1555–1564. Los Huancas, aliados de la conquista. Tres informaciones inéditas sobre la participación indígena en la conquista del Perú. 1559–1560–1561. Waldemar Espinoza Soriano. *Anales Científicos de la Universidad de1 Centro del Perú* (Huancayo) 1:393.

Fernández, Diego (vecino de Palencia)

1963a *Primera y Segunda Parte de la Historia del Perú*. Biblioteca de Autores Españoles desde la formación del lenguaje hasta nuestros días. Crónicas del Perú I, pp. CXIII + 398. Tomo CLIV, Ediciones Atlas, Madrid, Spain.

1963b *Historia del Perú*. Biblioteca de Autores Españoles desde la formación del lenguaje hasta nuestros días. Crónicas del Perú II, pp. 1–131. Tomo CLV, Ediciones Atlas, Madrid, Spain.

Fernández, Jorge

1983–1985 Cronología y paleoambiente del intervalo 12.550–520 A.P. (Pleistoceno-Holoceno) de la puna jujeña: Los derrames petrolíferos de barro negro y su contenido arqueológico y paleofaunístico (insectos, pájaros y mamíferos). *Anales de Arqueología y Etnología* (Universidad Nacional de Cuyo, Facultad de Filosofia y Letras, Mendoza) 38–40 (Pt. 1): 29– 42.

Fernández Baca A., Saúl

1971 La alpaca: Reproducción y crianza. *Boletín de divulgación* (Instituto Veterinario de Investigaciones Tropicales y de Altura, Universidad Nacional Mayor de San Marcos, Lima) 7: 1–43.

Fernández de Oviedo [y Valdéz], Gonzalo

1959a (1535) *Historia General y Natural de las Indias. II*. Biblioteca de Autores Españoles desde la formación del lenguaje hasta nuestros días. Edición y estudio preliminar de Juan Pérez de Tudela Bueso. Tomo CXVIII, Ediciones Atlas, Madrid, Spain.

1959b *Historia General y Natural de las Indias. III*. Biblioteca de Autores Españoles desde la formación del lenguaje hasta nuestros días. Edición y estudio preliminar de Juan Pérez de Tudela Bueso. Tomo CXIX, Ediciones Atlas, Madrid, Spain.

1959c *Historia General y Natural de las Indias. V*. Biblioteca de Autores Españoles desde la formación del lenguaje hasta nuestros días. Edición y estudio preliminar de Juan Pérez de Tudela Bueso. Tomo CXXI, Ediciones Atlas, Madrid, Spain.

Fernández Distel, Alicia A.

1974 Excavaciones arqueológicas en las cuevas de Huachichocana, Dep. de Tumbaya, prov. de Jujuy, Argentina. *Relaciones* (Buenos Aires) 8 (n.s.):101–134.

1975 Restos vegetales de etapas arcaicas en yacimientos del N. O. de la República Argentina (Pcia. de Jujuy). *Etnia* 22 (artículo 86):11– 24.

1986 La Cueva de Huachichocana, su posición dentro del Precerámico con agricultura incipiente del Noroeste argentino. In *Beiträge zur Allgemeinen und Vergleichenden Archäeologie*, Band 8, pp. 353– 430. Verlag Phillip von Zabern, Mainz am Rhein, Germany.

Ferreyrra, Ramón

1987 *Estudio sistemático de los algarrobos de la costa Norte del Perú*. Dirección de Investigación Forestal y de Fauna, Lima, Peru.

Flannery, Kent V., Joyce Marcus, and Robert G. Reynolds

1989 *The Flocks of the Wamani. A Study of Llama Herders on the Punas of Ayacucho, Peru*. Academic Press, New York.

Flores Ochoa, Jorge A.

1967 Los pastores de Paratía: Una introducción a su estudio. *Anales del Instituto de Estudios Socio-Económicos* (Universidad Técnica de Altiplano, Puno) 1(1):9–106.

1968 *Los pastores de Paratía: Una introducción a su estudio*. Instituto Indigenista Interamericano, series Antropología Social, 10. Mexico City, Mexico. [Author's Note: There is a 1979(b) English edition.]

1970 Notas sobre rebaños en la visita de Gutiérrez Flores. *Historia y Cultura* (Lima) 4:63–70.

1974–1976 Enqa, enqaychu, illa y khuya rumi. Aspectos mágico-religiosos entre pastores. *Journal de la Société des Américanistes* (Paris) 63:245–262.

1975a Pastores de alpacas. *Allpanchis Phuturinqa* (Cuzco) 8:5–23.

1975b Sociedad y cultura en la puna alta de los Andes. *América Indígena* (Mexico City) 35 (2):297–319.

1976 Pastores de la puna Andina. *Revista Universitaria* (Universidad Nacional San Antonio Abad del Cuzco) 63(130):56–83.

(Compiler)

1977a *Pastores de puna: Uywamichiq Punarunakuna*. Instituto de Estudios Peruanos, Lima, Peru.

1977b Pastores de alpacas de los Andes. In *Pastores de Puna*, pp. 15–49.

1977c Aspectos mágicos del pastoreo: *Enqa, enqaychu, illa y khuya rumi*. In *Pastores de Puna*, pp. 211–237.

1977d Pastoreo, tejido e intercambio. In *Pastores de Puna*, pp. 133–154.

1978 Classification et dénomination des camélidés sud-américaines. *Anales* (Paris) 33(5–6): 1006–1016.

1979a Desarrollo de las culturas humanas en las altas montañas tropicales (estrategias adaptativas). In *El Medio Ambiente, Páramo*, pp. 225–234. M. L. Salgado, editor. CEA-IVIC-UNESCO, CIFCA, Caracas, Venezuela.

1979b *Pastoralists of the Andes: The Alpaca Herders of Paratía*. Institute for the Study of Human Issues, Philadelphia, PA.

1982   Causas que originaron la actual distribución espacial de las alpacas y llamas. In *El Hombre y su Ambiente en los Andes Centrales*, pp. 63– 92. Luis Millones and Hiroyasu Tomoeda, editors. Senri Ethnological Studies, No. 10. National Museum of Ethnology, Osaka, Japan. [Author's Note: It was published again in 1984 (but the publication reads 1980) with the same title and text, but with the addition of a few bibliographical references, in *Revista del Museo e Instituto de Arqueología* (Universidad Nacional de San Antonio Abad del Cuzco) 23:223–250.]

1983   Pastoreo de llamas y alpacas en los Andes: Balance bibliográfico. *Revista Andina* (Cuzco) 1(1):175–218.

1987   Los corrales del ganado del Sol. *Kuntur. Perú en la Cultura*, No. 5, pp. 2–10. May–June. Lima, Peru.

(Compiler)

1988a *Llamichos y paqocheros: Pastores de llamas y alpacas*. CONCYTEC, CEAC y Editorial UNSAAC, Cuzco, Peru.

1988b Cambios en la puna. In *Llamichos y paqocheros*, pp. 273–293.

1988c Mitos y canciones ceremoniales en comunidades de puna. In *Llamichos y paqocheros*, pp. 237–251.

(Coordinator)

1990a Trabajos presentados al Simposio "RUR 6. El pastoreo altoandino: Orígen, desarrollo y situación actual." 46 Congreso Internacional de Americanistas, 4–8 de Junio de 1988, Amsterdam. Publicación auspiciada por la Comisión Ejecutiva del 46 Congreso Internacional de Americanistas en colaboración con el Centro de Estudios Andinos, Cuzco. CEAC, Cuzco, Peru.

1990b Posibilidades del pastoreo altoandino. Trabajos presentados al Simposio "RUR 6. El pastoreo altoandino: Orígen, desarrollo y situación actual," pp. 84–98. 46 Congreso Internacional de Americanistas, 4–8 de Junio de 1988, Amsterdam. Publicación auspiciada por la Comisión Ejecutiva del 46 Congreso Internacional de Americanistas en colaboración con el Centro de Estudios Andinos, Cuzco. CEAC, Cuzco, Peru.

Flores Ochoa, Jorge, Kim Mac Quarrie, and Javier Portus

1994   *Gold of the Andes. The Llamas, Alpacas, Vicuñas and Guanacos of South America*, Vol. I. Julio Soto, Impresores S.A., Barcelona, Spain.

Flores Ochoa, Jorge A., and Félix Palacios Rios

1978   La protesta de 1901: Un movimiento de pastores de la puna alta a comienzos del siglo XX. *Actes du XLIIe Congrès International des Américanistes*, Vol. III, pp. 83–94. Paris, France.

Fonseca Z., Oscar, and James B. Richardson III

1978   South American and Mayan cultural contacts at the Las Huacas site, Costa Rica. *Annals of the Carnegie Museum* 47, Article 13, pp. 299–317. Pittsburgh, PA.

Fornee, Niculoso de

1965 (1881–1897) Santiago Hamancay. Santa Catalina de Curavaci. Sant Pedro de Sayvita. Descripción de la tierra del Corregimiento de Abancay, de que es Corregidor Niculoso de Fornee. In *Relaciones Geográficas de Indias. Perú. II*. Marcos Jiménez de la Espada. Biblioteca de Autores Españoles desde la formación del lenguaje hasta nuestros días. Vol. II, pp. 26–29. Tomo CLXXXIV, Ediciones Atlas, Madrid, Spain.

Fornee, Niculoso de, Juan de Luque, Pedro de Placencia, Francisco Gallegos, and Martín de Cevereche

1965 (1881–1897) Zurite. Guarocondor. Anta, Puquibra. Descripción de la tierra del Corregimiento de Abancay, de que es Corregidor Niculoso de Fornee. In *Relaciones Geográficas de Indias. Perú. II*. Marcos Jiménez de la Espada. Biblioteca de Autores Españoles desde la formación del lenguaje hasta nuestros días. Vol. II, pp. 16–20. Tomo CLXXXIV, Ediciones Atlas, Madrid, Spain.

Fornee, Niculoso de, Francisco de Murcia, Martín de Salazar, Gerónimo Quipquin, and Martín de Cevereche

1965 (1881–1897) San Antón de Chincaypuquio. La Visitación de Nuestra Señora de Zumaro. La Encarnación Pantipata. Santiago Pivil. Descripción de la tierra del Corregimiento de Abancay, de que es Corregidor Niculoso de Fornee. In *Relaciones Geográficas de Indias. Perú. II*. Marcos Jiménez de la Espada. Biblioteca de Autores Españoles desde la formación del lenguaje hasta nuestros días. Vol. II, pp. 20–23. Tomo CLXXXIV, Ediciones Atlas, Madrid, Spain.

Fornee, Niculoso de, Alonso Vaez, Juan Velazquez, and Martín de Cevereche

1965 (1881–1897) Sant Sebastian Pampaconga. Sant Juan de Patallata. Santa Ana Chonta. Santiago Mollepata. Descripción de la tierra del Corregimiento de Abancay, de que es Corregidor Niculoso de Fornee. In *Relaciones Geográficas de Indias. Perú. II*. Marcos

Jiménez de la Espada. Biblioteca de Autores Españoles desde la formación del lenguaje hasta nuestros días. Vol. II, pp. 23–26. Tomo CLXXXIV, Ediciones Atlas, Madrid, Spain.

Frank, Eduardo N., and V. E. Wehbe

1994 Primer informe de avance del componente "camélidos domésticos." Acuerdo República Argentina-Unión Europea para el Programa de Apoyo para la Mejora en la producción de Pelos Finos de Camélidos Argentinos. Córdoba, Argentina.

Franklin, William L.

1973 High, wild world of the vicuña. *National Geographic* 143(1):76–91.

1975 Guanacos in Peru. *Oryx* 13(2):191–202.

1981 Living with guanacos: Wild camels of South America. *National Geographic* 160(1): 62–75.

1982 Biology, ecology, and relationship to man of the South American camelids. In *Mammalian Biology in South America*. Michael A. Mares and Hugh H. Genoways, editors. The Pymatuning Symposia in Ecology, Vol. 6, pp. 457–489. Special Publication Series, Pymatuning Laboratory of Ecology, University of Pittsburgh, Linesville, PA.

1983 Contrasting socioecologies of South America's wild camelids: The vicuña and the guanaco. In *Advances in the Study of Mammalian Behavior*, pp. 573–629. John F. Eisenberg and Devra G. Kleiman, editors. Special Publication 7. The American Society of Mammologists, Stillwater, OK.

Frézier, Amédée François

1716 *Relation du voyage de la mer du Sud aux côtes du Chili, du Pérou et du Brésil, fait pendant les années 1712, 1713 & 1714*. Paris, France.

FUCASUD. Pro Lama et Vicugna (Fundación Camélidos Sudamericanos).

1994 *Excerta Anatomica Camelidae. I.* Jorge M. Galotta and Silvia G. Márquez, compilers. Talleres Gráficos de Weben S.A., Buenos Aires, Argentina.

1995 *Excerta Camelidae. II.* J. M. Galotta, C. M. Nuevo, and M. D. Ghezzi, editors. Quetal S.A., Buenos Aires, Argentina.

1996 *Excerta Camelidae. III.* J. M. Galotta, C. M. Nuevo, and M. D. Ghezzi, editors. Quetal S.A., Buenos Aires, Argentina.

Fuji, Tatsuhiko, and Hiroyasu Tomoeda

1981 Chacra, Laime y Auquénidos, pp. 33–63. *Estudios Etnográficos del Perú Meridional*. Shozo Masuda, editor. University of Tokyo, Tokyo, Japan.

Fung, Rosa

1969 Las Aldas: Su ubicación dentro del proceso histórico del Perú antiguo. Dédalo. *Revista de Arte e Arqueología* (Museu de Arte e Arqueología, Universidade de Sao Paulo) 5 (9–10):5–208

Futuyma, Douglas J.

1979 *Evolutionary Biology*. Sinauer Associates, Sunderland, MA.

Gade, Daniel W.

1969 The llama, alpaca and vicuña: Fact vs. fiction. *Journal of Geography* 68(6):339–343.

1977 Llama, alpaca y vicuña: Ficción y realidad. In *Pastores de puna: Uywamichiq punarunakuna*, pp. 113–120. Jorge Flores Ochoa, compiler. Instituto de Estudios Peruanos, Lima, Peru. [Author's Note: This is a Spanish translation of the 1969 article, with no changes.]

1993 Leche y civilización andina: En torno a la ausencia del ordeño de la llama y alpaca. *Yearbook, Conference of Latin Americanist Geographers*, Vol. 19, pp. 3–14. [Author's Note: An English version of this article was included as Chapter 5, The Andes as a Dairyless Civilization, pp. 102–117, in his book, *Nature and Culture in the Andes* (1999, University of Wisconsin Press, Madison).]

García Cook, Angel

1974 El origen del sedentarismo en el área de Ayacucho, Perú. *Boletín*, INAH, Epoca 2 (11):15–30.

1981 The stratigraphy of Jaywamachay. Ac 335. In Richard S. MacNeish, Angel García Cook, Luis G. Lumbreras, Robert K. Vierra, and Antoinette Nelken-Terner, *Prehistory of the Ayacucho Basin, Peru*, Vol. II, *Excavations and Chronology*, pp. 57–79. University of Michigan Press, Ann Arbor.

Garcilaso Inca de la Vega

1959 *Comentarios Reales de los Incas*. 3 vols. Universidad Nacional Mayor de San Marcos, Lima, Peru.

1966 *Royal Commentaries of the Incas and General History of Peru [1604]. Part One.* H. V. Livermore, translator. University of Texas Press, Austin, TX.

Gasca, Pedro de La. *See* La Gasca, Pedro de

Gasparini, Graziano, and Luise Margolies

1977 *Arquitectura Inka*. Centro de Investigaciones Históricas y Estéticas, Facultad de Arquitectura y Urbanismo, Universidad Central de Venezuela, Caracas, Venezuela.

Gastiaburu, Teresa

1979 *El arte y la vida de Vicús. Colección del Banco Popular del Perú*. Lima, Peru.

Gaviria, Martín de
1965   Sancto Domingo de Chunchi. Relación que
        enbio a mandar Su Magestad se hiziese des-
        ta ciudad de Cuenca y de toda su provincia.
        In *Relaciones Geográficas de Indias. Perú. III.*
        Marcos Jiménez de la Espada. Biblioteca de
        Autores Españoles desde la formación del
        lenguaje hasta nuestros días. Vol. II, pp.
        285–287. Tomo CLXXXIV, Ediciones Atlas,
        Madrid, Spain.

Geismar, Joan H., and Sydne B. Marshall,
1973   Fauna of the 1952–1953 Ica-Nazca Expedi-
        tion. Anthro. G9493x. Mimeograph.

Gil, Juan
1989   *Mitos y utopías del descubrimiento. III. El Do-
        rado.* Alianza Editorial, Madrid, Spain.

Gilmore, Raymond M.
1950   Fauna and Ethnozoology of South America.
        In *Handbook of South American Indians*, Vol.
        6, *Physical Anthropology, Linguistics and Cul-
        tural Geography of South American Indians*,
        pp. 345–464. Julian H. Steward, editor. Bul-
        letin 143, Bureau of American Ethnology,
        Smithsonian Institution, Washington, DC.

Glade, Alfonso, and Claudio Cunazza
1992   Chile. In *South American Camelids. Action
        Plan for Their Conservation / Camélidos Sil-
        vestres Sudamericanos. Un Plan de Acción para
        su Conservación*, pp. 14–18, 46–50. Hernán
        Torres, compiler-editor. IUCN, Gland,
        Switzerland.

Goldstein, Paul S.
1990   La cultura Tiwanaku y las relaciones de sus
        fases cerámicas en Moquegua. In *Trabajos
        Arqueológicos en Moquegua, Perú*, Vol. 2, pp.
        31–58. L. K. Watanabe, M. E. Moseley, and
        F. Cabieses, compilers. Programa Contisuyo
        del Museo Peruano de Ciencias de la Salud,
        Southern Peru Copper Corporation, Lima,
        Peru.

Gómara, Francisco López de
1946 (1552) *Hispania Vitrix: Primera y Segunda Parte
        de la Historia General de las Indias.* Biblioteca
        de Autores Españoles desde la formación del
        lenguaje hasta nuestros días. Historiadores
        primitivos de Indias, Vol. I, pp. 157–294.
        Tomo XXII, Ediciones Atlas, Madrid, Spain.

Gómez, Juan, Juan Velez Benavente, and Alvaro Gar-
        cía de Balcazar
1965 (1881–1897)   Canaribamba. Relación que
        enbio a mandar Su Magestad se hiziese
        desta ciudad de Cuenca y de toda provincia.
        In *Relaciones Geográficas de Indias. Perú. III.*
        Marcos Jiménez de la Espada. Biblioteca de
        Autores Españoles desde la formación del

lenguaje hasta nuestros días. Vol. II, pp.
        281–285. Tomo CLXXXIV, Ediciones Atlas,
        Madrid, Spain.

Gomez-Tabanera, José Manuel
1988   Presentación de circunstancias a las llamadas
        "Noticias Secretas de América de Jorge Juan
        y Antonio de Ulloa." In *Noticias Secretas de
        América*, pp. v–xxv. Jorge Juan y Antonio de
        Ulloa. Colegio Universitario, Ediciones
        Istmo, Mundus Novus 9, Madrid, Spain.

Gonçalez Holguin, Diego
1952   *Vocabulario de la Lengva General de todo el
        Perv llamada Lengua Qquichua o del Inca.*
        Edición del Instituto de Historia, Universi-
        dad Nacional Mayor de San Marcos, Lima,
        Peru.

Gonzáles de Cuenca, Gregorio
Ms.   (1566) Archivo de Indias, Sevilla (AGI). Jus-
        ticia. Leg. 457, F.

González, Alberto Rex
1960   La estratigrafía de la Gruta de Intihuasi
        (Prov. de San Luis, R.A.) y sus relaciones con
        otros sitios precerámicos de Sudamérica. *Re-
        vista del Instituto de Antropología* 1:5–296.

Greenfield, Haskel J.
1988   The origin of milk and wool production in
        the Old World. *Current Anthropology* 29(4):
        573–587.

Grimwood, I. R.
1969   *Notes on the Distribution and Status of Some
        Peruvian Mammals. 1968.* American Com-
        mitee for International Wild Life Protection
        and New York Zoological Society, Special
        Publication 21. New York.

Grossman, Joel W.
1983   Demographic change and economic trans-
        formation in the south-central highlands of
        pre-Huari Peru. *Ñawpa Pacha* 21:45–126.

Guacorapacora (Guacrapaucar), Jerónimo
1972 (1558)   Memorias de los auxilios proporciona-
        dos por las tres parcialidades de Lurin-
        guanca, Ananguanca y Xauxa, la primera
        desde la llegada de Pizarro a Caxamarca y
        las dos segundas durante la pacificación de
        Hernández Girón, Lima, 1558. Los Huan-
        cas, aliados de la conquista. Tres informa-
        ciones inéditas sobre la participación indí-
        gena en la conquista del Perú.
        1558–1560–1561. Waldemar Espinoza Sori-
        ano. *Anales Científicos de la Universidad del
        Centro del Perú* 1:201–259.

Guaman Poma de Ayala, Phelipe
1936   *Nueva Corónica y Buen Gobierno.* Travaux et
        Mémoires de 1'lnstitut d'Ethnologie, Vol.
        XXIII. Institut d'Ethnologie, Paris, France.

Guaman Poma de Ayala, Felipe
1969 *Nueva Crónica y Buen Gobierno. Selección.* Casa de la Cultura del Perú, Lima, Peru.

Guffroy, Jean
1987 Les debuts de la sedentarisation et de l'agriculture dans les Andes Meridionales de l'Equateur. *L'Anthropologie* (Paris) 91(4): 873– 888.
1992 Las tradiciones culturales formativas en el alto Piura. In *Estudios de Arqueología Peruana*, pp. 99–121. Duccio Bonavia, editor. Fomciencias, Lima, Peru.

Guillen Guillen, Edmundo
1978 El testimonio inca de la conquista del Perú. *Bulletin de l'Institut Français d'Études Andines* (Lima) 7(3–4):33–57.

Gundermann K., Hans
1988 Ganadería aymara, ecología y forrajes. (Chile). In *Llamichos y paqocheros: Pastores de llamas y alpacas*, pp. 101–112. Jorge Flores Ochoa, compiler. Centro de Estudios Andinos, Cuzco, Peru.

Gurmendi, Jorge
1966 Fragilidad osmótica y mecánica en hematíes de auquénidos. *Archivos del Instituto de Biología Andina* (Lima) l(5):299–304.

Gutierrez de Santa Clara, Pedro
1963 *Quinquenarios o Historia de las Guerras Civiles del Perú.* Biblioteca de Autores Españoles desde la formación del lenguaje hasta nuestros días. Crónicas del Perú, III. Tomo CLXVI, Ediciones Atlas, Madrid, Spain.

Gutierrez Flores, Fray Pedro
1964 (1574) *Padron de los mil indios ricos de la provincia de Chucuito y de los pueblos, parcialidades y ayllos que son y la cantidad de ganado de la tierra que cada uno tiene.* Documentos Regionales para la Etnología y Etnohistoria Andinas, Tomo I, pp. 301–363. Adjunto a la Visita hecha a la Provincia de Chucuito por Garci Diez de San Miguel en el año de 1567. Ediciones de la Casa de la Cultura del Perú, Lima, Peru.

Gutierrez Flores, Pedro, and Juan Ramirez Segarra
1970 (1572) Resultas de la visita secreta lega que hizieron en la Provincia de Chucuito. . . . Documentos sobre Chucuito. *Historia y Cultura* (Lima) 4:5–48.

Guzman Ladron de Guevara, Carlos, and José Casafranca Noriega
1964 *Vicús. Informaciones Arqueológicas* No. 1. Ediciones de la Comisión Nacional de Cultura, Lima, Peru. [Author's Note: Few copies of this publication appeared, for the edition was withdrawn and destroyed by the Instituto Nacional de Cultura.]

Hall, Frank G.
1936 The effect of altitude on the affinity of hemoglobin for oxygen. *Journal of Biological Chemistry* 115:485–490.
1937 Adaptations of mammals to high altitude. *Journal of Mammalogy* 18:468–472.

Hall, Frank G., David B. Dill, and E. S. Guzmán Barrón
1936 Comparative physiology in high altitude. *Journal of Cellular Physiology* 8:301–313.

Haas, Jonathan
1985 Excavations on Huaca Grande: An initial view of the elite of Pampa Grande, Peru. *Journal of Field Archaeology* 12:391–409.

Hamann Carrillo, Sara María
1955–1956 Auquénidos del Perú para Madame Bonaparte. *Revista Histórica* (Lima) 22:216–221.

Hammond, Norman, and Karen Olsen Bruhns
1987 The Paute Valley Project in Ecuador, 1984. *Antiquity* 61(231):50–56.

Hansen, Barbara C. S., H. E. Wright, Jr., and J. P. Bradbury
1984 Pollen studies in the Junin area, central Peruvian Andes. *Geological Society of America Bulletin* 95:1454–1465.

Harcourt, Raoul D'
1950 L'impératrice Joséphine et les lamas. *Journal de la Société des Américanistes* (Paris) 39:259–262.
1962 *Textiles of Ancient Peru and Their Techniques.* University of Washington Press, Seattle, WA.

Harris, Olivia
1985 Ecological duality and the role of the center: Northern Potosi. In *Andean Ecology and Civilization. An Interdisciplinary Perspective on Andean Ecological Complementarity*, pp. 311– 335. Shozo Masuda, Izumi Shimada, and Craig Morris, editors. University of Tokyo Press, Tokyo, Japan.

Harris, Peter, Donald Heath, Paul Smith, David R.William, A. Ramirez, Hever Kruger, and D. M. Jones
1982 Pulmonary circulation of the llama at high and low altitudes. *Thorax* 37:38–45.

Hastenrath, Stefan L.
1967 Observations on the snow line in the Peruvian Andes. *Journal of Glaciology* 6(46):541–550.

Heath, Donald, Yolanda Castillo, Javier Arias-Stella, and Peter Harris
1969 The small pulmonary arteries of the llama and other domestic animals native to high altitude. *Cardiovascular Research* 3:75–78.

Heath, Donald, Peter Harris, Yolanda Castillo, and Javier Arias-Stella
1968   Histology, extensibility and chemical composition of the pulmonary trunk of dogs, sheep, cattle and llamas living at high altitude. *Journal of Pathology and Bacteriology* 96: 161–167.

Heath, Donald, Paul Smith, and Peter Harris
1976   Clear cells in the llama. *Experimental Cell Biology* 44:73–82.

Heath, Donald, Paul Smith, David R.Williams, Peter Harris, Javier Arias-Stella, and Hever Kruger
1974   The heart and pulmonary vasculature of the llama (*Lama glama*). *Thorax* 29:463–471.

Helmer, Marie
1956   Petite histoire de la Malmaison. *Journal de la Société des Américanistes* (Paris) 45:240–241.

Hemmer, Helmut
1976   Zum Problem der Herkunft des Alpakas (*Lama* sp. f. *pacos*). (The problem of the origin of the alpaca [*Lama* sp. f. *pacos*]). In *Säugetierkundliche Mitteilungen* (Mammalogical Information), Vol. 24, No. 3, pp. 193–200. (Edited by Arbeits Gemeinschaft für Säugetierforschung e. v. an der Zoologischen Staatssammlung). Munich, Germany.

Hemming, Francis, and Diana Noakes (editors)
1958a   *Official List of Works Approved as Available for Zoological Nomenclature*. First installment: Names 1–38. Printed by order of the International Trust for Zoological Nomenclature, London, UK.
1958b   *Official Index of Rejected and Invalid Works in Zoological Nomenclature*. Printed by order of the International Trust for Zoological Nomenclature, London, UK.

Hernández Sánchez-Barba, Mario
1968   Estudio preliminar. In *Descripción breve de toda la tierra del Perú, Tucumán, Río de La Plata y Chile*, pp. v–xvi. Reginaldo de Lizarraga. Biblioteca de Autores Españoles desde la formación del lenguaje hasta nuestros días. Tomo CCXVI, Ediciones Atlas, Madrid, Spain.

Herre, Wolf
1953a   Studien am Skelett des Mittelohres wilder und domestizierter Formen der Gattung *Lama* Frisch. *Acta Anatomica* 19:271–289.
1968   Zur Geschichte des Vorkolumbischen Haustiere Amerikas. In *El proyecto México de la Fundación Alemana para la Investigación Científica. I. Informe sobre los trabajos iniciados y proyectados*, pp. 90–97. Z. von Veselowsky and H. Dathe, editors. Franz Steinel Verlag, Wiesbaden, Germany.
1969   Los animales domésticos autóctonos sudamericanos y sus relaciones con el hombre. In *Mesa Redonda de Ciencias Prehistóricas y Antropológicas*, Vol. I, pp. 116–119. Pontificia Universidad Católica del Perú, Instituto Riva-Agüero, Seminario de Antropología, Lima, Peru.

Herre, Wolf, and M. Rohrs
1977   Zoological considerations on the origins of farming and domestication. In *Origins of Agriculture*, pp. 245–279. Charles A. Reed, editor. Mouton, The Hague and Paris.

Herrera [y Tordesillas], Antonio de
1945a (1601–1615) *Historia General de los hechos de los Castellanos, en las Islas, y Tierra-firme de el Mar Oceano*. Tomo II. Editorial Guarania, Buenos Aires, Argentina.
1945b   *Historia General de los hechos de los Castellanos, en las Islas, y Tierra-firme de el Mar Oceano*. Tomo V. Editorial Guarania, Buenos Aires, Argentina.
1945c (1738) *Historia General de los hechos de los Castellanos, en las Islas, y Tierra-firme de el Mar Oceano*. Tomo VI. Editorial Guarania, Buenos Aires, Argentina.
1945d   *Historia General de los hechos de los Castellanos, en las Islas, y Tierra-firme de el Mar Oceano*. Tomo VII. Editorial Guarania, Buenos Aires, Argentina.
1946a   *Historia General de los hechos de los Castellanos, en las Islas, y Tierra-firme de el Mar Oceano*. Tomo VIII. Editorial Guarania, Buenos Aires, Argentina.
1946b   *Historia General de los hechos de los Castellanos, en las Islas, y Tierra-firme de el Mar Oceano*. Tomo IX. Editorial Guarania, Buenos Aires, Argentina.
1947 (1738) *Historia General de los hechos de los Castellanos, en las Islas, y Tierra-firme de el Mar Oceano*. Tomo X. Editorial Guarania, Buenos Aires, Argentina.

Herrero, Miguel
1940   Las viñas y los vinos del Perú. *Revista de Indias* 1(1):111–116.

Hershkovitz, Philip
1969   The evolution of mammals on southern continents. VI. The recent mammals of Neotropical Region: A zoogeographic review. *Quaternary Review of Biology* 44(1):1–70.

Hesse, Brian
1980   Archaeological Evidence for Muscovy Duck in Ecuador. *Current Anthropology* 21(1):139–140.
1981   The association of animal bones with burial features. In *The Ayalán Cemetery. A Late Integration Period Burial on the South Coast of*

Ecuador, Appendix I, pp. 134–138. Douglas H. Ubelaker, editor. Smithsonian Contributions to Anthropology, No. 29. Smithsonian Institution Press, Washington, DC.

1982a Archaeological evidence for camelid exploitation in the Chilean Andes. *Säugetierkundliche Mitteilungen* 30(3):201–211.

Heucke, J., B. Dörges, and H. Klingel

1992 Ecology of feral camels in central Australia. In *Proceedings of the First International Camel Conference. 2nd–6th February 1992*, pp. 313–316. Edited by W. R. Allen, A. J. Higgins, I. G. Mayhew, D. H. Snow, and F. Wade, editors. R.&W. Publications (Newmarket), Amersham, Buckinghamshire, UK.

Hick, Michel H. V., and Eduardo Frank

1996 Preliminary results from introduction of South American domestic camelids (SADC) to Argentine Patagonia. In *Proceedings, 2nd European Symposium on South American Camelids*, pp. 203–213. Martina Gerken and C. Renieri, Universität Göttingen, Università degli Studi di Camerino, Matelica, Italy.

Hoces, Domingo

1992 Perú. In *South American Camelids. Action Plan for Their Conservation / Camélidos Silvestres Sudamericanos. Un Plan de Acción para su conservación*, pp. 19–22, 51–54. Hernán Torres, compiler-editor. IUCN, Gland, Switzerland.

Hochachka, Peter W., T. P. Momsen, James H. Jones, and C. Richard Taylor

1987 Substrate and O$_2$ fluxes during rest and exercise in a high-altitude-adapted animal, the llama. *American Journal of Physiology* 253: R298–R305.

Hocquenghem, Anne Marie

1987 *Iconografía Mochica*. Pontificia Universidad Católica del Perú, Fondo Editorial, Lima, Peru.

n.d. [1989?] *Los Guayacundos de Caxas y la Sierra Piurana. Siglos XV y XVI*. CIPCA (Centro de Investigación y Promoción del Campesinado), IFEA (Instituto Francés de Estudios Andinos). Tomo XLVIII, Travaux de l'IFEA, Lima, Peru.

Hocquenghem, Anne Marie, and Luc Ortlieb

1990 Pizarre n'est pas arrivé au Pérou durant une année El Niño. *Bulletin de l'Institut Français d'Études Andines* (Lima) 19(2):327–334.

1992 Eventos El Niño y lluvias anormales en la costa del Perú: Siglos XVI–XIX. *Bulletin de l'Institut Français d'Études Andines* (Paris) 21 (1):197–278.

Hoebel, E. Adamson

1961 *El hombre en el mundo primitivo*. Ediciones Omega, Barcelona, Spain.

Hoffstetter, Robert

1952 *Les Mammifères Pléistocènes de la République de l'Ecuateur*. Mémoires de la Société Géologique de France, Nouvelle Serie, Tome XXXI, Fasc. 1–4, Feuilles 1–49, Pl. I–VIII. Mémoire No. 66. Paris, France.

Hofmann, Rudolf K., Kai-Ch. Otte, Carlos F. Ponce del Prado, and M. A. Rios

1983 *El Manejo de la Vicuña Silvestre*, Vols. 1 and 2. Deutsche Gesellschaft für Technische Zusammenarbeit, Eschborn, Germany.

Holdridge, Leslie R.

1967 *Life Zone Ecology*. Tropical Science Center, San José, Costa Rica.

Holm, Olaf

1985 *Cultura Manteña-Huancavilca*. Museo Antropológico y Pinacoteca and Banco Central del Ecuador, Guayaquil. [Author's Note: This is a second edition. The first edition is identical and was published in 1982.]

Horkheimer, Hans

1958 *La alimentación en el Perú prehispánico y su interdependencia con la agricultura*. UNESCO, Programa de Estudios de la Zona Arida Peruana, Lima, Peru. Mimeograph.

1960 *Nahrung und Nahrungsgewinnung im vorspanischen Peru*. Bibliotheca Ibero-Americana. Colloquium Verlag, Berlin, Germany.

1961 *La Cultura Mochica*. Peruano Suiza, Lima, Peru.

1973 *Alimentación y obtención de alimentos en el Perú prehispánico*. Comentarios del Perú, 13. Universidad Nacional Mayor de San Marcos, Dirección Universitaria de Biblioteca y Publicaciones, Lima, Peru.

Huaman Poma de Ayala, Felipe. *See* Guaman Poma de Ayala, Phelipe

Humboldt, Alexander von

1849 *Ansichten der Natur, mit wissenschaftlichen Erlaeuterungen*. 2 vols. Cotta, Stuttgart and Tübingen, Germany.

1876 *Cuadros de la naturaleza*. Imprenta y Librería de Gaspar, Madrid, Spain. [Author's Note: "Los auquénidos" are discussed on pp. 172–174. *See* Humboldt 1971.]

1971 (1876) Los auquénidos. In *El Perú en la obra de Alejandro de Humboldt*, pp. 145–146. E. Núñez y G. Petersen, editors. Librería Studium, Lima, Peru.

Hurtado de Mendoza Avendaño, Andrés

1975 (1557) Provisión del gobierno superior del 20 de septiembre de 1557 para que por tiempo de

5 años no se hagan chacos de guanacos, vicuñas en los parajes que en ella se expresan. Apéndice documental. Economía y Ritual en los Condesuyos de Arequipa: Pastores y Tejedores del Siglo XIX. Luis Millones Santa Gadea. *Allpanchis Phuturinqa* (Cuzco) 8:65–66.

Hurtado de Mendoza, Andrés (Marqués de Cañete)
  1978 (1593) Carta a S. M. In *Los sucesos de Toledo en las dos últimas décadas del siglo XVI. El Servicio personal de los indios en el Perú (extractos del siglo XVI)*. Silvio Zavala. Tomo I, pp. 183–184. El Colegio de México, Mexico City, Mexico.

Hurtado de Mendoza, Luis
  1987 Cazadores de las Punas de Junín y Cerro de Pasco. Perú. *Estudios Atacameños*, No. 8. *Investigaciones Paleoindias al Sur de la línea ecuatorial*, pp. 198–243. Special issue, Lautaro Núñez and Betty Meggers, editors. Instituto de Investigaciones Arqueológicas, Universidad del Norte, San Pedro de Atacama, Chile.

Hurtado de Mendoza, Luis, and Carlos Chaud Gutiérrez
  1984 Algunos datos adicionales acerca del sitio de Callavallauri (Abrigo Rocoso No. 1 de Tschopik). *Arqueología y Sociedad* (Museo de Arqueología y Etnología, Universidad Nacional Mayor de San Marcos, Lima) 10:32–62.

Hyslop, John
  1984 *The Inka Road System*. Academic Press, New York.

Imbelloni, José
  1959 I cacciatori australi e i marginali del Pacifico. In *Le Razze e i Popoli della Terra*, Vol. IV, pp. 678–702. Renato Biasutti. Unione Tipografico-Editrice Torinese, Turin, Italy.

Inamura, Tetsuya
  1981a Adaptación ambiental de los pastores altoandinos en el Sur del Perú. In *Estudios Etnográficos del Perú Meridional*, pp. 65–83. S. Masuda, editor. University of Tokyo Press, Tokyo, Japan.

Instituto de Arte Peruano
  1938 *Muestrario de Arte Peruano Precolombino. I. Cerámica*. Publicaciones del Museo Nacional de Lima, Lima, Peru.

Irving, Washington
  1854 *Viajes y Descubrimientos de los Compañeros de Colón*. Imprenta de Gaspar y Roig, Madrid, Spain.

Isbell, William H.
  1986 Early ceremonial monuments of the Andes. A review essay by. . . . *Archaeoastronomy* 9:134–156.

Istituto Geografico De Agostini
  1993 *Calendario Atlante De Agostini*. Officine Grafiche de Agostini, Novara, Italy.

Italiano, Hernando
  1965 (1881–1897) Alusi. Relación que enbio a mandar Su Magestad se hiziese desta ciudad de Cuenca y de toda provincia. In *Relaciones Geográficas de Indias. Perú. III*. Marcos Jiménez de la Espada. Biblioteca de Autores Españoles desde la formación del lenguaje hasta nuestros días. Vol. II, pp. 287–289. Tomo CLX–XXIV, Ediciones Atlas, Madrid, Spain.

Izumi, Seiichi, and Toshihiko Sono
  1963 *Andes 2. Excavations at Kotosh, Peru. 1960*. Kadokawa Publishing Co., Tokyo, Japan.

Izumi, Seiichi, and Kazuo Terada
  1972 *Andes 4. Excavations at Kotosh, Peru. 1963 and 1966*. University of Tokyo Press, Tokyo, Japan.

Jacobs, Melville, and Bernhard J. Stern
  1962 *General Anthropology*. Barnes & Noble, New York.

Jain, N. C., and K. S. Keeton
  1974 Morphology of camel and llama erythrocytes as viewed with the scanning electron microscope. *British Veterinary Journal* 130:288–291.

Jensen, Peter M.
  1974 Un nuevo punto de vista sobre el problema de la adaptación a las grandes alturas en los Andes Peruanos. *Relaciones* (Sociedad Argentina de Antropología, Buenos Aires) 8:11–25.

Jensen, Peter M., and Robert R. Kautz
  1974 Preceramic transhumance and Andean food production. *Economic Botany* 28(1):43–55.

Jerez, Francisco de. *See* Xerez, Francisco de

Jessup, David
  1990 Rescate arqueológico en el Museo de Sitio de San Gerónimo, Ilo. In *Trabajos Arqueológicos en Moquegua, Perú*, vol. 3, pp. 151–165. L. K. Watanabe, M. E. Moseley, and F. Cabieses, compilers. Programa Contisuyo del Museo Peruano de Ciencias de la Salud, Southern Peru Copper Corporation, Lima, Peru.

Jiménez de la Espada, Marcos
  1895 La jornada del capitán Alonso Mercadillo a los indios Chupaichos e Iscaicingas. *Boletín de la Sociedad Geográfica de Madrid* 37:197–230.

  1938 (Note.) La Cibdad de Sant Francisco del Quito. In *Quito a través de los siglos*, p. 9. Recopilación y notas bio-bibliográficas de Eliecer Enríquez B. Imprenta Municipal, Quito, Ecuador.

  1965a *Relaciones Geográficas de Indias*. Biblioteca de Autores Españoles desde la formación del lenguaje hasta nuestros días. Tomo CLXXXIII, Ediciones Atlas, Madrid, Spain. [Author's Note: I believe that this citation is incorrect. Miller and Gill (1990) give this reference and I include it to avoid confusing the reader.]

1965b Relaciones Geográficas de Indias. In *Relaciones Geográficas de Indias, Perú. I.* Biblioteca de Autores Españoles desde la formación del lenguaje hasta nuestros días. Vol. I, pp. 3–117. Tomo CLXXXIII, Ediciones Atlas, Madrid, Spain.

1965c *Relaciones Geográficas de Indias. Perú. IV.* Biblioteca de Autores Españoles desde la formación del lenguaje hasta nuestros días. Vol. III. Tomo CLXXXV, Ediciones Atlas, Madrid, Spain.

Johnson, George R.
1930 *Peru from the Air.* American Geographical Society Special Publications, No. 12. New York.

Jones, John G.
1990 Results of the analysis of palynological investigation of camelid feces from Peru. Manuscript. Department of Anthropology, Texas A&M University, College Station.

Jones, John G., and Duccio Bonavia
1992 Análisis de coprolitos de llama (*Lama glama*) del Precerámico tardío de la Costa Nor-central del Perú. *Bulletin de l'Institut Français d'Études Andines* (Lima) 21(3):835–852.

Juan, Jorge, and Antonio de Ulloa
1988 (1826) *Noticias Secretas de América.* Colegio Universitario, Ediciones Istmo. Mundus Novus 9. Madrid, Spain. [Author's Note: This is a facsimile edition of the first edition, which was published in 1826.]

Juicio de Límites entre Perú y Bolivia
1906 *Prueba peruana presentada al Gobierno de la República Argentina* por Víctor M. Maúrtua. 15 vols. Barcelona, Spain.

Julien, Catherine J.
1985 Guano and resource control in sixteenth-century Arequipa. In *Andean Ecology and Civilization. An Interdisciplinary Perspective in Andean Ecological Complementarity,* pp. 185–231. Shozo Masuda, Izumi Shimada, and Craig Morris, editors. University of Tokyo Press, Tokyo, Japan.

Julien, Daniel C.
1981 Late prehispanic maritime adaptation on the North Coast of Peru. Paper presented at the International Andean Archaeology. Colloquium in Honor of Clifford Evans, April 28–29, University of Texas at Austin. Typescript. [Author's Note: The proceedings of this meeting were published in 1986, but this paper was not included.]

Jürgens, Klaus D.
1989 Sauerstoff-transport System von Säugetieren an das Leben in groben Hühen. *Naturwissenschaften* 76:410–415.

Jürgens, Klaus D., Manfred Pietschmann, Kazuhiro Yamaguchi, and Traute Kleinschmidt
1988 Oxygen binding properties, capillary densities and heart weights in high altitude camelids. *Journal of Comparative Physiology* [*B*] 158:469–477.

Kadwell, Miranda, Matilde Fernandez, Helen Stanley, Ricardo Baldi, Jane C. Wheeler, Raul Rosario, and Michael W. Bruford
2001 Genetic analysis reveals the wild ancestors of the llama and the alpaca. *Proceedings of the Royal Society London,* 268:3675–2584.

Katz, Lois (editor)
1983 *Art of the Andes. Pre-Columbian Sculpture and Painted Ceramics from the Arthur M. Sackler Collection.* The Arthur M. Sackler Foundation and The Arts Sciences and Humanities, Washington, DC.

Kaulicke, Peter
1974–1975 Reflexiones sobre la arqueología de la sierra de Lima. *Boletín del Seminario de Arqueología.* Arqueología PUC 15–16:29–36. Instituto Riva-Agüero, Pontificia Universidad Católica del Perú, Lima, Peru.

1979 Algunas consideraciones acerca del material óseo de Uchkumachay. In *Arqueología Peruana: Investigaciones arqueológicas en el Perú 1976,* pp. 103–111. Seminario. R. Matos M., compiler. Universidad Nacional Mayor de San Marcos y con el auspicio de la Comisión para Intercambio Educativo entre los Estados Unidos y el Perú, Lima, Peru.

1980 Der abri Uchkumachay und seine zeitliche Stellung innerhalb der lithischen Perioden Perus. *Materialen zur Allgemeinen und Vergleichenden Archaeologie. KAVA. Beiträge.* Band 2, pp. 429–458. Munich, Germany.

1989 La fauna osteológica de Cochasquí. In *Excavaciones en Cochasquí, Ecuador, 1964–1965,* pp. 242–246. Udo Oberem and Wolfgang W. Wurster, editors. Verlag Philipp von Zabern, Mainz am Rhein, Germany.

1991 El Periodo Intermedio Temprano en el Alto Piura: Avances del Proyecto Arqueológico "Alto Piura" (1987–1990). *Bulletin de l'Institut Français d'Études Andines* (Lima) 20(2): 381–422.

1993 La cultura Mochica: Arqueología, historia y ficción. *Histórica* (Lima) 17(1):93–107.

Kautz, Robert R., and Richard W. Keatinge
  1977   Determining site function: A north Peruvian
          coastal example. *American Antiquity* 42(1):
          86–97.
Keatinge, Richard W.
  1975   Urban settlement systems and rural sustain-
          ing communities: An example from Chan
          Chan's hinterland. *Journal of Field Archaeolo-
          gy* 2(3):215–227.
Kent, Jonathan D.
  1986   Introduction. In *The Osteology of South
          American Camelids*, pp. 1–2. V. R. Pacheco
          Torres, A. Altamirano Enciso, and E. Guerra
          Porras. Archaeological Research Tools, Vol.
          3. Institute of Archaeology, University of
          California, Los Angeles.
  1987   The most ancient south: A review of the
          domestication of andean camelids. In *Stud-
          ies in the Neolithic and Urban Revolution: The
          V. Gordon Childe Colloquium Mexico, 1986*,
          pp. 169–184. Linda Manzanilla, editor.
          BAR International Series 349. London,
          UK.
  1988a   El Sur más antiguo: Revisión de la domesti-
          cación de camélidos andinos. In *Llamichos y
          Paqocheros. Pastores de llamas y alpacas*, pp. 23–
          35. Jorge Flores Ochoa, compiler. CON-
          CYTEC, CEAC y Editorial UNSAAC,
          Cuzco, Peru. [Author's Note: This is a trans-
          lation of the 1987 article, but it is not en-
          tirely accurate.]
  1988b   Del cazador al pastor en los Andes Cen-
          trales. In *Rituales y Fiestas de las Américas*,
          pp. 127–145. Memorias del 45th Congreso
          Internacional de Americanistas. Ediciones
          Uniandes, Bogotá, Colombia.
Kessler, M., M. Gauly, C. Frese, and S. Hiendleder
  1996   DNA studies on South American camelids.
          In *Proceedings, 2nd European Symposium on
          South American Camelids*, pp. 269–278. Mar-
          tina Gerken and C. Renieri, editors. Univer-
          sität Göttingen and Università degli Studi di
          Camerino, Matelica, Italy.
Kleeman, Otto
  1975   Excavaciones en los valles del Caplina y
          Sama. In *Investigaciones Arqueológicas en los
          valles del Caplina y Sama (Dep. Tacna, Perú)*,
          pp. 85–118. H. Trimborn, O. Kleeman, H. J.
          Narr, and W. Wurster. Studia Instituti An-
          thropos, Vol. 25. Editorial Verbo Divino, Es-
          tella (Navarra), Spain.
Koford, Carl B.
  1957   The vicuña and the puna. *Ecological Mono-
          graphs* 27(2):153–219.

Kosok, Paul
  1965   *Life, Land and Water in Ancient Peru*. Long
          Island University Press, New York.
Kraglievich, Lucas J.
  1946   Sobre Camélidos chapadmalenses. Instituto
          del Museo de la Universidad Nacional de La
          Plata, *Notas del Museo de La Plata*, XI.
          Paleontología (93):317–328.
Kroeber, Alfred L.
  1963   *Anthropology: Culture Patterns and Processes*.
          Harcourt Brace & World, New York.
Kubler, George
  1946   The Quechua in the colonial world. In
          Handbook of South American Indians, Vol. 2,
          The Andean Civilizations, pp. 331–410. Julian
          H. Steward, editor. Bulletin 143, Bureau of
          American Ethnology, Smithsonian Institu-
          tion, Washington, DC.
Kumar, D., P. M. Raisinghani, and G. S. Manohar
  1992   Sarcoptic mange in camels: A review. In *Pro-
          ceedings of the First International Camel Confer-
          ence, 2nd–6th February 1992*, pp. 79–82. W. R.
          Allen, A. J. Higgins, I. G. Mayhew, D. H. Snow,
          and F. Wade, editors. R.&W. Publications
          (Newmarket), Ltd., Amersham, Switzerland.
La Gama, Sebastián de
  1975 (1540) Visita hecha en el valle de Jayanca por
          . . . El Valle de Jayanca y el reino de los
          mochica. Siglos XV y XVI, Waldemar Es-
          pinoza Soriano. *Bulletin de l'Institut Français
          d'Études Andines* (Lima) 4 (3–4):260–272.
La Gasca, Pedro de
  1976   Descripción del Perú. *Revista del Archivo
          Histórico del Guayas* 9: 37–57.
  1983–1984 (1549) Relación de algunos repartimi-
          entos de los reinos del Piru, de la cantidad
          de yndios que tienen y tributos que dan,
          sacado de la visita y tasa primera que se hizo
          por mandato del presidente lizenciado Gas-
          ca. La Tasa ordenada por el Licenciado
          Pedro de La Gasca (1549), María Rostwor-
          owski de Diez Canseco. *Revista Histórica*
          (Lima) 34:63–102.
Laming-Emperaire, Annette
  1968   Mission archéologique française au Chili
          austral et au Brésil Méridional. *Journal de la
          Société des Américanistes* 57:76–99.
Lange Topic, Theresa, Thomas H. McGreevy, and
  John R. Topic
  1987   A comment on the breeding and herding of
          llamas and alpacas on the North Coast of
          Peru. *American Antiquity* 52(4):832–835.
Lanning, Edward P.
  1960   Chronological and cultural relationship of
          early pottery styles in ancient Peru. Ph.D.

dissertation, Department of Anthropology, University of California, Berkeley.

1967a *Peru before the Incas*. Prentice-Hall, Englewood Cliffs, NJ

1967b *Archaeological Investigations on the Santa Elena Peninsula, Ecuador*. Report to the National Science Foundation on research carried out under grant GS-402, 1964–1965.

1970 Pleistocene man in South America. *World Archaeology* 2(1):90–111.

Larco Hoyle, Rafael

1938 *Los Mochicas*. Tomo I. Casa Editora "La Crónica" y "Variedades," Lima, Peru. [Author's Note: The complete study by Larco Hoyle was published in 2001 under the title *Los Mochicas* (Museo Arqueológico Rafael Larco Herrera, Fundación Telefónica, Lima), in two volumes. The 1938 text was included in volume I with changes.]

1941 *Los Cupisniques*. Casa Editora "La Crónica" y "Variedades," Lima, Peru.

1945 *Los Mochicas (Pre-Chimú, de Uhle y Early Chimú de Kroeber)*. Buenos Aires, Argentina.

1946 A culture sequence for the North Coast of Peru. In *Handbook of South American Indians*, Vol. 2, *The Andean Civilizations*, pp. 149–175. Julian H. Steward, editor. Bulletin 143, Bureau of American Ethnology, Smithsonian Institution, Washington, DC.

1948 *Cronología Arqueológica del Norte del Perú*. Hacienda Chiclín, Trujillo and Buenos Aires, Argentina.

Larrain Barros, Horacio

1974 Análisis de las causas de despoblamiento entre las comunidades indígenas del norte de Chile, con especial referencia a las hoyas hidrográficas de las quebradas Aroma y Tarapacá. *Norte Grande* 1(2):125–154. Instituto de Geografía, Universidad Católica de Chile, Santiago, Chile.

1978–1979 Identidad cultural e indicadores eco-culturales del grupo étnico Chango. *Norte Grande* 6:63–76. Instituto de Geografía, Universidad Católica de Chile, Santiago, Chile.

1980a Pedro Cieza de León. Visión geográfica e histórica del norte ecuatoriano (La Crónica del Perú). In *Cronistas de Raigambre Indígena*, pp. 107–376. Horacio Larrain Barros. Colección Pendoneros, 14. Instituto Otavaleño de Antropología, Otavalo, Ecuador.

1980b Felipe Guaman Poma de Ayala. In *Cronistas de Raigambre Indígena*, pp. 129–250. Horacio Larrain Barros. Colección Pendoneros, 15. Instituto Otavaleño de Antropología, Otavalo, Ecuador.

Larramendy, M., R. L Vidal, M. Bianchi, and N. Bianchi

1984 Camélidos Sudamericanos: Estudios genéticos. *Boletín de Lima* 6(35):92–96.

Latcham, Ricardo E.

1922 Los Animales domésticos de la América precolombina. *Publicaciones del Museo de Etnología y Antropología de Chile* (Santiago de Chile) 3(1):1–199.

1936 Atacameño Archaeology. *American Anthropologist*, n.s., 38(4):609–619.

Lathrap, Donald W.

1970 *The Upper Amazon*. Thames and Hudson, London.

1973 The antiquity and importance of long-distance trade relationships in the moist tropics of pre-Columbian South America. *World Archaeology* 5:170–185.

Lavallée, Danièle

1970 *Les représentations animales dans la céramique mochica*. Université de Paris, Mémoires de l'Institut d'Ethnologie-IV. Institut d'Ethnologie, Musée de l'Homme, Paris, France.

1978 Pasteurs préhistoriques des hauts plateaux andins. *L'Histoire* 5:33–42.

1979 Prehistoria de San Pedro de Cajas. In *Arqueología Peruana: Investigaciones arqueológicas en el Perú 1976*, pp. 113–132. Seminario. R. Matos M., compiler. Universidad Nacional Mayor de San Marcos y con el auspicio de la Comisión para Intercambio Educativo entre los Estados Unidos y el Perú, Lima, Peru.

1988 L'occupation préhistorique de 1'abri de Telarmachay (Pérou): Un aspect original de la néolithisation andine. *Académie des Inscriptions & Belles-Lettres*. Compte rendus des séances de l'année 1988. Avril–juin, pp. 266–288. Diffusion de Boccard, Paris, France.

1989 Un homme en Amerique il y a 300 000 ans? *Les Nouvelles de l'Archéologie* 36:14–16.

1990 La domestication animales en Amérique du Sud: Le point des connaisance. *Bulletin de l'Institut Français d'Études Andines* (Lima) 19 (1):25–44.

Lavallée, Danièle, and Michèle Julien

1975 El hábitat prehistórico en la zona de San Pedro de Cajas, Junín. *Revista del Museo Nacional* (Lima) 41:81–127.

1980–1981 Un aspect de la préhistoire andine: L'exploitation des camélidés et des cervidés au Formatif dans 1'abri de Telarmachay (Junín, Pérou). *Journal de la Société des Américanistes* (Paris) 67:97–124.

Lavallée, Danièle, Michèle Julien, and Jane Wheeler

1982 Telarmachay: Niveles precerámicos de ocupación. *Revista del Museo Nacional* (Lima) 46:

55–133. [Author's Note: Includes an appendix by P. Vaughan, pp. 128–133.)

Lavallée, Danièle, Michèle Julien, Jane Wheeler, and Claudine Karlin
1985   *Telarmachay: Chasseurs et pasteurs préhistoriques des Andes. I*, Vols. 1 and 2. Institut Français d'Études Andines. Editions Recherche sur les Civilizations, Synthèse No. 20. Paris, France. [Author's Note: A Spanish translation was published in 1995: *Telarmachay, cazadores y pastores prehistóricos de los Andes*, Vols. I and II. (Institut Français d'Études Andines, Lima). It corresponds to volume 88 of the *Travaux de l'Institut Français d'Études Andines.*]

Lavallée, Danièle, and Luis Guillermo Lumbreras
1986   *Le Ande dalla preistoria agli Incas*. Rizzoli, Milan, Italy.

Lecoq, Patrice
1987   Caravanes de lamas, sel et échanges dans une communauté de Potosí, en Bolivie. *Bulletin de l'Institut Français d'Études Andines* (Lima) 16(3–4):1–38.

Lemon, R. R. H., and C. S. Churcher
1961   Pleistocene geology and paleontology of the Talara region, Northwest Peru. *American Journal of Science* 259(6):410–429.

León, J. Alberto
1939   Algunas consideraciones sobre los camélidos de los Andes. *Boletín del Museo de Historia Natural "Javier Prado"* 3(11):95–105.

León Pinelo, Antonio
1943   *El Paraíso en el Nuevo Mundo*. Tomo II. Publícalo Raúl Porras Barrenechea, Bajo los Auspicios del Comité del IV Centenario del Descubrimiento del Amazonas. Lima, Peru.

León-Velarde, Fabiola, Daniel Espinoza, Carlos Monge C., and Christian de Muizon
1991   A genetic response to high altitude hypoxia: High hemoglobin-oxygen affinity in chicken (*Gallus gallus*) from the Peruvian Andes. *Comptes Rendus de l'Academie des Sciences* 313 (III):401–406.

Levillier, Roberto
1948   *América la bien llamada. II. Bajo la Cruz del Sur*. Editorial Guillermo Kraft, Buenos Aires, Argentina.

Lhote, Henri
1987   *Chameau et dromedaire en Afrique du Nord et au Sahara. Recherches sur leurs origines*, pp. 1–161. Office National des Approvisionnements et des Services Agricoles (ONAP-SA), Paris, France.

Lizana Salvatierra, Samuel
1993   Peligra la vicuña por aumento de caza furtiva. *El Comercio*, Año 154. No. 81,029. 28 August, p. A1. Lima, Peru.
1994   Cazadores han matado a miles de vicuñas: Masivo crimen ecológico se perpetró en Pampa Galeras, entre 1989 y marzo de este año. *El Comercio*, Año 155, No. 81,317. 12 June, p. A1. Lima, Peru.

Lizarraga, Reginaldo de
1968   *Descripción breve de toda la tierra del Perú, Tucumán, Río de la Plata y Chile*, pp. 1–213. Biblioteca de Autores Españoles desde la formación del lenguaje hasta nuestros días. Tomo CCXVI, Ediciones Atlas, Madrid, Spain. (Libro I: "Descripción breve de toda la tierra del Perú, Tucumán, Río de la Plata y Chile," pp. 1–102. Libro II: "De los prelados eclesiásticos del Reino del Perú, desde el reverendísimo Don Jerónimo de Loaisa, de buena memoria, y de los Virreyes que lo han gobernado, y cosas sucedidas desde Don Antonio de Mendoza hasta el Conde de Monterrey, y de los gobernadores de Tucumán y Chile," pp. 103–213.)

Lohmann Villena, Guillermo
1967   Una incógnita despejada: La identidad del Judío portugués, autor de la "Descripción general del Perú." *Revista Histórica* (Lima) 30:26–93.

López Aranguren, Dolores J.
1930a   Camélidos fósiles argentinos. *Anales de la Sociedad Científica Argentina*, 1st Semestre, 109:15–35.
1930b   Camélidos fósiles argentinos: Conclusión. *Anales de la Sociedad Científica Argentina*, 2nd Semestre, 109:97–126.

Lorandi, Ana María
1980   La frontera oriental del Tawantinsuyu: El Umasuyu y el Tucumán. Una hipótesis de trabajo. *Relaciones* (Buenos Aires), n.s., 14(1): 147–164.
1983   Mitayos y mitmaqkunas en el tawantinsuyu meridional. *Histórica* (Lima) 7(1):3–50.

Lorandi de Gieco, Ana María and Delia Magda Lovera
1972   Economía y patrón de asentamiento en la provincia de Santiago del Estero. *Relaciones* (Buenos Aires), n.s., 6:173–191.

Lothrop, Samuel K.
1946   Indians of the Paraná delta and La Plata littoral. In *Handbook of South American Indians*, Vol. l, *The Marginal Tribes*, pp. 177–190. Julian H. Steward, editor. Bulletin 143, Bureau

of American Ethnology, Smithsonian Institution, Washington, DC.

Lozano Machuca, Juan

1965 (1881–1897) Carta del factor de Potosí Juan Lozano Machuca al Virrey del Perú, en donde se describe la provincia de los Lipes. In *Relaciones Geográficas de Indias. Perú. II.* Apéndice III. Marcos Jiménez de la Espada. Biblioteca de Autores Españoles desde la formación del lenguaje hasta nuestros días. Vol. II, pp. 59–63. Tomo CLXXXIV, Ediciones Atlas, Madrid, Spain.

Lumbreras, Luis Guillermo

1967 La alimentación vegetal en los orígenes de la civilización andina. *Perú Indígena* (Lima) 26:254–273.

1974 *The People and Cultures of Ancient Peru.* Smithsonian Institution Press, Washington, DC. [Author's Note: There is a Spanish edition, *De los Pueblos, las Culturas y las Artes del Antiguo Perú*, Ediciones Moncloa-Campodonico, Lima, 1969.)

1980 El Imperio Wari. In *Historia del Perú,* Tomo II, pp. 9–91. Editorial Juan Mejía-Baca, Lima, Peru.

1989 *Chavín de Huántar en el nacimiento de la Civilización Andina.* Ediciones Indea. Instituto de Estudios Arqueológicos, Lima, Peru.

1993 *Chavín de Huántar: Excavaciones en la Galería de las Ofrendas.* Materialen zur Allgemeinen und Vergleichenden Archäologie, Band 51. KAWA (Kommision für Allgemeinen und Vergleichenden Archäologie). Verlag Philipp von Zabern, Mainz am Rhein, Germany.

Lynch, Thomas F.

1967 *The Nature of the Central Andean Preceramic.* Occasional Papers of the Idaho State University Museum, No. 21. Pocatello, ID.

1971 Preceramic transhumance in the Callejón de Huaylas, Peru. *American Antiquity* 36(2): 139–148.

1978 The South American Paleo-Indians. In *Ancient Native Americans*, pp. 455–489. Jesse D. Jennings, editor. W. H. Freeman, San Francisco, CA. [Author's Note: *See* Lynch 1983b.]

(Editor)

1980a *Guitarrero Cave: Early Man in the Andes.* Academic Press, New York.

1980b Setting and excavations. In *Guitarrero Cave,* pp. 3–28.

1983a Camelid pastoralism and the emergence of Tiwanaku. Civilizations in the South Central Andes. *World Archaeology* 15(1):1–14.

1983b The Paleo-Indians. In *Ancient South Americans*, pp. 87–137. Jesse D. Jennings, editor. W. F. Freeman, San Francisco, CA. [Author's Note: This is the study published in 1978, without changes.]

1986 Climate change and human settlement around the Late-Glacial Laguna de Punta Negra, northern Chile: The preliminary results. *Geoarchaeology* 1:43–70.

1990a Glacial-age man in South America? A critical review. *American Antiquity* 55(1):12–36.

1990b El hombre de la edad glacial en Suramérica: Una perspectiva europea. *Revista de Arqueología Americana* 1:141–185.

1991 Paleoindians in South America: A discrete and identifiable cultural stage? In *Clovis: Origins and Adaptations*, pp. 225–259. R. Bonnichsen and K. Turmire, editors. Center for the Study of First Americans, Corvallis, OR.

Lynch, Thomas F., and Susan Pollock

1981 La Arqueología de la Cueva Negra de Chobshi. Miscelánea Antropológica Ecuatoriana. *Boletín de los Museos del Banco Central del Ecuador* 1(1):92–119.

Lyon, Patricia J.

1984 An imaginary frontier: Prehistoric highland-lowland interchange in the Southern Peruvian Andes. In *Networks of the Past: Regional Interaction in Archaeology*, pp. 3–18. Peter D. Francis, F. J. Kense, and P. G. Duke, editors. The Archaeological Association of the University of Calgary, Calgary, BC.

Llagostera Martínez, Agustín

1979 9,700 years of maritime subsistence on the Pacific: An analysis by means of bioindicators in the north of Chile. *American Antiquity* 44(2):309–324.

Llerena Landa, J. Enrique

1957 *Cálculos, reducciones y equivalencias.* International Petroleum Co., Lima, Peru.

MacBride, J. Francis

1937 *Flora of Peru.* Botanical Series, XIII. Field Museum of Natural History, Chicago.

Macedo, Hernando de

1968 Osteológicos. In *Las Ruinas del Abiseo*, pp. 53–58. Duccio Bonavia. Universidad Peruana de Ciencias y Tecnología, Lima, Peru.

1979 Identification of animal bones. In *Excavations at La Pampa in the Northern highlands of Peru, 1975: Report 1 of the Japanese Scientific Expedition to Nuclear America*, pp. 97–98. K. Terada, editor. University of Tokyo Press, Tokyo, Japan.

Maccagno, Luis
1912   La raza de alpaca Suri. *Anales de la Dirección de Fomento* 11:1–6.
MacNeish, Richard S.
1969   *First Annual Report of the Ayacucho Archaeological Botanical Project*. Robert S. Peabody Foundation for Archaeology, Phillips Academy, Andover, MA.
1971   Early man in the Andes. *Scientific American* 224(4):36–46. [Author's Note: The same article was published under the same title in 1974 in *New World Archaeology, 1974: Theoretical and Cultural Transformations*, pp. 143– 153. Selected and with introduction by Ezra B. W. Zubrow, Margaret C. Fritz, and John M. Fritz. W. H. Freeman, San Francisco, CA.]
1981a   Seasonality of the components. In *Prehistory of the Ayacucho Basin, Peru*, Vol. II, *Excavations and Chronology*, pp. 149–166. Richard S. MacNeish, Angel García Cook, Luis G. Lumbreras, Robert K. Vierra, and Antoinette Nelken-Terner. University of Michigan Press, Ann Arbor, MI.
1981b   Ayamachay, Ac 102. In The stratigraphy of the other cave excavation. Robert K. Vierra and Richard S. MacNeish. In *Prehistory of the Ayacucho Basin, Peru*, Vol. II, *Excavations and Chronology*, pp. 114–121. Richard S. MacNeish, Angel García Cook, Luis G. Lumbreras, Robert K. Vierra, and Antoinette Nelken-Terner. University of Michigan Press, Ann Arbor, MI.
1981c   Synthesis and conclusions. In *Prehistory of the Ayacucho Basin, Peru*, Vol. II, *Excavations and Chronology*, pp. 199–257. Richard S. MacNeish, Angel García Cook, Luis G. Lumbreras, Robert K. Vierra, and Antoinette Nelken-Terner. University of Michigan Press, Ann Arbor, MI
1983   The Ayacucho Preceramic as a sequence of cultural energy-flow systems. In *Prehistory of the Ayacucho Basin, Peru*, Vol. IV, *The Preceramic Way of Life*, pp. 236–280. Richard S. MacNeish, Robert K. Vierra, Antoinette Nelken-Terner, Rochelle Lurie, and Angel García Cook. University of Michigan Press, Ann Arbor, MI.
MacNeish, Richard S., Rainer Berger, and Reiner Protsch
1970   Megafauna and man from Ayacucho, Highland Peru. *Science* 168:975–977.
MacNeish, Richard S., and Angel García Cook
1981a   Rosamachay, Ac 117. In The stratigraphy of the other cave excavations. Robert K. Vierra and Richard S. MacNeish. In *Prehistory*

of the Ayacucho Basin, Peru, Vol. II, *Excavations and Chronology*, pp. 121–124. Richard S. MacNeish, Angel García Cook, Luis G. Lumbreras, Robert K. Vierra, and Antoinette Nelken-Terner. University of Michigan Press, Ann Arbor, MI.
1981b   Ruyru Rumi, Ac 300. In The stratigraphy of the other cave excavations. Robert K. Vierra and Richard S. MacNeish. In *Prehistory of the Ayacucho Basin, Peru*. Vol. II, *Excavations and Chronology*, pp. 124–128. Richard S. MacNeish, Angel García Cook, Luis G. Lumbreras, Robert K. Vierra, and Antoinette Nelken-Terner. University of Michigan Press, Ann Arbor, MI.
MacNeish, Richard S., Angel García Cook, Luis G. Lumbreras, Robert K.Vierra, and Antoinette Nelken-Terner
1981   *Prehistory of the Ayacucho Basin, Peru*, Vol. II, *Excavations and Chronology*. University of Michigan Press, Ann Arbor, MI.
MacNeish, Richard S., and Antoinette Nelken-Terner
1983   Introduction to Preceramic contextual studies. In *Prehistory of the Ayacucho Basin, Peru*, Vol. IV, *The Preceramic Way of Life*, pp. 1–15. Richard S. MacNeish, Robert K. Vierra, Antoinette Nelken-Terner, Rochelle Lurie, and Angel García Cook. University of Michigan Press, Ann Arbor, MI.
MacNeish, Richard S., Antoinette Nelken-Terner, and Angel García Cook
1970   *Second Annual Report of the Ayacucho Archaeological-Botanical Project*. Robert S. Peabody Foundation for Archaeology, Phillips Academy, Publication No. 2. Andover, MA.
MacNeish, Richard S., Antoinette Nelken-Terner, and Robert K.Vierra
1980   Introduction. In *Prehistory of the Ayacucho Basin, Peru*, Vol. III, *Nonceramic Artifacts*, pp. 1–34. Richard S. MacNeish, Robert K. Vierra, Antoinette Nelken-Terner, and Carl J. Phagan. University of Michigan Press, Ann Arbor, MI.
MacNeish, Richard S., Thomas C. Patterson, and David L. Browman
1975   *The Central Peruvian Prehistoric Interaction Sphere*. Papers of the Robert S. Peabody Foundation for Archaeology, Vol. 7. Phillips Academy, Andover, MA.
MacNeish, Richard S., and Robert K.Vierra
1983a   The Preceramic way of life in the thorn forest riverine ecozone. In *Prehistory of the Ayacucho Basin, Peru*, Vol. IV, *The Preceramic Way of Life*, pp. 130–187. Richard S. MacNeish, Robert K. Vierra, Antoinette Nelken-Terner,

Rochelle Lurie, and Angel García Cook. University of Michigan Press, Ann Arbor, MI

1983b The Preceramic way of life in the puna ecozone. In *Prehistory of the Ayacucho Basin, Peru*, Vol. IV, *The Preceramic Way of Life*, pp. 225–235. Richard S. MacNeish, Robert K. Vierra, Antoinette Nelken-Terner, Rochelle Lurie, and Angel García Cook. University of Michigan Press, Ann Arbor, MI.

MacNeish, Richard S., Robert K. Vierra, Antoinette Nelken-Terner, Rochelle Lurie, and Angel García Cook

1983 *Prehistory of the Ayacucho Basin, Peru*, Vol. IV, *The Preceramic Way of Life*. University of Michigan Press, Ann Arbor, MI.

Maldonado, Angel

1952a Los antiguos habitantes de Nazca comieron frutos de algarroba o huarango y a sus excretos se deben los algarrobales pre-colombinos de esa zona. *Revista de la Facultad de Farmacia y Bioquímica* 14(55–56):69–78.

1952b Demostrando que en las lomas de Lachay y adyacentes hubo pastoreo de auquénidos por los antiguos peruanos. *Revista de la Facultad de Farmacia y Bioquímica* 14(55–56):53–68. [Author's Note: This article has not been mentioned in this book. I include it only for bibliographical purposes, but it has no scientific value whatsoever.]

Malpass, Michael A.

1983 Two Preceramic and Formative period occupations in the Cordillera Negra: Preliminary report. Paper presented at the Second Northeast Conference on Andean Archaeology and Ethnohistory, New York. Mimeograph.

Mandrini, Raul José

1989–1990 Notas sobre el desarrollo de la economía pastoril entre los indígenas del suroeste bonaerense (fines del siglo XVIII y comienzos del XIX). *Etnía* 34–35:67–87.

Marcus, Joyce

1985 Informe anual de las excavaciones en Cerro Azul, Valle de Cañete. Temporada de 1984. 15 de abril. Peru. Mimeograph.

1987 Prehistoric fishermen in the kingdom of Huarco. *American Scientist* 75(4):393–401.

Marcus, Joyce, Jeffrey D. Sommer, and Christopher P. Glew

1999 Fish and mammals in the economy of an ancient Peruvian kingdom. *Proceedings of the National Academy of Sciences* 96:6564–6570.

Markgraf, Vera

1988 Fell's Cave: 11,000 years of changes in paleoenvironment, fauna, and human occupation. In *Travels and Archaeology in South Chile*, pp. 196–201. Junius Bird and Margaret Bird; John Hyslop, editor. University of Iowa Press, Iowa City, IA.

Markgraf, Vera, and J. P. Bradbury

1982 Holocene climatic history of South America. Chronostratigraphic subdivisions of the Holocene. J. Mangerud, H. J. B. Birks, and D. Jager, editors. *Striae* 16:40–45.

Markham, Clements R.

1941 *Historia del Perú*. Editores Librería e Imprenta "Guia Lascano," Lima, Peru.

Márquez Miranda, Fernando

1946 The Chaco-Santiagueño Culture. In *Handbook of South American Indians*, Vol. 2, *The Andean Civilizations*, pp. 655–660. Julian H. Steward, editor. Bulletin 143, Bureau of American Ethnology, Smithsonian Institution, Washington, DC.

Marshall, Larry G.

1985 Geochronology and land-mammal biochronology of the Transamerican Faunal Interchange. In *The Great American Biotic Interchange*, pp. 49–85. F. G. Stehli and S. O. Webb, editors. Plenum Press. New York.

1988 Land mammals and the Great American Interchange. *American Scientist* 76(4): 380–388.

Marshall, Larry G., Annalisa Berta, Robert Hoffstetter, Rosendo Pascual, Osvaldo A. Reig, Miguel Bombin, and Alvaro Mones

1984 Mammals and stratigraphy: Geochronology of the continental mammal-bearing quaternary of South America. *Palaeovertebrata*, Mémoires Extraordinaires, pp. 1–76. Montpellier, France.

Marshall, Larry G., Robert F. Butler, Robert E. Drake, Garniss H. Curtis, and Richard H. Tedford

1979 Calibration of the Great American Interchange. *Science* 204(4390):272–279.

Masferrer Kan, Elio R.

1984 Criterios de organización andina: Recuay siglo XVII. *Bulletin de l'Institut Français d'Études Andines* (Lima) 13(1–2):47–61.

Mason, J. Alden

1962 *Las Antiguas Culturas del Perú*. Fondo de Cultura Económica. México. [Author's Note: There are English-language editions: *The Ancient Civilization of Peru*, Penguin Books, Middlesex, UK, 1957, 1961, 1964.]

Masuda, Shozo

1981 Cochayuyo, macha, camarón y higos charqueados. In *Estudios Etnográficos del Perú Meridional*, pp. 173–192. Shozo Masuda, editor. University of Tokyo, Tokyo, Japan.

1985 Algae collectors and *lomas*. In *Andean Ecology and Civilizations. An Interdisciplinary Perspective on Andean Ecological Complementarity*, pp. 233–250. Shozo Masuda, Izumi Shimada, and Craig Morris, editors. University of Tokyo Press, Tokyo, Japan.

1986 Las algas en la etnografía andina de ayer y de hoy. In *Etnografía e Historia del Mundo Andino. Continuidad y Cambio*, pp. 223–268. Shozo Masuda, editor. University of Tokyo, Tokyo, Japan.

Masuda, Shozo, Izumi Shimada, and Craig Morris (editors)

1985 *Andean Ecology and Civilization. An Interdisciplinary Perspective on Andean Ecological Complementarity*. University of Tokyo Press, Tokyo, Japan.

Mateos, P. Francisco

1964 Estudio preliminar y edición del . . . *Obras del P. Bernabé Cobo. I*. Biblioteca de Autores Españoles desde la formación del lenguaje hasta nuestros días. Tomo XCI, Ediciones Atlas, Madrid, Spain.

Matienzo, Juan de

1967 *Gobierno del Perú (1567)*. Edition et Étude préliminaire, Guillermo Lohmann Villena. *Travaux de l'Institut Français d'Études Andines* 11. Lima, Peru, and Paris, France.

Matos M., Ramiro

1980 La agricultura prehispánica en las punas de Junín. *Allpanchis Phuturinga* (Cuzco) 15: 91–108.

Matos M., Ramiro, and Rogger Ravines

1980 Período Arcaico (5,000–1,800 A.C.). In *Historia del Perú*, Vol. I, *Perú Antiguo*, pp. 157–250. Editorial Juan Mejía Baca, Lima, Peru.

Matos M., Ramiro, and John W. Rick

1978–1980 Los recursos naturales y el poblamiento precerámico de la puna de Junín. *Revista del Museo Nacional* (Lima) 44:23–68.

Matsuzawa, Tsugio

1978 The Formative site of Las Aldas, Perú: Architecture, chronology and economy. *American Antiquity* 43(4):652–673.

McGreevy, Tom, and Roxanne Shaughnessy

1983 High altitude land use in the Huamachuco area. In *Investigations of the Andean Past: Papers from the First Annual Northeast Conference on Andean Archaeology and Ethnohistory*, pp. 226–242. Daniel H. Sandweiss, editor. Cornell University Latin American Studies Program, Ithaca, New York.

Mejía Xesspe, M. Toribio

1931 Kausay. Alimentación de los indios. *Wira Kocha* (Lima) 1(1):9–24.

Melgar, Ernesto, Aura Gil, and Jorge Huaman

1971 DNA de alpaca: Composición de bases. *Archivos del Instituto de Biología Andina* (Lima) 4(2–4):82–86.

Mellafe, Rolando

1967 Consideraciones históricas sobre la Visita de Iñigo Ortiz de Zúñiga. In *Visita de la Provincia de León de Huánuco en 1562*, Iñigo Ortiz de Zúñiga. Tomo I, pp. 323–344. Universidad Nacional Hermilio Valdizán, Huánuco, Peru.

Mena, Cristóbal de

1968 (1534) *La Conquista del Perú*. In Biblioteca Peruana, Primera Serie, Tomo I, pp. 133–169. Editores Técnicos Asociados, Lima, Peru.

Mendiburu, Manuel de

1934 *Diccionario Histórico Biográfico del Perú*. Tomo X, pp. 259–262. Librería e Imprenta Gil, Lima, Peru.

Menghin, Oswaldo, and Alberto Rex González

1954 Excavaciones arqueológicas en el yacimiento de Ongamira, Córdoba (Rep. Arg.). Universidad Nacional de Eva Perón, Facultad de Ciencias Naturales y Museo. *Notas del Museo*, Tomo XVII, Antropología No. 67, pp. 213–274. Buenos Aires, Argentina.

Menzel, Dorothy

1968 *La Cultura Huari*. Las Grandes Civilizaciones del Antiguo Perú, Tomo VI. Cía de Seguros y Reaseguros Peruano-Suiza S.A., Lima, Peru. [Author's Note: This is a Spanish translation of "Style and Time in the Middle Horizon," published in 1964 in *Ñawpa Pacha* 2:1–105.]

1976 *Art as a Mirror of History in the Ica Valley, 1350–1570. Pottery Style and Society in Ancient Peru*. University of California Press, Berkeley and Los Angeles, CA.

1977 *The Archaeology of Ancient Peru and the Work of Max Uhle*. R. H. Lowie Museum of Anthropology, University of California, Berkeley.

Mercado de Peñalosa, Pedro de, Agustín Sánchez, Gabriel Gonzáles, Francisco de Uceda, and Melchior Molina

1965 Relación de la provincia de los Pacajes. In *Relaciones Geográficas de Indias. Perú. II*. Marcos Jiménez de la Espada. Biblioteca de Autores Españoles desde la formación del lenguaje hasta nuestros días. Vol. I, pp. 334–341. Tomo CLXXXIV, Ediciones Atlas, Madrid, Spain.

Meschia, Giacomo, Harry Prystowsky, André Hellegers, William Huckabee, James Metcalfe, and Donald H. Barron
1960   Observations on the oxygen supply to the fetal llama. *Quarterly Journal of Experimental Physiology* 45:284–291.

Métraux, Alfred
1946   Ethnography of the Chaco. In *Handbook of South American Indians*, Vol. l, *The Marginal Tribes*, pp. 197–370. Julian H. Steward, editor. Bulletin 143, Bureau of American Ethnology, Smithsonian Institution, Washington, DC.
1948a   Tribes of eastern Bolivia and the Madeira headwaters. In *Handbook of South American Indians*, Vol. 3, *The Tropical Forest Tribes*, pp. 381–463. Julian H. Steward, editor. Bulletin 143, Bureau of American Ethnology, Smithsonian Institution, Washington, DC.
1948b   Tribes of the middle and upper Amazon River. In *Handbook of South American Indians*, Vol. 3, *The Tropical Forest Tribes*, pp. 687–712. Julian H. Steward, editor. Bulletin 143, Bureau of American Ethnology, Smithsonian Institution, Washington, DC.
1948c   Tribes of the eastern slopes of the Bolivian Andes. In *Handbook of South American Indians*, Vol. 3, *The Tropical Forest Tribes*, pp. 465–506. Julian H. Steward, editor. Bulletin 143, Bureau of American Ethnology, Smithsonian Institution, Washington, DC.
1948d   The Guaraní. In *Handbook of South American Indians*, Vol. 3, *The Tropical Forest Tribes*, pp. 69–94. Julian H. Steward, editor. Bulletin 143, Bureau of American Ethnology, Smithsonian Institution, Washington, DC.

Meyers, Albert
1989   Análisis de la ceramica tosca. In *Excavaciones en Cochasquí, Ecuador: 1964–1965*. Udo Oberem and Wolfgang W. Wurster, editors. Materialen zur Allgemeinen und Vergleichenden Archäologie, Band 42, pp. 180–197. KAVA. Verlag Phillip von Zabern, Mainz am Rhein, Germany.

Miasta Gutiérrez, Jaime
1979   *El Alto Amazonas: Arqueología de Jaén y San Ignacio, Perú*. Universidad Nacional Mayor de San Marcos, Dirección de Proyección Social, Seminario de Historia Rural Andina. Lima, Peru. Mimeograph.

Miller, George R.
1984   Deer hunters and llama herders: Animal species selection at Chavín. In *The Prehistoric Occupation of Chavín de Huántar, Peru*. Appendix H., pp. 282–287. Richard L. Burg-

er. University of California Publications, Anthropology, Vol. 14. University of California Press, Berkeley and Los Angeles, CA.

Miller, George R., and Richard L. Burger
1995   Our father the cayman, our dinner the llama: Animal utilization at Chavín de Huántar, Peru. *American Antiquity* 60(3): 421–458.
1998   Ideología religiosa y utilización de animales en Chavín de Huántar. In *Excavaciones en Chavín de Huántar*. Appendix H, pp. 262–302. Richard L. Burger. Pontificia Universidad Católica del Perú, Fondo Editorial, Lima, Peru.

Miller, George R., and Anne L. Gill
1990   Zooarcheology at Pirincay, a Formative Period site in highland Ecuador. *Journal of Field Archaeology* 17(1):49–68.

Miller, Patricia D., and Natalio Banchero
1971   Hematology of the resting llama. *Acta Physiologica Latinoamericana* 21:81–86.

Miller, W. J., P. J. Hollander, and William L. Franklin
1985   Blood typing South American camelids. *The Journal of Heredity* 76:369–371.

Mogrovejo, Juan Domingo, and Cristóbal Makowski H.
1999   Cajamarquilla y los Mega Niños en el pasado prehispánico. *Iconos* 1:46–57.

Molina, Cristóbal de ("el Chileno")
1968 (1842)   *Conquista y población del Perú o destrucción del Perú*. Biblioteca Peruana, Tomo III, pp. 297–372. Editores Técnicos Asociados, Lima, Peru. [Author's Note: Porras (1986: 316–317) left open the possibility that this anonymous chronicle was written by Bartolomé de Segovia, also a priest. Both he and Molina were with Almagro in Chile.]

Molina, Giovanni Ignazio
1782   *Saggio sulla Storia Naturale del Chili*. Stamperia di S. Tommaso di Aquino, Bologna, Italy.

Monge C., Carlos
1989   Adaptación animal al gradiente hipóxico andino. *Interciencia* 14(1):8. [Author's Note: An English translation was published in this same issue, titled "Animal adaptation to low-oxygen Andean gradient," p. 7.]

Monge C., Carlos, and Fabiola León-Velarde
1991   Physiological adaptation to high altitude: Oxygen transportation in mammals and birds. *Physiological Reviews* 71(4):1135–1172.

Monge M., Carlos, and Carlos Monge C.
1968a   Adaptation to high altitude. In *Adaptation of Domestic Animals*, pp. 194–201. E. S. E. Hafez, editor. Washington State University, Pullman, WA; Lea and Febiger, Philadelphia, PA. [Author's Note: This article was published

in 1988 without any changes in *Carlos Monge. Obras. 4*. Estudios sobre altura, pp. 2171–2178. Consejo Nacional de Ciencias y Tecnología (CONCYTEC) and Universidad Peruana Cayetano Heredia, Lima, Peru.]

1968b Adaptación de los animales domésticos: Adaptación a las grandes alturas. *Archivos del Instituto de Biología Andina* 2(6):276–291. [Author's Note: This is a slightly modified Spanish translation of Monge M. and Monge C. 1968a.]

Monge C., Carlos, and José Whittembury

1976 High altitude adaptations in the whole animal. In *Environmental Physiology of Animals*, pp. 289–308. J. Bligh, J. L. Cloudsley-Thompson, and A. G. Mac Donald, editors. Wiley, New York.

Monod, Jacques

1971 *El azar y la necesidad. Ensayos sobre la filosofía natural de la biología moderna*. Breve Biblioteca de Respuestas. Barral Editores, Barcelona, Spain.

Montané, Julio

1968 Paleo-Indian remains from Laguna de Tagua-Tagua, central Chile. *Science* 161:1137–1138.

Montesinos, Fernando de

(1552) Anales del Perú Ológrafo inédito. (Footnote by F. Francisco Mateos in volume I of Cobo [1956: 421]. The original annotation is by Jiménez de la Espada.) [Author's Note: According to Romero (1936) there is a 1642 edition; however, this does not appear in Porras (1986: 493), who only mentions the 1906 edition.]

Montesinos, Fernando

1906 *Anales del Perú* (publicado por Víctor M. Maúrtua). Tomo I. 6472. Imp. de Gabriel L. y del Horno, Madrid, Spain.

Monzon, Luis de, Pedro Gonzáles, and Jhuan de Arbe

1965 (1881–1897) Descripción de la tierra del repartimiento de San Francisco de Atunrucana y Laramati, encomendado en Don Pedro de Córdova, jurisdicción de la Ciudad de Guamanga. Año de 1586. In *Relaciones Geográficas de Indias. Perú. I*. Marcos Jiménez de la Espada. Biblioteca de Autores Españoles desde la formación del lenguaje hasta nuestros días. Vol. I, pp. 226–236. Tomo CLXXXIII, Ediciones Atlas, Madrid, Spain.

Monzon, Luis de, Juan de Quesada, Gregorio Sánchez de Haedo, Juan Gutiérrez de Benavides, and Pedro Taipemarca

1965 (1881–1897) Descripción de la tierra del repartimiento de los Rucanas Antamarca de la corona real, jurisdicción de la ciudad de Guamanga. Año de 1586. In *Relaciones Geográficas de Indias. Perú. I*. Marcos Jiménez de la Espada. Biblioteca de Autores Españoles desde la formación del lenguaje hasta nuestros días. Vol. I, pp. 237–248. Tomo CLXXXIII, Ediciones Atlas, Madrid, Spain.

Monzon, Luis de, Beltrán Saravia, Pedro de Frias, and Pedro Taypimarca

1965 (1881–1897) Descripción de la tierra del repartimiento de Atunsora, encomendado en Hernando Palomino, jurisdicción de la ciudad de Guamanga. Año de 1586. In *Relaciones Geográficas de Indias. Perú. I*. Marcos Jiménez de la Espada. Biblioteca de Autores Españoles desde la formación del lenguaje hasta nuestros días. Vol. I, pp. 220–225. Tomo CLXXXIII, Ediciones Atlas, Madrid, Spain.

Moore, Katherine M.

1988a Hunting and herding economies on the Junin puna. In *Economic Prehistory of the Central Andes*, pp. 154–166. E. S. Wing and J. C. Wheeler, editors. BAR International Series 427. Oxford, UK.

1993 Animal management and health in the transition to domestication. Paper presented at the 16th Conference of the Society of Ethnobiology, Boston.

1998 Measure of mobility and occupational intensity in highland Peru. In *Seasonality and Sedentism. Archaeological Perspective from Old and New World Sites*, pp. 181–197. Thomas R. Rocek and Ofer Bar-Yosef, editors. Peabody Museum of Archaeology and Ethnology, Harvard University, Cambridge, MA.

Morales Soria D., Jorge, and Emiliano Aguirre

1980 Camélido finimioceno en Venta del Moro. Primera cita para Europa occidental. *Estudios Geológicos* 36:139–142.

Mori, Juan de, and Hernando Alonso Malpartida

1967 La visitación de los pueblos de los indios (1549). Archivo General de Indias, Sevilla. Justicia, legajo No. 397. Published in *Visita de la Provincia de León de Huánuco en 1562*, Iñigo Ortiz de Zúñiga, visitador. Tomo I, pp. 289–310. Universidad Nacional Hermilio Valdizán, Huánuco, Peru. [Author's Note: This "Visitación" was first published in *Travaux de l'Institut Français d'Études Andines* (Lima), 1955–1956.]

Morris, Craig, and Donald E. Thompson

1985 *Huánuco Pampa. An Inca City and Its Hinterland*. Thames and Hudson, London, UK.

Moseley, Michael Edward

1972 Subsistence and demography: An example of interaction from prehistoric Peru. *Southwestern Journal of Anthropology* 28(1):25–49.

Mujica, Elías
1985   Altiplano-coast relationship in the south-central Andes: From indirect to direct complementarity. In *Andean Ecology and Civilization. An Interdisciplinary Perspective on Andean Ecological Complementarity*, pp. 103–140. Shozo Masuda, Izumi Shimada, and Craig Morris, editors. University of Tokyo Press, Tokyo, Japan.

Muñóz de Linares, Elba, and Margarita A. García Sánchez
1986   *Los Camélidos sudamericanos y su literatura socio-cultural.* Ministerio de Trabajo, Instituto Indigenista Peruano. Universidad Nacional Mayor de San Marcos, Instituto Veterinario de Investigaciones Tropicales y de Altura (IVITA). Centro de Información Científica de Camélidos Sudamericanos (CICCS), Lima, Peru.

Murphy, Robert C.
1926   Oceanic and climatic phenomena along the West coast of South America during 1925. *Geographical Review* 16:26–54.

Murra, John V.
1946   The historic tribes of Ecuador. In *Handbook of South American Indians*, Vol. 2, *The Andean Civilizations*, pp. 785–821. Julian H. Steward, editor. Bulletin 143, Bureau of American Ethnology, Smithsonian Institution, Washington, DC.
1964a   Rebaños y pastores en la Economía del Tawantinsuyu. *Revista Peruana de Cultura* 2:76–101. [Author's Note: This was translated with slight modifications in 1965. It was also republished with the same title and text in 1975 in *Formaciones económicas y políticas del mundo andino*, pp. 117–144, Instituto de Estudios Peruanos, Lima, Peru.]
1964b   Una apreciación etnológica de la visita. Documentos Regionales para la Etnología y Etnohistoria Andinas. In *Visita hecha a la provincia de Chuquito por Garci Diez de San Miguel en el año 1567*. Tomo I, pp. 419–442. Ediciones de la Casa de la Cultura del Perú, Lima, Peru.
1965   Herds and herders in the Inca state. In *Man, Culture, and Animals: The Role of Animals in Human Ecological Adjustments*, pp. 185–215. A. Leeds and A. P. Vayda, editors. American Association for the Advancement of Science, Publication No. 78. Washington, DC.
1970   Información etnológica e histórica adicional sobre el reino Lupaqa. *Historia y Cultura* 4: 49–61.
1975   *See* Murra 1964a.

1978   *La organización económica del estado inca.* Siglo Veintiuno. América Nuestra, Mexico City, Mexico.

Murúa, Fr. Martín de
1922 (1911)   *Historia de los Incas, Reyes del Perú. Crónica del siglo XVI.* Colección de libros y documentos referentes a la historia del Perú. Tomo IV (2a Serie). Imprenta y Librería Sanmarti y Cia, Lima, Peru (includes books I, II, and III).
1925   *Historia de los Incas, Reyes del Perú. Crónica del siglo XVI* (Segunda parte). Colección de Libros y documentos referentes a la Historia del Perú. Tomo V (2a Serie), pp. 1–72. Imprenta y Librería Sanmarti y Cia, Lima, Peru (includes book IV).
1964   *Historia General del Perú.* Colección Joyas Bibliográficas. Bibliotheca Americana Vetus. II. Madrid, Spain.

Museo Amano
n.d. [1972]   *Huacos y tejidos precolombinos del Perú.* Edicion de "La Sociedad Latino-Americana," Lima, Peru.

Nachtigall, Horst
1966a   Ofrendas de llamas en la vida ceremonial de los pastores de la puna de Moquegua (Perú) y de la puna de Atacama (Argentina), y consideraciones histórico-culturales sobre la ganadería indígena. In *Actas y Memorias*, XXXVI Congreso Internacional de Americanistas, España. Vol. 3, pp. 193–198. Seville, Spain.

National Academy of Sciences
1979   *Tropical Legumes: Resources for the Future.* National Academies Press, Washington, DC.

Neveu-Lemaire, M., and G. Grandidier
1911   *Notes sur les mammiferes des hauts plateaux de l'Amerique du Sud.* Librairie H. Le Soudier, Paris, France.

Niemeyer F., Hans
1985–1986   La ocupación incaica de la cuenca alta del río Copiapó (IIIa Región de Atacama, Chile). El Imperio Inka. Actualización y perspectivas por registros arqueológicos y etno-históricos. *Comechigonia* 4 (Número especial):165–294.

Niemeyer F., Hans, and Virgilio Schiappacasse F.
1963   Investigaciones arqueológicas en las terrazas de Conanoxa, valle de Camarones (Provincia de Tarapacá). *Revista Universitaria* 48:101–166.

Noort, Olivier van
1926   *De Reis om de Wereld, 1598–1601.* Met inleiding en aanteekeningen uitgegeven door Dr. J. W. Ijzerman. Martinus Nijhoff, S-Gravenhage, Netherlands.

Novoa, C., and Jane C.Wheeler
  1984   Llama and alpaca. In *Evolution of Domesti-
          cated Animals*, pp. 116–128. Ian L. Mason,
          editor. Longman, London and New York.
          [Author's Note: There is a 1981 edition, is-
          sued by the same publisher, pp. 1–25.]

Nuevo Freire, Carlos M.
  1994   Sinopsis de historia natural de los Caméli-
          dos Sudamericanos. In *Excerta Anatomica
          Camelidae, 1*, pp. 4–9. Jorge M. Galotta and
          Silvia G. Márquez, compilers. FUCASUD,
          Pro lama et vicugna. Talleres Gráficos de
          Weben, S.A., Buenos Aires, Argentina.

Núñez, Alvaro
  1965   Relación de Zamora de los Alcaides, dirigida
          a la Audiencia de Quito. In *Relaciones Geográ-
          ficas de Indias, IV*, Marcos Jiménez de la Espa-
          da. Biblioteca de Autores Españoles desde la
          formación del lenguaje hasta nuestros días.
          Vol. III, pp. 136–138. Tomo CLXXXV, Edi-
          ciones Atlas, Madrid, Spain.

Núñez Atencio, Lautaro
  1969   Panorama arqueológico del Norte de Chile.
          In *Mesa Redonda de Ciencias Prehistóricas y
          Antropológicas*. Tomo II, pp. 197–217. Ponti-
          ficia Universidad Católica del Perú. Instituto
          Riva Agüero, Seminario de Antropología.
          Lima, Peru.
  1976   Geoglifos y tráfico de caravanas en el desier-
          to chileno. *Anales de la Universidad del Norte*
          10:147–201. (Tomo en homenaje al R. P.
          Gustavo Le Paige.)

Núñez Atencio, Lautaro, Juan Varela, and Rodolfo
          Casamiquela
  1983   *Ocupación Paleoindio en Quereo. Reconstrucción
          multidisciplinaria en el territorio semiárido de
          Chile*. Universidad del Norte, Antofagasta,
          Chile.
  1987   Ocupación Paleoindio en el Centro-Norte
          de Chile: Adaptación circunlacustre en las
          tierras bajas. In *Estudios Atacameños: Investi-
          gaciones Paleoindias al Sur de la Línea Ecuato-
          rial*, No. 8 (special issue), pp. 142–185. Lau-
          taro Núñez and Betty Meggers, editors.
          Instituto de Investigaciones Arqueológicas,
          Universidad del Norte, San Pedro de Ataca-
          ma, Chile.

Núñez Del Prado C., Oscar
  1968   Una cultura como respuesta de adaptación
          al medio ambiente. In *Actas y Memorias*,
          XXXVII Congreso Internacional de Ameri-
          canistas, República Argentina. Vol. IV, pp.
          241–260. Buenos Aires, Argentina.

Ocaña, Diego de
  1987 (1605?) *A través de la América del Sur*. Edición
          de Arturo Alvarez. Crónicas de América 33,
          Historia 16. Madrid, Spain.

Ochatoma Paravicino, José Alberto
  1992   Acerca del Formativo en Ayacucho. In *Estu-
          dios de Arqueología Peruana*, pp. 193–213. Duc-
          cio Bonavia, editor. Fomciencias, Lima, Peru.

Oliva, Anello
  1895 (1598) *Historia del Reino y Provincias del Perú de
          sus Incas Reyes, Descubrimiento y Conquista por
          los Españoles, de la Corona de Castilla con otras
          singularidades concernientes á la Historia*. Im-
          prenta y Librería de San Pedro, Lima, Peru.

Olivera, Daniel E.
  1991   El formativo en Antofagasta de la Sierra (Puna
          meridional Argentina): Análisis de sus posi-
          bles relaciones con contextos arqueológicos
          Agro-alfareros Tempranos del Noroeste Ar-
          gentino y Norte de Chile. In *Actas del XI
          Congreso Nacional de Arqueología Chilena*, pp.
          61–78. Sociedad Chilena de Arqueología,
          Santiago, Chile.

ONERN
  1976   *Mapa ecológico. Guía explicativa*. República
          del Perú, Oficina Nacional de Evaluación de
          Recursos Naturales, Lima, Peru.
  1985   *Los Recursos Naturales del Perú*. Lima, Peru.
          [Author's Note: Reprinted in 1988.]

Onuki, Yoshio
  1985   The *Yunga* zone in the prehistory of the cen-
          tral Andes: Vertical and horizontal dimensions
          in Andean ecological and cultural processes.
          In *Andean Ecology and Civilizations. An Interdis-
          ciplinary Perspective on Andean Ecological Com-
          plementarity*, pp. 339–356. Shozo Masuda, Izu-
          mi Shimada, and Craig Morris, editors.
          University of Tokyo Press, Tokyo, Japan.

Oré, Fray Luis Jerónimo de
  1992   *Símbolo católico indiano*. Edición facsimilar
          dirigida por Antonine Tibesar, O.F.M. Aus-
          tralis, Lima, Peru.

Orefici, Giuseppe
  1992   *Nasca. Archeologia per una riconstruzione stori-
          ca*. Jaca Book. Milan, Italy.

Oricain, Pablo José
  1906 (1790) Compendio Breve de Discursos varios
          sobre diferentes materias y noticias geográfi-
          cas comprehensivas á este Obispado del
          Cuzco que claman remedios espirituales. In
          *Juicio de Límites entre el Perú y Bolivia. Prue-
          ba Peruana presentada al Gobierno de la
          República Argentina*, Tomo undécimo, pp.

318–377. Víctor Maúrtua. Obispado y Audiencia del Cuzco, Imprenta de Heinrich y Co., Madrid, Spain.

Ortega Morejon, Di(ego) de, and Fray Chr(is)toual de Castro

1974 (1558) Relaçion y declaraçion del modo que este valle de Chincha y sus comarcanos se governavan antes que oviese yngas y despues q(ue) los vuo hasta q(ue) los cristianos entraron en esta tierra. La Relación de Chincha (1558), Juan Carlos Crespo. *Historia y Cultura* 8:93–104.

Ortiguera, Toribio de

1968 *Jornada del río Marañón con todo lo acaecido en ella, y otras cosas notables dignas de ser sabidas, acaecidas en las islas occidentales*, pp. 215–358. Biblioteca de Autores Españoles desde la formación del lenguaje hasta nuestros días. Tomo CCXVI, Ediciones Atlas, Madrid, Spain.

Ortiz de Zúñiga, Iñigo

1967 *Visita de la Provincia de León de Huánuco en 1562*. Tomo I. Universidad Nacional Hermilio Valdizán, Facultad de Letras y Educación, Huánuco, Peru.

1972 *Visita de la Provincia de León de Huánuco en 1562*. Tomo II. Universidad Nacional Hermilio Valdizán, Facultad de Letras y Educación, Huánuco, Peru.

Otte, K. C., and Venero, J. L.

1979a Analysis of the differential craniometry between the vicuña (*Vicugna vicugna*) and the alpaca (*Lama guanicoe pacos*). *Studies on the Neotropical Fauna and Environment* 14(2–3): 125–152.

Oviedo [y Valdéz], Gonzalo Fernández de
    See Fernández de Oviedo [y Valdéz], Gonzalo.

Pacheco Torres, Víctor R., Alfredo Altamirano Enciso , and Emma Guerra Porras

1986 *The Osteology of South American Camelids*. Archaeological Research Tools, Vol. 3. Institute of Archaeology. University of California, Los Angeles.

Padrón de los mil indios ricos . . .
    See Gutiérrez Flores 1964.

Palacios Rios, Félix

1977b Pastizales de regadío para alpacas. In *Pastores de puna. Uywamichiq punarunakuna*, pp. 155–191. Jorge Flores Ochoa, compiler. Instituto de Estudios Peruanos, Lima, Peru.

1981 Tecnología del Pastoreo. In *Runakunap Kawsayninkupaq Rurasqankunaqq. La tecnología en el mundo andino*, Tomo I, pp. 217–232. Selected and prepared by H. Lecht-

man and A. M. Soldi. Universidad Nacional Autónoma de México, Mexico City, Mexico.

1990 El simbolismo de la casa de los pastores aymaras. In *Trabajos presentados al simposio "RUR 6. El pastoreo altoandino: Origen, desarrollo y situación actual,"* pp. 63–83. Jorge Flores Ochoa, coordinator. 46 Congreso Internacional de Americanistas, 4–8 de junio de 1988. Amsterdam. Publicación auspiciada por la Comisión Ejecutiva del 46 Congreso Internacional de Americanistas, en colaboración con el Centro de Estudios Andinos, Cuzco. CEAC, Cuzco, Peru.

Palermo, Miguel Angel

1986 Reflexiones acerca del llamado "complejo ecuestre" en la Argentina. *Runa* (Buenos Aires) 16:157–178.

1986–1987 La expansión meridional de los camélidos domésticos: El caso del hueque de Chile. *Relaciones de la Sociedad Argentina de Antropología* (Buenos Aires), n.s., 17(1):67–79.

Palomino, Diego

1965 (1881–1897) Relación de las provincias que hay en la conquista del Chuquimayo que yo el Capitan Diego Palomino tengo por su Magestad y por el muy ilustre señor Pedro Gasca presidente de la Audiencia Real destos Reynos del Perú por su Magestad. In *Relaciones Geográficas de Indias. Perú. IV*. Ultimo Apéndice. Marcos Jiménez de la Espada. Biblioteca de Autores Españoles desde la formación del lenguaje hasta nuestros días. Vol. III, pp. 185–188. Tomo CLXXXV, Ediciones Atlas, Madrid, Spain.

Pascual, Rosendo

1954 Restos de vertebrados hallados en el abrigo de Ongamira (Córdoba). Excavaciones arqueológicas en el yacimiento de Ongamira, Córdoba (Rep. Arg.). Oswaldo Menghin and Alberto Rex Gonzáles. Universidad Nacional de Eva Perón, Facultad de Ciencias Naturales y Museo. *Notas del Museo*, Tomo XVII, Antropología, No. 67, pp. 269–274. Buenos Aires, Argentina.

1960 Informe sobre los restos de vertebrados hallados en la Caverna de Intihuasi y "Paraderos" vecinos de San Luis. *Revista del Instituto de Antropología* (Córdoba) 1:299–302.

1963 Determinación específica de fragmentos de huesos provenientes del sitio Cxa W(a)-Niveles 2 y 3. A) Huesos de mamiferos. Investigaciones arqueológicas en las terrazas de Conanoxa, valle de Camarones (Provincia de Tarapacá), Appendix 5. Hans Niemeyer F.

and Virgilio Schiappacasse F. *Revista Universitaria* (Universidad Católica de Chile, Santiago) 48:165–166.

Pascual, Rosendo, and Oscar E. Odreman Rivas
1973   Estudio del material osteológico extraido de la caverna de Huargo, Departamento de Huánuco, Perú. Excavaciones en la caverna de Huargo, Perú, Appendix 2. Augusto Cardich. *Revista del Museo Nacional* (Lima) 39:31–39.

Patterson, Bryan, and Rosendo Pascual
1968   Evolution of mammals on southern continents. V. The fossil mammal fauna of South America. *The Quarterly Review of Biology* 43 (4):409–451.

Patterson, Thomas C., John P. McCarthy, and Robert A. Dunn
1982   Polities in the Lurin Valley, Peru, during the Early Intermediate period. *Ñawpa Pacha* 20:61–82.

Paz Flores, M. Percy
1988   Ceremonias y pinturas rupestres. In *Llamichos y Paqocheros. Pastores de llamas y alpacas*, pp. 217–223. Jorge A. Flores Ochoa, compiler. CONCYTEC, CEAC y Editorial Universitaria UNSAAC, Cuzco, Peru.

Paz Maldonado, Jhoan de
1965 (1881–1897) Relación del pueblo de Sant-Andres Xunxi para el muy ilustre señor Licenciado Francisco de Auncibay, del consejo de Su Magestad y su oydor en la Real Audiencia de Quito. In *Relaciones Geográficas de Indias. Perú. III*. Marcos Jiménez de la Espada. Biblioteca de Autores Españoles desde la formación del lenguaje hasta nuestros días. Vol. II, pp. 261–264. Tomo CLXXXIV, Ediciones Atlas, Madrid, Spain.

Paz Ponce de León, Sancho de
1965 (1881–1897) Relación y descripción de los pueblos del Partido de Otavalo. 1582. In *Relaciones Geográficas de Indias. Perú. II*. Marcos Jiménez de la Espada. Biblioteca de Autores Españoles desde la formación del lenguaje hasta nuestros días. Vol. II, pp. 233–242. Tomo CLXXXIII, Ediciones Atlas, Madrid, Spain.

Pearsall, Deborah M.
1978   Paleoethnobotany in western South America: Progress and problems. In *The Nature and Status of Ethnobotany*, pp. 389–416. Richard I. Ford, editor. Anthropological Papers No. 67, Museum of Anthropology, University of Michigan, Ann Arbor, MI.

Pease G. Y., Franklin
1981   Las Relaciones entre las Tierras Altas y la Costa del Sur del Perú: Fuentes documentales. In *Estudios Etnográficos del Perú Meridional*, pp. 193–208. Shozo Masuda, editor. University of Tokyo, Tokyo, Japan.
1982   Relaciones entre los grupos étnicos de la Sierra Sur y la costa: Continuidades y cambios. El hombre y su ambiente en los Andes Centrales. *Senri Ethnological Studies* 10:107–122.
1989   La conquista española y la percepción andina del otro. *Histórica* (Lima) 13(2):171–196.
1991   *Los Incas. Una introducción*. Biblioteca "Lo que debo saber," Vol. I. Pontificia Universidad Católica del Perú. Fondo Editorial, Lima, Peru.
1992   *Perú: Hombre e Historia. Entre el siglo XVI y el XVIII*. Fundación del Banco Continental para el Fomento de la Educación y la Cultura. Ediciones Edubanco, Lima, Peru.

Pegram, R. G., and A. J. Higgins
1992   Camel ectoparasites: A review. In *Proceedings of the First International Camel Conference, 2nd–6th February 1992*, pp. 69–78. W. R. Allen, A. J. Higgins, I. G. Mayhew, D. H. Snow, and J. F. Wade, editors. R.&W. Publications (Newmarket), Amersham, Switzerland.

Penedo, M. C. T., M. E. Fowler, A. T. Bowling, D. L. Anderson, and L.Gordon
1988   Genetic variation in the blood of llamas, *Llama* [sic] *glama*, and alpaca, *Llama* [sic] *pacos. Animal Genetics* 19:267–276.

Peñaherrera del Aguila, Carlos (Director)
1963–1970 *Atlas Histórico Geográfico y de Paisajes Peruanos*. Presidencia de la República, Instituto Nacional de Planificación, Lima, Peru.
1969   *Geografía general del Perú*. Tomo I. *Aspectos físicos*. Lima, Peru.

Pereyra, Carlos
1988   Las "Noticias Secretas" de América y el enigma de su publicación. In *Noticias Secretas de América*, Appendix I, pp. 711–739 (5–33). Jorge Juan and Antonio de Ulloa. Colegio Universitario, Ediciones Istmo. Mundus Novus 9. Madrid, Spain. [Author's Note: This is a facsimile edition of the article published in 1940 in *Revista de Indias 2*.]

Pérez de Guevara, Juan
1965 (1881–1897) Carta de Juan Pérez de Guevara a Gonzalo Pizarro sobre su jornada de Rupa-Rupa. In *Relaciones Geográficas de Indias. Perú. IV*. Ultimo Apéndice. Marcos Jiménez de la Espada. Biblioteca de Autores Españoles desde la formación del lenguaje hasta nuestros días. Vol. III, pp. 168–170. Tomo CLXXXV, Ediciones Atlas, Madrid, Spain.

Pérez de Tudela Bueso, Juan
1963   El cronista Diego Fernández. In *Crónicas del Perú. I*. Estudio Preliminar y Edición por Juan Pérez de Tudela Bueso, pp. lxxvii–lxxxv. Biblioteca de Autores Españoles desde la formación del lenguaje hasta nuestros días. Tomo CLIV, Ediciones Atlas, Madrid, Spain.

Petersen G., Georg
1962   Las primeras operaciones militares de Francisco Pizarro en el Perú. In *Actas y Trabajos del II Congreso Nacional de Historia del Perú. Epoca Prehispánica*, Vol. II, pp. 359–383. Lima, Peru.

Petschow, D., Rosemarie Wurdinger, Johen Duhm, Gerhard Braunitzer, and Christian Bauer
1977   Causes of high blood O_2 affinity of animals living at high altitude. *Journal of Applied Physiology* 42:139–143.

Pigafetta, Antonio (Francisco)
1927   *Viaje alrededor del mundo (con notas de Carlos Amoretti)*, pp. 35–217. Primer Viaje en torno del globo, versión castellana de Federico Ruis Morcuende. Edición del IV Centenario, Espasa-Calpe, Madrid, Spain.
1970 (1519–1522)   *Primer Viaje en torno del Globo*, pp. 7–207. Translated by José Toribio Medina. Ed. Francisco de Aguirre, Buenos Aires, Argentina.

Pires-Ferreira, Edgardo
1979   Nomenclatura y nueva clasificación de los Camélidos Sudamericanos. *Politécnica* 4(2) [unpaginated]. (Escuela Politécnica Nacional, Quito, Ecuador.)
1981–1982   Nomenclatura y nueva clasificación de los Camélidos sudamericanos. *Revista do Museu Paulista*, n.s., 28:203–219. Universidade de Sao Paulo, Sao Paulo, Brazil. [Author's Note: This is the same text published in Ecuador in 1979. *See* Pires-Ferreira 1979.]

Pires-Ferreira, Edgardo, Jane Wheeler Pires-Ferreira, and Peter Kaulicke
1976   Utilización de animales durante el período precerámico en la Cueva de Uchcumachay y otros sitios de los Andes Centrales del Perú. Paper presented at the XLII International Congress of Americanists, Paris, France. [Author's Note: In 1977 it was published under the same title in *Journal de la Société des Américanistes* 64:149–154.]

Pizarro, Pedro
1968 (1844)   *Relación del descubrimiento y conquista de los reinos del Perú*. In Biblioteca Peruana. Tomo I, pp. 449–586. Editores Técnicos Asociados, Lima, Peru.

1978 (1571)   *Relación del descubrimiento y conquista de los Reinos del Pirú y del gouierno y horden que los naturales tenían y tesoros que en ellos se hallaron y de las demás cosas que en el an çubçedido hasta en dia desta fecha. Fecha por Pedro Piçarro conquistador y poblador destos dichos rreynos y vezino de la ciudad de Arequipa. Año de mill e quinientos e setenta y un años*. Relación del Descubrimiento y Conquista de los Reinos del Perú, pp. 1–262. Pontificia Universidad Católica del Perú. Fondo Editorial, Lima, Peru.

Plischke, Hans
1962   Ulrich Schmidel. In *Wahrhaftige Historien einer wunderbaren Schiffart.*, Ulrich Schmidel, pp. VII–XIX. Akademische Druck-u Verlagsanstalt. (Vol. 1 of "Frühe Reisen und Seefahrten in original Berichten"). Graz, Austria.

Pollard, Gordon C.
1971   Cultural change and adaptation in the central Atacama desert. *Ñawpa Pacha* 9:41–64.
1976   Identification of domestic *Lama* sp. from prehispanic northern Chile using microscopy. *El Dorado* 1(2):16–19.

Pollard, Gordon C., and Isabella M. Drew
1975   Llama herding and settlement in prehispanic northern Chile: Application of an analysis for determining domestication. *American Antiquity* 40(3):296–305.

Polo de Ondegardo, Juan
1916   Relación de los fundamentos acerca del notable daño que resulta de no guardar a los indios sus fueros. Junio 26 de 1571. In *Informaciones Acerca de la Religión y Gobierno de los Incas, seguidas de las Instrucciones de los Concilios de Lima*. Horacio H. Urteaga and Carlos Romero, editors. Colección de Libros y Documentos Referentes a la Historia del Perú, Tomo III, pp. 45–188. Imprenta y Librería Sanmarti y Cia, Lima, Peru.

(Juan) El Licenciado
1917 (1571)   *Informaciones Acerca de la Religión y Gobierno de los Incas*. Horacio H. Urteaga and Carlos Romero, editors. Colección de Libros y Documentos Referentes a la Historia del Perú, Tomo IV, pp. 45–94. Imprenta y Librería Sanmarti y Cia, 2nd Series, Lima, Peru.
1940 (1561)   Informe del Licenciado Juan Polo de Ondegardo al Licenciado Briviesca de Muñatones sobre la perpetuidad de las encomiendas en el Perú. *Revista Histórica* 13:128–196.

Ponce Del Prado, Carlos, and Kai Ch. Otte
  1984   Diseño de la política sobre Camélidos Sud-
          americanos silvestres en el Perú. In *La Vi-
          cuña*, pp. 1–11. Fernando Villiger, editor.
          Editorial Los Pinos, Lima, Peru.

Porras Barrenechea, Raúl
  1937   *Las relaciones primitivas de la conquista del
          Perú*. Imprimeries Les Presses Modernes,
          Paris, France.
  1949–1950 Crónicas perdidas, presentes y olvi-
          dadas sobre la Conquista del Perú. *Docu-
          menta*, Año II, No. l, pp. 179–243. Lima,
          Peru. [Author's Note: This article was in-
          cluded in Porras 1986:685–734.]
  1955   *Fuentes Históricas Peruanas*. Juan Mejía Baca
          and P. L. Villanueva, editors. Lima, Peru.
          [Author's Note: The cover says 1955, where-
          as inside it reads 1954.]
  (published by)
  1959   *Cartas del Perú (1524–1543)*. Colección de
          Documentos inéditos para la Historia del
          Perú. Edición de la Sociedad de Bibliófilos
          Peruanos, Lima, Peru.
  1967   *Las Relaciones Primitivas de la Conquista del
          Perú*. Instituto Raúl Porras Barrenechea, Es-
          cuela de Altos Estudios y de Investigaciones
          Peruanistas de la Universidad Nacional Ma-
          yor de San Marcos, Lima, Peru. [Author's
          Note: The first edition of this study was
          published in 1937.]
  1968   La Relación Sámano-Xerez. (Introduction
          and footnotes.) *Relación de Sámano-Xerez.
          1527*. Biblioteca Peruana. Primera Serie,
          Tomo I, pp. 5–6. Editores Técnicos Asocia-
          dos, Lima, Peru.
  1978   *Pizarro*. Editorial Pizarro, S.A., Lima, Peru.
  1986   *Los Cronistas del Perú (1528–1650) y otros
          Ensayos*. Biblioteca Clásicos del Peru/2. Edi-
          ciones del Centenario. Banco de Crédito del
          Perú, Lima, Peru. [Author's Note: The first,
          posthumous edition of *Los Cronistas del Perú
          (1528–1650)*, Sanmartí y Cia. Impresores,
          was published in Lima in 1962. This new
          edition is expanded and revised.]
  1992   Nota introductoria a Pedro Sancho (1534).
          Antología de1 Cuzco, p. 3. Raúl Porras Barre-
          nechea. Fundación M. J. Bustamante De la
          Fuente, Lima, Peru.

Poyart, Claude, Henri Wajcman, and Jean Kister
  1992   Molecular adaptation of hemoglobin function
          in mammals. *Respiration Physiology* 90: 3–17.

Pozorski, Shelia Griffis
  1976   Prehistoric subsistence patterns and site
          economics in the Moche Valley, Peru. Ph.D.
          dissertation, University of Texas, Austin.
  1979a   Prehistoric diet and subsistence of the
          Moche Valley, Peru. *World Archaeology* 2(2):
          163–184.
  1979b   Late prehistoric llama remains from the
          Moche Valley, Peru. *Annals of Carnegie Mu-
          seum* 48, Article 9:139–170.
  1980   Subsistencia Chimú en Chanchan. In *Chan-
          chan: Metrópoli Chimú*, pp. 181–193. Rogger
          Ravines, compiler. Instituto de Estudios Pe-
          ruanos, Instituto de Investigación Tecnológi-
          ca Industrial y de Normas Técnicas, Lima,
          Peru.
  1982   Subsistence systems in the Chimu state. In
          Chan Chan: Andean Desert City, pp. 177–196.
          Michael E. Moseley and Kent C. Day, edi-
          tors. University of New Mexico Press, Albu-
          querque, NM.
  1983   Changing subsistence priorities and early
          settlement patterns on the north coast of
          Peru. *Journal of Ethnobiology* 3(1):15–38.

Pozorski, Shelia G., and Thomas Pozorski
  1979   An early subsistence exchange system in the
          Moche Valley, Peru. *Journal of Field Archaeol-
          ogy* 6:413–432.
  1987   *Early Settlement and Subsistence in the Casma
          Valley, Peru*. University of Iowa Press, Iowa
          City, IA.

Pozorski, Thomas
  1979   The Las Avispas Burial Platform at Chan
          Chan, Peru. *Annals of the Carnegie Museum*
          48, Article 8:119–137.
  1980   Las Avispas: Plataforma funeraria. In *Chan-
          chan. Metrópoli Chimú*, pp. 231–242. Rogger
          Ravines, compiler. Instituto de Estudios Peru-
          anos, Instituto de Investigación Tecnológica
          Industrial y de Normas Técnicas, Lima, Peru.
  1982   Early Social stratification and subsistence
          systems: The Caballo Muerto Complex. In
          Chan Chan: Andean Desert City, pp. 225–
          253. Michael E. Moseley and Kent C. Day,
          editors. University of New Mexico Press,
          Albuquerque, NM.

Pozzi-Escot, Denise
  1985   Conchopata: Un poblado de especialistas
          durante el Horizonte Medio. *Bulletin de l'In-
          stitut Français d'Études Andines* (Lima) 14
          (3–4):115–129.

Pozzi-Escot, Denise, and Carmen Cardoza
  1986   *El consumo de Camélidos entre el Formativo y
          Wari en Ayacucho*. Instituto Andino de Estu-
          dios Arqueológicos, Universidad Nacional
          San Cristóbal de Huamanga, Ayacucho, Peru.

Prescott, Guillermo H.
  1955   *Historia de la Conquista del Perú*. Ediciones
          Imán, Buenos Aires, Argentina.

2000 *History of the Conquest of Mexico & History of the Conquest of Perú.* Cooper Square Press, New York.

Presta, Ana María, and Mercedes del Río

1993 Reflexiones sobre los Churumatas del Sur de Bolivia, Siglos XVI–XVII. *Histórica* (Lima) 17 (2):223–237.

Proulx, Donald A.

1968 *An Archaeological Survey of the Nepeña Valley, Peru.* Research Reports Number 2. Department of Anthropology, University of Massachusetts, Amherst, MA.

Pulgar Vidal, Javier

1950 Los auquénidos en Colombia. *Naturaleza y Técnica* 1(3):17–19 (83–85).

n.d. [1980?] *Geografía del Perú: Las ocho regiones naturales del Perú.* Editorial Universo, Lima, Peru. [Author's Note: There are several editions, all of them in Spanish.]

Quilter, Jeffrey

1991 Late Preceramic Peru. *Journal of World Prehistory* 5(4):387–438.

Quilter, Jeffrey, Bernardino Ojeda E., Deborah M. Pearsall, Daniel H. Sandweiss, John G. Jones, and Elizabeth S. Wing

1991 Subsistence economy of El Paraiso, an early Peruvian site. *Science* 251:277–283.

Radicati di Primeglio, Carlos

1967 Azarosa vida y obra de Jerónimo Benzoni. In *La Historia del Mundo Nuevo* de M. Jerónimo Benzoni, Milanés, pp. v–lviii. Universidad de San Marcos, Lima, Peru.

1985 Azarosa vida y obra de Jerónimo Benzoni. In *La Historia del Nuevo Mundo.* (*Relatos de su viaje por el Ecuador, 1547–1550)*, pp. 1–54. Girolamo Benzoni. Museo Antropológico y Pinacoteca del Banco Central del Ecuador, Guayaquil, Ecuador. [Author's Note: This is the same text as was published in 1967. *See* Radicati di Primeglio 1967.]

Raedeke M., Kenneth J.

1976 *El guanaco de Magallanes, Chile. Distribución y biología.* Corporación Nacional Forestal de Chile, Publicación No. 4. Santiago, Chile.

Raffino, Rodolfo A., Eduardo P. Tonni, and Alberto L. Cione

1977 Recursos alimentarios y economía en la región de la Quebrada del Toro, Provincia de Salta, Argentina. *Relaciones de la Sociedad Argentina de Antropología*, n.s., 11:9–30.

Raimondi, Antonio

1874 *El Perú. Tomo I. Parte preliminar.* Imprenta del Estado, Calle de la Rifa, Num. 58 por J. Enrique del Campo. Lima, Peru.

1943 *Notas de Viajes para su obra "El Perú."* Vol. 2, pp. 95–172. *Viaje al Departamento de Ancash, 1860–1861.* Publicado por el Ing. Alberto Jochamovitz. Imprenta Torres Aguirre, Lima, Peru.

1945 *Notas de Viajes para su obra "El Perú."* Vol. 3, pp. 2–174. *Viaje de Lima a Ayacucho, visitando las Quebradas de Lurín, del río Rímac y Huarochirí, 1862.* Publicado por el Ing. Alberto Jochamovitz. Imprenta Torres Aguirre, Lima, Peru.

1948 *Notas de Viajes para su obra "El Perú."* Vol. 4, pp. 3–83. *Viaje de Ica a Areguipa, visitando Nasca, Sun Juan de Lucanas, Puquio, Coracora, Chala, Caravelí y Ocoña, 1863.* Publicado por el Ing. Alberto Jochamovitz. Imprenta Torres Aguirre, Lima, Peru.

Ramírez, Balthasar

1936 (1597) *Description del Reyno del Pirú del sitio temple. Prouincias, obispados y ciudades, de los Naturales de sus lenguas y trage.* . . . Quellen zur Kulturgeschichte der präkolumbischen Amerika, Vol. III, pp. 1–122. H. Trimborn, editor. Veröffentlichung des Forschungsinstituts für Kulturmorphologie e. V., Frankfurt a. M. Strecker und Schröeder Verlag, Stuttgart, Germany.

Ramírez de Velasco, Juan

1965 (1881–1897) Información acerca de la provincia de Telan y Zuraca. *Relaciones Geográficas de Indias. Perú. II.* Apéndice III. Marcos Jiménez de la Espada. Biblioteca de Autores Españoles desde la formación del lenguaje hasta nuestros días. Vol. II, pp. 75– 78. Tomo CLXXXIV, Ediciones Atlas, Madrid, Spain.

Ramírez-Horton, Susan

1982 Retainers of the lords or merchants: A case of mistaken identity? In *El Hombre y su Ambiente en los Andes Centrales*, pp. 123–136. Senri Ethnological Studies, No.10. National Museum of Ethnology, Osaka, Japan.

Ramos de Cox, Josefina

1971 ¿Transporte pre-hispánico con llamas? Pando-Lima (P. V. 47-I, II y III) (1,260 D.C.–1,535 D.C.?). *Boletín del Seminario de Arqueología* (Pontificia Universidad Católica del Perú, Instituto Riva-Agüero, Lima) 10:68–71.

Ramos de Cox, Josefina, Cirilo Huapaya, Scarlett O'Phelan, Carlos Giesecke, Eric Stromber, Henriette Knoester, and Flor Villena

1974–1975 Informe preliminar sobre el proyecto de arqueología y computación del material del Complejo Pando. *Boletín del Seminario de Arqueología* (Pontificia Universidad Católica del Perú, Instituto Riva-Agüero, Lima), Arqueología PUC, 15–16:7–12.

Real Academia Española
1970   *Diccionario de la Lengua Español*, 19th ed. Editorial Espasa-Calpe, Madrid, Spain.
1992   *Diccionario de la Lengua Española*, 21st ed., Vols. I and II. Editorial Espasa-Calpe, Madrid, Spain.

Reitz, Elizabeth J.
1976   Virú 434: An agricultural site in coastal Peru. Preliminary report, University of Georgia Museum of Natural History, Athens, GA. Unpublished manuscript, in author's possession.
1979   Faunal materials from Viru 434: An Early Intermediate Period site from coastal Peru. *Florida Journal of Anthropology* 4(2):76–92.
1988a Faunal remains from Paloma, an Archaic site in Peru. *American Anthropologist* 90(2):310–322.
1988b Preceramic animal use on the central coast. In *Economic Prehistory of the Central Andes*, pp. 31–55. Elizabeth S. Wing and Jane C. Wheeler, editors. BAR International Series 427, Oxford, UK.
1990   *Vertebrate Fauna from El Azucar, Ecuador*. Museum of Natural History, University of Georgia, Athens, GA. Mimeograph.

Reynafarge, César
1966   Iron metabolism during and after altitude exposure in man and adapted animals (camelids). *Federation Proceedings* 25:1240–1242.

Reynafarge, César, José Faura, Alfredo Paredes, and Doris Villavicencio
1968   Erythrokinetics in high-altitude-adapted animals (llama, alpaca and vicuña). *Journal of Applied Physiology* 24:93–97.

Reynafarge Dávila, Baltazar
1971   Mecanismos moleculares de la adaptación a la hipoxia de las grandes alturas. *Archivos del Instituto de Biología Andina* 4(1):1–14.

Reynafarge Dávila, Baltazar, and Mario Rosenmann
1971   Niveles de 2, 3 difosfoglicerato en el eritrocito de mamíferos. *Archivo del Instituto de Biología Andina* 4:67–73.

Ribera (Rivera), Pedro de, and Antonio de Chávez y de Guevara
1965 (1881–1897) Relación de la ciudad de Guamanga y sus términos. Año de 1586. In *Relaciones Geográficas de Indias. Perú. I*. Marcos Jiménez de la Espada. Biblioteca de Autores Españoles desde la formación del lenguaje hasta nuestros días. Vol. I, pp. 181–204. Tomo CLXXXIII, Ediciones Atlas, Madrid, Spain. [Author's Note: *See* Rivera 1974.]

Richardson, James B. III
1978   Early man on the Peruvian north coast, early maritime exploitation and the Pleistocene and Holocene environment. In *Early Man in America, From a Circum-Pacific Perspective*, pp. 274–289. A. L. Bryan, editor. Occasional Papers No. 1 of the Department of Anthropology, University of Alberta, Edmonton, AB. Archaeological Research International.

Rick, John W.
1980   *Prehistoric Hunters of the High Andes*. Academic Press, New York.
1983   *Cronología, Clima y Subsistencia en el Precerámico Peruano*. Ediciones Indea, Instituto Andino de Estudios Arqueológicos, Lima, Peru.

Ridell, Francis A., and Lidio M.Valdez
1987–1988 Hacha y la ocupación temprana del valle de Acarí. *Gaceta Arqueológica Andina* 4(16):6–10.

Ridout, César A.
1942   La vicuña (*Auchenia vicugna*). Necesidad de conservar y propagar esta especie autóctona. *Boletín del Museo de Historia Natural "Javier Prado"* 6(22–23):400–409.

Rivera, Mario
1991   The prehistory of northern Chile: A synthesis. *Journal of World Prehistory* 5(1):1–47.

Rivera, Pedro de, and Antonio de Chávez y de Guevara
1974   Relación de la Ciudad de Guamanga y sus términos. Año de 1586. In *Huamanga: Una larga historia*, pp. 147–182. Consejo Nacional de la Universidad Peruana, Lima, Peru. [Author's Note: This was taken from de Ribera y Chávez y de Guevara (1881) 1965.]

Rivero, Mariano Eduardo de
1828   Sobre las vicuñas, huanacos, llamas y alpacas del Perú. Unpublished manuscript.

Rojas, P. Alonso de (?)
1986 (1889) Relación del descubrimiento del río de las Amazonas, hoy S. Francisco del Quito, y declaración del mapa donde esta pintado. In *La Aventura del Amazonas*, pp. 231–252. G. de Carvajal, P. de Almesto, and Alonso de Rojas. Rafael Díaz, editor. Crónicas de América 19, Historia 16. Madrid, Spain.

Rollins, Harold B., James B. Richardson III, and Daniel H. Sandweiss
1986   The birth of El Niño: Geoarcheological evidence and implications. *Geoarchaeology* 1(1):3–15.

Romero, Carlos A.
1936   El Camello en el Perú. *Revista Histórica* (Lima) 10 (Entrega 3):364–372.

Romero, Emilio
n.d.   Perú: Una Nueva Geografía. Tomo I. Librería Studium, Lima, Peru.

Rosas, Virginia
1990 Francia importa 300 camélidos de Chile para crianza oficial. *El Comercio*, Año 151, No. 79,981. 15 October, p. B2. Lima, Peru.

Rosas Lanoire, Hermilio
1970 La Secuencia Cultural del Período Formativo en Ancón. Bachelor's thesis, Archaeology, Universidad Nacional Mayor de San Marcos, Lima, Peru.

Rostworowski de Diez Canseco, María
1970 Mercaderes del valle de Chincha en la época prehispánica. Un documento y unos comentarios. *Revista Española de Antropología Americana* (Madrid) 5:135–178.

1972 Las etnias del valle del Chillón. *Revista del Museo Nacional* (Lima) 38:250–314.

1977a La estratificación social y el hatun curaca en el mundo andino. *Histórica* (Lima) 1(2):249–286.

1977b *Etnia y sociedad. Costa Peruana Prehispánica.* Instituto de Estudios Peruanos, Lima, Peru. [Author's Note: A second edition was published in 1989 by the same institution.]

1978 *Señoríos indígenas de Lima y Canta.* Instituto de Estudios Peruanos, Lima, Peru.

1981 *Recursos Naturales Renovables y Pesca, siglos XVI y XVII.* Instituto de Estudios Peruanos, Lima, Peru.

1982 Comentarios a la visita de Acarí de 1593. *Histórica* (Lima) 6(2):227–254.

1983–1984 La Tasa ordenada por el Licenciado Pedro de la Gasca (1549). *Revista Histórica* (Lima) 34:53–102.

1988a *Historia del Tahuantinsuyu.* Instituto de Estudios Peruanos, Consejo Nacional de Ciencia y Tecnología, Lima, Peru.

1988b Prólogo. In *Conflicts Over Coca Fields in XVIth-Century Peru*, pp. 69–81. M. Rostworowski de Diez Canseco. Studies in Latin American Ethnohistory & Archaeology, Joyce Marcus, series editor. Memoirs of the Museum of Anthropology, University of Michigan, No. 21. Ann Arbor, MI.

Rouse, Irving, and José María Cruxent
1963 Some recent radiocarbon dates for western Venezuela. *American Antiquity* 28(4):537–540.

Rowe, Ann P.
1977 *Warp-Patterned Weaves of the Andes.* The Textile Museum, Washington, DC.

Rowe, John Howland
1946 Inca culture at the time of the Spanish conquest. In *Handbook of South American Indians*, Vol. 2, *The Andean Civilizations*, pp. 183–330. Julian H. Steward, editor. Bulletin 143, Bureau of Ethnology, Smithsonian Institution, Washington, DC.

Ruiz de Arce, Juan
1968 (1933) *Advertencias.* Biblioteca Peruana, Primera serie. Tomo I, pp. 405–437. Editores Técnicos Asociados, Lima, Peru.

Sáez-Godoy, Leopoldo
1979 Introducción y notas. In *Crónica y relación copiosa y verdadera de los Reinos de Chile (1558).* Edición de Leopoldo Sáez-Godoy. Introducción: pp. v–xix; Notas: pp. 258–324. Bibliotheca Ibero-Americana, Colloquium Verlag, Berlin, Germany.

Sagástegui Alva, Abundio
1973 *Manual de las malezas de la costa norperuana.* Talleres Gráficos de la Universidad Nacional de Trujillo, Trujillo, Peru.

Salas, Alberto Mario
1950 *Las Armas de la Conquista.* Emecé Editores, Buenos Aires, Argentina.

Salas, Carla, and Sandro Cruz
1998 Alpacas peruanas en Suiza. *Revista Dominical de Expreso. Día Siete.* 8 November, pp. 3–5. Lima, Peru.

Salinas Loyola, Juan de
1965a (1881–1897) Relación de la ciudad de Sant Miguel de Piura. In *Relaciones Geográficas de Indias. Perú. II.* Marcos Jiménez de la Espada. Biblioteca de Autores Españoles desde la formación del lenguaje hasta nuestros días. Vol. II, pp. 33–45. Tomo CLXXXIV, Ediciones Atlas, Madrid, Spain. [Author's Note: In 1925 it was published as Anonymous and the date given was 1577, which is wrong, as it is 1571. It was published in *Relaciones geográfico-estadísticas del Perú fechas por el orden de las instrucciones y memoriales que mando despachar Su Majesta en 1577.* Colección de Libros y Documentos referentes a la Historia del Perú. Tomo V (2a parte), pp. 73–98 Imprenta y Librería Sanmartí y Cia., Lima, Peru.]

1965b (1881–1897) Relación y descripción de la ciudad de Loxa. In *Relaciones Geográficas de Indias. Perú. III.* Marcos Jiménez de la Espada. Biblioteca de Autores Españoles desde la formación del lenguaje hasta nuestros días. Vol. II, pp. 291–306. Tomo CLXXXIV, Ediciones Atlas, Madrid, Spain.

1965c (1881–1897) Descubrimiento, conquistas y poblaciones de Juan de Salinas Loyola. In *Relaciones Geográficas de Indias. Perú. IV.* Ultimo Apéndice. Marcos Jiménez de la Espada. Biblioteca de Autores Españoles desde la formación del lenguaje hasta nuestros días.

Vol. III, pp. 197–218. Tomo CLXXXV, Ediciones Atlas, Madrid, Spain. [Author's Note: It includes several accounts that have here been considered as a unit.]

1965d (1881–1897) Relación de la ciudad de Zamora de los Alcaides. In *Relaciones Geográficas de Indias. Perú. IV.* Marcos Jiménez de la Espada. Biblioteca de Autores Españoles desde la formación del lenguaje hasta nuestros días. Vol. III, pp. 125–135. Tomo CLXXXV, Ediciones Atlas, Madrid, Spain.

Salomon, Frank
1980   *Los señores étnicos de Quito en la época de los Incas.* Colección Pendoneros, No. 10, pp. 9–370. Instituto Otavaleño de Antropología, Otavalo, Ecuador. [Author's Note: An English version was published in 1986: *Native Lords of Quito in the Age of the Incas. The Political Economy of the North Andean Chiefdoms.* Cambridge Studies in Social Anthropology 59. Cambridge University Press, Cambridge, UK.]

Sámano-Xerez
1968 (1528) *La Relación Sámano-Xerez.* Biblioteca Peruana. Primera Serie. Tomo I, pp. 7–14. Editores Técnicos Asociados, Lima, Peru.

Sancho de la Hoz, Pedro
1968 (1534) *Relación para Su Majestad.* Biblioteca Peruana. Primera Serie. Tomo I, pp. 275– 343. Editores Técnicos Asociados, Lima, Peru.

Sandoval, Wilfredo
2001   Logran criar alpacas en la costa. ¡Que San José las acoja! *El Comercio*, Año 162, No. 83,916. 23 July, pp. A1, A10. Lima, Peru.

San Martín, Felipe, and Fred C. Bryant
1987   *Nutrición de los Camélidos Sudamericanos: Estudio de nuestro conocimiento.* Programa Colaborativo de Apoyo a la Investigación en Rumiantes Menores. Facultad de Medicina Veterinaria, Instituto Veterinario de Investigaciones Tropicales y de Altura, Universidad Nacional Mayor de San Marcos; Department of Range and Wildlife Management, College of Agricultural Sciences, Texas Tech University, Lubbock. Technical report T-9-505.
1988   Comparación de las tasas de pasaje de la fase líquida y de la fase sólida en el tracto digestivo de llama y ovino. *Investigaciones sobre pastos y forrajes de Texas Tech University en el Perú.* Tomo IV, pp. 84–95. Technical report T-9-550 of the College of Agricultural Sciences, Texas Tech University, Lubbock.

San Martín, Felipe, J. A. Pfister, L. Rosales, R. Farfán, and T. Huisa
1989   Comportamiento alimenticio al pastoreo de llamas, alpacas y ovinos en los Andes del

Perú. In *Investigaciones sobre pastos y forrajes de Texas Tech University en el Perú.* Vol . V, pp. 97–111. Felipe San Martín and Fred C. Bryant, editors. Technical report T-9-584 of the College of Agricultural Sciences, Texas Tech University, Lubbock.

San Martín, Gaspar de, and Juan Mosquera
1990   Visita al pueblo del Ynga. 1559. In *Visita y numeración de los pueblos del valle de los Chillos, 1551–1559*, pp. 159–168. Cristóbal Landázuri N., compiler. Fuentes para la Historia Andina. Marka. Abyala, Quito, Ecuador.

San Román C., Jorge
1993   Domestic camelids in Bolivia. In *European Symposium on South American Camelids*, pp. 249–254. Martina Gerken and C. Renieri, editors. Rheinische Friedrich-Wilhelms Universität, Bonn; Università degli Studi di Camerino, Facoltà di Medicina Veterinaria, Matelica, Italy.

Santa Cruz, Alcibíades
1942   La alimentación de los Mapuches antes de la Conquista. *Boletín de la Sociedad de Biología de Concepción* 16:5–10.

Santillán, Hernando de
1968 (1879) *Relación.* Biblioteca Peruana, Tomo III, pp. 365–463. Editores Técnicos Asociados, Lima, Peru. [Author's Note: In other editions it was published with the title *Relación del origen, descendencia, politica y gobierno de los Incas.*]

Santisteban Manrique, Ernesto
1963   Identificación de cuero y fanereos de Conanoxa. Investigaciones arqueológicas en las terrazas de Conanoxa, valle de Camarones (Provincia de Tarapacá), Appendix 3. Hans Niemeyer F. and Virgilio Schiappacasse F. *Revista Universitaria* (Santiago) 48:160–161.

Scaro, José Leonardo, and Mario Carlos Aggio
1966   Erythropoietin in high altitude resident animals. *Revue Canadienne de Biologie* 25:209–211.

Schaedel, Richard P.
1972   The city and the origin of the state in America. In *Actas y Memorias*, XXXIX Congreso Internacional de Americanistas, Tomo 2, pp. 15–33. Lima, Peru.
1985   Coast-highland interrelationship and ethnic groups in northern Peru (500 B.C.–A.D. 1980). In *Andean Ecology and Civilization. An Interdisciplinary Perspective on Andean Ecological Complementarity*, pp. 443–474. Shozo Masuda, Izumi Shimada, and Craig Morris, editors. University of Tokyo Press, Tokyo, Japan.

Schjellerup, Inge
1992   Patrones de asentamiento en las faldas orientales de los Andes de la región de

Chachapoyas. In *Estudios de Arqueología Peruana*, pp. 355–373. Duccio Bonavia, editor. Fomciencias, Lima, Peru.

Schmidel, Hulderico (Ulrico)

1749 (1531–1534) *Historia y descubrimiento del Río de la Plata y Paraguay por Hulderico Schmidel*. Historiadores primitivos de las Indias Occidentales, Vol. 3. Andrés Gonzáles Barcia, editor. Imprenta de Juan Zúñiga, Madrid, Spain. [Author's Note: This author is variously called Schmidt, Schmidts, and Schmidl.]

(Ulrich)

1962 *Wahrhafftige Historien einer wunderbaren Schiffart*. Akademische Druck-u. Verlagsanstalt. (Vol. 1 de "Frühe Reisen und Seefahrten in Original berichten.") Graz, Austria. [Author's Note: This is a facsimile edition of the one printed by Levinius Hulsius, Nuremberg, Germany, in 1602.]

(Ulrico)

1986 *Relatos de la conquista del Río de 1a Plata y Paraguay, 1534–1554*. El Libro de Bolsillo, Alianza Editorial, Madrid, Spain.

Schmidt, Max

1929 *Kunst und Kultur von Peru*. Impropyläen-Verlag zu Berlin, Germany.

Schonfelder, Uwe

1981 Cerámica fina y hallazgos menores. In *Cochasquí: Estudios Arqueológicos*, pp. 153–282. Serie: Arqueológica. Udo Oberem, compiler. Colección Pendoneros, 4. Otavalo, Ecuador.

Scossiroli, Renzo E.

1984 *L'Uomo e l'agricultura. Il problema delle origini*. Edagricola, Bologna, Italy.

Seibert, P.

1983 Human impact on landscape and vegetation in the central high Andes. In *Man's Impact on Vegetation*, pp. 261–276. W. Holzner, M. J. A. Werger, and I. Ikusima, editors. W. Junk, The Hague, Boston, London.

Seifert, H. S. H., H. Böhnel, S. Heiterfuss, J. Rengel, R. Schaper, U. Sukup, and U. Wernery

1992 Isolation of *C. perfringens* Type A from enterotoxaemia in camels and production of a locality-specific vaccine. In *Proceedings of the First International Camel Conference, 2nd–6th February 1992*, pp. 65–68. W. R. Allen, A. J. Higgins, I. G. Mayhew, D. H. Snow, and J. F. Wade, editors. R.&W. Publications (Newmarket), Amersham, Switzerland.

Shea, Daniel E.

n.d. [1985?] Preliminary discussion of prehistoric settlement patterns and relation to terracing in Achoma, Colca Valley, Arequipa, Peru. Mimeograph.

Shimada, Izumi

1978 Economy of a prehistoric urban context: Commodity and labor flow at Moche V Pampa Grande. *American Antiquity* 43(4): 569–592.

1979 Primer informe sobre el Proyecto Arqueológico Batán Grande-La Leche. Presentado al INC. Mimeograph.

1982 Horizontal archipelago and coast-highland interaction in North Peru: Archaeological models. In *El Hombre y su Ambiente en los Andes Centrales*, pp. 137–210. Luis Millones and Hiroyasu Tomoeda, editors. Senri Ethnological Studies, No. 10. National Museum of Ethnology, Osaka, Japan. [Author's Note: Inside the publication it reads 1980, whereas the cover says 1982.]

1985 Introduction. In *Andean Ecology and Civilization. An Interdisciplinary Perspective on Andean Ecological Complementarity*, pp. xi–xxxii. Shozo Masuda, Izumi Shimada, and Craig Morris, editors. University of Tokyo Press, Tokyo, Japan.

Shimada, Izumi, Carlos G. Elera, and Melody Shimada

1982 Excavaciones efectuadas en el centro ceremonial de Huaca Lucía-Cholope, del Horizonte Temprano, Batán Grande, Costa Norte del Perú. *Arqueológicas* (Museo Nacional de Antropología y Arqueología, Instituto Nacional de Cultura, Lima) 19:109–210.

Shimada, Izumi, Stephen Epstein, and Alan K. Craig

1982 Batán Grande: A prehistoric metallurgical center in Peru. *Science* 216:952–959.

Shimada, Melody

1982 Zooarchaeology of Huacaloma: Behavioral and cultural implications. In *Excavations at Huacaloma in the Cajamarca Valley, Peru, 1979*, pp. 303–336. Report 2 of the Japanese Scientific Expedition to Nuclear America. Kazuo Terada and Yoshio Onuki, editors. University of Tokyo Press, Tokyo, Japan.

1985 Continuities and change in patterns of faunal resource utilization: Formative through Cajamarca periods. In *The Formative Period in the Cajamarca Basin, Peru. Excavations at Huacaloma and Layzon, 1982*, pp. 289–310. Report of the Japanese Scientific Expedition to Nuclear America. Kazuo Terada and Yoshio Onuki, editors. University of Tokyo Press, Tokyo, Japan.

1988 Prehistoric subsistence in the north highlanda of Peru: Early Horizon to Late Intermediate [sic]. In *Economic Prehistory of the Andes*, pp. 131–145. Elizabeth S. Wing and Jane C.

Wheeler, editors. BAR International Series 427. Oxford, UK.

Shimada, Melody, and Izumi Shimada

1981 Explotación y manejo de los recursos naturales en Pampa Grande, sitio Moche V. Significado del análisis orgánico. *Revista del Museo Nacional* (Lima) 45:19–73.

1985 Prehistoric llama breeding and herding on the north coast of Peru. *American Antiquity* 50(1):3–26.

1987 Comment on the functions and husbandry of alpaca. *American Antiquity* 52(4):836–839.

Silva Santisteban, Fernando

1964 *Los Obrajes en el Virreinato del Perú*. Publicaciones del Museo Nacional de Historia, Lima, Peru.

Silverman, Helaine

1985 Cahuachi: Simplemente monumental. *Boletín de Lima* 7(41):85–95.

1988 Cahuachi: Non-urban cultural complexity on the south coast of Peru. *Journal of Field Archaeology* 15(4):403–430.

1993 *Cahuachi in the Ancient Nasca World*. University of Iowa Press, Iowa City, IA.

Sillau A., Hugo, Sergio Cueva, A. Valenzuela, and A. E. Candela

1976 O$_2$ transport in the alpaca (*Lama pacos*) at sea level and at 3,300 m. *Respiration Physiology* 27:147–155.

Simoens da Silva, Antonio Carlos

1980 Los camélidos andinos. In *El Perú visto por viajeros brasileños*, pp. 53–67. Estuardo Núñez. Centro de Estudios Brasileños. Gráfica Morsom, Lima, Peru. [Author's Note: This article was published in 1930 under the title *Viageros etnographicos Sul americanos: Peru*. Imprensa Nacional, Rio de Janeiro, Brazil.]

Simpson, George Gaylord

1950 History of the fauna of Latin America. *American Scientist* 38(3):361–389.

Simpson, George Gaylord, and Carlos de Paula Couto

1981 Fossil mammals from the Cenozoic of Acre, Brazil. 3. Pleistocene Edentata Pilosa, Proboscidea, Sirenia, Perissodactyla and Artiodactyla. *Iheringia Service of Geology* 6:11–74. [Author's Note: This was taken from *Biological Abstracts* 74(3):1638, ref. 15771, 1982.]

Simpson Vuilleumier, Beryl

1971 Pleistocene changes in the fauna and flora of South America. *Science* 173(3999):771–780.

Smith, Joseph E., Narla Mohandas, and Stephen B. Shoet

1979 Variability in erythrocyte deformability among various mammals. *American Journal of Physiology* 236:725–730.

Snarskis, Michael J.

1976 Stratigraphic excavations in the eastern lowlands of Costa Rica. *American Antiquity* 41(3):342–353.

Sotelo Narváez, Pedro

1965 (1881–1897) Relación de las provincias de Tucumán que dio Pedro Sotelo Narvaez, vecino de aquellas provincias, al muy ilustre señor licenciado Cepeda, presidente desta Real Audiencia de La Plata. In *Relaciones Geográficas de Indias. Perú. II.* Marcos Jiménez de la Espada. Biblioteca de Autores Españoles desde la formación del lenguaje hasta nuestros días. Vol. I, pp. 390–396. Tomo CLXXXIII, Ediciones Atlas, Madrid, Spain.

Soto, Isabel de

1992 Notes sur l'edition originale de *Nouvelle certaines des Isles du Peru. Nouvelles certaines des Isles du Peru, suivi de Nouvelles du Pérou de Miguel de Estete*, p. 11. Amiot, Lenganey. Thaon, France.

Soukup SDB, Jaroslav

n.d. [1987] *Vocabulario de los nombres vulgares de la flora peruana y catálogo de los géneros*. Editorial Salesiana, Lima, Peru.

Spunticchia, Roberto

1989–1990 *Modelli di insediamento e di sussistenza proponibili per i gruppi di cacciatori e raccoglitori della "puna" di Junín (Ande Centrali peruviane)*. Thesis in paleontology, Università degli Studi di Roma "La Sapienza," Faccoltà di Lettere e Filosofia, Roma, Italy.

Squier, E. George

1877 *Incidents of Travel and Exploration in the Land of the Incas*. Harper Brothers, New York.

1974 *Un viaje por tierras incaicas: Crónica de una expedición arqueológica (1863–1865)*. Edición auspiciada por la Universidad Nacional Mayor de San Marcos, con la colaboración de la Embajada de los Estados Unidos de América, para conmemorar el sesquicentenario de la Independencia del Perú. Buenos Aires, Argentina.

Stahl, Peter W.

1988 Prehistoric camelids in the lowlands of western Ecuador. *Journal of Archaeological Science* 15(4):355–366.

Stahl, Peter W., and Presley Norton

1987 Precolumbian animal domesticates from Salango, Ecuador. *American Antiquity* 52(2):382–391.

Stanish, Charles S.

1990 Economías agrarias post-Tiwanaku en la cuenca del río Moquegua. In *Trabajos Arqueológicos en Moquegua, Perú*, Vol. 2, pp. 115–160. L. K.

Watanabe, M. E. Moseley, and F. Cabieses, compilers. Programa Contisuyo del Museo Peruano de Ciencias de la Salud. Southern Peru Copper Corporation, Lima, Peru.

Stanley, Helene F., Miranda Kadwell, and Jane C. Wheeler
1994 Molecular evolution of the family Camelidae: A mitochondrial DNA study. *Proceedings of the Royal Society of London*, Series B, Biological Sciences 256(1345):1–6.

Steele, Arthur Robert
1964 *Flowers for the King (The Expedition of the Ruiz and Pavon and the Flora of Peru)*. Duke University Press, Durham, NC.

Stehberg, Rubén
1980 Aproximación metodológica al estudio del poblamiento humano de los Andes de Santiago (Chile). *Boletín del Museo de Historia Natural* (Santiago) 37:9–39.

Stehberg, Rubén, and Tom D. Dillehay
1988 Prehistoric human occupation in the arid Chacabuco-Colina ecotone in central Chile. *Journal of Anthropological Archaeology* 7:136–162.

Stevens, D. H., G. H. Burton, Julio Sumar Kalinowski, and Peter W. Nathanielsz
1980 Ultrastructural observations on the placenta of the alpaca (*Lama pacos*). *Placenta* 1:21–32.

Steward, Julian H.
1948 Tribes of the montana [sic]: An introduction. In *Handbook of South American Indians*, Vol. 3, *The Tropical Forest Tribes*, pp. 507–533. Julian H. Steward, editor. Bulletin 143, Bureau of American Ethnology, Smithsonian Institution, Washington, DC.
1949 South American cultures: An interpretative summary. In *Handbook of South American Indians*, Vol. 5, *The Comparative Ethnology of South American Indians*, pp. 669–772. Julian H. Steward, editor. Bulletin 143, Bureau of American Ethnology, Smithsonian Institution, Washington, DC.

Steward, Julian H., and Alfred Métraux
1948 Tribes of the Peruvian and Ecuadorian montana [sic]. In *Handbook of South American Indians*, Vol. 3, *The Tropical Forest Tribes*, pp. 535–656. Julian H. Steward, editor. Bulletin 143, Bureau of American Ethnology, Smithsonian Institution, Washington, DC.

Stiglich, Germán
1922 *Diccionario geográfico del Perú*. Imp. Torres Aguirre, Lima, Peru.

Stone, Doris
1972 *Pre-Columbian Man Finds Central America. The Archaeological Bridge*. A Peabody Museum Press Book. Harvard University, Cambridge, MA.

Stothert, Karen E., and Jeffrey Quilter
1991 Archaic adaptations of the Andean region, 9000 to 5000 B.P. *Revista de Arqueología Americana* 4:25–53.

Strong, William Duncan
1957 Paracas, Nazca, and Tiahuanacoid cultural relationships in south coastal Peru. *American Antiquity* 22(4, Pt. 2). Memoirs of the Society for American Archaeology, No. 13.

Strong, William Duncan, and John M. Corbett
1943 *A Ceramic Sequence at Pachacamac*. Columbia Studies in Archaeology and Ethnology, Vol. I, pp. 27–122. Columbia University Press, New York.

Strong, William Duncan, and Clifford Evans Jr.
1952 *Cultural Stratigraphy in the Virú Valley, Northern Peru. The Formative and Florescent Epochs*. Columbia Studies in Archaeology and Ethnology, Vol. IV. Columbia University Press, New York.

Stübel, Alphons
1885 (Letter sent to Johann Jakob von Tschudi). Das Lama (*Auchenia Lama* Fisch) in seinen Beziehungen zum altperuanischen Volksleben. Johann Jakob von Tschudi. Footnote 1, p. 96. *Zeitschrift für Ethnologie*. Organ der Berliner Gesellschaft für Anthropologie, Ethnologie und Urgeschichte, Band 17. Verlag von A. Asher & Co., Berlin, Germany.
1891 (1885) (Letter sent to Johann Jakob von Tschudi). In *Culturhistorische und Sprachliche Beiträge zur Kenntniss des Alten Perú*. Johan Jakob von Tschudi. Footnote 1. Denkschriften Kaiserlichen Akademie der Wissenschaften in Wien. Philosophisch-Historische Classe, Band 39-1, pp. 98. In Comission bei F. Tempsky, Buchhändler der Akademie der Wissenschaften, Wien, Austria.
1918 (1885) (Letter sent to Johann Jakob von Tschudi). In *Contribuciones a la historia, civilización y lingüistica del Perú Antiguo*, Tomo I, footnote 127, p. 209. Johann Jakob von Tschudi. Colección de Libros y Documentos referentes a la Historia del Perú. Translated by Germán Torres Calderón. Imprenta y Librería Sanmarti y Cía., Lima, Peru.
1969 (1885) (Letter sent to Johann Jakob von Tschudi). La Llama (*Auchenia Lama* Fisch) [sic]. Johann Jakob von Tschudi. Footnote 10. J. von Tschudi y la ethnohistoria peruana. Prólogo a la traducción de "La Llama" de Von Tschudi. Glynn Custred. In *Mesa Redonda de Ciencias Prehistóricas y Antropológicas*

Tomo I, pp. 126–127. Pontificia Universidad Católica del Perú. Instituto Riva-Agüero, Seminario de Antropología, Lima, Peru.

Sumar Kalinovski, Julio

1985 Introducción. In *Bibliografía de los Camélidos Sudamericanos*, pp. 11–13. Universidad Nacional Mayor de San Marcos. C. I. Instituto Veterinario de Investigaciones Tropicales y de Altura (IVITA). Convenio CIID-Canadá, Lima, Peru.

1988 Present and potential role of South American camelids in the high Andes. *Outlook on Agriculture* 17(1):23–29.

1992 Los camélidos domésticos en el Perú. *Boletín de Lima* 14(79):81–95.

Szaszdi, Adam

1988 *D. Diego Tomala, cacique de la isla de la Puná. Un caso de aculturación socioeconómica*. Museo Antropológico, Banco Central del Ecuador, Guayaquil, Ecuador.

Tabío, Ernesto E.

1977 *Prehistoria de la costa del Perú*. Academia de Ciencias de Cuba. Havana, Cuba.

Tarragó, Myriam N.

1978 Paleoecology of the Calchaqui Valley, Salta Province, Argentina. In *Advances in Andean Archaeology*, pp. 487–512. David L. Browman, editor. Mouton, The Hague and Paris.

Tauro, Alberto

1988 *Enciclopedia Ilustrada del Perú* (Tomo 4:1572–1573 and Tomo 5:1838–1839). Peisa, Lima, Peru.

Taylor, Anne-Christine, and Philippe Descola

1981 El conjunto jívaro en los comienzos de la conquista española del alto Amazonas. *Bulletin de l'Institut Français d'Études Andines* (Lima) 10(3–4):7–54.

Téllez Lúgaro, Eduardo, and Osvaldo Silva Galdames

1989 Atacama en el siglo XVI. La conquista hispana en la periferia de los Andes Meridionales. *Cuadernos de Historia* (Santiago de Chile) 9:45–69.

Tello, Julio C.

1942 *Origen y Desarrollo de las Civilizaciones Prehistóricas Andinas*. Lima, Peru.

1956 *Arqueología del Valle de Casma. Culturas: Chavín, Santa o Huaylas Yunga y Sub-Chimú*. Publicación Antropológica del Archivo "Julio C. Tello" de la Universidad Nacional Mayor de San Marcos, Vol. I. Editorial San Marcos, Lima, Peru.

Tello, Julio C., and Toribio Mejía Xesspe

1979 *Paracas. II Parte. Cavernas y Necrópolis*. Publicación Antropológica del Archivo "Julio C. Tello." Universidad Nacional Mayor de San Marcos, Lima, Peru; Institute of Andean Research, New York.

Tenney, S. Marsh, and John E. Remmers

1966 Alveolar dimensions in the lungs of animals raised at high altitude. *Journal of Applied Physiology* 21:1328–1330.

Terada, Kazuo

1979 *Excavations at La Pampa in the North Highlands of Peru, 1975. Report 1 of the Japanese Scientific Expedition to Nuclear America*. University of Tokyo Press, Tokyo, Japan.

Terada, Kazuo, and Yoshio Onuki

1982 *Excavations at Huacaloma in the Cajamarca Valley, Peru, 1979. Report 2 of the Japanese Scientific Expedition to Nuclear America*. University of Tokyo Press, Tokyo, Japan.

1985 *The Formative Period in the Cajamarca Basin, Peru: Excavations at Huacaloma and Layzón, 1982. Report 3 of the Japanese Scientific Expedition to Nuclear America*. University of Tokyo Press, Tokyo, Japan.

Thévenin, René

1961 *El origen de los animales domésticos*. Eudeba, Editorial Universitaria de Buenos Aires, Buenos Aires, Argentina.

Thomas, R. Brooke

1977 Adaptación humana y ecología de la puna. In *Uywamichiq punarunakuna. Pastores de puna*, pp. 87–111. Jorge A. Flores Ochoa, compiler. Instituto de Estudios Peruanos, Lima, Peru.

Toledo, Francisco de

1975 *Tasa de la Visita General de Francisco de Toledo*. Introducción y versión paleográfica de Noble David Cook. Universidad Nacional Mayor de San Marcos, Dirección Universitaria de Biblioteca y Publicaciones, Lima, Peru.

1978 (1575) Ordenanzas para los indios de todos los repartimientos y pueblos de este reino del Perú. IV. La legislación del Virrey Toledo. In *El Servicio personal de los indios en el Perú (extractos del siglo XVI)*. Silvio Zavala. Tomo I, pp. 142–145. El Colegio de México, Mexico City, Mexico. [Author's Note: This version is summarized and annotated.]

Tomoeda, Hiroyasu

1988 "La llama es mi chacra": El mundo metafórico del pastor andino. In *Llamichos y Paqocheros. Pastores de llamas y alpacas*, pp. 225–235. Jorge A. Flores Ochoa, compiler. CONCYTEC, CEAC y Editorial UNSAAC, Cuzco, Peru.

1993 Los Ritos Contemporáneos de Camélidos y la Ceremonia de la Citua. El Mundo Ceremonial Andino. Luis Millones and Yoshio Onuki, editors. *Senri Ethnological Studies* No. 37, pp. 289–306. National Museum of Ethnology, Osaka, Japan.

Tonni, E. P., and Gustavo G. Politis

1980   La distribución del guanaco (Mammalia, Camelidae) en la Provincia de Buenos Aires durante el Pleistoceno Tardío y Holoceno: Factores climáticos como causa de su retracción. *Ameghiniana* 17(1):53–66.

Topic, John R., Jr.

1978   Growth and development: Chan Chan during the Middle Horizon and Late Intermediate period. Trent University, Peterborough, UK. Mimeograph.

1980   Excavaciones en los barrios populares de Chanchan. In *Chanchan. Metrópoli Chimú*, pp. 267–282. Rogger Ravines, editor. Instituto de Estudios Peruanos. Instituto de Investigación Tecnológica, Industrial y de Normas Técnicas, Lima, Peru.

1982   Lower-class social and economic organization at Chan Chan. In *Chan Chan: Andean Desert City*, pp. 145–175. Michael E. Moseley and Kent C. Day, editors. A School of American Research Book. University of New Mexico Press, Albuquerque, NM.

1990   Craft production in the kingdom of Chimor. In *The Northern Dynasties: Kingship and Statecraft in Chimor*, pp. 145–176. Michael E. Moseley and Alana Cordy-Collins, editors. Dumbarton Oaks Research Library and Collection, Washington, DC.

Topic, Theresa Lange, Thomas H. McGreevy, and John R.Topic

1987   A comment on the breeding and herding of llamas and alpacas on the north coast of Peru. *American Antiquity* 52 (4):832–835.

Torres, Hernán

1992a   Antecedentes, Objetivos y Limitaciones del Plan. In *South American Camelids. An Action Plan for Their Conservation / Camélidos Silvestres Sudamericanos. Un Plan de Acción para su Conservación*, pp. 32–36. Hernán Torres, compiler-editor. IUCN, Gland, Switzerland.

1992b   Resumen ejecutivo. In *South American Camelids. An Action Plan for Their Conservation / Camélidos Silvestres Sudamericanos. Un Plan de Acción para su Conservación*, p. 31. Hernán Torres, compiler-editor. IUCN, Gland, Switzerland.

(Compiler-editor)

1992c   *South American Camelids. An Action Plan for Their Conservation / Camélidos Silvestres Su-damericanos. Un Plan de Acción para su Conservación*. IUCN, Gland, Switzerland.

Torres Saldamando, Enrique

1888   *Libro Primero de Cabildos de Lima*. 3 vols. Paris, France.

Tosi, Joseph A.

1960   *Zonas de vida natural en el Perú*. Instituto Interamericano de Ciencias Agrícolas de la OEA, Zona Andina. Boletín Técnico, No. 5, pp. 1–271. Lima, Peru.

Tovar S., Augusto

1973   El Parque Nacional del Huascarán. *El Serrano* (Lima) 22(279):10–15.

Tovar, Oscar

1973   Comunidades vegetales de la Reserva Nacional de Vicuñas de Pampa Galeras, Ayacucho, Perú. *Museo de Historia Natural Javier Prado*, Vol. 27(B):1–32.

Trimborn, Hermann

1936   Prólogo. In *Fuentes de la historia cultural de la América precolombina*, pp. xii–xiv. *Quellen zur Kulturgeschichte des präkolumbischen Amerika*. Published under the direction of H. Trimborn. Strecker und Schröder Verlag, Stuttgart,Germany

1954   *Pascual de Andagoya: Ein Mensch erlebt die Conquista*. Universität Hamburg. Abhandlungen aus dem Gebiet des Auslandskunde, Band 59, Reihe B. Völkerkunde, Kulturgeschichte und Sprachen, Band 33, pp. 1–154. Cram, de Gruyter & Co., Hamburg, Germany.

1975   Los valles de Caplina y Sama. In *Investigaciones Arqueológicas en los valles del Caplina y Sama (Dep. Tacna, Perú)*. H. Trimborn, O. Kleemann K., J. Narry, and W. Wurster. Studia Instituti Anthropos. Vol. 25, pp. 13–60. Editorial Verbo Divino, Estella (Navarre), Spain.

Troll, Carl

1931   Die geographischen Grundlagen der Andinen-Kulturen und des Inkareiches. *Iberoamericanisches Archiv* 5:258–294. [Author's Note: There is a Spanish translation. *See* Troll 1935.]

1935   Los fundamentos geográficos de las civilizaciones Andinas y del Imperio Incaico. *Revista Universitaria* 8(9):127–183.

1943   Die Stellung der Indianer-Hochkulturen im Landschaftsaufbau der Tropischen Anden. Zeitschrift d. Gesellshaft f. Erdkunde, No. 3–4. Berlin, Germany. [Author's Note: There is a Spanish translation. *See* Troll 1958.]

1958   *Las Culturas Superiores Andinas y el medio geográfico*. Publicaciones del Instituto de

Geografía, Facultad de Letras, Universidad Nacional Mayor de San Marcos, Lima, Peru.

1968 The cordilleras of the tropical Americas: Aspects of climatic phytogeographical and agrarian ecology. In *Geo-Ecology of the Mountainous Regions of the Tropical Americas/Geo-Ecología de las Regiones Montañosas de las Américas Tropicales*, pp. 15–56. Carl Troll. Colloquium Geographicum herausgegeben von Geographischen Institut der Universität Bonn. Ferd. Dümmlers Verlag, Bonn, Germany.

Trucco, Giovanni (a cura di)
1936 *Grande Dizionario Enciclopedico*, Vol. VII, p. 83. Unione Tipografico-Editrice Torinese, Turin, Italy.

1937 *Grande Dizionario Enciclopedico*, Vol. VIII, p. 946. Unione Tipografico-Editrice Torinese, Turin, Italy.

Trujillo, Diego de
1968 *Relación del descubrimiento del reyno del Perú.* Biblioteca Peruana, Primera Serie, Tomo II, pp. 9–29. Editores Técnicos Asociados, Lima, Peru.

Tschopik, Harry
1946 The Aymara. In *Handbook of South American Indians*, Vol. 2, *The Andean Civilizations*, pp. 501–573. Julian Steward, editor. Bulletin 143, Bureau of American Ethnology, Smithsonian Institution, Washington, DC.

Tschudi, Johann Jakob von
1846 *Perú. Reiseskizzen aus den Jahren 1838–1842.* St. Gallen, Switzerland. [Author's Note: There is a Spanish translation. *See* Tschudi 1966.]

1885 Das Lama (*Auchenia Lama* Fisch) in seinen Beziehungen zum altperuanischen Volksleben. In *Zeitschrift für Ethnologie*, pp. 93–109. Organ der Berliner Gesellschaft für Anthropologie, Ethnologie und Urgeschichte, Band 17. Verlag von A. Asher & Co., Berlin, Germany. [Author's Note: The same text was included in the 1891 edition with slight changes in some notes and additions in others. The section on the llama has not changed.]

1891 *Culturhistorische und Sprachliche Beiträge zur Kenntniss des Alten Perú.* Denkschriften Kaiserlichen Akademie der Wissenschaften in Wien. Philosophisch-Historische Classe, Band 39-1. In Commision bei F. Tempsky, Buchhändler der Akademie der Wissenschaften, Wien, Austria. [Author's Note: There is a Spanish translation. *See* Tschudi 1918.]

1918 *Contribuciones a la historia, civilización y lingüistica del Perú Antiguo.* Tomo I, pp. 1–236. Colección de Libros y Documentos referentes a la Historia del Perú. Germán Torres Calderón, translator. Imprenta y Librería Sanmarti y Cía., Lima, Peru.

1966 *Testimonio del Perú. 1838–1842.* Consejo Económico Consultivo Suiza-Perú, Lima, Peru.

1969 La Llama (*Auchenia Lama* Fisch) [sic] J. J. von Tschudi y la etnohistoria peruana. Prólogo a la traducción de "La Llama" de Von (Sic) Tschudi. Glynn Custred. In *Mesa Redonda de Ciencias Prehistóricas y Antropológicas*, Tomo I, pp. 123–138. Pontificia Universidad Católica del Perú, Instituto Riva-Agüero, Seminario de Antropología. Lima, Peru. [Author's Note: This version is part of the studies published by Tschudi in 1885 and 1891. The first Spanish translation was done in 1918; it is a poor, incomplete translation with mistakes.]

Ubbelohde Doering, Heinrich
1959 Bericht über archäologische Feldarbeiten in Perú. II. *Ethnos* 24(1–2):1–32.

Uhle, Max
1906 Los kjoekkenmöeddings del Perú. *Revista Histórica* I, Trimestre I:3–23.

Ulloa Mogollón, Juan de, Diego Hernández Talavera, Hernando Medel de la Feria, Amador Gonzáles, Gonzalo Gómez de Butrón, Miguel Nina Taypi, Juan Caquia, Francisco Inga Pacta, Diego Chacha, and Diego Chuqui Anco
1965 (1881–1897) Relación de la provincia de los Collagua para la discrepción de las Indias que Su Magestad manda hacer. In *Relaciones Geográficas de Indias. Perú. II.* Marcos Jiménez de la Espada. Biblioteca de Autores Españoles desde la formación del lenguaje hasta nuestros días. Vol. I, pp. 326–333. Tomo CLXXXIII, Ediciones Atlas, Madrid, Spain. [Author's Note: This account was published under the name of its first author alone in 1988, as part of *The Cultural Ecology, and History of Terracing and Terrace Abandonment in the Colca Valley of Southern Peru*, Vol. II, pp. 79–86. William M. Denevan, editor, Madison, WI.]

(Ulrich). See Schmidel, Hulderico.

(Ulrico). See Schmidel, Hulderico.

Unciga, Miguel
1981 (1575) Presentación. Primera información hecha por Don Juan Colque Guarache, cerca de sus predecesores y subcesión en el cacicazgo mayor de los Quillacas, Asanaques, Sivaroyos, Uruqillas y Haracapis, y de sus seruicios a fauor de Su Magestad en la

conquista, allanamiento y pacificación deste reino del Pirú. Año 1575. Juan Colque Guarache. El Reino Aymara de Quillaca-Asanaque, siglos XV y XVI. Waldemar Espinoza Soriano. *Revista del Museo Nacional* (Lima) 45:245–248.

Universidad Nacional Mayor de San Marcos. C. I. Instituto Veterinario de Investigaciones Tropicales y de Altura-IVITA. Centro de Información Científica de Camélidos Sudamericanos.

1985   *Bibliografía de los Camélidos Sudamericanos.* Convenio CIID-Canadá. Lima, Peru.

Uribe, María Victoria

1977–1978 Asentamientos prehispánicos en el altiplano de Ipiales, Colombia. *Revista Colombiana de Antropología* (Bogota) 21:57–195.

Usher, George

1974   *A Dictionary of Plants Used by Man.* Hafner Press/Macmillan, New York.

Usselmann, Pierre

1987   Un acercamiento a las modificaciones del medio físico latinoamericano durante la colonización: Consideraciones generales y algunos ejemplos en las Montañas Tropicales. *Bulletin de l'Institut Français d'Études Andines* (Lima) 16(3–4):127–135.

Valcarcel, Luis E.

1967   *Etnohistoria del Perú Antiguo. Historia del Perú (Incas).* Universidad Nacional Mayor de San Marcos, Lima, Peru.

Valderrama Fernández, Ricardo, and Carmen Escalante Gutiérrez

1983   Arrieros, troperos y llameros en Huancavelica. *Allpanchis Phuturinga* 13, 18(21):65–88.

Valdez Cárdenas, Lidio M.

1988b   Los camélidos en la subsistencia Nasca: El caso de Kawachi. *Boletín de Lima* 10(57):31–35.

1994   Cahuachi: New evidence for an early Nasca ceremonial role. *Current Anthropology* 35 (5):675–679.

1996   The pictographs from Hacha, Peru. In *Proceedings and Abstracts: 6th Annual Apala Student Conference,* January 12–14, 1996, pp. 98–104. Bruce D. Low and Rob Wondrasek, editors. University of Saskatchewan, Saskatoon, Canada.

Valverde, Vicente de (Fray)

1969a   *Carta Relación de Fray Vicente Valverde a Carlos V, sobre la conquista del Perú (1539).* Serie A, No. 3, Ediciones Universidad Nacional de Educación (La Cantuta). Lima, Peru. [Author's Note: The cover reads Fray Vicente de Valverde, *La Conquista del Perú.* I prefer to use the title that appears inside, as it is more complete.]

1969b   20 de Marzo de 1539. Fray Vicente de Valverde al Emperador. Cuzco. In *Carta Relación de Fray Vicente de Valverde a Carlos V sobre la conquista del Perú (1539),* pp. 19–62. Serie A, No. 3. Ediciones Universidad Nacional de Educación (La Cantuta). Lima, Peru.

Vallenas Pantigoso, Augusto

1970a   Fisiología de la digestión de los Auquénidos. In *Anales de la Primera Convención sobre Camélidos Sudamericanos (Auquénidos),* pp. 69–78. Universidad Nacional Técnica del Altiplano, Puno, Peru.

Van Der Hammen, Th., and G. W. Noldus

1985   Pollen analysis of the Telarmachay rock-shelter (Peru). In *Telarmachay. Chasseurs et Pasteurs Préhistoriques des Andes,* pp. 379–387. D. Lavallée, M. Julien, J. Wheeler, and C. Karlin, editors. Institut Français d'Études Andines, Editions Recherches sur les Civilizations. Synthèse No. 20. Paris, France.

Van Nice, P., Craig P. Black, and S. Marsh Tenney

1980   A comparative study of ventilatory response to hypoxia with reference to hemoglobin O_2-affinity in llama, cat, rat, duck and goose. *Comparative Biochemistry and Physiology* [A] 66:347–350.

Vargas Ugarte, Rubén, S. J.

1966   *Historia General del Perú. Virreinato (1551–1596). Tomo II.* Editor Carlos Milla Batres, Lima, Peru.

Varón Gabai, Rafael

1980   *Curacas y encomenderos. Acomodamiento nativo en Huaraz. Siglos XVI y XVII.* P. L. Villanueva, Editor. Lima, Peru.

Vásquez S., Segundo, and Víctor Vásquez S.

1986   Estudio de coprolitos de Camélidos. Anexo en Informe final del Proyecto "El Período Wari 3–4 y un establecimiento de pescadores en Chan Chan," Segundo Vásquez Sánchez. Presented to Fomciencias. Trujillo, Peru.

Vásquez Sánchez, Víctor F., and Teresa E. Rosales Tham

1998   Zooarqueología de la zona urbana de Moche. In *Investigaciones en la Huaca de la Luna 1996,* pp. 173–193. S. Uceda, E. Mujica, and R. Morales, editors. Proyecto Arqueológico Huacas del Sol y de la Luna. Facultad de Ciencias Sociales, Universidad Nacional de La Libertad [sic], Trujillo, Peru.

Vázquez de Espinosa, Antonio

1948   *Compendio y Descripción de las Indias Occidentales.* Smithsonian Miscellaneous Collections, Vol. 108. Washington, DC.

Vega, Andrés de
1965   La descripción que se hizo en la provincia de Xauxa por la instrucción de S. M. que a la dicha provincia se invio de molde. In *Relaciones Geográficas de Indias. Perú. I.* Marcos Jiménez de la Espada. Biblioteca de Autores Españoles desde la formación del lenguaje hasta nuestros días. Vol. 1, pp. 166–175. Tomo CLXXXIII, Ediciones Atlas, Madrid, Spain.

Viault, François Gilbert
1890   Phisiologie experimentale: Sur l'aumentation considérable du nombre de globules rouges dans le sang chez les habitants des hauts plateaux de l'Amerique du Sud. *Comptes Rendue de l'Academie des Sciences* 111:917–918.

Vierra, Robert K.
1981a   Tukumachay, Ac 351. In The stratigraphy of the other cave excavations. Robert K. Vierra and Richard S. MacNeish. In *Prehistory of the Ayacucho Basin, Peru*, Vol. II, *Excavations and Chronology*, pp. 129–133. Richard S. MacNeish, Angel García Cook, Luis G. Lumbreras, Robert K. Vierra, and Antoinette Nelken-Terner. University of Michigan Press, Ann Arbor, MI.
1981b   Chupas, Ac 500. In The stratigraphy of the other cave excavations. Robert K. Vierra and Richard S. MacNeish. In *Prehistory of the Ayacucho Basin, Peru*, Vol. II, *Excavations and Chronology*, pp. 138–144. Richard S. MacNeish, Angel García Cook, Luis G. Lumbreras, Robert K. Vierra, and Antoinette Nelken-Terner. University of Michigan Press, Ann Arbor, MI.

Villalba, Lilián
1992   Bolivia. In *South American Camelids. An Action Plan for Their Conservation / Camélidos Silvestres Sudamericanos. Un Plan de Acción para su Conservación*, pp. 10–13; 42–45. Hernán Torres, compiler-editor. IUCN, Gland, Switzerland.

Villavicencio de Izquierdo, Doris, César Reynafarge, A. Zavaleta, and J. Ramos
1970   Características electroforéticas de la hemoglobina de las alpacas. *Archivos del Instituto de Biología Andina* 3(5–6):193–199.

Villee, Claude A., Warren F. Walker, Jr., and Frederick E. Smith
1968   *General Zoology*. W. B. Saunders, Philadelphia, PA.

Visita de Acari
1973 (1593) *Historia y Cultura* 7:131–209.

Vivar, Gerónimo de
1979   *Crónica y relación copiosa y verdadera de los Reinos de Chile (1558)*. Edición de Leopoldo Sáez-Godoy. Bibliotheca Ibero-Americana. Colloquium Verlag, Berlin, Germany. [Author's Note: *See* Bibar 1966.]

Wagner, Claus
1986   Prólogo. Ulrico Schmidel y sus "historias verdaderas" y Notas al Prólogo y Notas al Texto. In *Relatos de la conquista del Río de la Plata y Paraguay, 1534–1554*, pp. 9–20, 119–120. Ulrico Schmidel. El Libro del Bolsillo, Alianza Editorial, Madrid, Spain.

Wallace, Dwight T.
1962   Cerrillos, an early Paracas site in Ica, Peru. *American Antiquity* 27(3):303–314.

Walton, William, Jr.
1811   *An historical and descriptive account of the four species of Peruvian Sheep, called Carneros de la Tierra; to which are added particulars respecting the domestication of the wild species, and the experiments hitherto made by the Spaniards, to cross the respective breeds to improve their wools*, pp. 7–183. W. Glindon, London, UK.

Watanabe, Luis K., Michael E. Moseley, and Fernando Cabieses (compilers)
1990   *Trabajos Arqueológicos en Moquegua, Perú*. 3 vols. Programa Contisuyo del Museo Peruano de Ciencias de la Salud. Southern Peru Copper Corporation, Lima, Peru.

Webb, S. David
1974   Pleistocene llamas of Florida, with a brief review of the Lamini. In *Pleistocene Mammals of Florida*, pp. 170–213. S. D. Webb, editor. University Press of Florida, Gainesville, FL.
1978   A history of savanna vertebrate in the New World. Part II. South America and the Great Interchange. *Annual Review of Ecology and Systematics* 9:393–426.
1985   Late Cenozoic mammal dispersals between the Americas. In *The Great American Biotic Interchange*, pp. 357–386. Francis G. Stehli and S. David Webb, editors. Plenum Press, New York.
1991   Ecogeography and the Great American Interchange. *Paleobiology* 17(3):266–280.

Webb, S. David, and Larry G. Marshall
1982   Historical biogeography of Recent South American mammals. In *Mammalian Biology in South America*, Vol. 6, pp. 39–52. Michael A. Mares and Hugh H. Genoways, editors. Special Publication Series, Pymatuning Laboratory of Ecology, University of Pittsburgh, Linesville, PA.

Webb, S. David, and Frank G. Stehli
1995 Selenodont Artiodactyla (Camelidae and Cervidae) from the Leisey Shell Pit, Hillsborough County, Florida. *Bulletin of the Florida Museum Natural History*. Paleontology and Geology of the Leisey Shell Pits, Early Pleistocene of Florida, Part II. Richard C. Hulbert Jr., Gary S. Morgan, and S. David Webb, volume editors. Vol. 37, Pt. II, Nos. 11–20, pp. 621–643. Gainesville, FL.

Weberbauer, Augusto
1945 *El Mundo Vegetal de los Andes Peruanos. Estudio Fitogeográfico*. Estación Experimental Agrícola de La Molina. Dirección de Agricultura, Ministerio de Agricultura, Lima, Peru.

Webster, Steven S.
1971a An indigenous Quechua community in exploitation of multiple ecological zones. In *Actas y Memorias*, Congreso Internacional de Americanistas, Vol. 3, pp. 174–183. Lima, Peru. [Author's Note: This was republished this same year with no changes. *See* Webster 1971b.]
1971b An indigenous Quechua community in exploitation of multiple ecological zones. *Revista del Museo Nacional* 37:174–183.

Wendt, W. E.
1976 El asentamiento precerámico en Río Seco, Perú. *Lecturas en Arqueología*, 3. Museo de Arqueología y Etnología, Universidad Nacional Mayor de San Marcos, Lima, Peru. [Author's Note: This is a translation of the 1963 article, Die Präkeramische Siedlung am Río Seco, Peru. *Baessler Archiv*, n.s., 11: 225–257.]

Wensvoort, J.
1992 Desert plants and diet for race camels. In *Proceedings of the First International Camel Conference, 2nd–6th February 1992*, pp. 323–326. W. R. Allen, A. J. Higgins, I. G. Mayhew, D. H. Snow, and F. Wade, editors. R.& W. Publications (Newmarket), Amersham, Switzerland.

West, Michael
1971a Prehistoric Human Ecology in the Virú Valley. *California Anthropologist* 1(1):47–56.
1971b Recent archeological studies in the Virú Valley, Peru. *California Anthropologist* 1(2):22–32.

West, Terry L.
1981a Llama caravans of the Andes. *Natural History* 90(12):62–73.

Wheeler, Jane C.
1982a On the origin and early development of camelid pastoralism in the Andes. In *Animals and Archaeology*. J. Clutton-Brock and Caro-

line Grigson, editors. British Archaeological Reports, Oxford, UK. Mimeograph.
1982b Aging llamas and alpacas by their teeth. *Llama World* 1(2):12–17.
1984a On the origin and early development of camelid pastoralism in the Andes. In *Animals and Archaeology*, Vol. 3, *Early Herders and Their Flocks*, pp. 395–410. Juliet Clutton-Brock and Caroline Grigson, editors. BAR International Series 202. British Archaeological Reports, Oxford, UK. [Author's Note: This is the same as the 1982a mimeographed text.]
1984b On the origin and the early development of camelid pastoralism in the Andes. In *Animals and Archaeology*, Vol. 4, *Husbandry and the Emergence of Breeds*, pp. 1–13. Juliet Clutton-Brock and Caroline Grigson, editors. British Archaeological Reports, Oxford, UK. [Author's Note: This is the same text as was published in 1982a and 1984a.]
1984c La domesticación de la alpaca (*Lama pacos* L.) y la llama (*Lama glama* L.) y el desarrollo temprano de la Ganadería Autóctona en los Andes Centrales. *Boletín de Lima* 6(36):74–84.
1984d Review of *Prehistoric Hunters of the High Andes* by John W. Rick. *American Antiquity* 49(1):196–198.
1985a La faune. In *Telarmachay: Chasseurs et Pasteurs Préhistoriques des Andes I*, Vol. 1, pp. 28–32. D. Lavallé, M. Julien, J. Wheeler, and C. Karlin. Institut Français d'Études Andines, Editions Recherches sur les Civilizations. Synthèse No. 20. Paris, France.
1985b De la chasse a l'élevage. In *Telarmachay: Chasseurs et Pasteurs Préhistoriques des Andes I*, Vol. 1, pp. 61–79. D. Lavallé, M. Julien, J. Wheeler, and C. Karlin. Institut Français d'Études Andines, Editions Recherches sur les Civilizations. Synthèse No. 20. Paris, France.
1986 Faunal remains from archaeological excavations at Coporaque. The Cultural Ecology, Archaeology, and History of Terracing and Terrace Abandonment in the Colca Valley of Southern Peru, pp. 291–312. Technical report to the National Science Foundation and the National Geographic Society. William M. Denevan, editor. Madison, WI.
1988a Nuevas evidencias arqueozoológicas acerca de la domesticación de la alpaca, la llama y el desarrollo de la ganadería autóctona. In *Llamichos y paqocheros. Pastores de llamas y alpacas*, pp. 45–57. Jorge A. Flores Ochoa, compiler. CONCYTEC, CEAC y Editorial UNSAAC, Cuzco, Peru.

1991   Origen, evolución y status actual. In *Avances y perspectivas del conocimiento de los Camélidos Sud Americanos*, pp. 11–48. Saúl Fernández Baca, editor. FAO, Organización de las Naciones Unidas para la Agricultura y la Alimentación, Oficina Regional de la FAO para América Latina y el Caribe, Santiago, Chile.

1993   The domestic South American Camelidae: Past, Present and Future. In *European Symposium on South American Camelids*, pp. 13–28. Martina Gerken and C. Renieri, editors. Rheinische Friederich-Wilhelms-Universität Bonn; Università degli Studi di Camerino, Facoltà di Medicina Veterinaria, Matelica, Italy.

1995   Evolution and present situation of the South American Camelidae. *Biological Journal of the Linnean Society* 54:271–295.

Wheeler, Jane, Carmen Rosa Cardoza, and Denise Pozzi-Escot

1977   Estudio provisional de la fauna de las Capas II y III de Telarmachay. *Revista del Museo Nacional* (Lima) 43:97–102.

Wheeler, Jane C., A. J. F. Russel, and Hilary Redden

1995   Llamas and alpacas: Pre-conquest breeds and post-conquest hybrids. *Journal of Archaeological Science* 22(6):833–840.

Wheeler, Jane C., A. J. F. Russel, and H. F. Stanley

1992   A measure of loss: Prehispanic llama and alpaca breeds. Razas prehispánicas de llamas y alpacas; la medida de lo que se ha perdido. *Archivos de Zootecnia* 41 (Suppl.): 467–475.

Wheeler Pires-Ferreira, Jane C.

1975   La fauna de Cuchimachay, Acomachay A, Acomachay B, Tellarmachay [sic] y Utco 1. El hábitat prehistórico en la zona de San Pedro de Cajas, Junín. Danièle Lavallée and Michèle Julien. *Revista del Museo Nacional* 41:120–127.

1977   Camelid domestication and subsistence strategies in the Central Andes. Research proposal submitted to the National Science Foundation. December 12.

Wheeler Pires-Ferreira, Jane C., Edgardo Pires-Ferreira, and Peter Kaulicke

1976   Preceramic animal utilization in the central Peruvian Andes. *Science* 194:483–490.

1977   Domesticación de los Camélidos en los Andes Centrales durante el Período Precerámico: Un modelo. *Journal de la Société des Américanistes* 64:155–165.

Whitehouse, Ruth D. (editor)

1983   *The Facts on File. Dictionary of Archaeology.* Fact On File Publications, New York. [Author's Note: There is an 1988 edition.]

Whittembury, José, Rodolfo Lozano, César Torres, and Carlos Monge C.

1968   Blood viscosity in high altitude polycythemia. *Acta Physiologica Latinoamericana* 18:355– 359.

Wilhelm Grob, Ottmar E.

1978   The Precolumbian Araucanian chicken (*Gallus inauris*) of the Mapuche Indians. In *Advances in Andean Archaeology*, pp. 189–196. David L. Browman, editor. Mouton, The Hague and Paris.

Wilson, David J.

1988   *Prehispanic Settlement Patterns in the Lower Santa Valley, Peru*. Smithsonian Institution Press, Washington, DC.

Willey, Gordon R.

1953   *Prehistoric Settlement Patterns in the Virú Valley, Peru*. Bulletin 155, Bureau of American Ethnology, Smithsonian Institution, Washington, DC.

1971   *An Introduction to American Archaeology*, Vol. II, *South America*. Prentice Hall, Englewood Cliffs, NJ.

Willey, Gordon R., and John M.Corbett

1954   *Early Ancón and Early Supe Culture. Chavín Horizon Sites of the Central Peruvian Coast.* Columbia University Press, New York.

Williams, Andrew, Donald Heath, Peter Harris, David Williams, and Paul Smith

1981   Pulmonary mast cells in cattle and llamas at high altitude. *Journal of Pathology* 134:1–6.

Wing, Elizabeth S.

1972   Utilization of animal resources in the Peruvian Andes. In *Andes 4. Excavations at Kotosh, Peru 1963 and 1966*, pp. 327–351. S. Izumi and K. Terada, editors. University of Tokyo Press, Tokyo, Japan.

1975a   Informe preliminar acerca de los restos de fauna de la cueva de Pachamachay, en Junín, Perú. Prehistoria y ecología humana en las punas de Junín. Ramiro Matos Mendieta. *Revista del Museo Nacional* (Lima) 41:79–80.

1975b   La domesticación de animales en los Andes. *Allpanchis Phuturinga* (Cuzco) 8(8):25–44.

1977a   Animal domestication in the Andes. In *Origins of Agriculture*, pp. 837–859. Charles Reed, editor. Mouton, The Hague and Paris.

1977b   Prehistoric subsistence patterns of the Central Andes and adjacent coast and spread on the use of domestic animals. Report to the

National Science Foundation, Soc 74-20634. Florida.

1977c Caza y pastoreo tradicionales en los Andes Peruanos. In *Pastores de puna. Uywamichiq punarunakuna*, pp. 121–130. Jorge Flores Ochoa, compiler. Instituto de Estudios Peruanos, Lima, Peru.

1978 Animal domestication in the Andes. In *Advances in Andean Archaeology*, pp. 167–188. David L. Browman, editor. Mouton, The Hague and Paris.

1980 Faunal remains. In *Guitarrero Cave: Early Man in the Andes*, pp. 149–172. Thomas F. Lynch, editor. Academic Press. New York.

1986 Domestication of Andean mammals. In *High Altitude Tropical Biogeography*, pp. 246–264. F. Villeumier and M. Monasterio, editors. Oxford University Press, New York.

1988 Use of animals by the Inca as seen at Huánuco Pampa. In *Economic Prehistory of the Central Andes*, pp. 167–179. Elizabeth S. Wing and Jane C. Wheeler, editors. BAR International Series 427. British Archaeological Reports, Oxford, UK.

1989 Human use of canids in the Central Andes. *Advances in Neotropical Mammalogy*, pp. 265–278. Gainesville, FL.

Wing, Elizabeth S., and Elizabeth J. Reitz
1982 Pisces, Reptilia, Aves, Mammalia. In *Los Gavilanes. Mar, desierto y oasis en la historia del hombre*, pp. 191–200. Duccio Bonavia. Corporación Financiera de Desarrollo, S.A., COFIDE, Instituto Arqueológico Alemán, Lima, Peru.

Winslow, Robert M., and Carlos Monge C.
1987 *Hypoxia, Polycythemia and Chronic Mountain Sickness*. Johns Hopkins University Press, Baltimore, MD.

Wright, H. E., Jr.
1980 Environmental history of the Junín plain and the nearby mountains. In *Prehistoric Hunters of the High Andes*, pp. 253–256. John W. Rick, editor. Academic Press, New York.

Xerez, Francisco de
1968 (1534) *Verdadera relación de la conquista del Perú y provincia del Cuzco llamada la Nueva Castilla*. Biblioteca peruana. Tomo I, pp. 191–272. Editores Técnicos Asociados, Lima, Peru.

Yacobaccio, Hugo Daniel
1984–1985 Una adaptación regional de cazadores-recolectores en los Andes Centro-Sur. *Relaciones de la Sociedad Argentina de Antropología*, n.s., 16:165–173.

1986 Must hunters walk so much? Adaptive strategies of south Andean hunter-gatherers (10,800–7,500 BP). In *Proceedings of the World Archaelogical Congress*, Vol. 2, *The Pleistocene Perspective*, p. 16. Allen & Unwin, London, UK.

Yamamoto, Norio
1985 The ecological complementarity of agro-pastoralism: Some comments. In *Andean Ecology and Civilization. An Interdisciplinary Perspective on Andean Ecological Complementarity*, pp. 85–99. Shozo Masuda, Izumi Shimada, and Craig Morris, editors. University of Tokyo Press, Tokyo, Japan.

Zárate, Agustín de
1968 (1555) *Historia del descubrimiento y conquista del Perú*. Biblioteca Peruana, Tomo II, pp. 105–413. Editores Técnicos Asociados, Lima, Peru.

Zavala, Silvio
1978a *El servicio personal de los indios en el Perú (Extractos del siglo XVI)*, Tomo I. El Colegio de México, Mexico City, Mexico.

1978b Estudios del Dr. Juan de Matienzo. In *El servicio personal de los indios en el Perú (Extractos del siglo XVI)*, Tomo I, pp. 51–61. Silvio Zavala. El Colegio de México, Mexico City, Mexico

Zeuner, F. E.
1963 *A History of Domesticated Animals*. Hutchinson, London, UK.

Zevallos Quiñones, Jorge
1984 La fundación de Trujillo. *El Comercio*, Año 145. No. 77,629. 7 de mayo, p. A2. Lima, Peru.

1989 *Los Cacicazgos de Lambayeque*. Trujillo, Peru.

1991 Notas introductorias. In *Etnohistoria del Area Virú-Santa: Un avance documental (Siglos XVI–XIX). Documentos para la Etnohistoria de la Zona Norte/1*, pp. 1–17. Instituto Departamental de Cultura, La Libertad, Dirección de Patrimonio Cultural Monumental de la Nación. Proyecto Especial de Irrigación Chavimochic. Trujillo, Peru.

15.2. SOURCES CITED BY OTHERS

Acosta-Solís, M.
1965 *Los recursos Naturales del Ecuador y su Conservación*. Instituto Panamericano de Geografía e Historia, México.

Actas del Cabildo de Santiago de 1541 a 1557.
1861 Colección . . . Vol. I., pp. 65–604.

Aguilar, T.
  1985   Arqueología de Sihuinacocha. Convención
         Internacional sobre Camélidos Sudameri-
         canos. V. *Libro de Resúmenes.* Cuzco, Perú,
         Junio 16–21. Universidad Nacional San An-
         tonio Abad, Cuzco, Peru.
Aguilar y Córdova, Diego de
  1938   El Marañón. *El Apogeo de la literatura Colonial.*
         Las poetisas anónimas. El Lunarejo-Caviedes.
         Biblioteca de Cultura Peruana, primera serie,
         No. 5, pp. 321–345. Paris, France.
Agustinos
  1865   Relación de la religión y ritos del Perú. In
         Colección de documentos inéditos relativos al des-
         cubrimiento, conquista y colonización de las pose-
         siones españolas en América y Oceanía. Direc-
         ción: Joaquín F. Pacheco y Francisco de
         Cárdenas. Serie 1, Tomo III, pp. 5–58. Ma-
         drid, Spain.
Albes, E.
  1918   The South American relatives of the camels.
         Bulletin of Panamerican Union 46:598–617.
Albrittom, E. C. (editor)
  1954   *Standard Values in Nutrition and Metabolism.*
         Philadelphia, PA.
Allen, G. M.
  1942   *Extinct and Vanishing Mammals of the Western*
         Hemisphere. American Committee for Inter-
         national Wild Life Protection, Special Pub-
         lication 11. New York.
Almeida Reyes, Eduardo
  1982   Los Camélidos en el Pucara de Rumicucho
         durante la época Incásica. Trabajo presenta-
         do en el Coloquio Internacional "Carlos Ze-
         vallos M." Guayaquil, Ecuador.
Altamirano, Alfredo
  n.d.   Sacrificio de Camélidos en Pacatnamú. MS on
         file in the Archive of Moche Art, University of
         California, Los Angeles. [Author's Note: I as-
         sume this is the same report as the 1984 MS.]
  1984   Sacrificios de Camélidos en Pacatnamú.
         Programa Académico de Arqueología. Uni-
         versidad Nacional Mayor de San Marcos,
         Lima, Peru. MS.
Altmann, S. A.
  1974   Baboons, space, time and energy. *American*
         Zoologist 14:221–248.
Ameghino, Florentino
  1884   Excursiones geológicas y paleontológicas en
         la provincia de Buenos Aires. *Boletín de la*
         Academia de Ciencias (Córdoba) 6:161–257.
  1889   Los mamíferos fósiles de la República Ar-
         gentina. *Academia Nacional de Ciencias de*
         Córdoba, 6, I–XXXII, pp. 1–1027. [Author's
         Note: I took this reference from Webb

         (1974). López Aranguren (1930b) gives
         (with the same date) *Contribuciones al cono-*
         cimiento de los Mamíferos Fósiles de la Repúbli-
         ca Argentina. Buenos Aires. I assume it is the
         same study.]
  1891   Mamíferos y aves fósiles argentinos: Espe-
         cies nuevas, adiciones, y correcciones. *Revista*
         Argentina de Historia Natural 1:240–259.
Andrade Marin, L.
  1936   *Llanganati. Expedición Italo-Ecuatoriana*
         Boschetti-Andrade Marín, 1933–1934. Imp.
         Mercantil.
Angulo, Fray Domingo
  1920   Diario de la segunda visita pastoral del arzo-
         bispo de los Reyes Don Toribio Alfonso de
         Mogrovejo. Libro de Visitas, 1539. *Revista*
         del Archivo Nacional l(1).
Anonymous
  1582a Relación de la tierra de Jaén. RGI: 1897, pp.
         28–33.
Anonymous
  1582b Relación de la doctrina a beneficio de Nam-
         bija y Yaguarsongo. RGI: IV, pp. 21–27.
Araníbar, J.
  1927   *Llamas y Alpacas, vicuñas y guanacos.* Romero
         E. C. Montevideo.
Arias, Peter
  1525   *Lettere di Pietro Arias Capitano Generale, della*
         conquista del paese del Mar Oceano Scripte alla
         Maesta Cesarea dalla Cipta di Panama delle
         cose ultimamente scoperte nel Mar Meridiano
         decto el Mar Sur. MDXXV fig. e. b. au titre
         16mo. s. 1 (en vers).
Armitage, G. M.
  1975   The extraction and identification of opal
         phytoliths from the teeth of ungulates. *Jour-*
         nal of Archaeological Science 2(3):187–197.
Arriaga, J. de
  1920   *La extirpación de la idolatría en el Perú.* Lima,
         Peru.
Ascasubi, Miguel
  1846   Informe cronológico de las misiones del reino
         de Chile, hasta 1789. In *Historia física y política*
         de Chile. Documentos sobre la historia, la estadís-
         tica y la geografía. Claudio Gay, 1846–1852,
         Vol. 1, pp. 300–400 (1846). Paris, France.
Athens, J. S.
  1980   *El proceso Evolutivo en las Sociedades Complejas*
         y la Ocupación del Período Tardío-Cara en los
         Andes Septentrionales del Ecuador. Colección
         Pendoneros, No. 2. Instituto Otavaleño de
         Antropología, Otavalo, Ecuador.
Atienza, Lope de
  1931   Compendio Historial del estado de los In-
         dios del Perú con mucha doctrina y cosas

notables de ritos, costumbres e inclinaciones que tienen, con docta doctrina y avisos para los que viven entre estos neófitos; nuevamente compuesto por . . . Clérigo Presbitero, criado de la Serenísima Reina de Portugal, Bachiller en Cánones, dirigido al Honorabilísimo señor Licenciado Juan de Obando, del Consejo de Estado, Presidente del Real Consejo de Indias. In *La Religión del Imperio de los Incas*. Vol. I. Apéndice, pp. 2–307. Jijón y Caamaño, editor. Escuela Tipográfica Salesiana, Quito, Ecuador.

Bader, R. S.
1957 Two Pleistocene mammalian faunas from Alchua County, Florida. *Bulletin of the Florida State Museum* 2:53–75.

Bahamonde, N., F. San Martin, and A. S. Pelliza
1986 Diet of guanaco and red deer in Neuquen Province, Argentina. *Journal of Range Management* 39:22–24.

Baker, Edward W., and G. W. Wharton
1952 *An Introduction to Acarology*. Macmillan, New York.

Bandera, Damián de la
1881 Relación General de la Disposición y Calidad de la Provincia de Guamanga, llamada San Joan de la Frontera, y de la Vivienda y Costumbres de los Naturales Della. Año de 1557. In *Relaciones Geográficas de Indias*. Tomo I, pp. 96–103. Ministerio de Fomento del Perú. Edit. Tip. de Manuel Hernández, Madrid, Spain. [Author's Note: *See* Bandera 1974.]

Bankes, George H. A.
1971 Some aspects of the Moche culture. Ph.D. dissertation, Institute of Archaeology, University of London, UK.

Banks, Nathan
1915 *The Acarina or Mites*. U.S. Department of Agriculture, Report No. 108. Washington, DC.

Barros Arana, Diego
1872 *Elementos de Historia Natural*, 2a edición. Santiago, Chile.

Barros Valenzuela, Rafael
1963 Anotaciones sobre los Lamidos en Chile. *Revista Universitaria. Anales de la Academia Chilena de Ciencias Naturales* 48(26):57–67.

Behrendt, G.
1960 Estudio sobre la creación de formas del *Odocoileus Peruvianus* Gray. *Pesca y Caza* 10:149–166.

Belding, David L.
1942 *Textbook of Clinical Parasitology*. Appleton-Century-Crofts, New York.

Benirschke, K.
1967 *Comparative Aspects of Reproduction Failures.* K. Benirschke, editor. Springler-Verlag, New York.

Benjamin, J.
1947 Breve reseña histórica y descriptiva de los auchénidos. *Campo* l(3):25–33.

Bennett, Wendell C.
1946 *Excavations in the Cuenca Region*. Yale Publications in Anthropology, No. 35. New Haven, CT.

Berdichewsky, Bernardo
1963 El precerámico de Taltal y sus correlaciones. *Centro de Estudios Antropológicos de la Universidad de Chile*, 16.

Bergen, W. Von (editor)
1963 *Wool Handbook*. John Wiley & Sons Interscience Publishers, New York.

Bermann, Marc P., Paul Goldstein, Charles Stanish, and Luis Watanabe
1989 The collapse of the Tiwanaku State. A view from the Osmore drainage. In *Ecology, Settlement and History in the Osmore Drainage, Peru*, pp. 269–285. D. S. Rice, C. Stanish, and P. R. Scarr, editors. BAR International Series, No. 545. British Archaeological Reports, Oxford, UK.

Berninger, O.
1919 Bosque y Tierra descubierta en el sur de Chile. *Geographische Abhandlungen* 3(1).

Bird, Junius
1938 Antiquity and migrations of the early inhabitants of Patagonia. *The Geographical Review* 28(2):250–275.
1943 *Excavations in Northern Chile*. Anthropological Papers of the American Museum of Natural History, New York.

Bird, Junius, and Louisa Bellinger
1954 *Paracas Fabrics and Nasca Needlework*. National Publishing Co., Washington, DC. [Author's Note: The reference is wrong. *See* Bird 1954.]

Boman, Eric
1908 *Antiquités de la Région andine de la République Argentine et du desert d'Atacama*. Paris, France.

Bombin, Miguel
1976 Modelo Paleoecológico Evolutivo para o Neoquaternario da Regiao de Campanha-Oeste do Rio Grande do Sul (Brasil). A formaçao Touro Passo, seu Conteúdo Fossilífero e a Pedogênese Pós-Deposicional. *Comunicações do Museu de Ciências da PUCRUGS*, No. 15, pp. 1–90. Porto Alegre, Brazil.

Borrero, Luis A.
1976   El hombre temprano y la extinción de la megafauna. Paper presented at the 5th Congreso de Arqueología del Uruguay, Atlántida. Mimeograph.
1977   La extinción de la megafauna: Su explicación por factores concurrentes. *Anales del Instituto de la Patagonia* 8:81–93.

Boyd, Robert, and Peter J. Richerson
1985   *Culture and the Evolutionary Process.* University of Chicago Press, Chicago.

Bradbury, J. P.
1982   Holocene chronostratigraphy of Mexico and Central America. Chronostratigraphic subdivision of the Holocene. J. Magerud, H. J. B. Birks and K. D. Jager, editors. *Striae* 16: 46–48.

Brack Egg, A.
1981   Situación actual de la vicuña en el Perú. In *Comunicaciones de la Vicuña*, No. 3, pp. 9–10. J. Alarcón M., editor. Instituto Nacional de Fomento Lanero, Centro de Documentación del Convenio Multinacional de Conservación y Manejo de la Vicuña, La Paz, Bolivia.

Branco, W.
1883   Ueber eine fossile Säugethier-Fauna von Punin bei Riobamba in Ecuador. II: Beschreibung der Fauna. *Paleo. Abh.* 1:57– 204.

Brandt
1841 [?}   *Academia de San Petersburgo*, IV, entrega V.

Braun, R.
1982   The Formative as seen from the southern Ecuadorian highlands. In *Primer Simposio de Correlaciones Antropológicas Andino-Meso-americanas*, pp. 41–99. Jorge G. Marcos and Presley Norton, editors. Escuela Superior Politécnica del Litoral, Guayaquil, Ecuador.

Brennan, Curtiss
1978   *Investigations at Cerro Arena: Incipient Urbanism on the Peruvian North Coast.* University Microfilms International, Ann Arbor, MI.

Bridges, E. L.
1949   *Uttermost Part of the Earth.* E. P. Dutton and Co., New York.

Brougère, Anne Marie
1980   Traditions, changements et écologie dans des communautés paysannes andines. Thèse. Paris, France.

Browman, David L.
1973   Pastoral models among the Huanca of Peru prior to the Spanish Conquest. Relaciones Antropológicas. A newsletter. *Bulletin on South American Anthropology* 1(1):40–44.

1974b   Precolumbian Llama caravan trade network. Paper presented at the International Congress of Americanists, Mexico City, Mexico. [Author's Note: This paper was not published in the proceedings of the aforementioned congress.]
1975   Trade patterns in the Central Highlands of Peru in the first millennium B.C. *World Archaeology* 6(3).

Bry, Theodor
1592   *Los Grandes Viajes.* Francofurti ad Moenum [Frankfort-on-Main], Germany.

Bryan, A. L.
1973   Paleoenvironments and cultural diversity in Late Pleistocene of South America. *Quaternary Research* 3(2):237–256.

Bunch, T. D., and W. C. Foote
?   Chromosome banding patterns. Relationship of the llama (*Lama glama*) and the Bactrian camel (*Camelus bactrianus*). *ADSA Annual Meeting and Divisional Abstracts.* Supplement 1. *JDS*, Vol. 66. University of Wisconsin, Madison, WI.

Bustinza, V.
1991   Mejoramiento genético. In *Producción de rumiantes menores: Alpaca*, pp. 113–128. C. Novoa and A. Flores, editors. Rerumen. Lima, Peru.

Bustinza Méndez, M. Julio, and F. R. Plaza Peres
1971   Recopilación bibliográfica de trabajos de Investigación y Monografías realizadas en el mundo sobre el género *Lama* de la familia Camelidae. Especies: *L. glama, L. paco* y *L. guanicoe. Revista de la Universidad* 3(4):219–249.

Busto Duthurburu, José Antonio del
1953   La Casa de Peralta en el Perú. Tesis Universitaria. Lima, Peru.
1966   *Francisco Pizarro, el Marqués Gobernador.* Artes Gráficas Marisal, Madrid, Spain.

Byrd, K. M.
1976   Changing animal utilization and their implications: Southwest Ecuador (6500 B.C.– A.D. 1400). Ph.D. dissertation, University of Florida, Gainesville, FL.

Cabrera, Angel
1935   Sobre la osteología de *Palaeolama. Anales del Museo Nacional* (Buenos Aires) 38:283–312. [Author's Note: This is the reference given by David S. Webb, 1974. However, for this same article Armando Cardozo gives *Museo Argentino de Ciencias Naturales*, Buenos Aires, Vol. 37, pp. 283–312.]

1957 Catálogo de los mamíferos de América del Sur, Vol. 1. *Revista del Museo Argentino de Ciencias Naturales Zoológicas* 4:1–307.

1961 Catálogo de los mamíferos de América del Sur, Vol. II. Sirenia-Perissodactyla-Artiodactyla-Rodentia-Cetacea. *Revista del Museo Argentino de Ciencias Naturales "Bernardino Rivadavia" e Instituto Nacional de Investigación en Ciencias Naturales y Ciencias Zoológicas* 4(2):309–732.

Cabrera, Angel, and José Yepes
1940 *Mamíferos sud-americanos: Vida, costumbres y descripción*. Historia Natural Ediar. Comp. Argentina de Editores, Buenos Aires, Argentina. [Author's Note: There is a later edition. *See* Cabrera and Yepes 1960.]

Cajal, J.
1981 Situación del guanaco en la República Argentina. Trabajo presentado a la VI Convención Internacional sobre Camélidos Sudamericanos. Corporación Nacional Forestal. Instituto de la Patagonia, Punta Arenas, Chile.

Calkin, V. I.
1970 Drevnejsie domašnie životnye Srednej Azii, Soobščenie I. *Bjulleten Moskovskogo Obščestva Ispytatelej Prirody-Otdel Biologiceškij* 76(1): 145–159.

Camargo, R. and A. G. Cardozo
1971 Ensayo comparativo de la Capacidad de Digestión en la llama y la oveja. In *Actas y Trabajos de la III Reunión Latinoamericana de la Producción de Animales*. Bogotá, Colombia.

Canals Frau, Salvador
1953 *Las poblaciones indígenas de la Argentina. Su origen. Su pasado. Su presente*. Buenos Aires, Argentina.

Cañadas Cruz, Luis
1983 *El mapa bioclimático y ecológico del Ecuador*. Quito, Ecuador.

Capdeville, Augusto
1921 *Notas acerca de la arqueología de Taltal*. Ecuador.

Capurro, S., L. F. Silva, and F. C. Silva
1960 Estudios cromatográficos y electroforéticos en camélidos sudamericanos. *Investigaciones Zoológicas Chilenas* 6:49–64.

Carbajal, O. P., Fray Gaspar de
1955 *Relación del Nuevo descubrimiento del famoso Río Grande de las Amazonas*. Fondo de Cultura Económica, Mexico City, Mexico. [Author's Note: *See* Carvajal 1894, 1986.]

Cardozo (Gonzáles), Armando
1966 *Bibliography of Camelidae 1–6*. Utah State University, Department of Animal Science, Logan, UT.

1968 Bibliografía de los Camélidos. *Boletín Experimental*, No. 32. Estación Experimental Ganadera de Patacamaya, Bolivia.

1970 Bibliografía de los Camélidos Sudamericanos. Quito. Instituto Interamericano de Ciencias Agrícolas, Zona Andina. Programa de Investigaciones. *Suplemento al Boletín Experimental*, No. 32, de la División de Investigaciones (Bolivia) y presentada a la I Convención de Auquénidos, Puno, Peru.

1973 *Preliminares sobre la presencia y trascendencia de Camélidos en el Ecuador*. Curso corto intensivo sobre producción animal. Guayaquil, Ecuador, Noviembre. Guayaquil, IICA, Ecuador.

1975b *Sobre el origen y filogenia de los camélidos sudamericanos*. IICA-CITA, Bogotá, Colombia.

1976–1977 Presencia y trascendencia de los camélidos sudamericanos. *El Agro* 21(4):24–26; 22(1):20–22. Ecuador.

1977a *Bibliografía de los camélidos sudamericanos*. Universidad Nacional de Jujuy, Jujuy, Argentina.

1977b *Bibliografía de los Camélidos Sudamericanos-A*. Universidad Nacional de Jujuy, Dirección de Publicaciones, Jujuy, Argentina.

1978 *Bibliografía de los Camélidos Sudamericanos-B*. Policrom. Artes Gráficas, La Paz. (?), Bolivia.

1980 Población y geografía de la vicuña. *Comunicaciones de la Vicuña*, No. 1. A. Cardozo, editor. Instituto Nacional de Fomento Lanero, Centro de documentación del convenio multinacional de Conservación y Manejo de la Vicuña, La Paz, Bolivia.

1982 Avances en el conocimiento de la fibra de llama. *Cuadernos de la Academia Nacional de Ciencias de Bolivia* 60:25–36.

Casamiquela, Rodolfo
1969 Enumeración crítica de los mamíferos continentales pleistocenos de Chile. *Rehue* (Concepción) 2:143–172.

Casas, P. P., fray Bartolomé de las
1951 *Historia de las Indias*. Gráfica Panamericana, Mexico City, Mexico. [Author's Note: *See* Casas 1981.]

(Casas, Bartolomé de las)
1961 *Obras escogidas de Fray Bartolomé de las Casas. Historia de las Indias. Tomo 2*. Biblioteca de Autores Españoles, Tomo 96, pp. 1–447. Madrid, Spain. [Author's Note: *See* Casas 1981.]

Charnot, Y.
1959 A propos de l'ecologie des camélidés [Concerning the ecology of the camel family]. *Bull. Soc. Sc. Nat. et Dhys. Maroc*. 39(1):29–39.

Chávez, J.
1991 Mejoramiento genético de alpacas y llamas.
      In *Avances y Perspectivas del Conocimiento de los*
      Camélidos Sudamericanos, pp. 151–190. S.
      Fernández Baca, editor. FAO, Santiago,
      Chile.

Chauca, Denise, Antonieta Valenzuela, and Hugo Sillan
1970 Estudio comparativo de digestibilidad in
      vitro de fibra cruda entre la Alpaca y el
      Ovino. In *Cuarto Boletín Extraordinario*, pp.
      100–104. Centro de Investigación. Instituto
      Veterinario de Investigaciones Tropicales y
      de Altura (IVITA), Lima, Peru.

Churcher, C. S.
1965 Camelid material of the genus *Palaeolama*
      Gervais from Talara Tar-Seeps, Peru, with a
      description of a new subgenus, *Astylolama*.
      Proceedings of the Zoological Society of London
      145(Pt. 2):161–205.

Cieza de León, Pedro de
1864 *The Travels of Pedro de Cieza de León, A.D.*
      1532–50. C. R. Markham. translator-editor.
      Hakluyt Society, London, UK. [Author's
      Note: *See* Cieza de León 1973.]
1966 (1881) *Guerra de Chupas* [=] *Guerra Siviles del*
      Perú, II. Marqués de la Fuensanta del Valle y
      D. José Sancho Rayón. Colección de Docu-
      mentos Inéditos para la Historia de España,
      Tomo LXXVI. Madrid. Kraus Reprint, Va-
      duz, Switzerland. [Author's Note: *See* Cieza
      de León 1881.]
1984 *Obras completas de Pedro Cieza de León.* C.
      Sáenz de Santa María, Editor. La Crónica
      del Perú, Partes 1–3. Monumento Hispano-
      Indiana 2, Tomo 1. Instituto "Gonzalo Fer-
      nández de Oviedo," Madrid, Spain. [Author's
      Note: *See* Cieza de León 1984, 1985, 1987.]

Collier, Donald, and John V. Murra
1943 *Survey and Excavations in South Ecuador.* An-
      thropological Series, Vol. 35. Field Museum
      of Natural History, Chicago, IL.

Condorena, N.
1980 Algunos índices de producción en alpaca
      bajo el sistema de esquila anual establecido
      en La Raya. *Rev. Inv. Pec.* 5:50–55. ([IVITA]
      Universidad Nacional Mayor de San Mar-
      cos.)

Cook, Noble David
1982 *The People of the Colca Valley: A Population*
      Study. Westview Press, Boulder, CO.

Cope, E. D.
1878 Description of new extinct Vertebrata from
      the upper Tertiary and Dakota formations.
      Bulletin of the U.S. Geological and Geographi-
      cal Survey, Territories 4:379, 396.

1893 A preliminary report on the vertebrate pale-
      ontology of the Llano Estacado. *Geological*
      Survey Texas, Annual Report 4:1–136.

Cordero, Luis
1955 *Diccionario quichua-español; Español-quichua.*
      Casa de la Cultura Ecuatoriana, Quito,
      Ecuador.

Córdoba y Figueroa, P.
1862 *Historia de Chile, por el maestre de campo don . . .*
      (1492–1717). Colección . . . Vol. II.

Córdova Valenzuela, Gustavo
1977 *Los auquénidos en el Antiguo Perú según Tschu-*
      di. Carácter Sagrado, importancia económica y
      utilidad de la especie en el Tahuantinsuyo. Grá-
      fica Inclán, Lima, Peru.

Coreal, François
1722 *Voyages de François Coreal aux Indes Occiden-*
      tales, contenant ce qu'il a vû de plus remar-
      quable pendant son séjour depuis 1666 jusqu'en
      1697, traduit de l'Espagnol. Avec une Rélation
      de la Guiane de Walter Raleigh et le voyage de
      Narbrough à la Mer du Sud par le Détroit de
      Magellan. 3 vols. Chez Robert-Marc d'E-
      spilly, Paris, France.

Correal, U., and Thomas van der Hammen
1971 *Investigaciones Arqueológicas en los Abrigos Ro-*
      cosos del Tequendama. Biblioteca del Banco
      Popular, Bogotá, Colombia.

Craig, Alan
1984 On the persistence of error in paleoenviron-
      mental studies of Western South America.
      In *Simposio Culturas Atacameñas*, pp. 45–56.
      B. Bittmann, editor. 44 International Con-
      gress of Americanists, Manchester; Universi-
      dad del Norte, Antofagasta, Chile.

Crawford, R. M., N. Wishart, and R. M. Campbell
1970 A numerical analysis of high-altitude vegeta-
      tion in relation to soil erosion in the Eastern
      Cordillera of Peru. *Journal of Ecology* 58(1):
      173–191.

Cronin, T. M.
1981 Rates and possible causes of neotectonic
      vertical crustal movements of the emerged
      southeastern United States Atlantic Coastal
      Plain. *Geological Society of America Bulletin*
      92:812–833.

Crosby, Alfred W.
1986 *Ecological Imperialism: The Biological Expan-*
      sion of Europe, 900–1900. Cambridge Uni-
      versity Press, Cambridge, UK.

Cruzat Añaños, A.
1967 Ocupación aldeana en la altiplanicie de
      Chupas. Tesis de grado para Bachiller en
      Antropología, Universidad Nacional San
      Cristóbal de Huamanga, Ayacucho, Peru.

Curtain, C. C., G. Pascoe, and R. Hayman
1973 [?] *Biochem. Genet.* 10(3):253–262.

Custred, G.
1974 Llameros y Comercio Interregional. In *Reciprocidad e Intercambio en los Andes Peruanos*, pp. 252–289. G. Alberti and E. Mayer, editors. Instituto de Estudios Peruanos, Lima, Peru.

Dalquest, W. W., and O. Mooser
1980 Late Hemphillian mammals of the Ocote local fauna, Guanajato, Mexico. Pearce-Sellard Series, *Texas Memorial Museum* 32:1–25.

Darwin, Charles
1845 *Journal of Researches into the natural history and geology of the countries visited during the voyage of HMS* Beagle *round the world.* Ward Lock and Co., London, UK.

Davids, M.
1987 La llama. *Fauna Argentina*, 129. Centro Editor de América Latina, Buenos Aires, Argentina.

Del Río, Mercedes, and Ana María Presta
1984 Un estudio etnohistórico en los corregimientos de Tomina y Yamparaes: Casos de multietnicidad. *Runa* 14.

Dennler de la Tour, G.
1954 The guanaco. *Oryx* 2(5):273–279.
1955 The vicuña. *Nature* 175(4451):332–333.

Descripción del Virreinato del Perú
1965 *Crónica inédita de comienzos del siglo XVII.* Universidad Nacional del Litoral (Colección de Textos y Documentos, serie B, No. 1). Rosario, Argentina.

Dianderas, A.
1954 Contribución al estudio histórico descriptivo y zoométrico de la vicuña (*Lama vicugna*). Tesis Med. Vet. Universidad Nacional Mayor de San Marcos, Facultad de Medicina Veterinaria y Zootecnia, Lima, Peru.

Diccionario de Autoridades
1976 (1726–1737) *Diccionario de la lengua castellana ... por la Real Academia Española.* 3 vols. Biblioteca Románica Hispánica, 5. Diccionarios, 3. Facsímile. Madrid, Spain.

Dillehay, Tom D.
n.d. A brief note on *Lama glama*: An economic link between coastal and highland groups on the central coast. University of Kentucky, Lexington, KY. Unpublished manuscript.

Dollfus, Olivier
1982 Development of land-use patterns in the central Andes. *Mountain Research and Development* 2(1):39–48.

Dorsey, G. A.
1901 *Archaeological Investigations on the Island of La Plata, Ecuador.* Publication 54. Field Museum of Natural History, Chicago, IL.

Dostal, W.
1958 Zur Frage der Entwicklung des Beduinentums. *Archiv für Völkerkunde* 13:1–14.

Dransart, Penny Z.
1991b Fibre to fabric: The role of fibre in camelid economies in prehispanic and contemporary Chile. Ph.D. dissertation, Linacre College, Oxford University, Oxford, UK.

Driesch, A. van den
1976 A guide to the measurement of animal bones from archaeological sites. *Peabody Museum Bulletin*, No. 1. Harvard University, Cambridge, MA.

Druss, M.
1978 Environment, subsistence, and settlement patterns of the Chiuchiu complex (ca. 2700 to 1600 B.C.) of the Atacama Desert, Northern Chile. Dissertation, Columbia University, New York.
1980 The oasis theory revisited: Variation in the subsistence-settlement system of the final Preceramic Chiuchiu Complex, northern Chile, and its implications for the process of Andean camelid domestication. Paper present at the 33rd Northwest Anthropological Conference, 27–29 March, Bellingham, Washington.
1981 Environmental change and camelid domestication in the mid Loa region, northern Chile. To be published in *Ñawpa Pacha*. [Unpublished manuscript.] [Author's Note: To date it has not been published in the abovementioned journal.]

Eckerlin, R. H., and C. E. Stevens
1973 Bicarbonate secretion by the glandular saccules of the llama stomach. *The Cornell Veterinarian* 63:436–445.

Edmund, A. Gordon
1965 A late Pleistocene fauna from the Santa Elena Peninsula, Ecuador. *Royal Ontario Museum, Life Series.* Contribution No. 63. Toronto, Canada.

Eisenberg, J. F. (editor)
1979 *Vertebrate Ecology in the northern Neotropics.* Smithsonian Institution Press, Washington, DC.

Ellenberg, Heinz
1958 Wald oder Steppe? Die natürliche Planzaudecke der Anden Perus. I und II. *Die Umschau in Wissenschaften und Tecknik* 21:645–648; 22:679–801.

Engelhardt, W., and K. Rubsamen
1979   Digestive physiology of camelids. In *The Workshop on Camels*, pp. 307–346. Khartoum, Sudan.

Engelhardt, von W., K. Rubsamen, and R. Heller
1984   *The Camelid, An All-purpose Animal*. Vol. I. Proceedings of the Khartoum Workshop on Camels, December 1979. W. R. Cockrill, editor. Scandinavian Institute of African Studies, Uppsala, Sweden.

England, B. G., W. C. Foote, D. G. Matthes, Armando G. Cardozo, and S. Riera
1969   Ovulation and the *corpus luteum* function in the llama (*Lama glama*). *Journal of Endocrinology* 45:505–513.

Epstein, H.
1971   *The Origin of Domestic Animals of Africa*. Vol. 2 (pp. 545–584). New York.

Erize, E.
1960   *Diccionario comentado mapuche-español*. Cuadernos del Sur. Universidad Nacional del Sur, Buenos Aires, Argentina.

Estrada, Emilio
1954   *Ensayo preliminar sobre arqueología del Milagro*. Publicaciones del Archivo Histórico del Guayas, Guayaquil, Ecuador.
1957b   *Ultimas Civilizaciones Pre-históricas de la Cuenca del Río Guayas*. Publicación del Museo Víctor Emilio Estrada, No. 2. Guayaquil, Ecuador.
1962   *Arqueología de Manabí Central*. Publicación del Museo Víctor Emilio Estrada, No. 7. Guayaquil, Ecuador.

Falkner, P. Tomás
1974   *Descripción de la Patagonia y de las partes contiguas de la América del Sur*, 2nd ed. Hachette, Buenos Aires, Argentina.

Fallet, M.
1961   Vergleichende Untersuchungen zur Wollbildung südamerikanischer Tylopoden. *Zeitschrift für Tierzuchtung und Zuchtungsbiologie* 75:34–56.

Febres, A.
1882   *Diccionario araucano-español o sea Calepino chileno-hispano por el P . . . de la Compañía de Jesús*. Juan A. Alsina. Buenos Aires, Argentina.

Fernández Baca, S.
1975   Alpaca raising in the high Andes. *World Animal Review*, FAO, 14:1–8.
1978   Llamoids or New World Camelidae. In *An Introduction to Animal Husbandry in the Tropics*, pp. 499–518. G. Williamson and W. J. A. Payne, editors. Longman, London.
1961–1962 Algunos aspectos del desarrollo dentario de la Alpaca (*Lama pacos*). *Revista de la Facultad de Medicina Veterinaria* 16–17:88–103.

1971   La alpaca: Reproducción y crianza. IVITA. *Boletín de Divulgación* 7. Lima, Peru.

Fernández Baca, Saúl, and César Novoa M.
1963–1966 Estudio comparativo de la digestibilidad de los forrajes en ovinos y alpacas. *Revista de la Facultad de Medicina Veterinaria* 18–20:88–96.
1968   Primer ensayo de inseminación artificial de alpacas (*Lama pacos*) con semen de vicuña (*Vicugna vicugna*). *Revista de la Facultad de Medicina Veterinaria* 22:9–18.

Fernández de Oviedo y Valdéz, G.
1851–1855 *Historia General y Natural de las Indias, Islas y Tierra Firme del Mar Océano por el capitán . . . primer cronista del Nuevo Mundo*. 5 vols. Imprenta de la Real Academia de la Historia. Madrid, Spain. [Author's Note: *See* Fernández de Oviedo 1959a, 1959b, and 1959c.]

Fernández de Oviedo, Gonzalo
1959   *Historia General y Natural de las Indias*. IV. Biblioteca de Autores Españoles desde la formación del lenguaje hasta nuestros días. Edición y estudios preliminares de Juan Pérez de Tudela Bueso. Tomo CXX, Ediciones Atlas, Madrid, Spain.

Fischer, H.
1882   Bericht über eine Anzahl Steinsculpturen aus Costa Rica. *Abhandlungen herausgegeben von Naturwissenchaftlichen vereine zu Bremen* 7:153–175.

Flores, J. A.
1973   Velocidad de pasaje de la ingesta y digestibilidad en alpacas y ovinos. Tesis. Prog. Acad. Med. Vet., Universidad Nacional Mayor de San Marcos, Lima, Peru.

Flores Ochoa, Jorge A.
1981   Factores que posiblemente han intervenido para definir la presente distribución espacial de llamas y alpacas en los Andes Centrales. Resumenes. Convención Internacional sobre Camélidos Sudamericanos. IV. Punta Arena, Chile.
1984   Origen, distribución y aspectos socioeconómicos del pastoreo. In *Pastoreo y pastizales de los Andes del Sur del Perú*, pp. 13–47. Mario Tapia and Jorge Flores Ochoa. Instituto Nacional de Investigación y Promoción Agropecuaria (Programa Colaborativo de Apoyo a la Investigación en rumiantes menores), Lima, Peru.

Forbes, D.
1870   On the Ayamara Indians of Bolivia. *Journal of the Ethnology Society of London* 2.

Franklin, W. L.
1974   The social behavior of the Vicuña. In *The Behavior of Ungulates and Its Relationship to Man-*

agement. V. Geist and F. Watther, editors. International Union on Conservation of Nature and Natural Resources publications, New Series, 24, 1:477–487. Morges, Switzerland.

1978   *Socioecology of the Vicuña*. Ph.D dissertation, Utah State University. Logan. University Microfilms International, Ann Arbor, MI.

1980   Territorial marking behavior by the South American vicuña. In *Chemical Signals; Vertebrates and Aquatic Invertebrates*, pp. 53–66. D. Müller-Schwarze and S. Silverstein, editors. Plenum Press, New York.

Free, J. P.

1944   Abraham's Camel. *Journal of Near Eastern Studies* 3:187–197.

Frick, C.

1929   History of the earth. *Natural History* 24: 106–108.

Fritz, U., and U. Schönfelder

1987   New results concerning the integration period in the southern highland of Ecuador. *Institute of Archaeology Bulletin* 23:127–150. [Author's Note: Stahl (1988: Table l, 358) misspells the last name of the second author as Schoenfelder; furthermore, this reference is not included in his bibliography.]

Fuenzalida, H.

1936   Noticia sobre los fósiles encontrados en la Hacienda Chacabuco, en abril de 1929. *Revista Chilena de Historia Natural* 40:96–99.

Galdos Rodríguez, Guillermo

1977   Visita a Atico y Caravelí (1549). *Revista del Archivo General de la Nación* 4–5:55–80.

Gandia, Enrique de

1935   *Historia de la Provincia de Santa Cruz*. Buenos Aires, Argentina.

García, Pedro A.

1836 (1822)   *Diario de la Expedición de 1822 á los campos del Sud de Buenos Aires, desde Morón hasta la Sierra de la Ventana. . . .* Imprenta del estado. Colección de Obras y Documentos relativos . . . Tomo Cuarto. Buenos Aires, Argentina.

García Márquez, Manuel Enrique

1988   Excavaciones en dos viviendas Chiribaya en el Yaral, valle de Moquegua. Tesis para optar el grado de Bachiller en Ciencias Históricas Arqueológicas. Universidad Católica Santa María, Arequipa, Peru.

Garmendia, A. E., G. H. Palmer, J. C. de Marrtini, and T. C. McGuire

1957   Failure of passive immunoglobulin transfer: A major determinant of mortality in newborn alpacas (*Lama pacos*). *American Journal of Veterinary Research* 48(10):1472–1476.

Gauthier-Philthers, H., and A. I. Dagg

1981   *The Camel: Its Evolution, Ecology, Behavior, and Relationship to Man*. University of Chicago Press, Chicago, IL.

Gay, Claudio

1847   *Zoología. Historia Física y Política de Chile*, Vol. I (Mamíferos), pp. 19–182. Santiago, Chile.

Gazin, C. L.

1942   The Late Cenozoic vertebrate faunas from the San Pecho Valley, Arizona. *Proceedings, U.S. National Museum* 92:475–518.

George, W.

1962   *Animal Geography*. Heinemann, London, UK.

Gervais, Henri, and Florentino Ameghino

1880   *Les mamiferes fossiles de l'Amérique du Sud*. Librairie F. Sary, Paris, France.

Gervais, Paul

1855   Recherches sur les Mamifères de l´Amérique méridionale. In *Zoologie de l´Expedition de Castelnau*, pp. 1–63. P. Bertrand, Paris, France.

1869 (1867)   Nouvelles recherches sur les animaux vertébrés vivants et fossils. In *Zoologie et paleontologie generales*, pp. 1–263. P. Bertrand, Paris, France.

Godoy, J. C.

1948   Glosas históricas sobre los camélidos. *Campo y suelo argentino* 32(285):64–65, 67.

1948   Glosas históricas sobre los camélidos andinos. Ministerio de Agricultura. *Almanaque* 23:273–277.

1952   Glosas históricas sobre los camélidos andinos. *Res*. (Argentina), 22:26755–26756.

Goiçueta (Goizueta), Miguel de

1852   Viaje del capitán Juan Ladrillero al descubrimiento del estrecho de Magallanes, 1558. In *Historia, física y política de Chile. Documentos sobre la historia, la estadística y la geografía*. Claudio Gay, 1846–1852, Vol. 2, pp. 55–98. (1852). Paris, France.

Gongora Marmolejo, A. de

1862   *Historia de Chile desde su descubrimiento hasta el año de 1575, compuesta por el capitán . . . y seguida de varios documentos*. Colección. . . . Vol. II.

Gonzáles Suárez, Federico

1890–1903   *Historia general de la República del Ecuador*. 9 vols. Quito, Ecuador.

1969   *Historia General de la República del Ecuador*. Vol. I. Casa de la Cultura Ecuatoriana, Quito, Ecuador.

1973   *Historia General de la República del Ecuador*. 5 vols. Clásicos Ariel, Quito, Ecuador.

González de Nájera, A.

1889   *Desengaño y reparo de la guerra del Reino de Chile*. Colección de Historiadores de Chile, Vol. XVI. Santiago, Chile.

Goodland, R.
1966 On the savanna vegetation of Calabozo, Venezuela and Rupunumi, British Guiana. *Boletín de la Sociedad Venezolana de Ciencias Naturales* 26:341–359.

Graham, R. W.
1992 *Palaeolama mirifica* from the central Mississippi River Valley; Paleoecological and evolutionary implications. *Journal of Vertebrate Paleontology* 12(3):31A.

Gray, Annie P.
1954 *Mammalian Hybrids; A Check-list with Bibliography.* Commonwealth Bureau of Animal Breeding and Genetics, Technical communication No. 10. Farnham Royal, Bucks Commonwealth Agricultural Bureau, Edinburgh, UK. [Author's Note: Novoa and Wheeler [1984] give the date of publication as 1972, and Wheeler [1991] says 1954. This is the correct date. The title is incomplete in both cases, and the bibliographical reference is mistaken.]

Gregory, J. T.
1939 Two new camels from the late Lower Pliocene of South Dakota. *Journal of Mammalogy* 20:366–368.
1942 Pliocene vertebrate from Big Spring Canyon, South Dakota. *Univ. Publ. Dept. Geol. Sci.* 26:307–446.

Grimmwood, I. R.
1968 Endangered mammals in Peru. *Oryx* 9(6): 411–421.

Grohs, W.
1974 Los Indios del Alto Amazonas del siglo XVI al XVIII. *Bonner Amerikanistiche Studien*, II. Bonn, Germany.

Groot, Ana María, L. P. Correa, and Eva Hooikas
1976 Informe presentado al Banco de la República sobre el proyecto Nariño. Unpublished manuscript. Biblioteca del Museo del Oro del Banco de la República, Bogotá, Colombia.

Guerrero, C.
1971 Enfermedades parasitarias de las alpacas. IVITA, *Boletín de Divulgación* 8:38–61.

Guerrero Vergara, Ramón
1880 Los descubrimientos del estrecho de Magallanes. In *Anuario hidrográfico de la Marina*, Vol. VI. Santiago, Chile.

Güttler, Eva
1986 *Untersuchungen über die Haltung Zucht Physiologie und Pathologie der Fortpflanzung und Krankheiten von Lamas in den Andean Argentiniens* [Investigations on Handling, Breeding, Physiology and Pathology of Reproduction and Diseases of Llamas in the Andes of Argentina, in German]. Inaugural-Dissertation zur Erlangun des Doktorgrades beim Fachbereich Veterinärmedizin der Justus-Liebig-Universität Giessen, Germany.

Guzmán, S. F.
1980 La conservación y propagación de la vicuña en Bolivia. *Comunicaciones de la vicuña*, A. Cardozo, editor. No. 2, pp. 15–16. Instituto Nacional de Fomento Lanero, Centro de Documentación del Convenio Multinacional de Conservación y Manejo de la Vicuña, La Paz, Bolivia.

Haffer, J.
1974 Avian speciation in tropical South America. *Pub. Nuttall Ornith. Club* 14:1–390.

Hahn, Eduard
1896 *Die Haustiere und ihre Beziehungen zur Wirtschaft des Menschen.* Verlag von Duncker & Humboldt, Leipzig, Germany.

Harisse, Henry
1866 *Bibliotheca Americana Vetustisima. A Description of works relating to America. Published Between the Years 1492 and 1551.* Geo. P. Philes, New York.

Harrison, J. A.
1985 Giant camels from the Cenozoic of North America. *Smithsonian Contributions in Paleobiology* 57:1–29.

Harrison, John V.
1944 The Geology of the Central Andes in part of the Province of Junín. *The Quarterly Journal of the Geological Society of London* 99:1–36.

Hartman, C. V.
1907 Archaeological research on the Pacific Coast of Costa Rica. *Memoirs of the Carnegie Museum* 3:1–188.

Hay, O. P.
1921 Descriptions of species of Pleistocene Vertebrata types or specimens most of which are preserved in the United States National Museum. *Proceedings, U.S. National Museum* 59:599–642.

Hay, O. P., and H. J. Cook
1930 Fossil vertebrates near or in association with human artifacts at localities near Colorado, Texas; Frederick, Oklahoma; and Folsom, New Mexico. *Proceedings, Colorado Museum of Natural History* 9:4–40.

Hellmich, W.
1940 Die Bedeutung des Andenraumes in biogeographischen Bild Südamerikas. In *Tier und Umwelt in Südamerika*, hrsg. v. H. Krieg. Iberoamerik Studien, Band 13. Hamburg, Germany.

Hemmer, H.
1975 Zur Herkunft des Alpakas. *Zeitschrift des Kölner Zoo* 18(2):59–66.
Hernández, E.
1981 Estudio breve de la llama en el Ecuador. *Resúmenes, Convención Internacional sobre Camélidos Sudamericanos*, Vol. IV. Punta Arenas, Chile.
Hernández Principe, Rodrigo
1923 (1622) Idolatrías en Recuay. *Inca* 1:25–49.
Herre, Wolf
1952 Studien über die Wilden und domestizierten Tylopoden Südamerikas. *Der Zoologisches Garten*, n.s. 19(2–4):70–98.
1953b Die Herkunft des alpaka. *Säugetierkundliche Mitteilungen* 1:176–177.
1961 Tiergeographische Betrachtungen an vorkolumbianischen Haussäugetieren Südamerikas. *Schriften des Geographischen Instituts der Universität Kiel* 20:289–304.
1973 Results of modern zoological research (Domestication research and history of domestic animals). In *Domestikationsforschung und Geschichte Der Haustiere, International Symposium*, pp. 57–68. J. Matolcsi, editor. Akademiai Kiado, Budapest, Hungary.
Herre, Wolf, and Lothar Kaup
1969 Uber Reste Tylopoden aus Mexiko. *Zeitschrift für zoologische Systematik und Evolutionsforschung* 7:243–254. Hamburg and Berlin, Germany.
Herrera, Antonio de
1726–1730 *Historia general de los hechos de los castellanos, en las islas y tierra-firme de el Mar Océano* (9 tomes in 4 volumes). Oficina Real de Nicolás Franco e Imprenta de Fco. Martínez Abad, Madrid, Spain. [Author's Note: *See* Herrera 1945a, 1945b, 1945c, 1945d, 1946a, 1946b, 1947.]
1730 *Descripción de las Indias Occidentales*. Madrid, Spain.
Herrera y Tordesillas, A. de
1545–1624 *Historia general de los hechos de los Castellanos en las Islas y Tierra firme del Mar Océano*. Madrid, Spain.
Herrera, Rafael
1895 Informe a Pedro Fernández Concha sobre la administración de la Hacienda Los Condes. 29 de Junio 1895. Typewritten copy; manuscript.
Hershkovitz, P.
1958 A geographical classification of neotropical mammals. *Fieldiana: Zoology* 36(6):581–620.
1972 The recent mammals of the neotropical region: A zoogeographic and ecological review. In Evolution, Mammals, and Southern

Continents, pp. 311–432. A. Keast, F. C. Erk, and B. Glass, editors. State University of New York Press, Albany, New York.
Hesse, Brian
1982b Animal domestication and oscillating climates. *Journal of Ethnobiology* 2:1–15.
1984 Archaic exploitation of small mammals and birds in northern Chile. *Estudios Atacameños* 7:42–61.
1986 Buffer resources and animal domestication in prehistoric northern Chile. *Archaezoologica* (Melanges), pp. 73–85.
n.d. Harvest profiles for small and large camelids from archaic sites in northern Chile. Manuscript.
Hibbard, C. W.
1963 *Tanupolama vera* (Matthew) from the late Hemphillian of Beaver County, Oklahoma. *Transactions of the Kansas Academic Sciences* 66:267–269.
Hibbard, C. W., and Walter W. Dalquest
1962 Artiodactyls from the Seymour Formation of Knox County, Texas. *Papers of the Michigan Academy of Science, Arts and Letters* 47: 83–99.
Hibbard, C. W., and Elmer S. Riggs.
1949 Upper Pliocene vertebrates from Keefe Canyon, Meade County, Kansas. *Bulletin of the Geological Society of America* 60:829–860.
Hidalgo Lehuede, J.
1981 Culturas y etnias protohistóricas: Area andina meridional. (1). *Chungará* 8:209–253.
Hoffmann, R.
1975 Trabajo presentado en la Segunda Mesa Redonda Internacional sobre la Conservación de la Vicuña. Nasca, Peru.
Hofmann, R., K. Otte, and M. Rios
1983 El manejo de la vicuña silvestre. *Deutsche Gesellschaft für Tecnische Zusammenarbeit (GTZ) Eschoborn*, Tomo I, pp. 66–80.
Hoffstetter, R.
1948 Sobre la presencia de un camélido en el Pleistoceno superior de la Costa Ecuatoriana. *Boletín Informativo de Ciencias Naturales* 2(5):23–25.
1970 Evolución de comunidades, cambios faunísticos e integraciones biocenóticas de los vertebrados del Cenozoico de América del Sur. In *Actas del IV Congreso Latinoamericano de Zoología*, Vol. II, Quito, Ecuador: pp. 955– 969; Perú: pp. 971–990. Caracas, Venezuela.
Holm, Olaf
1953 El tatuaje entre los aborígenes prepizarrianos de la costa ecuatoriana. *Cuadernos de Historia y Arqueología* 7, 8: 56–92.

Horn, D.
  1977   Animal utilization in the Lake Titicaca Basin
          (Chiripa). Paper presented at the 42nd An-
          nual Meeting of the Society for American
          Archaeology.
Housse, Rafael R. P.
  1930   Estudios sobre el guanaco. *Revista Chilena de
          Historia Natural* 34:38–48.
  1953   *Animales salvajes de Chile en su clasificación
          moderna*. Ediciones de la Universidad de
          Chile, Santiago, Chile.
Howard, Walter E.
  1969   Relationship of wildlife to sheep husbandry
          in Patagonia, Argentina. University of Cali-
          fornia, Davis. Unpublished report.
  1970   *Relationship of wildlife to sheep husbandry in
          Patagonia, Argentina*. Sheep husbandry Re-
          search Project UNDP/SF/FAO, Vol. 14, pp.
          1–31.
Howell, A. B.
  1944   *Speed in Animals*. University of Chicago
          Press, Chicago, IL.
Hsu, T. C., and K. Benirschke
  1967, 1974   *An Atlas of Mammalian Chromosomes*.
          Vol. 1, p. 40; Vol. 8, p. 389. Springer-Verlag,
          New York.
Humboldt, Alexander, and Aimé (Goujaud) Bonpland
  1815–1826   *Reise in die Aequinoctial-Gegenden des
          menen Continents in den Jahren 1799, 1800,
          1801, 1802, 1803 und 1804 verfasst von Hum-
          boldt und A. Bonpland*. 5 vols. Cotta, Stutt-
          gart, Tübingen, Germany.
Ignacio de Armas, Juan
  1888   *La zoología de Colón y de los primeros explo-
          radores de América*. Havana, Cuba.
Illiger, Carolus
  1811   *Prodromus systematis mammalium et avium*.
          1–302 pp. Berlin, Germany.
Inamura, Tetsuya
  1981b   Llama and alpaca: The life of the Andean
          pastoralist. *Minzokugaku* 16:100–113.
International Llama Association
  1982   The ILA Breeders and Owners Directory:
          Profile of a comunity. *Llama World* 1:22.
IVITA (Instituto de Investigaciones Tropicales y de
          Altura)
  1971   *Bibliografía de los camélidos sudamericanos*. Au-
          thor, Lima, Peru.
Jarman, P. J., and A. R.Sinclair
  1979   Feeding strategy and the patterns of re-
          source partitioning in ungulates. In *Seren-
          geti: Dynamics of an Ecosystem*, pp. 130–162.
          A. R. Sinclair and M. Norton-Griffiths, edi-
          tors. University of Chicago Press, Chicago,
          IL.

Jensen, Peter M.
  1972   High altitude adaptations in Andean South
          America. University of California, Davis,
          Department of Anthropology. Manuscript.
Jijón y Caamaño, Jacinto
  1914   Contribución al conocimiento de los aborí-
          genes de la Provincia de Imbabura en la
          República del Ecuador. In *Estudios de Prehis-
          toria Americana, II*. Blass y Cía. Impresores,
          Madrid, Spain.
  1920   Nueva contribución al conocimiento de los
          aborígenes de la provincia de Imbabura. *Bo-
          letín de la Sociedad Ecuatoriana de Estudios
          Históricos* 4:1–120, 183–245.
  1927   Puruhá. Contribución al conocimiento de
          los aborígenes de la provincia del Chimbo-
          razo de la República del Ecuador. *Boletín de
          la Academia Nacional de Historia* 6, 12, and
          14. Edición separada.
  1941   *El Ecuador Interandino y Occidental antes de la
          Conquista Castellana*, Vol. 2. Editorial Ecua-
          toriana, Quito, Ecuador.
  1951   *Antropología prehispánica del Ecuador*. La
          Prensa Católica, Quito, Ecuador.
Jiménez de la Espada, Marcos
  1892–1894   La traición de un tuerto. *La Ilustración Es-
          pañola*, 22 de Agosto de 1892, continúa el 22 de
          Agosto de 1894, 30 de Agosto de 1894, 8 de
          Setiembre de 1894. [Author's Note: The refer-
          ence was taken from another source by Frank-
          lin Pease, who was unable to verify their relia-
          bility. (Franklin Pease, pers. commun., 1990).]
Jonson, L. W.
  1983   The llama stomach: Structure and function.
          Llama World 1(4):33–34.
Juan, Jorge, and Antonio de Ulloa
  1918   *Noticias secretas de América*. 2 vols. Editorial
          América, Madrid, Spain. [Author's Note: *See*
          Juan and Ulloa 1988.]
Jungius, H.
  1971   The vicuña in Bolivia: The status of an en-
          dangered species, and recommendations for
          its conservation. *Zeitschrift für Saügetier-
          kunde* 36:129–146.
Kautz, Robert Raymond
  1976   *Late-Pleistocene Paleoclimates and Human
          Adaptation on the Western Flank of the Peru-
          vian Andes*. Ph.D. dissertation, University of
          California, Davis. University Microfilms,
          Ann Arbor, MI, 1980.
Keast, A.
  1972   Comparisons of contemporary mammal
          faunas of southern continents. In *Evolution,
          Mammals, and Southern Continents*, pp. 433–
          501. A. Keast, F. C. Erk, and B. Glass, edi-

tors. State University of New York Press, Albany, NY.

Kent, Jonathan D.
1982a *The Domestication and Exploitation of the South American Camelids: Methods of Analysis and Their Application to Circum-Lacustrine Archaeological Sites in Bolivia and Peru.* Ph.D. dissertation, Washington University, St. Louis. University Microfilms International, Ann Arbor, MI.
1982b Osteon population density and age in South American camelids. Paper presented at the 47th Annual Meeting of the Society for American Archaeology, Minneapolis, MN.

Kostritsky, B., and S.Vilches
1974 *Informe en Extenso, Proyecto Santuario Nacional del Guanaco, Calipuy.* Dirección General de Forestal y Caza. Ministerio de Agricultura. Lima, Peru.

Kraglievich, Lucas J.
1946 Sobre camélidos Chapadmalenses. *Notas del Museo de La Plata. Paleo.,* 11: 317–329.

Kreuzer, F.
1966 Transport of oxygen and carbon dioxide at altitude. In *Excercise at Altitude.* R. Margaria, editor. Excerpa Medica Foundation, Amsterdam, Netherlands.

Krickeberg, Walter
1922 *Amerika.* George Buschan, Illustrierte Völkerkunde. Band 1. Stuttgart, Germany.

Krumbiegel, I.
1944 Die neuweltlichen tylopoden. *Zoologischer Anzeiger* 145:45–70.
1952 *Lamas.* Neue Brehmbücherer, Leipzig, Germany.

Kurent, D. M., B. R. Shafit, and J. J. Maio
1973 [?] *Molecular Biology* 81:272–284.

Ladrillero, Juan de (1557)
1880 Descripción de la costa del mar océano desde el sur de Valdivia hasta el Estrecho de Magallanes. In *Anuario hidrográfico de la Marina de Chile.* Tomo VI. Santiago, Chile.

Langer, P.
1974 Stomach evolution in the Artiodactyla. *Mammalia* 2(38):295–314.

Larrea, Carlos Manuel
1919 *Relación de "El Descubrimiento y Conquista del Perú."* Quito, Ecuador.
1971 *Notas de Prehistoria e Historia Ecuatoriana.* Corporación de estudios y publicaciones, Quito, Ecuador.
1972 *Prehistoria de la región andina del Ecuador.* Corporación de estudios y publicaciones, Quito, Ecuador.

Latcham, Ricardo
1923 *La existencia de la propiedad en el antiguo Imperio de los Incas.* Santiago de Chile, Chile.
1938 *Arqueología de la región atacameña.* Prensas de la Universidad de Chile, Santiago de Chile, Chile.

Lavallée, Danièle, and J. Wheeler
1985 Pastores tempranos en las Punas de Junín, Perú. Paper presented at the XLV International Congress of Americanists, Bogotá, Ecuador.

Lears, A. E.
1895 The coast desert of Peru. *Bulletin of the American Geographical Society* 28.

Leguia, G.
1991 Enfermedades parasitarias. In *Avances y perspectivas del conocimiento de los Camélidos Sudamericanos,* pp. 325–362. S. Fernández Baca, S., editor. FAO, Santiago, Chile.

Lenz, R.
1905 *Diccionario Etimológico de las Voces Chilenas derivadas de lenguas indígenas americanas.* Santiago, Chile.

León, A.
1932a *Los Auquénidos.* UNTA, Puno, Peru.
1932b Les Auchenides: Notes phylogeniques et zoologiques, études zoo-technique. Ph.D. dissertation, Escuela de Medicina Veterinaria de Alfort.
n.d. *Les Auchenides: Notes Phylogeniques et Zoologiques.* Vigot Frères, Editeurs, Paris, France.

León (Solís), L.
1983 Expansión inca y resistencia indígena en Chile, 1470–1536. *Chungará* 10:95–115.

Lewis, C.
1947 *Golden fleece of the Andes.* The Grace Log, Grace Steamship Company, May–June.

Liais, Emmanuel
1872 *Climats, géologie, faune et géographie botanique du Brésil.* 640 pp. Paris, France.

Link, P.
1949 *Alpaca, llama, vicuña and guanaco.* Imprenta Ferrari Hermanos, Buenos Aires, Argentina.

Lönnberg, Einar
1913 Notes on guanacos. *Arkiv für Zoology* 8(19): 1–8.

Lothrop, Samuel K.
1929 *Polychrome Guanaco Cloaks of Patagonia.* Museum of the American Indian, Heye Foundation, Vol. 7, No. 6, pp. 1–30. New York.

Lozano, Pedro
1733 *Descripción chorográfica del terreno, ríos, árboles y animales de las dilatadísimas provincias del Gran Chaco Gualamba [. . .]* Córdova, Argentina. [Author's Note: Espinoza Soriano (1985–

1986) dates it in 1733. Lorandi (1980) uses another edition but gives the date as 1756. The reference is as follows: *Descripción Corográfica del Gran Chaco Gualamba* por Pedro Lozano, 1873 (1756). Publicación dirigida por A. Lamas, Colección de Obras, Documentos y Noticias inéditas o poco conocidas, Buenos Aires, Argentina.]

Ludeña de la Vega, Guillermo
1975 *La obra del Cronista Indio Felipe Guamán Poma de Ayala. Biografía del Cronista con reproducciones de sus dibujos*. Tomo I. Impreso en Talleres de la Editorial "Nueva Educacion," Lima, Peru.

Lumbreras, Luis Guillermo
1970 Proyecto de investigaciones arqueológicas en Puno, sobre el papel de la domesticación y el uso de los auquénidos en el desarrollo de las culturas altiplánicas andinas. Informe anual de las Actividades del Museo de Arqueología y Etnología, Anexo 9. Universidad Nacional Mayor de San Marcos, Lima, Peru.
1971 Proyecto de Investigaciones Arqueológicas en Puno. *Pumapunku* (La Paz) 3:58–67.

Lumley, H. de, M. A. de Lumley, M. Beltrao, Y. Yokoyama, J. Labeyrie, J. Danon, G. Delibrias, C. Falgueres, and J. L. Bischoff
1988 Découvertes d'outils taillés associés à des faunes du Pleistocène moyen dans la Toca da Esperança, Etat de Bahia. Brésil. *Comptes Rendues Acad. Sci.* 306, ser. II:241–247.

Lund, Peter W.
1837a Om Huler i Kalksteen det Indre af Brasilien, der Tildeels Indeholde Fossile Knokler. Förste Afhandling: Lappa Nova de Maniquine. *Kong. Danske Vidensk. Selsk. Naturv. Math. Afh.* 6:207–249.
1837b Ander Afhanding: Lappa de Cerca Grande. *Kong. Danske Vidensk. Selsk. Naturv. Math. Afh.* 6:307–332.
1842 Blik paa Brasiliens. Dyreverden for sidste Jordomvaeltuing. Fjerde Afhandling, Fortsaettelse af Patterdyrene. *Danske Vidensk. Selsk. Skrifter* 9:137–208.

Lydekker, R. (editor)
1901 *Library of Natural History*. Vol. 2, pp. 990–1006. Saalfield Publ. Co.

Llagostera Martínez, Agustín
1977 Ocupación humana en la costa norte de Chile asociada a peces local-extintos y a litos geométricos. *Actas del VII Congreso de Arqueología de Chile*, Vol. I, pp. 93–114.

M. y R.
1557 La visita que hizo Diego Mendez e frai Pedro Rengel (de los Puruháes encomendados en Juan de Padilla). AGI/S Justicia 671: f. 242r–257r.

M. y S.M.
1559 Visita de la encomienda de Francisco Ruiz hecha por Juan Mosquera y Cristóbal de San Martín por mandado del gobernador Gil Ramírez Dávalos. AGI/S Justicia 683:f. 798r–874v.

Maccagno, Luis
1932 *Los Auquénidos Peruanos*. Publicación del Ministerio de Fomento, Dirección de Agricultura y Ganadería, Sección de Defensa y Propaganda, Lima, Peru. [Author's Note: There is a second 1956 edition, published by the Ministerio de Agricultura, Lima, Peru.]

MacDonagh, E. J.
1940 Observaciones sobre Guanacos Cruzados con Llamas en Barreto (Córdova). Instituto del Museo de la Universidad de La Plata, Argentina. *Revista del Museo de La Plata*, Tomo II, Zoología, No. 10. [Author's Note: This is the citation made by Sumar (1988). However, Wheeler (1991) gives the reference as follows: Observaciones sobre Guanacos Cruzados con Llamas. *Revista del Museo de La Plata. Zoología* 2(10):5–84.]

Macera, Pablo et al.
1968 *Mapas Coloniales de Haciendas Cuzqueñas*. Universidad Nacional Mayor de San Marcos, Seminario de Historia Rural Andina, Lima, Peru.

Macfadden, B. J., O. Siles, P. Zeitler, N. M. Johnson, and K. E. Campbell
1983 Magnetic polarity. Stratigraphy of the Middle Pleistocene (Ensenadan) Tarija Formation of southern Bolivia. *Quaternary Research* 19:172–187.

Macfadden, B. J., and R. G. Wolf
1981 Geological investigations of late Cenozoic vertebrate bearing deposits in southern Bolivia. In *Actas del II Congreso Latino-Americano de Paleontología*, pp. 765–778. Porto Alegre, Brazil.

Machoni, Antonio
1936 (1700) *Descripción de las provincias del Chaco, y/confinantes según las relaciones modernas, y noticias adquiridas por diversas entradas de los missioneros de la Compañía de Jesús que se ha hecho en este siglo de 1700*/Po. Petroschi Sculp. Cf Furlong, No. XVIII.

Mackey, Carol J.
1977 Llama herding in the Chimu state. Paper presented at the 42nd Annual Meeting of the Society for American Archaeology, New Orleans, LA.

McGreevy, Thomas
1984 The role of pastoralism in Prehispanic and modern Huamachuco. Master's thesis, Department of Antropology, Trent University, Peterborough, Ontario, Canada.

Macquera, E.
1991 Persistencia fenotípica y caracterización de los tipos de llama Kara y Lanuda. Tesis, Universidad Nacional Agraria La Molina, Lima, Peru.

Madueño, C. G.
1912 Monografía de la vicuña. *Pan-Amer. Sci. Cong.*, Vol. 16, No. 2, pp. 5–30. Santiago de Chile, Chile.

Malaga Medina, Alejandro
1977 Los Collaguas en la historia de Arequipa en el siglo XVI. In *Collaguas I*, pp. 93–129. Franklin Pease, editor. Pontificia Universidad Católica del Perú, Lima, Peru.

Malpartida, E. and A. Florez
1980 *Estudio de la selectividad y consumo de la vicuña en Pampa Galeras.* Universidad Nacional Agraria, No. 23. Lima, Peru.

Mann, W. M.
1930 Wild animals in and out of the zoo. *Smithsonian Scientific Series* 6:302–303.

Mann, G. (et al.?)
1953 Colonias de guanacos—*Lama guanicoe*—en el desierto septentrional de Chile. *Investigaciones Zoológicas Chilenas* 1(1)[10?]:11–13. [Author's Note: Larrain Barros (1978–1979) only gives Mann as the author, and for him the issue number of the journal is 10, whereas for Bittmann (1986) the authors are Mann et al. [sic], and the issue number is 1.]

Manrique, N.
1985 *Colonialismo y pobreza campesina, Caylloma y el valle del Colca, siglos XVI–XX.* Desco. Centro de Estudios y Promoción del Desarrollo, Lima, Peru.

Marcos, J. G., and P. Norton
1981 Interpretación sobre la arqueología de la Isla de la Plata. *Miscelánea Antropológica Ecuatoriana* 1:136–154.

Marelli, Carlos A.
1931 Los vertebrados exhibidos en los zoológicos del Plata. *Memorias del Jardín Zoológico* 4:1–274.

Mariño de Lovera, P.
1865 [1595] *Crónica del Reino de Chile (. . .).* Colección de Historiadores de Chile y documentos relativos a la historia nacional. Vol. VI. Imprenta del Ferrocarril, Santiago, Chile.

Markgraf, M.
1983 Late and postglacial vegetational and paleo-climatic changes in subantarctic, temperate, and arid environments in Argentina. *Palynology* 7:43–70.

Markgraf, Vera
1985 Late Pleistocene faunal extinction in southern Patagonia. *Science* 228:1110–1112.

Marqués de Cañete
1953 (1556) Provisión del gobierno superior, . . . de 1557, para que . . . no se hagan chacos de guanacos, vicuñas, etc. Documentos del siglo XVI, Virrey Marqués de Cañete. *Revista del Archivo Histórico* (Cuzco) 4(4):61–64. [Author's Note: It should have been cited as Andrés Hurtado de Mendoza.]

Márquez Miranda, Fernando, and Eduardo Mario Cigliano
1961 Problemas arqueológicos en la zona de Ingenio del Arenal (Pcia. de Catamarca, Rep. Argentina). *Revista del Museo de la Plata*, n.s., *Sección Antropología*, V.

Marshall, L. G., S. D. Webb, J. J. Sepkoski, Jr., and D. M. Raup
1982 Mammalian evolution and the Great American Interchange. *Science* 215:1351–1357.

Martin, P. S., and H. E. Wright
1967 *Pleistocene Extinction: The Search for a Cause.* Yale University Press, New Haven, CT.

Martínez, R.
1567 Memorial sobre la encomienda de Bagua. AGI, Justicia, 416.

Matthew, W. D., and H. F. Osborn
1909 Faunal list of the Tertiary Mammalia of the West. *Bulletin of the U. S. Geological Service* 361:91–138.

Matthews, L. H.
1971 *The Life of Mammals*, Vol. 2. Universe Book, New York.

Meade, G. E.
1945 *The Blanco Fauna.* University of Texas Publication No. 4401, pp. 509–556.
1953 An early Pleistocene vertebrate fauna from Frederick, Oklahoma. *Journal of Geology* 61:452–460.

Means, Philip Ainsworth
1928 Biblioteca Andina. Part One: The chroniclers, or the writers of the sixteenth and seventeenth centuries who treated of the pre-Hispanic history and culture of the Andean countries. *Transactions of the Connecticut Academy of Arts and Sciences*, Vol. 29. New Haven, CT.

Medina, José Toribio
1882 *Los aborígenes de Chile.* Imprenta Gutenberg, Santiago, Chile. [Author's Note: There is a

1952 edition with the same title published by Fondo Histórico y Bibliográfico José Toribio Medina, Santiago, Chile.]

Meggers, Betty J.
1966   *Ecuador*. Thames & Hudson, London, UK.

Meighan, Clement
1970   Excavations at Guatacondo, Chile, 1969. A preliminary report on the field activities. Grant G.S. 2652. University of California, Los Angeles, CA.

Meisch, Lynn
1987   Alpacas back in Ecuador. *Llamas: The International Camelid Journal*, November, December, pp. 24–26.

Menegaz, A. N., F. J. Goin, and E. Ortiz Janreguizar
1989   Análisis morfológico y morfométrico multivariado de los representantes fósiles y vivientes del género *Lama* (Artiodactyla, Camelidae). Sus implicancias sistemáticas, biogeográficas, ecológicas y biocronológicas. *Ameghiniana* 26(3–4):153–172.

Mengoni Goñalons, Guillermo L.
1978   El aprovechamiento de los recursos faunísticos en el interior de Patagonia meridional continental: Hipótesis y modelos. Trabajo presentado al VI Congreso Nacional de Arqueología del Uruguay, Salto, Uruguay.

1979   La domesticación prehistórica de camélidos en el área andina: técnicas, métodos y modelos. In *Actas de las Jornadas de Arqueología del Noroeste Argentino. Antoquitas-Asociación de Amigos del Instituto de Arqueología. "Profesor J. M. Suetta,"* Publicación No. 2, pp. 190–198. Buenos Aires, Argentina.

(Mengoni, G.)
1981   Visibilidad arqueológica de clases de edad en conjuntos arqueofaunísticos de Camélidos. In *Resúmenes Convención Internacional sobre Camélidos Sudamericanos*, IV, pp. 33–34. Punta Arenas, Chile. [Author's Note: This clearly is a mistake. It should read Mengoni Goñalons, Guillermo L.]

Mengoni Goñalons, Guillermo L.
1982   El hombre prehistórico en la puna argentina: Su relación con la Fauna. Primera Reunión Nacional de Ciencias del Hombre en Zonas Aridas. Mendoza, Argentina.

1983   Prehistoric utilization of faunal resources in arid Argentina. In *Animals and Archaeology. 1. Hunters and Their Prey*. Caroline Grigson, editor. BAR International Series 163. British Archaeological Reports, Oxford, UK. [Author's Note: The editors are J. Clutton-Brock and C. Grigson. Also, the page numbers have been omitted: 325–335.]

(Mengoni, G.)
1986   Patagonian prehistory: Early exploitation of faunal resources (13,500–8,500 B.P.). In *New Evidence for the Pleistocene Peopling of the Americas*, pp. 271–279. A. L. Bryan, editor. Center for the study of Early Man, University of Maine, Orono, ME. [Author's Note: This is a mistake. The author is Guillermo L. Mengoni Goñalons.]

Meriam, J. C., and C. Stock
1925   A Llama from the Pleistocene of McKittrick, California. *Carnegie Institute of Washington Publication* 347:38–44.

Mester, A. M.
1985   Un taller Manteño de la concha madre perla del sitio Los Frailes, Manabí. *Miscelánea Antropológica Ecuatoriana* 5:101–111.

Middleton, William D.
1992   The extraction of phytoliths from prehispanic and contemporary camelid dental calculus. Paper presented at the 57th Annual Meeting of the Society for American Archaeology.

Mikesel, N. W.
1955   Notes on the dispersal of the dromedary. *Southwestern Journal of Anthropology* 9:231–245.

Miller, George R.
1979   An introduction to the ethnoarchaeology of the Andean camelids. Ph.D. dissertation in Anthropology, Graduate Division of the University of California, Berkeley, CA.

1981   Subsistence and social differentiation at Chavín de Huántar: Some insights from the preliminary analysis of the faunal remains. Paper presented at the 46th Annual Meeting of the Society of American Archaeology, San Diego, CA.

1983   Ch'arki y Chavín: Análisis preliminar de los restos faunísticos de las zonas de ocupación. In *Xauxa* I. R. Matos M., editor. Lima, Peru.

n.d.   Preliminary report on the mammalian remains from Cardal. Unpublished report (1989) in possession of Richard L. Burger and Lucy Salazar-Burger.

Miller, G. S.
1924   A second instance of the development of rodent-like incisors in an artiodactyl. *Proceedings of the United States National Museum* 66 (8):1–4.

Miller, S., J. Rottmann, and R. D. Taber
1973   Dwindling and endagered ungulates of Chile: *Vicugna, Lama, Hippocamelus,* and *Pudu*. In *North American Wildlife and Natural Resource Conference*, Vol. 38, pp. 55–68.

Miller, W. E.
1968 Occurrence of a giant bison, *Bison latifrons*, and a slender-limbed camel, *Tanupolama*, at Rancho La Brea. *Los Angeles County Museum, Contributions in Science* 147:1–9.

Millones, J. O.
1982 Patterns of land-use and associate environmental problems of the Central Andes. An integrated summary. In *Mountain Research and Development*, Vol. 1, No. l, pp. 49–61. Special issue. United Nations University, International Mountain Society. UNEP-MAB, Boulder, CO.

Millones Santa Gadea, Luis
1975 Economía y ritual en los Condesuyos de Arequipa: Pastores y tejedores del siglo XIX. *Allpanchis* 8(8):45–66.

Ministerio de Agricultura
1978 *El guanaco de Magallanes. Chile. Su distribución y biología.* Publicación Técnica, No. 4. Corporación Forestal, Chile.

Ministerio de Agricultura, Ganadería y Minería
1977 Estudio integral del guanaco. *Boletín Serie Técnica*, No. l. Río Negro, Argentina.

Miotti, L., M. Salemme, and A. Menegaz
1989 El Manejo de los Recursos Faunísticos Durante el Pleistoceno Final y Holoceno Temprano en la Pampa y Patagonia. Precirculados del IX Congreso Nacional de Arqueología Argentina, pp. 102–118. Universidad de Buenos Aires, Buenos Aires, Argentina.

Molina, Juan Ignacio.
1878a *Compendio de la historia jeográfica natural i civil del Reino de Chile.* Publicado anónimo en Bolonia en 1776 (. . .). Colección . . . Vol. XI, pp. 185–304. [Author's Note: *See* Molina 1782.]

1878b *Compendio de la historia jeográfica, natural i civil del Reino de Chile.* Escrito en italiano por el abate. . . . Colección . . . Vol. XV, pp. 305–522. [Author's Note: *See* Molina 1782.]

1901 *Compendio de la historia civil del reino de Chile.* Colección de Historiadores de Chile, Tomo XXVI. Santiago, Chile. [Note: *See* Molina 1782.]

Mones, A.
1988 Nuevos registros de Mamíferos fósiles de la Formación San José (Plioceno-Pleistoceno inferior) (Mammalia: Xenarthra; Artiodactyla; Rodentia). *Comm. Paleont. Mus. Hist. Nat.* 1(20):255–277.

Montagu, I.
1965 Communication on the current survival in Mongolia of the wild horse (*Equus przewaleski*), wild camel (*Camelus bactrianus ferus*), and wild ass (*Equus hemionus*). *Proceedings of the Zoological Society* 144:425–428.

Montellano, M.
1989 Pliocene Camelidae of Rancho E. Ocote, Central Mexico. *Journal of Mammalogy* 70(2): 359–369.

Montoya, Alfredo J.
1984 *Como evolucionó la ganadería en la época del Virreinato.* Plus Ultra, Buenos Aires, Argentina.

Moore, Katherine M.
1988b Animal procurement and use in prehistoric highland Peru. Paper presented at the 49th Annual Meeting of the Society for American Archaeology, Portland, OR.

1989 Hunting and the origins of herding in Peru. Ph.D. dissertation, Department of Anthropology, University of Michigan, Ann Arbor, MI. [Author's Note: Baied y Wheeler (1993) cite this dissertation in this way. However, Flannery et al. (1989) indicate that the title is "Evolution of Specialized Animal Economies in Prehistoric Highland Peru." Also, no date is given for the dissertation. In the 1998 article the author gave the title as "Hunting and the Origin of Herding in Prehistoric Highland Peru," thus confirming that the date is 1989 (*see* Moore 1998:197).]

Morán, Emilio F.
1979 *Human Adaptability: An Introduction to Ecological Anthropology.* Wadsworth Publishing Co., Inc., Belmont, CA.

Mori, Juan de
1889 (1535) *Relación hecha por Juan de Mori de la expedición de Simón de Alcazaba.* In Medina, 1889, Vol. 3, pp. 316–330 (Medina, José Toribio, 1889. *Colección de documentos inéditos para la historia de Chile* . . . Vol. 3. Santiago, Chile).

Moro, S., and C. Guerrero
1971 *La alpaca, enfermedades infecciosas y parasitarias.* Instituto Veterinario de Investigaciones Tropicales y de Altura, *Boletín de Divulgación*, No. 8. Lima, Peru.

Moro S., Manuel, Augusto P. Vallenas, and collaborators
1985 Informe sobre estudios realizados en alpacas en el año 1957. Ministerio de Agricultura, Dirección de Ganadería. *Informaciones*, No. 6. Lima, Perú.

Morrison, P.
1966 Insulative flexibility in the guanaco. *Journal of Mammalogy* 47:18–23.

Mostny, G., and F. Hans Niemeyer
1983 *Arte rupestre chileno.* Departamento de Extensión Cultural del Ministerio de Educación. Colección de Arte Chileno, Santiago, Chile.

Müller, Otto F.
1776   *Zoologiae Danicae prodromus, seu animalium Daniae et Norvegiae indigenarum characteres, nomina, et synonymia imprimis popularium.* 1–282 pp. Havnial, Typis, Hallagerus.

Murphy, R. C.
1926   Oceanic and climatologic phenomenon along the West Coast of South America during 1925. *The Geographical Review* 16(1).

Musters, G. C.
1871   *At Home with the Patagonians.* John Murray, London, UK.

Nachtigall, Horst
1964   Woher stammt das Nomadentum? *Die Umschau in Wissenschaft und Technik*, 15 Januar, Year 64. Frankfurt, Germany.
1966b *Indianische Fischer Feldbauer und Viehzüchter.* Beiträge zur peruanischen Völkerkunde, Berlin, Germany.

Namnandorj, O.
1970   *Conservation and Wild Life in Mongolia.* Field Research Project. Miami, FL.

Neira, M.
1970   Avances sobre la arqueología peruana y la domesticación de los auquénidos. In *Convención sobre camélidos sudamericanos (Auquénidos).* Universidad Nacional Técnica del Altiplano, Puno, Peru.

Netherly, Patricia J.
1977   On defining the North Coast of Peru. Paper presented at the 42nd Annual Meeting of the Society for American Archaeology, New Orleans, LA.
1978   *Local Level Lords on the North Coast of Peru.* University Microfilms, Ann Arbor, MI.

Niemeyer, Hans
1963   Excavaciones en un cementerio incaico en la hacienda Camarones (Prov. de Tarapacá). *Revista Universitaria.* Universidad Católica, Santiago, Chile.

Nordenskiöld, E.
1908   Einfundort für säugethier-fossilien in Perú. *Arkiv. Zool.* 4(11):22.

Noriega B., Humberto
1947   Ganado lanar en el Ecuador. Tesis Ing. Agr. Universidad Central, Quito, Ecuador.

Norton, Presley
1986   El señorío Salangone y la liga de mercaderes. El cartel spondylus-balsa. *Miscelánea Antropológica Ecuatoriana* 6:131–143.

Norton, Presley, R. Lunnis, and N. Nayling
1983   Excavaciones en Salango, provincia de Manabí, Ecuador. *Miscelánea Antropológica Ecuatoriana* 3:9–80.

Novoa, C.
1980   *La conservación de especies nativas en América Latina.* FAO, Consulta técnica sobre la conservación de los recursos de genética animal, Rome, Italy.
1981   La conservación de especies nativas en América Latina. In *Animal Genetic Resources Conservation and Management.* FAO Animal Production and Health, Paper No. 24, pp. 349–363. FAO, Rome, Italy

Núñez, Diego (Diego, Núñez)
1858   Apontamiento de que V. A. quer saber. *Revista Trimestral del Instituto Brasileiro* 2: 365–369.

Núñez Atencio, Lautaro
1981   Paleoindian and archaic cultural periods in the arid and semiarid regions of north Chile. Manuscript.
1983   *Paleoindio y Arcaico en Chile: Diversidad, Secuencia y Proceso.* Ediciones Cuicuilco, Serie Monografías, Escuela Nacional de Antropología e Historia, Mexico City, Mexico.

Núñez, Lautaro, and C. Santoro
1988   Cazadores de la Puna Seca y Salada del área centro-sur andina (norte de Chile). *Estudios Atacameños* 9:11–60.

Núñez A., Lautaro, J. Varela, and R. Casamiquela
1987   Ocupación paleoindia en el Centro-Norte de Chile: Adaptación circunlacustre en las tierras bajas. *Estudios Atacameños* 8:142–185. [Author's Note: Also published in *Investigaciones paleoindias al sur de la línea ecuatorial*, edited by L. Núñez and Betty Meggers. Taraxacum, Washington, DC.]

Núñez de Pineda y Bascuñán, F. de
1863   *Cautiverio feliz del Maestro de Campo General don . . . y razón individual de las guerras dilatadas del Reino de Chile. . . .* Colección. . . Vol. III.

Olivares, Miguel de
1864   *Historia militar, civil y sagrada de lo acaecido en la conquista y pacificación del Reino de Chile, desde la primera entrada de los españoles hasta la mitad del siglo décimo octavo de nuestra Redención.* Colección. . . Vol. IV.

Oporto, N. R.
1977   *Proyecto Provincial para el Manejo de la Fauna Silvestre. Subproyecto Integral del Guanaco (Lama guanicoe). I. Estudios Preliminares.* Provincia de Río Negro, Ministerio de Agricultura y Minería, Serie Técnica, 1.

Oré, Lvis Hieronymo
1598   *Symbolo catholico indiano.* Antonio Ricardo, Lima, Peru. [Author's Note: *See* Oré 1992.]

Orlove, Benjamin S.
1977a *Alpacas, Sheep and Man*. Academic Press, New York.
1977b Integration through production: The use of zonation. *American Ethnologist* 4(1):84–101.
1982 Native Andean pastoralist: Traditional adaptations and recent changes. In *Contemporary Nomadic and Pastoral Peoples: Africa and Latin America*, pp. 95–136. P. C. Saltzman, editor. Studies in Third World Societies, Publication No. 17. Williamsburg.

Ortega, I. M.
1985 Social organization and ecology of a migratory guanaco population in southern Patagonia. Master's thesis, Iowa State University, Ames, IA.

Osgood, W. H.
1916 Mammals of the Collins-Day South American Expedition. *Field Museum of Natural History: Zoological Series*, Publication No. 189, 10(4):199–217. Chicago, IL.
1943 The mammals of Chile. *Field Museum of Natural History: Zoological Series*, No. 30. Chicago, IL.

Otte, K. C., and Hofman, R. K.
1979 Utilización racional de la vicuña silvestre en el Perú. *Acta Zoológica Lilloana* 34:141–152.

Otte, K. C., and Venero, J. L.
1979b *Análisis de la craneometría diferencial entre la vicuña* (Vicugna vicugna) *y la alpaca* (Lama guanicoe pacos). Proyecto para la utilización racional de la vicuña silvestre, Lima, Peru.

Ovaglie, Alonso d'
1646 *Istoria relazione del regno di Chile*. Rome, Italy.

Ovalle, Alonso de
1888 *Histórica relación del Reyno de Chile y de las misiones y ministerios que ejercita en él la Compañía de Jesús (. . .)*. 2 vols. Colección . . . , Vols. XII–XIII.
1969 (1646) *Histórica relación del reino de Chile*. Instituto de Literatura Chilena, Santiago, Chile.

Palacios Rios, Felix
1977a Hiwasha Uywa Uywataña, Uka Uywaha Hiwasaru Uyusitu: Los pastores de Chichillapi. Tesis de maestría, Programa de Perfeccionamiento en Ciencias Sociales, Pontificia Universidad Católica del Perú, Lima, Peru.

Palomino, Diego
1549 Relación de las provincias que hay en la conquista del Chuquimayo que yo el capitán . . . tengo por Su Majestad y por el muy Ilustre Señor Pedro Gasca presidente de la Audiencia Real destos Reynos del Perú por su Majestad. RGI. 1897; IV, pp. xlvi–lii.

Parodi, R.
1947 La presencia del Género *Paleolama* en los túmulos indígenas de Santiago del Estero con algunas consideraciones sobre *Lama cordubensis* y otras especies. Sociedad Científica, *Anales* 143(1):3–9.

Parsons, Jeffrey R.
1988 Changing regional settlement in the Sierra Central: Implications for changing relationships between cultivators and herders, AD 500–1500. Paper presented at the symposium, "The Late Prehispanic and Early Colonial Periods in Perú." Ann Kendall and Brian S. Bauer, coordinators. 46th International Congress of Americanists. Amsterdam, Netherlands.

Pascual, R. et al. [sic]
1966 *Paleontografía bonaerense: IV. Vertebrata*. 1–202 pp. Comm. Invest. Cienc., La Plata, Argentina. [Author's Note: This is the reference given by Webb 1974.]

Pascual, R. E., D. Ortega Hinojosa, and E. Tonni
1965 Las edades del Cenozoico mamalífero de la Argentina con especial atención a aquellas del territorio bonaerense. *Anales de la Comisión de Investigación Científica*, Provincia de Buenos Aires, VI:165 ff.

Paso y Troncoso, Francisco del (editor)
1939–1942 *Epistolario de Nueva España, 1505–1818 . . .* (16 vols.). Mexico City, Mexico.

Patterson, Bryan, and Rosendo Pascual
1969 Evolution of mammals on southern continents: V. The fossil mammal fauna of South America. *Quarterly Review of Biology* 43:409–451.
1972 The fossil mammal fauna of South America. In *Evolution, Mammals, and Southern Continents*, pp. 247–309. A. Keast, F. C. Erk, and B. Glass, editors. State University of New York Press, Albany, NY.

Patzelt, Erwin
1979 *Fauna del Ecuador*. Editorial Las Casas, Quito, Ecuador.

Paucar, A., J. Tellez, L. Neyra, and J. Rodríguez
1984 Estudio Tecnológico del Beneficio de Vicuñas. In *La Vicuña*, pp. 33–48. F. Villiger, compiler. Editorial Los Pinos, Lima, Peru.

Pearson, Oliver P.
1943 The status of the vicuña in Southern Peru. *Journal of Mammalogy* 24(1):97.
1951 Mammals in the highlands of Southern Peru. *Bulletin of the Museum of Comparative Zoology* 106(3):117–174.

Pease, Franklin (editor)
1977 *Collaguas. I.* Pontificia Universidad Católica del Perú, Lima, Peru.

Pernoud, R. (editor)
1942   Un journal de voyage inédit au long des côtes du Chili et du Pérou. *L'Amérique du Sud au XVIIIe siecle. Mélanges anecdotiques et bibliographiques.* Imprimerie du "Petit Nantais," Nantes, France.

Perkins, Dexter, and Patricia Daly
1968   A hunter's village in Neolithic Turkey. *Scientific American* 219(5):96–106.

Petersen, Georg
1935   Estudios climatológicos en el Noroeste Peruano. *Boletín de la Sociedad Geológica del Perú* 7(2).
1941   La marcha de Francisco Pizarro. *Chaski* 1(3): 8–10.

Philippi, Rodolfo Armando
1860   *Viaje al Desierto de Atacama hecho de orden del Gobierno de Chile en el verano 1853–1854.* Librería de Eduardo Anton, Halle in Saxony, Germany.

(Philippi, Rodulfo)
1872   *Elementos de Historia Natural,* 2nd ed. Santiago, Chile.

(Philippi, R.)
1893   Noticias preliminares sobre los huesos fósiles de Ulloma. *Anales de la Universidad de Chile,* Vol. 82. Santiago, Chile. [Author's Note: Dillehay et al. (1992) write Phillipi. It must be a mistake.]

(Philippi, R. A.)
1966   El llamado desierto de Atacama y las grandes formaciones de altiplano de los Andes del sur de 19° latitud sur. *Boletín de la Academia Nacional de Ciencias Naturales* 45: 283– 322.

Pieters, H.
1954 [?]   2. *Tierpsychology* 2:213–303.

Pigafetta, Antonio
1906   *Magellan's Voyage around the World.* 2 vols. and index. J. A. Robertson, translator-editor. Cleveland, OH. [Author's Note: *See* Pigafetta 1927, 1970.]

Pizarro, Francisco
1540   Título de Pedro Monge, por . . . *CDIHCH* 6:168–169.

Pizarro, Pedro
1965   *Relación del descubrimiento y conquista de los reinos del Perú.* Biblioteca de Autores Españoles desde la formación del lenguaje hasta nuestros días, 168. CLXVIII, pp. 159–242. Ediciones Atlas, Madrid, Spain. [Author's Note: *See* Pizarro Pedro 1968, 1978.]

Plaza, J.
1973   Sarna ovina. *Agro informativo,* No. 189. Servicio Agrícola Ganadero, Chile.

Pocock, R. I.
1923   The external characters of the Pigmy Hippopotamus (*Choeropsis liberiensis*) and the Suidae and Camelidae. *Proceedings of the Zoological Society of London,* pp. 531–549. London, UK.

Politis, G., and M. Salemme
1989   Prehispanic mammal exploitation and hunting strategies in the Eastern Pampa Sub-Region of Argentina. In *Hunters of the Recent Past,* pp. 234–245. L. Davis and B. O. K. Reeves, editors. One World Archaeology Series. Unwin Hyman, London, UK.

Ponce Paz, Ramón
1951   Para una bibliografía de auquénidos. In *Contribución a la bibliografía de veterinaria americana,* pp. 15–18. Congreso Panamericano de Medicina Veterinaria, Lima, Peru.

Prichard, H.
1902   Field-notes upon some of the larger mammals of Patagonia, made between September 1900 and June 1901. *Proceedings of Zoological Society* 1:272–277.

Quinn, W. H., V. T. Neals, and S. Antúnez de Mayolo
1986   Preliminary report on El Niño occurrences over the past four and a half centuries. National Science Foundation, ATM-85 15014, Corvallis, Oregon. Oregon State University, Report 86-16. Manuscript.
1987   EI Niño occurrences over the past four and a half centuries. *Journal of Geophysical Research* 93(C13):14449–14461.

Radtke, U.
1987   Paleo sea levels and discrimination of the last and the penultimate interglacial fossiliferous deposits by absolute methods and geomorphological investigations; illustrated from marine terraces in Chile. *Berliner Geographische Studien* 25:313–342.

Raedeke, Kenneth J.
1979   *Population Dynamics and Socioecology of the Guanaco* (Lama guanicoe) *of Magallanes.* Chile. Ph.D. dissertation, College of Forest Resources, University of Washington, Seattle. University Microfilms International, Ann Arbor, MI.
1980   Food habits of the guanaco (*Lama guanicoe*) of Tierra del Fuego, Chile. *Turrialba* 30:177–181.

Ramírez, A.
1987   *Alpaca Clostridium perfringens Type A Enterotoxemia: Purification and Assays of the Enterotoxin.* University Microfilms International, Ann Arbor, MI.
1991   Enfermedades infecciosas. In *Avances y perspectivas del conocimiento de los Camélidos Sud-*

americanos, pp. 263–324. S. Fernández Baca, editor. FAO, Santiago, Chile.

Ramírez-Horton, Susan
1979 The economic organization of the valleys of the North Coast: A preliminary analysis of the late prehispanic era. Paper presented at the Segunda Jornada de Etnía, Ayllu y Parcialidad. Museo Nacional de Historia, Lima, Peru.

Raven, Peter H., and D. I. Axelrod.
1975 History of the flora and fauna of Latin America. *American Scientist* 63:420–429.

Reiner, R. J., F. C. Bryant, R. D. Farfán, and B. F. Craddock
1987 Forage intake of alpacas grazing Andean rangeland in Peru. *Journal of Animal Science* 64:868–871.

Reynafarge S., J. Faura, D. Villavicencio, A. Curaca, L. Reynafarge, L. Oyola, L. Contreras, E. Vallenas, and A. Faura
1975 Oxygen transport of hemoglobin in high-altitude animals (Camelidae). *Journal of Applied Physiology* 38:806–810.

Ribeiro, Darcy
1985 *Las Américas y la civilización.* Centro Editor de América Latina, Colección Bibliotecas Universitarias, Buenos Aires, Argentina.

Rice, Don S., Charles Stanish, and Phillip R. Scarr (editors)
1989 *Ecology, Settlement, and History in the Osmore Drainage, Peru.* Oxford BAR International Series 545. Oxford, UK.

Riddell, Francis A., and Dorothy Menzel
1954 Fieldnotes on excavation at Acarí, 1954. University of California, Berkeley, CA. Manuscript.

Ripinsky, M. M.
n.d. The occurrence of the camel in the ancient Egypt and the Sahara. Mimeograph.
1975 The camel in ancient Arabia. *Antiquity* 69: 295–298.

Rivera, Pedro de, and Antonio de Chávez y Guevara
1881 Relación de la Ciudad de Guamanga y Términos. Año de 1586. In *Relaciones Geográficas de Indias.* Ministerio de Fomento del Perú, Tomo I, pp. 105–139. Edit. Tip. de Manuel Hernández, Madrid, Spain. [Author's Note: *See* Rivera and Chávez y de Guevara 1974.]

Rodríguez, R., and H. Torres
1981 El hábitat de la vicuña en Chile. In *Resúmenes Convención Internacional sobre Camélidos Sudamericanos*, IV. Punta Arenas, Chile.

Román y Zamora, Gerónimo
1897 (1575) Repúblicas del mundo. *Colección de libros raros o curiosos que tratan de América*, Tomos 14–15. Madrid, Spain.

Romero, Carlos A.
1913 Noticias cronológicas de la gran ciudad del Cuzco por el Dr. Dn. Diego de Esquivel y Navia. *Revista Histórica* 5:209–224.

Romero, Elías C.
1927 *Llamas y alpacas, vicuñas y guanacos.* Buenos Aires, Argentina.

Romero, E.
1936 La llama en la Argentina. *La Chacra* 8(69).

Romero, Fernando, and Emilia Romero Del Valle
1943 Probable Itinerario de los Tres Primeros Viajes Marítimos para la Conquista del Perú. Sobretiro del No. 16 de la *Revista de América*. Mexico City, Mexico. [Author's Note: This article was apparently published with the same title and in the same year in *Boletín de la Sociedad Geográfica de Lima* 60(1).]

Rosales, D. de
1877–1878 *Historia general del Reyno de Chile.* 3 vols. Imprenta del Mercurio, Valparaiso, Chile.

Rosemblat, A.
1954 *La población indígena y el mestizaje en América, 1.* Nova, Buenos Aires, Argentina.

Rosenmann, A. M., and P. Morrison
1963 Physiological response to heat and dehydration in the guanaco. *Physiological Zoology* 36: 45–51.

Roth, J. A., and J. Laerm
1990 A late Pleistocene vertebrate assemblage from Edisto Island, South Carolina. *Brimleyana* 3:1–29.

Rottmann, J.
1981 Situación de los camélidos en Chile. In *Resúmenes Convención Internacional sobre Camélidos Sudamericanos*, IV, pp. 13–14. Punta Arenas. Chile.

Royo y Gómez, José
1960a Características paleontológicas y geológicas del yacimiento de vertebrados de Muaco, Estado Falcón, con industria lítica humana. Memoria del III Congreso Geológico Venezolano, Tomo 2. *Boletín de Geología*, Publicación Especial, 3:501–505.
1960b El yacimiento de vertebrados pleistocenos de Muaco, Estado Falcón, Venezuela, con industria lítica humana. In *Report of the International Geological Congress*, XXI Session, Norden, Pt. 4, pp. 154–157. Copenhagen, Denmark.

Rübsamen, K., and W. V. Engelhardt
1975 Water metabolism in the llama. *Comparative Biochemistry and Physiology* 52A:595–598.

Ruiz, Hipólito
1952 *Relacion histórica del viage, que hizo a los reynos del Perú y Chile el botánico d. Hipólito Ruiz en*

el año 1777 hasta el de 1778, en cuya época regresó a Madrid. Segunda edición, enmendada y completada . . . por . . . Jaime Jaramillo-Arango. Tomo Primero. Madrid, Spain.

Rusconi, Carlos

1930a La presencia del general "Platygonus" en túmulos indígenas de época prehispánica. Addenda a "Las especies fósiles argentinas de pecaríes (Tayassuidae) y sus relaciones con las del Brasil y Norte América." *Anales del Museo Nacional de Historia Natural* 36:228–340.

1930b Nueva especie de "Palaeolama" del Pleistoceno argentino (*P. brevirostris*). *Revista Chilena de Historia Natural* 34:208–219, 338–348.

1931 La dentadura de *Palaeolama* en relación a la de otros camélidos. *Revista de Medicina Veterinaria* 13:250–273. [Author's Note: The reference comes from Cardozo (1975a), but in his bibliography the date is given as 1931, and in the text as 1930. The correct date is 1931.]

1936 El carpincho más grande del mundo. *Revista Geográfica Amer.* 5(29):131–135.

Salomon, Frank

1978 Systémes politiques verticaux aux marches de l'Empire Inca. *Annales* (Paris) 33:5–6.

1986 *Native Lords of Quito in the Age of the Incas. The Political Economy of the North Andean Chiefdom.* Cambridge Studies in Social Anthropology, 59. Cambridge University Press, Cambridge, UK.

Salvador Lara, Jorge

1971 *Esquema para el estudio de la prehistoria en el Ecuador.* Editorial Ecuatoriana, Quito, Ecuador.

San Martín, Felipe

1987 Comparative forage selectivity and nutrition of South American camelids and sheep. Ph.D. dissertation, Texas Tech University, Lubbock, TX.

Santillán, Diego Abad de, et al. [sic]

1965 *Historia argentina.* Tomo I. Tipografía editora Argentina, Buenos Aires, Argentina.

Santoro, C., and L. Nuñez

1987 Hunters of the dry puna and salt puna in northern Chile. *Andean Past* 1:57–109.

Sawyer, Michael J.

1985 An analysis of mammalian faunal remains from the site of Huaricoto, PAN3-35. Master's thesis, Department of Anthropology, California State University, Hayward, CA.

Saxon, Earl C.

1978 Natural prehistory, archaeology and ecology at the uttermost part of the earth. Department of Anthropology. University of Durham, UK.

Schmidel, Ulrich

1903 *Viaje al Río de Lá Plata (1534–1554).* Notas bibliográficas y biográficas por Bartolomé Mitre. Prólogo, traducción y anotaciones por Samuel A. Lafone Quevedo. Buenos Aires, Argentina. [Author's Note: *See* Schmidel Hulderico 1749.]

Schmidt, Christian R.

1975 Breeding seasons and notes on some other aspects of reproduction in captive camelids. *International Zoo Yearbook* 13:387–390. [Author's Note: Raedeke (1976) gives the date for this citation as 1973 in the text but 1975 in the bibliography.]

Schmidt-Nielsen, K., B. Schmidt-Nielsen, S. A. Jarnum, and T. R. Houpt

1957 Body temperature of the camel and its relation to water economy. *American Journal of Physiology* 188:103–112.

Shimada, Izumi

1977 Socioeconomic organization at Moche V Pampa Grande, Peru: Prelude to a major transformation to come. Ph.D. dissertation, University of Arizona, Tucson, AZ.

Shimada, Izumi, and colleagues [sic]

1981 The Batan Grande-La Leche Archaeological Project: The first two seasons. *Journal of Field Archaeology* 8:405–446.

Shimada, Melody, and Izumi Shimada

1979 Herding and the economic role of llama on the pre-Incaic Peruvian coast: New data and reconsideration. Unpublished manuscript.

Sick, W. D.

1963 *Wirtschaftsgeographie von Ecuador.* Stuttgart Geographische Studien, Stuttgart, Germany.

Sillau, H., S. Cueva, D. Chauca, A.Valenzuela, and W. Cárdenas

1972 Observaciones sobre el transporte de oxígeno en la alpaca en la altura y a nivel del mar. *Revista de Investigaciones Pecuarias* 1:129–136.

Simpson, G. G.

1928 Pleistocene mammals from a cave in Citrus County, Florida. *American Museum Novitates* 328:1–16.

1929 Pleistocene mammalian fauna of the Seminole Field, Pinillas County, Florida. *Bulletin, American Museum of Natural History* 56:561–599.

1932 Miocene land mammals from Florida (note on the Pleistocene). *Florida Geological Survey Bulletin* 10:11–41.

1940 Review of the mammal-bearing Tertiary of South America. *Proceedings of the American Philosophical Society* 83:649–709.

1945 *The Principles of Classification and Classifications of Mammals. Bulletin.* American Museum Natural History 85:1–350.

1962 *Evolution and Geography: An Essay on Historical Biogeography with Special Reference to Mammals.* University of Oregon Press, Eugene, OR.

1980 *Splendid Isolation. The Curious History of South American Mammals.* Yale University Press, New Haven, CT.

Snarskis, Michael J.
1975 Excavaciones estratigráficas en la Vertiente Atlántica de Costa Rica. *Vínculos* 1(1):2–17.

Spalding, Karen
1984 *Huarochirí. An Andean Society under Inca and Spanish Rule.* Stanford, CA.

Stanish, Charles
1989 An archaeological evaluation of an ethnohistorical model in Moquegua. In *Ecology, Settlement and History in the Osmore Drainage, Peru*, pp. 303–320. Don S. Rice, Charles Stanish, and Phillip R. Scarr, editors. BAR International Series 545. British Archaeological Reports, Oxford, UK.

Stanley, H. F., and Jane C. Wheeler
1992 Molecular evolution and genetic diversity of the Camelidae. International Conference on the Molecular Evolution. University Park, PA.

Steele, Zulma
1946 How the camel come to America. *Fauna* 8 (3):78–79.

Stehberg, Rubén
1967 Un sitio habitacional alfarero temprano en el interior de la Quinta Normal, Santiago. In *Tomo Homenaje G. Le Paige.* Universidad del Norte. Antofagasta.Chile.

1981 El complejo prehispánico Aconcagua en la Rinconada de Huechún. *Publicaciones Ocasionales del Museo Nacional de Historia Natural* 35:3–87.

Steinbacher, G.
1953 Zur Abstammung des Alpaka, *Lama pacos* (Linne, 1785). *Säugetierkundliche Mitteilungen* 1(2):78–79.

Steward, Julian, and L. C. Faron
1959 *Native Peoples of South America.* McGraw-Hill Book Co., New York.

Stock, Chester
1928 *Tanupolama*, a new genus of llama from the Pleistocene of California. *Carnegie Institute of Washington Publication* 393:29–37.

Stouse, Pierre A. D., Jr.
1970 The distribution of llamas in Bolivia. *Proceedings of the Association of American Geographers* 2:136–140.

Stroock, S. I.
1937 *Llamas and Llamaland.* S. Stroock & Co., New York.

Sumar Kalinovski, Julio
1974 Los camélidos sudamericanos como factor de producción en los Andes Altos. In *Reunión del Programa Regional Cooperativo de los Andes Altos*, 4a. Pasto. Informe Bogotá, IICA (Informes de Conferencias, Cursos y Reuniones), No. 54, pp. 311–322. Colombia.

1983 *Studies on Reproductive Pathology in Alpacas.* Faculty of Veterinary Medicine, Swedish University of Agriculture Science, Uppsala, Sweden, and IVITA, San Marcos University, Lima, Peru.

Tambussi, C. P., and E. P. Tonni
1985 Aves del sitio arqueológico Los Toldos, Cañadón de las Cuevas, provincia de Santa Cruz (República Argentina). *Ameghiniana* 22:69–74.

Tapia, M.
1971 Pastos naturales del Perú y Bolivia. *Publicación Miscelánea*, No. 85. IICA.

Tapia, M., and J. A. Flores
1984 *Pastoreo y pastizales de los Andes del Sur del Perú.* Inst. Nac. de Inv. y Promoción Agropecuaria. Prog. Col. de Apoyo de la Investigación en Rumiantes Menores. Lima, Peru.

Tapia, M., and J. L.Lascano
1970 Contribución al conocimiento de alpacas pastoreando. *I Convención Internacional sobre Camélidos Sudamericanos* (Summary). Universidad Nacional Técnico del Altiplano, Puno, Peru.

Taylor, K. M., D. A.Hungerford, R. L. Snyder, and F. A.Ulmer
1968 [?] *Cytogenetics* 7 (7).

Teruel, Luis de
1918 (1613) Idolatrías de los indios Huachos y Yauyos. *Revista Histórica* (Lima) 6:180–197.

Thomas, Oldfield
1891 Note on some ungulate mammals. *Proceedings of the Zoological Society of London*, pp. 384–389.

Thomas, R. Brooke
1973 *Human Adaptations to a High Andean Energy Flow System.* Occasional Papers in Anthropology. Department of Anthropology, Pennsylvania State University, University Park, PA.

Thorvildsen, K.
1964 Gravrøser på Umm an-Nar. *KUML*, pp. 191–219.

Tillman, A.
1981 *Speechless Brothers: The History and Care of Llamas.* Early Winters Press, Seattle, WA.

Tonni, E. P., and J. H. Laza
1976   Paleoetnozoología del área de la Quebrada del Toro, Provincia de Salta. *Relaciones de la Sociedad Argentina de Antropología*, n.s., 10: 131–140.

Tonni, Eduardo P., Gustavo G. Politis, and Luis M. Meo Guzman
1980   La presencia de *Megatherium* en un sitio arqueológico de la pampa bonaerense (Rep. Argentina): Su relación con la problemática de las extinciones pleistocénicas. In *Actas del VII Congreso Nacional de Argueología*. Colonia del Sacramento, Uruguay.

Torres, Hernán
1984   *Distribución y conservación de la vicuña*. Informe Especial No. l. UICN/CSE. Grupo especialista en Camélidos Silvestres Sudamericanos, Gland, Switzerland.

Tosi, M.
1974   Some data for the study of prehistoric cultural areas on the Persian Gulf. *Proceedings of the Arabian Seminar* Vol. 4, pp. 145–171.

Tribaldos de Toledo, Luis
1864 (1735)   *Vista General de las Continuadas guerras, difícil conquista del gran reino, provincia de Chile*. Colección de Historiadores de Chile, Santiago, Chile.

Trimborn, Hermann
1928   *Die Kultur historische Stellung der Lama zucht in der Wirtschaft der peruanischen Erntevölker*. Anthropos, Vienna, Austria.

Troll, Carl
1980   Las culturas superiores andinas y el medio geográfico. *Allpanchis*, 15. *La Agricultura Andina* 2:3–55. [Author's Note: *See* Troll 1958.]

Troncoso, R.
1982   Evaluación de la capacidad de carga animal del Parque Nacional Lanca. Informe de Consultoría. Corporación Nacional Forestal, Arica, Chile.

Trues, D. L., and Harvey Crew
1972   Archaeological Investigations in northern Chile. University of California, Davis, Department of Anthropology, Davis, CA. Unpublished manuscript.

Trujillo, Diego de
1948   *Relación del descubrimiento del reyno del Perú*. Escuela de Estudios Hispanoamericanos, Seville, Spain. [Author's Note: *See* Trujillo 1968.]

Tschudi, J. J. von
1844–1846   *Untersuchungen über die Fauna Peruana*. St. Gallen, Switzerland.

1847   *Travels in Peru, During the Years 1838–1842*. David Bogue, London, UK. [Author's Note: *See* Tschudi 1966.]

Ubelaker, Douglas H.
1981   *The Ayalán Cemetery. A Late Integration Period Burial Site on the South Coast of Ecuador*. Smithsonian Contributions to Anthropology, No. 29. Smithsonian Institution, Washington, DC.

Uerpmann, Hans-Peter
1973   Animal bone finds and economic archaeology: A critical study of "osteoarchaeological method." *World Archaeology* 4(3):307–322.

Uhle, Max
1919   *Fundamentos étnicos de la arqueología de Tacna y Arica*. Ecuador.

Ulloa, J. J.
1748   La Llama. *Relación histórica del viaje de América Meridional*. I Parte, Vol. 2, pp. 589–590.

Valcarcel, Luis E.
1922   Glosario de la vida incaica. *Revista Universitaria* (Cuzco) 11(39):3–19.

Valdéz Cárdenas, Lidio
1988a   Patrones de subsistencia Nasca. Una perspectiva desde Kawachi y Tambo Viejo. Tesis presentada para optar la Licenciatura en Arqueología. Facultad de Ciencias Sociales. Universidad Nacional de San Cristóbal de Huamanga. Ayacucho, Peru.

Valdivia, P(edro). de
1861   *Cartas de don . . . al emperador Carlos V.* Colección. . . [de Historiadores de Chile], Vol. 1, pp. 1–62. Santiago, Chile.

1970 (1551)   *Cartas de Relación de la conquista de Chile*. Editorial Universitaria, Santiago, Chile.

Valenzuela, Francisco de
1906   Información de servicios de Francisco de Valenzuela (1578). In *Juicio de Límites Perú-Bolivia*, Vol. VII, *Vilcabamba*, pp. 98–116.

Vallenas, A.
1965   Some physiological aspects of digestion in the alpaca (*Lama pacos*). In *Physiology of Digestion in the Ruminants*, pp. 147–158. R. W. Dougherty, editor. Butterworths, Washington, DC.

Vallenas, Augusto
1970b   Comentarios sobre la posición de los camélidos sud-americanos en la sistemática. *Boletín Extraordinario Instituto Veterinario de Investigaciones Tropicales y de Altura* 4:128–141.

Vallenas, A. P., J. F. Cummings, and J. F. Munnell
1971   A gross study of the compartmentalized stomach of two New World Camelids, the llama and guanaco. *The Journal of Morphology* 134:399–407.

Vallenas, A. P., and C. E. Stevens
1971a Motility of the llama and guanaco stomach. *American Journal of Physiology* 220:275–282.
1971b Volatile fatty acid and concentration and pH of llama and guanaco forestomach digesta. *Cornell Veterinarian* 61:239–252.

Van der Hammen, T.
1974 The Pleistocene changes of vegetation and climate in tropical South America. *Journal of Biogeography* 1:3–26.

Van Soest, P. J.
1982 *Nutritional Ecology of the Ruminant*. O & B Books, Inc., Corvallis, OR.

Vargas Ugarte, S. J.
1949 *Historia del Perú*. Imprenta Baiocco, Lima, Peru.

Vázquez de Espinosa, Antonio
1942 *Compendium and Description of the West Indies*. Smithsonian Institution Miscellaneous Collection, Vol. 102, Publication 3646. Washington, DC. [Author's Note: *See* Vázquez de Espinosa 1948.]

Vassberg, D. E.
1978 Concerning pigs, the Pizarros and the agropastoral background of conquerors of Peru. *Latin American Research Review* 13:47–61.

Vau, E.
1936 [?] *Kühn-Arch* 40.

Vaughan, T. A.
1978 *Mammalogy*. W. B. Saunders Co., Philadelphia, PA.

Velasco, Juan
1946 *Historia del Reino de Quito. I. La Historia Natural*. Empresa Editorial "El Comercio," Quito, Ecuador.

Viault, F.
1895 *Ultramar*. Société d'editions littéraires, Paris, France.

Vidal Gormaz, F.
1879 *Anuario Hidrográfico de la Marina*, Tomo V. Santiago, Chile.

Vietmeyer, N. D.
1978 Incredible odyssey of a visionary Victorian peddlar. *Smithsonian* 9:91–102.
1979 A visionary adventurer ruined by bureaucracy (history). *The Bulletin*, January 16.

Von Pilters, Hilde
1954 Untersuchungen über angeborene Verhaltensweisen bei Tylopoden, unter besonderer Berücksichtigung der Neuweltlichen Formen. *Zeitschrift für Tierpsychologie* 11(2):213–303.

Walker, Ernest P., et al. [sic]
1964 *Mammals of the World*. 3 vols. Johns Hopkins University Press, Baltimore, MD. [Author's Note: This is the reference given by Rae-
deke (1976), whereas for Larramendy et al. (1984) only Walker is the author. This agrees with the reference Reitz (1979) gives for Walker (1975) and which apparently is another edition.]

Walker, Ernest P.
1975 *Mammals of the World*. Johns Hopkins University Press, Baltimore, MD.

Walton, W.
1918 (1811?) *An historical and descriptive account of the Peruvian sheep, called "carneros de la tierra"; and of the experiments made by Spaniards to improve and respective breeds. To which is added, an account of a successful attempt to domesticate the vicuña in the England, and a recomendation to this species to cross with our native flocks.* L. Boot. Printed for John Harding, London, UK.

Waterhouse, G. R., and Charles Darwin
1839 *The Zoology of the Voyage of the* Beagle. *Part II. Mammalia*. London, UK.

Webb, S. David
1965 The osteology of Camelops. *Bulletin of the Los Angeles County Museum* 1:1–54.
1972 Locomotor evolution in camels. *Forma et Function* 5:99–112.
1977 A history of savanna vertebrates in the New World. Part I: North America. *Annual Review of Ecology and Systematics* 8:355–380.

Weberbauer, Augusto
1922 Vegetationskarte der peruanischen Anden. [No name of publication given.]

Webster, Steven S.
1973 Native pastoralism in the South Andes. *Ethnology* 12(2):115–133.

Weischet, Wolfgang
1966 Zur Klimatologie der Nordchilenischen Wüste. *Meteorologische Rundschau* 19(1).

West, Michael
1977 Late Formative period subsistence strategies on the north coast of Peru. Paper presented at the Annual Meeting of the Institute for Andean Studies, Berkeley, CA.
1978 Ethnographic observation in the Virú Valley, Peru, and their implications for archaeological interpretation. Paper presented at the 18th Annual Meeting of the Institute for Andean Studies, Berkeley, CA.

West, Terry
1981b *Alpaca Production in Puno, Peru*. Small Ruminant Collaborative Research Support Program, Sociology Technical Report Series, Publication 3. Department of Rural Sociology, University of Missouri, Columbia, MO.

Wheeler, Jane

1981   Camelid domestication at Telarmachay. Paper presented at the Ethnobiology Conference IV, Columbia, MO.

1982c Camelid domestication and the early development of pastoralism in the Andes. Paper presented at the 4th International Conference of the International Council for Archaeozoology. British Archaeological Reports, Oxford, UK.

1983   Nuevas evidencias arqueozoológicas sobre domesticación de alpacas (*Lama pacos*) y llamas (*Lama glama*) y el desarrollo temprano de la ganadería autóctona andina. In *Abstracts of the Ninth Latin American Congress of Zoology*, Arequipa.

1988b Origin and evolution of the South American Camelidae. Paper presented at the Western Veterinary Conference 6th Anual Meeting, Las Vegas.

1988c Llamas and alpacas of South America. In *Selected Papers, Western Veterinary Conference*, pp. 301–310. Las Vegas.

Wheeler, Jane, and Elías J. Mujica

1981   Prehistoric Pastoralism in the Lake Titicaca Basin, Peru: 1979–1980 Field Season. Final project report submitted to the National Science Foundation.

Wheeler Pires-Ferreira, Jane, Ramiro Matos, and Edgardo Pires-Ferreira

1975   *Informe del Programa de Arqueología.* Seminario de Historia Rural Andina. Serie Huamán Poma I. Lima, Peru.

Wilbert, J[ohannes]

1974   *The Thread of Life. Symbolism of Miniature Art from Ecuador.* Dumbarton Oaks Studies in Pre-Columbian Art and Archaeology, Vol. 12. Trustees for Harvard University, Washington, DC.

Wilson, Paul, and William L. Franklin

1985   Male group dynamics and inter-male aggression of guanaco in southern Chile. *Zeitschrift für Tierpsychologie* 69:305–328.

Wing, Elizabeth S.

1973   Utilization of animal resources in the Andes. Report to the National Science Foundation (GS 3021).

1975c Hunting and herding in the Peruvian Andes. In *Archaeozoological Studies*, pp. 302–308. A. T. Clason, editor. American Elsevier, New York.

1985   Provisional preliminary report on the faunal remains from El Paraiso (PV46-35). Peru. MS on file, Florida State Museum, University of Florida, Gainesville, FL.

Winge, Herluf

1906   Jordfundne og nulevende Hovdyr (Ungulata) fra Lagoa Santa, Minas Geraes, Brasilien. *E Museo Lundii* 4 (III /1):1–239.

Winsor, Justin

1886   *Narrative and Critical History of America*, Vol. II. Boston and New York.

Winterhalder, Bruce, Robert R. Larsen, and Thomas Brooke

1974   Dung as an essential resource in a highland Peruvian community. *Human Ecology* 2(2): 43–55.

Winterhalder, Bruce P., and R. B. Thomas

1978   Geoecology of southern Highland Peru: A human adaptation perspective. *Institute of Arctic Alpine Research Paper* No. 27, pp. 1–91.

Wolf, Teodoro

1892   *Geografía y geología del Ecuador.* Tip. de F. A. Brokhaus, Leipzig, Germany.

1975   *Geografía y geología del Ecuador.* Casa de la Cultura Ecuatoriana, Quito, Ecuador.

Wortman, J. L.

1898   The extinct Camelidae of North America and some associated forms. *Bulletin of the American Museum of Natural History* 10: 93–142.

Wright, H. E., Jr.

1983   Late Pleistocene glaciation and climate around the Junín plain, central Peruvian highlands. *Geografiska Annaler* 65A:35–43.

Yacobaccio, H.

1989   Close to the edge: Early adaptations in the South Andean Highlands. In *The Pleistocene Perspective: Innovation, Adaptation and Human Survival*, pp. 32–54. A. Simon and S. Joyce, editors. One World Archaeology Series. Unwin, London, UK.

Yagil, R.

1985   *The Desert Camel. Comparative Physiological Adaptation.* S. Karger, Basel, Switzerland.

Zárate, Agustín de

1853   *Historia del descubrimiento de la provincia del Perú . . .* [1555]. Madrid, Spain. [Author's Note: See Zárate 1968.]

Zlatar, V.

1983   Replanteamiento sobre el problema Caleta Huelén 42. *Chungará* 10:21–28.

ACKNOWLEDGMENTS

In expanding and revising the bibliography for the English edition I had the help of Jimi Espinoza.

A Note on
Appendices A and B

In the Spanish edition of this book, *Camélidos Sudamericanos*, Duccio Bonavia expressed frustration over the fact that, since Volume 1 of Richard S. MacNeish's *Prehistory of the Ayacucho Basin, Peru* had still not been published, the data on faunal remains (including those of camelids) were unavailable to him. In response to this problem, Elizabeth S. Wing and Kent V. Flannery, who analyzed the fauna from MacNeish's Ayacucho sites, have provided brief summaries of their findings.

603

Animal Remains from Archaeological Sites in the Ayacucho Basin Below 3000 m Elevation

Elizabeth S. Wing

Excavations in the Ayacucho Basin recovered large samples of identifiable animal bones and teeth. The total sample from the lower-elevation sites is 12,821 identified fragments, representing 37 taxa. These remains were recovered from six rockshelters and two open-air sites. The larger shelters were Pikimachay (Ac 100), Ayamachay (Ac 102), Puente (Ac 158), and Rosamachay (Ac 117), and the two very small rockshelters were Ac 244 and Ac 245. The two open-air sites were the important cultural centers of Wari (As 4) and Wichqana (As 18).

People occupied the lower-elevation sites in the basin for a long span of time. The earliest deposits in Pikimachay include remains of extinct animals and date as early as 12,500 BC. One deposit from Rosamachay extends into Spanish Colonial times, as indicated by the presence of the Old World domestic animals, pig and cow.

The locations of the sites are at elevations between 2500 m and 3000 m. The higher rockshelters within this range are located today in arid thorn scrub vegetation. Those sites closer to the river are in humid thorn forest. The locations of the sites and the time periods of their occupation played a critical role in the animals that were used.

The data derived from the identification of the faunal remains from the lower-elevation sites can be divided into four components and discussed separately. The earliest is the assemblage of extinct animals from Pikimachay (Ac 100), and the most recent includes the remains of introduced Old World domestic animals buried in the Rosamachay (Ac 117) rockshelter. Two of the rockshelters, Pikimachay (Ac 100) and Puente (Ac 158), have strata with abundant remains of small rodents and perching birds. These are viewed as incidental to the refuse produced by people living in the rockshelters. The fourth component is composed of the remains of extant animals used by the inhabitants of the Ayacucho Basin. Identification and analysis of these remains are the focus of the zooarchaeological research in the basin and will be summarized in greater detail than the other groups of remains.

The extinct animal remains are represented by teeth and bones and were identified by S. David Webb, Distinguished Curator Emeritus at the Florida Museum of Natural History. These remains were excavated from Pikimachay (Ac 100) deposits dating from 12,500 to 7100 BC. These strata also include human remains and a variety of extant species. The extinct animals are primarily sloths (*Erimotherium* sp., family Mylodontiidae). One mastodon bone and seven horse bones (*Equus* sp.) were also identified.

At the other end of the time scale represented by the Ayacucho Project are remains of introduced European domesticates. These were excavated from Rosamachay (Ac 117). The deposits in this rockshelter range in time from the Preceramic Period to Colonial times, as attested by remains of pigs (*Sus scrofa*) and one cow (*Bos taurus*). The pig remains represent five young individuals, including one close to newborn, which were all buried. In addition, four dogs (*Canis familiaris*), one of which was a newborn, were buried. It is not possible to distinguish whether these dogs were native or introduced. Burials of two guinea pigs (*Cavia porcellus*) clearly represent native animals. One of these guinea pigs was mummified and wrapped in woven fabric. The other is a partial skeleton and associated with a piece of fabric. The intrusive burials, some clearly Spanish Colonial, shed doubt on the cultural affiliation from this site.

The faunal component viewed here as incidental is composed of small rodents (such as *Akodon* sp. and *Phyllotis* sp.) and perching birds (Passeriformes). These animals are found primarily in early levels (5750–3000 BC) in Pikimachay (Ac 100) and to a lesser extent in Puente (Ac 158). Though these animals could conceivably be a human food resource, the burials include unburned bones and mummified specimens. Furthermore, they are associated with birds of prey (Acciptridae and Falconidae) and a barn owl (*Tyto alba*), which may account for their presence. Small rodents still occupied the rockshelters at the time excavations were being conducted.

The excavations of the Ayacucho Project recovered a large faunal sample associated with deposits laid down during a long period of human occupation. This time period encompasses cultural changes in manipulation of the environment, domestication of animals, and development of agriculture. The economic species are the focus of this research. Vertebrates that predominate in the lower-elevation sites are guinea pigs (*Cavia* spp., wild and domestic), viscacha (*Lagidium peruanum*), dog (*Canis familiaris*), fox (*Pseudalopex culpaeus*), hog-nosed skunk (*Conepatus rex*), camelids (Camelidae, wild and domestic), deer (*Odocoileus virginianus* and *Hippocamelus antisensis*), tinamou (Tinamidae), and pigeon (Columbidae). Evidence for distant connections includes the presence of jaguar (*Felis onca*) and parrot (*Amazona* sp.).

Many of the animals, such as the carnivores and birds, are present in low numbers throughout the sites, while others, such as guinea pigs, camelids, and deer, are abundantly represented during certain time periods. Relatively great abundance can indicate a successful hunting strategy or control of an animal through domestication. Familiarity gained through hunting can sometimes lead to management and then domestication.

Guinea pig remains are abundant, representing between 50% and 92% of the identified faunal remains prior to 5500 BC. The guinea pig burial wrapped in fabric recovered from Rosamachay implies special regard for this animal which endured after the relative abundance of this animal declined. Their remains gradually diminished from 44% to 20% of the fauna during the period from 5500 to 2500 BC. Gade (1967) has suggested that guinea pigs may have sought the warmth and protection of human shelter with the added enticement of food refuse. Little effort may have been required to tame and ultimately domesticate guinea pigs.

During the time period from 2500 to 1750 BC, remains of deer are relatively abundant in the lower-elevation sites. This suggests intensive hunting. During this time period, remains of domestic dogs are still scarce. Dogs may not have taken part in the hunting enterprise. However, dog remains become common and consistently present after 1750 BC.

At this same time, camelids predominate in the lower-elevation faunas. Camelids make up 85% of the identified faunal remains at both the Wichqana (As 18) and Wari (As 4) sites. Evidence for both llama (*Lama glama*) and alpaca (*Vicugna pacos*), based on the distinctive form of the incisors, is present among the camelid remains at Wari (As 4) dating from AD 650–850. Camelids are multifaceted animals, providing meat, bone, skins, wool, and use as beasts of burden. In a landscape where important resources are at different elevations and where the most productive grazing lands are at high elevations, all of the products and services provided by the camelids were of great value and made other cultural development possible.

REFERENCES

Gade, Daniel W.
   1967 The guinea pig in Andean folk culture. *Geographical Review* 57:213–224.
Wing, Elizabeth S.
   n.d.   Animal remains from the Ayacucho Basin. Unpublished manuscript.
   1986 Domestication of Andean mammals. In *High Altitude Tropical Biogeography*, edited by F. Vuillemier and M. Monasterio, Chap. 10, pp. 246–264. Oxford University Press, Oxford.

Animal Remains from Archaeological Sites in the Ayacucho Basin Above 3000 m Elevation

Kent V. Flannery

More than 75,000 bones and teeth were recovered from sites above 3000 m during MacNeish's Ayacucho Project. At least 18 taxa were represented. The faunal remains came from four caves: Jaywamachay (Ac 335), Chupas Cave (Ac 500), Ruyru Rumi Cave (Ac 300), and Tukumachay (Ac 351). Of all these sites, it was Jaywamachay that produced by far the greatest faunal sample: more than 72,000 fragments of bones and teeth, of which 5,764 were from deer, 2,236 were from camelids, and 63,637 were identifiable only as from ungulates.

The sites above 3000 m were located in one of two environmental zones: humid woodland or low puna. The zone ranging from roughly 3000 to 4000 m would once have been a woodland with trees 10–15 m high, perhaps dominated by *quiñua* (*Polylepis* sp.), *Berberis*, *Escallonia*, and *mutuy* (*Cassia* sp.). Receiving 500–1000 mm of rain and snow, it would have been perhaps the richest environment for hunting in the entire Ayacucho Basin, especially for guanaco, white-tailed deer, and huemal deer. The humid woodland also teems with viscacha (*Lagidium peruanum*), white-eared opossum (*Didelphis albiventris*), fox (*Pseudalopex culpaeus*), hog-nosed skunk (*Conepatus rex*), tinamou (*Nothoprocta ornata*), pigeons (*Columba maculosa*), and doves (*Metriopelia melanoptera* and *Zenaidura auriculata*). Wild guinea pigs (*Cavia* cf. *tschudii*) live in huge colonies in the *Scirpus* marshes of this zone. Other small rodents diagnostic of the *Polylepis* woodland are the grass mouse, *Akodon boliviensis*, and the vesper mouse, *Calomys sorellus*. Jaywamachay (3350 m) and Chupas Cave (3496 m) both lie in the humid woodland.

Ascending to 4000 m, one enters the treeless zone that Quechua speakers call *sallqa*. This zone is divided by Spanish speakers into *puna baja* and *puna alta*. Ruyru Rumi Cave (4033 m) lies at the transition from humid woodland to low puna, while Tukumachay (4350 m) is squarely in the low puna. This zone consists of rolling meadows of bunch grass (*Stipa ichu*) that overshadow the more delicate grasses (e.g., *Festuca* and *Calamagrostis*), which are among the preferred foods of the vicuña. Chuño potatoes, oca, and masua can be grown in the low puna if conditions are right. Vicuña (and occasionally huemal deer) are the main prey. The hog-nosed skunk is well represented, but opossums cannot tolerate the nine months of frost. The bird life of the puna includes several species of tinamou, the puna flicker (*Colaptes rupicola*), Ridgway's ibis (*Plegadis ridgwayi*), the Andean goose (*Chloephaga melanoptera*), the Andean gull (*Larus serranus*), and several species of teal (*Anas flavirostris* and *A. versicolor*). Among the small rodents characteristic of this zone are vesper mice (*Calomys lepidus*) and leaf-eared mice (*Phyllotis pictus* and *P. sublimus*).

Elizabeth Wing (Appendix A) reports that prior to 5500 BC, guinea pigs contributed 50%–92% of the identified remains from sites below 3000 m. The remains from the higher-altitude caves could not have been more different. Prior to 5500 BC, 98% of the identified bones from sites like Jaywamachay were from deer and camelids. Puente phase levels (9000–7100 BC), however, produced three specimens of *Cavia* cf. *tschudii*, the wild guinea pig living today in the *Scirpus* marsh at Totorabamba, near Jaywamachay.

Not until Rancha phase levels (500–200 BC) did domestic guinea pigs show up at Jaywamachay, but this may only be because of a hiatus in occupation between 5800 and 500 BC. Owing to this hiatus, we must look to lower-elevation sites such as the Puente Rock Shelter (Ac 158) for data on early guinea pig domestication.

One similarity between the higher-elevation and lower-elevation sites is the scarcity of domestic dog remains. Wing (Appendix A) reports that dogs were scarce at the lower-elevation sites until roughly 1750 BC. The same was true at the higher-elevation sites. A few dogs showed up in Jaywa phase levels (7100–5800 BC), but they were never common until after sedentary life had been established.

Because of its staggering ungulate bone samples, Jaywamachay Cave deserves special consideration. It was occupied on at least 15–16 occasions, from the Puente phase (9000–7100 BC) through the Jaywa phase (7100–5800 BC), after which it was abandoned until Rancha phase times (500–200 BC).

From 9000 to 5800 BC, Jaywamachay seems mainly to have served as a camp site for hunting white-tailed deer (*Odocoileus virginianus*), huemal deer (*Hippocamelus antisensis*), and guanaco (*Lama guanicoë*). Camelid fragments small enough to represent vicuña (*Vicugna vicugna*) were rare. Vicuñas should have been available on the low puna some 700 m above Jaywamachay, but it appears that hunters shifted their camps to the *sallqa* to hunt them.

Stratigraphic Zones I–H (Puente phase) and G–C (Jaywa phase) were major encampments. Some of these levels had 8,000–11,000 fragments of ungulate bone each, with duplicate skeletal elements indicating a minimum of 30–35 individual animals per level. Given the low probability that duplicate elements were saved from every animal, the true number of deer and guanaco killed during some of these encampments could be as high as 100. Antlers, both shed and unshed, and the remains of late-term deer fetuses indicate that several seasons of the year were involved.

An interesting aspect of Jaywamachay is that virtually all the ungulate bone had been burned. Cooking was often done in a *pachamanca*, or Andean earth-oven, but even that fact is insufficient to explain the high levels of burning; thus, it may be that green bone was used as fuel.

Unfortunately, the Puente and Jaywa phases seem to have antedated the domestication of camelids in the Ayacucho Basin. I took measurements of all the camelid bones and evaluated them, based on the "decision rules" for large and small camelids proposed by Wing as the result of her discriminant analysis of guanaco, vicuña, llama, and alpaca skeletons (Wing 1972 and personal communication). I also separated fused and unfused epiphyses to see whether there was a sudden increase in juvenile camelids at any point in the sequence. Despite the taking of many measurements, and the observing of many epiphyses, neither the Puente phase nor the Jaywa phase yielded either a significant increase in camelids smaller than the guanaco or a significant increase in juvenile camelids.

Let us move on now to the Piki phase (5800–4400 BC). For data from this phase, we must turn to Zone E of Chupas Cave (Ac 500), a modest occupation during which at least two deer and two camelids were killed. One of the camelids was large, in the guanaco size range, while the other was smaller. The sample of bones is too small to support a claim of domestication for the small camelid.

Next comes the Chihua phase (4400–3100 BC). This would seem to be a crucial period for camelid domestication, since it is coeval with Telarmachay Cave near Junín, where Wheeler (1984) found evidence for early camelid domestication. Her evidence consisted not only of animals in the llama size range but also high percentages of very young camelids at ca. 4300 BC.

Unfortunately, our Chihua phase data come from Zone C of Ruyru Rumi Cave (Ac 300), which produced only 10 bones identifiable as camelid. Only one minimum individual is represented, and the evidence consists mainly of elements from one forelimb (scapula, radius, ulna, and metacarpals). It would be tempting to suggest that the hunters of Zone C arrived at Ruyru Rumi carrying one forelimb of camelid (perhaps preserved as charqui) as part of their provisions for a hunting trip. Unfortunately, the evidence is too inconclusive to support that suggestion. They also killed at least two deer while at the cave.

Finally, we come to the Cachi phase (3100–1750 BC), represented by Zone C1 of Tukumachay (Ac 351). This cave occupation yielded a stone corral of precisely the type that might have been used to control domestic camelids. This structure reminds us that some of the most compelling evidence for early camelid domestication in Peru and Argentina is provided by corrals, both in caves and preceramic open-air sites (Mengoni Goñalons and Yacobaccio 2006). Unfortunately, Tukumachay was a very damp cave in which bone preservation was poor; Zone C1 yielded only one splinter identifiable as coming from an ungulate.

In sum, the oldest evidence of camelid domestication in Peru still comes from the period 4500–3500 BC (Wheeler 1984), and the evidence becomes even stronger by 2000 BC (Kent 1982, Moore 1989). The Ayacucho Project reinforced the evidence from Junín, but did not extend domestication farther back in time. In fact, the Ayacucho results remind us that caves do not necessarily hold all the answers to what happened in a region. Jaywamachay had great bone preservation but was not occupied during the crucial Chihua phase; Tukumachay was occupied in Chihua times but had poor preservation.

References

Kent, Jonathan D.
1982 The domestication and exploitation of the South American camelids: Methods of analysis and their application to circum-lacustrine archaeological sites in Bolivia and Peru. Ph.D. dissertation, Department of Anthropology, Washington University, St. Louis.

Mengoni Goñalons, Guillermo L., and Hugo D. Yacobaccio
2006 The domestication of South American camelids: A view from the south-central Andes. In *Documenting Domestication: New Genetic and Archaeological Paradigms*, edited by Melinda A. Zeder, Daniel G. Bradley, Eve Emshwiller, and Bruce D. Smith, pp. 228–244. University of California Press, Berkeley, CA.

Moore, Katherine M.
1989 Hunting and the origins of herding in Peru. Ph.D. dissertation, Department of Anthropology, University of Michigan, Ann Arbor.

Wheeler, Jane C.
1984 La domesticación de la alpaca (*Lama pacos* L.) y la llama (*Lama glama* L.) y el desarrollo temprano de la ganadería autóctona en los Andes Centrales. *Boletín de Lima* 36:74–84.

Wing, Elizabeth S.
1972 Utilization of animal resources in the Peruvian Andes. In *Andes 4: Excavations at Kotosh, Peru, 1963 and 1966*, edited by Seiichi Izumi and Kazuo Terada, pp. 327–351. University of Tokyo Press, Tokyo.

INDEX

Note: Page numbers followed by "f" indicate a figure drawing; by a "t" a table; and in italics, a photograph.